Lecture Notes in Computer Science 2658

Edited by G. Goos, J. Hartmanis, and J. van Leeuwen

Springer
Berlin
Heidelberg
New York
Hong Kong
London
Milan
Paris
Tokyo

Peter M.A. Sloot David Abramson
Alexander V. Bogdanov Jack J. Dongarra
Albert Y. Zomaya Yuriy E. Gorbachev (Eds.)

Computational Science – ICCS 2003

International Conference
Melbourne, Australia and St. Petersburg, Russia
June 2-4, 2003
Proceedings, Part II

Springer

Volume Editors

Peter M.A. Sloot
University of Amsterdam, Informatics Institute, Section of Computational Science
Kruislaan 403, 1098 SJ Amsterdam, The Netherlands
E-mail: sloot@science.uva.nl

David Abramson
Monash University, School of Computer Science and Software Engineering
Wellington Road, Clayton, VIC 3800, Australia
E-mail: davida@csse.monash.edu.au

Alexander V. Bogdanov
Yuriy E. Gorbachev
Institute for High-Performance Computing and Information Systems
Fontanka emb. 6, St. Petersburg 191187, Russia
E-mail: bogdanov,gorbachev @hm.csa.ru

Jack J. Dongarra
University of Tennessee and Oak Ridge National Laboratory, Computer Science Dept.
1122 Volunteer Blvd., Knoxville, TN 37996-3450, USA
E-mail: dongarra@cs.utk.edu

Albert Y. Zomaya
The University of Sydney, School of Information Technologies, CISCO Systems
Madsen Building F09, Sydney, NSW 2006, Australia
E-mail: zomaya@it.usyd.edu.au

Cataloging-in-Publication Data applied for

A catalog record for this book is available from the Library of Congress

Bibliographic information published by Die Deutsche Bibliothek
Die Deutsche Bibliothek lists this publication in the Deutsche Nationalbibliographie; detailed bibliographic data is available in the Internet at <http://dnb.ddb.de>.

CR Subject Classification (1998): D, F, G, H, I, J, C.2-3

ISSN 0302-9743
ISBN 3-540-40195-4 Springer-Verlag Berlin Heidelberg New York

Springer-Verlag Berlin Heidelberg New York
a member of BertelsmannSpringer Science+Business Media GmbH

http://www.springer.de

Printed in Germany

Typesetting: Camera-ready by author, data conversion by PTP-Berlin GmbH
Printed on acid-free paper SPIN: 10931165 06/3142 5 4 3 2 1 0

Preface

Some of the most challenging problems in science and engineering are being addressed by the integration of computation and science, a research field known as computational science.

Computational science plays a vital role in fundamental advances in biology, physics, chemistry, astronomy, and a host of other disciplines. This is through the coordination of computation, data management, access to instrumentation, knowledge synthesis, and the use of new devices. It has an impact on researchers and practitioners in the sciences and beyond. The sheer size of many challenges in computational science dictates the use of supercomputing, parallel and distributed processing, grid-based processing, advanced visualization and sophisticated algorithms.

At the dawn of the 21st century the series of International Conferences on Computational Science (ICCS) was initiated with a first meeting in May 2001 in San Francisco. The success of that meeting motivated the organization of the second meeting held in Amsterdam April 21–24, 2002, where over 500 participants pushed the research field further.

The International Conference on Computational Science 2003 (ICCS 2003) is the follow-up to these earlier conferences. ICCS 2003 is unique, in that it was a single event held at two different sites almost opposite each other on the globe – Melbourne, Australia and St. Petersburg, Russian Federation. The conference ran on the same dates at both locations and all the presented work was published in a single set of proceedings, which you hold in your hands right now.

ICCS 2003 brought together experts from a range of disciplines: mathematicians and computer scientists providing basic computing expertise, and researchers and scientists from various application areas who are pioneering advanced applications of computational methods in sciences such as physics, chemistry, life sciences, engineering, arts and humanities; along with software developers and vendors. The intent was to discuss problems and solutions in these areas, identify new issues, and shape future directions for research, as well as help industrial users apply advanced computational techniques.

Many of the advances in computational science are related to Grid Computing. The Grid has provided a way to link computation, data, networking, instruments and other resources together to solve today's complex and critical problems. As such, it is becoming a natural environment for the computational sciences. In these proceedings you will find original research in this new era of computational science and the challenges involved in building the information infrastructure needed to enable science and engineering discoveries of the future.

These four volumes, LNCS 2657, 2658, 2659 and 2660, contain the proceedings of the ICCS 2003 meeting. The volumes consist of over 460 peer-reviewed, contributed and invited papers presented at the conference in Melbourne, Australia and St. Petersburg, Russian Federation. The acceptance rate for oral presentations was 40% of the submitted papers. The papers presented reflect the aim of the scientific organization to bring together major players in the emerging field of computational science.

The conference included 27 workshops (10 in St. Petersburg and 17 in Australia), 6 presentations by Keynote speakers, and over 460 contributed papers selected for oral presentations and posters. Each paper/poster was refereed by at least two referees.

We are deeply indebted to all the authors who submitted high-quality papers to the conference, without this depth of support and commitment there would have been no conference at all. We acknowledge the members of the program committee and all those involved in the refereeing process, and the workshop organizers and all those in the community who helped us to convene a successful conference. Special thanks go to Dick van Albada, Martin Lack, Zhiming Zhao and Yan Xu for preparation of the proceedings; they did a marvelous job! Amitava Datta, Denis Shamonin, Mila Chevalier, Alexander Boukhanovsky and Elena Stankova are acknowledged for their assistance in the organization and all those 1001 things that need to be done to make a large (distributed!) conference like this a success!

Of course ICCS 2003 would not have been possible without the support of our sponsors, and we therefore gratefully acknowledge their help in the realization of this conference.

Amsterdam, June 2003

Peter M.A. Sloot,

on behalf of the co-editors:
David Abramson
Alexander Bogdanov
Jack J. Dongarra
Albert Zomaya
Yuriy Gorbachev

Organization

The conference was organized by the Section Computational Science, The University of Amsterdam, The Netherlands; the Innovative Computing Laboratory at The University of Tennessee, USA; the School of Computer Science and Software Engineering, Monash University, Victoria, Australia; the School of Information Technologies, The University of Sydney, New South Wales, Australia; and the Institute for High Performance Computing and Information Systems, St. Petersburg, Russian Federation.

Conference Chairs

Alexander Bogdanov, Chair of the St. Petersburg ICCS 2003 conference site
David Abramson, Chair of the Melbourne ICCS 2003 conference site
Jack J. Dongarra, Scientific and Overall Co-chair (The University of Tennessee, Knoxville, USA)
Peter M.A. Sloot, Scientific and Overall Chair (The University of Amsterdam, The Netherlands)

Workshops Organization and Program Chairs

Yuriy Gorbachev (IHPCIS, St. Petersburg, Russian Federation)
Albert Zomaya (The University of Sydney, Australia)

Local Organizing Committees

Martin Lack & Associates Pty. Ltd. (Australia)
Elena Stankova (IHPCIS, Russian Federation)
Alexander Boukhanovsky (IHPCIS, Russian Federation)
Mila Chevalier (NIP, Russian Federation)

Program Committee

Albert Y. Zomaya (The University of Sydney, Australia)
Alexander Bogdanov (IHPCIS, Russia)
Alexander Zhmakin (PhTI RAS, Russian Federation)
Alfons Hoekstra (The University of Amsterdam, The Netherlands)
Alistair Rendell (Australian National University, Australia)

Andrzej M. Goscinski (Deakin University, Australia)
Antonio Lagana (University of Perugia, Italy)
Azzedine Boukerche (University of North Texas, USA)
Bastien Chopard (University of Geneva, Switzerland)
Beniamino Di Martino (Seconda Universita' di Napoli, Italy)
Bernard Pailthorpe (The University of Queensland, Australia)
Dale Shires (US Army Research Laboratory, USA)
David A. Bader (University of New Mexico, USA)
Dick van Albada (The University of Amsterdam, The Netherlands)
Dieter Kranzlmueller (Johannes Kepler University Linz, Austria)
Edward Moreno (Euripides Foundation of Marilia, Brazil)
Elena Zudilova (The University of Amsterdam, The Netherlands)
Francis Lau (The University of Hong Kong, Hong Kong)
Geoffrey Fox (Indiana University, USA)
Graham Megson (The University of Reading, UK)
Greg Watson (LANL, USA)
Hai Jin (Huazhong University of Science and Technology, China)
Hassan Diab (American University of Beirut, Lebanon)
Hong Shen (Japan Advanced Institute of Science and Technology, Japan)
James Glimm (Stony Brook University, USA)
Jemal H. Abawajy (Carleton University, Canada)
Jerzy Wasniewski (UNI-C Danish IT Center for Education and Research, Denmark)
Jesús Vigo-Aguiar (University of Salamanca, Spain)
Jose Laginha Palma (University of Porto, Portugal)
Kevin Burrage (The University of Queensland, Australia)
Koichi Wada (University of Tsukuba, Japan)
Marian Bubak (AGH, Cracow, Poland)
Matthias Müller (University of Stuttgart, Germany)
Michael Johnson (The University of Sydney, Australia)
Michael Mascagni (Florida State University, USA)
Nikolay Borisov (SPbSU, Russian Federation)
Paul Coddington (University of Adelaide, Australia)
Paul Roe (Queensland University of Technology, Australia)
Peter Kacsuk (MTA SZTAKI Research Institute, Hungary)
Peter M.A. Sloot (The University of Amsterdam, The Netherlands)
Putchong Uthayopas (Kasetsart University, Thailand)
Rajkumar Buyya (Melbourne University, Australia)
Richard Ramaroson (ONERA, France)
Robert Evarestov (SPbSU, Russian Federation)
Rod Blais (University of Calgary, Canada)
Ron Perrott (Queen's University of Belfast, UK)
Rosie Renaut (Arizona State University, USA)
Srinivas Aluru (Iowa State University, USA)
Stephan Olariu (Old Dominion University, USA)

Tarek El-Ghazawil (George Washington University, USA)
Vaidy Sunderam (Emory University, USA)
Valery Zolotarev (SPbSU, Russian Federation)
Vasil Alexandrov (The University of Reading, UK)
Vladimir P. Nechiporenko (Ministry of Industry, Science and Technologies, Russian Federation)
Xiaodong Zhang (National Science Foundation, USA)
Yong Xue (Chinese Academy of Sciences, China)
Yuriy Gorbachev (IHPCIS, Russian Federation)
Zdzislaw Meglicki (Indiana University, USA)

Workshop Organizers

Computer Algebra Systems and Their Applications
A. Iglesias (University of Cantabria, Spain)
A. Galvez (University of Cantabria, Spain)

Computer Graphics
A. Iglesias (University of Cantabria, Spain)

Computational Science of Lattice Boltzmann Modeling
B. Chopard (University of Geneva, Switzerland)
A.G. Hoekstra (The University of Amsterdam , The Netherlands)

Computational Finance and Economics
X. Deng (City University of Hongkong, Hongkong)
S. Wang (Chinese Academy of Sciences, China)

Numerical Methods for Structured Systems
N. Del Buono (University of Bari, Italy)
L. Lopez (University of Bari, Italy)
T. Politi (Politecnico di Bari, Italy)

High-Performance Environmental Computations
E. Stankova (Institute for High Performance Computing and Information Systems, Russian Federation)
A. Boukhanovsky (Institute for High Performance Computing and Information Systems, Russian Federation)

Grid Computing for Computational Science
M. Müller (University of Stuttgart, Germany)
C. Lee (Aerospace Corporation, USA)

Computational Chemistry and Molecular Dynamics
A. Lagana (Perugia University, Italy)

Recursive and Adaptive Signal/Image Processing (RASIP)
I.V. Semoushin (Ulyanovsk State University, Russian Federation)

Numerical Methods for Singular Differential and Differential-Algebraic Equations
V.K. Gorbunov (Ulyanovsk State University, Russian Federation)

Workshop on Parallel Linear Algebra (WoPLA03)
M. Hegland, (Australian National University, Australia)
P. Strazdins (Australian National University, Australia)

Java in Computational Science
A. Wendelborn (University of Adelaide, Australia)
P. Coddington (University of Adelaide, Australia)

Computational Earthquake Physics and Solid Earth System Simulation
P. Mora (Australian Computational Earth Systems Simulator)
H. Muhlhaus (Australian Computational Earth Systems Simulator)
S. Abe (Australian Computational Earth Systems Simulator)
D. Weatherley (QUAKES, Australia)

Performance Evaluation, Modeling and Analysis of Scientific Applications on Large-Scale Systems
A. Hoisie, (LANL, USA)
D.J. Kerbyson, (LANL, USA)
A. Snavely (SDSC, University of California, USA)
J. Vetter, (LLNL, USA)

Scientific Visualization and Human-Machine Interaction in a Problem Solving Environment
E. Zudilova (The University of Amsterdam, The Netherlands)
T. Adriaansen (Telecommunications & Industrial Physics, CSIRO)

Innovative Solutions for Grid Computing
J.J. Dongarra (The University of Tennessee, USA)
F. Desprez (LIP ENS, France)
T. Priol (INRIA/IRISA)

Terascale Performance Analysis
D.A. Reed (NCSA, USA)
R. Nandkumar (NCSA, USA)
R. Pennington (NCSA, USA)
J. Towns (NCSA, USA)
C.L. Mendes (University of Illinois, USA)

Computational Chemistry in the 21st Century: Applications and Methods
T.H. Dunning, Jr. (JICS, ORNL, USA)
R.J. Harrison (ORNL, USA)
L. Radom (Australian National University, Australia)
A. Rendell (Australian National University, Australia)

Tools for Program Development and Analysis in Computational Science
D. Kranzlmueller (Johannes Kepler University, Austria)
R. Wismüller (University of Vienna, Austria)
A. Bode (Technische Universität München, Germany)
J. Volkert (Johannes Kepler University, Austria)

Parallel Input/Output Management Techniques (PIOMT2003)
J.H. Abawajy (Carleton University, Canada)
Dynamic Data Driven Application Systems
F. Darema (NSF/CISE, USA)
Complex Problem-Solving Environments for Grid Computing (WCPSE02)
D. Walker (Cardiff University, UK)
Modeling and Simulation in Supercomputing and Telecommunications
Y. Mun (Soongsil University, Korea)
Modeling of Multimedia Sychronization in Mobile Information Systems
D.C. Lee (Howon University, Korea)
K.J. Kim (Kyonggi University, Korea)
OpenMP for Large Scale Applications
B. Chapman (University of Houston, USA)
M. Bull (EPCC, UK)
Modelling Morphogenesis and Pattern Formation in Biology
J.A. Kaandorp (The University of Amsterdam, The Netherlands)
Adaptive Algorithms for Parallel and Distributed Computing Environments
S. Moore (University of Tennessee, USA)
V. Eijkhout (University of Tennessee, USA)

Sponsoring Organizations

The University of Amsterdam, The Netherlands
Hewlett-Packard
Springer-Verlag, Germany
Netherlands Institute in St. Petersburg, (NIP)
Ministry of Industry, Science and Technologies of the Russian Federation
Committee of Science and High Education of the Government of St. Petersburg
St. Petersburg State Technical University
Institute for High Performance Computing and Information Systems, St. Petersburg
IBM Australia
Microsoft
Cray Inc.
Dolphin Interconnect
Microway
Etnus
ceanet
NAG
Pallas GmbH

Table of Contents, Part II

Russian Track

Workshop on Computational Finance and Economics

Workshop on Numerical Methods for Structured Systems

Workshop on High-Performance Environmental Computations

Workshop on Computational Chemistry and Molecular Dynamics

Workshop on Recursive and Adaptive Signal/Image Processing (RASIP)

Workshop on Numerical Methods for Singular Differential and Differential-Algebraic Equations

Poster Papers

Table of Contents, Part I

Russian Track

Keynote Paper

Track on Scientific Computing in Mathematics, Mechanis, Physics, and Chemistry

Track on New Large-Scale Simulation Application Areas

Track on Parallel and Distributed Computing

Track on Grid Computing and Hybrid Computational Methods

Track on New Algorithmic Approaches to Existing Application Areas

Track on Advanced Numerical Algorithms

Track on Problem Solving Environments (Including: Visualisation Technologies, Web Technologies, and Software Component Technologies

Track on Computer Algebra Systems and Their Applications

Workshop on Computer Graphics

Workshop on Computational Science of Lattice Boltzmann Modeling

Table of Contents, Part III

Australian Track

Track on Applications

Track on Clusters and Grids

Track on Models and Algorithms

Track on Web Engineering

Track on Networking

Track on Parallel Methods and Systems

Track on Data Mining

Workshop on Parallel Linear Algebra (WoPLA03)

Workshop on Java in Computational Science

Workshop on Computational Earthquake Physics and Solid Earth System Simulation

Workshop on Performance Evaluation, Modeling, and Analysis of Scientific Applications on Large-Scale Systems

Workshop on Scientific Visualization and Human-Machine Interaction in a Problem Solving Environment

Workshop on Innovative Solutions for Grid Computing

Table of Contents, Part IV

Australian Track

Workshop on Terascale Performance Analysis

Workshop on Computational Chemistry in the 21st Century: Applications and Methods

Workshop on Tools for Program Development and Analysis in Computational Science

Workshop on Parallel Input/Output Management Techniques (PIOMT2003)

Workshop on Dynamic Data Driven Application Systems

Workshop on Complex Problem-Solving Environments for Grid Computing (WCPSE02)

Workshop on Modeling and Simulation in Supercomputing and Telecommunications

Workshop on Modeling of Multimedia Synchronization in Mobile Information System

Workshop on OpenMP for Large Scale Applications

Workshop on Modeling Morphogenesis and Pattern Formation in Biology

Workshop on Adaptive Algorithms for Parallel and Distributed Computing Environments

Poster Papers

Workshop on Computational Finance and Economics

Parallel Computing Method of Valuing for Multi-asset European Option*

Weimin Zheng[1], Jiwu Shu[1], Xiatie Deng[2], and Yonggen Gu[3]

[1] Department of Computer Science and Technology
Tsinghua University., Beijing, 100084, China
zwm-dcs@tsinghua.edu.cn,shujw@tsinghua.edu.cn

[2] Department of Computer Science
City University of Hong Kong,Hong Kong SAR,China
CSDENG@cityu.edu.hk

[3] Institute of Systems Science Academia Sinica
Beijing, 100080, China

Abstract. A critical problem in Finance Engineering is to value the option and other derivatives securities correctly. The Monte Carlo method (MC) is an important one in the computation for the valuation of multi-asset European option. But its convergence rate is very slow. So various quasi Monte Carlo methods and there relative parallel computing method are becoming an important approach to the valuing of multi-asset European option. In this paper, we use a number-theoretic method, which is a H-W method, to generate identical distributed point set in order to compute the value of the multi-asset European option. It turns out to be very effective, and the time of computing is greatly shortened. Comparing with other methods, the method computes less points and it is especially suitable for high dimension problem.

1 Introduction

The benefit and the risk of derivatives tools are not only influenced by the self relationship between demands and services, but also rely on the balance of demand and serve of underlying asset. The key problem of the financial project is how to estimate the value of option and other portfolio properly. Black and Scholes concluded the precise formula to estimate the value of European call option and put option[1]. But now the problem of estimating the value of European option, which is relying on several underlying asset price, is not solved preferably[1]. At the present time, there are three methods to estimate the value of option[2]: formulas, deterministic numerical methods and Monte Carlo simulation. MC method is an important one for estimating the value of European option. Random numbers are used on lots of disparate routes for sampling, and

* Weimin Zheng,Prof., Present research project: computer architecture, parallel and distribute process and math finance and so on. This research is supported by a joint research grant (No:60131160743, N_CityU102/01) of NSFC/RGC.

P.M.A. Sloot et al. (Eds.): ICCS 2003, LNCS 2658, pp. 3–9, 2003.

the variables of underlying asset in the world of riskless follow these tracks. We can calculate the benefit of each route and such benefit discounts due to riskless rate. This lose and benefit are discount according to riskless rate. Then we use the arithmetical average value of all benefit after discount as the evaluation of option. Compared with other methods, MC method is effective if there are several variables, because MC method leads to an approximately linear increase in the computation time when the number of variables grows, while most of other methods lead to an exponential increase. In the MC method, samples must be done on the route of every variable in each simulation. For example, in a simulation, N samples are obtained from multi-dimension standard normal school. Then on one of simulative routes for the estimating the value of option relying on n variables, there are N samples needed. To estimate accurately, both the number of execution of simulation and the number of samples N are large. Because the convergence rate of MC is quite slow, when the number of variables n is also large, some methods, such as tens and hundreds, $\mathbf{O}(n^{-1/2})$ are needed to deal with N sample routes, and the compute load goes beyond the capacity of a single computer. Therefore Parallel computing method and quasi Monte Carlo method are used widely to solve the problems[2][3][4][5][6].

Many researches have been done aboard. Paskov brought forward the method of generating quasi MC sequence of Soblo and Halton to estimate the value of European option relying on several variables, and they compared the performance of these two methods[7]; Pagageorgiou and Traub selected a quasi MC method, which is faster than MC method and uses less sample points, to solve a problem of European option relying on 360 variables[8].In addition, Acworth compared several MC methods and quasi MC methods detailedly, and concluded that quasi MC method is better than normal MC method[4]. In [3], a quasi MC method called (t,m,s)-net method is selected to estimate the value of European option of underlying asset. In this paper, we introduce NTM and select H-W (HUA Luogeng - WANG Yuan) method to generate consistent distributed point set to estimate the value of European option of several underlying asset, and make out satisfying result in little time.

2 Model of Estimating The Value of European Option of Multiple Assets

We consider how to evaluate the value of European option with multiple statuses. The value of its derivatives securities relies on the N statuses variable, such as the price and elapsed time τ of the discount of ventual asset. Suppose S_i $(i = 1, 2, \cdots, n)$ as the value of underlying asset_I and $V(S_1, S_2, \cdots, S_n, \tau)$ as the value of derivatives securities. According to the Black-Scholes equation, we can acquire the patiel diferebtial coefficients equation of estimating the derivatives securities value of European option of several variables[1]:

$$\frac{\partial V}{\partial \tau}=\frac{1}{2}\sum_{i=1}^{n}\sum_{j=1}^{n}\rho_{ij}\sigma_i\sigma_j S_i S_j\frac{\partial^2 V}{\partial S_i \partial S_j}+r\sum_{i=1}^{n}\frac{\partial V}{\partial S_i}-rV, 0<S_1,\cdots,S_n<\infty, \tau\in(0,T) \tag{1}$$

and ρ_{ij}, $(i,j=1,2,\cdots,n)$ is the relative coefficient, a known constant. $\sigma_i>0$ is volatility of underlying asset. $r>0$ is riskless rate. Both σ_i and r are constants. Suppose $S=(S_1,S_2,\cdots,S_n)^T$ as the asset price vector at current time t. $V(S,\tau)$ is meant to represent $V(S_1,S_2,\cdots,S_n,\tau)$ and S_T is meant to represent the asset price vector at expire time T. The boundary condition is:

$$V(S_T,0)=\max(X-\max(S_1,\cdots,S_n),0)$$

Here X is the strike price. For European call option, the boundary condition $V(S_T,0)$ is

$$\begin{cases} C_{\max}(S,0)=\max(\max(S_1,\cdots,S_n)-X,0)\\ C_{\min}(S,0)=\max(\min(S_1,\cdots,S_n)-X,0)\end{cases} \tag{2}$$

And for European put option, the boundary condition $V(S_T,0)$ is

$$\begin{cases} P_{\max}(S,0)=\max(X-\max(S_1,\cdots,S_n),0)\\ P_{\min}(S,0)=\max(X-\min(S_1,\cdots,S_n),0)\end{cases} \tag{3}$$

To estimate the value of European option of several status variables, the equation(1) can be induced to multi-integral problem[5]:

$$V(S,\tau)=e^{-r\tau}\int_0^\infty\int_0^\infty\cdots\int_0^\infty V(S,0)\Psi(S_T;S,\tau)dS_T \tag{4}$$

Here:

$$\Psi(S_T;S,\tau)=\frac{1}{(2\pi\tau)^{\frac{n}{2}}\sqrt{det\ R}\tilde{\sigma}\tilde{S_T}}\exp(-\frac{1}{2}W_T^T R^{-1}W_T) \tag{5}$$

is the transform density function of several variables, where

$$W_T=\left(\frac{\ln S_{T_1}-\hat{S}_1}{\sigma_1\sqrt{\tau}},\cdots,\frac{\ln S_{T_n}-\hat{S}_n}{\sigma_n\sqrt{\tau}}\right) \tag{6}$$

$$\hat{S}_i=\ln S_i+\left(r-\frac{\sigma_i}{2}\right)\tau, i=1,2,\cdots,n,\ \tilde{\sigma}=\prod_{i=1}^{n}\sigma_i, \tilde{S}_i=\prod_{i=1}^{n}S_{T_i} \tag{7}$$

$R=(\rho_{ij})_{n\times n}, \rho_{ii}=1$ and when $i\neq j, \rho_{ij}\in(0,1)$

3 The Parallel Strategy and Algorithm

3.1 NTM Method

NTM is the derivation of numeric theory and proximate analysis. In fact it is also a kind of quasi MC method. The key problem of computing approximately the multi-integral on S- dimension unit cube C^s using NTM method is how to obtain the symmetrically distributed points set on C^s. Assume $P_n = \{c_k^{(n)}, k = 1, \cdots, n\}$ as a points set on C^s. If it is a NT-nets on C^s, in the other word, it has low difference[3], $I(f)$ can be approached by :

$$I(f, P_n) = \frac{1}{n} \sum_{k=1}^{n} f(c_k^{(n)}) \tag{8}$$

Therefore, how to conclude the best quadrature formula is equivalence to how to find the best consistent distributed point set. In the reference[9], Korobov put forward the method to find the best consistent distributed point set, and the error rate is $\mathbf{O}(n^{-1}(\log n)^s)$, Considered at the point of view of approximation, the result of Korobov method is a existence theorem, so it is difficult to solve real problems using this method. Therefore HUA Luogeng and WANG Yuan (called H-W method) brought up a method that obtains net point aggregation using partition round region[9], which is called H-W method, and the error rate is $\mathbf{O}(n^{-\frac{1}{2}-\frac{1}{2(s-1)}+\varepsilon})$, H-W method obtains symmetrically distributed points set by this way:

$$\gamma = \left\{\left(\left\{2\cos\frac{2\pi}{p}\right\}, \left\{2\cos\frac{4\pi}{p}\right\}, \cdots, \left\{2\cos\frac{2\pi n}{p}\right\}\right), k = 1, 2, \cdots\right\} \tag{9}$$

Here, p is a prime number and $p \geq 2n+3$, $\{x\}$ is meant to represent the fraction part of x. By the means of $\gamma_i(1 \leq i \leq s$ rational number approach defined at (9), H-W method brought forward a method obtaining net point aggregation, which is called partition round region method[3]. Here we use the algorithm of parallel computing, combining the method of numeric theory, to resolve the high dimension integral problem in estimate the value of European option.

3.2 Method Comparison

For the European option of several assets, the number of assets is normally to be tens, or even hundreds. Therefore the multi-integral to compute is tens-integral, or even hundreds-integral. At present the method to compute multi-integral is approximate enumeration, and the quality of solution relies on the large numbers of points set. As the scale of problem and the quality of solution increase, the computing time increases. Sometimes, because the increase of computing dimension or the scale of problem often overwhelms the time restriction easily, the method would loose its feasibility. For example, sometimes we change the multi-integral to overlapped integral of single integral on [0,1], then apply the formula on single integral in turns. But this traditional method is not feasible sometimes. For example, Simpson formula, the error rate is $\mathbf{O}(n^{-2/s})$, and the convergent rate is $\mathbf{O}(n^{-1/2})$, When s is some of large, the number of approximate computing point increases quickly. Another MC method of formatting quadrature formula is to transform the analytic problem to a probability problem with the same solution,

$$I(f,n) = \frac{1}{n}\sum_{i=1}^{n} f(x_i)$$

then study this probability problem by statistical simulation. The error rate of this method is $\mathbf{O}(n^{-1/2})$ and is better than $\mathbf{O}(n^{-1/s})$ The convergent rate is irrespective to the number of dimensions, but it is very slow, just $\mathbf{O}\sqrt{\ln(ln(n))/n}$. The efficiency of MC method is not nice, and only when n is very large, $I(f,n)$ can obtain the satisfied approached result. So in order to increase the convergent rate and reduce the scale of computing, a lot of quasi MC methods emerge as the times require[3][4]. Especially as the NTM develops, the method of computing multi-integral develops quickly. The points set of C^s obtained by NTM is more even, less point number and less computation than by MC method.

3.3 Parallel Computing of NTM

When we use parallel method to compute equation (4), we first make the problem discrete, then divide the compute task into several parts averagely and distribute them to corresponding processor to do. Equation (4) changes to:

$$V(S,\tau) = e^{-rt}\int_0^{\infty}\cdots\int_0^{\infty} Q(S_T;S,\tau)dS_T \tag{10}$$

where

$$Q(S_T;S,\tau) = V(S_T,0)\Psi(S_T;S,\tau) = V(S_T,0)\frac{\exp\left(-\frac{1}{2}W_T^T R^{-1} W_T\right)}{(2\pi\tau)^{\frac{n}{2}}\sqrt{det\ \tilde{R}\tilde{\sigma}\tilde{S}_T}} \tag{11}$$

$$V(S_T,0) = \max(X - \max(S_{T_1},\cdots,S_{T_n}),0) \tag{12}$$

Suppose $\{\theta_j\} = \{(\theta_1,\cdots,\theta_n)\}$, $\Delta\tau = T/M$, $\Delta S = a/N$, (T is the expire time, a is strike price) After making discrete,equation (8) changes to

$$V_{i,k} = V(i\Delta S,k\Delta t) = V(i_1\Delta S,\cdots,i_n\Delta S,k\Delta\tau) = \frac{\exp(rk\Delta\tau)}{N}\sum_{j=1}^{n} Q(a\theta_j;i\Delta S,k\Delta\tau) \tag{13}$$

Here N is the number of sample points in each status. The value of derivatives securities at some time for different asset can be obtained by equation (13). The cluster can deal with the problem of dividing the compute grid point easily and apply on parallel compute of equation (13).

4 Experiment Result and Conclusion

We use MPI environment in cluster system. When the number of dimension is certain, the grid point can be generated ahead, be stored in a file and be read out when needed. But when the number of dimension is some of large, the file generated is very large, so it must be divided into several parts in the parallel environment of cluster system. Therefore each processor generates NT-net grid points parallel and deals with the computation of data generated by itself. After each processor finished the computation of itself, we collect the result. In the process of parallel computation, there is nearly not data communication. We take the computation of estimating the value of 50 assets option for an example, the parameter is selected as [3]. The computation result is also similar with this paper. Table 1 lists the speedup using different number of processors on "Tsinghua TongFang Explorer 108 Cluster System".

Table 1. Speedup in different number of processors

Number of Processor	1	2	4	6	8	12
Speedup(S_p)	/	1.89	3.67	5.34	6.83	9.96

At present, when the number of assets, which is relied on by European option, is very large, such as tens or hundreds, if we need to get a precise result, the number of execution and the number of sample N are some of large. Common MC methods can not match the time restriction. In this paper, NTM is selected. H-W method generates consistent distributed points set to estimate the value of European option of several underlying assets, and obtains satisfied result, with advanced algorithm and short computing time. We conclude that the method is suited for high dimension computation.

References

1. Kwok Y.K., Mathematical Models of Financial Derivatives. Spring-Verlag Singapore Pte. Ltd(1998)
2. Stavros A. Z., High-Performance Computing in Finance: The Last 10 Years and The Next. Parallel Computing, 25(1999)2149-2175
3. Jenny X L, Gary L M. Parallel Computing of A Quasi-Monte Carlo Algorithm for Valuing Derivatives. Parallel Computing, 26(2000) 641-653
4. Acworth P., Broadie M., Glasserman P., A Comparison of Some Monte Carlo and Quasi Monte Carlo Techniques for Option Pricing, in: Niederreiter H., Hellekalek P., Larcher G., Zinterhof P.(Eds.) Monte Carlo and Quasi-Monte Carlo Methods 1996, Lecture Notes in Statistics, Springer,Berlin,127(1998)1-18
5. Perry S.C., Grimwood R.H., Kerbyson D. J, et al. Performance Optimization of Financial Option Calculations. Parallel Computing, 26(2000)623-639
6. Morokoff W., Caflish R.E., Quasi-Random Sequences and Their Discrepancies. SIAM J.Sci.Stat.Computing, 15(1994)1251-1279
7. Paskov S.H. New Methodologies for Valuing Derivatives, in: Mathematics of Derivatives Securities. Isaac Newton Inst., Cambridge Univ.Press, Cambridge(1996)

8. Papageorgiou H.A., Traub J.F.. New Results on Deterministic Pricing of Financial Derivatives, Technical Report CUCS-028-96, Columbia: Department of computer Science, Columbia University(1996)
9. Fang K.T., Wang Y.. Applications of Number-theoretic Method in Statistics, The Science Press, Beijing, P.R.China (1996)

A Fuzzy Approach to Portfolio Rebalancing with Transaction Costs*

Yong Fang[1], K.K. Lai[2]**, and Shou-Yang Wang[3]

[1] Institute of Systems Science, Academy of Mathematics and Systems Sciences, Chinese Academy of Sciences, Beijing 100080, China
yfang@amss.ac.cn

[2] Department of Management Sciences, City University of Hong Kong, Kowloon, Hong Kong
mskklai@cityu.edu.hk

[3] Institute of Systems Science, Academy of Mathematics and Systems Sciences, Chinese Academy of Sciences, Beijing 100080, China
swang@mail.iss.ac.cn

Abstract. The fuzzy set is a powerful tool used to describe an uncertain financial environment in which not only the financial markets but also the financial managers' decisions are subject to vagueness, ambiguity or some other kind of fuzziness. Based on fuzzy decision theory, two portfolio rebalancing models with transaction costs are proposed. An example is given to illustrate that the two linear programming models based on fuzzy decisions can be used efficiently to solve portfolio rebalancing problems by using real data from the Shanghai Stock Exchange.

1 Introduction

In 1952, Markowitz [8] published his pioneering work which laid the foundation of modern portfolio analysis. It combines probability theory and optimization theory to model the behavior of economic agents under uncertainty. Konno and Yamazika [5] used the absolute deviation risk function, to replace the risk function in Markowitz's model thus formulated a mean absolute deviation portfolio optimization model. It turns out that the mean absolute deviation model maintains the nice properties of Markowitz's model and removes most of the principal difficulties in solving Markowitz's model.

Transaction cost is one of the main sources of concern to portfolio managers. Arnott and Wagner [2] found that ignoring transaction costs would result in an inefficient portfolio. Yoshimoto's emperical analysis [12] also drew the same conclusion. Due to changes of situation in financial markets and investors' preferences towards risk, most of the applications of portfolio optimization involve a revision of an existing portfolio, *i.e.*, portfolio rebalancing.

Usually, expected return and risk are two fundamental factors which investors consider. Sometimes, investors may consider other factors besides the expected

* Supported by NSFC, CAS, City University of Hong Kong and MADIS.

** Corresponding author

P.M.A. Sloot et al. (Eds.): ICCS 2003, LNCS 2658, pp. 10–19, 2003.

return and risk, such as liquidity. Liquidity has been measured as the degree of probability involved in the conversion of an investment into cash without any significant loss in value. Arenas, Bilbao and Rodriguez [1] took into account three criteria: return, risk and liquidity and used a fuzzy goal programming approach to solve the portfolio selection problem.

In 1970, Bellman and Zadeh [3] proposed the fuzzy decision theory. Ramaswamy [10] presented a portfolio selection method using the fuzzy decision theory. A similar approach for portfolio selection using the fuzzy decision theory was proposed by León *et al.* [6]. Using the fuzzy decision principle, Östermark [9] proposed a dynamic portfolio management model by fuzzifying the objective and the constraints. Watada [11] presented another type of portfolio selection model using the fuzzy decision principle. The model is directly related to the mean-variance model, where the goal rate (or the satisfaction degree) for an expected return and the corresponding risk are described by logistic membership functions.

This paper is organized as follows. In Section 2, a bi-objective linear programming model for portfolio rebalancing with transaction costs is proposed. In Section 3, based on the fuzzy decision theory, two linear programming models for portfolio rebalancing with transaction costs are proposed. In Section 4, an example is given to illustrate that the two linear programming models based on fuzzy decisions can be used efficiently to solve portfolio rebalancing problems by using real data from the Shanghai Stock Exchange. A few concluding remarks are finally given in Section 5.

2 Linear Programming Model for Portfolio Rebalancing

Due to changes of situation in financial markets and investors' preferences towards risk, most of the applications of portfolio optimization involve a revision of an existing portfolio. The transaction costs associated with purchasing a new portfolio or rebalancing an existing portfolio have a significant effect on the investment strategy. Suppose an investor allocates his wealth among n securities offering random rates of returns. The investor starts with an existing portfolio and decides how to reconstruct a new portfolio.

The expected net return on the portfolio after paying transaction costs is given by

$$\sum_{j=1}^{n} r_j(x_j^0 + x_j^+ - x_j^-) - \sum_{j=1}^{n} p(x_j^+ + x_j^-) \tag{1}$$

where r_j is the expected return of security j, x_j^0 is the proportion of the security j owned by the investor before portfolio reblancing, x_j^+ is the proportion of the security j bought by the investor, x_j^- is the proportion of the security j sold by the investor during the portfolio rebalancing process and p is the rate of transaction costs.

Denote $x_j = x_j^0 + x_j^+ - x_j^-, j = 1, 2, \cdots, n$. The semi-absolute deviation of return on the portfolio $x = (x_1, x_2, \cdots, x_n)$ below the expected return over the past period t, $t = 1, 2, \cdots, T$ can be represented as

$$w_t(x) = |\min\{0, \sum_{j=1}^{n}(r_{jt} - r_j)x_j\}|. \tag{2}$$

where r_{jt} can be determined by historical or forecast data.

The expected semi-absolute deviation of the return on the portfolio $x = (x_1, x_2, \cdots, x_n)$ below the expected return can be represented as

$$w(x) = \frac{1}{T}\sum_{t=1}^{T} w_t(x) = \frac{1}{T}\sum_{t=1}^{T}|\min\{0, \sum_{j=1}^{n}(r_{jt} - r_j)x_j\}|. \tag{3}$$

Usually, the anticipation of certain levels of expected return and risk are two fundamental factors which investors consider. Sometimes, investors may wish to consider other factors besides expected return rate and risk, such as liquidity. Liquidity has been measured as the degree of probability of being able to convert an investment into cash without any significant loss in value. Generally, investors prefer greater liquidity, especially since in a bull market for securities, returns on securities with high liquidity tend to increase with time. The turnover rate of a security is the proportion of turnover volumes to tradable volumes of the security, and is a factor which may reflect the liquidity of the security. In this paper, we assume that the turnover rates of securities are modelled by possibility distributions rather than probability distributions.

Carlsson and Fullér [4] introduced the notation of crisp possibilistic mean (expected) value and crisp possibilistic variance of continuous possibility distributions, which are consistent with the extension principle. Denote the turnover rate of the security j by the trapezoidal fuzzy number $\hat{l}_j = (la_j, lb_j, \alpha_j, \beta_j)$. Then the turnover rate of the portfolio $x = (x_1, x_2, \cdots, x_n)$ is $\sum_{j=1}^{n} \hat{l}_j$. By the definition, the crisp possibilistic mean (expected) value of the turnover rate of the portfolio $x = (x_1, x_2, \cdots, x_n)$ can be represented as

$$E(\hat{l}(x)) = E(\sum_{j=1}^{n}\hat{l}_j x_j) = \sum_{j=1}^{n}(\frac{la_j + lb_j}{2} + \frac{\beta_j - \alpha_j}{6})x_j. \tag{4}$$

Assume that the investor does not invest the additional capital during the portfolio rebalancing process. We use $w(x)$ to measure the risk of the portfolio and use the crisp possibilistic mean (expected) value of the turnover rate to measure the liquidity of the portfolio. Assume the investor wants to maximize return on and minimize the risk to the portfolio after paying transaction costs. At the same time, he requires that the liquidity of the portfolio is not less than a given constant through rebalancing the existing portfolio. Based on the above discussions, the portfolio rebalancing problem is formulated as follows:

$$
\text{(P1)}\begin{cases}
\max \sum_{j=1}^{n} r_j(x_j^0 + x_j^+ - x_j^-) - \sum_{j=1}^{n} p(x_j^+ + x_j^-) \\
\min \sum_{t=1}^{T} \frac{|\sum_{j=1}^{n}(r_{jt}-r_j)x_j| + \sum_{j=1}^{n}(r_j - r_{jt})x_j}{2T} \\
s.t. \ \sum_{j=1}^{n} (\frac{la_j+lb_j}{2} + \frac{\beta_j-\alpha_j}{6})x_j \geq l, \\
\sum_{j=1}^{n} x_j = 1, \\
x_j = x_j^0 + x_j^+ - x_j^-, j = 1, 2, \cdots, n, \\
0 \leq x_j^+ \leq u_j, j = 1, 2, \cdots, n, \\
0 \leq x_j^- \leq x_j^0, j = 1, 2, \cdots, n.
\end{cases}
$$

where l is a given constant by the investor and u_j represents the maximum proportion of the total amount of money devoted to security $j, j \in S$.

Eliminating the absolute function of the second objective function, the above problem can be transformed into the following problem:

$$
\text{(P2)}\begin{cases}
\max \sum_{j=1}^{n} r_j(x_j^0 + x_j^+ - x_j^-) - \sum_{j=1}^{n} p(x_j^+ + x_j^-) \\
\min \frac{1}{T} \sum_{t=1}^{T} y_t \\
s.t. \ \sum_{j=1}^{n} (\frac{la_j+lb_j}{2} + \frac{\beta_j-\alpha_j}{6})x_j \geq l, \\
y_t + \sum_{j=1}^{n}(r_{jt} - r_j)x_j \geq 0, t = 1, 2, \cdots, T, \\
\sum_{j=1}^{n} x_j = 1, \\
x_j = x_j^0 + x_j^+ - x_j^-, j = 1, 2, \cdots, n, \\
0 \leq x_j^+ \leq u_j, j = 1, 2, \cdots, n, \\
0 \leq x_j^- \leq x_j^0, j = 1, 2, \cdots, n. \\
y_t \geq 0, t = 1, 2, \cdots, T.
\end{cases}
$$

where l is a given constant by the investor.

The above problem is a bi-objective linear programming problem. One can use several algorithms of multiple objective linear programming to solve it efficiently.

3 Portfolio Rebalancing Models Based on Fuzzy Decision

In the portfolio rebalancing model proposed in above section, the return, the risk and the liquidity of the portfolio are considered. However, investor's satisfactory degree is not considered. In financial management, the knowledge and experience of an expert are very important in decision-making. Through comparing the present problem with their past experience and evaluating the whole portfolio in terms of risk and liquidity in the decision-making process, the experts may estimate the objective values concerning the expected return, the risk and the liquidity. Based on experts' knowledge, the investor may decide his levels of aspiration for the expected return, the risk and the liquidity of the portfolio.

3.1 Portfolio Rebalancing Model with Linear Membership Function

During the portfolio rebalancing process, an investor considers three factors (the expected return, the risk and the liquidity of the portfolio). Each of the factors is transformed using a membership function so as to characterize the aspiration level. In this section, the three factors are considered as the fuzzy numbers with linear membership function.

a) Membership function for the expected return on the portfolio

$$\mu_r(x) = \begin{cases} 0 & \text{if } E(r(x)) < r_0 \\ \frac{E(r(x))-r_0}{r_1-r_0} & \text{if } r_0 \le E(r(x)) \le r_1 \\ 1 & \text{if } E(r(x)) > r_1 \end{cases}$$

where r_0 represents the necessity aspiration level for the expected return on the portfolio, r_1 represents the sufficient aspiration level for the expected return of the portfolio.

b) Membership function for the risk of the portfolio

$$\mu_w(x) = \begin{cases} 1 & \text{if } w(x) < w_0 \\ \frac{w_1-w(x)}{w_1-w_0} & \text{if } w_0 \le w(x) \le w_1 \\ 0 & \text{if } w(x) > w_1 \end{cases}$$

where w_0 represents the necessity aspiration level for the risk of the portfolio, w_1 represents the sufficient aspiration level for the risk of the portfolio.

c) Membership function for the liquidity of the portfolio

$$\mu_{\hat{l}}(x) = \begin{cases} 0 & \text{if } E(\hat{l}(x)) < l_0 \\ \frac{E(\hat{l}(x))-l_0}{l_1-l_0} & \text{if } l_0 \le E(\hat{l}(x)) \le l_1 \\ 1 & \text{if } E(\hat{l}(x)) > l_1 \end{cases}$$

where l_0 represents the necessity aspiration level for the liquidity of the portfolio, l_1 represents the sufficient aspiration level for the liquidity of the portfolio.

The values of r_0, r_1, w_0, w_1, l_0 and l_1 can be given by the investor based on the experts' knowledge or past experience. According to Bellman and Zadeh's maximization principle, we can define $\lambda = \min\{\mu_r(x), \mu_w(x), \mu_{\hat{l}}(x)\}$.

The fuzzy portfolio rebalancing problem can be formulated as follows:

$$(\text{P3}) \begin{cases} \max \lambda \\ s.t. \;\; \mu_r(x) \ge \lambda, \\ \quad\quad \mu_w(x) \ge \lambda, \\ \quad\quad \mu_{\hat{l}}(x) \ge \lambda, \\ \quad\quad \sum_{j=1}^{n} x_j = 1, \\ \quad\quad x_j = x_j^0 + x_j^+ - x_j^-, j = 1, 2, \cdots, n, \\ \quad\quad 0 \le x_j^+ \le u_j, j = 1, 2, \cdots, n, \\ \quad\quad 0 \le x_j^- \le x_j^0, j = 1, 2, \cdots, n, \\ \quad\quad 0 \le \lambda \le 1. \end{cases}$$

Furthermore, the fuzzy portfolio rebalancing problem can be rewritten as follows:

$$\text{(P4)} \begin{cases} \max \lambda \\ s.t. \ \sum_{j=1}^{n} r_j x_j - \sum_{j=1}^{n} p(x_j^+ + x_j^-) \geq \lambda(r_1 - r_0) + r_0, \\ \quad \frac{1}{T}\sum_{t=1}^{T} y_t \leq w_1 - \lambda(w_1 - w_0), \\ \quad \sum_{j=1}^{n} (\frac{la_j + lb_j}{2} + \frac{\beta_j - \alpha_j}{6}) x_j \geq \lambda(l_1 - l_0) + l_0, \\ \quad y_t + \sum_{j=1}^{n} (r_{jt} - r_j) x_j \geq 0, t = 1, 2, \cdots, T, \\ \quad \sum_{j=1}^{n} x_j = 1, \\ \quad x_j = x_j^0 + x_j^+ - x_j^-, j = 1, 2, \cdots, n, \\ \quad 0 \leq x_j^+ \leq u_j, j = 1, 2, \cdots, n, \\ \quad 0 \leq x_j^- \leq x_j^0, j = 1, 2, \cdots, n, \\ \quad y_t \geq 0, t = 1, 2, \cdots, T, \\ \quad 0 \leq \lambda \leq 1. \end{cases}$$

where r_0, r_1, l_0, l_1, w_0 and w_1 are constants given by the investor based on the experts' knowledge or past experience.

The above problem is a standard linear programming problem. One can use several algorithms of linear programming to solve it efficiently, for example, the simplex method.

3.2 Portfolio Rebalancing Model with Non-linear Membership Function

Watada [11] employed a logistic function for a non-linear membership function $f(x) = \frac{1}{1+exp(-\alpha)}$. We can find that a trapezoidal membership function is an approximation from a logistic function. Therefore, the logistic function is considered much more appropriate to denote a vague goal level, which an investor considers.

Membership functions $\mu_r(x), \mu_w(x)$ and $\mu_{\hat{l}}(x)$ for the expected return, the risk and the liquidity on the portfolio are represented respectively as follows:

$$\mu_r(x) = \frac{1}{1 + exp(-\alpha_r(E(r(x)) - r_M))}, \tag{5}$$

$$\mu_w(x) = \frac{1}{1 + exp(\alpha_w(w(x) - w_M))}, \tag{6}$$

$$\mu_{\hat{l}}(x) = \frac{1}{1 + exp(-\alpha_l(E(\hat{l}(x)) - l_M))} \tag{7}$$

where α_r, α_w and α_l can be given respectively by the investor based on his own degree of satisfaction for the expected return, the level of risk and the liquidity. r_M, w_M and l_M represent the middle aspiration levels for the expected return, the level of risk and the liquidity of the portfolio respectively. The value of r_M, w_M and l_M can be gotten approximately by the values of r_0, r_1, w_0, w_1, l_0 and l_1, i.e. $r_M = \frac{r_0+r_1}{2}$, $w_M = \frac{w_0+w_1}{2}$ and $l_M = \frac{l_0+l_1}{2}$.

Remark: α_r, α_w and α_l determine respectively the shapes of membership functions $\mu_r(x)$, $\mu_w(x)$ and $\mu_{\hat{l}}(x)$ respectively, where $\alpha_r > 0, \alpha_w > 0$ and $\alpha_l > 0$. The larger parameters α_r, α_w and α_l get, the less their vagueness becomes.

The fuzzy portfolio rebalancing problem can be formulated as follows:

$$(\mathrm{P5}) \begin{cases} \max \eta \\ s.t. \quad \mu_r(x) \geq \eta, \\ \qquad \mu_w(x) \geq \eta, \\ \qquad \mu_{\hat{l}}(x) \geq \eta, \\ \qquad \sum_{j=1}^{n} x_j = 1, \\ \qquad x_j = x_j^0 + x_j^+ - x_j^-, j = 1, 2, \cdots, n, \\ \qquad 0 \leq x_j^+ \leq u_j, j = 1, 2, \cdots, n, \\ \qquad 0 \leq x_j^- \leq x_j^0, j = 1, 2, \cdots, n, \\ \qquad 0 \leq \eta \leq 1. \end{cases}$$

Let $\theta = \log\frac{1}{1-\eta}$, then $\eta = \frac{1}{1+exp(-\theta)}$. The logistic function is monotonously increasing, so maximizing η makes θ maximize. Therefore, the above problem may be transformed to an equivalent problem as follows:

$$(\mathrm{P6}) \begin{cases} \max \theta \\ s.t. \quad \alpha_r(\sum_{j=1}^{n} r_j x_j - \sum_{j=1}^{n} p(x_j^+ + x_j^-)) - \theta \geq \alpha_r r_M, \\ \qquad \theta + \frac{\alpha_w}{T} \sum_{t=1}^{T} y_t \leq \alpha_w w_M, \\ \qquad \alpha_l \sum_{j=1}^{n} (\frac{la_j + lb_j}{2} + \frac{\beta_j - \alpha_j}{6}) x_j - \theta \geq \alpha_l l_M, \\ \qquad y_t + \sum_{j=1}^{n} (r_{jt} - r_j) x_j \geq 0, t = 1, 2, \cdots, T, \\ \qquad \sum_{j=1}^{n} x_j = 1, \\ \qquad x_j = x_j^0 + x_j^+ - x_j^-, j = 1, 2, \cdots, n, \\ \qquad 0 \leq x_j^+ \leq u_j, j = 1, 2, \cdots, n, \\ \qquad 0 \leq x_j^- \leq x_j^0, j = 1, 2, \cdots, n, \\ \qquad y_t \geq 0, t = 1, 2, \cdots, T, \\ \qquad \theta \geq 0. \end{cases}$$

where α_r, α_w and α_l are parameters which can be given by the investor based on his own degree of satisfaction regarding the three factors.

The above problem is also a standard linear programming problem. One can use several algorithms of linear programming to solve it efficiently, for example, the simplex method.

Remark: The non-linear membership functions of the three factors may change their shape according to the parameters α_r, α_w and α_l. Through selecting the values of these parameters, the aspiration levels of the three factors may be described accurately. On the other hand, deferent parameter values may reflect deferent investors' aspiration levels. Therefore, it is convenient for deferent investors to formulate investment strategies using the above portfolio rebalancing model with non-linear membership functions.

4 An Example

In this section, we give an example to illustrate the models for portfolio rebalancing based on fuzzy decision as proposed in this paper. We suppose that an investor wants to choose thirty different types of stocks from the Shanghai Stock Exchange for his investment.

The rate of transaction costs for stocks is 0.0055 in the two securities markets on the Chinese mainland. Assume that the investor has already owned an existing portfolio and he will not invest the additional capital during the portfolio rebalancing process. The proportions of the stocks are listed in Table 1.

Table 1. The proportions of stocks in the existing portfolio

Stock	1	2	3	4	5	6	7
Proportions	0.05	0.08	0.05	0.35	0.10	0.12	0.25

Suddenly, the financial market situation changes, and the investor needs to change his investment strategy. In the example, we assume that the upper bound of the proportions of Stock j owned by the investor is 1. Now we use the fuzzy portfolio rebalancing models in this paper to re-allocate his assets. At first, we collect historical data of the thirty kinds of stocks from January, 1999 to January, 2002. The data are downloaded from the website www.stockstar.com. Then we use one month as a period to get the historical rates of returns of thirty-six periods. Using historical data of the turnover rates of the securities, we can estimate the turnover rates of the securities as the trapezoidal fuzzy numbers.

In the following, we will give two kinds computational results according to whether the investor has a conservative or an aggressive approach.

At first, we assume that the investor has a conservative and pessimistic mind. Then the values of r_0, r_1, l_0, l_1, w_0, and w_1 which are given by the investor may be small. They are as follows: $r_0 = 0.028$, $r_1 = 0.030$, $l_0 = 0.020$, $l_1 = 0.025$, $w_0 = 0.025$ and $w_1 = 0.035$.

Considering the three factors (the return, the risk and liquidity) as fuzzy numbers with trapezoidal membership function, we get a portfolio rebalancing strategy by solving (P4). The membership grade λ, the obtained risk, the obtained return and obtained liquidity are listed in Table 2.

Table 2. Membership grade λ, obtained risk, obtained return and obtained liquidity when $r_0 = 0.028$, $r_1 = 0.030$, $l_0 = 0.020$, $l_1 = 0.025$, $w_0 = 0.025$ and $w_1 = 0.035$.

λ	obtained risk	obtained return	obtained liquidity
0.835	0.0266	0.0297	0.0301

Considering the three factors (the return, the risk and liquidity) as fuzzy numbers with non-linear membership function, we get a portfolio rebalancing strategy by solving (P6).

In the example, we give three deferent values of parameters α_r, α_w and α_l. The membership grade η, the obtained risk, the obtained return and obtained liquidity are listed in Table 3.

Table 3. Membership grade η, obtained risk, obtained return and obtained liquidity when $r_M = 0.029$, $w_M = 0.030$ and $l_M = 0.0225$.

η	θ	α_r	α_w	α_l	obtained risk	obtained return	obtained liquidity
0.811	1.454	600	800	600	0.0282	0.0314	0.0304
0.806	1.425	500	1000	500	0.0286	0.0319	0.0303
0.785	1.295	400	1200	400	0.0289	0.0322	0.0302

Secondly, we assume that the investor has an aggressive and optimistic mind. Then the values of r_0, r_1, l_0, l_1, w_0, and w_1 which are given by the investor are big. They are as follows: $r_0 = 0.028$, $r_1 = 0.036$, $l_0 = 0.021$, $l_1 = 0.031$, $w_0 = 0.032$ and $w_1 = 0.036$.

Considering the three factors (the return, the risk and liquidity) as fuzzy numbers with trapezoidal membership function, we get a portfolio rebalancing strategy by solving (P4). The membership grade λ, the obtained risk, the obtained return and obtained liquidity are listed in Table 4.

Table 4. Membership grade λ, obtained risk, obtained return and obtained liquidity when $r_0 = 0.028$, $r_1 = 0.036$, $l_0 = 0.021$, $l_1 = 0.031$, $w_0 = 0.032$ and $w_1 = 0.036$.

λ	obtained risk	obtained return	obtained liquidity
0.890	0.0324	0.0351	0.0298

Considering the three factors (the return, the risk and liquidity) as fuzzy numbers with non-linear membership function, we get a portfolio rebalancing strategy by solving (P6).

In the example, we give three deferent values of parameters α_r, α_w and α_l. The membership grade η, the obtained risk, the obtained return and obtained liquidity are listed in Table 5.

Table 5. Membership grade η, obtained risk, obtained return and obtained liquidity when $r_M = 0.032$, $w_M = 0.034$ and $l_M = 0.026$.

η	θ	α_r	α_w	α_l	obtained risk	obtained return	obtained liquidity
0.849	1.726	600	800	600	0.0318	0.0349	0.0295
0.836	1.630	500	1000	500	0.0324	0.0353	0.0293
0.802	1.396	400	1200	400	0.0328	0.0355	0.0295

From the above results, we can find that we get the different portfolio rebalancing strategies by solving (P6) in which the different values of the parameters (α_r, α_w and α_l) are given. Through choosing the values of the parameters α_r, α_w and α_l according to the investor's frame of mind, the investor may get a favorite portfolio rebalancing strategy. The portfolio rebanlancing model with the non-linear membership function is much more convenient than the one with the linear membership function.

5 Conclusion

Considering the expected return, the risk and liquidity, a linear programming model for portfolio rebalancing with transaction costs is proposed. Based on fuzzy decision theory, two fuzzy portfolio rebalancing models with transaction costs are proposed. An example is given to illustrate that the two linear programming models based on fuzzy decision-making can be used efficiently to solve portfolio rebalancing problems by using real data from the Shanghai Stock Exchange. The computation results show that the portfolio rebanlancing model with the non-linear membership function is much more convenient than the one with the linear membership function. The portfolio rebalaning model with non-linear membership function can generate a favorite portfolio rebalancing strategy according to the investor's satisfactory degree.

References

1. Arenas, M., Bilbao, A., Rodriguez, M.V.: A Fuzzy Goal Programming Approach to Portfolio Selection. European Journal of Operational Research 133 (2001) 287–297.
2. Arnott, R.D., Wanger, W.H.: The Measurement and Control of Trading Costs. Financial Analysts Journal 46(6) (1990) 73–80.
3. Bellman, R., Zadeh, L.A.: Decision Making in a Fuzzy Environment. Management Science 17 (1970) 141–164.
4. Carlsson, C., Fullér, R.: On Possibilistic Mean Value and Variance of Fuzzy Numbers. Fuzzy Sets and Systems 122 (2001) 315–326.
5. Konno, H., Yamazaki, H.: Mean Absolute Portfolio Optimization Model and Its Application to Tokyo Stock Market. Management Science 37(5) (1991) 519–531.
6. León, T., Liern, V., Vercher, E.: Viability of Infeasible Portfolio Selection Problems: a Fuzzy Approach. European Journal of Operational Research 139 (2002) 178–189.
7. Mansini, R., Speranza, M.G.: Heuristic Algorithms for the Portfolio Selection Problem with Minimum Transaction Lots. European Journal of Operational Research 114 (1999) 219–233.
8. Markowitz, H.M.: Portfolio Selection. Journal of Finance 7 (1952) 77–91.
9. Östermark, R.: A Fuzzy Control Model (FCM) for Dynamic Portfolio Management. Fuzzy Sets and Systems 78 (1996) 243–254.
10. Ramaswamy, S.: Portfolio Selection Using Fuzzy Decision Theory, Working Paper of Bank for International Settlements, No.59, 1998.
11. Watada, J.: Fuzzy Portfolio Model for Decision Making in Investment. In: Yoshida, Y. (eds.): Dynamical Asspects in Fuzzy Decision Making. Physica-Verlag, Heidelberg (2001) 141–162.
12. Yoshimoto, A.: The Mean-Variance Approach to Portfolio Optimization Subject to Transaction Costs. Journal of the Operational Research Society of Japan 39 (1996) 99–117.

Mining Investment Venture Rules from Insurance Data Based on Decision Tree

Jinlan Tian, Suqin Zhang, Lin Zhu, and Ben Li

Department of Computer Science and Technology
Tsinghua University., Beijing, 100084, PR China

Abstract. Classification is a basic method of Data Mining. In this paper, we first introduce the basic concept of classifier and how to evaluate the precision of the classifier in this paper. Then we expatiate that how to use the Decision Tree Classifier to search the factors which will bring more venture at the guarantee slip, on the basis of the guarantee slip and compensation information database established by insurance agents. As a result, we gain some useful rules which will be useful to control investment venture.

1 Introduction

Data Mining, which is also called Knowledge Discovery in Databases(KDD), is an advanced process of finding and extracting reliable, novel, effective and comprehensible patterns hidden in a large amount of data. Data Mining technologies have brought significant effects to industries and other domains in the recent years. It is only four or five years from theoretic research to developing Data Mining products abroad. Data Mining technology is more and more often utilized in large companies, business, bank, insurance and telecommunication departments. It just puts up a great power of developing potential.

Insurance is a kind of operation with great venture. Venture evaluation has a significant effect to insurance company. Whether an insurance company could be successful depends on choosing a balance between competitive insurance premium and the venture of insurance. Insurance premium is always confirmed by analyzing and estimating some important factors such as individual health of policy-holders at health-insurance, car style at automobile-insurance, and so on. The situation of insurance market is always changing, so insurance companies should establish insurance premium on the basis of analyzing data of former years. At the present time, professionals of insurance companies adopt only curt analytical methods, analysts make decisions by their experience with a large number of data statistics. These curt methods are very difficult to use and affected by subjective factors.

Data Mining provides a circumstance to analyze insurance investment database. There are many methods of Data Mining which can be applied to venture analysis. We will emphasize on Decision Tree Classifier method in this paper, gain some helpful rule of controlling insurance venture by finding more venturesome area from guarantee slip and compensation information database.

P.M.A. Sloot et al. (Eds.): ICCS 2003, LNCS 2658, pp. 20–27, 2003.

2 The Basic Concept of Classifier

Classification is a very important method of Data Mining. Classification is the task of assigning a discrete label value to an unlabeled record. In doing so, records are divided into predefined groups. A classifier is a model that predicts one attribute of a set of data when given other attributes. A training set is needed to construct a classifier. The training set consists of records in the data for which the label has been supplied. An attribute is an inherent characteristic in the dataset. The attribute being predicted is called the label, and the attributes used for prediction are called the descriptive at-tributes. A concrete form of stylebook can be represented as $(v_1, v_2, \cdots, v_n; c)$. The v_i expresses as the value of each field, and the c expresses as a class.

The training set is the base of constructing a classifier. An attribute at the training set is defined as the classification label. The type of label attribute must be discrete, and if the number of the label attribute value is fewer(2 or 3 values is the best), the error-rate is much lower. An algorithm that automatically builds a classifier from a train-ing set is called an inducer. After generating an inducer, unlabeled records in the data-set could be built into such specific classes. Classifier also can predict the value of label attribute. There are several basic classifiers as rendered below.

1) *Decision Tree Classifiers.* A Decision Tree Classifier classifies data from attribute set by predicting the label for each record to make a series of decision. For example, a Decision Tree generated from a training set may predict a man with a family, a car which costs from \$15000 to \$23000 and two children, will have a good credit. Such Decision Tree classifier could be used to judge the credit degree of a person. MineSet, as a Data Mining tool provided by SGI, generates a Tree Visualization to display the structure of the Decision Tree. Each decision is represent as a node at the tree.

2) *Option Tree Classifiers.* Like Decision Tree classifiers, Option Tree classifiers also assign each record to a class. Instead of picking an attribute to split on for the root node at Decision Tree, Option Tree contain special Option Node, the Option Node may split into several branches. For example, an Option Node in a car-producing-area Option Tree may chooses kilometers per gallon, horsepower, number of cylinder, or weight of a car as the attributes. However, one node just can choose only one at-tribute at most at one time in Decision Tree. We could consider more situations synthetically when using Option Tree. Option Tree is generally more accurate than Decision Tree, but larger.

3) *Evidence Classifiers.* An Evidence Classifier classifies data through checking probability of some specific results of an attribute. For instance, it may estimate a man with a car which costs \$15000 to \$23000 has a probability of 70% to have a good credit, but the remain 30% person may have unreliable credit. Evidence Classifier predicts the classification result with the maximum probability on the basis of a simple probability model. MineSet Evidence Visualizer displays the result of evidence classification. It gives answers to users' questions such as "if $\cdots$ how about $\cdots$".

3 How to Evaluate the Precision of Classifiers

When a classifier is built, it is useful to know how well you can expect it to perform in the future (what is the classifier's error-rate). Factors affecting classification error-rate include:

1) *The number of records in the training set.* Since the inducer must learn from the training set, the larger the training set, the more reliable the classifier should be; how-ever, the larger the training set, the longer it takes the inducer to build a classifier. The improvement to the error-rate decreases as the size of the training set increases.

2) *The number of attributes.* More attributes mean more combinations for the inducer to compute, making the problem more difficult for the inducer and requiring longer time. Note that sometimes random correlations can lead the inducer astray; consequently, it might build less accurate classifiers (technically, this is known as "over fitting").

3) *The information in the attributes.* Sometimes there is not enough information in the attributes to correctly predict the label with a low error-rate (for example, trying to determine someone's salary based on their eye color). Adding other attributes (such as profession, hours per week, and age) might reduce the error-rate.

4) *The distribution of future unlabeled records.* If future records come from a distribution different from that of the training set, the error-rate probably will be high. For example, if you build a classifier from a training set containing family cars, it might not be useful when attempting to classify records containing many sport cars, because the distribution of attribute values might be very different.

There are two common methods of estimating the error-rate of a classifier as de-scribed below. Both of these assume that future records will be sampled from the same distribution as the training set.

1) *Holdout.* A portion of the records (commonly two-thirds) is used as the training set, while the rest is kept as a test set. The inducer is shown only two-thirds of the data and builds a classifier. The test set is then classified using the induced classifier, and the error-rate or loss on this test set is the estimated error-rate or estimated loss. This method is fast, but since it uses only two-thirds of the data for building the classifier, it does not make efficient use of the data for learning. If all the data were used, it is possible that a more accurate classifier could be built.

2) *Cross-Validation.* The dataset is splitted into k mutually exclusive subsets of approximately equal size. The inducer is trained and tested k times; each time, it is trained on all the data minus a different fold, then tested on that holdout fold. The estimated error-rate is then the average of the errors obtained. Cross-Validation can be repeated multiple times (t). For a t times k-fold cross-validation, $k \times t$ classifiers are built and evaluated. This means the time for cross-validation is $k \times t$ times longer. Increasing the number of repetitions (t) increases the running time and improves the error estimate and the corresponding confidence interval.

Generally, a holdout estimate should be used at the exploratory stage, as well as on dataset over 5,000 records. Cross-validation should be used for the final classifier building phase, as well as on small datasets.

4 Application of Decision Tree Classifier at Insurance Operations

Decision Tree method comes from Concept Learning System (CLS), and then ID3 method emerged as a peak of Decision Tree algorithm. The method has evolved to C4.5 at last which can deal with continuous attributes. Other famous Decision Tree methods include CART and Assistant.

The input of Decision Tree construction is a set of data with class-label, and the result of the construction is a binary tree or a multiple tree. The inner nodes (non-leaf nodes) of the binary tree generally represent as a logical judgment, such as $a_i = v_j$. (a_i is an attribute of some class, v_j is the possible value of the attribute.) The branches of the node are the result of the logical judgments. Each inner node of the multiple tree represents as the attribute of some class, the branches of the node represent all values of the attribute. The number of the branches equal to the number of possible values of the attribute. The labels on Leaf nodes are class for some instance.

Decision Tree starts from the root of the tree, and taking appropriate branches ac-cording to the attribute or question asked about at each branch node. One eventually comes to a leaf node. For example, multiple tree, if all of the data in training set belong to the same class, they will form a leaf node, and the content of the leaf node is the label of that class. Otherwise, the method will choose an attribute with some strategy, dividing the dataset into several subsets according to possible values of the attribute, making the data of each subset have the same attribute value, and then handling each subset in the same way recursively. The binary tree also follows this method except for choosing a reasonable logical judgment.

We will introduce how to utilize MineSet classifiers, the product of SGI, to mine hospitalization insurance data of some city. The hospitalization insurance database consists of individual information table, company information table, periodical (in a month) compensation table and so on. The concrete contents of each table are rendered below:

Table 1. Individual information table

Individual Insurance No.	Name	Sex	Date of Birth	Company No.	Total Salary per Year	Insured Date
3504274308250011	X	male	19430825	0000000663	19411	19970701
3502116405101511	Y	female	19640510	0000000663	15529	19970701
3502115409043551	Z	male	19540904	0000000664	7051	19970901
...	...	...	...	...	...	...

Table 2. Company information table

Company NO.	Company Name	Area Code	Type of Company	Insured Date
0000000330	computer corporation	05	03(enterprise)	19971101
0000000331	tade informatino center	03	03(enterprise)	19970901
0000000352	maternity hospital	01	02(public institution)	19970701
...	...	...	...	...

Table 3. Compensation table in one month

Compensation Bill No.	Compensatory Clerk NO.	Individual Insurance NO.	Compensatory Money	Compensatory Date
424300	01	3526017202011021	17.78	19980101
424190	06	3502056009140011	78.2	19980101
424191	19	3502047201172011	274.5	19980101
...	...	...	...	...

The procedures of Data Mining are discussed below:

1)*Preparing the Data.* We should prepare the data before data mining. For example, we should remove redundant information in the dataset, such as individual name, company name, insured date and so on. We also should make a statistic of compensation times of hospitalization insurance in a period of time. There is an individual compensation information table rendered below after preparing the data.

Table 4. Individual compensation information table

Individual Insurance NO.	Age	Total Salary per Year	Type of Company	Area No.	Compensation Times	If Compensating
3502043808264031	60	7051	03(enterprise)	03	0	0(no)
3502114704291511	51	14287	02(public institution)	01	8	1(yes)
3502042604134011	72	6376	09	01	21	1(yes)
...	...	...	...	...	...	...

2) *Analyzing the Data.* MineSet can build a classifier to predict one particular attribute when given some attributes in a set of data. The attribute being predicted is called the label, and the attributes used for prediction are called the descriptive attributes. MineSet can build a classifier automatically from a training set. The training set is consists of records whose labels are already given on the basis of existent attributes. After the generation, the classifier could be used to classify the records which have no label attribute in the data set. The value of the label can be predicted by the classifier.

Whether policy-holders claim for compensation is the most concerned information when analyzing insurance operation. Towards the dataset mentioned above, we define the attribute "if compensating" as the label attribute. Other information such as "individual insurance NO." belongs to irrelevant information. The attribute "if compensating" is derived from the attribute "compensation times", so "compensation times" can be removed because of the repetition. The remains of the attributes include "age", "total salary per year", "type of company" and "area code". The training set consists of all of the compensation information of that month.

3) *Data Mining.* We firstly apply "column weightiness" method of MineSet to find the columns which are more effective to label attribute than other columns, so we will avoid subjectiveness based on our experience in this way. The results of "column weightiness" method are three attributes, "age", "total salary per year" and "type of company", which are most effective to label attribute.

Select the "Decision Tree" mining tool, select the mode as "Classifier and Error", and set some options of that mode, then push "go!" button to run the inducer. At last we get a Decision Tree on the insurance dataset. Fig. 1 illustrates the Decision Tree.

4) *Analyzing and Comprehend the Data.* MineSet provides us a binary tree, and it can make a decision at each node according to descriptive attributes. Pointing to a node causes the specific information of the node to be displayed. All possible out-comes are marked on the horizontal lines emanating from each decision node. Each line indicates the value against which the attribute of that mode was tested. Analyzing the specific information of the root node, we can see that there are 6401 records in the training set. The number of customers who had not claimed for compensation is 5377, at the rate of 84.00%. The number of customers who had claimed for compensation is 1024, at the rate of 16.00%.

Note that in this tree the root split on the age of the policy-holders, the age is the most important factor, this result matches our daily experience that older person may not be in a good health condition. However, it is hard to distinguish accurately how old a person can be regarded as an "aged person". MineSet mining tools could give an accurate quantitative conclusion. In our example, we can see that the root node split into two branches by the age of 56. The left branch ($age < 56$) contains 4140 records, and the number of customers in the left branch who had not claimed for compensation is 3742, at the rate of 90.39%. The number of customers who had claimed for compensation is 398, at the rate of 9.61%. The right branch ($age > 56$) contains 2261 records. The number of customers at the right branch who had claimed for compensation is 626, at the rate of 27.69%. The compensation rate increases notable at the right branch. Applying the mining tools to hospitalization insurance dataset, we just gain a rule of the venture of insurance investment that "There is a higher compensation probability when a policy-holder is older than 56." If we apply database query method to such dataset, some condition must be given beforehand, and it will be very difficult and over work loaded by analyzing data statistics artificially.

We can get some other rules about compensation at the right branch of the root node. For example, next factor is "total salary per year". Considering that policy-holders with high salary may pay more money on taking exercises and health care, on the other hand, policy-holders with low salary may pay less. So

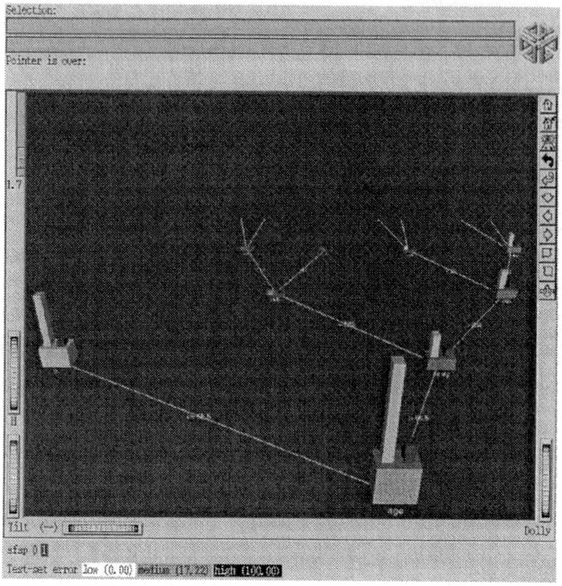

Fig. 1. The Decision Tree on the insurance dataset

it is credible that salary has an obvious influence of compensation situation. The factor "type of company" is another factor on the right branch. We can see from the tree that the compensation probability of the policy-holders who work at enterprise is much lower than that of the policy-holders who work at public institution. Combined with the concrete circumstance of hospitalization insurance domestically, we can explain such result in this way: The payment of fee-for-service is related to the style of company. The policy-holders who work at enterprise will pay more of the total fee, and insurance company will pay lesser. But the policy-holders who work at public institution will pay much less fee of the total and insurance company will pay most of it. Under this circum-stances, the policy-holders who work at enterprise will not go to see the doctor if he or she has a light sickness.

We can predict the compensation probability in the future according to the Decision Tree and detailed information of policy-holders, and then adjust the fee criterion of some kinds of policy-holders on the basis of compensation probability which has been predicted. Just for example, considering a policy-holder at the age of 58, working in enterprise and the total salary of 12000 per year, we follow the binary tree from root to leaf and predict that the compensation probability of that person is 9.84%, lower than the average probability. So the insurance company may decrease the insurance premium of such policy-holders. However, considering a policy-holder at the age of 59, working in public institution and the total salary of 9500 per year, the Decision Tree predicts that the compensation probability of that person is 37.56%, much higher than the average probability.

So the insurance company may increase the insurance premium of such policyholders.

If users want to gain some more detailed rules such as classifying policyholders under 56 years old, MineSet will provide data filtration function. Using such function, you can get the requisite training set by setting "$age < 56$" as the filtrating condition, then apply the Decision Tree method on this training set to get the requisite Decision Tree.

The Option Tree Visualizer's functionality is the same as for Decision Tree except that the Option Tree extends a regular Decision Tree classifier by allowing Option Nodes. An Option Node shows several options that can be chosen at a decision node in the tree. For example, we can choose one of the four branches from the root node. They are "age", "total salary per year", "type of company" and "area node". Instead of using a single attribute at a node in Decision Tree, the option node provides you with several options. However, the time necessary to build an Option Tree under the default setting is much longer than that needed to build a Decision Tree. The Option Tree has two notable advantages:

1) *Higher Comprehensibility.* The option nodes enhance comprehensibility of the factors affecting the class label by showing several choices that can be made. When flying over the tree, you can choose an option that you believe is easier to understand, or better for predictions.

2) *Higher Precision.* The option nodes reduce the risk of making a mistake by averaging the votes made by the options below. Every option leads to a sub tree that can be thought of as an "expert". The option node averages these experts' votes. Such averaging can lead to a better classifier with a lower error rate.

5 Conclusions

In conclusion, the classification method of Data Mining builds Decision Tree or Option Tree based on training sets accumulated in database, and then predicts new data according to the classifier. Classification methods can be applied not only at insurance field, but also at other investment field such as banking and stockjobbing or other trades. It will bring helpful policy supports to managers. Data Mining, as a new technical field, will be applied far and wide in China.

References

1. Heikki Mannnila, Hannu Toivonen and A. Inkeri. Verkamo, "Efficient algorithms for discov-ering association rules," AAAI Workshop on Knowledge Discovery in Databases, pages 181–192, July 1994
2. K.Decker and S.Focardi, "Technology Overview: A Report on Data Mining," ftp://ftp.cscs.ch/pub/CSCS/techreports
3. Tony Xiaohua Hu, "Knowledge Discovery in Databases: An Attribute-Oriented Rough Set Approach," http://www.cs.bham.ac.uk/ anpdm_docs
4. SGI Company, MineSet2.0 Tutorial
5. Gao Wen, "KDD: Knowledge Discovery in Databases," Computer World, vol. 37, 1998

Market-Based Interest Rates: Deterministic Volatility Case *

Guibin Lu[1] and Qiying Hu[2]

[1]School of Economics & Management,Xidian University, Xi'an,710071,China
guibinlu@163.com

[2]College of International Business & Management, Shanghai University, Shanghai 201800, China.
qyhu@mail.shu.edu.cn

Abstract. Central banks issue often many kinds of bonds to guide their benchmark interest rates. Their market data are thought of to reflect current state of the countries financial system. Then at least how much data is needed? Based on the framework of HJM model, We prove that the amount of the data needed is related to the form of the volatility function of forward rates, and then the initial forward rate curve is not essential.

1 Introduction

In some emergent financial markets, benchmark interest rates are policy-based and determined by their governments. While their central banks issue often many kinds of interest rate instruments to make their benchmark interest rates market-based.

Spot rate models are popular in theory and in practice. However, spot rates are not observable on the market. So we have to recalibrate the term structure of interest rate, given initial market data. Then forward rate models, such as Heath-Jarrow-Morton model(HJMM)[1], arise, which eliminate the disadvantages of spot rate models and fit initial forward rate curve naturally. However, there is one natural problem, that is, finite amount of data, but one initial curve, is presented in the market. Does the initial curve is essential for curve fitting problem, or finite amount of data is sufficient? This problem can be also stated as follows: can some group of market data be treated as state variable as forward rate curve does?

From a geometric view of the interest rate theory, [2] proved that in deterministic volatility models of HJM framework, forward rate curves can be represented as a function of a state variable of finite dimension. [3] presented detailed steps to construct this function. [4] and [5] discussed some cases of stochastic volatility and proved similar results as deterministic cases. This paper will prove that forward rate curve is not essential to describe the system state, while several distinct groups of market data can represent the same system state.

2 Musiela Parameterized HJM Model

Zero-coupon bonds are the representative interest rate instruments in the bond market, which is sold in the price less than their face values and bought back in the price of face values, and no interests is paid in the duration of their existence. Without loss of generality, their face values are assumed to be one unit.

*This project was supported by National Natural Science Foundation of China.

P.M.A. Sloot et al. (Eds.): ICCS 2003, LNCS 2658, pp. 28–33, 2003.

First of all, some important definitions are given as follows.

$P(t,x)$: the price at time t of a zero-coupon bond maturing at time $t+x$, $t,x \geq 0$;

$r(t,x)$: the forward rate at time t of a riskless loan which is contracted at time $t+x$ and matured instantaneously, $t,x \geq 0$;

$R(t)$: spot rate at time t, $t \geq 0$.

We assume that the market is frictionless and the bonds are perfectly divisible. There are some explicit relationships among the bonds prices, forward rates and spot rates:

$$r(t,x) = -\frac{\partial \log P(t,x)}{\partial x} \tag{1}$$

$$P(t,x) = \exp\{-\int_0^x r(t,s)ds\} \tag{2}$$

$$R(t) = r(t,0) \tag{3}$$

We assume that the bond market is a filtered probability space $(\Omega, \mathcal{F}, Q, \{\mathcal{F}_t\}_{t\geq 0})$, where Ω is the sample space, Q is the martingale measure defined on Ω, $\mathcal{F}$ is the filtration, and $\{\mathcal{F}_t\}$ is the time t filtration. Let $W(t)$ be a m-dimensional standard Wiener process. Assume that $R(t)$ is $\{\mathcal{F}_t\}$-measurable, that is , it is known at time t and $R(t)$ is a one-to-one mapping between $\{\mathcal{F}_t\}$ and its value set. In probability theory, $\{\mathcal{F}_t\}$ is the information set of the probability system at time t and its element means system event. Above all, the value of stochastic variable $R(t)$ reflects the system state. $R(t)$ is called the state variable of the system, and the evolution of the process $\{R(t), \forall t \geq 0\}$ reflects the transformation of the system state in the state space. The forward rates $r_t(\cdot)$ and bond prices $P(t,\cdot)$ are also state variables ([6]). Then we can use several distinct state variables to reflect the same system state and there must be some homeomorphic mapping between corresponding state spaces.

In the HJM model, the evolution of the forward rates is a stochastic differential equation:

$$\begin{aligned} dr(t,x) &= \mu(t,x)dt + \sigma(t,x)dW_t, \\ r(0,x) &= r_0^*(x), \forall x \geq 0. \end{aligned} \tag{4}$$

where, $\mu(t,x)$ and $\sigma(t,x)$ are 1-dimensional drift coefficient and d-dimensional volatility coefficient, respectively, and $r_0^*(\cdot)$ is the initial forward curve. Under the arbitrage-free condition, HJM proves that the drift coefficient must satisfy

$$\mu(t,x) = \frac{\partial}{\partial x} r(t,x) + \sigma(t,x)\int_0^x \sigma(t,s)ds. \tag{5}$$

So, given the initial forward rate curve $r_0^*(\cdot)$ and particular forms of volatility coefficient $\sigma(t,x)$, we can get the forward rates in the future with help of dynamics of the forward rate.

3 Finite-Dimensional Realization (FDR)

Given real numbers $\beta > 1$ and $\gamma > 0$, let $\mathcal{H}_{\beta,\gamma}$ be the Hilbert space of all differential functions $r : \mathbf{R}_+ \rightarrow \mathbf{R}$ satisfying the norm condition

$$\| r \|_\gamma < \infty.$$

Here the norm is defined by

$$\| r \|_{\beta,\gamma}^2 = \sum_{n=0}^{\infty} \beta^{-n} \int_0^{\infty} (\frac{d^n r(x)}{dx^n})^2 \exp\{-\gamma x\} dx.$$

Because the following results are independent of the values of β, γ, we write $\mathcal{H}$ for $\mathcal{H}_\gamma$. Surely the term forward rate curve simply refers to a point in $\mathcal{H}$.

We assume that the volatility function σ has the form

$$\sigma : \mathcal{H} \times \mathbf{R}_+ \to \mathbf{R}^m.$$

Here, each component of $\sigma(r, x) = (\sigma_1(r, x), \ldots, \sigma_m(r, x))$ is a functional of the infinite dimensional r-variable, and a function of the real variable x. We can rewrite the forward rate curve $\{r(t, x), \forall x \geq 0\}$ at time t as the form of r_t, which is a function-valued stochastic variable. Thus r_t satisfies the following Stratonovich differential equation

$$\begin{aligned} dr_t &= \mu(r_t)dt + \sigma(r_t) \circ dW_t, \\ r_0 &= r_0^*, \end{aligned} \tag{6}$$

where

$$\mu(r) = \frac{\partial}{\partial x} r(t, x) + \sigma(t, x) \int_0^x \sigma(t, s) ds - \frac{1}{2} \sigma_r^{'} [\sigma(r)], \tag{7}$$

and $\sigma_r^{'}$ is a Frechet derivative of the volatility function $\sigma(r)$ with respect to r, $\circ$ means the Stratonovich integral.

Due to the special properties of Stratonovich integral, there is no second-order items here and then we can rewrite the above Stratonovich stochastic differential equation by

$$\frac{dr_t}{dt} = \mu(r_t) + \sigma(r_t) \cdot \nu_t, \tag{8}$$

where ν_t denotes noise.

Given the dynamics of the above forward rate curves, and a special form of the functional σ, we refer $\{\mu, \sigma\}_{LA}$ to the Lie algebra generated by $\{\mu, \sigma\}$.

Assume f is a smooth mapping on the space $\mathcal{H}$, and x is a fixed point in $\mathcal{H}$, we write the solution of the following equation as $x_t = \exp\{ft\}x$:

$$\frac{dx_t}{dt} = f(x_t), \quad x_0 = x. \tag{9}$$

When the dimension of $\{\mu, \sigma\}_{LA} = d < \infty$ is finite, this Lie algebra can be spanned by a set of d smooth mappings $f_1, \ldots, f_d$. Let

$$G(z) = \exp\{f_1 z_1\} \cdots \exp\{f_d z_d\} r_0^*, \; z = (z_1, \ldots, z_d) \in \mathbf{R}^d \tag{10}$$

[2] proves that when the Lie algebra $\{\mu, \sigma\}_{LA}$ has a finite dimension d, there exists an invariant submanifold ς near the initial point r_0^*, such that $\{\mu, \sigma\}_{LA}$ is contained in a tangent space of the manifold ς , and r_t can be realized as

$$\begin{cases} r_t = G(z_t), \\ dz_t = a(z_t)dt + b(z_t) \circ dW_t, \\ z_0 = z_0^*, \end{cases} \tag{11}$$

where $a(z_t), b(z_t)$ and z_0^* are respectively drift coefficient, volatility coefficient of the process z_t and its initial point, and G is a diffeomorphism. They can be deduced by the relationship G between r_t and z_t. In other words, the infinite-dimensional forward rate curve r_t can be realized as an invertible function G of a finite-dimensional diffusion process z_t, which is called the finite-dimensional realization of the forward rate curve r_t.

4 Minimal State Variables

The value of the forward rate curve r_t reflects the system state at time t. However, infinite amount of data is needed to obtain the forward rate curve, while finite amount of data is available in the market. Then a natural problem arises: Does there exist a set of finite data which is sufficient to reflect the system state?

4.1 Main Results

$\{G(z), \forall z \in \mathbf{R}^d\}$ is a parameterized form of the manifold ς, there exists an open neighborhood U of the point 0 in the space $\mathbf{R}^d$, and an open neighborhood V of the point r_0^* in the manifold ς, such that

$$V = G(U),$$

where $G^{-1}(V)$ denotes the coordinate system of the manifold ς at r_0^*, or G is a homeomorphic mapping from $\mathbf{R}^d$ to ς. The inverse mapping G^{-1} of G exists, and there exists one-to-one relationship between $\mathbf{R}^d$ and ς. It means that z_t is also a state variable as r_t does. We can refer to the stochastic variables r_t, z_t and $R(t)$ as homeomorphic mappings between the filtration $\mathcal{F}_t$ and $\mathbf{R}^\infty$($\mathbf{R}^d$,$\mathbf{R}^1$, respectively), their state spaces are of the same structures with infinite-dimension, d-dimension, and 1-dimension, respectively.

Theorem 1. The total forward rate curve is not essential to describe the current system state, while a set of market data with d dimension is sufficient.

Proof. A homeomorphic mapping is of full order, so the order of the homeomorphic mapping G^{-1} equals to the order of the space of z_t, which is exactly d.

Given a set of maturity $\{x_1, x_2, \cdots, x_d\}$, we have from equation (11) that

$$r_t(x_i) = G(x_i, z_t), \quad i = 1, 2, \cdots, d \tag{12}$$

The Jaccobi matrix of the function G has a same order as G, which equals to d. Then the above equations has a unique solution.

Above all, if we have d units of market forward rates $(r_t(x_1), r_t(x_2), \ldots, r_t(x_d))$, the value of z_t can be computed. It means that in the space of the forward rates, d units of separate points is sufficient to describe the current system state. □

Theorem 1 shows that the total forward rate curve is not essential to describe the current system state, we need only d units of market data, which is identical to the dimension of the Lie algebra $\{\mu, \sigma\}_{LA}$, and depend on the form of the volatility function σ.

Theorem 2. At time t, two sets of market data will generate the same state of system.

Proof. we have proved in theorem 1 that the function $G : \mathbf{R}^d \to \mathcal{H}$ is a homeomorphic mapping, which implies that there is a one-to-one relationship between the space

$\mathbf{R}^d$ and $\mathcal{H}$. From equation (11), we conclude that given a forward rate curve, there is only one unique value of z_t corresponding to it. On the basis of theorem 1, the same system state will be obtained, given any two sets of market data. □

When we choose a model for the real world, or the form of the volatility function, the dimension of the Lie algebra $\{\mu, \sigma\}_{LA}$ can be deduced with the help of the theory of algebra, then acquire the form of the function G, at last, choosing a set of d units market data, the current system state can be well described.

Proposition 4.4 in [2] shows that we can also directly choose d units data of the forward rates as the state of system. [2] also proves that, when $d \leq 2$, the forward rate process can be realized as a spot rate process. In this case, we can say that the market data reflect the benchmark interest rates of the central bank.

4.2 Construction of the State Variable

[2] gives the construction of the state variable z_t as follows.

(1) choose a collection $f_1, f_2, \ldots, f_d$ of smooth mapping which spans $\{\mu, \sigma\}_{LA}$;

(2) compute the form of $G(z_t) = \exp\{f_1 z_1\} \cdots \exp\{f_d z_d\} r_0^*$;

(3) Due to the Stratonovich differential equations of r_t and z_t, the following relationships hold,

$$G' a = \mu, G' b = \sigma. \tag{13}$$

Thus a, b can be from the above equations.

4.3 Identification of Error-Priced Market Data

When error-priced data exists in the market, we can use the following method to identify them based on the above results.

(1) When there is only one error data, from theorem 2, we arrange all market data available into three sets, and from equation (11), three units value of z_t is obtained, then the only one which differs from other two means that this set contains the error data. Second, arranging this set into two subsets, together with other true-priced data, we form two new sets of d units data, and find the set containing the error data. Go on until the error data is determined.

(2) The case of two error data: the method is similar to the above case.

As long as the market data available is enough,error data should be identified by using the above method.

5 Conclusion

From a geometric view of interest rate theory, we study the existence problem of minimal state variables by using some important results of FDR problem of [2]. In the framework of HJM model, we find that in the case of deterministic volatility, the system state can be well described based on the available information in the current market, and any set of market data generates the unique state of system.

References

[1] Heath,D., Jarrow,R., Morton,A.: Bond Pricing and the Term Structure of Interest Rates: A New Methodology for Contingent Claims Valuation. Econometrica **60** (1992) 77–105

[2] Bjork, T., Svensson, L.: On the existence of finite dimensional nonlinear realizations for nonlinear forward rate rate models. Math. Fina.**11** (2001) 205–243

[3] Bjork, T., Landen C.: On the construction of finite dimensional nonlinear realizations for nonlinear forward rate models. To appear in Fina. and Stoch. (2003)

[4] Bjork, T., Landen C., Svenssom, L.: On finite Markovian realizations for stochastic volatility for- ward rate models. Working paper. Stockholm School of Economics (2002)

[5] Filipovic, D. Teichmann, J.: Finite dimensional realizations for stochastic equations in the HJM framework. To appear in J. of Func. Anal. (2003)

[6] Bjork, T., Christensen, B.J.: Interest rate dynamics and consistent forward rate curves. Math. Fina. **9** (1997) 323–348

Double Auction in Two-Level Markets*

Ning Chen[1], Xiaotie Deng[2], and Hong Zhu[1]

[1] Dept. of Computer Science, Fudan University, China
{012021113, hzhu}@fudan.edu.cn
[2] Dept. of Computer Science, City University of Hong Kong
csdeng@cityu.edu.hk

Abstract. In the general discussion of double auction, three properties are mainly considered: incentive compatibility, budget balance, and economic efficiency. In this paper, we introduce another property of double auction: semi-independence, from which we are trying to reveal the essential relation between incentive compatibility and economic efficiency.

Babaioff and Nisan [1] studied supply chain of markets and corresponding protocols that solve the transaction and price issues in markets chain. In the second part of the paper, we extend their model to two-level markets, in which all markets in the supply chain are independent and controlled by different owners. Beyond this basic markets chain, there is a communication network (among all owners and another global manager) that instructs the transaction and price issues of the basic markets. Then we discuss incentive compatible problems of owners in the middle level of the markets in terms of semi-independence.

1 Introduction

With the rapid progresses of e-commerce over the Internet, a number of economic concepts have been integrated with computer science extensively, such as Game Theory [9,10], Mechanism Design [7,8] and Auction Theory [6]. One of the most remarkable combinations is that of electronic market, which is on the basis of double auctions.

Double auction, a classic economic concept, specifies that multiple sellers and buyers submit bids to ask for transactions for some well-defined goods [4]. For different purposes, the designed protocols should satisfy required properties over the Internet. Three properties are mainly concerned: incentive compatibility, budget balance, and economic efficiency. The first one emphasize the truthful behavior of agents (sellers and buyers), whereas the last two are macro requirements to the outcome. It's well known that the three properties can not be hold simultaneously under mild assumptions [7]. Hence we are trying to seek for the more essential relations among these properties. Specifically, we present

* This work was supported by grants from the Research Grants Council of the Hong Kong SAR, China [CityU 1081/02E], and the National Natural Science Foundation of China [60273045].

P.M.A. Sloot et al. (Eds.): ICCS 2003, LNCS 2658, pp. 34–45, 2003.

another property, semi-independence, from which we study the relation between incentive compatibility and economic efficiency deeply.

Note that such electronic market, on which the double auction applies, can not reflect the information of related commodities exactly. Hence Babaioff and Nisan introduced the model of supply chain of markets [1] to solve this problem. For example, in lemonade-stand industry, a lemon market sells lemons, a squeezing market offers squeezing services, and lemonade are provided by a juice market. These three markets are composed of the supply chain of lemonade. In addition, Babaioff and Nisan introduced symmetric (pivot) protocol for exchanging information along markets. These protocols determine the supply/demand curve for each market, and allow each market to function independently.

However, markets of different goods in that model are controlled by a single person. Sometimes, a supply chain of markets may be very difficult to construct and manage for a single one. Hence we propose to study a model that allows the individual market to be fully independent. Specifically, we consider two-level markets, where each market in the original supply chain is controlled by a different owner. Moreover, there is another higher level market among owners. The transaction and price issues of the basic markets chain are determined through the interaction of owners at the upper level market.

Many appealing problems arise in such two-level markets model. For example, as participants of the markets, all owners may have their own targets and utilities. Thus we need consider the truthful behaviors of owners, *i.e.*, incentive compatibility. Observing that as a link between two-level markets, such behaviors include two directions: to agents and to manager. That is, the owner may lie to each of them or both. Therefore in this paper, we study the relation between the mechanisms of the markets chain and the truthful behavior of owners. Specifically, we show several sufficient and necessary conditions that guarantee incentive compatibility for different types of owners.

Note that the study of truthful behavior of owners is more closer to practical (electronic) markets, since intermediaries (owners) play an important role in the transactions of the markets and the performance of the whole markets is mostly determined by behaviors of these owners. As we have seen in reality, the loss of efficiency and budget deficit are mainly due to the selfish behaviors of such owners. Therefore we believe that this paper is an important step in the quest for the mechanisms that promote economic efficiency and market revenue.

In section 2, we review the basic concepts of double auction and introduce the property of semi-independence. Next, we briefly review the supply chain markets model of [1] and introduce the model of two-level markets. In section 4, we study the issue of incentive compatibility of owners in two-level markets.

2 Double Auction

2.1 The Model

In a market of a kind of goods, there are n sellers and buyers, respectively. Each seller has one indivisible goods to sell and each buyer plans to buy at most one item. Each agent (seller or buyer) has a privately known non-negative real, termed as *type*, representing the true valuation that the seller/buyer wants to charge/pay.

To win the auction, each agent submits a value to the auctioneer. Let s_i be the i-th supply bid (in the non-decreasing order) of sellers and d_j the j-th demand bid (in the non-increasing order) of buyers, *i.e.*, $0 \leq s_1 \leq s_2 \leq \cdots \leq s_n$ and $d_1 \geq d_2 \geq \cdots \geq d_n \geq 0$. We denote $S = (s_1, \ldots, s_n)$ as the *supply curve* and $D = (d_1, \ldots, d_n)$ as the *demand curve* of market. The *utility* for trading agent (*i.e.*, *winner*) is the absolute value of the difference between price and his true type. For non-trading agent, the utility is zero. We assume that all agents are self-interested, that is, they aim to maximize their own utilities. Note that to maximize the utilities, agents may not submit their types, the strategy is determined according to different double auction rules.

Double auction (DA) rule R is a mechanism that, upon receiving input: supply and demand curves S and D, specifies the quantity $q = R_q(S, D)$ of transactions to be conducted and the price $p_s = R_s(S, D)$, $p_d = R_d(S, D)$ that the trading sellers/buyers receive/pay. Note that in all DA discussed in this paper, non-trading agents receive/pay zero, and once the trade quantity q is fixed, the winners are the first q sellers (with lowest supply bids) and buyers (with highest demand bids), respectively. Moreover, we only concern *non-discriminating* DA, *i.e.*, the price paid by all buyers is same and the price paid to all sellers is same too, but the two values are not necessarily equal.

Let l, *optimal trade quantity*, be the maximal index such that $s_l \leq d_l$. Assume without loss of generality that there always exist the $(l+1)$-th supply and demand bid, s_{l+1} and d_{l+1}, such that $s_{l+1} > d_{l+1}$. Otherwise, we may add sellers with bids ∞ and buyers with bids zero[1]. Followings are some classic DA rules:

- **k-DA:** ([12,2]) $q = l$, and $p_s = p_d = k \cdot s_l + (1 - k) \cdot d_l$, where $k \in [0, 1]$.
- **VCG DA:** ([11,3,5]) $q = l$, and $p_s = \min\{s_{l+1}, d_l\}$, $p_d = \max\{s_l, d_{l+1}\}$.
- **Trade Reduction (TR) DA:** ([1]) $q = l - 1$, and $p_s = s_l$, $p_d = d_l$.

2.2 Semi-independence of Double Auction

We denote $(a_1, \ldots, a_{i-1}, a_{i+1}, \ldots, a_n)$ as a_{-i} and (a_{-i}, a_i) as the tuple $(a_1, \ldots, a_n)$. In this paper, we mainly consider the following properties of DA: (i) *incentive compatibility*: The DA motivates self-interested agents to submit their types to maximize the utility values. It's easy to see that for any incentive compatible DA, there must be $p_s \geq s_q$ and $p_d \leq d_q$, where q units of goods traded. (ii) *budget balance*: The total payment of buyers should be at least the total amount given to sellers, *i.e.*, the revenue of the mechanism is non-negative. (iii) *economic efficiency*: The desired outcome should maximize the total types of all agents. That is, all sellers with types below the market clearing price should trade with all buyers whose types are above the clearing price [1].

[1] Note that for any efficient DA, the optimal trade quantity l may not be identical under different supply and demand curves.

Note that trading l units of goods maximizes efficiency if all agents submit their true types. We denote such DA that trading l units of goods as *efficient DA*. Following let's first look at another property of DA, semi-independence.

Definition 1 (Semi-Independence): Given efficient DA R, supply and demand curves S and D, let $p_s = R_s(S, D)$, $p_d = R_d(S, D)$, and $l = R_q(S, D)$ be optimal trade quantity. We say price p_s is *semi-independent* of s_i under S and D, if

– $1 \leq i \leq l$, $R_s((s_{-i}, s'_i), D) = p_s$, for $\forall$ $s'_i \in [s_{i-1}, s_i]$ (if i=1, $s'_1 \in [0, s_1]$).

– $l < i \leq n$, $R_s((s_{-i}, s'_i), D) = p_s$, for $\forall$ $s'_i \in [s_i, s_{i+1}]$ (if $i = n$, $s'_n \in [s_n, \infty)$).

That is, when s_i changes to its neighbor bid continuously, the price p_s will not change. To sellers, the price is *semi-independent* of i-th supply bid if for $\forall$ S and D, price $R_s(S, D)$ is semi-independent of the i-th supply bid s_i under S and D. If the price is semi-independent of all supply bids under any supply and demand curves, we say DA R is *semi-independent* of supply curve. Similarly, we can define R is semi-independent of demand curve.

Following lemma reveals the essential connections between incentive compatibility and economic efficiency on the basis of semi-independence.

Lemma 1 For any efficient and incentive compatible DA R, let l be the optimal trade quantity, then the price to sellers/buyers is semi-independent of i-th supply/demand bid, for all $1 \leq i \leq l$. And if the $(l+1)$-th supply/demand bid is strictly smaller/larger than the $(l+2)$-th bid, the price is not semi-independent of the $(l+1)$-th bid.

Proof. We only prove the case to sellers and supply bids, the other one is similar. We first prove the first part, that is, the price to sellers is semi-independent of i-th supply bid, for all $1 \leq i \leq l$.

For any supply and demand curves $S = (s_1, \dots, s_n)$ and D, let $p_s = R_s(S, D)$. Then we need to show that p_s is semi-independent of $s_1, \dots, s_l$. Suppose otherwise, that there exists i, $1 \leq i \leq l$, such that modifying s_i to s'_i will change p_s to p'_s, where s'_i satisfies Definition 2.1. If $p_s < p'_s$, the seller who bids s_i can increase his utility simply by submitting s'_i untruthfully. If $p_s > p'_s$, let $(s_1, \dots, s_{i-1}, s'_i, s_{i+1}, \dots, s_n)$ be true types of all sellers respectively. Then same as above, the seller who bids s'_i can increase his utility by reporting s_i. A contradiction to incentive compatibility.

It remains to prove that the price to sellers is not semi-independent of the $(l+1)$-th supply bid s_{l+1} if $s_{l+1} < s_{l+2}$. The idea is to construct a pair of supply and demand curves $S = (s_1, \dots, s_n)$ and $D = (d_1, \dots, d_n)$, *s.t.* $d_{l+1} < s_l < s_{l+1} < d_l$, to show that the price $p_s = R_s(S, D)$ is not semi-independent of s_{l+1}. It's easy to see that such pair of curves does exist.

Claim 1. Price $p'_s = R_s(S', D) = R_s(S, D)$, where $S' = (s_{-l}, s_{l+1}) = (s_1, \dots, s_{l-1}, s_{l+1},$
$s_{l+1}, s_{l+2}, \dots, s_n)$, *i.e.*, $p'_s = p_s$.

Proof of the Claim. Note that the l-th supply bid equals the $(l+1)$-th supply bid in S', since $d_{l+1} < s_{l+1} < d_l$, the number of trading goods under S' and D is also l according to the efficiency of R, *i.e.*, $R_q(S', D) = l$. Thus from above discussion, price p'_s is semi-independent of the l-th supply bid s_{l+1}. Hence, when reducing s_{l+1} to s_l (note that $s_l \in [s_{l-1}, s_{l+1}]$), the price does not change, *i.e.*, $p'_s = p_s$. □

Our following discussions are based on S' and D. Note that $R_q(S', D) = l$, and the l-th supply bid is s_{l+1} in S', thus there must be $p'_s \geq s_{l+1}$. Next we consider two cases.

Case 1. $p'_s > s_{l+1}$. In this case, let s_{l+1} be the true type of the $(l+1)$-th seller who bids s_{l+1}. Then if he submits the bid truthfully, he won't get any utility. But if he submits an sufficiently small value $\varepsilon \geq 0$, the supply curve will be $\widetilde{S} = (\varepsilon, s_1, \dots, s_{l-1}, s_{l+1}, s_{l+2}, \dots,$
$s_n)$ (assuming the submitted bids of other sellers don't change). According to R, he will be a trading seller since the number of trading goods is still l ($s_{l-1} \leq s_l < d_l$ and $s_{l+1} > d_{l+1}$). Following we consider the price $\widetilde{p}_s = R_s(\widetilde{S}, D)$. Because $\widetilde{p}_s$ is semi-independent of all the first l supply bids, $(\varepsilon, s_1, \dots, s_{l-1})$, then the following supply curves sequence share the same price:

$$\begin{aligned} S &= (s_1, s_2, \dots, s_l, s_{l+1}, \dots, s_n) \rightarrow \\ &\quad (\varepsilon, s_2, s_3, \dots, s_l, s_{l+1}, \dots, s_n) \rightarrow \\ &\quad (\varepsilon, s_1, s_3, \dots, s_l, s_{l+1}, \dots, s_n) \rightarrow \cdots \rightarrow \\ &\quad (\varepsilon, s_1, s_2, \dots, s_{l-1}, s_{l+1}, \dots, s_n) = \widetilde{S}. \end{aligned}$$

That is, we have $R_s(\widetilde{S}, D) = R_s(S, D)$, *i.e.*, $\widetilde{p}_s = p_s = p'_s$. Hence the seller who bids ε untruthfully will get utility $\widetilde{p}_s - s_{l+1} = p'_s - s_{l+1} > 0$, a contradiction.

Case 2. $p'_s = s_{l+1}$. It's easy to see there exists $\delta > 0$ such that $d_{l+1} < s_{l+1} + \delta < d_l$ and $s_{l+1} + \delta < s_{l+2}$. Thus when the $(l+1)$-th supply bid changes from s_{l+1} to $s_{l+1} + \delta$, if price p'_s is changed, then the lemma follows. Otherwise, we have

$$R_s((s_1, \dots, s_{l-1}, s_{l+1}, s_{l+1} + \delta, s_{l+2}, \dots, s_n), D) = p'_s.$$

Similar as claim 1, we can ensure the same price p'_s even increasing the l-th supply bid from s_{l+1} to $s_{l+1} + \delta$. That is,

$$R_s((s_1, \dots, s_{l-1}, s_{l+1} + \delta, s_{l+1} + \delta, s_{l+2}, \dots, s_n), D) = p'_s.$$

Let $s_{l+1} + \delta$ be the true type of the l-th seller who bids $s_{l+1} + \delta$. Then although he is a trading seller, but the utility of him is negative, *i.e.*, $p'_s - (s_{l+1} + \delta) < 0$, which contradicts the property of incentive compatibility.

Therefore, we have constructed specified supply and demand curves such that the price is not semi-independent of the $(l+1)$-th supply bid, hence the lemma follows. □

Note that if $s_{l+1} = s_{l+2}$, the lemma also works for the first index i, where $l + 1 < i < n$ and $s_{l+1} = \cdots = s_i < s_{i+1}$, such that the price to sellers is semi-independent of s_i. If $s_{l+1} = \cdots = s_n$, then the price to sellers is semi-independent of s_n.

3 Supply Chain of Markets

Supply chain of markets is a sequence of markets, where the first one M^0 provides resources, the last one M^t consumes the final desired goods and all middle markets $M^1, \ldots, M^{t-1}$ convert the previous goods to the following one sequentially. Without loss of generality, assume the number of agent in each market is equal to n.

3.1 Symmetric Protocol

Note that only the *agents* (buyers) in the last consume market submit demand bids, whereas other *agents* (sellers) in other markets submit supply bids. For any market M^i, denote $S^i = (s_1^i, \ldots, s_n^i)$ as the *supply curve* and $D^i = (d_1^i, \ldots, d_n^i)$ the *demand curve* of M^i, where $0 \leq s_1^i \leq \ldots \leq s_n^i$ and $d_1^i \geq \ldots \geq d_n^i \geq 0$. Here $S^0, S^1, \ldots, S^{t-1}, D^t$ are composed of the submitted bids of agents of each market respectively, other curves are computed in terms of the following symmetric protocol by Babaioff and Nisan [1].

Symmetric Protocol:

Input: Supply/demand curves $S^0, S^1, \ldots, S^{t-1}, D^t$.

Algorithm:

(1) $s_j^t = \sum_{i=0}^{t-1} s_j^i$, for $1 \leq j \leq n$.

(2) $e_j = d_j^t - s_j^t$.

(3) $d_j^i = s_j^i + e_j$, for $0 \leq i \leq t-1$, $1 \leq j \leq n$.

Output: Demand/supply curves $D^0, D^1, \ldots, D^{t-1}, S^t$.

We stress here that above definition is just in terms of its mathematical sense, the original one (exchanging information along the sequence of markets) is referred to [1]. One of the most attractive properties of symmetric protocol is the following lemma.

Lemma 2 The supply curve S^i of market M^i is independent of its demand curve D^i under symmetric protocol, for all $0 \leq i \leq t$.

Proof. We only prove the lemma for supply market M^0, others are similar. Let $E = (e_1, \ldots, e_n)$, According to symmetric protocol, the demand curve of M^0 is:

$$D^0 = S^0 + E = S^0 + (D^t - \sum_{i=0}^{t-1} S^i) = D^t - \sum_{i=1}^{t-1} S^i$$

which implies that S^0 is independent of D^0. □

3.2 Two-Level Markets

Babaioff and Nisan [1] studied the supply chain of markets based on that the auctioneer, who creates the markets chain, conducts all affairs among them. Sometimes, however, it may be very difficult to deal with such a huge markets network for a single one.

Therefore, we consider the following *two-level markets* model, in which all basic markets in original supply chain are independent and controlled by different *owners*, rather than a single one. We denote the owner of market M^i as O^i. Among all owners and another (global) *manager*, the previous auctioneer, there is a communication network (market) that instruct the transaction and price issues of all basic markets. That is, all owners submit supply/demand curves to the manager, who specifies affairs of basic markets in terms of the global mechanism. Formally,

Definition 2 (Global Mechanism) A *global mechanism* G of manager, upon receiving supply/demand curves of markets owners, specify the following two issues:

1. Computing demand/supply curve for each market in terms of symmetric protocol.
2. Deciding the number of transactions to be conducted and the price that the sellers/buyers should receive/pay among all markets.

In this paper we only consider the case that take DA (with symmetric protocol) as global mechanism, *i.e.*, *global DA mechanism*. That is, the number of transactions and price in all markets are determined independently by (the unique) DA. Hence, all discussed properties of DA also apply in the global case.

Example 1 (Lemonade stand industry [1]) Assume there are three communicating electronic markets that produce lemonade: a lemon market M^0, sells lemons; a squeezing market M^1, offers squeezing services; a juice market M^2, from which buyers buy lemonade. There're three agents in each market with valuations showed in the left columns of each market in Graph 1. Then all participants work as follows:

1. Each agent submits his supply/demand bid to the corresponding owner. (Assume all agents know the protocol of global mechanism G at first).
2. Each owner submits the received bids (curve) to the manager.
3. The manager performs the global mechanism G, and returns the results to owners.
4. Each owner returns the above results to agents in his market.
5. Each market acts independently in terms of mechanism G and corresponding curves, then agents get their awards from the owner. For example, if we use global VCG DA, two juices are sold in juice market at price $8 = \max\{8, 7\}$ each. □

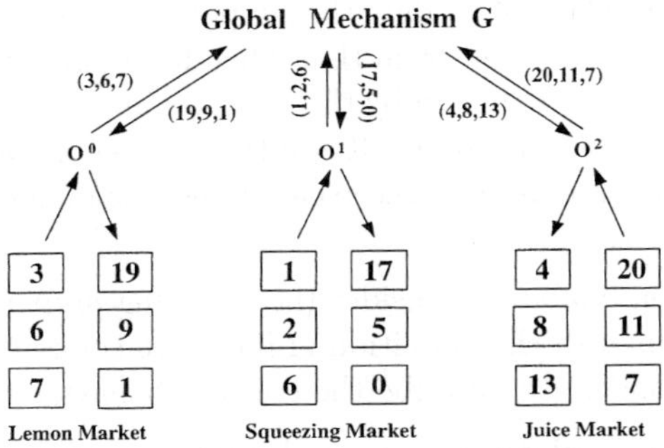

Graph 1: Operation of Lemonade Stand Industry

Note all agents participate step 1, 4, 5, whereas the manager participates step 2, 3. That is, agents and manager do not communicate directly. For simplicity, following we denote M as the supply market M^0 and omit the index 0 of all bids and curves of M.

4 Selfish Owners

Observing that besides performing the protocol of two-level markets, each owner may has his own goal and to optimize it, he may not execute the protocol correctly. Thus we need take their reactions into consideration. In this paper, we assume the *utility* of each owner is the currency he obtained from the market. Hence in this case, the function of the selfish owner is, on one hand, controlling the market, on the other hand, maximizing his utility. Following we only consider owner O of supply market M. Other owners of the conversion markets and demand market are similar.

Example 2 (Selfish owner O of lemon market under global VCG DA) Consider owner O of lemon market showed in above section, the supply bids that he received are $(3, 6, 7)$. If O submits the supply curve truthfully, the manager will return corresponding demand curve $(19, 9, 1)$ to him. According to VCG DA, the first two sellers are traded at price 7 each, and O will pay them 14, which is obtained from the manager. Then we regard that O does not get any utility. However, if O submits $(3, 6, 8)$ untruthfully, the manager will handout 16 to him but at the inner lemon market, O only needs to pay 14 to the two trading sellers. Thus O gets 2 dollars successfully from his lying. (Note that agents do not know what the supply curve that O submits to the manager). □

Remark 1. In fact, the trick of O is that the manager only knows the submitted bid 8, whereas agents only know the true bid 7. Similarly, O may get utility by lying to agents. Moreover, this phenomenon also arises in global k-DA and TR DA.

Remark 2. For non-incentive compatible DA, the submitted bids of agents may not be their true types. But here, we only consider the conduct of owners, *i.e.*, whether he can get positive utility from his lying.

Remark 3. Note that when O submits the supply curve untruthfully, the number of transactions may be different from that when he submits the curve truthfully. Thus, to avoid from being detected by the manager and agents, all owners should be restricted to *quantity consistent* (QC): if the number of transactions is q when all owners truthfully submit their received curves, the quantity should also be q even if any owner lies.

Therefore it's reasonable to consider the following property of global mechanism (here, to distinguish the definition of incentive compatibility of agents, we use the term *immaculate* to describe the truthful behavior of owners):

Definition 3 (Immaculate) In two-level markets, a global mechanism is said to be *immaculate* if for any market owner O^i, $0 \leq i \leq t$, and supply/demand curves submitted/returned by other owners, O^i won't get any positive utility under the restriction of QC, no matter what the supply/demand curve he submits/returns.

4.1 Both-Side Untruthful

In Example 2, O may get positive utility simply by deceiving all agents in M that the demand curve was $(19, 6, 1)$ instead of the true one $(19, 9, 1)$. Thus O (who obtains 14 from the manager) gets 2 dollars (he only pays 12 to the first two sellers). That is, owners may lie to both the manager and agents in his market, *i.e.*, both-side untruthful.

Definition 4 (Pseudo-Constant DA Families) DA rule R belongs to *pseudo-constant DA families* if trading quantity is the unique variable of price function to trading sellers. That is, for any supply and demand bids, there exists a function f, *s.t.* the price to all trading sellers is $f(q)$, where q is the trading quantity according to R.

It's easy to see that for any global DA mechanism that is contained in pseudo-constant DA families, the owner won't get any utility under QC constraint.

Theorem 1 Any global DA mechanism R that does not belong to pseudo-constant DA families is not immaculate when the owner is both-side untruthful.

Proof. Because R is not contained in pseudo-constant DA families, there exist two pairs of supply/demand curves (S, D) and (S', D'), such that $q = R_q(S, D) = R_q(S', D')$, and $R_s(S, D) \neq R_s(S', D')$. Let $p_s = R_s(S, D)$ and $p'_s = R_s(S', D')$. Assume without loss of generality that $p_s < p'_s$. Thus we can regard S as the true bids of sellers, whereas owner O submits S' to the manager. Assume the demand curve of market M is D' according to other markets' curves. Note that under symmetric protocol, there always exist the submitted curves of other markets owners such that the demand curve of M is D' (Lemma 2). Hence in terms of S' and D', the manager decides q units of goods to be trade and the price to trading sellers is p'_s. When getting all payments, $q \cdot p'_s$, O returns D to all agents in M. In such situation, O only needs to pay trading sellers p_s rather other the true price p'_s (since $p_s = R_s(S, D)$). Therefore, O defalcates $q \cdot (p'_s - p_s) > 0$ from his lying. □

4.2 One-Side Untruthful – To Manager

Although pseudo-constant DA families avoid corruptions of owners, such mechanisms perform infeasibly because they can not reflect the supply curves efficiently. That is, the price may deviate from the values of bids arbitrarily, since it only depends on the number of transactions. Therefore in this section, we restrict the power of owners that assume they are always truthful to agents in their own markets (we denote such owners as *type* 1). That is, the demand/supply curve that the owner returns to agents is always the truth. Hence, the utilities of owners are only from the manager, and all trading agents indeed get their original payment.

Lemma 3 If all owners are type 1 and efficient global DA mechanism R is not semi-independent of supply curve, then R is not immaculate.

Proof. Trivially, there exist supply/demand curves S and D and i, $1 \leq i \leq n$, such that p_s is not semi-independent of s_i, where $p_s = R_s(S, D)$. Note that Lemma 2 implies that the demand curve D is independent of $S = (s_{-i}, s_i)$. Let l be optimal trade quantity.

Since p_s is not semi-independent of s_i, there exists $S' = (s_{-i}, s_i')$ such that $p_s' \neq p_s$, where $p_s' = R_s(S', D)$, and s_i' satisfies Definition 1. Following we consider two possible cases. If $p_s < p_s'$, let S be submitted bids of agents, then O can get $l \cdot (p_s' - p_s)$ utilities by submitting S' untruthfully to the manager (note that untruthfulness of the owner is restricted to QC constraint). If $p_s' < p_s$, let S' be submitted bids of agents. Similar as above, O can get $l \cdot (p_s - p_s')$ utilities by submitting S to the manager.

Thus O may always get positive utility as long as he does not truthfully submit supply curve. Therefore the global DA mechanism R is not immaculate. □

Whereas if p_s is semi-independent of all supply bids, intuitively, no matter what the supply curve that O submits, he won't get any utility from the manager. Formally,

Lemma 4 If all owners are type 1 and efficient global DA mechanism R is semi-independent of supply curve, then R is immaculate.

Proof. Assume submitted bids of agents in M are $s_1, \ldots, s_n$ and $s_1 \leq s_2 \leq \cdots \leq s_n$. Let $D = (d_1, \ldots, d_n)$ be demand curve of market M, which is computed in terms of symmetric protocol. Let $p_s = R(S, D)$. To prove R is immaculate, we only need to show that for any supply curve $S' = (s_1', \ldots, s_n')$ that O submits, $p_s' = p_s$, where $p_s' = R_s(S', D)$. Note that QC constraint implies that $s_l' \leq d_l$ and $s_{l+1}' > d_{l+1}$. Let $\widetilde{S} = (s_1', \ldots, s_l', s_{l+1}, \ldots, s_n)$ and $\widetilde{p}_s = R_s(\widetilde{S}, D)$. We observe the following two facts.

Claim 1. $\widetilde{p}_s = p_s$.

Proof of the Claim. We use mathematical induction on the number k that the last $n-k$ elements of S and $\widetilde{S}$ are all identical, *i.e.*,

$$(s_{k+1}', \ldots, s_l', s_{l+1}, \ldots, s_n) = (s_{k+1}, \ldots, s_l, s_{l+1}, \ldots, s_n),$$

where $0 \leq k \leq l$. The cases of $k = 0, 1$ are trivial. Suppose the claim is correct when $k \leq j-1$, $2 \leq j < l$. Then if $k = j$, we have

$$(s'_{j+1}, \cdots, s'_l, s_{l+1}, \cdots, s_n) = (s_{j+1}, \cdots, s_l, s_{l+1}, \cdots, s_n) \;\; \& \;\; s'_j \neq s_j.$$

Assume without loss of generality that $s'_j < s_j$ (the other part is similar as above). We consider two cases about the relations between s'_j and $(s_1, \cdots, s_j)$ as follows.

Case 1. There exists i, $1 \leq i \leq j-1$, such that $s_i \leq s'_j < s_{i+1}$. Then the following supply curves sequence share the same price:

$$\begin{aligned} S &= (s_1, \cdots, s_i, s_{i+1}, \cdots, s_j, s_{j+1}, \cdots, s_n) \to \\ & (s_1, \cdots, s_i, s'_j, s_{i+2}, \cdots, s_j, s_{j+1}, \cdots, s_n) \to \\ & (s_1, \cdots, s_i, s'_j, s'_j, s_{i+3}, \cdots, s_j, s_{j+1}, \cdots, s_n) \to \cdots \to \\ & (s_1, \cdots, s_i, s'_j, \cdots, s'_j, s_{j+1}, \cdots s_n) \to \\ & (s'_1, \cdots, s'_i, s'_{i+1}, \cdots, s'_j, s_{j+1}, \cdots, s_n) = \widetilde{S}. \end{aligned}$$

Note that the last step is by induction hypothesis and others are by semi-independence.

Case 2. $s'_j < s_1 \leq \cdots \leq s_j$. Similar as above case, supply curve S and $\widetilde{S}$ share the same price. Therefore, by induction, we have $\widetilde{p}_s = p_s$. □

Claim 2. $p'_s = \widetilde{p}_s$.

Proof of the Claim. (sketch) Similar as the proof of Claim 1, but here the induction is on the number k that the first $n-k$ elements of S' and $\widetilde{S}$ are identical, where $0 \leq k \leq n-l$. We omit details here. □

Therefore the payment p'_s, $\widetilde{p}_s$ and p_s, on submitted supply curves S', $\widetilde{S}$ and S respectively, are all identical, *i.e.*, $p'_s = \widetilde{p}_s = p_s$. Thus the lemma follows. □

Theorem 2 When all owners are type 1, efficient global DA mechanism R is immaculate if and only if it's semi-independent of supply curve.

4.3 One-Side Untruthful – To Agents

Contrary to above subsection, we may assume all owners are always truthful to manager (*e.g.*, they may be appointed by the manager) and they can return deliberately constructed demand/supply curves to agents to get utility (we denote such owners as *type* 2). Note that the utilities of owners in this case are from their lying to agents. Similarly, we have the following theorem.

Theorem 3 When all owners are type 2, efficient global DA mechanism R is immaculate if and only if it's semi-independent of demand curve.

From above discussion and Lemma 1, we have the following corollary.

Corollary 1 If global DA mechanism R is efficient and incentive compatible, then it's not immaculate, regardless the types of agents.

Thus global VCG mechanism is not immaculate. Note that although global k-DA and TR mechanisms do not satisfy the conditions of above corollary (efficiency and incentive compatibility), they are not immaculate mechanisms too. But as to global TR mechanism, we can modify it to immaculate one simply by exchanging the price functions to trading sellers and buyers.

5 Conclusion and Further Research

A more general model in practice, two-level markets, is studied in this paper. Note that our discussions are based on symmetric protocol, but all results also apply to other protocols that share Lemma 2.

In addition, we study incentive compatible problem for selfish owners who aim to maximize their own utilities. Contrary to this direction, we may consider another one that the goal of each owner is to optimize some common values, such as the total utility of trading agents and income/outcome of the agents in his market. Actually, this work (on selfless owners) is under investment right now.

As we have seen in this paper and many other works, the implementation of incentive compatibility is on the cost of decreasing the revenue of the markets. Thus, whether there exists a weaker feasible notion of incentive compatibility (such as approximation or average case) is a very meaningful direction in the future work.

References

1. M. Babaioff, N. Nisan, *Concurrent Auctions Across the Supply Chain*, EC 2001.
2. K. Chatterjee, W. Samuelson, *Bargaining under Incomplete Information*, Operations Research, 31:835–851, 1983.
3. E. H. Clarke, *Multipart Pricing of Public Goods*, Public Choice, 17–33, 1971.
4. D. Friedman, J. Rust, *The Double Auction Market Institutions, Theories, and Evidence*, Addison-Wesley Publishing Company, 1991.
5. T. Groves, *Incentives in Teams*, Econometrica, 617–631, 1973.
6. P. Klemperer, *Auctin Theory: A guide to the Literature*, Journal of Economics Surveys, 13(3):227–286, 1999.
7. A. Mas-Collel, W. Whinston, J. Green, *Microeconomic Theory*, Oxford University Press, 1995.
8. N. Nisan, A. Ronen, *Algorithmic Mechanism Design (Extended Abstract)*, STOC 1999, 129–140.
9. M. J. Osborne, A. Rubistein, *A Course in Game Theory*, MIT Press, 1994.
10. C. H. Papadimitriou, *Algorithms, Games, and the Internet*, STOC 2001, 749–753.
11. W. Vickrey, *Counterspeculation, Auctions and Competitive Sealed Tenders*, Journal of Finance, 8–37, 1961.
12. R. Wilson, *Incentive Efficiency of Double Auctions*, Econometrica, 53:1101–1115, 1985.

Community Network with Integrated Services

ZhiMei Wu, Jun Wang, and HuanQiang Zhang

Institute of Software, Chinese Academy of Sciences

Abstract. This paper first introduces the general problems met when building Broadband Community Access Network which supports Integrated Services, and then presents some solutions and achievements on multicast service, QoS provision, security, network accounting and billing. In the third section, this paper gives the architecture of a community network, which is a broadband access network based on high-speed switching Ethernet, and provides integrated services, such as digital TV, IP telephone, WWW, Email, and so on. This paper also gives an implementation of this broadband access system, which can afford the special requests of classified transfer, QoS and security, and at the same time provides the accounting and billing function. Lastly, this paper indicates the unfinished part of this broadband access system, and describes the future development.

Keywords: Broadband Access, Community Network, Integrated Services, QoS, Multicast, Security

1 Introduction

With the rapid improvement in communication and computer technology, broadband community network access to Internet and enterprise network has been the consequent request, which can give home users voice service and more video services. In the earlier of Internet, the data exchange is the main service provided by business based on TCP/IP architecture. The Internet of 1990s has not only a great improvement in scale and range, but also a extension to audio and video. During this period, researchers have tried every method to improve the quality of service of transferring audio and video streams in Internet, and also tried to provide the security for data transfer and some reasonable accounting policies. Lately, people have got great achievement in the transfer quality of audio and video in IP network. These improvements expand the application range of IP Network, and now pure IP telephone is quite popular, and TV service on IP is at the beginning. Except the traditional data service and the replacement with telephone and TV service, the broadband IP network[1] can provide some new-style applications, such as interactive TV, video conference, E-commerce[4], e-learning and so on. “All is on in IP” is the future of network development.

2 Issues in Broadband Access Network

2.1 Broadband Access Technologies

There are several typical technologies used for community broadband access network, which are xDSL, Cabel Modem, FTTx+LAN and Wireless access.

P.M.A. Sloot et al. (Eds.): ICCS 2003, LNCS 2658, pp. 46–53, 2003.

DSL stands for "Digital Subscriber Line", a broadband technology that uses telephone lines and digital coding to create a connection to the internet from computer. That link can carry vast amounts of voice, video and data information at very high speeds. The Asynchronous Digital Subscriber Line(ADSL) can provide 6~8Mbps data rates in the downstream direction while the upstream is relatively slow, 144~384Kbps. The Very-High-Bit-Rate Digital Subscriber Line(VDSL)[2] can reach speeds up to 52Mbps downstream, and up to 13Mbps upstream, but it operates only under distances not longer than 1.5 kilometers while ADSL can operate over 3.5 Kilometers. It is xDSL's main advantage that it can utilize the user's existed telephone line, and its disadvantage is too expensive device. If using xDSL, the telecommunication office must keep the same number of xDSL device with the users'. So xDSL is only suitable for the residential area with low density.

Cable Modem is another broadband technology used for community access network. Cable systems were originally designed to deliver broadcast television signals efficiently to subscribers' homes. To deliver data services over a cable network, one television channel (in the 50-750MHz) is typically allocated for downstream traffic to homes and another channel (5-42MHz) is used to carry upstream signals. It have a very high-speed downstream bandwidth (frequency can reach 1GHz), and is a very good access technology if only single-direction access. But there are several big problems when it is been reconstructed to accept two-way streams, they are the expensive reconstruction price, very limited upstream bandwidth, and the amplified noise by upstream amplifiers.

FTTx+LAN is the extension of Internet, which is widely used in Enterprise network and office network[5][6]. It carries several benefits for residential users: very cheap access, easily to upgrade more advanced system, great extensibility. It will need to solve more technology problems such as QoS, security and network accounting, if FTTx+LAN are provided for internet access, IP telephone, LAN TV of residential users. Its disadvantage is that this solution has very high technical threshold to be crossed.

Wireless access is also an interesting broadband technology, especially in the area which can't be easily wired. Bluetooth and 802.11b are the main standard of wireless access. They are not generally used because their short distance and slow bandwidth.

2.2 QoS Provision

Our object is to implement an IP network which can carry many services. In this network, the traditional burst-out traffic can share the same network device (router, switch and links) with the traffic which request more strict delay, delay jitter, bandwidth and lost-packet rate. So it can be named a QoS network, in which there are many kinds of service, every of which have many traffic. The QoS means that in the network it must be sure to provide the corresponding, expectable bandwidth for the special class of service traffic without the knowledge of other traffics. IP packets travel from source to one link, then across switch and router, at the last reach to another link of destination. For QoS, these packets' delay and jitter must be in a restrict range defined by service. In broadband access network, main influence came from the IP packet's transfer policy of switch and the allocated bandwidth of links for the service.

We setup a rate shaping and a rate control policy which based on dual leaky bucket algorithm[3] for the service provider, especially for the video service provider which have very high-speed output, to prevent the overflow of data link's bandwidth. We also add the guarantee for quality of service by an access control algorithm in network device for the classes of service by the current available bandwidth.

In switches, we add classify, queue and scheduler policy for traffics. The classify policy is admitted to specify a class for a destination address, which class is allocate a single transmit queue. A packet is dispatched to a specified class of queue by switch after it is accepted. The packets in high priority of queue are sent prior to lower priority of queue. Switch can provide classified services by classifying, queue and scheduler function. For example, switch can transmit the data by the defined priority of control packets, key service packets, voice packets, video packets and general packets.

2.3 Security of Home Network

The home network's security includes computer's security and network's security. In this paper, we just only concentrate on the network security, which existed in anywhere have network applications. The typical network security is the security of business data by transfer in E-commerce. The content of network security ranges from application layer to physics layer of OSI, and mainly is in network layer, data link layer and physics layer when talked about in broadband access network.

The content of security in physics layer is to prevent wire tapping. Traditional Ethernet have a basic idea of carrier detect, in which way signal is broadcast in a shared line from which every one can listen others' signal. This is not suitable to the request of physics security. In our broadband access system, we use the switched Ethernet as the basic transfer link, every user have and only have a separated line with one of switch port, the signals of which can't be listened by others.

To provide the security in data link layer, we must prevent the user's own information to be broadcast to other users. In general, VLAN can divide the whole network to several broadcast area, so as to separate the data communication form each other. But there are some problems in community network, in which many home of community share a video stream of one video server and they should be in one and the same VLAN with video server. And at the other times when they have a business or shopping by network, they hope to be separated with others. So we implemented an "asymmetric VLAN" technology to support the two requests.

PPPOE is a security technology of network layer on Ethernet. It basic idea is that by PPPOE, IP packet is put into a PPP frame, which is encapsulated by Ethernet frame. Every PPPOE user must login on PPPOE server, and establish a virtual PPPOE channel with PPPOE server on Ethernet before transferring data. When PPPOE user want to communicate with other user of a same LAN or outer network, PPPOE user must send or receive data to PPPOE server firstly, and then PPPOE server will transmit the data to their destination address or PPPOE user. Under PPPOE, everyone can't see others' IP packet directly and they must get admits to network by authentication, so user's security and data's security can be provided well.

2.4 Network Accounting

An accounting system should include both user authentication and billing. In a community network, user authentication can be based on PPPOE, or other connection properties like switch port number, MAC address of home gateway, or IP addresses; And the following price schemes can be used: (1) Flat fee pricing, this is probably the most simple and easy pricing scheme. That means subscribers pay each month certain fixed amount of money, independent of what services they received or how much traffic they used; (2) Usage-based pricing: The idea behind this scheme is that the costs should reflect the amount of usage. The switches will record the information of how much a user used the network, i.e. the amount of data a user has moved, or the quantity of service he has enjoyed. That way anybody who thinks he needs a bigger share of the resources of the network can have it by paying more. This pricing scheme is superior to the flat fee scheme, for it can utilize the information of how a user use the network; (3) Policy and content based pricing: This is a far more complicated pricing scheme, it takes more considerations into the pricing decision, such as the category of services, the traffic amount and time period of the services, even the discount policy for those big customers can be included. This is a more decent and integral scheme than the above twos.

3 Broadband Access System

3.1 Basic Ideas

Community broadband access network for integrated services is a very complicated system, which involves many software and hardware technologies (Figure 1), such as embedded real-time operating systems, multicast systems, QoS, network management, communication protocols and routing algorithms.

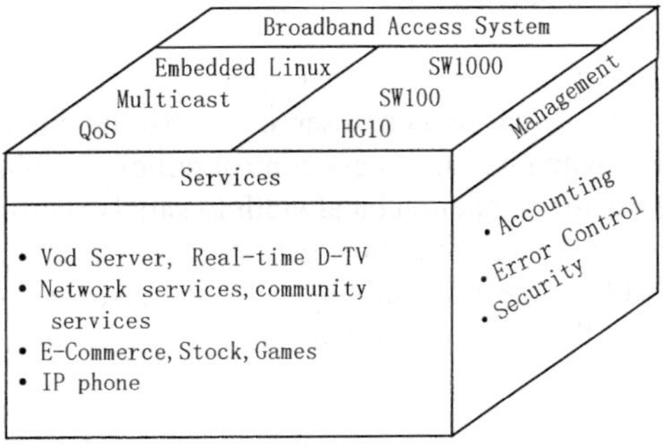

Fig. 1. Technologies of community broadband access system

3.1.1 Embedded Real-Time Operating Systems

Embedded Real-Time Linux is our choice of Embedded Real-Time operating system. Its kernel has all public source code, which made it easily to develop in the special hardware environment, and easily to improve the system's performance. Another advantage is that its cost is very cheap than the business system with same performance. Now, Embedded Real-Time Linux has been used in switches and home gateway devices.

The software system of our Gbit access switch, named SW1000, is developed based on the Linux system. We have tailored the Linux kernel, TCP/IP protocol stack and some network application modules, such as ftp, telnet and so on, and add install/uninstall module, self-protective module, flow accounting module, VLAN-support module, IGMP Snooping module, interrupt and debug module to make up of a new kernel. In user's space, we add the system configure interface, VLAN management interface, network management interface and accounting proxy module.

3.1.2 Multicast Systems

The multicast source streams come from digital TV server, video server and other video provider of WAN, which are selected to decode and playback in home gateway by residential users. The multicast systems involve the server, switch and home gateway device. Server can control multicast streams' output rate by rate shaping and rate control algorithm, at the same time the receivers decrease the influence of delay jitter on the playback quality by pre-buffering video data. The IGMP modules in home gateway cooperate with IGMP Snooping modules in switch to implement the dynamic join and exit of a multicast group. This dynamic group management is the base of dynamic selection of channel. A demand of one channel is in fact that a user joins the multicast group corresponding to this channel, the exit of one channel is correspond to the exit of the multicast group, the switch from one channel to another channel is made up of the two operations.

3.1.3 QoS Provisioning Systems

The QoS system exists in the server, switch and home gateway. It is implemented based on the RSVP, by which switch reserve the bandwidth that server or home gateway requests. The switch has an access control policy, by which switch will deny the new streams if there are no enough bandwidth to satisfy their requests, so to avoid the network's congestion. Server and home gateway transfer and receive data streams only if their bandwidth request is satisfied. Switch also can classify, queue and scheduler the packets by the defined priority.

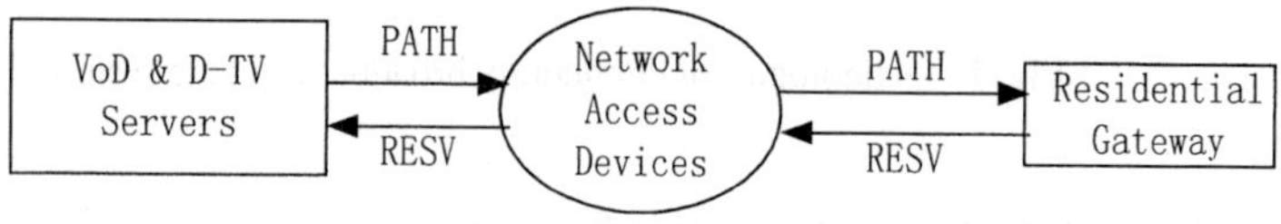

Fig. 2. The mechanism of RSVP

3.1.4 Network Management Systems

Network management system of a broadband community network consists of error management, configuration management, performance management, security management and account management. It bears the same design principle as the network management system in telephone networks, while they have very different focuses. The telephone network focuses on the management of network elements, while a community network puts more emphasis on the category of services and QoS management.

3.2 Implementation

The idea of our IP-based broadband community network is offering convergent services of Internet, IP phone and Digital TV through combined network of fiber and LAN. The system is composed of Digital TV receivers, video servers, TV conference servers, accounting servers, network management servers, broadband switches, home gateways, PCs, analog or digital TV sets, and IP phone(Fig. 3).

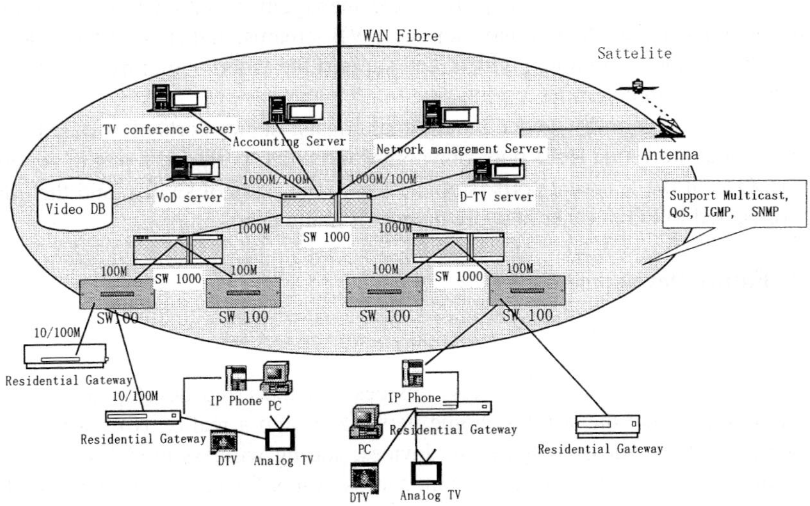

Fig. 3. IP-based broadband community network

Digital TV server's main function is to receive the signals from satellites, demodulate the DVB streams from them, demux the specified streams and put them into the IP packets, lastly send them to users by network or store them in the video database. This digital TV server consists of a high-performance computer, a receiving card for satellite TV, network card and the software system.

Video server's main function is to store, edit and send audio or video program. It mainly consists of a high-performance computer, a network card and a video database, which usually is a large scale disk-array.

3.2.1 Switches

The Gigabit switch SW1000 and the 100M switch SW100 are the main network access devices for community network. SW1000 supports 8~48 ports of 10/100Mbps or 1~6 Gigabit ports, each of them has the line-speed data switch capability. SW1000 consists of four modules: switch fabric modules, fast Ethernet modules, Gigabit Ethernet modules, supervisor modules. SW100 support 16~32 ports of 10/100Mbps, each of which has the line-speed data switch capability. SW100 consists of three modules: switch fabric modules, fast Ethernet modules, supervisor modules.

3.2.2 Home Gateway

Home gateway is the network access device for home network, and it is also a playback device for digital TV and other video programs. Its main function include: provide the playback of digital TV with SDTV quality, and provide IP telephone service, and provide internet data access such as www, email or other, and provide video-on-demand service, and provide the management automation for community. This device can decode and playback the DVB streams, and support the video output of RGB, S-Video, composite Video, and support the video output by PAL/NTSC, and support stereo audio output.

The home gateway device's interfaces consist of two ports of 10BaseT Ethernet, one S-Video port, one RGB port, one composite video port, and a pair of stereo audio ports. One of the Ethernet ports is connected to the ports of switch, and the other one is connected to PC or other network device of home user.

3.3 Future Development

3.3.1 Wireless Access

The coming down price of wireless devices and the rapid development of wireless technologies make the wireless home network design a reality. Now we are making enhancements to our home gateway devices, adding wireless modules to the device, so that the gateway can function as an access points for household wireless devices. (Figure 4).

3.3.2 Transition to IPv6

The community broadband network is the extension of Internet, which connects every home into a whole network. But the scale of this network is limited because the limited number of IP address. Today's IP address is based on IPv4, of which available address will exhaust in the very short future. The NAT technology can only slow down this process and can't radically solve the lacking-address problem because more and more applications are based on the bidirectional UDP communication. Another problem is the QoS and security which can't be taken great improvement by IPv4. IPv6 is designed to address these problems[7], so it becomes another focus of our future research. The content of IPv6 project consists of how to transfer from IPv4 to IPv6 smoothly and the new feature of broadband access and integrated service under IPv6.

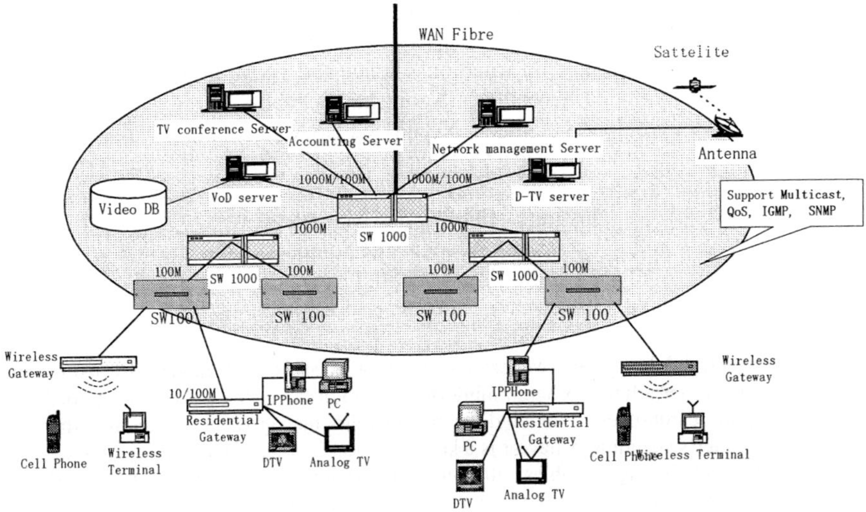

Fig. 4. Enhancement of Broadband Community Network

4 Conclusion

The broadband access network system for integrated services which this paper design and implement is based on IP, and is suitable to not only the community network with dense residential , but also the campus and enterprise network. This system's success will greatly drive the information process of community.

References

[1] Rajiv Jain, "Building the New Broadband Access Network—Considerations & Experience," APRICOT 2002, Bangkok, March, 2002,

[2] M. Schenk, D. Schmucking, A. Worner, I.Ruge, "VDSL (Very high bit-rate digital subscriber line) - A bridge and alternative to FTTH," Proc. Europ. Conference on Networks and Optical Communications (NOC'96), Heidelberg, June 1996.

[3] P. Pancha and M. E. Zarki, "Leaky bucket access control for VBR MPEG video," in Proc IEEE INFOCOM, pp. 796–803, 1995

[4] D. Messerschmitt and J. P. Hubaux, "Opportunities for E-Commerce in Networking," IEEE Commun. Mag., Sept. 1999.

[5] P. Wang, M. Moh, "Ethernet Growing up: The Extensible Broadband Access Infrastructure," ISSLS 2000, Stockholm Sweden, June 2000.

[6] T. Ishihara, M. Okuda, K. lguchi, "Broadband Access Network Providing Reliable Ethernet Connectivity," Fujitsu Scientific & Technical Journal 2001-6, vol37–1, June, 2001

[7] J. Yamada, K. Wakayama, E. Sato, "IPv6 Broadband Access Network Systems" Hitachi Review 2002-6 Vol51–2, June 2002.

A Set of Data Mining Models to Classify Credit Cardholder Behavior*

Gang Kou[1], Yi Peng[1], Yong Shi[1], and Weixuan Xu[2]

[1]College of Information Science and Technology
University of Nebraska at Omaha
Omaha, NE 68182, USA
yshi@unomaha.edu

[2]Institute of Policy and Management
Chinese Academy of Sciences
Beijing, 100080, China
wxu@mail.casipm.ac.cn

Abstract. In this paper, we present a set of classification models by using multiple criteria linear programming (MCLP) to discover the various behaviors of credit cardholders. In credit card portfolio management, predicting the cardholder's spending behavior is a key to reduce the risk of bankruptcy. Given a set of predicting variables (attributes) that describes all possible aspects of credit cardholders, we first present a set of general classification models that can theoretically handle any size of multiple-group cardholders' behavior problems. Then, we implement the algorithm of the classification models by using SAS and Linux platforms. Finally, we test the models on a special case where the cardholders' behaviors are predefined as five classes: (i) bankrupt charge-off; (ii) non-bankrupt charge-off; (iii) delinquent; (iv) current and (v) outstanding on a real-life credit card data warehouse. As a part of the performance analysis, a data testing comparison between the MCLP and induction decision tree approaches is demonstrated. These findings suggest that the MCLP-data mining techniques have a great potential in discovering knowledge patterns from a large-scale real-life database or data warehouse.

Keywords: Data Mining, Multi-criteria Linear Programming, Classification, Algorithm, SAS and Linux platforms

1 Introduction

The history of credit card can be traced back to 1951 when the Diners' Club issued the first credit card in the US to 200 customers who could use it at 27 restaurants in New York [1]. At the end of fiscal 1999, there are 1.3 billion payment cards in circulation and Americans made $1.1 trillion credit purchases [2]. These statistics show that credit card business becomes a major power to stimulate the US economy growth in the last fifty years. However, the increasing credit card delinquencies and

* This research has been partially supported by a grant under (DUE-9796243), the National Science Foundation of USA, a National Excellent Youth Fund under (#70028101), National Natural Science Foundation of China and a grant from the K.C. Wong Education Foundation, Chinese Academy of Sciences.

P.M.A. Sloot et al. (Eds.): ICCS 2003, LNCS 2658, pp. 54–63, 2003.

personal bankruptcy rates are causing plenty of headaches for banks and other credit issuers. The increase in personal bankruptcy rates was substantial. From 1980 to 2000, the number of individual bankruptcy filings in the US increased approximately 500% [3]. How to predict bankruptcy in advance and avoid huge charge-off losses becomes a critical issue of credit card issuers. Since a credit card database can contain hundreds and thousands of credit transactions, it is impossible to discover or predict the cardholders' behaviors without using mathematical tools. In fact, the practitioners have tried a number of quantitative techniques to conduct the credit card portfolio management. Some examples of known approaches are (1) Behavior Score developed by Fair Isaac Corporation (FICO) [4]; (2) Credit Bureau Score also developed by FICO [4]; (3) First Data Resource (FDR)'s Proprietary Bankruptcy Score [5]; (4) Multiple-criteria Score [6,7] and (5) Dual-model Score [8]. A basic characteristic of these models is that they first consider the behaviors of the cardholders as two predefined classes: bankrupt accounts and non-bankrupt accounts according to their historical records. Then they use either statistical methods or neural networks to compute the Kolmogorov-Smirnov (KS) value that measures the largest separation of these two cumulative distributions of bankrupt accounts and non-bankrupt accounts in a training set [9]. The resulting KS values from the learning process are applied to the real-life credit data warehouse to predict the percentage of bankrupt accounts in the future. Thus, these methods can be generally regarded as two-group classification models in credit card portfolio management.

In order to discover more knowledge for advanced credit card portfolio management, multi-group (group number is larger than two) data mining methods are needed. Comparing with two-group classification, the multi-group classifier enlarges the difference between bankrupt accounts and non-bankrupt accounts. This enlargement increases not only the accuracy of separation, but also provides more useful information for credit card issuers or banks. From theoretical point of view, a general model of multi-group classifier is easy to construct. From practical point of view, the best control parameters (such as class boundaries) have to be identified through a learning process on a training data set. Therefore, finding the practical technology with certain size of multi-group classifier is not trivial task. Peng *et al* [10] and Shi *et al* [7] have explored a three-group classifier considering the number of months where the account has been overlimit. This model produces the prediction distribution for each behavior class and inner relationship between these classes so that credit card issuers can establish their credit limit policies for various cardholders.

The purpose of this paper is to build a set of data mining models to classify credit cardholders' behavior. Then these models are tested on the real-life data for five groups of credit cardholders who are predefined as: Bankrupt charge-off accounts, Non-bankrupt charge-off accounts, Delinquent accounts, Current accounts and Outstanding accounts. Bankrupt charge-off accounts are accounts that have been written off by credit card issuers because of cardholders' bankrupt claims. Non-bankrupt charge-off accounts are accounts that have been written off by credit card issuers due to reasons other than bankrupt claims. The charge-off policy may vary among authorized institutions.

2 Models of Multiple Criteria Linear Programming Classification

A general problem of data classification by using multiple criteria linear programming can be described as [11]:
Given a set of r variables or attributes in database $\boldsymbol{a} = (a_1, \ldots, a_r)$, let $\boldsymbol{A}_i = (A_{i1}, \ldots, A_{ir}) \in \boldsymbol{R}^r$ be the sample observations of data for the variables, where $i = 1, \ldots, n$ and n is the sample size. If a given problem can be predefined as s different classes, $G_1, \ldots, G_s$, then the boundary between the jth and $j+1th$ classes can be b_j, $j = 1, \ldots, s-1$. We want to determine the coefficients for an appropriate subset of the variables, denoted by $\boldsymbol{X} = (x_1, \ldots, x_r)^T \in \boldsymbol{R}^r$ and scalars b_j such that the separation of these classes can be described as follows:

$\boldsymbol{A}_i\boldsymbol{X} \le b_1$, $\boldsymbol{A}_i \in G_1$; $b_{k-1} \le \boldsymbol{A}_i\boldsymbol{X} \le b_k$, $\boldsymbol{A}_i \in G_k$, $k = 2, \ldots, s-1$; $\boldsymbol{A}_i\boldsymbol{X} \ge b_{s-1}$, $\boldsymbol{A}_i \in G_s$; where $\boldsymbol{A}_i \in G_j$, $j = 1, \ldots, s$, means that the data case $\boldsymbol{A}_i$ belongs to the class G_j.

The quality of classification is measured by minimizing the total overlapping of data and maximizing the distances of every data to its class boundary simultaneously. Let α_i^j be the overlapping degree with respect of data case $\boldsymbol{A}_i$ within G_j and G_{j+1}, and β_i^j be the distance from $\boldsymbol{A}_i$ within G_j and G_{j+1} to its adjusted boundaries. A multiple criteria linear programming (MCLP) classification model can be defined as:

(M1) *Minimize* $\Sigma_i \Sigma_j \alpha_i^j$ and *Maximize* $\Sigma_i \Sigma_j \beta_i^j$

Subject to: $\boldsymbol{A}_i\boldsymbol{X} \le b_1 + \alpha_i^1$, $\boldsymbol{A}_i \in G_1$; $b_{k-1} - \alpha_i^{k-1} \le \boldsymbol{A}_i\boldsymbol{X} \le b_k + \alpha_i^k$, $\boldsymbol{A}_i \in G_k$, $k = 2, \ldots, s-1$; $\boldsymbol{A}_i\boldsymbol{X} \ge b_{s-1} - \alpha_i^{s-1}$, $\boldsymbol{A}_i \in G_s$; $b_{k-1} + \alpha_i^{k-1} \le b_k - \alpha_i^k$, $k = 2, \ldots, s-1$,

where $\boldsymbol{A}_i$ are given; $\boldsymbol{X}$ and b_j are unrestricted; and $\alpha_i^j, \beta_i^j \ge 0$, for $j = 1, \ldots, s-1$, $i = 1, \ldots, n$.

Note that the constraints $b_{k-1} + \alpha_i^{k-1} \le b_k - \alpha_i^k$ ensure the existence of the boundaries. If minimizing the total overlapping of data, maximizing the distances of every data to its class boundary, or a given combination of both criteria is considered separately, model (M1) is reduced to linear programming (LP) classification (known as linear discriminant analysis), which is initiated by Freed and Glover [12]. However, the single criterion LP could not determine the "best tradeoff" of two misclassification measurements. Therefore, the model (M1) is potentially better than LP classification in identifying the best tradeoff of the misclassifications for data separation. To facilitate the computation on the real-life data, a compromise solution approach [13,14,15] is employed to reform model (M1) for the "best tradeoff" between $\Sigma_i \Sigma_j \alpha_i^j$ and $\Sigma_i \Sigma_j \beta_i^j$. Let us assume the "ideal values" for $s-1$ classes overlapping $(-\Sigma_i \alpha_i^1, \ldots, -\Sigma_i \alpha_i^{s-1})$ be $(\alpha_*^1, \ldots, \alpha_*^{s-1}) > 0$, and the "ideal values" of

$(\Sigma_i \beta_i^1, \ldots, \Sigma_i \beta_i^{s-1})$ be $(\beta_*^1, \ldots, \beta_*^{s-1})$. When $-\Sigma_i \alpha_i^j > \alpha_*^j$, we define the regret measure as $-d_{\alpha j}^+ = \alpha_*^j + \Sigma_i \alpha_i^j$; otherwise, it is 0, where $j = 1, \ldots, s\text{-}1$. When $-\Sigma_i \alpha_i^j < \alpha_*^j$, we define the regret measure as $d_{\alpha j}^- = \alpha_*^j + \Sigma_i \alpha_i^j$; otherwise, it is 0, where $j = 1, \ldots, s\text{-}1$. Thus, we have:

Theorem 1. $\alpha_*^j + \Sigma_i \alpha_i^j = d_{\alpha j}^- - d_{\alpha j}^+$, $|\alpha_*^j + \Sigma_i \alpha_i^j| = d_{\alpha j}^- + d_{\alpha j}^+$, and $d_{\alpha j}^-, d_{\alpha j}^+ \geq 0, j = 1, \ldots, s\text{-}1$.

Similarly, we can derive:

Corollary 1. $\beta_*^j - \Sigma_i \beta_i^j = d_{\beta j}^- - d_{\beta j}^+$, $|\beta_*^j - \Sigma_i \beta_i^j| = d_{\beta j}^- + d_{\beta j}^+$, and $d_{\beta j}^-, d_{\beta j}^+ \geq 0$, $j = 1, \ldots, s\text{-}1$.

Incorporating the above results into model (M1), it is reformulated as:

$$\text{(M2)} \quad Minimize \sum_{j=1}^{s-1} (d_{\alpha j}^- + d_{\alpha j}^+ + d_{\beta j}^- + d_{\beta j}^+)$$

Subject to: $\alpha_*^j + \Sigma_i \alpha_i^j = d_{\alpha j}^- - d_{\alpha j}^+, j = 1, \ldots, s\text{-}1; \beta_*^j - \Sigma_i \beta_i^j = d_{\beta j}^- - d_{\beta j}^+, j = 1, \ldots, s\text{-}1;$

$\mathbf{A}_i X \leq b_1 + \alpha_i^1, \mathbf{A}_i \in G_1; b_{k-1} - \alpha_i^{k-1} \leq \mathbf{A}_i X \leq b_k + \alpha_i^k, \mathbf{A}_i \in G_k, k = 2, \ldots, s\text{-}1;$

$\mathbf{A}_i X \geq b_{s-1} - \alpha_i^{s-1}, \mathbf{A}_i \in G_s; b_{k-1} + \alpha_i^{k-1} \leq b_k - \alpha_i^k, k = 2, \ldots, s\text{-}1,$

where $\mathbf{A}_i$, α_*^j, and β_*^j are given; X and b_j are unrestricted; and $\alpha_i^j, \beta_i^j, d_{\alpha j}^-, d_{\alpha j}^+, d_{\beta j}^-, d_{\beta j}^+ \geq 0$, for $j = 1, \ldots, s\text{-}1, i = 1, \ldots, n$.

We can call model (M1) or (M2) is a "weak separation formula" since it considers as many overlapping data as possible. We can build a "medium separation formula" on the absolute class boundaries in (M3) and a "strong separation formula" which contains as few overlapping data as possible in (M4).

$$\text{(M3)} \quad Minimize \sum_{j=1}^{s-1} (d_{\alpha j}^- + d_{\alpha j}^+ + d_{\beta j}^- + d_{\beta j}^+)$$

Subject to: $\alpha_*^j + \Sigma_i \alpha_i^j = d_{\alpha j}^- - d_{\alpha j}^+, j = 1, \ldots, s\text{-}1; \beta_*^j - \Sigma_i \beta_i^j = d_{\beta j}^- - d_{\beta j}^+, j = 1, \ldots, s\text{-}1;$

$\mathbf{A}_i X \leq b_1, \mathbf{A}_i \in G_1; b_{k-1} \leq \mathbf{A}_i X \leq b_k, \mathbf{A}_i \in G_k, k = 2, \ldots, s\text{-}1;$

$\mathbf{A}_i X \geq b_{s-1}, \mathbf{A}_i \in G_s; b_{k-1} + \varepsilon \leq b_k - \alpha_i^k, k = 2, \ldots, s\text{-}1,$

where A_i, ε, α^j_*, and β^j_* are given; X and b_j are unrestricted; and α^j_i, β^j_i, $d^-_{\alpha j}$, $d^+_{\alpha j}$, $d^-_{\beta j}$, $d^+_{\beta j} \geq 0$, for $j = 1, \ldots, s\text{-}1$, $i = 1, \ldots, n$.

$$\text{(M4)} \quad \text{Minimize} \sum_{j=1}^{s-1} (d^-_{\alpha j} + d^+_{\alpha j} + d^-_{\beta j} + d^+_{\beta j})$$

Subject to: $\alpha^j_* + \Sigma_i \alpha^j_i = d^-_{\alpha j} - d^+_{\alpha j}$, $j = 1, \ldots, s\text{-}1$; $\beta^j_* - \Sigma_i \beta^j_i = d^-_{\beta j} - d^+_{\beta j}$, $j = 1, \ldots, s\text{-}1$;

$A_i X \leq b_1 - \alpha^1_i$, $A_i \in G_1$; $b_{k-1} + \alpha^{k-1}_i \leq A_i X \leq b_k - \alpha^k_i$, $A_i \in G_k$, $k = 2, \ldots, s\text{-}1$;

$A_i X \geq b_{s-1} + \alpha^{s-1}_i$, $A_i \in G_s$; $b_{k-1} + \alpha^{k-1}_i \leq b_k - \alpha^k_i$, $k = 2, \ldots, s\text{-}1$,

where A_i, α^j_*, and β^j_* are given; X and b_j are unrestricted; and α^j_i, β^j_i, $d^-_{\alpha j}$, $d^+_{\alpha j}$, $d^-_{\beta j}$, $d^+_{\beta j} \geq 0$, for $j = 1, \ldots, s\text{-}1$, $i = 1, \ldots, n$.

A loosing relationship of models (M2), (M3), and (M4) is given as:

Theorem 2. If a data case A_i is classified in a given class G_j by model (M4), then it may be in G_j by using models (M3) and (M2). If a data case A_i is classified in a given class G_j by model (M3), then it may be in G_j by using models (M2).

Example 1. As an illustration, we use a small training data set adapted from [16,17] in Table 1 (Column 1-6) to show how the two-class model works.
Suppose whether or not a customer buys computer relates to the attribute set {Age, Income, Student and Credit_rating}. We first define the variables Age, Income, Student and Credit_rating by numeric numbers as follows:
For Age: "≤30" assigned to be "3"; "31…40" to be "2"; and ">40" to be "1".
For Income: "high" assigned to be "3"; "medium" to be "2"; and "low" to be "1".
For Student: "yes" assigned to be "2" and "no" to be "1".
For Credit_rating: "excellent" assigned to be "2" and "fair" to be "1".
G1 = {yes to buys_computer} and G2 = {no to buys_computer}
Then, let $j = 1, 2$ and $i = 1, \ldots, 14$, model (M2) for this problem to classify the customer's status for {buys_computer} is formulated by

Minimize $d_\alpha^- + d_\alpha^+ + d_\beta^- + d_\beta^+$
Subject to:
$\alpha^* + \Sigma_i \alpha_i = d_\alpha^- - d_\alpha^+, \beta^* - \Sigma_i \beta_i = d_\beta^- - d_\beta^+$,
$2x_1 + 3x_2 + x_3 + x_4 = b + \alpha_1 - \beta_1$, $x_1 + 2x_2 + x_3 + x_4 = b + \alpha_2 - \beta_2$
$x_1 + x_2 + 2x_3 + x_4 = b + \alpha_3 - \beta_3$, $2x_1 + x_2 + 2x_3 + 2x_4 = b + \alpha_4 - \beta_4$
$3x_1 + x_2 + 2x_3 + x_4 = b + \alpha_5 - \beta_5$, $x_1 + 2x_2 + 2x_3 + x_4 = b + \alpha_6 - \beta_6$
$3x_1 + 2x_2 + 2x_3 + 2x_4 = b + \alpha_7 - \beta_7$, $2x_1 + 2x_2 + x_3 + 2x_4 = b + \alpha_8 - \beta_8$
$2x_1 + 3x_2 + 2x_3 + x_4 = b + \alpha_9 - \beta_9$, $3x_1 + 3x_2 + x_3 + x_4 = b + \alpha_{10} - \beta_{10}$

$3x_1 + 3x_2 + x_3 + 2x_4 = b + \alpha_{11} - \beta_{11}$, $x_1 + x_2 + 2x_3 + 2x_4 = b + \alpha_{12} - \beta_{12}$
$3x_1 + 2x_2 + x_3 + x_4 = b + \alpha_{13} - \beta_{13}$, $x_1 + 2x_2 + x_3 + 2x_4 = b + \alpha_{14} - \beta_{14}$
where α^*, and β^* are given, x_1, x_2, x_3, x_4 and b are unrestricted, and α_i, β_i, d_α^-, d_α^+, d_β^-, $d_\beta^+ \geq 0$, $i = 1,\ldots, 14$.

Before solving the above problem for data classification, we have to choose the values for the control parameters α^*, β^* and b. Suppose we use $\alpha^* = 0.1$, $\beta^* = 30000$ and $b = 1$. Then, the optimal solution of this linear program for the classifier is obtained as Column 7 of Table 1, where only cases A_8 and A_{12} are misclassified. In other words, cases $\{A_1, A_2, A_3, A_4, A_5, A_6, A_7, A_9\}$ are correctly classified in G1, while cases $\{A_{10}, A_{11}, A_{13}, A_{14}\}$are found in G2. Similarly, when we apply models (M3) and (M4) with $\varepsilon = 0$, one of learning processes provides the same results where cases $\{A_1, A_2, A_3, A_5, A_8\}$ are correctly classified in G1, while cases $\{A_{10}, A_{11}, A_{12}, A_{14}\}$ are correctly found in G2. Then, we see that cases $\{A_1, A_2, A_3, A_5\}$ classified in G1 by model (M4) are also in G1 by models (M3) and (M2), and cases $\{A_{10}, A_{11}, A_{14}\}$ classified in G2 by model (M4) are in G2 by model (M3) and (M2). This is consistent to Theorem 2.

Table 1. A two-class data set of customer status

Cases	Age	Income	Student	Credit Rating	Class: buys_ computer	Training results
A_1	31… 40	high	no	fair	yes	success
A_2	>40	medium	no	fair	yes	success
A_3	> 40	low	yes	fair	yes	success
A_4	31… 40	low	yes	excellent	yes	success
A_5	≤30	low	yes	fair	yes	success
A_6	>40	medium	yes	fair	yes	success
A_7	≤30	medium	yes	excellent	yes	success
A_8	31… 40	medium	no	excellent	yes	*failure*
A_9	31… 40	high	yes	fair	yes	success
A_{10}	≤30	high	no	fair	no	success
A_{11}	≤30	high	no	excellent	no	success
A_{12}	>40	low	yes	excellent	no	*failure*
A_{13}	≤30	medium	no	fair	no	success
A_{14}	>40	medium	no	excellent	no	success

3 Algorithm Implementation and Software Development

A general algorithm to execute the MCLP classification method can be outlined as:

Algorithm 1.
Step 1 Build a data mart for task data mining project.
Step 2 Generate a set of relevant attributes or dimensions from a data mart, transform the scales of the data mart into the same numerical measurement and determine predefined classes, classification threshold, training set and verifying set.
Step 3 Use the MCLP model to learn and compute the best overall score (***X****) of the relevant attributes or dimensions over all observations.
Step 4 Discover the interested patterns that can best match the original classes under the threshold by choosing the proper control parameters (α^*, β^* and b). If the patterns are found, go to Step 5. Otherwise, go back to Step 3.
Step 5 Apply the final learned score (***X*****) to predict the unknown data cases.

Two versions of actual software have been developed for the MCLP classification method. The first version is based on the well-known commercial SAS platform. In this software, we have applied SAS codes to execute Algorithm 1 in which the MCLP models (M2)-(M4) utilize SAS linear programming procedure. The second version of the software is written by C++ language running on Linux platform. The reason for developing Linux version of the MCLP classification software is that the majority of database vendors, such as IBM are aggressively moving to Linux-based system development. Our Linux version goes along with the trend of information technology. Because many large companies currently use SAS system for data analysis, our SAS version is also useful to conduct data mining analysis under SAS environment.

4 Experimental Results from a Real-Life Database

Given a set of attributes, such as monthly payment, balance, purchase, and cash advance and the criteria about "bankruptcy", the purpose of data mining in credit card portfolio management is to find the better classifier through a training set and use the classifier to predict all other customer's spending behaviors [18]. The frequently used data-mining model in the business is still two-class separation technique. The key of two-class separation is to separate the "bankruptcy" accounts from the "current" accounts and identify as many bankruptcy accounts as possible. This is also known as the method of "making black list." The examples of popular methods are Behavior Score, Credit Bureau Score, FDC Bankruptcy Score, and Set Enumeration Decision Tree Score [6]. These methods were developed by either statistics or decision tree. Using a terabyte real credit database of the major US Bank, the SAS version of two-class MCLP model (as in Example) has demonstrated a better predication power (e.g., higher KS values) than these popular business methods [6].

4.1 Testing on Five-Class MCLP Models

We demonstrate the experimental results of 5-class MCLP model in Linux version on a real-life credit data warehouse from a major US bank. Since this database contains 64 attributes with a lot of overlapping, we employed the weak separation model (M2).

In the five-class MCLP model, we defined five classes as Bankrupt charge-off accounts (the number of over-limits ≥ 13), Non-bankrupt charge-off accounts (7 ≤ the number of over-limits ≤ 12), Delinquent accounts (3 ≤ the number of over-limits ≤ 6), Current accounts (1 ≤ the number of over-limits ≤ 2), and Outstanding accounts (no over limit). We selected 200 samples with 40 accounts in each class as the training set. The same 5,000 samples with 53 in G1; 165 in G2; 379 in G3, 482 in G4 and 3921 in G5 were used as the verifying set. We found, in the training process, that G1 has been correctly identified 47.5% (19/40), G2 55% (22/40), G3 47.5% (19/40), G4 42.5% (17/40), and KS values 42.5 for G1 vs. G2, G2 vs. G3 and G3 vs. G4, but 67.5 for G4 vs. G5 in Figure 1. When we used this as the better classifier, we predicted the verifying set as G1 for 50.9% (27/53), G2 for 49.7% (82/165), G3 for 40% (150/379), G4 for 31.1% (150/482) and G5 for 54.6% (2139/3921). The predicted KS values are 36.08 for G1 vs. G2, 23.3 for G2 vs. G3, 27.82 for G3 vs.

G4, and 42.17 for G4 vs. G5 in Figure 2. This indicates that the separation between G4 and G5 is better than other situations. In other words, the classifier is favorable to G4 vs. G5. We note that many real-life applications do not require more than five classes separations. This claim is partially supported by psychological studies. According to [19], Human attention span is "seven plus or minus two". Therefore, for the practical purpose of classification in data mining, classifying five interesting classes in a terabyte database can be very meaningful.

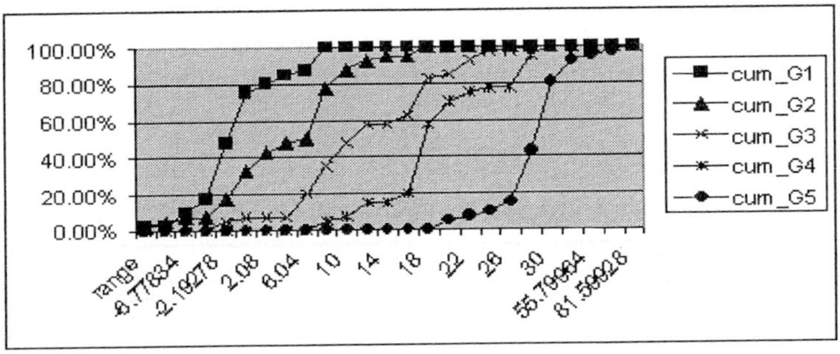

Fig. 1. Five-Class Training Data Set

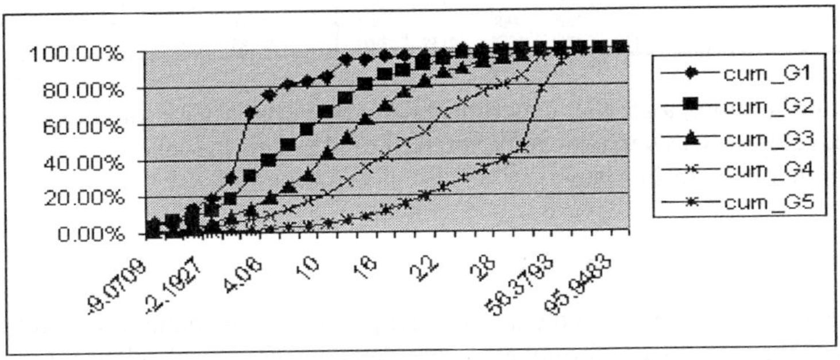

Fig. 2. Five-Class Verifying Data Set

4.2 Comparison of MCLP Method and Decision Tree Method

A commercial Decision Tree software C5.0 (the newly updated version of C4.5) was used to test the classification accuracy of the multiple classes (from two to five) against MCLP methods [17, 20]. Given a terabyte credit card database of the major US Bank, the number samples for the training sets were 600 for two-class problem, 300 for three-class problem, 160 for four-class problem, and 200 for five-class problem. The verifying sets were used the same 5,000 credit card records as in Section 4.1. Table 2 summarizes these comparisons of two methods. As we see, generally the decision tree method does a better job than the MCLP method on training sets when the sample size is small. When applying the classifier from the training process on the larger verifying sets, the MCLP method outperforms the decision tree method. Two issues may cause this evidence. One is that the MCLP method as a linear model may miss some nonlinear nature of data, while the decision tree is nonlinear model. This could be the reason why the latter is better than the former in the training process. However, the robustness and stability of the MCLP is better than the decision tree when the classifier is applied to predict the classification of the verifying sets. This may be due to the fact that the MCLP method employs optimization to find the optimal factors from all feasible factors for the scores while the decision tree method just selects the better tree from a limited built trees, which is not a best tree. In addition, when the decision tree gets big (i.e., the size of the verifying sets increase), the pruning procedure may further eliminate some better branches.

Table 2. Five-class Comparison of MCLP and Decision Tree

5 Classes	Multi Criteria Linear Programming							Decision Tree						
	Class	(a)	(b)	(c)	(d)	(e)	Total	Class	(a)	(b)	(c)	(d)	(e)	Total
T-Set	1	**19**	21	0	0	0	40	1	**35**	4	0	1	0	40
	2	3	**22**	15	0	0	40	2	0	**37**	1	2	0	40
	3	0	8	**19**	13	0	40	3	0	1	**38**	0	1	40
	4	0	0	6	**17**	17	40	4	0	0	5	**29**	6	40
	5	0	0	0	10	**30**	40	5	0	0	0	2	**38**	40
	Class	(a)	(b)	(c)	(d)	(e)	Total	Class	(a)	(b)	(c)	(d)	(e)	Total
V-Set	1	**27**	18	6	2	0	53	1	**41**	6	4	2	0	53
	2	26	**82**	42	14	1	165	2	30	**82**	31	16	6	165
	3	20	145	**150**	41	13	379	3	38	104	**159**	63	14	379
	4	13	70	182	**150**	32	482	4	26	73	178	**162**	44	482
	5	28	110	470	1074	**2139**	3921	5	85	95	1418	428	**1895**	3921

Furthermore, a parallel experimental study on the MCLP classifications through the developed SAS version can be referred to [18]. For the sake of space, we will not elaborate on the results here.

5 Concluding Remarks

There are some research and experimental problems remaining to be explored. From the structure of MCLP formulation, the detailed theoretical relationship of models (M2), (M3), and (M4) need to be further investigated in terms of classification separation accuracy and predictive power. In addition, in the proposed MCLP models, the penalties to measure the "cost" of misclassifications (or coefficients of $\Sigma_i \Sigma_j \alpha_i^j$ and $\Sigma_i \Sigma_j \beta_i^j$) were fixed as 1. If they are allowed to change, their influence on the classification results can be studied, and a theoretical sensitivity analysis of the misclassification in the MCLP models will be also conducted. From mathematical structure point of view, a multiple criteria non-linear classification model may be generalized if the hyper-plane X becomes non-linear cases, say X^p, p >1. The possible connection of the MCLP classification with the known Support Vector Machine (SVM) method in pattern recognition can be researched. In the empirical tests, we have noticed that identifying the optimal solution for model (M2), (M3), or (M3) in the training process may be time-consuming. Instead, we can apply the concept of fuzzy multiple criteria linear programming to seek a satisfying solution that may lead to a better data separation. Other well-known methods, such as neural networks, rough set, and fuzzy set should be considered into part of the extensive comparison study against the MCLP method so that the MCLP method can be known in the data mining community for both researchers and practitioners. We will report any significant results from these ongoing projects in the near future.

References

1 The First Credit Card Was Issued in 1951. http://www.didyouknow.cd/creditcards.htm.
2 Debt statistics. http://www.nodebt.org/debt.htm.
3 Stavins, J.: Credit Card Borrowing, Delinquency, And Personal Bankruptcy. New England Economic Review. July/August 2000, http://www.bos.frb.org/economic/neer/neer2000/neer400b.pdf.
4 www.fairisaac.com
5 http://www.firstdatacorp.com
6 Shi, Y., Wise, M., Luo, M., Lin, Y.: Data Mining in Credit Card Portfolio Management: A Multiple Criteria Decision Making Approach. In: Koksalan, M., Zionts, S. (eds.): Multiple Criteria Decision Making in the New Millennium. Springer, Berlin (2001) 427-436.
7 Shi, Y., Peng, Y., Xu, W., Tang, X.: Data Mining Via Multiple Criteria Linear Programming: Applications in Credit Card Portfolio Management. International Journal of Information Technology and Decision Making. **1** (2002) 131-151.
8 Lin, Y.: Improvement On Behavior Scores by Dual-Model Scoring System. International
9 Journal of Information Technology and Decision Making. **1** (2002) 153-164.
10 Conover, W.J.: Practical Nonparametric Statistics. Wiley (1999).
11 Peng, Y., Shi, Y., Xu, W: Classification for Three-Group Of Credit Cardholders' Behavior Via A Multiple Criteria Approach. Advanced Modeling and Optimization. **4** (2002) 39-56.
12 Kou, G., Peng, Y., Shi, Y., Wise M., Xu, W.: Discovering Credit Cardholders' Behavior by Multiple Criteria Linear Programming. Working Paper, College of Information Science and Technology, University of Nebraska at Omaha (2002).
13 Freed, N. , Glover, F.: Simple but Powerful Goal Programming Models for Discriminant Problems. European Journal of Operational Research. **7** (1981) 44–60.
14 Shi, Y., Yu, P.L.: Goal Setting and Compromise Solutions. In Karpak, B., Zionts, S. (eds.): Multiple Criteria Decision Making and Risk Analysis Using Microcomputers. Springer-Verlag, Berlin (1989) 165–204.
15 Shi, Y.: Multiple Criteria Multiple Constraint-levels Linear Programming: Concepts, Techniques and Applications. World Scientific Publishing, River Edge, New Jersey (2001).
16 Yu, P.L.: Multiple Criteria Decision Making: Concepts, Techniques and Extensions. Plenum, New York (1985).
17 Han, J., Kamber, M.: Data Mining: Concepts and Techniques. Morgan Kaufmann Publishers, San Francisco, California (2001).
18 Quinlan, J.: Induction of Decision Trees. Machine Learning. **1** (1986) 81–106.
19 Peng, Y.: Data Mining in Credit Card Portfolio Management: Classifications for Card Holder Behavior. Master Thesis, College of Information Science and Technology, University of Nebraska at Omaha (2002).
20 Miller, G.A.: The Magical Number Seven, Plus or Minus Two: Some Limits on Our Capacity for Processing Information. The Psychological Review. **63** (1956) 81–97.
21 http://www.rulequest.com/see5-info.html

Continuous Time Markov Decision Processes with Expected Discounted Total Rewards*

Qiying Hu[1], Jianyong Liu[2], and Wuyi Yue[3]

[1] College of International Business & Management, Shanghai University, Shanghai 201800, China.
qyhu@mail.shu.edu.cn
[2] Institute of Applied Mathematics, Academia Sinica, Beijing 100080, China.
[3] Dept. of Information Science and Systems Engineering Konan University, Kobe 658-8501, JAPAN
yue@konan-u.ac.jp

Abstract. This paper discusses continuous time Markov decision processes with criterion of expected discounted total rewards, where the state space is countable, the reward rate function is extended real-valued and the discount rate is a real number. Under necessary conditions that the model is well defined, the state space is partitioned into three subsets, on which the optimal value function is positive infinity, negative infinity, or finite, respectively. Correspondingly, the model is reduced into three submodels, by generalizing policies and eliminating some worst actions. Then for the submodel with finite optimal value, the validity of the optimality equation is shown and some its properties are obtained.

1 Introduction

Markov decision processes (MDP) have been studied well since its beginning in 1960s. While continuous time MDP (CTMDP) [1], as one of its three basic models, was also studied well though has some delay with respect to the other two basic models, discrete time MDP (DTMDP) [2], and semi-Markov decision processes (SMDP) [3]. A new area is the hybrid system which combines event-driven dynamics and time-driven dynamics, e.g., see [4]. The criterions include discounted criterion, average criterion, expected total rewards and mixed criterion, etc. The standard results in MDP with discounted criterion include the following three aspects. 1) The model is well defined. 2) The optimality equation holds. 3) A stationary policy achieving the supermum of the optimality equation will be optimal. In order to obtain these standard results, some conditions should be required. The general and the most usual method to study a MDP model is first to present a set of conditions for the model, and then, based on the conditions, show the standard results 1), 2) and 3) successively.

There are various conditions presented in literature, especially for DTMDP and SMDP. As for CTMDP with discounted criterion, [5] studied it with unbounded transition rates by using the general method. In [6], the author studied the CTMDP also with unbounded transition rates but by using a transformation method, which can transform the CTMDP into a DTMDP under the discounted criterion. Under this transformation, the corresponding optimality equations and discounted objective functions for the stationary

* This research was supported by the National Natural Science Foundation of China, and by Institute of Applied Mathematics, Academia Sinica and by GRANT-IN-AID FOR SCIENTIFIC RESEARCH (No.13650440), Japan.

P.M.A. Sloot et al. (Eds.): ICCS 2003, LNCS 2658, pp. 64–73, 2003.

policies in the CTMDP model and the DTMDP model are equivalent. So the results for CTMDP can be obtained directly from that for DTMDP. In [7], the author studied CTMDP with bounded transition rates also by a transformation, but under which only the discounted objectives for stationary policies in the CTMDP model and the DTMDP model are equivalent. On the other hand, in [8] the author presented a set of conditions for unbounded reward rate. Recently, in [9] the authors discussed a denumerable-state CTMDP with unbounded transition and reward rates. But the method they used is a combination of that of [6] and [8]. In [10], the authors discussed the same model and same conditions but on average criterion.

But there are few studies about expected total rewards criterion. Though the methods presented in [6] and [7] may be used to study it with nonpositive or nonnegative rewards, it is restricted to the stationary policies and can not deal with the general reward rate function or the negative discount rate.

The various conditions presented in literature are only sufficient for MDP. On the contrary, we try to study the necessary conditions, i.e., we want to see what results can be obtained under the condition that the MDP model is well defined. This condition is only the standard result 1), and is obviously the precondition for studying MDP. It is interesting to see if the standard results 2) and 3) can be implied by it. In [11], we studied it for DTMDP with expected discounted total rewards.

This paper is a subsequent one to [11] for CTMDP, where the state space is countable, the reward rate function is extended real-valued and the discount rate may be any real number. The criterion is the discounted expected total rewards with no limits on the discount factor. So it includes the traditional discounted criterion and the expected total rewards criterion. We first generalize the general Markov policies into piecewise semi-Markov policies. Then under the condition that the model is well defined, we show that after eliminating some worst actions, the state space S can be partitioned into three subsets, on which the optimal value function equals $+\infty$, $-\infty$, or is finite, respectively. According to it, the original MDP model can be decomposed into three corresponding sub-models. In the one with finite optimal value, the reward rate function is finite and bounded above at each state, and the validity of the optimality equation is discussed.

The remainder of the paper is organized as follows. Section 2 gives the formulation of the model and presents two conditions, under which Section 3 decomposes the state space and the MDP model. Section 4 discusses some properties of the CTMDP model. In Section 5, the validity of the optimality equation with finite optimal value is shown and several its properties are discussed. While Section 6 is a concluding section.

2 Model and Conditions

The CTMDP model discussed here is

$$\{S, A(i), q_{ij}(a), r(i,a), U_\alpha\} \tag{1}$$

where the state space S and the action set $A(i)$, available at state i, are countable; $\{q_{ij}(a)|i,\ j\in S, a\in A(i)\}$ is the state transition rate family satisfying $q_{ij}(a) \geq 0$ for $i \neq j$ and $\sum_j q_{ij}(a) = 0$ for $(i,a)\in \Gamma := \{(i,a)|i\in S, a\in A(i)\}$, and it is assumed that $\lambda(i) := \sup\{-q_{ii}(a)|a\in A(i)\} < \infty$ for $i\in S$; the reward rate function $r(i,a)$ is extended real-valued; U_α is the objective function for the criterion of expected discounted total rewards with discount factor $\alpha\in(-\infty, +\infty)$, and will be defined below.

We suppose that the measure about the time variable t is the Lebesgue measure.

We define the following policies as in literature, a Markov policy $\pi = (\pi_t, t \geq 0) \in \Pi_m$, a stochastic stationary policy $\pi_0 \in \Pi_s$, a stationary policy $f \in F = \times_i A(i)$. For a policy $\pi = (\pi_t)$ and $s \geq 0$, we define a policy $\pi^s = (\pi_t^*) \in \Pi_m$ by $\pi_t^* = \pi_{s+t}$ for $t \geq 0$. For any policy $\pi = (\pi_t) \in \Pi_m$ and $t \geq 0$, we define a matrix $Q(\pi, t) = (q_{ij}(\pi, t))$ with $q_{ij}(\pi, t) = \sum_{a \in A(i)} q_{ij}(a) \pi_t(a|i)$ and a vector $r(\pi, t) = (r_i(\pi, t))$ with $r_i(\pi, t) = \sum_{a \in A(i)} r(i, a) \pi_t(a|i)$. Thus, $q_{ij}(\pi, t)$ and $r_i(\pi, t)$ are respectively the state transition rate family and the reward rate function under policy π. If $\pi = \pi_0 \in \Pi_s$, then both $Q(\pi_0, t)$ and $r(\pi_0, t)$ are independent of t, and will be denoted respectively by $Q(\pi_0) = (q_{ij}(\pi_0))$ and $r(\pi_0) = (r_i(\pi_0))$.

CONDITION A: For any policy $\pi \in \Pi_m$, the $Q(\pi, t)$-process $\{P(\pi, s, t), 0 \leq s \leq t < \infty\}$ exists uniquely and is the minimal one; moreover, for any $0 \leq s \leq t \leq u < \infty$,

$$\frac{\partial}{\partial t} P(\pi, s, t) = P(\pi, s, t) Q(\pi, t), \ P(\pi, s, u) = P(\pi, s, t) P(\pi, t, u),$$

$$\sum_j P_{ij}(\pi, s, t) = 1, \ P_{ij}(\pi, s, s) = \delta_{ij}, \ i, j \in S.$$

One can find the constructing algorithm for the minimal Q-process in [12] (II. 17) for stationary case and in [1] for nonstationary case. Condition A is true when $q_{ij}(a)$ is bounded, or under the assumptions presented in [5] when $q_{ij}(a)$ is unbounded.

Now, we generalize the concept of policies. Let $Y(t)$ be the state of the process at time t. Given any integer N, real numbers $\{t_i, i = 1, 2, \ldots, N\}$ with $0 = t_0 < t_1 < \ldots < t_N < t_{N+1} = \infty$, and Markov policies $\{\pi^{n,i}, n = 0, 1, 2, \ldots, N, i \in S\} \subset \Pi_m$, we define a policy $\pi = (\pi^{n,i}; n = 0, 1, 2, \ldots, N, i \in S)$ as follows: for $n = 0, 1, 2, \ldots, N$, if $Y(t_n) = i$, then $\pi^{n,i}$ is used in time interval $[t_n, t_{n+1})$, i.e., the action is chosen according to $\pi_{t-t_n}^{n,i}(\cdot|j)$ at time $t \in [t_n, t_{n+1}]$ if $Y(t) = j \in S$. Such a policy, denoted by $\pi = (\pi^{n,i})$ for short, is called a (finite) *piecewise semi-Markov policy*, the set of which is denoted by $\Pi_m(s)$. If all $\pi^{n,i} = f^{n,i} \in F$, then $\pi = (f^{n,i})$ is called a piecewise semi-stationary policy, the set of which is denoted by $\Pi_s^d(s)$.

For such a policy π, if $Y(t_n) = i$, then the system in $[t_n, t_{n+1}]$ is a Markov process with transition probability matrix $P(\pi^{n,i}, s, t)$. So, the system under it is a special case of piecewise Markov process (see [13]). In details, for each s and t with $0 \leq s \leq t$ and $i, j \in S$, suppose that $s \in [t_m, t_{m+1}]$ and $t \in [t_n, t_{n+1}]$ for some $m \leq n$, then the state transition probability that the system will be in state j at time t provided that the system is in state i at time s and in state k at time t_m is

$$\begin{aligned} P_{ij}^k(\pi, s, t) &:= P_\pi\{Y(t) = j | Y(s) = i, Y(t_m) = k\} \\ &= \sum_{j_1} P_{ij_1}(\pi^{m,k}, s - t_m, t_{m+1} - t_m) \\ &\quad \cdot \sum_{j_{n-m}} P_{j_{n-m-1} j_{n-m}}(\pi^{n-1, j_{n-m-1}}, 0, t_n - t_{n-1}) \\ &\quad \cdot P_{j_{n-m} j}(\pi^{n, j_{n-m}}, t_n, t). \end{aligned} \tag{2}$$

For $i, j \in S$, let $P_{ij}(\pi, t) = P_{ij}^i(\pi, 0, t)$.

Now, we define the objective function, for a Markov policy $\pi \in \Pi_m$, by

$$U_\alpha(\pi) = \int_0^\infty \exp(-\alpha t) P(\pi, t) r(\pi, t) dt \tag{3}$$

where the integral is the Lebesgue integral. It is the expected discounted total rewards on the whole time axis under π. Let $U_\alpha(\pi, t) := U_\alpha(\pi^t)$, for $t \geq 0$. Obviously,

$$U_\alpha(\pi, t) = \int_t^\infty \exp(-\alpha(s-t))P(\pi, t, s)r(\pi, s)ds \tag{4}$$

is the expected discounted, to time t, total rewards on the time axis $[t, \infty)$ under π. And for $\pi = (\pi^{n,i}) \in \Pi_m(s)$ with $\{t_n, n = 1, 2, \cdots, N\}$ and $t \geq 0$, we define inductively

$$\begin{aligned} U_\alpha^{n,k}(\pi, t, i) = & \int_t^{t_{n+1}} \exp(-\alpha(s-t)) \sum_j P_{ij}(\pi^{n,k}, t, s) r_j(\pi^{n,k}, s) ds \\ & + \exp(-\alpha(t_{n+1} - t)) \sum_j P_{ij}(\pi^{n,k}, t, t_{n+1}) U_\alpha^{n+1,j}(\pi, t_{n+1}, j), \\ & \qquad t \in [t_n, t_{n+1}),\ n = 0, 1, \ldots, N-1,\ k, i \in S, \\ U_\alpha^{N,k}(\pi, t, i) = & U_\alpha(\pi^{N,k}, t - t_N, i),\ t \geq t_N,\ k, i \in S. \end{aligned} \tag{5}$$

Let $U_\alpha^{n,k}(\pi, t_n, i) = 0$ for $t = t_n$ and $k \neq i$, and $U_\alpha(\pi, i) = U_\alpha^{0,i}(\pi, 0, i)$. Let $U_\alpha(\pi)$ be the vector with its i-th component $U_\alpha(\pi, i)$.

CONDITION B: $U_\alpha(\pi)$ is well defined (may be infinity) for each $\pi \in \Pi_m(s)$.

This condition means that :1) $\sum_j P_{ij}(\pi, t) r_j(\pi, t)$, and furthermore, the integral in Eq. (3), are well-defined for each $\pi \in \Pi_m$; 2) $\sum_j P_{ij}(\pi, t, s) U_\alpha(\pi', s, j)$ is well-defined for every policies $\pi \in \Pi_m$ and $\pi' \in \Pi_m(s)$; 3) the sum in Eq. (5) is well-defined. The condition is necessary to discuss CTMDP. It is well known that it is true whenever $\alpha > 0$ and $r(i, a)$ is bounded above or below; or $\alpha \geq 0$ and $r(i, a)$ is nonnegative or nonpositive. Condition B is assumed to be true throughout the paper.

Conditions A and B hold mean that the CTMDP model (Eq. (1)) is well defined.

Because a policy $\pi \in \Pi_m$ is also a piecewise semi-Markov policy with arbitrary N and $t_1, t_2, \ldots, t_N$, it follows from Eq. (5) that for $\pi \in \Pi_m, t \geq 0$,

$$U_\alpha(\pi) = \int_0^t \exp(-\alpha s)P(\pi, s)r(\pi, s)ds + \exp(-\alpha t)P(\pi, t)U_\alpha(\pi, t), \tag{6}$$

which means that $P(\pi, t)$ can be put out of the integral $\int_t^\infty$, that is,

$$\begin{aligned} & \int_t^\infty \exp(-\alpha(s-t))P(\pi, s)r(\pi, s)ds \\ & = P(\pi, t) \int_t^\infty \exp(-\alpha(s-t))P(\pi, t, s)r(\pi, s)ds \\ & = P(\pi, t)U_\alpha(\pi, t). \end{aligned}$$

Eq. (6) is still true for policies $\pi \in \Pi_m(s)$ by defining $r(\pi, s)$ adequately.

Let the optimal value function be $U_\alpha^*(i) = \sup\{U_\alpha(\pi, i) | \pi \in \Pi_m(s)\}$ for $i \in S$. For $\varepsilon \geq 0, \pi^* \in \Pi_m(s)$, if $U_\alpha(\pi^*, i) \geq U_\alpha^*(i) - \varepsilon$ (if $U_\alpha^*(i) < +\infty$) or $\geq 1/\varepsilon$ (if $U_\alpha^*(i) = +\infty$), then π^* is called ε-optimal. Here, $1/0 = +\infty$ is assumed. 0-optimal is simply called optimal.

3 Eliminating the Worst Actions

First, we introduce some concepts. State j can be reached from state i (and write $i \to j$) if there are a policy $\pi \in \Pi_m(s)$ and $t \geq 0$ such that $P_{ij}(\pi, t) > 0$. It is easy to see that

$i \to j$ iff there are $\pi \in \Pi_m$ and $t \geq 0$ such that $P_{ij}(\pi, t) > 0$, or equivalently there are $n \geq 0$, states $j_1, j_2, \ldots, j_n \in S$ and $f \in F$ such that $q_{ij_1}(f) q_{j_1 j_2}(f) \ldots q_{j_n j}(f) > 0$. It is apparent that if $i \to j$ and $j \to k$, then $i \to k$. For a subset $S_0 \subset S$ and a state i, if there is a state $j \in S_0$ such that $i \to j$, then we say that S_0 can be reached from state i, which is denoted by $i \to S_0$. Let $S_0^* = \{i | i \to S_0\}$ be a set of states that can reach S_0. Because $i \to i$, so $S_0 \subset S_0^*$. A subset S_0 of S is called a closed (state) set if $q_{ij}(a) = 0$ for all $i \in S_0, a \in A(i)$ and $j \bar{\in} S_0$, or equivalently, $(S - S_0)^* = S - S_0$. Similarly as above, S_0 is closed iff $P_{ij}(\pi, t) = 0$ for all $i \in S_0, \pi \in \Pi_m(s), j \bar{\in} S_0$ and $t \geq 0$.

For any closed subset S_0, if the system's initial state $i \in S_0$, then the system will remain in S_0 irrespective of the policies used. Thus, the restriction of CTMDP to S_0,

$$S_0\text{-CTMDP} := \{S_0, (A(i), i \in S_0), p_{ij}(a), r(i, a), U_\alpha\}$$

is also a CTMDP, which is called an induced sub-CTMDP by S_0. Its policies are restriction of the original policies to S_0. It is clear that Condition A and B are also true for S_0-CTMDP. Let its objective function be $U_\alpha^{S_0}(\pi)$.

THEOREM 1: For any closed subset $S_0 \subset S, \pi \in \Pi_m(s)$ and $i \in S_0$, $U_\alpha(\pi, i) = U_\alpha^{S_0}(\pi, i)$.

The theorem says that the induced sub-CTMDP by a closed set S_0 is equivalent to the original CTMDP in subset S_0. So, if both S_0 and $S - S_0$ are closed, then CTMDP can be partitioned into two smaller parts:S_0-CTMDP and $(S - S_0)$-CTMDP. On the other hand, if S_0 is closed while $U_\alpha^*(i)$ for $i \in S - S_0$ is known, or a (ε-)optimal policy can be obtained in $S - S_0$, then one need to discuss only S_0-CTMDP. Thus the state space is partitioned and reduced.

On the other hand, some actions may be eliminated with no influence on the essential of the model.

DEFINITION 1: Suppose that $A_1(i) \subset A(i)$ for $i \in S$. We denote by CTMDP$'$ the CTMDP with $A(i)$ being replaced by $A_1(i)$ (a symbol " $'$ " is added). If for any policy π of the (original) CTMDP there is a policy π' of the CTMDP$'$ such that $U_\alpha(\pi, i) \leq U'_\alpha(\pi', i)$ for all i, then the CTMDP is equivalent to the CTMDP$'$, and we say that $A(i)$ can be reduced as $A_1(i)$ for $i \in S$, or $a \in A(i) - A_1(i)$ can be eliminated for $i \in S$.

Now, we denote by $U(i) = \sup\{r(i, a) | a \in A(i)\}$ and $L(i) = \inf\{r(i, a) | a \in A(i)\}$ respectively the supremum and infimum of the reward rate function $r(i, a)$ over the action set $A(i)$ for $i \in S$. Let $S_U = \{i | U(i) = +\infty\}, S_{=\infty} = \{i |$there is $\pi \in \Pi_m(s)$ such that $U_\alpha(\pi, i) = +\infty\}, S_\infty = \{i | U_\alpha^*(i) = +\infty\} - S_{=\infty}, S_{-\infty} = \{i | U_\alpha^*(i) = -\infty\}, S_0 = S - S_{=\infty} - S_\infty - S_{-\infty} = \{i | -\infty < U_\alpha^*(i) < \infty\}$. These state subsets have obvious meanings.

LEMMA 1:

1) For $i \in S_U$, there is a policy $\pi_0 \in \Pi_s$ such that $r_i(\pi_0) = +\infty$. So $U_\alpha(\pi_0, i) = +\infty$ and $S_U \subset S_{=\infty}$.
2) For $i \in S$ with $L(i) = -\infty$, there is a policy $\pi_0 \in \Pi_s$ such that $r_i(\pi_0) = -\infty$ and then $U_\alpha(\pi_0, i) = -\infty$.
3) For $i \in S, L(i) = -\infty$ and $U(i) = +\infty$ can not be true simultaneously.

THEOREM 2:

1) $S_{=\infty}^* = S_{=\infty}$ and so $S' := S - S_{=\infty}$ is closed.

2) For $i{\in}S' - S_{-\infty}$, $A(i)$ can be reduced as

$$A_1(i) = \{a{\in}A(i)|r(i,a) > -\infty \quad \text{and} \sum_{j\in S_{-\infty}} q_{ij}(a) = 0\}. \tag{7}$$

After the reduction, $S^*_{-\infty} = S_{-\infty}$ and so $S'' := S' - S_{-\infty}$ becomes closed.
3) For $i{\in}S''$, $A_1(i)$ can further be reduced as

$$A_2(i) = \{a{\in}A_1(i)|\exists\pi{\in}\Pi_m \text{ with } U_\alpha(\pi,i) > -\infty \text{ and the Lebesgue measure of } \{s\in[0,t]|\pi_s(a|i) > 0\} \text{ is positive for each } t > 0\}. \tag{8}$$

After this reduction, $S^*_\infty = S_\infty$, and so $S_0 := S'' - S_\infty$ is closed.

By Theorem 1 and 2, S can be partitioned into four subsets: $S_{-\infty}, S_{=\infty}, S_\infty, S_0$. In $S_{-\infty}$, each policy is optimal; in $S_{=\infty}$, there is an optimal policy (in fact, there is a stochastic stationary optimal policy in S_U); in $S_\infty, U_\alpha(\pi,i) < \infty$ for each π, while $U^*_\alpha(i) = \infty$, and thus there is no optimal policy, and in $S_0, U^*_\alpha(i)$ is finite, and S_0 is closed after eliminating some worst actions. So, one can consider only the following CTMDP:

$$S_0\text{-CTMDP} = \{S_0, A_2(i), q_{ij}(a), r(i,a), U_\alpha\}. \tag{9}$$

Because $i{\in}S_{-\infty}$ when $A_2(i) = \emptyset$, Eq. (9) is a CTMDP; furthermore, we have

$$-\infty < U^*_\alpha(i) < +\infty, -\infty < r(i,a){\le}U(i) < +\infty, \qquad \forall i,a. \tag{10}$$

It is easy to see that all the above results restricted to $\Pi_s(s)$ are also true.

In the remaining of this paper, we discuss mainly the S_0-CTMDP, and so will write S_0 and $A_2(i)$ by S and $A(i)$ respectively for convenience.

4 Some Properties

This section discusses some properties of S_0-CTMDP (Eq. (9)) and simplifies the expression of $A_2(i)$. First, the following lemma is from [12] (II. 15-17).
LEMMA 2: Suppose that $P(t) = (p_{ij}(t))$ is a homogeneous state transition probability matrix on a countable state space S with a finite transition rate family $Q = (q_{ij})$. Let $q_i = -q_{ii}$. Then there are nonnegative continuous functions $g_{ij}(t)$ for $i,j{\in}S$, on $[0,\infty)$, such that

$$p_{ij}(t) = \exp(-q_i t)\int_0^t \exp(q_i s) q_i g_{ij}(s)ds + \exp(-q_i t)\delta_{ij}, \qquad i,j{\in}S, t\ge 0$$

where δ_{ij} denotes the Kronecker delta function, and for $s > 0, t \ge 0$,

$$\lim_{s\to 0+} g_{ij}(s) = (1-\delta_{ij})q_{ij}/q_i, \sum_j g_{ij}(s) = 1, \; g_{ij}(s+t) = \sum_k g_{ik}(s)p_{kj}(t).$$

Based on the above lemma, one can proved the following two lemmas.

LEMMA 3: Suppose that $P(t) = (p_{ij}(t)), Q$ and $g_{ij}(t)$ are as in Lemma 2, $\sup_i q_i < \infty, u$ is a finite nonnegative function in $S, Z \subset S, i \in S$. If $\sum_{j \in Z} p_{ij}(t^*)u_j$ is finite for some $t^* > 0$, then $h_i(t) := q_i \exp(q_i t) \sum_{j \in Z} g_{ij}(t)u_j$ is finite and continuous in $[0, t^*)$, and $\sum_{j \in Z} q_{ij}u_j < \infty$; otherwise, $h_i(t) = +\infty$ for all $t > 0$.

LEMMA 4: Using the symbols in Lemma 2, Suppose that $\sup_i q_i < \infty, u$ is a finite function in $S, t^* > 0$ and $i \in S$. If $\sum_j p_{ij}(t)u_j$ is finite in $[0, t^*]$, then its derivative is well-defined and continuous in $[0, t^*)$, and

$$\frac{d}{dt}\Big\{\sum_j p_{ij}(t)u_j\Big\} = \sum_j \frac{d}{dt}p_{ij}(t)u_j = \sum_j \{-q_i p_{ij}(t) + q_i g_{ij}(t)\}u_j.$$

Having the above several lemmas for pre preparation, now we can prove the following theorem, where $S^+ := \{i \in S_0 | U^*_\alpha(i) \geq 0\}$ and $S^- := \{i \in S_0 | U^*_\alpha(i) < 0\}$.

THEOREM 3: 1) $P(\pi, t)U^*_\alpha < \infty$ is well defined for each $\pi \in \Pi_m(s)$ and $t > 0$.

2) For $\pi \in \Pi_m(s), t > 0$ and $i \in S$, if $\sum_j P_{ij}(\pi, t)U^*_\alpha(j) = -\infty$, then $U_\alpha(\pi^*, i) = -\infty$ for any piecewise semi-Markov policy $\pi^* = (\pi^{n,j}) \in \Pi_m(s)$ with $\pi^{0,i} = \pi$ and $t_1 = t$, especially, $U_\alpha(\pi, i) = -\infty$.

COROLLARY 1: Suppose that there is $f^* \in F$ such that $Q(f^*)$ is bounded, then $\sum_j q_{ij}(a)U^*_\alpha(j) < \infty$ is well defined for any $i \in S$ and $a \in A(i)$.

COROLLARY 2: Suppose that $f \in F$ with bounded $Q(f), i \in S, t^* > 0, [P(f,t)U^*_\alpha]_i$ is finite in $[0, t^*]$, then $[P(f,t)U^*_\alpha]_i$ is differentiable in $[0, t^*)$, its derivative is continuous and

$$\frac{d}{dt}\Big\{\sum_j P_{ij}(f,t)U^*_\alpha(j)\Big\} = \sum_j \frac{d}{dt}P_{ij}(f,t)U^*_\alpha(j),$$

$$\sum_j [P(f,t)Q(f)]_{ij}U^*_\alpha(j) = \sum_j P_{ij}(f,t)[Q(f)U^*_\alpha(j)]_j, \qquad t \in [0, t^*). \tag{11}$$

We conjecture that the result in Corollary 2 is also true for $\pi \in \Pi_m$, but it needs that Lemma 2 holds for a nonhomogeneous Markov process, which is not known to us.

To conclude equations in this section, we give the following theorem on a simplified expression for $A_2(i)$.

THEOREM 4: If $q_{ij}(a)$ is uniformly bounded, then

$$A_2(i) \subset \{a \in A_1(i) | \sum_j q_{ij}(a)U^*_\alpha(j) > -\infty\}, \qquad i \in S; \tag{12}$$

moreover, if $h_i(t)$ is finite and continuous whenever $\sum_{j \in Z} q_{ij}u_j$ is finite in Lemma 3, then

$$A_2(i) = \{a \in A_1(i) | \sum_j q_{ij}(a)U^*_\alpha(j) > -\infty\}, \qquad i \in S. \tag{13}$$

Remark 1: 1) By the above theorem, if S^- is finite, or $U^*_\alpha(i)$ is bounded below, then Eq. (13) is true when $q_{ij}(a)$ is uniformly bounded; 2) S^- is empty if U^*_α is nonnegative, especially, if the reward function is nonnegative; 3) $U^*_\alpha(i)$ is bounded below if $\alpha > 0$ and the reward function is bounded below.

5 Optimality Equation

This section shall deal with the standard results 2) and 3) (see Section 1), that is, we shall show the optimality equation and the optimality of policies achieving the optimality equation for S_0-CTMDP, under the assumption that $\{q_{ij}(a)\}$ is uniformly bounded, i.e., $\lambda = \sup\{-q_{ii}(a)|i \in S, a \in A(i)\} < \infty$. For $\pi \in \Pi_m(s), t \geq 0$ and a finite function $u = (u(i))$ on S, we define

$$U_\alpha(\pi, t, u) = \int_0^t \exp(-\alpha s)P(\pi, s)r(\pi, s)ds + \exp(-\alpha t)P(\pi, t)u$$

whenever the right hand side is well-defined. Denote $U_\alpha^*(\pi, t) = U_\alpha(\pi, t, U_\alpha^*)$ for short, which is well-defined by Theorem 3. Certainly, $U_\alpha^*(\pi, t)$ is the expected discounted total rewards if π is used in $[0, t]$ and then an optimal policy is used from t.

LEMMA 5: $U_\alpha^* = \sup\{U_\alpha^*(\pi, t)|\pi \in \Pi_m(s)\}$ for $t \geq 0$, and $U_\alpha^*(\pi, t)$ is nonincreasing in t for any $\pi \in \Pi_m(s)$.

Now, we introduce our third condition.

CONDITION C: For each $i \in S$ and $a \in A(i)$, there is f and $t > 0$ such that $f(i) = a$ and $U_\alpha^*(f, t, i) > -\infty$.

Remark 2: Two sufficient conditions for Condition C are as follows: 1) the conditions given in Theorem 4, especially, when S^- is finite or U_α^* is bounded below (see Remark 1); 2) for each $i \in S$, $A(i)$ can be reduced as

$$A'(i) = \{a \in A(i)| \sup_{f \in F: f(i)=a} U_\alpha(f, i) > -\infty\},$$

which means that any action $a \in A(i)$ should be eliminated if any stationary policy f using it will have negative infinite objective value. In fact, if $A(i)$ can be reduced as $A'(i)$, then it follows Lemma 5 that $U_\alpha^* \geq U_\alpha^*(f, t) \geq U_\alpha(f) > -\infty$ for each $f \in F$ and $t > 0$.

THEOREM 5: Under Condition C, U_α^* satisfies the following optimality equation:

$$\alpha U_\alpha^*(i) = \sup_{a \in A(i)} \{r(i, a) + \sum_j q_{ij}(a)U_\alpha^*(j)\}, \qquad i \in S. \tag{14}$$

The policy set is generalized here, but it is often our pleasure to restrict an ($\varepsilon \geq 0$) optimal policy to a smaller and simpler policy set. To do this, our first result is the following theorem, which says that the optimality can be restricted to Π_m, the set of Markov policies, iff the optimal value function restricted to Π_m also satisfies the optimality Eq. (14). Let $U_\alpha^m = \sup\{U_\alpha(\pi)|\pi \in \Pi_m\}$. We affirm that U_α^m is finite. In fact, if $U_\alpha^m(i_0) = -\infty$ for some $i_0 \in S$, then it is easy to see from Eq. (5) that $U_\alpha(\pi, i_0) = -\infty$ for each $\pi \in \Pi_m(s)$. Thus $U_\alpha^*(i_0) = -\infty$, which is a contradiction. But $U_\alpha^m \leq U_\alpha^* < \infty$, so, U_α^m is finite.

THEOREM 6: $U_\alpha^* = U_\alpha^m$ iff U_α^m is a solution of the optimality Eq. (14).

In order to obtain some properties for the optimality Eq. (14), we define a set, denoted by W, of finite functions $u = (u(i))$ on S satisfying the following conditions: for each $\pi \in \Pi_m(s)$ and $i \in S$, $\sum_j P_{ij}(\pi, t)u(j) < \infty$ is well-defined for all $t \geq 0$. Moreover, $\sum_j P_{ij}(\pi, t)u(j) > -\infty$ whenever $\sum_j P_{ij}(\pi, t)U_\alpha^*(j) > -\infty$ for each $t \geq 0$ and $i \in S$. W is nonempty for $U_\alpha^* \in W$. It is clear that $U_\alpha(\pi, t, u) < \infty$ is well-defined for each $u \in W$.

LEMMA 6: Suppose that $\varepsilon \geq 0, \beta + \alpha \geq 0, u \in W, \pi \in \Pi_m$ and $i \in S$. If π and u satisfy the following two conditions, then $u(i) \leq U_\alpha(\pi, i) + (\beta + \alpha)^{-1}\varepsilon$.

$$\alpha u \leq r(\pi, t) + Q(\pi, t)u + \exp(-\beta t)\varepsilon e, \text{ a.e. } t \geq 0, \tag{15}$$

$$\liminf_{t \to \infty} \exp(-\alpha t) \sum_j P_{ij}(\pi, t)u(j) \leq 0. \tag{16}$$

THEOREM 7: Suppose that $u \in W$ is a solution of the optimality Eq. (14) and $i \in S$.

1) if for some $\beta > -\alpha$ and each $\varepsilon > 0$, there is a policy $\pi \in \Pi_m(s)$ with $U_\alpha(\pi, i) > -\infty$ satisfying Eq. (15) and Eq. (16), then $u(i) \leq U_\alpha^*(i)$;
2) if u satisfies the following (23) for each $\pi \in \Pi_m(s)$ with $U_\alpha(\pi, i) > -\infty$, then $u(i) \geq U_\alpha^*(i)$,

$$\limsup_{t \to \infty} \exp(-\alpha t) \sum_j P_{ij}(\pi, t)u(j) \geq 0. \tag{17}$$

It is clear that there is often a policy $\pi = (f_t) \in \Pi_m^d$ satisfying Eq. (15), but it may be not true that $U_\alpha(\pi, i) > -\infty$. On the other hand, U_α^* often satisfies Eq. (17) for $\pi \in \Pi_m(s)$ with $U_\alpha(\pi, i) > -\infty$. In fact, by Eq. (6) we know that if $U_\alpha(\pi, i) > -\infty$, then $\sum_j P_{ij}(\pi, t)U_\alpha^*(j)$ is also finite for each $t \geq 0$, and

$$\limsup_{t \to \infty} \exp(-\alpha t) \sum_j P_{ij}(\pi, t)U_\alpha^*(j) \geq \limsup_{t \to \infty} \exp(-\alpha t) \sum_j P_{ij}(\pi, t)U_\alpha(\pi, t, j) = 0.$$

The following corollary can be proved easily by Theorem 7 and Lemma 6.

COROLLARY 3: Provided that Eq. (14) holds,

1) for any given $f \in F$, if f attains supremum of Eq. (14), f and U_α^* satisfy Eq. (16) and $U_\alpha(f) > -\infty$, then f is optimal;
2) for some $\pi^* \in \Pi_m(s)$, if $U_\alpha(\pi^*)$ is a solution Eq. (14), then π^* is optimal;
3) if for any $\varepsilon > 0$, there is a Markov policy $\pi \in \Pi_m^d$ with $U_\alpha(\pi) > -\infty, \pi$ and U_α^* satisfy Eq. (15) and Eq. (16) for each $i \in S$, then $U_\alpha^* = \sup\{U_\alpha(\pi) | \pi \in \Pi_m^d\}$;
4) if $\alpha > 0, \varepsilon \geq 0, f \in F$ attains the ε-supremum of Eq. (14), f and U_α^* satisfy Eq. (16), $U_\alpha(f) > -\infty$, then f is $\alpha^{-1}\varepsilon$-optimal; moreover, if such f exists for each $\varepsilon > 0$, then $U_\alpha^* = \sup\{U_\alpha(f) | f \in F\}$;
5) if $U_\alpha^* \leq 0$, then U_α^* is the largest solution of Eq. (14) in W satisfying conditions given in 1) of Theorem 7;
6) U_α^* is the smallest solution of Eq. (14) in W satisfying Eq. (17) for $\pi \in \Pi_m(s)$ and $i \in S$ with $U_\alpha(\pi, i) > -\infty$.

COROLLARY 4: For $f \in F$ and $i \in S$ with $U_\alpha(f, i) > -\infty, \sum_j q_{ij}(f)U_\alpha(f, j)$ is finite and

$$\alpha U_\alpha(f, i) = r(i, f) + \sum_j q_{ij}(f)U_\alpha(f, j). \tag{18}$$

To conclude equations in this section, we discuss the CTMDP model (see Eq. (1)) restricted to $\Pi_s^d(s)$, the set of piecewise semi-stationary policies. In this case, Theorem 2 is still true except that " $\leq U(i)$" should be deleted in Eq. (10) and $A_2(i)$, defined by Eq. (8), should be redefined by

$$A_2(i) = \{a \in A_1(i) | \text{there is } f \in F \text{ such that } f(i) = a \text{ and } U_\alpha(f, i) > -\infty\}.$$

Thus, Condition C is trivial. By noting that Corollary 2 and 4 also hold for f. The following theorem can be proved similarly as Theorem 5 and 6.

THEOREM 8: Restricted to $\Pi_s^d(s), U_\alpha^{*d} := \sup\{U_\alpha(\pi)|\pi \in \Pi_s^d(s)\}$ satisfies Eq. (14), moreover, $U_\alpha^s := \sup\{U_\alpha(f)|f \in F\}$ satisfies Eq. (14) iff $U_\alpha^{*d} = U_\alpha^s$.

6 Conclusions

This paper discussed CTMDP with expected discounted total rewards under the necessary conditions that the model is well defined. We partitioned the state space into three subsets, on which the optimal value is negative infinity, positive infinity and finite respectively. Thus the discussion on the CTMDP could be restricted in the sub-state space with finite optimal value (we call it a sub-CTMDP). In fact, the reward rate function of this sub-CTMDP is finite and is bounded above in the action. Finally, we showed, for this sub-CTMDP, its optimality equation and the optimality of policies achieving the optimality equation.

Further research may include if we can deal with the state partition and action elimination directly on the optimality equation such that the optimality equation can be obtained whenever it is well defined. Also, Condition C may be proved.

References

1. Kakumanu, P.V.: Continuous Time Markov Decision Models with Applications to Optimization Problems. Technical Report **63**, Dept. of Oper. Res., Cornell Univ. (1969)
2. Lewis, M.E. and Puterman, M.L.: A Probabilistic Analysis of Bias Optimality in Unichain Markov Decision Processes. IEEE Trans. on Autom. Contr. **46** (2001) 96–100
3. Lippman, S.A.: On Dynamic Programming with Unbounded Rewards, Mgt. Sci. **21** (1975) 1225–1233
4. Cassandras, C.G., Pepyne, D.L. and Wardi, Y.: Optimal control of A Class of Hybrid Systems. IEEE Trans. on AC **46** (2001) 398–415
5. Song, J.: Continuous Time Markov Decision Processes with Nonuniformly Bounded Transition Rate Family, Scientia Sinica Series A, **11** (1988) 1281–1290
6. Hu, Q.: CTMDP and Its Relationship with DTMDP. Chinese Sci. Bull. **35** (1990) 710–714
7. Serfozo, R.F.: An Equivalence Between Continuous and Discrete Time Markov Decision Processes, J. Oper. Res. **27** (1979) 60–70
8. Hou, B.: Continuous-time Markov Decision Processes Programming with Polynomical Reward, Thesis, Institute of Appl. Math. Academic Sinica, Bejing (1986).
9. Guo, X.P. and Zhu, W.P.: Denumerable-state Continuous-time Markov Decision Processes with Unbounded Transition and Reward Rates under the Discounted Criterion. J. Appl. Prob. **39** (2002) 233–250
10. Guo, X.P. and Zhu, W.P.: Denumerable-state Continuous-time Markov Decision Processes with Unbounded Cost and Transition Rates under Average Criterion. ANZIAM J. **43** (2002) 541–557
11. Hu, Q. and Xu, C.: The Finiteness of the Reward Function and the Optimal Value Function in Markov Decision Processes. J. Math. Methods in Ope. Res. **49** (1999) 255–266
12. Chung, K.L.: Markov Chains with Stationary Transition Probabilities. Springer-Verlag (1960)
13. Kuczura, A.: Piecewise Markov Processes. SIAM J. Appl. Math. **24** (1973) 169–181

Core Equivalence in Economy for Modal Logic*

Takashi Matsuhisa**

Department of Liberal Arts and Sciences, Ibaraki National College of Technology
Nakane 866, Hitachinaka-shi, Ibaraki 312-8508, Japan.
mathisa@ge.ibaraki-ct.ac.jp

Abstract. We investigate a pure exchange economy under uncertainty with emphasis on the logical point of view; the traders are assumed to have a multi-modal logic with non-partitional information structures. We propose a generalized notion of rational expectations equilibrium for the economy and we show the core equivalence theorem: The ex-post core for the economy coincides with the set of all its rational expectations equilibria.

Keywords: Multi-modal logic, Pure exchange economy under reflexive information structure, Ex-post core, Rational expectations equilibrium, Core equivalence theorem.
Journal of Economic Literature Classification: D51, D84, D52, C72.

1 Introduction

This article relates economies and multi-agent modal logic. Let us consider a pure exchange atomless economy under uncertainty. As far as the standard notion of economy either with complete information or with incomplete information, the role of traders' knowledge and beliefs remains obscured: The economy has not been investigated from the epistemic point of view. Here this article aims to fill that gap. Specifically, we seek epistemic conditions of traders' knowledge for the equivalence between equilibria and core in a pure exchange atomless economy under uncertainty.

The purposes of this article are: First to propose the multi-modal logic **KT** by which the traders use making their decision, secondly to establish the extended notion of rational expectations equilibrium for the economy, and finally to investigate the relationship between the ex-post core and the rational expectations equilibrium allocations with emphasis on modal logical point of view.

The stage is set by the following: Suppose that the trader have the multi-agent modal logic **KT**: It is an extension of the propositional logic with many modal operators requiring only the axiom (T) " each traders does not know a sentence whenever it is not true." The logic have non-partitional information structures, each of which gives an interpretation of the logic. Each trader has

* This is an extended abstract and the final form will be published elsewhere.

** Partially supported by the Grant-in-Aid for Scientific Research(C)(2)(No.14540145) in the Japan Society for the Promotion of Sciences.

P.M.A. Sloot et al. (Eds.): ICCS 2003, LNCS 2658, pp. 74–83, 2003.

own utility function which is measurable, but he/she is not assumed to know the function completely. It is shown that

Main Theorem. (Core equivalence theorem). *In a pure exchange atomless economy under generalized information, assume that the traders have the multi-modal logic* **KT** *and they are risk averse. Then the ex-post core coincides with the set of all rational expectations equilibrium allocations for the economy.*

Many authors have investigated several notions of core in an economy under asymmetric information (e.g., Wilson [11], Volij [10], Einy et al [5] and others). The serious limitations of the analysis in these researches are its use of the 'partition' structure by which the traders receive information. The structure is the Kripke semantics for the modal logic **S5**;[1] it is obtained if each trader t's possibility operator $P_t : \Omega \to 2^{\Omega}$ assigning to each state ω in a state space Ω the information set $P_t(\omega)$ that t possesses in ω is reflexive, transitive and symmetric. From the epistemic point of view, this entails t's knowledge operator $K_t : 2^{\Omega} \to 2^{\Omega}$ that satisfies 'Truth' axiom **T**: $K_t(E) \subseteqq E$ (what is known is true), the 'positive introspection' axiom **4**: $K_t(E) \subseteqq K_t(K_t(E))$ (we know what we do) and the 'negative introspection' axiom **5**: $\Omega \setminus K_t(E) \subseteqq K_t(\Omega \setminus K_t(E))$ (we know what we do not know).[2]

One of these requirements, symmetry (or the equivalent axiom **5**), is indeed so strong that describes the hyper-rationality of traders, and thus it is particularly objectionable. The recent idea of 'bounded rationality' suggests dropping such assumption since real people are not complete reasoners. In this article we weaken both transitivity and symmetry imposing only reflexivity. As has already been pointed out in the literature, this relaxation can potentially yield important results in a world with imperfectly Bayesian agents (e.g. Geanakoplos [7]).

The idea has been performed in different settings. Among other things Geanakoplos [7] showed the no speculation theorem in the extended rational expectations equilibrium under the assumption that the information structure is reflexive, transitive and *nested* (Corollary 3.2 in Geanakoplos [7]). The condition 'nestedness' is interpreted as a requisite on the 'memory' of the trader.

However all those researches have been lacked the logics that represents the traders' knowledge. This article proposes the multi-modal logic of the traders and the economies under generalized information structure as the models for the logic. In the structure we shall relax the transitivity, and we extend the ex-post core equivalence theorem of Einy et al [5] into the models with removing out transitivity and symmetry.

This article is organized as follows: In Section 2 we present the multi-modal logic **KT** and give its finite model property. Further we introduce the notion "economy for logic **KT**", a generalized notion of rational expectations equilibrium and ex-post core for the economy. Section 3 gives the existence theorem of rational expectations equilibrium. In Section 4 we give an outline of the proof of

[1] C.f.: Chellas [3], Fagin, Halpern et al [6].

[2] C.f.: Bacharach [2], Fagin, Halpern et al [6].

Main theorem. Finally we conclude by giving some remarks about the assumptions of the theorem. The detailed proofs except for Theorem 1 and discussions are given in Matsuhisa, Ishikawa and Hoshino [9].

2 Economy for Multi-modal Logic

2.1 Logic of Knowledge KT

Let T be a set of *traders* and $t \in T$. Let us modal logics for traders as folows: The *sentences* of the language form the least set containing each *atomic* sentence $\mathbf{P}_m (m = 0, 1, 2, \dots)$ closed under the following operations:

- nullary operators for *falsity* $\bot$ and for *truth* $\top$;
- unary and binary syntactic operations for *negation* $\neg$, *conditionality* $\rightarrow$ and *conjunction* $\wedge$, respectively;
- unary operation for *modality* $\Box_t$ with $t \in T$.

Other such operations are defined in terms of those in usual ways. The intended interpretation of $\Box_t\varphi$ is the sentence that 'trader t knows a sentence φ.'

A *modal logic* L is a set of sentences containing all truth-functional tautologies and closed under substitution and modus ponens. A modal logic L' is an *extension* of L if $L \subseteqq L'$. A sentence φ in a modal logic L is a *theorem* of L, written by $\vdash_L \varphi$. Other proof-theoretical notions such as *L-deducibility, L-consistency, L-maximality* are defined in usual ways. (See, Chellas [3].)

A *system of traders' knowledge* is a modal logic L closed under the $2n + 3$ rules of inference ($\mathrm{RE}_\Box$) and containing the schema (N), (M), (C), and (T): For every $t \in T$,

$$(\mathrm{RE}_\Box) \qquad \frac{\varphi \longleftrightarrow \psi}{\Box_t\varphi \longleftrightarrow \Box_t\psi}$$

$$(\mathrm{N}) \qquad \Box_t\top;$$

$$(\mathrm{M}) \qquad \Box_t(\varphi \wedge \psi) \longrightarrow (\Box_t\varphi \wedge \Box_t\psi);$$

$$(\mathrm{C}) \qquad (\Box_t\varphi \wedge \Box_t\psi) \longrightarrow \Box_t(\varphi \wedge \psi);$$

$$(\mathrm{T}) \qquad \Box_t\varphi \longrightarrow \varphi.$$

Definition 1. The *multi-modal logic* **KT** is the minimal system of trades' knowledge.

2.2 Information and Knowledge[3]

Trader t's information structure is a couple $\langle \Omega, P_t \rangle$, in which Ω be a non-empty set and P_t is a mapping of Ω into 2^Ω. It is said to be *reflexive* if

Ref $\quad \omega \in P_t(\omega) \quad$ for every $\omega \in \Omega$,

[3] See Fagin, Halpern et al [6].

and it is said to be *transitive* if

Trn $\xi \in P_t(\omega)$ implies $P_t(\xi) \subseteqq P_t(\omega)$ for any $\xi, \omega \in \Omega$.

An *information structure* is a structure $\langle \Omega, (P_t)_{t \in T} \rangle$ where Ω is common for all trader, and it is called an *RT-information structure* if each P_t is reflexive and transitive.

Given our interpretation, a trader t for whom $P_t(\omega) \subseteqq E$ knows, in the state ω, that some state in the event E has occurred. In this case we say that at the state ω the trader t knows E. i's *knowledge operator* K_t on 2^Ω is defined by $K_t(E) = \{\omega \in \Omega | P_t(\omega) \subseteqq E\}$. The set $P_t(\omega)$ will be interpreted as the set of all the states of nature that t knows to be possible at ω, and $K_t E$ will be interpreted as the set of states of nature for which t knows E to be possible. We will therefore call P_t t's *possibility operator* on Ω and also will call $P_t(\omega)$ t's *possibility set* at ω. A possibility operator P_t is determined by the knowledge operator K_t such as $P_t(\omega) = \bigcap_{K_t E \ni \omega} E$. However it is also noted that the operator P_t cannot be uniquely determined by the knowledge operator K_t when P_t does not satisfy the both conditions **Ref** and **Trn**.

A *partitional* information structure is an RT-information structure $\langle \Omega, (P_t)_{t \in T} \rangle$ with the additional condition: For each $t \in T$ and every $\omega \in \Omega$,

Sym $\xi \in P_t(\omega)$ implies $P_t(\xi) \ni \omega$.

2.3 Finite Model Property

A *model on* an information structure is a triple $\mathcal{M} = \langle \Omega, (P_t)_{t \in T}, V \rangle$, in which $\langle \Omega, (P_t)_{t \in T} \rangle$ is an information structure and a mapping V assigns either **true** or **false** to every $\omega \in \Omega$ and to every atomic sentence $\mathbf{P}_m$. The model $\mathcal{M}$ is called *finite* if Ω is a finite set.

Definition 2. By $\models^{\mathcal{M}}_{\omega} \varphi$, we mean that a sentence φ is *true* at a state ω in a model $\mathcal{M}$. *Truth at a state* ω in $\mathcal{M}$ is defined by the inductive way as follows:

1. $\models^{\mathcal{M}}_{\omega} \mathbf{P}_m$ if and only if $V(\omega, \mathbf{P}_m) = \mathbf{true}$, for $m = 0, 1, 2, \ldots$;
2. $\models^{\mathcal{M}}_{\omega} \top$, and not $\models^{\mathcal{M}}_{\omega} \bot$;
3. $\models^{\mathcal{M}}_{\omega} \neg\varphi$ if and only if not $\models^{\mathcal{M}}_{\omega} \varphi$;
4. $\models^{\mathcal{M}}_{\omega} \varphi \longrightarrow \psi$ if and only if $\models^{\mathcal{M}}_{\omega} \varphi$ implies $\models^{\mathcal{M}}_{\omega} \psi$;
5. $\models^{\mathcal{M}}_{\omega} \varphi \wedge \psi$ if and only if $\models^{\mathcal{M}}_{\omega} \varphi$ and $\models^{\mathcal{M}}_{\omega} \psi$;
6. $\models^{\mathcal{M}}_{\omega} \Box_t \varphi$ if and only if $P_t(\omega) \subseteq ||\varphi||^{\mathcal{M}}$, for $t \in T$;

Where $||\varphi||^{\mathcal{M}}$ denotes the set of all the states in $\mathcal{M}$ at which φ is true; this is called the *truth set of* φ. We say that a sentence φ is *true in the model* $\mathcal{M}$ and write $\models^{\mathcal{M}} \varphi$ if $\models^{\mathcal{M}}_{\omega} \varphi$ for every state ω in $\mathcal{M}$. A sentence is said to be *valid in* an information structure if it is true in every model on the information structure.

Let Σ be a set of sentences. We say that $\mathcal{M}$ is a *model for* Σ if every member of Σ is true in $\mathcal{M}$. An information structure is said to be *for* Σ if every member of Σ is valid in it. Let $\boldsymbol{R}$ be a class of models on a *reflexive* information structure. A modal logic L is *sound with respect to* $\boldsymbol{R}$ if every member of $\boldsymbol{R}$ is a model for L. It is *complete with respect to* $\boldsymbol{R}$ if every sentence valid in all members of $\boldsymbol{R}$ is a theorem of L. We say that L is *determined by* $\boldsymbol{R}$ if L is sound and complete with respect to $\boldsymbol{R}$.

A modal logic L is said to have the *finite model property* if it is determined by the class of all finite models in $\boldsymbol{R}$. The following theorem can be shown by the same way described in Chellas [3].

Theorem 1. *The multi-modal logic* **KT** *has the finite model property.*

From now on we consider t's information structure $\langle \Omega, (P_t)_{t\in T}, V\rangle$ as a finite model for $\mathbf{KT}_t$.

2.4 Economy for Logic KT

Let Ω be a non-empty *finite* set called a *state space*, and let 2^{Ω} denote the field of all subsets of Ω. Each member of 2^{Ω} is called an *event* and each element of Ω a *state*. The space of the traders is a measurable space (T, Σ, μ) in which T is a set of traders, Σ is a σ-field of subsets of T whose elements are called *coalitions*, and μ is a measure on Σ.

A *pure exchange economy under uncertainty* is a tuple $\langle T, \Sigma, \mu, \Omega, \mathbf{e}, (U_t)_{t\in T}, (\pi_t)_{t\in T}\rangle$ consisting of the following structure and interpretations: There are l commodities in each state of the state space Ω , and it is assumed that Ω is *finite* and that the consumption set of trader t is $\mathbb{R}^l_+$;

- (T, Σ, μ) is the measure space of the traders;
- $\mathbf{e} : T \times \Omega \to \mathbb{R}^l_+$ is t's *initial endowment* such that $\mathbf{e}(\cdot, \omega)$ is μ-measurable for each $\omega \in \Omega$;
- $U_t : \mathbb{R}^l_+ \times \Omega \to \mathbb{R}$ is t's von-Neumann and Morgenstern utility function;
- π_t is a subjective prior on Ω for a trader $t \in T$.

For simplicity it is assumed that (Ω, π_t) is a finite probability space with π_t *full support*[4]for almost all $t \in T$.

Definition 3. An *pure exchange economy for logic* **KT** is a structure $\mathcal{E}^{KT} = \langle \mathcal{E}, (P_t)_{t\in T}, V\rangle$, in which $\mathcal{E}$ is a pure exchange economy such that $\langle \Omega, (P_t)_{t\in T}, V\rangle$ is a finite model for the logic **KT**. Furthermore it is called an economy *under RT-information structure* if each P_t is a reflexive and transitive information structure.

Remark 1. An economy under asymmetric information is an economy $\mathcal{E}^{KT}$ under partitional information structure (i.e., each P_t satisfies the three conditions **Ref**, **Trn** and **Sym**.)

[4] I.e., $\pi_t(\omega) \gneqq 0$ for every $\omega \in \Omega$.

Let $\mathcal{E}^{KT}$ be a pure exchange economy for logic **KT**. We denote by $\mathcal{F}_t$ the field generated by $\{P_t(\omega) \mid \omega \in \Omega\}$ and by $\mathcal{F}$ the join of all $\mathcal{F}_t (t \in T)$; i.e. $\mathcal{F} = \vee_{t \in T} \mathcal{F}_t$. We denote by $\{A(\omega) \mid \omega \in \Omega\}$ the set of all atoms $A(\omega)$ containing ω of the field $\mathcal{F} = \vee_{t \in T} \mathcal{F}_t$.

Remark 2. The set of atoms $\{A_t(\omega) \mid \omega \in \Omega\}$ of $\mathcal{F}_t$ does not necessarily coincide with the partition induced from P_t.

We shall often refer to the following conditions: For every $t \in T$,

A-1 $\int_T \boldsymbol{e}(t,\omega) d\mu \gneqq 0$ for all $\omega \in \Omega$.
A-2 $\boldsymbol{e}(t,\cdot)$ is $\mathcal{F}_t$-measurable
A-3 For each $x \in \mathbb{R}^l_+$, the function $U_t(x,\cdot)$ is $\mathcal{F}_t$-measurable, and the function: $T \times \mathbb{R}^l_+ \to \mathbb{R}, (t,x) \mapsto U_t(x,\omega)$ is $\Sigma \times \mathcal{B}$-measurable where $\mathcal{B}$ is the σ-field of all Borel subsets of $\mathbb{R}^l_+$.
A-4 For each $\omega \in \Omega$, the function $U_t(\cdot,\omega)$ is continuous, strictly increasing on $\mathbb{R}^l_+$.
A-5 For each $\omega \in \Omega$, the function $U_t(\cdot,\omega)$ is continuous, increasing, strictly quasi-concave and *non-saturated*[5] on $\mathbb{R}^l_+$.

Remark 3. It is plainly observed that **A-5** implies **A-4**. We note also that **A-3** does not mean that trader t knows his/her utility function $U_t(\cdot,\omega)$.[6]

2.5 Ex-post Core

An *assignment* $\boldsymbol{x}$ is a mapping from $T \times \Omega$ into $\mathbb{R}^l_+$ such that for every $\omega \in \Omega$, the function $\boldsymbol{x}(\cdot,\omega)$ is μ-measurable, and for each $t \in T$, the function $\boldsymbol{x}(t,\cdot)$ is at most $\mathcal{F}$-measurable. We denote by $\mathcal{A}ss(\mathcal{E}^{KT})$ the set of all assignments for the economy $\mathcal{E}^{KT}$.

By an *allocation* we mean an assignment $\boldsymbol{a}$ such that for every $\omega \in \Omega$,

$$\int_T \boldsymbol{a}(t,\omega) d\mu \leqq \int_T \boldsymbol{e}(t,\omega) d\mu.$$

We denote by $\mathcal{A}lc(\mathcal{E}^{KT})$ the set of all allocations, and for each $t \in T$ we denote by $\mathcal{A}lc(\mathcal{E}^{KT})_t$ the set of all the functions $\boldsymbol{a}(t,\cdot)$ for $\boldsymbol{a} \in \mathcal{A}lc(\mathcal{E}^{KT})$.

An assignment $\boldsymbol{y}$ is called an *ex-post improvement* of a coalition $S \in \Sigma$ on an assignment $\boldsymbol{x}$ at a state $\omega \in \Omega$ if

Imp1 $\mu(S) \gneqq 0$;
Imp2 $\int_S \boldsymbol{y}(t,\omega) d\mu \leqq \int_S \boldsymbol{e}(t,\omega) d\mu$; and
Imp3 $U_t(\boldsymbol{y}(t,\omega),\omega) \gneqq U_t(\boldsymbol{x}(t,\omega),\omega)$ for almost all $t \in S$.

We shall present the notion of core in an economy $\mathcal{E}^{KT}$ for logic **KT**.

---

[5] I.e.; For any $x \in \mathbb{R}^l_+$ there exists an $x' \in \mathbb{R}^l_+$ such that $U_t(x',\omega) \gneqq U_t(x,\omega)$.
[6] That is, $\omega \notin K_t([U_t(\cdot,\omega)])$ for some $\omega \in \Omega$, where $[U_t(\cdot,\omega)] := \{ \xi \in \Omega \mid U_t(\cdot,\xi) = U_t(\cdot,\omega)\}$. This is because the information structure is not a partitional structure.

Definition 4. An allocation $\boldsymbol{x}$ is said to be an *ex-post core* allocation of a pure exchange economy for logic **KT** if there is no coalition having an ex-post improvement on $\boldsymbol{x}$ at any state $\omega \in \Omega$. The *ex-post core* denoted by $\mathcal{C}^{ExP}(\mathcal{E}^{KT})$ is the set of all the ex-post core allocations of $\mathcal{E}^{KT}$.

Let $\mathcal{E}^{KT}$ be the pure exchange economy for logic **KT** and $\mathcal{E}^{KT}(\omega)$ the economy with complete information $\langle T, \Sigma, \mu, \mathbf{e}(\cdot, \omega), (U_t(\cdot, \omega))_{t \in T} \rangle$ for each $\omega \in \Omega$. We denote by $C(\mathcal{E}^{KT}(\omega))$ the set of all core allocations for $\mathcal{E}^{KT}(\omega)$.

Proposition 1. *Let $\mathcal{E}^{KT}$ be a pure exchange economy for logic* **KT** *satisfying the conditions* **A-1**, **A-2** *and* **A-3**. *Suppose that the economy is atomless (that is, (T, Σ, μ) is non-atomic measurable space.) The ex-post core of $\mathcal{E}^{KT}$ is non-empty (i.e., $\mathcal{C}^{ExP}(\mathcal{E}^{KT}) \neq \emptyset$). Moreover, $\mathcal{C}^{ExP}(\mathcal{E}^{KT})$ coincides with the set of all assignments $\boldsymbol{x}$ such that $\boldsymbol{x}(\cdot, \omega)$ is a core allocation for the economy $\mathcal{E}^{KT}(\omega)$ for all $\omega \in \Omega$: i.e.,*

$$\mathcal{C}^{ExP}(\mathcal{E}^{KT}) = \{\boldsymbol{x} \in \mathcal{A}lc(\mathcal{E}^{KT}) \mid \boldsymbol{x}(\cdot, \omega)) \in \mathcal{C}(\mathcal{E}^{KT}(\omega)) \text{ for all } \omega \in \Omega\}.$$

□

2.6 Expectation and Pareto Optimality

Let $\mathcal{E}^{KT}$ be the pure exchange economy for logic **KT**. We denote by $\mathbf{E}_t[U_t(\boldsymbol{x}(t, \cdot)]$ the *ex-ante* expectation defined by

$$\mathbf{E}_t[U_t(\boldsymbol{x}(t, \cdot)] := \sum_{\omega \in \Omega} U_t(\boldsymbol{x}(t, \omega), \omega)\pi_t(\omega)$$

for each $\boldsymbol{x} \in \mathcal{A}ss(\mathcal{E}^{KT})$. We denote by $\mathbf{E}_t[U_t(\boldsymbol{x}(t, \cdot))|P_t](\omega)$ the *interim* expectation defined by

$$\mathbf{E}_t[U_t(\boldsymbol{x}(t, \cdot)|P_t](\omega) := \sum_{\xi \in \Omega} U_t(\boldsymbol{x}(t, \xi), \xi)\pi_t(\xi|P_t(\omega)).$$

Definition 5. An allocation $\boldsymbol{x}$ in an economy $\mathcal{E}^{KT}$ is said to be *ex-ante Pareto-optimal* if there is no allocation $\boldsymbol{a}$ with the two properties as follows:

PO-1 For almost all $t \in T$, $\mathbf{E}_t[U_t(\boldsymbol{a}(t, \cdot)] \geqq \mathbf{E}_t[U_t(\boldsymbol{x}(t, \cdot)]$.
PO-2 The set of all the traders $s \in T$ such that

$$\mathbf{E}_s[U_s(\boldsymbol{a}(t, \cdot)] \gneqq \mathbf{E}_s[U_s(\boldsymbol{x}(t, \cdot)].$$

is not a μ-null set.

2.7 Rational Expectations Equilibrium

Let $\mathcal{E}^{KT} = \langle N, \Omega, (e_t)_{t \in T}, (U_t)_{t \in T}, (\pi_t)_{t \in T}, (P_t)_{t \in T} \rangle$ be a pure exchange economy for logic **KT**. A *price system* is a non-zero $\mathcal{F}$-measurable function $p : \Omega \to \mathbb{R}^l_+$. We denote by $\sigma(p)$ the smallest σ-field that p is measurable, and by $\Delta(p)(\omega)$

the atom containing ω of the field $\sigma(p)$. The *budget set* of a trader t at a state ω for a price system p is defined by

$$B_t(\omega, p) := \{ x \in \mathbb{R}^l_+ \mid p(\omega) \cdot x \leqq p(\omega) \cdot \boldsymbol{e}(t, \omega) \}.$$

Let $\Delta(p) \cap P_t : \Omega \to 2^\Omega$ be defined by $(\Delta(p) \cap P_t)(\omega) := \Delta(p)(\omega) \cap P_t(\omega)$; it is plainly observed that the mapping $\Delta(p) \cap P_t$ satisfies **Ref**. We denote by $\sigma(p) \vee \mathcal{F}_t$ the smallest σ-field containing both the fields $\sigma(p)$ and $\mathcal{F}_t$, and by $A_t(p)(\omega)$ the atom containing ω. It is noted that

$$A_t(p)(\omega) = (\Delta(p) \cap A_t)(\omega).$$

Remark 4. If P_t satisfies **Ref** and **Trn** then $\sigma(p) \vee \mathcal{F}_t$ coincides with the field generated by $\Delta(p) \cap P_t$.

We shall propose the extended notion of rational expectations equilibrium for an economy $\mathcal{E}^{KT}$.

Definition 6. A *rational expectations equilibrium* for an economy $\mathcal{E}^{KT}$ under reflexive information structure is a pair $(p, \boldsymbol{x})$, in which p is a price system and $\boldsymbol{x}$ is an allocation satisfying the following conditions:

RE 1 For almost all $t \in T$, $\boldsymbol{x}(t, \cdot)$ is $\sigma(p) \vee \mathcal{F}_t$-measurable.
RE 2 For almost all $t \in T$ and for every $\omega \in \Omega$, $\boldsymbol{x}(t, \omega) \in B_t(\omega, p)$.
RE 3 For almost all $t \in T$, if $\boldsymbol{y}(t, \cdot) : \Omega \to \mathbb{R}^l_+$ is $\sigma(p) \vee \mathcal{F}_t$-measurable with $\boldsymbol{y}(t, \omega) \in B_t(\omega, p)$ for all $\omega \in \Omega$, then

$$\mathbf{E}_t[U_t(\boldsymbol{x}(t, \cdot))|\Delta(p) \cap P_t](\omega) \geqq \mathbf{E}_t[U_t(\boldsymbol{y}(t, \cdot))|\Delta(p) \cap P_t](\omega)$$

pointwise on Ω.
RE 4 For every $\omega \in \Omega$, $\int_T \boldsymbol{x}(t, \omega) d\mu = \int_T \boldsymbol{e}(t, \omega) d\mu$.

The allocation $\boldsymbol{x}$ in $\mathcal{E}^{KT}$ is called a *rational expectations equilibrium allocation.*

We denote by $RE(\mathcal{E}^{KT})$ the set of all the rational expectations equilibria of a pure exchange economy $\mathcal{E}^{KT}$ for logic **KT**, and denote by $\mathcal{R}(\mathcal{E}^{KT})$ the set of all the rational expectations equilibrium allocations for the economy

3 Existence Theorem

We can prove the existence theorem of the generalized rational expectations equilibrium for a pure exchange economy $\mathcal{E}^{KT}$ for logic **KT**. Let $\mathcal{E}^{KT}(\omega)$ be the economy with complete information for each $\omega \in \Omega$. We set by $W(\mathcal{E}^{KT}(\omega))$ the set of all the competitive equilibria for $\mathcal{E}^{KT}(\omega)$, and we denote by $\mathcal{W}(\mathcal{E}^{KT}(\omega))$ the set of all the competitive equilibrium allocations for $\mathcal{E}^{KT}(\omega)$.

Theorem 2. *Let $\mathcal{E}^{KT}$ be a pure exchange economy for logic* **KT** *satisfying the conditions* **A-1**, **A-2**, **A-3** *and* **A-4**. *Suppose that the economy is atomless (that is, (T, Σ, μ) is non-atomic measurable space.) Then there exists a rational expectations equilibrium for the economy; i.e., $RE(\mathcal{E}^{KT}) \neq \emptyset$.*

Proof. See Appendix.

Remark 5. Matsuhisa and Ishikawa [8] shows Theorem 2 for an economy under RT-information structure.

4 Proof of Main Theorem

We can now state explicitly Main theorem in Section 1 as follows:

Theorem 3. *Let $\mathcal{E}^{KT}$ be a pure exchange economy for logic* **KT** *satisfying the conditions* **A-1**, **A-2**, **A-3** *and* **A-4**. *Suppose that the economy is atomless (that is, (T, Σ, μ) is non-atomic measurable space.) Then the ex-post core coincides with the set of all rational expectations equilibrium allocations; i.e., $\mathcal{C}^{ExP}(\mathcal{E}^{KT}) = \mathcal{R}(\mathcal{E}^{KT})$.*

In view of Theorem 2 it is first noted that $\mathcal{R}(\mathcal{E}^{KT}) \neq \emptyset$. Because $\mathcal{E}^{KT}(\omega)$ is an atomless pure exchange economy for each $\omega \in \Omega$, it follows from the core equivalence theorem of Aumann [1] that $\mathcal{C}(\mathcal{E}^{KT}(\omega)) = \mathcal{W}(\mathcal{E}^{KT}(\omega))$ for any $\omega \in \Omega$. We shall observe that Main theorem immediately follows from the above Proposition 1 together with the below Proposition 2:

Proposition 2. *Let $\mathcal{E}^{KT}$ be a pure exchange economy for logic* **KT** *satisfying the conditions* **A-1**, **A-2**, **A-3** *and* **A-4**. *Then the set of all rational expectations equilibrium allocations $\mathcal{R}(\mathcal{E}^{KT})$ coincides with the set of all the assignments $\boldsymbol{x}$ such that $\boldsymbol{x}(\cdot, \omega)$ is a competitive equilibrium allocation for the economy with complete information $\mathcal{E}^{KT}(\omega)$ for all $\omega \in \Omega$; i.e.,*

$$\mathcal{R}(\mathcal{E}^{KT}) = \{\boldsymbol{x} \in \mathcal{A}lc(\mathcal{E}^{KT}) \mid \textit{There is a price system } p \textit{ such that } (p(\omega), \boldsymbol{x}(\cdot, \omega)) \in W(\mathcal{E}^{KT}(\omega)) \textit{ for all } \omega \in \Omega\}.$$

□

5 Concluding Remarks

We shall give a remark about the ancillary assumptions in results in this article. Could we prove the theorems under the generalized information structure removing out the reflexivity? The answer is no vein. If trader t's possibility operator does not satisfy **Ref** then his/her expectation with respect to a price cannot be defined at a state because it is possible that $\Delta(p)(\omega) \cap P_t(\omega) = \emptyset$ for some $\omega \in \Omega$.

Could we prove the theorems without four conditions **A-1**, **A-2**, **A-3** and **A-4** together with **A-5**. The answer is no again. The suppression of any of these assumptions renders the existence theorem of rational expectations equilibrium (Theorem 2) vulnerable to the discussion and the example proposed in Remarks 4.6 of Matsuhisa and Ishikawa (2002).

Appendix

Proof of Theorem 2

In view of the conditions **A-1**,**A-2**, **A-3** and **A-4**, it follows from the existence theorem of a competitive equilibrium for an atomless economy with complete information[7] that for each $\omega \in \Omega$, there exists a competitive equilibrium $(p^*(\omega), \boldsymbol{x}^*(\cdot, \omega)) \in W(\mathcal{E}^{KT}(\omega))$. We take a set of strictly positive numbers $\{k_\omega\}_{\omega \in \Omega}$ such that $k_\omega p^*(\omega) \neq k_\xi p^*(\xi)$ for any $\omega \neq \xi$. We define the pair $(p, \boldsymbol{x})$ as follows: For each $\omega \in \Omega$ and for all $\xi \in A(\omega)$, $p(\xi) := k_\omega p^*(\omega)$ and $\boldsymbol{x}(t, \xi) := \boldsymbol{x}^*(t, \omega)$. It is noted that $\boldsymbol{x}(\cdot, \xi) \in W(\mathcal{E}^{KT}(\omega))$ because $\mathcal{E}^{KT}(\xi) = \mathcal{E}^{KT}(\omega)$, and we note that $\Delta(p)(\omega) = A(\omega)$.

We shall verify that $(p, \boldsymbol{x})$ is a rational expectations equilibrium for $\mathcal{E}^{KT}$: In fact, it is easily seen that p is $\mathcal{F}$-measurable with $\Delta(p)(\omega) = A(\omega)$ and that $\boldsymbol{x}(t, \cdot)$ is $\sigma(p) \vee \mathcal{F}_t$-measurable, so **RE 1** is valid. Because $(\Delta(p) \cap P_t)(\omega) = A(\omega)$ for every $\omega \in \Omega$, it can be plainly observed that $\boldsymbol{x}(t, \cdot)$ satisfies **RE 2**, and it follows from **A-3** that for almost all $t \in T$, $\mathbf{E}_t[U_t(\boldsymbol{x}(t, \cdot))|\Delta(p) \cap P_t](\omega) = U_t(\boldsymbol{x}(t, \omega), \omega)$ On noting that $\mathcal{E}^{KT}(\xi) = \mathcal{E}^{KT}(\omega)$ for any $\xi \in A(\omega)$, it is plainly observed that $(p(\omega), \boldsymbol{x}(t, \omega)) = (k_\omega p^*(\omega), \boldsymbol{x}^*(t, \omega))$ is also a competitive equilibrium for $\mathcal{E}^{KT}(\omega)$ for every $\omega \in \Omega$, and it can be observed by the above equation that **RE 3** is valid for $(p, \boldsymbol{x})$, in completing the proof. □

References

1. Aumann, R. J.: Markets with a continuum of traders, Econometrica 32 (1964) 39–50.
2. Bacharach, M. O.: Some extensions of a claim of Aumann in an axiomatic model of knowledge, Journal of Economic Theory 37 (1985) 167–190.
3. Chellas, B. F.: Modal Logic: An introduction. Cambridge University Press, Cambridge, London, New York, New Rochelle, Melbourne, Sydney (1980)
4. Debreu, G.: Existence of competitive equilibrium. In Arrow, K.J. and Intriligator, M.D., (eds): Handbook of Mathematical Economics, Volume 2. North-Holland Publishing Company, Amsterdam (1982) 697–744.
5. Einy, E., Moreno, D. and Shitovitz, B.: Rational expectations equilibria and the ex-post core of an economy with asymmetric information, Journal of Mathematical Economics 34 (2000) 527–535.
6. Fagin, R., Halpern, J.Y., Moses, Y., and Vardi, M.Y. Reasoning about Knowledge. The MIT Press, Cambridge, Massachusetts, London, England (1995)
7. Geanakoplos, J.: Game theory without partitions, and applications to speculation and consensus, Cowles Foundation Discussion Paper No.914 (1989) (Available in http://cowles.econ.yale.edu)
8. Matsuhisa, T. and Ishikawa, R.: Rational expectations can preclude trades. Working paper. Hitotsubashi Discussion Paper Series No.2002-1 (2002) (Available in http://wakame.econ.hit-u.ac.jp/)
9. Matsuhisa, T., Ishikawa, R. and Hoshino, Y.: Core equivalence in economy under generalized information. Working paper. Hitotsubashi Discussion Paper Series No.2002-12 (2002) (Available in http://wakame.econ.hit-u.ac.jp/)
10. Volij, O.,: Communication, credible improvements and the core of an economy with asymmetric information, International Journal of Game Theory 29 (2000) 63–79.
11. Wilson, R.: Information, efficiency, and the core of an economy, Econometrica 40 (1978) 807–816.

[7] C.f., Theorem 9 in Debreu [4]

Model on Analysis of Industrial Relation Based on the Binary Relation Theory

Kai-ya Wu[1], Xiao-jian Chen[1], Jia-zhong Qian[2], and Ru-zhong Li[2]

[1] School of Business, University of Science and Technology of China, Hefei, 230026, China
wuky2000@vip.sina.com

[2] School of Natural Resources and Environmental Engineering, Hefei University of Technology, Hefei, 230009, China
qjzy@hfut.edu.cn

Abstract. Based on the binary relation theory and Warshall's algorithm, a model on the connected incidence relation of the industry system is set up, and a handy method of quantitative analysis is provided for understanding the industrial structure and the relation. The model is applied to analyze the connected incidence relation of the industrial system in Anhui Province. The results show the model is effective with simple principle and handy operation. And it provides accurate reference for the analysis of the connected effects among different industries as well as the adjustment and optimism of the industrial structure.

1 Introduction

The binary relation indicates certain relevance between two elements in the set, which largely exist in the economic phenomenon. There are various relation and affections of different levels in the industrial system. By means of the input-output table, the analysis of the industrial relation aim to analyze the relation happening in the process of producing, distributing and exchanging among industries (Zhi-biao Liu, 2001).

In recent years, the analysis methods have been improved a lot. For instance, the optimal theory, as well as the Graph Model (Bing-xin Zhao, 1996.) and Degree-Hierarchy Structure Model (Liu and Zhou, 1999) are used in the analysis of the industrial relation. This essay utilizes exploringly the binary relation theory to establish the incidence relation of the industry system and to calculate the connected incidence relation through Warshall's algorithm. From this, it defines the influential industrial set under analyzing the industrial relation of industry system and the evolving regular of industrial structure.

2 Principle & Method Base

2.1 Establish the Incidence Relation r among Industries in the Industrial System

Take the whole industrial system as a set S: take the industries divided according to a certain principle as $1, 2, 3, \cdots n$-the elements of S, then it comes that $S = 1, 2, 3, \cdots n$. Establish the incidence relation r of the set S:

P.M.A. Sloot et al. (Eds.): ICCS 2003, LNCS 2658, pp. 84–89, 2003.

$$r = \{(i,j)|i \in s, j \in s, d_{i,j} \geq \alpha\} \tag{1}$$

In this expression, $\widetilde{d}_{ij}$ is determined by W.leontief's inverse matrix $\widetilde{D} = (\widetilde{d})_{ij}$ and α is the incidence critical value given previously. The compound relation between the incidence relation r and r :

$$r \circ r = \{(p,q)|p \in s, q \in s, \exists t \in s((p,t) \in r, (t,q) \in r)\} \tag{2}$$

If $(p,q)|p \in r \circ r$,that is, an industry t exists in the whole system S, the incidence relation r then happens between p and t. Meanwhile, anther r happens between t and q. therefore, by the compound calculation of the incidence relation r, a larger incidence relation can be obtained in the system S.

We mark the relational matrix of r as $M_r = (\alpha_{ij})_{n \times n}$, in which

$$y = \begin{cases} 1 & (i,j) \in r \\ 0 & (i,j) \notin r \end{cases} \tag{3}$$

We marked the relational matrix of $r \circ r$ as M_{ror}. If there's at least one industry t in S, which makes a relation between p and t, namely $(p,t) \in r$.and a relation between t and q, namely$(t,q) \in r$, then a relation between p and q must form $(p,q) \in r$. However, there may not only one industry t that can meet the system S, another one t can also satisfy the quest $(p,t') \in r$,$(t',q) \in r$. In all these situation,$(p,q) \in r \circ r$ will be available.

In this way, when we scan row p and column q of M_r , if we find at least one t which makes the number of t and q is 1, then the number of the site of row t and column q is also 1. Otherwise, it will be 0. Scan one row of M_r and each column of M_r, we will know M_{ror} and all the other rows of M_{ror} can also be obtained in the similar way.

Thus, we'll get M_{ror} through Boolean's plus method of matrix. That is, $M_{ror} = M_r \circ M_r = (b_{pq})_{n \times n}$,in which $b_{pq} = \bigvee\limits_{t=1}^{n} (\alpha_{pt} \wedge \alpha_{tq})$. where, $\vee$ is Boolean's add, conforming to $0 \vee 0 = 0$, $0 \vee 1 = 1$, $1 \vee 0 = 1$, $1 \vee 1 = 1$; $\wedge$ is Boolean's multiplication, conforming to $0 \wedge 0 = 0$, $0 \wedge 1 = 0$,$1 \wedge 0 = 0$,$1 \wedge 1 = 1$.

According to the incidence relation r generated from the System S, a larger incidence relation $r \circ r$ among the industries of S will be easily got.

2.2 Establish Connected Incidence Relation

In the industrial system, the connotation of the relation is abundant while the affections among the industries are complex. From the perspective of input-output analysis, W.leontief's anti-coefficient can only reflect the strong and the weak of industrial relation, but not the overall relation.

Besides, any kind of industrial development has double effects. On one side, it increases the supply and improves other industries' developments. On the other side, it stimulates the needs which drive other industries' developments. For this reason, we draw into the connected incidence relation R of the system S.

If $r^k = r^{k+1}$ establishes in r, then we name $R = r \cup r^2 \cup r^3 \cup \cdots \cup r^k$ as the connected incidence relation in S. In this expression r^k indicates that there are k compound calculation in r of S, that is $r^k = r \circ r \circ \cdots \circ r$ (here, k equals to the number of r).

Because the industrial system $S = \{1, 2, 3, \cdots, n\}$ is a limited set, the transitive closure of r is $t(r) = r \cup r^2 \cup r^3 \cup \cdots \cup r^n$.When $r^k = r^{k+1} (k \leq n)$,we have

$$\begin{aligned} r^{k+2} &= r^{k+1} \circ r = r^k \circ r = r^{k+1}, \\ r^{k+3} &= r^{k+2} \circ r = r^{k+1} \circ r = r^k \circ r = r^{k+1}, \\ &\cdots, \\ r^n &= r^{k+1}. \end{aligned}$$

In other words, $r^k = r^{k+1} = r^{k+2} = \cdots = r^n$. Therefore, the connected incidence relation $R = r \cup r^2 \cup r^3 \cup \cdots \cup r^n$.

Thus by adopting the famous Warshall's algorithm to calculate the transitive closure, we can quickly get the incidence relational matrix M_R of the connected incidence relation R.

If matrix $M_R = (c_{ij})_{m \times n}$ already got through Warshall's algorithm (Kolman, 2001), the set $\{j | c_{ij} = 1\}$ is called the put industry set of j in the system, while $\{j | c_{ji} = 1\}$ is called the push industry set. Till now, a quite clear quantitative result about the industrial relation in S will emerge.

2.3 Warshall's Algorithm for Computing Transitive Closure

Warshall's algorithm is an efficient method for computing the transitive closure of a relation, which based on the construction of a sequence of zero-one matrices. These matrices are $w_0, w_1, \cdots, w_n$, where $w_0 = M_R$ is the zero-one matrix of this relation, and $w_k = [w_{ij}^{[k]}]$.It can computes $M_{t(R)}$ by efficiently computing $w_0 = M_R, w_1, w_2, \cdots, w_n = M_{t(R)}$.

LEMMA: Let $w_k = [w_{ij}^{[k]}]$ be the zero-one matrix that has a 1 in its $(i, j)th$ position if and only if there is a path from v_i to v_j with interior vertices from the set $\{v_1, v_2, \cdots, v_k\}$. Then $w_{ij}^k = w_{ij}^{[k-1]} \vee (w_{ik}^{[k-1]} \wedge w_{kj}^{[k-1]})$, whenever i,$j$and k are positive integers not exceeding n.

The Lemma give us the means efficiently to compute the matrices $w_k (k = 1, 2, \cdots, n)$. We display psendocode for Warshall's algorithm.

```
Procedure Warshall (M_R : n × n zero-one matrix)
  begin
    w := M_R
    for k:=1 to n
    begin
      for j:=1 to n
        w_ij = w_ij ∨ (w_ik ∧ w_kj)
    end
  end.{w = [w_ij] is M_t(R)}
```

The computational complexity of Warshall's Algorithm can easily be computed in term of bit operations. To find the entry $w_{ij}^{[k]}$ from the entries $w_{ij}^{[k-1]}, w_{ik}^{[k-1]}$ and $w_{kj}^{[k-1]}$ using Lemma above requires two bit operations. To find all n^2 entries of w_k from those of w_{k-1} requires $2n^2$ bit operations. Since warshall's algorithm begins with $w_0 = M_R$ and computes the sequences of $0-1$ matrices $w_0, w_1, \cdots, w_n = M_{t(R)}$, the total number of bit operations is $n \cdot 2n^2 = 2n^3$.

3 Analysis of Relation for Anhui Industrial System as a Case

According to the 40 industrial departments classified in Anhui's Input-output Table in 1997 (Anhui's Input-Output Office,1999) and the complete consuming coefficient table, we choose an appropriate communicative critical value α (here $\alpha = 0.1$) to establish the incidence relational matrix M_r, and to calculate M_R by Warshall operation is below(next page).

Based on the M_R,We can figure out the connected incidence relation of Industrial system directly in Anhui Province. The main results are attained as followings.

(1) There's a group of connected incidence relation in Anhui's system. $S_0 = \{2, 3, 4, 5, 11, 12, 14, 15, 16, 24, 28, 33, 35\}$,that is {Coal mining and selection industry,Petroleum and natural gas mining industry, Metal and nonmetal mining and selection industry, Crude oil product and coke purification industry, Chemical industry, Metal purification and metal products, machine-building industry, Electricity and Vapour products and supply, Freight transportation and preservation, Financial insurance, Social service}.

The internal development and variation of this group can bring about the relative changes in other industries and at the same time improve those industries outside this group. For their strong overall functions, these industries can be the motive of economic development, which serves as the pivot to maintain economy's stability, compatibility and its high pace.

(2)The group S_0 includes resource,energy,service and other basic industries.In developing point of view,these industries must have the privilege be developed,which will pave the way for the continual development of Anhui's economy.

(3)Agriculture (industry code 1) is the basic industry that proposes a particular status in Anhui.On one hand,it can improve the development of food,textile,costume and other groups (code 6,7,8,10,22,29) and meanwhile pull the development of the communicative connected group S_0,so it fully displays that Anhui is a big and strong agriculture province.

(4) Social service (industry code 35),especially the tourist industry,is very important in Anhui.It can not only improve the development of S_0,but also all the other industry's developments.

(5)Financial industry (industry code 35) has impact on all the industries including S_0.It will do good to the whole nation's economy.

$$M_R=\begin{bmatrix}
1&0&0&0&0&1&1&1&0&1&0&0&0&0&0&0&0&0&0&0&0&1&0&0&0&0&0&0&1&0&0&0&0&0&0&0&0&0&0&0\\
1&1&1&1&1&1&1&1&1&1&1&1&1&1&1&1&1&1&1&1&1&1&0&1&1&1&1&1&1&1&0&1&0&0&1&1&0&1&1&0\\
1&1&1&1&1&1&1&1&1&1&1&1&1&1&1&1&1&1&1&1&1&1&0&1&1&1&1&1&1&1&0&1&0&0&1&1&0&1&1&0\\
1&1&1&1&1&1&1&1&1&1&1&1&1&1&1&1&1&1&1&1&1&1&0&1&1&1&1&0&1&1&0&1&0&0&1&1&0&1&1&0\\
1&1&1&1&1&1&1&1&1&1&1&1&1&1&1&1&1&1&1&1&1&1&0&1&1&1&1&1&1&1&0&1&0&0&1&1&0&1&1&0\\
0&0&0&0&0&1&0&0&0&0&0&0&0&0&0&0&0&0&0&0&0&0&0&0&0&0&0&0&0&0&0&0&0&0&0&0&0&0&0&0\\
0&0&0&0&0&0&1&1&0&1&0&0&0&0&0&0&0&0&0&0&0&1&0&0&0&0&0&0&1&0&0&0&0&0&0&0&0&0&0&0\\
0&0&0&0&0&0&0&1&0&0&0&0&0&0&0&0&0&0&0&0&0&0&0&0&0&0&0&0&0&0&0&0&0&0&0&0&0&0&0&0\\
0&0&0&0&0&0&0&0&1&0&0&0&0&0&0&0&0&0&0&0&0&0&0&0&0&0&0&0&0&0&0&0&0&0&0&0&0&0&0&0\\
0&0&0&0&0&0&0&0&0&1&0&0&0&0&0&0&0&0&0&0&0&0&0&0&0&0&0&0&1&0&0&0&0&0&0&0&0&0&0&0\\
1&1&1&1&1&1&1&1&1&1&1&1&1&1&1&1&1&1&1&1&1&1&0&1&1&1&1&1&1&1&0&1&0&0&1&1&0&1&1&0\\
1&1&1&1&1&1&1&1&1&1&1&1&1&1&1&1&1&1&1&1&1&1&0&1&1&1&1&0&1&1&0&1&0&0&1&1&0&1&1&0\\
0&0&0&0&0&0&0&0&0&0&0&0&0&0&0&0&0&0&0&0&0&0&0&0&0&0&0&0&0&0&0&0&0&0&0&1&0&0&0&0\\
1&1&1&1&1&1&1&1&1&1&1&1&1&1&1&1&1&1&1&1&1&1&0&1&1&1&1&0&1&1&0&1&0&0&1&1&0&1&1&0\\
1&1&1&1&1&1&1&1&1&1&1&1&1&1&1&1&1&1&1&1&1&1&0&1&1&1&1&1&1&1&0&1&0&0&1&1&0&1&1&0\\
1&1&1&1&1&1&1&1&1&1&1&1&1&1&1&1&1&1&1&1&1&1&0&1&1&1&1&1&1&1&0&1&0&0&1&1&0&1&1&0\\
0&0&0&0&0&0&0&0&0&0&0&0&0&0&0&0&1&0&0&0&0&0&0&0&0&0&0&0&1&0&0&0&0&0&0&0&0&0&0&0\\
0&0&0&0&0&0&0&0&0&0&0&0&0&0&0&0&0&1&0&0&0&0&0&0&0&0&0&0&0&0&0&0&0&0&0&0&0&0&0&0\\
0&0&0&0&0&0&0&0&0&0&0&0&0&0&0&0&0&0&1&1&0&0&0&0&0&0&0&0&0&0&0&0&0&0&0&0&0&1&0&0\\
0&0&0&0&0&0&0&0&0&0&0&0&0&0&0&0&0&0&0&0&0&0&0&0&0&0&0&0&1&0&0&0&0&0&0&0&0&1&0&0\\
0&0&0&0&0&0&0&0&0&0&0&0&0&0&0&0&0&0&0&0&0&0&0&0&0&0&0&0&0&0&0&0&0&0&0&0&0&0&0&0\\
0&0&0&0&0&0&0&0&0&0&0&0&0&0&0&0&0&0&0&0&0&0&0&0&0&0&0&0&0&0&0&0&0&0&0&0&0&0&0&0\\
0&0&0&0&0&0&0&0&0&0&0&0&0&0&0&0&0&0&0&0&0&0&0&0&0&0&0&0&0&0&0&0&0&0&0&0&0&0&0&0\\
1&1&1&1&1&1&1&1&1&1&1&1&1&1&1&1&1&1&1&1&1&1&0&1&1&1&1&1&1&1&1&1&0&0&1&1&0&1&1&0\\
0&0&0&0&0&0&0&0&0&0&0&0&0&0&0&0&0&0&0&0&0&0&0&1&0&0&0&0&0&0&0&0&0&0&0&0&0&0&0&0\\
0&0&0&0&0&0&0&0&0&0&0&0&0&0&0&0&0&0&0&0&0&0&0&0&0&0&0&0&0&0&0&0&0&0&0&0&0&0&0&0\\
0&0&0&0&0&0&0&0&0&0&0&0&0&0&0&0&0&0&0&0&0&0&0&0&0&0&0&0&0&0&0&0&0&0&0&0&0&0&0&1\\
1&1&1&1&1&1&1&1&1&1&1&1&1&1&1&1&1&1&1&1&1&1&0&1&1&1&1&1&1&1&0&1&0&0&1&1&0&1&1&0\\
0&0&0&0&0&0&0&0&1&0&0&0&0&0&0&0&0&0&0&0&0&0&0&0&0&0&0&0&0&0&0&0&0&0&0&0&0&0&0&0\\
1&1&1&1&1&1&1&1&1&1&1&1&1&1&1&1&1&1&1&1&1&1&0&1&1&1&1&1&1&1&0&1&0&0&1&1&0&1&1&0\\
0&0&0&0&0&0&0&0&0&0&0&0&0&0&0&0&0&0&0&0&0&0&0&0&0&0&0&0&0&0&0&0&0&0&0&0&0&0&0&0\\
0&0&0&0&0&0&0&0&0&0&0&0&0&0&0&0&0&0&0&0&0&0&0&0&0&0&0&0&0&0&0&0&0&0&0&0&0&0&0&0\\
1&1&1&1&1&1&1&1&1&1&1&1&1&1&1&1&1&1&1&1&1&1&1&1&1&1&1&1&1&1&1&1&1&0&1&1&0&1&1&0\\
0&0&0&0&0&0&0&0&0&0&0&0&0&0&0&0&0&0&0&0&0&0&0&0&0&0&0&0&0&0&0&0&0&0&0&0&0&0&0&0\\
1&1&1&1&1&1&1&1&1&1&1&1&1&1&1&1&1&1&1&1&1&1&0&1&1&1&1&1&1&1&0&1&0&0&1&1&0&1&1&0\\
0&0&0&0&0&0&0&0&0&0&0&0&0&0&0&0&0&0&0&0&0&0&0&0&0&0&0&0&0&0&0&0&0&0&0&0&0&0&0&0\\
0&0&0&0&0&0&0&0&0&0&0&0&0&0&0&0&0&0&0&0&0&0&0&0&0&0&0&0&0&0&0&0&0&0&0&0&0&0&0&0\\
0&0&0&0&0&0&0&0&0&0&0&0&0&0&0&0&0&0&0&0&0&0&0&0&0&0&0&0&0&0&0&0&0&0&0&0&0&0&0&0\\
0&0&0&0&0&0&0&0&1&0&0&0&0&0&0&0&0&0&0&0&0&0&0&0&0&0&0&0&0&0&0&0&0&0&0&0&0&0&0&0\\
0&0&0&0&0&0&0&0&0&0&0&0&0&0&0&0&0&0&0&0&0&0&0&0&0&0&0&0&0&0&0&0&0&0&0&0&0&0&0&0
\end{bmatrix}$$

(6)The essence of the dominant industry lies in its leading effect in the industrial structure system.Anhui's practical situation should be taken into account in selecting the dominant industry.But on addition,the agriculture and social service which have special status should also be included.In this way,we can realize the quick increase on the basis of the optimism of the industrial structure.

4 Conclusion

From the above-mentioned analysis, conclusions can be reached, the model based on the binary relation theory and Warshall's algorithm is effective with simple sufficient mathematical theories and convenient operation. And the model is applied. By adopting the statistics of the input-output graph, through the computer programming, the connected incidence relation can be obtained quickly under different incidence critical value, which will offer accurate reference for the analysis of the connected effects among different industries as well as the adjustment and optimism of the industrial structure.

Acknowledgements. This research was supported by the National Social Science Foundation of China (02BJY043) and the National Natural Science Foundation of China(40202027). The authors would like to thank the anonymous reviewer for their help.

References

1. Zhi-biao Liu:Analysis of Modern Industry Economy.Nanjing University Press(2001)
2. Bing-xin Zhao:Researching on the graph model applied to the industrial relation.Systems Engineering-Theory & Practice.2 (1997):39–42
3. Yong-qing Liu,Zhou Chan-shi:Degree-hierarchy structure model for the industry system of Guangdong Province and Its application. Systems Engineering -Theory & Practice.3(1999):116–131
4. Bernard Kolman:Discrete Mathematical Structure.Prentice Hall,Inc(2001)
5. Anhui's Input-Output Office:Anhui's Input-Output Table in 1997

Has Chinese Stock Market Become Efficient? Evidence from a New Approach

Max Chen[1] and Yongmiao Hong[2]

[1] Peking University, Beijing, 100871, PR China,
maxchen@gsm.pku.edu.cn

[2] Cornell University, 424 Uris Hall, Ithaca, NY 14850, U.S.A.
yh20@cornell.edu

Abstract. Using a new statistical procedure suitable to test efficient market hypothesis in presence of volatility clustering, we find significant evidence against the weak form of efficient market hypothesis for both Shanghai and Shenzhen stock markets, although they have become more efficient at the later stage. We also find that Share A markets are more efficient than Share B markets, but there is no clear evidence on which stock market, Shanghai or Shenzhen, is more efficient. These findings are robust to volatility clustering, a key feature of high-frequency financial time series. They have important implications on predictability of stock returns and on efficacy of capital asset pricing and allocation in Chinese economy.

1 Chinese Stock Market: An Emerging Global Market

With In this section, we briefly describe the development of Chinese stock market, which will be helpful for readers to understand our empirical findings. Thanks to the gradual but continuous and massive economic reform initiated in 1978, Chinese economy has been growing rapidly and steadily for more than two decades, with an average of annual GDP growth rate of around 9%. The reforms in restructuring state-own enterprises and pricing commodity markets have recorded remarkable success (see [1], [2] and [3]). Although still different from a competitive market economy, Chinese economy has been growing out of plan and become market-driven in most sectors.

Compared to other aspects of Chinese economic reform, the development of Chinese financial market has lagged behind. The stock markets were introduced in late 1990, 13 years after the economic reform started in 1978. Although Chinese state-owned enterprises did improve productivity in 1980s, non-performing loans to state-owned enterprises by Chinese commercial banks had been accumulating. As a consequence, commercial banks are unwilling and unable to provide further loans to poor-performing state-owned enterprises. On the other hand, private savings in Chinese banking sector have been increasing, with an annual saving rate between 30% to 40% over the last two decades. Chinese government had to find a solution to this dilemma, and the setup of a stock market was apparently a natural choice. There are two stock exchanges markets in China.

P.M.A. Sloot et al. (Eds.): ICCS 2003, LNCS 2658, pp. 90–98, 2003.

One is located in Shenzhen, a southern city next to Hong Kong and one of the five special economic zones in China where market mechanisms were first introduced in China. The other stock exchange is located in Shanghai, the largest industrial and financial center in China. Both Shenzhen and Shanghai stock exchanges are regulated by the official China Securities Regulatory Commission (CSRC), which was set up in 1992, and has been responsible for regulating and monitoring Chinese stock market. CSRC offers two types of stocks—Share A and Share B, for domestic and foreign investors respectively. Share A markets, traded in local (i.e., Chinese) currency, are designed for domestic investors, and Share B markets, introduced in 1992 and traded in major foreign currencies, are designed for foreign investors. The introduction of Share B markets aims at attracting foreign capitals, but Share A and B markets were segmented to avoid adverse impact of financial turmoil from international financial markets. Effort has been made to merge these two markets. Starting from 02/2001, CSRC allows domestic investors holding foreign currencies to invest in Share B markets, and starting from 12/2002, qualified foreign institutional investors—mutual fund management institutions, insurance companies, securities companies, commercial and investment banks are allowed to invest in Share A markets.

Although starting relatively late, Chinese stock market, equipped with most advanced hardware facilities in the world, has developed rather rapidly. Its information delivering system, via its trading network, Internet and Reuters terminals, provides Chinese market updates in a timely manner to all domestic stock trading branches across China and to 150 counties and regions over the world. A great deal of effort has also been devoted to improving market structures, legal systems, and institutional arrangements. The stock market has been playing an increasingly important role in Chinese economy. By the end of October 2002, 1,215 companies, the majority of them stated-own, are listed in Shanghai and Shenzhen stock exchanges markets. By the end of 2001, the total market capitalization was 4,352.2 billion yuan, or 45.37% of Chinese GDP, and the tradable capitalization was 1,446.3 billion yuan, or 15.08% of Chinese GDP.

This later ratio of capitalization is still low compared to mature stock markets, but it is 10 times more than the ratio in 1991. Chinese stock market has apparently become the largest emerging stock market in the world.

Unlike the mature stock markets in developed economies, the majority of investors in Chinese stock market are individuals, accounting for more than 99% in 2002 in terms of the number of opened accounts. Institutional investors only accounts for about 0.5% in 2002, but they contribute about 60% of investment in Chinese stock markets. Also, compared to mature stock markets, the turnover rates and P/E ratios in Chinese stock market have been very high, indicating over-speculative activity.

Given its potential in scale, rapid growth, increasing openness to and integration with international financial markets, it is important and interesting to investigate how Chinese stock market has been performing over the past decade. In this paper, we shall examine whether Chinese stock market has achieved the weak form of market efficiency in the sense of Fama (1970,1991). On one hand, although Chinese government still has certain direct or indirect control of financial resources such as regulating the number of new listed companies, as well as the quota and initial prices of new listed stocks, Chinese stock market has had the most flexible market mechanism in Chinese economy. With the improved legal systems and institutional arrangements, one may expect that Chinese stock market has achieved some form of market efficiency. On the other hand, the existence of government policy intervention, insider trading, revealment of misleading information on listed companies, and some irrational behavior of Chinese individual investors might lead one to expect that Chinese stock market has not been efficient. Market (in)efficiency has immediate interest to both domestic and foreign investors. For example, the predictability of stock returns is closely related to the market timing ability of mutual fund managers for active fund management. More importantly, market (in)efficiency has far-reaching implications for the efficacy of capital asset allocation and pricing. The study of EMH can also shed some light on the development of other emerging financial markets, particularly those of transitional economies (e.g., Russia and Eastern Europe). With the availability of a decade of daily observations, we now have a sufficiently large sample to provide reliable statistical inference on EMH for Chinese stock market.

2 Efficient Market Hypothesis

We now discuss the definition of the weak form of market efficiency, its implication and a suitable and powerful test.

2.1 Weak Form of Efficient Market Hypothesis

Let P_t be a stock price at time t. We define the stock return at time t in terms of the relative percentage change in the stock price P_t from time $t-1$ to time t:

$$Y_t = 100 \ln(P_t/P_{t-1}) \tag{1}$$

Let I_{t-1} denotes the collection of stock returns available at time $t-1$; i.e., $I_{t-1} = \{Y_{t-1}, Y_{t-2}, ...\}$. In exploring the dynamics of how stock return Y_t changes over time, an important and interesting hypothesis is the weak form of EMH, which can be formally stated as

$$\mathbf{H}_0: \qquad E(Y_t|I_{t-1}) = E(Y_t) \text{ almost surely (a.s.).} \tag{2}$$

intuitively, the unconditional mean $E(Y_t)$ is the long-run average stock return, and the conditional mean $E(Y_t|I_{t-1})$ is the best one-step-ahead stock return one can expect to obtain by fully and efficiently utilizing I_{t-1}, the information on the entire past history of stock returns. When $\mathbf{H}_0$ in (2) holds, no trading strategy can beat the market systematically by earning an extra return higher than the market average over the long run while bearing the same risk. Of course, investors may still be able to beat the market in the short-run by sheer luck.

When I_{t-1} is extended to include other information, we could define the semi-strong form or strong form of market efficiency. See [4], [5] and [6] for more discussion. We emphasize that EMH in form of (2) can be derived from the stochastic Euler equation associated with an intertemporal utility maximization of a representative investor subject to a budget constraint (e.g., [7]).

2.2 A Generalized Spectral Derivative Approach

Hong and Lee [8] propose a specification test for the adequacy of a time series conditional mean model with estimated parameters. It is based on the generalized spectrum proposed in [9], which is an analytic tool for nonlinear time series, just as power spectrum is an analytic tool for linear time series (cf. [10], [11]).

Suppose $\{Y_t\}$ is a strictly stationary process with marginal characteristic function $\varphi(u) \equiv E(e^{iuY_t})$ and pairwise joint characteristic function $\varphi_j(u,v) \equiv E(e^{iuY_t + ivY_{t-|j|}})$, where $i \equiv \sqrt{-1}$, $u, v \in (-\infty, \infty)$ and $j = 0, \pm 1, \cdots$. The basic idea of the generalized spectrum is to first transform the data via an exponential function

$$Y_t \longrightarrow \exp(iuY_t) \tag{3}$$

and then consider the spectrum of the transformed series $\{e^{iuY_t}\}$:

$$f(\omega, u, v) \equiv \frac{1}{2\pi} \sum_{j=-\infty}^{\infty} \sigma_j(u,v) e^{-ij\omega}, \qquad \omega \in [-\pi, \pi] \tag{4}$$

where ω is the frequency, $\sigma_j(u,v)$ is the autocovariance function of the transformed series:

$$\sigma_j(u,v) \equiv \mathrm{cov}(e^{iuY_t}, e^{ivY_{t-|j|}}), \qquad j = 0, \pm 1, \cdots \tag{5}$$

The function $f(\omega, u, v)$ can capture any type of pairwise serial dependence in $\{Y_t\}$, i.e., dependence between Y_t and Y_{t-j} for any nonzero lag j, including the dependent processes with zero autocorrelation. The generalized spectrum $f(\omega, u, v)$ itself is not suitable for testing EMH, because it can capture serial dependence not only in mean but also in higher order conditional moments. For example, the generalized spectrum $f(\omega, u, v)$ can capture the ARCH process, which is an *m.d.s.*

The resulting test statistic in [8] is

$$M_1(p) \equiv \left[\sum_{j=1}^{T-1} k^2(j/p)(T-j) \int \left| \hat{\sigma}_j^{(1,0)}(0,v) \right|^2 dW(v) - \hat{C}_1(p) \right] / \sqrt{\hat{D}_1(p)} \tag{6}$$

Where, $p \equiv p(T)$ is a smoothing parameter called bandwidth, and $k(\cdot)$ is a symmetric kernel function that assigns a weight to each lag j. Examples of $k(\cdot)$ include the Bartlett kernel and the Parzen kernel. And $W : \mathbf{R} \longrightarrow \mathbf{R}^+$ is a nondecreasing function that weights set about zero equally,

$$\hat{C}_1(p) = \sum_{j=1}^{T-1} k^2(j/p) \frac{1}{T-j} \sum_{t=j+1}^{T-1} (Y_t - \bar{Y})^2 \int \left| \hat{\psi}_{t-j}(v) \right|^2 dW(v) \tag{7}$$

$$\hat{D}_1(p) = 2\sum_{j=1}^{T-2}\sum_{l=1}^{T-2} k^2(j/p)k^2(l/p)\times$$

$$\int\int\left|\frac{1}{T-\max(j,l)}\sum_{t=\max(j,l)+1}^{T}(Y_t-\bar{Y})^2\hat{\psi}_{t-j}(v)\hat{\psi}_{t-l}(v')\right|^2 dW(v)dW(v') \quad (8)$$

where $\hat{\psi}_t(v) = e^{ivY_t} - \hat{\varphi}(v)$, and $\hat{\varphi}(v) = T^{-1}\sum_{t=1}^{T} e^{ivY_t}$. Under EMH, we have

$$M_1(p) \to N(0,1) \textit{ in distribution} \quad (9)$$

provided $p \equiv p(T) \to \infty$ as $T \to \infty$ and certain regularity conditions hold.[3] Upper-tailed N(0,1) critical values (e.g., 1.645 at the 5% level) should be used.

The $M_1(p)$ test has other appealing features. As explained earlier, the generalized spectral derivative $f^{(0,1,0)}(\omega,0,v)$ only focuses on checking serial dependence in conditional mean, and thus is suitable to test EMH. It will not falsely reject EMH when there exists volatility clustering and serial dependence in higher order conditional moment. On the other hand, $f^{(0,1,0)}(\omega,0,v)$ can detect both linear and nonlinear departures from EMH. Thus, it has more power against a wider range of departures from EMH than any autocorrelation-based tests, even if the latter were applicable in the presence of conditional heteroskedasticity of unknown form. Moreover, the $M_1(p)$ test can check a large number of lags. This will ensure power against departures from EMH at an unknown high lag order. Usually, the use of a large number of lags might be not powerful against many practical alternatives, due to the loss of a large number of degrees of freedom. Fortunately, this is not the case with the $M_1(p)$ test, which discounts higher order lags via the kernel $k(\cdot)$. The downward weighting by $k(\cdot)$ ensures good power of $M_1(p)$ in practice because it is consistent with the stylized fact that financial markets are usually more affected by the recent past events than by the remote past events. This is one of the advantage of frequency domain analysis over time domain analysis. The latter usually gives equal weighting to each lag, which is obviously not efficient when a large number of lags are used.

3 Data

Historically the majority of studies on EMH have focused on the predictability of common stock returns. Likewise, we will consider eight major stock indices in both Shanghai and Shenzhen stock markets from 12/1990 to 10/2002, obtained

[3] In Hong and Lee (2002), the observed raw data $\{Y_t\}_{t=1}^{T}$ is replaced with a sample of estimated residuals $\{\hat{\varepsilon}_t\}_{t=1}^{T}$, where $\hat{\varepsilon}_t = Y_t - g(I_{t-1},\hat{\theta}), g(\cdot,\cdot)$ is a conditional mean model, and $\hat{\theta}$ is a $\sqrt{T}$-consistent finite-dimensional parameter estimator. Hong and Lee (2002) show that the limit distribution of $M_1(p)$ does not depend on parameter estimation uncertainty (i.e., one can treat estimated parameters as if it were equal to the true parameter values). As a consequence, one can use $M_1(p)$ to test EMH without any modification. The EMH hypothesis involves no parameter estimation.

from China Stock Market & Accounting Research (CSMAR) Database and WISe from Shanghai Wind Information Corporation. These indices are Shanghai Composite (SHC) index, Shanghai Share A (SHA) index, Shanghai Share B (SHB) index, Shanghai 180 (SH180) index, Shenzhen Composite (SZC) index, Shenzhen Share A (SZA) index, Shenzhen Share B (SZB) index, and Shenzhen Constituent (SZCS) index. They are most representative of the overall performance of Shanghai and Shenzhen stock markets. Their starting dates are between 1990 and 1992, except for SH180, which starts from 07/01/1996. All eight indices have the same ending date, 10/31/2002. There exists rather strong volatility clustering in both Shanghai and Shenzhen markets, and in both Share A and Share B markets. However, it appears that Share A and Share B indices have different volatility clustering patterns. Except for Share B indices, there were more variations in the early part of the sample period (corresponding to the subsample before 12/16/1996) than in the later part of the sample period (corresponding to the period after 12/16/1996). This was perhaps due to the implementation of a 10% band limit on daily stock price changes with a $T+1$ settlement rule. In contrast, for Share B indices, there were more variations in the later part of the sample than in the early part of the sample period. This indicates the booming of Share B markets, which might be due to the introduction of a legal regularion by Chinese government on 01/1996 to encourage foreign investment in Share B markets.

The histograms compare the unconditional distributions of stock returns with a normal distribution having the same sample mean and sample variance. All stock returns are apparently nonnormal. They all have a higher peak around zero and heavier tails than the normal distribution, implying a large excess kurtosis. There are some extreme large stock returns, both positive and negative. For Share A and Composite indices, Shanghai market has a higher average return (sample mean) than Shenzhen market for the whole sample and for the subsamples respectively, although the sample standard deviations are not always larger. For Share B indices, Shanghai market has a smaller average return and a smaller standard deviation than Shenzhen market for the whole sample and the first subsample, but it has a higher return with a smaller standard deviation than Shenzhen market for the second subsample.

For indices rather than Share B indices, the standard deviations over the whole sample and over the first subsample are larger than the standard deviations over the second subsample. For Share B indices, however, the standard deviations are larger for the second subsample than for the whole sample and for the first subsample.

There is no strong evidence for skewness except for Shanghai Composite and Share A indices, and Shenzhen Constintuent Index. For all indices, skewness over the whole sample and over the first subsample is all positive; for the second subsample, all indices have a smaller skewness and some of them have a negative skewness. All stock returns have a larger kurtosis than implied by the normal distribution over the whole sample and over the two subsamples. The kurtosis is much larger over the whole sample and over the first subsample than over the second subsample. Both Shanghai Composite and Share A indices have an extremely large kurtosis over the whole sample and over the first subsample. Shenzhen Constituent index also has a very large kurtosis over the first subsample. This is apparently due to one or a few extreme market movements, such as the one on 05/21/1992, when a daily return of 105%, or a 72% return in log-price difference for Shanghai Composite index is recorded.

4 Empirical Evidence

Because of the existence of persistent volatility clustering in Chinese stock returns, we should use a statistical procedure that is robust to conditional heteroskedasticity of unknown form in testing EMH. Except for SH180 with $c \leq 5$, all indices are firmly rejected at the 5% significance level for all c. The strong rejection of EMH might be due to imperfect institutional arrangement, frequent government policy intervention, and the irrational investors' behavior in the stock market at the early stage. To examine whether Chinese stock market has achieved EMH, or whether Chinese stock market has become more effiicient at the later stage, we also examine the subsamples before and after 12/16/1996, when Chinese stock market began to implement a 10% band limit on daily stock price changes. Before this date, transaction rules were changed several times by governments according to stock market conditions. Since 1997, CSRC has taken charge of both Shenzhen and Shanghai stock markets, marking the end of the experimental period and the beginning of a new development period for a more unified Chinese stock market.

From the first subsample, the results indicate the rejection of EMH for all indices except Shenzhen Share A and Shanghai 180 indices, for most choices of c at the 1% level. Note that SH180 only has 115 observations before 12/16/1996, so the insingifinat result with SH180 may be due to the small sample problem.

For most choices of c, the values of $M_1(\hat{p}_0)$ based on the second subsample reject EMH for all eight indices at the 5% significance level, including Shenzhen Share A and SH180 indices. There appears some evidence that there exists departures from EMH at higher order lags (which implies a longer-run mean reverting) for the Composite and Share A indices, because the $M_1(\hat{p}_0)$ statistics become larger and significant when c becomes larger.

Why has Chinese stock market not become efficient after a decade? Because this is an important issue, we are tempted to provide some speculations here. First, Chinese stock market, although one decade has passed since its setup in 1990, remains far away from a mature capitalist stock market. Unlike commodity markets in Chinese economy, where prices are basically determined by market forces, Chinese government still has direct and indirect control of financial resources, including the number of new listed companies as well as the quota and initial prices of new listed stocks. Stock prices do not fully incorporate all available public information. Investors have to respond to non-price signals. In particular, they have to speculate somewhat irregular government policy intervention, although the scope and frequency of policy intervention have been diminished since 1997. Second, for a capital market to function well, it is important for investors to receive high-quality information on listed companies in a timly manner. However, there have been many instances that listed companies may deliver misleading information on their financial accounting data. In addition, the existence of price limit (e.g., the 10% band limit), market segmentation, and insider trading may have slowed down information flow to investors. Of course, more careful empirical examination is needed to sort out possible sources of market inefficiency, but this is beyond the scope of the present paper.

A comparison between the two subsamples suggests that except for Shenzhen Share A and SH180 indices, the $M_1(\hat{p}_0)$ statistic is generally larger in the first subsample than in the second subsample. Given that the two subsamples have similar sample sizes for all indices except SH180, this indicates that both Shanghai (in terms of Composite, Share A and Share B indices) and Shenzhen (in terms of Composite, Constituent, and Share B indices) stock markets have become more efficient at the later stage. It appears that Chinese government's regulation of Chinese stock market has apparently been improving and working in the right direction.

A comparison of the whole and the two subsamples also reveals that for both Shanghai and Shenzhen markets, the values of $M_1(\hat{p}_0)$ are much larger for Share B indices than Share A indices. This indicates that for both Shanghai and Shenzhen stock markets, Share A markets are more efficient than Share B markets. We note that there is no clear systematic pattern on which market, Shanghai or Shenzhen, is more efficient. Both Shanghai and Shenzhen stock exchanges have been competing in building up Chinese stock market.

Why are Share B markets less efficient than Share A markets? A well-function financial market requires a high volume of transactions, which will quickly wipe out any arbitrage opportunity, thus achieving market efficiency. Share B markets have been featured with a low level of transactions for most of time over the last decade. This is so even after Chinese government opened Share B markets to domestic Chinese investors in 02/2001. It may be expected with the integration between the segmented Share A and Share B markets, Share B markets will become more efficient.

The empirical findings documented here have important implications. For example, the violation of EMH implies that stock returns in Chinese stock market are predictable using the past history of stock returns. This has immediate interest to both domestic and foreign investors. In particular, it implies that it is possible to beat the market by using a suitable trading strategy. A relevant question is how one can predict stock returns in Chinese stock market. Do there exist trading strategies that have superior out-of-sample predictable ability? On the other hand, the violation of EMH has important implications on the efficacy of capital asset pricing for the listed companies in Chinese stock market. It would be interesting and important to develop an appropriate asset pricing model to examine what factors (e.g., macroeconomic factors) will help determine asset prices of listed companies in Chinese stock market. All these issues are left for subsequent research.

5 Conclusion

Using a generalized spectral derivative test suitable and powerful to test EMH in presence of volatility clustering, we find significant evidence against the weak form of efficient market hypothesis for both Shanghai and Shenzhen stock markets, although there exists some evidence that they have become more efficient at the later stage. We also find that Share A markets are more efficient than Share B markets. Our findings are robust to volatility clustering of unknown form. Some speculations are given to explain the empirical findings.

We thank the seminar participants at China Center for Economic Research (CCER), Peking University, Antai School of Management, Shanghai Jiao Tong University, and School of Management, Xiamen University for comments, and the China Postdoctoral Science Foundation, the National Science Foundation of United States and School of Economics and Management, Tsinghua University for support. We thank the Research Center for Financial Mathematics and Financial Engineering, Peking University for the powerful and valuable CPU time.

References

1. Groves, T., Hong, Y., McMillan, J., Naughton, B.: Incentives in Chinese State-owned Enterprices, Quarterly Journal of Economics. 1 (1994) 183-209.
2. Naughton, B.: Growing out of Plan, Oxford University Press: Oxford. (1994)
3. Lin, Y.F., Cai, F., Zhou, L.: China Miracle, Chinese University of Hong Kong Publisher: Hong Kong. (1996)
4. Fama, E.F.: A Review of Theory and Empirical Work, Journal of Finance. 25 (1970) 383-417.
5. Fama, E.F.: Efficient Markets II. Journal of Finance. 46 (1991) 1575-1618.
6. Campbell, C., Lo, A., MacKinlay A.C.: Econometrics of Financial Markets, Princeton University Press: Princeton, New Jersey. (1997)
7. Sargent, T., Ljungqvist, L.: Recursive Macroeconomic Theory. MIT Press: Cambridge, MA. (2002)
8. Hong, Y., Lee Y.: Generalized Spectral Tests for Conditional Mean Specification in Time Series with Conditional Heteroskedasticity of Unknown Form. Working paper, Department of Economics & Department of Statistical Science, Cornell University. (2002)
9. Hong, Y.: Hypothesis Testing in Time Series via the Empirical Characteristic Function: A Generalized Spectral Density Approach. Journal of the American Statistical Association. (1999) 84, 1201-1220.
10. Priestley, M.B.: Spectral Analysis and Time Series. Academic press: London. (1981)
11. Hamilton, J.: Time Series Analysis, Princeton University Press: Princeton, New Jersey. (1994)

Workshop on Numerical Methods for Structured Systems

Multi-symplectic Spectral Methods for the Sine-Gordon Equation

A.L. Islas and C.M. Schober

Department of Mathematics and Statistics, Old Dominion University
Department of Mathematics, University of Central Florida
cschober@mail.ucf.edu

Abstract. Recently it has been shown that spectral discretizations provide another class of multi-symplectic integrators for Hamiltonian wave equations with periodic boundary conditions. In this note we develop multi-symplectic spectral discretizations for the sine-Gordon equation. We discuss the preservation of its phase space geometry, as measured by the associated nonlinear spectrum, by the multi-symplectic spectral methods.

1 Introduction

One approach to generalizing the concept of symplecticity to encompass partial differential equations (PDEs) is to develop a local concept of symplecticity that treats space and time equally [1,2,3,4,5]. A Hamiltonian PDE (in the "1+1" case of one spatial and one temporal dimension) is said to be *multi-symplectic* (MS) if it can be written as

$$\boldsymbol{M} z_t + \boldsymbol{K} z_x = \boldsymbol{\nabla}_z S(z), \qquad z \in \mathbb{R}^d, \tag{1}$$

where $\boldsymbol{M},\ \boldsymbol{K} \in \mathbb{R}^{d\times d}$ are skew-symmetric matrices and $S:\ \mathbb{R}^d \to \mathbb{R}$ is a smooth function. The term MS is applied to system (1) in the sense that associated with $\boldsymbol{M}$ and $\boldsymbol{K}$ are the 2-forms

$$\omega(U,V) = V^T \boldsymbol{M} U, \quad \kappa(U,V) = V^T \boldsymbol{K} U, \quad U,\ V \in \mathbb{R}^d,$$

which define a symplectic space-time structure (symplectic with respect to more than one independent variable).

Symplecticity is a global property for Hamiltonian ODEs. In contrast, an important aspect of the MS structure is that symplecticity is now a local property, i.e., symplecticity may vary over the spatial domain and from time to time. This local feature is expressed through the following MS conservation law (MSCL):

$$\partial_t \omega + \partial_x \kappa = 0, \tag{2}$$

where U, V are any two solutions of the variational equation associated with (1)

$$\boldsymbol{M} dz_t + \boldsymbol{K} dz_x = \boldsymbol{S}_{\boldsymbol{zz}}(z) dz.$$

P.M.A. Sloot et al. (Eds.): ICCS 2003, LNCS 2658, pp. 101–110, 2003.

One consequence of multi-symplecticity is that when the Hamiltonian $S(z)$ is independent of x and t, the PDE has an energy conservation law (ECL) [2]

$$\frac{\partial E}{\partial t} + \frac{\partial F}{\partial x} = 0, \quad E = S(z) - \frac{1}{2} z^T \boldsymbol{K}\, z_x, \quad F = \frac{1}{2} z^T \boldsymbol{K}\, z_t, \tag{3}$$

as well as a momentum conservation law

$$\frac{\partial I}{\partial t} + \frac{\partial G}{\partial x} = 0, \quad G = S(z) - \frac{1}{2} z^T \boldsymbol{M}\, z_t, \quad I = \frac{1}{2} z^T \boldsymbol{M}\, z_x. \tag{4}$$

When the local conservation laws are integrated in x, using periodic boundary conditions, we obtain the global conservation of the total energy and total momentum.

MS integrators are discretizations of the PDE which preserve exactly a discrete version of the MSCL (2). In other words, MS integrators have been designed to preserve the MS structure, but not necessarily the local conservation laws or global invariants. Even so, numerical experiments using MS integrators, e.g., for the nonlinear Schrodinger (NLS) equation, have demonstrated that MS methods have remarkable conservation properties (cf. [4,5]). For example, we showed that the local and global energy and momentum are preserved far better than expected, given the order of the scheme. In addition, the global norm and momentum were preserved within roundoff.

However, the numerical experiments for the NLS equation demonstrated that MS finite difference schemes can have difficulty in resolving spatial structures in very sensitive regimes [4]. On the other hand, spectral methods have proven to be highly effective methods for solving evolution equations with simple boundary conditions. As the number N of space grid points increases, errors typically decay at an exponential rate rather than at polynomial rates obtained with finite difference approximations [7]. For the NLS equation we showed that a significant improvement in the resolution of the qualitative features of the solution is obtained with a MS spectral method [6].

In this note we focus on the question of preservation of the phase space geometry of nonlinear wave equations by MS spectral methods. We use the sine-Gordon equation (SG) as our model equation. In the next section we present the MS formulation of the SG equation and a description of the phase space geometry in terms of the associated nonlinear spectrum. In section 3 we provide the MS spectral discretization of the SG equation. In section 4 we implement both MS and nonsymplectic spectral methods and use the nonlinear spectrum of the SG equation as a basis for comparing the effectiveness of the integrators. The relevant quantities to monitor are the periodic/antiperiodic eigenvalues of the associated spectral problem. These eigenvalues are the spectral representation of the action variables and are directly related to the geometry of the SG phase space. Significantly, we show that the MS spectral methods provide an improved resolution of phase space structures, as measured by the nonlinear spectrum, when compared with non-symplectic spectral integrators.

2 The MS and Integrable Structure of the SG Equation

The MS form of the SG equation,

$$u_{tt} - u_{xx} + \sin u = 0, \tag{5}$$

is obtained by introducing the new variables $v = u_t$, $w = u_x$. This results in the system of equations

$$\begin{aligned} -v_t + w_x &= \sin u \\ u_t \qquad &= v \\ - u_x &= -w, \end{aligned} \tag{6}$$

which can be written in standard MS form (1) with

$$z = \begin{pmatrix} u \\ v \\ w \end{pmatrix}, \qquad \boldsymbol{M} = \begin{pmatrix} 0 & -1 & 0 \\ 1 & 0 & 0 \\ 0 & 0 & 0 \end{pmatrix}, \qquad \boldsymbol{K} = \begin{pmatrix} 0 & 0 & 1 \\ 0 & 0 & 0 \\ -1 & 0 & 0 \end{pmatrix}$$

and Hamiltonian $S(z) = -\cos u + \frac{1}{2}\left(v^2 - w^2\right)$.

The energy and flux are given by

$$\begin{aligned} E &= S - \tfrac{1}{2} z^T \boldsymbol{K} z_x = -\cos u + \tfrac{1}{2}\left(v^2 - w^2 - u w_x + u_x w\right), \\ F &= \tfrac{1}{2} z^T \boldsymbol{K} z_t = \tfrac{1}{2}\left(u w_t - u_t w\right), \end{aligned}$$

respectively. Deriving relations (3)-(4) for the SG equation, the (ECL) can be simplified to

$$E_t + F_x = \left(-\cos u + \tfrac{1}{2}(v^2 + w^2)\right)_t - (vw)_x = 0.$$

Similarly, the momentum conservation law is given by

$$I_t + G_x = \left(\cos u + \tfrac{1}{2}(v^2 + w^2)\right)_x - (vw)_t = 0.$$

2.1 Integrable Structure of the SG Equation

The phase space of the SG equation (5) with periodic boundary conditions can be described in terms of the Floquet spectrum of the following linear operator (the spatial part of the associated Lax pair [9]):

$$\mathcal{L}(u, \lambda) = \left[\boldsymbol{A}\frac{d}{dx} + \frac{i}{4}\boldsymbol{B}(u_x + u_t) + \frac{1}{16\lambda}\boldsymbol{C} - \lambda \boldsymbol{I}\right],$$

where

$$\boldsymbol{A} = \begin{pmatrix} 0 & -1 \\ 1 & 0 \end{pmatrix}, \qquad \boldsymbol{B} = \begin{pmatrix} 0 & 1 \\ 1 & 0 \end{pmatrix} \qquad \boldsymbol{C} = \begin{pmatrix} e^{iu} & 0 \\ 0 & e^{-iu} \end{pmatrix}, \qquad \boldsymbol{I} = \begin{pmatrix} 1 & 0 \\ 0 & 1 \end{pmatrix},$$

u is the potential and $\lambda \in \mathbb{C}$ denotes the spectral parameter.

The fundamental solution matrix M, defined by the conditions $\mathcal{L}(u,\lambda)M = 0$ and $M(x,x;u,\lambda) = I$, is used to introduce the Floquet discriminant $\Delta(u;\lambda) = Tr[M(x+L,x;u,\lambda)]$. The Floquet discriminant is analytic in both its arguments. Moreover, for a fixed λ, Δ is invariant along solutions of the SG equation: $\frac{d}{dt}\Delta(u,\lambda) = 0$. Since Δ is invariant and the functionals $\Delta(u,\lambda)$, $\Delta(u,\lambda')$ are pairwise in involution, Δ provides an infinite number of commuting invariants for the SG equation.

The spectrum of $\mathcal{L}$ is given by the following condition on the discriminant: $\sigma(\mathcal{L}) = \{\lambda \in \mathbb{C} | \Delta(u;\lambda) \in \mathbb{R}, -2 \leq \Delta(u;\lambda) \leq 2\}$. When discussing the numerical experiments, we monitor the following elements of the spectrum which determine the phase space geometry of the SG equation: (a) Critical points λ^c, specified by the condition $d/d\lambda \Delta(q;\lambda)|_{\lambda=\lambda^c} = 0$, and (b) Double points λ^d, which are critical points that satisfy the additional constraints $\Delta(q;\lambda)|_{\lambda=\lambda^d} = \pm 2$, $d^2/d\lambda^2 \Delta|_{\lambda=\lambda^d} \neq 0$. Complex double points correspond, in general, to critical saddle-like level sets and can be used to label their associated homoclinic orbits.

The periodic/antiperiodic spectrum provides the actions in an action-angle description of the system. The values of these actions fix a particular level set. Let λ denote the spectrum associated with the potential u. The level set defined by u is then given by, $\mathcal{M}_u \equiv \{v \in \mathcal{F} | \Delta(v,\lambda) = \Delta(u,\lambda), \lambda \in \mathbb{C}\}$. Typically, $\mathcal{M}_u$ is an infinite dimensional stable torus. However, the SG phase space also contains degenerate tori which may be unstable. If a torus is unstable, its invariant level set consists of the torus and an orbit homoclinic to the torus. These invariant level sets, consisting of an unstable component, are represented, in general, by complex double points in the spectrum. A complete and detailed description of the SG phase space structure is provided in [9].

3 MS Spectral PDEs

Bridges and Reich have shown that Fourier transforms leave the MS nature of a PDE unchanged [3]. A MS discretization of (1) is then obtained by truncating the Fourier expansion. This produces a system of Hamiltonian ODEs which can be discretized in time using symplectic methods. We briefly summarize these results.

Consider the space $L_2(I)$ of L-periodic, square integrable functions in $I = [-L/2, L/2]$ and let $U = \mathcal{F}u$ denote the discrete Fourier transform of $u \in L_2(I)$. Here $\mathcal{F} : L_2 \to l_2$ denotes the Fourier operator which gives the complex-valued Fourier coefficients $U_k \in \mathbb{C},\ k = -\infty, \ldots, -1, 0, 1, \ldots, \infty$, which we collect in the infinite-dimensional vector $U = (\ldots, U_{-1}, U_0, U_1, \ldots) \in l_2$. Note that $U_{-k} = U_k^*$. We also introduce the L_2 inner product, which we denote by (u, v) and the l_2 inner product, which we denote by $\langle U, V \rangle$. The inverse Fourier operator $\mathcal{F}^{-1} : l^2 \to L^2$ is defined by $\langle V, \mathcal{F}u \rangle = (\mathcal{F}^{-1}V, u)$. Furthermore, partial differentiation with respect to $x \in I$ simply reduces to $\partial_x u = \mathcal{F}^{-1}\Theta U$ where $\Theta : l_2 \to l_2$ is the diagonal spectral operator with entries $\theta_k = i2\pi k/L$.

These definitions can be generalized to vector-valued functions $z \in L_2^d(I)$. Let $\hat{\mathcal{F}} : L_2^d(I) \to l_2^d$ be defined such that $Z = (Z^1, \ldots, Z^d) = \hat{\mathcal{F}}z = (\mathcal{F}z^1, \ldots, \mathcal{F}z^d)$.

Thus with a slight abuse of notations and after dropping the hats, we have $Z = \mathcal{F}z$, $z = \mathcal{F}^{-1}Z$, and $\partial_x z = (\partial_x z^1, \ldots, \partial_x z^d) = (\mathcal{F}^{-1}\Theta Z^1, \ldots, \mathcal{F}^{-1}\Theta Z^d) = \mathcal{F}^{-1}\Theta Z$.

Applying these operators to (1), one obtains an infinite dimensional system of ODEs

$$\boldsymbol{M}\partial_t Z + \boldsymbol{K}\Theta Z = \boldsymbol{\nabla}_Z \bar{S}(Z), \quad \bar{S}(Z) = \int_{-L}^{L} S(\mathcal{F}^{-1}Z)\, dx.$$

This equation can appropriately be called a MS spectral PDE with associated MSCL

$$\partial_t \Omega + \Theta \mathcal{K} = 0, \qquad \Omega = \mathcal{F}\omega, \quad \mathcal{K} = \mathcal{F}\kappa, \tag{7}$$

and ECL

$$\partial_t E + \Theta F = 0, \quad E = \mathcal{F}e, \quad F = \mathcal{F}f,$$

in Fourier space.

3.1 MS Spectral Schemes for the SG Equation

A MS spatial discretization of the PDE is given by the truncated Fourier series,

$$U_k = \frac{1}{\sqrt{N}} \sum_{l=1}^{N} u_l\, e^{-\theta_k (l-1)\Delta x}, \; u_l = u(x_l), \; x_l = -\frac{L}{2} + (l-1)\Delta x, \; \Delta x = \frac{L}{N},$$

with

$$\theta_k = \begin{cases} i\frac{2\pi}{L}(k-1) & \text{for } k = 1, \ldots, N/2, \\ 0 & \text{for } k = N/2+1, \\ -\theta_{N-k+2} & \text{for } k = N/2+2, \ldots, N. \end{cases}$$

A family of discrete MSCLs exists which resemble the conservation law (7) in Fourier space [3].

Using the discrete Fourier transform, system (6) becomes

$$\begin{aligned} -\partial_t V_k + \theta_k W_k &= \boldsymbol{F}_k(\sin u) \\ \partial_t U_k \qquad\quad &= V_k \\ - \theta_k U_k &= -W_k. \end{aligned} \tag{8}$$

This system of equations can be recombined into a single equation. In Fourier space we have that (8) becomes

$$\ddot{U}_k = -\theta_k^2 U_k - (\boldsymbol{F} \sin u)_k\,, \tag{9}$$

where

$$(\boldsymbol{F} \sin u)_k = \frac{1}{N} \sum_{l=0}^{N-1} \sin u_l\, e^{-\theta_k l \Delta x},$$

and the discrete Hamiltonian is given by

$$H = \frac{1}{2} \sum_{l=0}^{N-1} \left[|\dot{U}_l|^2 + \mu_l^2 |U_l|^2 \right] - \frac{1}{N} \sum_{l=0}^{N-1} \cos u_l. \tag{10}$$

To maintain the multi-symplecticity, a symplectic integrator in time should be used. To discretize (9) in time we note that the Hamiltonian (10) is separable which allows one to use explicit symplectic integrators. A general form of explicit higher order symplectic schemes is given by (see e.g., [10]),

$$\begin{aligned} &a_1 = p_j, \quad b_1 = q_j, \\ &a_{i+1} = a_i - C_i\, k\, V'(b_i), \quad b_{i+1} = b_i + D_i\, k\, T'(a_{i+1}), \quad i = 1, \ldots, m, \\ &p_{j+1} = a_m, \quad q_{j+1} = b_m, \end{aligned}$$

where the coefficients C_i and D_i are determined so that the scheme is symplectic and of order $O(k^m)$. For example, a first order scheme is given by $m = 1$ and $C_1 = 1, \;\; D_1 = 1$. Similarly, a second-order scheme is given by $m = 2$ and $C_1 = 0, \;\; C_2 = 1, \;\; D_1 = \frac{1}{2} = D_2$. A fourth-order integrator S_4 can be obtained by forming the following product of second-order integrators S_2

$$S_4(k) = S_2(\beta k)\, S_2(\alpha k)\, S_2(\beta k),$$

where $\alpha = -2^{1/3}\beta$ and $\beta = 1/(2 - 2^{1/3})$.

4 Numerical Experiments

The MS property can be lost in a discretization by using a non-symplectic discretization either in space or in time. Here we examine the loss of multi-symplecticity due to the time discretization. We compare the performance of the spectral discretization implemented in time with second- and fourth-order symplectic methods (and thus MS) versus non-symplectic Runge-Kutta methods of the same order. In the numerical experiments we focus on determining whether the MS integrators preserve the structure of the SG phase space appreciably better than the nonsymplectic methods.

As derived using the inverse scattering theory [12], the SG equation has an infinite number of local conservation laws and global invariants. In this example, rather than monitor the local energy and momentum conservation laws, we choose to examine the preservation of the nonlinear spectrum, which incorporates all of the global invariants and determines the phase space structure.

Under the SG flow, the spectrum remains invariant. However, due to perturbations induced by the numerical discretization, the spectrum evolves in time. The evolution of the spectrum under the numerical flow is primarily due to the time discretization. To determine the effectiveness of MS spectral integrators in capturing the phase space structure, we compute the spectral content of the initial data and monitor its evolution under the different schemes.

The following initial data is used in the numerical experiments:

$$u(x,0) = \pi + 0.1\cos(\mu x), u_t(x,0) = 0,$$

with parameters $\mu = 2\pi/L$ and $L = 2\sqrt{2}\pi$. This initial data is for solutions in the unstable regime as the zeroth double point remains closed, i.e. the initial data is on the level set containing the homoclinic manifold. (Closed double points cannot be preserved by the numerical schemes and in the following experiments one observes that the zeroth mode is immediately split into a gap state by the numerical scheme.)

To interpret the evolution of spectrum plots, note that under perturbations the complex double points can split in two ways – either into a gap along an arc of the circle, or into a cross along the radius (Figure 1). For each set of

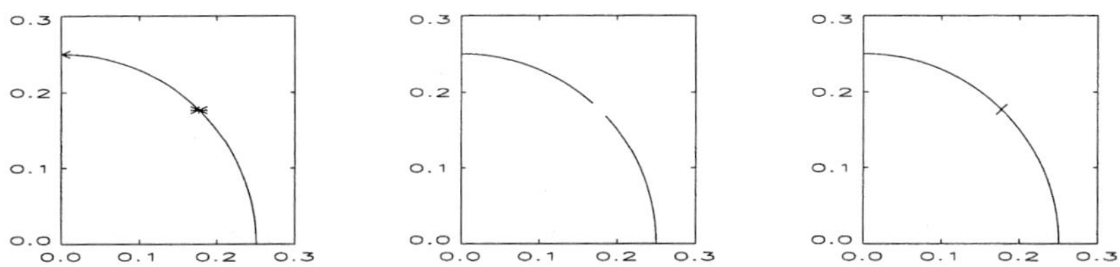

Fig. 1. The nonlinear spectrum. (a) Homoclinic orbit, (b) Inside the homoclinic orbit ('gap state'), (c) Outside the homoclinic orbit ('cross state').

experiments, we show a signed measure of the splitting distance for the zeroth mode as a function of time. Positive and negative values represent gap and cross states, respectively. When the splitting distance passes through zero, the double points coalesce and homoclinic crossings occur.

We consider the exponentially accurate spectral scheme (9) implemented in time with either Runge-Kutta (2nd and 4th-order) or with symplectic (1st, 2nd and 4th-order) integrators. These integrators will be denoted by S-2RK, S-4RK and S-1SY, S-2SY, S-4SY. In the spectral experiments we use $N = 32$ Fourier modes and a fixed time step $\Delta t = L/512$.

The splitting distance obtained with S-1SY is not shown but it is worth noting that even with the first-order symplectic integrator, *bounded* oscillations are observed. The splitting distance for both modes obtained with S-1SY is $O(10^{-2})$. Using S-2RK and S-2SY (Figure 2), the spectrum for the first mode does not execute any homoclinic crossings for $0 < t < 500$ and so the torus component is accurately preserved. However, the zeroth mode does display homoclinic crossings which occur earlier than with the lower order S-1SY.

Since the initial data is chosen on the homoclinic manifold, it is to be expected that there will be an earlier onset and higher density of homoclinic crossings when a more accurate scheme is used. Refinement can accentuate the frequency of homoclinic crossings as the numerical trajectory is trapped in a narrower band about the homoclinic manifold. The main observation is that with the nonsymplectic S-2RK there is a $O(10^{-3})$ linear drift in the error in the nonlin-

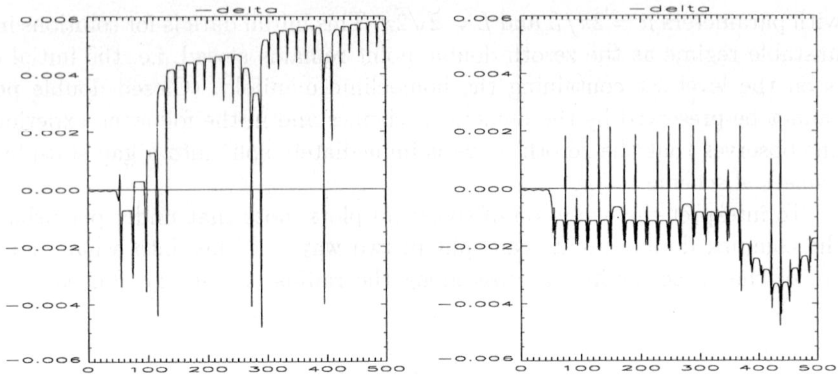

Fig. 2. Left: S-2RK: $u(x,0) = \pi + 0.1\cos\mu x$, $u_t(x,0) = 0$, $N = 32$, $t = 0 - 500$. Right: S-2SY: $u(x,0) = \pi + 0.1\cos\mu x$, $u_t(x,0) = 0$, $N = 32$, $t = 0 - 500$.

ear spectrum. The error in the nonlinear spectrum is smaller with S-2SY and further, it doesn't drift. The drift in the nonlinear spectrum obtained with S-2RK can be eliminated on the timescale $0 < t < 500$ by increasing the accuracy of the integrator and using S-4RK. In this case the nonlinear spectral deviations are $O(10^{-4})$ for S-4RK and S-4SY (Figure 3). There does not seem to be an appreciable difference.

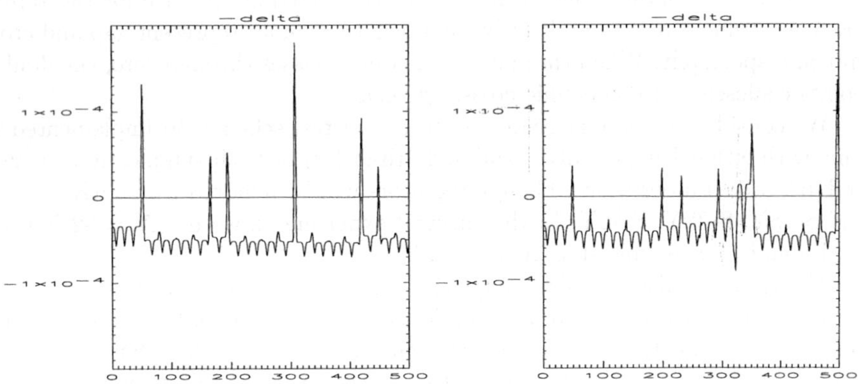

Fig. 3. Left: S-4RK: $u(x,0) = \pi + 0.1\cos\mu x$, $u_t(x,0) = 0$, $N = 32$, $t = 0 - 500$. Right: S-4SY: $u(x,0) = \pi + 0.1\cos\mu x$, $u_t(x,0) = 0$, $N = 32$, $t = 0 - 500$.

In long time studies of low dimensional Hamiltonian systems, symplectic integrators have been reported as superior in capturing global phase space structures since standard integrators may allow the actions to drift [8,11]. We continue the integration to $t = 10,000$ and examine the time slice $10,000 \leq t \leq 10,500$. For

S-4RK (Figure 4) a drift has occured. The deviations in the actions associated with the zeroth mode oscillates about 1.2×10^{-4} whereas for S-4SY (Figure 4) it oscillates about 5×10^{-5}. Although the drift observed with nonsymplectic schemes can be reduced by using a higher order integrator, it is not eliminated and simply occurs on a longer timescale. This problem is avoided using the MS integrator.

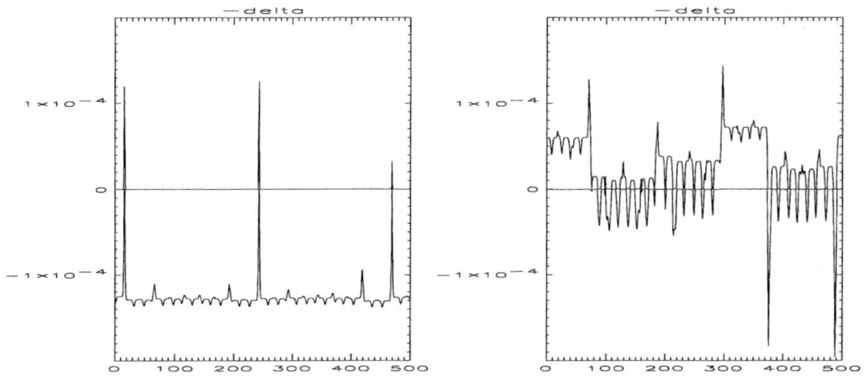

Fig. 4. Left: S-4RK: $u(x,0) = \pi + 0.1 \cos \mu x$, $u_t(x,0) = 0$, $N = 32$, $t = 10000 - 10500$. Right: S-4SY: $u(x,0) = \pi + 0.1 \cos \mu x$, $u_t(x,0) = 0$, $N = 32$, $t = 10000 - 10500$.

5 Conclusions

MS integrators are discretizations of the PDE which preserve exactly a discrete version of the MSCL. In this note, we have developed MS spectral integrators for the sine-Gordon equation. The benefits of these integrators are greater qualitative fidelity and superior preservation of local conservation laws and global invariants. The numerical experiments show that the MS spectral methods provide an improved resolution of phase space structures when compared with nonsymplectic spectral integrators.

References

1. Reich, S.: Multi-Symplectic Runge-Kutta Collocation Methods for Hamiltonian Wave Equations. J. of Comput. Phys. **157** (2000) 473–499
2. Bridges, T.J., Reich, S.: Multi-Symplectic Integrators: numerical schemes for Hamiltonian PDEs that conserve symplecticity. University of Surrey, Technical Report (1999)
3. Bridges, T.J., Reich, S.: Multi-Symplectic Spectral Discretizations for the Zakharov-Kuznetsov and shallow water equations. University of Surrey, Technical Report (2000)

4. Islas, A.L., Karpeev, D.A., Schober,C.M.: Geometric integrators for the nonlinear Schrödinger equation. J. of Comp. Phys. **173** (2001) 116–148
5. Islas, A.L., Schober,C.M.: Multi-symplectic spectral methods for the Gross-Pitaevski equation. Lect. Notes Comp. Sci. **2331** (2002) 486–495
6. Islas, A.L., Schober,C.M.: Multi-symplectic methods for generalized Schrödinger equations. Fut. Gen. Comput. Sys. **950** (2003)
7. Fornberg, B.: A practical guide to pseudospectral methods. Cambridge University Press (1998)
8. Channell, P.J., Scovel, C.: Symplectic integration of Hamiltonian systems. Nonlinearity **3** (1990) 1–13
9. Ercolani, N., Forest, M.G., McLaughlin, D.W.: Geometry of the modulational instability, Part III. Physica D **43** (1990) 349–360
10. Yoshida, H.: Construction of higher order symplectic integrators. Phys Lett. A **150** (1990) 262–268
11. Sanz-Serna, J., Calvo, M.: Numerical Hamiltonian problems. Chapman and Hall, London (1994)
12. Ablowitz, M.J., Segur, H.: Solitons and the inverse scattering transform. SIAM Studies in Applied Math., SIAM, Philadelphia (1981)

A Survey on Methods for Computing Matrix Exponentials in Numerical Schemes for ODEs

Nicoletta Del Buono and Luciano Lopez

Dipartimento Interuniversitario di Matematica,
Università degli Studi di Bari
Via E. Orabona, 4 - I-70125 Bari ITALY
{delbuono,lopezl}@dm.uniba.it

Abstract. This paper takes a look at numerical procedures for computing approximation of the exponential of a matrix of large dimension. Existing approximation methods to evaluate the exponentiation of a matrix will be reviewed paying more attention to Krylov subspace methods and Schur factorization techniques. Some theoretical results on the bounds for the entries of the exponential matrix and some implementation details will be also discussed.

1 Introduction

Several problems in mathematics and physics can be formulated in terms of finding a suitable approximations to certain matrix functions. Particularly, the issue of computing the matrix exponential $f(A) = e^{-tA}$, $t \geq t_0$, is one of the most frequently encountered tasks in matrix function approximation and it has received a renew attention from the numerical analysis community. This problem arises in many areas of applications, as for instance the solution of linear parabolic partial differential equations which needs the numerical solution of n dimensional systems of ODEs $\dot{y} = -Ay + b(t)$, $y(0) = y_0$, $t > 0$ [3,15,16], or recently in the field of geometric integration. In fact, most Lie group methods, as Runge-Kutta/Munthe-Kaas schemes, Magnus expansions and Fer expansions [19,20,28], require to suitable approximate the matrix exponential from a Lie algebra $g \subset \mathbb{R}^{n\times n}$ once (and often repeatedly) at each time step. This can be a very challenging task for large dimension matrix. Moreover, the context of Lie-group method imposes a crucial extra requirement on the approximant: it has to reside in the Lie group $G \subset GL(n, \mathbb{R})$ associated to the Lie algebra g.

In general, this property is not fulfilled by many standard approximations unless the exponential is evaluated exactly.

This can be done for instance for a 3×3 skew symmetric matrices, whose exponential is given exactly by the *Euler Rodriguez* formula

$$\exp(A) = I + \frac{\sin(\alpha)}{\alpha}A + \frac{1}{2}\left(\frac{\sin(\alpha/2)}{\alpha/2}\right)^2 A^2,$$

where

P.M.A. Sloot et al. (Eds.): ICCS 2003, LNCS 2658, pp. 111–120, 2003.

$$A = \begin{pmatrix} 0 & a_3 & -a_2 \\ -a_3 & 0 & a_1 \\ a_2 & -a_1 & 0 \end{pmatrix}, \quad \alpha = (a_1^2 + a_2^2 + a_3^2)^{1/2}.$$

Other exact formulas for exponentials of skew symmetric matrices can be obtained making use of the *Cayley-Hamilton* theorem which for every $A \in GL(n, \mathbb{R})$ allows us to have the basic decomposition $\exp(tA) = \sum_{k=0}^{m-1} f_k(t)A^k$, where m is the degree of the minimal polynomial of A and $f_0, \dots, f_{m-1}$ are some analytic functions that depend on the characteristic polynomial of A. However, these algorithms are practical only up to dimension eight. In fact, for large scale matrices, they require the computation of high powers of the matrix in question and hence high computational costs due to the matrix-matrix multiplication; moreover they suffer of some computational instabilities due to the direct use of the characteristic polynomial.

Based on the above remarks, it appears clear that exponential of large matrices cannot be evaluated analytically and that an algorithm which is simple and efficient and satisfies some geometric properties is highly desiderable.

In the following pages, we will review some existing approximation methods, highlighting also some theoretical aspects. In the last section we will give also a brief overview on a new methodology with can be consider an hybrid scheme for exponential approximation.

2 Review of Existing Approximation Methods

Classical methods for the evaluation of a matrix exponential can be classified into three main categories:

(i) rational approximants;
(ii) Krylov subspace methods;
(iii) techniques based on numerical linear algebra;

further, within the context of Lie-group theory other methods for the approximation of the exponential have been recently introduced:

(iv) the splitting methods.

2.1 Rational Approximants

A rational approximation of the exponential function replaces it by a rational function, $\exp(z) \approx r(z) := p_\alpha(z)/q_\beta(z)$, where p and q are polynomials of degree α and β respectively, $q(0) = 1$ and the error term $e^z - p(z)/q(z)$ is small in a chosen norm. Thus for the matrix $\exp(A)$ one has to compute two matrix-value polynomials, $p(A)$ and $q(A)$, and invert the latter to obtain $r(A)$.

Probably, the most popular approximant of this kind is the *diagonal* (ν, ν) *Padé approximations* where

$$p_\nu(z) = \sum_{k=1}^{\nu} \frac{(2\nu - k)!\nu!}{2\nu!k!(\nu - k)!} z^k \quad \text{and} \quad q_\nu(z) = p_\nu(-z).$$

Unfortunately, the Padé approximants are good only near the origin (see [13]). However, this problem can be overcome with the so-called scaling and squaring technique, which exploits the identity

$$\exp(A) = (\exp(A/2^k))^{2^k}$$

as follows. First, a sufficiently large k is chosen so that $A/2^k$ is close to 0, then a diagonal Padé approximant is used to calculate $\exp(A/2^k)$, and finally the result is squared k times to obtain the required approximation to $\exp(A)$. This basic approach is implemented in the Matlab function `expm` to evaluate the exponential of a matrix with a computational cost between $20n^3$ and $30n^3$ operations.

About the behavior of rational approximants when applied to matrix belonging to a Lie algebra, the following theorem states that an important category of Lie group leads itself to suitable rational approximations.

Theorem 1. *(see [5]) Let* $G = \{Y \in GL(n, \mathbb{R}) : \; Y^T P Y = P\}$, *where* P *is a non-singular* $n \times n$ *matrix and let* $g = \{X \in gl(n, \mathbb{R}) : \; XP + PX^T = 0\}$ *be the corresponding Lie algebra. Let* ϕ *be a function analytic in a neighborhood* U_0 *of* 0, *with* $\phi(0) = 1$ *and* $\phi'(0) = 1$. *If*

$$\phi(z)\phi(-z) = 1, \qquad \forall z \in U_0, \tag{1}$$

then $\forall X \in g$, $\Phi(tX) \in G$ *for all* $t \in \mathbb{R}^+$ *sufficiently small.*

Examples of Lie group verifying the above theorem are the orthogonal and the symplectic group. Moreover, the diagonal Padé approximants are analytic functions that fulfill (1), and this guarantees that their approximation of the exponential of a matrix in the Lie-algebra of a quadratic Lie group lies in the corresponding group.

Another class of approximations of $\exp(A)$ are based on the Chebyshev series expansion of the function $\exp(z)$ on the spectral interval of A. Now let A be an Hermitian matrix and suppose that e^z is analytic in a neighborhood of the spectral interval $[\lambda_1, \lambda_n]$ of A. The approximation of $\exp(A)v$ has the form

$$\exp(A)v \approx \sum_{k=0}^{m} a_k C_k(\frac{1}{l_1}A - \frac{1}{l_2}I)v,$$

where $l_1 = (\lambda_n - \lambda_1)/2$ and $l_2 = (\lambda_n + \lambda_1)/2$, C_k is the kth Chebyshev polynomial of the first kind and the a_k are the Chebyshev coefficients of $e^{(l_1 t + l_2)}$, $t \in [-1, 1]$. Chebyshev polynomial approximation for the exponential of symmetric large sparse matrices was considered in [34], while in [2] a lower bound for $\| \exp(-\tau A)v\|_2$ based on the computation of an eigenvector associated with λ_1 was obtained. The exponential of a non-symmetric matrix has also been approximated in [26,27] using the Faber polynomial and the Faber series.

2.2 Methods of Numerical Linear Algebra

A simple technique to evaluate $\exp(A)$ is by spectral decomposition: if $A = VDV^{-1}$ where D is diagonal matrix with diagonal elements λ_i, $i = 1, \ldots, n$, then

$$\exp(A) = V \exp(D) V^{-1} = V \begin{pmatrix} e^{\lambda_1} & 0 & \ldots & 0 \\ 0 & e^{\lambda_2} & & \\ \vdots & & \ddots & \\ & & \ldots & e^{\lambda_n} \end{pmatrix} V^{-1}$$

This, however, is not a viable techniques for general matrices. In place of spectral decomposition, one can factorize A in a different form, e.g. into Schur decomposition (see [13]) $A = QTQ^T$, where Q is an orthogonal $n \times n$ matrix and T is an $n \times n$ upper triangular matrix or a block upper triangular matrix (when the eigenvalue of A can be clustered in blocks) and therefore $\exp(A) = Q \exp(T) Q^T$.

2.3 Krylov Subspace Methods

The basic idea of the Krylov subspace techniques is to project the exponential of the large matrix onto a small Krylov subspace. Particularly, an approximation to the matrix exponential operation $\exp(A)v$ of the form

$$\exp(A)v \approx p_{m-1}(A)v,$$

where $A \in GL(n, \mathbb{R})$, v is an arbitrary nonzero vector, and p_{m-1} is a polynomial of degree $m-1$. Since this approximation is an element of the Krylov subspace

$$K_m(A, v) = \text{span}\{v, Av, \cdots, A^{m-1}v\},$$

the problem can be reformulated as that of finding an element of $K_m(A, v)$. To find such an approximation can be used the Arnoldi and nonsymmetric Lanczos algorithms, respectively. The Arnoldi algorithm generates an orthogonal basis of the Krylov subspace using $v_1 = v/\|v\|$ as initial vector.

Arnoldi algorithm
1. Compute $v_1 = v/\|v\|$.
2. Do $j = 1, 2, \ldots, m$.
   $w := Av_j$
   Do $i = 1, 2, \ldots, j$
      $h_{i,j} := (w, v_j)$ $\qquad w := w - h_{i,j} v_j$
   Compute $h_{j+1} = \|w\|_2$ and $v_{j+1} = w/h_{j+1,j}$.

Arnoldi algorithm produces an orthonormal basis $V_m = [v_1, v_2, \cdots, v_m]$ of the Krylov subspace $K_m(A, v)$ and a $m \times m$ upper Hessemberg matrix H_m with elements $h_{i,j}$ such that

$$AV_m = V_m H_m + h_{m+1,m} v_{m+1} e_m^T$$

from which we get $H_m = V_m^T A V_m$. Therefore H_m represents the projection of the linear transformation A onto the subspace K_m, with respect to the basis V_m. Based on this we can also write $\exp(A)v \approx \beta V_m \exp(H_m) e_1$ (where $\beta = \|v\|_2$). Now, the problem of computing $\exp(A)v$ has been reduced to the task of computing the lower-dimensional expression $\exp(H_m)e_1$, which for $m \ll n$ is usually much easier, e.g., by diagonalization of H_m.

The computational cost required by the Krylov subspace methods with the Arnoldi iteration is sum of the following partial amounts of operations, counting both multiplications and additions: $2mn^2 + 2nm(m-1)$ operation to compute the Krylov subspace $K_m(A, v)$ of dimension m; Cm^3 computations for the evaluation of the exponential of the Hessemberg matrix H_m; $2nm$ operations arising from the multiplication of $\exp(H_m)$ with the orthogonal basis. However, when n is large and $m \ll n$, these costs are subsumed in that of the Arnoldi iteration, and the leading factor is $2mn^2$ ($2mn^3$ for matrices).

Some error bounds for the Arnoldi approximation of $\exp(A)v$ can be given for various classes of matrices.

Theorem 2. *(see [17]) Let A be a complex square matrix of large dimension n, and v a given n-dimensional vector of unit length ($\|v\| = 1$) and $err_m = \|\exp(\tau A)v - V_m \exp(\tau H_m) e_1\|$ be the error in the Arnoldi approximation of $\exp(\tau A)$. Then err_m satisfies the following bounds:*

(1) if A is Hermitian negative semidefinite matrix with eigenvalues in the interval $[-4\rho, 0]$ then:

$$\begin{aligned} err_m &\le 10 e^{-m^2/5\rho\tau}, & \sqrt{4\rho\tau} \le m \le 2\rho\tau, \\ err_m &\le 10(\rho\tau)^{-1} e^{-\rho\tau}(e\rho\tau/m)^m, & m \ge 2\rho\tau; \end{aligned}$$

(2) if A is skew-Hermitian matrix with eigenvalues in an interval on the imaginary axis of length 4ρ, then

$$err_m \le 12 e^{-(\rho\tau)^2/m}(e\rho\tau/m)^m, \quad m \ge 2\rho\tau;$$

(3) if A has numerical range contained in the disk $|z + \rho| \le\lt \rho$, then:

$$err_m \le 12 e^{-\rho\tau}(e\rho\tau/m)^m, \quad m \ge 2\rho\tau.$$

Another well known algorithm for constructing a convenient basis of K_m is the Lanczos algorithm which, starting from two vectors v_1 and w_1, generates a biorthogonal basis of the subspaces $K_m(A, v_1)$ and $K_m(A^T, w_1)$.

Lanczos algorithm

1. Compute $v_1 = v/\|v\|$ and select w_1 so that $(v_1, w_1) = 1$.
2. Do $j = 1, 2, \ldots, m$.

   $\alpha_j := (Av_j, w_j)$
   $\hat{v}_{j+1} := Av_j - \alpha_j v_j - \beta_j v_{j-1}$
   $\hat{w}_{j+1} := A^T w_j - \alpha_j w_j - \delta_j w_{j-1}$
   $\beta_{j+1} := \sqrt{|(\hat{v}_{j+1}, \hat{w}_{j+1})|}$, $\delta_{j+1} := \beta_{j+1} sign[(\hat{v}_{j+1}, \hat{w}_{j+1})]$
   $v_{j+1} = \hat{v}_{j+1}/\delta_{j+1}$ and $w_{j+1} = \hat{w}_{j+1}/\beta_{j+1}$.

Setting $V_m = [v_1, v_2, \ldots, v_m]$, $W_m = [w_1, w_2, \ldots, w_m]$ and $T_m = tridiag(\delta_i, \alpha_i, \beta_i)$ then $W_m^T V_m = V_m W_m = I$, with I the identity matrix and

$$AV_m = V_m T_m + \delta_{m+1} v_{m+1} e_m^T,$$

from which we can derive the approximation $\exp(A)v \approx \beta V_m \exp(T_m)e_1$. The exponential matrix $\exp(T_m)$ can be computed again from the eigendecomposition $T_m = Q_m D_m Q_m^T$, with diagonal D_m, via $\exp(T_m) = Q_m \exp(D_m) Q^T$.

Observe, that both these methods reduce to the same technique when the matrix is symmetric.

For the Lanczos method it can also be proved analogous error bounds (except for different constants) as that given for the Arnoldi method in Theorem 2 (see [17]).

2.4 Splitting Methods

Splitting methods have been considered by various authors in different contexts: for constructing symplectic methods or volume preserving algorithms, and in PDE context [5,7,15]. A good survey can be found in [24].

The idea these methods are based on is the following: a given $n \times n$ matrix A in a Lie algebra g is split in the form

$$A = \sum_{k=1}^{s} A_k, \tag{2}$$

where $A_k \in g$, $k = 1, 2, \ldots, s$, has an exponential that can be easily evaluated exactly so that

$$\exp(tA_1)\exp(tA_2)\cdots\exp(tA_s) = \exp(tA) + O(t^{p+1})$$

for sufficiently large value of $p \geq 1$. Of course this procedure is competitive with direct evaluations of $\exp(tA)$ only when each $\exp(tA_i)$ is easy to evaluate exactly and their products are cheap.

These requirements are satisfies when each $C = A_i$ is a low rank matrix, i.e. $C = \sum_{l=1}^{p} \alpha_l \beta_l^T = \alpha\beta^T$, with $\alpha_l, \beta_l \in \mathbb{R}^n$, where $p \geq 1$ is a small integer and $\alpha = [\alpha_1, \ldots, \alpha_p]$, $\beta = [\beta_1, \ldots, \beta_p]$ are $n \times p$ matrices. In this case, the function $\exp(tC)$ can be calculated explicitely via the formula

$$\exp(tC) = I + \alpha D^{-1}(\exp(tD) - I)\beta^T. \tag{3}$$

where $D = \beta^T \alpha$ is nonsingular (see [5]). Observe that each $\exp(tD)$ can be approximated with a diagonal (ν, ν)-Padé approximant at the cost of νp^3 flops.

This low-rank splitting can be easy generalized splitting A into matrices having one raw and one column (or a few rows and a few columns) (see [5] for more details). Recently, it has been demonstrated also that the splitting (2) can be improved when A belongs to a Lie algebra g by letting

$$A = \sum_{k=1}^{r} a_k Q_k, \tag{4}$$

where $r = dim(g)$ and $Q_1, Q_2, \ldots, Q_r$ is a basis of the algebra, and

$$\exp(tA) \approx \exp(g_1(t)Q_1)\exp(g_2(t)Q_2)\ldots\exp(g_r(t)Q_r),$$

and $g_1, g_2, \ldots, g_r$ are polynomials ([6]). In this case, the right choice of the basis and the use of certain features of the Lie algebra assume a fundamental importance [6].

3 A Recent Approach Based on SVD Techniques

Recently, other approaches for evaluating both $\exp(A)$ and $\exp(\tau A)y$ where A is a sparse skew symmetric matrix of large dimension n, y is a given vector, and τ is a scaling factor, have been proposed which takes advantage from some of the approaches discussed before. In [9] a procedure based on an effective Schur decomposition of the skew symmetric matrix A consisting into two main steps is presented. We will describe briefly this method. In the first step the skew-symmetric matrix A is tranformed via the Arnoldi procedure into its Hessenberg form H, e.g. $A = QHQ^T$ with Q $n \times n$ orthogonal matrix and H an Hessenberg $n \times n$ matrix possessing a tridiagonal structure. Then in the second step a Schur decomposition of H is obtained by using the singular value decomposition of a bidiagonal matrix of half size of H. The proposed procedure allows to take full advantage of the sparsity of A and of the tridiagonal form of H. In fact, the main cost for evaluating $\exp(A)$ is $\frac{9}{2}n^3$ flops which reduces to $2n^3$ flops when the banded structure of H is exploited; further, kn^2 flops are needed for evaluating $\exp(\tau A)y$.

Remark 1. In the implementation of a method to approximate the exponential of a matrix it could be important to know whether the entries of the matrix function $\exp(A)$ exhibit some kind of decay behavior away from the main diagonal, and to be able to estimate the rate of decay. Generally, even A is a sparse, $\exp(A)$ is usually a dense matrix. However when the matrix in study presents a banded structure it can be proved that the elements of the exponential decay very rapidly away from the diagonal and in practical computation they can be set to zero away from some bandwidth which depends on the required accuracy threshold (see [1,21]. Let A be a banded matrix of bandwidth $s \geq 1$ then the extra diagonal entries of $\exp(A)$ rapidly decay, as

$$|[\exp(A)]_{k,l}| < \exp(\rho) I_{|k-l|}(2\rho), \qquad \text{for } k, l = 1, 2, \ldots, n, \tag{5}$$

where $[\exp(A)]_{k,l}$ denotes the (k, l) entry of the matrix $\exp(A)$, $\rho = \max_{2 \leq j \leq n} h_j$ and $I_r(z)$ is the modified Bessel function with $I_r(2\rho) \cong \frac{\rho^r}{r!}$ for $r >> 1$ (see [21] for details). It not always clear at all how, for a bounded A, the nearness of $\exp(A)$ to a banded matrix can be exploited for improving Krylov subspace approximation or into rational approximation, but it has been used successfully in the context of splitting methods and in linear algebra methods to save some computational costs.

4 Exponentiation in Numerical Methods for ODEs

As observed above, the use of exponential of matrices for the numerical integration of ordinary differential equations on manifolds has been successfully adopted for designing a large number of geometrical methods which behave more favorably than standard integrators (see for instance [14]). In this final section, we will report some geometric integrators for linear and nonlinear differential equations.

Second and fourth order Magnus methods for $y' = A(t)y$

MG2	MG4
$A_1 = A(t_n + h/2);$	$A_1 = A(t_n + (\frac{1}{2} - \frac{\sqrt{3}}{6})h)$
$\omega_n = A_1;$	$A_2 = A(t_n + (\frac{1}{2} + \frac{\sqrt{3}}{6})h)$
$y_{n+1} = \exp(h\omega_n)y_n$	$\omega_n = \frac{1}{2}(A_1 + A_2) + \frac{\sqrt{3}}{12}h[A_2, A_1];$
	$y_{n+1} = \exp(h\omega_n)y_n$

Third order Crouch-Grossman method for $y' = A(t, y)y$:

$$A_1 = A(t_n, y_n);$$

$$A_2 = A(t_n + \tfrac{3}{4}h, \exp(\tfrac{3}{4}hA_1)y_n);$$

$$A_3 = A(t_n + \tfrac{3}{4}h, \exp(\tfrac{1}{108}hA_2)\exp(\tfrac{119}{216}hA_1)y_n);$$

$$y_{n+1} = \exp(\tfrac{13}{51}hA_3)\exp(-\tfrac{2}{3}hA_2)\exp(\tfrac{24}{17}hA_1)y_n$$

Exponential methods for nonlinear initial value problem $y' = f(y)$

- The exponential fitted Euler method:

$$y_{n+1} = y_n + h\varphi(hA)f(y_n),$$

- A symmetric exponential method:

$$y_{n+1} - y_n = \exp(hA)(y_{n-1} - y_n) + 2h\varphi(hA)f(y_n),$$

where with $A = f'(y_n)$ is Jacobian matrix of f and $\varphi = \frac{e^z - 1}{z}$.

References

1. Benzi M., Golub G.H.: Bounds fro the entries of matrix functions with applications to preconditioning, BIT 39 (3) (1999) pp 417–438.
2. Bergamaschi, L., Vianello, M.: Efficient computation of the exponential operator for large, sparse, symmetric matrices. Numer. Linear Algebra Appl. 7 (2000), pp. 27–45.
3. Calvetti D., Reichel L.: Exponential Integration methods for large stiff systems of differential equations. Iterative Methods in Scientific Computation II (1999), pp 1–7.

4. Castillo P., Saad Y.: Preconditioning the matrix exponential operator with applications. J. Sci. Comput. 13(3) (1998), pp. 275–302.
5. Celledoni E., Iserles A.: Approximating the exponential from a Lie algebra to a Lie group. Math. Comp. 69 (2000), pp. 1457–1480.
6. Celledoni E., Iserles A.: Numerical calculation of the exponential based on Wei-Norman equation Tech. Report, University of Cambridge.
7. Celledoni E., Iserles A.: Methods for the approximation of the matrix exponential in a Lie-algebraic setting. IMA J. Numer. Anal., No. 21, (2001) pp. 463–488.
8. Cheng, H.-W., Yau S.S.T.: More explicit formulas for the matrix exponential. Linear Algebra Appl. 262 (1997), pp. 131–163.
9. Del Buono N., Lopez L., Peluso R.: Computation of Exponentials of Real Skew Symmetric Matrices by SVD Techniques, Tech. Report 7/03 (2003).
10. Dieci L., Papini A.: Padé approximation for the exponential of a block triangular matrix. Linear Algebra Appl. 308 (2000), pp. 183–202.
11. Dieci L., Papini A.: Conditioning of the exponential of a block triangular matrix. Numer. Algorithms 28, (2001), pp. 137–150.
12. Gallopoulos E., Saad Y.: Efficient solution of parabolic equations by Krylov approximation methods, SIAM J. Sci. Stat. Comput. 13, No. 5, pp. 1236–1264, 1992.
13. Golub G.H., van Loan C.F.: Matrix Computation. 3th edn, John Hopikns, Baltimore.
14. Hairer E., Lubich C., Wanner. G.: Geometric Numerical Integration: Structure-Preserving Algorithms for Ordinary Differential Equations. Springer-Verlag, Berlin, 2002.
15. Hochbruck M., Lubich C.: Exponential integrators for quantum-classical molecular dynamics, BIT 39, No.4, pp. 620–645, 1999.
16. Hochbruck M., Lubich C.: Exponential integrators for large systems of differential equations, SIAM J. Sci. Comput. 19, No.5, pp. 1552–1574, 1998.
17. Hochbruck M., Lubich C.: On Krylov subspace approximations to the matrix exponential operator SIAM J. Numer. Anal. 34, No.5, pp. 1911–1925, (1997).
18. Horn R., Johnson C.: Topics in Matrix Analysis. Cambridge University Press, New York, 1995.
19. Iserles, A., Nørsett S.P., Rasmussen A.F.: Time symmetry and high-order Magnus methods. Appl. Numer. Math. 39, (2001) 379–401.
20. Iserles A., Munthe-Kaas H., Nørsett S., Zanna A.: Lie-group methods, Acta numerica. Cambridge: Cambridge University Press. Acta Numer. 9, 215–365 (2000).
21. Iserles A.: How large is the exponential of a banded matrix? N. Z. J. Math. 29, No.2, pp. 177–192 (2000).
22. Iserles A., Zanna A.: Efficient computation of the matrix exponential by Generalized Polar Decompositions. DAMTP Technical Report NA2002/09, University of Cambridge, UK (2002).
23. Leite F.S., Crouch, P.: Closed forms for the exponential mapping on matrix Lie groups based on Putzer's method. J. Math. Phys. 40, No.7, pp. 3561–3568, 1999.
24. McLachlan, R. I., Quispel G.R.W.: Splitting methods. Acta Numerica 2002.
25. Moler C.B., Van Loan C.F.: Nineteen dubious ways to compute matrix exponential. SIAM Review 20 (1978), pp. 801–836 .
26. Moret I., Novati P.: An interpolatory approximation of the matrix exponential based on Faber polynomials. J. Comput. Appl. Math. 131 (2001) pp. 361–380.
27. Moret I., Novati P.: The computation of functions of matrices by truncated Faber series. Numer. Funct. Anal. Optimization 22, (2001), pp. 697–719.
28. Munthe-Kaas, H.: High order Runge Kutta methods on manifolds. Appl. Numer. Math. 29(1) (1999), pp. 115–127.

29. Saad Y.: Analysis of some Krylov subspace approximation to the matrix exponential operator, SIAM J. Numer. Anal. 29, No. 1, pp. 209–228, 1992.
30. Van Loan C.: The sensitivity of the matrix exponential. SIAM J. Matrix Anal. Appl. 14 (1977), pp. 971–981.
31. Ben Taher R., Rachidi, M.: Some explicit formulas for the polynomial decomposition of the matrix exponential and applications. Linear Algebra Appl. 350, (2002) pp. 171–184.
32. Politi T.: A formula for the exponential of a real skew-symmetric matrix of order 4. BIT 41, No.4, (2001) pp. 842–845.
33. Tal-Ezer H.: Polynomial approximation of functions of matrices and applications. J. Sci. Comput. 4 (1989), pp. 25–60.
34. Tal-Ezer H.: Spectral methods in time for parabolic problems. SIAM J. Numer. Anal. 26 (1989), pp. 1–11.
35. Zanna A., Munthe-Kaas H.Z.: Generalized polar decompositions for the approximation of the matrix exponential. SIAM J. Matrix Anal. Appl. 23(3) (2002), pp. 840–862.

A Discrete Approach for the Inverse Singular Value Problem in Some Quadratic Group

T. Politi

Dipartimento Interuniversitario di Matematica, Politecnico di Bari, Via Orabona 4, I-70125 Bari (Italy). pptt@dm.uniba.it

Abstract. In this paper the solution of an inverse singular value problem is considered. First the decomposition of a real square matrix $A = U\Sigma V$ is introduced, where U and V are real square matrices orthogonal with respect to a particular inner product defined through a real diagonal matrix G of order n having all the elements equal to ± 1, and Σ is a real diagonal matrix with nonnegative elements, called $G-$singular values. When G is the identity matrix this decomposition is the usual SVD and Σ is the diagonal matrix of singular values. Given a set $\{\sigma_1, \ldots, \sigma_n\}$ of n real positive numbers we consider the problem to find a real matrix A having them as $G-$singular values. Neglecting theoretical aspects of the problem, we discuss only an algorithmic issue, trying to apply a Newton type algorithm already considered for the usual inverse singular value problem.

1 Introduction

Inverse problems are an important topic in applied mathematics (statistics, data analysis, applied linear algebra), since some times it is important to recovery some general structures (for instance matrices) starting from known (e.g. experimentally) data (eigenvalues, singular values, some prescribed entries). In this work we consider a special inverse problem related to the already studied inverse singular value problem (see [2]). In particular, firstly we consider the decomposition of a real square matrix A of order n

$$A = U\Sigma V \tag{1}$$

where U and V are $G-$orthogonal (or hypernormal) matrices and Σ is a nonnegative diagonal matrix. When G is the Minkowski matrix, (1) has some interesting applications in the study of polarized light (see [9,10]). Hence, we consider the inverse singular value problem for the decomposition (1). At the present, we do not know if this problem has a practical interest, but we observe that in the last years there has been a growing interest towards other kinds of SVD (for instance the Hyperbolic Singular Value Decomposition having some applications in signal processing and other fields of the engineering [1,7,11]). The paper is organized as follows: in the Section 2 we describe some algebraic properties of a product between vectors of $\mathbb{R}^n$ and introduce the singular value decomposition (1), in

P.M.A. Sloot et al. (Eds.): ICCS 2003, LNCS 2658, pp. 121–130, 2003.

Section 3 we present a discrete approach for the inverse singular value problem in the groups defined in Section 2 and finally in Section 4 we show a numerical example.

2 Singular Value Decomposition in Quadratic Groups

Let us denote by $\mathcal{D}_p$ the set of real diagonal matrices of dimension n, with elements equal to ± 1 and with p elements equal to $+1$. Let $G \in \mathcal{D}_p, 1 \leq p < n$, and let us define the following inner product:

$$\mathbf{x}, \mathbf{y} \in \mathbb{R}^n : \qquad (\mathbf{x}, \mathbf{y}) = \mathbf{x}^T G \mathbf{y}. \tag{2}$$

Given a vector $\mathbf{x} \in \mathbb{R}^n$, the number

$$\mathbf{x}^T G \mathbf{x} = \sum_{i=1}^{n} g_{ii} x_i^2$$

is called *Hypernormal Norm* even if it does not define a norm since it could be nonpositive. If G is the identity matrix of order n, then (2) defines the usual Euclidean scalar product. Using the hypernormal norm, the following classification among the vectors of $\mathbb{R}^n$ can be introduced.

Definition 1. *A vector* $\mathbf{x} \in \mathbb{R}^n$ *is called* Timelike *(*Strictly Timelike*) if* $(\mathbf{x}, \mathbf{x}) \geq 0$ *(if* $(\mathbf{x}, \mathbf{x}) > 0$*).*

Definition 2. *A vector* $\mathbf{x} \in \mathbb{R}^n$ *is called* Spacelike *(*Strictly Spacelike*) if* $(\mathbf{x}, \mathbf{x}) \leq 0$, *(if* $(\mathbf{x}, \mathbf{x}) < 0$*).*

Definition 3. *The set* $\{\mathbf{x} \in \mathbb{R}^n \mid (\mathbf{x}, \mathbf{x}) = 0\}$ *is called* Light Cone.

Definition 4. *The vectors* $\mathbf{x}, \mathbf{y} \in \mathbb{R}^n$ *are* $G-$orthogonal *with respect to (2) if*

$$(\mathbf{x}, \mathbf{y}) = \mathbf{x}^T G \mathbf{y} = 0.$$

We observe that the $G-$orthogonality does not imply the linear independence. In fact, if $\mathbf{x}$ is in the light cone, it is $G-$orthogonal to all the vectors $\mathbf{y} = \alpha \mathbf{x}$. Among the most used matrices G, we have the *Lorentz matrix* when

$$G = \begin{pmatrix} 1 & \\ & -I_{n-1} \end{pmatrix} \tag{3}$$

and the *Minkowski matrix* if

$$G = \begin{pmatrix} I_3 & \\ & -1 \end{pmatrix}$$

which has important applications in the mathematics of the relativity theory. Now we give some definition about real matrices with respect to (2).

Definition 5. *A real square matrix A of order n is said G-*adjoint *of $B \in \mathbb{R}^{n\times n}$ if*

$$(A\mathbf{x}, \mathbf{y}) = (\mathbf{x}, B\mathbf{y})$$

for all $\mathbf{x}, \mathbf{y} \in \mathbb{R}^n$.

The matrix $G-$adjoint of A is usually denoted by A^+. From (2) it follows that

$$A^+ = GA^TG.$$

Definition 6. *A real square matrix A of order n is called $G-$*selfadjoint *if*

$$(A\mathbf{x}, \mathbf{y}) = (\mathbf{x}, A\mathbf{y}).$$

If A is $G-$selfadjoint then $A = A^+ = GA^TG$. The concept of $G-$orthogonality can be extended also to real matrix.

Definition 7. *A real square matrix A of order n is said G-*orthogonal *(or* hypernormal*) if*

$$A^{-1} = A^+ = GA^TG.$$

See [8] for further properties and applications of hypernormal matrices. If A is a $G-$orthogonal matrix then

$$AGA^T = G.$$

If we consider the quadratic group related to matrix G, i.e. the set

$$\mathcal{H}_G(\mathbb{R}) = \{Y \in \mathbb{R}^{n\times n} \mid \det(Y) \neq 0,\ YGY^T = G\},$$

it coincides with the set of $G-$orthogonal matrices. In [6] the conservative solution of differential systems in these groups has been considered. Related to $\mathcal{H}_G(\mathbb{R})$ it is possible to define the its algebra as the set

$$\mathbf{h}_G = \{A \in \mathbb{R}^{n\times n} \mid GA + A^TG = 0\}.$$

If $A \in \mathbf{h}_G$ then

$$A^T = -GAG.$$

If G is the identity matrix then $\mathbf{h}_G$ is the set of real skew-symmetric matrices (sometimes matrices in $\mathbf{h}_G$ are also called $G-$skew-symmetric). If

$$G = \begin{pmatrix} I_p & O \\ O & -I_{n-p} \end{pmatrix}$$

$A \in \mathbf{h}_G$ can be characterized in the following way

$$A = \begin{pmatrix} A_{11} & A_{12} \\ A_{12}^T & A_{22} \end{pmatrix}$$

where $A_{11} \in \mathbb{R}^{p \times p}$ and $A_{22} \in \mathbb{R}^{(n-p)\times(n-p)}$ are real skew-symmetric matrices and $A_{12} \in \mathbb{R}^{p\times(n-p)}$. Given $A \in \mathbb{R}^{n\times n}$ if there exist two real G–orthogonal matrices U and V, and a diagonal matrix Σ with nonnegative elements such that

$$A = U\Sigma V$$

then it is called G–Singular Value Decomposition of A (or $G - SVD$). In particular it is called Singular Value Decomposition in Minkowski Space if G is the matrix (3). For the existence of this decomposition the following theorem holds (see [4,9]).

Theorem 1. *The G–SVD of a real matrix A exists iff*

1. *Matrix A^+A is diagonalizable and has real nonnegative eigenvalues;*
2. *$\mathcal{N}(A^+A) = \mathcal{N}(A)$*

where $\mathcal{N}(A)$ denotes the null space of A.

The elements of Σ are called G–singular values of A. They are the square roots of the eigenvalues of A^+A. In fact

$$A^+A = V^+\Sigma U^+U\Sigma V = V^+\Sigma^2 V = V^{-1}\Sigma^2 V.$$

3 Inverse Singular Value Problem in Quadratic Groups

Inverse problems involving eigenvalues and singular values have been extensively studied in last years (see [2,3,5]). The classical inverse singular value problem is stated as follows: given real general matrices $B_0, B_1, \ldots, B_n \in \mathbb{R}^{n\times n}$, and a set of nonnegative numbers $\sigma_1, \ldots, \sigma_n$ find the real values $c_1, \ldots, c_n$ such that the singular values of the matrix

$$B(\mathbf{c}) = B_0 + \sum_{i=1}^{n} c_i B_i \tag{4}$$

are $\sigma_1, \ldots, \sigma_n$. We suppose $\sigma_i \neq \sigma_j$ if $i \neq j$. Starting from the same premise it is obvious to define an analogous inverse problem also for other kinds of singular value decomposition. It is possible to ask that matrix (4) has the G–singular values equal to $\sigma_1, \ldots, \sigma_n$. In [2] two different approaches are analyzed for the inverse singular value problem: the first is continuous and the other one is discrete.

The first approach exploits the property that the eigenvalues of the symmetric matrix

$$\begin{pmatrix} 0 & A \\ A^T & 0 \end{pmatrix}$$

are plus and minus of the singular values of A, then a continuous flow is derived having the diagonal matrix of the singular values as limit point. In our case this approach cannot be applied since the G–singular values are the eigenvalues, with signs plus and minus, of the non-symmetric matrix

$$\begin{pmatrix} 0 & A \\ GA^TG & 0 \end{pmatrix}$$

then it cannot treated as a symmetric inverse eigenvalue problem. The second approach is a Newton-type algorithm having quadratic, but not global convergence. We try to apply this second approach to our problem. Following [2] we define

$$M(\Sigma) = \{U^+ \Sigma V \in \mathbb{R}^{n \times n} \mid U, V \in \mathcal{H}_G(\mathbb{R})\} \tag{5}$$

the set of all real matrices in $\mathbb{R}^{n\times n}$ whose G–singular values are $\sigma_1, \ldots, \sigma_n$. Denoting by

$$B = \{B(\boldsymbol{c}) \mid \boldsymbol{c} \in \mathbb{R}^n\}$$

the set of matrices of the form (4), the inverse G-singular value problems is equivalent to find an intersection of the two sets $M(\Sigma)$ and B. If $X^{(m)} \in M(\Sigma)$ there exist two matrices $U^{(m)}, V^{(m)} \in \mathcal{H}_G(\mathbb{R})$ such that

$$U^{(m)^+} X^{(m)} V^{(m)} = \Sigma. \tag{6}$$

We recall that

$$U^{(m)^{-1}} = U^{(m)^+} = GU^{(m)^T} G, \qquad V^{(m)^{-1}} = V^{(m)^+} = GV^{(m)^T} G.$$

Now we need to find an intercept of the line tangent to the manifold $M(\Sigma)$ at $X^{(m)}$ with the set B. Since a tangent vector $T(X)$ to $M(\Sigma)$ at a point $X \in M(\Sigma)$ has the form

$$T(X) = XK - HX$$

where K, H are G–skew-symmetric matrices and it is well-known that

$$X + T(X) = X + XK - HX$$

represents the line tangent to $M(\Sigma)$ emanating from X then we need to find two matrices $H^{(m)}, K^{(m)} \in \mathrm{h}_G$ such that

$$X^{(m)} + X^{(m)} K^{(m)} - H^{(m)} X^{(m)} = B(\boldsymbol{c}^{(m+1)}). \tag{7}$$

From (6) it is also:

$$X^{(m)} = U^{(m)} \Sigma V^{(m)^{-1}},$$

then (7) becomes:

$$U^{(m)} \Sigma V^{(m)^{-1}} + U^{(m)} \Sigma V^{(m)^{-1}} K^{(m)} - H^{(m)} U^{(m)} \Sigma V^{(m)^{-1}} = B(\boldsymbol{c}^{(m+1)})$$

$$\Sigma V^{(m)^{-1}} + \Sigma V^{(m)^{-1}} K^{(m)} - U^{(m)^{-1}} H^{(m)} U^{(m)} \Sigma V^{(m)^{-1}} = U^{(m)^{-1}} B(\boldsymbol{c}^{(m+1)})$$

$$\Sigma + \Sigma V^{(m)^{-1}} K^{(m)} V^{(m)} - U^{(m)^{-1}} H^{(m)} U^{(m)} \Sigma = U^{(m)^{-1}} B(\mathbf{c}^{(m+1)}) V^{(m)}$$

$$\Sigma + \Sigma G V^{(m)^T} G K^{(m)} V^{(m)} - G U^{(m)^T} G H^{(m)} U^{(m)} \Sigma = G U^{(m)^T} G B(\mathbf{c}^{(m+1)}) V^{(m)}.$$

Setting

$$\tilde{K}^{(m)} = G V^{(m)^T} G K^{(m)} V^{(m)}$$

$$\tilde{H}^{(m)} = G U^{(m)^T} G H^{(m)} U^{(m)}$$

$$\tilde{B}^{(m)} = G U^{(m)^T} G B(\mathbf{c}^{(m+1)}) V^{(m)}.$$

the equation becomes

$$\Sigma + \Sigma \tilde{K}^{(m)} - \tilde{H}^{(m)} \Sigma = \tilde{B}^{(m)}. \tag{8}$$

The matrices $\tilde{K}^{(m)}$ and $\tilde{H}^{(m)}$ are in the tangent space h_G, in fact

$$G\tilde{K}^{(m)} + \tilde{K}^{(m)^T} G = G\left[G V^{(m)^T} G K^{(m)} V^{(m)}\right] + \left[G V^{(m)^T} G K^{(m)} V^{(m)}\right]^T G =$$

$$= V^{(m)^T} G K^{(m)} V^{(m)} + V^{(m)^T} K^{(m)^T} G V^{(m)} =$$

$$= V^{(m)^T} G K^{(m)} V^{(m)} - V^{(m)^T} G K^{(m)} V^{(m)} = 0.$$

The proof for $\tilde{H}^{(m)}$ is similar. We can exploit the n^2 scalar equations (8) to evaluate the elements of unknown matrices $\tilde{K}^{(m)}$ and $\tilde{H}^{(m)}$ and the unknown vector $\mathbf{c}^{(m+1)}$. Let us consider if they are enough to compute the elements unknown. Matrices $\tilde{K}^{(m)}$ and $\tilde{H}^{(m)}$ are in h_G, whose dimension is $[n(n-1)]/2$ hence they are characterized by $n^2 - n$ parameters plus n elements of vector $\mathbf{c}^{(m+1)}$ we have exactly n^2 unknowns, which can be computed from the equations. For $i \neq j$ (8) gives

$$\tilde{b}_{ij}^{(m)} = \sigma_i \tilde{k}_{ij}^{(m)} - \tilde{h}_{ij}^{(m)} \sigma_j \tag{9}$$

while the equations for diagonal elements are

$$\tilde{b}_{ii}^{(m)} = \sigma_i, \qquad i = 1, \ldots, n. \tag{10}$$

Using (10) it is possible to compute the elements of the vector $\mathbf{c}^{(m+1)}$. In fact

$$\sigma_i = \tilde{b}_{ii}^{(m)} = \mathbf{e}_i^T \tilde{B}^{(m)} \mathbf{e}_i$$

where $\mathbf{e}_i$ is the i–th unit vector of $\mathbb{R}^n$, and from (4)

$$\tilde{B}^{(m)} = G U^{(m)^T} G B(\mathbf{c}^{(m+1)}) V^{(m)} =$$

$$= G U^{(m)^T} G \left[B_0 + \sum_{j=1}^{n} c_j B_j \right] V^{(m)} =$$

$$= G U^{(m)^T} G B_0 V^{(m)} + \sum_{j=1}^{n} c_j G U^{(m)^T} G B_j V^{(m)}.$$

Hence

$$\tilde{b}_{ii}^{(m)} = \boldsymbol{e}_i^T \left[GU^{(m)^T} GB_0 V^{(m)} + \sum_{j=1}^{n} c_j GU^{(m)^T} GB_j V^{(m)} \right] \boldsymbol{e}_i =$$

$$= \boldsymbol{e}_i^T GU^{(m)^T} GB_0 V^{(m)} \boldsymbol{e}_i + \sum_{j=1}^{n} c_j \boldsymbol{e}_i^T GU^{(m)^T} GB_j V^{(m)} \boldsymbol{e}_i.$$

The vector $\boldsymbol{c}^{(m+1)}$ is the solution of the linear system

$$A^{(m)} \boldsymbol{c}^{(m+1)} = \boldsymbol{\sigma} - \boldsymbol{b}, \tag{11}$$

where $A \in \mathbb{R}^{n \times n}$ is the matrix with elements

$$a_{ij}^{(m)} = \boldsymbol{e}_i^T GU^{(m)^T} GB_j V^{(m)} \boldsymbol{e}_i$$

and

$$b_i = \boldsymbol{e}_i^T GU^{(m)^T} GB_0 V^{(m)} \boldsymbol{e}_i$$

and $\boldsymbol{\sigma}$ denotes the vector $(\sigma_1, \ldots, \sigma_n)$. If (11) has a unique solution then vector $\boldsymbol{c}^{(m+1)}$ is known. To compute matrices $\tilde{K}^{(m)}$ and $\tilde{H}^{(m)}$ we exploit (9). We recall the structure of matrices in $\mathbf{h}_G$

$$\tilde{K}^{(m)} = \begin{pmatrix} K_{11}^{(m)} & K_{12}^{(m)} \\ K_{12}^{(m)^T} & K_{22}^{(m)} \end{pmatrix}$$

where $K_{11}^{(m)} \in \mathbb{R}^{p \times p}$ and $K_{22}^{(m)} \in \mathbb{R}^{(n-p)\times(n-p)}$ are real skew-symmetric matrices. First we consider the equations (i, j), with $1 \le i < j \le p$ and $p+1 \le i < j \le n$:

$$\tilde{b}_{ij}^{(m)} = \sigma_i \tilde{k}_{ij}^{(m)} - \tilde{h}_{ij}^{(m)} \sigma_j \tag{12}$$

and

$$\tilde{b}_{ji}^{(m)} = \sigma_j \tilde{k}_{ji}^{(m)} - \tilde{h}_{ji}^{(m)} \sigma_i = -\sigma_j \tilde{k}_{ij}^{(m)} + \sigma_i \tilde{h}_{ij}^{(m)}. \tag{13}$$

Solving (12) and (13) we obtain:

$$\tilde{k}_{ij}^{(m)} = \frac{\sigma_i \tilde{b}_{ij}^{(m)} - \sigma_j \tilde{b}_{ji}^{(m)}}{\sigma_i^2 - \sigma_j^2}$$

$$\tilde{h}_{ij}^{(m)} = \frac{\sigma_i \tilde{b}_{ji}^{(m)} + \sigma_j \tilde{b}_{ij}^{(m)}}{\sigma_i^2 - \sigma_j^2}.$$

From the equations (i, j), $1 \le i \le p$ and $p+1 \le j \le n$, it is:

$$\tilde{b}_{ij}^{(m)} = \sigma_i \tilde{k}_{ij}^{(m)} - \tilde{h}_{ij}^{(m)} \sigma_j \tag{14}$$

and

$$\tilde{b}_{ji}^{(m)} = \sigma_j \tilde{k}_{ji}^{(m)} - \tilde{h}_{ji}^{(m)} \sigma_i = \sigma_j \tilde{k}_{ij}^{(m)} - \sigma_i \tilde{h}_{ij}^{(m)}. \tag{15}$$

Solving (14) and (15) we obtain:

$$\tilde{k}_{ij}^{(m)} = \frac{\sigma_j \tilde{b}_{ji}^{(m)} - \sigma_i \tilde{b}_{ij}^{(m)}}{\sigma_j^2 - \sigma_i^2}$$

$$\tilde{h}_{ij}^{(m)} = \frac{\sigma_i \tilde{b}_{ji}^{(m)} - \sigma_j \tilde{b}_{ij}^{(m)}}{\sigma_j^2 - \sigma_i^2}.$$

We use $\tilde{K}^{(m)}$ and $\tilde{H}^{(m)}$ to compute $K^{(m)}$ and $H^{(m)}$, in fact

$$K^{(m)} = G(V^{(m)^{-1}})^T G \tilde{K}^{(m)} V^{(m)^{-1}} = V^{(m)} \tilde{K}^{(m)} G V^{(m)^T} G$$

$$H^{(m)} = G(U^{(m)^{-1}})^T G \tilde{H}^{(m)} U^{(m)^{-1}} = U^{(m)} \tilde{K}^{(m)} G U^{(m)^T} G.$$

The next step is to project $B(\mathbf{c}^{(m+1)})$ to $M(\Sigma)$. It is possible to use the exponential map, since, given a matrix H in $\mathbf{h}_G$, then e^H belongs to the quadratic group. The computation of the exponential of a matrix is too expensive, then the Cayley transform can be used, if

$$D = \left(I + \frac{H^{(m)}}{2}\right)\left(I - \frac{H^{(m)}}{2}\right)^{-1}, \qquad F = \left(I + \frac{K^{(m)}}{2}\right)\left(I - \frac{K^{(m)}}{2}\right)^{-1},$$

then

$$X^{(m+1)} = D^+ X^{(m)} F.$$

The computation of the matrix can be avoided since only G–orthogonal matrices $U^{(m)}$ and $V^{(m)}$ are needed:

$$U^{(m+1)} = D^+ U^{(m)}, \qquad V^{(m+1)} = F V^{(m)}.$$

The convergence of the method will not be considered in this work.

4 A Numerical Example and Conclusions

In this last section we present a numerical example concerning the discrete method introduced in the previous section.

Example 1. In this case we have considered a problem of dimension 4, the matrix G is

$$G = \begin{pmatrix} I_2 & 0 \\ 0 & -I_2 \end{pmatrix}$$

and we have chosen the following random symmetric matrices B_i:

$$B_0 = \begin{pmatrix} 1.4353 & 1.1345 & 0.7833 & 0.5754 \\ 1.1345 & 0.7065 & 0.8811 & 1.1264 \\ 0.7833 & 0.8811 & 0.9568 & 1.2707 \\ 0.5754 & 1.1264 & 1.2707 & 1.7857 \end{pmatrix}$$

$$B_1 = \begin{pmatrix} 0.5462 & 1.0596 & 0.9154 & 1.0762 \\ 1.0596 & 1.8168 & 0.3103 & 0.4132 \\ 0.9154 & 0.3103 & 1.2816 & 0.3617 \\ 1.0762 & 0.4132 & 0.3617 & 1.9886 \end{pmatrix}$$

$$B_2 = \begin{pmatrix} 0.8796 & 0.7333 & 0.7728 & 0.5254 \\ 0.7333 & 1.1831 & 0.9896 & 0.9110 \\ 0.7728 & 0.9896 & 1.1885 & 0.5023 \\ 0.5254 & 0.9110 & 0.5023 & 1.2917 \end{pmatrix}$$

$$B_3 = \begin{pmatrix} 1.9338 & 0.8019 & 1.6053 & 0.1655 \\ 0.8019 & 1.6375 & 1.1775 & 1.0814 \\ 1.6053 & 1.1775 & 0.6922 & 0.5885 \\ 0.1655 & 1.0814 & 0.5885 & 1.7120 \end{pmatrix}$$

$$B_4 = \begin{pmatrix} 0.9805 & 1.2666 & 0.7582 & 1.4403 \\ 1.2666 & 0.8244 & 0.9508 & 0.5583 \\ 0.7582 & 0.9508 & 1.3864 & 1.0502 \\ 1.4403 & 0.5583 & 1.0502 & 0.3976 \end{pmatrix}.$$

Then we have taken the following vector $\boldsymbol{c}$:

$$\overline{\boldsymbol{c}} = (6.2520E-01, 7.3336E-01, 3.7589E-01, 9.8765E-03)^T$$

and have considered as G–singular values those of matrix $B(\overline{\boldsymbol{c}})$:

$$\boldsymbol{\sigma} = (0.39364, 1.1338, 2.5057, 10.4847)^T.$$

Hence we have perturbed each entry of the vector $\overline{\boldsymbol{c}}$ with random quantities between 0 and 0.5, and have considered this one as the initial guess for the iterations. In fact the algorithm, when it converges, has only a local convergence (see [2]) then it is necessary to start from a point near the solution. In Table 1 the error, measured in the 2-norm, on the G–singular values at each iteration are shown.

Table 1. Example 1

Iteration	Error
0	$2.5289e+0$
1	$1.1410e+0$
2	$7.2304e-1$
3	$3.2589e-2$
4	$8.9911e-3$
5	$3.0011e-4$
6	$9.8812e-5$

In the paper we have presented a discrete approach to the inverse singular value problem in quadratic groups. The algorithm is similar to the one already introduced for the usual inverse singular value problem. The algorithm described needs to be investigated more carefully, in fact it breaks down very often (the matrix $A^{(m)}$ is some times singular or the matrix $B(\boldsymbol{c}^{(m)})$ does not satisfy the hypotheses on the existence of the G-singular value decomposition. A possible solution to these troubles could be a different parameterization of the manifold (5). This features will be analyzed in future together with the solvability of the problem. Moreover it should be considered the analogous problem for the Hyperbolic Singular Value Decomposition (see [1,7,11]) which has some applications in signal processing and other fields of engineering. Some other aspects of the problem need further investigations, for instance the application of continuous techniques (see [2]).

References

1. Bojanczyk A.W., Onn R., Steinhardt A.O.: Existence of The Hyperbolic Singular Value Decomposition. Linear Algebra Appl. **185** (1993) 21–30
2. Chu M.T.: Numerical methods for inverse singular value problems. SIAM J. Numer. Anal. **29** (3) (1992) 885–903
3. Chu M.T.: Inverse eigenvalue problems. SIAM Rev. **40** (1) (1998) 1–39
4. Di Lena G., Piazza G., Politi T.: *An algorithm for the computation of the G−singular values of a matrix.* Manuscript.
5. Friedland S.: Inverse eigenvalue problems. Linear Algebra Appl. **17** (1977) 15–51
6. Lopez L., Politi T.: Applications of the Cayley Approach in the Numerical Solution of Matrix Differential Systems on Quadratic Groups. Appl. Num. Math. **36** (2001) 35–55
7. Onn R., Steinhardt A.O., Bojanczyk A.W.: The Hyperbolic Singular Value Decomposition and applications. IEEE Trans. Sign. Proc. **39** (7) (1991) 1575–1588
8. Rader C.M., Steinhardt A.O.: Hyperbolic Householder Transforms. SIAM J. Matrix Anal. **9** (1988) 269–290
9. Renardy M.: Singular values decomposition in Minkowski space. Linear Algebra Appl. **236** (1996) 53–58
10. Xing Z.: On the deterministic and nondeterministic Mueller matrix. J. ModernOpt. **39** (1992) 461–484
11. Zha H.: A note on the existence of The Hyperbolic Singular Value Decomposition. Linear Algebra Appl. **240** (1996) 199–205

Two Step Runge-Kutta-Nyström Methods for Oscillatory Problems Based on Mixed Polynomials

Beatrice Paternoster

Dipartimento di Matematica e Informatica
Universitá di Salerno, Italy
beapat@unisa.it

Abstract. We consider two step Runge-Kutta-Nyström methods for the numerical integration of $y'' = f(x, y)$ having periodic or oscillatory solutions. We assume that the frequency ω can be estimated in advance. Using the linear stage representation, we describe how to derive two step Runge-Kutta-Nyström methods which integrate trigonometric and mixed polynomials exactly. The resulting methods depend on the parameter $\nu = \omega h$, where h is the stepsize.

1 Introduction

We are concerned with the second order initial value problem

$$y''(t) = f(t, y(t)), \quad y(t_0) = y_0, \quad y'(t_0) = y'_0, \qquad y(t), f(t, y) \in R^n, \qquad (1)$$

having periodic or oscillatory solutions, which describes many processes in technical sciences. Examples are given in celestial mechanics, molecular dynamics, seismology, and so on.

For ODEs of type (1), in which the first derivative does not appear explicitly, it is preferable to use a direct numerical method, instead of reducing the ODEs (1) into a first order system. An interesting and important class of initial value problems (1) which can arise in practice consists of problems whose solutions are known to be periodic, or to oscillate with a known frequency. Classical methods require a very small stepsize to track the oscillations and only methods which take advantage of some previous knowledge about the solution are able to integrate the system using a reasonable large stepsize. Therefore, efficiency can be improved by using numerical methods in which *a priori* information on the solution (as for instance, a good estimate of the period or of the dominant frequency) can be embedded.

In the following let us assume that a good estimate of the dominant frequency ω is known in advance. The aim is to exploit this extra information and to modify a given integration method in such a way that the method parameters are 'tuned' to the behavior of the solution. Such an approach has already been proposed by Gautschi in 1961 [6] for linear multistep methods for first-order differential equations in which the dominant frequencies ω_j are a priori known.

P.M.A. Sloot et al. (Eds.): ICCS 2003, LNCS 2658, pp. 131–138, 2003.

Exploiting this idea, many numerical methods with coefficients depending on the predicted frequency are available in literature. Paternoster introduced Runge–Kutta–Nyström methods based on trigonometric polynomials [10], and methods resulting exact in phase when the high frequency component is produced by a linear part [11]. Coleman et al. considered methods based on mixed collocation [4]. In the class of *exponential–fitted* methods many methods are available in literature. Ixaru in [7] focused on the numerical formulae associated with operations on oscillatory functions. Then many papers followed, we only cite here some of them [8,14,15,16]; see also the references therin.

In this paper we consider two step Runge–Kutta–Nyström methods for (1) having periodic or oscillatory solutions, for which a good estimate of the frequency is known in advance. We treat the TSRKN method as a composite linear multistep scheme, as done in Albrecht's approach [1,2]. Following the approach of [6,10,12], we define the *trigonometric order* of the methods. In section 2 we recall the conditions to obtain two step Runge–Kutta–Nyström methods which integrate algebraic polynomials exactly. In section 3 we give the definition of *trigonometric order* of the TSRKN method, and state the conditions to satisfy for integrating trigonometric polynomials exactly.

2 Two Step Runge-Kutta-Nyström Methods Based on Algebraic Polynomials

We consider the two step Runge–Kutta–Nyström methods (TSRK)

$$
\begin{aligned}
Y_{i-1}^j &= y_{i-1} + hc_j y'_{i-1} + h^2 \textstyle\sum_{s=1}^m a_{js} f(x_{i-1} + c_s h, Y_{i-1}^s), \qquad j = 1,\dots,m \\
Y_i^j &= y_i + hc_j y'_i + h^2 \textstyle\sum_{s=1}^m a_{js} f(x_i + c_s h, Y_i^s), \qquad j = 1,\dots,m, \\
y_{i+1} &= (1-\theta) y_i + \theta y_{i-1} + h \textstyle\sum_{j=1}^m v'_j y'_{i-1} + h \sum_{j=1}^m w'_j y'_i + \\
&\quad h^2 \textstyle\sum_{j=1}^m (v_j f(x_{i-1} + c_j h, Y_{i-1}^j) + w_j f(x_i + c_j h, Y_i^j)), \\
y'_{i+1} &= (1-\theta) y'_i + \theta y'_{i-1} + h \textstyle\sum_{j=1}^m (v'_j f(x_{i-1} + c_j h, Y_{i-1}^j) + w'_j f(x_i + c_j h, Y_i^j)).
\end{aligned}
\tag{2}
$$

for the initial value problem (1). θ, v_j, w_j, v'_j, w'_j, a_{js}, $j, s, = 1,\dots,m$ are the coefficients of the methods, which can be represented by the Butcher array

$$
\begin{array}{c|c}
\mathbf{c} & \mathbf{A} \\
\hline
 & \mathbf{v} \\
 & \mathbf{w} \\
\theta & \\
 & \mathbf{v}' \\
 & \mathbf{w}'
\end{array}
=
\begin{array}{c|cccc}
c_1 & a_{11} & a_{12} & \cdots & a_{1m} \\
c_2 & a_{21} & a_{22} & \cdots & a_{2m} \\
\vdots & \vdots & \vdots & & \vdots \\
c_m & a_{m1} & a_{m2} & \cdots & a_{mm} \\
\hline
\theta & v_1 & v_2 & \cdots & v_m \\
 & w_1 & w_2 & \cdots & w_m \\
 & v'_1 & v'_2 & \cdots & v'_m \\
 & w'_1 & w'_2 & \cdots & w'_m
\end{array}
\tag{3}
$$

The TSRKN method (2), introduced in [13], was derived as an indirect method from the two step Runge–Kutta method presented in [9]. In comparison with classical one step Runge–Kutta–Nyström methods, TSRKN methods need a lower number of stages to rise to a given order of convergence. Indeed, advancing from x_i to x_{i+1}, we only have to compute Y_i, because Y_{i-1} have already been evaluated in the previous step. Therefore the computational cost of the method depends on the matrix A, while the vector v adds extra degrees of freedom.

It is known that the method (2) is zero–stable if [13]

$$-1 < \theta \leq 1 \tag{4.}$$

We treat formulas (2) by extending Albrecht's technique [1,2] to the numerical method we considered, as in [10,12]. According to this approach, we regard the TSRKN method (2) as a composite linear multistep scheme on a not equidistant grid.
Y_{i-1}^j and Y_i^j in (2) are called *internal* stages; y_{i+1} and y'_{i+1} are the *final* stages, which give the approximation of the solution and its derivative of the solution in the step point x_i.

We associate a linear difference operator with each internal stage Y_i^J of (2), in the following way:

$$\mathcal{L}_j[z(x);h] = z(x+c_jh) - z(x) - hc_jz'(x) - h^2\sum_{s=1}^{m}(a_{js}z''(x+c_sh), \tag{5}$$

for $j = 1,\dots,m$, While the operator

$$\begin{aligned}\bar{\mathcal{L}}[z(x);h] \quad &= z(x+h) - (1-\theta)z(x) - \theta z(x-h) - h(\textstyle\sum_{j=1}^m v'_jz'(x-h)+ \\ &\textstyle\sum_{j=1}^m w'_jz'(x)) - h^2\sum_{j=1}^m(v_jz''(x+(c_j-1)h) + w_jz''(x+c_jh)),\end{aligned} \tag{6}$$

is associated with the stage y_{i+1} in (2). Finally

$$\begin{aligned}\bar{\mathcal{L}}'[z(x);h] &= hz'(x+h) - h(1-\theta)z'(x) - \theta hz'(x-h) - \\ &h^2\textstyle\sum_{j=1}^m(v'_jz''(x+(c_j-1)h) + w'_jz''(x+c_jh))\end{aligned} \tag{7}$$

is associated with the final stage y'_{i+1} in (2). It follows that

$$\mathcal{L}_j[1;h] = \mathcal{L}_j[x;h] = 0, \quad j=1,\dots,m,$$

which implies that $y(x_i+c_jh) - Y_i^j = O(h)$ for $h \to 0$. Moreover

$$\bar{\mathcal{L}}[1;h] = \bar{\mathcal{L}}'[1;h] = \bar{\mathcal{L}}'[x;h] = 0, \quad j=1,\dots,m,$$

If we annihilate (6) on the function $z(x) = x$, then from
$\bar{\mathcal{L}}[x;h] = 0$, it follows that

$$\sum_{j=1}^{m}(v'_j + w'_j) = 1 + \theta \tag{8}$$

which represents the consistency condition already derived in [9,13] which, together with (4), ensures that the TSRKN is convergent with order at least one.

If (5) is identically equal to zero when $z(x) = x^p$, i.e. if $\mathcal{L}_j[x^p; h] = 0$, then

$$\sum_{s=1}^{m} a_{js} c_s^{p-2} = \frac{c_j^p}{p(p-1)}, \quad j = 1, \ldots, m. \tag{9}$$

Moreover, if (6) is equal to zero when $z(x) = x^p$, i.e. $\bar{\mathcal{L}}[x^p; h] = 0$, then

$$\sum_{j=1}^{m} (v_j (c_j - 1)^{p-2} + w_j c_j^{p-2}) = \frac{1 - (-1)^p \theta}{p(p-1)} - \frac{(-1)^{p-1}}{p-1} \sum_{j=1}^{m} v'_j. \tag{10}$$

Finally, if we annihilate (7) on the function $z(x) = x^p$, then from $\bar{\mathcal{L}}'[x^p; h] = 0$, it follows that

$$\sum_{j=1}^{m} (v'_j (c_j - 1)^{p-2} + w'_j c_j^{p-2}) = \frac{1 - (-1)^{p-1} \theta}{(p-1)}. \tag{11}$$

We can now give the following definitions:

Definition 1. *An m–stage TSRKN method is said to satisfy the simplifying conditions $C_2(p)$ if its parameters satisfy*

$$\sum_{s=1}^{m} a_{js} c_s^{k-2} = \frac{c_j^k}{k(k-1)}, \qquad j = 1, \ldots, m, \ k = 1, \ldots, p.$$

Definition 2. *An m–stage TSRKN method (2) is said to satisfy the simplifying conditions $B_2(p)$ if its parameters satisfy*

$$\sum_{j=1}^{m} (v_j (c_j - 1)^{k-2} + w_j c_j^{k-2}) = \frac{1 - (-1)^k \theta}{k(k-1)} - \frac{(-1)^{k-1}}{k-1} \sum_{j=1}^{m} v'_j,$$

$$j = 1, \ldots, m, \ k = 1, \ldots, p.$$

Definition 3. *An m–stage TSRKN method is said to satisfy the simplifying conditions $B'_2(p)$ if its parameters satisfy*

$$\sum_{j=1}^{m} (v'_j (c_j - 1)^{k-2} + w'_j c_j^{k-2}) = \frac{1 - (-1)^{k-1} \theta}{(k-1)}, \qquad k = 1, \ldots, p.$$

$C_2(p)$, $B_2(p)$ and $B'_2(p)$ allow the reduction of order conditions of trees in the theory of two step RKN methods, which is under development by the author of this paper; moreover they also mean that all the quadrature formulas represented by the TSRKN method have order at least p, similarly as it happens in the theory of Runge–Kutta methods [3].

The following theorem can be obviously proved by using Albrecht's theory [1,2]:

Theorem 1. *If $C_2(p)$, $B_2(p)$ and $B_2'(p)$ hold, then the m–stage TSRKN method (1.2) has order of convergence p.*

Proof. $C_2(p), B_2(p)$ and $B_2'(p)$ imply that all the stages of the method have order p or, in Albrecht's terminology, that each stage in (2) has order of *consistency* p, so that the method has order of *consistency* p. In this case the method converges with order at least p.

It is worth mentioning that the conditions $C_2(p)$, $B_2(p)$ and $B_2'(p)$ are only sufficient conditions for the TSRKN method to have order p, but not necessary. Indeed the final stage must have order of *consistency* p, which is the condition $B_2'(p)$, but it is not necessary that also the internal stages have order of *consistency* p. If all the stages have order of *consistency* p, then they are exact on any linear combination of the power set $\{1, x, x^2, \ldots, x^p\}$, and this implies that the TSRKN method results exact when the solutions of the system of ODEs(1) are algebraic polynomials. Moreover the simplifying conditions $C_2(p)$, $B_2(p)$ and $B_2'(p)$ are a constructive help for the derivation of new numerical methods within the class of TSRKN methods having a high order stage, and will be a useful basis for characterizing collocation TSRKN methods.

3 Two Step Runge-Kutta-Nyström Methods Based on Trigonometric Polynomials

Now, we can consider TSRKN methods which integrate ODEs (1) having periodic or oscillatory solutions, which can be expressed through trigonometric polynomials.

Let us suppose that a good approximation of the dominant frequency ω is known in advance, and that (4) and (8) hold, so that the method is convergent with order at least one.

Following Gautschi [6] and [10,12], we state now the definition of *trigonometric* order.

Definition 4. *The two step RKN method*

$$\begin{array}{c|c} \mathbf{c}(\nu) & \mathbf{A}(\nu) \\ \hline & \mathbf{v}(\nu) \\ & \mathbf{w}(\nu) \\ \theta & \mathbf{v}'(\nu) \\ & \mathbf{w}'(\nu) \end{array}$$

is said to be of trigonometric order q, relative to the frequency ω, if the associated linear difference operators (5)–(7) satisfy

$$\begin{aligned}
&\mathcal{L}_j[1;h] = \bar{\mathcal{L}}[1;h] = \bar{\mathcal{L}}'[1;h] = 0, && j = 1,\dots,m;\\
&\mathcal{L}_j[\cos r\omega x;h] = \mathcal{L}_j[\sin r\omega x;h] = 0, \quad i = 1,\dots,m, && r = 1,\dots,q;\\
&\bar{\mathcal{L}}[\cos r\omega x;h] = \bar{\mathcal{L}}[\sin r\omega x;h] = 0, && r = 1,\dots,q;\\
&\bar{\mathcal{L}}'[\cos r\omega x;h] = \bar{\mathcal{L}}'[\sin r\omega x;h] = 0, && r = 1,\dots,q,
\end{aligned}$$

with $\nu = \omega h$.

It is already known that methods with *trigonometric order* q have *algebraic order* $2q$ (see [6,10] for the definition of algebraic order) and therefore have order of convergence $2q$.

It is easy to verify that a TSRKN method has *trigonometric* order q, according to Definition 4, if its parameters satisfy the following systems:

$$A(q) = \begin{cases} \sum_{s=1}^{m} a_{js}\cos(rc_s\nu) = \dfrac{1-\cos(rc_j\nu)}{r^2\nu^2} \\ \qquad\qquad\qquad\qquad\qquad\qquad\qquad j = 1,\dots,s \\ \sum_{s=1}^{m} a_{js}\sin(rc_s\nu) = \dfrac{c_j}{r\nu} - \dfrac{\sin(rc_j\nu)}{r^2\nu^2} \end{cases}$$

$$VW\theta(q) = \begin{cases} \sum_{j=1}^{m}(v_j\cos(r\,(c_j-1)\nu) + w_j\cos(rc_j\nu)) = \\ \qquad \dfrac{(1-\cos(r\nu))(1-\theta)}{r^2\nu^2} + \dfrac{\sum_{j=1}^{m} v'_j\sin(r\nu)}{r\nu} \\ \sum_{j=1}^{m}(v_j\sin(r\,(c_j-1)\nu) + w_j\sin(rc_j\nu)) = \\ \qquad \dfrac{-\sin(r\nu)(1+\theta)}{r^2\nu^2} + \dfrac{\sum_{j=1}^{m} v'_j\cos(r\nu) + \sum_{j=1}^{m} w'_j}{r\nu} \end{cases}$$

$$V'W'\theta(q) = \begin{cases} \sum_{j=1}^{m}(v'_j\cos(r(c_j-1)\nu) + w'_j\cos(rc_j\nu) = \dfrac{\sin(r\nu)(1+\theta)}{r\nu} \\ \sum_{j=1}^{m}(v'_j\sin(r(c_j-1)\nu) + w'_j\sin(rc_j\nu) = \dfrac{(1-\theta)(1-\cos(r\nu)}{r\nu} \end{cases}$$

for $r = 1,\dots,q$.

The following theorem states some constructive conditions to derive TSRKN methods giving exact solution (within the roundoff) error when the solution of ODEs (1) is a mixed polynomial which oscillates with frequency ω.

Theorem 2. *If the coefficients of the TSRKN method satisfy the conditions* $C_2(p), B_2(p), B'_2(p)$ *and* $A(q), VW\theta(q), V'W'\theta(q)$, *then the TSRKN method integrates any linear combination of* $\{1, x, x^2, \dots, x^p, \cos\omega x, \sin\omega x, \cos 2\omega x, \sin 2\omega x, \dots, \cos q\omega x, \sin q\omega x\}$ *exactly.*

The construction of the methods requires the solutions of the linear systems $C_2(p), B_2(p), B'_2(p)$ and $A(q), VW\theta(q), V'W'\theta(q)$, which is underdetermined. It is possible to solve the uncoupled linear systems after fixing some free parameters. Through symbolic computation it is possible to determine the analytical expressions of the remaining parameters of the method; for a high number of stages the involved systems have to be solved numerically.

4 Conclusions

Numerical methods for (1) having frequency–dependent parameters are quite widely used methods [4,5,6,8,10,11,14,15,16], when a good approximation of the frequency ω to be fitted is a priori available. In this paper we design the approach to be used in the derivation of two step Runge–Kutta–Nyström methods in the case that only one frequency is fitted, but the development of TSRKN methods in which more frequencies are fitted can be considered as well. The linear stability analysis of these methods has not be considered in this paper, and has to follow the lines drawn in [5,8,4].

Recently some authors [16] addressed the problem of how choosing the optimal value of the frequency to predict, and this new perspective enlarges the sphere of application of methods with ν–dependent parameters, where ν is given by the product of the fitted frequency and the stepsize.

The coefficients of methods which are frequency–dependent involve combinations of trigonometric functions.

It is known that a common feature of this type of methods is that heavy cancelations occur during the evaluation of the coefficients from their closed form, which becomes increasingly severe as ν tends to 0. Therefore, in the application, the expansions of the coefficients in powers of ν, generated for example by Maple or Mathematica, is strongly recommended.

References

1. Albrecht, P.: Elements of a general theory of composite integration methods, Appl. Math. Comp. **31** (1989) 1–17.
2. Albrecht, P.: A new theoretical approach to RK methods, SIAM J. Numer. Anal. **24**(2) (1987) 391–406.
3. Butcher, J.C.: The Numerical Analysis of Ordinary Differential Equations: Runge–Kutta and General Linear Methods, Wiley, New York (1987).
4. Coleman, J.P., Duxbury, S.C.: Mixed collocation methods for $y'' = f(x,y)$, J. Comput. Appl. Math. **126** (2000) 47–75.
5. Coleman J.P., Ixaru, L.Gr.: P–stability and exponential–fitting methods for $y'' = f(x,y)$, IMA J. Numer. Anal. **16** (1996) 179–199.
6. Gautschi, W.: Numerical integration of ordinary differential equations based on trigonometric polynomials, Numer. Math. **3** (1961) 381–397.
7. Ixaru, L. Gr.: Operations on oscillatory functions, Comput. Phys. Comm. **105** (1997) 1–19.
8. Ixaru, L.Gr., Paternoster, B.: A conditionally P–stable fourth–order exponential–fitting method for $y'' = f(x,y)$, J. Comput. Appl. Math. **106** (1999) 87–98.
9. Jackiewicz, Z., Renaut, R., Feldstein, A.: Two–step Runge–Kutta methods, SIAM J. Numer. Anal. **28**(4) (1991) 1165–1182.

10. Paternoster, B.: Runge–Kutta(–Nyström) methods for ODEs with periodic solutions based on trigonometric polynomials, Appl. Numer. Math. **28**(2–4) (1998) 401–412.
11. Paternoster, B.: A phase–fitted collocation–based Runge–Kutta–Nyström method, Appl. Numer. Math. **35**(4) (2000) 239–355.
12. Paternoster, B.: General two–step Runge–Kutta methods based on algebraic and trigonometric polynomials, Int. J. Appl. Math. **6**(4) (2001) 347–362.
13. Paternoster, B.: Two step Runge-Kutta-Nyström methods for $y'' = f(x, y)$ and P-stability, *Computational Science - ICCS 2002, Lecture Notes in Computer Science 2331, Part III*, P.M.A.Sloot, C.J.K.Tan, J.J.Dongarra, A.G.Hoekstra Eds., 459–466, Springer Verlag, Amsterdam (2002).
14. Simos, T.E.: An exponentially–fitted Runge–Kutta method for the numerical integration of initial–value problems with periodic or oscillating solutions, Comput. Phys. Comm. **115** (1998) 1–8.
15. Vanden Berghe, G., De Meyer, H., Van Daele, M., Van Hecke, T., Exponentially–fitted Runge–Kutta methods, J. Comput. Appl. Math. **125** (2000) 107–115.
16. Vanden Berghe, G., Ixaru, L. Gr., De Meyer, H.: Frequency determination and step–length control for exponentially–fitted Runge–Kutta methods, J. Comput. Appl. Math. **132** (2001) 95–105.

A Symplectic Lanczos-Type Algorithm to Compute the Eigenvalues of Positive Definite Hamiltonian Matrices

Pierluigi Amodio*

Dipartimento di Matematica, Università di Bari,
Via E. Orabona 4, I-70125 Bari, Italy,
amodio@dm.uniba.it

Abstract. The Lanczos algorithm is a well known procedure to compute few eigenvalues of large symmetric matrices. We slightly modify this algorithm in order to obtain the eigenvalues of Hamiltonian matrices $H = JS$ with S symmetric and positive definite. These matrices represent a significant subclass of Hamiltonian matrices since their eigenvalues lie on the imaginary axis. An implicitly restarted procedure is also considered in order to speed-up the convergence of the algorithm.

1 Introduction

Many applications require the numerical approximation of the eigenvalues of a $2n \times 2n$ real Hamiltonian matrix $H = JS$ where

$$J = \begin{pmatrix} O & I_n \\ -I_n & 0 \end{pmatrix}$$

and S is a large and sparse symmetric matrix. As an example, we cite the solution of the continuous-time algebraic Riccati equations [9,11] of the form

$$Q + A^T X + XA - XGX = 0,$$

where A, G and Q are known $n \times n$ matrices, G and Q symmetric, and the solution X is also symmetric.

It is well known that the considered eigenvalues are symmetric with respect to the real and imaginary axes. In this paper we are particularly interested in a subclass of Hamiltonian matrices where S is symmetric and positive definite. In this case the eigenvalues lie (two by two symmetrically with respect to the origin) along the imaginary axis. These matrices arise in many application fields that deal with evolutionary problems whose solutions satisfy a certain conservation law, typically the energy of the system (see, for example, [5]). In the rest of the paper we refer to these matrices as positive definite Hamiltonian matrices [2].

All the numerical methods for the eigenvalues computation of Hamiltonian matrices use symplectic transformations in order to maintain the Hamiltonian

* Work supported by GNCS.

P.M.A. Sloot et al. (Eds.): ICCS 2003, LNCS 2658, pp. 139–148, 2003.

structure of the matrix. In fact, if P is symplectic (that is, $P^TJP = J$) and H is Hamiltonian, then also $P^{-1}HP$ is Hamiltonian. For this reason, most of the existing algorithms are symplectic modifications of more general approaches as, for instance, the symplectic QR method in [6,7,13] and the symplectic iterative methods [3,4].

In particular, the symplectic Lanczos algorithm in [4] where the directions generated by the procedure are symplectic rather than orthogonal, is strictly related to the method introduced in this paper. The idea inherited this algorithm is the following: let

$$S = \begin{pmatrix} S_{11} & S_{12} \\ S_{12}^T & S_{22} \end{pmatrix}$$

with S_{11} and S_{22} symmetric, the matrix H may be represented in the form

$$H = \begin{pmatrix} S_{12}^T & S_{22} \\ -S_{11} & -S_{12} \end{pmatrix} \equiv \begin{pmatrix} V & W \\ Z & -V^T \end{pmatrix}.$$

The approach in [4], as well as those in [1,10] which define canonical forms for the Hamiltonian matrices, transform the above matrix in

$$\begin{pmatrix} V_1 & W_1 \\ & -V_1^T \end{pmatrix}$$

with V_1 triangular and V_1 and W_1 having few nonnull diagonals. This is called a Hamiltonian triangular form for the matrix H.

Conversely, based on a theoretical result in [2], for positive definite Hamiltonian matrices we derive a canonical form (W_2 tridiagonal)

$$\begin{pmatrix} & W_2 \\ -I & \end{pmatrix}$$

by means of a transformation that we prove to be symplectic. We will explain how to derive this form in Section 2. Then in Section 3, following the results presented in [4,8], we derive an implicit restarting technique that can be easily applied to the considered class of matrices.

2 Tridiagonal Canonical Form

In [2] existence results on a diagonal form for Hamiltonian matrices $H = JS$ with S positive definite is given. In the same paper, this result is also rearranged in order to obtain the following symplectic transformation in the simplest canonical form.

Theorem 1. *Given a real Hamiltonian matrix $H = JS$ with S symmetric and positive definite, a real symplectic matrix Z exists such that*

$$HZ = Z\begin{pmatrix} & D_1 \\ -D_1 & \end{pmatrix} \equiv ZJ\begin{pmatrix} D_1 & \\ & D_1 \end{pmatrix} \qquad (1)$$

with $D_1 > 0$ diagonal.

Even if the proof of Theorem 1 is presented in [2], it is interesting to sketch its main steps. As a remark, we recall that:

- if A is a symmetric matrix, then all the eigenvalues are real and an orthogonal matrix U exists such that
$$AU = U\Lambda;$$
- if A is a skew-symmetric matrix, then all the eigenvalues are pure imaginary and a unitary matrix $U = [U_1\ \overline{U}_1]$ exists such that
$$AU = iU\begin{pmatrix}\Lambda_1 & \\ & -\Lambda_1\end{pmatrix}.$$
Moreover, from $V = \sqrt{2}[Re(U_1)\ Im(U_1)]$ one has
$$AV = VJ\begin{pmatrix}\Lambda_1 & \\ & \Lambda_1\end{pmatrix}, \quad V^TV = I,$$
that is, a skew-symmetric matrix is similar to a positive definite Hamiltonian matrix by means of an orthogonal matrix.

Proof. Let us start from the decomposition of the symmetric and positive definite matrix S in diagonal form, that is $S = Q\Lambda Q^T$, with Q orthogonal and Λ positive. Then

$$(JS)JQ = (JQ\Lambda Q^T)JQ = JQ(\Lambda Q^TJQ)$$

shows that JS is similar to ΛQ^TJQ by means of the transformation matrix JQ. Now, since the matrix $\Lambda^{1/2}Q^TJQ\Lambda^{1/2}$ is skew-symmetric, it will admit the following similarity transformation

$$\Lambda^{1/2}Q^TJQ\Lambda^{1/2} = VJ\begin{pmatrix}D_1 & \\ & D_1\end{pmatrix}V^T, \tag{2}$$

and hence

$$\Lambda Q^TJQ = \Lambda^{1/2}\left(\Lambda^{1/2}Q^TJQ\Lambda^{1/2}\right)\Lambda^{-1/2} = (\Lambda^{1/2}V)J\begin{pmatrix}D_1 & \\ & D_1\end{pmatrix}(\Lambda^{1/2}V)^{-1}.$$

Let $U = JQ\Lambda^{1/2}V$. One has that

$$\begin{aligned} U^TJU &= -(V^T\Lambda^{1/2}Q^TJ)J(JQ\Lambda^{1/2}V) = V^T\Lambda^{1/2}Q^TJQ\Lambda^{1/2}V \\ &= V^TVJ\begin{pmatrix}D_1 & \\ & D_1\end{pmatrix}V^TV = J\begin{pmatrix}D_1 & \\ & D_1\end{pmatrix} \end{aligned}$$

and therefore the matrix

$$Z = U\begin{pmatrix}D_1^{-1/2} & \\ & D_1^{-1/2}\end{pmatrix} = JQ\Lambda^{1/2}V\begin{pmatrix}D_1^{-1/2} & \\ & D_1^{-1/2}\end{pmatrix}$$

is symplectic and satisfies (1). □

Equation (1) allows us to easily obtain the eigenvalues of JS as $i\lambda_j$ and $-i\lambda_j$, where λ_j is a diagonal element of D_1. On the other hand, as it occurs in the case of symmetric matrices, the above theorem cannot be used to compute the eigenvalues numerically. For this reason, different approaches need to be introduced.

We now observe that $-S^{1/2}JSJS^{1/2}$ is symmetric and positive definite, then it is possible to apply to this matrix the Lanczos algorithm, that is one should compute a matrix $Z_{k+1} = [Z_k \; z_{k+1}]$ which is defined by means of $k+1$ orthogonal columns and a tridiagonal matrix T_k such that

$$-S^{1/2}JSJS^{1/2}Z_k = Z_kT_k + \beta_{k+1}z_{k+1}e_k^T,$$

where e_k is the last unit vector of $\mathbb{R}^k$. Then, from $V_{k+1} = S^{-1/2}Z_{k+1} = [V_k \; v_{k+1}]$, one has

$$-H^2V_k = V_kT_k + \beta_{k+1}v_{k+1}e_k^T \tag{3}$$

with V_k of size $2n \times k$ and such that

$$V_k^TSV_k = I_k. \tag{4}$$

If $-i\lambda_j$, for $j = 1, \ldots, n$, are the eigenvalues of H, then the eigenvalues of H^2 are $-\lambda_1^2, -\lambda_1^2, -\lambda_2^2, -\lambda_2^2, \ldots$, that is, H^2 has double eigenvalues with a subspace of size 2 associated. This means that, if no breakdown has previously occurred, this algorithm theoretically will stop after n steps, providing a tridiagonal symmetric and positive definite matrix T. In fact, from

$$-H^2V = VT, \qquad V^TSV = I_n, \tag{5}$$

one has

$$-V^TSH^2V = V^TSVT = T$$

and $-SH^2$ is symmetric and positive definite.

Hence the eigenvalues of H^2 may be obtained applying an algorithm for symmetric and positive definite matrices to a half sized matrix T. Unfortunately, the decomposition (3)-(4) is not stable and can be only used to compute a limited number of extremal eigenvalues. Otherwise, it needs continue reorthogonalizations which make necessary to store all the columns of V.

Anyway, from a theoretical point of view, we derive the following symplectic transformation which is numerically more reliable than that defined in Theorem 1.

Theorem 2. *Given a real Hamiltonian matrix $H = JS$ with S symmetric and positive definite, a real symplectic matrix U exists such that*

$$HU = U\begin{pmatrix} & T \\ -I & \end{pmatrix} \equiv UJ\begin{pmatrix} I & \\ & T \end{pmatrix}$$

with T symmetric and positive definite tridiagonal matrix.

Proof. Let

$$W = -HV, \tag{6}$$

from (5) one has

$$HW = VT \tag{7}$$

and, consequently,

$$H[V\ W] = [-W\ VT] = [V\ W] \begin{pmatrix} & T \\ -I & \end{pmatrix}. \tag{8}$$

The matrix $U = [V\ W]$ is symplectic, that is

$$U^T JU = \begin{pmatrix} V^T JV & V^T JW \\ W^T JV & W^T JW \end{pmatrix} = J.$$

In fact, from (6) one has $JW = SV$ and

$$V^T JW = V^T SV = I.$$

Moreover, $V^T JV = O$ since each column v_i of V may be expressed in terms of an even polynomial of degree $2(i-1)$ in H, that is $v_i = p_{i-1}(H^2)v_1$, and for any integers i and l

$$v_i^T Jv_l = v_1^T(p_{i-1}(H^2))^T Jp_{l-1}(H^2)v_1 = v_1^T(p_{i+l-1}(H))^T Jp_{i+l-1}(H)v_1 = 0$$

being, for any vector v of appropriate length, $v^T Jv = 0$. Finally, $W^T JW = O$ for a similar reasoning. □

The above theorem states that U transforms the matrix H in a matrix

$$J \begin{pmatrix} I & \\ & T \end{pmatrix}$$

which corresponds to the tridiagonal form of symmetric matrices. This is the simplest canonical transformation which can be computed for Hamiltonian matrices and proves the existence of a canonical form for positive definite Hamiltonian matrices which is slightly different from the one obtained in Theorem 1. In fact, from $T = Q_1 D_1^2 Q_1^T$, with D_1 diagonal and Q_1 orthogonal, it results that $\mathrm{diag}(Q_1, Q_1)$ is symplectic and

$$\begin{pmatrix} Q_1^T & \\ & Q_1^T \end{pmatrix} \begin{pmatrix} & T \\ -I & \end{pmatrix} \begin{pmatrix} Q_1 & \\ & Q_1 \end{pmatrix} = \begin{pmatrix} & D_1^2 \\ -I & \end{pmatrix} \equiv J \begin{pmatrix} I & \\ & D_1^2 \end{pmatrix}.$$

Now we derive an algorithm for the computation of matrices U and T in Theorem 2. Let $V_k = [v_1, \ldots, v_k]$ and

$$T_k = \begin{pmatrix} \alpha_1 & \beta_2 & & \\ \beta_2 & \alpha_2 & \ddots & \\ & \ddots & \ddots & \beta_k \\ & & \beta_k & \alpha_k \end{pmatrix}, \tag{9}$$

the generic row of (3) is

$$-H^2 v_j = \beta_j v_{j-1} + \alpha_j v_j + \beta_{j+1} v_{j+1}$$

and allows us to compute v_{j+1}. The coefficient β_j is computed in order to have $v_j^T S v_j = 1$, while α_j derives from

$$0 = v_j^T S v_{j+1} = \frac{1}{\beta_{j+1}}(v_j^T S H w_j - \alpha_j) = \frac{1}{\beta_{j+1}}(-v_j^T H^T S w_j - \alpha_j),$$

where $w_j = -Hv_j$ in analogy with (6).

The complete algorithm to obtain the tridiagonal form is the following:

$\tilde{v}_1$ arbitrary
$\beta_1 = (\tilde{v}_1^T S \tilde{v}_1)^{1/2}$
for $j = 1, 2, \ldots, n$
  $v_j = \tilde{v}_j / \beta_j$
  $w_j = -Hv_j$
  $\alpha_j = w_j^T S w_j$
  if $j = 1$
    $\tilde{v}_{j+1} = Hw_j - \alpha_j v_j$
  elseif $j < n$
    $\tilde{v}_{j+1} = Hw_j - \alpha_j v_j - \beta_j v_{j-1}$
  end
  $\beta_{j+1} = (\tilde{v}_{j+1}^T S \tilde{v}_{j+1})^{1/2}$
  if $\beta_{j+1} = 0$, stop
end

We observe that $\alpha_j > 0$ and $\beta_j \geq 0$. The algorithm stops before n steps if $\beta_j = 0$ (i.e., $\tilde{v}_j = 0$), that is, when $\tilde{v}_1$ belongs to an invariant subspace of H^2. The vectors w_j need not to be stored, while the v_j are required for any reorthogonalization. If the above algorithm is applied to matrices JS with S nondefinite, then $\tilde{v}_j^T S \tilde{v}_j$ should be less than zero and the procedure breaks. A modification of the algorithm in order to overcome this problem will be investigated in future.

3 Implicit Restarting

As observed previously, the Lanczos method is widely used to compute only a small subset of the eigenvalues. In this case, the algorithm is stopped after a fixed number k of iterations, then it restarts with a different initial vector v_1 obtained by the performed iterations. The idea is just to compute the new starting vector in order to obtain, after k steps, $\beta_{k+1} = 0$, that is an invariant subspace which allows us to compute k eigenvalues.

The implicit restarted Lanczos algorithm shows several advantages since the storage requirement is fixed, there are no spurious eigenvalues, and deflation techniques similar to those applied to the QR algorithm may be applied [8].

As an example, we now analyze how it is possible to apply this technique to the Lanczos algorithm described in the previous section. Let us start from $k+2$ steps of equation (3)

$$-H^2 V_{k+1} = V_{k+1} T_{k+1} + \beta_{k+2} v_{k+2} e_{k+1}^T,$$

where e_{k+1} represents the last unit vector of $\mathbb{R}^{k+1}$. Then, for a given real parameter μ and from $T_{k+1} - \mu I = QR$ we have the following equalities

$$\begin{aligned} (-H^2 - \mu I) V_{k+1} &= V_{k+1}(T_{k+1} - \mu I) + \beta_{k+2} v_{k+2} e_{k+1}^T \\ (-H^2 - \mu I) V_{k+1} &= V_{k+1} QR + \beta_{k+2} v_{k+2} e_{k+1}^T \\ (-H^2 - \mu I)(V_{k+1} Q) &= (V_{k+1} Q) RQ + \beta_{k+2} v_{k+2} e_{k+1}^T Q \\ -H^2 (V_{k+1} Q) &= (V_{k+1} Q)(RQ + \mu I) + \beta_{k+2} v_{k+2} e_{k+1}^T Q \end{aligned} \tag{10}$$

where $RQ + \mu I$ is a symmetric and positive definite tridiagonal matrix. The last equation cannot be however considered as obtained by a Lanczos procedure since

$$\beta_{k+2} e_{k+1}^T Q = [\widehat{\beta}_{k+1} e_k^T \;\; \widetilde{\beta}_{k+2}], \tag{11}$$

that is, it has two elements different from zero (instead of one, see (9)). Anyway, equation (10) is useful to define the vector $v_1^+ = (V_{k+1} Q) e_1$ (e_1 is the first unit vector of $\mathbb{R}^{k+1}$) as the starting vector of the new Lanczos iterations. Its relation with v_1 is

$$(-H^2 - \mu I) v_1 = \rho_1 v_1^+$$

where ρ_1 is the (1,1) element of R, is obtained by applying the second equation of (10) to e_1 and from $Re_1 = \rho_1 e_1$.

Let us now partition the matrices $V_{k+1} Q = [V_k^+ \; \tilde{v}_{k+1}]$ and

$$RQ + \mu I = Q^T T_{k+1} Q = \begin{pmatrix} T_k^+ & \widetilde{\beta}_{k+1} e_k \\ \widetilde{\beta}_{k+1} e_k^T & \widetilde{\alpha}_{k+1} \end{pmatrix}. \tag{12}$$

Substitution of (11) and (12) in the last equation of (10) gives

$$-H^2 [V_k^+ \; \tilde{v}_{k+1}] = [V_k^+ \; \tilde{v}_{k+1} \; v_{k+2}] \begin{pmatrix} T_k^+ & \widetilde{\beta}_{k+1} e_k \\ \widetilde{\beta}_{k+1} e_k^T & \widetilde{\alpha}_{k+1} \\ \widehat{\beta}_{k+1} e_k^T & \widetilde{\beta}_{k+2} \end{pmatrix}. \tag{13}$$

The first k columns of (13) may be rewritten in the form

$$-H^2V_k^+ = V_k^+T_k^+ + \beta_{k+1}^+ v_{k+1}^+ e_k^T \tag{14}$$

where

$$v_{k+1}^+ = \frac{1}{\beta_{k+1}^+}(\widetilde{\beta}_{k+1}\tilde{v}_{k+1} + \widehat{\beta}_{k+1}v_{k+2})$$

and β_{k+1}^+ is such that $(v_{k+1}^+)^T S v_{k+1}^+ = 1$. Hence equation (14) represents the implicit application of k steps of the Lanczos algorithm to the starting vector v_1^+.

This technique may be iterated. Starting from (14), we repeat one additional step of the Lanczos iteration and then again the implicit method until $\beta_{k+1} \approx 0$.

The application of p shifts $\mu_1, \mu_2, \ldots, \mu_p$ is straightforward and is used to speed-up the computation of k eigenvalues simultaneously. We summarize its use in the following algorithm.

```
compute v_1,...,v_k, ṽ_{k+1}, α_1,...,α_k, β_2,...,β_{k+1} by
  means of k steps of the Lanczos algorithm
while |β_{k+1}| is greater than a fixed tolerance
    compute v_{k+1},...,v_{k+p}, ṽ_{k+p+1}, α_{k+1},...,α_{k+p},
      β_{k+2},...,β_{k+p+1} by means of p additional steps
      of the Lanczos algorithm
    v_{k+p+1} = ṽ_{k+p+1}/β_{k+p+1}
    construct the matrix T_{k+p} as in (9)
    choose the parameters μ_1, μ_2,...,μ_p
    Q = I_{k+p}; T̃ = T_{k+p}
    for i = 1,...,p
        Q_i R_i = T̃ − μ_i I
        T̃ = Q_i^T T̃ Q_i
        Q = Q Q_i
    end
    define q_{k+p,k} as the (k+p, k) element of Q
    define β̃_{k+1} as the (k+1, k) element of T̃
    β̂_{k+p+1} = β_{k+p+1} q_{k+p,k}
    [V_k^+ V̂_p] = V_{k+p} Q
    define v̂_{k+1} as the first column of V̂_p
    define v_1,...,v_k as the k columns of V_k^+
    ṽ_{k+1} = β̃_{k+1} v̂_{k+1} + β̂_{k+p+1} v_{k+p+1}
    define α_1,...,α_k and β_2,...,β_k as the main diagonal
      and the lower diagonal of T̃
    β_{k+1} = (ṽ_{k+1} S ṽ_{k+1})^{1/2}
end
```

The matrix-by-matrix operations are not expensive since k and p are small with respect to n. This means that the computational cost of the overall algorithm depends only on the number of iterates in the Lanczos method.

The choice of the parameters μ_j gives rise to different strategies. For example, if we set μ_j, $j = 1, \ldots, p$, as p of the eigenvalues of T_{k+p}, that is, if we use the exact shift selection strategy [12], then $\widetilde{T}$ has the following structure

$$\begin{pmatrix} T_k^+ & \\ & D_p \end{pmatrix},$$

where D_p is diagonal with $\mu_1, \ldots \mu_p$ as main diagonal entries. This strategy gives good results, especially when used to compute the largest eigenvalues in modulus of the positive definite Hamiltonian matrix.

4 Conclusions

The Lanczos process has been modified in order to compute the eigenvalues of positive definite Hamiltonian matrices. The obtained algorithm is symplectic and requires half of the workspace of the original algorithm. Moreover, since the procedure gives a symmetric and positive definite matrix, known techniques for this class of matrices (for example, the implicit restarting) can be used to improve the computation.

References

1. Ammar, G., Mehrmann, V.: On Hamiltonian and symplectic Hessenberg forms. Linear Algebra Appl. **149** (1991), 55–72
2. Amodio, P., Iavernaro, F., Trigiante, D.: Conservative perturbations of positive definite Hamiltonian matrices, Numer. Linear Algebra Appl., (2003), in press
3. Benner, P.: Symplectic balancing of Hamiltonian matrices, SIAM J. Sci. Comput., **22 (5)** (2001), 1885–1904
4. Benner, P., Faßbender, H.: An implicit restarted symplectic Lanczos method for the Hamiltonian eigenvalue problem, Linear Algebra Appl., **263** (1997), 75–111
5. Brugnano, L., Trigiante, D.: Solving ODEs by Linear Multistep Initial and Boundary Value Methods. Gordon & Breach, Amsterdam, 1998
6. Bunse-Gerstner, A.: Matrix factorization for symplectic QR–like methods, Linear Algebra Appl., **83** (1986), 49–77
7. Byers, R.: A Hamiltonian QR–algorithm, SIAM J. Sci. Stat. Comput., **7** (1986), 212–229
8. Calvetti, D., Reichel, L., Sorensen, D.C.: An implicitly restarted Lanczos method for large symmetric eigenvalue problems, ETNA, Electron. Trans. Numer. Anal., **2** (1994), 1–21
9. Lancaster, P., Rodman, L.: The algebraic Riccati equation. Oxford University Press, Oxford, 1995
10. Lin, W., Mehrmann, V., Xu, H.: Canonical forms for Hamiltonian and symplectic matrices and pencils, Linear Algebra Appl. **302-303** (1999), 469–533

11. Rosen, I., Wang, C.: A multilevel technique for the approximate solution of operator Lyapunov and algebraic Riccati equations. SIAM J. Matrix Anal. Appl. **32** (1992), 514–541
12. Sorensen, D.C.: Implicit application of polynomial filters in a k-step Arnoldi method, SIAM J. Matrix Anal. Appl. **13** (1992), 357–385
13. Van Loan, C.: A symplectic method for approximating all the eigenvalues of a Hamiltonian matrix, Linear Algebra Appl., **16** (1984), 233–251

Applying Stabilization Techniques to Orthogonal Gradient Flows

C. Mastroserio[1] and T. Politi[2]

[1] Dipartimento Interuniversitario di Matematica, Università di Bari
Via Orabona 4, I-70125 Bari (Italy). carmen@dm.uniba.it
[2] Dipartimento Interuniversitario di Matematica, Politecnico di Bari,
Via Orabona 4, I-70125 Bari (Italy). pptt@dm.uniba.it

Abstract. The solution of ordinary differential systems on manifolds could be treated as differential algebraic equation. In this paper we consider the solution of orthogonal differential systems deriving from the application of the gradient flow techniques to minimization problems. Neglecting the constraints for the solution a differential system is derived. Hence the problem is modified introducing a stabilization technique which is a function of the constrain. The advantage of this approach is that it is possible to apply non conservative numerical methods which are cheaper. Some numerical examples are shown.

1 Introduction

Many problem of practical interest can be modeled by systems of differential equations whose solutions satisfy some invariants, usually defined explicitely by algebraic constraints. In recent years particular attention has been paid to the development of numerical methods which approximate the solution of such a system while preserving the invariant to machine precision. These methods usually need the computational of matrix exponential once (and often repeatedly) or the solution of linear systems at each time step (see [9,6,8]) and this highly increases their computational costs.

In this paper we consider a class of differential systems derived - via a gradient flow technique - from a constraint minimization problem (on the orthogonal manifold) of a particular objective function. In this class of problems is not crucial to preserve the orthogonality of the solution, but the main interest is to get, as soon as possible, the minimum point.

Hence, we will show how it is possible to solve this differential system with invariants applying explicit methods with a splitting technique. To do this, we review the regularization technique described in [1,2] and applied to the Stiefel manifold in [4].

In [2] an important difference between the stabilization and the regularization techniques for differential system is pointed out. In regularization methods the problem is perturbed in order to obtain another system easier to solve. In this case, the solution we obtain is not the same of the initial system, hence this

P.M.A. Sloot et al. (Eds.): ICCS 2003, LNCS 2658, pp. 149–157, 2003.

implies that the perturbation to introduce must be small. On the other hand, in the stabilization techniques, the solution of the two systems (the given problem and the perturbed one) is the same, since the perturbation term vanishes when a function satisfies the constraint. Hence, the perturbation parameter does not need to be small (or large).

The paper is organized as follows: in Section 2 some stabilization techniques are recalled, in Section 3 the problem to be solved is stated and a modified gradient flow is derived introducing a perturbation term on the minimization function, in Section 4 a method for the perturbed orthogonal flow is described and finally in Section 5 some numerical tests are shown.

2 A Survey on Stabilization Techniques

Let us consider the differential system

$$Y'(t) = F(t, Y(t)), \tag{1}$$

with initial condition $Y(0) = Y_0$. For sake of simplicity we assume that there is only a unique function $Y(t)$ satisfying (1). Together with (1) we suppose that there is an invariant set (or a constraint) defined by an algebraic equation:

$$H(Y) = 0 \tag{2}$$

such that if $H(Y_0) = 0$ then $H(Y(t)) = 0$ for all $t > 0$. It is not very important to distinguish between the vector or the matrix case (i.e. Y_0 is a square real matrix and so is $Y(t)$). The important question is how to design a numerical method which preserves the properties given by the invariant set or to exploit the information on the solution given by (2) to improve the quality of the approximate solution in terms, for instance, of stability or of global error. Applications of (1)-(2) include some mechanical systems with quadratic constraints, preserving the orthogonality of a solution matrix and isospectral flows, preserving the set of the eigenvalues of the initial condition matrix. One of the most popular techniques for the stabilization of invariants is the one described by Baumgarte [3] applied to a Differential Algebraic Equation

$$\dot{y}(t) = F(y) - B(y)x$$

with invariant

$$0 = g(y).$$

Let $G = g_y(y)$ be a full rank matrix, the stabilization technique consists in replacing the invariant with a linear combination of its derivatives

$$\ddot{g}(y) + \gamma_1 \dot{g}(y) + \gamma_2 g(y) = 0$$

or

$$\dot{g}(y) + \gamma g(y) = 0.$$

In this case the approach is equivalent to replace the differential system by

$$\dot{y} = F(y) - \gamma B(GB)^{-1} g(y).$$

In [2] the following result is proved.

Theorem 1. *Consider the differential system (1) and the invariant defined by (2) and apply the stabilization*

$$Y'(t) = F(Y(t)) - \gamma F_Y(Y) H(Y) \tag{3}$$

where $F_Y = D(H_Y D)^{-1}$, *if there exists a constant* γ_0 *such that*

$$\|H_Y(Y) F(Y)\|_2 \leq \gamma_0 \|H(Y)\|_2$$

for all Y *belonging to a neighborhood of the invariant set, then it is asymptotically stable invariant manifold of (3) for* $\gamma > \gamma_0$. *In particular if* H *is an integral invariant of (1), i.e.*

$$H_Y(Y) F(Y) = 0, \qquad \forall Y,$$

then the invariant set is an asymptotically stable invariant manifold of (3) for any $\gamma > 0$.

It should be pointed out that the numerical solution can be now obtained discretizing (3) using also nonstiff integrators. This approach seems to be better than the Baumgarte and the projected invariants methods.
If we consider the differential system

$$Y'(t) = F(t, Y) Y, \qquad Y(0) = Y_0 \tag{4}$$

with $F(t, Y)$ continuous skew-symmetric function for all $Y \in \mathcal{O}_n(\mathbb{R})$ and Y_0 orthogonal square matrix of order n, it is well known that the solution $Y(t)$ is an orthogonal matrix for all $t \geq 0$. The system (4) can be considered as a system on the set of square real matrices with the nonlinear constrain

$$H(Y) = Y(t) Y^T(t) - I_n = 0 \tag{5}$$

where I_n is the identity matrix of order n. In [4] the following modified equation is proposed in order to make the manifold of orthogonal matrices attractive for the system (4):

$$Y'(t) = F(t, Y) Y - \gamma Y(t) P(Y) \tag{6}$$

where γ is a real positive number.

3 Gradient Flow Approach for the Regularization

In this section we consider the class of differential systems on the manifold of orthogonal matrices and we apply a gradient flow technique in order to minimize a given matrix function. Consider the isospectral manifold

$$M(\Lambda) = \{ A \in \mathbb{R}^{n \times n} \mid A = Q^T \Lambda Q, Q \in \mathcal{O}(n) \}$$

where Λ is a given diagonal real matrix and $\mathcal{O}(n)$ is the set of $n \times n$ orthogonal matrices. The problem we are interesting in can be formulated as follows:

$$\begin{aligned}&\text{Minimize } \frac{1}{2}\|Q^T \Lambda Q - P(Q^T \Lambda Q)\|_F^2 \\ &\text{subject to } Q^T Q = I\end{aligned} \tag{7}$$

where $\|\cdot\|_F$ denotes the Frobenius norm of a matrix and $P(\cdot)$ is the projection function. Introducing the Frobenius inner product of two matrices $A, B \in \mathbb{R}^{n\times n}$, defined by

$$\langle A, B\rangle = \text{trace}(AB^T) = \sum_{i,j=1}^{n} a_{ij} b_{ij},$$

the minimization problem (7) becomes:

$$\begin{aligned}&\text{Minimize } \frac{1}{2}\langle f(Q), f(Q)\rangle \\ &\text{subject to } g(Q) = 0\end{aligned} \tag{8}$$

where,

$$f(Q) = Q^T \Lambda Q - P(Q^T \Lambda Q), \qquad g(Q) = Q^T Q - I.$$

Introducing a real nonnegative parameter ε it is possible to transform (8) into an unconstrained minimization problem:

$$\text{Minimize } \frac{1}{2}\langle f(Q), f(Q)\rangle + \frac{\varepsilon}{2}\langle g(Q), g(Q)\rangle. \tag{9}$$

Again we put

$$\varphi(Q, \varepsilon) = \frac{1}{2}\left[\langle f(Q), f(Q)\rangle + \varepsilon\langle g(Q), g(Q)\rangle\right]$$

and compute the Fréchet derivative of φ at Q acting on H:

$$\begin{aligned}\varphi'(Q,\varepsilon)H &= \langle f'(Q)H, f(Q)\rangle + \varepsilon\langle g'(Q)H, g(Q)\rangle = \\ &= \langle H^T \Lambda Q + Q^T \Lambda H - P'(Q^T \Lambda Q)(H^T \Lambda Q + Q^T \Lambda H), f(Q)\rangle + \\ &\quad + \varepsilon\langle H^T Q + Q^T H, g(Q)\rangle = \\ &= \langle H^T \Lambda Q + Q^T \Lambda H - P(H^T \Lambda Q + Q^T \Lambda H), f(Q)\rangle + \\ &\quad + \varepsilon\langle H^T Q + Q^T H, g(Q)\rangle.\end{aligned}$$

Since the projection $P(\cdot)$ is orthogonal to function $f(Q)$, we get

$$\varphi'(Q,\varepsilon)H = \langle H^T \Lambda Q + Q^T \Lambda H, f(Q)\rangle + \varepsilon\langle H^T Q + Q^T H, g(Q)\rangle =$$

$$= \langle H^T \Lambda Q, f(Q)\rangle + \langle Q^T \Lambda H, f(Q)\rangle + \varepsilon\langle H^T Q, g(Q)\rangle +$$

$$+\varepsilon\langle Q^T H, g(Q)\rangle =$$

$$= \langle H, \Lambda Q f(Q)^T\rangle + \langle H, \Lambda Q f(Q)\rangle + \varepsilon\langle H, Qg(Q)^T\rangle + \varepsilon\langle H, Qg(Q)\rangle =$$

$$= \langle H, \Lambda Q f(Q)^T + \Lambda Q f(Q) + \varepsilon\left(Qg(Q)^T + Qg(Q)\right)\rangle.$$

This last equation suggests that the gradient of φ at a general matrix H, with respect to the Frobenius inner product, can be interpreted as the matrix:

$$\varphi'(Q,\varepsilon) = \Lambda Q f(Q)^T + \Lambda Q f(Q) + \varepsilon\left(Qg(Q)^T + Qg(Q)\right) =$$

$$= \Lambda Q\left(f(Q)^T + f(Q)\right) + \varepsilon Q\left(g(Q)^T + g(Q)\right) =$$

$$= \Lambda Q\left(2Q^T \Lambda Q - 2P(Q^T \Lambda Q)\right) + \varepsilon Q\left(2Q^T Q - 2I\right) =$$

$$= 2\Lambda Q\left[Q^T \Lambda Q - P(Q^T \Lambda Q)\right] + 2\varepsilon Q(Q^T Q - I).$$

The choice of the projection function $P(\cdot)$ leads to different classes of problem on the isospectral manifold $M(\Lambda)$. It should be observed that the obtained unconstrained differential systems obtained considering the gradient of function φ can be seen as a stabilized differential system (it is easy to observe that the last term is just the stabilized term introduced in the previous section).

4 A Numerical Method for Regularized System

Once the regularization term has been added to the given differential system to solve it some numerical aspects are to be taken into account. A possible technique is the post-stabilization of the numerical solution (see [2]). It consists in the application of a stabilization step with respect to the manifold at the end of each time step. In the case of orthogonal differential systems we can consider the following method introduced in [2] and used in [4] to solve differential systems on the Stiefel manifold in order to compute a subset of Lyapunov exponents of a dynamical system. First the system (4) (or the system (6) choosing $\gamma = 0$) is integrated from t_k to t_{k+1} using an explicit method and obtaining an approximate solution $\tilde{Y}_{k+1}$. Then the regularization term

$$Y'(t) = -\gamma Y(t)P(Y) \qquad (10)$$

is integrated using, for instance, forward Euler with the same stepsize h but with γ chosen so that $\gamma h = \frac{1}{2}$ and taking as approximation at the previous step $\tilde{Y}_{k+1}$:

$$Y_{k+1} = \tilde{Y}_{k+1} - \gamma h \tilde{Y}_{k+1} P(\tilde{Y}_{k+1}) \tag{11}$$

and then

$$Y_{k+1} = \tilde{Y}_{k+1} \left(I_n - \frac{1}{2} P(\tilde{Y}_{k+1}) \right).$$

This last step can be seen as the application of one step of the Schulz method to compute the polar decomposition (see [7]).

5 Numerical Tests

Example 1. Let us consider the following gradient flow introduced in [5]. If L is a real symmetric matrix then the orthogonal flow defined by the differential system:

$$Y'(t) = Y[Y^T LY, P(Y^T LY)], \qquad Y(0) = Y_0 \tag{12}$$

with Y_0 a random orthogonal matrix, converges to the eigenvector matrix related to L. In this case the projection function is

$$P(X) = \text{diag}(X)$$

while the objective function to minimize is (7). We have solved (12) using the numerical method introduced in the previous section. Figure 1 shows the orthogonal error of the numerical approximation Y_n, given by

$$E_n = \|Y_n^T Y_n - I\|_F$$

while in Figure 2 are reported the values of objective function. In Figure 3 is shown the orthogonal error using the numerical method but applying two iterates of (11). In this case the orthogonal error is smaller but considering the values of the objective function they are the same shown in Figure 2. In this case it seems that a better integration in the manifolds does not imply a speeder descent toward the minimum point of the objective function.

Example 2. Given a real symmetric matrix A we want to find a least squares approximation of A that is still symmetric but has a prescribed set of eigenvalues $\{\lambda_1, \ldots, \lambda_n\}$. In this case the orthogonal flow defined by the differential system:

$$Y'(t) = Y[Y^T LY, A], \qquad Y(0) = Y_0 \tag{13}$$

with Y_0 a random orthogonal matrix. Figure 1 shows the orthogonal error of the numerical approximation while in Figure 2 are sketched the values of objective function.

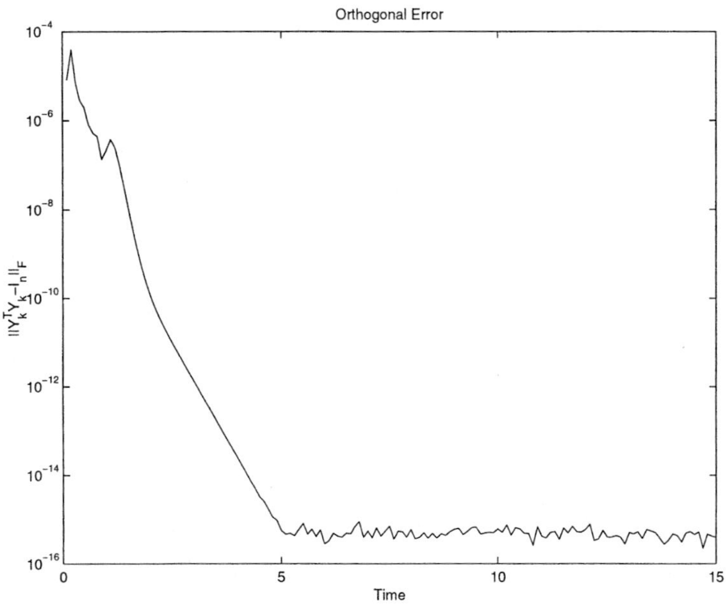

Fig. 1. Orthogonal error for Example 1.

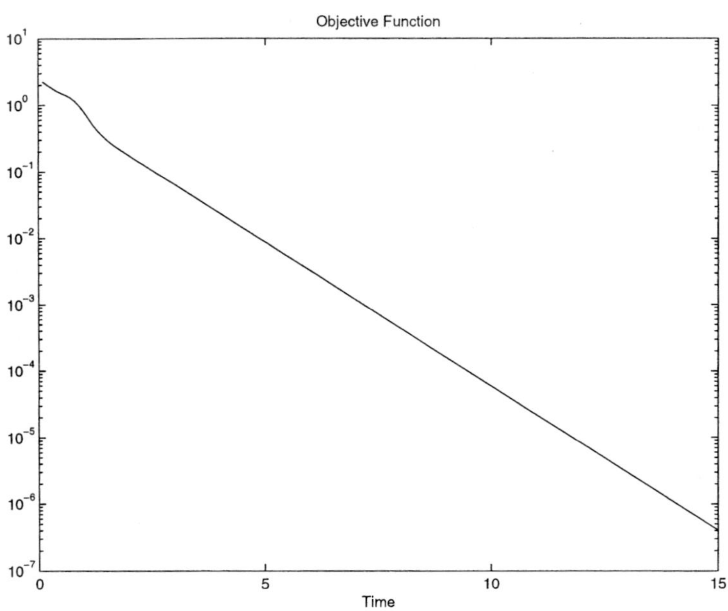

Fig. 2. Objective function for Example 1.

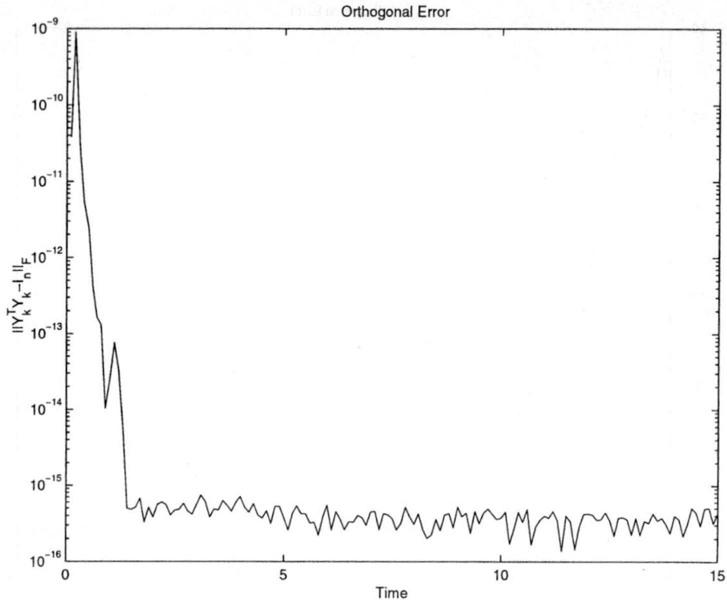

Fig. 3. Orthogonal error for Example 1.

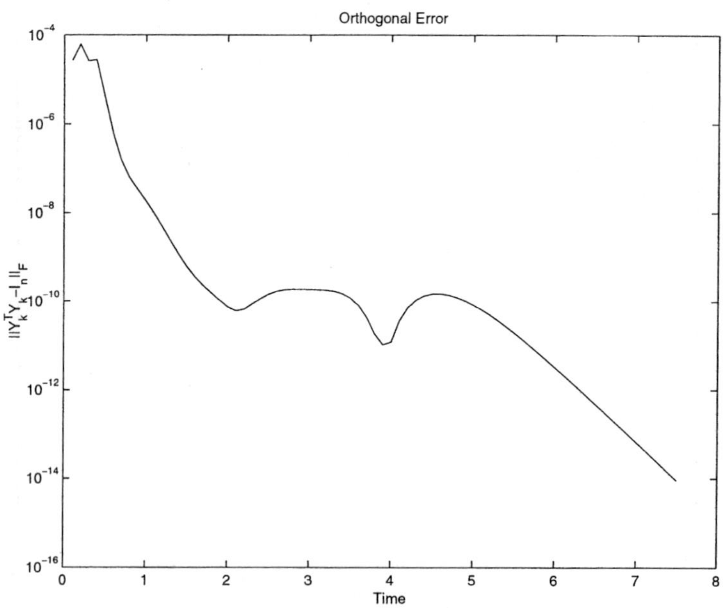

Fig. 4. Orthogonal errors for Example 2.

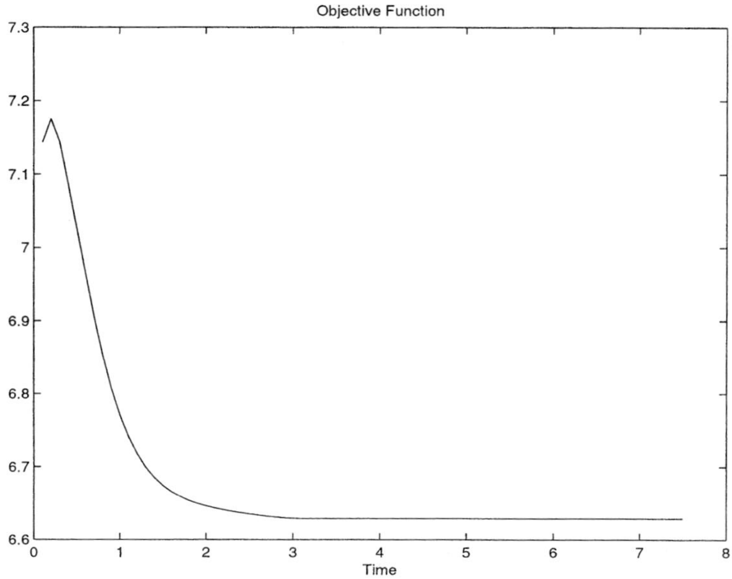

Fig. 5. Objective function for Example 2.

References

1. Ascher U.M.; Stabilization of invariants of discretized differential systems. Numer. Algorithms **14** (1997) 1–24
2. Ascher U.M., Chin H., Reich S.: Stabilization of DAEs and invariant manifolds. Numer. Math. **14** (1994) 131–149
3. Baumgarte J.: Stabilization of constraints and integrals of motion in dynamical systems. Comp. Math. Appl. Mech. Eng. **1** (1972) 1–16
4. Bridges T.J., Reich S.: Computing Lyapunov exponents on a Stiefel manifold. Physica D **156** (2001) 219–238
5. Chu M.T., Driessel K.R.: The projected gradient method for least squares matrix approximations with spectral constraints. SIAM J. Numer. Anal. **47** (1990) 1050–1060
6. Diele F., Lopez L., Peluso R.: The Cayley transform in the numerical solution of unitary differential systems. Adv. in Comp. Math. **8** (1998) 317–334
7. Higham D.J.: Time-stepping and preserving orthonormality. BIT **37** (1997) 24–36
8. Lopez L., Politi T.: Applications of the Cayley approach in the numerical solution of matrix differential systems on quadratic groups. Appl. Num. Math. **36** (2001) 35–55
9. Munthe-Kaas H.: Runge-Kutta methods on Lie groups. BIT **38** (1998) 92–111

Workshop on High-Performance Environmental Computations

Coupling General Circulation Models on a Meta-computer

Wolfgang Joppich[1] and Johannes Quaas[2]

[1] Fraunhofer Institute for Algorithms and Scientific Computing (FhG–SCAI), Schloss Birlinghoven, D–53754 Sankt Augustin, Germany
[2] Laboratoire de Météorologie Dynamique du CNRS, 4 place Jussieu, F–75252 Paris, France

Abstract. Sophisticated climate models, describing the circulation in atmosphere and oceans, respectively, have been coupled together. This has been done using two different remote supercomputers in an efficient way for the first time.
Two state–of–the–art climate models have been chosen, a coupling tool has been adapted, and a new developed message–passing library has been used.

1 Introduction

One of the most important research topics at the moment is the understanding and the prediction of the climate change due to anthropogenic influences. The most useful instrument therefore are comprehensive models, which describe the climate system. At least the circulation in it's two main subsystems, the atmosphere and the ocean, has to be considered in climate research.
The most efficient way to take into account several climate subsystems is to couple together models of the respective systems. Models of different parts of the climate system generally are developed independently by different research groups. Coupling the existing codes of subsystems is reasonable, because the only interaction between two subsystems occurs at their common surface. Only small changes within the approved models then have to be made.
The integration of climate models requires extremely large computing capacities. To become more and more reliable, the resolution of the model grids has to be increased and more physical processes in the subsystems and their interaction have to be taken into account. Both approaches lead top the need of even larger computing power. Beside the development of more efficient numerical solutions for the integration it is the application and adaption of new technical resources that promises to satisfy the necessary computing requirements.
In the latter area massively parallel systems have been introduced very successfully during the last years. In climate modeling, the integration of the physical equations is distributed to a large number of processors using domain decomposition technique.
If the models of two or more climate subsystems are coupled, one can consider

P.M.A. Sloot et al. (Eds.): ICCS 2003, LNCS 2658, pp. 161–170, 2003.

to distribute the computation of the two models, not only on different subsets of the processors of one computer, but use different computers for the different models. In the general case, each of the models should be allowed to run in parallel. This technique is called "meta–computing". This could have at least three advantages: First, in the ideal case the overall integration time will be reduced to that of the most time–consuming component. Secondly, for each model the suitable hardware can be chosen. This is can improve largely the efficiency, if e.g. one model runs more efficient on a vector machine, and the other one e.g. on a parallel system. If codes are written for different hardware, a transfer of the codes to a common hardware can be avoided. Thirdly it would be easier for different research centers, which develop models of different parts of the climate system, to work together. While coupling the codes, their respective models can reside on their own machines.
The goal of our work presented here is to test the hardware needed and to develop or adapt, respectively, the suitable software needed for the coupling of climate models on a meta–computer. We developed a new coupled model using two approved general circulation models of the atmosphere and the ocean to gain the experience with the coupling of climate models on a meta–computer.
We did this in the framework of a project called "Distributed Computation of Weather– and Climate Models", carried out by the Institute for Algorithms and Scientific Computing of the German National Research Center for Information Technology (GMD–SCAI, St. Augustin) together with the Alfred Wegener–Institute for Polar and Marine Research (AWI, Bremerhaven). This project was part of the "Gigabit Testbed West"–project (GTBW) funded by the German Research Network (DFN) [1][2]. The GTBW provided an ATM–network with a bandwidth of 2.4 GBit/s and a latency of 7ms, connecting the GMD and the Forschungszentrum Jülich (FZJ) crossing a distance of about 100km [3]. Beside our project, a variety of other scientific applications have been tested on the testbed. Connected to the gigabit network were an IBM SP2 at GMD's site and a Cray T3E at the FZJ.

Known to the authors are three other works on the topic of the distributed computation of coupled climate models.
The first was the work of C. R. Mechoso et al. 1993 at the University of California, Los Angeles [4]. Not only the atmosphere– and the ocean model ran distributed, but the atmosphere model was also splitted in two parts. Although this work was a big step concerning the idea of meta–computing in climate research, the results have been disappointing because of the insufficient performance of the used technique. The respective programs ran each sequential, and the model needed 193s for the computation of one day. The time for the communication between the models consumed 1930s, exactly the ten–fold time. The standard deviation for the communication time had a value of 1220s (10 measurements) and therewith the same order of magnitude as the mean value.
1998 a coupling of French climate models has been carried out with a similar coupling technique as used by us [5]. Also in this work the models ran each se-

quential. An integration over one year has been carried out. But as the network was not reliable, some phases of the integration had to be done several times. Inthe third work concerning this topic by Météo–France an integration of the coupled model had only been simulated. This work focused on the network security [6].

2 Description of the Coupled Models

We have chosen two well–known climate models. As atmosphere model we used the Integrated Forecast System (IFS) of the European Centre for Medium–Range Weather Forecasts (ECMWF), and as ocean model the Modular Ocean Model (MOM) of the Geophysical Fluid Dynamics Laboratory (GFDL), in a version which is maintained by the AWI.

2.1 The Atmosphere Model IFS

The IFS has been developed by the ECMWF jointly with Météo–France from 1993 on [7]. In the present work we used the model version "CY16R2".
The governing equations of the IFS are the so–called "primitive equations" which describe the general circulation in the atmosphere. Prognostic variables are the horizontal velocity, the temperature, the humidity and the surface pressure. The land surface parametrization of the IFS contains a bulk model, which represents processes such as evapotranspiration, snow formation and melting, precipitation and runoff using four layers below the earth's surface [8]. As boundary condition the IFS needs the sea surface temperature (SST) for climate runs.
The IFS applies a spectral technique for the solution of it's differential equations of dynamics. Several computations, e.g. the radiation processes, are carried out in grid–point space. Also the coupling to the ocean model takes place in the grid–point space. We used the model in a spectral resolution with the triangular truncation T63, which corresponds to a horizontal mesh of $1.9° \times 1.9°$. To avoid a very high resolution at high latitudes, the IFS uses a so–called "reduced grid". As schematically shown in figure 1, near the poles less grid points along one latitude are used compared to the low latitudes.

In the vertical, the model has 31 layers, and a time–step of 1 hour has been used. The version used in our experiments is parallelized using the message–passing–interface (MPI)–library.
IFS is the operational model for the numerical weather prediction of the ECMWF. It has also been used for several climate runs, e.g. in the AMIP [9]. The IFS has been coupled with some ocean models, e.g. by the ECMWF for it's seasonal forecast [10][11]. With MOM it hasn't been coupled so far.

2.2 The Ocean Model MOM

Based on works of K. Bryan, the MOM is developed by the Geophysical Fluid Dynamics Laboratory (GFDL, Princeton) since 1969 [12]. The version MOM 2

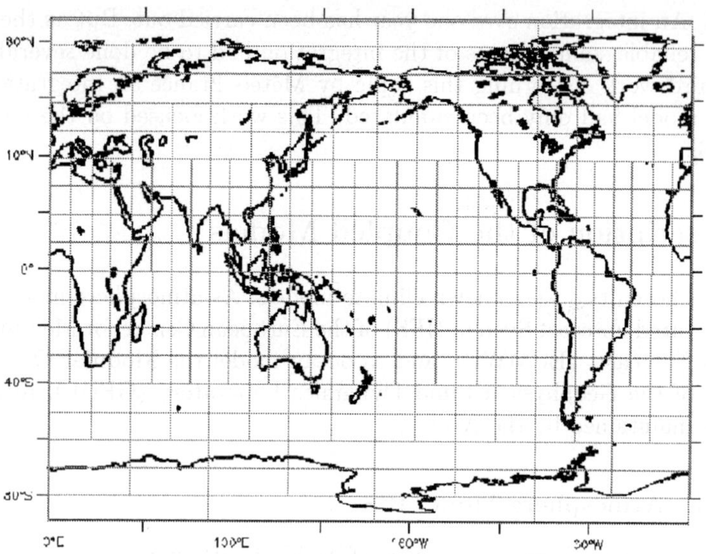

Fig. 1. Scheme of the "reduced grid".

used by us has been released by the GFDL 1995 [13]. Our MOM–version has been further developed by AWI, in particular a sea–ice model has been included. MOM uses the primitive equations to describe the circulation in the oceans. It's prognostic variables are the horizontal velocity, the temperature and the salinity. The sea–ice model simulates the formation and melting of ice, it's drift and the radiative balance at it's surface [14]. The MOM needs input of horizontal wind stress (τ_x and τ_y), the wind speed (V), the temperature (T), the dew point temperature (T_d, the total cloud cover (N), and the precipitation (P) at it's surface as boundary conditions.
We used the MOM in a horizontal resolution of 194×92 grid points with 29 layers in the vertical. The time step has been 4 hours. The code is parallelized using the Cray–SHMEM–library.
MOM is widely used in the climate modeling community, and has been coupled to several atmosphere models [4][15].

3 The Coupling Method

We used the so–called "external coupling" approach, in which the respective models are kept almost unchanged [16]. In each of the models coupling interfaces are defined, in which the data is sent and received, respectively. The data transfer between the models is managed by an additional program, the so–called "coupling tool". We wanted to use the message passing technique to exchange the data between the models, in particular the MPI library to achieve an efficient and portable way of coupling.

3.1 The Adapted Coupling Tool: The NCAR CSM–Coupler

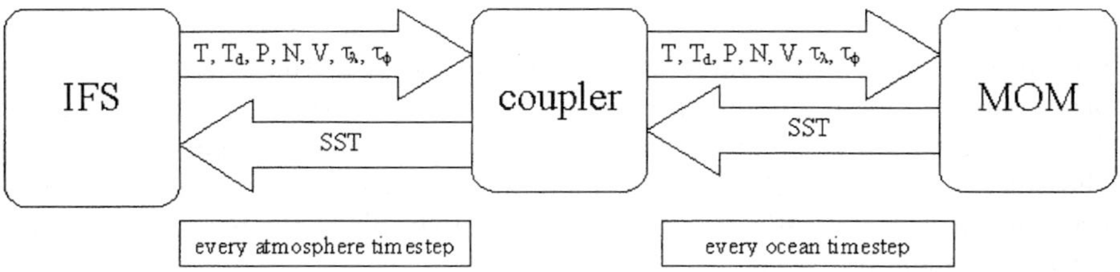

Fig. 2. Exchange of data between IFS and MOM

As coupling tool, the CSM–coupler of the National Center for Atmospheric Research (NCAR, Boulder) has been adapted.
In general, such a coupling tool receives the data from one model, performs computations, if required, interpolates the fields between the different grids, and sends it to the other model. In NCAR's use for it's Climate System Model (CSM), the CSM–coupler performs the data exchange between four models describing atmosphere, ocean, sea–ice and land surface, respectively [17]. As in our work the sea–ice model was coupled internally to the ocean model and we decided to keep the land surface parametrizations of the atmosphere model, the communication constructs for the land and the sea–ice in the coupler had to be eliminated. Further, at NCAR the coupler computes the fluxes between the models of NCAR's CSM. The computation of the input fluxes is a relevant part of a particular model. As the formulas used by the CSM–coupler are very different form those used by IFS and MOM, this parts of the coupler had to be removed.
The remaining tasks to be handled by the coupler are receiving data from one model, interpolating it between the grids, and send it to the other model, as schematically illustrated in figure 2. The atmosphere time step is longer than the ocean time step, and so the coupler averages the data from the atmosphere's time step to the ocean's time step. As the CSM–coupler needs regular grids for the interpolation routine and the interfaces are designed for the use at NCAR, the coupler had to be adapted according to the needs of our models.
However, the CSM–coupler had the advantage to use MPI. Other coupling tools like the in the climate modeling community well–known coupler "OASIS" of the Centre Européen de Recherche et de Formation en Calcul Scientifique (CERFACS, Toulouse) didn't support the use of the MPI–library at the time we started our work (OASIS does so from the version 3.0 on, released in July 2000) [18].

3.2 The New MPI-Library "Meta-MPI"

Supercomputer vendors supply efficient MPI–libraries only for their own hardware. The use of a public–domain implementation of MPI, such as MPICH, doesn't provide an efficient library for massively parallel systems [2]. In the framework of the GTWB–project therefore a new MPI–implementation, the "Meta–MPI", has been developed. Meta–MPI uses for the communication between several programs router–nodes. If one program sends data to another one, the data is sent by the router node of this program to the router node of the other one. So for the use of n programs $n \times (n+1)$ router–nodes are necessary. The configuration of Meta–MPI is done by an input–file, comparable with the "hostfile" used by MPICH. The configuration can be done using a graphical user interface. Meta–MPI has an efficiency comparable to the hardware vendor's MPI–implementation [19].

3.3 Interfaces in the Models

In both models modular interfaces for the coupling have been defined. The routines for sending and receiving data have been implemented similar to the writing of output and the reading of input by the models, respectively. As the coupling is managed by the serial coupler, the data is received by one processor and distributed to all the others. To send data to the coupler, it has to be gathered by one processor, which sends the data to the coupler. The conversion of different units and, in IFS, the calculation of current values of accumulated precipitation fields is carried out by the interface routines. In the IFS an additional interpolation between the "reduced grid" and the regular grid had to be introduced, and in the MOM the different grids of ocean and sea–ice had to be handled.

4 Results

To test the coupled model and our coupling technique, we carried out some test runs. In these runs, the IFS has been computed on the IBM SP2, the MOM and the coupler on the Cray T3E. The SP2 is the less powerful machine, so there all available nodes have been used. From the 34 nodes of the SP2, two were necessary for the communication managed by Meta–MPI, so 32 were left for the integration of the model. To achieve load–balance between the two models, we used eight nodes of the Cray T3E for the integration of the MOM. With this configuration the integration of the MOM on the T3E consumed slightly less time than the IFS on the SP2. The coupler ran on one processor of the T3E (see figure 3).

The coupled model has been integrated for eight months. For these integrations, the IFS has been initialized using analysis data of the ECMWF for January 1, 1997. For the MOM, a "spinup"–run in uncoupled mode has been carried out

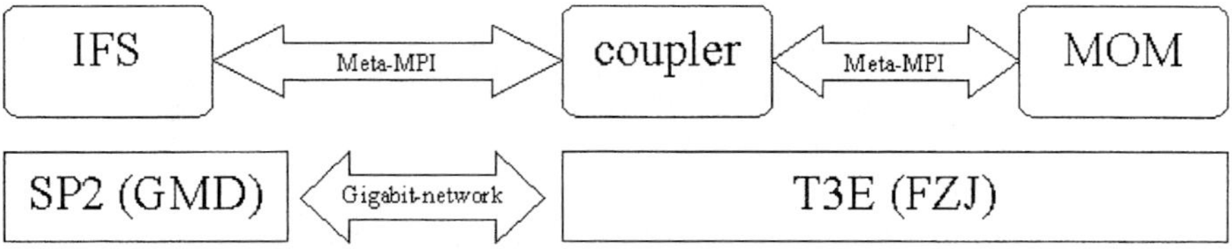

Fig. 3. The models and the underlying hardware.

over 100 years integrated time to achieve a climatological equilibrium. In this spinup–run, the model has been forced using climatological boundary values for the wind forcing. For the initialization for the temperature– and salinity–fields climatological data has been used. Starting with the begin of January, the models have been coupled together.

4.1 The Results of the Coupled Model

The main goal of our work was the development the tools and the technique for the coupling of climate models, and the results of the coupled model shall only be reported briefly in a qualitative manner.
The two models have been coupled together for the first time, and a first goal was therefore to develop a stable running coupled model. This has successfully been reached.
In figure 4 the sea surface temperature is plotted, as it is seen by the IFS. The treatment of the land–/sea–mask suppresses the use of the SST provided by the ocean model in the IFS, if the IFS assumes any land below a mesh box. This leads to the particular land–distribution shown in figure 4.
Note that the temperature over sea ice is too cold over the arctic. The reason for this is the special representation of the ice temperature in the sea ice model, which makes it difficult to get the real ice surface temperature. Both facts mentioned require further development for climate scenario runs. Figure 5 shows the temperature as diagnosed in 2m above the surface in the atmosphere, averaged over July. In general, it agrees with the expected temperature distribution.

4.2 Technical Results

Every time step, eight physical fields in the grid–point space had to be sent – seven from the atmosphere model to the coupler, and one in the opposite direction. The horizontal grid consisted of 96 points in latitude and 192 points in longitude, so for a 8–Byte–representation of the data results a data volume of 1.3 MB. To send such a volume of data through the used network, a time in the range of less than $10ms$ were necessary.The computing time consumed for the integration of one time step of the atmosphere model has been on average 11s.
During the test–runs of the model never any failure of the network has been occurred.

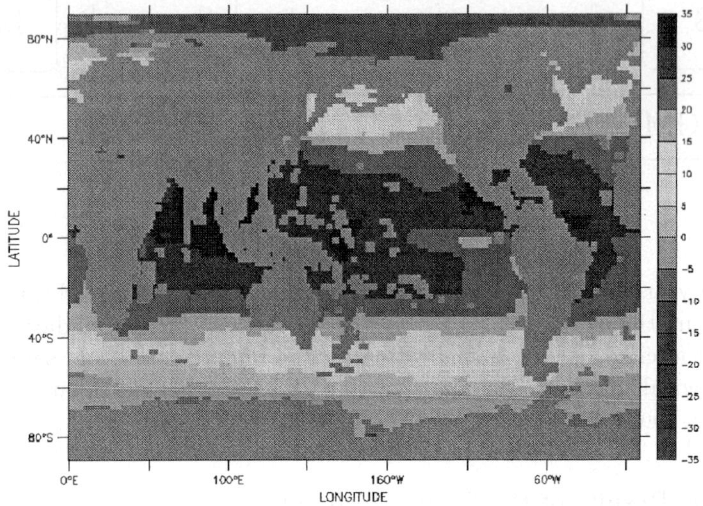

Fig. 4. Average of the sea surface temperature in July.

The comparison with test runs in uncoupled mode showed, that in our case it was even much faster to run the ocean model on a remote computer and send the boundary values via the network to the atmosphere model than to read it simply from the SP2's hard disk.

5 Summary and Conclusion

Using state–of–the–art general circulation models for atmosphere and ocean, we developed a new coupled climate model. The two models, the atmosphere model IFS of the ECMWF and the ocean model MOM in a version extended by an internally coupled sea–ice model by the AWI, have been coupled together for the first time. The coupled model runs stable and produces in general reliable results. For scenario climate runs further development would be necessary.
We coupled the models using an adapted version of NCAR's CSM–coupler. Because many changes had to be introduced to adapt this tool, probably the choice of a more general coupling library would be reasonable for future works. The MpCCI–library for example provides the facilities needed, and supports additionally a coupling in parallel, so that the data does not need to be gathered by one processor for sending, but processors of the two models working on the same domain can directly communicate together [20].
The models ran on two remote computers, one Cray T3E in Jülich, the other one on an IBM SP2 at the GMD St. Augustin. The data exchange took place on a

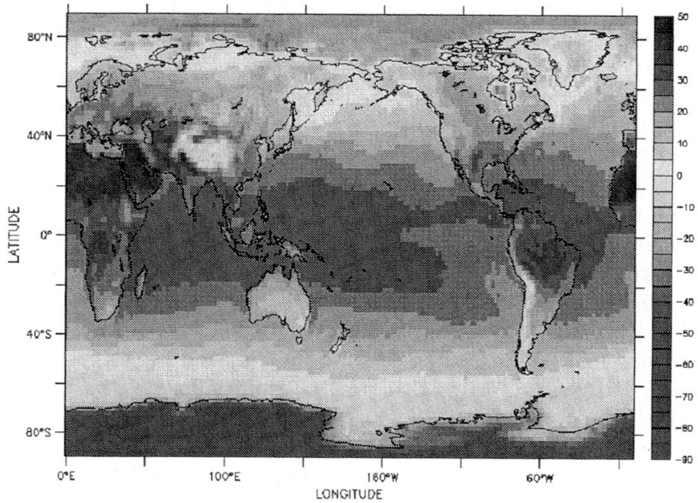

Fig. 5. Average of the temperature in 2m in july.

gigabit test–network supplied by the DFN, crossing the distance of about 100km. We used a new implemented message passing–library, "Meta–MPI", which has the advantage of a high–performance connection of different supercomputers and a quite easy handling. Neither bandwidth of the network used nor it's latency have been a constraint for our simulation. The data volume of 1.3MB has been sent in some ms, a factor of 1000 faster than the computation time. The network has been perfectly reliable. In comparison to former studies of distributed computation of climate models our work has been the first which does it in an efficient way. As networks of the quality used by us will become more and more available to every research center, we hope that our coupling technique can be used by scientists to couple their climate models together.
Projects concerning this topic are actually carried out by our partners of the AWI, and are planned by the GMD together with several climate research centers in Germany.

References

1. T. Eickermann, P. Wunderling, R. Niederberger and R. Völpel, Aufbruch ins Jahr 2000 – Start des Gigabit Testbed West, DFN–Mitteilungen 45 (1998) 13–15.
2. Eickermann, T., J. Henrichs, M. Resch, R. Stoy and R. Völpel, Metacomputing in gigabit environments: networks, tools, and applications, Parallel Comp. 24 (1998) 1847–1873.
3. F. Hommes and R. Niederberger, Gigabit Testbed West Technik, in: *DFN–Symposium*, DFN, Berlin, 1999.

4. C. Mechoso, C.–C. Ma, J. Ferrara and J. Spahr, Parallelization and distribution of a coupled atmosphere–ocean general circulation model, Mon. Wea. Rev. 121 (1993) 2026–2076.
5. C. Cassou, P. Noyret, E. Savault, O. Thual, L. Terray, D. Beaucourt and M. Imbard, Distributed ocean–atmosphere modeling and sensitivity to the coupling flux precision: the CATHODe project, Mon. Wea. Rev. 126 (1998) 1035–1053.
6. TEN34, Final report on advanced application monitoring, Deliverable D13.2, DANTE Ltd., Cambridge, 1998.
7. M. Déqué, C. Dreveton, A. Braun and D. Cariolle, THe ARPEGE/IFS atmosphere model: A contribution to the french community climate modelling, Clim. Dyn. 10 (1994) 249–266.
8. ECMWF, Documentation of the ECMWF Integrated Forecast System (IFS), Technical Report, ECMWF, Reading, U.K., 1999
9. W. Gates, J. Byle, C. Covey, C. Dease, C. Doutriaux, R. Drach, M. Fiorino, P. Gleckler, J. Hnilo, S. Marlais, T. Phillips, G. Potter, B. Santer, K. Sperber, K. Taylor and D. Williams, An overview of the results of the atmospheric models intercomarison project (AMIP), Technical Report 45, PCMDI, Livermore, California, USA, 1998
10. T. Stockdale, M. Latif, G. Burgers and J.–O. Wolff, Some sensitivities of a coupled ocean–atmosphere GCM, Tellus 46A (1994) 367–380
11. T. Stockdale, Coupled ocean–atmosphere forecasts in the presence of climate drift, Mon. Wea. Rev. 125 (1997) 809–818
12. K. Bryan, A numerical method for the study of the circulation of the world ocean, J. Comp. Phy. 4 (1969), 347–376
13. R. Pacanowski, MOM 2 documentation, user's guide and reference manual, Technical Report 3, GFDL, Princeton, USA, 1995
14. H. Fischer, Vergleichende Untersuchungen eines optimierten dynamisch–thermodynamischen Meereismodells mit Beobachtungen im Weddelmeer, Technical Report 166, Alfred–Wegener–Institut for Marine and Polar Research, Bremerhaven, Germany, 1995
15. M. Fischer and A. Navarra, GIOTTO – A coupled atmosphere–ocean general circulation model: The tropical pacific, Q.J.R. Met. Soc. 126 (2000) 1991–2012
16. K. Cassirer, T. Horie and W. Joppich, On the use of coupling interfaces for the transfer of data from the global model GME to the local model LM of the German Weather Service, Technical Report, GMD–SCAI, St. Augustin, Germany, 1998
17. F. Bryan, B. Kauffman, W. Large and P. Gent, The NCAR CSM flux coupler, Technical Report, NCAR, Boulder, Colorado, USA, 1997
18. L. Terray, S. Valcke and A. Piacentini, OASIS 2.3 user's guide and reference manual, Technical Report TR/CGMC/99–37, CERFACS, TOulouse, France, 1999
19. J. Henrichs, Meta–MPI, Technical Report, Pallas GmbH, Brühl, Germany, 1998
20. R. Ahrem, M. G. Hackenberg, P. Post, R. Redler and J. Roggenbuck, Specification of MpCCI Version 1.0, GMD–SCAI, St. Augustin, 2000

Numerical Simulation of Cloud Dynamics and Microphysics

Elena N. Stankova and Mikhail A. Zatevakhin

Institute for High Performance Computing and Information Systems, Fontanka, 6, 191186, St. Petersburg, Russia
{misha,lena}@fn.csa.ru

Abstract. It is well known that convective clouds play an essential role in the evolution of the aerosol particles in the atmosphere by providing spatial redistribution and transformation of aerosol due to in cloud chemical reactions, dynamical and microphysical processes. The influence of the clouds upon the aerosol properties increase greatly in extreme conditions, such as volcano eruptions, explosions, large fires, when enormous number of aerosol particles of different size and chemical compound are injected in the atmosphere. In this paper the overview of the main results, obtained by the authors in the field of numerical modeling of the convective clouds, developing in natural and extreme conditions and the investigation of their role in the evolution of the atmospheric aerosol is presented.

1 Introduction

Computational fluid dynamics and meteorology in particular are among the major consumers of high performance computer technology. Effective numerical modeling of convective flows developing in natural and extreme conditions allows to calculate online transport and transformation of the atmospheric particles and effluents in real time thus making possible to use the model for the operational forecast of the local pollution propagation. A detailed analysis of the processes of such kind requires development of the effective algorithms of its numerical realization including high-performance algorithms for modern supercomputer calculations. The effective numerical algorithms will allow to incorporate the model into in the operational regional or global climate models for online forecasting of the pollutant propagation in the certain area and possible prevention of undesirable effects of catastrophic phenomenon frequently accompanied by the convective cloud development.

A role of clouds in evolution of atmospheric pollutant is quite significant. Cloud dynamics strongly modifies the distribution of pollutant through advection and more local eddy motions over the edges of the cloud. Cloud microphysics participates in pollutant evolution by means of condensation/evaporation and collection processes transferring them to the drops during precipitation formation and reinjecting back into the air by evaporation of droplets and raindrops. On the one hand, the processes in the clouds result in cleaning the atmosphere by wet removal of the pollution. On the other hand they are responsible for so called acid rains, damaging the material surfaces of aquatic and forest ecosystems. In addition, clouds participate actively in trace

P.M.A. Sloot et al. (Eds.): ICCS 2003, LNCS 2658, pp. 171–178, 2003.

chemical redistribution, affect the global radiation budget and thus influence global background atmosphere.

Though clouds of all types contribute significantly to atmospheric processes, the case of so called deep convection, when the upper boundary of the cloud can reach the tropopause level, is the most interesting one from the view point of pollutant delivering to the upper levels of troposphere. Besides, deep convection plays an exclusively important role as pollutant converter and transport mean in various extreme situations when enormous amount of chemically active aerosol particles are being injected into the atmosphere.

Convective clouds are very complex natural objects, with a wide spectrum of interacting processes of different time/space scales (convective transport, microphysical interactions, chemical reactions, etc.). A brief review of the resent cloud models where the most complete set of the in-cloud processes are considered [1]-[6] shows, that in each model only a few features of cloud processes interaction are being considered rigorously while the rest ones are treated approximately, with the use of strong simplifying assumptions and crude parameterizations. Thus, an urgent need does exist in a numerical model capable of representing properly all the essential interactions between dynamical, microphysical, chemical and aerosol processes. This model implementation will apparently demand supercomputer facilities and the effective numerical algorithms including special techniques for supercomputer applications.

This paper is dedicated to the description of the numerical algorithms used for simulation dynamical, turbulent, microphysical and chemical processes in convective clouds developing in natural and extreme conditions. Algorithms were realized in the frame of one and two-dimensional numerical models. The first one is intended for natural clouds investigation, the second – primarily for investigation of the clouds developing in extreme conditions. The second model was also used for describing the polluted cloud formation that occurs during buoyant thermal rising.

2 Algorithms Used in the One–Dimensional Model

A time-dependent, one-dimensional model is used for an isolated warm cumulus cloud life cycle simulation. In this model the cloud is assumed to have a cylindrical form with constant radius R in an environment at rest. All equations expressed in cylindrical coordinates and all cloud variables are represented with mean values averaged over the horizontal cross section of the cloud [7]. The system of equations combines the equations of vertical motion, the equation of continuity, the first law of the thermodynamics, and the equations of water vapor and liquid hydrometeors transport. The dynamic interaction between the cloud and its environment is modeled by two entrainment terms: turbulent entrainment representing lateral mixing at the side boundaries of the cloud and dynamic entrainment, representing the systematic inflow or outflow of air required to satisfy mass continuity. Time splitting method is used for numerical solution of the equations. The formation and growth of drops by nucleation, condensation and stochastic collection are modeled in detail with the help of stochastic collection equation. For calculation of the terms describing processes of condensation/evaporation the Lagrangian scheme was chosen [8], which guarantee the conservation of cloud drops number. For the calculation of coalescence process the modification of the well-known Kovetz and Olund method [9] is used, which allows to obtain good solution of stochastic collection equation at the relatively low number of drop size intervals and provides precise fulfillment of the mass conservation law.

In such kind of problems the major part of computation time is consumed during the solution of stochastic coalescence equation especially when the real collection kernel is calculated directly in calculation process. To reduce the run time the collection kernel of such equation, being once calculated must be stored in computer memory. This demands the array of the size $n^2 m$, where n is the number of droplet size classes and m is the number of space grid points. It is evidently needs great amount of computer memory, but allows reducing run time essentially.

To investigate the role of convection in aerosol pollutant redistribution a very effective numerical algorithms are needed which could be incorporated into the dynamical block of the model without a significant growth of the computational expenses. Development and implementation of such algorithms will allow assessing more accurately the effect of the aerosol particles size distribution on the pollutant concentrations.

For the problem solution the model was modified in nucleation block description. Nucleus vertical transport and reinjecting back into the air by evaporation of cloud drops were taken into account. The particle was assumed to be activated when achieving critical supersaturation for given size. The results of numerical investigation of the cloud effect upon aerosol pollution redistribution in the atmosphere revealed strong feedback between cloud and aerosol parameters. Vertical distribution of the aerosol particles in different outer atmospheric conditions with unstable temperature distribution (case 1) and the other (case 2) with the small layer of the temperature inversion on the 50 min of cloud development is represented in Fig.1.

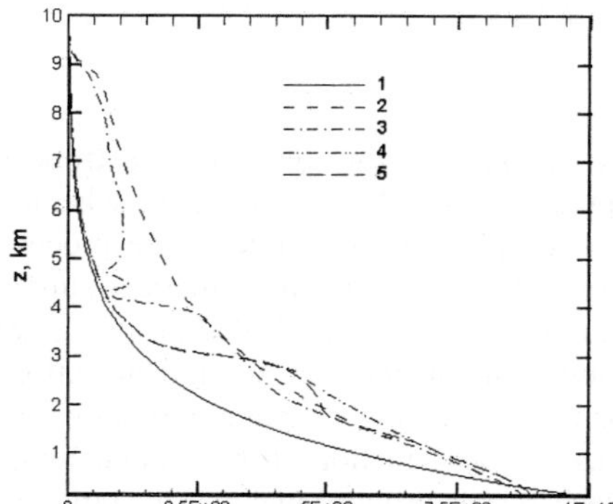

Fig. 1. Vertical profiles of aerosol particles obtained on the 50min of cloud development, where 1- initial aerosol distribution, 2 - hydrophobic particles (case 1), 3 - cloud condensation nucleus (case 1), 4 - hydrophobic particles (case 2), 5 - cloud condensation nucleus (case 2).

For the analyses of the calculation results a special viewer was elaborated, which allowed looking through the output data which structure was not prescribed in advance. This was obtained by special output procedure, which created data base with all necessary information about the structure of the output data fields and its content. The viewer utilizes this data to form the user's menu system and provide all features necessary for visualization.
Calculations with the model gave the results, which were in good agreement with observations of the dynamical and microphysical properties of warm cumulus cloud (Fig.2). The development of the droplet spectrum during cloud evolution up to the precipitation formation was particularly well simulated.

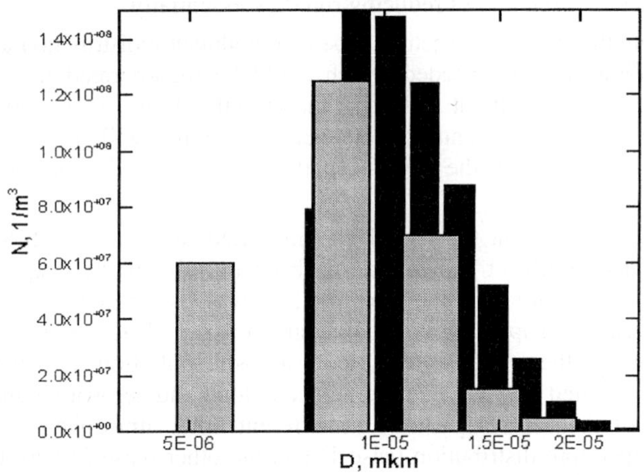

Fig. 2. Numerical and experimental data comparison. (drop size distribution) Dark columns - numerical data, light columns - numerical data.

One-dimensional model was used for investigation of the effect of the convective cloud droplets on the chemical composition of the troposphere. For this purpose effective and computationally chip numerical scheme of chemical species redistribution along the droplet spectrum due to the processes of condensation and evaporation was elaborated [8]. In general, numerical simulation of cloud droplet – chemical species interaction is very complicated problem, demanding solution of the kinetic equation for multicomponent composition distribution function [10]. Proposed approach is based upon the following simple assumptions: if the droplet spectrum is divided into several size classes, fraction of the chemical species redistributed due to the microphysical processes is proportional to the ratio of the droplet mass flux between classes to the total droplet mass class contains. Nevertheless it allows obtaining satisfactory results using relatively small computer resources.
Calculation showed that the main effect of the convective cloud on the chemical composition of the air is the depletion of the well-dissolved species concentrations and their removal by precipitation. The transformation of aqueous species in different size droplets due to their coagulation is very important for the cloud and precipitation

chemistry. The vertical transport in the convective cloud may affect essentially the vertical distribution of chemical species of the troposphere, especially if there is a surface pollution source.

Test calculations showed that the method application for simulation intensive collection at high aerosol particle concentrations required essential reduction of the time step for obtaining positive solution. For overcoming this constraint the following modification of the initial scheme was proposed. A solution is obtaining in two stages as the time step is splitting. On the first stage the following equation is solved (see [9]):

$$\frac{M_j^{n+1/2} - M_j^n}{\Delta t} = -S_j^- , \tag{1}$$

where M_j – aerosol particle mass in the interval $m_{j-1/2}$, $m_{j+1/2}$, m_j, j=0,1,...,N – coordinates of the grid points in the mass space.

$$S_j^- = \sum_{l=1}^{N-1} S_{jl}^- \tag{2}$$

$$S_{jl}^- = K(m_j, m_l)\frac{M_j^{n+1/2} M_l^n}{m_l} \tag{3}$$

$K(m_j,m_l)$ – is collection kernel.

Solution of the equation (1) is as follows:

$$M_j^{n+1/2} = \frac{M_j^n}{1 + \Delta t \sum_{l=1}^{N-1} K(m_j, m_l) M_l^n / m_l} \tag{4}$$

The further calculations are conducted according to the formulas:

$$\frac{M_j^{n+1} - M_j^n}{\Delta t} = S_j^+ - S_j^- \tag{5}$$

$$S_j^+ = \sum_{k=1}^{N-1}\sum_{l=1}^{N-1} S_{jkl}^+ \tag{6}$$

$$S_{gjl}^+ = \alpha_{gjl} S_{jl}^- \tag{7}$$

$$\alpha_{gjl} = \begin{cases} \dfrac{m_{g+1} - m_j - m_l}{m_{g+1} - m_g} & at\ m_g < m_j + m_l \le m_{g+1} \\ \dfrac{m_j + m_l - m_{g-1}}{m_g - m_{g-1}} & at\ m_{g-1} < m_j + m_l \le m_g \\ 0 & \text{in the other cases} \end{cases} \tag{8}$$

It is easy to show that

$$M_j^{n+1} = M_j^n + \Delta t\left(S_j^+ - S_j^-\right) = M_j^n - \Delta t S_j^- + \Delta t S_j^+ = M_j^{n+1/2} + \Delta t S_j^+, \tag{9}$$

Test calculations showed that at relatively small time step the solution obtained by explicit and implicit schemes practically coincided. At large time step implicit scheme produces a certain error. Nevertheless it provides positive solution at any time step (as it is clear from [8]) and guarantees mass conservation similar to the original scheme.

3 Algorithms Used in the Two–Dimensional Model

The numerical study of the convective flows developing in extreme conditions was based upon the system of equations of the turbulent convection, described in detail in [11] and [12]. In case of deep convection, especially associated with the extreme conditions, the problem is strengthen by the presence of crucial variations of the system thermodynamic parameters. At such conditions the conventional Boussinesq approximation becomes unjustified and one has to use the full compressible Navier - Stokes equations. However at low Mach numbers typical for clouds convection, these equations becomes acoustically "stiff" and their solution presents an extremely complicated computational problem [13]. Besides, one of the most difficult physical issues associated with the cloud dynamics modeling is an appropriate treatment of turbulent transport. A turbulence model used for this purpose should be capable of providing an adequate description of the strong stream-line curvature, flow rotation, and buoyancy effects being crucial in deep clouds convection.

Two-dimensional model is the model based upon the solution of equations, describing the axisymmetric turbulent flow of compressible two-phase two-component medium considering condensation of water vapour and precipitation formation. The model is appropriate for the description of the flows with arbitrary vertical scales and temperature variations. The description of the processes of turbulent mixing is provided with the help of *k*-ε model modified by taking into account the influence of the streamline curvature and buoyancy forces. It was shown in [11], that for the correct description of the stratified flows one of the constants in the *k*-ε turbulence model should have different values in the stable and unstable conditions. The alternative directions scheme is used for the numerical solution of the system of equations. The suppression of the nonphysical oscillations is provided by means of the specially elaborated monotonization procedures [14] Calculation of the processes of turbulent energy generation and dissipation is separated in the additional step of the splitting procedure, where fully explicit scheme is used. As the dimensions of the disturbed region of calculation vary greatly, the simulations were performed in the domain with movable external boundaries where movable grid with fixed number of points was inserted. The model is intended for the investigations of the flows developing at the presence of high-energy sources on the ground surface or in the air. As a matter of fact this problem is converged to the simulation of the buoyant thermal rising in the stratified medium.

The numerical simulation provided with this model allowed to obtain the description of all stages of the thermal evolution, including the stage of its transformation to the buoyant vortex ring, which is characterized by the fully coincidence of the regions of maximum temperature and maximum vorticity. Such structure is preserved till the end of the thermal rising when its speed begins to decrease due to the generation of the secondary vortex system in the stable stratified atmosphere.

For taking into account the phase transitions of the water vapor the main model equations are written for the general case of two-phase two-component flow movement. In this case the processes of cloud droplets evaporation-condensation were described in the equilibrium approach, for the description of precipitation formation the Kessler parameterization scheme was used. Moist convection numerical experiments results showed, that while thermal ascending the condensed water once appeared in the updraft region near the axis of symmetry propagates then to the whole cloud volume. The further cloud evolution depends essentially upon the environmental conditions. When the relative humidity of the air is more than 50% in the lower 5km atmospheric layer release of the latent heat of condensation resulted in the development of the convective flow similar to that in the natural convective clouds. In the regions of high water content the formation of the precipitation particles capable to reach the ground takes place.

This model appended with the equations describing the aerosol-droplet interaction [15] was used for the investigations of the vertical transport of the aerosol particle injected in the atmosphere as a result of the powerful explosion. The results of numerical experiments showed that the process of so called «nucleation coalescence» [16] is responsible for the rate of the capture of the large aerosol particles by the cloud drops. Regions with low concentration of dry aerosol particles locate near cloud lower boundary and under the vortex core, where the ambient air is involved into the cloud.

4 Conclusions

Several numerical algorithms intended for simulation of dynamical, microphysical and chemical processes in convective clouds are presented. The algorithms provide development and operation of adequate cloud models and effective codes of their numerical realization. Special technique elaborated for supercomputer applications allows conducting a wide set of numerical experiments with reasonable/affordable computation resources. Wide set of numerical experiments was conducted aimed to investigate the processes of pollutant source formation transport and transformation.

References

1. Trembley A. and Leighton H.: A three-dimensional cloud chemistry model. J. Clim. Appl. Meteorol. **25** (1986) 652–671
2. Roelofs G.J.: Drop size dependent sulphate distribution in a growing cloud. J.Atm.Chem. **14** (1992) 109–118
3. Wang C. and Chang J.S.: A three-dimensional numerical model of cloud dynamics, microphysics, and chemistry, 1. Concepts and Formulation. J. Geophys. Res. **98** (1993) 14827–14844
4. Gregoire P.J. et al.: Impact of cloud dynamics on tropospheric chemistry: advances in modeling the interactions between microphysical and chemical processes. J.Atmosph.Chem. **18** (1994) 247–266
5. Audiffren N. et al.: Effects of a polidiaperse cloud on tropospheric chemistry. J.Geophys.Res. **101** (1996) 25949–25966
6. Guiciullo C.S. and Pandis S.N.: Effect of composition variations in cloud droplet populations on aqueous-phase chemistry. J.Geophys.Res. **102** (1997) 9375–9385
7. Shiino, J.: A numerical study of precipitation development in cumulus clouds. Pap.Met.Geophys. **29** (4) (1978) 157–194
8. Karol I. L., Zatevakhin M. A., Ozhigina N. A., Ozolin Y. E., Ramaroson R., Rozanov E.V., Stankova E.N.: A model of dynamical, microphysical and photochemical processes in a convective cloud. (In Russian) Izvestia *FAO* **36** (6) (2000) 1–16
9. Stankova E.N., Zatevakhin M.A.: The modified Kovetz and Olund method for the numerical solution of stochastic coalescence equation. Proceedings 12th International Conference on Clouds and Precipitation. Zurich (1996) 921–923
10. Gelbard F., Seinfeld J.H.: Simulation of Multicomponent Aerosol Dynamics. Journal of Colloidal and Interface Science **78** (9) (1980)
11. Dovgalyuk Yu.A., Zatevakhin M.A., Stankova E.N.: Numerical simulation of buoyant thermal using k-e model. J.Appl. Met. **33** (1994) 1118–1126.
12. Zatevakhin M.A.: Turbulent Thermal in a Humid Atmosphere High Temperature. **39** (4) (2001) 532–539
13. Kuznetsov A., Niculin D., Strelets M., Zatevakhin M.: 3D Simulation of the Interaction between Large Thermally Buoyant Clouds in the Earth Atmosphere. In Proceedings of the Second European Computational Fluid Dynamics Conference Stuttgart (1994) 1010–1017
14. Zatevakhin M.A., Stankova E.N.: Monotonization of finite-difference schemes of numerical solution of hydrodynamics equations. Tr.Gl.Geofiz.Obs. **534** (1991) 73–86. (in Russian)
15. Stankova E.N., Zatevakhin M.A.: Investigation of aerosol-droplet interaction in the mature convective clouds using the two-dimensional model. Proceedings 14th Int. Conf. Nucleation and Atm.Aeros. Helsinki (1996) 901–903.
16. Bradley, M.M.: Numerical simulation of nucleation scavenging within smoke plume above large fires. Proceedings Int. Conf. Energy Transform. Lausanne .(1987)

Optimal Numerical Realization of the Energy Balance Equation for Wind Wave Models

Igor V. Lavrenov

State Research Centre of the Russian Federation
Arctic and Antarctic Research Institute
lavren@aari.nw.ru

Abstract. The optimal numerical realization of the energy balance equation in wind wave models is proposed. The scheme is separated into two parts: the numerical source term integration and the energy propagation numerical realization. The first one is based on a spreading numerical method. Semi-analytical solution is used for integration of source term, which includes the wind wave input, dissipation term and exact non-linear energy transfer function. The energy propagation numerical realization is based on the utilisation of the diffusive operator, which is implemented in the semi-Lagrangian numerical method. It gives opportunity to remove the garden-sprinkler effect of the energy balance equation solution for the case of wind wave propagation in the ocean. The new technique yields accurate results and has the additional advantage of being numerically stable, thereby allowing the use of a large time step (three hours and more). The method could be regarded as a general alternative for the numerical realization of energy balance equation in the models.

1 Introduction

The successful solution of the hindcast or forecast wind wave problem depends on the quality of physical model and on the equation numerical realization in the mathematical modelling. The errors of the solution depend on the accuracy of wind speed calculation as well. Nevertheless the errors of the energy balance equation numerical realization can be no less than the errors coming from the wind value accuracy and from the unsatisfactory implementation of the wind wave physics in the model [1,2,3,4,6,7]. In the present study the optimal numerical realization of the energy balance equation in wind wave models is proposed, which gives an opportunity to get reliable results by using sufficient large time step of numerical integration.

2 Formulation of the Problem

The evolution of a two-dimensional ocean wave spectrum $S(\omega,\beta,\varphi,\theta,t)$ with respect to frequency ω and direction β (measured here counter clockwise from the parallel) as a function of latitude φ and longitude θ is described by the transport equation [3,4]:

P.M.A. Sloot et al. (Eds.): ICCS 2003, LNCS 2658, pp. 179–187, 2003.

$$\frac{\partial S}{\partial t}+\frac{1}{\cos\varphi}\frac{\partial(\dot{\varphi}\cos\varphi S)}{\partial\varphi}+\frac{\partial(\dot{\theta} S)}{\partial\theta}+\frac{\partial(\dot{\beta} S)}{\partial\beta}=G\ , \tag{1}$$

where $G=G(\omega,\beta,\varphi,\theta,t)$ is the source function, describing wind input, dissipation and non-linear wave-wave interaction. The functions $\dot{\varphi},\dot{\theta},\dot{\beta}$ represent the rates of change of location (φ,θ) and direction of propagation β of a wave packet travelling along a ray on spherical surface.

$$\dot{\varphi}=C_g\sin\beta/R\,,\dot{\theta}=C_g\cos\beta/R\cos\varphi\,,\dot{\beta}=-\,C_g\tan\varphi\cos\beta/R \tag{2}$$

where C_g is a group velocity and R is an Earth radius.

The scheme of numerical realization of the energy balance equation (1) with (2) is separated into two parts: the energy propagation numerical realization and the numerical source term integration.

3 Numerical Scheme of the Wave Energy Propagation

The numerical solution of the energy balance equation (1) requires the discretization of the continuous wave spectrum $S(\omega,\beta)$ in the frequency-direction space. The mean energy contained in the frequency $(\omega_k-0.5\Delta\omega\leq\omega\leq\omega_k+0.5\omega)$ and in the direction bands $(\beta_l-0.5\Delta\beta\leq\beta\leq\beta_l+0.5\Delta\beta)$ can be determined as:

$$\overline{S}(\omega_k,\beta_l)=\frac{1}{\Delta\beta\Delta\omega}\int_{\omega_k-\frac{\Delta\omega}{2}}^{\omega_k+\frac{\Delta\omega}{2}}\int_{\beta_l-\frac{\Delta\beta}{2}}^{\beta_l+\frac{\Delta\beta}{2}}S(\omega,\beta)\,d\beta d\omega\quad . \tag{3}$$

In order to derive the mean energy approximation in the bands the bi-quadratic interpolation is applied between points $\omega_{k-1},\omega_k,\omega_{k+1}$, β_{l-1}, β_l and β_{l+1}.

$$\overline{S}(\omega_k,\beta_l)=\sum_{i,j=-1}^{1}a_ib_jS(\omega_{k+i},\beta_{l+j})\quad , \tag{4}$$

where $a_1=a_{-1}=b_{-1}=b_1=1/24;\ a_0=b_0=11/12$. In a similar manner the energy balance equation (1) can be adapted to incorporate the effects of a finite frequency-angular resolution:

$$\frac{\partial\bar{S}}{\partial t}+\frac{\partial\bar{S}}{\partial\varphi}(1+\varepsilon/2+\delta)\dot{\varphi}+\frac{\partial\bar{S}}{\partial\theta}(1+\varepsilon/2+\delta)\dot{\theta}+\frac{\partial\bar{S}}{\partial\beta}(1-\varepsilon/2+\delta)\dot{\beta}+$$
$$+\varepsilon\frac{C_g}{R}\left[\frac{\partial}{\partial\beta}\left\{\cos\beta\frac{\partial\bar{S}}{\partial\varphi}-\frac{\sin\beta}{\cos\varphi}\frac{\partial\bar{S}}{\partial\theta}\right\}+\tan\varphi\sin\beta\frac{\partial^2\bar{S}}{\partial\beta^2}\right]-$$
$$-\delta\frac{C_g\omega}{R}\frac{\partial}{\partial\omega}\left[\left\{\sin\beta\frac{\partial\bar{S}}{\partial\varphi}+\frac{\cos\beta}{\cos\varphi}\frac{\partial\bar{S}}{\partial\theta}\right\}-\tan\varphi\frac{\partial}{\partial\beta}\left\{\cos\beta\bar{S}\right\}\right]+$$
$$+O(\varepsilon^2)+O(\delta^2)=G(\bar{S}) \quad , \tag{5}$$

where $\varepsilon=(\Delta\beta)^2/12<<1$, $\delta=(\Delta\omega/\omega)^2/12<<1$. It is supposed that $\varepsilon\sim\delta$. Unlike to (1) the equation (5) contains correction terms of first order of small parameters ε and δ, which allow taking into account the fine value of the angular and frequency discretization of spectrum.

Using (2) and (5) one can derive the energy balance equation, which includes the main correction term in the form:

$$\frac{\partial\bar{S}}{\partial t}+\frac{\partial\bar{S}}{\partial\varphi}\tilde{\dot{\varphi}}+\frac{\partial\bar{S}}{\partial\theta}\tilde{\dot{\theta}}+\frac{\partial\bar{S}}{\partial\beta}\tilde{\dot{\beta}}-A\frac{\varepsilon C_g}{R}\frac{\partial^2\bar{S}}{\partial\beta^2}=G(\bar{S}) \quad , \tag{6}$$

where $A=R/L-\tan\varphi\sin\beta$, $\tilde{\dot{\varphi}}=\dot{\varphi}(1+\varepsilon/2)$, $\tilde{\dot{\theta}}=\dot{\theta}(1+\varepsilon/2)$, $\tilde{\dot{\beta}}=\dot{\beta}(1-\varepsilon/2)$,

L is scale of wave propagation. The term correcting the left side of the energy transfer equation (6) is approximated by a simple angular diffusive operator, which describes a slight "exchange" of the energy between the nearest angular components.

The numerical realization of the energy propagation part of the equation (1) results in numerical solution of the equation (5) or (6). The numerical scheme is based on semi-Lagrangian method in combination with utilisation of angular diffusive operator (INTERPOL method [4,5]).

Let us consider the swell propagation to the South from the initial disturbance, which centre is located at point $\theta_0=0^0, \varphi_0=72^0$. The wave height partial distribution is approximated by the function $\exp(-\delta r/L_{\max})$, where δr is the distance from centre, $L_{\max}$ is a correlation radius. The significant wave height at the centre is 10m, mean wave period is 15 s. The propagation problem is solved by different methods: analytically, using the first order upwind scheme (FOUS), which is used in the WAM model [3], and by the INTERPOL method as well.

The initial partial distribution of significant wave height is shown in fig.1a. The wave height distributions after 24 h of propagation obtained by the analytical solution, the FOUS and the INTERPOL method (the latter two both using 12 directions and a time step of 30 min) are presented in fig.1b, 1c and 1d, respectively). In the FOUS and, to a lesser extent, the INTERPOL results, most of the wave energy tends to concentrate along the model direction nearest to the initial direction of propagation. In this direction wave heights are overestimated, while they are underpredicted in other directions. This is "garden-sprinkler" effect [1,4,5] caused by the crude angular resolution of the model. For FOUS solution with 12 directions the local errors can rise to 40% in the direction of wave propagation, and to -35% in other directions. During time the errors along the model direction nearest to the path of propagation grow to +25% within 12 h, and at certain points exceed the +90% level at t=48 h. The negative error levels involve in a less spectacular manner, from peak values of 15-20% at t=12 to -40% at t=48 h of propagation. Increasing the number of directions in the FOUS to 24 reduces the error levels by roughly half. The INTERPOL method with 12 directions is in better agreement with the analytical solution than its FOUS counterpart; its accuracy is comparable to that of the FOUS with 24 directions.

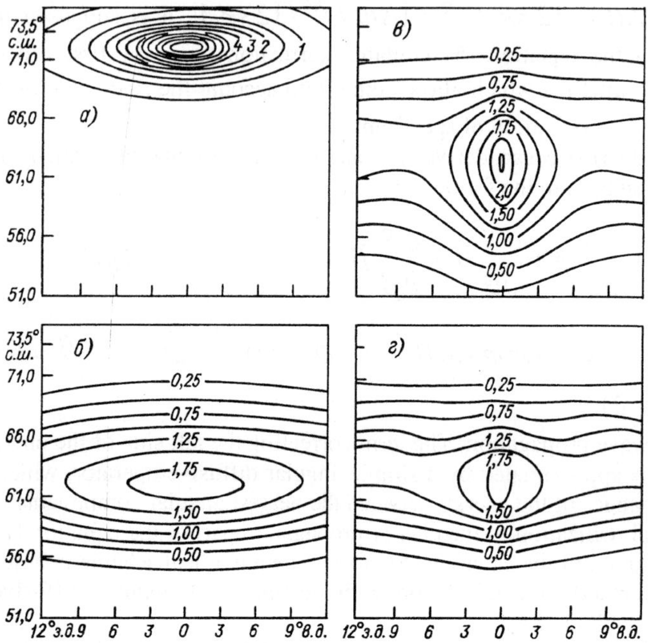

Fig. 1. The spatial distribution of wave heights: 1a - initial distribution at t=0 h; 1b - analytical solution at t=24 h, 1c - solution by FOUS with 12 direction at t=24 h, 1d - INTERPOL solution (with 12 directions and a time step of 30 minutes) at t=24h.

The INTERPOL method with 24 directions also reduces the error levels by half.

Fig.2 shows the temporal evolution of the total wave energy $\Sigma(t)$, its mean location $<\varphi(t)>$, the spatial spreading function $\theta(t)$ and root mean square error *RMSE*. Both the FOUS and INTERPOL schemes are able to reproduce the behaviour of the total energy and mean position of the wave disturbance fairly well (fig.2a, 2b). At the beginning stages of wave propagation all methods give a growing up *RMSE* level (fig.2d). The error achieves 8% level at 12 h and 20% at 40 h for FOUS with 12 directions. The double increasing of the number of directions diminishes errors by halves for the final stage of wave propagation, but for the beginning stage the error level remains approximately the same. The INTERPOL method with 12 directions and the same time step of 20 minutes gives 5% error at *t*=12.5 hours and 12% at 40 hours. The increasing of the time step up to 3 h gives 3% *RMSE* at t=12 h and 11% error at 40 h and 10% for 6-h time step. So, the increasing of the time step diminishes the error level.

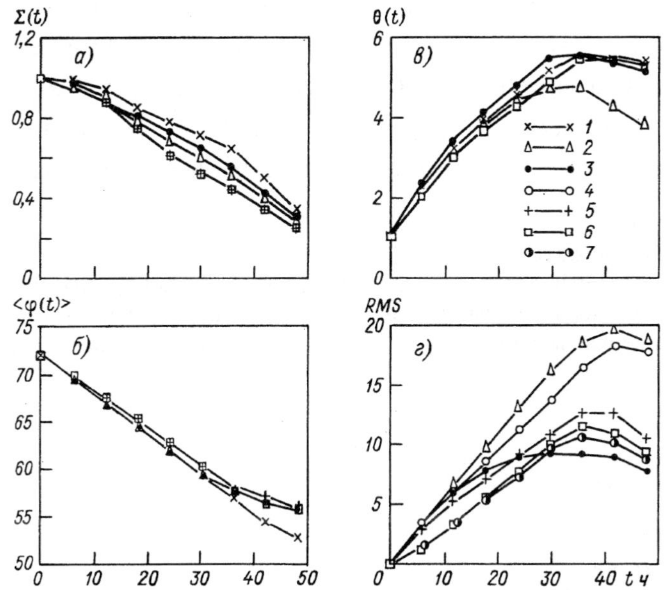

Fig. 2. The evolution of the numerical integral parameters in time. 2a - the normalized total energy $\sum(t)$; 2b - the latitude of the mean energy location $<\varphi(t)>$, 2c - the spatial spreading function $\theta(t)$; 2d - the root mean square error $RMSE$ (t). 1 - analytical solution; 2 - FOUS solution by 12 directions, Δ t= 20 min., 3 - FOUS solution by 24 directions, Δ t = 20 min.; 4 - FOUS solution by 12 directions, Δ t= 20 min. with the angle shift at $\Delta\beta/2$. 5- INTERPOL solution by 12 directions, Δ *t*= 20 min. 6 - INTERPOL solution by 12 directions, Δ *t*= 3 h, 7- INTERPOL solution by 12 directions, Δ *t*= 6 h.

4 Numerical Realization of Source Term Numerical Integration

The second step in total numerical scheme is the numerical realization of the source term integration. The scheme for the next source term is developed:

$$G = G_{in} + G_{nl} + G_{ds} \quad , \tag{7}$$

where G_{in} is the function, describing a wind input, it depends linearly on spectrum $G_{in} = A\,S$ [3], where A is the function of frequency ω, angle of spectral component propagation β and wind speed $\vec{U}$. G_{nl} is the exact form of the non-linear energy transfer computed by optimal algorithms based on the numerical method of integration of the highest precision [4]; G_{ds} is the function of energy dissipation, it consists of two components $G_{ds} = G_{ds}^1 + G_{ds}^2$. The first G_{ds}^1 is a dissipation mechanism due to white capping, it is the non-linear function of the wave spectrum, it depends on wind speed as well: $G_{ds}^1 = -A\,S^{\gamma+1}$ (where $\gamma = \gamma(\omega)$). The second component of dissipation is the linear function of the wave spectrum. It doesn't depend on the wind: $G_{ds}^2 = -B\,S$, B is the function of frequency ω, angle of spectral component propagation β. It can depend on slowly varying integral parameters of the wave spectrum [3].

For the numerical integrating of the source term a spreading algorithm is used. It consists of two steps. At first the simple first order Euler's scheme (or it can be semi-implicit numerical scheme [3] as well) is applied for the numerical integration of equation

$$\frac{S^{n+\rho} - S^n}{\Delta t} = G_{nl} + G_{ds}^2 \quad , \tag{8}$$

where $0 \le \rho \le 1$.

At the second step the next finite difference equation is solved

$$\frac{S^{n+1} - S^{n+\rho}}{\Delta t} = G_{in} + G_{ds}^1 \quad . \tag{9}$$

The solution of the equation (9) could be approximated analytically as

$$S^{n+1} = S^{n+\rho}\left(A\,(S^{n+\rho})^{\gamma} \pm \left|1 - A(S^{n+\rho})^{\gamma}\right| \mathrm{e}^{-\gamma B \Delta t}\right)^{-1/\gamma} . \tag{10}$$

The numerical experiments show that numerical scheme (8)-(10) is able to produce stable numerical results by using time step 1, 3, and 6 hours instead of time step in 20 minutes used in the WAM model. That is why in the present numerical scheme there is no need to introduce any restriction on the wave spectrum and source function numerical values. Such restriction used to include in traditional numerical scheme, because of instability of numerical results [2,3].

Fig.3 presents the test results of numerical simulations for the frequency spectrum at $t = 30$ h duration. The same computations were carried out by the second-order numerical scheme (Adam's method) for the different time step of numerical integration. The comparison results show that results of the second-order numerical scheme for time step in 3 min practically coincide with those of spreading method

(8)-(10) for time step in 1 h. By increasing the time step up to 3 and 6 hours the numerical value of frequency spectrum remains stable. It is less only by 10% for time step in 3 h and by 17% for time step in 6 h than appropriate value for time step in 1 h. Fig.4 shows the results of changing the total energy during time for the same methods. The comparison results present the high accuracy of the method (8)-(10) for the time step in 1 h and 3 h. For large time step the total increasing of energy becomes smaller because of underestimation of non-linear energy transfer contribution.

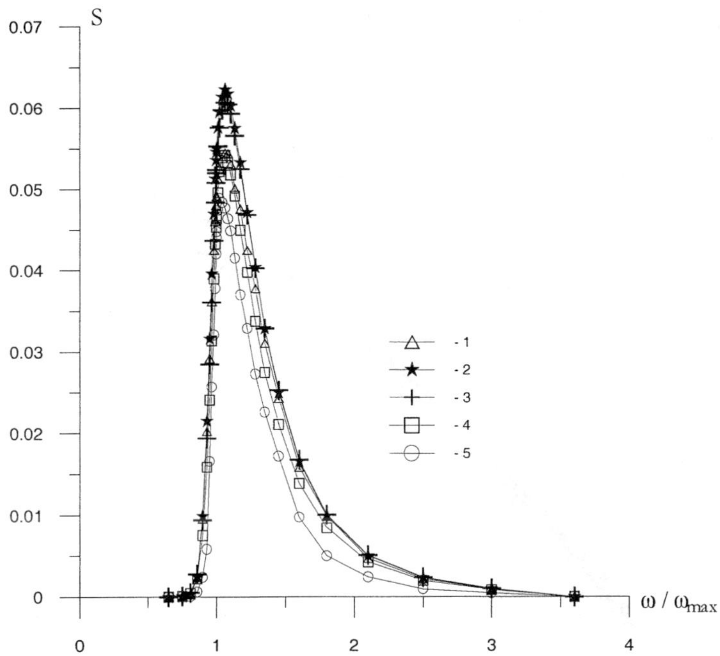

Fig. 3. Numerical value of frequency spectrum at t=30 h, 1-solution without non-linear energy transfer action; 2 – the second order method, time step $\Delta t=$ 3 min; 3- spreading method $\Delta t=$ 1 h; 4 - $\Delta t=3$ h; 5 - $\Delta t=$ 6 h.

5 Conclusions

The optimal numerical realization of the energy balance equation in wind wave models is proposed. The scheme is separated into two parts: the energy propagation numerical realization and the numerical source term integration. The energy propagation numerical realization is based on the semi-Lagrangian numerical method. The original correction term is derived to remove the garden-sprinkler effect of the solution of energy balance equation for the case of wave propagation on spherical

surface. The method results proved to be much closer to the analytical solution than the finite difference schemes. The method is absolute stabile. Its higher accuracy is achieved by using larger propagation time steps (three hours and more). The main feature of the method is that it may be applicable to solve the energy balance equation with a fine space grid by using larger time step than it is allowed by the traditional finite-difference schemes.
The second part of the total scheme is based on a spreading numerical method. It uses the semi-analytical solution for integration source term, which includes the wind wave input, dissipation term, and exact non-linear energy transfer function. The method gives reliable and stable results for time steps up to three hours and more. The new technique yields

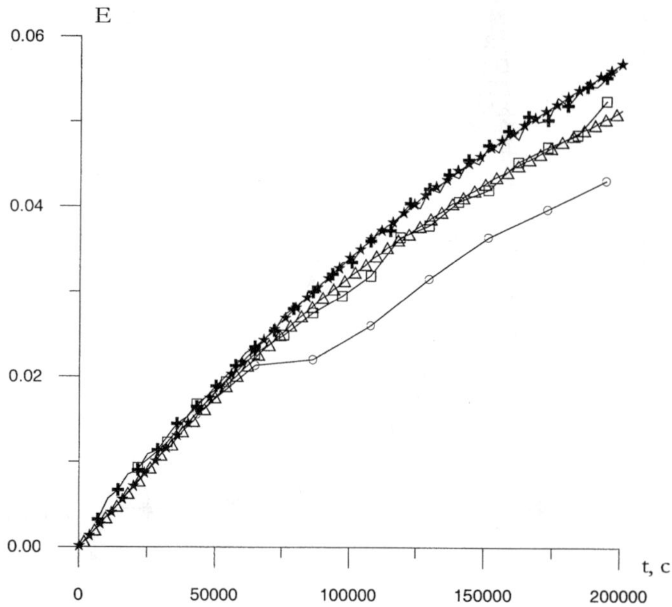

Fig. 4. Total energy evolution in time (designations are the same as to Fig.3)

accurate results and has the additional advantage of being numerically stable, thereby allowing to use a large time step (three hours and more).
The method can be regarded as a general alternative for the numerical realization of energy balance equation in the models. It gives opportunity to approach the time step of numerical integration of the energy balance equation to the time step of the wind speed information input in the numerical wind wave simulation to solve operational problem in the optimal way.

Acknowledgements. The study is supported by the Grants: RFBR 01-05-64846, INTAS-99-666, INTAS-01-025, INTAS-01-234, INTAS-01-2156.

References

1. Booij, N. and Holthuijsen, L.H.: Propagation of Ocean Waves in Discrete spectral wave model. Journal of Comput. Phys. 68, (1987), 307–326
2. Burgers, G.: A guide to the Nedwam wave model. Scient.Rep, WR-90-04, KNMI, De Bilt.
3. Komen G.J., Cavaleri, L., Donelan, M., Hasselmann, K., Hasselman, S., Janssen, P.A.E.M Dynamics and modeling of ocean waves, Cambridge University press, (1994), 532
4. Lavrenov, I.V.: Wind Waves in Ocean. Dynamics and numerical simulation. Springer, (2003), 375
5. Lavrenov, I.V., Onvlee, J.: New approach for reduction of the garden-sprinkler effect in third generation wave models, Scient.Rep. KNMI, De Bilt. (1993)
6. Tolman, H.L.: Effects of numeric on the physics in a third-generation wind-wave model, J.Phys.Ocean.22, (1992), 1095–1111
7. Zambresky, L.F.: A verification study of global WAM model December 1987 - November 1988, ECMWF Technical Report 63, ECMWF, Reading, (1989), 86

Dynamic Modelling of Environment-Industry Systems

Igor Kantardgi

Moscow State Civil Engineering University, Dept. of Water Management and Sea Ports, Novokosinskaya Street, 14-1-4, 111672, Moscow, Russia
kantardgi@rambler.ru, Igor_Kantardgi@stankin.ru

Abstract. The various kinds of modelling of environment-industry systems in environmental management are considered. The system dynamics models of interaction between industrial production and environment are developed. The basic model of the main factors and links is upgraded to the detailed models, which may be applied directly in the management process. The models are applied in the system of various levels, including model of Moscow city and machine-building enterprise. It's shown that the modelling provides the necessary information for the support the decision-making.

1 Introduction

The modern approach to supply the sustainable development of industry relates to the concept of the environmental management which is presented in the international standards serial ISO 14000 [1, as an example of standard from serial]. The practical application of the environmental management needs the information support. This information may be obtained from the mathematical modelling of various systems.

International standard ISO 14001 is focused specifically onto environmental management. The Life Cycle Assessment methodology [1] gives the inventory panel of the sources of environmental load. To reach the integrated indicator of the environmental load in the terms of environmental damage, the next stages of modelling have to follow the first inventory stage – Fig.1 [2]:

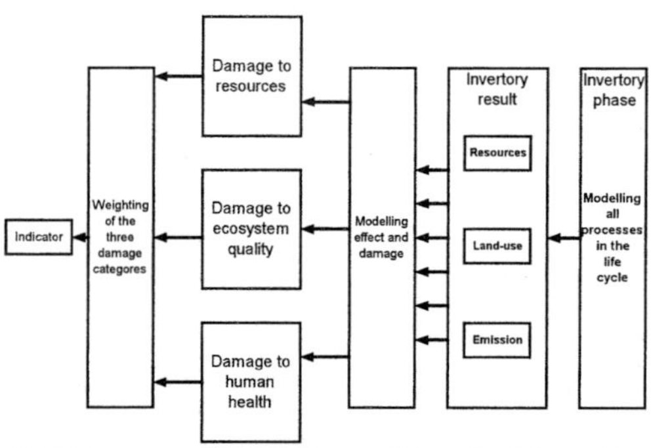

Fig. 1. Main steps of Eco-indicator 99 methodology (from [2])

P.M.A. Sloot et al. (Eds.): ICCS 2003, LNCS 2658, pp. 188–194, 2003.

- Fate analysis (links from emissions to concentrations);
- Effect analysis (links from concentrations to health effects, resources and ecosystems changes);
- Damage analysis (from effects to assessment of damages of human health, resources, and ecosystems);
- LC eco-indicator determination through damages normalization and weighting.

Fate, effects, and damage analysis is supported by specific kinds of modelling. But the management procedures need the special system models, what can provide the information for the factual management [3]. These models are developed in the report.

The system dynamics modelling software VENSIM PLE by Ventana Systems Inc. (academic version) [4] is applied for model computer realization.

2 Basic Model of Environment-Industry Interaction

The basic model (model 3) was designed to study the main factors and links between them. It's also used for development of the more realistic model versions.

The model is presented in Fig.2. It includes two parts: environmental damage and economy of enterprise. In this model the environmental damage is presented by the total pollution flow directed into environment. The enterprise pays the environmental taxes what are proportional to the difference between real pollutant flow and normative quota of pollution. The coefficient of proportionality is the environmental taxes rate.

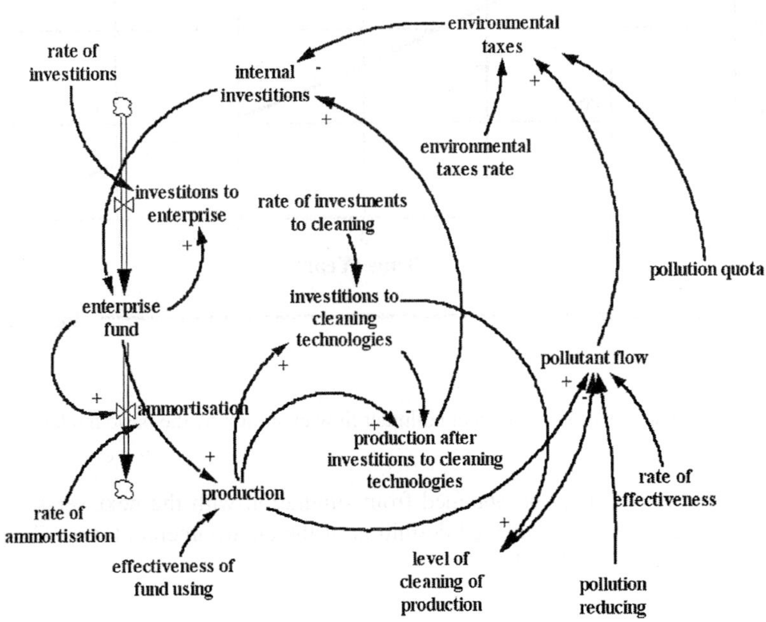

Fig. 2. The basic model simulating environment-industry main links

The charge of the benefit after paying the environmental taxes may be spent for the internal investments in the production. To decrease the pollution flow and environmental taxes the part of the benefit is directed to the cleaning technologies. So, the major feedback loop of the model is the following one:

Enterprise fund – production – investments in cleaning technologies – level of cleaning production – pollutant flow – environmental taxes – internal investments – enterprise fund .

This feedback loop is positive (Fig.2), and to control the system the following negative feedback loop is included:

Production – investments in cleaning technologies – production after investments in cleaning technologies – internal investments – enterprise fund – production .

The example of simulation with the basic model is shown in Fig.3. The time step of the model is one year, the model period is 6 years, and the simulated enterprise is successful in decreasing of the environmental pollution.

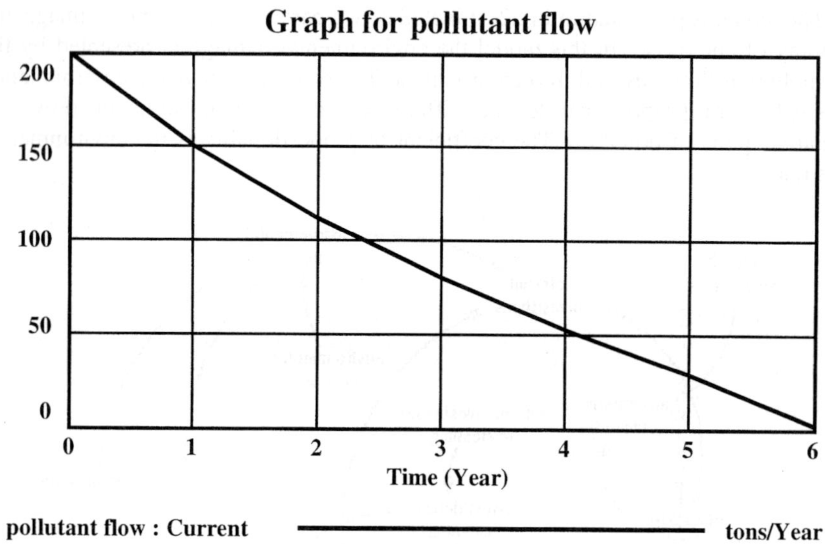

Fig. 3. Simulation of the total pollutant flow evolution by the basic model

The nontrivial results may be obtained from simulation with the next version of the basic model with more complicated definition of the environmental taxes. The model (model 4) is presented in Fig.4.

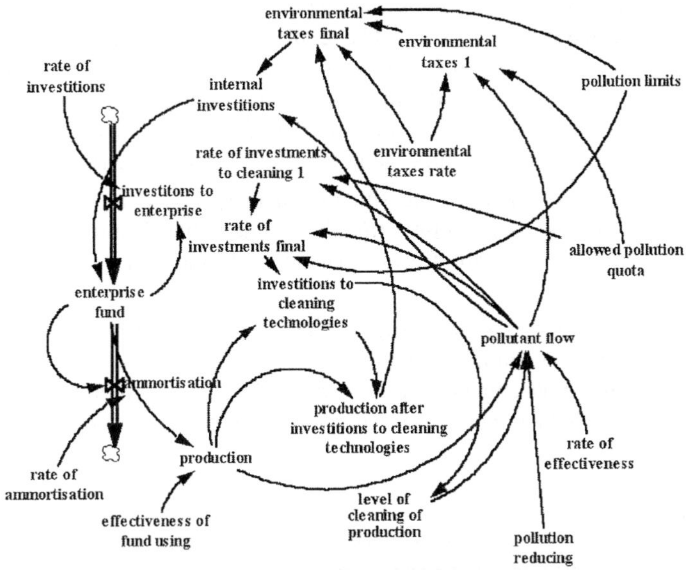

Fig. 4. Model 4 with two-steps limits of pollution

In model 4 the environmental taxes depend on the level of the factual pollutants flow in comparison with the allowed quota or normative limit. If the level of the factual pollutants flow exceeds the normative limit the environmental taxes rate five times more than if not.

Model also utilizes the more complicated strategy of enterprise: the investments in the new technologies increase depending on the results of decrease of pollutant flow in the previous year.

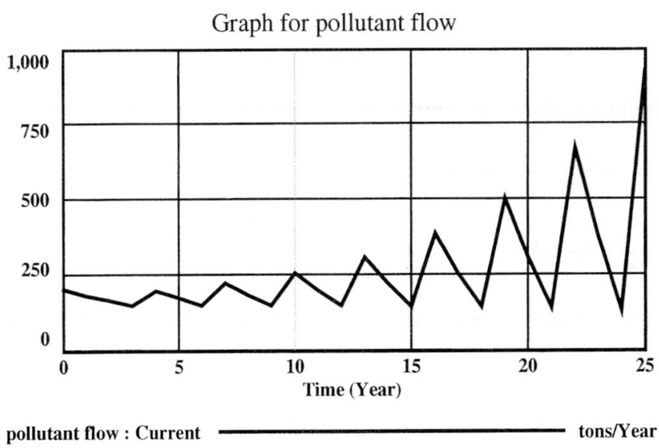

Fig. 5. Simulated oscillations of pollutant flow

As the simulation result the oscillations of the factors were obtained – fig. 5. The period of oscillations is 3 years. The explanation of the pollutant flow periodic variations is clear. If in the previous year the pollutant flow is lower than quota, the management of enterprise cut the investments in cleaning technologies in the current year what result in the increase of the pollutant flow in the next year. The amplitude of the oscillations increases with time in relation with the increase of production.

The approach to improve the situation may be found by simulation also. The problem is to find the minimal investments volume in the cleaning technologies in the years with allowed pollutant flows what supports the sustainable development of enterprise.

The found optimum is shown in Fig.6. The simulation has been carried out for the same boundary and initial conditions. The other factors like enterprise fund, production, and environmental taxes show the similar dynamic behavior. The oscillations are obtained in all cases when the initial pollutant flow is close to limit.

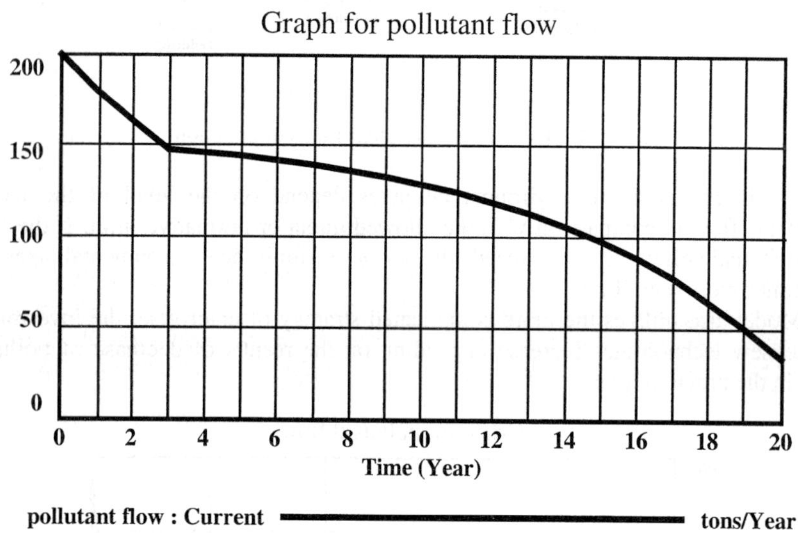

Fig. 6. Improved environmental management of enterprise (simulation result)

3 Working Model

The working model is designed for selected pollutant. The example is shown below for the dangerous Chromium VI compounds emitted into air. The causal loop diagram of the model (model 8) is shown in Fig.7.

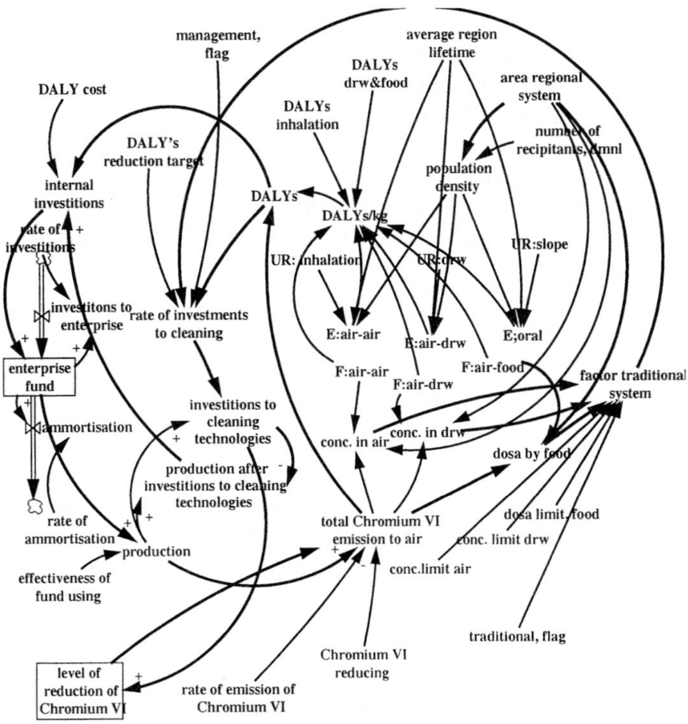

Fig. 7. Model of Chromium VI emission to air environmental damage

The environmental damage to human health is measured in DALYs (Disability Adjusted life Years) scale. The model provides two possibilities of control. The first approach is basing on the simulation the traditional control when factual concentration of the substance is compared with normative limits. The concentration of Ch VI oxides in air, drinking water (drw) and food may be applied. To define the concentrations in water and food the fate factors (F) are used [2].

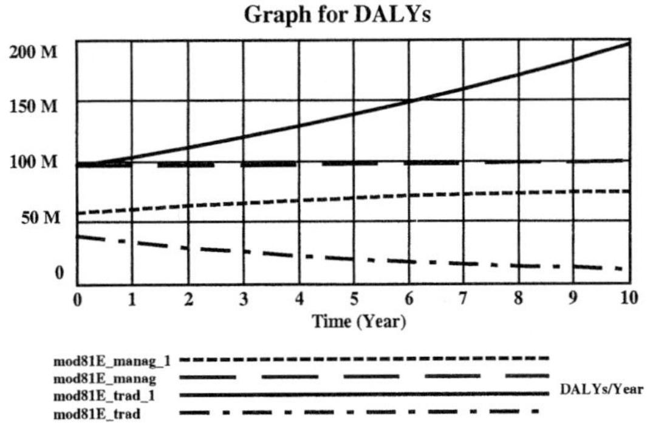

Fig. 8. Simulation of the environmental damage caused by emission of Ch VI to air of "close Moscow"

The second approach is the management basing on the system simulation with the main objective to reach the adopted decrease of the environmental damage. The human health damage is estimated using the unit-risk concept.

The results of simulation for the model of "close Moscow" (example) are shown in Fig.8. Only in one simulated scenario (mod_81E_trad) the environmental damage decreases effectively with time.

The model includes 75 variables and factors, and the optimal values of the variables with initial values and constant factors for the system management may be obtained by simulation.

4 Conclusions

The system dynamics modelling of the environment-industry systems can provide the information what is necessary for the management procedures. The modelling approach may be applied for the system of various dimensions in space and time. The simulation of scenarios supports the finding of the optimal decision.

References

1. International standard ISO 14040. Environmental Management – Life Cycle Assessment – Principles and framework. ISO (1997)
2. Goedkoop, M., Spriensma, R`: The Eco-indicator 99. A Damage Oriented Method for Life Cycle Impact Assessment. PRe Consultants B.V., Amersfoort, The Netherlands, Internet version (1999)
3. Kantardgi, I., Kaliagin, V., Kharif, Ch., Purvis, M. (eds.): Management of Natural Resources. Teaching manual, Nyzhny Novgorod State Technical University, Nyzhny Novgorod (2002)
4. Ford, A. Modelling the Environment. An Introduction to system Dynamics Modelling of Environmental Systems. Island Press (1999)

Simulation of Water Exchange in Enclosed Water Bodies

Erdal Özhan[1] and Lale Balas[2]

[1]Middle East Technical University, Civil Engineering Department, Coastal Engineering Laboratory, 06531 Ankara, Turkey
ozhan@metu.edu.tr
[2]Gazi University, Faculty of Engineering and Architecture, Civil Engineering Department, 06570 Ankara, Turkey
lalebal@gazi.edu.tr.

Abstract. A 0-D (box type) mathematical flushing model and a three-dimensional baroclinic numerical model have been presented that are used to simulate transport processes in coastal waters. The numerical model consists of hydrodynamic, transport and turbulence model components. In the hydrodynamic model component, the Navier-Stokes equations are solved with the Boussinesq approximation. The transport model component consists of the pollutant transport model and the water temperature and salinity transport models. In this component, the three dimensional convective diffusion equations are solved for each of the three quantities. In the turbulence model, a two-equation k-ε formulation is solved to calculate the kinetic energy of the turbulence and its rate of dissipation, which provides the variable vertical turbulent eddy viscosity. Horizontal eddy viscosity can be simulated by the Smagorinsky algebraic sub-grid scale turbulence model. The solution method is a composite finite difference-finite element method. In the horizontal plane finite difference approximations and in the vertical plane finite element shape functions are used. The governing equations are solved implicitly in the Cartesian coordinate system. The horizontal mesh sizes can be variable. To increase the vertical resolution, grid clustering can be applied. In the treatment of coastal land boundaries, the flooding and drying processes can be considered. The developed numerical model predictions are verified by the hydraulic model studies conducted on the forced flushing of marinas in enclosed seas.

1 Introduction

Tide, wind, fresh water inflow, gravitation, macro instabilities and waves are the six basic mechanisms which produce motion and mixing in water bodies. Along Turkish shores the tidal ranges are typically from 0.1 m. to 0.3 m., which are quite low values for sufficient levels of water exchanges. Therefore, lower frequency motions due to wind or fresh water inflow are dominant throughout the Turkish shores. Gravitational movement superposes over the effects of wind and the fresh water inflow.

In recent years, there have been a lot of studies performed on 0-Dimensional (box type water exchange, 2-Dimensional (depth averaged) and 3-Dimensional (full picture) transport models. 0-D numerical models do not pay attention to the internal hydraulics within the water body [1],[2]. The most simplifying assumption of the model

P.M.A. Sloot et al. (Eds.): ICCS 2003, LNCS 2658, pp. 195–204, 2003.

is the complete and uniform mixing of the pollutant at any time with water present in the water body. 2-D numerical models, where the flow is depth averaged, can be successfully used to simulate circulation, when density differences do not play a significant role. Tidal circulation in coastal areas is a typical example. Such models are also used for designing optimal layout of marinas to assure sufficient flushing. The use of 3-D models is unavoidable in all cases where the influence of density distribution, or vertical velocity variations can not be neglected. Inflow of lower salinity river plumes into coastal sea is a typical example, but often also in the simulation of tidal flows the influence of density differences can not be neglected. In cases of wind driven flows in smaller areas near the shore, the depth-averaged 2-D models can not simulate the 3-D character of the flow. With wind-driven flows the density stratification causes an additional effect: an important diminishing of the shear stress between the horizontal layers, the consequence being that the surface layer circulation can differ considerably from the depth averaged circulation. Strong stratification in water bodies usually also demands the use of 3-D models [3]. In this paper a 0-D flushing model proposed for enclosed water bodies and a 3-D fully implicit transport model have been presented.

2 Zero Dimensional, Box Type Flushing Model

A non-conservative substance with a first order decay reaction is considered. Its concentration is C_o at the start of the computations. The intrusion of the pollutant into the coastal water body continues at a constant rate P. As the model is zero dimensional, the pollutant concentration is assumed not to vary spatially. This requires complete mixing inside the water body at all times. The model equation, stating the conservation of the pollutant mass in the enclosed water body is written as [1],[2],[4]:

$$\frac{dC}{dt}=-(k+\frac{Q}{V})C+\frac{P}{V} \tag{1}$$

in which, C: the instantaneous pollutant concentration; k: decay coefficient; Q: entering discharge; V: water volume inside the marina.

When the flushing is due to the tidal discharge alone, the bulk conservation equation becomes:

$$\frac{dC}{dt}=-kC+\frac{P}{V}\text{ for } nT\le t\le(n+\frac{1}{2})T \text{ (ebb tide)} \tag{2}$$

$$\frac{dC}{dt}=-(k+\frac{Q}{V})C+\frac{P}{V}\text{ for } (n+\frac{1}{2})T\le t\le(n+1)T \text{ (flood tide)} \tag{3}$$

in which, T: tidal period, n: a positive integer (“zero” included) and t=0, T, 2T,…, are the times of mean high tide level (i.e. the onset of the ebb tide). The timely variations of the tidal discharge Q and the water volume V are used as:

$$Q=-\frac{1}{2}A_S R\, w \sin(wt) \tag{4}$$

$$V=V_S+\frac{1}{2}A_S R\cos(wt) \tag{5}$$

where, A_s: Surface area at mean sea level; V_s: Mean tide level volume; R: Tidal range (from mean low level to mean high level) and w: $2\pi/T$.

The solution giving the pollutant concentration at the time of high tide after the n- th tidal cycle is:

$$C_n=a^n C_o+\frac{b(1-a^n)}{1-a}C_a \tag{6}$$

in which,

$$a=\frac{M-1}{M+1}e^{-kT} \tag{7}$$

$$b=\frac{2}{kT}\left[\frac{M}{M+1}(1-e^{-k\frac{T}{2}})-a(1-\frac{k^2}{M(k^2+w^2)})+ae^{k\frac{T}{2}}(1+\frac{k^2}{M(k^2+w^2)})\right] \tag{8}$$

$$M=\frac{2V_s}{A_sR}=2\frac{\bar{h}}{R};C_a=\frac{PT}{2V_s} \tag{9}$$

where, h: mean water depth, M: flushing parameter.

If the substance is conservative then k=0 and equation for C_n still holds if the parameters a and b are redefined as:

$$a=\frac{M-1}{M+1}\ ;\ b=\frac{M(1+\frac{M-1}{M+1})}{M+1} \tag{10}$$

The zero dimensional model presented above is applied to Ölüdeniz Lagoon. Ölüdeniz Lagoon is an example of choked lagoons [8]. The lagoon is connected to the sea by a single narrow and shallow channel. This entrance channel serves as a dynamic filter and tidal currents are damped out. In the Mediterranean Sea tidal ranges are typically in the order of 0.2 to 0.3 meters. Despite the smallness of the tidal amplitude, a considerable level of flushing is induced by the tidal motion [6].

Flushing of Ölüdeniz lagoon is controlled primarily by the two processes: firstly the tidal action and secondly the wind driven currents. Density driven currents are rather weak compared to the tide and wind induced currents [7]. Tidal motion at the site of Ölüdeniz Lagoon is typically semi-diurnal. The mean tidal range is about 0.15m and the tidal period is 12 hours and 25 minutes. Lagoon is rather deep with a mean water depth of 17 m. In the middle of the lagoon, the water depth reaches about 40m.

The model provides a quick assessment of the degree of tidal flushing to be expected. The mean depth of the lagoon is 17 m. and the surface area is about 406000 m^2. The tidal motion at the site of Ölüdeniz Lagoon is semi-diurnal type. The mean tidal range is about 0.15 m., which is typical around the Turkish Mediterranean coastline. The flushing parameter has a value of 227. The C_n/C_o ratios are computed for a conservative pollutant and results are plotted in Figure 1. This figure depicts the situation that if no pollutant is introduced into the lagoon (i.e. for $C_a/C_o=0$), it takes 261days for the pollutant amount in the lagoon to be flushed out by the tidal currents alone up to level of 99 %, i.e. to have the value of $C_n/C_o=1\%$. If the pollutant continues *entering the lagoon waters, the flushing period of the lagoon gets much longer. Furthermore, it is computed that $C_n = C_o$ for all times if the pollutant addition rate is such that $C_a/C_o=0.0044$. As it is observed from Figure 1, the cleansing of the lagoon can not be realized at all if $C_a/C_o=0$.*

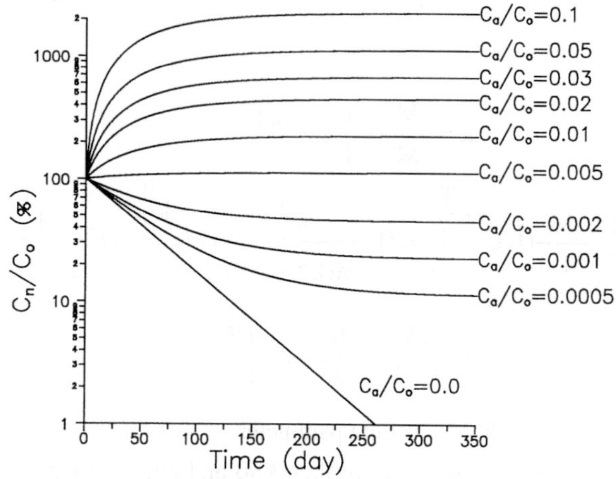

Fig. 1. The changes of pollutant concentration with time for various rates of pollutant addition.

3 Three Dimensional Numerical Model

An unsteady three-dimensional baroclinic circulation model (HIDROTAM3) has been developed to simulate the transport processes in coastal water bodies [3]. The model predictions were verified by using several experimental and analytical results published in the literature and its successful use for a variety of real life cases was demonstrated [3][8][10].

The governing hydrodynamic equations in the three dimensional Cartesian coordinate system are as follows [3]:

$$\frac{\partial u}{\partial x}+\frac{\partial v}{\partial y}+\frac{\partial w}{\partial z}=0 \tag{11}$$

The momentum equations in the orthogonal horizontal directions x and y:

$$\frac{\partial u}{\partial t}+u\frac{\partial u}{\partial x}+v\frac{\partial u}{\partial y}+w\frac{\partial u}{\partial z}=fv-\frac{1}{\rho_o}\frac{\partial p}{\partial x}+2\frac{\partial}{\partial x}(\nu_x\frac{\partial u}{\partial x})+\frac{\partial}{\partial y}(\nu_y(\frac{\partial u}{\partial y}+\frac{\partial v}{\partial x}))+\frac{\partial}{\partial z}(\nu_z\frac{\partial u}{\partial z}) \tag{12}$$

$$\frac{\partial v}{\partial t}+u\frac{\partial v}{\partial x}+v\frac{\partial v}{\partial y}+w\frac{\partial v}{\partial z}=-fu-\frac{1}{\rho_o}\frac{\partial p}{\partial y}+2\frac{\partial}{\partial y}(\nu_y\frac{\partial v}{\partial y})+\frac{\partial}{\partial x}(\nu_x(\frac{\partial v}{\partial x}+\frac{\partial u}{\partial y}))+\frac{\partial}{\partial z}(\nu_z\frac{\partial v}{\partial z}) \tag{13}$$

The momentum equation in the vertical direction z:

$$\frac{\partial w}{\partial t}+u\frac{\partial w}{\partial x}+v\frac{\partial w}{\partial y}+w\frac{\partial w}{\partial z}=-\frac{1}{\rho_o}\frac{\partial p}{\partial z}-g+\frac{\partial}{\partial y}(\nu_y(\frac{\partial w}{\partial y}+\frac{\partial v}{\partial z}))+\frac{\partial}{\partial x}(\nu_x(\frac{\partial w}{\partial x}+\frac{\partial u}{\partial z}))+\frac{\partial}{\partial z}(\nu_z\frac{\partial w}{\partial z}) \tag{14}$$

where, x,y: Horizontal coordinates; z: Vertical coordinate; t: Time; u,v,w: Velocity components in x,y,z directions at any grid locations in space; v_x, v_y, v_z : Eddy viscosity coefficients in x,y and z directions respectively; f: Corriolis coefficient; ρ(x,y,z,t): In situ water density; ρ_o: Reference density; g: Gravitational acceleration; p: Pressure.

The temperature and salinity variations are calculated by solving the three dimensional convection-diffusion equation which is written as:

$$\frac{\partial Q}{\partial t}+u\frac{\partial Q}{\partial x}+v\frac{\partial Q}{\partial y}+w\frac{\partial Q}{\partial z}=\frac{\partial}{\partial x}(D_x\frac{\partial Q}{\partial x})+\frac{\partial}{\partial x}(D_y\frac{\partial Q}{\partial y})+\frac{\partial}{\partial z}(D_z\frac{\partial Q}{\partial z}) \tag{15}$$

where, D_x, D_y and D_z: Turbulent diffusion coefficient in x,y and z directions respectively; Q: Temperature (T) or salinity (S). The conservation equation for a pollutant constituent is:

$$\frac{\partial C}{\partial t}+u\frac{\partial C}{\partial x}+v\frac{\partial C}{\partial y}+w\frac{\partial C}{\partial z}=\frac{\partial}{\partial x}(D_x\frac{\partial C}{\partial x})+\frac{\partial}{\partial x}(D_y\frac{\partial C}{\partial y})+\frac{\partial}{\partial z}(D_z\frac{\partial C}{\partial z})-k_pC \tag{16}$$

where, C: Pollutant concentration; k_p: Decay rate of the pollutant; D_x, D_y and D_z: Turbulent diffusion coefficient in x,y and z directions respectively.

The two-equation k-ε turbulence model is used for turbulence modeling. The model equations for the kinetic energy and dissipation of the kinetic energy are:

$$\frac{\partial k}{\partial t}+u\frac{\partial k}{\partial x}+v\frac{\partial k}{\partial y}+w\frac{\partial k}{\partial z}=\frac{\partial}{\partial z}(\frac{v_z}{\sigma_k}\frac{\partial k}{\partial z})+P+B-\varepsilon+F_k \tag{17}$$

$$\frac{\partial \varepsilon}{\partial t}+u\frac{\partial \varepsilon}{\partial x}+v\frac{\partial \varepsilon}{\partial y}+w\frac{\partial \varepsilon}{\partial z}=\frac{\partial}{\partial z}(\frac{v_z}{\sigma_\varepsilon}\frac{\partial \varepsilon}{\partial z})+c_{1\varepsilon}\frac{\varepsilon}{k}(P+c_{3\varepsilon}B)-c_{2\varepsilon}\frac{\varepsilon^2}{k}+F_\varepsilon \tag{18}$$

where, k: Kinetic energy; ε: Rate of dissipation of kinetic energy; F_k: Horizontal diffusion terms for the kinetic energy; F_ε: Horizontal diffusion terms for the dissipation of kinetic energy; P: Stress production of the kinetic energy; B: Buoyancy production of the kinetic energy.The stress production of the kinetic energy is defined by:

$$P=v_h\left[2\left(\frac{\partial u}{\partial x}\right)^2+2\left(\frac{\partial v}{\partial y}\right)^2+\left(\frac{\partial u}{\partial y}+\frac{\partial v}{\partial x}\right)^2\right]+v_z\left[\left(\frac{\partial u}{\partial z}\right)^2+\left(\frac{\partial v}{\partial z}\right)^2\right] \tag{19}$$

where, v_h is the horizontal eddy viscosity and u,v are the horizontal water particle velocities in x and y directions respectively.

The vertical eddy viscosity is calculated by:

$$v_z=C_\mu\frac{k^2}{\varepsilon} \tag{20}$$

To account for large-scale turbulence generated by the horizontal shear, horizontal eddy viscosity can be simulated by the Smagorinsky algebraic sub-grid scale turbulence model.

4 Model Application to a Laboratory Flume

The predictions of transport model component are compared with the results of a laboratory model harbor study using dye tracer [9]. The model harbor was set to be a square, with prototype dimensions of 432mx432m. The tidal range was 0.1 m and tidal period was 708 s. Dye was injected into the enclosed basin to provide an initial uniform concentration across the basin. Dye experimental results and numerical model predictions at low tide for the first cycle and for the second cycle are presented in Figure (2) and in Figure (3) respectively.

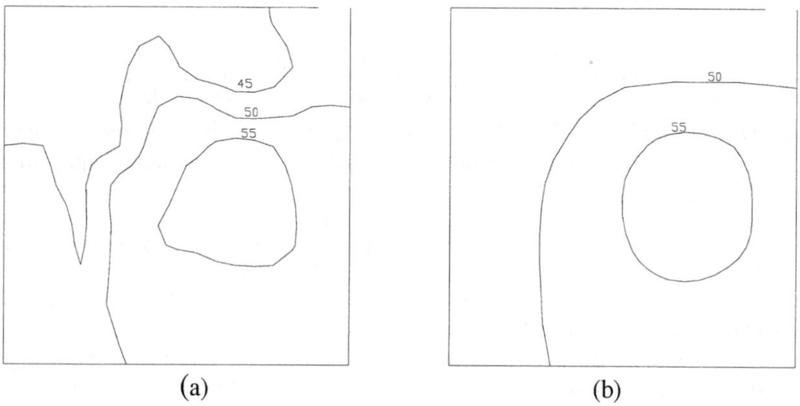

Fig. 2. a) Dye experimental results at low tide for the first tidal cycle, b) Numerical simulations at low tide for the first tidal cycle (% of initial concentration contours).

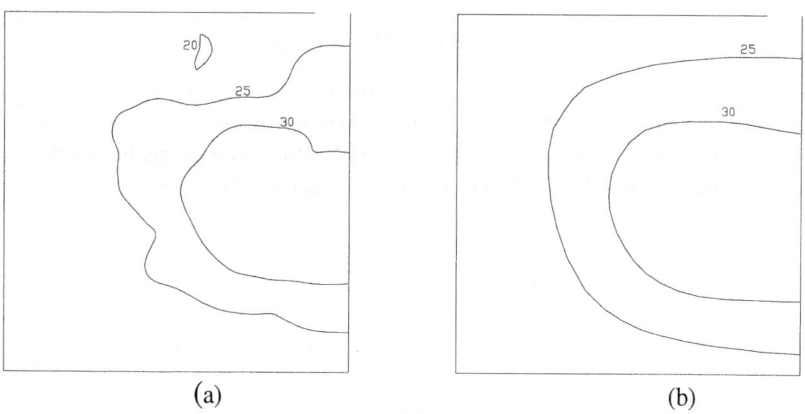

Fig. 3.. a) Dye experimental results at low tide for the second tidal cycle, b) Numerical simulations at low tide for the second tidal cycle (% of initial concentration contours).

Centroidal distribution is well predicted by the model. However, close to the harbor entrance, the predicted concentrations are about 17% higher than the measurements. The main reason for this may be the estimation of rather weak flow and small eddies close to the open boundary. Overall the numerical predictions are in good agreement with the experimental results.

5 Forced Flushing of Marinas in Enclosed Seas

Construction of a marina disturbs the natural flow patterns and normally deteriorates the water quality in and around the project site [3]. Water enclosed in a marina basin is normally tranquil and has a restricted contact with the outside sea. Water exchange takes place only through the entrance. The cross sectional area of the entrance is often small and the exchange is low especially in areas where the tidal range is small. Water, which enters into the basin, can not freely circulate. Limited water circulation may result in poor water quality levels in the marina. It is often necessary to apply some special design features to enhance flushing of marinas. The use of a "Morning-Glory Spillway" like structure to pump out water from the marina for improving the flushing performance is investigated. A physical model study was performed in the Coastal Engineering Laboratory of the Middle East Technical University. The measured velocities in the physical model at the grid points neighboring the intake are used as the boundary conditions in the mathematical model. The length scale of the model was 1/50. The length and width of the rectangular model marina basin were 5.80 m. and 2.8 m., respectively. The average water depth was 0.2 m. Using the length scale of 1/50, these dimensions correspond to a prototype marina of 290 m by 140 m., with a water depth of 10 m. Surface water was withdrawn from the marina by installing a structure similar to the morning-glory spillway. The water taken from the basin was pumped into the open sea. The discharge point was selected so that, the motion resulting from the pumped water does not affect the circulation in the basin. Velocity measurements were taken with a 'Minilab SD-12' microscale 3-axis ultrasonic current meter. The morning glory shaped intake structure was placed at various locations in the basin. Two of them, which are at the corners, are presented in this paper. The choice of the corner locations is due to the concern for affecting the maneuvering space of the crafts inside the marina. The grid system used has a square mesh size of 10x10 m. The vertical eddy viscosity is calculated by the k-ε model and horizontal eddy viscosities are predicted by the sub-grid scale turbulence model. At t=0, the pump is started, so that the water begins to flow in the intake, whereas the remaining part of the water body is assumed to be at rest and the water surface is assumed to be horizontal. The land boundaries are taken as fixed boundaries. Velocities measured in the physical model at the grid points neighboring the intake are used as the boundary conditions of the intake location in the mathematical model. Steady state conditions are reached approximately 1.5 hours after the start of pumping. The paths followed by the floats in the physical model are compared with the results obtained from the mathematical model in Figures (4-5). The velocity distributions at the surface layer as obtained from the mathematical model are also shown. The average velocities along the paths followed by the floats in both physical and mathematical models are compared in Table (1). In Case I, the intake structure is placed on the right end corner of the marina (135 m. x 285 m.). The best agreement is obtained for the path followed by the float number 2 released at location (100 m. x 0 m.) with an average velocity relative error of 0.55%. The highest relative error in the average velocity is calculated as 24.4% for the float number 7 which released at location (30 m. x 10 m.). For this path, the velocities predicted by the numerical model are less than the velocities observed in the physical model.

Table 1. Average velocities along the float paths.

Float number	Release location (x,y) in m.	Physical model average velocity (cm/s)		Mathematical model average velocity (cm/s)	
		Case I	Case II	Case I	Case II
1	90 x 0	**	4.04	**	4.15
2	100 x 0	3.64	5.27	3.66	4.51
3	110 x 0	3.61	5.16	3.88	5.04
4	120 x 0	3.72	3.88	3.91	4.01
5	130 x 0	**	**	**	**
6	70 x 10	2.24	**	1.98	**
7	30 x 10	0.93	**	0.70	**
8	130 x 240	**	2.72	**	3.51

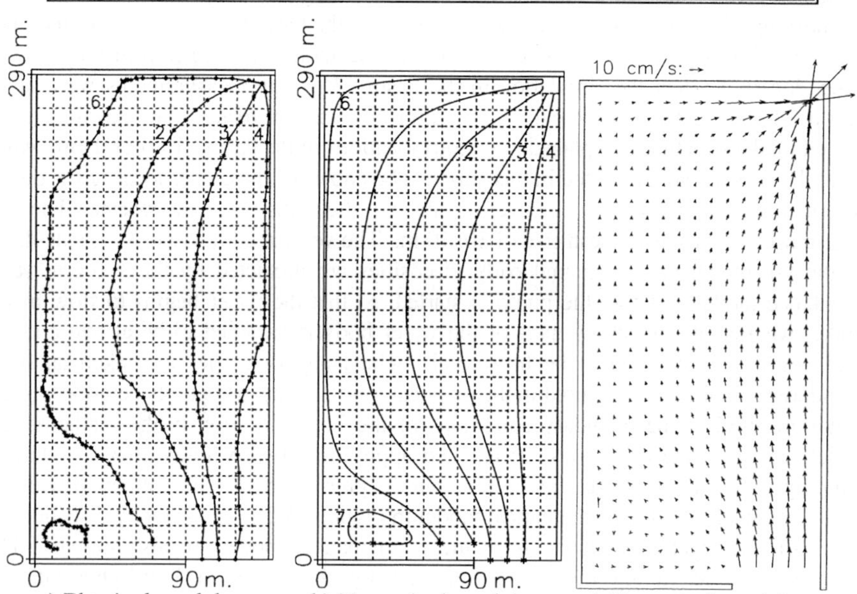

a) Physical model b) Numerical model c) Numerical model

Fig. 4. Float paths and velocity distributions at the surface layer, when the morning glory is placed on the right end corner of the marina.

There is a dead zone where the released float does not move through the intake, since there occurs a gyre. Therefore, the float number 7 follows a counter clockwise elliptical route. The location of dead zone is almost the same for physical and numerical models.

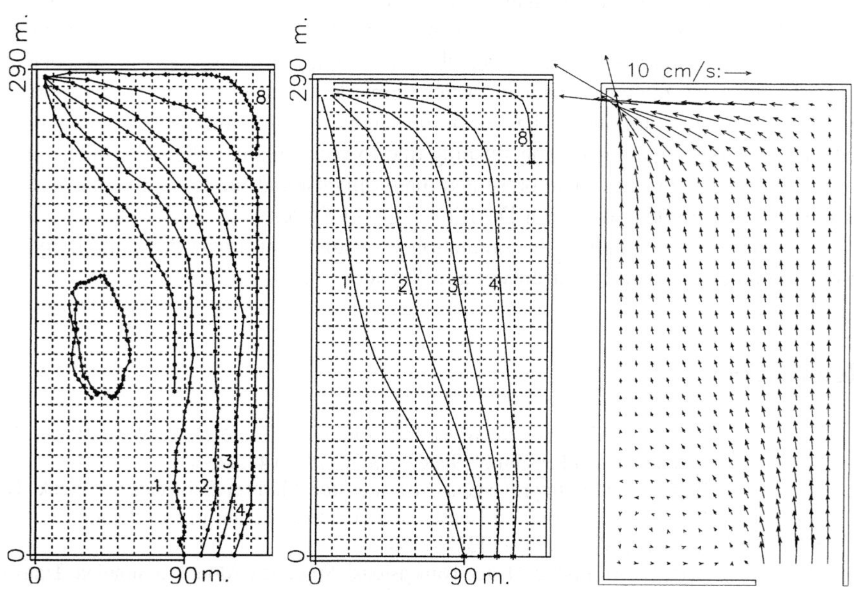

a) Physical model b) Numerical model c) Numerical model

Fig. 5. Float paths and velocity distributions at the surface layer, when the morning glory is placed on the left end corner of the marina.

In Case II, the intake structure is placed on the left end corner of the marina (5 m. x 285 m.). The closest agreement in the path trajectory is obtained for the float number 8, which is released at location (130 m. x 240 m.). However, the relative error in its average velocity is the highest, 29 %. In the physical model, a dead zone is observed in the left-hand side of the basin towards the middle of the y-axis. On the other hand, the dead zone predicted by the numerical model is on the left entrance corner of the basin and it occupies a rather restricted area. The pumping operation should be carried out during flood tide only to derive the greatest efficiency from such a system. The best location of the intake for removal of water and the optimum discharge rate are two crucial issues for designing such a scheme. The first question becomes especially important in the case of complex basins where the problems of "dead regions" are normally more significant. Such features have traditionally been investigated in physical models.

6 Conclusions

By assuming complete and uniform mixing of the pollutant at any time with the water present in the lagoon, zero dimensional model is applied to the Ölüdeniz Lagoon. This assumption is a crude assumption for the Ölüdeniz lagoon, which is rather deep.

However the zero dimensional mathematical model provides a quick assessment of the degree of tidal flushing to be expected. It shows that flushing due to tidal motion alone is not sufficient for maintaining an acceptable water quality in the lagoon. Secondly, a three dimensional numerical model which consists of hydrodynamic, transport and turbulence model components has been presented. Three-dimensional modeling is necessary for investigating lagoon flushing due to mechanisms other than tide, such as wind and density currents. The presented three dimensional model is a valuable design tool and can be implemented in a Decision Support System. It may be used in diverse coastal applications including the prediction of natural flushing rates caused by tidal motion, wind effect, density currents or a fresh water inflow.

References

1. Özhan, E.: Flushing of Marinas with Weak Tidal Motion. In: W.R. Blain, W.R., Weber, N.B. (eds): Marinas: Planing and Feasibility. Southampton, (1989) 485–498.
2. Özhan, E.: Water Quality Improvement Measures for Marinas Subjected to Weak Tidal Motion. Proc. Second Int. Conf. on Coastal and Port Engineering in Developing Countries, COPEDEC 3, Kenya, Vol.2, (1991) 1337–1350.
3. Balas, L., Özhan, E.: An Implicit Three Dimensional Numerical Model to Simulate Transport Processes in Coastal Water Bodies, Int. Journal for Numerical Methods in Fluids, Vol. 34 (2000) 307–339.
4. Balas, L., Özhan, E.: Flushing of Ölüdeniz Lagoon, Proc. of the Joint Conference MEDCOAST'99-EMECS'99, Antalya, Turkey, Vol.3 (1999) 1873–1884.
5. Balas, L., Özhan, E.: Three Dimensional Modelling of Transport Processes in Göksu Lagoon System, Proc.of the Second International Conference on the Mediterranean Coastal Environment, MEDCOAST'95, Tarragona, Spain, Vol. 3 (1995) 1661–1672.
6. Balas, L., Özhan, E.: Three Dimensional Modelling of Transport Processes in Stratified Coastal Waters, Proc.of International Conference on Hydroinformatics, IAHR, Copenhagen, Denmark, (1998) 97–104.
7. Balas, L., Özhan, E., Öztürk C.: Three Dimensional Modelling of Hydrodynamic and Transport Processes in Ölüdeniz Lagoon. Proc. of the Third International Conference on the Mediterranean Coastal Environment, MEDCOAST'97, Qawra, Malta, Vol. 2, (1997) 1097–1109.
8. Balas, L., Özhan E.: Applications of a 3-D Numerical Model to Circulations in Coastal Waters, Coastal Engineering Journal, Vol. 43, No 2 (2001) 99–120.
9. Balas, L.: Simulation of Pollutant Transport in Marmaris Bay, China Ocean Engineering., Vol.15, No.4 (2001) 565–578.
10. Balas, L., Özhan, E.: Three Dimensional Modelling of Stratified Coastal Waters. Estuarine, Coastal and Shelf Science, Vol..54 (2002) 75–87.

A Baroclinic Three Dimensional Numerical Model Applied to Coastal Lagoons

Lale Balas[1] and Erdal Özhan[2]

[1] Gazi University, Faculty of Engineering and Architecture, Civil Engineering Department, 06570 Ankara, Turkey,
lalebal@gazi.edu.tr.
[2]Middle East Technical University, Civil Engineering Department, Coastal Engineering Laboratory, 06531 Ankara, Turkey
ozhan@metu.edu.tr

Abstract. An implicit baroclinic unsteady three-dimensional model (HIDROTAM3) which consists of hydrodynamic, transport and turbulence model components, has been implemented to two real coastal water bodies namely, Ölüdeniz Lagoon located at the Mediterranean coast and Bodrum Bay located at the Aegean Sea coast of Turkey. M2 tide is the dominant tidal constituent for the coastal areas. The flow patterns in the coastal areas are mainly driven by the wind force. Model predictions are highly encouraging and provide favorable results.

1 Introduction

Coastal water bodies are generally stratified which may cause significant vertical and lateral density gradients. In most of the real coastal water body applications, it is necessary to predict the vertical structure of the flow due to these density gradients. Similarly, the wind induced currents have a strong 3-D character, which require a comprehensive 3-D numerical model. A three dimensional numerical model has been developed to simulate the transport processes in coastal waters by the authors [1]. In coastal water bodies such as estuaries, lagoons or bays, the horizontal length scale is much larger than the vertical scale. The flow is predominantly horizontal and the vertical acceleration is hardly compared with the gravity acceleration. The tidal range of the Turkish Mediterranean Sea coast is small, the spring ranges being typically 20-30 cm., hence the dominant forcing for circulation and water exchange is due to the wind action. In the literature, it is possible to find a large number of one, two and three dimensional numerical models which have been used for the estimation of circulation patterns and water quality in coastal water bodies [2], [3].

2 Application to Ölüdeniz Lagoon

With the developed three dimensional numerical model, the tidal circulation patterns are investigated in Ölüdeniz Lagoon by using an input tidal wave with an amplitude of a=0.15 m. and a period of T=12 hours 25 min [4]. Ölüdeniz Lagoon is located at the

P.M.A. Sloot et al. (Eds.): ICCS 2003, LNCS 2658, pp. 205–212, 2003.

Mediterranean coast of Turkey (Figure 1). Turkish coastline along the western Mediterranean is increasingly being popular. Öludeniz Lagoon which is strategically located along the sailing route, is busy almost all over the year. The town of Fethiye located inland of the lagoon is one of the most developed coastal resorts along the Turkish coastline. In order to prevent the excessive pollution, it is forbidden to enter the lagoon waters by any motorboats. Simulation is started from the state of no motion at the low tide when the water level is minimum everywhere in the lagoon, and the tidal currents are zero. Then water level along the open sea boundary increases with respect to a sinusoidal wave. Flow velocities are computed at 6 levels along the water depth. The density of the water is assumed to be constant. The shoreline boundary is assumed to be fixed. The horizontal square mesh size is constant, $\Delta x = \Delta y = 20$ m. To simulate the pollutant transport an inner mesh is fitted to the area as shown in Figure 2.

The propagation of the tidal wave is mainly affected by the geometric and topographic conditions. The mean water depth at the entrance channel is around 3.5 m. and the width is around 60 m. Due to this narrow and shallow entrance, tidal currents are damped out. Therefore the tidal velocities inside the lagoon are hardly compared with the velocities around the entrance channel. The period of the tidal wave, semi diurnal type, is so long that, the wave length is much larger than the length of the coastal water body considered. Therefore the associated displacement of the waters is essentially horizontal, resulting in nearly horizontal flows. The flow directions at all layers are nearly the same.

Using the estimated tidal current patterns, the transport of a pollutant is also simulated. For the simulation of the transport of pollutant discharged at the point shown on Figure 2, the coliform count is used as the tracer. The initial concentration is taken as 10^6 bac/ml at the discharge point and is assumed to be 1 bac/ml in the rest of the lagoon waters. The discharge is assumed to take place steadily into the surface layer and to start at the same time as the tidal action. It is assumed that the value of T_{90} is 2 hours. Vertical eddy viscosity is calculated by the k-ε turbulence model. Since the horizontal motion has an intensity, the Smagorinsky algebraic sub-grid scale turbulence model is used to estimate the horizontal eddy viscosities. The distribution of pollutant concentrations at the surface (from inner to outer contour in bac/ml is: 900000, 300000, 100000, 50000, 25000, 10000, 5000, 2500, 1000, 500, 250, 10) and at the bottom (from inner to outer contour in bac/ml is: 10000, 5000, 2500, 1000, 500, 250, 10) layers are given hourly in Figure 3. It is seen that the progress of pollutant is stabilized almost at 6 hours, and it covers the computational grid shown in the Fig. 2. The pollutant concentration level at the bottom layers is about 10 times less than the level at the surface layers.

3 Application to Bodrum Bay

Model has been applied to Bodrum Bay to simulate the circulation pattern and the transport of pollutant. Bodrum Bay is located at the Aegean Sea coast of Turkey (Figure 1) where the dominant forcing for the water exchange is due to the wind action. The town of Bodrum located inland of the Bay is one of the most developed coastal resorts along the Agean Sea coast of Turkey. In Bodrum Bay, there are two sea outfalls discharging directly into the Bay. The sea outfall at Kızılburun has a length of 1500 m.

and the one at İnceburun has a length of 900 m. They discharge nearly at water depths of 40 m.

The sea outfalls are shown in Figure 4 with a dotted line. The grid system used has a square mesh size of 350x350 m. Near the sea outfall a more refined mesh with a grid size of 50 m x 50 m. is fitted. The water depth is divided into 6 layers of equal thickness. In the simulations of pollutant transport the coliform count is used as the tracer. The rate of disappearance of pathogenic bacteria and viruses due to die-off approximately follows first order kinetics. The die of constant k_p, is computed in terms of T_{90}, the time required for 90 percent of the initial bacteria to die, and equals $2.3/T_{90}$.

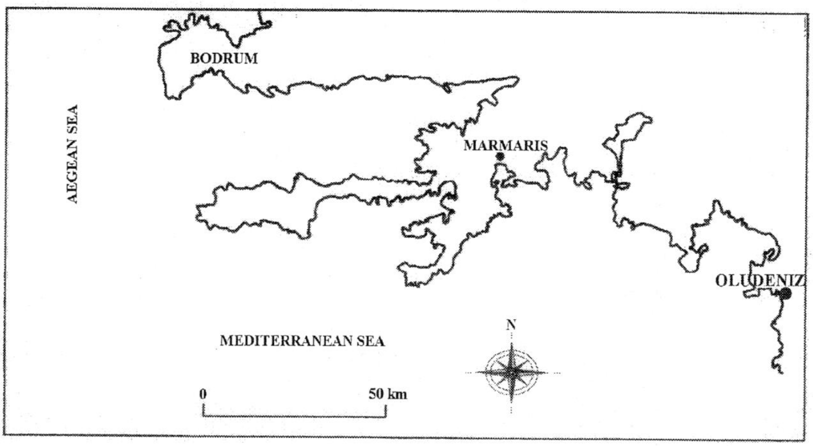

Fig. 1. Location of Bodrum Bay and Ölüdeniz Lagoon.

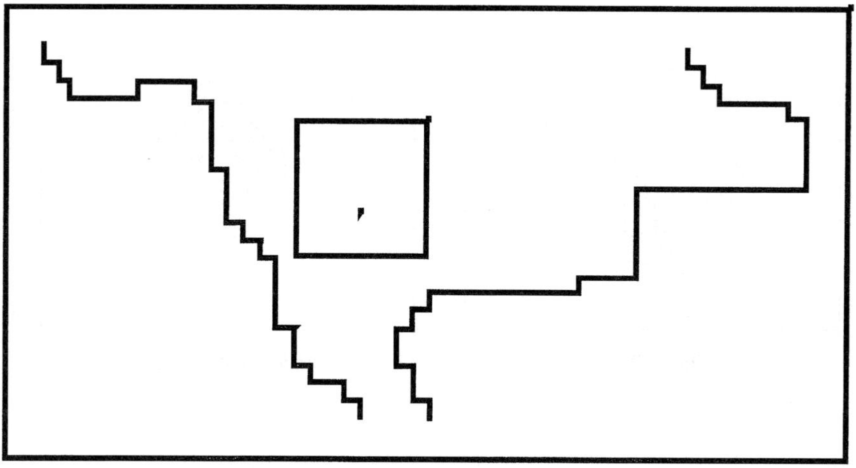

Fig. 2. Pollutant discharge location and the computational grid of Ölüdeniz Lagoon.

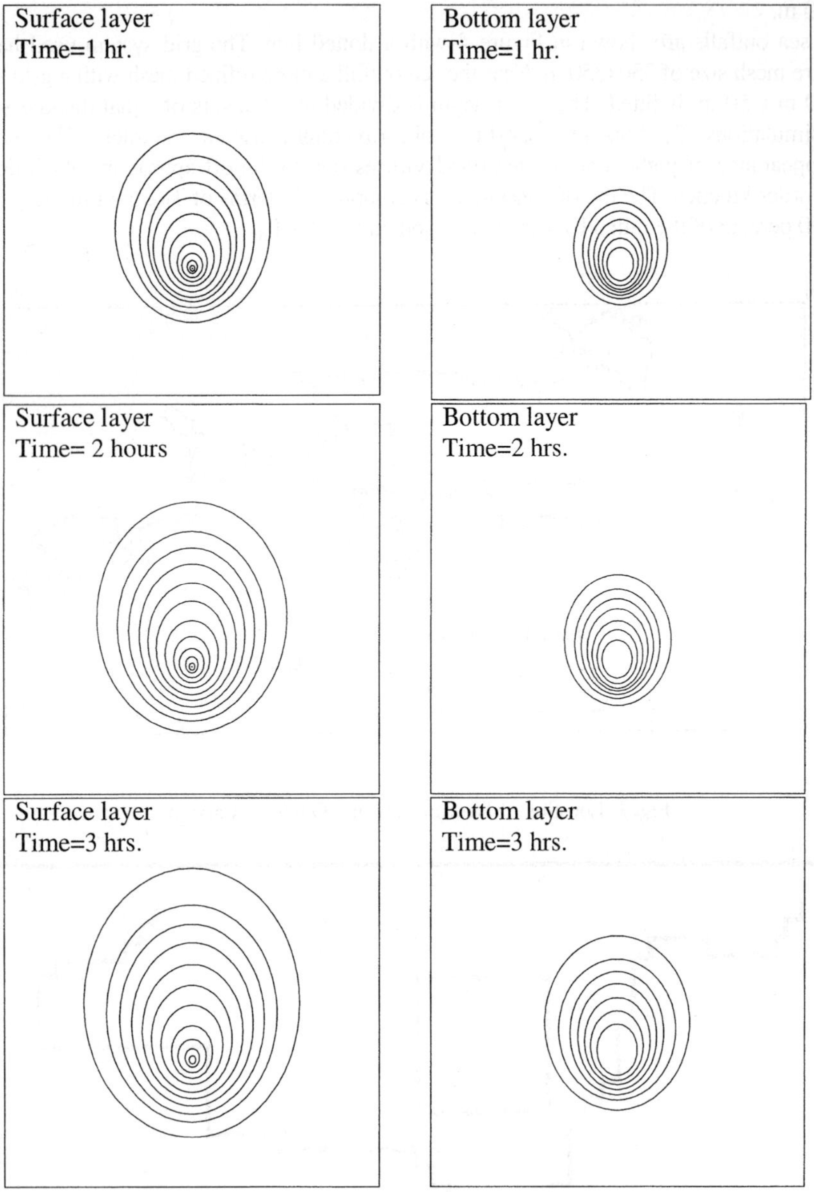

Fig. 3. Progress of concentration contours under tidal action.

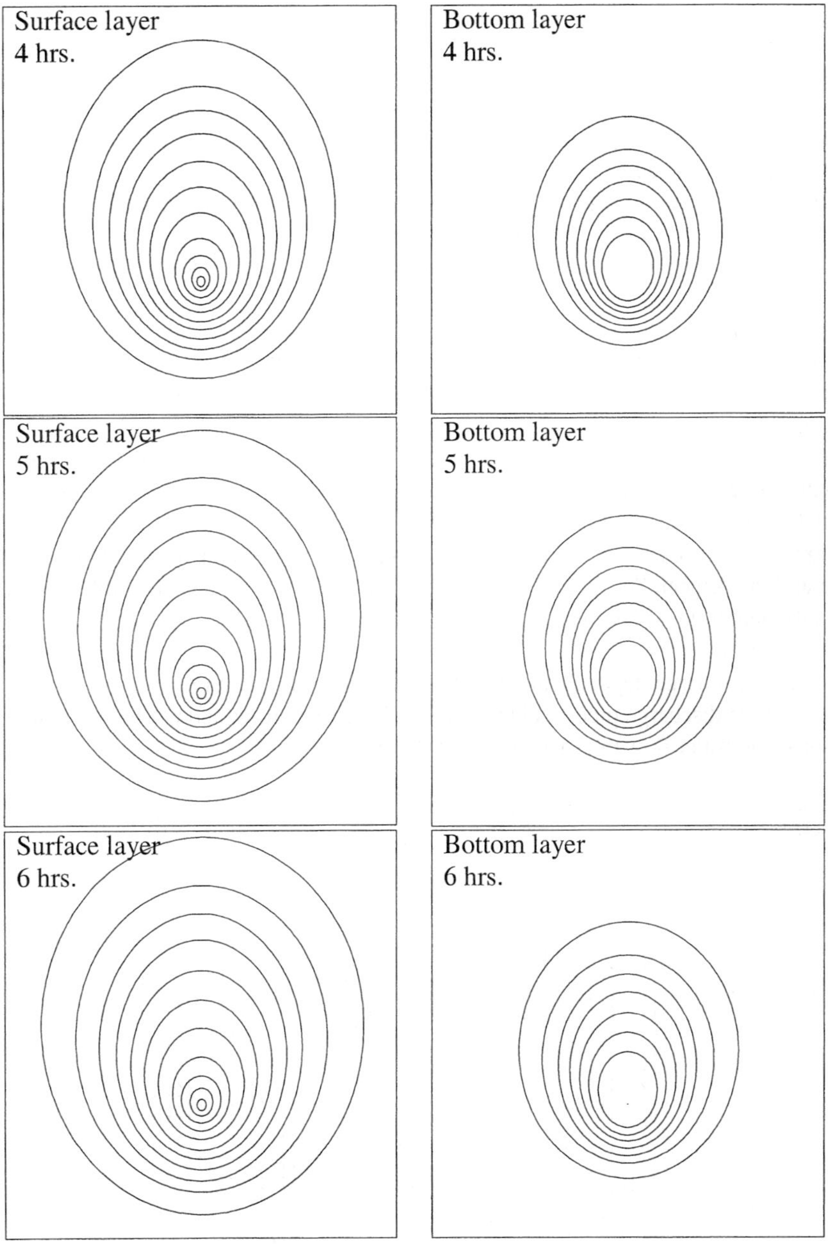

Fig. 4. (cont'd). Progress of concentration contours under tidal action.

For the coastal area wind shear is the dominant forcing that induce circulation. Wind characteristics are obtained from the measurements of the meteorological station in Bodrum for the period of 1990-2001. The wind analysis shows that the most critical wind direction for wind speeds greater than 5 m/s, is the SSW direction. In the model simulations, Bodrum Bay is subjected to a wind action with a speed of 8 m/s. blowing from SSW. Simulation is started from a state of rest that, there is no water motion and water surface is horizontal. The steady state is reached approximately 4 hours after from the start of the wind action. The water density is taken constant. Horizontal eddy viscosities are calculated by the sub-grid scale turbulence model and the vertical eddy viscosity is calculated by the k-ε turbulence model.

The steady state flow patterns near the surface and the bottom are sketched in Figure 4 and Figure 5 respectively. The transport of a pollutant is simulated using the coliform count as the tracer. The initial concentration is taken as 10^6 bac/ml at the discharge point. The discharge is assumed to take place steadily. T_{90} value is measured as 1.5 hours. Distribution of pollutant concentrations at the surface and at the bottom layers are showen in Figure 4 and in Figure 5 respectively. The area which has pollutant concentrations over 10^3 bac/100 ml is a considerable area that includes the coastal band width of 200 m.

Sea outfalls in Turkey must obey the regulations stated by the Turkish Water Pollution Control Regulations [5]. Some of the regulatory norms are summarized here; a)The initial dilution should not be less than 40 and preferably be more than 100; b)During the summer season, T_{90} value should be taken at least equal to 2 hours for the Mediterranean Sea and the Aegean Sea, and to 1.5 hours for the Black Sea; c) The total dilution should be sufficient to yield less than 1000 total coliform per 100 ml for 90% of the time when sampled in preserved areas of human contact within 200 m from the shoreline; d) The minimum discharge depth should be 20m. If reaching this depth is impractical, then the length of the discharge pipe, excluding the diffusor section, should not be less than 1300 m for waste water flow rates larger than 200 m^3/day.

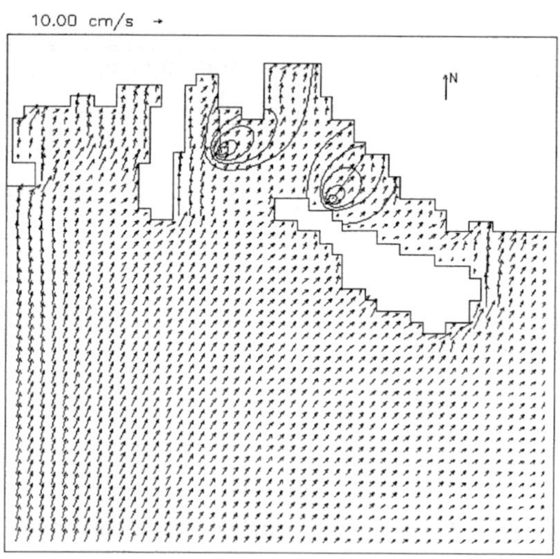

Fig. 5. Circulation pattern at the sea surface and pollutant contours (Pollutant contours from inner to outer: 500000 bac/100ml, 100000 bac/100ml, 25000 bac/100ml, 5000 bac/100ml, 1000 bac/100ml)

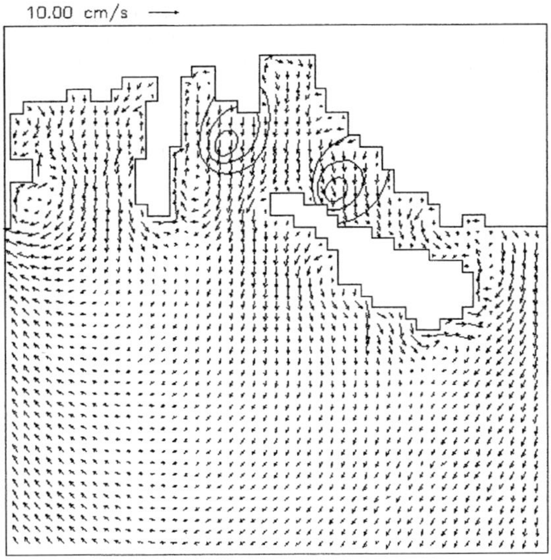

Fig. 6. Circulation pattern at the sea bottom and pollutant contours (Pollutant contours from inner to outer: 100000 bac/100ml, 5000 bac/100ml, 1000 bac/100ml).

4 Conclusions

An implicit baroclinic 3-D numerical model developed to simulate the transport processes in coastal waters is presented. Model is applied to Ölüdeniz Lagoon to simulate the tidal circulation pattern and the dispersion of a pollutant discharged near the lagoon entrance. Model predictions are encouraging for Ölüdeniz Lagoon that is a type of chocked lagoon. Model predictions are also compared with the site measurements, which are in good agreement with each other [6]. Model has been applied to Bodrum Bay that there exist two sea outfalls discharging into the Bay. The model with a great success simulates wind induced circulation pattern and the pollutant diffusion. Presented 3-D model HIDROTAM3, can serve as a powerful design tool and can be implemented in Decision Support Systems.

References

1. Balas, L., Özhan E. : An Implicit Three Dimensional Numerical Model to Simulate Transport Processes in Coastal Water Bodies, *International Journal for Numerical Methods in Fluids (SCI-Core)*, 34, (2000) 307–339.
2. Huang, W., Spaulding, M. : 3D model of estuarine circulation and water quality induced by surface discharges, *Journal of Hydraulic Engineering, ASCE,* **121**, (1995) 300–311.
3. Roberts, P.J.W. : Modeling Mamala Bay outfall plumes, II: Far field", *Journal of Hydraulic Engineering, ASCE*, 126(6), (1999) 574–583.

4. Balas L., Özhan E. : Flushing of Ölüdeniz Lagoon, Proceedings of the Joint Conference MEDCOAST'99-EMECS'99, Antalya, Turkey, Vol.3, (1999) 1873–1884.
5. Balas, L. : Simulation of Pollutant Transport in Marmaris Bay", *China Ocean Engineering*, 15 (4), (2001) 565–578.
6. Balas L., Özhan E.: Applications of a 3-D Numerical Model to Circulations in Coastal Waters", *Coastal Engineering Journal, CEJ,* 43, No 2., (2001) 99–120.

Stochastic Simulation of Inhomogeneous Metocean Fields. Part I: Annual Variability

Alexander V. Boukhanovsky[1], Harald E. Krogstad[2],
Leonid J. Lopatoukhin[3], and Valentin A. Rozhkov[3]

[1]Institute for High Performance Computing and Information Systems,
St. Petersburg, Russia
avb@fn.csa.ru, http://www.csa.ru

[2] Dept. Mathematical Sciences, NTNU, Trondheim, Norway
harald.krogstad@math.ntnu.no

[3] Oceanology Dept., State University, St. Petersburg, Russia
leonid@LL1587.spb.edu

Abstract. The paper discusses stochastic models of scalar and vector metocean fields based on time varying Empirical Orthogonal Functions in space, and autoregressive time series models for the coefficients in the expansions. The models are fitted to an extensive data set from the Barents Sea and verified by studying field extreme value properties.

1 Introduction

Annual variations of spatial and temporal properties are characteristic features of most metocean fields, and today several approaches for statistical analyses and modeling of such fields have been developed. For example, in the papers of Dragan *et al.* [4] and Lopatoukhin *el al.* [13] the time series of air and water temperature, river run-off, wind speed, wave heights and sea level variations have been treated as Periodically Correlated Stochastic Processes (PCSP). For spatial fields, periodical representations based on Empirical Orthogonal Functions (EOFs) have also been proposed by Kim&Wu [10] and Kim&North, [11].

In the present paper we shall study the annual variations of metocean events as inhomogeneous random fields by constructing probabilistic description of their spatial and annual variabilities. The aims of the current research have been to

- develop computationally efficient procedures for stochastic simulation of ensembles of annual variations of metocean fields;
- apply the above mentioned procedures to numerical investigations of the field extremes.

P.M.A. Sloot et al. (Eds.): ICCS 2003, LNCS 2658, pp. 213–222, 2003.

2 The Data Set

All computations carried out by the stochastic models have been tested against an extensive data set from the Barents Sea region. The input data set consists of fields of atmospheric sea level pressure (SLP), wind speed (WS) and air temperature (AT) from the NCEP/NCAR reanalysis data [9]. The input wave fields are numerically obtained fields calculated by Wave Watch model [15]. A summary of sampling and duration of the data sets are presented in Table 1.

Table 1. Sampling and duration characteristics of the data set. A Gaussian grid with latitude variable step is used for air temperature and wind speed.

	Field	Time step	Spatial Sampling	Duration	Source
1	Sea level pressure	6h	2.5°×2.5°	1948-1999	NCEP/ NCAR
2	Air temperature	24h	$\Delta\lambda$=1.875°	1958-1997	
3	Wind speed	6h	$\Delta\varphi \approx 1.9°$	1948-1999	
4	Wind waves	6h	0.5°×1.5°	1971-1997	Wave Watch III

3 Analysis of Inhomogeneous Metocean Fields

The second order inhomogeneity of a metocean field $\zeta(\mathbf{r},t)$ is defined in terms of the mean value $m_\zeta(\mathbf{r},t) = \mathrm{E}\zeta(\mathbf{r},t)$, and the covariance function

$$K_\zeta(\mathbf{r}_1,\mathbf{r}_2,t,\tau) = \mathrm{E}\left\{\zeta^0(\mathbf{r}_1,t)\zeta^0(\mathbf{r}_2,t+\tau)\right\},$$

where ^{0} denotes the *centering* operation, $X^0 = X - \mathrm{E}(X)$. Alternatively, the covariance function may be defined as $K_\zeta(\mathbf{r}_1,\boldsymbol{\rho},t,\tau)$, where $\mathbf{r} = (x, y)$ are the geographical coordinates and $\boldsymbol{\rho} = \mathbf{r}_2 - \mathbf{r}_1$ the spatial lag. The large scale inhomogeneity is due to variations in the spatial distribution of atmospheric or ocean forcing.

A preliminary analysis of the data shows that during winter, a low mean SLP, $m_{SLP}(\mathbf{r},t)$, in the western part of the Barents sea is prevailing with an increasing pressure towards north, south and east. In summer, the situation is the opposite with the highest pressure in the center of the sea, and dropping in all directions away from the center. However, during all the year, the sea level pressure is below standard (1013.2hPa). The standard deviation $\sigma_{SLP} = K_{SLP}(\mathbf{r},\mathbf{r},t,0)^{1/2}$ of the SLP variations is about 8 hPa in winter, and only 2.5–3.5 hPa in summer.

The standard deviation of the air temperature in the center of sea in winter is increasing from 2° to 6° toward north, whereas the variability in summer is fairly constant around 2.5–3.5°.

We may conclude, based on the variations of $m_\zeta(\mathbf{r},t)$ and $\sigma_\zeta(\mathbf{r},t)$, that these metocean fields are inhomogeneous in space as well as non-stationary in time. In fact, seasonal changes in the spatial patterns is a typical feature of the spatio-temporal variability of metocean fields. Since the physical interpretation of the full covariance function is harder than for the functions $m_\zeta(\mathbf{r},t)$, $\sigma_\zeta(\mathbf{r},t)$, this has lead us to consider EOFs, which are calculated by means of the integral equation

$$\int_\Omega K_\zeta(\mathbf{r}_1,\mathbf{r}_2)_t \varphi_{kt}(\mathbf{r}_2)d\mathbf{r}_2 = \lambda_{kt}\varphi_{kt}(\mathbf{r}_1)\,, \tag{1}$$

where λ_{kt} are eigenvalues and $K_\zeta(\mathbf{r}_1,\mathbf{r}_2)_t = K_\zeta(\mathbf{r}_1,\mathbf{r}_2,t,0)$ [12]. The first and second EOFs for the SLP are shown in Fig. 1.

The covariance function for vector fields $\mathbf{V}(\mathbf{r},t)$ like wind, currents, and temperature gradients, is defined

$$\mathbf{K}_\mathbf{V}(\mathbf{r},\boldsymbol{\rho},t,\tau) = \mathrm{E}\left\{\mathbf{V}^0(\mathbf{r},t)\otimes\mathbf{V}^0(\boldsymbol{\rho},\tau)\right\}, \tag{2}$$

where $\otimes$ is the tensor (or outer) product of $\mathbf{V}^0 = \mathbf{V}-\mathbf{m}_\mathbf{V}$ [1]. Note that $\mathbf{K}_\mathbf{V}(\mathbf{r},\boldsymbol{\rho})_t$ is a tensor,

$$\begin{pmatrix} K_u & K_{uv} \\ K_{vu} & K_v \end{pmatrix} \tag{3}$$

where K_u ,K_{uv}, K_v are auto- and cross-covariance functions of the components $\left(\mathbf{V}_x,\mathbf{V}_y\right)$ of $\mathbf{V}$ in geographical coordinates. Similarly to Eqn. (2), vector EOFs (VEOFs) may be computed by a *tensor integral equation*

$$\int_\Omega \mathbf{K}_\mathbf{V}(\mathbf{r}_1,\mathbf{r}_2)_t \boldsymbol{\Psi}_{kt}(\mathbf{r}_2)d\mathbf{r}_2 = \lambda_{kt}\boldsymbol{\Psi}_{kt}(\mathbf{r}_1) \tag{4}$$

where $\boldsymbol{\Psi}_k = \varphi_k\mathbf{i}+\psi_k\mathbf{j}$ and $(\mathbf{i},\mathbf{j})$ are the unit vectors in geographical coordinates [3].

The first VEOFs of wind speed for January and July are shown in Fig. 1. We observe that not only the EOFs of sea level pressure, but also the VEOFs for the wind change quite significantly from winter to summer.

The EOFs and VEOFs are both orthogonal with respect to the inner product defined by the covariance function, and may therefore be used as basis functions for an orthogonal expansion of the fields. Thus, the field $\zeta(\mathbf{r},t)$ may be expanded as

$$\zeta(\mathbf{r},t)-m(\mathbf{r},t) = \sum_k a_k(t)\varphi_{kt}(\mathbf{r},t) \tag{5}$$

The basis functions $\varphi_{kt}(\mathbf{r},t)$ are periodic in time, and may thus be considered as generalizations of the component representation of a field discussed in Dragan *et al.*[4],

$$\zeta(t) = \sum_k \xi_k(t)\exp(-i\Lambda_k t) \tag{6}$$

where $\xi_k(t)$ are the stationary random functions, $\Lambda_k = 2\pi k/T$ are frequencies, and T the fundamental period of variation (1 year).

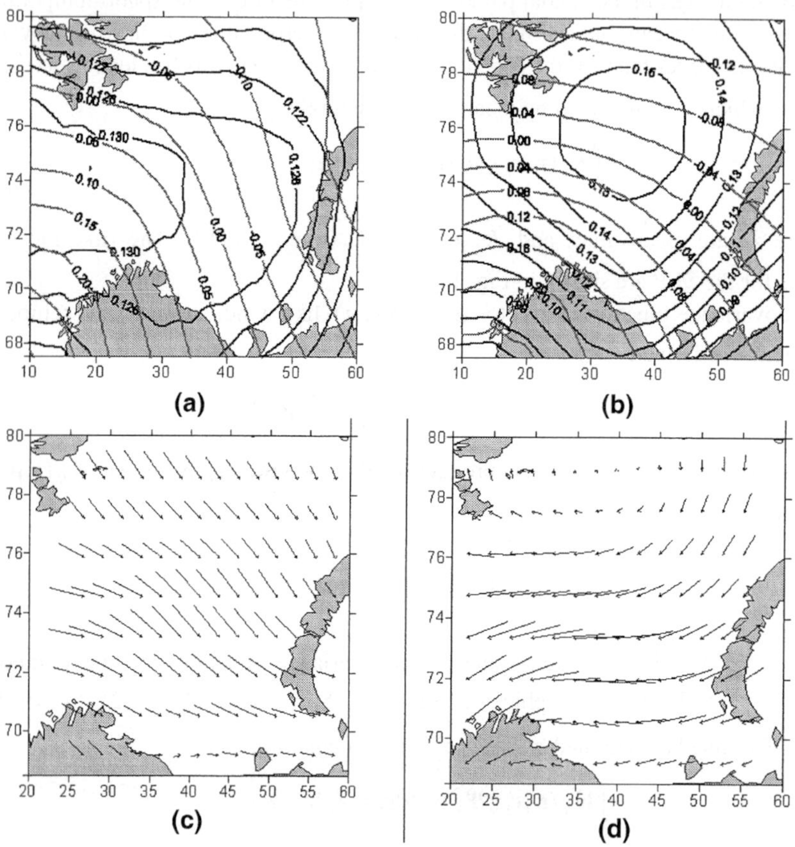

Fig. 1. Periodical EOFs of metocean fields: SLP: (*a,b*) and WS (*c,d*) for different months: January (*a,c*) and July (*c,d*). (Black – first EOF of SLP, Gray – second one)

If we estimate EOFs $\varphi_{k\tau}(\mathbf{r})$ on the basis of the *averaged* spatial covariance function over the year, we obtain an *all months basis* $\varphi_k(\mathbf{r})$. The annual variations will then only be manifested in the coefficients $a_k(t)$ of Eqn. (5), as illustrated for the SLP auto- and cross correlation functions $K_{a_k}(\tau)$ in Fig. 2. It is seen that the autocorrelation in $a_k(t)$ is much smaller when we use the seasonal EOFs. Moreover, in that case, the cross correlation between the different time functions $a_k(t)$ become negligible.

An estimate of the convergence of the series in Eqn. (5) for SLP, air temperature, wave heights and wind speed, calculated by the mean of the ratio $\lambda_i / \sum_k \lambda_k$ is shown in Table 2.

Table 2. Estimates of the speed of convergence (in %) of the orthogonal expansions in Eqn. (5) of various metocean fields by EOFs. Representative months (I,IV,VII and XI) for the periodic basis functions.

k	Periodic basis $\varphi_{kt}(\vec{r})$				Time-independent basis $\varphi_k(\vec{r})$
	I	IV	VII	XI	
Sea level pressure					
1	84.0	81.8	67.7	77.5	82.0
2	8.3	8.8	19.6	12.0	7.7
3	5.3	6.3	8.0	6.9	7.3
Air temperature					
1	57.5	60.9	43.8	64.7	89.8
2	17.6	17.1	21.8	14.6	3.3
3	8.8	6.3	10.7	7.0	3.2
Wind wave heights					
1	75.4	70.0	65.6	75.7	70.6
2	10.0	13.3	13.2	9.6	22.3
3	2.5	7.5	8.3	10.9	5.3
Wind speed (at 10m level)					
1	39.1	42.6	53.5	46.8	28.2
2	19.7	25.3	17.0	28.9	26.0
3	17.1	12.1	11.0	9.0	20.8

It seen from Table 2, that the first term in Eqn. (5) holds 40–85% of the total variance of the field values. A sum of the first three terms in (5) holds about 85–95% of the variance. Let us also note that different orthogonal polynomial, e.g., Chebyshev, Legendre, spherical harmonics etc. may be used as basis functions in Eqn. (5) [2]. However, the convergence of the EOFs expansion is the best in terms of the second-order metric defined by the covariance function.

4 The Stochastic Simulation Procedure

Taking into account Table 2, the relation in Eqn. (5) may be written as

$$\zeta(\mathbf{r},t) = m(\mathbf{r},t) + \sum_k a_k(t)\varphi_{kt}(\mathbf{r},t) + \varepsilon(\mathbf{r},t) \tag{7}$$

where N is a representative number of terms, and $\varepsilon(\mathbf{r},t)$ is residual white noise (both in space and time). The noise variance is typically not more than 20% of σ_ζ^2. Bearing in mind the type of autocovariance functions $K_{a_k}(\tau)$ shown in Fig. 3, and ignoring the cross correlations defined by $K_{a_k a_j}(\tau)$, the stochastic processes $a_k(t)$ may be presented in the form of uncorrelated scalar autoregressive models [7]:

$$a_k(t) = \sum_{j=1}^{p} \Phi_j^{(k)} a_k(t-j) + \delta^{(k)}(t) \tag{8}$$

Here the $\Phi_j^{(k)}$ coefficients are calculated from the covariance function $K_{a_k}(\tau)$, and $\delta^{(k)}(t)$ is a constant variance Gaussian white noise, only dependent on k. The relations (7) and (8) constitute a stochastic model of the inhomogeneous (by **r**) and periodically correlated (by t) random field $\zeta(\mathbf{r},t)$.

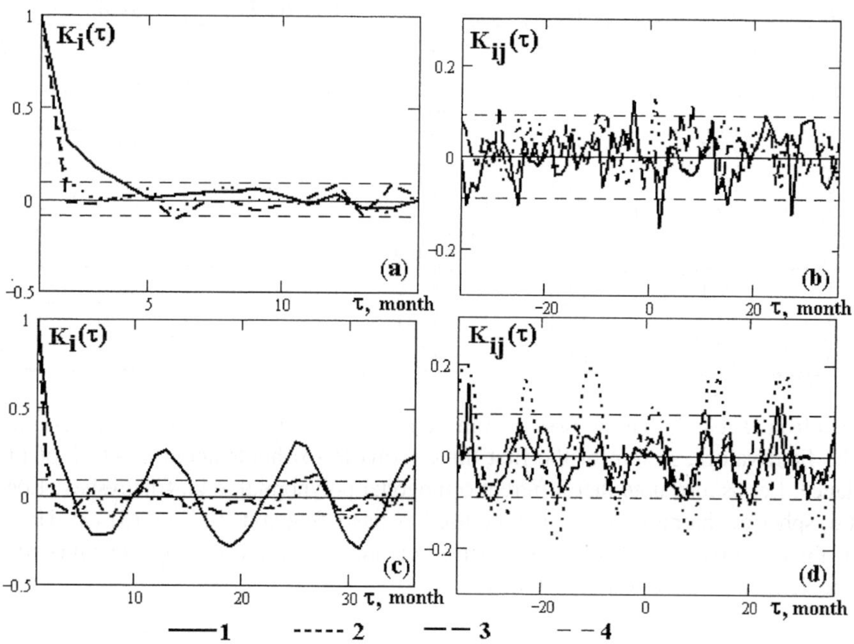

Fig. 2. Auto- (a,b) and cross (b,d) covariance functions (AKF, CKF) for the expansion coefficients in Eqn. (7) of SLP field by periodic basis (a,b) and averaged by months basis (c,d). For the AKF, 1–3 denotes the serial number of the coefficient. For the CKF, 1 corresponds to $(i=2, j=1)$, 2 to $(i=3,j=1)$, and 3 to $(i=3,j=2)$. Moreover, 4 are the 90% limits confidence interval for the estimates.

The functions $a_k(t)$ and $\varepsilon(\mathbf{r},t)$ in Eqn. (7) enable us to regard it both as a regression and a factor model [8]. The terminology of multidimensional statistical analysis is not always suitable for the analysis of random processes and fields. E.g., the model in Eqn. (7) is known as a regression, but not as a factor model, although $a_k(t)$ are random functions, and, moreover, $\{\Phi_i^{(k)}\}$ are calculated by means of $K_{a_k}(\tau)$. The functions

$\varphi_{kt}(\mathbf{r})$, calculated by Eqns. (1) and (3), are known as EOFs, and not as the factor loadings.

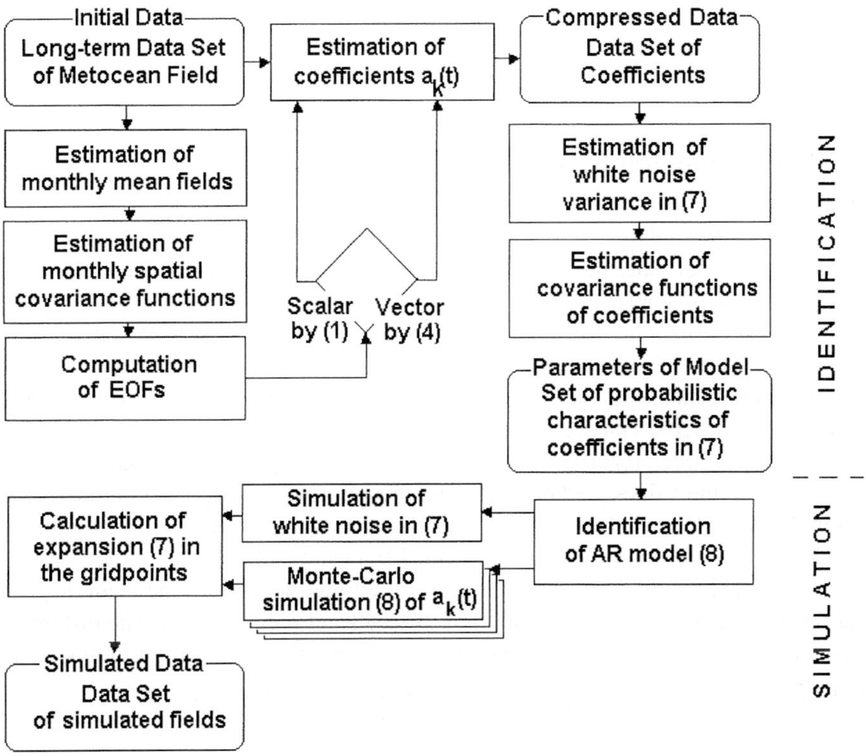

Fig. 3. Stochastic simulation of annual variation of inhomogeneous metocean fields: general algorithm.

The general algorithm of the stochastic model identification and simulation is shown in Fig. 3. The model identification starts by estimating $\{m(\mathbf{r},t)\}$ and $\{\varphi_{kt}(\mathbf{r})\}$ for a discrete set of grid points on a monthly basis. By expanding the fields into EOF series by means of $\{\varphi_{kt}(\mathbf{r})\}$, the remaining time variations is reduced to the time series of scalar coefficients $\{a_k(t)\}$. The coefficient time series, which now are stationary, are then in turn fitted to suitable autoregressive models by invoking the covariance functions $K_{a_k}(\tau)$. The estimation of autoregressive coefficients $\Phi_j^{(k)}$ in Eqn. (8) for each time series $a_k(t)$ is discussed in numerous places, e.g. in **[7]**. Finally, the variance of the white noise field $\varepsilon(\mathbf{r},t)$ is calculated.

When using the model for simulations, an ensemble of independent realizations of $a_k(t)$ for the time span $[0,T]$ of the computations, is first produced by Eqn. (8). Then, the realizations of the full spatiotemporal metocean fields are obtained by means substituting of the simulated series of $a_k(t)$ into Eqn. (7) and adding the white noise term.

The result of the simulation procedure is the set of the simulated metocean fields $\zeta(\mathbf{r},t)$ or $\mathbf{V}(\mathbf{r},t)$ in a discrete set of grid points $\mathbf{r}_{ij} = (x_i, y_j)$ and at discrete times, $\{t_n\}$.

5 Verification and Extreme Value Analysis

The developed simulation procedure may be verified and tested by considering the problem of spatial extremes. In practice, due to inhomogeneity of the fields, there arise questions of joint extremes in two or more spatial points. A Gaussian kernel estimate [14] of the joint distribution of mean monthly minima of SLP in the SW and E parts of Barents Sea is presented in Table 3 (left). This estimation was obtained from the initial data set given in Table 1. It is seen that although the two parts of the sea are more than 1000 km from each other, the correlation is rather high (about 0.7). Since extreme value properties were not considered as a part of the model identification, the extreme values predictions by the model provide independent model verification. In the right part of the Table 3, a similar Gaussian kernel estimate of joint SLP minima in the same points have been obtained from simulations using Eqns. 7 and 8. It is observed that the maxima of the joint probability are located on the main diagonal, and this reflects the overall correlation of the sea level pressure field. The differences between probabilities are insignificant. Even the largest difference between $p_{i,j}^{(1)} = 31.5\%$ and $p_{ij}^{(2)} = 17.1\%$ is inside the 90% confidence interval.

Table 3. Kernel estimates of joint distributions of sea level pressure minima in the points (φ=70° N, λ= 15° E) and (φ= 76° N, λ= 45° E).

φ= 70° N, λ= 15° E (hPa)	Sample data					Probability model (Eqns. 7,8)				
	φ= 76° N, λ= 45° E (hPa)									
	5 0	0 -5	-5 -10	-10 -15	-15 -20	5 0	0 -5	-5 -10	-10 -15	-15 -20
2 –2	10.9	7.5	0.5	–	–	15.9	10.1	0.1	–	–
-2 –6	6.2	17.1	4.1	0.1	–	6.7	31.5	1.8	–	–
-6 –10	1.8	15.2	10.1	0.3	–	0.6	14.3	8.4	1.1	–
-10 –14	0.2	4.6	9.6	1.1	0.3	–	0.7	3.6	1.7	0.2
-14 –18	–	0.9	5.2	2.3	1.4	–	0.0	0.9	1.7	0.7

Annual extremes, calculated from monthly mean values of water temperature and salinity, sea level, wave heights, may in different years be observed in different months. Thus, the distribution for the annual extreme, $F_Y(x)$, will then a mixture of distributions for the monthly mean $F_i(x)$ [13] with the weights ρ_i – the probabilities of appearance of X_{min} or X_{max} in the month i.

We may now pose two questions:

Does the distribution in Eqn. (9) extend from a single point extreme to a field extreme?
Does the model in Eqns. (7,8) reproduce both the distribution (9) and its field modification?
Table 4 answers these questions in the affirmative. We observe a quite good agreement between sample and simulated extremes over the entire field. E.g., from the data only in 7.5% of the mean monthly temperatures in the Barents Sea is below –39º. By the model simulations, this value has changed to 5.5%, which, however, is well inside 90% confidence limits. A similar agreement is observed for the probability ρ_i of monthly extremes. E.g., the annual lowest air temperature over the Barents Sea occurs in January 40% of the time, whereas this occurs 45% of the time in the simulations, again well inside the 90% confidence limits.

Table 4. Probability characteristics of annually varying metocean fields

Minimum of sea level pressure (hPa-1000)										
Data	Probability by months ρ_i (%)							Integral probability (%)		
	IX	X	XI	XII	I	II	III	<0	<-5	<-15
Sample	2.5	10	20	20	20	15	12.5	95	75	1.5
90% Confidence interval	0 7.4	0.5 9.5	7.3 36.2	7.3 36.2	7.3 36.2	3.7 26.3	2.0 23.0	88.1 100	61.3 88.7	0 5.3
Model (7,8)	–	7.5	15	30	20	15	12.5	95	63	<1
Minima of air temperature (°Ñ)										
Data	Probability by months ρ_i (%)							Integral probability (%)		
	IX	X	XI	XII	I	II	III	<-30	<-33	<-39
Sample	–	–	2.5	15	40	27.5	15	90	47.5	7.5
90% Confidence interval	–	–	0 7.4	3.7 26.3	24.5 55.5	13.4 41.6	3.7 26.3	80.5 99.5	31.7 63.3	0 15.8
Model (7,8)	–	–	5	12.5	45	27.5	10	95	70	5.5

6 Conclusions

This paper has shown that computational multivariate statistics of spatio-temporal fields may be used for describing the metocean regime in seas and oceans.
By applying field expansions in terms of periodical EOFs and VEOFs, it has been possible to develop stochastic models which are field generalizations of PCSPs.
The models may be applied to simulate ensembles of time varying metocean fields like SLP, WS, AT, and sea waves.
Investigation of spatial and temporal features of annual extremes of measured and simulated metocean fields may be used for independent model verification. Further applications will be discussed in a separate paper in these proceedings.

Acknowledgment. This research has been partly founded by INTAS 99-0666 Project: "Estimation of extreme metocean events".

References

1. Belyshev A.P., Klevantsov Yu.P., Rozhkov V.A. Probabilistic analysis of sea currents. Leningrad, Gymet P.H., (1983) 264 p. (in Russian)
2. Blais J.A.R. Estimation and spectral analysis. Univ. of Calgary Press, (1988) 132 p.
3. Boukhanovsky A.V., Degtyarev A.B., Rozhkov V.A. Peculiarities of computer simulation and statistical representation of time–spatial metocean fields. LNCS #2073, Springer–Verlag, (2001) 463–472
4. Dragan Ya. P., Rozhkov V.A., Yavorsky I.N. Methods of probabilistic analysis of rhythms of oceanological processes. Leningrad, Gymet. P.H., (1987) 320 p. (in Russian)
5. Dubrovin B.A., Novikov S.P., Fomenko A.T. Modern geometry: methods and applications. Moscow, "Nauka", (1979) 760 p. (in Russian)
6. Isihara A. Statistical physics. Acad. Press, New York-London, (1971) 472 p.
7. Jenkins G.M., Watts D.G. Spectral analysis and its application. Holden-Day, San-Francisco (1969)
8. Johnson R.A., Wichern D.W. Applied multivariate statistical analysis. Prentice-Hall Internalional, Inc., London (1992) 642 p.
9. Kalnay E., M. Kanamitsu, R. Kistler, W. Collins, D. Deaven, L. Gandin, M. Iredell, S. Saha, G. White, J. Woollen, Y. Zhu, A. Leetmaa, R. Reynolds, M. Chelliah, W. Ebisuzaki, W.Higgins, J. Janowiak, K. C. Mo, C. Ropelewski, J. Wang, R. Jenne, D. Joseph. The NCEP/NCAR 40-Year Reanalysis Project. Bulletin of the American Meteorological Society, ¹3, March (1996)
10. Kim, K.-Y., and G. R. North, EOFs of harmonizable cyclostationary processes, J. Atmos. Sci., 54 (1997) 2416–2427
11. Kim, K.-Y., and Q. Wu, A comparison study of EOF techniques: Analysis of nonstationary data with periodic statistics, J. Clim. 12 (1999) 185–199
12. Loeve M. Fonctions aleatories de second odre. C.R. Acad. Sci. 220, (1945)
13. Lopatoukhin L.J., Rozhkov V.A., Ryabinin V.E., Swail V.R., Boukhanovsky A.V., Degtyarev A.B. Estimation of extreme wave heights. JCOMM Technical Report, WMO/TD #1041 (2000)
14. Silverman B.W. Density estimation for statistics and data analysis. London, Chapman & Hall, (1986)
15. Tolman H. User manual and system documentation of WAVE WATCH-III. NOAA technical note. (1999) 124 p.

Stochastic Simulation of Inhomogeneous Metocean Fields. Part II: Synoptic Variability and Rare Events

Alexander V. Boukhanovsky[1], Harald E. Krogstad[2], Leonid J. Lopatoukhin[3], Valentin A. Rozhkov[3], Gerassimos A. Athanassoulis[4], and Christos N. Stephanakos[4]

[1] Institute for High Performance Computing and Information Systems, St. Petersburg, Russia
avb@fn.csa.ru, http://www.csa.ru

[2] Dept. Mathematical Sciences, NTNU, Trondheim, Norway
harald.krogstad@math.ntnu.no

[3] Oceanology Dept., State University, St. Petersburg, Russia
leonid@LL1587.spb.edu

[4] Technical University of Athens, Athens, Greeece,
{mathan,chstef}@central.ntua.gr

Abstract. The paper discusses stochastic models for the synoptic variability of metocean fields using spatio-temporal impulse representations combined with Markov walks of storms in space. The models are fitted to an extensive data set of ocean wave fields from the Barents Sea and verified by studying the fields duration and extreme value properties.

1 Introduction

Metocean fields, like atmospheric pressure, wind speed, ocean waves etc., have a complex spatial and temporal variability. Traditionally, the approach for statistical formalization of such phenomena has been based on a *multiscale* hypothesis proposed by Andrey S. Monin [17]. The hypothesis suggests modeling the total variability by means of a set of stochastic models for each temporal scale separately, and with the interdependence taken into account parametrically.

Stochastic simulations on the scale of annual and inter-annual variations were considered in the first part of this paper [9], but for metocean fields such as ocean waves, the synoptic variability has more interest, since its impact on the total variation and the extreme events is higher.

The synoptic variability corresponds to temporal scales from a few hours to some days. For atmospheric processes, the associated spatial scales are 2000–3000 km, as opposed to 500–700 km for the oceanic ones. The nature of the synoptic variability may be explained as a stochastic alternation between *storms* and *calms*, that is, cyclones and anticyclones in the atmosphere [1], and storms and calms for wind and wave fields [15].

P.M.A. Sloot et al. (Eds.): ICCS 2003, LNCS 2658, pp. 223–233, 2003.

There have been many papers devoted to the analysis and statistical description of the marginal [2,3,4,10,15,20,21] and even joint [7,22] synoptic variability of metocean processes in terms of stochastic series at fixed points in space. However, these methods become complicated if we consider the variability of the full spatio-temporal fields. Currently, there are at least two different approaches:

- *The Euler approach*, considering the joint variability of the field in a set of fixed points.
- *The Lagrange approach*, considering the motion of spatial structures in the fields.

The Euler approach adopts multivariate time series models $\zeta_k(t) = \zeta(\mathbf{r}_k, t)$ for description and simulation [19]. In the Lagrangian approach, the alternation of storm and calm periods may be described in terms of *spatio-temporal* random events, or *impulses* [13,15,20]. In general, it is difficult to say which approach is best. However, for the analysis of extreme events over the whole field, the Lagrange's approach may be more reliable because it is independent of choice of particular grid points. In the present paper, we describe a computationally efficient method for stochastic simulation of the synoptic variations of metocean fields based on the Lagrange approach, and fit the model to an extensive data set from the Barents Sea. The model is verified against the data for T-years field extremes.

2 The Data Set

The ocean wave field is described in terms of the directional wave spectrum and today, advanced numerical models, based on the numerical integration of the wave action equations, produce directional spectra in a fine grid, using the wind field as input. In this paper, the NCEP/NCAR reanalyzed wind fields 1971–99 [12] and the 4th generation wave model *WaveWatch-v1.18* [23] have been applied for a simulation of a long-term set of ocean wave fields for the Barents Sea region (Fig.1). The time step in the simulations is 6 hours, and wind and wave measurements from Barents Sea buoys were used for verification [16].

Instead of using the full ensemble of calculated spectra [15], we consider here only the most basic parameter, the significant wave height, $h(\mathbf{r},t)$, defined as four times the standard deviation of the surface height. The synoptic variability is summarized in the long-term distribution of $h(t)$ in fixed points, and Table 1 shows seasonal averages of the median $h_{0.5}$ and the r.m.s. value σ_h of h, for three locations in the Barents Sea, as marked in Fig. 1.

Table 1. Average wave height distribution parameters

Region	Parameters of long-term wave heights distributions							
	1		2		3		Total area	
	$h_{0.5}$ [m]	σ_h [m]	$h_{0.5}$ [m]	σ_h [m]	$h_{0.5}$ [m]	σ_h [m]	$h_{0.5}$ [m]	σ_h [m]
Winter	2.2	1.5	2.0	1.4	1.9	1.5	2.2	0.9
Spring	1.4	1.2	1.2	1.1	1.1	1.3	1.4	0.8
Summer	0.9	0.6	0.8	0.6	0.7	0.6	0.8	0.3
Autumn	1.6	1.1	1.6	1.2	1.5	1.2	1.6	0.7

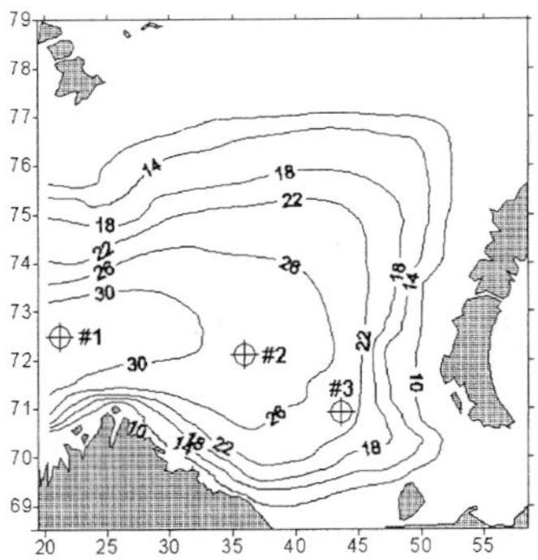

Fig. 1. Spatial distribution of 100-years maximal (0.1%) wave heights in the Barents Sea.

The data set is clearly spatially inhomogeneous [14] and the interval analysis, in accordance to [15], shows that the difference is significant, basically due to fetch limitations. The inhomogeneity of the wave field is even more clearly seen from a map of the spatial distribution of extreme events, *e.g.* waves encountered once 100 years (see Fig. 1). These results were obtained by the BOLIVAR method [15] applied to the same 28-years data. The highest waves are observed in the western part, and the intensity of the waves decreases towards SE.

3 The Spatial Parametrization of Storms

It is obvious from Table 1 and Fig. 1 that the wave fields are inhomogeneous in space and have the seasonal periodicity. We are going to describe these features in terms of spatial structures – *storms* and *calms*. For atmospheric phenomena this procedure is developed rather well, see, *e.g.* [1]. For ocean fields (such as ocean waves) it is possible to generalize Angelides' [2] definition of storms to the spatio-temporal domain,

$$\Omega(t) = \{\mathbf{r}: \quad h(\mathbf{r},t) \geq z\}, \tag{1}$$

where z is the level of the storm and additional parameters are defined in Table 2. Note that $\{h^+,\mathbf{r}^+\}$ characterize the extreme and $\{\bar{h},\mathbf{r}_0\}$ the general behavior of the storm in space.

Table 3 shows seasonally averaged storm statistics. The probability of storm occurrence is P_{total}, and the conditional probability for the numbers of storms is P_N. Thus, the level z = 2m is exceeded in at least one point in 43.7% of the cases during winter. However, in only 1.4% of those cases the number of storms is greater than two.

The mean, r.m.s. and 95% quantiles for h^+, $\bar{h}$, and L show that the parameters of the storms are strongly dependent on z and have a clear annual variability. It is interesting to observe that the smallest storm diameter L occurs in spring, and this may be caused by the seasonal variations of fetch related to the ice cover of the Barents Sea. The maximal ice cover (more that 60% of the sea area) is observed in March-April, and the minimal cover (10-15%) in September-October [8].

Geospatial vector statistics are displayed for vectors quantities [6,7,9]. These are the modulus and direction of the mean vector, $\{|\mathbf{m}|,\varphi_{\mathrm{m}}\}$, the principal axes and rotation angle of the r.m.s. ellipse $\{\lambda_1,\lambda_2,\alpha\}$, and the coefficient of variation, $\rho=(\lambda_1^2+\lambda_2^2)^{1/2}/|\mathbf{m}|$. It is seen that $\mathbf{r}^+$ is shifted from towards west $\mathbf{r}_0$, but less than about 50 km. The coefficient of variation ρ is quite high, however.

Table 2. Characterizing parameters of the storms

Description	Notation	Definition
Area	$S_\Omega(t)$	$\int_{\Omega(t)} d\mathbf{r}$
Equivalent diameter	$L(t)$	$2\sqrt{S_\Omega(t)/\pi}$
Averaging wave height	$\bar{h}(t)$	$\int_{\Omega(t)} h(\mathbf{r},t)d\mathbf{r} \Big/ S_\Omega(t)$
Geometric center ("center of gravity")	$\mathbf{r}_0(t)$	$\int_\Omega h(\mathbf{r},t)\mathbf{r}d\mathbf{r} \Big/ \int_\Omega h(\mathbf{r},t)d\mathbf{r}$
Maximum wave height	$h^+(t)$	$\max_{\mathbf{r}\in\Omega(t)}\left[h(\mathbf{r},t)\right]$
Location of the maximal wave height	$\mathbf{r}^+(t)$	$\left\{\mathbf{r}: h(\mathbf{r},t)=h^+(t)\right\}$

Fig. 2 shows two examples of typical storms, six hours apart, with some of the above parameters marked on the map. The storm velocity, $\mathbf{W}=\partial\mathbf{r}_0/\partial t$, and although the mean velocity of the storms is only 3.4-11.9 km/h, the variations may reach 50 km/h with large variability, as indicated by the relatively high values of both λ_1 and λ_2.

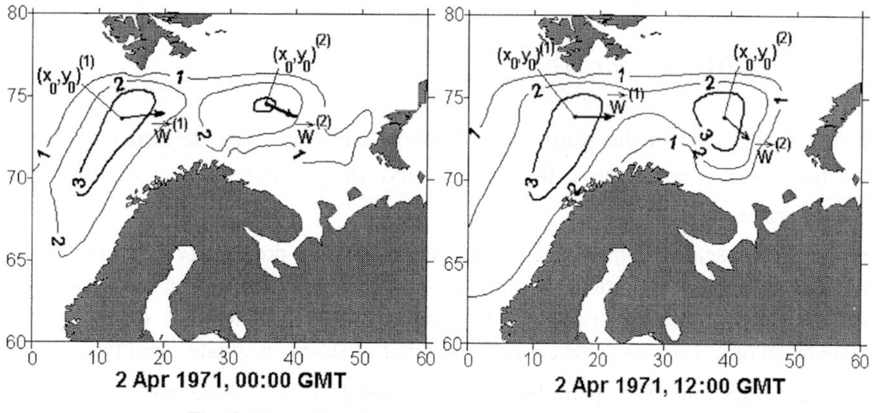

Fig. 2. Examples of storm movement in synoptic terms.

The analysis of Table 3 shows that, in general, only one storm occurs at a time, and the position of highest wave is not far from the storm's geometrical center, $\mathbf{r}_0$. Moreover, $\{h^+,L\}$ may be applied for defining the storm's spatio-temporal behavior.

Table 3. Seasonal estimates of parameters of the storms in the Barents sea. W – Winter, SP – Spring, SU – Summer, AU – Autumn. See text for further explanations.

Parameter		Level z = 2.0 m				Level z = 4.0 m			
		W	SP	SU	A	W	SP	SU	A
P_{total}		43.7	30.6	23.2	47.3	20.5	6.2	1.0	12.5
P_N (%)	N=1	87.3	94.6	93.5	85.7	95.3	96.9	100	92.6
	N=2	11.3	5.0	6.0	13.1	4.6	3.1	–	6.9
	N≥3	1.4	0.4	0.5	1.2	0.1	–	–	0.5
h^+ [m]	Mean	2.4	2.4	2.3	2.4	4.5	4.4	4.3	4.4
	r.m.s	0.5	0.4	0.3	0.4	0.5	0.4	0.3	0.4
	95%	4.0	3.7	3.2	3.6	6.2	5.3	5.1	5.8
$\bar{h}$ [m]	Mean	3.0	2.8	2.6	2.9	5.1	4.9	4.7	5.0
	r.m.s	1.1	0.8	0.6	0.9	1.1	0.8	0.6	0.9
	95%	7.0	5.7	4.8	6.2	8.8	7.1	6.5	8.2
L [km]	Mean	460	440	450	543	460	380	360	445
	r.m.s.	335	260	290	372	270	216	195	269
	95%	1460	1200	1317	1592	1170	1000	910	1243
$\mathbf{r}_0 - \mathbf{r}^+$ [km]	$\lvert\mathbf{m}\rvert$	45	49	18	47	31	35	16	13
	φ_{m}	276	258	314	284	255	238	354	286
	λ_1	152	137	113	164	122	107	56	108
	λ_2	96	78	81	120	84	65	46	82
	α	145	139	153	155	133	137	109	137
	ρ	4.0	3.2	7.6	4.3	4.8	3.5	4.3	10.6
Storm Center Velocity $\mathbf{W}$ [km/h]	$\lvert\mathbf{m}\rvert$	3.4	5.7	5.8	4.8	7.0	11.9	11.4	8.2
	φ_{m}	297	275	253	277	280	277	272	288
	λ_1	28.0	20.3	21.1	29.4	24.1	19.6	20.1	22.3
	λ_2	17.0	13.8	15.5	18.1	10.5	8.1	8.5	12.1
	α	180	150	153	180	137	130	136	134
	ρ	9.7	4.33	4.5	7.11	3.7	1.8	1.9	3.09

4 A Stochastic Model for the Synoptic Variability

Traditional models are based on stochastic differential equations of the form

$$\frac{\partial h}{\partial t} + \mathbf{W} \cdot \nabla h = G(\mathbf{r},t) \,, \tag{2}$$

where $G(\mathbf{r},t)$ is a source function. For wind waves, Eqn. (2) follows from the wave action equation [23] after integrating over the spectrum and linearizing the derivative term. If we expand the solution of Eqn. (2) using Galerkin techniques we obtain

$$h(\mathbf{r},t) = \sum\nolimits_k a_k(t)\Phi_k(\mathbf{r},t \mid \Xi_k) \,, \tag{3}$$

where $\{\Phi_k\}$ are spatio-temporal basis functions, depending on a set of parameters Ξ_k, and $\{a_k\}$ the corresponding coefficients. The coefficients $\{a_k\}$ will be obtained as solutions of a system of algebraic equations after a suitable parametrization of G.

In contrast to this Eulerian approach, the Lagrange approach considered below writes the field as moving *spatio-temporal impulse structures*, and estimates the characteristics directly from the initial data set, without considering the source function.

4.1 Storm Motion

It is possible to model the temporal sequence of storm centres $\mathbf{r}_0(t)$ as a Markov chain with a finite number of spatial locations (regions). Fig. 3c shows a map with nine regions in the Barents Sea, and Table 4 estimates of the transition and limit probabilities of the Markov chain for the level z = 2 (for all seasons). The additional ground state "C" (*calm*) signifies no storms.

Table 4. Transition and limit probabilities of the storm sequence. Here "+" is the probabilities below 1%, blank space – zero probabilities.

State	Transition probabilities (%)										Limit Prob.
	SW	S	SE	W	0	E	NW	N	NE	C	
SW	**11.7**	17.2	2.2	15.6	20.6	2.2	+	+		29.4	1.2
S	1.6	**33.8**	8.4	3.6	9.3	2.8	+	+	+	39.2	6.7
SE	+	9.8	**27.3**	3.3	4.3	5.5	+	1.6	2.7	44.7	3.3
W	3.5	8.1	+	**31.3**	23.2	1.4	1.2	+	+	28.6	5.4
O	+	11.2	3.9	5.8	**44.2**	10.1	+	2.4	1.8	19.9	9.3
E	+	4.9	10.2	2.8	8.6	**25.3**	+	1.4	7.2	38.8	2.9
NW	1.1	2.2	3.3	11.1	2.2		**14.4**	16.7	1.1	47.8	0.6
N	+		2.4	2.4	10.2	3.6	3.0	**20.5**	22.9	33.7	1.1
NE	+	2.1	1.4	2.8	4.5	3.1		3.8	**31.7**	49.8	1.9
C	+	3.3	1.5	3.5	3.7	0.9	0.5	0.5	0.7	**84.4**	67.6

The diagonal elements dominate, showing that the storms tend to remain in the same region during one synoptic term. The limit probabilities show strong spatial inhomogeneity with the maximal occurrence of storms (9.3%) observed in the central region (O), and lowest in the NW, N, and SW regions. The Markov chain produces an alternating sequence of *storms* (of duration $\Im$) and *calms* (of duration Θ).

4.2 Parametrization of Spatio-Temporal Impulses

Let us consider a storm $\{\mathbf{r}_0(t), h^+(t), S_\Omega(t)\}$, $t \in [t_0, t_0 + \Im]$. The size of the storm area $S_\Omega(t)$ may in fact be derived from a quantile diagram (Fig. 3a) of wave heights at a specific synoptic term t: $S_\Omega(t)$ is equal to the fraction of wave heights larger than z times the total area of the region. It is also possible to apply a regression between h^+ and $S_\Omega(t)$. Some information about the *shape* of the storm area could be obtained from moments of inertia parameters $(\lambda_1, \lambda_2, \alpha)$, similar to those in Table 2.

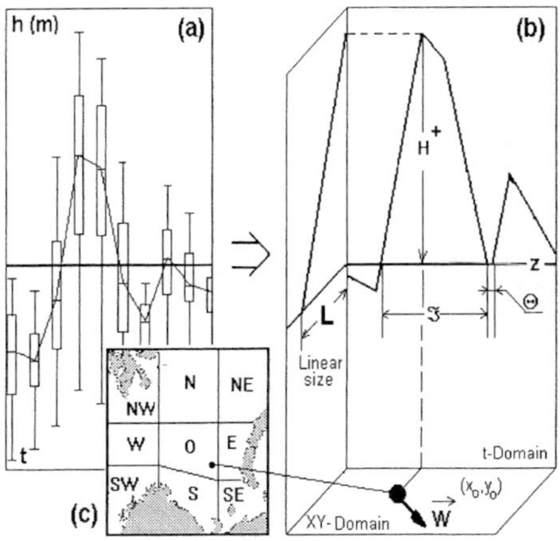

Fig. 3. The spatio-temporal parametrization of storm variability.

Our aim is to parametrize the storm impulses in terms of the overall maximum wave height H^+ and the associated storm area S^+. This parametrization, given in terms of a set of parameters $\{H^+, S^+, \Im, \Theta\}$ and the Markov process for $\mathbf{r}_0(t)$, generalizes the BOLIVAR approach [15,20] from time series to spatio-temporal fields. The data have shown that as the values $(H^+, \Im)$ as (H^+, L^+) are highly dependent: (couple correlation are 0.7–0.9).

Since H^+ is an extreme value, the distribution, conditional on $\Im$, is approximately Gumbel,

$$F\left(H^+ \mid \Im\right) = \begin{cases} \exp\left[-\exp\left(-\left(H^+ - A(\Im)\right)/B(\Im)\right)\right], & H^+ \geq z, \\ 0, & H^+ < z, \end{cases} \tag{4}$$

and this has been validated in [20].

It now remains to specify a time function for $h^+(t)$ and a spatial field function $\Phi(\mathbf{r})$ as in Eqn. 3. It is shown in [15] that a piecewise linear function in time is sufficient, whereas for $\Phi(\mathbf{r})$ it is possible to use an elliptic cone (1st order), or a elliptic paraboloid (2nd order), see also Section 5 below.

4.3 Simulation Procedure

The simulation procedure is shown in Fig. 4. In first step is to estimate the parameters in Table 2 for each synoptic term *t*. The level z may depend on the season and obtained from models of the annual variability, as described in [9]. The second step is the spatio-temporal impulse parametrization of the time series $\{h^+(t), L(t)\}$, the parameters of Eqn. (4), and the parameters (transition and limit probabilities) of the Markov chain model for $\{\mathbf{r}_0(t)\}$.

The simulation starts with the Monte-Carlo simulation. Based on the realizations, the durations $\{\Im_k\}$ are found and the value of H^+ are simulated from Eqn. (4). The diameter L^+ is obtained regression on H^+, and finally, all synthesized parameters are substituted into Eqn. (3), thus generating a synthetic field for all points $(\mathbf{r}, t)$.

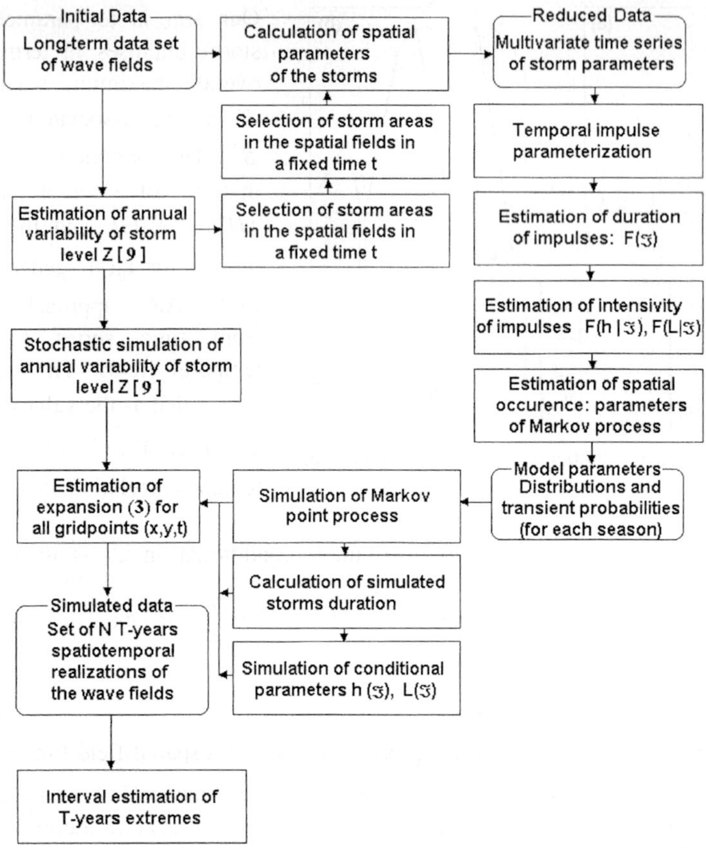

Fig. 4. Scheme of stochastic simulation procedure of synoptic variability

5 Verification and Extreme Value Analysis

There arise at least three questions for verification of the simulation procedure:

- Does the empirical storm duration statistics fit the Markov model statistics?
- Does the Lagrangian model described above reproduce the annual spatial extremes seen in the data?
- Does the Lagrangian model reproduce estimates of T-years extremes in the fixed points estimated by approaches like BOLIVAR or AMS [15]?

Figure 5 answers the first and second questions. Figure 5a shows the *quantile biplot* (Q-Q-plot) of simulated (h^*) and sample (h) values of annual maxima of significant wave height of over the whole sea. In Fig. 5b the same biplot is shown for simulated ($\Im^*$) and sample ($\Im$) values of storm durations. We observe good agreement between sample and simulated characteristics over the entire field. For answering the third question, the extremes for return period of 1, 10, and 100 years were calculated for the points 1-3 in Fig. 1 by the AMS approach (with parametrical confidence intervals [15]) and by the Lagrangian stochastic simulation, see Table 5.

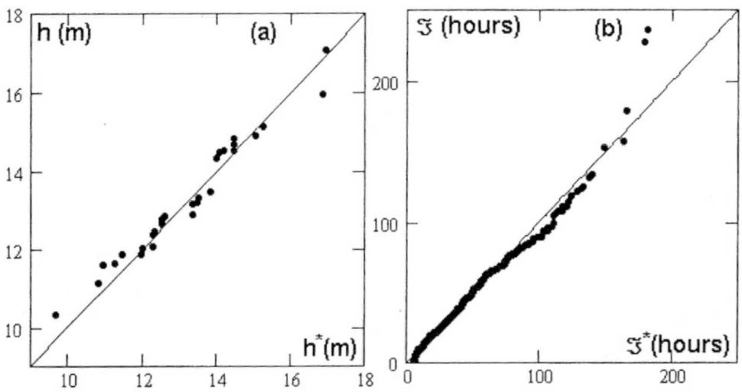

Fig. 5. Verification of the simulation model

All simulations were carried out for two types of the spatial impulse approximation $\Phi(\mathbf{r})$. The 1[sr] order shape is an elliptic cone, and in this case the Lagrangian model underestimates the *T*-years extremes. By assuming the more reasonable form of an elliptical paraboloid for $\Phi(\mathbf{r})$, the values become in much better agreement with the AMS method.

Table 5. Sample and simulated values of *T*-year wave heights in Barents sea.

Point		1			2			3		
T		1	10	100	1	10	100	1	10	100
Sample estimates	Point data	10.3	13.2	16.7	9.6	11.4	13.7	9.0	10.8	13.2
	90% conf. interval	9.3 11.8	11.9 15.1	15.0 19.2	8.6 11.0	10.3 13.1	12.3 15.8	8.1 10.4	9.7 12.4	11.9 15.2
Simulation	1st order	8.0	10.3	13.0	6.9	7.8	9.0	6.7	7.5	8.5
	2nd order	9.8	12.2	15.1	9.2	11.6	14.5	8.4	10.6	13.4

The results of the model verification are satisfying and give confidence to use the Lagrangian model for analysis and numerical studies of the spatio-temporal variations of extreme synoptic events. Nevertheless, adaptation of the model to more complex sea basins is problematic and requires further development of the storm impulse approximations and the random walks model.

6 Conclusions

This paper has demonstrated that computational multivariate statistics of spatio-temporal fields may be used for describing the synoptic variability of metocean fields. The synoptic variability is modelled by a Lagrangian approach, which describes the storm motion by a spatial Markov random walk and the storm's extension and field properties as spatio-temporal impulses, characterized by a limited set of parameters.

Model properties not considered in the fitting have been used for verification, and the field extremes predicted by the model are shown to fit very well to estimates of extreme values obtained by other approaches.

Acknowledgment. This research has been partly founded by INTAS 99-0666 Project: "Estimation of extreme metocean events".

References

1. Agee E.M. Trends in cyclone and anticyclone frequency and comparison with periods of warming and cooling over the Northern Hemisphere. J. Clim., 4, 1991, pp. 263–267.
2. Angelides D.C., Veneciano D., Shyam Sunder S. Random sea and reliability of offshore foundations. J. Eng. Mech. Div., 1981, vol. 107, #1, pp. 131–148.
3. Athanassoulis G.A., Vranas P.B., Soukissian T.H., 1992. A new model for long-term stochastic analysis and prediction. Journ. Ship Res., v 36, N1, pp. 1–16.
4. Athanassoulis G.A., Stephanakos Ch.N., 1995. A nonstationary stochastic model for lon g-term time series of significant wave height, Journ. Geoph. Res., v 100 (C8), pp. 16149–16162.
5. Bharucha-Reid A.T. Elements of the theory of Markov Processes and their applications. MC Graw-Hill Book Company Inc., New-York, Toronto, Tokyo, 1960.
6. Belyshev A.P., Klevantsov Yu.P., Rozhkov V.A. Probabilistic analysis of sea currents. Leningrad, Gymet P.H., 1983, 264 pp. (in Russian)
7. Boukhanovsky A.V., Degtyarev A.B., Rozhkov V.A. Peculiarities of computer simulation and statistical representation of time–spatial metocean fields. LNCS #2073, Springer–Verlag, 2001, pp.463–472.
8. Boukhanovsky A.V., Mironov E.U., Rozhkov V.A. Annual rhythms and extremes of Barents sea iceness. Rev. of Russian Geographical Society, vol. 134, 3, 2002, pp. 6–16 (in Russian)
9. Boukhanovsky A.V., Krogstad H.E., Lopatoukhin L.J., Rozhkov V.A. Stochastic simulation of inhomogeneous metocean fields: Part I: Annual variability. Proc. ICCS'03, LNCS, Springer-Verlag, 2003 *(This volume)*.
10. Jardine T.P., Latham F.R. An analysis of wave heights records for the NE Atlantic. Quarterly Journal of Roy. Met. Soc., 1981, vol. 107, pp. 415–426.
11. Kallenberg O.. Random Measures. Berlin, Academie-Verlag, 1983
12. Kalnay E. et al. The NCEP/NCAR 40-Year Reanalysis Project. Bulletin of the American Meteorological Society, No. 3, March, 1996.
13. Levin B.R. Theoretical background of statistical radiophysics. Moscow, Soviet Radio, 1969
14. Loeve M. Probability theory. D. van Nostrand Company Inc., London, 1955.
15. Lopatoukhin L.J., Rozhkov V.A., Ryabinin V.E., Swail V.R., Boukhanovsky A.V., Degtyarev A.B. Estimation of extreme wave heights. JCOMM Technical Report, WMO/TD #1041, 2000.
16. Lopatoukhin L., Rozhkov V., Boukhanovsky A., Degtyarev A., Sas'kov K., Athanassoulis G., Stefanakos C., Krogstad H., The spectral wave climate in the Barents sea. Proc. OMAE 2002, paper 28397.
17. Monin A..S. An Introduction to the Theory of Climate. D. Reidel, 1986.

18. Ogorodnikov V.A., Protasov A.V. Dynamic probabilistic model of atmospheric processes and the variational methods of data assimilation. Russ. J. Numer. Anal. Math. Mod., 12, 1997, pp. 461–479.
19. Ripley B.D. Spatial statistics. NY, Chichester, Brisbane, Toronto: J. Willey & Sons, 1981.
20. Rozhkov V., Lopatoukhin L., Lavrenov I., Dymov V, Boukhanovsky A., Simulation of storm waves. Physics of Atmosphere and Ocean, 2000, 36, N5. (Translation from Russian)
21. Stefanakos Ch.N., Athanassoulis G.A., A unified methodology for the analysis, completion and simulation of nonstationary time series with missing-values, with application to wave data, Applied Ocean Research, Vol. 23/4, pp. 207–220, 2001.
22. Stefanakos Ch.N., Athanassoulis G.A., Bivariate stochastic simulation based on nonstationary time series modeling, ", 13rd International Offshore and Polar Engineering Conference, ISOPE'2003, Honolulu, Hawaii, May 25–30, 2003.
23. Tolman H. User manual and system documentation of WAVE WATCH-III. NOAA technical note. 1999.

Stochastic Simulation of Inhomogeneous Metocean Fields. Part III: High-Performance Parallel Algorithms

Alexander V. Boukhanovsky and Sergey V. Ivanov

Institute for High Performance Computing and Information Systems,
St. Petersburg, Russia
avb@fn.csa.ru, http://www.csa.ru

Abstract. The paper discusses the high-performance parallel algorithms for stochastic simulation of metocean processes and fields. The approaches for parallel representation of sample estimation procedures, linear stochastic systems, Markov chains, periodically correlated processes and inhomogeneous fields are proposed. The speedup of the proposing algorithms is studied in respect to parameters of the models.

1 Introduction

Metocean data fields, like atmospheric pressure, wind speed, ocean waves etc. have a complex spatial and temporal variability. Recently the huge databases of metocean data in the irregular gridpoints are collected (see e.g. [14]). Development of environmental models and use them for data assimilation and reanalysis [10], has allowed to create global information arrays of metocean fields in points of a regular spatial-temporal grid. For the analysis and synthesis of these data the special models, considers in the first [4] and second [5] parts of this paper, has been developed. The modeling procedures often require large amounts of computational resources and are therefore executed on parallel computer systems.

Generally, *parallelization* of statistical computational procedures (including Monte-Carlo techniques) is based on decomposition of sample on the equal sub-volumes (see e.g. [16]). This approach is valid only for independent random values (RV), because in terms of time series (TS) or stochastic fields (SF) the elements of sample are interdependent. Therefore, for the dependent data models, the sophisticated reformulation of the sequential algorithm (and correspondent code) is requires. There are at least two *extensive* ways to solve this problem.

The first way is the automatic translation of sequential code by means of the loop parallelization tools [18]. But if the stochastic algorithm has explicit formalization of interdependence (e.g. – parametrical regression), then the efficiency of this procedure is rather low. The second way concerns the using of the free parallel scientific libraries (as ATLAS, PBLAS, PLAPACK, ScaLAPACK etc. [19]) for the compiling of the code. But the majority of the computational procedures orients on the elected computational tacks (matrix algebra, PDE solving and optimization). Moreover, sometimes

P.M.A. Sloot et al. (Eds.): ICCS 2003, LNCS 2658, pp. 234–244, 2003.

the most labor-consuming part of the algorithm is not respect to any standard procedure.

Thus, the development of the parallel stochastic algorithms is the *creative* problem. The best solution may be obtained using the paradigm of problem "reflection" to parallel architecture of computer [3], take in mind the specifics of data. The main goals of this paper are the follows:

- To illustrate the principles of parallelization for stochastic simulation of metocean processes and fields.
- To study the computational efficiency of the proposed parallel algorithms in respect to parameters of stochastic models.

2 Theoretical Model of Parallel Program

The design of scalable and portable algorithms requires the previous formalization of the theoretical model of parallel program. One of the simple models for computational applications is the BSP (bulk-synchronous parallel) model, associated with simultaneous computation of *p* parallel threads, with barrier synchronization [8]. It allows consider any parallel program in terms of cortege $\langle p, C, \eta \rangle$. Here $C = (V, E)$ is the communication graph, (where V are the vertexes and E are the edges), and $\eta = \langle L_i, g, f_p \rangle$ are the characteristic of processors loading. The values L_i (associated with V_i) are the times of parallel computations of thread *i,* and g is the communication time. The value f_p is the part of sequential operations.

There are a few indexes characterizing the performance of parallel algorithms [7]. For statistical application we consider the speedup index $S_p = T_1 / T_p$ (where T_i are the measured time of computations with *i* parallel threads) as the measure of efficiency $\varepsilon_p = S_p / p$. For maximization of S_p the follows is requires: (1) graph *C* includes at least *p* parallel threads, (2) the loadings of the processors are balanced $L_i / L_j \approx 1$, (3) and value g is minimized.

In practice these requirements are not enough for absolute maximization of S_p, because BSP model ignores the platform-dependent features, e.g. cashing. Therefore, the validation of proposed algorithms would be controlled by means of computational experiment. We use the on-shell cluster "*Paritet*" (4x2-processor nodes), designed in Institute for High Performance Computing and Data Bases (Russia). In spite of the low number of nodes, this cluster reproduced in scale the *Beowulf* architecture and may be used for *qualitative* analysis of parallel algorithms.

3 Parallelization Principles for Stochastic Models

The paradigm of parallel algorithms design requires the formalization of principles for parallelism detection, based on the features of the stochastic models. Below the three general principles are considered.

3.1 Parallel Algorithms, Based on the Ensemble Decomposition

The principle of ensemble decomposition is based on the data parallelism. For metocean fields the first level of decomposition technique is result of the *multiscale* hypothesis proposed by Andrey S. Monin [15]. The hypothesis suggests modelling the total variability of process ζ by means of a set of stochastic models for each temporal scale separately, and with the interdependence taken into account parametrically. It allows present the total distribution function $P_\zeta(x)$ over the probability space Ω in terms of combined distribution

$$P_\zeta(x) = \int_\Omega G_\zeta(x,y)dF_\xi(y)dy = \sum_{k=1}^{p}\int_{\Omega_k} G_\zeta(x,y)dF_\xi^{(k)}(y)dy\,. \tag{1}$$

Here G_ζ is the main scale distribution function and F_ξ is the distribution function of above-scale driving process ξ. Eqn. (1) allows consider the probabilistic characteristic of ζ for each sub-volume $\Omega_k = \{\xi_k\}, \bigcup_{k=1}^{p}\Omega_k = \Omega$ in parallel. In practice for the best balancing of the processor loadings it is easy to consider the initial dataset in terms of natural metocean scales (e.g. day, month, year etc.). For example, in the papers [4,5] the function G_ζ from Eqn. (1) is associated with synoptic variability and F_ξ - with annual and year-to-year variability. Therefore, in simple case (continuous data analysis) the processing of the synoptic data may be carrying out for 12 months in parallel. For the irregular data (series with the data missing) the dynamical balancing is required.

The second level of decomposition is based only on statistical properties of ensemble in terms of RV model. Let us consider the sample estimate Ξ^* of parameter Ξ as any statistical sum $\langle \Im_\xi(x)\rangle_{\Omega^*}$ over the sample Ω^* [22] (here $\langle\ldots\rangle$ is the operation of sample averaging). Such definition allows compute the $\Xi^*_{(k)}$ over the Ω^*_k in the p parallel threads. The obtained sample of estimates $\left\{\Xi^*_{(k)}\right\}_{k=1}^{p}$ may be used for estimation of the total value Ξ^* and its sample variability, roughly if $N >> p$:

$$\Xi^* = p^{-1}\sum_{k=1}^{p}\Xi^*_{(k)},\quad |\Xi - \Xi^*| \le t_\alpha(p)\frac{\sigma^*_\Xi}{\sqrt{p}}\,. \tag{2}$$

Here $t_\alpha(p)$ is the $1-\alpha\%$ quantile of Student' distribution, and σ^*_{Ξ} is the estimate of r.m.s., calculated over the sample $\left\{\Xi^*_{(k)}\right\}_{k=1}^{p}$. In the Fig. 1(a) the communication graph of the parallel algorithm is shown. The vertex ***A*** is the data preparation and storage on the local nodes, vertex ***B*** – parallel computation the $\Xi^*_{(k)}$, and vertex ***D*** – computation of total estimate (2).

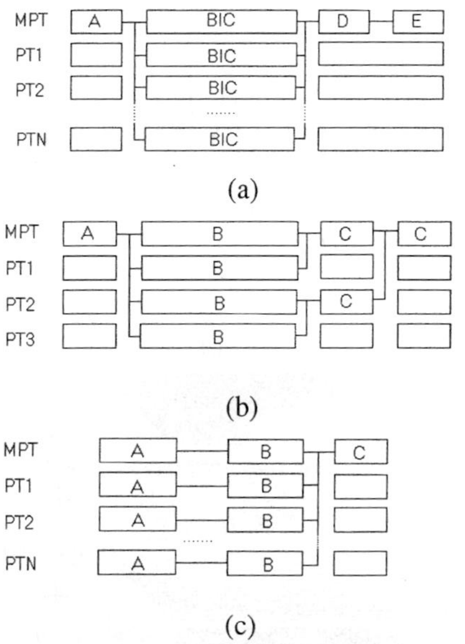

Fig. 1. Communication graphs of the parallel algorithms
(Here PT – Parallel thread, MPT – Main parallel thread)

For some statistical parameters another operators are uses instead of $\langle \ldots \rangle$. For example, for univariate RV the $q\%$-quantiles may be estimated as order statistics $\xi^*_q = sort(\Omega^*)_{[qN]+1}$. Such definition is clear for parallelism detection: the sample estimates of $x_q^{*(k)} = sort(\Omega^{(k)})_{[qN_k]+1}$ for all threads $k=\overline{1,p}$ are considers in parallel (see Fig. 1(a), vertex ***C***). For the computation of total estimate the Eqn. (2) may be used also; but for last terms of sample such estimates are biased. Therefore, the total sorting of previously sorted Ω^*_k is required (vertex ***E***). In the Fig 1(a) seen, that the most labor-consuming operations: estimation of sums (***B***) and sorting (***C***) are parallel.

In the Fig 2(a) the speedup indexes S_p vs. p for statistical estimation of univariate RV are shown for different volumes N of the sample Ω^*. It is seen, that the scalability of the algorithm is rather good, especially – for high value of N.
Even for RV model, the inverse problem – stochastic simulation by Monte-Carlo technique sometimes require more computational resources, than the estimation. The simulation of random numbers is traditional parallel problem (see e.g. [21]). The parallel scheme of RV simulation is close to Fig. 1(a), when vertex ***B*** contains the random number generation instead of statistical summation.

3.2 Parallel Algorithms, Based on the Strong Mixing Principle

The ensemble decomposition principle is the best mainly for the RV model. For the TS and SF modeling more sophisticated approaches are require. Here we consider the class of the parallel algorithms, based on the strong mixing principle [11]. In terms of

TS model this expressed as $F_{\zeta(t)\zeta(s)}(x,y)\xrightarrow[|t-s|\to\infty]{}F_{\zeta(t)}(x)F_{\zeta(s)}(y)$, where t and s are the time moments. This fact allows design the algorithms on the base of the parallel simulation of p independent time series, and its further sewing.

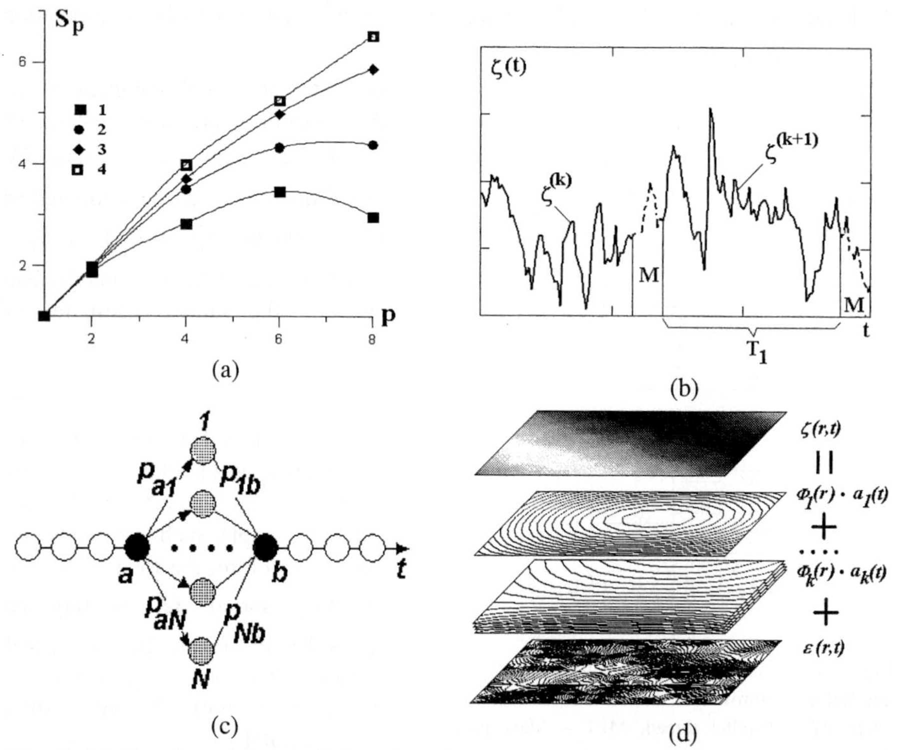

Fig. 2. **(a)** Speedup of the parallel statistical estimation. Here 1-N=2000, 2-N=4000, 3-N=10000, 4-N=20000.
(b-d) Illustration of the main parallelization principles: (linear stochastic systems **(b)**, Markov chains **(c)** and models of inhomogeneous stochastic fields **(d)**).

3.2.1 Parallel Models of Linear Stochastic Systems

For simulation of Gaussian stationary TS and homogeneous SF $\zeta(\mathbf{u})$ the model of linear stochastic system in terms of partial differential equation may be adopted [1]:

$$\left[\mathop{*}_{k=1}^{N} L_k\right]\zeta(\mathbf{u})=\left[\mathop{*}_{k=1}^{N} E_k\right]\varepsilon(\mathbf{u})+\left[\mathop{*}_{k=1}^{N} Q_k\right]\eta(\mathbf{u}). \tag{3}$$

Here $\mathbf{u}=(t,\mathbf{r})$, where $\mathbf{r}$ is vector of spatial coordinates and t is the time, ε is the white noise field, η is the driving stochastic process (input signal) and $L_k=\sum_{i=0}^{n_k} l_{ki}\frac{\partial^i}{\partial u_k^i}$, $E_k=\sum_{i=0}^{m_k} e_{ki}\frac{\partial^i}{\partial u_k^i}$, $Q_k=\sum_{i=0}^{s_k} q_{ki}\frac{\partial^i}{\partial u_k^i}$ are the partial differential operators with constant coefficients ($*$ is the operator composition). When $N = 1, q_{ki} = 0, e_{ki} = 0, k \geq 1$, the Eqn. (3) reduces to the well-known autoregressive model [9]:

$$\zeta_t = \sum_{i=1}^{M} \Phi_i \zeta_{t-i} + \sigma_\varepsilon \varepsilon_t, \quad t = \overline{M+1,T} . \tag{4}$$

The parameters $\Phi_i, \sigma_\varepsilon$ are obtains by means of linear equation system

$$\sum_k \Phi_k K_{|i-k|} = K_i, \quad K_i = E\left[(\zeta_t - m_\zeta)(\zeta_{t-i} - m_\zeta)\right] . \tag{5}$$

Here K_i is the value of covariance function of ζ_t. It may be estimated by means of parallel algorithm, described in Chapter 3.1. It is seen, that the Eqn. (4) has an explicit recurrence, thus, the direct loop parallelism is impossible here. Taking to account, that $M \ll T$, let us consider the parallelization principle, shown in Fig. 2(b). If we decompose the length of simulated series as $T = pT_1 + (p-1)M$, the follows steps may be carry out:

- Estimation of the model parameters by Eqn. (5) (vertex ***A***)
- Parallel simulation of T_1-length independent time series $\zeta_t^{(k)}, k = \overline{1,p}$ (vertex ***B***)
- *Couple sewing* of the simulated series by Eqn. (4) (vertex ***C***).

Table 1. Speedup indexes for parallel simulation of linear dynamic systems (left part) and Markov chains (right part)

Linear dynamic system (4-6)					Markov chain (7-8)				
T	M	*Number of processors p*			T	M	*Number of processors p*		
		2	4	8			2	4	8
2000	10	1.7	1.3	0.8	10^3	10	1.0	0.9	0.7
2000	40	1.8	2.9	3.8	10^3	100	0.9	0.7	0.4
4000	10	1.7	2.6	3.2	10^4	10	1.7	2.3	2.7
4000	40	1.9	3.3	4.8	10^4	100	1.8	2.5	2.7
8000	10	1.8	2.9	3.9	10^5	10	1.6	2.5	3.3
8000	40	1.9	3.4	5.6	10^5	100	1.9	3.5	5.8

The algorithm of couple sewing is close to approach for environmental data missing recovering from the paper [20]. It allows present the values of $\tilde{\zeta}_t, t = T_1 + 1, ..., T_1 + M$ by means of recurrent equation

$$\tilde{\zeta}_t = \sum_{i=1}^{M} \Theta_i^{(1)} \zeta_{t-i} + \sum_{j=1}^{M} \Theta_j^{(2)} \zeta_{t+j} + \sigma_\delta \delta_t . \tag{6}$$

Here $\Theta_i^{(\bullet)}, \sigma_\delta$ are the model parameters, and δ_t is the white noise. The Eqn. (6) is the double-side generalization of Eqn. (5), and the values $\left\{\zeta_{T_1-M+1}^{(k)}, \ldots, \zeta_{T_1}^{(k)}\right\}$ and $\left\{\zeta_1^{(k+1)}, \ldots, \zeta_M^{(k+1)}\right\}$ of the TS from two parallel threads *(k)* and *(k+1)* may be consider as the boundary values. The procedure like (5) is used for the estimation of $\Theta_i^{(\bullet)}$. In the Table 1 (left part) the speedup indexes S_p vs. p are present for different values of T

and M. For the short data series the S_p became less 1, but with the increasing of T the speedup became rather better (e.g. for $T = 8000$, $M = 40$ the $S_p = 5.6$ and $\varepsilon = 70\%$ for 8 processors).

3.2.2 Parallel Markov Models

For stationary Markov chains and processes the strong mixing principle is also valid. Taking to account, that for numerical computations the discrete representation of the continuous Markov processes are traditionally uses, we consider only discrete M-states Markov chains with parameters [2]:

$$\pi = \begin{bmatrix} \pi_1 \\ \vdots \\ \pi_M \end{bmatrix}, \quad \mathbf{P} = \begin{bmatrix} p_{11} & & p_{1M} \\ & \ddots & \\ p_{M1} & & p_{MM} \end{bmatrix}, \quad \sum_{k=1}^{M} p_{ik} = \sum_{k=1}^{M} \pi_k = 1, \quad i = \overline{1,M}. \tag{7}$$

Here π is the vector of limit probabilities and $\mathbf{P}$ is the matrix of transient probabilities. The general principle of parallelism is shown in the Fig. 2(c) for first order Markov chain. The parallelization technique is close to the same for linear stochastic systems (the communication graph in the Fig. 1(b)). It consists of three stages:

- Estimation of parameters (7) and data preparation (vertex ***A***)
- Parallel simulation of independent sub-chains (vertex ***B***)
- *Couple sewing* of the simulated sub-chains (vertex ***C***).

The sewing procedure uses the states in the tail of first sub-chain (a) and head of second sub-chain (b) as the boundary conditions for simulation of the sewing Markov state k, with the conditional probability:

$$p_k^{(a,b)} = \frac{p_{ak}\, p_{kb}}{\sum_{l=1}^{M} p_{al}\, p_{lb}}, \quad a,b,k = \overline{1,M}. \tag{8}$$

In the Table 1 (right part) the speedup indexes S_p vs. p for Markov chains with different T and M are shown. For low T=10^3 the speedup may be less 1, due to high communication expenses for sewing of the sub-chains. But for high length of the chain (T=10^5) the values of S_p increases in respect to M and p. E.g. for Markov chain with the 100 states the S_4=3.5 and S_8=5.8 times.

3.3 Parallel Algorithms, Based on the Functional Approximation Principle

The principle of functional approximation is based on the classical representation of TS or spatio-temporal SF $\zeta(\mathbf{r},t)$ in terms of the deterministic function, dependent from the set of random arguments Ξ [12]:

$$\zeta(\mathbf{r},t) = \zeta(\mathbf{r},t \mid \Xi). \tag{9}$$

It allows decompose the spatial (or spatio-temporal) domain on the set of equal sub-volumes. The main advantages of this principle are obvious for the nonstationary TS and inhomogeneous SF, where the strong mixing principle is not valid.

3.3.1 Periodically Correlated Time Series

One of the simplest examples of nonstationary TS is the model of periodically correlated stochastic process (PCSP) $\zeta(t)$, where the mathematical expectation $m_{\zeta}(t)=m_{\zeta}(t+\tau)$ and covariance function $K_{\zeta}(t,s)=K_{\zeta}(t+\tau,s+\tau)$, τ is the period of correlation (e.g. – one year). The PCSP model is widely uses for simulation of the annual variability of different metocean processes [13], e.g. sea waves, wind speed, atmospheric pressure, ice cover, air and water temperature [6] etc. In the book [17] the simulation algorithm for PCSP with explicit formalization of dependence is proposed. Instead of this, let us consider the alternative *parametrical* model of PCSP as expansion [4]:

$$\zeta(t)=\sum_{k}\alpha_k(t)exp(i\Lambda_k t). \tag{10}$$

Here $\Xi=\{\alpha_k(t)\}$ - the set of parameters, $\Lambda_k=2\pi k/\tau$. The inverse transformation of Eqn. (10) allows obtain the explicit expression for TS $\alpha_k(t)$:

$$\alpha_k(t)=\int_0^t \zeta(s)H(t,s)exp(-i\Lambda_k s)ds. \tag{11}$$

Its covariance function is:

$$K_{\alpha_k\alpha_j}(t,s)=\int_0^t\int_0^s R(t,s,x,y)K_{\zeta}(x,y)exp\left[-i(\Lambda_k x+\Lambda_j y)\right]dxdy. \tag{12}$$

Here $H(t,s)$ is the kernel function, $R(t,s,x,y)=H(t,s)H(x,y)$. When $H(t,s)$ is the step function for $s\in\left[t,t-\tau\right]$, the time series $\alpha_k(t)$ became stationary and Gaussian [6]. The Eqns. (10,12) allows the domain parallelization of PCSP computation. Communication graph of this algorithm is shown in the Fig. 1(c), where vertex ***A*** respects to simulation of TS (11) with covariance function (12) by means of multivariate autoregressive model (4), and vertex ***B*** is the parallel computation of the Eqn. (10) for equal time intervals $\left[t_k,t_{k+1}\right]_{k=0}^{p}$. After the sub-volumes computations all the data send to the main computational thread (vertex ***C***).

3.3.2 Inhomogeneous Spatio-Temporal Fields

The principle of functional approximation is applied for simulation of the inhomogeneous spatio-temporal metocean fields. In [4,5] such models are presented as the expansion:

$$\zeta(\mathbf{r},t)-m(\mathbf{r},t)=\sum_{k=1}^{M}a_k(t)\Phi_k(\mathbf{r},t)+\varepsilon(\mathbf{r},t). \tag{13}$$

Here $a_k(t)$ are the coefficients, $m(\mathbf{r},t)$ is the mathematical expectation, $\Phi_k(\mathbf{r},t)$ is the spatio-temporal basis, $\varepsilon(\mathbf{r},t)$ is the inhomogeneous white noise. The estimation of $a_k(t)$ is fully discussed in the paper [4]. The general principle of parallelism in Eqn. (13) is shown in the Fig. 2(d). It is seen, that there are at least two alternative ways for parallelization. The first *(horizontal)* way is the domain decomposition on the equal spatial areas, and calculation the Eqn. (13) for each area in parallel. The communication graph of this algorithm is shown in the Fig. 1(c) and discussed in Section 3.3.1. The second *(vertical)* way is based on the parallel computation of the coefficients (vertex *A* in Fig 1(c)) and the terms $a_k(t)\Phi_k(\mathbf{r},t)$ for all the gridpoints $\{\mathbf{r}_k\}_{k=1}^{N}$, and finally – summation of Eqn. (13) in the main parallel thread (vertex *C*). Theoretically both schemes are valid, but in practice the real speedup depends from the total numbers of gridpoints N and numbers of basic functions *M*. In the table 2 the speedup indexes S_p vs. *p* are present for different values of *N* and *M*. Take in mind, that the parameters a_k may be considered as RV, stationary SF or PCSP [4], we carry out all the computations for three classes of complexity (associated with loadings L_A for simulation of coefficients).

Table 2. Speedup indexes for parallel simulation of inhomogeneous metocean fields

Complexity	*Horizontal* parallelization					*Vertical* parallelization				
	M	*N*	*Processors p*			*M*	*N*	*Processors p*		
			2	4	8			2	4	8
I	4	500	1	1.2	1.2	10	100	0.7	0.4	0.4
	8	1000	1.5	2.1	2.3	100	100	1.0	0.9	0.7
II	4	500	1.5	2.0	1.6	10	100	1.9	1.6	1.8
	8	1000	1.9	3.4	5.7	100	100	~2	1.7	3.2
III	8	500	~2	3.7	6.1	10	100	1.9	3.4	1.9
	16	1000	~2	3.9	7.7	100	100	~2	3.9	5.9

From the table 2 seen, that for horizontal (domain) parallelization the speedup is the highest for the high-complexity models. E.g., for third class of complexity (PCSP model of coefficients) the S_8=7.7 (efficiency ε_8=96%) when *M*=16 and *N*=1000 spatial points. For the complexity class I (RV model, describing only spatial variability), the speedup is low. The vertical (sum) parallelization allows obtain the high speedup only if the number of basic functions is close to number of spatial points. Therefore, this way is not adopts for the reduction of data dimensionality in stochastic models and may be considered only for the specific problems, as the simulation of the fields with very complicated spectral structure.

4 Conclusions

This paper has demonstrated the main principles of parallel algorithms design for computational multivariate statistics of spatio-temporal metocean fields. The principles of ensemble decomposition, strong mixing and functional approximation allow develop the parallel stochastic models for dependent time series and fields (including autoregressive TS, Markov chains, PCSP and inhomogeneous spatio-temporal SF). The analysis of speedup sensitivity to the model parameters shown, that the efficiency of the proposed algorithms is the best for huge model datasets, in practice applying for numerical study of extreme metocean events [5].

Acknowledgment. This research has been partly founded by INTAS 99-0666 Project: "Estimation of extreme metocean events".

References

1. Adomian G. Stochastic systems. *Academic Press*, NY (1983)
2. Bharucha-Reid A.T. Elements of the theory of Markov Processes and their applications. *MC Graw-Hill Book Company Inc., New-York, Toronto, Tokyo* (1960).
3. Bogdanov A.V., Gevorkyan A.S., Stankova E.N., Pavlova M.I. Deterministic computation towards indeterminism. *Proceedings of ICCS'02, LNCS, 2331, Springer-Verlag* (2002), pp. 1176–1183.
4. Boukhanovsky A.V., Krogstad H.E., Lopatoukhin L.J., Rozhkov V.A. Stochastic simulation of inhomogeneous metocean fields: Part I: Annual variability. *Proc. ICCS'03, LNCS* (2003) (*This volume*).
5. Boukhanovsky A.V., Krogstad H.E., Lopatoukhin L.J., Rozhkov V.A., Athanassoulis G.A., Stephanakos Ch.N. Stochastic simulation of inhomogeneous metocean fields: Part II: Synoptic variability and rare events. *Proc. ICCS'03, LNCS* (2003) (*This volume*).
6. Boukhanovsky A.V., Mironov E.U., Rozhkov V.A. Annual rhythms and extremes of Barents sea iceness. *Rev. of Russian Geographical Society, vol. 134, 3* (2002) pp. 6–16 *(in Russian)*
7. Cosnard M., Trystan D. Parallel algorithms and architectures. *Int. Thomson Publishing Company* (1995)
8. Gerbessiotis A.V. Architecture independent parallel algorithm design: theory vs practice. *Future Generation Computer Systems*, 18 (2002), pp. 573–593.
9. Jenkins G.M., Watts D.G. Spectral analysis and its application. *Holden-Day, San-Francisco* (1969)
10. Kalnay E., M. Kanamitsu, R. Kistler, W. Collins, D. Deaven, L. Gandin, M. Iredell, S. Saha, G. White, J. Woollen, Y. Zhu, A. Leetmaa, R. Reynolds, M. Chelliah, W. Ebisuzaki, W.Higgins, J. Janowiak, K. C. Mo, C. Ropelewski, J. Wang, R. Jenne, D. Joseph. The NCEP/NCAR 40-Year Reanalysis Project. *Bulletin of the American Meteorological Society*, №3, March, (1996).
11. Leadbetter M., Lindgren G., Rootzen H. Extremes and related properties of random sequences and processes. *Springer-Verlag, NY,* (1986)
12. Loeve M. Probability theory. *D. van Nostrand Company Inc., London,* (1955).
13. Lopatoukhin L.J., Rozhkov V.A., Ryabinin V.E., Swail V.R., Boukhanovsky A.V., Degtyarev A.B. Estimation of extreme wave heights. *JCOMM Technical Report, WMO/TD #1041* (2000).
14. Mikhailov N.N. Vyazilov E.D., Lomanov V.I., Studyonov N.S., Shairmardonov M.Z.: Russian Marine Expeditionary Investigations of the World Ocean. R. Tatusko and S. Levitus (eds.), World Data Center for Oceanography, International Ocean Atlas and Information Series, Vol. 5, *NOAA Atlas NESDIS 56, U.S. Government Printing Office, Washington, D.C.* (2002) 184 pp.
15. Monin A.S. An Introduction to the Theory of Climate. *D. Reidel* (1986)
16. Musial G., Debski L. Monte Carlo method with parallel computation of phase transitions in the three-dimensional Ashkin-Teller mode. *Proc. PPAM'01, LNCS, 2328, Springer-Verlag,* (2002), pp. 535–543.

17. Ogorodnikov V.A., Prigarin S.M. Numerical modelling of random processes and fields: algorithms and applications. *VSP, Utrecht, the Netherlands* (1996) 240 p.
18. Pandle S., Agrawal D.P. (Eds.) Compiler Optimization for Scalable PC, *LNCS 1808* (2001).
19. Parallel scientific libraries. *On the site http://parallel.ru/cluster* (2003)
20. Stefanakos Ch.N., Athanassoulis G.A., A unified methodology for the analysis, completion and simulation of nonstationary time series with missing-values, with application to wave data, *Applied Ocean Research, Vol. 23/4*, (2001) pp. 207–220.
21. Tan C. J. K. The PLFG parallel pseudo-random number generator. *Future Generation Computer Systems*, 18 (2002), pp. 693–698.
22. Zacks Sh. The theory of statistical inference. John Wiley & Sons., Inc. (1971)

Workshop on Grid Computing for Computational Science

Performance Comparison of Process Allocation Schemes Depending upon Resource Availability on Grid Computing Environment

Hiroshi Yamamoto[1], Kenji Kawahara[1], Tetsuya Takine[2], and Yuji Oie[1]

[1] Dept. of Computer Science and Electronics, Kyusyu Institute of Technology, Kawazu 680-4, Iizuka, 820-8502 Japan
yamamoto@infonet.cse.kyutech.ac.jp
{kawahara, oie}@cse.kyutech.ac.jp

[2] Dept. of Applied Mathematics and Physics Graduate School of Informatics, Kyoto University, Yoshidahonmachi, Sakyou–ku, Kyoto, 606–8501 Japan
takine@amp.i.kyoto-u.ac.jp

Abstract. Improvements in the performance of end-computers and networks have recently made the construction of a grid system over the Internet feasible. A grid environment comprises many computers, each having a set of components and distinct performance, that are shared among many users and managed in a distributed manner. Thus, it is important to focus on a situation in which the computers are used unevenly due to decentralized management by different process schedulers. In the present study, as a preliminary theoretical investigation of the effect of such features on the performance of process schedulers, the average execution time of a long-lived process in the process allocation scheme employed in the decentralized environment is analytically derived using M/G/1–PS queues. The impact of the distribution of CPU utilization on the performance of these schemes is also investigated assuming CPU utilization for each computers in a grid environment follows a probability distribution function.

1 Introduction

Recent improvements in the performance of end-computers and networks have made it feasible to construct a grid system over the Internet. A grid system [1] is constructed by connecting geographically distributed computers over the Internet to secure greater computing power. Several fundamental services are indispensable for sharing computational resources (e.g., CPU time, memory, storage, etc.) in a grid environment, including security management, resource management, and process execution on remote hosts, which are run as a low-level middleware [2]. The Globus Toolkit [3] produced by the Globus project [4] is well-known middleware providing these important services. Using these services, a computer (User) can securely submit a set of process executions to the

P.M.A. Sloot et al. (Eds.): ICCS 2003, LNCS 2658, pp. 247–256, 2003.

grid environment. Users require a scheduling system linked to the middleware to appropriately select computers to which the processes are to be allocated.

The grid environment consists of many computers, each with distinct processing performance, and the resources of each computer are shared among many users who are geographically distributed in a decentralized manner. The resource status of each computer is managed in a distributed manner rather than a centralized manner, most likely by many different organizations [5]. Although the number of processes executed on each computer and its utilization cannot be controlled strictly, the resources can be used unevenly and dynamically. To achieve efficient grid computing, it is necessary to develop a process allocation scheme that considers a distributed resource management architecture. Process allocation schemes based upon the resource status of available computers have already been proposed [6][7]. However, such schemes do not cater for decentralized resource management, instead assuming that users are able to perceive all processes, like in cluster computing.

In the present research, we focus particular attention on a situation in which computers are very likely to be used unevenly and utilization will change dynamically because of decentralized management by different process schedulers. Specifically, in this paper, the effect of such features on the performance of process schedulers is preliminarily investigated based on theoretical considerations. The process allocation schemes treated here are implemented in a decentralized manner in which users initially investigate resource status such as CPU performance and utilization, and then divide local long-lived process into a number of *sub-processes* to be allocated to computers with good resource availability.

The computers are modeled as M/G/1–PS (processor sharing) queues, and the average execution time of long-lived process in each scheme are derived analytically. Adopting M/G/1–PS queues allows the average sub-process sojourn time as a function of resource status alone. As the grid computing environment is constantly evolving, it is very difficult to precisely describe the CPU utilization. Hence, a probability distribution function associated with CPU utilization is introduced to describe the uneven and dynamic utilization of computers, allowing the impact of these features on the performance of proposed process allocation schemes to be evaluated.

The rest of this paper is organized as follows: In Section 2, a resource management model of grid computing is described, and the process allocation schemes are proposed. Section 3 outlines the analytical evaluation of the average process execution time, which is used in section 4 to evaluate the characteristics of the proposed process allocation schemes. Finally, this paper is concluded in section 5.

2 Grid Computing Architecture

2.1 Resource Management Architecture

The resource management architecture employed here is shown in Fig. 1, consisting of computers and directories.

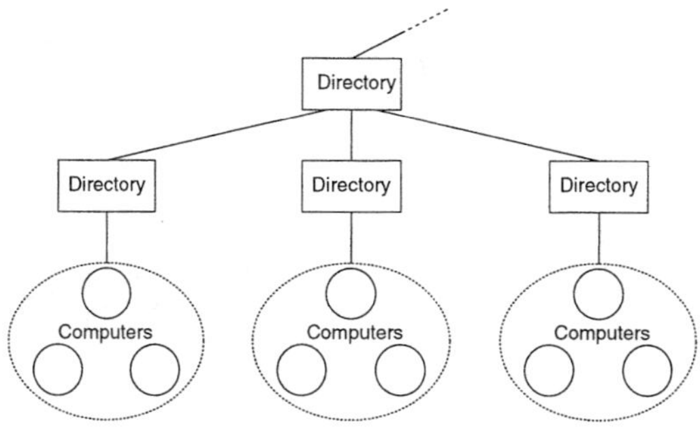

Fig. 1. Resource Management Architecture.

Computers Every computer is assumed to manage its own performance and capability using a database. This information is measured periodically using a resource management toolkit such as the Network Weather Service [8]. In this context, computers execute processes allocated by other computers involved in the grid, while simultaneously executing necessary local processes.

Here, CPU performance is denoted C_i, and CPU utilization ρ_i $(0 \leq \rho_i < 1)$ is defined as the load imposed on the computer R_i. The computer can execute processes allocated to it at the rate of C_i per unit time, and ρ_i of the total capability of the CPU is consumed when the user allocates new sub-process.

Directory The directory manages the static computer information as a database, including the operating system, processing capability, memory capacity, and IP addresses. The database is updated periodically when computers notify the directory. According to requests from users, the directory asks computers under its control to send back information related to the resource status, which is cached in the directory database, and then replies to users with the obtained information. Each entry registered with a database has a lifetime [9]: if the entry has not been updated during the lifetime, the directory determines that the relevant computer is no longer available.

2.2 Grid Computing

The grid computing process is outlined in Fig. 2. After a process arrives, the user divides long-lived processes into n independent small-scale sub-processes. The long-lived process has a computational complexity of X. Here, as overhead in each sub-process due to segmentation is neglected for simplicity, the computational complexity of each sub-process x can be defined as X/n. Next, the user requests the directory to investigate the resource status of computers, and to reply with the relevant information. The user selects n computers to which sub-processes are allocated based on the information from the directory. The computer resource information is presented in detail in section 3. In this study,

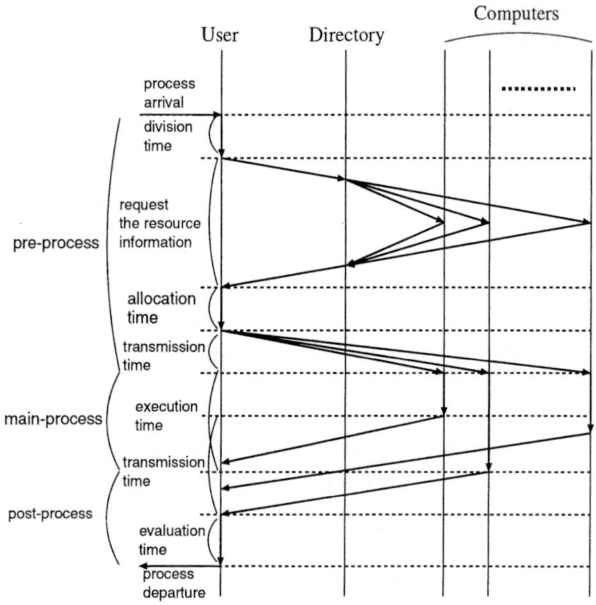

Fig. 2. Grid Computing.

it is reasonably assumed that all processes are independent, have the same priority, and are fairly scheduled in a round-robin manner on all computers. The overall process execution time is determined by the time when the result of the latest sub-process is returned.

As shown in Fig. 2, the interval from the allocation of sub-processes to the completion of all sub-processes is defined as the *main-process*. Processes to be executed before the main-process begins are defined as the *pre-process*, and those executed after the main-process are defined as the *post-process*. In this paper, the performance of each sub-process allocation scheme is evaluated based on the process execution time for the main-process.

2.3 Sub-process Allocation Scheme

Some adaptive sub-process allocation schemes are described below. Here, the limited case of at most one sub-process allocated to each computer is considered.

Scheme 1 (Random Selection). The user randomly selects n (number of sub-processes) computers from all the available computers, then allocates all n sub-processes as selected. In this scheme, the user does not have to consider the computer resource information, thereby minimizing the pre-process time.

Scheme 2 (Resource Availability First Selection). The user selects n least-loaded computers and allocates sub-processes to these computers. In this scheme,

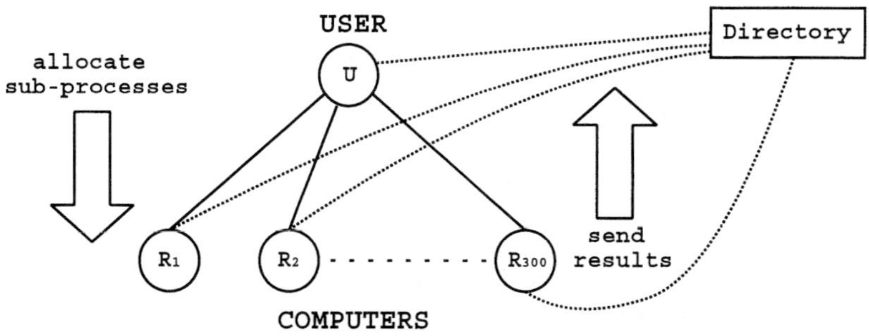

Fig. 3. Analytical Model.

the user needs to receive the current resource information for all computers and compare the available processing capability. This results in a lengthy pre-process time.

Scheme 3 (Random Selection with Resource Availability Threshold). The user requests the directory to ask all computers to send back a reply if the computer satisfies the requirement that the current utilization is lower than some threshold T. The user then randomly selects n computers among those responding to the inquiry, and allocates sub-processes accordingly. When T is set relatively low, fewer than n computers may satisfy the requirement, in which case the user increases T by some amount and re-investigates the resource status. This process is repeated until n or more computers satisfy the requirement.

The pre-process time of Scheme 3 is expected to be shorter than that of Scheme 2 because the user does not need to receive resource information from all available computers, and also does not need to compare the information.

3 Analytical Model

As illustrated in Fig. 3, the analytical model is defined such that the user is directly connected to 300 computers over the Internet. Transmission delay between users is neglected. The performance of each of the schemes described in section 2.3 is evaluated using this model. Computers are evenly divided into three groups in terms of CPU performance; class A, B and C, corresponding to CPU performance C_A, C_B and C_C of 50, 100 and 150, respectively. A CPU performance of 1.0 indicates that the computer takes a unit time of 1.0 to calculate a process of unit workload. Thus a CPU with a performance of 50 can calculate a process with a workload of 100 in 2 unit periods. As only the main-process is considered here, the average process execution time is adopted as a performance measure, and is obtained analytically as follows.

Each computer is modeled using M/G/1–PS queues, where all processes in the system are fairly served with a round-robin scheduler. Thus, this model allows for the coexistence of other processes, which may be allocated from other users and/or generated by the local computer. If a sub-process with computational complexity x is allocated on computer R_i with CPU performance C_i and utilization ρ_i, the sub-process mean sojourn time is given by

$$E[W_i] = \frac{x/C_i}{1-\rho_i}. \tag{1}$$

Considering that the average process execution time is defined as the interval between the allocation and receipt of all sub-process, if one process is divided by the user into n sub-processes, which are allocated to computers $R_1, \cdots, R_n$ the average process execution time is defined as the maximum sub-process mean sojourn time as follows.

$$E[W] = \max\{E[W_1], \cdots, E[W_n]\}. \tag{2}$$

The current resource capability I_i of computer R_i, which indicates the average processing time for a process with unit workload, is defined as

$$I_i = \frac{1/C_i}{1-\rho_i}. \tag{3}$$

Even if the CPU performance of the three groups differ, the user can select computers in order of the value I_i to allocate sub-processes in Scheme 2. In Scheme 3, the user randomly selects computers with current resource capability I_i higher than the predetermined threshold T. Thus, each scheme mentioned in section 2.3 is valid, and is evaluated as follows:

1. Computer resource information is investigated using Eq. (3).
2. Sub-processes are allocated to n computers according to the current resource capability.
3. The average process execution time is evaluated using Eq. (2).

However in general, the CPU utilization changes dynamically. Therefore, 10,000 experiments were conducted under the condition that the average CPU utilization of available computers $\bar{\rho}$ $(0 \leq \bar{\rho} < 1)$ is fixed while the CPU utilization of each computer is changed. The mean of the average process execution time for all experiments is then employed as the performance measure.

The grid computing environment is constantly evolving, which makes it very difficult to precisely describe the CPU utilization of computers involved. Thus, the CPU utilizations is assumed to follow a power distribution in which the CPU utilization ρ_i of computer R_i is given as follows.

$$F(\rho_i) = \rho_i^c. \quad \left(i = 1, \cdots, 300, \quad c = \frac{\bar{\rho}}{1-\bar{\rho}}\right) \tag{4}$$

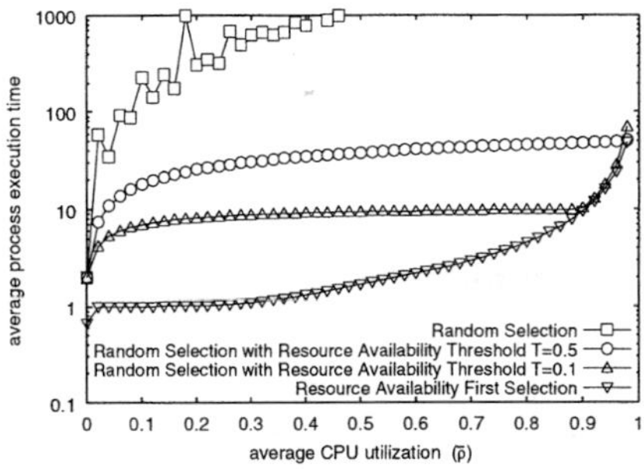

Fig. 4. Average Process Execution Time vs. Average CPU Utilization.

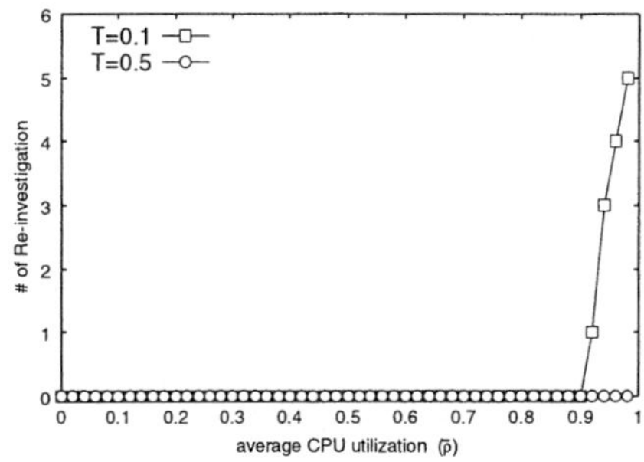

Fig. 5. Number of Re-investigations vs. Average CPU Utilization.

4 Results and Discussion

4.1 Average Process Execution Time vs. Average CPU Utilization

The workload of each process X (i.e., computational complexity) was set to 10,000, and the number of sub-processes n was set to 100. Therefore, the computational complexity x of each sub-process is 100 (=10000/100).

Figure 4 shows the average process execution time as a function of average CPU utilization. In random selection (Scheme 1), 100 computers are randomly chosen independent of CPU utilization or class. Therefore, some of the sub-processes will be assigned to computers of class A, resulting in an execution time of as long as 2 (=100/50) even when the average CPU utilization is 0. With the power distribution, allocation under scheme 1 may result in assignment to a highly loaded computer. As a result, the average process execution time increases markedly with average CPU utilization.

In Scheme 2, the user selects the n least-loaded computers and allocates sub-processes accordingly. Consequently, the average process execution time of Scheme 2 outperforms the other schemes treated here. The performance of Scheme 3 falls between schemes 1 and 2 in that n computers are selected randomly, but from computers that are not highly loaded. From Fig. 4, as the threshold T is reduced, the average process execution time becomes smaller over a wide range of $\bar{\rho}$ and approaches that of Scheme 2. However, reducing T may also result in re-investigation if too few computers are available. Although this is a concern, Fig. 5 shows that no re-investigation occurs under the present model until $\bar{\rho}$ becomes more than 0.9, even in the case of $T = 0.1$.

4.2 Ratio of Sub-processes Allocated to Computers in Each Class

Given the present method of updating computer information, it may occur that many users will select similar sets of computers based on information that may not be the latest, which could lead to critical performance degradation.

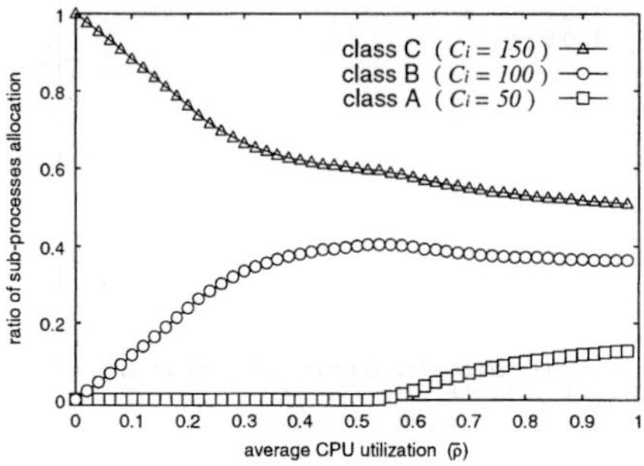

Fig. 6. Ratio of Sub-Processes Allocated to Each Class (Resource Availability First Selection).

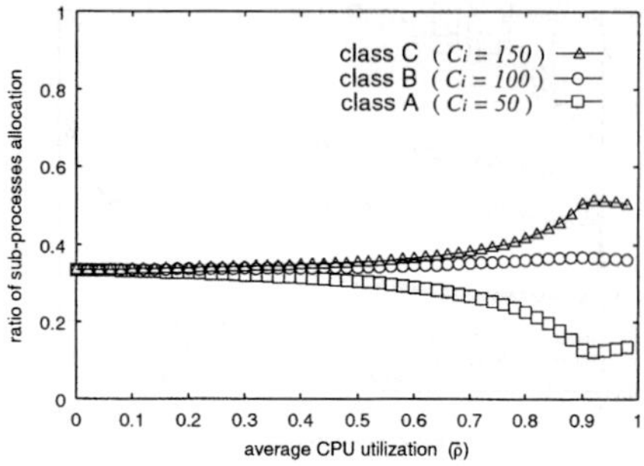

Fig. 7. Ratio of Sub-Processes Allocated to Each Class (Random Selection with Resource Availability Threshold: $T = 0.1$).

To examine this in more detail, the ratio of sub-processes allocated to computers belonging to each class is investigated. As shown in Fig. 6, Scheme 2 allocates many sub-processes to computers with higher CPU performance (i.e., class C). As the average CPU utilization increases, computers with relatively low CPU performance are also selected. However, even if the average CPU utilization is 0.5, 60% of all sub-processes are assigned to computers in class C. Therefore, it is very likely under Scheme 2 that many users in the grid environment will select a few computers with high performance or that were lightly loaded at the same time, thereby leading to critical performance degradation if processes are allocated independently by several schedulers.

Figure 7 shows that Scheme 3 selects computers evenly from all classes. For example, in the case that the average CPU utilization is smaller than 0.5, the ratio of sub-processes allocated to each class is around 0.33. Therefore, Scheme 3 will use all types of computers fairly, while providing moderate performance as shown in Fig. 4.

4.3 Effect of the Number of Sub-processes on the Average Process Execution Time

The effect of the number of sub-processes is illustrated in Fig. 8 in terms of the average process execution time for average CPU utilization $\bar{\rho}$ of 0.6. Normally, as the number of sub-processes increases, the complexity of each sub-process x decreases, resulting in a comparable decrease in average process execution time for Scheme 2 down to some minimum value. As shown in Fig. 8, as the number of sub-processes is increased past 130, the average process execution time begins to increase again because the user must allocate sub-processes to highly loaded computers with higher probability. The average process execution time for Scheme 3 improves with the number of sub-processes as long as all the selected computers have I_i smaller than T. From Fig. 8, the average process

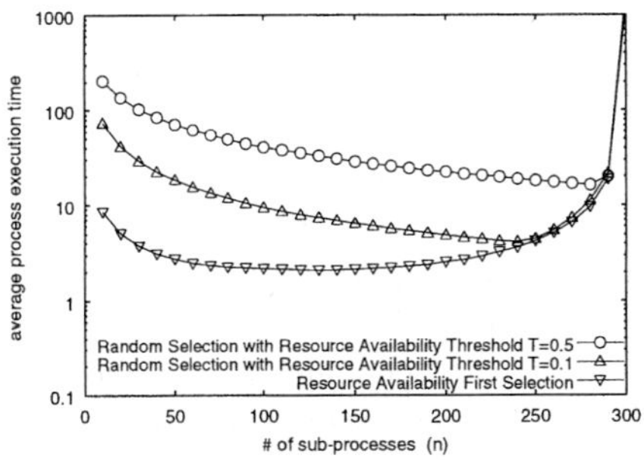

Fig. 8. Average Process Execution Time vs. the Number of Sub-processes.

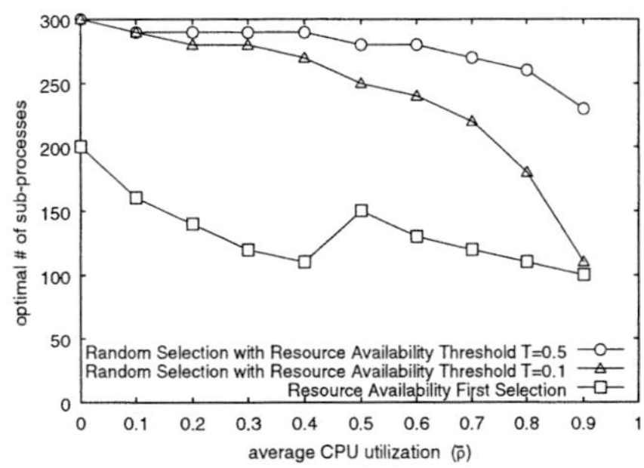

Fig. 9. Optimal Number of Sub-Processes.

execution time decreases with increasing n for $T = 0.1$ as long as n is smaller than 240. In this case, no re-investigation occurs.

As mentioned above, there exists an optimal number of sub-processes that minimizes the average execution time for some value of $\bar{\rho}$. Figure 9 illustrates the optimal number of sub-processes for each $\bar{\rho}$. The optimal number of sub-processes is dependent on $\bar{\rho}$, demonstrating that a larger number of sub-processes does not always lead to an improvement in the process execution time. Therefore, the number of sub-processes should be determined carefully in consideration of $\bar{\rho}$, that is, the average load imposed on the grid computing system. This figure also shows that the optimal number of sub-processes is larger for higher threshold T.

5 Conclusion

The performance of sub-process allocation schemes was analytically evaluated using a model grid computing environment in which computers were modeled as M/G/1–PS queues. Three process allocation schemes were examined; random selection of computers (Scheme 1), selection of n least-loaded computers based on the current resource availability (Scheme 2), and random selection of n computers from among those with processing capability greater than a predefined threshold (Scheme 3). The latter two schemes employ current resource availability information for computers in the grid, requiring a model of CPU utilization. To achieve this in a grid computing environment, a power distribution was used to describe an environment in which computers are very likely to be used unevenly and to have dynamically changing utilization.

The performance of each scheme was investigated preliminarily in terms of the average process execution time expressed as a function of average CPU utilization. It was also shown that there exists an optimal number of sub-processes at which grid computing is most effective. Scheme 1 is very simple to implement, but the user may unknowingly choose computers that are highly loaded. Scheme 1 is therefore considered to be inadequate even if the average CPU utilization is relatively low. Scheme 2 outperforms both of the other schemes treated here, but may concentrate sub-process allocation to a few computers, possibly resulting in sever performance degradation if processes are independently allocated by several process schedulers at a similar time. Scheme 3 achieves the most moderate performance over a wide range of average CPU utilization without concentrating process allocation. Therefore, Scheme 3 appears to be greatest practical interest in terms of ease of implementation and moderate performance provided. The optimal number of sub-processes was shown to vary dependent on average CPU utilization, indicating that a larger number of sub-processes does not always lead to an improvement in the process execution time.

Acknowledgments This work was supported in part by a Grant-in Aid for Scientific Research on Priority Areas (2) (14019074) of the Ministry of Education, Culture, Sports, Science and Technology, Japan.

References

1. I. Foster and C. Kesselman, *The GRID Blueprint for a New Computing Infrastructure,* Morgan Kaufmann Publishers, 1998.
2. I. Foster, C. Kesselman, and S. Tuecke, "The Anatomy of the Grid", *International Journal of Supercomputer Applications*, Vol. 15, Num. 3, 2001.
3. I. Foster and C. Kesselman, "The Globus Project: A Status Report", *Proc. of IPPS/SPDP '98 Heterogeneous Computing Workshop*, pp. 4–18, 1998.
4. The Globus Project, *http://www.globus.org/.*
5. K. Czajkowski, S. Fitzgerald, I. Foster, and C. Kesselman, "Grid Information Services for Distributed Resource Sharing", *Proc. of the Tenth IEEE International Symposium on High-Performance Distributed Computing (HPDC-10)*, IEEE Press, August 2001.
6. E. Heymann, M. A. Senar, E. Luque, and M. Livny, "Adaptive Scheduling for Master-Worker Applications on the Computational Grid", *Proc. of the First International Workshop on Grid Computing (GRID 2000)*, Bangalore, India, December 2000.
7. G. Shao, F. Berman, and R. Wolski, "Master/Slave Computing on the Grid", *Proc. of 9th Heterogeneous Computing Workshop*, Cancun, Mexico, pp. 3–16, May 2000.
8. R. Wolski, N. T. Spring, and J. Hayes, "The Network Weather Service: A Distributed Resource Performance Forecasting Service for Metacomputing", *Journal of Future Generation Computing Systems*, Vol. 15, Num. 5–6, pp. 757–768, October 1999.
9. S. Gullapalli, K. Czajkowski, C. Kesselman, and S. Fitzgerald, "Grid Notification Framework (VB0.1)", *Grid Forum Working Draft GWD-GIS-019, www.gridforum.org*, June 2001.

Efficient Load Balancing by Adaptive Bypasses for the Migration on the Internet

Yukio Hayashi

Japan Advanced Institute of Science and Technology, Ishikawa, Japan

Abstract. We study a dynamic load balancing problem for servers on the Internet as Grid, based on the differences of distributed computing to parallel computing. We propose an adaptive method according to initially assigned load; the optimal migration flow is directly obtained without any iterations, and the conditions of migration for the bottleneck edges are relaxed by the bypasses on a cactus adaptively extended from a spanning tree. Simulation results show that the number of rounds for the migration is decreased under 2/3 for the conventional method, and that the cost with respect to the instability of migration is also decreased in about the half. Thus, the adaptively constructed cactus is considered as a practically efficient topology.

1 Introduction

On the rapid growth of computer networks, distributed computing has attracted considerable attention. In the efficient distributed computing, load balancing is one of the important issues. The problem is to balance the load of servers (computers in general) on a network such as the Internet. In other words, it is to suppress the idling or over-load states as few as possible by using only local communications between servers. Thus, we should remark that it is different from the load balancer for Web-accesses at the end of switch level (Layer 4 or 5) based on the DNS round robin [1] or for a centralized master/slave system.

Load balancing algorithms have been so far developed for parallel computers, whose processor is corresponding to the server in a distributed system. For example, to only locally balance the differences of load between neighbors, initiation methods are triggered by a processor with load level under or over a certain threshold. The former is called "receiver-initiation," and the later "sender-initiation" [7]. They are very simple but ad hoc. In addition, the global balancing is not guaranteed, and the setting of threshold is generally difficult.

On the other hand, the most popular approach to avoid the returns of migration is based on two phases: 1) calculation of the migration flow on each connection, 2) migration of the load between the nearest-neighbors in run-time. As the typical strategies [10], there are dimension exchange (DE) and diffusion (DF) methods. With the DE method, a processor communicates its nearest-neighbor one at a time. It is suitable for the parallel computer on a special homogeneous architecture such as the hyper-cube structure. With the DF method, a processor communicates all of the nearest-neighbors. It is superior for asynchronous

P.M.A. Sloot et al. (Eds.): ICCS 2003, LNCS 2658, pp. 257–266, 2003.

processings in multi-port communications [10] associated with a heterogeneous distributed system. Therefore, we focus the DF method for load balancing in distributed systems on a general topology.

Recently, the optimal polynomial scheme (OPS) within a finite number of iterations has been proposed [3]. The key-idea is to iteratively control polynomials for the calculation of diffusion flow by using the eigenvalues of a Laplacian. For parallel computers, it is an established method as a generalization of the conventional first order, second order, and Chebyshev schemes, and extended to the connection structure of a cartesian product of graphs [4]. However, there is a problem for the ordering of migration. It may cause a chain of load requirements for mutually related cycles, then a processor must be waited for a long time until the arrival of much load. Moreover, the calculations of all eigenvalues are necessary in advance for a topology. However, it is not negligible in distributed systems because the topology may be changed on the Internet, while to find the best topology is still open [2] (may be intractable).

In this paper, we propose **an adaptive method according to initially assigned load**. In this method induced from a quadratic programming (QP) problem [3][6] equivalent to solving the DF method, **the optimal flow is obtained by a variation or perturbation on a cactus adaptively extended from a spanning tree** (as a least connected component). The merits in computation and migration are that it can be directly calculated by the independency of each cycle in a cactus and the efficient message passings on a tree without cycles, and that **the conditions of migration for the bottleneck edges are relaxed by bypasses on a cactus**. Indeed, we show the proposed method is more efficient than the conventional tree walking algorithm (TWA) [7] through simulations for one-port and multi-port cases; **The number of rounds is decreased** under a certain ratio by the relaxation of the conditions for migration.

2 Dynamic Load Balancing Problem

In this section, we present a new problem setting for a dynamic load balancing in distributed computing on the Internet. Then, we point out that the DF method is equivalent to a QP problem, whose form is applied to the derivation of a variational or perturbative method on a cactus in the next section.

2.1 Distributed System on the Internet

Considering the differences of distributed computing to parallel computing: loose coupling, independence of processing elements, and heterogeneity [8], we define the load balancing problem in a network based on IP packet routings. To simplify the discussion, we assume that the performances of servers are the same.

- The amount of load is defined by the ratio of processes in ready state to the performance, which is measurable between the adjacent servers. We assume it is divisible as same in many literatures [2][3][4][6][10], because our application considered in this paper has very much data or many combination of parameters in independent processes.

- The connection between servers are logical on dynamic IP packet routings. We consider a computer-system on the Internet or in a WAN as Grid. However, the proposed method in this paper is applicable to a smaller system in a LAN.
- The connection structure is not restricted to regular ones, such as a ring, lattice, or homogeneous Cayley graph, but a heterogeneous general topology with both dense and sparse connections.
- Since long-distance connections may exists, we basically consider asynchronous local communications between servers.

Thus, processings are dominant than communication overheads in our application. Strictly speaking, the definition of load index is impossible without actually running, because it depends on many factors: not only on the response times for CPU, memory and I/O resources, but also on their communications or process granularity. It is known as "Key Issues in Dynamic Load Balancing" [10]. However, it is natural to consider the estimated values in a large problem with very much data or many combination of parameters.

2.2 Equivalent QP Problem to the DF Method

Although many schemes have been proposed for the DF method [2][3][4][6], it essentially result in solving a diffusion equation by a discrete Laplacian L on a simple undirected graph (V, E), whose vertices and edges are corresponding to the servers and connections, respectively.

The matrix-vector representation is

$$\mathbf{Lf} = \begin{bmatrix} \cdots & \cdots & -w_e & \cdots\cdots \\ \vdots & \ddots & 0 & \vdots \;\; \vdots \\ -w_e & 0 & \sum w_e & \vdots \;\; \vdots \\ \vdots & \vdots & \vdots & \ddots \;\; \vdots \\ \cdots & \cdots & \cdots & \cdots\cdots \end{bmatrix} \begin{pmatrix} f(1) \\ \vdots \\ f(u) \\ \vdots \\ f(|V|) \end{pmatrix},$$

the u th element is

$$Lf(u) = -\sum_{v \sim u} w_e(f(v) - f(u)), \tag{1}$$

where $v \sim u$ denotes a set of the adjacent vertices to $u \in V$, $f(u)$ the amount of load at server u, $w_e = w_{\bar{e}} > 0$ the weight of edge (e; $u \to v$ or $\bar{e}:\ v \to u \in E$, which depends on only choosing the direction).

In this paper, w_e is defined as "the stability of traffic for the migration between servers" (e.g. time-invariance measured by ping), not as an accelerative parameter in numerical schemes for the DF method. To discuss the characteristics of traffic, the conventional queuing theory based on independent Poisson arrival is insufficient, in even macroscopic phase transition, a statistical analysis such as Contact Process is needed for the self-similarity of packet density fluctuations [9]. However, it is beyond our present scope. Thus, we don't consider the weights of communication efficiency for data transfer speed with delay or band-width, because it is microscopically unstable and indefinite in the dynamic routings.

Let us consider a diffusion equation:

$$\frac{\partial \mathbf{f}}{\partial t} = -L\mathbf{f}. \tag{2}$$

It is converged to the solution

$$\bar{f} \stackrel{\text{def}}{=} \frac{\sum_{u \in V} f(u)}{|V|}, \tag{3}$$

with monotone decreasing $\sum_{u \in V}(f(u) - \bar{f})^2$. It is also known that the total load $\sum_{u \in V} f(u)$ is conserved in (2) at any time.

Intuitively, the flow for migration is accumulated by the right side of (2) according to the difference of load between adjacent servers, and each flow on e is enhanced to be much migration as larger w_e (more stable) in the right side of (1).

On the other hand, as first order scheme, the difference equation for (2) is given by

$$\mathbf{f}^k = (I - \Delta t L)\mathbf{f}^{k-1} = \underbrace{F \times \ldots \times F}_{k} \mathbf{f}^0, \tag{4}$$

where $F \stackrel{\text{def}}{=} I - \Delta t L$, I denotes an unit matrix, $\mathbf{f}^k$ the load vector at k th iteration, $\mathbf{f}^0$ initial load vector. We assume that the step-width Δt satisfies $1 \leq \Delta t \times \sum_{e \in E_u} w_e$, where E_u denotes a set of edges connected to $u \in V$ [6]. Hereafter, we rewrite w_e by $\Delta t \times w_e$.

Since the convergence of (4) is very slow, other schemes have been developed [3][4][6], in which we use that the equation (4) is equivalent to

$$y_e^{k-1} = -w_e(f^{k-1}(v) - f^{k-1}(u)),$$

$$z_e^k = z_e^{k-1} + y_e^{k-1}, \quad z_e^0 = 0,$$

$$f^k(u) = f^{k-1}(u) - \sum_{e \in E_u} y_e^{k-1},$$

and also to the following QP problem [3][6],

$$\min \ \tfrac{1}{2}\mathbf{z}^T W^{-1} \mathbf{z}, \tag{5}$$

$$s.t. \ B\mathbf{z} = \mathbf{f}^0 - \bar{\mathbf{f}}, \tag{6}$$

where $W \stackrel{\text{def}}{=} diag(w_e)$, B denotes the incidence matrix, $\mathbf{z}$ flow vector for the migration (z_e the flow on an edge $e \in E$, the positive value corresponds to the direction), $\bar{\mathbf{f}} \stackrel{\text{def}}{=} (\bar{f}, \ldots, \bar{f})$ the balancing solution vector of (3).

In the QP problem, the balancing condition (6) includes a feasible solution with wastful flows on cycles, therefore the minimization (5) is needed. In other words, the equivalent DF method implicitly gives us the optimal solution as the global balancing. Note that (5) represents a cost with respect to the instability of migration. If we don't consider the weight (as all $w_e = 1$), it becomes a simple cost for the no-wastful flow.

2.3 Efficient Message Passing on a Tree

When the connection structure is a spanning tree, the flow z_e is directly obtained by efficient message passings, instead of the iterative calculations in such [2][3][4][6]. It is known as TWA [7], in which a feasible solution for (6) is the optimal, then z_e is recursively calculated from leaves to the root as follows.

For a leaf u: $z_e = f(u) - \bar{f}$,
where $e \in E$ is an edge to the parent of $u \in V$.
For others: $z_{e'} = f(v) - \bar{f} + \sum_e z_e$,
where $e' \in E$ is an edge to the parent of $v \in V$, and $\{e\}$ in the summation is a set of edges from the children of v.

Although TWA is seemed to be a centralized control method, the communications are composed of the sending/receiving values of z_e or the accumulated load (the individual values of load are not needed) to each parent and the broadcasting the value $\bar{f}$ from the root, its additional computation is only (3) at the root. Therefore, they can be performed by locally asynchronous communications and processings in a distributed manner.

3 Adaptively Constructed Cactus

In this section, for the load balancing, we propose an adaptive method on a cactus. A simple graph, whose each edge is contained in an exactly one cycle, is called "cactus." In the proposed method, we show that the optimal flow is directly obtained by a variation or perturbation for the QP problem, in which the flows of bottleneck edges are bypassed on a cactus. Consequently, the conditions of migration are relaxed, and the number of rounds is decreased as shown in the next section.

First, we consider a spanning tree as the least connection structure among servers. After applying TWA to it, at each vertex u, we find a bottleneck edge:

$$e = \arg \max_{e \in E_u} \left\{ \frac{|z_e|^2}{w_e} \right\},$$

which depends on the initial load.

Next, in the connected edges to u, we find the pair e' which minimizes the cost for the extended QP problem by adding a bypass e'' as shown in Figure 1. The flow $z_{e'}$ has the inverse direction of z_e at u (for input or output).

We call the following approach "variation or perturbation," because the problem is extended to the minimization (5) on a cactus according to the initial load. Note that the balancing condition (6) still holds due to the bypass modifications.

For the bypass flow Δz on e'', the variation of cost in (5) is given by

$$\delta C(\Delta z) \stackrel{\text{def}}{=} \frac{(z_e - \Delta z)^2}{w_e} + \frac{(z_{e'} - \Delta z)^2}{w_{e'}} + \frac{\Delta z^2}{w_{e''}} - \left(\frac{z_e^2}{w_e} + \frac{z_{e'}^2}{w_{e'}} \right).$$

The most decreasing is obtained at the extreme point $\frac{\partial(\delta C)}{\partial(\Delta z)} = 0$. Since each cycle on a cactus is independent, we can easily derive the solution

$$\Delta z_{opt} = \frac{w_{e'} w_{e''} z_e + w_e w_{e''} z_{e'}}{w_{e'} w_{e''} + w_e w_{e''} + w_e w_{e'}} > 0. \tag{7}$$

From this, we also derive $\delta C(\Delta z_{opt}) < 0$; the cost is always decreased by adding bypasses on a cactus.

Thus, we can obtain the optimal solution for the larger QP problem (equivalent to solving the DF method) on an adaptively extended cactus by applying only local modifications, because each cycle is independent; If there are common edges on cycles, iterative calculations are necessary. We summarize the adaptive method on a cactus as follows. The communication complexity is $O(|V|)$.

< Distributed Algorithm on a Cactus >

Step 0. At the trriger from a server with heavy load, the following processes for the global balancing are initiated in run-time.

Step 1. Construct a minimum spanning tree (MST) by applying a distributed algorithm e.g. [5]. Here, the distance may be based on geographical locations of servers, or on the averaged communication delay.

Step 2. Calculate the flow by applying TWA to it.

Step 3. Find a bottleneck edge and the candidate of bypass at each vertex, independently. To construct a cactus (Figure 2), the candidates of bypass are mutually excluded by time-stamps or other appropriate criterions.
For the fixed bypass, calculate the modified flows $z_e - \Delta z_{opt}$, $z_{e'} - \Delta z_{opt}$, and Δz_{opt} by (7).

Step 4. Asynchronously migrate it between servers, as soon as possible (first come, first served) in a locally distributed manner.

Altough the above discussion is devoted to a ternary cycle, it is quite same for other cycles: quaternary or longer cycles. However, the ternary is practically better in the following reasons.

- The mutual exclusion is restricted in the alternative combination of triangles. If we consider longer cycles, it may be intractable that many edges are complicatedly related.
- Each server can directly communicate to the nearest-neighbors. While, for longer cycles, it must pass the information z_e or w_e through intermediators.
- Both ends of a bypass edge are probably close in the geographical locations.

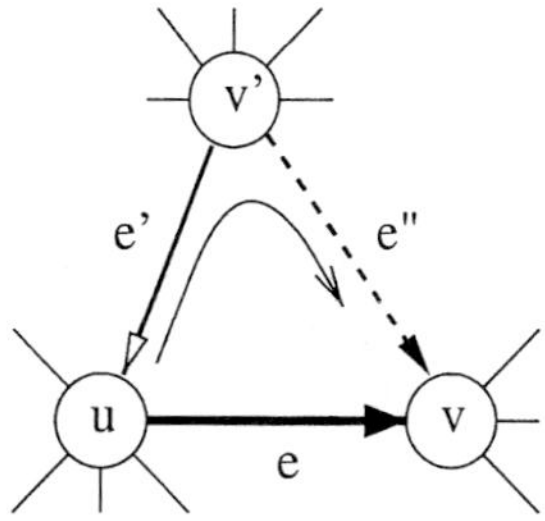

Fig. 1. Bypass for a bottleneck edge e.

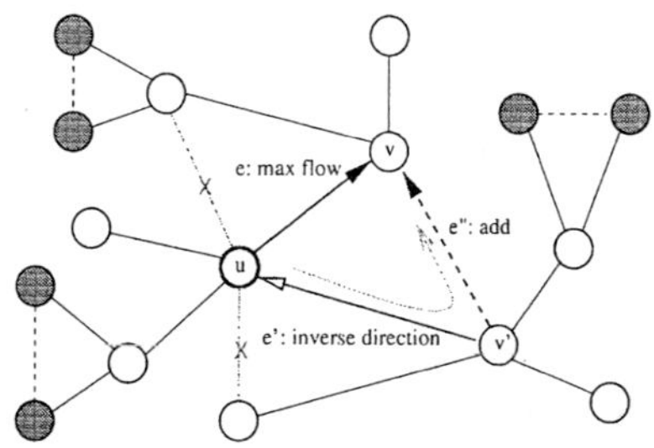

Fig. 2. Mutual exclusion among three triangles for constructing a cactus. The other triangles far from them have no-relation.

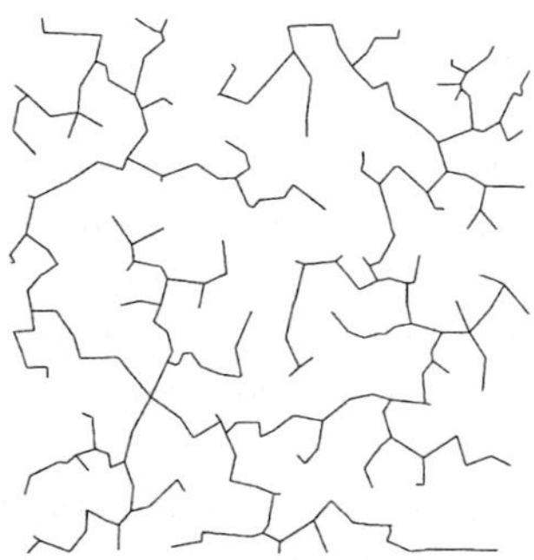

Fig. 3. MST ($|V| = 300$).

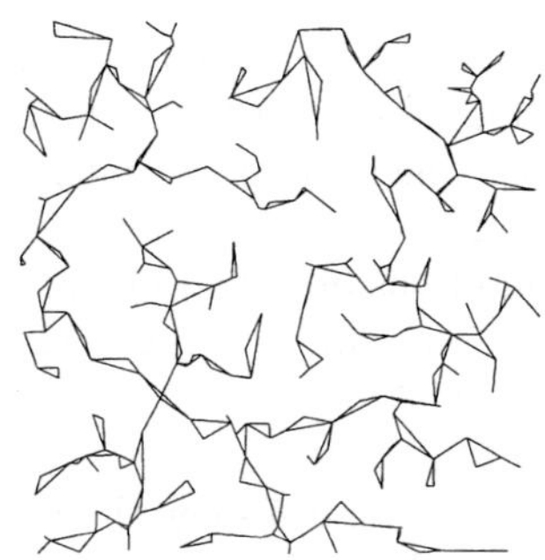

Fig. 4. Ternary cactus ($|V| = 300$).

4 Simulation Results

We investigate the effect of the relaxed conditions for migration by bypasses on an adaptively constructed cactus according to the initial load. Since the migration is a process in Step 4 after the calculation of flow, the effect is common and independent of any schemes (even if we apply other schemes such as OPS [3]), but depends on the initial load and the topology.

To set a topology among servers and investigate the general properties in our method, we first construct graphs with sizes $|V| = 100, 200, 300, 400, 500$, whose vertices are randomly located. In addition, we set the weight w_e on the complete graph and the initial load $f(u)$ at each vetrex by $(0.5, 1)$ and $(0, 200)$ uniformly random numbers, respectively. In this first setting, we consider the complete graph to only prepare for the extension to a cactus from the MST. After fixing a topology, the vertex and edge are corresponding to a server and connection for the communication. Remenber that w_e represents the stability of migration between servers. We also set the time-stamps by (0, 1) uniformly random numbers. An example of the MST is shown in Figure 3, and a cactus constructed by the previous algorithm is shown in Figure 4.

We verify the effect of bypasses on a cactus. In our simulations, to simply the discussion, we assume that there exist a constant interval between the activations for the migration. Because a server with less load than z_e must be waited until the required load is arrived, even if each server can asynchronously activate it. We call the timing for migration "round," which is not depended on the amount of flow and the routing. At each round, the edges which satisfy the condition of migration become active. Note that the asynchronous property is included in the random time-stamps. Strictly speaking, since the activation depends on the data transfer speed, the delay by traffic congestion, and the application (the amount of data for migration), it is an underestimation. In other words, a real system takes much more times than the number of rounds $\times$ a constant interval.

We investigate the minimum, maximum, and average values of the rounds in ten trials for each size of graphs. In the multi-ports, the migrations can be simultaneously activated at a vertex. The difference between the minimum and maximum is due to whether fitting or unfitting of the initial load for the topology; in a case, the load may be migrated through many and long paths on the MST or cactus. The average rounds are shown in Figure 5. By the bypasses on a cactus, the conditions of migration are relaxed, then the rounds are decreased; the ratio (of the rounds on a cactus to that on the MST) is about 2/3 as shown in Figures 6 and 7. The result for the multi-ports is about 10 % improved than that for the one-port.

Figures 8 and 9 show the ratio of the costs with respect to the instability of migration. By the bypasses on a cactus, it is also decreased to about the half ($0.4 \sim 0.5$). Moreover, the average efficiency: active edges for the migration per rounds is about twice of that in the MST. Through the simulations, we have found that the bypasses on a cactus are constantly 36 % added for all of sizes $|V|$ and initial load in ten trials. It suggests that there exist a something of general law in constructing a ternary cactus from any spanning tree.

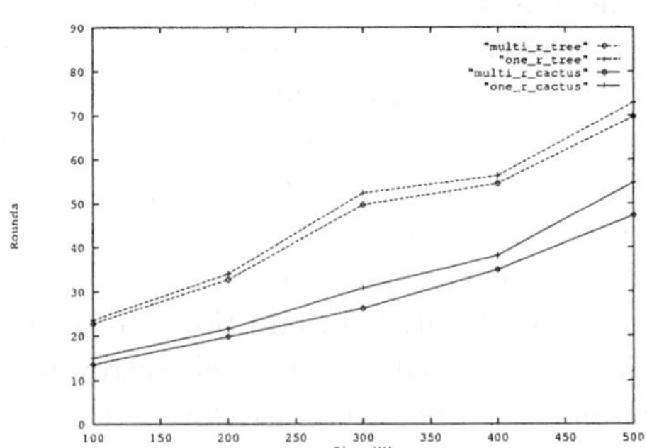

Fig. 5. Average rounds for the multi-ports $\diamond$ and one-port $+$ cases. The solid and dashed lines are corresponding to the results for a cactus and the MST, respectively.

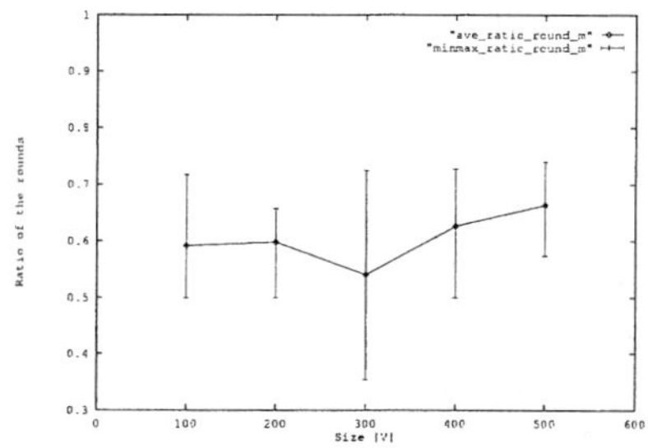

Fig. 6. Ratio of the rounds (multi-ports).

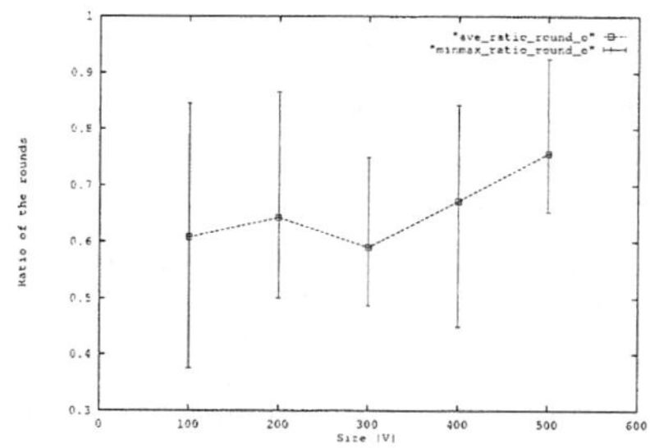

Fig. 7. Ratio of the rounds (one-port).

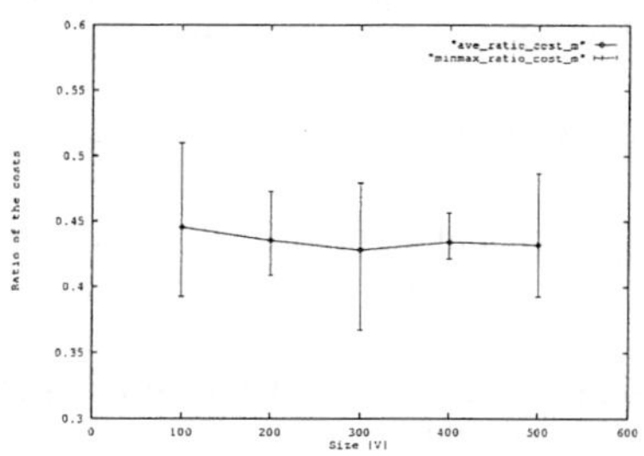

Fig. 8. Ratio of the costs (multi-ports).

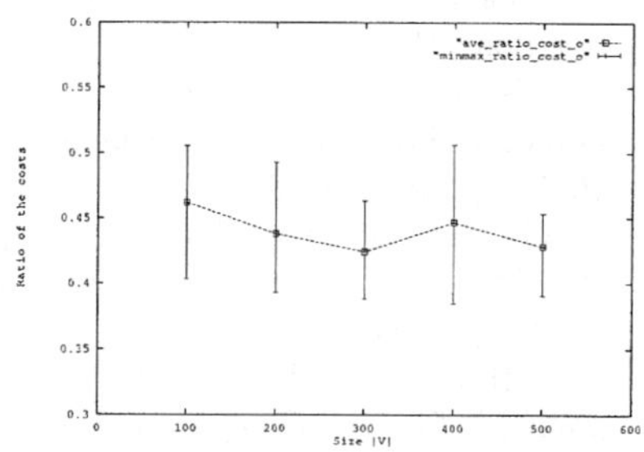

Fig. 9. Ratio of the costs (one-port).

5 Conclusion

We have considered the dynamic load balancing problem for servers on the Internet as Grid, based on the differences of distributed computing to parallel computing: loose coupling, independence of processings, and heterogeneity. In general, the efficiency for load balancing depends not only on calculation schemes for the migration flow but also on heterogeneous topologies of the connection between servers. However, to find the optimal topology is intractable in the state-of-the-art.

In this paper, we have proposed an adaptive method according to initially assigned load, and presented it as a distributed algorithm based on only local communications and asynchronous processings. The method has been induced from a QP problem equivalent to solving the DF method, in which the optimal flow is obtained by using local variational or perturbative computations, and the conditions of migration for the bottleneck edges are relaxed by the bypasses on a cactus adaptively extended from the MST.

Simulation results have shown that the number of rounds for the migration is decreased under 2/3 for the conventional TWA, and that the cost with respect to the instability of migration is also decreased in about the half. These results are underestimations, however, not depended on any schemes and implementation technologies. Thus, the adaptively constructed cactus is considered as a practically efficient topology, in addition, suitable for dynamically reconstructive distributed systems (with flexible connections and changes of servers). The implementation and evaluation on real systems are further studies, e.g. for parameter search in scientific computations or cooperative Web crawlers.

References

1. T. Bourke. *Server Load Balancing*, O'Reilly, 2001.
2. T. Decker, B. Monien, and R. Preis. "Towards Optimal Load Balancing Topologies," A. Bode et al. (Eds): Euro-Par2000, *LNCS* 1900, pp. 277–287, 2000.
3. R. Diekmann, A. Frommer, and B. Monien. "Efficient Schemes for Nearest Neighbor Load Balancing," *Parallel Computing*, Vol. 25, pp. 789–812, 1999.
4. R. Elsässer, A. Frommer, B. Monien, and R. Preis. "Optimal and Alternating-Direction Load Balancing Schemes," P. Amestoy et al. (Eds.): Euro-Par'99, *LNCS*, 1685, pp. 280–290, 1999.
5. R. Gallager, P. Humblet, and P. Spira. "A Distributed Algorithm for Minimun Weight Spanning Trees," *ACM Trans. on Prog. Lang. and Systems*, Vol. 5, No. 1, pp. 66–77, 1983.
6. Y.F. Hu, R.J. Blake. "An Improved Diffusion Algorithm for Dynamic Load Balancing," *Parallel Computing*, Vol. 25, pp. 417–444, 1999.
7. W. Shu, and M.Y. Wu. "Runtime Incremental Parallel Scheduling on Distributed Memory Computers," *IEEE Trans. on Parallel and Distributed Sysytems*, vol. 7, no. 6, pp. 637–649, 1996.
8. V.S. Sunderam, and G.A. Geist. "Heterogeneous Parallel and Distributed Computing," *Parallel Computing*, Vol. 25, pp. 1699–1721, 1999.
9. M. Takayasu, A.Yu. Tretyakov, K. Fukuda, and H. Takayasu. "Phase Transition and $1/f$ noise in the Internet Packet Transport," D.E. Wolf et al. (Eds): *Traffic and Granular Flow '97*, Springer Singapore, pp. 57–74, 1998.
10. C. Xu, and F.C.M. Lau. Load Balancing in Parallel Computers -Theory and Practice-, Kluwer Academic Publishers, 1997.

A Distributed Data Storage Architecture for Event Processing by Using the Globus Grid Toolkit

Han Fei[1], Nuno Almeida[1], Paulo Trezentos[1], Jaime E. Villate[2], and Antonio Amorim[3]

[1] ADETTI, Edificio ISCTE, University of Lisbon,
Avenida das Forças Armadas,
1600-082 Lisbon, Portugal
{Han.Fei, Nuno.Almeida, Paulo.Trezentos}@iscte.pt

[2] Department of Physics
School of Engineering,
University of Porto
villate@fe.up.pt

[3] Faculdade de Ciencias, University of Lisbon,
Campo Grande, Edificio C8, sala 8.3.05,
1749-016 Lisbon, Portugal
Antonio.Amorim@fc.ul.pt

Abstract. In this paper we discuss a Grid-based Event Processing System (GEPS). Data intensive problems broadly exist in many scientific computational areas; usually their needs for super storage and computing capacities are difficult to be fully satisfied. Meanwhile the Globus Toolkit has become the de facto standard of building high performance distributed computing environments. Event processing and filtering is a kind of data intensive problem in high-energy physics area. Using the Globus grid toolkit, we have constructed the GEPS system, which provides web-based access to grid computing environments for event processing. Performance result indicates that event processing and filtering can be effectively implemented on GEPS.

1 Introduction

In many scientific disciplines, the need for terabyte data storage, processing and transferring is emerging as a crucial problem; nevertheless large computing and storage facilities are always scarce resources. The storage and computing capabilities are often temporarily and geographically distributed unevenly, sometimes redundant in one place meanwhile scarce in other places. With the scientific and technical applications becoming more and more complicated and sophisticated, many researchers, working and living in different places, have to not only cooperate in the same research project but must also access distributed computing resources.

It is unlikely that conventional methods can meet the demands of providing and sharing these resources. A blueprint of computational grids leveled at addressing these difficulties has been proposed. [1]

P.M.A. Sloot et al. (Eds.): ICCS 2003, LNCS 2658, pp. 267–274, 2003.

A Grid [2] is super-computing net, which can connect distributed mainframe computers, super-computers, as well as large numbers of desk top computing devices into easy-to-use computing facilities.

2 The LHC Computing Problem

In the Large Hadron Collider (LHC) accelerator at CERN, there are 10^9 collisions per second taking place per second. Each collision contains about 1 MB of information. One single collision is called an "event". Each event is recorded by surrounding particle detectors for later processing and filtering to pick out the physically interesting ones. Events are recorded at a typical rate of 100 Hz. Considering the data intensive aspect of event processing, computational grids are a possible solution.

2.1 Related Work

Gfarm (Grid Data Farm) is an event processing project [3][4] at KEK (High Energy Accelerator Research Organization) and ICEPP (International Center for Particle Physics, the University of Tokyo). A large scale distributed Gfarm file is divided into several fragments and distributed across the disks in the Gfarm file system. A Gfarm file is a logical aggregation of physical file fragments distributed over many CPU nodes. The processing jobs access the Gfarm files through the Gfarm parallel I/O library, and the job executes in parallel at each node where the physical file fragments reside. The Gfarm file system daemon runs on each node to facilitate remote file operation with access control. When a job is submitted into the Gfarm server, it is redistributed to nodes, which contain the fragment database files. When the job is finished, the results will be retrieved across the network.

Parallel ROOT (PROOF) [5] is another event processing system. The ROOT client session creates a master server on a remote cluster, and then the master server in turn creates slave servers on all the nodes in the cluster. All the slave servers execute user job in parallel. The master server distributes the event data packets to every slave server, carefully adjusting the packet size such that the slower slave servers get smaller data packets than faster slave servers. PROOF uses a TChain object to provide a single logical view of many geographically distributed physical files. The master server keeps a list of all generated packets per slave, so in case a slave failed then remaining slaves can reprocess its packets.

An application in Gfarm system need to use Gfarm I/O library to access Gfarm file, so already existing applications need to be changed and recompiled, and this change means the application will be Gfarm Only. In the case of PROOF, because the PROOF toolkit is relatively reliant on specific grid techniques, the application can't always utilize the latest grid feature, which can be available only after PROOF provides a realization of that feature. For solving these inconveniences, we propose a distributed Grid-based approach, which facilitates intensive event raw data storage and processing, while providing a uniform application staging interface.

3 GEPS Prototype

In GEPS system we make use of the Grid infrastructure, a back-end database, LDAP directory query, and PHP script web interface. The scalability of GEPS can be easily obtained through freely adding into or picking out any grid computing and storage node. GEPS works like a portal. Behind the friendly appearance of GEPS, many Grid related details are well hidden. Geographically distributed physicists can easily cooperate over the same event processing project, share dispersed events data file, stage jobs, query job status, share computing resources, transfer data file, and visualize events filtering results.

3.1 Introduction of the Events Application

Event processing application is programmed in C++ by using the Root Toolkits [6]. Root is an object-oriented framework, aimed at solving the data analysis challenges of the high-energy physics discipline. It provides a large collection of specific utilities to manage information in an efficient way. Root provides not only an application programming interface (API), but a integrated Root tree class data file visualization environment. The creation of the Root data file has several steps. The first step is to create a structure to store all the raw information of the events. This process consists of the creation of a shared library, which contains all the variables of the event, track, vertices, as well as relation objects.

If the shared library produced in the first step works well, then the next step is to create a Root tree to storage all the objects presented in the raw information file. The Root tree class is optimized to reduce storage space usage and enhance accession speed. Inside the Root tree there is one branch with all events, inside this branch are all event variables that include the tracks, vertices, and relations.

After all the information appears in the Root tree, now it is the time for scrutinizing, one by one, which event will be the candidate that meets the processing standard. The calibration procedure based on the processing standard will be done on each event, then the result will be stored in a new tree with the same structure.

Based on the Grid-enabled computing net, we can divide event raw data into different storage parts, which can be stored in geographically distributed Grid resource nodes. After that, we can stage processing and filtering procedures in a parallel manner, monitor the application running status, collect results, merge the different results data into final data file, and visualize the final data.

3.2 The GEPS Architecture

Figure 1 describes the GEPS architecture. GEPS has an easy-to-use and friendly interface, which is programmed in the PHP script language. No matter where the end user is, the services of GEPS can be easily approached through Internet. After logging into the main-page, several optional functions can be chosen by the end user. The user can request the summarized or detailed information of the available Grid resources. The user can simply fill in a job description form to express some needed information about the job, such as: what is the executable, where does the executable reside, to which Grid nodes is the executable going to be submitted, where is the raw data file, where does the end user want the output to be stored. After filling in the specification of one job, the user can continue to describe another job. The next step is simply push the "submit" button, then jobs will be submitted to grid nodes. After submitting all jobs, the user can continuely monitor the running status of submitted jobs.

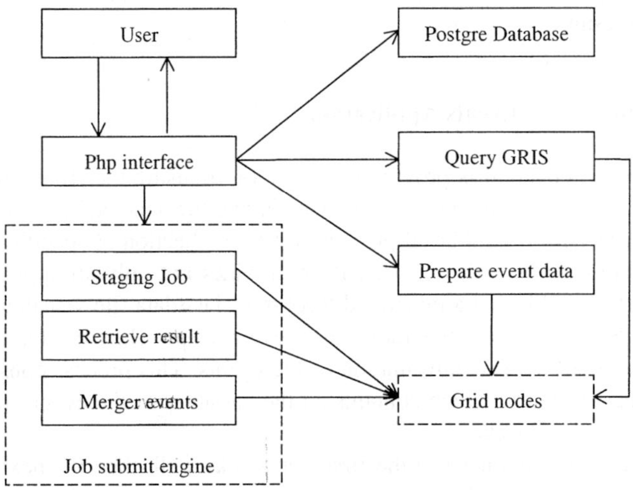

Fig. 1. The GEPS architecture. The user can use GEPS through a PHP scripted interface, which hides many realization details from the outside user.

When the jobs have done, the distributed event result data files will be automatically merged to form the final result, which will be stored in the user specified site. Finally the user can utilize the Root visualization tool to see the event processing and filtering result.

The end user interface is programmed in the PHP script language. It will receive the job descriptions provided by the end user, and it will insert the new job into the PostgreSQL database. If the end user wants to query information about the grid computing environment, the PHP scripts will call a function to query a GRIS ldap server.

Before submitting the executables, the event raw data needs to be copied to the grid nodes, according to the demands of the job specification.

The grid job submission engine will parse the job specification tuple in the PostgreSQL database, analyze the job executing environment and raw event data distribution demands, synthesize the RSL sentences, submit the jobs, monitor the status of submitted job.

4 The GPES Prototype Implementation

4.1 The Computational Grids Environment

The Globus Grid toolkit evolved out of the I-WAY high performance distributed computing experiment [7]. Before the Globus Grid became the de facto high performance computing environment, there were other candidate grid architectures, include using object-based technology and web technology [8].

Table 1. Globus components in GEPS.

Component	Usage
GRAM	Executable staging
GRIS in MDS	Query Grid node information
GASS	Transfer raw data, retrieve remote results

Table 1 lists the grid components used in GEPS. In the Grid job submission engine, the new job specification tuples are selected from the back-end PostgreSQL database. For each new job, by parsing the job specification tuple, a job Resource Specification Language (RSL) sentence is formulated, then a raw data file is transferred (by using GASS components) in accordance with the setting of relevant resources, and then the GRAM component (globus-gram-client) is used for remotely submitting and managing job. The run time stdout and stderr is defined in the RLS sentence. After all submitted jobs having finished, GASS file access functions are used for retrieving distributed event results.

4.2 Query GRIS LDAP Server

Monitoring Discovering Service (MDS). The Globus Toolkit has provided an information Monitoring and Discovery Service (MDS)[9], which acts as a resource information registry and discovery agent. The MDS includes a standard, configurable information provider framework called a Grid Resource Information Service (GRIS). GRIS is implemented as an OpenLDAP[10][11] server. Each Grid node can run a local GRIS.

Through GEPS, the end user can query properties of the grid nodes, such as how many processors are available at this moment, what bandwidth is provided, etc. The MDS provides two interfaces: interactive and programmatic. By default, a GRIS service is automatically configured to port 2135. In our GEPS, the grid-info routine obtains the overall Grid node information by querying this port through the LDAP protocol. The PHP script will call the grid-info routine to get the results.

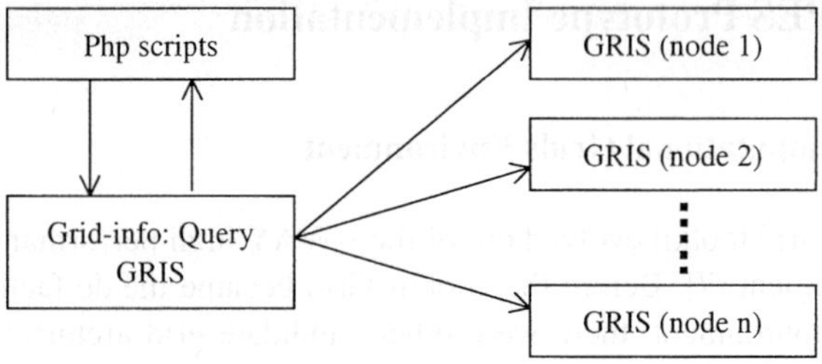

Fig. 2. Querying Grid node resource information through the LDAP protocol. In GEPS the end user selects interesting objects or just simply makes a choice of default objects. The PHP scripts then call Grid-info. Grid-info will send LDAP queries to the GRIS in each Grid node and get the available resources list.

5 Experiment and Results

From August to October in 2002, we tested 13 groups of raw event data, and with a total of 130 experiment executions (for decreasing the effect of system and network latency in executable staging and data transfer). The current GEPS demonstration prototype temporarily consists of two server, gandalf and hobbit. Because the GEPS topology structure has the feature of scalability, in the future more nodes can be easily incorporated. One of the advantages of computational grids is that any part can be easily changed without any global effect.

Different granularities of event data will dramatically affect the overall performance of the GEPS system. This is reasonable, because with many smaller files of raw event data, the portion of system cost dedicated to raw data transfer will become larger in total execution time. Based on the event data file size, Figure 3 gives the relation between running only on hobbit and running in parallel between gandalf and hobbit. The unit on Y-axis is time cost in second, and the unit in X-axis is the number of events in raw event data file. In raw event file each event is about 1M bytes in size. From the illustration we can easily see that the data file size of approximate 2000 events is a watershed. Data files consisting of less than 2000 events run in tightly coupled computing environments will have better performance. But usually our event raw data files can be easily much larger than 2000 events. From the results illustrated in Figure 3 we know that to some extent our GEPS has provided better performance.

The GEPS network connection is fast Ethernet. The user defines the raw event data distribution by using the RSL sentence. Before a job can be submitted to the grid gatekeeper through the grid client API, the raw event data will firstly be transferred to the grid nodes in accordance with the raw event data distribution specification. GEPS currently uses Globus GASS file access API for transferring raw data and result file between gird nodes. Figure 3 only gives the comparison of processing time cost between GEPS and hobbit.

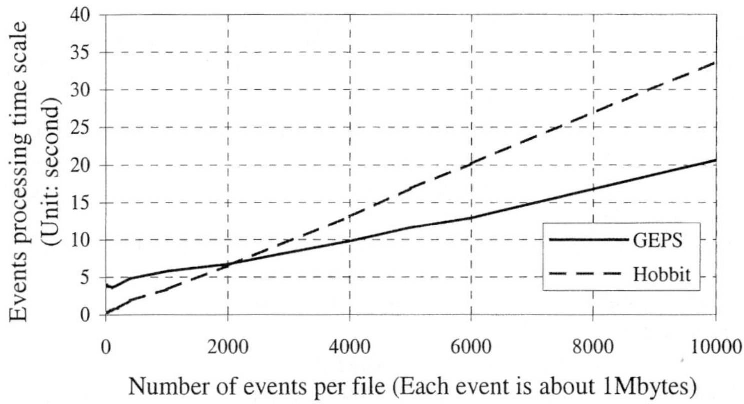

Fig. 3. Performance in GEPS & hobbit with different event raw data file sizes.

6 Conclusions and Future Work

We have described the GEPS prototype, which provides an integrated meta computing environment for event processing and filtering. In GEPS, Grid related detail and relevant middleware specifics have been hidden from the end user. GEPS facilitates the scalability of intensive event data storage. Using GEPS, physicists can easily administer and share distributed data and take advantage of distributed computing resources. This prototype has incorporated to date innovative Grid concepts and mechanisms.

The smaller bandwidth and the larger latency due to the geographical distribution of the Grid computational resources are the main reason of parallel inefficiency. We are working on adding GridFTP into our prototype. Because multiple TCP streams and proper TCP buffer sizes are very important to obtaining better performance in TCP wide area links [12], we are trying to add this feature into the GEPS prototype. We are also exploring the feasibility of solving other physics problems in the GEPS prototype environment.

Acknowledgements. This work was supported by Fundação da Ciência e Técnologia under the grant CERN/P/FIS/43719/2001. The first author gratefully acknowledges the postdoctoral fellowship by the FCT. The third author would like to thank ADETTI (Associação para o Desenvolvimento das Telecomunicações e Técnicas de Informática) for their support to this work.

References

1. I.Foster and C. Kesselman: The Grid: Blueprint for a New Computing Infrastructure. Morgan Kaufmann (1999)
2. S. Barnard, R. Biswas, S. Saini, R. Van der Wijngaart, M. Yarrow, L. Zechter, I. Foster, O. Larsson: Large-Scale Distributed Computational Fluid Dynamics on the Information Power Grid using Globus. Proc. of Frontiers '99 (1999)
3. Y.Morita, O.Tatebe, S.Matsuoka, N.Soda, H.Sato, Y.Tanaka, S.Sekiguchi, S.Kawabata, Y.Watase, M.Imori, T.kobayashi: Grid Data Farm for Atlas Simulation Data Challenges, Proceedings of CHEP 2001 (International Conference on Computing in High Energy and Nuclear Physics) (2001) 699–701
4. Osamu Tatebe, Youhei Morita, Satoshi Matsuoka, Noriyuki Soda, Hiroyuki Sato, Yoshio Tanaka, Satoshi Sekiguchi, Yoshiyuki Watase, Masatoshi Imori, Tomio Kobayashi: Grid Data Farm for Petascale Data Intensive. Electrotechnical Laboratory, Techinical Report, TR-2001-4. http://datafarm.apgrid.org
5. René Brun, Fons Rademakers: Distributed Parallel Interactive Data Analysis Using the Proof System. Proceedings of CHEP 2001 (International Conference on Computing in High Energy and Nuclear Physics) (2001) 704–707
6. http://root.cern.ch
7. I. Foster, J. Geisler, W. Nickless, W. Smith, S. Tuecke: Software Infrastructure for the I-WAY High Performance Distributed Computing Experiment. Proc. 5th IEEE Symposium on High Performance Distributed Computing (1997) 562–571
8. S. Brunett, K. Czajkowski, S. Fitzgerald, I. Foster, A. Johnson, C. Kesselman, J. Leigh, S. Tuecke: Application Experiences with the Globus Toolkit. Proceedings of 7th IEEE Symp. on High Performance Distributed Computing, July 1998
9. S. Fitzgerald, I. Foster, C. Kesselman, G. von Laszewski, W. Smith, S. Tuecke: A Directory Service for Configuring High-Performance Distributed Computations. Proc. 6th IEEE Symposium on High-Performance Distributed Computing (1997) 365–375
10. Heinz Johner, Michel Melot, Harri Stranden, Permana Widhiasta: LDAP Implementation Cookbook. SG24-5110-00, IBM. International Technical Support Organization, http://www.redbooks.ibm.com
11. Heinz Johner, Larry Brown, Franz-Stefan Hinner, Wolfgang Reis, Johan Westman. Understanding LDAP. SG24-4986-00, IBM. International Technical Support Organization, http://www.redbooks.ibm.com
12. J. Lee, D. Gunter, B. Tierney, B, Allcock, J. Bester, J. Bresnahan, S. Tuecke: Applied Techniques for High Bandwidth Data Transfers Across Wide Area Networks. Proceedings of International Conference on Computing in High Energy and Nuclear Physics, Beijing, China, September (2001)

Generalization of the Fast Consistency Algorithm to a Grid with Multiple High Demand Zones

Jesús Acosta-Elias and Leandro Navarro-Moldes
Polytechnic University of Catalonia, Spain[1]
{jacosta, leandro}@ac.upc.es

Abstract. One of the main challenges of grid systems of large scale and data intensive is that of providing high availability and performance, in spite of the unreliability and delay occasioned by the size of Internet. Replication enables us to meet such a challenge with success. In the context of weak consistency, the fast consistency algorithm prioritizes replicas with high demand. Nevertheless, the fast consistency algorithm only works well in a single zone of high demand, whereas in multiple high demand zones its performance is poor. In this paper, we propose an algorithm chosen according to demand, whereby the replicas in each zone of high demand select leader replicas that subsequently construct a logical topology, linking all the replicas together. In this way, changes are able to reach all the high demand replicas without the low demand zones forming a barrier to prevent this from happening.

1 Introduction

Replication in a grid system should provide scalability [20], apart from increasing availability and performance – factors of great importance when working on Internet scale. Replicas in a grid system are made up of servers with the same content and providing the same service. If the system can withstand failures in both the links and the servers, availability increases. If a server cannot be reached by a group of users, these users are able to contact another server who can be reached at that time.

If the latency produced in the links is decreased, performance improves. Users will be able to contact the nearest server, thus avoiding having gone through links and routers in order to reach more distant servers. A further advantage is found in the saving of bandwidth, which is a resource in low supply. Considerable improvement in performance is also achieved, since there are many servers with the same data and services that can all attend user's requests simultaneously.

Replication algorithms have traditionally been classified into strong and weak consistency. The strong consistency algorithm ensures that all the replicas have exactly the same content (synchronous systems) before any transaction is carried out, where in an unreliable network like Internet, with a large amount of replicas, latency can become high so that it becomes impossible for using the system. The strong consistency

[1] This work has been partially supported by the Mexican Ministry of Education under contract PROMEP-57, the Spanish MCYT project COSACO (TIC2000-1054), and the Catnet EU project.

P.M.A. Sloot et al. (Eds.): ICCS 2003, LNCS 2658, pp. 275–284, 2003.

algorithm [5, 8, 9] is suitable for systems with few replicas, and on a reliable network where a large amount of bandwidth is available.

However, weak consistency algorithms, i.e. Golding's Refdbms [12], Thesis's Adya [4], Usenet [14, 22], Coda [23], Bayou [21], Ficus [13], Grapevine [24], generate very little traffic, low latency, and are more scalable. They do not sacrifice either availability or reply time in order to guarantee strong consistency, but only need to ensure that the replicas eventually converge to a consistent state in a finite, but not bounded, period of time. They are very useful in systems where it is not necessary for all the replicas to be totally consistent for carrying out transactions (systems that withstand a certain degree of asynchrony).

Golding's research into weak consistency prompted us to develop the "Fast Consistency" algorithm [1], which prioritizes the high demand replicas, and in which we see that with very few additional signaling bytes considerable improvement is obtained, since with a low number of anti-entropy sessions it is possible to deliver consistent content to a greater number of clients. Nevertheless, in [2] it is demonstrated that this algorithm has a very good performance in a zone with one high demand region, whereas in multiple regions of high demand its performance becomes poor due to the formation of islands of locally consistent replicas. To tackle this problem successfully, we propose a combined mechanism for converting multiple zones of high demand into a single zone, thus obtaining the best possible performance from the fast consistency algorithm.

2 The Model

The model of our distributed grid system consists of a number of N nodes that communicate via message passing. By simplicity we assume a fully replicated system, i.e, all nodes must have exactly the same content. But in real applications, while all nodes may want to be notified of changes, only a set of nodes may decide to keep a replica.

Every node is a server that gives services to local clients. Clients make requests to a server, and every requested service is a "read" operation, a "write" operation, or both. When a client invokes a "write" operation in a server, this operation (change) must be propagated to all servers (replicas) in order to guarantee the consistency of the replicas. An update is a message that carries a "write" operation to the replica in other neighboring nodes.

In this model, the demand of a server is measured as the number of service requests by their clients per time unit.

3 The Problem

The problem consists in that when there exist multiple zones of high demand, surrounded by zones of low demand, the propagation of the changes by means of the fast consistency algorithm is slowed up by the zones of low demand. In other words, the fast consistency algorithm is unable to carry the changes across the zones of low demand to the zones of high demand, since it has no overall knowledge of where the zones with high demand replicas are located. Each node that carries into effect both the weak consistency algorithm and the fast consistency algorithm is aware only of its immediate neighbors; that is, all those replicas at distance one.

The existence of low demand regions, such as that we can see in Fig 1, zone II, and in which the updates arrive at a relatively low speed, gives rise to areas that the fast consistency algorithm finds difficult to overcome. It also causes the high demand regions (Fig. 1, zone I) to be isolated one from the other, and therefore their content undergoes delay before being mutually consistent.

These highly consistent regions surrounded by low consistency zones appear as islands formed by locally consistent replicas.

To illustrate this, a change produced in A, for example, quickly reaches B, but rises until C at a relatively slow speed, then falls rapidly to D. Ideally, the change produced in A should reach D just as rapidly as B, but in fact this is not the case since Zone II stands in the way.

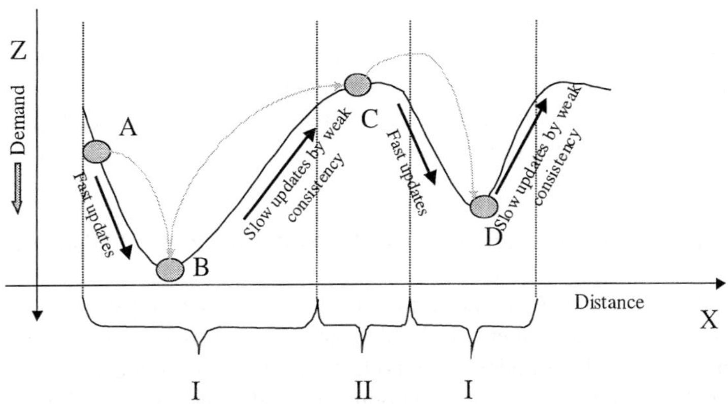

Fig. 1. Shows a cross section of the distribution of nodes. The demand is on the Z axis, and on the X axis we have the distribution in the X-Y plane, where Y = k. The nodes are shown in such a way that those with the most demand are towards the bottom, and those with the least demand are found towards the top. Two different kinds of regions can be seen, valleys (I) and mountains (II).

4 Proposal

Using an election algorithm [16, 6, 11], the replicas in each region of high demand will choose a leader node. The leader node in each region of high demand (island) must be one of the nodes of greater demand (most requests) in the zone. The changes brought about within the zone by the fast consistency algorithm reach the nodes of greater demand in each zone. This algorithm ensures that the changes reach these zones in very few sessions.

When every island has a representative, a superset of representatives is formed, which make up a logical topology connecting all the islands together. Its main task will be to take all the changes brought about in each region to the rest of the regions, without having to cross the zones of low demand.

4.1 Election Algorithm

A choice begins at a configuration in which all the nodes are in the same state and reach a configuration exactly where one node is the leader. It is necessary to make a choice; either when a centralized algorithm is put into effect in a distributed system, or after a system crashes and it is not known which nodes are functioning.

Definition of the election algorithm: Each node has the same local algorithm. The algorithm is decentralized when it is able to start an execution by an arbitrary non-empty subset of nodes.

It is called a leader network, or it is said to contain a leader if there is exactly one node, which knows that it is "the leader". The availability of a leader can be exploited by providing it with a distinguished local algorithm, while all other nodes execute the same local algorithm (which is different from the leader algorithm).

One may now deduce that the model that adapts to the problem of finding the node with most demand is that belonging to a leader network. Nevertheless, this algorithm is more general. Our problem is more specific, since the leader node is the one with most demand and is found at the bottom of a high demand zone.

All the nodes in a zone of high demand cooperate together to select the node, which will carry out the processes of the leader node.

4.2 The Algorithm for Electing the Leader Node on a Zone of High Demand

For the development of this algorithm, we consider a peer-to-peer network with N nodes connected to each other via E links and with a distinct and finite weight (demand) on each one of the nodes. Each node executes the same local algorithm, which consists in first sending messages (announcing its demand) to its neighbors via the corresponding (adjacent) links, awaiting the arrival of the messages (neighboring demand) and processing them. The messages are transmitted in all directions and arrive after an unpredictable but finite delay.

Each node at a random time (fig. 3) will cast its vote for the neighboring node having the greatest demand, and will send it a message notifying it that the vote has been cast. Each vote is unique and unrepeated; it has the ID of the node casting it, a time stamp, and a time to live (in real applications it is an empirical number) necessary for avoiding loops or for preventing the vote from circulating infinitely around the network.

Each node that receives a vote passes the vote on to whichever of its neighbors has the largest number of requests, and so on successively, until after an unbounded but finite period of time the majority of votes cast on an island (high demand zone) have only one node, which will be the node selected (the leader). It is not possible to ensure that all the votes of the nodes on an island reach the node of greatest demand, since the number of nodes that make up a zone of high demand is not known. Neither do we know how many votes are still traveling without having arrived at the node of greatest demand. However, it is possible to ensure that the votes in a high demand zone will not travel to other zones, since only replicas of higher demand are propagated, and never toward the zones of lower demand. In order for a node to take on the role of a leader node, it is sufficient that, after a period of time, the number of accumulated votes is different from zero.

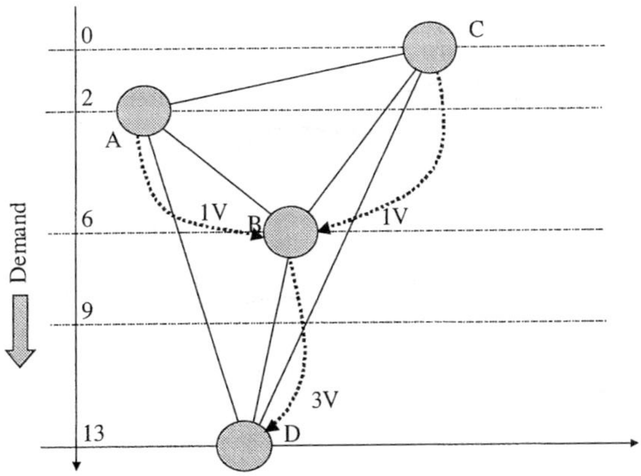

Fig 2. An example of a demand based leader election algorithm.

Table 1. Accumulated votes in the nodes of figure 2.

Node	Votes
A	0
B	0
C	0
D	3

Informally speaking this algorithm works in the following way (fig. 2):

- Node C votes for its neighbor having most demand (B) and sends it its vote.
- Node B receives a vote from C, and since B has a neighbor with greater demand, it sends C's vote to this neighbor (D).
- Node D receives C's vote from B. D looks to see if it has a neighbor with greater demand. Since it does not, it keeps the vote. So D now has one vote.
- Node A votes for its neighbor with greatest demand (B), and sends it its vote.
- Node B receives a vote from A, and as B has a neighbor with greater demand, it sends A's vote to this neighbor (D).
- Node D receives A's vote from B. D looks to see if it has a neighbor with greater demand. It does not, so it keeps the vote. D now has two votes
- Node B votes for its neighbor having greatest demand (D), and sends it its vote.
- Node D receives B's vote. D looks to see if it has a neighbor with greater demand. It does not, so it keeps the vote. D now has three votes.

If node D sees no nodes with more demand than itself, it does not send the votes it has received to any other node; neither does it cast its own vote. Therefore it retains all the votes, and thus becomes the leader node in this high demand zone.

Every time it is necessary to construct the superset of leaders, it is enough for each node having a number of votes greater than zero to announce itself.

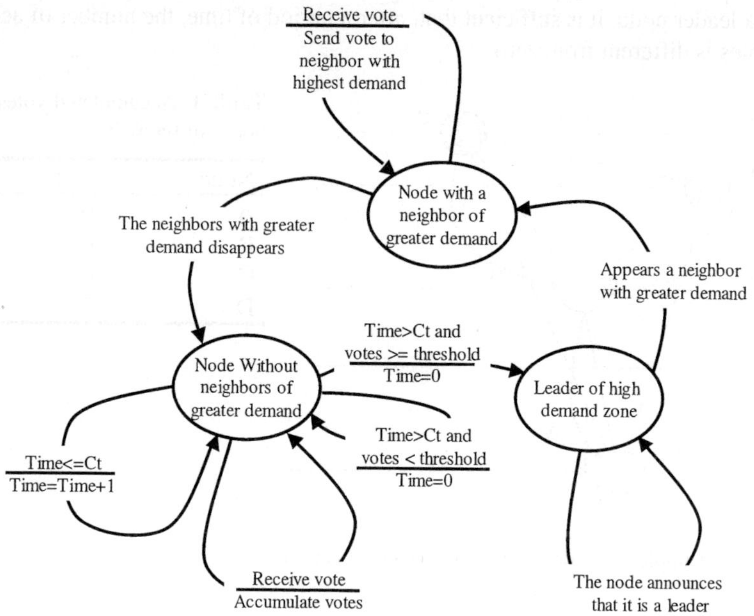

Fig.3. The state machine of leader election algorithm.

4.3 The Building of the Leader Network

The node that knows it is the leader now has the task of finding other leader nodes, if they exist. It must also announce itself so that other nodes become aware of its existence. The mechanism by which a leader node announces itself is by sending a message as part of the weak consistency protocol. This protocol ensures that the message arrives to all the nodes in a finite, but unbounded, period of time, and therefore to the leader nodes as well, assuming they exist.

When a leader node receives a message from another leader node, it keeps the ID of the node sending the message in a table. Each leader node has a table containing the data of the other leader nodes that know of its existence. This table is replicated in each leader node and is reconstructed dynamically. The ID of a leader node is included in the table on arrival of a message of announcement. It is not necessary to remove a node from the table of leaders because the table is dynamically reconstructed periodically, the period of time being at least equal to the time (expressed in sessions) necessary for the message to cross the entire network of replicas.

5 Validation

In order to validate our proposal, a simulator based on NS2 [19] has been built. We simulate the behavior of the algorithms on a grid network with synthetic demand. Our fast consistency algorithm has been tested in [1] using a generator of random representative Internet topologies [10,17,18], now in this paper we use as scenarios for applying our algorithms, surfaces of 17*17 nodes on which the different levels, representing the demands, are synthetically generated by the diamond-square algorithm, which is a classic algorithm for generating fractal surfaces that resemble landscapes with scaling properties; that is to say, self-similar. In this way, we achieve a scenario sufficiently general to ensure that the results obtained in the simulations do not depend on the particular or local conditions of a specific scenario. To reduce the effects of randomness, and to prevent the results from depending on the characteristics of a particular fractal surface, one hundred different scenarios have been generated, on each one of which a thousand experiments have been carried out.

5.1 Performance Metric

The purpose of the "fast consistency" algorithm with high demand zones interconnected is to improve the performance of the weak consistency algorithms, with particular emphasis on increasing the speed with which these algorithms carry the changes to the zones of greatest demand, so that a greater number of clients may have access to fresh content in a shorter period of time. It is for that reason that our experiments are centered on measuring these speeds. The performance (speed) is measured in terms of the anti-entropy sessions needed for all the zones to receive the messages with the changes generated en the rest of the nodes that make up the grid.

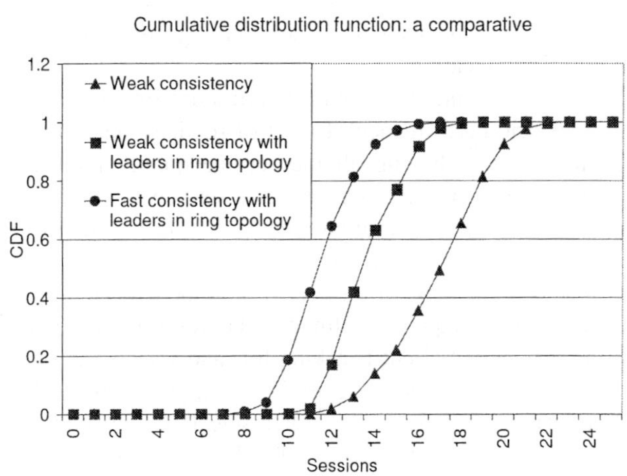

Fig. 4

5.2 Simulation Results

The leader nodes having a number of votes greater than or equal to a threshold (i.e. average of all node leaders) are selected and interconnected in a ring topology, which joins together the zones of high demand. In this topology, each node sees the same network diameter. The results can be seen in Figure 4(CDF); we see that Fast Consistency with leaders interconnected has a better performance than Weak Consistency, and also the effects of the change on the topology when the leader nodes are connected in a ring topology.

6 Related Work

Work whose aim is to adapt replication systems to clients' needs, in TACT [25] for example, propose that the consistency between replicas vary in a continuous way between weak and strong. Three metrics are used to limit consistency – numerical error, order error, and staleness. In TACT, it is the client who specifies the level of consistency required. This level may take any value between weak and strong. With fast consistency, however, updated content is carried to the high demand replicas without incurring the huge costs involved with the strong consistency.

In Fluid Replication [7], clients can create replicas dynamically, whenever and wherever they are necessary. When a client sees that performance is low, either because of increasing demand on the network, or because of client mobility, he can create replicas on a waystation, which is a node on which replicas can be created. This system seeks to provide replicas when and where they are required. Our proposal concerns algorithms for keeping replicas consistent. Our algorithms can be used to complement the fluid replication system. In order that replicas in fluid replication permanently updated by fast consistency could be created wherever there was a great concentration of clients. It is in such cases where our algorithm could act as an excellent complement to fluid replication.

In [15], S. Krishnamurthy and others put forward a dynamic selection replica algorithm based on the historic performance of a set of replicas, and which is capable of dynamically selecting replicas having all the service quality requirements a client might request. For example, a client who needs a service that responds to requests within a specific time expresses these requirements as a service quality specification. The client can specify his QoS requirements at the beginning or during execution. The specification includes the name of the service and the expected response time. A scheduler intercepts clients' requests, estimates the response time of the different replicas offering the service the client is asking for, and selects the subset of replicas that have what the client requires. We propose is giving preferential treatment to large client subsets. Furthermore, the dynamic replica selection algorithm has been tested on AquA; a middleware based on CORBA in local area networks, although it works very well in large scale distributed systems such as Internet.

7 Conclusions and Future Work

The main challenge to distributed systems on Internet scale is that of maintaining availability and performance, in spite of the low reliability and the delay associated with large size broad networks. To meet this challenge successfully, the most valuable tool is replication. However, it is not a magical solution; it involves the problem of keeping replicas consistent. Weak consistency is an algorithm that maintains replicas consistent, sacrificing data freshness in a controlled way in order to improve availability and performance. In the context of weak consistency, the fast consistency algorithm prioritizes replicas with high demand in such a way that a large number of clients receive fresh content. This algorithm gives high performance in one zone of high demand, but in multiple zones this performance becomes poor. To improve this poor performance, we propose an election algorithm based on demand. After an interval of time t, the replica that accumulates a great number of votes in a zone of high demand assumes the role of leader in that zone. The leaders in each zone together make up a logical hierarchy [3] whose task is to convey the changes made in each zone to the high demand replicas, preventing the zones of low demand from forming a barrier to the fast consistency algorithm. To validate this proposal we have constructed a simulator on Network Simulator [19]; we have feed it with a set of replicas in which there exist multiple zones of high demand, connecting these zones together by means of a logical topology. The result is a considerable improvement in the algorithm performance compared with the same topology without the interconnected high demand zones. This allows us to conclude that with very few additional bytes in signaling, and increasing the complexity of the fast consistency algorithm a little – since periodic election must be carried out so that the hierarchy of leader replicas corresponds to the dynamic state of the system – demand is changeable, and the leader nodes and the links between them must be kept up to date. In this work we have not considered the effect of the local minimums, which can slow down fast consistency and/or create false leaders. It may be necessary for the replicas to be sensitive to field effects, as it occurs in the relativity theory of gravity, and thus the replicas would have information beyond their neighbors lying at distance one.

References

1. Jesús Acosta Elias, Leandro Navarro Moldes, "A Demand Based Algorithm for Rapid Updating of Replicas ", IEEE Workshop on Resource Sharing in Massively Distributed Systems (RESH'02), July 2002.
2. Jesús Acosta Elias, Leandro Navarro Moldes, "Behaviour of the fast consistency algorithm in the set of replicas with multiple zones with high demand", Symposium in Informatics and Telecommunications, SIT2002.
3. N. Adly, M. Nagi and J. Bacon. 'A Hierarchical Asynchronous Replication Protocol for Large Scale Systems'. Proc. of the IEEE Workshop on Parallel and Distributed Systems, Princeton, New Jersey, October 1993, pp.152-157.
4. Adya, "Weak Consistency: A Generalized Theory and Optimistic Implementations for Distributed Transactions", PhD thesis MIT, Department of Electrical Engineering and Computer Science, March 1999.
5. K. P. Birman, "The process group approach to reliable distributed computing", Communications of ACM, December 1993/Vol. 36, No. 12
6. Ernest Chang, Rosemary Roberts, "An improved algorithm for decentralized extrema-finding in circular configurations" of processes Communications of the ACM May 1979 Volume 22 Issue 5

7. L. P. Cox, and Noble, B. D. 2001."Fast Reconciliations in Fluid Replications", Int. Conf. On Dist. Comp. Syst. (ICDCS) (April 2001). http://mobility.eecs.umich.edu/papers/icdcs01.pdf
8. D. J. Dietterich, "DEC data distributor: for data replication and data warehousing". In Int. Conf. On Management of data, pp 468, ACM, May 1994.
9. V. Duvvuri, P. Shenoy and R. Tewari, "Adaptative Leases: A Strong Consistency Mechanism for the World Wide Web", IEEE INFOCOM 2000, p 834-843.
10. M. Faloutsos, P. Faloutsos, and C. Faloutsos, "On Power-Law Relationships of the Internet Topology", ACM SIGCOMM, Cambridge, MA, September 1999
11. R. G. Gallager, P. A. Humblet, P. M. Spira, "A Distributed Algorithm for Minimum-Weight Spanning Trees" ACM Transactions on Programming Languages and Systems (TOPLAS) January 1983 Volume 5 Issue 1
12. R. A. Golding, "Weak-Consistency Group Communication and Membership", PhD thesis, University of California, Santa Cruz, Computer and Information Sciences Technical Report UCSC-CRL-92-52, December 1992.
13. R. Guy, J. Heidemann, W. Mak, T. Page, Jr., G. Popek and D. Rothmeier. Implementation of the Ficus Replicated File Sys. Proc. USENIX Conf, June 1990.
14. Brian Kantor, Phil Lapsley RFC0977, http://www.ietf.org/rfc/rfc0977.txt, 1986
15. S. Krishnamurtyhy, W. Sanders, and M. Cukier, "A Dynamic Replica Selection Algorithm for Tolerating Timing Faults", In Proc. Of the The International Conference on Dependable Systems and Networks, pages 107-116, July 2001.
16. G. LeLann, "*Distributed systems - towards a formal approach*," in Information Processing 77, B. Gilchrist, Ed. North-Holland, 1977.
17. Medina, A. Lakhina, I. Matta, and J. Byers, "BRITE: Universal Topology Generation from a User's Perspective",
18. Medina, I. Matta, and J. Byers, "On the Origin of Power Laws in Internet Topologies", ACM Computer Communication Review, p. 160-163, April 2000.
19. The Network Simulator: http://www.isi.edu/nsnam/ns/
20. C. Neuman, "Scale in Distributed Systems. In Readings in Distributed Computing Systems", IEEE Computer Society Press, 1994
21. K. Petersen, M. J. Spreitzer, D. B. Terry, M. M. Theimer, and Demers, "Flexible Update Propagation for Weakly Consistent Replication", Proc. of the 16th ACM Symp. on Op. Syst. Prin. (SOSP-16), S. Malo, France, Oct. 5-8,97, p. 288-301.
22. Yasushi Saito, Jeffrey C. Mogul, and B. Verghese. A Usenet Performance Study, http://www.research.digital.com/wrl/projects/newsbench/usenet.ps. Nov 98."
23. M. Satyanarayanan, Scalable, Secure, and Highly Available Distributed File IEEE Computer May 1990, Vol. 23, No. 5
24. Michel D. Schroeder, Andrew D. Birrel, and Roger M. Needham, Experience with Grapevine: The Growth of a Distributed System, ACM Transactions of Computer Systems, Vol. 2, No. 1, February 1984, Pages 3-23.
25. Haifeng Yu, Amin Vahdat, "Desing and Evaluation of a Continuous Consistency Model for Replicated Services", Proceedings of the Fourth Symposium on Operating Systems Design and Implementation (OSDI), October 2000.

Performance Analysis of a Parallel Application in the GRID

Holger Brunst[2], Edgar Gabriel[1,3], Marc Lange[1], Matthias S. Müller[1], Wolfgang E. Nagel[2], and Michael M. Resch[1]

[1] High Performance Computing Center Stuttgart
Allmandring 30, D-70550 Stuttgart, Germany
{gabriel,lange,mueller,resch}@hlrs.de

[2] ZHR, Dresden University of Technology
D-01062 Dresden, Germany
{brunst,nagel}@zhr.tu-dresden.de

[3] Innovative Computing Laboratories,
Computer Science Department,
University of Tennessee, Knoxville, TN, USA

Abstract. Performance analysis of real applications in clusters and GRID like environments is essential to fully exploit the performance of new architectures. The key problem is the deepening hierarchy of latencies and bandwidths and the growing heterogeneity of systems. This paper discusses the basic problems of performance analysis in such clustered and heterogeneous environments. It further presents a software environment that supports the user in running codes and getting more insight into the performance of the application. In order to give a proof of the concept a code for direct numerical simulation of reactive flows is run in a GRID like hardware environment, and the performance analysis is presented.

1 Introduction

For some time, parallel programming has been mainly a problem of splitting the work of a code into a number of equally sized pieces and distributing them correctly to all available processors. In addition, communication issues had to be taken care of. This required a substantial amount of work but at least the programmer could safely assume that the code would run on a homogeneous architecture. The speed of processors would be the same as well as the speed of communication between individual processors. Moving from single processor systems to parallel computers was thus a cumbersome task but added just one additional level of complexity.

With the advent of clusters [1] and the concept of the GRID [2] the situation has changed. The new hardware concept is to put together basic building blocks to form a new system. Such basic building blocks can be anything from a single processor PC or a multi processor SMP to a full cluster. In that sense what we

P.M.A. Sloot et al. (Eds.): ICCS 2003, LNCS 2658, pp. 285–294, 2003.

get are systems that are more and more dynamic in composition. In addition, the programmer is faced with a hierarchy of bandwidths, latencies and protocols that is getting deeper and more heterogeneous [3]. From the point of view of performance analysis for real applications this architectural trend represents a challenge both for the programmer and the tool provider.

This paper deals with the performance analysis of a parallel application from the area of combustion simulation, which was ported to distributed GRID environments due to its tremendous demand on resources. During this procedure two problems had to be solved: first, we had to port the application to a GRID environment. Second, we wanted to evaluate the potential of this application in a heterogeneous GRID environment. Both steps required detailed analysis of the communication patterns of the application, using a performance analysis tool.

The structure of the paper is like follows: the problem of performance analysis in GRID environments and the performance analysis tool Vampir is presented in Sect. 2. Section 3 presents the application used throughout the rest of the paper. Section 4 presents then the performance analysis of this application and how we solved the identified problems. Finally, in Sect. 5 we summarize the paper and present the ongoing work in this area.

2 How to Adapt Existing Analysis Technology

Today, clustered parallel computers are available to a large user comunity that deals with numerical simulation. These systems mostly manufactured from standard components off the shelf (COTS) often provide a very good price/performance ratio. Therefore, parallel computing more and more also becomes attractive for commercial applications and thus widens the community of parallel application developers who need tool support.

For increased requirements of resources, clusters can be interconnected via a GRID middleware. Unfortunately, the different communication layers in such a hybrid system make it highly complicated to develop parallel applications performing well. Therefore, it is crucial to gain insight into the application's internal behavior when executed; as this is the only possibility to locate performance bottlenecks in many cases.

The performance evaluation and the subsequent optimization of a parallel program executed on a GRID imposes special requirements on the generation and visualization of performance data [4]. In order to allow for a largely transparent view of the application's behavior, there is a need for some background mechanism that collects and merges the performance data generated on the distributed nodes of the GRID. This task is not a simple one as one has to cope with the synchronization problem of multiple non-synchronous system clocks. Current performance tools are familiar with the analysis and visualization of performance data generated on a homogeneous parallel machine. So far, there is no graphical support for the specifics introduced by GRID computing. In our work we identified the following key issues towards the performance analysis of parallel GRID applications:

- Appropriate data base formats for parallel distributed trace data
- Unified and transparent access to trace data
- Post synchronization of asynchronous trace data
- Visualization of trace data with focus on additional communication layer

Conceptionally, a great deal of these issues has been veryfied with the following software framework that was the basis for this paper. For the communication part we used a GRID-aware implementation of the MPI standard called PACX-MPI. The performance optimization was carried out with an adapted version of the performance analysis tool Vampir [5–7] which was one of the results of the german UNICORE [8] project. The porting of the associated tracing library Vampirtrace onto PACX-MPI has been done in the scope of the European project METODIS [9] and is currently further enhanced in the DAMIEN [10] project. The following properties can be analyzed with this framework:

- Program state changes
- Collective operations
- Peer to peer communication
- Hardware performance monitors
- OpenMP parallel regions
- File I/O
- Environmental information
- Source code locators

Based on event traces, the framework provides timeline representations and accumulated performance measures for selected time intervals of a program run. The timelines show the parallel program's states for an arbitrary period of time. Performance problems that are related to imbalanced code or bad synchronization can easily be detected with the latter as they cause irregular patterns. When it comes to performance bottlenecks caused by bad cache access or weak usage of the floating point units, the timelines are less useful. Hardware counters can be found in almost every processor now. Support of these counters in the form of hardware performance monitors is available in combination with process timeline diagrams.

The different performance categories named above need to be related to each other in order to create a complete view of an application's runtime behavior. Flexible grouping allows performance analysis from different perspectives. When it comes to the visualization task of a multi level GRID application, the grouping capability can be used to navigate through the trace data on different levels of abstraction. This approach works fine for both event trace oriented displays (timelines) and statistic displays for accumulated data.

3 A Real World Application Test Case

This section provides some background on the application from the field of reactive flow simulation which has been analysed in an GRID-like environment. In Sect. 3.1, we briefly discuss why one is interested in the use of clustered systems for this type of application. The basic properties of the code and the physical problem to which it is applied are described in Sect. 3.2.

The performance problems we were faced executing the first runs and the following performance analysis are described then in the subsequent sections.

3.1 Direct Numerical Simulation of Reactive Flows

Chemical reactions occuring in turbulent flows play an essential role in a variety of technical applications like e.g. combustion in automotive engines and gas turbines. There is obviously a large interest in an improved energy efficiency and reduced emissions of these processes. Better and more accurate models for turbulent combustion are needed to improve the quality of numerical simulations of such technical combustion processes. The information about the basic interaction mechanisms of turbulent transport and chemical reactions which is needed for the development of improved turbulent combustion models can be obtained by direct numerical simulations (DNS), in which turbulent reactive flows are computed with a great level of detail.

The set of coupled partial differential equations to be solved in DNS are the Navier-Stokes equations for reactive flows as given e.g. in [11]. These equations express the conservation of mass, momentum, and energy, which are also needed for the DNS of non-reacting flows, as well as the conservation of the mass fractions of all the chemical species. The number of species to be considered varies between nine for the hydrogen-air system and several hundreds in the case of higher hydrocarbons. The computation of the chemical source-terms and of the multicomponent diffusion velocities are the most time consuming parts of such DNS.

Due to the nature of turbulence and the chemical kinetics, a very high resolution is needed in space and time, which leads to a large computational effort. E.g., the DNS of the temporal evolution of a premixed methane flame over a physical time of a few milliseconds in a twodimensional domain with an area of about 3 cm^2 took about 60 hours using 256 PEs of a Cray T3E-900 [12]. As turbulence is an inherently three-dimensional phenomenon, some of it's aspects could only be reproduced precisely by performing similar DNS in 3D requiring at least a hundred times as many grid points. Making efficient usage of large distributed computing environments would be an essential contribution to reach this goal.

3.2 The Application Program PCM

PCM is a code developed for the direct numerical simulation (DNS) of reactive flows on parallel computers with distributed memory using message-passing communication [12, 13]. In favor of being able to utilize detailed models for the computation of chemical reaction kinetics and the molecular transport, it is currently restricted to the simulation of two-dimensional flow fields. While some features of turbulent flows are thus not present in the simulations, it has been shown that many other aspects of turbulent combustion processes are missing if oversimplified models for chemistry or molecular diffusion are used [12].

The spatial discretization in PCM is performed using a finite-difference scheme with sixth-order central-derivatives, avoiding numerical dissipation and leading to very high accuracy. The integration in time is carried out using a fourth-order fully explicit Runge-Kutta method with adaptive control of the time step. This fully explicit formulation leads to a parallelization strategy, which is based on a regular two-dimensional domain decomposition of the physical space, projected onto a corresponding two-dimensional processor topology. For a given computational grid and number of processors it is tried to minimize the length of the subdomain boundaries and thus the amount of communication. However, the different communication speeds in a GRID-computing environment are not considered in the initial distribution up to now. After this initial decomposition, each processor node controls a rectangular subdomain of the global computational domain. In addition to the grid points belonging to a node's subdomain, a three points wide halo region is stored on each node. Using the values at the grid points of this halo region an integration step on the subdomain is carried out independently from the other nodes. After each integration step, the new values of the domain boundaries are exchanged between neighboring nodes using MPI.

The application case which has been used for the analysed runs is the DNS of a turbulent mixing-layer with a cold H_2/N_2 mixture on one side and heated air on the other. Periodic boundary conditions are used perpendicular to the mixing layer. This is a simple model configuration to study the influence of turbulence on autoignition processes, which is important in combustion systems like Diesel engines where fuel autoignites after being released into a turbulent oxidant of elevated temperature. More details on this configuration and some results regarding the physics can be found in [12, 13].

Recently, our main platform for production runs have been Cray T3E systems and large PC-clusters on which a high performance and a very good scaling behavior has been achieved [14]. While our first experiences with PCM in a metacomputing environment showed a great potential of such distributed platforms for this type of application, it became also obvious that a considerable optimization effort will be needed to fully exploit this potential [15].

4 Analysis of PCM with Vampir

For the tests presented in this paper, 24 processors on a Cray T3E at the High Performance Computing Center Stuttgart were used. For the distributed tests, we used two partitions of the same machine, creating a virtual metacomputer. The first 12 processes were running on the first partition (T3E-A), while the other 12 processes were running on the second partition (T3E-B). The internal latency between two processes on the same machine was 7 μs, the achievable bandwidth was around 300 MBytes/s, while the external latency between the two partitions was in the range of 4 ms, the achievable external bandwidth was approximately 10 MBytes/s. The latter values are typical in local area networks (LAN). Thus, the communication characteristics of this virtual metacomputer are similar to GRID environments with respect to the fact, that we have to deal with the same hierarchies in the quality of communication. Coupling two partitions of the same machine has additionally one major advantage for our analysis compared to using two different machines: we could exclude most effects due to network traffic of other users, which made our results more easily comparable to each other. The communication library used to make PCM run across distributed environments is PACX-MPI [16], an implementation of the MPI standard optimized for GRID environments.

The following analysis is centered around two questions: During the initial porting of the application onto PACX-MPI, we were facing the problem, that the unexpected message queue in PACX-MPI had from time to time an unexpected overrun, which neither the developers of the library, nor the application developers could explain. Second, we wanted to evaluate the potential of this application for future use in distributed environments, since the computational demand of realistic simulations is tremendous.

4.1 Processing Part

The application can be split into three parts: preprocessing, processing, and application checkpointing. The preprocessing part is dominated by communication, since all file-operations are handled by the process with rank zero. The execution time for the preprocessing is increasing from approximately 3.5 seconds for the Cray-T3E run to about 6 seconds for the metacomputing run. This is due to the fact that the broadcast-operations require external communication. However, the preprocessing-part does not have any dramatic influence on the performance of a real production run.

In a real production run, the overall execution time is dominated by the processing part, since the application will execute several thousand iterations, instead of the 20 iterations shown in this analysis. Zooming into this section shows, that the ratio of communication to computation for this part of the application is a lot better. Using the summary chart, the exact times spent in communication and in computation can be determined. For the processing part, approximately 20 % of the execution time is just spent in communication. Taking into account, that we have to deal with a tightly coupled application using an environment with latencies and bandwidths typical for local-area networks, this result is quite good.

The message statistic display in Fig. 1 shows the communication pattern in the processing part. Two different patterns can be identified in the left part of the figure:

1. The penta-diagonal structure reflects the neighborhood communication in the application, which again is the result of the 2D domain decomposition.
2. Additionally, a much lower amount of communication is visible between the MPI-process with the rank zero and all other processes. A part of this additional communication is due to the load-balancing integrated in the application. Node zero collects the execution time of each process for a certain number of iterations. Depending on the fact, whether the load imbalance between processes is higher than a user defined limit, work will be redistributed between the nodes.

The key-issue for an application to achieve high performance in a clustered environment is high data locality. The assignment of sub-domains to processors

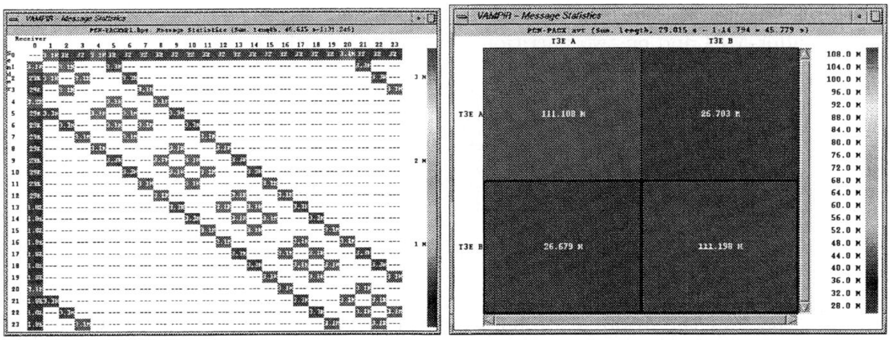

Fig. 1. Communication pattern in the processing part of PCM for a **4** × 6 process topology, process view (left) and cluster-view (right).

should reflect the hardware topology. One possible approach is to use an optimized MPI_Cart_create. However, in our case the application performs the mapping itself, because it also supports other message passing libraries. Nevertheless an optimal result may be achieved, because the assumption that processes with ranks that have small differences are close to each other is fulfilled by PACX-MPI. But in general the configuration has to be optimized concerning the communication pattern with respect to periodic boundary conditions, orientation of domain and others. Here, further analysis with Vampir is needed. The result of one configuration can be seen in in the right part of Fig. 1, where the total amount of data is shown, which has been sent between processes inside each machine, and the total amount of data transfered between the machines. The communication inside each cluster-node is dominating compared to the communication between both cluster-nodes. While totally 109 MBytes of data have been sent inside of each cluster-node, just around 26 MBytes of data have crossed the boundaries of the machines.

The good performance behavior of this part of the application is also visible in Table 1, where we compare the execution time for five iterations on the virtual metacomputer with the same number of iterations on the Cray T3E. While the execution time increases slightly, it is still in the same range as on the Cray T3E. This is mainly the result of the regular communication pattern shown in Fig. 1 (left) and of the high data locality shown in Fig. 1 (right).

4.2 Application Checkpointing

Long production runs have to save regularly the current status of the run in order to avoid the loss of results in case of a machine crash. In case several machines and networks are involved in such a simulation, like given in GRID-environments, the probability of loosing a subsystem even increases. For the application under examination here, a regular checkpointing is in addition required for tracking the temporal evolution of the solution.

Table 1. Comparison of the average execution time for 5 time steps on a single machine with vendor-MPI and on two partitions with PACX-MPI.

Number of nodes	Execution time with Cray-MPI	Execution time with PACX-MPI
12	17.8 sec	19.1 sec
24	10.3 sec	12.4 sec

This part of the application shows a very different behavior on a single parallel machine (see left part of Fig. 2) compared to a cluster of T3E's (see right part of Fig. 2). The analysis with Vampir revealed the reason for this very different behavior and in addition explained the problem of gathering thousands of messages on the process with rank zero mentioned above.

On the Cray T3E, the execution of this part is dominated by the process with rank zero. As Fig. 2 shows all other processes start sending their part of the global data to the process zero approximately at the same time, and block then in a wait-operation. The reason for this blocking is, that for the used message length, Cray-MPI switches to the synchronous-send mode. Thus, the wait-operation can only return, when the according receive-operation has been started. This behavior causes all processes to be more or less synchronized with process zero.

With PACX-MPI, the processes sending their data to the process zero behave differently. They send all their messages in a row to the destination process without synchronizations between the individual sends. The reason for this is that PACX-MPI does not switch to a synchronous send-mode for external operations, unless explicitly required by the application by using *MPI_Ssend*. Additionally, the communication-daemons involved in the external operations can receive and transfer the messages faster than the process zero is able to handle the file-operations. Since the wait-operation to each of the initiated send-routines can

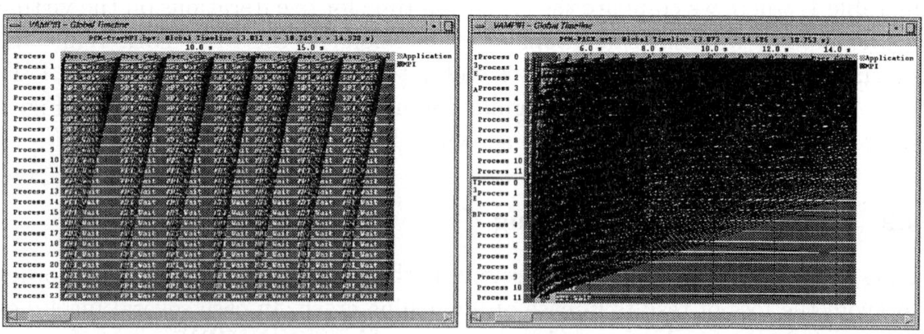

Fig. 2. Communication behavior of PCM during writing of results-files with Cray-MPI (left) and PACX-MPI (right)

return as soon as '*data is safely stored away*'[17], they are returning, as soon as the data is received by the communication-daemon handling all outgoing messages. Thus, the processes can proceed immediately with the next communication, without being synchronized with process zero. This behavior explains the gathering of much higher number of messages at the process doing the file I/O operations than in the single system case. To achieve a similar behavior with PACX-MPI like with Cray-MPI, the application will have to use the synchronous versions of the send-operations of MPI in this part of the application.

5 Conclusions

In this paper, the problems for performance analysis of parallel applications arising in clustered systems are briefly discussed. The components of a software environment have been presented which enables the user to run and analyse applications in GRID-like environments. The functionality of this software environment has been demonstrated for a real world application. In addition to an optimized, topology-aware application and an optimized, GRID-aware MPI library a thorough understanding of the MPI implementation was required to achieve good performance.

The visualization capabilities of Vampir enabled us to analyse the communication pattern of the application with respect to the different levels of communication, which helps to find the best application settings for a given GRID environment. The analysis with Vampir was not only useful for application tuning but also revealed the reason for the buffering of an unexpectedly high number of messages with PACX-MPI. The reason for this problem could not be found without the help of a runtime analysis tool. Once found, it was relatively easy to implement a solution for this problem.

Future extension of Vampir will include further features that are necessary to understand possible performance problems. Examples are a visualization of the protocol types which have been used for the individual messages, or the gathering of environment variables on the different platforms in the Grid that influence the behavior of the MPI library, like MPI_BUFFER_MAX.

Acknowledgments

The authors would like to thank their home institutions for providing their machines for this study. We would also like to thank Hans-Christian Hoppe from PALLAS GmbH for the support for VAMPIRtrace.

References

1. R. Buyya. *High Performance Cluster Computing, Volume 1: Architectures and Systems, Volume 2: Programming and Applications* . Prentice Hall, 1999.
2. Ian Foster and Carl Kesselman. *The Grid: Blueprint for a New Computing Infrastructure.* Morgan Kaufmann, 1999.

3. S. Pickles, F. Costen, J. Brooke, E. Gabriel, M. Müller, M. Resch and S. Ord. *Metacomputing Across Intercontinental Networks.* Future Generation Computer Systems, 7, 2001, 911–918.
4. H. Brunst, W. E. Nagel, and H.-C. Hoppe. Group-Based Performance Analysis for Multithreaded SMP Cluster Applications. In R. Sakellariou, J. Keane, J. Gurd, and L. Freeman, Eds., *Euro-Par 2001 Parallel Processing*, LNCS 2150, 148–153, Springer, 2001.
5. H. Brunst, M. Winkler, W. E. Nagel, and H.-C. Hoppe. Performance Optimization for Large Scale Computing: The Scalable VAMPIR Approach. In V. N. Alexandrov, J. J. Dongarra, B. A. Juliano, R. S. Renner, and C. K. Tan, Eds., *Computational Science – ICCS 2001, Part II*, LNCS 2074, 751–760, Springer, 2001.
6. S. Moore, D. Cronk, K. London, and J. Dongarra. Review of Performance Analysis Tools for MPI Parallel Programs. In Y. Cotronis and J. Dongarra, Eds., *Recent Advances in Parallel Virtual Machine and Message Passing Interface, 8th European PVM/MPI Users' Group Meeting, Proceedings*, LNCS 2131, 241–248, Springer, 2001.
7. W. E. Nagel, A. Arnold, M. Weber, H.-C. Hoppe, and K. Solchenbach. *VAMPIR: Visualization and Analysis of MPI Resources.* Supercomputer, 12(1), 1996, 69–80. http://www.pallas.de/pages/vampir.htm.
8. J. Almond, D. Snelling. *UNICORE: Secure and Uniform Access to Ditributed Resources via the World Wide Web.* 1998. http://www.fz-juelich.de/unicore/whitepaper.ps.
9. *METODIS – Metacomputing Tools for Distributed Systems.* http://www.hlrs.de/organization/pds/projects/metodis.
10. *DAMIEN – Distributed Applications and Middleware for Industrial Use of European Networks.* http://www.hlrs.de/organization/pds/projects/damien.
11. F. A. Williams. *Combustion Theory.* second edition, Benjamin/Cummings, 1985.
12. M. Lange and J. Warnatz. Investigation of Chemistry-Turbulence Interactions Using DNS on the Cray T3E. In E. Krause and W. Jäger, Eds., *High Performance Computing in Science and Engineering '99*, 333–343, Springer, 2000.
13. M. Lange. Parallel DNS of Autoignition Processes with Adaptive Computation of Chemical Source Terms. In C. B. Jenssen, T. Kvamsdal, H. I. Andersson, B. Pettersen, A. Ecer, J. Periaux, N. Satofuka, and P. Fox, Eds., *Parallel Computational Fluid Dynamics: Trends and Applications*, 551–558, Elsevier Science, 2001.
14. M. Lange. Massively Parallel DNS of Flame Kernel Evolution in Spark-Ignited Turbulent Mixtures. In E. Krause and W. Jäger, Eds., *High Performance Computing in Science and Engineering '02*, 425–438, Springer, 2002.
15. E. Gabriel, M. Lange and R. Rühle. Direct Numerical Simulation of Turbulent Reactive Flows in a Metacomputing Environment. In *Proceedings of the 2001 ICPP Workshops*, 237–244, IEEE Computer Society, 2001.
16. E. Gabriel, M. Resch and R. Rühle. Implementing MPI with Optimized Algorithms for Metacomputing. In A. Skjellum, P. V. Bangalore, Y. S. Dandass, Eds., *Proceedings of the Third MPI Developer's and User's Conference*, MPI Software Technology Press, 1999.
17. Message Passing Interface Forum. *MPI: A Message-Passing Interface Standard (Version 1.1).* 1995. http://www.mpi-forum.org.

Workshop on Computational Chemistry and Molecular Dynamics

Linear Algebra Computation Benchmarks on a Model Grid Platform

Loriano Storchi[1], Carlo Manuali[2], Osvaldo Gervasi[3], Giuseppe Vitillaro[4], Antonio Laganà[1], and Francesco Tarantelli[1,4]

[1] Department of Chemistry, University of Perugia,
via Elce di Sotto, 8, I-06123 Perugia, Italy
redo@thch.unipg.it, lag@unipg.it, franc@thch.unipg.it

[2] CASI, University of Perugia,
via G. Duranti 1/A, I-06125 Perugia, Italy
carlo@unipg.it

[3] Department of Mathematics and Informatics, University of Perugia,
via Vanvitelli, 1, I-06123 Perugia, Italy
osvaldo@unipg.it

[4] Istituto di Scienze e Tecnologie Molecolari, CNR,
via Elce di Sotto, 8, I-06123 Perugia, Italy
peppe@thch.unipg.it

Abstract. The interest of the scientific community in Beowulf clusters and Grid computing infrastructures is continuously increasing. The present work reports on a customization of Globus Software Toolkit 2 for a Grid infrastructure based on Beowulf clusters, aimed at analyzing and optimizing its performance. We illustrate the platform topology and the strategy we adopted to implement the various levels of process communication based on Globus and MPI. Communication benchmarks and computational tests based on parallel linear algebra routines widely used in computational chemistry applications have been carried out on a model Grid infrastructure composed of three 3 Beowulf clusters connected through an ATM WAN (16 Mbps).

1 Introduction

Grid computing [1] is promising to establish itself as a revolutionary approach to the use of computing resources. At the same time, Beowulf-type clusters (BC) [2] have become very popular as computing platforms for the academic and scientific communities, showing extraordinary stability and fault tolerance at very attractive cost/benefits ratios. The availability of Gigabit Ethernet further facilitates the assembly of high throughput workstation clusters of the Beowulf type. It is therefore a natural and smooth development to explore the possibility of integration between the Grid and the cluster paradigms. By connecting together, at the hardware and software levels, several clusters one can in principle build a Grid computing platform very useful for scientific applications. While the Grid model is often viewed as a cooperative collection of individual computers, it is clear that, in order to make such a model efficient for scientific applications, and we think in particular of computational chemistry ones, much effort must be

P.M.A. Sloot et al. (Eds.): ICCS 2003, LNCS 2658, pp. 297–306, 2003.

devoted toward re-designing the scheduling and communication software (and possibly the application themselves) so that the Grid topology and interconnections are explicitly taken into account. As we mentioned, this refers especially to local clustering. Some work along these lines was for example carried out by developing a MPI-based library of collective communication operations explicitly designed to take into account two layers (wide-area and local-area) of message passing [3].

The present work presents a study of the performance of a model platform composed of three Beowulf clusters connected via Internet and assembled as a Grid based on Globus Toolkit 2 [4], with a view on using it for quantum chemistry applications. The aspects of the computational environment and the optimizations we have focused on concern in particular:

1. a centralized installation of the Globus software into a NFS shared directory and the definition of a method to specify the cluster's hosts parameters for MDS.
2. An implementation of Globus and MPICH-G2 [5] for Grid management taking explicitly into account the local tightly-coupled MPI level (LAM/MPI in our case) for the intra-cluster communication among the nodes.
3. A modification of the LAM/MPI *broadcast* implementation in order to optimize LAM/MPI throughput on the Grid.

The resulting computing model shows very promising features and supports the existing wide interest in the Grid approach to computing. In Sec. 2 we give some details of the platform we have set up. In Sections 3 and 4 we illustrate the need for *topology-awareness* by discussing the performance of some broadcast schemes. In Sec. 5 the performance of some parallel linear algebra kernels very important for computational chemistry applications is analyzed.

2 A Model Computing Grid

The work described in the present paper has been carried out on a computing platform composed of three Beowulf-type workstation clusters, named *GRID*, *HPC* and *GIZA*. The nodes within each cluster are interconnected through a dedicated switched network, with bandwidth of 100 Mbps for *HPC* (Fast-Ethernet), 200 Mbps for *GIZA* (Fast-Ethernet with 2 NICs in Channel Bonding [6]) and 1 Gbps (Gigabit-Ethernet) for *GRID*. *HPC* and *GRID* are interconnected through a Fast-Ethernet local area network (LAN) at 100 Mbps, while they are connected to *GIZA* over a 16Mbps ATM wide area network (WAN). The connection among the three clusters is schematically shown in Fig. 1, where the the effective average communication bandwidth is indicated. Table 1 summarizes the main characteristics of each cluster.

On our model Grid we have installed Globus Toolkit 2, introducing some modifications of the reference installation procedure [4] which are of general relevance for Grids based on Beowulf clusters. In a Beowulf cluster environment it is common practice to adopt a configuration where a specialized node, called

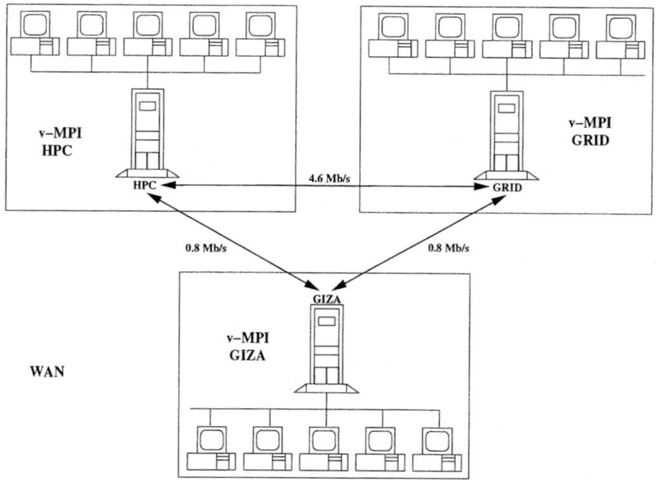

Fig. 1. Representation of the model Grid.

frontend, acts as a firewall for the other nodes and hosts all the Internet services needed by the cluster. In particular, this node usually hosts the NIS to allow one-time login of users, the shared (NFS exported) filesystems and the Automount service. In particular, the directory `/usr/local` is exported via NFS and contains the software packages shared by the other cluster nodes. In our case, all nodes access Globus in `/usr/local/globus`. For security reasons, we have restricted the range of port numbers Globus may use. The information about the nodes used by MDS is kept in the custom directory `/usr/local/globus/etc/nodes`, where a subdirectory for each node of the cluster is created to keep the GRIS node configuration files. This directory must be named as the value of the `HOSTNAME` environment variable. On each node, the GRIS subsystem is activated at boot time, invoking the `SXXgris` command customized in order to point to the right configuration directory defined by the environment variable `sysconfdir`. In this way all nodes refer to the GIIS server running on the frontend of the cluster and all the resources of the cluster may be inspected trough the LDAP server. To implement such configuration, some modifications to the MDS configuration files are necessary. In particular, in each node the following line must appear in `Grid-info-site-policy.conf` so that all MDS operations from any node of the cluster are accepted:

`policydata: (&(Mds-Service-hn=*.`*cluster-IP-domain*`)(Mds-Service-port=2135))`

and the file `Grid-info-resource-register.conf` in each node of the cluster, in order to register the local GRIS server on the GIIS server, must contain:

`reghn:` *GIIS-server.cluster-IP-domain*
`hn:` *GRIS-server.cluster-IP-domain*

Table 1. Characteristics of the three Beowulf clusters used for the model Grid.

Cluster:	*GRID*	*HPC*	*GIZA*
Number of nodes:	9 SMP working nodes, 18 working CPU	8 SMP working nodes, 16 working CPU	8 SMP working nodes
Processor type:	Intel Pentium III double-processor	Intel Pentium II double-processor	Intel Pentium III single-processor
Clock:	18 CPUs at 1GHz	4 CPUs at 550 Mhz, 1 CPU at 450 Mhz, 3 CPUs at 400 Mhz	8 CPUs at 800 Mhz
RAM:	18 Gbyte	4 Gbyte	4 Gbyte
HDISK:	EIDE/SCSI	EIDE/SCSI	EIDE
Type of switched Network:	Gigabit-Ethernet	Fast-Ethernet	Fast-Ethernet, 2 channels bonded
Linux Kernel type:	RedHat 7.2 2.4.13 SMP MOSIX	RedHat 6.2 2.2.16-3 SMP	RedHat 7.0 2.2.19

One very important aspect of our exercise was that we wanted to retain, in the Globus Grid, the ability to exploit the local level of MPI parallelism on each cluster in a very general way. As the local level of MPI environment, we have chosen to adopt LAM/MPI (version 6.5.6). To use the cluster's local MPI implementation one needs to compile the *Globus Resource Management* SDK with the flavor `mpi`. As has already been noted [7], this operation fails on Beowulf nodes (only some vendor MPI are supported). To overcome this problem, we have recompiled the package `globus_core` according to the following sequence of commands:

```
# export GLOBUS_CC=gcc;
# export LDFLAGS='-L/usr/local/globus/lib'
# /usr/local/globus/BUILD/globus_core-2.1/configure
  --with-flavor=gcc32mpi --enable-debug --with-mpi
  --with-mpi-includes=-I/usr/include/
  --with-mpi-libs="-L/usr/lib -lmpi -llam"
# make all
# make install
```

After successful compilation of `globus_core`, the installation of the *Globus Resource Management* SDK can be accomplished without problems. In order to be able to build the MPICH-G2 package correctly, it is further necessary to introduce some defines which are missing in the include file `mpi.h` of LAM/MPI.

```
#define MPI_CHARACTER          ((MPI_Datatype) &lam_mpi_character)
#define MPI_COMPLEX            ((MPI_Datatype) &lam_mpi_cplex)
#define MPI_DOUBLE_COMPLEX     ((MPI_Datatype) &lam_mpi_dblcplex)
#define MPI_LOGICAL            ((MPI_Datatype) &lam_mpi_logic)
#define MPI_REAL               ((MPI_Datatype) &lam_mpi_real)
#define MPI_DOUBLE_PRECISION   ((MPI_Datatype) &lam_mpi_dblprec)
#define MPI_INTEGER            ((MPI_Datatype) &lam_mpi_integer)
#define MPI_2INTEGER           ((MPI_Datatype) &lam_mpi_2integer)
#define MPI_2REAL              ((MPI_Datatype) &lam_mpi_2real)
#define MPI_2DOUBLE_PRECISION  ((MPI_Datatype) &lam_mpi_2dblprec)
```

To enable the spawning of MPICH-G2 jobs and to simplify the addition of nodes to the cluster we let the $GLOBUS_GRAM_JOB_MANAGER_MPIRUN macro point to the following script replacing the standard mpirun command:

```
export LAMRSH=rsh
export LAM_MPI_SOCKET_SUFFIX="GJOB"$$
/usr/bin/lamboot /usr/local/globus/hosts >> /dev/null 2>&1
/usr/bin/mpirun -c2c -O -x `/usr/local/globus/bin/glob_env` $*
rc=$?
/usr/bin/lamhalt >> /dev/null 2>&1
exit $rc
```

Here glob_env is a small program which returns the list of current environment variables so that they can be exported by mpirun (-x flag).

3 Topology-Aware Functions: Broadcast Models

Since a Grid is made up of many individual machines connected in a Wide Area Network (WAN), two generic processes may be connected by links of different kind, resulting in widely varying point-to-point communication performance. Therefore, the availability of topology-aware collective functionalities, and more generally of topology discovery mechanisms, appears to be of fundamental importance for the optimization of communication and performance over the Grid. The aim of such topology-aware strategies would be to minimize point-to-point communications over slow links in favor of that over high bandwidth and/or low latency links. In some cases it may also be possible to schedule data exchange so that as much slow-bandwidth communication as possible is hidden behind fast traffic and process activity. Knowledge of the topology of a Grid is ultimately knowledge of the characteristics of the communication link between any two processes at any given time. A first useful approximation to such detailed map is provided by the classification scheme described in the MPICH-G2 specifications [5, 8]. In this scheme there are four levels of communication: levels 1 and 2 account for TCP/IP communications over a WAN and a LAN, respectively. Level 3 concerns processes running on the same machine and communicating via TCP/IP, while level 4 entails communication through the methods provided by the local MPI implementations. Thus, the MPICH-G2 layer implements topology awareness over levels 1, 2 and 3, while level 4 operations are left to the vendor-supplied implementation of MPI.

We have tried to obtain a first demonstration of the benefits of topology-aware collectives by comparing the performance on our Grid of three different broadcast methods: *(i)* the broadcast operation provided by MPICH-G2, *(ii)* an optimized topology-aware broadcast of our own implementation, and *(iii)* a non-topology-aware broadcast based on a flat binomial-tree algorithm. As is clear, all processes can communicate at Level 1, because all machines are in the same WAN, but only processes that run on machines belonging to the same cluster (*HPC*, *GRID* or *GIZA*) can communicate using the locally supplied MPI (l-MPI) methods. The l-MPI we have used is LAM/MPI [9] and provides for communication via TCP/IP among nodes in a dedicated network or via shared-memory for processes running on the same machine. In accord with the MPICH-G2 communication hierarchy, we can thus essentially distinguish between two point-to-point communication levels: inter-cluster communication (Level 1) and intra-cluster communication (Level 4). As already mentioned, we notice that the effective bandwidth connecting *GIZA* to *HPC* and *GRID* is slower than that between *HPC* and *GRID*. This asymmetry may be thought of as simulating the communication inhomogeneity of a general Grid.

Consider now a typical broadcast, where one has to propagate some data to 24 machines, 8 in each cluster. For convenience, the machines in cluster *HPC* will be denoted $p_0, p_1, \ldots, p_7$, those in *GRID* as $p_8, p_9, \ldots, p_{15}$, and those in *GIZA* as $p_{16}, p_{17}, \ldots, p_{23}$. The MPICH-G2 `MPI_Bcast` operation over the communicator `MPI_COMM_WORLD`, rooted at p_0, produces a cascade of broadcasts, one for each communication level. Thus, in this case, there will be a broadcast at the WAN inter-cluster level, involving processes p_0, $p8$ and p_{16}, followed by three intra-cluster propagations, where l-MPI will take over. So we have just two inter-cluster point-to-point communication steps, one from p_0 to p_8 and another from p_0 to p_{16}, and then a number of intra-cluster communications. The crucial point to be made here is that communication over the slow links is minimized, while the three fast local (intra-cluster) broadcast propagations can take place in parallel.

In this prototype situation, the strategy adopted in our own implementation of the broadcast is essentially identical, but we have optimized the broadcast operation at the local level. The essential difference between our implementation of the broadcast and the LAM/MPI one is that in the latter, when a node propagates its data via TCP/IP, non-blocking (asynchronous) send operations over simultaneously open sockets are issued, while we opted for blocking operations. The local broadcast tree is depicted in Fig. 2. The LAM/MPI choice appears to be optimal on a high bandwidth network where each node is connected inde-

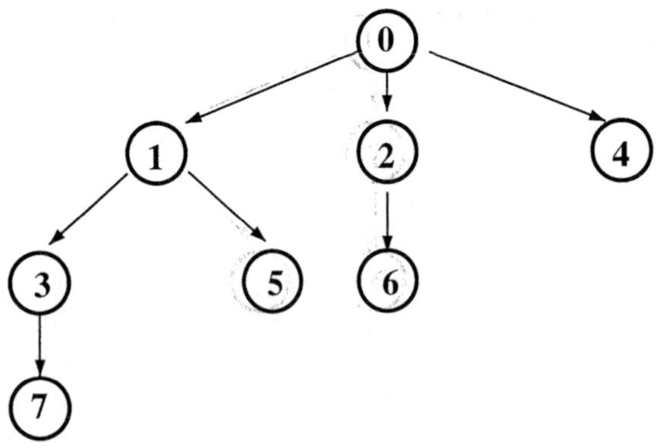

Fig. 2. Local broadcast tree in a 8-node cluster.

pendently to all the others, but it clearly loses efficiency on a switched-ethernet cluster, where simultaneous send operations issued by a node necessarily share the available bandwidth. It is not difficult to see that on small clusters this translates in a remarkable efficiency loss. In particular, in our 8-node case, with the tree of Fig. 2, a factor of two in broadcast time is observed: if τ is the basic transmission *time step*, i.e., the time required for data exchange between any two nodes in an isolated send/receive operation (for large data transfers this is roughly the amount of data divided by the bandwidth), then the synchronous broadcast completes in about 6τ, while our version takes just 3τ.

4 Broadcast Tests

In all our tests we have propagated ~38 Mb of data ($5 \cdot 10^6$ double precision real numbers). Thus, on a local switched 100 Mbit ethernet such as that of the *HPC* cluster the measured time-step is $\tau = 3.4$ s and the optimized broadcast takes about 10 seconds. On *GIZA*, where each node multiplexes over two ethernet cards, the time is exactly halved. It is instructive to compare the performance of topology-aware broadcasts with the no-topology-aware one. The latter, as previously mentioned, is executed using a flat binomial tree algorithm involving all nodes of the Grid, without consideration for the different link speeds. This procedure of course results in a larger amount of slow inter-cluster communication at the expenses of the fast local transfers. An example is shown in Fig. 3 where a broadcast is propagated again from node p_0. As can be seen, we have here only 7

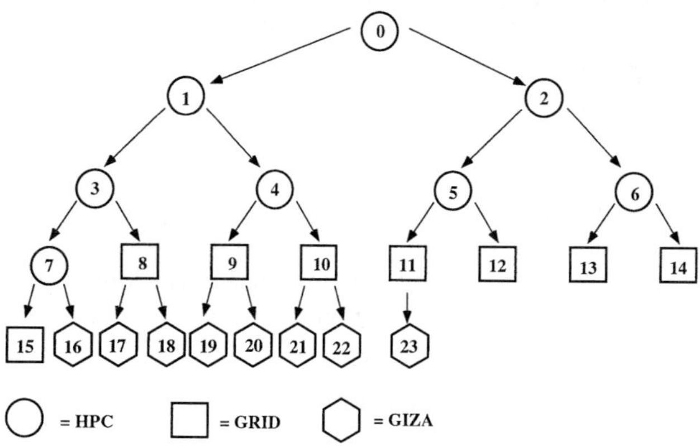

Fig. 3. Flat broadcast tree over the whole Grid.

intra-cluster data transfers (the same number we actually have in a single local broadcast) and as many as 16 inter-cluster data transfers instead of the two we have in the topology-aware algorithm. By taking into account the various link speeds it is possible to give a rough estimate of the overall times required for the three broadcast algorithms. For a 38Mb data propagation, as before, these are compared with the measured ones in Table 2. It should be noted that the

Table 2. Performance of different broadcast algorithms over the Grid.

Broadcast type	Measured time (s)	Estimated time (s)
MPICH-G2	70	69
Locally optimized	60	59
No-topology aware	448	440

serial link between either *HPC* or *GRID* and *GIZA* is not dedicated. Thus, the reported link speed of 0.8 Mbyte/s (see Fig. 1) is an average value obtained over various measurements at the time of our study. In the first case of Table 2 the broadcast operation is executed by MPICH-G2 in two steps. First, a broadcast over the WAN takes place, consisting in practice of two non-blocking sends from p_0 (*HPC*) to p_8 (*GRID*) and p_{16} (*GIZA*). The rate of these data transfers falls much below the available local bandwidth and they go through different routes, thus they should overlap quite effectively: the overall time required coincides with the time of the *HPC-GIZA* transfer, i.e. about 49 s. After the inter-cluster broadcast, three parallel local broadcasts, one on each cluster. The time for these operations is the largest of the three, i.e. ~20 seconds, required on *HPC* (LAM/MPI is used). Thus, the total time for a MPICH-G2 global broadcast is estimated in 69 seconds, matching closely the observed time. The second case of Table 2 differs from the first only for the optimization of the local intra-cluster broadcasts. As we have shown, this halves the local time from 20 to 10 seconds.

The last case of broadcast reported in Table 2 is the flat no-topology-aware structure whose tree is sketched in Fig. 3. All data transfers originating from the same node are synchronous (blocking). First of all, a local broadcast within *HPC* takes place. After 3 time steps (~10 s) all local nodes except p_6 have been reached and p_3, p_4 and p_5 simultaneously start their sends toward *GRID*. The time for these would be ~32 s, but it is in fact longer because after the first 3.4 s also p_6 starts sending through the same channel toward *GRID*. A better estimate yields 38 s. After this the 8 transfers from *HPC* and *GRID* toward *GIZA* start taking place and the time required for these, sharing the same channel, is of course of the order of 400 s. In total, the observed broadcast time of 448 seconds is therefore very well explained and the dominance of the long distance transfers, which in a topology-aware implementation is suitably minimized, is evident.

5 Linear Algebra Benchmarks

Since the intended use of our cluster Grid is for quantum chemistry and nuclear dynamics applications, we have also performed some preliminary tests on its performance with standard linear algebra routines. To this purpose, we have installed the ScaLAPACK package [10] on top of MPICH-G2. The resulting software hierarchy is depicted in Fig. 4. As the figure shows, we have installed, besides the standard BLAS and LAPACK software packages at the local level, the BLACS software (Basic Linear Algebra Communication Subprograms) [11] on top of MPICH-G2, PBLAS (a parallel BLAS library) and ScaLAPACK, a

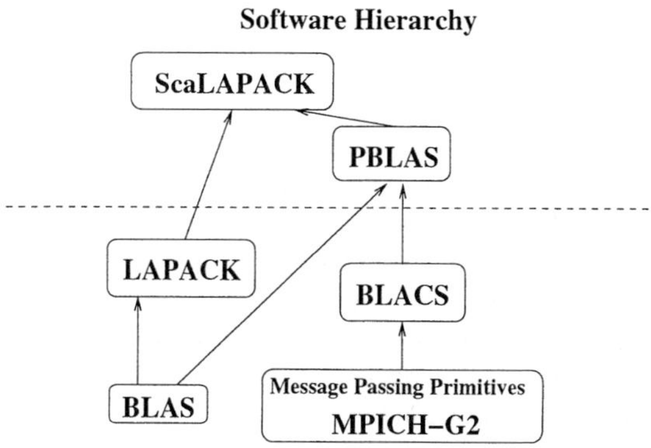

Fig. 4. ScaLAPACK software hierarchy on the Grid. The broken line separates the local environment (below) from the global one (above).

library of high-performance parallel linear algebra routines. BLACS is a message-passing library designed for linear algebra and it implements a data distribution model consisting of a one- or two-dimensional array of processes, each process storing a portion of a given array in block-cyclic decomposition. On this data distribution scheme PBLAS and ScaLAPACK work. Besides parallel execution, the data distribution model enables the handling of much larger arrays than would be possible in a data replication scheme.

The initial tests we have made use one of the fundamental PBLAS routines, PDGEMM, which performs matrix multiplication on double precision floating point arrays. The algorithm used in PDGEMM is essentially the one described in ref. [12] where a two-dimensional block-cyclic decomposition of the arrays is distributed and data transfers occur mostly along one dimension. Thus, our Grid of workstation clusters lends itself quite naturally to such decomposition if we arrange the machines in each cluster along a different row (say) of the BLACS array and ensure that data communication (the exchange of array blocks) takes place predominantly along rows of the array (i.e., intra-cluster).

We have measured PDGEMM performance for such an arrangement for the multiplication of two 20000 by 20000 matrices observing an effective speed of about 2.5 Gflops. Comparing this to the theoretical aggregate speed of the Grid in the absence of communication, it is clear that data transfers dominate the Grid activity. This is confirmed also by the fact that the global performance is almost completely independent of the size of the array blocks used in the block-cyclic decomposition: the measured speed varies from 2.52 Gflops to 2.55 Gflops for block sizes 64 to 256 (but the top speed is reached already for a block size of 160). The block size substantially affects local matrix multiply performance through the extent of cache re-use and a larger block size probably also improves data communication slightly by decreasing the impact of network latency. It should finally be noted that by exchanging rows with columns of the processor array, i.e., increasing the amount of inter-cluster data transfers, the expected performance deterioration reaches 70%.

6 Conclusions

In this work we have discussed how to realize, on a model Grid made up of three workstation cluster, two Globus communication levels, inter-cluster and intra-cluster, using MPI. The benefits of topology-aware communication mechanisms have been demonstrated by comparing a two-level implementation of broadcast with a flat binary-tree one. The model Grid has been further tested by measuring the performance of parallel linear algebra kernels of common use in theoretical chemistry applications, arranged to exploit the two communication levels. The tests have provided useful indications on how to make efficient use of a Grid platform composed of workstation clusters.

Acknowledgments

This research has been financially supported by MIUR, ASI and CNR.

References

1. Foster, I., Kesselman, C. (eds.): The Grid: Blueprint for a Future Computing Infrastructure. Morgan Kaufmann Publishers, USA (1999)
2. See e.g.:`http://www.beowulf.org`
3. Kielmann, T., Hofman, R.F.H., Bal, H.E., Plaat, A., Bhoedjang, R.A.F.: MAGPIE: MPI's Collective Communication Operations for Clustered Wide Area Systems. ACM Sigplan Notices, Vol. 34 (1999) 131–140
4. The Globus Project:`http://www.globus.org`
5. Foster, I., Karonis, N.: A Grid-Enabled MPI: Message Passing in Heterogeneous Distributed Computing Systems. SC'98, Orlando, Florida (1998). See also `http://www3.niu.edu/mpi/`
6. See e.g.:`http://sourceforge.net/projects/bonding/`
7. See the Globus mailing lists, e.g.: `http://www-unix.globus.org/mail_archive/mpich-g/2002/04/msg00007.html`
8. Karonis, N., de Supinski, B., Foster, I., Gropp, W., Lusk, E., Bresnahan, J.: Exploiting hierarchy in parallel computer networks to optimize collective operation performance. Fourteenth International Parallel and Distributed Processing Symposium, Cancun, Mexico (2000).
9. `http://www.lam-mpi.org/`
10. `http://www.netlib.org/scalapack/slug/scalapack_slug.html`
11. (a) Dongarra, J., Van de Geijn, R.: Two dimensional basic linear algebra communication subprograms. Computer Science Dept. Technical Report CS-91-138, University of Tennessee, Knoxville, USA (1991) (also LAPACK Working Note #37); (b) Dongarra, J., Van de Geijn, R., Walker, D.: Scalability issues in the design of a library for dense linear algebra. Journal of Parallel and Distributed Computing, Vol. 22, N. 3 (1994) 523-537 (also LAPACK Working Note #43).
12. Fox, G., Otto, S., Hey, A.: Matrix algorithm on a hypercube I: matrix multiplication. Parallel Computing, Vol. 3 (1987) 17–31

Uniform Access to the Distributed Resources for the Computational Chemistry Using UNICORE

Jarosław Pytliński, Lukasz Skorwider
ICM Warsaw University
Pawińskiego 5a, 02-106 Warsaw, Poland
{pyciu,luksoft}@icm.edu.pl

Krzysztof Benedyczak, Michał Wroński, Piotr Bała
Faculty of Mathematics and Computer Science
Nicolaus Copernicus University
Chopina 12/18, 87-100 Toruń, Poland,
{golbi,wrona,bala}@mat.uni.torun.pl

Valentina Huber
Research Center Jülich
D-52425 Jülich, Germany
v.huber@fz-juelich.de

Abstract. In this paper we describe deployment of qunatum chemical and biomolecular applications on the grid. We have used UNICORE infrastructure as framework for development dedicated user interface to the number of existing computational chemistry codes as well as molecular bilogy databases. The user interface is integrated with the UNICORE client based on plugin mechanism which provides general grid functionality such as single login, job submission and control mechanism.

1 Introduction

Research in the area of computational chemistry requires access to the computer resources usually not available at the user workstation. This includes various types of resources such as computational servers, visualization workstations as well as number of databases. Current desktop computers provide significant CPU speed, disk space as well as visualization capabilities and therefore become important computational platforms. More often they are used for graphical representation and analysis of results both theoretical and experimental. However, nontrivial problems still require large memory, disk space and CPU time, available only at remote systems or sites. Also large and frequently updated databases are available remotely and cannot be hosted locally, also because of the license restrictions.

Distributed computational resources cannot be effectively utilized using typical unix tools based on the remote login. Users sharing external as well as departmental or even local computers have to solve number of practical problems.

P.M.A. Sloot et al. (Eds.): ICCS 2003, LNCS 2658, pp. 307–315, 2003.

They are faced with different site polices and practices such as different security, different user identification, different priorities or scheduling mechanisms and so on. Additionally, users adopt latest developments in the graphical user interfaces and simply do not want to use traditional, command line based tools. Number of applications exists but none of them can cover full range of functionality. In result user has to work with different programs dedicated to computations, visualization or database access. Each of them has specific user interface and requires dedicated and often non compatible file formats.

Currently various computational techniques become standard tools for a large number of users. One should note that increasing number of them have no long-standing experience in large scale computing and data mining as well as advanced visualization and expects intuitive tools which are easy to learn and use. Such tools cannot be just another application, but rather should allow for integration of existing tools and programs within one uniform framework. For example, user has to be able to use visualization software he is used to and should be able to replace it with new version or even another application.

Recent advances in computer technology, especially grid tools make them good candidate for development of uniform user interface to distributed resources [1]. Computational grids enable sharing a wide variety of geographically distributed resources and allow selection and aggregation of distributed resources across multiple organizations for solving large scale computational and data intensive problems in chemistry. This includes also access to various databases as well as unique experimental facilities. In order to attract users, grid interfaces must be easy to install, simple to use and able to work with different operating system and hardware platforms.

Most of the computational chemistry programs which provide graphical user interface run locally and access to the remote resources is limited. User can however use them for input preparation for applications running at the remote systems but this requires additional effort to transfer input and output files. On another hand, computational intensive applications in general have no dedicated user interfaces. Some expensive commercial graphical tools are available as well as academic software but even sophisticated tools such as GaussView [5], Tripos [6] or Cerius [7] are able to run user applications only on local workstations and do not take advantage of grid environment. These and many other applications can be used as aid for input preparation, but in most cases job must be prepared and submitted manually by the user.

The existing web technology can provide powerful user interface to computational resources and databases. It cannot however achieve high numerical efficiency required for most simulation problems. This leads us to solutions consisting on web interface for the user and traditional, usually batch type, numerical simulation engine.

This approach is presented by WebMo [8] which is web based submission system for quantum chemistry codes such as Gaussian [9], Gamess [10] and Mopac [11]. This tool is limited to the local batch systems and has no grid capabilities. The web submission to the geographically distributed systems is possible within BioCore [12] which is web interface to the molecular dynamics code NAMD [13]. Currently this system is limited to the particular MD code and single visualization package (VMD). NPACI Gamess Portal presents analogous approach for quantum mechanical code.

Presented problems can be solved with the help of grid middleware which provides access to remote high performance resources. However most of the development goes to the design and production of general tools which can be used in different situations and provide simple access to remote resources. The existing solutions like globus [2], Legion [3], LSF [4] address most important issues. Unfortunately, these tools require an advanced computer skills for installation and usage. Moreover, in order to provide required functionality user application must be modified, sometimes significantly. This is not possible for commercial or legacy codes which cannot be easily adopted by the user. Most grid solutions have limited user interface not addressing application specific issues. The significant exception is UNICORE [14] middleware which provides uniform interface to the distributed computer resources of different type.

2 UNICORE

The details of the UNICORE can be found elsewhere [19] and we will summarize here its most important features.
UNICORE is uniform interface to the computer resources which allows users to prepare, submit and control application specific jobs and file transfers. Jobs to be run on the same system can be grouped in job-groups. The user specifies target system for job group as well as resource requirements for CPU time, number of CPUs amount of memory and disk space for job. Jobs can have complicated structure with various job subgroups which can be run at different systems. Subgroup jobs and file transfers can be ordered with user defined dependencies. This includes conditional execution and loops with control which can be used as environment variables in the user's jobs. The user input is mapped to the target system specific commands and options by the UNICORE infrastructure. Compared to other tools UNICORE has wider functionality, is more flexible and allows for much easier integration of the user interface with external applications.

The UNICORE architecture is based, like other grid middleware, on the three tier model. It consists of user, server and target system tier. The user tier consists of the graphical user interface, the UNICORE client, written as Java application. It offers the functions to prepare and control jobs and to set up and maintain the user's security environment. The UNICORE job can be build from multiple parts which can be executed asynchronously or dependently on different systems at different UNICORE sites. Within the UNICORE environment user has a comfortable way to use distributed computing resources without having to learn about site or system specifics.

In order to overcome this disadvantage the software is introduced to the UNICORE through the Incarnation DataBase (IDB) entries which describe details of the local installation. IDB allows for definition of the environment variables for the script executed by the user. In this way we can define variables required for particular application or define path to the program.

The UNICORE infrastructure allows also for registration of the applications available on the target machine through SOFTWARE_RESOURCE entries in the IDB. The program name and program version together with shell commands required to set up execution environment are stored in the dedicated section of the IDB file. The UNICORE client has build-in capabilities to check software resource entries on the particular target systems and use it for job construction.

The UNICORE client provides also the plugin mechanism which become very attractive and efficient method for integration of applications with grid middleware. Plugin is written in Java and is loaded to the UNICORE client during start, or on request. Once it is available in the client, in addition to the standard futures such as preparation of the script job, user gets access to the menu which allows preparation application specific jobs. Dedicated plugins can be used for database access or postprocessing. Since plugins are written in Java, they can be easily integrated with external applications or existing Java applications or applets.

The UNICORE security is based on the Secure Socket Layer (SSL) protocol and the X.509 certificates. SSL uses public key cryptography to establish connections between client and server. Therefore each component of the UNICORE infrastructure has a public-private key pair with the public part known by the others. The keys have to be certified by a certification authority. The user's X.509 certificate is his identification and is maintained by the UNICORE client application in a encrypted data base. The client has to know about the CA which signs the user and gateway certificates. The SSL is used for the connection coming over insecure internet to the UNICORE gateway. The user is authenticated by the gateway when presenting certificate. The user certificate is translated to the user login under which all tasks are executed on the target system. The user can use number of certificates to access different remote resources. The authentication scheme supports also project concept and jobs can be executed using different projects with single login.

The UNICORE client provides users with tools for certificate management such as certificate signature request generation, keystroke editor or public key export.

3 Computational Chemistry Applications

The UNICORE software provides general framework for running user applications. This includes input preparation, job submission and control and finally postprocessing of the results. Advanced visualization and access to the databases and other distributed sources of information should alse be considered here. The wide functionality and especially flexibility obtained with plugin concept makes UNICORE good candidate for general framework for access to the various, not only computational, resources. In the next section we will describe extensions which covers wide range of computational chemistry scenarios.

3.1 Application Specific Interface

In most cases input for the CPU intensive chemistry application has text form and is prepared with standard text editor. This approach requires significant experience from the user and any mistake in the file format results an error and extends time in which results will be obtained. The preparation of the input file requires knowledge of many keywords to describe atom coordinates, molecular symmetry, geometry optimization, electronic properties, etc.

This disadvantages can be removed by the application specific interface which assist users and generates parameters automatically to the text area, where it can be modified by experienced users by hand, e.g. they can add new keywords. It checks the input data and their interdependencies and prevents users from entering incorrect values while requesting to set mandatory parameters. Once generated input can be stored format to be reused and modified in the future.

Using UNICORE plugin concept such interfaces has been developed by us for the most popular CPU intensive quantum chemistry application such as Gaussian98, Car-Parrinello Molecular Dynamics calculations [20] and molecular dynamics code Amber [16].

In addition to the standard input preparation features plugins can provide user with additional functionality. For example Amber plugin contains additional help on keywords and parameters together with information on default values, Gaussian plugin estimates required CPU time.

More detailed description of the plugin functionality can be found in [17].

One should note that plugin development requires knowledge of Java programming as well as UNICORE Client interface, but once technology is learned the development is fast.

3.2 Job Submission

After successful generation of the input files user has to construct job and submit it to the target system. Traditionally, user develops shell scripts and adopts them to the target system. UNICORE can use well established model and job can be constructed using script tasks. We have developed templates for most popular biomolecular applications including Gaussian98, Gromos96 [15], Amber and Charmm [18]. The templates can be easily modified by the user in the part including application input, exactly as it is done for batch jobs. Significant difference comes from the fact using UNICORE middleware scripts can be run on any target system without modifications taking advantage of the environment variables defined in the IDB.

The job configuration includes information on the input files which has to be transfered form user workstation to the system where particular application will be executed as well as information on the output files which must be transferred back. Once this information is provided by the user all required file transfer will be performed automatically during job submission or retrieval of the results.

As described above, the UNICORE allows user for advanced workflow management. This includes creation subtasks which can be run on the different target systems. For example the structure minimalization with Amber was run on one target system than results were transferred to another system where molecular dynamics simulations were performed.

While application specific input is prepared using dedicated plugin the proper job task is generated automatically. In this case user avoids script programming and job can be created even by non-experienced user. In addition, the plugin checks the availability of the software on sites and allows job submission only to sites, where the particular program is installed. Furthermore the site administrators can specify in the Incarnation DataBase several software versions with appropriated environments, e.g. different paths to binaries, and the plugin displays available versions in a choice box to be selected by a user.

As part of the job preparation, user specifies resources required by the job. This includes CPU time, required memory size and disk space. UNICORE client provides user with dedicated resource editor which displays information on the target system resources and allows for their adjustment to the particular job. Currently resources must be specified by the user, however the Gaussian plugin has build in estimation of the CPU time required for the job execution. This is based on the known algorithm scaling as well as on stored in the IDB target system performance and is calculated according to the particular input. Such estimate is of course not accurate and rather provides user with the upper limit of the CPU time.

Recent developments of the UNICORE middleware allows for easy discovery which target system has particular software installed. Dedicated plugin lists target systems either from the particular site, or from all sites user has access to.

3.3 Job Control

Once job is submitted user can check job status, monitor execution and retrieve output to the local workstation. All these functions can be performed from the UNICORE client with a single login during client startup. User can monitor job status and retrieve output to any computer connected to the network with the UNICORE client installed, in particular other than one used for job submission.

Using graphical interface provided by the UNICORE Client user can obtain information on the jobs status and is allowed to hold, resume or cancel executing job. With the dedicated plugin status of the job can be checked periodically. Since molecular dynamics applications require some specific job control mechanism we have developed plugin which display status of all users job known to the UNICORE Client. All available sites are checked and known jobs are listed and displayed in the table which can be differently ordered. User has also possibility to select particular job and get more details on it using standard UNICORE Client functionality.

Standard UNICORE monitoring mechanism provides only general information on the job or subjob status and lacks details such as estimation of the job progress or information on the intermediate results. This feature is required especially for long running molecular dynamics simulations.

We have removed this disadvantage by designing plugin which allows to select running job from the UNICORE Client and monitor its execution on the target system. The Plugin prepares service job which based on the provided UNICORE job number enters this job working directory on the target system and gets access to the generated files. Selected files can be retrieved to the users workstation as they exists in the working directory or can be filtered to transfer only most significant information. Plugin provides user with typical filters such as retrieval of the last lines of selected file, but more advanced filters can be written. The postprocessed file is transfered to the UNICORE client and can be displayed in the text window or used for visualization.

Designed plugins extends UNICORE client functionality and adopts it to the computational chemistry requirements for job control and monitoring. Especially these tools significantly help user to to handle number of long running jobs.

3.4 Analysis and Visualization

The important part of most computational chemistry applications is visualization and analysis of the results. In the past this task has been on the dedicated workstations, but recent development in computer hardware and architecture allows to use common workstations. Chemistry users adopt these changes and number of visualization packages has been developed and is used in both Windows an Unix environment. Since UNICORE provides user with easy to use, uniform interface to the resources we decided to use it also for visualization. This was performed by introducing graphics capabilities to the selected plugins. For example CPMD plugin which takes care on input preparation and retrieval of results is able to plot graphs using data transfered files during the calculation. The trajectory of the molecular structure can be displayed from the plugin used external java package JMol. The graphics can be updated on a user's request or a user can specify the timestep for the automatic update. This feature allow to monitor the calculation of long running jobs or to store the computing resources by aborting jobs with wrong bahaviour.

Similar functionality is available in Gaussian and Amber plugins.

We have decided to use existing visualization packages because users are used to them and there is no need to develop just another visualization tool. As result we were able to integrate UNICORE plugins with such packages as JMol, JMV or RasMol/RasWin as well as any other application available on the Client workstation.

3.5 Database Access

The Librarian plugin allows for access to the remote databases from the UNICORE client. It allows user to prepare query to the PDB database [22], widely used crystallographic database and nucleotides database Entrez [23]. While plugin is started user can build database query using dedicated interface: simple, based on the keyword search, or advanced one with full functionality provided by the database. The query is submitted to the databases web interface and result is returned to the UNICORE client. The search results can be browsed, visualized or saved for later processing. Since copies of both databases are available at different locations, the plugin checks automatically which mirror is currently available. For the user query the one which provides answer in the shortest time is used.

4 Conclusions

The UNICORE software was used to establish European computational grid - EUROGRID [24]. BioGRID is application oriented grid which adopt EUROGRID infrastructure to the specific area, namely molecular biology and quantum chemistry.

The UNICORE software was used as framework providing uniform access to the number of resources and applications important for computational chemistry and molecular biology. This includes seamless access to the computational resources, databases as well as framework for application specific interfaces development. Examples presented here demonstrates capabilities of the UNICORE middleware and shows directions for further development. Modular architecture based on the plugin concept and demonstrated visualization capabilities open number of possible applications in the different areas.

One should note, that UNICORE middleware can be used also together with other grid middleware, especially can be used as job preparation and submission tool for globus.

Acknowledgements This work is supported by European Commission under IST grant 20247. The software was used using EUROGRID facilities at ICM Warsaw University (Poland), Forschungszentrum Jülich (Germany), University of Manchester - CSAR (UK), IDRIS (France) and University of Bergen - Parallab (Norway).

References

1. C. Kesselman I. Foster, editor. *The Grid: Blueprint for a Future Computing Infrastructure.* Morgan Kaufman Publishers, USA, 1999.
2. I. Foster and C. Kesselman. Globus: A metacomputing infrastructure toolkit. *Int. J. Scientific Applications*, 11(2):115–128, 1997.
3. Legion. University of Virginia. Charlottesville. VA USA Http://legion.virginia.edu.
4. LSF. Platform Computing. http://www.platform.com.
5. GaussView. Gaussian Inc., Pittsburgh PA. USA. 2000.
6. Tripos. Tripos Inc. St. Louis. USA.
7. Cerius^2. Accelerys, San Diego. USA.
8. Webmo. http://www.webmo.net.
9. M. J. Frisch, G. W. Trucks, H. B. Schlegel, G. E. Scuseria, M. A. Robb, J. R. Cheeseman, V. G. Zakrzewski, J. A. Montgomery, Jr., R. E. Stratmann, J. C. Burant, S. Dapprich, J. M. Millam, A. D. Daniels, K. N. Kudin, M. C. Strain, O. Farkas, J. Tomasi, V. Barone, M. Cossi, R. Cammi, B. Mennucci, C. Pomelli, C. Adamo, S. Clifford, J. Ochterski, G. A. Petersson, P. Y. Ayala, Q. Cui,

K. Morokuma, P. Salvador, J. J. Dannenberg, D. K. Malick, A. D. Rabuck, K. Raghavachari, J. B. Foresman, J. Cioslowski, J. V. Ortiz, A. G. Baboul, B. B. Stefanov, G. Liu, A. Liashenko, P. Piskorz, I. Komaromi, R. Gomperts, R. L. Martin, D. J. Fox, T. Keith, M. A. Al-Laham, C. Y. Peng, A. Nanayakkara, M. Challacombe, P. M. W. Gill, B. Johnson, W. Chen, M. W. Wong, J. L. Andres, C. Gonzalez, M. Head-Gordon, E. S. Replogle, and J. A. Pople. Gaussian 98. 2001.

10. M. W. Schmidt, K. K. Baldridge, J. A. Boatz, M. S. Gordon S. T. Elbert, J. H. Jensen, S. Koseki, N. Matsunaga, K. A. Nguyen, S. J. Su, T. L. Windus, M. Dupuis, and J. A. Montgomery. General atomic and molecular electronic structure system. *J. Comput. Chem.*, 14:1347–1363, 1993.
11. J. J. P. Stewart, L. P. Davis, and L. W. Burggraf. Semi-empirical calculations of molecular trajectories: method and applications to some simple molecular systems. *J. Comp. Chem.*, 8(8):117–23, 1987.
12. M. Bhandarkar, G. Budescu, W. F. Humphrey, J. A. Izaguirre, S. Izrailev, L. V. Kal, D. Kosztin, F. Molnar, J. C. Phillips, and K. Schulten. Biocore: A collaboratory for structural biology. In A. G. Bruzzone, A. Uchrmacher, and E. H. Page, editors, *Proceedings of the SCS International Conference on Web-Based Modeling and Simulation.* San Francisco, California, 1999.
13. L. Kal, R. Skeel, M. Bhandarkar, R. Brunner, A. Gursoy, N. Krawetz, J. Phillips, A. Shinozaki, K. Varadarajan, and K. Schulten. Namd2: Greater scalability for parallel molecular dynamics. *J. Comp. Phys.*, 151:283–312, 1999.
14. UNICORE. Unicore Forum. http://www.unicore.org.
15. W. Van Gunsteren and H. J. C. Berendsen. *GROMOS (Groningen Molecular Simulation Computer Program Package).* Biomos, Laboratory of Physical Chemistry, ETH Zentrum, Zurich, 1996.
16. P. Kollman. *AMBER (Assisted Model Building with Energy Refinement).* University of California, San Francisco, USA, 2001.
17. J. Pytliński, L. Skorwider, V. Huber, K. Bednarczyk and P. Bala iUnicore- an uniform platform for chemistry on the grid,2002
18. B. R. Brooks, R. E. Bruccoleri, B. D. Olafson, D. J. States, S. Swaminathan, and M. Karplus. A program for macromolecular energy, minimization, and dynamics calculations. *J. Comp. Chem.*, 4:187–217, 1983.
19. P. Bala, B. Lesyng and D. Erwin Eurogrid-european computational grid testbed in press,2002
20. CPMD consortium. http://www.cpmd.org.
21. V. Huber. Supporting Car-Parrinello Molecular Dynamics with Unicore. In *Computational Science ICCS - San Francisco, USA, Proceedings, Pt. 1*, volume 2073, pages 560–567. Springer Verlag, 2001.
22. H. M. Berman, J. Westbrook, Z. Feng, G. Gilliland, T. N. Bhat, H. Weissig, I. N. Shindyalov and P. E. Bourne. The Protein Data Bank. *Nucleic Acids Research*, 28:235-242, 2000.
23. http://www.ncbi.nlm.nih.gov.
24. http://www.eurogrid.org.

Common Data Format for Program Sharing and Integration

Elda Rossi[1], Andrew Emerson[1], and Stefano Evangelisti[2]

[1]CINECA, via Magnanelli 6/3, 40033 Casalecchio di Reno (BO) – Italy
[2]Laboratoire de Physique Quantique, UMR 5626, Université Paul Sabatier, 118 Route de Narbonne, F-41062 Toulouse CEDEX – France

Abstract. This paper describes the design and implementation of a common data format within abiGrid, a grid-based project that connects different research groups of Quantum Chemists. The research activity of the partners is focused on orbital localization in a Multi-Reference context, since orbital localization is a necessary step towards the development of efficient methods for the treatment of large systems.The goal of abiGrid is to permit the use and interchange of home-made programs, while maintaining the individuality of the different codes that, as research tools, are subject to changes and evolution. Central points of the project are the design of a Common Data format and an extensive use of the Grid technology. In the present contribution the structure of the common data format is described. The format is based on XML, which has already used for chemical applications (CML). The problem of the large amount of data produced by ab-initio calculations is also addressed.

Keywords: Grid Technology, Meta-systems, Globus, XML-CML, ab-initio methods, Local Orbitals.

1 Introduction

The treatment of large systems is becoming a key issue in Quantum Chemistry (QC). Indeed it is hoped that in the near future it will be possible to apply QC techniques to fields involving macromolecules such as nano-sciences, biology, and so on. In this context, Linear-Scaling (LS) methods (algorithms whose computational complexity scales proportionally with the system size) are particularly interesting [1]. LS methods take advantage of the locality of most physical interactions to neglect distant contributions, thus achieving a linear growth. In order to use the locality of the interaction, however, it is essential to work with local orbitals, and this is the reason of a renewed interest on orbital-localization techniques in QC in the last years.

The applicability of local-orbital methods in a Multi-Reference (MR) context is presently studied by a group of researchers working on different aspects of Methodological and Applicative QC [2-8]. Since the different groups involved in the project are working on ab-initio codes developed in each single laboratory, program integration is a central point of this research, and a considerable effort is currently being devoted to this task. In order to work in this direction, two working group activities were proposed within COST in Chemistry [9].

P.M.A. Sloot et al. (Eds.): ICCS 2003, LNCS 2658, pp. 316–323, 2003.

In the present contribution, we will focus on one of these activities, "AbiGrid: a meta-laboratory for code integration in ab-initio methods" [10], whose goal is to design and implement a tool, based on grid technology, that will enhance scientific cooperation by making easier the use and interchange of home-made programs, while maintaining the individuality of the different codes that, as research tools, are subjected to changes and evolutions dictated by the research activities of the different groups. All the partners involved in the Meta-Laboratory activity have been developing quantum chemistry codes for internal use for many years. These codes are complementary and their combined use is very important for new collaborations. On the other hand program sharing is difficult since the codes have not been written with sharing considerations in mind and no standard format has been respected in input and output planning.

The solution, in perspective, is the integration of all the codes into a single meta-system for QC calculations. This meta-system is not expected to substitute the general large packages for Quantum Chemistry available today on the market, like Gaussian or MolCAS or others, which are obviously much more complete and reliable. This is instead a collaborative environment for researchers that need to share their own home-made codes for common interests.

At the moment, we are focusing our work on the communication between the different codes. Two strategies can be adopted: one possibility is to write a series of one-to-one interfaces, connecting pairs of codes. The other possibility is to adopt a single data format, and to write interfaces from each code to this standard format. The first solution can be in some cases more desirable, particularly when two programs share already some data format. If the number of codes to be integrated becomes larger, however, the number of one-to-one interfaces becomes unreasonably large, and the use of a single data format is certainly preferable. Therefore, it was this solution that was adopted in our project.

2 The XML Choice

The strategy therefore was to decide on a common data format and write a series of "wrapper" codes in order to convert the input and output files from each application to or from the common data format.

The design of file formats able to represent particular types of information is a common problem but is particularly pertinent to the area of computational chemistry. In general, no standard format has ever been adopted for any area of chemistry with most software providers tending to use their own proprietary solutions. However, in recent years developments in the evolution of the creation of documents destined for the WEB has led to a new formalism termed XML (eXstensible Mark-up Language) [11]. XML is not a new computer language but rather a "meta-language", a set of rules which can be used to generate a syntax and vocabulary appropriate to the data to be described. In this way many "flavors" of XML have been created to represent information from many diverse areas (e.g business, medicine, archaeology, etc.) as well as the successor to HTML for web-page design. The main advantage of an XML-based data format is that it is extensible, i.e. new types of data can be easily and seamlessly added to the language definition if the need arises – something which is not possible with the currently available file types. In addition, since XML documents contain just freely formatted text they are human readable and can be easily interpreted by text parsers and other programs and converting between them provides a lossless mechanism of data transfer. There are also many readily available utilities for processing XML documents which can be transferred over the internet using existing and commonly used protocols (e.g. hypertext transfer protocol or http).

To our knowledge there is only one mature implementation of an XML for chemistry; this has been named Chemical Mark-up Language or CML and has been designed by Peter Murray-Rust and Henry Rzepa [12]. There has yet to be a widespread adoption of CML by the computational chemistry community but interest is growing and the need for a more unified approach to storing chemical information is generally accepted. Although in principle any kind of data can be included in CML, it has been primarily designed for representing molecular structures and chemical reactions rather than for QC quantities. The authors of CML are also in the process of formulating an XML definition called CCML, or Computational Chemistry Mark-up language, which is expected to be able represent information relevant to computer simulation such as algorithms, parameterizations (e.g. force-fields) and job control. However, neither CML nor the forthcoming CCML was designed with quantum chemistry properties in mind so we have therefore decided to construct our own XML definition (or "XML schema") appropriate for handling data in QC calculations. We should point out that having our own XML (tentatively designated QCML, for Quantum Chemistry Mark-Up Language), does not imply abandoning CML or CCML completely because XML allows one to have multiple "namespaces"

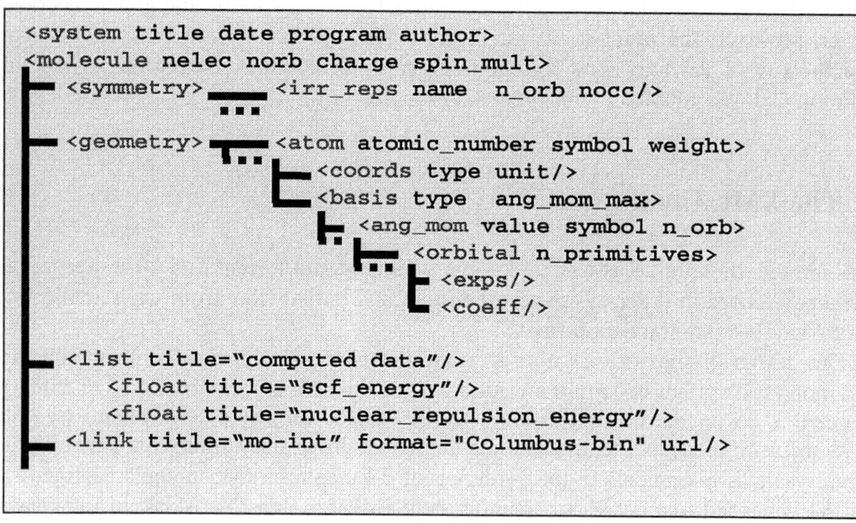

Fig. 1. A preliminary schema of QCML. A QC computation can be described with a limited number of "tags" and "attributes"

(i.e. mark-up language definitions) within the same document. Thus, although we expect to use mainly QCML in our files, we are free to use CML if it is more appropriate for the data item in question (e.g. for molecular structures). The design of

QCML is still at an early stage but there will be tags such as <symmetry>, <geometry>, <coefficient>, etc together with appropriate attributes for describing the quantities required or outputted by the QC applications. Eventually a schema will be published which will allow the validation of any QCML document. In fig. 1 a preliminary schema of QCML is shown.

We point out in particular the use of the general CML "list/float" tag to maintain computed quantities like different types of energy and the "link" tag that will be discussed in another section (XML and large data files).

3 XML Wrappers and Tools

All the data describing the chemical system of interest are kept in a "data repository" as showed in fig. 2. The data repository is based on the XML format and is somewhere on the network.

All the efforts for integration and conversion from/to the common format are handled by "wrappers", programs specifically designed for each single code that generate the input files starting from the repository (INPUT wrapper) and that, at the end, store important data from the output files into the repository (OUTPUT wrapper).

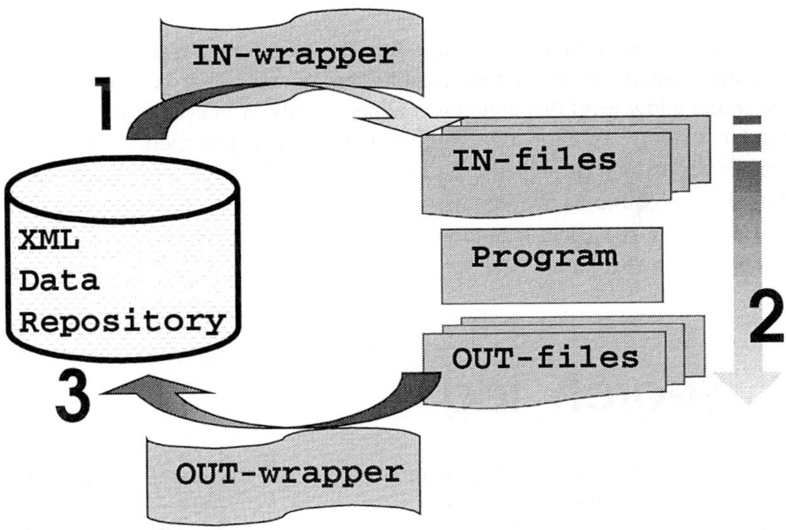

Fig. 2. The input wrapper converts data from the repository to the input files (1), the code runs as usual producing output files (2),at the end the output wrapper converts selected output data to the repository (3)

This strategy allows us to leave the program unchanged. If we modify the program or we add a new one, we only have to modify or to write the specific wrappers. So in theory the meta-system will not be affected by the change in any other way.One important task therefore is to write the wrapper programs which will perform the conversions to and from this common format which could also be used for documentation purposes and with appropriate utilities for results analysis. To do this we have adopted two strategies:

1. The writing of a number of PERL scripts to perform the conversions between the application files and the QCML data format.
2. The construction of a FORTRAN library of low-level and high-level subroutines designed to aid the reading and writing of XML documents.

PERL was chosen because it is one of the most powerful text processing languages and there is already an extensive library of modules for parsing, writing and validating XML documents [13]. In addition, PERL interpreters are generally found on all UNIX systems and since it does not need to be explicitly compiled, transferring the PERL wrappers between the different hardware and software architectures in a grid environment is straightforward.

The FORTRAN library was written at the request of the application program authors and maintainers to facilitate the integration of their codes into the meta-system. FORTRAN in fact is well known and widely used throughout this community.

Although FORTRAN is not usually recommended for text manipulation, it is possible (particularly with FORTRAN 90) and the routines can be validated with their PERL equivalents.

The routines in the library are organised in three levels: the top level doing macro tasks, the intermediate levels that are used by the previous ones but can also be used by the user and a low level that contains routines mainly of internal use.

In fig.3 an example of a top level routines is presented. The routine **`Search_tag`** searches the repository (trough the **`file`** argument) for a given tag (**`atom`** in the example). It is possible to specify what atom (**`id3`**, that is the atom no. 3, in the example) and what specific attribute we are interested in (**`symbol`**). The result will be the symbol of that atom (**`Ca`** for example) stored as a string in the **`value`** argument.

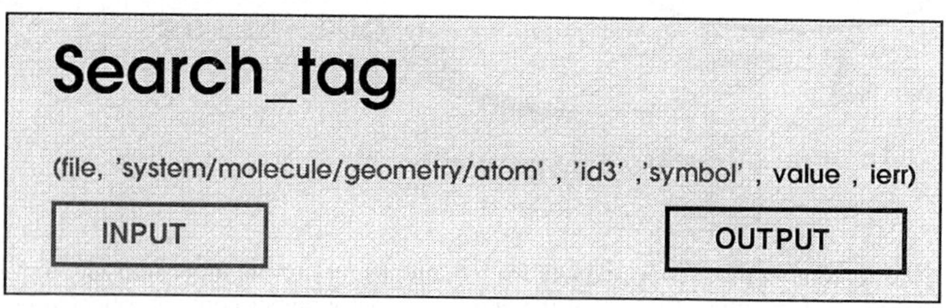

Fig. 3. The `search_tag` routine from the FORTRAN library

4 XML and Large Data Files

XML, as well as being text-based, is also rather verbose and requires more disk space than other data formats in order to store the equivalent quantity of data. This is not usually considered important bearing in mind the storage available on current computer systems but QC codes very often produce and access huge quantities of data, such as electron integrals or checkpoint files. For efficiency, both in terms of disk space and CPU time, these files are usually written in a binary format.

Due to the size of the stored data (perhaps even hundreds of Gbytes for very large calculations), it is unthinkable to keep these data in a formatted XML file. Although large XML files can be compressed quite efficiently with standard compression utilities at the end of a run, the increase in data access time during program execution would severely affect program performance. Thus, when large binary files are required it was decided to create an XML file containing a link to the location of the binary file, together with information describing its format and contents. Although not ideal there seems to be no other workaround without affecting program performance.

5 The Implementation

A number of existing programs have been used as building blocks for the prototypal meta-system. In most cases the codes are home-made, but sometimes they are general use packages, distributed freely over the internet (as is the case for Dalton and Columbus). The programs in question are as follows:

- COLUMBUS (General ab-initio electronic package) [14]
- DALTON (General ab-initio package) [15]
- CAS-DI (Multi-Reference Configuration Interaction) [16]
- EPCISO (Spin-orbit Configuration Interaction) [17]
- NEVPT (Multi-Reference Perturbation Theory, Perturbative-Variational approaches) [18]
- LOCNAT (Localized Multireference algorithm) [3,4]
- FCI (Full Configuration Interaction): [19]
- PROP (Property Calculation) [20]

Some of these codes can be considered "zero level" programs, as they do not require pre-computed structured data. On the other hand, a program like FCI needs molecular orbital integrals to start. It has been designed to take those data from a specific commercial program, but allowing the integrals to come from other sources will increase flexibility.

Another problem we are facing is that of "multiple instances" of the same input data. For example, all codes need information like "number of atoms", "electrons", "basis set", etc., each of them in a different input format. The solution is to collect all data at the beginning and transfer them to each program in a suitable format.

The first prototype of this implementation, was the interface between the COLUMBUS and FCI chains. This was because the FCI code requires a very limited set of data (essentially, one- and two-electron integrals), and contains a rather limited set of options. However, we notice that the strategy adopted will permit a future integration of the other codes within the same XML scheme. Work has already started on the PERL parsers needed to convert the various input and output files to their QCML equivalents. The main problem being encountered is not so much the writing or parsing of the XML code, the library routines make this a simple programming task, but instead gathering together all the quantities present in the many files required and generated by the applications.

6 Conclusion

We are well on the way to interfacing different ab-initio codes via an XML wrapper, which permits the data transfer from one code to the other. The problem of the large amount of data produced and/or needed by most of the QC codes (Orbital Coefficients, Molecular Integrals, Configuration Coefficients, etc) and for which a verbose solution like XML does not seem appropriate, was solved by writing in the XML file the locations and descriptions of the binary files containing the data. We are working on the extension of this approach to the different codes involved in the abiGrid project. We believe that the proposed solution, in which the different codes maintain their individuality, and are under the direct responsibility of the groups which wrote them, can represent an efficient way to enhance the collaboration between different researchers belonging to the Quantum Chemistry community.

References

[1.] S. Goedecker, Rev. Mod. Phys., 71 (1999) 1085
[2.] N. Guihéry, J.-P. Malrieu, S. Evangelisti, and D. Maynau, Chem. Phys. Lett. 349 (2001) 555
[3.] D. Maynau, S. Evangelisti, N. Guihéry, C.J. Calzado and J.-P. Malrieu, J. Chem. Phys. 116 (2002) 10060
[4.] C. Angeli, S. Evangelisti, R. Cimiraglia and D. Maynau, J. Chem. Phys. 117 (2002) 10525
[5.] C. Angeli, C.J. Calzado, R. Cimiraglia, S. Evangelisti, N. Guiéhry, J.-P. Malrieu, and D. Maynau, J. Comp. Meth. Sci. and Engen. 3 (2002) 1.
[6.] C.J. Calzado, S. Evangelisti, D. Maynau, J. Mol. Struct. (THEOCHEM), in press.
[7.] C. Angeli, C.J. Calzado, R. Cimiraglia, S. Evangelisti, N. Guihéry, T. Leininger, J.-P. Malrieu, D. Maynau, J.V. Pitarch, and M. Sparta, Mol. Phys. in press.
[8.] C. Angeli, C.J. Calzado, R. Cimiraglia, S. Evangelisti, and D. Maynau, Mol. Phys. in press.
[9.] http://cost.cordis.lu/src/home.cfm
[10.] http://cost.cordis.lu/src/extranet/publish/D23WGP/d23-0006-01.htm
[11.] see for example http://www.w3.org/XML
[12.] Murray-Rust, Henry S. Rzepa and Michael Wright, New J. Chem., 2001, 618–634. (http://www.xml-cml.org/
[13.] See, for example, the CPAN archive, http://www.cpan.org
[14.] http://www.itc.univie.ac.at/~hans/Columbus/columbus.html; H. Lischka, R. Shepard, R. M. Pitzer, I. Shavitt, M. Dallos, Th. Müller, P. G. Szalay, M. Seth, G. S. Kedziora, S. Yabushita and Z. Zhang, Phys. Chem. Chem. Phys. 3 (2001) 664.

[15.] http://www.kjemi.uio.no/software/dalton/dalton.html
[16.] N.Ben Amor, D. Maynau Chem. Phys. Lett. 286 (1998) 211.
[17.] V. Vallet, L. Maron, C. Teichteil et J.P. Flament, J. Chem. Phys. 113 (2000) 1391.
[18.] C.Angeli, R.Cimiraglia, S.Evangelisti, T,Leininger, J.-P. Malrieu, J.Chem. Phys. 114 (2001) 10252.
[19.] G.L. Bendazzoli, S.Evangelisti, J.Chem. Phys. 98 (1993) 31.
[20.] J. Pitarch-Ruiz, J. Sánchez-Marín, D. Maynau. J. Chem. Phys. 112 (2000) 1655.

A Multiscale Virtual Reality Approach to Chemical Experiments

Antonio Riganelli[1], Osvaldo Gervasi[2], Antonio Laganà[1], and Margarita Albertì[3]

[1] Department of Chemistry, University of Perugia, Via Elce di Sotto, 8, 06123 Perugia, Italy
{lag,auto}@dyn.unipg.it

[2] Department of Mathematics and Informatics, University of Perugia, Via Vanvitelli, 1, 06123 Perugia, Italy
osvaldo@unipg.it
http://ogervasi.unipg.it

[3] Department of Physical Chemistry, University of Barcelona, Marti i Franques, 1, 08028 Barcelona, Spain

Abstract. A multiscale virtual reality approach to chemical experiments has been implemented using VRML, CML and Java technologies. Three case studies of increasing difficulty are examined to the end of dealing with different combinations of human and molecular scale virtual reality approaches.

1 Introduction

Among natural sciences Chemistry is by definition a phenomenological discipline rationalized in terms of an invisible reality (atoms and molecules) whose properties and rules are based on the first principles of physics. This has fostered, in the past, the development of chemical models and, in recent times, of chemical simulations. More recently, thanks to the development of high performance concurrent platforms, the role of simulations in chemical investigation has rapidly increased and research has moved from simple to complex systems and from abstract to realistic models embracing full dimensional theoretical treatments which include both human (meter) and molecular (nanometer) scales. However, most often, the related computational apparatus is so heavy and the feedback so poor that the insight obtained from a simulation is very limited even if supported by traditional post-processing visual representations[1].

Analogues of natural phenomena and visual representations have always played a crucial role in the development of science, in general, and of chemistry, in particular[2, 3]. The most recent step of this ongoing process of making analogues of the real world is the use of virtual reality, VR (see for example refs. [4–7]), at both meter and nanometer level. The present computational capacity (in particular, massively parallel and grid computing) makes it possible, in fact,

P.M.A. Sloot et al. (Eds.): ICCS 2003, LNCS 2658, pp. 324–330, 2003.

to link the heavy number crunching activity of a simulation with realistic sensorial perceptions of the user at both levels in order to enhance his/her insight and intuition. To this end, the virtual experiment (human virtual reality, HVR) and the simulated interaction between atoms and molecules (molecular virtual reality, MVR) can be combined together in a multiscale approach.

Along this line we are building some virtual chemical experiments three of which are described in this paper. In Section 2 we analyze the computational tools used and the objects defined for assembling the HVR and the MVR environments. In Section 3 we describe in some detail the HVR components of the three chemical experiments. In Section 4 an MVR component of a virtual reality simulation of the proposed experiments is discussed.

2 Tools and Objects for the HVR and MVR

Contrary to conventional visualization paradigms, the virtual reality ones when applied to a chemical laboratory (see for examplet Ref.[8] for a multimedia product used for teaching laboratory practices) allow the user to visualize, manipulate and interact with the computer generated structures and materials of the experimental setup being used and of the processes being considered. This is obtained by introducing some elements of immersive VR. These elements make the user an active component of the virtual scene especially if supported by suitable immersive devices like the Head Mounted Display (HMD), the Tracking Devices and/or the Data Gloves.

In the work reported in this paper VR applications are implemented using a Window of World (WoW)[9] approach based on the following tools:

VRML:[10, 11] to represent the Molecular Virtual World and to enable its navigation using suitable VRML Web browsers[12]. A migration to X3D[13] is planned for the near future.

Java:[14] to automate the generation of the necessary VRML code from the data produced by the computational engines used in the simulation and to assemble a dynamical representation of the Molecular Virtual World. Along this line a set of specific Java classes has been created to associate the proper chemical and visual properties (like size and color) to each atom. The simulation provides a description of the system considered in terms of the number of the constituting atoms and bonds. For each atom the chemical symbol, the spatial coordinates of each considered arrangement, the number and the type of bonds are specified. The Java program recognizes the atoms from the chemical symbol, extracts its chemical and visual properties and generates the VRML code for the dynamical rendering of the Molecular Virtual World.

CML:[15] to deal properly, in conjunction with XML[16], with the representation of the physical and chemical properties of the intervening atoms and molecules.

Using the above mentioned tools a typical HVR chemical laboratory environment was built. A key component of such a virtual chemical laboratory is given in Fig. 1 where a laboratory bench provided with a fume-hood is shown. In the figure also some laboratory basic tools are shown. In this virtual laboratory the user can walk-by, use the various components and perform the implemented experiments (as described in Section 3).

3 The Chemical Experiments

The up-to-date implemented experiments are the following.

The first experiment deals with the measurement of the dependence of the volume of a gas from the pressure at constant temperature (Boyle law) [17]. The related HVR environment is the one in which the user adjusts the height of a mercury container connected by a rubber pipe to an air filled burette. By placing the container at various heights one obtains an imbalance in the observed levels of mercury (and therefore a larger or a smaller pressure) as well as a change in the air volume in the burette. By plotting the measured gas volume as a function of the pressure exerted on it, one can work out the proportionality constant that is related to the Boltzmann constant k_B [18]. The rationalization of the experiment at MVR level is given by a molecular dynamics application (the same used also for the other experiments and illustrated in Section 4). Using that application the behaviour of the gas + liquid Hg system is mimicked.

The second experiment deals with flame spectroscopy [19, 20]. The colour of the flame depends on the emission spectrum of the substance added to it. In the HVR environment the user first cleans a wire loop of platinum or nickel-chromium by dipping it into a vessel containing hydrochloric or nitric acid and then he/she immerges it into the powder or the solution of an ionic (metal) salt. The loop is finally placed in the clear or blue part of the flame and the resulting colour is observed. A spectroscopic data base allows the individualization of the colours associated with a given element and the processes occurring inside the flame is simulated using a VMR component based on molecular dynamics.

The third experiment is concerned with an expansion jet of a given gas. In the HVR environment a gas is allowed to flow through a valve from a high pressure chamber into a low pressure region. This generates an expansion regime within which several physical and chemical processes can occur including non reactive energy transfer, reactive processes, recombinations, electronic transitions, etc.. In this case too, the VMR component is simulated using a set of molecular dynamics programs.

Accordingly, three specific virtual experimental apparatuses were built. These are:

a) A mercury containing container connected to a burette through a rubber pipe;
b) A platinum wire loop;
c) A high pressure chamber connected to an expansion vessel through an air valve.

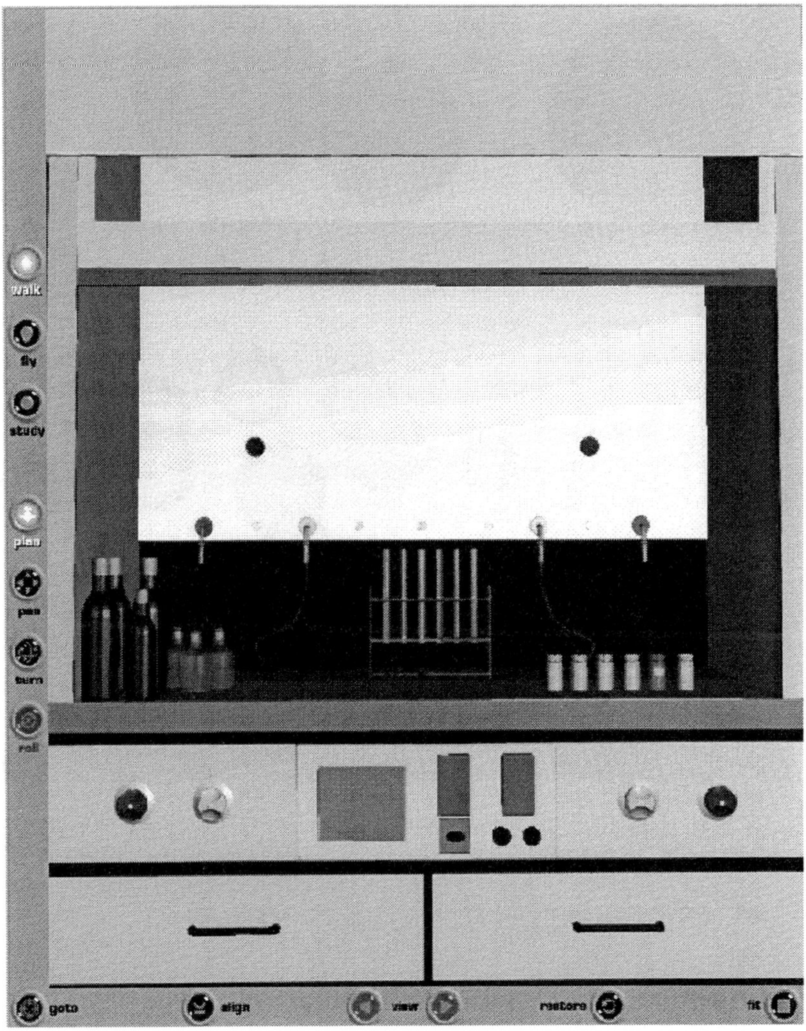

Fig. 1. An HVR illustration of a laboratory bench provided by a fume-hood and some basic tools

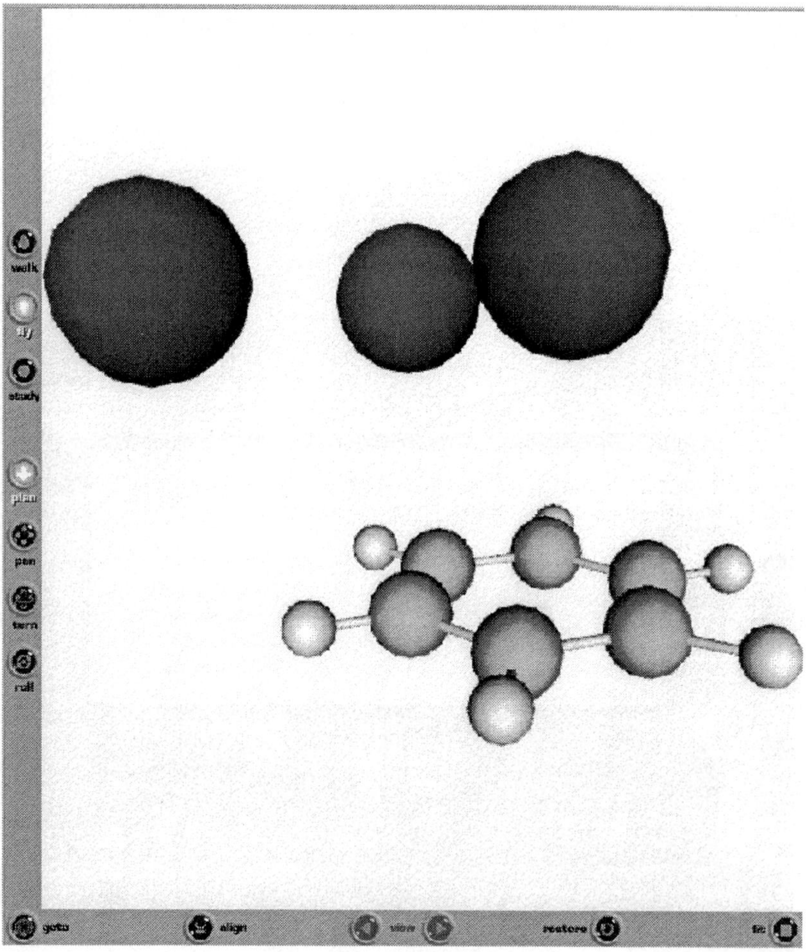

Fig. 2. An MVR representation of a typical benzene-Ar_3 cluster

4 The Molecular Dynamics Simulation

As already mentioned, the MVR component of the chemical experiments considered in the present paper is based on the molecular dynamics suite of codes DL_POLY [21]. DL_POLY is organized as a distributable molecular dynamics simulation package of subroutines, programs and data files allowing the study of macromolecules, polymers, ionic systems and solutions. We use for our purposes its most recent version DL_POLY_2 (version 2.13). This version of DL_POLY includes a Graphical User Interface, written in Java, offering additional functionalities enhancing its usage in an MVR context.

For illustrative purposes we discuss here the use of DL_POLY for the creation of an Ar seeded neutral benzene jet. This system is of particular interest since only a few benzene-rare gas_n isomers can coexist during the lifetime of the cluster in the expansion jet. In connection with the relative abundances measured in the experiment, the dynamical simulation should predict structural transitions occurring during the cooling process.

Dynamical calculations are, as usual, carried out by expressing the interaction in terms of components having simple two-body well-known functional forms (like, for instance, the Lennard-Jones 12-6 or the Morse ones [22]). To overcome the limitations of the pairwise additive formulation of the potential that, as an example, makes it difficult to describe properly the interaction for geometries other than the most stable one, especially near the in-plane configurations [23], we have recently adopted also a new-semiempirical potential describing in a more realistic way the interaction between the benzene and the rare gas atoms [24].

Up to date, most of the runs have been carried out by considering clusters with two and three Ar atoms. A picture of a typical clustering situation for a three Ar atom system is illustrated in Fig. 2. The figure shows that a suitable geometry is that of three Ar atoms sitting on the same (upper, in the figure) side of the benzene molecule in a quite asymmetric configuration (this cannot be properly perceived in a single shot picture like Fig. 2 while it can be fully appreciated in an MVR environment in which the observer can move around the molecule).

5 Conclusions

The aim of the paper is to show how is possible to use virtual reality approaches to describe a chemical experiment at both a human and a molecular level using a virtual reality approach. To this end a multiscale virtual reality approach has been adopted to deal with the description of the physical environment, HVR scale, and with the molecular environment, MVR scale. The main features of the three experiments of different complexity and of the common molecular dynamics engine have been outlined in the paper. The application is still in a prototype stage. In spite of that, however, it has already shown to be very effective. This has prompted us to plan its porting on a grid platform in order to exploit its full potentiality.

Acknowledgments

This research has been financially supported by MIUR, CNR and ASI as well as by the D23 Action (Metachem) of COST in Chemistry. M.A. also acknowledges the financial support from the Ministerio de Educacion, Cultura y Deporte: PR2002-0014), the Spanish DGICYT (PB97-0919 and BQU2001-3018) and the Generalitat de Catalunya (CUR 20001SGR-00041).

References

1. Wolff, R.S., Yaeger, L.: Visualization of Natural Phenomena. Springer-Verlag New-York (1993)
2. Maldonado, T.: Reale e Virtuale. Feltrinelli, Milano (1994)
3. Zare, N.R.: Visualizing Chemistry. Journal of Chemical Education, Vol. 79 (2002) 1290-1291
4. Casher, O., Leach, C., Page, C.S., Rezpa, H.S.: Virtual reality modelling language (VRML) in Chemistry. Chemistry in Britain, Vol. 34 (1998) 26-31
5. Krieger, J.H: Doing Chemistry in a virtual world. Chemical & Engineering News (1996) 35-41
6. Ruiz, I.R., Espinosa, E.L., Garcia, G.C., Gómez-Nieto, MÁ.: Computer-Assisted Learning of Chemical Experiments through a 3D Virtual Laboratory. Lecture Notes in Computer Science, Vol. 2329 (2002) 704-712
7. Garratt, J., Clow, D., Hodgson, A., and Tomlinson, A.: Computer Simulation and Chemical Education - A review of Project eLABorate. Chemistry Education Review, Vol. 14 (1999) 51-73 (ISSN 0972-0316).
8. Brattan, D. Jevons, O.M., and Rest, A.J. (Eds): Practical Laboratory Chemistry, Series of 19 CD ROM Programmes for teaching laboratory skills. Chemistry Video Consortium, Southampton, October 1999 (`http://www.soton.ac.uk/ chemweb/cvc/` and `http://www.emf-v.com`).
9. `http://vr.isdale.com/WhatIsVR.htm`
10. Ames, A., L., Nadeau, D., R., Moreland, J., L.: VRML 2.0 Sourcebook. Wiley Computer Publishing, New York Tokio (1997)
11. Hartman, J., Wernecke, J.: VRML 2.0 Handbook: building moving worlds on the web. Addison Wesley (1996)
12. *OpenSource cross-platform VRML browsers*:
    OpenVRML (`http://www.openvrml.org`); FreeWRL
    (`http://freewrl.sourceforge.net`);
    Vendor VRML browsers:
    Cortona from Parallel Graphics (`http://www.parallelgraphics.com/cortona/`); Contact from Blaxxun (`http://developer.blaxxun.com/`); Cosmo Software from Computer Associates (`http://www.cai.com/cosmo`)
13. X3D Consortium: `http://www.x3d.org`
14. Flanagan, D.: Java in a Nutshell. O'Reilly (1999)
15. Chemistry Markup Language: `http://www.xml-cml.org`
16. EXtensible Markup Language: `http://www.xml.org`
17. Riganelli, A., Pacifici, L., Gervasi, O., Laganà, A., Workshop on Multimedia in Chemistry Education, Perugia, (2002)
18. `http://www.chm.davidson.edu/ChemistryApplets/GasLaws/BoylesLaw.html`
19. McKelvey, G.M., Gilbert, G.L., McWherter, C.: Flame Tests that Are Portable, Storable, and Easy to Use. Journal of Chemical Education, Vol. 75 (1998) 55-56
20. `http://www.spectroscopynow.com`
21. Smith, W., Forester, T.R., J., Molecular Graphics, Vol. 14 (1996) 136-141; `http://www.dl.ac.uk/TCS/Software/DL_POLY/`
22. Stone, A.: The Theory of Intermolecular forces. Oxford University Press, Oxford (1996)
23. Amos, A.T., Palmer, T.F., Walters, A., Burrows, B.L., Chem. Phys. Letters, Vol. 172 (1990) 503-508
24. Pirani, F., Cappelletti, D., Liuti, G., Chem. Phys. Letters, Vol. 350 (2001) 286-289

Theoretical Analysis on Mechanisms Implied in Hybrid Integrated Circuit Building

Giacomo Giorgi[1], Filippo De Angelis[1], Nazzareno Re[2], and Antonio Sgamellotti[1]

[1] Dipartimento di Chimica e Istituto CNR di Scienze e Tecnologie Molecolari, Università di Perugia, via Elce di Sotto 8, I-06123 Perugia, Italy
giac@thch.unipg.it
[2] Facoltà di Farmacia, Università G. D'Annunzio, I-66100 Chieti, Italy

Abstract. Nowadays the relevance of silicon chemistry is increasing due to the applications of Si, in particular in semiconductors, in order to obtain an increasing number of performing materials. A theoretical investigation based on the Density Functional Theory has been done on the reaction paths implied in the Pt-catalysed alkene hydrosilylation, a process through which the selective grafting of organic molecules to silicon can be obtained. We studied the Si-H oxidative addition of SiH_4 to a diphosphine molecule, the C_2H_4 insertion on Pt-H and Pt-SiH_3 bonds, the isomerisation of the product of the C_2H_4 insertion, and the two following Si-C and C-H reductive eliminations. The set of these processes are known as the Chalk- Harrod mechanism and the modified Chalk-Harrod mechanism, respectively. The goal of this work has been to identify the rate determining step of both mechanisms. The dynamics of the oxidative addition step has been simulated.

1 Introduction

A theoretical investigation has been performed on the reaction pathways involved in the $Pt(PH_3)_2$- catalysed ethylene hydrosilylation by SiH_4. We have studied both mechanisms proposed to explain the formation of the vinyl-silane product. First of all we have analysed the Chalk-Harrod mechanism. It consists of three processes: (I) the Si-H oxidative addition of SiH_4 to $Pt(PH_3)_2$, (II) the C_2H_4 insertion into the Pt-H bond, (III) the Si-C reductive elimination. Next we analysed the modified Chalk-Harrod mechanism. It consists of the same preliminary oxidative addition of SiH_4 to the $Pt(PH_3)_2$, followed by the C_2H_4 insertion into the Pt-Si bond and, finally, by the C-H reductive elimination. We have investigated each step participating in both mechanisms by gradient-corrected DFT methods, and then performed *ab initio* molecular dynamics simulations on the initial oxidative addition step. One of the best known catalysts for hydrosilylation process is chloroplatinic acid (H_2PtCl_6, Speier's catalysts)[1], for which Chalk and Harrod have suggested the two reaction mechanisms [2] previously described in order to obtain the final organosilicon compound.(See Fig.1) Considering the ambiguous behavior of the Speier's catalyst, we used a model Pt(0)

P.M.A. Sloot et al. (Eds.): ICCS 2003, LNCS 2658, pp. 331–340, 2003.

diphosphine complex which has been demonstrated to be the active species in the well characterised homogeneous catalysis by $Pt(PR_3)_4$[3]. Sakaki and coworkers

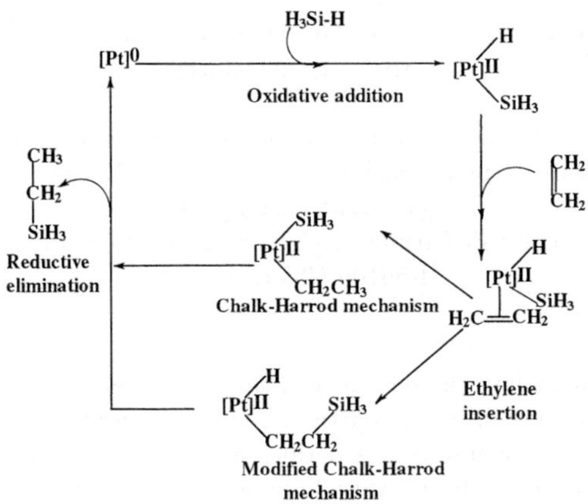

Fig. 1

Fig. 1. Chalk-Harrod and modified Chalk-Harrod mechanisms scheme

have theoretically investigated the Chalk-Harrod and modified Chalk-Harrod mechanisms at HF and post-HF level[4, 5], while Coussens et al.[6] have done a DFT analysis on the C_2H_4 insertion in the cationic systems $(H)Pt(PX_3)_2{}^+$. Since in the works by Sakaki the geometry optimisation of the C_2H_4 insertion step was done at the uncorrelated Hartree-Fock level, in our work we have investigated both hydrosilylation mechanisms performing geometry optimisation of each step at a correlated level of theory. It means that both the Chalk-Harrod and modified Chalk-Harrod mechanisms were treated by gradient-corrected DFT methods. Furthermore, *ab initio* molecular dynamics (AIMD) simulations were carried out by means of the Car- Parrinello method, to study the dynamical features of the initial Si-H oxidative addition to $Pt(PH_3)_2$, the initial step that is shared by both mechanisms.

2 Computational Details

All the static calculations have been performed using the Gaussian 98[7] program package. Reactants, transition states (TS) and products geometries of the two mechanisms analysed have been fully optimised using Density Functional Theory. A preliminary study on the oxidative addition process has been recently done [8] using the "pure" BPW91 [7] exchange-correlation functional to provide consistency between static and dynamic analyses. In this paper, we have used both the "pure" BPW91 and hybrid B3LYP[7] functionals, comparing the results achieved with the two different approaches for the activation barrier (E_a)

of the Rate Determining Step of the whole mechanism. We have made geometry optimisation using the relativistic pseudopotential LANL2DZ[7] for the Pt atom, the 6-31G**[7] basis set for Si, C and the reacting H atom, and the 6-31G[7] for P and for the remaining H atoms: we summarise these basis set as BSa. The optimised geometries have been subsequently employed to perform single-point calculations. Here we have still used the relativistic pseudopotential LANL2DZ for the Pt atom, while for all other atoms we have employed the 6-311++G(3df,3pd)[7] basis set: we summarise these basis set as BSb. Every TS structure for the key reaction steps has been checked by doing an Intrinsic Reaction Coordinate (IRC) analysis[7]. We have found an excellent agreement between the "pure" BPW91 and the hybrid B3LYP functionals. In fact, considering BSa, the activation energy for the Chalk-Harrod RDS is 32.2 kcal/mol and 32.1 kcal/mol for BPW91 and B3LYP, respectively. For the Car-Parrinello calculations gradient corrected AIMD simulations have been performed using the parallel version developed by F.De Angelis of the Car-Parrinello code [9] implementing Vanderbilt pseudopotentials[10–12]. The Perdew-Zunger[13] parameterisation has been employed for the LDA exchange-correlation functional, while the gradient-corrected functional is taken from Ref.[14]. Core states are projected out using pseudopotentials. For Pt atom "ultra-soft" pseudopotentials were generated following the Vanderbilt scheme, while the Hamann-Schluter-Chang (HSC) pseudopotential[15] has been used for P and Si atoms. The wavefunctions were expanded in plane waves up to an energy cutoff of 25 Ry. Periodic boundary conditions were used by placing the model molecule in a cubic box of 10.6Å, to avoid coupling between periodic images. The equations of motion were integrated using a time step of 6 a.u. (0.145 fs) with an electronic fictitious mass μ= 500 a.u.. Constrained dynamics simulations were performed by means of the SHAKE algorithm[16], employing the slow-growth method[17]. The thermal equilibrium was maintained checking the temperature of the nuclei by a Nosé thermostat[18, 19], which creates a canonical ensemble (NVT). All simulations were performed at 300 K.

3 Results and Discussion

3.1 Oxidative Addition

In Figure 2 we show the oxidative reduction reaction mechanism. Our results achieved at the BPW91/BSa level, show a structure for the TS ($TS_{R\rightarrow 1}$ where the Si-H bond does not lie in the P-Pt-P plane. Moreover, we have that the Si-H bond approaches the Pt atom in an out- of-plane fashion. This can justify the presence of a *trans*-$Pt(PH_3)(H)(SiH_3)$ product that is not obtainable through an in-plane approach.

The AIMD simulations we performed confirm the reaction mechanism of the oxidative addition process. Constrained dynamics simulations were made by varying the Si-Pt distance in the range 4.6-2.3Å (see Fig.3), the former value corresponding to the unbound free reactants, the latter being close to the Si-Pt distance in the oxidative addition product. Dynamics simulations were initialised

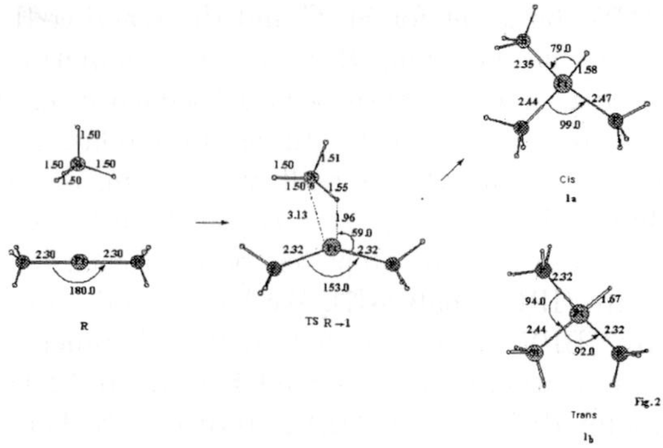

Fig. 2. Oxidative addition process: reactants, TS and products (1a and 1b). 1a is the most stable product (cis). (Å, degrees)

by fixing the distance Si-Pt at 4.6Å, minimising the corresponding structure, and thermalising the resulting system at 300 K for 1.4 ps. Subsequently, we decreased the Si-Pt distance in the cited range during a time span of 5.8 ps.

3.2 Chalk-Harrod Mechanism

We have considered the four-coordinated $PtH(SiH_3)(PH_3)(C_2H_4)$ complex. This species is the product of a reaction that involves the loss of a PH_3 group followed by an C_2H_4 coordination on the Pt-H bond (endothermic by 43.7 kcal/mol compared to free reactants R in Fig.2). We will hereafter refer to C_α and C_β as the C_2H_4 carbons in complex 2 facing the H and the PH_3 groups, respectively. We find a point of inflection instead of the minimum we should expected for the insertion products and, to clarify this point, we have done a linear transit calculation constraining the bond distance between C_β and H atom bound to Pt to vary in the range 2.5-1.0Å. A "true" TS has been found for the isomerisation step; the geometry optimisation of $TS_{2\rightarrow4}$ (see Fig.4), has led to an E_a=25.1 kcal/mol. We then have calculated that the more stable product is 4b because of the *trans*-effect given by the bulky group SiH_3. In fact, in the product 4a we have an agostic interaction between an hydrogen bound to C_α and the Pt atom that, being in the trans site to the SiH_3, does not allow the stabilising *trans*-effect. The 4a and 4b products were found 3.7 and 4.7 kcal/mol below 2.

3.3 Modified Chalk-Harrod Mechanism

Contrary to the "simple" Chalk-Harrod mechanism, in the modified mechanism we have distinguished an insertion process and an isomerisation step. The main

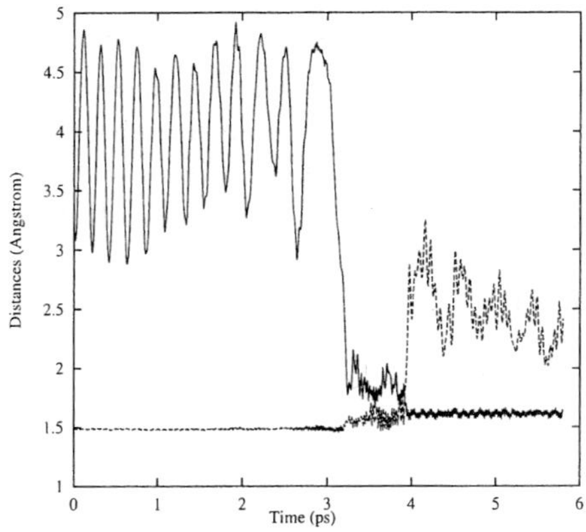

Fig. 3. Time evolution of the Pt-H (solid line) and Si-H (dashed line) distances for the time span going from 3 to 4 ps, corresponding to a Pt-Si distance decrease from 3.20 to 2.74 Å. Time in ps, distances in Å.

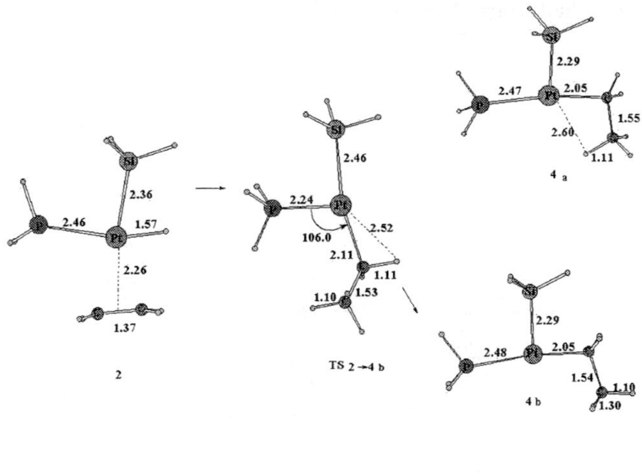

Fig. 4. Chalk-Harrod mechanism: Insertion plus Isomerisation step. Reactants, TS and products (4a and 4b). 4b is the most stable product.

difference between the Chalk-Harrod mechanism and its modified version is observable in the insertion process. In fact, while in the Chalk-Harrod mechanism the C_2H_4 insertion takes place on the Pt-H bond, in the modified version it takes place on the Pt-Si bond (see Fig.5). We have obtained an insertion product (3') characterised by an agostic interaction between an hydrogen atom bound to the Si and the Pt. 3' can simply isomerise and give the products 4'a and 4'b (see Fig.6). Here we have, contrary to the Chalk-Harrod mechanism, that the 4'a is the more stable product because in the product 4'b H group is too small to give *trans*-effect and the agostic interaction between an H bound to Si and the Pt atom stabilises slightly 4'a.

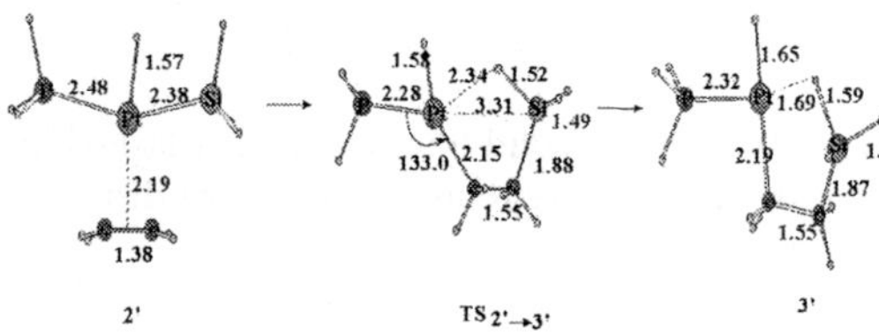

Fig. 5. Modified Chalk-Harrod mechanism: Insertion step. Reactants, TS, and product.

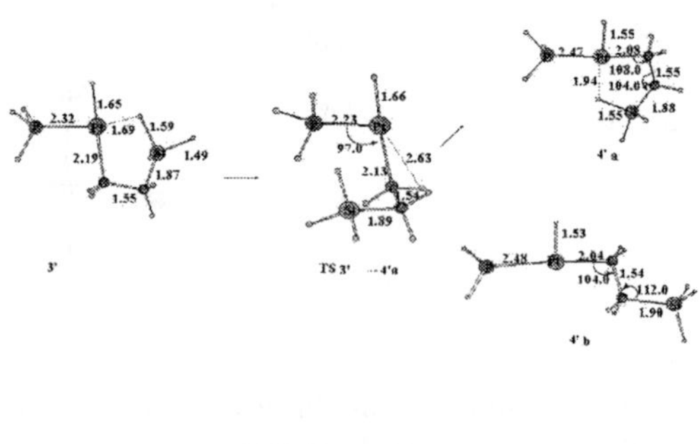

Fig. 6. Modified Chalk-Harrod mechanism: Isomerisation step. Reactant, TS and products (4'a and 4'b). 4'a is the most stable product.

3.4 Reductive Elimination

In the Chalk-Harrod mechanism, the C_2H_4 insertion and the isomerisation is followed by coordination of a PH_3 group, before of the Si-C reductive elimination. The 4b product has been considered as the starting reactant because of its bigger stability (see Fig.7). The energy gain of PH_3 coordination is computed to be 45.4 kcal/mol. The resulting diphosphinic complex 5 then undergoes the Si-C reductive elimination leading to the final $SiH_3CH_2CH_3$ reaction product P, restoring the initial catalyst. The main characteristic of this reaction is to find in the TS non-planar structure, contrary to the C-H reductive elimination. For this step we have computed an activation energy of 24.1 kcal/mol (referred to 5), with a small exothermicity of 3.7 kcal/mol.
In the modified Chalk-Harrod mechanism the C-H reductive elimination (see Fig.8) is again preceded by the coordination of a PH_3 group to the 4'a species leading to 5'. The TS connecting 5' to the final products P ($TS_{5' \rightarrow P}$) shows the concerted breaking of the Pt-C_α bond and the formation of the H-C_α bond. This process is kinetically favored over the Si-C reductive elimination. In fact, here we have an E_a for the C-H reductive elimination 5.2 kcal/mol lower than that for Si-C (18.9 vs. 24.1 kcal/mol) reflecting the directionality of the sp^3 valence orbital of SiH_3. Also thermodynamics confirms this trend: we have found the C-H reductive elimination favored over the Si-C one by 10.7 kcal/mol (-14.4 vs. -3.7 kcal/mol), reflecting the lower stability of 5' with respect to 5 (see Figs.9 and 10, respectively).

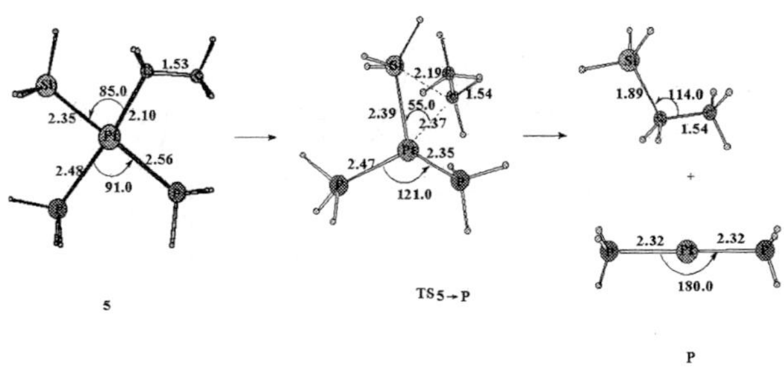

Fig. 7. Si-C Reductive Elimination: Reactant, TS, and Products

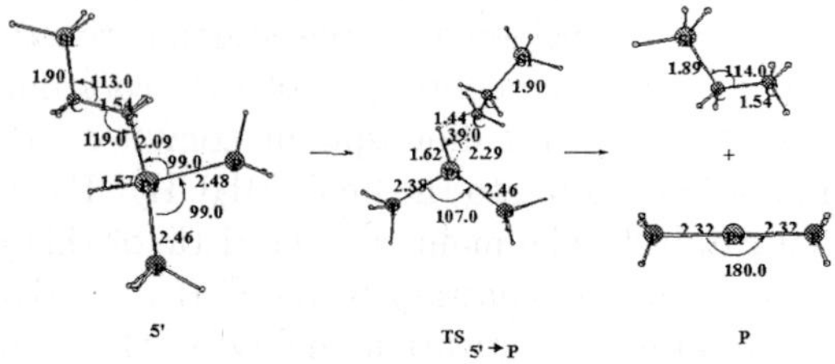

Fig. 8. C-H Reductive Elimination: Reactant, TS, and Products

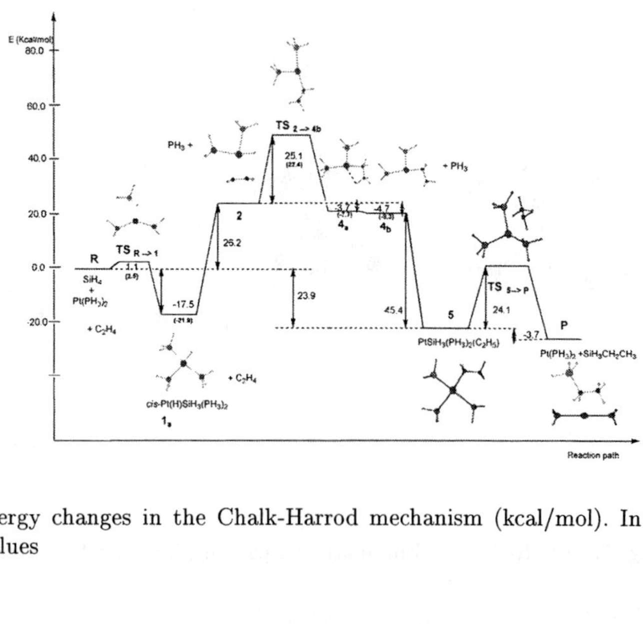

Fig. 9. Energy changes in the Chalk-Harrod mechanism (kcal/mol). In parentheses Sakaki's values

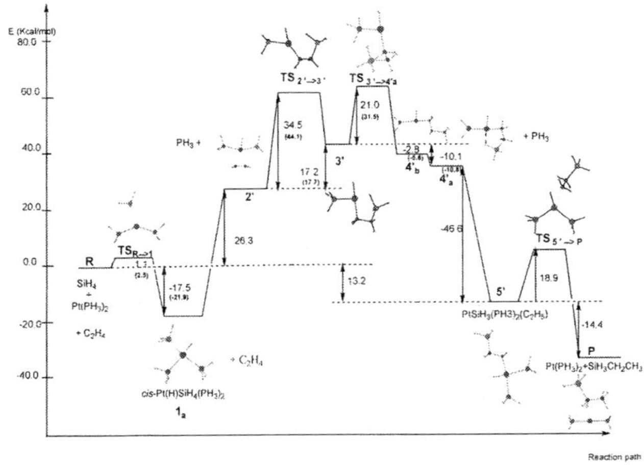

Fig. 10. Energy changes in the modified Chalk-Harrod mechanism (kcal/mol). In parentheses Sakaki's values

4 Conclusion

In this work we have analysed the mechanisms implicated in the Pt-catalysed hydrosilylation of C_2H_4 by combining static and dynamic calculations. The Chalk-Harrod mechanism and the modified Chalk-Harrod mechanism have been investigated by a DFT approach, checking the effect of different exchange-correlation functionals and basis set expansion on the energetics of both mechanisms; furthermore, dynamics simulations, by means of Car-Parrinello method, have been performed on the preliminary oxidative addition step. This analysis have revealed that the formation of the preliminary Pt-H bond takes place before that the Si-H bond breaks, going directly to the more stable *cis* isomer through a non planar TS.
In agreement with previous theoretical works, we compute a maximum energy barrier of 25.1 kcal/mol and 38.2 kcal/mol in correspondence of the C_2H_4 insertion and isomerisation step, for the Chalk-Harrod and modified Chalk-Harrod mechanisms, respectively.

References

1. Speier, J.L., Webster, J.A., Barnes, G.H,: The Addition of Silicon Hydrides to Olefinic Double Bonds. Part II. Use of Group VIII Metal Catalysts. J. Am. Chem. Soc.**79**, (1957), 974–979
2. Harrod, J.F., Chalk, A.J.,: Wender, I., Pino, P.(Eds.), Organic Synthesis via Metal Carbonyls, vol.2, Wiley, New York, 1977, p.673.
3. Prignano, A.L., Trogler, W.C.,: Silica-supported bis(trialkylphosphine)platinum oxalates. Photogenerated catalysts for hydrosilylation of olefins. J. Am. Chem. Soc. **109**, 1987, 3586–3595

4. Sakaki, S., Mizoe, N., Sugimoto, M.: Theoretical Study of Platinum(0)-catalyzed Hydrosilylation of Ethylene. Chalk-Harrod Mechanism or Modified Chalk-Harrod Mechanism. Organometallics, **17**, 1998, 2510–2523
5. Sakaki, S., Ieki, M.: Oxidative addition reactions of saturated Si-X bonds (X=H, F, C, or Si) to $Pt(PH_3)_2$. An ab initio MO/MP4 study. J. Am. Chem. Soc., **115**, 1993, 2373–2381
6. Coussens, B.B., Buda, F., Oevering, H., Meier, R.J.: Simulations of Ethylene Insertion in the Pt^{II} Bond of $(H)Pt(PX_3)_2{}^+$. Organometallics **17**, 1998, 795–801
7. Frisch, M.J., Trucks, G.W., *et al.*: Gaussian, Inc., Pittsbourgh PA, 1998 and references therein.
8. Giorgi, G., De Angelis, F., Re, N., Sgamellotti, A.: Oxidative addition of SiH_4 to $Pt(PH_3)_2$: a dynamical density functional study. Chem. Phys. Lett., **364**, 2002, 87–92
9. F. De Angelis, PhD thesis, University of Perugia, (1999).
10. The implementation that we use is described in: Pasquarello, A., Laasonen, K., Car, R., Lee, C., Vanderbilt, D.: Ab initio Molecular Dynamics for d-Electron Systems: Liquid Copper at 1500 K. Phys. Rev. Lett., **69**, 1992, 1982–1985
11. The implementation that we use is described in: Pasquarello, A., Laasonen, K., Car, R., Lee, C., Vanderbilt, D.: Car-Parrinello molecular dynamics with Vanderbilt ultrasoft pseudopotentials. Phys. Rev. B, **47**, 1993, 10142–10153
12. Vanderbilt, D.: Soft self-consistent pseudopotentials in generalized eigenvalue formalism. Phys. Rev. B, **41**, 1990, 7892–7895
13. Perdew, J.P., Zunger, A.: Self-interaction correction to density-functional approximations for many-electron systems. Phys. Rev. B, **23**, 1981, 5048–5079
14. Perdew, J.P., Chevary, J.A., Vosko, S.H., Jackson, K.A., Pederson, M.R., Singh,D.J., Fiolhais, C.: Atoms, molecules, solids and surfaces: Applications of the generalized gradient approximation for exchange and correlation. Phys. Rev. B, **46**, 1992, 6671–6687
15. Hamann, D.R., Schlüter, M., Chiang, C.: Norm-Conserving Pseudopotentials. Phys. Rev. Lett., **43**, 1979, 1494–1497
16. Ryckaert, J.P., Ciccotti, G,, Berendsen, H.J.: Numerical Integration of the Cartesian Equation of Motion of a System with Constraints: Molecular Dynamics of N-Alkanes. J. Comp. Phys., **23**, 1977, 327–341
17. Straatsma, T.P., Berendsen, H.J.C., Postma, J.P.M.: Free energy of hydrophobic hydration: A molecular dynamics of noble gas in water . J. Chem. Phys., **85**, 1986, 6720–6727
18. Nosé, S.: A molecular dynamics method for simulations in the canonical ensemble. Mol. Phys., **52**, 1984, 255–268
19. Hoover, W.G.: Canonical dynamics: phase-space distributions. Phys. Rev. A, **31**, 1985, 1695–1697

Parallel Models for a Discrete Variable Wavepacket Propagation

D. Bellucci[2], S. Tasso[2], and A. Laganà [1]

[1]Department of Chemistry, University of Perugia, Via Elce di Sotto, 8, 06123 Perugia, Italy
lag@dyn.unipg.it
[2] Department of Mathematics and Informatics, University of Perugia, Via Vanvitelli, 1, 06123 Perugia, Italy
daniele@dyn.unipg.it, sergio@unipg.it

Abstract. The parallelization of a code carrying out the discrete variable propagation of a wavepacket is discussed. The performances obtained from a Task Farm model are compared with those obtained from a Pipeline model.

1 Introduction

Atom diatom reactive scattering calculations are often carried out using wavepacket techniques. The difficulty of carrying out these calculations lies in the fact that as the total angular momentum increases the matrices involved become rapidly so large that they make the computer code quite inefficient. In fact, the algorithms usually utilized to carry out the time propagation of the wavepacket require at each time step the use of all the elements of the matrices. This makes it inconvenient to apply a fine grain parallelization of the code.

As an alternative, for our time dependent code (in which the wavepacket is propagated in the AV routine by dealing only with its real component[1]) we adopted a more localized Discrete Variable technique that at each time step (τ) performs a series of matrix operations that can be schematized as follows:

$$G = A \cdot C + C \cdot B^T + V \odot C. \tag{1}$$

Eq (1) is recursive since the value of C is taken at time $\tau - 1$ while that of G is taken at time τ (the value of C at time τ, in turns, depends on that of G at the same time τ even though, in order to simplify the notation, we have dropped the time subindex). For simplicity we also assume that all the matrices involved in the calculation are square matrices of order nr. Such a constraint, however, can be easily removed with no prejudice for the results.

The starting point of our study is the sequential version of AV (see Fig. [1] for its pseudocode). By adopting a domain representation *by rows* the j-th row of matrix G can be expressed as follows:

P.M.A. Sloot et al. (Eds.): ICCS 2003, LNCS 2658, pp. 341–349, 2003.

$$Row(j,G) = \sum_{k=1}^{nr} A(j,k) \cdot Row(k,C) + Row(j,C) \cdot B^T + Row(j,V) \odot Row(j,C). \tag{2}$$

To optimize the usage of space all the matrices were stored on the secondary memory minimizing the occupation of the main memory. When adopting a domain representation by rows it is easy to show that:

- Only a row vector is needed to carry out the calculations of Eq (2);
- The first term of Eq (2) ensures that accesses to I/O are optimized by storing each matrix *by row*;
- The *transpose* operation is avoided by multiplying $Row(j,C)$ by B rows since they are the columns of B^T [2].

```
Do j = 1,n
   Row(j,G) = 0_n
   ReadFromFile (Row(j,A))
   Do k = 1,n
      ReadFromFile (Row(k,C))
      Row(j,G) = Row(j,G) + A(j,k) · Row(k,C)
      If (k = j) then
         Keep in main memory Row(j,C)
      EndIf
   EndDo
   Do h = 1,n
      ReadFromFile (Row(h,B))
      G(j,h) = G(j,h) + Row(j,C) · Row(h,B)
   EndDo
   ReadFromFile (Row(j,V))
   Row(j,G) = Row(j,G) + Row(j,V) ⊙ Row(j,C)
   WriteToFile (Row(j,G))
EndDo
```

Fig. 1. Pseudocode of the sequential version of the AV routine.

2 The Task Farm Model

The first attempt to parallelize AV was performed using MPI and a Task Farm model [3]. In the startup phase the Master process distributes the rows of C using a cyclic policy. This implies that in the startup phase if $j \equiv i \bmod M$ the vector $Row(j,C)$ is sent to the Worker W_i (with M being the number of scheduled Workers). At the end of the startup phase each Worker has stored the rows received from the Master in a local (unshared) secondary space storage hereafter called $Dataset(W_i, C)$. In the same way, all the elements of the matrices A and B referred by the Worker W_i are stored in a similiar *Dataset* during the wavepacket inizialization (immediately before the first step $\tau = 0$).

To carry out the subsequent operational phase the Master process adopts a scheduling policy to *broadcast* $Row(j, C)$ (that is needed to perform the calculation of the second term of Eq (1)) to each Worker. Each scheduled Worker loads from a local (unshared) secondary memory space all the elements needed to carry out the calculation of:

- $w_i = \sum_{k \in D_i} A(j,k) \cdot Row(k, C)$
- $w_i(h) = w_i(h) + Row(j, C) \cdot Row(h, B) \qquad \forall h \in E_i$

where:

- i is the index of the Worker process;
- j is the index of the Row of G to be calculated;
- $D_i = [i]_M$ with M being the number of scheduled Workers;
- E_i is the set of contiguous indices assigned to Worker W_i allowing an optimum load balancing among the Workers.

Through a reduce operation the Master reassembles the resulting rows of matrix G. The pseudocodes of the Master and the Worker processes are given in Figs. [2],[3] respectively.

```
Process Master
Do i = 1, nr
   ReadFromFile (< Row(i,V), Row(i,C) >)
   MPI_Bcast (Every Worker, Row(i,C))
   T = Row(i,V) ⊙ Row(i,C)
   MPI_Reduce (Master, T, Row(i,H))
   WriteToFile (Row(i,H))
EndDo
```

Fig. 2. Pseudocode of the operational phase of the Master Process of the Task Farm model.

Several speedup measurements were performed using a Cluster of Linux Workstations on which the Gnu compiler and the Lam-MPI library were made available. The size of the matrices was varied from 512 to 1024. Speedups achieved when using the Task Farm Model are shown in Fig. [4]. They scale as the ideal speedup indicated for comparison in the same figure.

3 An Improved Task Farm Model

A first improvement of the simple Task Farm model (hereafter called STF) was introduced by changing the scheduling policy. The cyclic policy introduces,

```
Process W_i
Do j = 1, nr
   MPI_Bcast (From Master, Row(j, C))
   w_i = 0
   Read ({A(j, k)|k ∈ D_i} from Dataset(W_i, A))
   ForEach k ∈ D_i
     Read (Row(k, C) from Dataset(W_i, C))
     w_i = w_i + A(j, k) · Row(k, C)
   EndFor
   ForEach h ∈ E_i
     Read (Row(k, B) from Dataset(W_i, B))
     w_i(h) = w_i(h) + Row(j, C) · Row(h, B)
   EndFor
   MPI_Reduce (Master, w_i)
EndDo
```

Fig. 3. Pseudocode of the i-th Worker process operational phase.

in fact, a certain amount of overhead due to the fact that the communication system has to be initialized everytime a block of data is sent.

The improved Task Farm (ITF) minimizes such an overhead adopting a block scheduling policy. Each Worker has to receive at least a matrix partition during the startup phase. Thus the Master process needs at least M steps to complete the scheduling operation (remember that M is the number of Workers used). The block policy consists of sending a block of contiguos rows of C to each Worker.

To estimate the advantage gained in terms of a reduction of comunication time when using the block policy of the ITF model instead of the cyclic one we plot in Fig. [5] the amount $Dt = t_{STF} - t_{ITF}$ defined as the difference between the STF and ITF the elapsed times (t).

As apparent from Fig. [5] the *block* scheduling policy progressively outperforms the cyclic one on the four node Beowulf as the size of the matrices grows. Such an advantage slowly decreases as the number of processes increases although it remains substantially constant in percentage with respect to an increase of the size of the matrices and of the number of processes. This result, however, exploits also the fact that the A, B and V matrices remain unchanged during the calculation. Accordingly, in the ITF model there is not need to repeat their distribution to the Workers at every time step.

Related speedups are given in Fig. [6]. The figure shows that the ITF model, in addition to improving over the STF one, still scales as the ideal curve when the dimension of the matrices increases.

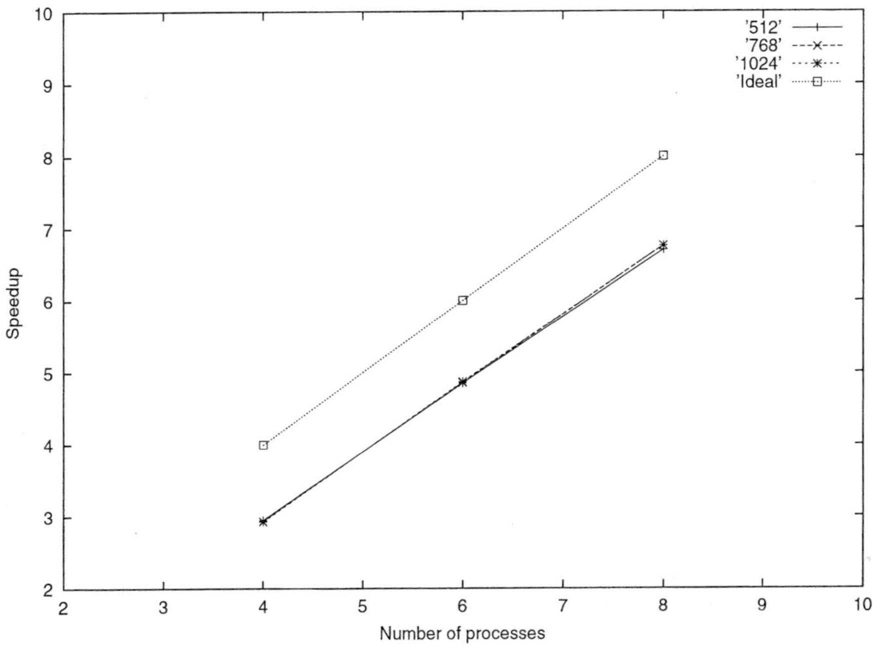

Fig. 4. Speedups obtained for the *STF* model on the 8 node Beowulf plotted as a function of processes.

4 The Pipeline Model

During the operational phase of the Task Farm the calculation of $A \cdot C + C \cdot B^T$ is performed by all available Worker processes in a way that overlaps the calculation of $V \odot C$ performed by the Master process. In this way related sequentiality is suppressed and possible limiting effects on the speedup are avoided.

The Pipeline solution differs from the Task Farm ones in that the calculation of G is partitioned among the scheduled processes. Therefore, in an M stage Pipeline the i-th process[1] W_i calculates the following vector:

- $w_i = \sum_{k \in D_i} A(j,k) \cdot Row(k,C)$
- $w_i(k) = w_i(k) + Row(j,C) \cdot Row(k,B) + V(j,k) \odot C(j,k) \qquad \forall k \in D_i$

where

- j represents the index of the row of G to be calculated;
- D_i is a set of contiguous indices assigned to process W_i allowing an optimum load balancing among the stages.

[1] Each process is now a stage of the pipe.

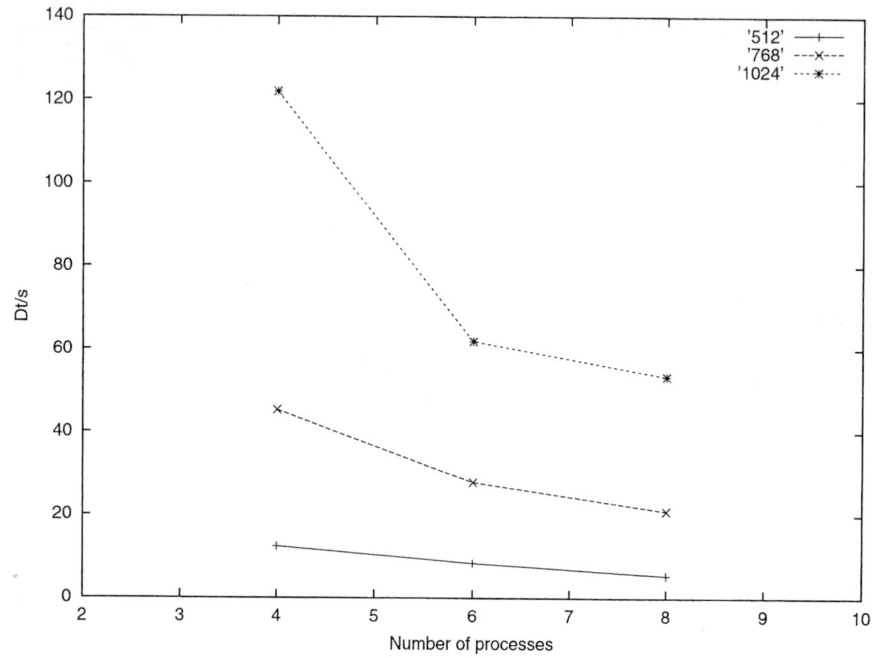

Fig. 5. Time difference between cyclic and block policy.

Distributed calculations of $Row(j, G)$ can be described as follow: the W_1 stage calculates w_1, loads $Row(j+1, C)$ and sends the t-uple $< Row(j+1, C), w_i >$ to W_2. For each $i = 2, \cdots, M-1$ stage W_i receives the t-uple from W_{i-1}, performs the sum $w_i = w_i + w_{i-1}$ and then sends the new t-uple $< Row(j+1, C), w_i >$ to W_{i+1}. The last stage carries out the final calculation of $Row(j, G)$ summing its vector (w_M) to the second term of the t-uple received from W_{M-1} which contains $\sum_{i=1}^{M-1} w_i$. Like in the Task Farm model each vector needed to calculate w_i by stage W_i is loaded from a local (unshared) secondary memory storage to avoid possible conflicts.

A possible stencil for stage W_i is illustrated in Fig. [7]. In our work we made use of the persistent comunication requests [4] to avoid overheads caused by the repeated initialization of the comunication system.

Speedups obtained for this approach are plotted in Fig. [8]. A comparison with the speedups calculated for the Task Farm model (see Fig. [6]) shows that the Pipeline model is less efficient though the size of the matrices does not affect the speedup obtained. The efficiency obtained for the Pipeline solution is constant (percentual efficiency is 46% irrespective of the number of processes used and of the size of the matrices employed) implying that the Pipeline model

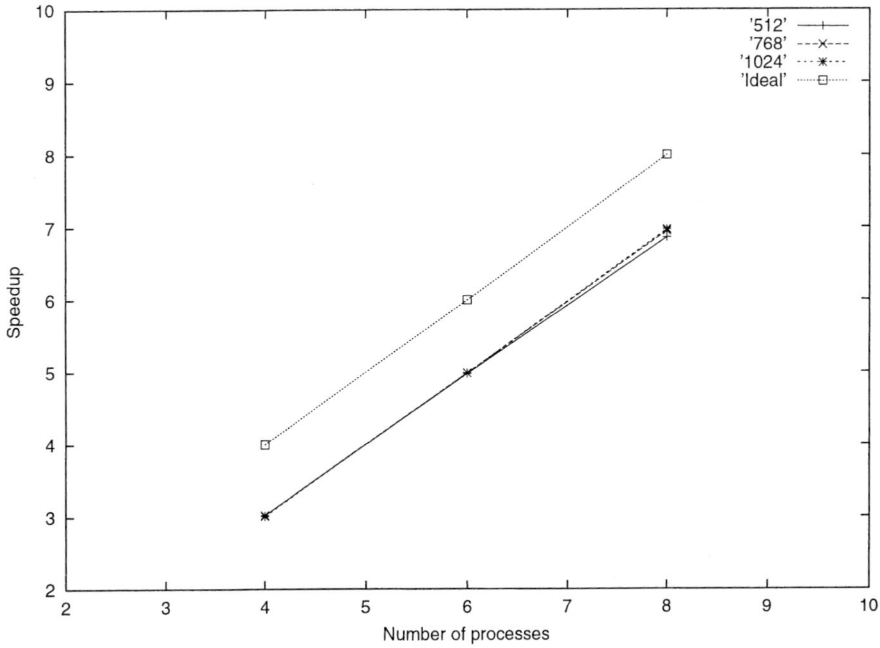

Fig. 6. Speedups obtained for the *ITF* model.

is well balanced. This suggests that a possible evolution of this model is a Pipeline made of Task Farms. Work on such a hybrid model is in progress.

5 Conclusion

Discrete variable approaches to wavepacket propagation techniques require to compute the following recursive matrix operation

$$G = A \cdot C + C \cdot B^T + V \odot C$$

where the value of the elements of C depend on the previous value of the elements of G.

This calculation engages a significant amount of computing resources. In this paper we have discussed how the request of computing resources can be reduced. To this end we have proposed a Task Farm model. Our study shows that a Task Farm model is appropriate to this end and that further reduction in both computing time and memory occupation can be obtained by approaches based on models accounting for the memory hierarchy of MIMD concurrent architectures.

$\cdots$ *calculation of* $Row(j,G)$ $\cdots$
$w_i = 0_n$
Read $(\{A(j,k)|k \in D_i\}$ from $Dataset(W_i, A))$
ForEach $k \in D_i$
  Read $(Row(k,C)$ from $Dataset(W_i, C))$
  $w_i = w_i + A(j,k) \cdot Row(k,C)$
  Read $(Row(k,B)$ from $Dataset(W_i, B))$
  $w_i(k) = w_i(k) + Row(j,C) \cdot Row(k,B) + V(j,k) \odot C(j,k)$
EndFor
If $(i == 1)$ **then**
  ReadFromFile $(Row(j+1,C))$
Else
  MPI_Recv (From W_{i-1}, $< Row(j+1,C), w_{i-1} >$)
  $w_i = w_i + w_{i-1}$
EndIf
If $(i \neq M)$ then
  MPI_Send $(W_{i+1}, < Row(j+1,C), w_i >)$
Else
  $\{Row(j,G) = w_i\}$
  WriteToFile $(Row(j,G))$
EndIf
$\cdots\ \cdots$

Fig. 7. Pseudocode of the *i*-th stage of the Pipeline model.

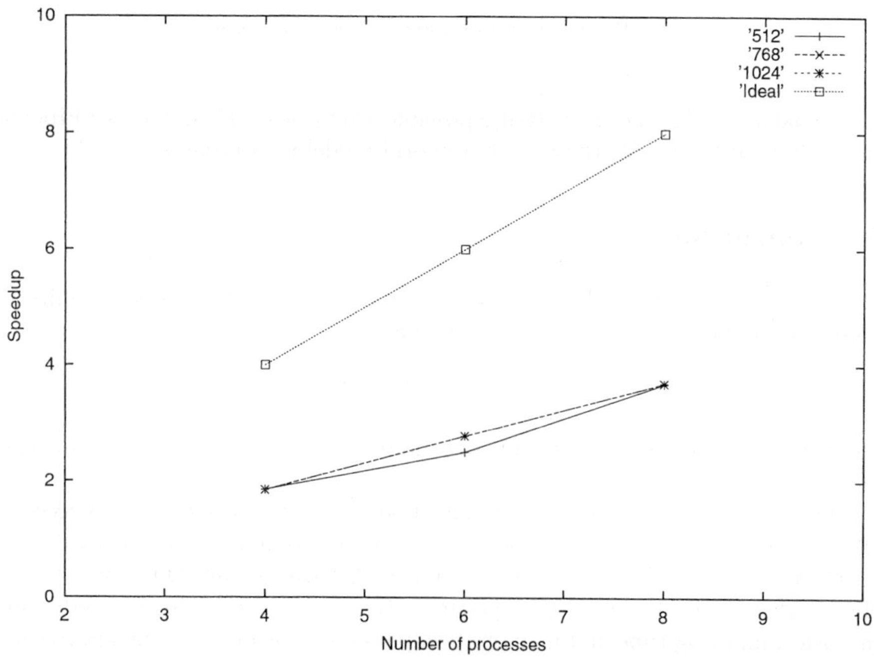

Fig. 8. Speedups obtained for the Pipeline model on the 8 node Beowulf plotted as a function of the number of processes.

This has led to an improved Task Farm model that adopts a block scheduling policy. Speedups reached by this solution are satisfactory and are not affected by an increase of the matrix size (in the size interval of our investigation). Further benefits could be ensured by the use of a *thread safe* communication library [5–7] that could reduce the need for collective communication.

We have also investigated the performances of a Pipeline model. These are lower than those of a Task Farm model though being characterized by a constant value of the efficiency with respect to the number of activated processes and to the size of the matrices. This suggests a possible use of a Hybrid Model in which each stage of the Pipeline is made of a Task Farm.

Acknowledgments.
This work has been financially supported by MIUR, ASI, CNR and COST in Chemistry (Action D23).

References

1. G. G. Balint-Kurti, Time dependent quantum approaches to chemical reactions, Lectures Notes in Chemistry, **75** (2000) 74 - 88.
2. G. D'Agosto, Parallel approaches to integrate the Schrödinger equation using a time dependent technique, Diploma thesis, (2000) Perugia.
3. D. Bellucci, S. Tasso, A. Laganà, Fine grain parallelism for discrete variable approaches to wavepacket calculations, Lecture Notes in Computer Science, **2331** (2002) 918 - 925.
4. M. Snir, S. Otto, S. Huss-Lederman, D. Walker, J. Dongarra, MPI: The Complete Reference, MIT Press, Cambridge Massachusetts, (1996).
5. A. Skjellum, B. Protopov, S. Hebert, A thread taxonomy for MPI, Proc. of MPIDC, (1996).
6. A. Chowdappa, A. Skjellum, N. Doss, Thread-safe message passing with P4 and MPI, Tech. Rep. TR-CS-941025, Computer Science Department and NSF Engineering Research Center, Mississipi State University, (1994).
7. M. Danelutto, C Pucci, A compact thread-safe communication library for efficient cluster computing, Lecture Notes in Computer Science, **1823** (2000) 407 - 416.

Calculation of Displacement Matrix Elements for Morse Oscillators

Zimei Rong, Dominique Cavagnat, and Laure Lespade

Laboratoire de Physico-Chimie Moléculaire, UMR 5803 CNRS
Université de Bordeaux I,
351 cours de la Libération, F-33405 Talence Cedex, France
zimeirong@hotmail.com

Abstract. Displacement matrix elements of the Morse oscillators are widely used in physical and chemical computation. Many papers have been published about the analytical expressions of the displacement matrix elements. People still use an integration method rather than the analytical expressions to calculate the displacement matrix elements. Analytical expressions of the diagonal and off diagonal displacement matrix elements $\langle v+m|q^n|m\rangle$ with intermediate variables are derived, where v, m and n are non-negative integers and n • 6. Calculation with these expressions is superior to that with numerical integration in precision, calculation speed and convenience.

1 Introduction

The Morse potential is the most popular anharmonic diatomic potential model in molecular spectroscopy and molecular dynamics. The matrix elements are widely used in calculation of transition energies, intensities and relative physical observables. Many physical quantities, such as dipole moment function, are functions of displacement, which gives to displacement matrix elements (abbreviated as matrix elements thereafter) extra importance. The Morse potential is expressed as [1]

$$V(q) = D(1 - e^{-aq})^2 \ , \tag{1}$$

where D is the dissociation energy and a is the scaling factor. The first advantage of the Morse oscillator is that its simple energy level equation represents very well the actual molecular energy levels and its parameters can easily be found from observed transition energies. Another advantage is the relatively simple expressions of its analytical wave functions, from which the analytical matrix elements are obtained [2 - 9]. Many previous papers have dealt with the analytical expressions of the Morse matrix elements. Heaps and Herzberg derived the analytical matrix elements for the n = 1 and 2 [2]. Herman and Rubin obtained general expressions for the matrix elements [3]. However, these expressions have never gain broad acceptance because of their cumbersome formulation [10]. The algorithm of the analytical expressions is poor [8]. The analytical expressions involve computational difficulties of significant rounding

P.M.A. Sloot et al. (Eds.): ICCS 2003, LNCS 2658, pp. 350–356, 2003.

off errors due to near cancellation terms in the summations [3, 5, 7]. Too much calculation is required for only some matrix elements of analytical expressions [5, 6, 8]. Only a few expressions for the diagonal [4, 5, 7] and off diagonal matrix elements [4, 8] among the published analytical matrix elements are more convenient than an integral method. Thus, a numerical integration method was thought more convenient than the analytical expressions to calculate the matrix elements [11]. On the other hand, accuracy of the numerical integration will deteriorate with increase of m, n and v [6]. We have previously derived compact expressions of the matrix elements $<v|q^n|0>$ for n • 6 without summations [12] and used in a simulation program of CH stretching methyl internal rotation [13]. But, in that paper, those expressions are limited to the calculation with the ground state.

In this paper, we derive general expressions for the diagonal and off diagonal matrix elements $<v+m|q^n|m>$ for n • 6, which are used in our spectroscopic calculations. To keep the expressions compact and calculation fast, intermediate variables are introduced. For verification, these analytical matrix elements are compared with those from numerical integrations of the analytical wave functions.

2 Morse Oscillator

The Schrödinger equation associated to the Morse oscillator is given by [1]

$$-\frac{\hbar^2}{2\mu}\frac{d^2\Psi(q)}{dq^2}+D(1-e^{-aq})^2\Psi(q)=E\Psi(q) \ . \tag{2}$$

The vibrational energy levels are given by [1]

$$E_v=(v+1/2)\hbar\omega-(v+1/2)^2\hbar x \ , \tag{3}$$

where v is the quantum number, ω is the frequency and x is the anharmonicity, which can be expressed with the parameters D and a. Either the parameter pair (ω, x) or (D, a) uniquely determines a Morse oscillator. The frequency and anharmonicity are usually obtained from the observed transitions fitted to the Birge-Sponer equation,

$$\nu_{v\leftarrow 0}/v=(\omega+x)+vx \ . \tag{4}$$

Three intermediate parameters, $k=\omega/x$, $\beta=k-2v-1$ and $y=k\exp(-aq)$ are used in the wave function formulation. The orthonomal wave functions are non-degenerate and are given by [1]

$$|v>=N_v e^{-y/2}y^{\beta/2}L_v^{\beta}(y) \ , \tag{5}$$

where the normalisation constant is given by

$$N_v=[a\beta\Gamma(v+1)/\Gamma(k-v)]^{1/2} \ , \tag{6}$$

where • is the gamma function. In the solution, a generalised Laguerre polynomial is used, which is defined as [14]

$$L_v^\beta(y) = \sum_{i=0}^{v} \frac{(-1)^i \Gamma(\beta+v+1)}{i!(v-i)!\Gamma(\beta+i+1)} y^i \ . \tag{7}$$

Generalised Laguerre polynomials can be calculated from the following recurrences [14]

$$L_0^\beta(y) = 1 \ , \tag{8}$$

$$L_1^\beta(y) = 1+\beta-y \ , \tag{9}$$

$$vL_v^\beta(y) = (2v-1+\beta-y)L_{v-1}^\beta(y) - (v-1+\beta)L_{v-2}^\beta(y) \ . \tag{10}$$

3 Displacement Matrix Elements

The Morse matrix elements $<v+m|q^n|m>$ can be calculated with numerical integrations of the wave functions in Eq. (5) or with the analytical expressions [see ref 2 - 9]. The deriving procedure of the analytical matrix elements can be divided into two parts [4]. The matrix elements of y^λ are calculated first and then, the matrix elements of q^n are calculated with the relationship:

$$q^n = \left(\frac{-1}{a}\right)^n \frac{d^n}{d\lambda^n}\left(\frac{y}{k}\right)^\lambda \Bigg|_{\lambda=0} \ . \tag{11}$$

The matrix elements of y^λ were given as:

$$<v+m|\, y^\lambda \,|m> = \frac{N_0 N_v}{a} \sum_{i=0}^{m} \frac{\Gamma(k+\lambda-m-v-i-1)\Gamma(i-\lambda+1)\Gamma(i-\lambda+v+1)}{\Gamma(i+v+1)\Gamma(m-i+1)\Gamma(i+1)\Gamma(1-\lambda+v)\Gamma(1-\lambda-v)} \ . \tag{12}$$

In evaluating the off diagonal matrix elements $<v+m|q^n|m>$, where $v > 0$, both gamma functions $\Gamma(1-\lambda-v)$ and polygamma functions $\psi^{(i)}(1-\lambda-v)$ are used, where i is a non negative integer. They have poles at • = 0, which can be solved with the following two equations:

$$\psi(1-\lambda-v)/\Gamma(1-\lambda-v)\,|_{\lambda\to 0} = (-1)^v \Gamma(v) \ , \tag{13}$$

$$\psi(1-\lambda-v) = \psi(\lambda+v) + \pi \cot[(\lambda+v)\pi] \ . \tag{14}$$

The matrix elements of q^n are calculated by combining Eqs. (11) and (12). The number of terms in the matrix elements increases dramatically as n increases. To keep the expressions compact, the following intermediate variables are introduced

$$y_0 = \frac{(-1)^v}{v}\sqrt{\frac{(k-2m-1)(k-2m-2v-1)\Gamma(m+1)\Gamma(m+v+1)}{\Gamma(k-m)\Gamma(k-m-v)}} \ , \tag{15}$$

$$y_c = \Gamma(k-m-v-i-1)/\Gamma(m-i+1) \ , \tag{16}$$

$$y_{i+1} = \psi^{(i)}(k-m-v-i-1) - \Gamma(i+1)[i^{-i-1} + \sum_{j=v+1}^{j=v+i} j^{-i-1}] - \delta_{i0}\ln k \ . \tag{17}$$

The off diagonal matrix elements can then be expressed as

$$<v+m|q|m> = -y_0 a^{-1}\Gamma(k-m-v)/(k-2m-v-1)\Gamma(m+1) \ , \tag{18}$$

$$<v+m|q^2|m> = y_0 a^{-2}\sum_{i=0}^{m} y_c(2y_1) \ , \tag{19}$$

$$<v+m|q^3|m> = -y_0 a^{-3}\sum_{i=0}^{m} y_c(3y_1^2 + 3y_2 - \pi^2) \ , \tag{20}$$

$$<v+m|q^4|m> = y_0 a^{-4}\sum_{i=0}^{m} y_c(4y_1^3 + 12y_1y_2 + 4y_3 - 4y_1\pi^2) \ , \tag{21}$$

$$\begin{aligned}<v+m|q^5|m> = -y_0 a^{-5}\sum_{i=0}^{m} y_c(&5y_1^4 + 30y_1^2y_2 + 20y_1y_3 + 15y_2^2 + 5y_4 \\ &+ 10y_1^2\pi^2 + 10y_2\pi^2 - \pi^4)\end{aligned} \ , \tag{22}$$

$$\begin{aligned}<v+m|q^6|m> = y_0 a^{-6}\sum_{i=0}^{m} y_c(&60y_1^3y_2 + 60y_1^2y_3 + 90y_1y_2^2 + 30y_1y_4 + 60y_2y_3 \\ &+ 6y_1^5 + 6y_5 - 20y_1^3\pi^2 - 60y_1y_2\pi^2 - 20y_3\pi^2 + 6y_1\pi^4)\end{aligned} \ . \tag{23}$$

For the diagonal matrix elements the following intermediate variables are used,

$$x_d = (k-2m-1)\Gamma(m+1)/\Gamma(k-m) \ , \tag{24}$$

$$x_c = \Gamma(k-m-i-1)/\Gamma(m-i+1) \ , \tag{25}$$

$$x_i = \psi^{(i)}(k-m-i-1) - 2\Gamma(i+1)\sum_{j=1}^{i} j^{-i-1} - \delta_{i0}\ln(k) \ . \tag{26}$$

The diagonal matrix elements $<m|q^n|m>$ can thus be written as

$$<m|q|m>=-x_d a^{-1}\sum_{i=0}^{m} x_c x_0 \ , \tag{27}$$

$$<m|q^2|m>=x_d a^{-2}\sum_{i=0}^{m} x_c (x_0^2+x_1) \ , \tag{28}$$

$$<m|q^3|m>=-x_d a^{-3}\sum_{i=0}^{m} x_c (x_0^3+3x_0x_1+x_2) \ , \tag{29}$$

$$<m|q^4|m>=x_d a^{-4}\sum_{i=0}^{m} x_c (x_0^4+6x_0^2x_1+4x_0x_2+3x_1^2+x_3) \ , \tag{30}$$

$$\begin{aligned}<m|q^5|m>=-x_d a^{-5}\sum_{i=0}^{m} x_c (&x_0^5+10x_0^3x_1+10x_0^2x_2+15x_0x_1^2+5x_0x_3 \\ &+10x_1x_2+x_4)\end{aligned} \ , \tag{31}$$

$$\begin{aligned}<m|q^6|m>=x_d a^{-6}\sum_{i=0}^{m} x_c (&x_0^6+15x_0^4x_1+20x_0^3x_2+45x_0^2x_1^2+15x_0^2x_3 \\ &+60x_0x_1x_2+6x_0x_4+15x_1^3+15x_1x_3+10x_2^2+x_5)\end{aligned} \ . \tag{32}$$

The matrix elements with the ground state $<v|q^n|0>$ were derived previously [12] and are listed for comparison. With the following intermediate variables

$$y_0=(-1)^v\sqrt{(k-2v-1)\Gamma(v)\Gamma(k-v-1)/v(k-v-1)\Gamma(k-1)} \ , \tag{33}$$

$$y_{i+1}=\psi^{(i)}(k-v-1)+\psi^{(i)}(v)-(-1)^{(i)}\psi^{(i)}(1)-\delta_{i0}\ln k \ , \tag{34}$$

the off diagonal matrix elements are expressed as

$$<v|q|0>=-y_0 a^{-1} \ , \tag{35}$$

$$<v|q^2|0>=y_0 a^{-2}(2y_1) \ , \tag{36}$$

$$<v|q^3|0>=-y_0 a^{-3}(3y_1^2+3y_2-\pi^2) \ , \tag{37}$$

$$<v|q^4|0>=y_0 a^{-4}(4y_1^3+12y_1y_2+4y_3-4y_1\pi^2) \ , \tag{38}$$

$$\begin{aligned}<v|q^5|0>=-y_0 a^{-5}(&5y_1^4+30y_1^2y_2+20y_1y_3+15y_2^2+5y_4 \\ &+10y_1^2\pi^2+10y_2\pi^2-\pi^4)\end{aligned} \ , \tag{39}$$

$$<v\,|\,q^6\,|\,0>= y_0 a^{-6}(60y_1^3y_2 + 60y_1^2y_3 + 90y_1y_2^2 + 30y_1y_4 + 60y_2y_3 \\ + 6y_1^5 + 6y_5 - 20y_1^3\pi^2 - 60y_1y_2\pi^2 - 20y_3\pi^2 + 6y_1\pi^4) \quad . \tag{40}$$

The diagonal matrix elements are expressed with intermediate variables

$$x_i = \psi^{(i)}(k-1) - \delta_{i0}\ln(k) \ , \tag{41}$$

as

$$<0\,|\,q\,|\,0>= -a^{-1}x_0 \ , \tag{42}$$

$$<0\,|\,q^2\,|\,0>= a^{-2}(x_0^2 + x_1) \ , \tag{43}$$

$$<0\,|\,q^3\,|\,0>= -a^{-3}(x_0^3 + 3x_0x_1 + x_2) \ , \tag{44}$$

$$<0\,|\,q^4\,|\,0>= a^{-4}(x_0^4 + 6x_0^2x_1 + 4x_0x_2 + 3x_1^2 + x_3) \ , \tag{45}$$

$$<0\,|\,q^5\,|\,0>= -a^{-5}(x_0^5 + 10x_0^3x_1 + 10x_0^2x_2 + 15x_0x_1^2 + 5x_0x_3 + 10x_1x_2 + x_4) \ , \tag{46}$$

$$<0\,|\,q^6\,|\,0>= a^{-6}(x_0^6 + 15x_0^4x_1 + 20x_0^3x_2 + 45x_0^2x_1^2 + 15x_0^2x_3 \\ + 60x_0x_1x_2 + 6x_0x_4 + 15x_1^3 + 15x_1x_3 + 10x_2^2 + x_5) \quad . \tag{47}$$

The general expressions of the off diagonal matrix elements $<v+m|q^n|m>$ in Eqs. (18) – (23) and the matrix elements $<v|q^n|0>$ in Eqs. (35) – (40) have the same coefficients in the corresponding expressions, the summation only appearing in the general expressions. The same pattern occurs by comparing the diagonal matrix elements $<m|q^n|m>$ in Eqs. (27) – (32) with the matrix elements $<0|q^n|0>$ in Eqs. (42) – (47). The n is limited to 6 in our work, and can be extended further simply, however the term of the expression increases with the n increasing. The matrix elements $<v+m|q^n|m>$ are verified by numerical comparison with those from the analytical matrix elements $<v|q^n|0>$ and those from numerical integration. For example, the same result $<6|q^6|0>$ = 0.00000123459177 $Å^6$ is obtained from the Eqs. (23), (40) and numerical integration (with reduced mass 0.97959254204827 amu, frequency 2988.84 cm^{-1} and anharmonicity 51.59 cm^{-1}). However, the calculation with the numerical integration needs much more time than that with the analytical expressions. High accuracy is achieved in our expressions without the sign alternation in the summation, which causes the round off errors in previous expressions [4, 5, 7].

4 Conclusion

The analytical expressions of the Morse matrix elements $<v+m|q^n|m>$ for n • 6 have been derived and programmed with the input of the reduced mass, frequency and anharmonicity. Intermediate variables are employed to keep the expressions compact and computation fast. Our expressions are exact and verified with a numerical integration method. Compared with the previous analytical matrix elements, great accuracy, simplicity and fast computation are achieved with these expressions.

References

1. Morse, P. M.: Phys. Rev. 34, (1929) 57–64
2. Heaps, H. S., Herzberg, G.: Z. Phys. 133, (1952) 48–64
3. Herman, R., Rubin, R. J.: Astrophys. J. 121, (1955) 533–540
4. Sage, M. L.: Chem. Phys. 35, (1978) 375–380
5. Gallas, J. A. C.: Phys. Rev. A 21, (1980) 1829–1834
6. Sovkov, V. B., Ivanov, V. S.: Opt. Spectrosc. 59, (1986) 733–735
7. Requena, A., Piñeiro, A. L., Moreno, B.: Phys. Rev. A 34, (1986) 4380–4386
8. Nagaoka, M., Yamabe, T.: Phys. Rev. A. 38, (1988) 5408–5411
9. López-Piñeiro, A., Moreno, B.: Phys. Rev. A 41, (1990) 1444–1449
10. Ogilvie, J. F.: The Vibrational and Rotational Spectroscopy of Diatomic Molecules. Academic Press, San Diego, (1998)
11. Truhlar, D. G., Onda, K.: Phys. Rev. A 23, (1981) 973–974
12. Rong, Z.: Ph.D. Thesis, Otago University, (2002)
13. Rong Z., Kjaergaard, H. G.: J. Phys. Chem. A 106, (2002) 6242–6253
14. Abramowitz; M., Stegun, I. A.: Handbook of Mathematical Functions. Dover, New York, (1965).
15. Herzberg, G.: Spectra of Diatomic Molecules. D. Van Nostrand Company, Inc., New York, (1951), 501–581

Appendix: Polygamma Function

The polygamma functions are extensively used in the expressions. The asymptotic formula is given as follows [14]

$$\psi^{(i)}(z) = (-1)^{(i-1)} \left(\frac{\Gamma(i)}{z^i} + \frac{\Gamma(i+1)}{2z^{i+1}} + \sum_{j=1}^{\infty} B_{2j} \frac{\Gamma(2j+i)}{\Gamma(2j+1) z^{2j+i}} \right). \tag{48}$$

where B_{2j} are the Bernoulli numbers. For most molecules [15], k value or the argument z is large enough, so that only very few terms are required to gain a high accuracy.

Initial Value Semiclassical Approaches to Reactive and non Reactive Transition Probabilities

N. Faginas Lago and A. Laganà

Dipartimento di Chimica, Università di Perugia, Via Elce di Sotto, 8, 06123 Perugia, Italy

Abstract. Semiclassical initial value representations (IVR) of the S matrix elements for state to state transitions provide a practical way for incorporating quantum mechanical effects into the classical estimate of the transition probability of elementary chemical processes. This semiclassical IVR approach applies to the study of the reactive and non-reactive transitions in atom diatom collisions. As a case study the collinear H + Cl_2 reaction is considered.

1 Introduction

Molecular dynamics calculations are becoming a kind of universal tool to rationalize the behaviour of chemical processes. However, most often, they are performed using classical mechanics that can not reproduce quantum effects. Accordingly, highly inaccurate results are obtained when interference, tunneling and resonance effects are important. On the other hand, quantum calculations are exact but difficult to carry out especially when dealing with large systems. As an alternative to quantum techniques semiclassical approaches are often used. Semiclassical approaches retain the simplicity of classical mechanics (they are based on the integration of classical trajectories) while introducing quantum-like corrections by exploiting the information carried by the classical phase.

A key problem of the traditional semiclassical formulations is that they require the solution of a two point boundary problem and the numerical search for related root trajectories. This makes the search for semiclassical contributions to the transition probabilty cumbersome and unsuitable for parallel and distributed computing. In this paper has been adopted the more recent initial value representation (IVR) approach [1] that avoids the search for root trajectories by properly mapping the formulation of the S matrix elements into the space of the initial variables.

In order to more easily tackle the problem of dealing with reactive transitions on which the literature (see ref. [1], and references therein) is rather poor our investigation was focused on atom diatom collinear systems and on the possibility of using a unified approach for both reactive and non reactive transitions.

The paper is articulated as follows: in Section 2 the IVR semiclassical method is outlined; in Section 3 the specialization of the method for collinear systems is given; in Section 4 the H + Cl_2 reactive case study is illustrated; in Section

P.M.A. Sloot et al. (Eds.): ICCS 2003, LNCS 2658, pp. 357–365, 2003.

5 a comparison of exact quantum and IVR semiclassical product vibrational distributions is carried out.

2 The IVR Semiclassical S Matrix

The basic semiclassical approach [1] pivots on the definition of the matrix element of the time evolution operator (propagator) $K_{1\to 2}(t)$ that for the atom diatom case reads

$$K_{1\to 2}(t) \equiv \langle\psi_2|e^{-iHt/\hbar}|\psi_1\rangle = \int d\mathbf{x}_1 \int d\mathbf{x}_2 \psi_2^*(\mathbf{x}_2)\psi_1(\mathbf{x}_1)\langle\mathbf{x}_2|e^{-iHt/\hbar}|\mathbf{x}_1\rangle \tag{1}$$

which is the probability amplitude for a transition from the bound initial state 1 (described by the wavefunction ψ_1 in the coordinate $\mathbf{x}_1$ at time zero) to the bound state 2 (described by the wavefunction ψ_2 in the coordinate $\mathbf{x}_2$ at time t) for a system whose Hamiltonian is H. Using the standard semiclassical approximation for the coordinate representation of the propagator one has

$$K_{1\to 2}(t) = \sum_{roots} \int d\mathbf{x}_1 \int d\mathbf{x}_2 \psi_2(\mathbf{x}_2)^* \psi_1(\mathbf{x}_1) \times \left[(2\pi i\hbar)^F \left|\frac{\partial \mathbf{x}_2}{\partial \mathbf{P}_1}\right|\right]^{-1/2} e^{iS_t(\mathbf{x}_2,\mathbf{x}_1)/\hbar} \tag{2}$$

where F is the number of degrees of freedom and, P_i is the momentum of the ith state. $S_t(\mathbf{x}_2, \mathbf{x}_1)$ is the classical action associated with the trajectory connecting $\mathbf{x}_1$ to $\mathbf{x}_2$ in time t that can be formulated as

$$S_t(\mathbf{x}_2, \mathbf{x}_1) = \int_0^t dt' \mathbf{P}(t')\dot{\mathbf{x}}(t') - H(\mathbf{P}(t'), \mathbf{x}(t')) \tag{3}$$

where primed quantities are related to products and dotted quantities are time derivatives.

The Jacobian factor in Eq. (2)

$$\left|\frac{\partial \mathbf{x}_2}{\partial \mathbf{P}_1}\right| = \left|\frac{\partial \mathbf{x}_t(\mathbf{P}_1, \mathbf{x}_1)}{\partial \mathbf{P}_1}\right|$$

is evaluated at the roots, that is at the solutions of the nonlinear two boundary value problem

$$\mathbf{x}_t(\mathbf{P}_1, \mathbf{x}_1) = \mathbf{x}_2. \tag{4}$$

As already mentioned, in general, there are multiple roots since $\mathbf{x}_t$ needs not to be a monotonic function of $\mathbf{P}_1$ and the summation in Eq. (2) is over all such roots. This process is often unstable (especially when both reactive and non reactive events can take place) and disrupts concurrency in trajectory calculations.

To eliminate the need for carrying out the root finding process, the IVR approach maps the final position independent variable $\mathbf{x}_2$ into the initial variable $\mathbf{P}_1$ and evaluates numerically the integrals of Eq. (2). In addition, the propagator in the Cartesian coordinates or momentum representation [3] is substituted by its semiclassical formulation. This leads to the expression

$$K_{1\to 2}(t) = \int d\mathbf{x}_1 \int d\mathbf{P}_1 \left[\left| \frac{\partial \mathbf{x}_t(\mathbf{x}_1, \mathbf{P}_1)}{\partial \mathbf{P}_1} \right| / (2\pi i\hbar)^F \right]^{1/2} e^{iS_t(\mathbf{x}_1,\mathbf{P}_1)/\hbar} \psi_2(\mathbf{x}_t)^* \psi_1(\mathbf{x}_1) \tag{5}$$

in which $S_t(\mathbf{x}_1, \mathbf{P}_1) \equiv S_t(\mathbf{x}_t(\mathbf{x}_1, \mathbf{P}_1), \mathbf{x}_1)$, $\mathbf{x}_2(\mathbf{x}_1, \mathbf{P}_1)$ is written in a more general form as $\mathbf{x}_t(\mathbf{x}_1, \mathbf{P}_1)$, the double ended boundary condition of Eq. (4) has been replaced by an integral over the phase space of the initial conditions and the Jacobian factor $(\partial \mathbf{x}_t/\partial \mathbf{P}_1)$ appears in the numerator (rather than in the denominator as is in Eq. (2)) avoiding so far possible singularities.

3 The Collinear Atom Diatom Case Study

To carry out the numerical evaluation of the semiclassical **S** matrix we have simplified the problem by considering the atom diatom collinear processes of the type A + BC $(\nu_i) \to$ AB (ν_f) + C (reactive) and A + BC $(\nu_i) \to$ A + BC (ν_f) (non reactive). The classical collinear atom diatom Hamiltonian function reads as

$$H = \frac{1}{2\mu_{A,BC}} P_R^2 + \frac{1}{2\mu_{BC}} P_r^2 + V(R, r) \tag{6}$$

where $V(R, r)$ is the potential of the system R and r the reactant Jacobi coordinates, P_R and P_r are the related momenta, μ_{BC} and $\mu_{A,BC}$ the reduced masses of the diatom and of the atom-diatom system. Classical trajectories were computed by integrating the following set of Hamilton equations

$$\dot{R} = P_R/\mu_{A,BC} \tag{7}$$

$$\dot{r} = P_r/\mu_{BC} \tag{8}$$

$$\dot{P}_R = -\partial V/\partial R \tag{9}$$

$$\dot{P}_r = -\partial V/\partial r \tag{10}$$

(where $\dot{R}$, $\dot{P}_R$, $\dot{r}$ and $\dot{P}_r$ are the time derivatives of R, P_R, r and P_r, respectively) using a fourth order Runge-Kutta method and adjusting the stepsize to guarantee energy conservation and collision time constance.

To compute the phase associated with each trajectory, the following fifth differential equation was integrated

$$\dot{S}_t = -\omega\dot{n} - R\dot{P}_R \tag{11}$$

where S_t is the already mentioned classical action formulated in terms of the action angle variables J and ω (defined as $\omega = \int \partial P_r(J)/\partial J dr$ and $2n\left(n+\frac{1}{2}\right)\hbar = J = \oint P_r dr$ [4] with n being the classical vibrational number continuously varying through the integer values of its discrete quantum analogue ν and ω being the oscillator phase angle).

The collision is started at a value of R sufficiently large (R_0) that makes the atom diatom interaction negligible. The quantum number ν_i for the transition $1 \rightarrow 2$ determines the initial value of the classical number n_i and hence the initial value of P_R (P_{R_0}) by energy conservation. The initial value of ω (ω_0), to which corresponds uniquely a value of r (r_0), is chosen on a regular grid in the range [-0.5 - 0.5]. Thus the initial boundary conditions are

$$\begin{aligned} R_0 &= Large \\ P_{R_0} &= -\{2\mu_{A,BC}[E-\varepsilon_{n_i}]\}^{1/2} \\ r_0 &= r(\omega_0, n_i) \\ P_{r_0} &= P_r(\omega_0, n_i) \end{aligned} \tag{12}$$

where E is the total energy of the system and ε_{n_i} the initial vibrational energy of the diatom in the state $n_i = \nu_i$.

In the procedure illustrated above, the use of action angle variables is motivated by the need of carrying out a proper selection of initial conditions that is more satisfactory if the oscillator phase is sampled uniformly. For this reason, our computational procedure first selects the initial values of the action angle variables of state 1 on a regular grid, then carries out the integration of the Hamilton equations using the Jacobi coordinates and finally transforms back all quantities into the action angle variables of state 2.

For our calculations we have considered the diatoms as a Morse oscillators, since this makes it possible to work out analytically the relationships linking the action angle variables to the other variables.

The IVR formulation of the semiclassical **S** matrix can then be obtained by expressing it in terms of the propagator and integrating over time t. The IVR S matrix element ($S^{IVR}_{1\rightarrow 2}(E)$) then reads [5]

$$\begin{aligned} S^{IVR}_{1\rightarrow 2}(E) = -e^{-i(k_1R_1+k_2R_2)} \int dP_{r_0} \int dr_0 \int dP_{R_0} \\ \times \left[\left|\frac{\partial(r_t, R_t)}{\partial(P_{r_0}, P_{R_0})}\right| / (2\pi i\hbar)^F\right]^{1/2} e^{i[Et+S_t(P_{r_0}, r_0, P_{R_0}, R_0)]/\hbar} \\ \psi_2(r_t)\psi_1(r_0)\hbar\sqrt{k_1k_2}/P_{R_t} \end{aligned} \tag{13}$$

with μ being the reduced mass of the system and R_t and r_t the value of the Jacobi coordinates obtained at time t by integrating Eqs. 12 starting from the initial conditions R_0, r_0,P_{R_0} and P_{r_0}. The two $k_j(j=1,2)$ factors are the initial and final wave numbers defined as

$$k_j = \sqrt{2\mu(E-\varepsilon_j)/\hbar^2}. \tag{14}$$

Finally, the detailed probability $P_{1\to2}(E)$ of the 1 to 2 transition is worked out as usual from the related $S_{1\to2}^{IVR}(E)$ element by taking its square modulus

$$P_{1\to2}(E) = |S_{1\to2}^{IVR}(E)|^2. \tag{15}$$

4 The H + Cl_2 Case Study

The potential V adopted for the H + Cl_2 system is of the LEPS type

$$V(r_{AB}, r_{BC}, r_{AC}) = \sum_{i=1}^{3} Q_i - \left\{ \sum_{i\le j}^{3} (J_i - J_j)^2/2 \right\}^{1/2}$$

with

$$Q_i = \left\{ {}^1E_i - \frac{1-\Delta_i}{1+\Delta_i} {}^3E_i \right\}/2$$

$$J_i = \left\{ {}^1E_i + \frac{1-\Delta_i}{1+\Delta_i} {}^3E_i \right\}/2$$

$${}^1E_i = D_i \exp[-\beta_i(r_i - r_i^e)]\{\exp[-\beta_i(r_i - r_i^e)] - 2\}$$

$${}^3E_i = \frac{1}{2} D_i \exp[-\beta_i(r_i - r_i^e)]\{\exp[-\beta_i(r_i - r_i^e)] + 2\}$$

where the index i extends over all diatomic pairs (AB, BC and AC) and r_i is the interatomic distance of the ith diatom. The value of the parameters D_i, r_i^e, β_i and Δ_i (for the H + Cl_2 system [6]) are given in Table 1.

Table 1. Value of the LEPS parameters for HCl and Cl_2

	$\beta_i(\AA^{-1})$	$r_i^e(\AA)$	Δ_i	D_i(eV)
HCl	1.86932	1.273200	0.067000	4.625970
Cl_2	2.008020	1.999800	-0.113000	2.516970

Among practical advantages of using the LEPS potential is the fact that the asymptotic diatom quantum eigenenergies have the closed form expression

$$\varepsilon_\nu = -D\left[1 - \frac{\hbar\beta}{(2\mu D)^{1/2}}\left(\nu + \frac{1}{2}\right)\right]^2 \tag{16}$$

where, as usual, ν is the vibrational quantum number and the angle variable ω is related to the internuclear distance r by the relationship

$$r - r^e = \frac{1}{\beta}\ln\left[\frac{1-(1+\varepsilon_\nu/D)^{1/2}\cos(2\pi\omega)}{(-\varepsilon_\nu/D)}\right]. \tag{17}$$

The vibrational wave function $\phi_\nu(r)$ has the following form

$$\phi_\nu(r) = N_\nu e^{-z/2} z^{c/2} L_\nu^c(z) \tag{18}$$

where

$$\begin{aligned} N_\nu &= [\beta c\nu!/\Gamma(k-\nu)]^{1/2}, \\ c &= k - 2\nu - 1, \\ k &= 2(2\mu_{BC}D)^{1/2}/\beta\hbar \\ z &= ke^{-\beta(r-r^e)} \end{aligned} \tag{19}$$

with $L_\nu^c(z)$ being the Laguerre polynomial and $\Gamma(i)$ the Gamma function [7].

5 Product Vibrational Distributions

As already mentioned, non reactive processes are the most consolidated case study for semiclassical approaches and the most relevant for Molecular Dynamics calculations. The peculiarity of the case considered here is that the LEPS potential includes the reactive channel and, therefore, it highly distorts the triatom even when reaction does not occur and when reactive processes coexist with non reactive ones. This makes of particular interest the study of the near the threshold behaviour. A preliminary study of the results obtained for the H + Cl_2 system is given in [8] where an analysis of the convergence with the number of integrated trajectories is performed. Here, an extended analysis of the accuracy of state to state IVR semiclassical probabilities for both reactive and non reactive processes is given.
The accuracy of semiclassical results is tested against quantum results by comparing the structure of the calculated product vibrational distribution (PVD). For illustrative purposes we report here also quasiclassical values determined by assigning the classical result to the closest quantum state and then taking the ratio between the number of trajectories assigned to the considered quantum final state and the total number of integrated trajectories.

In Fig. 1 the non reactive PVDs calculated for values of the collision energy (E_{tr}) extending up to 1.26 kcal/mol (the threshold energy) are given. As apparent from the figure quasiclassical probabilities are always 1 for the elastic transitions and 0 for the inelastic ones. On the contrary, semiclassical IVR values (empty squares connected by dashed dotted lines) reproduce much better the structure of quantum (solid diamonds connected by dashed lines) probabilities. The deviation of semiclassical IVR from quantum results of the PVD does not usually exceed 10% although there is no clear systematic for the related error.

At energies above the threshold the product channel becomes accessible and the probability of populating some product vibrational states differs from zero. Reactive state to state semiclassical IVR probabilities calculated at $\nu_i = 0$ (upper row) and $\nu_i = 1$ (lower row) are plotted (dashed dotted line) in Fig. 2 as product vibrational distributions (as a function of ((ν_f) at increasing value of the total energy (from the left hand side to the right hand side). For comparison also exact quantum (dashed line) and quasiclassical ones (solid line) are shown in the same figure.

The figure shows that differences between reactive probabilities calculated using different methods can be quite large. In particular, the deviation of quasiclassical results is significant when product vibrational distributions show a definite structure (like the bimodal shape obtained at $\nu_i = 1$). On the contrary, the agreement between semiclassical IVR and quantum PVDs is definitely better in particular because they reproduced the related structure is when there is more than one maximum.

6 Conclusions

The significant progress made by Molecular Dynamics is prompting the introduction of more accuracy in calculating state to state transition probabilities when classical mechanics dynamical techniques are used. In this paper numerical tests of the traditional way of building quantum corrections into classical mechanics approaches through the calculation of the classical phase associated with the trajectories described by the system and the evaluation from it of semiclassical S matrix elements have been discussed. In particular, the recently proposed IVR formulation of the semiclassical S matrix has been considered because of its straightfarword concurrent implementation. Results of the test are extremely encouraging and plans are being made for its implementation on metacomputer and grid platforms.

Acknowledgments.

This research has been financially supported by MIUR, ASI and CNR (Italy) and COST in Chemistry Action D23 (Europe).

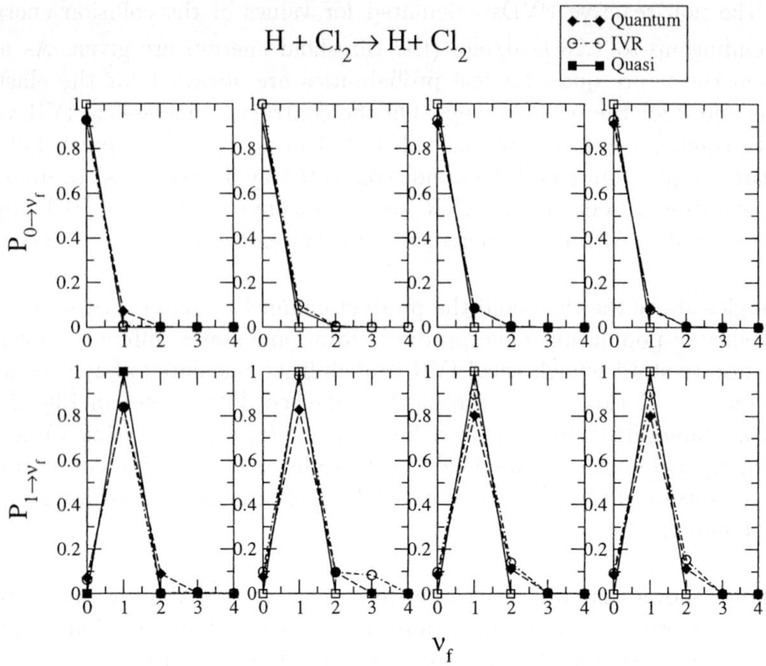

Fig. 1. PVDs of the semiclassical IVR inon reactive probabilities (empty squares connected by dashed-dotted lines $P_{0\rightarrow\nu'}(E_{tr})$ (upper panel) and $P_{1\rightarrow\nu'}(E_{tr})$ (lower panel) calculated at $\nu_i = 0$ (upper panel) and $\nu_i = 1$ (lower panel) at $E_{tr} = 0.02$, 0.03, 0.052 and 1.26 kcal mol^{-1} (from the left hand side to the right hand side)for the H + $Cl_2(\nu_i) \rightarrow HCl(\nu_f)$ + Cl. For comparison also quantum (diamonds connected by dashed lines) and quasiclassical (solid squares connected by solid lines) results are shown.

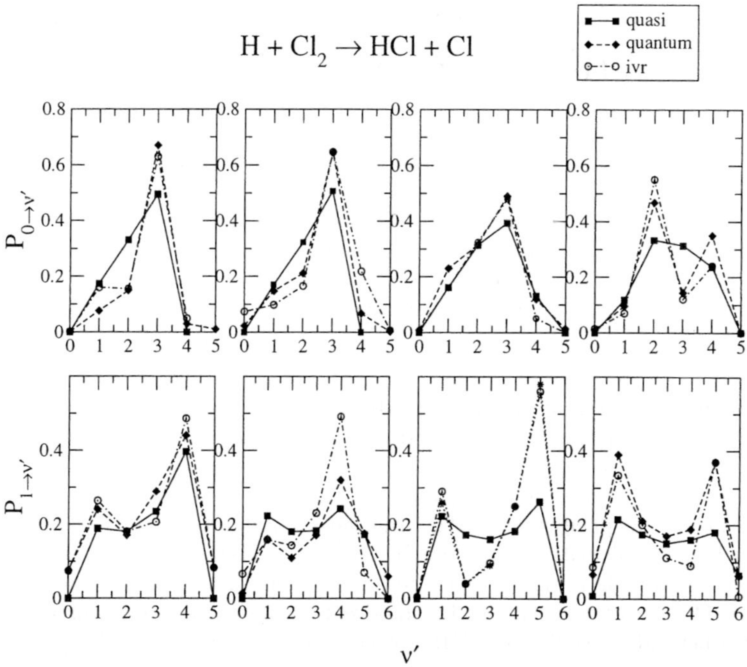

Fig. 2. PVDs of the semiclassical IVR reactive probabilities (empty squares connected by dashed-dotted lines $P_{0\to\nu'}(E_{tr})$ (upper panel) and $P_{1\to\nu'}(E_{tr})$ (lower panel) calculated at $\nu_i = 0$ (upper panel) and $\nu_i = 1$ (lower panel) at $E_{tr} = 3, 4, 5$ and 6 kcal mol^{-1} (from the left hand side to the right hand side) for the H + $Cl_2(\nu_i) \to HCl(\nu_f)$ + Cl. For comparison also quantum (diamonds connected by dashed lines) and quasiclassical (solid squares connected by solid lines) results are shown.

References

1. Miller, W. H., *J. Phys. Chem. A* **105**, 2942 (2001).
2. Miller, W. H., *J. Chem. Phys.* **53**, 3678 (1970).
3. Miller, W. H. and T. F. George, *J. Chem. Phys.* **56**, 5668 (1972).
4. Goldstein, H., *Classical Mechanics*, Addison-Wesley, Massachusetts, 1970.
5. Skinner, D. E and Miller, W. H., *Chem. Phys. Lett.* **300**, 20 (1999).
6. Laganà, A., *Gazzetta Chimica Italiana* **111**, 459 (1981).
7. Abramowitz, M. and Stegun, I. A., *Handbook of Mathematical Functions*, Dover, 1970.
8. Faginas Lago, N., *PhD Thesis, University of Perugia*, Perugia (2002)

Theoretical Investigations on the Reactions of $C_6H_5^+$ and $C_{10}H_7^+$ with D_2

Marco Di Stefano, Marzio Rosi, and Antonio Sgamellotti

CNR Institute of Molecular Sciences and Technologies, c/o
Department of Chemistry, University of Perugia, Via Elce di Sotto 8,
06123 Perugia, Italy
marco@thch.unipg.it

Abstract. The recent discovery of benzene and polycyclic aromatic hydrocarbons (PAHs) in the interstellar clouds has increased the interest in this entire class of molecules, which result to be the object of several theoretical and experimental investigations. In this work, we have studied the reactions of the phenylium, $C_6H_5^+$, and the naphtylium cation, $C_{10}H_7^+$, with D_2. All calculations have been carried out at B3LYP/6-31G* level of theory and they suggest that the addition products, $C_6H_5D_2^+$ and $C_{10}H_7D_2^+$, once formed, can be stabilized by unreactive collisions or lose a neutral molecule (H_2 or HD) or an atom (H or D), respectively.

1 Introduction

Polycyclic aromatic hydrocarbons (PAHs) have been postulated to be an important, ubiquitous component of the interstellar medium based on their high stability against UV radiation. Many members of this class have been already studied [2] [3] [4], even though most of the available information principally concerns the smallest terms of this class, like benzene [5] and naphtalene [6]. By experimental investigations, it has been demonstrated that PAHs are expected to be present as a mixture of free, neutral and ionized gas phase molecules [7] following a large size distribution from small systems (less than 25 carbon atoms) to large graphitic plateles. They can be present both in their neutral and/or ionized forms and are thought to be the origin of a variety of interstellar infrared emission features observed in the 700 to 3100 cm^{-1} range. Recent experimental studies of reactions of the PAHs cations $C_6H_6^+$ and $C_{10}H_8^+$ with H atoms [1] [8] [9] have shown that association channels are dominant and they lead to observable quantities of protonated PAHs. In the same studies, it was also noticed that the cations $C_6H_5^+$ and $C_{10}H_7^+$ did not react with atomic hydrogen whereas they showed particular reactivity towards the hydrogen molecules with a supposed association mechanism [1] [8] [9]. Such kind of reactions, in which PAHs and atomic or molecular hydrogen are involved, are generally postulated to be of fundamental importance in numerous processes which can take place in the interstellar medium [3].

In this paper, we will show the results of our theoretical investigations on the reactions $[C_6H_5^+ + D_2]$ and $[C_{10}H_7^+ + D_2]$. The direct addition of D_2 to

P.M.A. Sloot et al. (Eds.): ICCS 2003, LNCS 2658, pp. 366–375, 2003.

the cations of interest immediately leads to the formation of the intermediates $C_6H_5D_2^+$ and $C_{10}H_7D_2^+$ and both these processes result to be strongly exothermic. These addition complexes may either be stabilized by unreactive collisions or dissociate by losing H_2 (or HD) in the case of $C_6H_5D_2^+$ and H (or D) for the latter cation [10].

2 Theoretical Methods

Calculations have been carried out with the B3LYP hybrid exchange correlation functional [11] as implemented in GAUSSIAN98 [12] in conjunction with the 6-31G* basis set [13]. For all the species involved in the investigated reactions, we have optimized their geometries and computed their harmonic frequencies, in order to confirm the presence of either a minimum or a saddle point. All thermochemical calculations have been carried out at 298.15 K and 1 atm by adding, for each species, the zero point correction and the thermal corrections to enthalpy and Gibbs free energy to the calculated B3LYP values. The zero point correction to energy has been considered so that we could evaluate the thermochemistry of our reactions in conditions which are closer to the interstellar ones, since they are characterized by much lower values of temperature.

3 $C_6H_5^+ + D_2$

The relative abundant literature information on the existence and reactivity of the phenylium cation, $C_6H_5^+$, indicates that it can be generated by electron impact on benzene [14] and two isomeric structures are known to exist for this cation. They are thought to be the cyclic and linear isomers which are readily distinguished by their different reactivities. While linear $C_6H_5^+$ reacts neither with H_2 nor with H, a reaction between H_2 and cyclic $C_6H_5^+$ was observed in several studies [14] [15]. The reaction with D_2 was already experimentally investigated by Sablier et al. [16]. The cyclic $C_6H_5^+$ can exist both in a triplet and a singlet state and we calculated the latter to be more stable by 0.81 eV. In Figure 1, we report a schematic representation of the potential energy surface for the association of D_2 to the phenylium cation and two different mechanisms may be hypothesized.

According to the results on the $[C_6H_5 + H_2]$ reaction [17], the process of interest with D_2 should first proceed by the deuterium abstraction mechanism to form $[C_6H_5D^+ + D]$; successively, the deuterium addition to $C_6H_5D^+$ would give the intermediate of interest. The triplet state $C_6H_5^+$ can add the deuterium molecule with this proposed mechanism. For the abstraction step, the transition state found on the potential energy surface gives rise to a barrier height of 0.13 eV. The second D atom approaches the $C_6H_5D^+$ intermediate out of the plane of the ring and this addition is exothermic by 3.41 eV and barrierless [18] [19]. This result is understandable in terms of the electronic structures of the ions involved. The radical cations, with their highly reactive, open-shell structures, readily add a deuterium atom to produce the intermediate $C_6H_5D_2^+$ with a

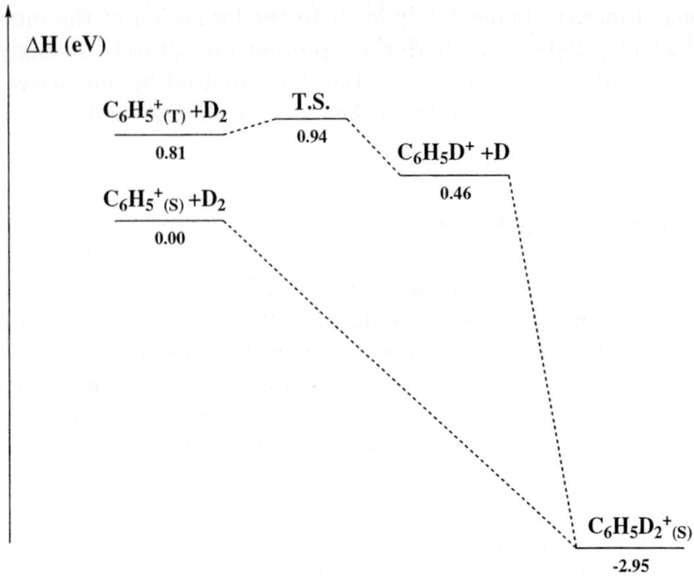

Fig. 1. Schematic representation of the [$C_6H_5^+$ + D_2 → $C_6H_5D_2^+$] reaction at B3LYP/6-31G* level of theory. (S) and (T) mean singlet and triplet state

more favourable closed-shell electron configuration (singlet state, see Fig. 1). The resultant cation is far less reactive with D (or H) atoms because it already has the preferred closed-shell electron structure and the addition of a second atom would brake this favourable configuration. The singlet isomer does not exibit the same behaviour and this result is rather foreseeable, since the abstraction step is endothermic of 0.46 eV. It will prefer adding D_2 in a direct 1,1-association mechanism. This addition can be rationalized as an insertion process involving the D-D bond and the vacant sp^2 orbital on the ipso carbon of the phenylium ion. Furthermore, it does not require any energetic barrier and this result is also supported by a similar investigation carried out on the 1,1-elimination process [18]. The exothermicity for this mechanism is computed to be 2.95 eV.

The intermediate $C_6H_5D_2^+$, once formed, can either be stabilized by unreactive collisions with D_2 or form $C_6H_3D_2^+$ and $C_6H_4D^+$ by losing either H_2 or HD, or, finally, it could reform the reactants by releasing D_2 [10]. So as to rationalize the experimental data, we focussed on the fate of the $C_6H_5D_2^+$ intermediate. In Figure 2, we show the potential energy surface of the $C_6H_5D_2^+$ intermediate at B3LYP/6-31G* level of theory.

By considering the cation $C_6H_5D_2{}^+{}_a$, which is the species directly formed from the phenylium ion according to the mechanism previously described, a 1-2 deuterium shift can be observed, with the consequent formation of the intermediate $C_6H_5D_2{}^+{}_b$. The barrier is 0.55 eV at B3LYP/6-31G* level of theory. $C_6H_5D_2{}^+{}_b$ evolves in different manners; firstly, it can lose a HD molecule in a barrierless process and form $C_6H_4D^+$ (ΔH= 2.92 eV). Alternatively, by a fur-

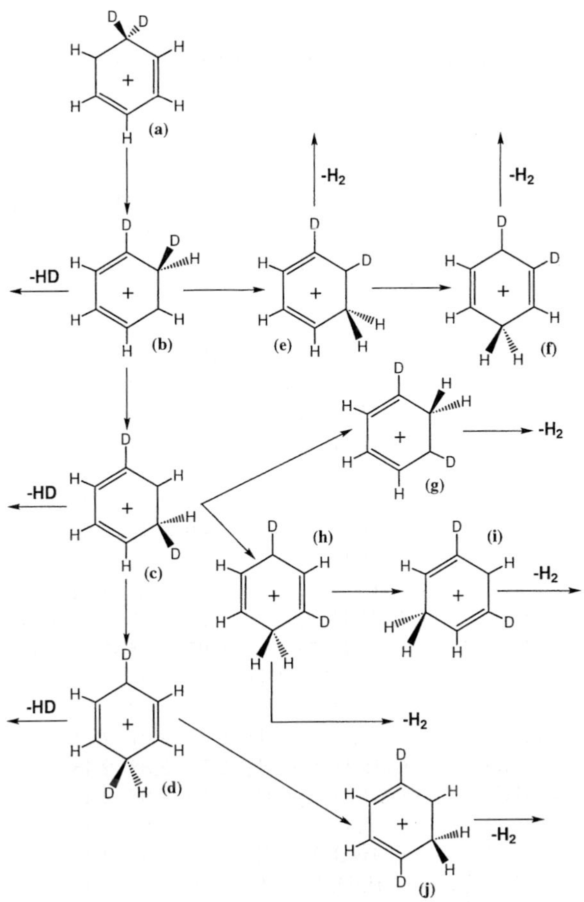

Fig. 2. Schematic representation of the $C_6H_5D_2^+$ potential energy surface at B3LYP/6-31G* level of theory

ther shift of a D atom, it can evolve into $C_6H_5D_2{}^+{}_c$ with an activation barrier of 0.55 eV. Finally, a hydrogen shift can be considered with the formation of $C_6H_5D_2{}^+{}_e$ and the energetic barrier of this third process is 0.52 eV. At this point, $C_6H_5D_2{}^+{}_e$ will be able either to lose a H_2 molecule forming $C_6H_3D_2^+$ (ΔH= 2.88 eV) or to shift a further H atom to give $C_6H_5D_2{}^+{}_f$. In Table 1, we present the energetics for all the reactions drawn in Figure 2, reporting their ΔE_0, ΔH, ΔG and E_a values at B3LYP/6-31G* level of theory.

By comparing the thermochemical values for the loss of HD and H_2 from the intermediate $C_6H_5D_2^+$ with the exothermicity of the association reaction [$C_6H_5^+ + D_2$] (2.95 eV), the production of $C_6H_4D^+$ and $C_6H_3D_2^+$ results to be exothermic by 0.03 eV and 0.07 eV with respect to the reactants, explaining, thus, their presence in the experimental apparatus [10]. However, it is also worth noticing that all the activation barriers for the isomerization processes lie much lower in energy than the release of a neutral molecule. Therefore our calculations can rationalize why in the experimental analysis large concentrations of $C_6H_5D_2^+$ were detected together with lower quantities of $C_6H_4D^+$ and $C_6H_3D_2^+$ [10].

Table 1. Thermochemical parameters (eV) calculated at B3LYP/6-31G* level of theory for the processes described in Figure 2. ΔH, ΔG and E_a values have been calculated at the temperature of 298.15 K. ΔE_0, ΔH and ΔG for the isomerization processes are in meV.

Reaction	ΔE_0	ΔH	ΔG	E_a
${C_6H_5D_2^+}_a \rightarrow {C_6H_5D_2^+}_b$	-5.09	-4.90	-5.42	0.55
${C_6H_5D_2^+}_b \rightarrow {C_6H_5D_2^+}_c$	-1.42	-1.61	-1.18	0.55
${C_6H_5D_2^+}_c \rightarrow {C_6H_5D_2^+}_d$	0.82	0.84	0.49	0.55
${C_6H_5D_2^+}_b \rightarrow {C_6H_5D_2^+}_e$	-7.54	-8.54	-6.20	0.52
${C_6H_5D_2^+}_e \rightarrow {C_6H_5D_2^+}_f$	-0.60	-0.76	-0.68	0.53
${C_6H_5D_2^+}_c \rightarrow {C_6H_5D_2^+}_g$	-4.71	-4.49	-3.84	0.52
${C_6H_5D_2^+}_c \rightarrow {C_6H_5D_2^+}_h$	-5.31	-5.25	-4.54	0.52
${C_6H_5D_2^+}_h \rightarrow {C_6H_5D_2^+}_i$	-2.20	-2.45	-1.66	0.53
${C_6H_5D_2^+}_d \rightarrow {C_6H_5D_2^+}_j$	-6.91	-6.94	-5.50	0.52
${C_6H_5D_2^+}_b \rightarrow C_6H_4D^+ + HD$	2.83	2.92	2.51	-
${C_6H_5D_2^+}_c \rightarrow C_6H_4D^+ + HD$	2.83	2.92	2.51	-
${C_6H_5D_2^+}_d \rightarrow C_6H_4D^+ + HD$	2.83	2.91	2.51	-
${C_6H_5D_2^+}_e \rightarrow C_6H_3D_2^+ + H_2$	2.79	2.88	2.48	-
${C_6H_5D_2^+}_f \rightarrow C_6H_3D_2^+ + H_2$	2.79	2.87	2.48	-
${C_6H_5D_2^+}_g \rightarrow C_6H_3D_2^+ + H_2$	2.79	2.88	2.48	-
${C_6H_5D_2^+}_h \rightarrow C_6H_3D_2^+ + H_2$	2.79	2.88	2.48	-
${C_6H_5D_2^+}_i \rightarrow C_6H_3D_2^+ + H_2$	2.79	2.88	2.48	-
${C_6H_5D_2^+}_j \rightarrow C_6H_3D_2^+ + H_2$	2.79	2.88	2.48	-

4 $C_{10}H_7^+ + D_2$

In analogy with $C_6H_5^+$, even the naphtylium cation, $C_{10}H_7^+$, is known to react with the H_2 molecule, whereas no reactions occur with atoms [8] [9]. Previous theoretical calculations on the naphtylium cation indicated this species is planar with the positive charge largely delocalized on the deprotonated ring [20] and our B3LYP/6-31G* calculations are in good agreement with these results. Then, we have been able to locate two different isomers of $C_{10}H_7^+$ and they can be indicated as $C_{10}H_7^+$-α and $C_{10}H_7^+$-β, where α and β denote the sites where the hydrogen atom has been abstracted (see Figure 3).

Although they result to be both planar, the abstraction of a H atom destroys the aromaticity of the naphtalene π system and they own only a C_S symmetry.

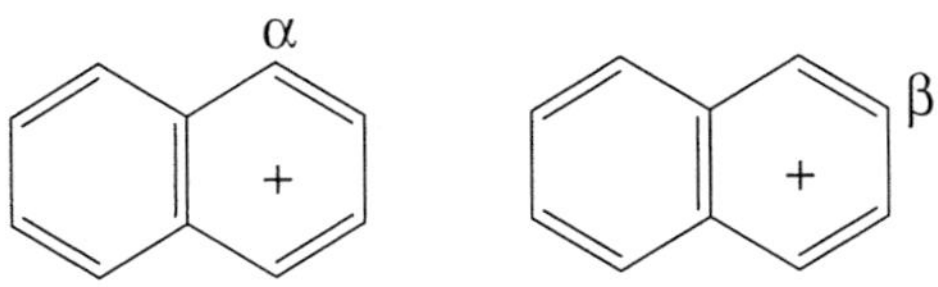

Fig. 3. Schematic representation of $C_{10}H_7{}^+$-α and $C_{10}H_7{}^+$-β

By a comparison of their energies, the $C_{10}H_7{}^+$-α isomer is more stable than β by nearly 0.058 eV and, by calculating the energies of the $C_{10}H_7D_2{}^+$-α and $C_{10}H_7D_2{}^+$-β intermediates, which can be directly originated from the previous reactants by D_2 addition, the α isomer results to be lower in energy by 0.13 eV. We can therefore assume that our process of interest will principally involve the α isomers. The association reaction [$C_{10}H_7{}^+$ + H_2] is known to occur under low density conditions [1] [8] [9]. Additional information is available on the reaction [$C_{10}H_8{}^+$ + H] [21]. Ho et al. [22] supposed the reaction [$C_{10}H_7{}^+ + H_2 \rightarrow C_{10}H_8{}^+$ + H] to be slightly endothermic, while our calculations suggest the same process to be exothermic by 0.34 eV. When D_2 reacts with the naphtylium cation, the intermediate $C_{10}H_7D_2{}^+$ is mainly produced with lower quantities of the cationic species $C_{10}H_7D^+$ and $C_{10}H_6D_2{}^+$ which could be formed by the further loss of a D or a H atom respectively [10].

On the D_2 addition to $C_{10}H_7{}^+$, we have optimized a transition state for the abstraction-addition mechanism (E_a= 0.17 eV), whereas the 1,1-addition is a barrierless process exothermic by 3.17 eV. In Figure 4, we report a schematic representation of the $C_{10}H_7D_2{}^+$ potential energy surface, where we concentrate only on the possible shifts of a deuterium atom around the molecular rings. Ten different structural isomers have been optimized and they all reveal comparable stabilities, except those cations in which the coordination of the deuterium atom involves the carbons in common with the two six-membered rings. These latest species have been indicated by the letters *e* and *f* and they can only lose the D atom to form the $C_{10}H_7D^+$ cations. At this point of the study, we focussed our attention only to the shifts involving the D atoms, without considering the same phenomenon in the case of the H atoms (see Figure 4).

As shown in Figure 4, by starting from $C_{10}H_7D_2{}^+{}_a$, which is our initial isomer, three different pathways can be followed. Firstly, we can consider the breaking of a C-D bond with the formation of $C_{10}H_7D^+$ and this reaction is endothermic by 2.82 eV. Alternatively, $C_{10}H_7D_2{}^+{}_b$ is formed: this reaction is slightly endothermic and it requires an activation barrier of 0.72 eV. The cation *b* can evolve by losing either a D (ΔH= 2.69 eV) or a H atom (ΔH= 2.61 eV) or give $C_{10}H_7D_2{}^+{}_c$. A third possible channel involves the isomerization $C_{10}H_7D_2{}^+{}_a \rightarrow C_{10}H_7D_2{}^+{}_f$ with an activation energy of 1.10 eV. In Table 2, we report the thermochemical parameters for all the processes drawn in Figure 4, considering both the activation barriers and the loss of the D and H atoms.

As we can observe in Table 2, the activation energies for the investigated isomerization processes result to be lower than the calculated values for the loss of the D and H atoms from the same intermediates. This result implies that $C_{10}H_7D_2^+$, once formed, can easily allow the atomic shifts all along its structure with the consequent formation of a remarkable number of isomers which differ one another for the position of the isotopically labelled atoms.

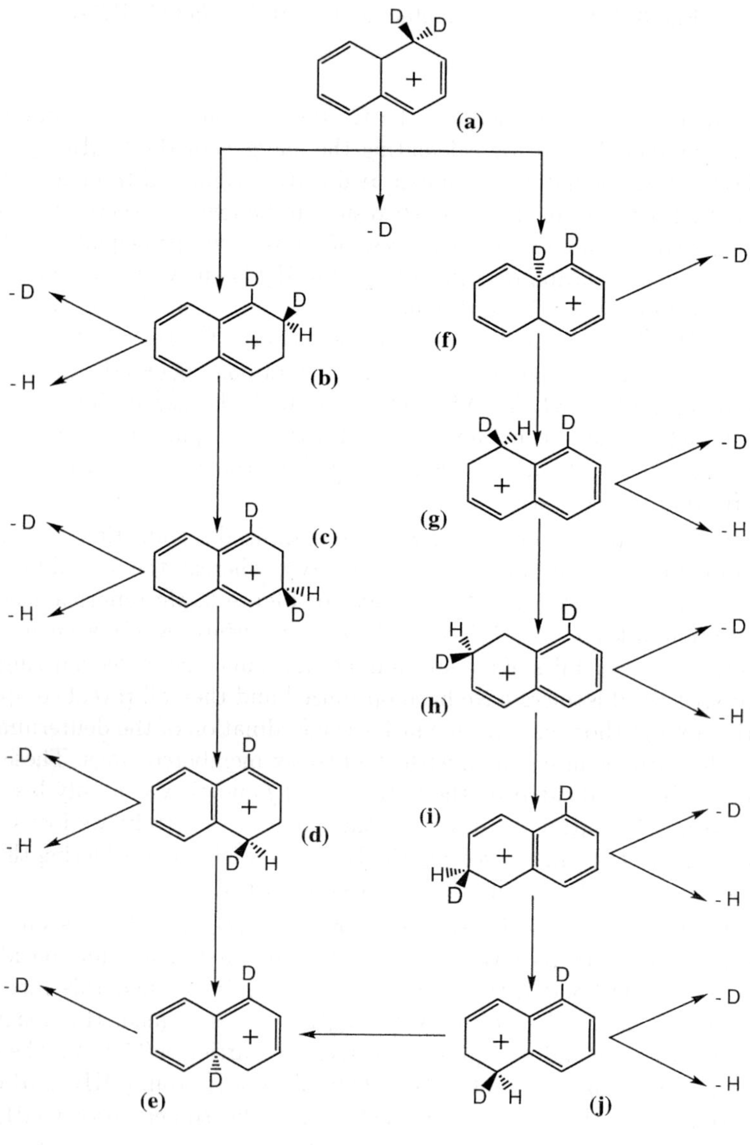

Fig. 4. Schematic representation of the $C_{10}H_7D_2^+$ potential energy surface at B3LYP/6-31G* level of theory

Table 2. Thermochemical parameters (eV) calculated at B3LYP/6-31G* level of theory for the processes described in Figure 4. ΔH, ΔG and E_a values have been calculated at the temperature of 298.15 K.

Reaction	ΔE_0	ΔH	ΔG	E_a
${C_{10}H_7D_2^+}_a \rightarrow {C_{10}H_7D_2^+}_b$	0.13	0.13	0.13	0.72
${C_{10}H_7D_2^+}_b \rightarrow {C_{10}H_7D_2^+}_c$	0.01	0.01	0.01	0.81
${C_{10}H_7D_2^+}_c \rightarrow {C_{10}H_7D_2^+}_d$	-0.13	-0.13	-0.14	0.59
${C_{10}H_7D_2^+}_d \rightarrow {C_{10}H_7D_2^+}_e$	0.88	0.88	0.89	1.11
${C_{10}H_7D_2^+}_a \rightarrow {C_{10}H_7D_2^+}_f$	0.88	0.88	0.89	1.10
${C_{10}H_7D_2^+}_f \rightarrow {C_{10}H_7D_2^+}_g$	-0.88	-0.88	-0.89	0.23
${C_{10}H_7D_2^+}_g \rightarrow {C_{10}H_7D_2^+}_h$	0.13	0.13	0.14	0.72
${C_{10}H_7D_2^+}_h \rightarrow {C_{10}H_7D_2^+}_i$	0.00	0.00	0.00	0.81
${C_{10}H_7D_2^+}_i \rightarrow {C_{10}H_7D_2^+}_j$	-0.13	-0.13	-0.14	0.59
${C_{10}H_7D_2^+}_j \rightarrow {C_{10}H_7D_2^+}_e$	0.88	0.88	0.89	1.11
${C_{10}H_7D_2^+}_a \rightarrow C_{10}H_7D^+ + D$	2.77	2.82	2.45	-
${C_{10}H_7D_2^+}_b \rightarrow C_{10}H_7D^+ + D$	2.64	2.69	2.32	-
${C_{10}H_7D_2^+}_b \rightarrow C_{10}H_6D_2^+ + H$	2.55	2.61	2.25	-
${C_{10}H_7D_2^+}_c \rightarrow C_{10}H_7D^+ + D$	2.64	2.69	2.32	-
${C_{10}H_7D_2^+}_c \rightarrow C_{10}H_6D_2^+ + H$	2.55	2.61	2.25	-
${C_{10}H_7D_2^+}_d \rightarrow C_{10}H_7D^+ + D$	2.78	2.83	2.46	-
${C_{10}H_7D_2^+}_d \rightarrow C_{10}H_6D_2^+ + H$	2.69	2.74	2.39	-
${C_{10}H_7D_2^+}_e \rightarrow C_{10}H_7D^+ + D$	1.89	1.95	1.57	-
${C_{10}H_7D_2^+}_f \rightarrow C_{10}H_7D^+ + D$	1.89	1.94	1.57	-
${C_{10}H_7D_2^+}_g \rightarrow C_{10}H_7D^+ + D$	2.77	2.83	2.46	-
${C_{10}H_7D_2^+}_g \rightarrow C_{10}H_6D_2^+ + H$	2.69	2.74	2.39	-
${C_{10}H_7D_2^+}_h \rightarrow C_{10}H_7D^+ + D$	2.64	2.69	2.32	-
${C_{10}H_7D_2^+}_h \rightarrow C_{10}H_6D_2^+ + H$	2.55	2.61	2.25	-
${C_{10}H_7D_2^+}_i \rightarrow C_{10}H_7D^+ + D$	2.64	2.69	2.32	-
${C_{10}H_7D_2^+}_i \rightarrow C_{10}H_6D_2^+ + H$	2.55	2.61	2.25	-
${C_{10}H_7D_2^+}_j \rightarrow C_{10}H_7D^+ + D$	2.77	2.83	2.46	-
${C_{10}H_7D_2^+}_j \rightarrow C_{10}H_6D_2^+ + H$	2.69	2.74	2.39	-

5 Conclusions

In this paper, we present a theoretical study at B3LYP/6-31G* level of theory of the reactions [$C_6H_5^+ + D_2$] and [$C_{10}H_7^+ + D_2$]. They both involve PAHs and the deuterium molecule, which are well-known constituents of the interstellar clouds. The 1,1-addition of the D_2 molecule to the phenylium and naphtylium cations does not require any activation energy and the products of this reactions are the experimentally revealed cations $C_6H_5D_2^+$ and $C_{10}H_7D_2^+$. The fate of these two species is different. From the experimental results on both systems [10], it was noticed that the intermediate $C_6H_5D_2^+$ could dissociate by losing either a HD or a H_2 molecule giving the cations $C_6H_4D^+$ and $C_6H_3D_2{}^+$, whereas the intermediate $C_{10}H_7D_2{}^+$ could lose a H atom forming $C_{10}H_6D_2{}^+$ or a D atom with the consequent synthesis of $C_{10}H_7D^+$. All our calculations rationalize these surveys and, in addiction, offer a more complete view of the possible reactive pathways which may be followed by both the systems of interest.

Acknowledgments.
The authors would like to thank Prof. Paolo Tosi and Prof. Davide Bassi of the University of Trento (Italy) for providing the experimental information prior to publication.

References

1. Scott, G.B.I., Fairley, D.A., Freeman, C.G., McEwan, M.J., Adams, N.G., Babcock, L.M.: $C_mH_n{}^+$ Reactions with H and H_2: An Experimental Study. J. Phys. Chem. A **101** (1997) 4973–4978
2. Anicich, V.G., McEwan, M.J.: Ion-molecule chemistry in Titan's ionosphere. Planet. Space Sci. **45** (1997) 897–921
3. Bohme, D.K.: PAH and Fullerene Ions and Ion/Molecule Reactions in Interstellar and Circumstellar Chemistry. Chem. Rev. **92** (1992) 1487–1508
4. Ehrenfreund, P., Charnley, S.B.: Organic Molecules in the Interstellar Medium, Comets and Meteorites: A Voyage from Dark Clouds to the Early Earth. Annu. Rev. Astron. Astrophys. **38** (2000) 427–483
5. Cernicharo, J., Heras, A.M., Tielens, G.G.M., Pardo, J.R., Herpin, F., Guélin, M., Waters, L.B.F.M.: Infrared Space Observatory's Discovery of C_4H_2, C_6H_2 and Benzene in CRL 618. The Astrophysical Journal **546** (2001) L123–L126
6. Messenger, S., Amari, S., Gao, X., Walker, R.M., Clemett, S.J., Chillier, X.D.F., Zare, R.N., Lewis, R.S.: Indigenous Polycyclic Aromatic Hydrocarbons in Circumstellar Graphite Grains from Primitive Meteorites. The Astrophysical Journal **502** (1998) 284–295
7. Salama, F., Joblin, C., Allamandola, L.J.: Neutral and Ionized PAHs: contribution to the interstellar extinction. Planet. Space Sci. **43** (1995) 1165–1173
8. Le Page, V., Keheyan, Y., Bierbaum, V.M., Snow, T.P.: Chemical Constraints on Organic Cations in the Interstellar Medium. J. Am. Chem. Soc. **119** (1997) 8373–8374
9. Snow, T.P., Le Page, V., Keheyan, Y., Bierbaum, V.M.: The interstellar chemistry of PAH cations. Nature **392** (1998) 259–261
10. Tosi, P., Bassi, D.: Private Communication.
11. Becke, A.D.: Density-functional thermochemistry. III. The role of exact exchange J. Chem. Phys. **98** (1993) 5648–5652
12. Frisch, M.J., Trucks, G.W., Schlegel, H.B., Scuseria, G.E., Robb, M.A., Cheeseman, J.R., Zakrzewski, V.G., Montogomery Jr., J.A., Stratman, R.E., Burant, J.C., Dapprich, S., Millam, J.M., Daniels, A.D., Kudin, K.N., Strain, M.C., Farkas, O., Tomasi, J., Barone, V., Cossi, M., Cammi, R., Mennucci, B., Pomelli, C., Adamo, C., Clifford, S., Ochterski, J., Petersson, G.A., Ayala, P.Y., Cui, Q., Morokuma, K., Malick, D.K., Rabuck, A.D., Raghavachari, K., Foresman, J.B., Cioslowski, J.,

Ortiz, J.V., Baboul, A.G., Stefanov, B.B., Liu, G., Liashenko, A., Piskorz, P., Komaromi, I., Gomperts, R., Martin, R.L., Fox, D.J., Keith, T., Al-Laham, M.A., Peng, C.Y., Nanayakkara, A., Gonzalez, C., Challacombe, M., Gill, P.M.W., Johnson, B., Chen, W., Wong, M.W., Andres, J.L., Gonzalez, C., Head-Gordon, M., Replogle, E.S., Pople, J.A.: GAUSSIAN 98, Revision A.7, Gaussian Inc., Pittsburg PA, (1998).

13. Frisch, M.J., Pople, J.A., Binkley, J.S.: Self-consistent molecular orbital methods 25. Supplementary functions for Gaussian basis sets. J. Chem. Phys. **80** (1984) 3265–3269
14. Knight, J.S., Freeman, C.G., McEwan, M.J., Anicich, V.G., Huntress, W.T.: A Flow Tube Study of Ion-Molecule Reactions of Acetylene. J. Phys. Chem. **91** (1987) 3898–3902
15. Petrie, S., Javahery, G., Bohme, D.K.: Gas-Phase Reactions of Benzenoid Hydrocarbon Ions with Hydrogen Atoms and Molecules: Uncommon Constraints to Reactivity. J. Am. Chem. Soc. **114** (1992) 9205–9206
16. Sablier, M., Capron, L., Mestdagh, H., Rolando, C.: Reactivity of phenylium ions with H_2 or D_2 in the gas phase: exchange versus addition. C. R. Acad. Sci. Paris. **t. 139** Série II (1994) 1313–1318
17. Mebel, A.M., Lin, M.C., Yu, T., Morokuma, K.: Theoretical Study of Potential Energy Surface and Thermal Rate Constants for the C_6H_5 + H_2 and C_6H_6 + H Reactions. J. Phys. Chem. A **101** (1997) 3189–3196
18. Bauschlicher, C.W., Ricca, A., Rosi, M.: Mechanism for the growth of polycyclic aromatic hydrocarbons (PAH) cations. Chem. Phys. Lett. **355** (2002) 159–163
19. Hudgins, D.M., Bauschlicher, C.W., Allamandola, L.J.: Closed-shell polycyclic aromatic hydrocarbons cations: a new category of interstellar polycyclic aromatic hydrocarbons. Spectrochimica Acta Part A **57** (2001) 907–930
20. Du, P., Salama, F., Loew, G.H.: Theoretical study of the electronic spectra of a polycyclic aromatic hydrocarbon, naphtalene, and its derivatives. Chem. Phys. **173** (1993) 421–437
21. Herbst, E., Le Page, V.: Do H Atoms Stick to PAH Cations In the Interstellar Medium? Astronomy and Astrophysics **344** (1998) 310–316
22. Ho, Y.P., Dunbar, R.C., Lifshitz, C.: C-H Bond Strenght of Naphtalene Ion. A Reevaluation Using New-Time-Resolved Photodissociation Results. J. Am. Chem. Soc. **117** (1995) 6504–6508

Density Functional Investigations on the C-C Bond Formation and Cleavage in Molecular Batteries

Paola Belanzoni, Marzio Rosi, and Antonio Sgamellotti

Istituto di Scienze e Tecnologie Molecolari del CNR, c/o Dipartimento di Chimica, Università di Perugia,
Via Elce di Sotto, 8 06123 Perugia, Italy
paola@thch.unipg.it

Abstract. Density functional calculations have been performed on titanium, nickel, molybdenum and niobium Schiff base complexes and titanium, nickel porphyrinogen complexes in order to understand the behaviour of these systems in redox processes. In titanium and nickel Schiff base complexes C-C σ bonds are formed upon reduction, while in titanium and nickel porphyrinogen complexes C-C σ bonds are formed upon oxidation. In both systems, the formation or the cleavage of C-C bonds avoids a variation in the oxidation state of the metal and these C-C bonds act not only as electron reservoirs, but also as a buffer for the oxidation state of the metal. In the molybdenum Schiff base complexes a preferential formation of metal-metal bonds upon reduction is calculated, while in the niobium analogues the formation of C-C bonds competes with that of M-M bonds, the latter being the first ones to be involved in electron-transfer reactions.

1 Introduction

The use of a chemical bond across two atoms for the storage and subsequent release of a pair of electrons seems to be quite an obvious concept. The reductive or oxidative coupling and the reverse decoupling can be the mechanism through which we must pass through [1]. Many mechanisms are known which operate in at least one direction [2, 3], while very often the reverse process does not even occur. A major question is under which conditions can we take advantage of a simple chemical bond for storing and releasing electrons. The major unquestionable goal is the facile reversibility associated with no overall change when the chemical system is returned to its original state. This is a normal event when the redox reactions involving formation and cleavage of a chemical bond are performed electrochemically [4]. However, reversibility to the original state is much more rarely observed when the redox system is involved in the exchange of electrons with chemical substrates [5]. As a matter of fact, the exchange of electrons with the substrate usually occurs at centres which function as electron reservoirs, so that the substrate remains bonded to them [6].

P.M.A. Sloot et al. (Eds.): ICCS 2003, LNCS 2658, pp. 376–385, 2003.

With the perspective of the use of chemical bonds as electron reservoirs, we selected the formation and cleavage of C-C single bonds as suitable candidates for storing and releasing a pair of electrons. Moreover, we tried to put metal-metal and carbon-carbon bonds in competition within the same molecular framework. The present paper is focused on density functional calculations on suitable model systems, performed in order to understand how C-C units can function as electron shuttles in redox processes, the role played by the metal and the requisites for the reversibility of the electron storage.

2 Computational Details

2.1 Model Systems

We can distinguish between two main classes of compounds: i) transition metal Schiff base complexes, in which the C-C σ bond arises from a reductive process, as depicted in Figure 1, and ii) porphyrinogen complexes, in which the C-C σ bond originates from an oxidative process, as shown in Figure 2.

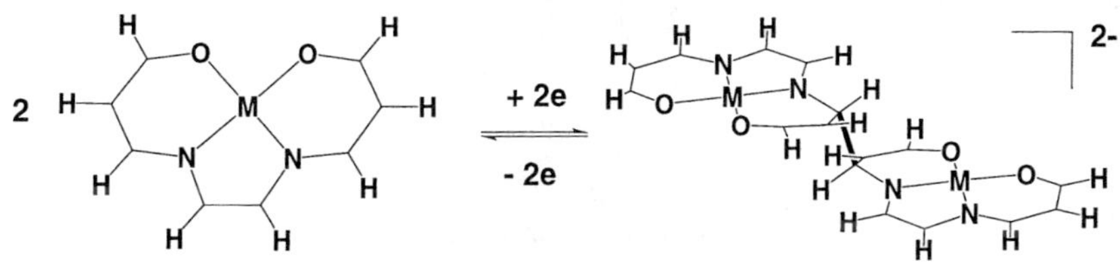

Fig. 1. C-C σ bond formation upon a reductive process

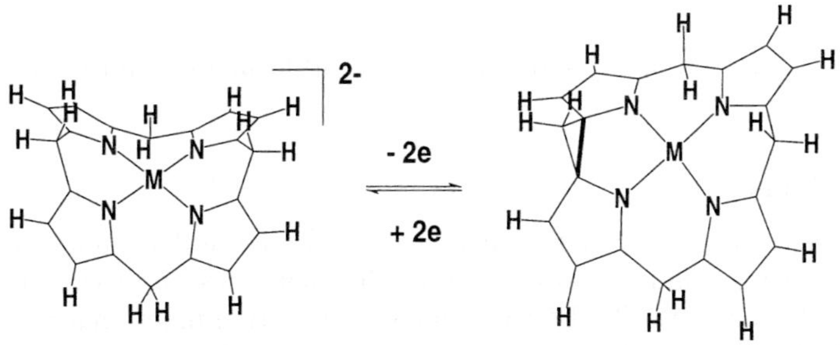

Fig. 2. C-C σ bond formation upon an oxidative process

For the first class of complexes, [Ni(salophen)] [salophen ≡ N,N' - phenylenebis-(salicylideneaminato)dianion], [Ti(salophen)Cl$_2$], $[Ti(salophen)]_2^{2+}$ and $[M(salophen)]_2$ (M=Mo,Nb) are considered as models for the oxidized form of transition metal Schiff base complexes, while $[Ti_2(*salophen_2*)]^{2-}$ and $[M_2(*salophen_2*)]$ (M=Mo,Nb), in which *salophen$_2$* is the octadentate, octaanionic ligand derived by a four-electron reduction of two salophen ligands, are considered as model of reduced, imino-coupled, transition metal Schiff base complexes. For the second class of complexes, $[TiH_8N_4]$ and $[Ni(porphyrinogen)]^{2-}$ are considered as model for the reduced form of a porphyrinogen complex.

In order to make the calculations feasible, some simplifications have been made in the investigated systems. In particular, in the salophen ligands the aromatic rings were substituted by C=C double bonds, while in porphyrinogen the ethyl *meso* groups were replaced by hydrogen atoms, as shown in Figure 3.

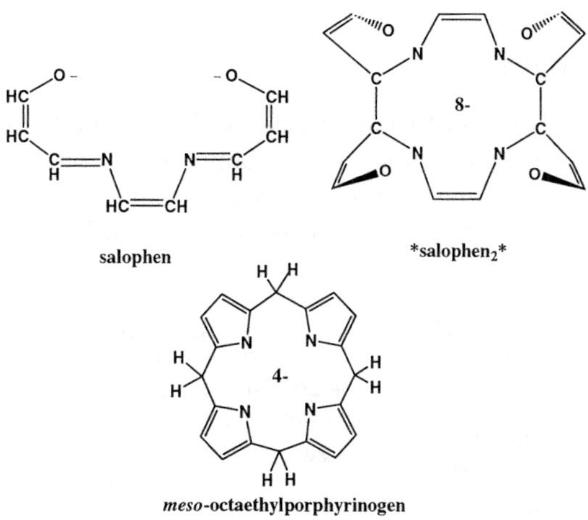

Fig. 3. Model representation of the salophen and *meso*-octaethylporphyrinogen ligands used in the calculations

These simplifications should not affect significantly the electronic distribution around the transition metal and the C-C σ bond formation and breaking mechanisms.

2.2 Methods

Density functional theory (DFT) has been used for the determination of equilibrium geometries and the evaluation of the energetics of all the investigated systems and processes. The BP86 exchange-correlation functional was used for all the calculations. This functional is based on the Becke's nonlocal exchange [7] and Perdew nonlocal correlation [8, 9] corrections to the local density approximation. Open shell systems have been calculated with the spin unrestricted approach. The calculations have been performed using the ADF (Amsterdam Density Functional) program package [10] (Gaussian 94 program package [11] for titanium systems) and were done on a cluster of IBM RISC/6000 workstations and on IBM SP3. The atomic charges have been obtained through a Mulliken population analysis.

2.3 Basis Sets

For the ADF calculations the molecular orbitals were expanded in an uncontracted double-ζ STO basis set for all atoms with the exception of 3d and 4s Ni orbitals, and of 4d and 5s Mo and Nb orbitals for which we used a triple-ζ STO basis set. This basis set was augmented by one 4p function for Ni, two 5p functions for Mo and Nb, one 2p polarization function for hydrogen and one 3d polarization function for the other elements. The cores (Ni: 1s-2p; Mo,Nb: 1s-3d; C,N,O: 1s) have been kept frozen. The basis set employed for the Gaussian 94 calculations on the titanium-Schiff base complexes is based on the Wachters-Hay set [12, 13] for the transition metal atom and on the 6-31G* set [14] for all the other atoms, except chlorine, which was described with a 6-31G set [14]. The same basis set, with the addition of diffuse functions [15] on nitrogen, was used for most of the calculations performed on the titanium-porphyrinogen complexes. Due to the size of these systems, however, some geometric optimizations of these complexes were performed using a smaller set, based on the Wachters-Hay set [12, 13] for the transition metal atom, the 6-31G set [14] for nitrogen and the 3-21G set [16] for carbon and hydrogen. Only the spherical harmonic components of the basis sets were used.

2.4 Geometry Optimization

The geometry of the model systems considered was fully optimized starting from parameters deduced from the available experimental X-ray structures [17–19]. We considered C_{2v} symmetry for [Ti(salophen)Cl_2] and [Ni(salophen)], C_i symmetry for $[\text{Ti(salophen)}]_2^{2+}$, $[\text{Ni(salophen)}]_2^{2-}$, $[\text{Ti}_2(\text{*salophen}_2\text{*})]^{2-}$ and -[M_2(*salophen$_2$*)] M=Mo,Nb, C_{2h} symmetry for $[\text{M(salophen)}]_2$ M=Mo,Nb, C_2 symmetry for $[\text{Ni(porphyrinogen)}]^{2-}$ and C_s symmetry for [Ni(porphyrinogen)(Δ)]. The counterions were not considered explicitely in the model systems. For the titanium-porphyrinogen complexes, first we fully optimized the geometry of $[H_8N_4]^{4-}$ and $[H_8N_4(\Delta)_2]$, where $[H_8N_4H_4]$ is the porphyrinogen and Δ denotes a cyclopropane unit, in order to compare the reduced and oxidized forms of the free ligand. Starting from this optimized geometry for the ligands, we subsequently optimized the geometry of the titanium complexes. We considered both the species $[\text{M(IV)}H_8N_4]$ and $[\text{M(0)}H_8N_4(\Delta)_2]$, in order to analyze the relative stability. The geometry optimizations of the free ligands were performed with the larger basis set, while those of the complexes were performed with the smaller basis set; for the titanium complexes, however, we computed the energies at the optimized geometries also with the larger basis set. The optimizations were performed considering an S_4 symmetry for $[MH_8N_4]$ and a C_2 symmetry for $[MH_8N_4(\Delta)_2]$.

3 Results and Discussion

3.1 Titanium Schiff Base and Titanium Porphyrinogen Complexes

The optimized structure of [Ti(salophen)Cl_2] shows an octahedral coordination around the transition metal, while the optimized geometry of the dimeric species $[Ti(salophen)]_2{}^{2+}$ shows a square pyramidal coordination around the metal. The main features of the optimized structure of $[Ti_2(*salophen_2*)]^{2-}$ are a clear rearrangement of the geometry of the salophen ligands, the presence of two C-C bonds (1.618 Å) between carbon atoms of imino groups of the two original salophen ligands, and the presence of an interaction between the two metal centers, the Ti-Ti distance being 2.452 Å. The ground state of [Ti(salophen)Cl_2] is the singlet $^1A_1$, and the highest occupied molecular orbital (HOMO) is mainly localized on the salophen ligand. The lowest unoccupied molecular orbital (LUMO) and the orbitals immediately at higher energy have mainly titanium d character, with small C=N π-antibonding components. The ground state of $[Ti(salophen)]_2{}^{2+}$ is the triplet ^3A_u, with the two singly occupied orbitals essentially metal in character. The LUMO and the orbitals immediately above are mainly bonding and antibonding combinations of d orbitals of the two Ti atoms, which do not interact at all between them. In both the model systems we have considered for the oxidized species, a reductive process should involve the metal, with a change in its oxidation state, since the LUMO is mainly metal d in character, at least in so far as the structure of the complex does not change. The ground state of the reduced species $[Ti_2(*salophen_2*)]^{2-}$ is the singlet ^1A_g, and the HOMO represents a σ bond between the two metal centers. The orbitals immediately lower in energy are mainly ligand in character and the C-C σ bonds of the imino groups of the two different salophen units lie at even lower energy. The LUMO is a bonding combination of metal d orbitals. An oxidative process of this system involves in the first place the metal centers and should cause a change in the oxidation state of the metal. However, the charge on the Ti atoms does not differ appreciably between the oxidized species $[Ti(salophen)]_2{}^{2+}$ and the reduced species $[Ti_2(*salophen_2*)]^{2-}$. Therefore, we can say that in the redox process the metal is the first species that acquires or loses electrons, but the geometry rearrangement of the system is able to buffer the variation in the number of the electrons and to maintain an unchanged oxidation state of the metal.

The optimized geometry of the free porphyrinogen ligand $[H_8N_4]^{4-}$ has a very distorted structure that allows the negatively charged N atoms to be as far away as possible from each other. The $[H_8N_4(\Delta)_2)]$ ligand has a calix structure, with the two cyclopropane rings pointing away from the cavity of the system. Energetically, the reduced species is strongly destabilized, so that the redox cycle shown in Figure 4 does not seem to be possible for the free ligands. The inclusion of a Ti^{4+} ion interacting with the ligands gives rise to an almost planar structure for $[TiH_8N_4]$ system and to a structure for $[TiH_8N_4(\Delta)_2]^{4+}$ very similar to that of the free ligand, with the metal located on top of the calix. The reduced species in this case is much more stable than the oxidized one, as expected.

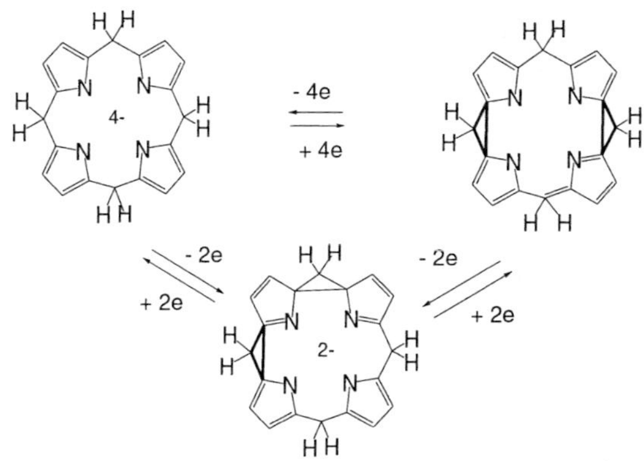

Fig. 4. C-C σ bond formation upon an oxidative process

From our calculations we can say that titanium prefers high oxidation states (IV) in the presence of ligands like porphyrinogen that are able to accept electrons. With a metal that needs electrons, porphyrinogen assumes the structure with two cyclopropane units and it is able to donate up to four electrons, while with a metal that can be easily oxidized porphyrinogen accepts up to four electrons. We conclude that the presence of the porphyrinogen coordinated to a metal has a buffering effect on the oxidation state of the metal. The only requirement is that electrons must be exchanged in couples; the removal of each couple corresponds to the formation of a cyclopropane unit on the porphyrinogen skeleton.

3.2 Nickel Schiff Base and Nickel Porphyrinogen Complexes

The optimized structure of the oxidized form of the nickel Schiff base model system [Ni(salophen)] shows a planar coordination around the transition metal. The reduction by two electrons of the monomeric species gives rise, through the reductive coupling of the imino groups, to the dimeric species $[Ni(salophen)]_2{}^{2-}$. The reduction does not affect significantly the coordination around the metal, which, as it is suggested by the Mulliken population analysis, does not change its oxidation state. The main variation upon reduction is the formation of a C-C σ bond of 1.617 Å between carbon atoms of imino groups of the two original salophen ligand. In the optimized structure of $[Ni(porphyrinogen)]^{2-}$ species, the presence of the metal in the middle of the N_4 core of the porphyrinogen gives rise to an almost planar structure. The oxidation by two electrons of this species gives rise to [Ni(porphyrinogen)(Δ)], whose optimized structure does not show significant variations in the coordination of the nickel, while shows a strong rearrangement in the ligand, leading to the formation of a cyclopropane ring in it. Also in this case we do not have a variation in the oxidation state of the metal, the oxidation being localized on the ligand through the formation of a C-C σ bond of 1.613 Å. [Ni(salophen)] and [Ni(porphyrinogen)(Δ)] show therefore a complementary behaviour upon redox processes: the first system indeed gives rise to the formation of C-C σ bonds upon reduction, while the second one shows a similar behaviour upon oxidation. Moreover, [Ni(salophen)] (**1**) and [Ni(porphyrinogen)]$^{2-}$ (**3**) can be seen as the partners of a redox reaction, whose products are [Ni(salophen)]$_2^{2-}$ (**2**) and [Ni(porphyrinogen)(Δ)] (**4**):

$$2 \text{ x } (\mathbf{1}) + (\mathbf{3}) \rightarrow (\mathbf{2}) + (\mathbf{4})$$

The reverse of this reaction is reported in Figure 5.

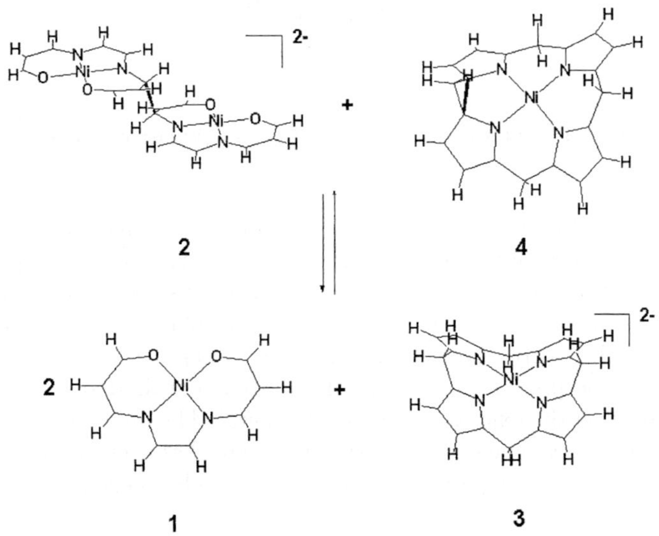

Fig. 5. An electrochemical cell based on C-C bonds breaking

From the calculated total energies of the systems, this reaction is computed to be exothermic and its ΔE is equal to -275.9 kJ mol^{-1}. Assuming $\Delta G \simeq \Delta E$, a rough estimate of the standard voltage of this redox reaction is $\Delta V \simeq 1.4$ V. The above reaction can be considered in an electrochemical cell, whose functioning is based on C-C bond breaking. Half-reaction (**4**) → (**3**) is the cathode of the cell, while half-reaction (**2**) → 2 x (**1**) is the anode of the cell. While the discharge of the cell implies C-C bonds cleavage, its recharge gives rise to formation of C-C bonds.

3.3 Molybdenum and Niobium Schiff Base Complexes

The optimized structure of [Mo(salophen)]$_2$ shows a Mo-Mo bond length of 2.195 Å which is typical of a Mo-Mo quadruple bond, that is expected for a Mo (II) complex. A metal-to-ligand four-electron transfer in the Mo (II) complex [Mo(salophen)]$_2$ would give rise to the Mo (IV) complex [Mo$_2$(*salophen$_2$*)], whose optimized geometry shows a Mo-Mo bond length of 2.538 Å, very close to a Mo-Mo double bond, and the presence of two C-C σ bonds between carbon atoms of the imino groups of the two salophen ligands. [Mo_2(*$salophen_2$*)] is computed to be less stable than [Mo(salophen)]$_2$ by 295.0 kJ mol^{-1}. In the optimized structure of [Nb(salophen)]$_2$, the calculated Nb-Nb bond length (2.433 Å) is typical of an Nb-Nb triple bond, as expected for an Nb (II) complex. A metal-to-ligand four-electron transfer in the Nb (II) complex [Nb(salophen)]$_2$ would give rise to the Nb (IV) complex [Nb_2(*$salophen_2$*)], whose optimized geometry shows a Nb-Nb bond length of 2.625 Å, very close to a Nb-Nb single bond, and the presence of two C-C σ bonds between carbon atoms of the imino groups of the two salophen ligands. [Nb_2(*$salophen_2$*)] is computed to be less stable than [Nb(salophen)]$_2$ by only 52.3 kJ mol^{-1}. For [Nb_2(*$salophen_2$*)] complex, the HOMO orbital essentially describes a Nb-Nb σ bond, while the LUMO orbital can be mainly characterized as a bonding δ orbital. The next orbital above the LUMO has mainly Nb-Nb π bonding character. The next valence orbitals at lower energies with respect to the HOMO are mainly composed of ligand atoms orbitals, while the orbitals describing the two C-C σ bonds between carbon atoms of imino groups of the two salophen ligands lie at lower energies. An oxidation of [Nb_2(*$salophen_2$*)], therefore, should affect only the Nb-Nb bond and not the two C-C σ bonds. Both [Nb_2(*$salophen_2$*)] and [Nb(salophen)]$_2$ can formally give rise to different levels of electron storage. The electron reservoirs are the C-C σ bonds between the carbon atoms of the imino groups of the two salophen ligands or the metal-metal bonds in the [Nb_2(*$salophen_2$*)] species, while only metal-metal bonds can act as electron reservoirs in [Nb(salophen)]$_2$ complexes. In order to analyze this point we investigated, in addition to the neutral species, the 2-, 2+, 4+ and 6+ species, which allow the storage or release of up to eight electrons (Fig. 6).

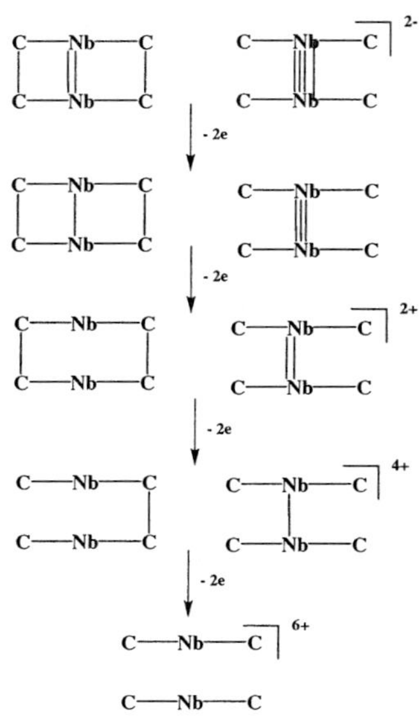

Fig. 6. Different levels of electrons storage

The optimized geometry of $[Nb_2(*salophen_2*)]^{2-}$ shows two C-C σ bonds between the carbon atoms of the imino groups of the two salophen ligands and the presence of a double (σ and δ) bond between the two metal centres. The neutral compound shows a single Nb-Nb bond and two C-C σ bonds. A two-electron oxidation of this species implies the cleavage of the Nb-Nb bond, and the removal of another two electrons implies the breaking of one C-C σ bond and, finally, the removal of the next two electrons gives rise to two units which do not interact at all. With regard to the $[Nb(salophen)]_2{}^x$ species (where x=2-,0,2+,4+,6+), we have a Nb-Nb quadruple bond in the dianion, a triple bond in the neutral species, a double bond in the dication, a single bond in the 4+ species and a lack of any metal-metal interaction in the 6+ species. In the two analysed series of compounds, the metal-metal bonds are always preferred with respect to carbon-carbon bonds.

4 Conclusion

The present study at density functional level has shown that titanium, nickel Schiff base complexes and titanium, nickel porphyrinogen complexes show similar, although complementary, behaviour in redox processes. In transition metal Schiff base complexes C-C σ bonds are formed upon acquisition of pairs of electrons, while in transition metal porphyrinogen complexes C-C σ bonds are formed upon loss of pairs of electrons. In both systems, the formation or the breaking of the C-C bonds avoids a variation in the oxidation state of the metal. These C-C bonds, therefore, act not only as electrons reservoirs, but also as a buffer of the oxidation state of the metal. The lack of variation in the oxidation state of the metal is the first step towards the reversibility of the redox process. The complementary behaviour in redox processes of nickel Schiff base complexes and nickel porphyrinogen complexes can be considered in a reversible electrochemical cell with the discharge process based on C-C bonds breaking and the recharge based on C-C bonds formation. The evaluation of the total energies of the investigated systems suggests that such an electrochemical cell presents an electrochemical potential suitable for practical applications. In molybdenum Schiff base complexes, a preferential formation of metal-metal bonds upon reduction is calculated, while in niobium analogues the formation of C-C bonds competes with that of M-M bonds, the latter being the first ones to be involved in electron-transfer reactions.

References

1. Savéant, J.-M.: Electron Transfer, Bond Breaking and Bond Formation. In: Tidwell, T.T. (eds.): Advances in Physical Organic Chemistry, Vol. 35. Academic Press, San Diego (2000) 117–192
2. Hartley, F.R., Patai, S. (eds.): Carbon-Carbon Bond Formation Using Organometallic Compounds. Wiley, Chichester (1985)
3. Cotton, F.A., Wilkinson, G., Murillo, C.A., Bochmann, M.: Advanced Inorganic Chemistry, 6th ed. Wiley, New York (1999) 957–962 and references therein

4. Baik, M.-H., Ziegler, T., Schauer, C.K.: Density Functional Theory Study of Redox Pairs. 1.Dinuclear Iron Complexes That Undergo Multielectron Redox Reactions Accompanied by a Reversible Structural Change. J. Am. Chem. Soc. **122** (2000) 9143–9154 and references therein
5. Rathore, R., Le Magueres, P., Lindeman, S.V., Kochi, J.K.: A Redox-Controlled Molecular Switch Based on the Reversible C-C Bond Formation in Octamethoxytetraphenylene. Angew. Chem. Int. Ed. **39** (2000) 809–812 and references therein
6. Rathore, R., Kochi, J.K.: Donor/Acceptor Associations and Electron-Transfer Paradigm in Organic Reactivity. In: Tidwell, T.T. (eds.): Advances in Physical Organic Chemistry, Vol. 35. Academic Press, San Diego (2000) 192–318 and references therein
7. Becke, A.D.: Density-Functional Exchange-Energy Approximation with Correct Asymptotic Behavior. Phys. Rev. A **38** (1988) 3098–3100
8. Perdew, J.P.: Density-Functional Approximation for the Correlation Energy of the Inhomogeneous Electron Gas. Phys. Rev. B **33** (1986) 8822–8824
9. Perdew, J.P.: Erratum Phys. Rev. B **34** (1986) 7406
10. ADF Program System Release 2.3.0 http://www.scm.com
11. Frisch, M.J., Trucks, G.W., Schlegel, H.B., Gill, P.M.W., Johnson, B.G., Robb, M.A., Cheeseman, J.R., Keith, T.A., Petersson, G.A., Montgomery, J.A., Raghavachari, K., Al-Laham, M.A., Zakrzewski, V.G., Ortiz, J.V., Foresman, J.B., Cioslowski, J., Stefanov, B.B., Nanayakkara, A., Challacombe, M., Peng, C.Y., Ayala, P.Y., Chen, W., Wong, M.W., Andres, J.L., Replogle, E.S., Gomperts, R., Martin, R.L., Fox, D.J., Binkley, J.S., Defrees, D.J., Baker, J., Stewart, J.P., Head-Gordon, M., Gonzalez, C., Pople, J.A.: Gaussian 94, Gaussian, Inc., Pittsburgh, PA, 1995
12. Wachters, A.J.H.: Gaussian Basis Set for Molecular Wavefunctions Containing Third-Row Atoms. J. Chem. Phys. **52** (1970) 1033–1036
13. Hay, P.J.: Gaussian Basis Sets for Molecular Calculations. The Representation of 3d Orbitals in Transition Metal Atoms. J. Chem. Phys. **66** (1977) 4377–4384
14. Frisch, M.J., Pople, J.A., Binkley, J.S.: Self-Consistent Molecular Orbital Methods. 25.Supplementary Functions for Gaussian Basis Sets. J. Chem. Phys. **80** (1984) 3265–3269 and references therein
15. Clark, T., Chandrasekhar, J., Spitznagel, G.W., Schleyer, P. von R.: Efficient Diffuse Function-Augmented Basis Sets for Anion Calculations. III.The 3-21G Basis Set for First-Row Elements, Lithium to Fluorine. J. Comput. Chem. **4** (1983) 294–301
16. Binkley, J.S., Pople, J.A., Hehre, W.J.: Self-Consistent Molecular Orbital Methods. 21.Small Split-Valence Basis Sets for First-Row Elements. J. Am. Chem. Soc. **102** (1980) 939–947
17. Franceschi, F., Solari, E., Floriani, C., Rosi, M., Chiesi-Villa A., Rizzoli, C.: Molecular Batteries Based on Carbon-Carbon Bond Formation and Cleavage in Titanium and Vanadium Schiff Base Complexes. Chem. Eur. J. **5** (1999) 708–721
18. Franceschi, F., Solari, E., Scopelliti, R., Floriani, C.: Metal-Mediated Transfer of Electrons between Two Different C-C Single Bonds That Function as Electron-Donor and Electron-Acceptor Units. Angew. Chem. Int. Ed. **39** (2000) 1685–1687
19. Floriani, C., Solari, E., Franceschi, F., Scopelliti, R., Belanzoni, P., Rosi, M.: Metal-Metal and Carbon-Carbon Bonds as Potential Components of Molecular Batteries. Chem. Eur. J. **7** (2001) 3052–3061

Violation of Covalent Bonding in Fullerenes

E.F. Sheka

Peoples' Friendship University of Russia
117923 Moscow, Russia
sheka@icp.ac.ru

Abstract. Electronic structure of X_{60} molecules (X=C, Si) is considered in terms of 60 odd electrons and spin-dependent interaction between them. Conditions for the electrons to be excluded from the covalent pairing are discussed. A computational spin-polarized quantum-chemical scheme is suggested to evaluate four parameters (energy of radicalization, exchange integral, atom spin density, and squared spin eigenvalue) to characterize the effect quantitatively. A polyradical character of the species, weak for C_{60} and strong for Si_{60}, is established.

1 Introduction

It cannot be said that fullerenes suffer from the lack of theoretical considerations. Both a basic molecule C_{60} and its homologues C_{70}, C_{84}, etc. as well as analogues Si_{60}, Ge_{60} have been repeatedly and thoroughly studied [see 1-7 and references therein]. In some sense, the molecule turned out to be a proving ground for testing different computational techniques, from a simplest to the most sophisticated. Constantly justifying the molecule stability, steadily repeated attempts of the molecule calculations are concentrated mainly on the reliability of reproducing the molecule structure and its possible distortion. There have been no doubts therewith concerning covalent bonding of atoms in the molecules. It has been taken for granted that all valence electrons participate in covalent pairing. That was the reason for the closed shell approximation to be exploited independently of whichever computational method has been used. The first breakdown of the assurance of the approach validity has been made by a comparative examination of the C_{60} and Si_{60} molecules [8-10] that has shown a strange feature in the high-spin states behavior of the molecules. As occurred, a sequence of spin-varying states, singlet (RHF)-triplet-quintet formed a progressively growing series by energy for the C_{60} molecule while for the Si_{60} molecule energy of the triplet and quintet states turned out to drop drastically with respect to the RHF singlet. Obviously, the peculiarity has clearly demonstrated the difference in the electronic structure of both molecules. However, as occurred, the observation is of much bigger importance since it concerns the basic properties of odd electrons behavior in fullerenic structures. The current paper is devoted to the phenomenon which is based on the violation of the odd electrons covalent coupling. The paper is arranged in the following way. Section 1 is devoted to conceptual grounds of the carried computational experiment. Computational techniques used in

P.M.A. Sloot et al. (Eds.): ICCS 2003, LNCS 2658, pp. 386–403, 2003.

the study is described in Section 2. Section 3 presents the results for lone pairs of odd electrons a well as for a set of pairs incorporated in the C_{60} and Si_{60} structures. The essentials of the study are discussed in Section 4.

2 Conceptual Grounds

Fullerenes are typical species with odd electrons that is why a concept on aromaticity has been expanded over the species since the very moment of their discovery [11]. However further examinations have highlighted that in spite of extreme conjugation, fullerenes behave chemically and physically as electron-deficient alkenes rather than electron-rich aromatic systems [12] so that the electrons pairing seems to be the main dominant of electronic structure. Conceptually, the problem of an electron pair is tightly connected with a fundamental problem of quantum theory related to the hydrogen molecule. According to the Heitler-London theory [13], two hydrogen atoms (electrons) retain their individuality (atomic orbitals, involving spin), and look like two individual radicals with spin S=1/2 when they are far from each other (weak interaction). When the distance approaches the interatomic chemical bond (strong interaction), the electrons, as well as their spins, become delocalized over both atoms, their properties are described by generalized molecular function (molecular orbital) and spins are aligned in an antiparallel way to provide tight covalent bonding between the atoms. As shown by forthcoming calculations [14], a continuos transition in the electron behavior from free radical-like to tightly coupled covalent bonding is observed indeed when the distance between the atoms changes from the infinity to the chemical bond length. By other words, the covalent bonding fades away when the electron interaction is weakening.

In the consequence of the topic of the current papers, two problems should be pointed out when this fundamental finding occurred to be of crucial importance. The first concerns diradicals in organic chemistry [15-22]. The phenomenon is caused by a pair of odd electrons connected with either C-C or C-N and N-N atom pairs and is common for species largely varying by composition. Generalizing its main aspects, the phenomenon essentials are caused by a violation of the above-mentioned atomic coupling from the covalent one in the part connected with odd electrons. Scheme in Fig.1 explains the main points of the diradical problem. Initially doubly degenerated atomic levels Ψ_A, and Ψ_B are splitted due to electron interaction with the energy difference $\Delta\varepsilon$. Two spins of the relevant electrons can be distributed over the splitted levels by five different ways. Configurations I, II, III, and IV are related to singlet state while the only configuration V describes the triplet one. As a result, the triplet state is spin-pure at any $\Delta\varepsilon$, while the singlet state is either purely covalent (configuration I) and, consequently, spin-pure at large $\Delta\varepsilon$, or is a mixture of configurations I-IV and becomes spin-mixed. The energy difference $\Delta\varepsilon$ turns out to be the main criterion for attributing the species to either covalently bound or diradical species and the analysis of carbenes [18,22] can be considered as the best example of this kind.

The other problem is related to molecular magnets presented by dimers composed of two transition metal atoms surrounded by extended molecular ligands [23-25]. Odd

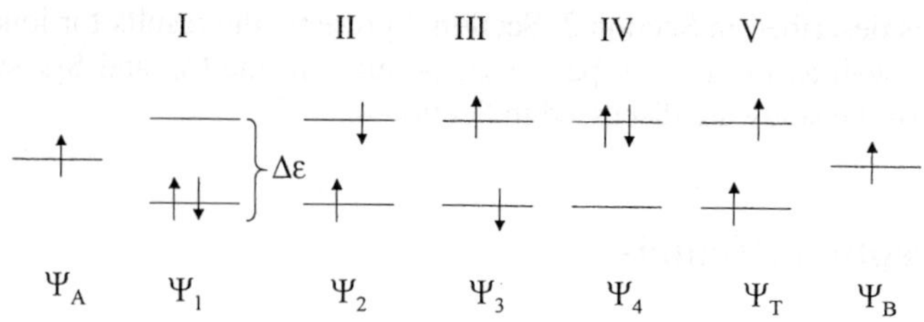

Fig. 1. Diagram of the energy level and spin configurations of a lone electron pair

electrons are associated with the metal atoms and their interaction is *a priori* weak. As previously, the triplet state is spin-pure while the singlet state is spin-mixed and is described by a combination of functions Ψ_1, Ψ_2, and Ψ_3.[1] First attempts of the electron interaction analysis have been based on the direct consideration of configuration interaction [23,24]. Later on Noodleman [25] suggested an alternative while practically feasible computational scheme based on quantum-chemical spin-polarized technique.

Following these general concepts, two fullerenes C_{60} and Si_{60} have been considered in the current study. The analysis has been done in due course of extended computational experiment fulfilled in the framework of spin-polarised Hartree-Fock calculations that has highlighted the main characteristics which are responsible for the molecule peculiar behavior.

3 Exchange Integrals

As shown for both diradicals [18,19] and molecular magnets [24], the criterion based on the quantity $\Delta\varepsilon$ lays the foundation of a qualitative analysis of the phenomenon, whilst important when tracing the odd electrons behavior when changing, say, structural parameters of the species. At the same time, as shown in the previous Section, the peculiarities of the odd electron pair behavior are caused by spin-mixing related to the singlet state of the pair. Therefore, the spin-dependent energy should be more appropriate quantitative characteristic of the phenomenon. Actually, as mentioned earlier, the value gradually decreases when weakening the electron interaction as shown for the hydrogen molecule [14]. In its turn, decreasing the value under controlled conditions will indicate growing the deviation of the electron coupling from the covalent bonding. Therefore, the problem is concentrated now around correct estimation of the value.

Let us consider two limit cases of strong and weak electron interaction. Obviously, diatomic molecules cover the former case. According to the Heitler-London theory [13], the energy of states of spin multiplicity $2S+1$, where S is full spin of two atoms with spins S_A and S_B each, ranging from 0 to $S_A + S_B$, can be expressed as

---

[1] The configuration Ψ_4 is usually omitted since the relevant state energy is high.

$$E^{2S+1} = E(0) - 4S_A S_B J_{AB} \sim E'(0) + S(S-1)J_{AB} . \tag{1}$$

Here *E(0)* and/or $E'(0)$ is the energy of the singlet state formed by covalently coupled electron pairs, S_A and S_B denote remained free spin of atoms *A* and *B*, integral J_{AB} describes the electron exchange. In case of homonuclear molecules, $S_A^{\max} = S_B^{\max} = n/2$, where *n* determines the number of unpaired electrons. The expression was inspired by the Heizenberg theory of ferromagnetism [26] and occurred to be quite useful practically in describing high-spin states of diatomic molecules [13]. Shown, the exchange integral is negative for the majority of molecules with only rare exclusion such as oxygen molecule and a few others. Silently implied therewith, the high-spin states are spin-pure that explains the appearance of spin-dependent part in Ex.(1) in form $S(S-1)J_{AB}$ where factor *S(S-1)* corresponds to the eigen value of operator $\hat{S}^2$. Applying to the general problem of odd electrons, Ex.(1) suggests the integral J_{AB} to be the main energetic criterion of the electron behavior in the limit of strong interaction. In what follows, the expression will be in use in the form

$$J = \frac{E^{2S_{\max}+1} - E^0}{S_{\max}(S_{\max}-1)}, \tag{2}$$

where $E^{2S_{\max}+1}$ and E^0 are related to the states of the highest and the lowest multiplicities, respectively. A practical usefulness of the expression is resulted from the fact that both needed energies can be quite accurately determined by using modern quantum-chemical tools. Obviously, spin-polarized techniques should be used. The value E^0, related to covalently bound singlet state, is well determined by a closed shell version of the technique while that simultaneously provides reliable determination of the $E^{2S_{\max}+1}$ value that corresponds to the ferromagnetic alignment of all spins. The ferromagnetic spin configuration is unique under any conditions (see Fig.1) so that the relevant solution is always true and the corresponding eigen functions satisfy both the Hamiltonian and $\hat{S}^2$ operator equations. Below a spin-polarized Hartree-Fock technique will be used for the values determination so that Ex.(2) can be rewritten in the following way

$$J = \frac{E^{UHF}(S_{\max}) - E^{RHF}(0)}{S_{\max}(S_{\max}-1)}. \tag{3}$$

Oppositely to covalently bound unique singlet state in the limit of strong interaction, the state becomes broken by both space and spin symmetry [27] when odd electron interaction weakens. As suggested in [25], the one-determinant singlet wave function in this case can be expressed as

$$|\psi_B\rangle = (N!)^{-1/2} M^{-1/2} \det[(a_1 + cb_1)\alpha, a_2\alpha ... a_n\alpha|(b_1 + ca_1)\beta, b_2\beta ... b_n\beta]$$
$$\approx M^{-1/2}(\phi_1 + c\phi_2 + c\phi_3). \tag{4}$$

The principal determinant ϕ_1 describes pure covalent coupling of *n* odd electrons[2] while small amounts of the charge transfer determinants ϕ_2 and ϕ_3, corresponding to $A^- - B^+$ and $A^+ - B^-$ configurations (see II and III in Fig.1) are mixed in due to nonorthogonality of atomic orbitals $\overline{a_1} = a_1 + cb_1$ and $\overline{b_1} = b_1 + ca_1$. The open shell manner for the function $|\psi_B\rangle$ expression is just appropriate to distinguish electron spins of atoms *A* and *B*. The function describes the singlet state and corresponds to the antiferromagnetic (AF) coupling of odd electrons.

As shown in [25], the energy of the above AF state is a specific weighted average of the energies of the pure spin multiplets. However, according to Ex.(4) it can be expressed as

$$E_B \equiv E_{AF}^{UHF} = E_{\rm cov} - E_{rad} , \tag{5}$$

where the latter is originated from the ionic contributions and is an independent *measure* of the deviation of the AF coupling energy from the covalent one. The term can be called as the *energy* of either *radicalization*, or *spin-mixing*, or *non-covalence* depending on which namely aspect is to be emphasized. In what follows the first nomination is preferred. Since both energies E_{AF}^{UHF} and $E_{\rm cov} = E_{AF}^{RHF}$ can be calculated within the same QCh approach by using the corresponding open shell and closed shell versions, the E_{rad} energy can be readily evaluated as the difference $E_{rad} = E_{AF}^{RHF} - E_{AF}^{UHF}$. Since ionic energies are always negative, $E_{rad} \geq 0$.

When odd electrons are covalently coupled, $E_{AF}^{UHF} = E_{AF}^{RHF}$ and, consequently, $E_{rad} = 0$. The corresponding exchange integral *J* which provides the high-spin series of the electron energies has to be determined by Ex.(3). In its turn, $E_{rad} \neq 0$ is an unambiguous indication that the odd electron coupling deviates from the covalent one. As suggested in [25], the *J* values can be determined therewith according to the following expressions

$$E_{AF}^{UHF} = E_F^{UHF} + S_{\max}^2 J , \tag{6}$$

and

[2] Doubly occupied canonical molecular orbitals that describe paired electrons are omitted [25].

$$J = \frac{E_{AF}^{UHF} - E_{F}^{UHF}}{S_{\max}^{2}}, \tag{7}$$

where E_{AF}^{UHF} and E_{F}^{UHF} correspond to the lowest (S=0) and highest (S=S_{max}) multiplicity of the n electron system and are determined by one of the spin-polarized UHF technique. As has been already mentioned, the ferromagnetic state always corresponds to a true solution of the relevant QCh equations. According to [25], energies of the series of high-spin-pure states are described as

$$E(S) = E_{AF}(0) - \frac{S(S+1)}{S_{\max}^{2}}\left(E_{AF}^{UHF} - E_{F}^{UHF}\right) = E_{AF}(0) - S(S+1)J\,, \tag{8}$$

where pure singlet state has the form

$$E_{AF}(0) = E_{F}^{UHF} + S_{\max}(S_{\max}+1)J = E_{AF}^{UHF} + S_{\max}J\,. \tag{9}$$

It is important to notice that Exs. (3), (6)-(9) are valid not only for lone pair of odd electrons. They retain their form in the case of n identical pairs with that difference that the exchange integral J is substituted by $\sim J/n$. In the weak interaction limit it is followed from the explicit expressions for the integral [25]. In the limit of strong interaction it was proved by a comparative study of the H_2 and H_6 systems [14, 28].

Expressions (3), (7) and (8) form the ground of the carried computational experiment which is aimed at analysis of the odd electron properties of two fullerene molecules C_{60} and Si_{60}. The computations have been performed by using semiempirical spin-polarized CLUSTER-Z1 sequential codes [29] in the version which is adequate to the AM1 technique [30]. Additionally to the mentioned, two other quantities were calculated, namely, eigenvalues of the $\hat{S}^2$ operator [31,32]

$$\langle S**2\rangle^{UHF} = 1/4\left(N_{\alpha} - N_{\beta}\right)^2 + 1/2\left|N_{\alpha} - N_{\beta}\right| - Sp\left(P^{\alpha}SP^{\beta}S\right) \tag{10}$$

and spin density at atom A

$$Sp_A = \sum_{i \in A} P_{ii}^{\alpha} - P_{ii}^{\beta}\,. \tag{11}$$

Here N_{α} and N_{β} ($N_{\alpha} \geq N_{\beta}$, N_{α} + N_{β} = N,) are the numbers of electron with spin up and down, respectively, N is the total number of electrons while P^{α} and

P^{β} present the relevant density matrices. A comparison of the $\langle S**2\rangle^{UHF}$ values[3] with the exact $\langle S**2\rangle = S(S-1)$ makes possible an analysis of the purity of the considered spin states[4].

4 Results

4.1 Electron Pair in the X_{60} Structure

In both organic and silicon chemistry the atom composition of pairs with odd electrons is rather variable (see, for example, [15,19]). Below we shall restrict ourselves by pairs of the >C-C< and >Si-Si< (below >X-X<) type only, where each atom is connected with three neighbors and which are characteristic for fullerenes X_{60}. Individual pairs in the fullerenes structure can be formed by a virtual dehydrogenation of the $X_{60}H_{60}$ molecules, as shown in Fig.2. Both basic molecules are tightly bonded covalently with $E_{rad}=0$ (see Table 1). Similar hexagon fragments were selected within the molecule structure which were then partially dehydrogenated that resulted in the formation of 1,2- and 1,4- pairs of odd electrons. The calculated values E^{RHF}, E_{AF}^{UHF}, and E_{F}^{UHF} $(S_{\max}=1)$ are listed in Table 1[5]. Hereinafter Ex.(3) was used when determining exchange parameter J for pairs with $E_{rad}=0$ while Ex.(7) was applied to determine J for pairs with $E_{rad}>0$.

---

[3] Easy to show, that the ferromagnetic limit of $\left\langle \hat{S}^2 \right\rangle^{UHF}$ at N_{α}-N_{β}=2n, that corresponds to $S_{\max}=n$, is equal to $\left\langle \hat{S}^2 \right\rangle^{UHF}_{\max} = 1/4N(N+2) = S_{\max}(S_{\max}+1)$.

[4] Application of the above semi-empirical technique is not crucial for the study from a conceptual viewpoint. Only its highly effective computational facilities has favored the choice. Spin-polarized DFT techniques works in similar situations absolutely analogously [33, 34], however their rather modest computational efficiency seriously prevents from carrying out an extended computational experiment which involves multiple studying of large systems at different spin multiplicity.

[5] Hereinafter in the paper energetic parameters are presented by heats of formation, $\Delta H = E_{tot} - \sum_{A} E_{elec}^{A} + EHEAT^{A}$, where $E_{tot} = E_{elec} + E_{nuc}$. E_{elec} and E_{nuc} are electronic and nuclear energies of the studied system, E_{elec}^{A} is the electronic energy of an isolated atom A and $EHEAT^{A}$ is the heat of formation for atom A. All values are calculated within the same computational session.

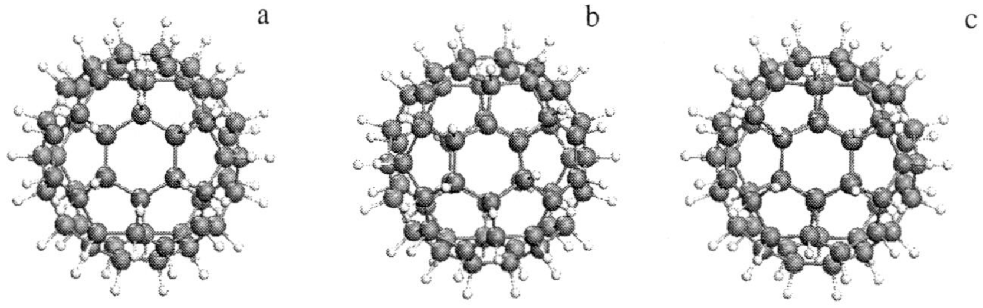

Fig. 2. Molecules $X_{60}H_{60}$ (a) and $X_{60}H_{58}$ with 1,2- (b) and 1,4- (c) pairs of odd electrons

According to Hoffman's classification [18], the first of the mentioned pairs is related to *via space* one while the other presents a *via bond* pair. As seen from the table, the formation of the 1,2-pair in the $C_{60}H_{58}$ molecule does not disturb the covalent bonding since, as previously, $E_{rad} = 0$ so that E_{AF}^{UHF} and E_{F}^{UHF} describe spin-pure states with spin density at atoms equal either to zero or to one in the singlet and triplet states, respectively. Exchange parameter J is rather big and similar to that one of the ethylene molecule (see Table 1). The other pair of the $C_{60}H_{58}$ molecule is characterized by a significant energy E_{rad}, small exchange parameter J and noticeable deviation of the calculated values $<S^{**}2>^{UHF}$ from exact. Taking together, the features doubtlessly show the deviation from the covalent coupling in the pair that forces to take it as a diradical as conventionally accepted.

Oppositely to the carbon species, the formation of any pair in the $Si_{60}H_{58}$ molecule is followed by well evident diradical effects. Thus, energy $E_{rad} > 0$ for both pairs; the values $<S^{**}2>^{UHF}$ differ form the exact ones; atomic spin density S_{at} at the pair atoms is large in the spin-mixed singlet state and considerably exceeds 1 in the triplet. As previously, the 1,2-pair and 1,4- pair differ rather drastically. The diradical character of both pairs is quite obvious. The discussed characteristics of the $Si_{60}H_{60}$ molecule pairs are similar to those of silicoethylene (see Table 1). As known [35], the latter does not exist in the gaseous state and is mentioned with respect to silicoethylene polymer that might be explained by its evidently diradical character.

4.2 Set of Odd Electron Pairs in the X_{60} Structures

If lone odd electron pairs have been considered at least qualitatively and semi-quantitatively [18, 19, 22], the only study of a cyclic H_6 cluster [14,28] can be attributed to the examination of the pair sets. At the same time sets of pairs >C-C< are >Si-Si< not a rarity for both organic and silicon chemistry. Enough to mention well extended class of aromatic compounds.

Since hexagon motive X_6 is deeply inherent in fullerenic structures, its exploitation as a model set of odd electron pairs seems quite natural. Additionally, X_{10} configuration attracts attention since there are strong arguments to consider the

Table 1. Energetic characteristics of one pair of odd electrons in the X_{60} structures[1]

Molecular species	Quantity	E^{RHF}	E_{AF}^{UHF}	$E_F^{UHF}(S_{max})$	E_{rad}
$C_{60}H_{60}$	ΔH, kcal/mol	334.161	334.161		**0**
	$\langle S**2\rangle^{UHF}$		0		
$C_{60}H_{58}$ 1,2-pair	ΔH, kcal/mol	316.319	316.315	352.457	**0.004**
	$\langle S**2\rangle^{UHF}$		0	2.021	
	Sp_A		0/0	1/1	
	J, kcal/mol		**-18.07**		
$C_{60}H_{58}$ 1,4-pair	ΔH, kcal/mol	389.304	334.422	334.623	**54.88**
	$\langle S**2\rangle^{UHF}$		1.031	2.027	
	Sp_A		-1.03/+1.03	+1.02/+1.02	
	J, kcal/mol			**-0.20**	
Ethylene	ΔH, kcal/mol	16.449	16.449	49.241	**0**
	$\langle S**2\rangle^{UHF}$		0	2.008	
	Sp_A		0/0	+1.02/+1.02	
	J, kcal/mol		**-16.40**		
$Si_{60}H_{60}$	ΔH, kcal/mol	441.599	441.597		**0.002**
	$\langle S**2\rangle^{UHF}$		0		
$Si_{60}H_{58}$ 1,2-pair	ΔH, kcal/mol	461.070	457.430	464.639	**3.64**
	$\langle S**2\rangle^{UHF}$		1.050	2.148	
	Sp_A		-1.5/+1.5	+1.31/+1.31	
	J, kcal/mol		**-7.21**		
$Si_{60}H_{58}$ 1,4-pair	ΔH, kcal/mol	504.097	453.606	454.510	**50.49**
	$\langle S**2\rangle^{UHF}$		1.383	2.269	
	Sp_A		-1.63/+1.63	+1.55/+1.55	
	J, kcal/mol		**-0.904**		
Silicoethylene	ΔH, kcal/mol	54.502	48.934	54.185	**5.57**
	$\langle S**2\rangle^{UHF}$		0.899	2.017	
	Sp_A		-1.33/+1.33	+1.17/+1.17	
	J, kcal/mol		**-5.25**		

[1]Ethylene and silicoethylene are calculated in the current study.

perdehydronaphthalene-C_{10} as a building stone of the C_{60} molecule [36-38]. The corresponding two fragments studied in the current paper are shown in Fig.3 in the form of $X_{60}H_{54}$ and $X_{60}H_{50}$ molecules. As previously, those are formed by a virtual dehydrogenation of the basic $X_{60}H_{60}$ species. Two molecular species X_6H_6 and $X_{10}H_8$ are added to provide a completed picture of the pair sets. X_{60} molecules complete the study. The calculated characteristics are given in Table 2.

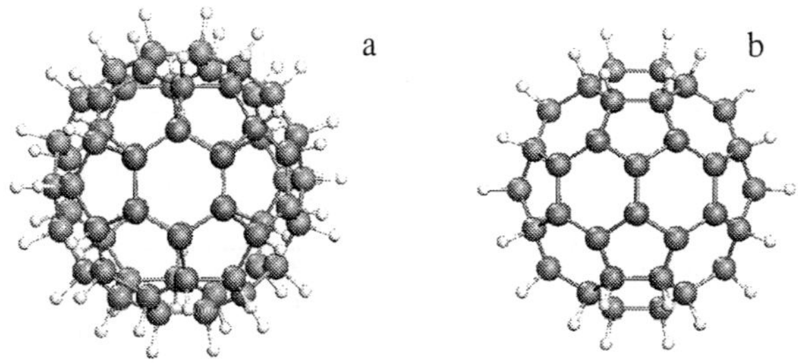

Fig. 3. Molecules $X_{60}H_{54}$ (a) and $X_{60}H_{50}$ (b) with X_6 and X_{10} fragments, respectively

X_6H_6 and X_6 fragment. There are three electron pairs in the molecular structures, $S_{max} = 3$, and the relevant state of the ferromagnetic aligning of six spins corresponds to septet. As seen from Table 2, the C_6H_6 molecule is tightly bound covalently, $E_{rad} = 0$. Both singlet and septet states are spin-pure, however, the singlet state spin density is slightly nonzero and is regularly distributed over the molecule atoms with values shown in the table. The exchange parameter J is still big whilst lower with respect to that of ethylene molecule. The benzene fragment C_6 of the $C_{60}H_{54}$ molecule behaves fully similarly so that its formation does not affect the covalent bonding. As in the case of the C_6H_6 molecule, singlet state spin density on the benzene fragment atoms is also nonzero and bigger than previously.

Absolutely another picture can be seen for the Si_6H_6 molecule while both the molecule itself and its Si_6 analogue in the $Si_{60}H_{54}$ species behave quite similarly. Those are characterized by large values of E_{rad}, by small parameter J and by atomic spin densities, similar in value in both singlet and septet state. The latter is spin-pure enough while the singlet state is evidently spin-mixed since the obtained $\langle S**2 \rangle^{UHF}$ value differs significantly from zero. The discussed features force to admit that both silicobenzene molecule and its analogue in the $Si_{60}H_{54}$ molecule should be attributed to *polyradicals*.

$X_{10}H_8$ and X_{10} fragment. There are five odd electron pairs in the molecular species, $S_{max} = 5$, and the multiplicity of the ferromagnetic state is 11. Analyzing data given in Table 2, an unexpected discovery can be made concerning the violation of the molecule covalent bonding in the naphthalene $C_{10}H_8$. As far as known, the stability of the covalent bound singlet state of the molecule has never been in doubt. However, the carried calculations show that not only E_{rad} noticeably differs from zero but $\langle S**2 \rangle^{UHF}$ is non-zero as well showing spin-mixed character of the singlet state. High values of atomic spin density are also remarkable. The findings evidence convincingly a polyradical behavior of the molecule, though not too strong As seen from Table 2, the tendency is not only kept but even strengthened for the C_{10}

Table 2. Energetic characteristics of sets of odd electrons pairs in the X_{60} structures[1]

Molecular species	Quantity	E^{RHF}	E_{AF}^{UHF}	$E_{F}^{UHF}(S_{max})$	E_{rad}
C_6H_6	ΔH, kcal/mol	21.954	21.952	162.350	**0.002**
	<S**2>	0	0	12.016	
	Sp_A	0	-0.15/+0.15	1.01	
	J, kcal/mol			**-11.70**	
C_6 fragment in **$C_{60}H_{54}$**	ΔH, kcal/mol	330.476	330.292	484.045	**0.18**
	<S**2>	0	0.207	12.027	
	Sp_A	0	-0.29/+0.29	1.09	
	J, kcal/mol			**-12.80**	
$C_{10}H_8$	ΔH, kcal/mol	40.466	38.619	291.512	**1.85**
	<S**2>	0	0.743	30.025	
	Sp_A	0	- (0.41-0.47)/ + (0.41-0.47)	(0.97-1.09)	
	J, kcal/mol			**-10.12**	
C_{10} fragment in **$C_{60}H_{50}$**	ΔH, kcal/mol	363.146	360.027	612.829	**3.12**
	<S**2>	0	1.009	30.035	
	Sp_A	0	- (0.50-0.52)/ + (0.50-0.52)	(0.96-0.97)	
	J, kcal/mol			**-10.11**	
C_{60}	ΔH, kcal/mol	972.697	955.380	2629.790	**17.32**
	<S**2>	0	4.937	930.386	
	Sp_A	0	±(0.61 – 0)	1.0-0.8	
	J, kcal/mol			**-1.86**	
Si_6H_6	ΔH, kcal/mol	144.509	121.246	158.973	**23.26**
	<S**2>	0	2.678	12.029	
	Sp_A	0	-1.51/+1.51	1.09	
	J, kcal/mol			**-4.19**	
Si_6 fragment in **$Si_{60}H_{54}$**	ΔH, kcal/mol	511.168	488.902	527.641	**22.27**
	<S**2>		3.174	12.164	
	Sp_A		-1.67/+1/67	1.14-1.09	
	J, kcal/mol			**-4.30**	
$Si_{10}H_8$	ΔH, kcal/mol	226.706	188.134	242.668	**38.57**
	<S**2>	0	4.609	30.457	
	Sp_A	0	- (1.53-1.71)/ + (1.53-1.71)	complicated distribution	
	J, kcal/mol			**-2.18**	
Si_{10} fragment in **$Si_{60}H_{50}$**	ΔH, kcal/mol	566.321	519.85	599.43	**46.47**
	<S**2>	0	5.585	31.347	
	Sp_A	0	- (1.68-1.93)/ + (1.68-1.93)	complicated distribution	
	J, kcal/mol			**-3.18**	
Si_{60}	ΔH, kcal/mol	1295.988	999.215	1513.208	**296.77**
	<S**2>	0	31.764	930.576	
	Sp_A	0	±(2.00-0.94)	(0.98-1.01)	
	J, kcal/mol			**-0.57**	

[1] Data dispersion is given in brackets

fragment in the $C_{60}H_{50}$ molecule. Supposing the fragment to be a building stone of the C_{60} molecule [36-38], its properties may genetically forecast a possible polyradical character of C_{60}[6].

The data in Table 2 related to siliconathtalene and Si_{10} fragment of the $Si_{60}H_{50}$ molecule leave no doubts concerning polyradical character of both molecules. Evidently, the effect is much bigger comparing with that of carbon species. For both silicon species E_{rad} are big, $\langle S**2\rangle^{UHF}$ drastically differs from zero for singlet states and even in the ferromagnetic states the $\langle S**2\rangle^{UHF}$ values do not coincide with the exact ones. The latter is followed by a non-regular distribution of the atomic spin density over atoms.

Fullerenes C_{60} and Si_{60}. There are 30 odd electron pairs in each molecule, $S_{\max} = 30$, and the multiplicity of the states with ferromagnetic alignment of all 60 spins is 61. As seen from Table 2, $E_{rad} > 0$ for both molecules, for the Si_{60} just drastically. The UHF singlet states are spin-mixed, and again, the mixing for silicon species is just enormous that is seen from the deviation of the $\langle S**2\rangle^{UHF}$ values from zero. In both cases atomic spin density is quite considerable and is distributed over atoms in a rather complicated way. Fig.4 presents the spin density distribution in a manner when the presented value gradually grows from the left to the right while the sums over the values in both cases are zero. White-and-black image of the X_{60} molecule shown in the figure insert highlights the space distribution of positive and negative spin density over atoms. Taking together, the presented data make a polyradical character of the odd electron bonding in both molecules completely evident.

5 Discussion

The carried analysis has convincingly shown that X_{60} fullerene composition of atoms with odd electrons results in weakening the covalent bonding that is just drastic in the case of silicon species. To describe the phenomenon concisely, the term *polyradicalization* has been suggested to emphasize that a rather peculiar chemical behavior should be expected from the species. As shown in the study, four parameters can be proposed to describe the effect quantitatively. The main parameter $E_{rad} = E^{RHF} - E_{AF}^{UHF}$ indicates straightforwardly the bonding weakening, when it is non-zero. Three other parameters, namely, exchange parameter *J*, $\langle S**2\rangle^{UHF}$, and Sp_A, describe quantitatively conditions under which the weakening occurs.

[6] The disclosed feature has forced us to check the tendency for a series of aromatic hydrocarbons. The calculations have convincingly shown the strengthening of the effect when going from naphthalene to pentacene for both carbon and silicon species. The results will be partially discussed in the next Section.

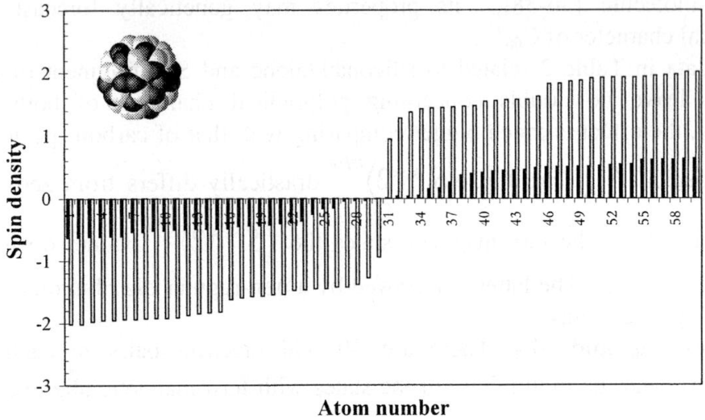

Fig. 4. Spin density distribution over atoms of molecule C_{60} (dense black bars) and Si_{60} (contour bars) in the UHF singlet state. Insert: space distribution of the density for the Si_{60} molecule

Taking together, they provide a complete picture of the phenomenon discussed. Additionally, the computational approach used makes it possible to determine the energy of high-spin states of the studied compounds. Fig.5 presents the data related to *E(S)* and $\langle S**2\rangle^{UHF}$ calculated in accordance with Exs. (8) and (10) for both C_{60} and Si_{60} species. The series of the $E^{UHF}(S)$ energies of spin-mixed states, calculated straightforwardly by using the applied tool, are included for comparison. As seen from the figure, the difference between the series of spin-pure and spin-mixed states is not too big as might be expected. This is much more surprising since the calculated $E^{UHF}(S)$ relate to the optimized structures which are different for different spin states, sometimes rather significantly, while the *E(S)* series is related to the same structure. A conclusion can be made that the applied spin-polarized QCh tool provides a quite reliable presentation of high-spin states. Curves 3 in the figure plots the ratio $\varsigma(\%) = \dfrac{\langle S**2\rangle^{UHF} - S(S-1)}{S(S-1)}$ which characterizes spin purity of the states. As seen from the figure, in the case of C_{60}, the high-spin states become spin-pure at rather low spin values while only at high spins the similar is observed for the Si_{60} molecule.

Even in the first studies of diradicals, Hoffman [16-18] and other authors [19] have tried to exhibit the criterion of the transition from covalent pairing to odd electron pair radicalization. However, only the energy difference $\Delta\varepsilon = \varepsilon_2 - \varepsilon_1$ between the energies of two orbitals of the pair was suggested that was not enough to formalize the criterion. A considerable extension of the number of quantitative parameters, readily accessible by the modern spin-polarized QCh techniques, makes now possible to suggest a formal criterion for the transition. Given in Fig. 6 presents

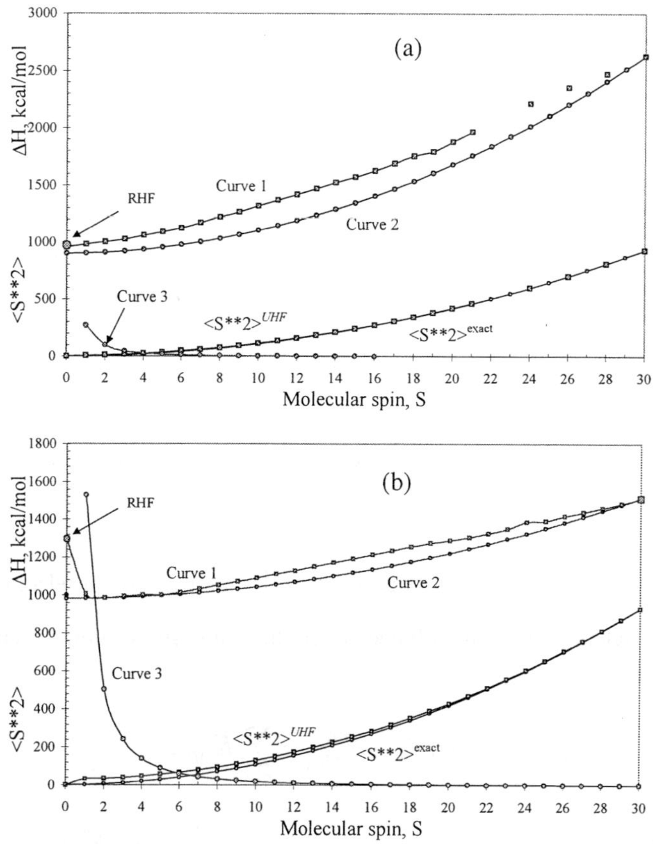

Fig. 5. Heat of formation of the UHF- (curves 1) and pure- (curves 2) spin states of the molecules C_{60} (a) and Si_{60} (b); RHF singlet states are shown by arrows; curves 3 present ζ values (see text)

the dependence of E_{rad} versus exchange parameter J on the basis of the data summarized in Table 3. As seen, the dependence for both carbon and silicon species is quite similar and exhibits a clearly seen quasi-threshold character. One may conclude that for the studied species the transition starts when J reaches ~10 kcal/mol.

Dependencies $E_{rad}(J)$, or more precise, the steepness of the curves after transition, well formalize the difference in the polyradicalization of different species. As seen in the figure, the steepness is a few times more for the silicon species in comparison with carbon molecules. The obvious preference shown by silicon atoms towards polyradicalization instead of double bond formation is well supported by high values of atomic spin densities (see Tables 1 and 2). The latter quantity, in its turn, is provided by electrons taken out of chemical bonding [39]. Actually, Fig.7 presents

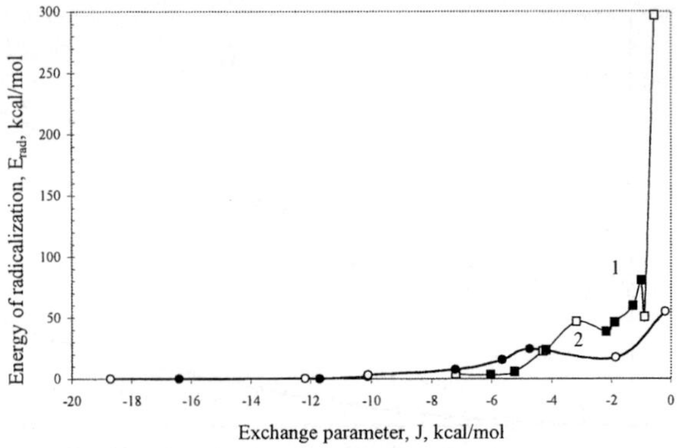

Fig. 6. Energy of polyradicalization versus exchange parameter *J* for odd electron pairs >Si-Si< (1) and >C-C< (2). Empty and filled points correspond to fullerenic and "aromatic" structures, respectively

absolute values of the atom spin density $|Sp_A|$ multiplied by an electron spin, and atom free valence V_A^{free} distributed over the molecules atoms. The latter is determined as

$$V_A^{free} = N_{val}^A - \sum_{B \neq A} K_{AB}\ , \qquad (12)$$

where N_{val}^A is the number of valent electrons of atom *A* while $\sum_{B \neq A} K_{AB}$ presents a

Table 3. Fundamental energetic parameters of >X-X< odd electron pairs, kcal/mol

Molecular species	Carbon		Silicon	
	E_{rad}	J	E_{rad}	J
X_2H_4	0.00	-16.40	5.57	-5.25
1,2 pair $X_{60}H_{58}$	0.004	-18.07	3.64	-7.21
1,4 pair $X_{60}H_{58}$	54.88	-0.20	50.49	-0.90
X_6H_6	0.002	-11.70	23.26	-4.19
X_6 fragment of $X_{60}H_{54}$	0.18	-12.80	22.27	-4.30
$X_{10}H_8$	1.85	-10.12	38.57	-2.18
X_{10} fragment of $X_{60}H_{50}$	3.12	-10.11	46.47	-3.18
$X_{14}H_{10}$	7.45	-7.22	45.97	-1.90
$X_{18}H_{12}$	15.32	-5.66	59.53	-1.29
$X_{22}H_{14}$	24.23	-4.74	80.22	-1.01
X_{60}	17.32	-1.86	296.77	-0.57

generalized bond index[7], summarized over all atoms excluding atom A. A close similarity should be noted between the two values, which are calculated independently. Taking together, the data present a quantitative explanation of the difference in bonding carbon and silicon atoms and answer the question why silicon atoms "dislike" sp_2 hybridization [9,10]. Qualitatively, this can be described in the following way. While carbon atom interaction forces odd electron to participate in the action thus strengthening it, silicon atoms prefer to leave the odd electrons free in a form of spin density, while the atom interaction is kept at much weaker level. The obtained findings throw light as well on why "...A comparison of the chemistry of tetravalent carbon and silicon reveals such gross differences that the pitfalls of casual analogies should be apparent" [42].

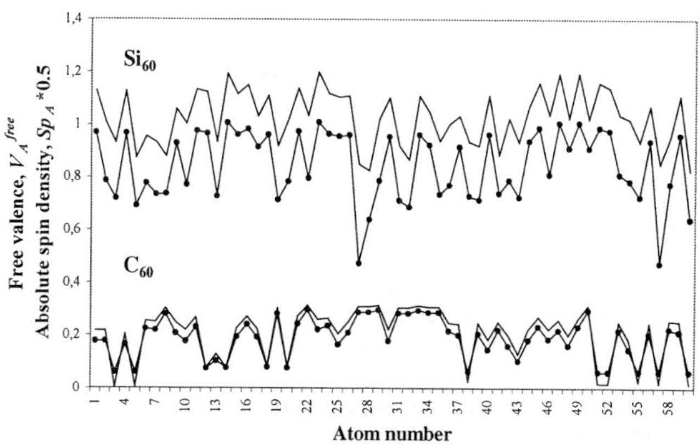

Fig. 7. Free valence (solid curves) and absolute spin density (dotted curves) distributed over atoms of C_{60} and Si_{60} molecules

Thus disclosed polyradical character of the basic fullerene molecule C_{60} is well consistent with extreme diversity of its properties which sometimes seem to be even contradictory. Enough to mention, for example, diamagnetism of a free molecule and pristine C_{60} crystals [43,44] and a ferromagnetic behavior of either carpet-like polymerized crystalline species [45] or TDAE- C_{60} charge complex [46]. However, even Salem and Rowland have already mentioned [19] that diradical electronic structure is readily amenable to any external effects caused by either intramolecular chemical substitution or by intermolecular interaction. Actually, recent studies have shown [47] that above mentioned peculiarities in the magnetic behavior are tightly connected with changing the main polyradical characteristics E_{rad}, J, and Sp_A. Another examples of "strange" behavior of the C_{60} admixtures such as effective

[7] A generalized bond index is determined as $K_{AB} = |P_{AB}|^2 + |Sp_{AB}|^2$ [40], where the first term is known as Wiberg's bond index [41] which was introduced to characterize the covalence of chemical bonds in molecules with closed electron shells while the second goes from the spin density matrix.

inhibition of radical destruction of polystyrene [48] as well as sensitization of penetration through cell membranes [49] should be obviously analyzed from this viewpoint.

Polyradical character of the Si_{60} molecule is supported experimentally even more ponderably. The first consequence provided by the phenomenon may be formulated as a prohibition of the existence of particular chemical entities and even whole classes such as alkenes, alkynes, aromatic compounds, and fullerenes, typical for its carbon analogue. Only polyradicalization suppression can provide chemical stability that in the latter case may be realized in form of the $Si_{60}H_{60}$ and $Si_{60}C_{60}$ species [9,10]. This very high susceptibility of silicon atoms to polyradicalization provides as well magnetism of its crystal surfaces [50] that has already become a physical reality oppositely to the absence of similar effects for diamond crystal.

Acknowledgements. The author is greatly indebted to V.A.Zayets for stimulating and valuable discussions.

References

1. Chang, A.H.H., Ermler, W.C., Pitzer R.M.: J. Phys. Chem. **95** (199) 9288.
2. Weaver, J.H.: Acc. Chem. Res. **25** (1992) 143.
3. Bulusheva, L.G.: PhD Thesis, Institute of Inorganic Chemistry SB RAS, Novosibirsk (1998).
4. Lee, B.X., Cao, P.L., Que, D.L.: Phys Rev **61B** (2000) 1685.
5. Nagase, S.: Pure Appl. Chem. **65** (1993) 675.
6. Slanina, Z., Lee, S.L.: Fullerene Sci.Technol.**2** (1994) 459.
7. Lee, B.X., Jiang, M., Cao, P.L.: J.Phys.: Condens. Matter **11** (1999) 8517.
8. Sheka, E.F., Nikitina, E.A., Zayets, V.A., Ginzburg, I.Ya.: JETP Letters **71** (2001) 177.
9. Sheka, E.F., Nikitina, E.A.: Doklady RAN **378** (2001) 1.
10. Sheka, E.F., Nikitina, E.A., Zayets, V.A., Ginzburg, I.Ya.: Int Journ Quant Chem **88** (2002) 441.
11. Kroto H.W., Heath J.R., O'Brien S.C., Curl R.F., Smalley, R.E.: Nature **318** (1985) 162
12. Fowler, P.W., Ceulemans, A.: J.Phys.Chem. **99** (1995) 508.
13. Herzberg, G.: Molecular Spectra and Molecular Structure. I.Spectra of Diatomic Molecules, 2nd edn., Van Nostrand, Prinston, N.J. (1950) p.353.
14. Gubanov, V.A., Likhtenstein, A.I., Postnikov, A.V.: Magnetism and Chemical Bonding in Crystals (in Russian), Nauka, Moskva (1985) Chapter 2.
15. Hay J.M., Thomson, R.H.: Organic Chemistry of Stable Free Radicals, Academic Pres, New York, (1955).
16. Hoffman, R., Imamura, A., Hehre, W.J.: J.Amer.Chem.Soc. **90** (1968) 1499.
17. Hoffman, R.: Chem.Commun. (1969) 240.
18. Hoffman, R.: Accounts Chem.Res. **4** (1971) 1.
19. Salem, L., Rowland, C.: Angew.Chem. Intern Edit. **11** (1972) 92.
20. Hoffman, R.: J.Chem.Phys. **39** (1963) 1397.
21. Hoffman, R.: J.Chem.Phys. **40** (1964) 2474, 2480, and 2745.
22. Harrison, J.W.: In W.Kirsme (Ed.): Carbene Chemistry, 2nd edn., Academic Press, New York (1971) Chapter 5, p.159.
23. Ellis, D.E., Freeman, A.J.: J.Appl.Phys.**39** (1968) 424.
24. Hay, P.J., Thibeault, J.C., Hoffman, R.: J Amer Chem Soc. **97** (1975) 4884.

25. Noodleman, L.: J Chem Phys **74** (1981) 5737.
26. Heizenberg, W.: Ztschr.Phys. (1928) 325.
27. Benard, M.: J Chem Phys **71** (1995) 2546.
28. Mattheis, L.F.: Phys.Rev. **123** (1961) 1219.
29. Zayets, V.A. CLUSTER-Z1: Quantum-Chemical Software for Calculations in the *s,p*-Basis: Institute of Surface Chemistry, Nat. Ac.Sci. of Ukraine: Kiev (1990).
30. Dewar, M.J.S., Zoebisch, E.G., Healey, E.F., Stewart, J.J.P.: J Amer Chem Soc **107** (1985) 3902.
31. Löwdin, P.O.: Phys Rev **97** (1955) 1509.
32. Zhogolev, D.A., Volkov, V.B.: Methods, Algorithms and Programs for Quantum-Chemical Calculations of Molecules (in Russian), Naukova Dumka: Kiev (1976).
33. Ginzberg, A.P.: J Amer Chem Soc. **102** (1980) 111.
34. Norman, J.G., Ryan, P.B., Noodleman, L.: J Amer Chem Soc. **102** (1980) 4279.
35. Brief Chemical Encyclopedia (in Russian), Sov. enziklopedia:Moscow, (1963) p.799.
36. Bulychev, B.M., Udod, I.A.: Russian Chem. Journ. **39** (1995) No.2, 9.
37. Alekseev, N.I., Dyuzhev,G.A. Techn.Phys. **46** (2001) 573, 577.
38. Tomilin, F.N., Avramov, P.V., Varganov, S.A., Kuzubov, A.A., Ovchinnikov, S.G. Phys.Sol.State, **43** (2001) 973.
39. Khavryutchenko, V., Sheka, E., Aono, M., Huang, D.-H. Phys. Low-Dim. Struct. (1998) No.3/4, 81.
40. Semenov, S.G.: In Evolution of the Valence Doctrine (in Russian), Moscow, Khimia (1977) p.148.
41. Wiberg, K.B.:Tetrahedron, **24** (1968) 1083.
42. Gaspar, P., Herold, B.J. In W.Kirsme (Ed.): Carbene Chemistry, 2nd edn., Academic Press, New York (1971) Chapter 13, p.504.
43. Haddon, R.C., Scheemeyer, L.F., Waszczak, J.V., Glarum, S.H., Tycko, R., Dabbah, G., Kortan, A.R., Muller, A.J., Mujsce, A.M., Rosseinsky, M.J., Zahurak, S.M., Makhija, A.V., Thiel, F.A., Raghavachari, K., Cockayne, E., Elser V. : Nature **350** (1991) 46.
44. Luo, W.L., Wang, H., Ruoff, R.C., Cioslowski, J., Phelps, S.: Phys. Rev. Lett. **73** (1994) 186.
45. Makarova, T.L., Sundqvist, B., Esquinazi, P., Höhne, R., Kopelevich, Y., Scharff, P., Davydov, V. A., Kashevarova, L. S., Rakhmanina, A. V.: Nature **413** (2001) 716.
46. Allemand P.-M., Khemani K.C., Koch A., Wudl F., Holczer K., Donovan S., Grüner G., Thompson J.D.: Science **253** (1991) 301.
47. Sheka, E.F.: In 6th Fock's School on Quantum Chemistry, V.Novgorod University, V.Novgorod (2003).
48. Volkova, N.N., Sumannen, E.V. 13th Symposium on Modern Physical Chemistry, Institute of Chemical Physics Problems of RAS, Tuapse, Russia (2001) p.122
49. Kotelnikova, R.A., Bogdanov, G.N., Romanova, V.S., Parnes, Z.N.: Mol. Mat. **11** (1998) 111.
50. Sheka, E.F., Nikitina, E.A., Zayets, V.A.: Surface Science (in print)

Workshop on Recursive and Adaptive Signal/Image Processing (RASIP)

Jointly Performed Computational Tasks in the Multi-mode System Identification*

Innokenti Semoushin

Ulyanovsk State University, 42 Leo Tolstoy Str., 432970 Ulyanovsk, Russia
SemoushinIV@ulsu.ru http://staff.ulsu.ru/semoushin/

Abstract. In this paper, we propose a new closed-loop multi-mode system identification setup that allows the jointly performed change point detection, fault diagnosis, supervised training of a bank of signature generators, unsupervised identification of the optimal filter, and monitored system accommodation on the basis of incomplete noisy data. We validate the setup by simulating numerical examples.

1 Introduction

In the literature on modern control systems, more emphasis is placed on detection [1] and identification [2] as separate problems. At the same time, reality provides evidence that in many cases fault detection-diagnosis must be performed together with system identification and accommodation. In adaptive multi-mode or hybrid stochastic systems with an uncertain mode switching mechanism, identification is to be launched repeatedly every time when a mode switch is detected. In the equivalent manner, identification must be stopped when the identification algorithm provides the highest (or tolerable) pitch of agreement with the observed data. Thus, FDD-SIA is a composite challenging problem needing not only a theoretical but also computational research.

In this work, we propose a unified approach to perform FDD-SIA computational tasks simultaneously in a single system. The outline of the paper is as follows: Section 2 describes the monitored system (MS). In Sect. 3 we formulate the Statistical Orthogonality Principle (SOP) and refer to the Accessible Indirect Performance Index (AIPI). The SOP-based FDD is described in Sect. 5. Section 6 shows one of identification algorithms (IA) intended to identify the optimal Steady-State Kalman Filter (SSKF) as a substitute for the Feedback Filter (FF). The MS accommodation is briefly described in Sect. 7. Some experimental results are shown in Sect. 8. Finally, Sect. 9 concludes the paper.

2 Monitored System

For the monitored system we consider an errors-in-variables (EIV) control system with the observed data $z(t_i) = \begin{bmatrix} y(t_i) \\ u(t_{i-1}) \end{bmatrix}$, $i = 1, 2, \ldots$, composed of the control

* This work was supported in part by the Russian Ministry of Education (grant No. T02-03.2-3427).

P.M.A. Sloot et al. (Eds.): ICCS 2003, LNCS 2658, pp. 407–416, 2003.

input $u \in \mathbb{R}^q$ and the measurement output $y \in \mathbb{R}^m$. The system is parameterized by an uncertainty vector $\theta \in \mathbb{R}^l$ and consists of a plant (P), a sensor (S) and a stabilizing feedback controller(FC) (Fig.1(a)). The P state equation is

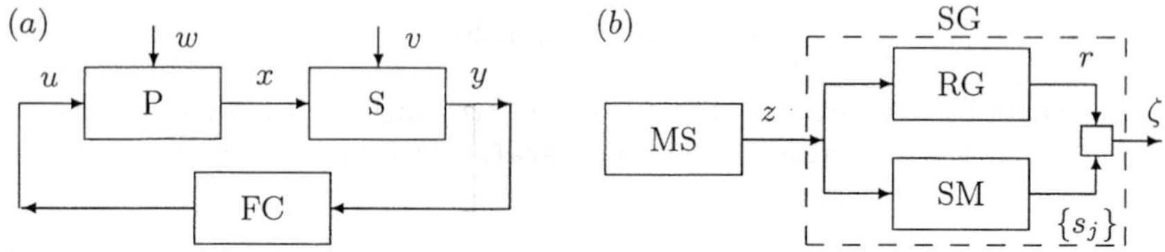

Fig. 1. (a) Monitored System (MS), and (b) Statistical orthogonality. Legend: P – plant, S – sensor, FC – feedback controller, RG – residual generator, SM – sensitivity model, SG – signature generator

$$x(t_{i+1}) = \Phi_\theta x(t_i) + \Psi_\theta u(t_i) + w(t_i), \quad i = 0, 1, \ldots \tag{1}$$

with $x \in \mathbb{R}^n$ and a noise $w(t_i)$ where the random initial $x(t_0)$ at some t_0 has a finite mean $\bar{x}_0$ and a finite covariance $P_0 \geq 0$ and $\{u(t_i)\}$ is wide-sense stationary and so $\mathrm{E}\{||u(t_i)||^2\} < \infty$ for all t_i. The S equation is

$$y(t_i) = Hx(t_i) + v(t_i), \quad i = 1, 2, \ldots \tag{2}$$

with $\mathrm{rank}(H) < n$ and a noise $v(t_i)$. Both $\{w(t_i)\}$ and $\{v(t_i)\}$ are i.i.d. zero mean mutually independent wide-sense stationary sequences whose covariances are $\mathrm{E}\{w(t_i)w(t_i)^T\} = Q_\theta \geq 0$ and $\mathrm{E}\{v(t_i)v(t_i)^T\} = R_\theta > 0$ for all t_i. The FC, in compliance with the well-known separation property, is formed by a feedback filter (FF) cascaded with a feedback regulator (FR). The FF equations are

$$\tilde{x}(t_{i+1}^-) = \bar{\Phi}\tilde{x}(t_i^+) + \Psi u(t_i), \quad \tilde{x}(t_i^+) = \tilde{x}(t_i^-) + \bar{K}[y(t_i) - H\tilde{x}(t_i^-)] \tag{3}$$

with some initials $\tilde{x}(t_0^+)$, $u(t_0)$. Constant transition matrix $\bar{\Phi}$ and constant gain $\bar{K}$ are chosen in a sense arbitrarily or designed to be optimal in a steady state (as $t_0 \to -\infty$) w.r.t. a *fixed* (fault-free) system mode specified by a nominal value θ_0 of θ. The FR equation is

$$u(t_i) = f_{\mathrm{FR}}[\tilde{x}(t_i^+)] = -\bar{G}_{\mathrm{FR}}\tilde{x}(t_i^+) \tag{4}$$

with a function $f_{\mathrm{FR}}[\cdot]$ which can be chosen to satisfy the second equality in (4) with a constant gain matrix $\bar{G}_{\mathrm{FR}}$. By (3), (4), it is predetermined that $u(t_i)$ depends on all available data $z(t_1^i) = [z(t_1), \cdots, z(t_i)]$, the latter is a notation for a composite (stackable) column vector. Note that $\begin{bmatrix} w(t_i) \\ v(t_j) \end{bmatrix}$ is independent of $u(t_k)$ and $x(t_k)$ for all $t_i \geq t_k$, $t_j > t_k$.

The overall closed-loop control system (1), (2), (3), (4) is assumed to be asymptotically stable in all modes of operating referred to by the subscript θ, and so all the processes within the system are wide-sense stationary at every t_i as $t_0 \to -\infty$ (*the main assumption*). The system is designed to operate with a minimum expected control cost

$$J_c = \lim_{j \to -\infty} \mathrm{E}\{x^T(t_{N+1})X_f x(t_{N+1}) + \sum_{i=j}^{N}[x^T(t_i)Xx(t_i) + u^T(t_i)Uu(t_i)]\} \quad (5)$$

for $N > 0$, $X_f \geq 0$, $U > 0$, $X \geq 0$, and in all modes holds the *main properties*:

- ⋆ $(\Phi_\theta, Q_\theta^{1/2})$ is stabilizable, $(\Phi_\theta, X^{1/2})$ is detectable, (Φ_θ, H) is completely observable, and $(\Phi_\theta, \Psi_\theta)$ is completely controllable.

to guarantee the existence of the optimal steady-state parameters for (3), (4).

The system itself can operate in several (*finitely* or *infinitely* many) modes. This brings the problem to the realm of hybrid or multi-mode systems. The mode characterized by θ_0 can be treated as the main one. The other modes characterized by some θ_1, θ_2, ..., $\theta_{\mathcal{M}}$ – for the case of *finitely* many modes, – can be viewed as some alternative modes not obligatory (albeit possibly) faulty. In this case we have to test the $\mathcal{M}$ pairs of hypotheses, in each pair between

$$\mathtt{H}_0 = \{\theta^\dagger = \theta_0\} \quad \text{and} \quad \mathtt{H}_\mu = \{\theta^\dagger = \theta_\mu = \theta_0 + \Upsilon_\mu\} \quad (6)$$

where $\theta^\dagger$ is the *true* value of θ, $\mu = 1, 2, \ldots, \mathcal{M}$, and Υ_μ is the μ-th change on θ (the μ-th alternative mode of operation). Allowing for arbitrary faults, which are modelled by any sizable change Υ on θ not violating the above main assumption, we come to the case of *infinitely* many modes. By this, we generalize the problem as we have to test the continuum set of pairs of hypotheses

$$\mathtt{H}_0 = \{\theta^\dagger = \theta_0\} \quad \text{and} \quad \mathtt{H}_\Upsilon = \{\theta^\dagger = \theta_\Upsilon = \theta_0 + \Upsilon\} \ . \quad (7)$$

In both cases (6) and (7) we do not assume a specific mode switching behavior (for example, a Markovian one) and view it as deterministic (albeit unknown to the observer) like controlled by an independent actor.

3 Statistical Orthogonality Principle

Definition 1. *Given l pairs $\{r, s_j\}$, $j = 1, 2, \ldots, L$, each formed by column vectors $r = r(t_i)$ and $s_j = s_j(t_k)$. Then $r(t_i)$ and $s_j(t_k)$ are said to be* statistically orthogonal *at distance $\Delta t = t_k - t_i$, that is denoted as $r(t_i) \perp s_j(t_k)$, if*

$$\forall t_i, t_k, \Delta t = t_k - t_i = \text{const} : \quad \mathrm{E}\{s_j^T(t_k)r(t_i)\} = 0 \quad (8)$$

and $r(t_i)$ is said to be statistically orthogonal to $S(t_k) = [s_1(t_k) \mid \cdots \mid s_L(t_k)]$ at $\Delta t = t_k - t_i$ if (8) holds for all given pairs, that is denoted as $r(t_i) \perp S(t_k)$.

Let r and S in Definition 1 be computed by L pairs of parallel blocks RG and SM as shown in Fig.1(*b*). By RG and SM are meant *Residual Generator* and *Sensitivity Model* both parameterized by a design parameter $\hat{\theta} \in \mathbb{R}^L$ not fully but partly being an exact replica of θ, the MS parameter, in its elements' numbering and sense. Thus, $r(t_i) = r(t_i, \hat{\theta})$ and $S(t_k) = S(t_k, \hat{\theta})$.

Definition 2. *Let* RG *and* SM *be designed in such a way that* in case (6) *for each* $\mu = 0, 1, 2, \ldots, \mathcal{M}$, *parameter* $\hat{\theta}$ *takes a specified value* $\hat{\theta}_\mu$ *such that*

$$\forall \mu : \quad \{\hat{\theta} = \hat{\theta}_\mu\} \equiv \{\, r(t_i, \hat{\theta}) \perp S(t_i, \hat{\theta}) \;\leftrightarrows\; \theta^\dagger = \theta_\mu \,\} \tag{9}$$

or in case (7) *for each* Υ, *parameter* $\hat{\theta}$ *takes a specified value* $\hat{\theta}_\Upsilon$ *such that*

$$\forall \Upsilon : \quad \{\hat{\theta} = \hat{\theta}_\Upsilon\} \equiv \{\, r(t_i, \hat{\theta}) \perp S(t_i, \hat{\theta}) \;\leftrightarrows\; \theta^\dagger = \theta_\Upsilon \,\} \tag{10}$$

where $\leftrightarrows$ *stands for "iff" (if and only if). Then (9) and (10) is termed* Statistical Orthogonality Principle.

The parallel blocks in Fig.1(*b*), RG and SM, may be variously designed to satisfy Definitions 1 and 2. However as system (1), (2), (3), (4) is stochastic, the presence of an *(adaptive) estimator* (AE), denote it by $g(t_{i|i-1}, \hat{\theta})$, is requisite for obtaining a residual $r(t_i, \hat{\theta})$. With it, writing $r(t_i, \hat{\theta}) \perp S(t_i, \hat{\theta})$ is straightforward as a necessary (and on occasion, sufficient) condition for minimum of

$$J(\hat{\theta}) = 1/2 \lim_{t_j \to -\infty} \mathrm{E}\{\|r(t_i, \hat{\theta})\|^2 \mid y(t_j^{i-1})\} \ . \tag{11}$$

Thus, SOP implementation boils down to the construction of (11) so as to obtain

$$S(t_i, \hat{\theta}) = \left[\frac{\partial r(t_i, \hat{\theta})}{\partial \hat{\theta}_1} \middle| \cdots \middle| \frac{\partial r(t_i, \hat{\theta})}{\partial \hat{\theta}_L} \right] \ . \tag{12}$$

4 Identification Performance Indices

Minimum Prediction Error methods [3] offer to predict by $g(t_{i|i-1}, \hat{\theta})$ the system output $y(t_i)$ and use $r(t_i, \hat{\theta}) = y(t_i) - g(t_{i|i-1}, \hat{\theta})$ as a residual. Minimizing (11) with such residual leads to the biased parameter estimates for the given EIV system [4], [5]. Had $r(t_i, \hat{\theta})$ in (11) instead been in one of the following forms

$$r(t_i, \hat{\theta}) = \{\, x(t_i) - g(t_{i|i-1}, \hat{\theta}) \;\text{ or }\; \hat{x}_{\mathrm{OPT}}(t_{i|i-1}) - g(t_{i|i-1}, \hat{\theta}) \,\} \tag{13}$$

where $\hat{x}_{\mathrm{OPT}}(t_{i|i-1})$ is the one-step SSKF predictor for $x(t_i)$, minimizing (11) would have helped to avoid the bias. Obviously, (13) provides two clear examples of *Direct* (but) *Inaccessible Residual* (DIR) and in this case (11) is an example of *Direct* (but) *Inaccessible Performance Index* (DIPI).

Remark 1. In the limit as $t_0 \to -\infty$, FF equations (3) assure coincidence

$$\tilde{x}(t_i^+) = \hat{x}_{\mathrm{OPT}}(t_{i|i}) \quad \text{and} \quad \tilde{x}(t_i^-) = \hat{x}_{\mathrm{OPT}}(t_{i|i-1}) \tag{14}$$

if optimally designed for the fault-free system mode specified by $\theta^\dagger = \theta_0$.

Remark 2. If a set of AEs $g(t_{i|i-1}, \hat{\theta})$ in (13) contains SSKF, then $g(t_{i|i-1}, \hat{\theta})$ minimizing (11) coincides with $\hat{x}_{\mathrm{OPT}}(t_{i|i-1})$. If a fault occurs, we consider SSKF a *hidden* object to be identified to substitute FF and be used to modify FR.

Problem 1. To implement both SOP-based FDD and SSKF identification, we need to construct the *Accessible* albeit *Indirect Residual* (AIR) as a substitute for the DIR in order to change from the DIPI to the *Accessible* albeit *Indirect Performance Index* (AIPI) in (11), the latter having the same minimizing estimator as the DIPI.

Solution 1. Being the system completely observable enables the solution to be found in the form:

$$\mathrm{AIR}(t_i,\hat{\theta}) = \mathrm{DIR}(t_i,\hat{\theta}) + \mathrm{SP}(t_i), \quad \mathrm{SP}(t_i) \perp \mathrm{DIR}(t_i,\hat{\theta}),\ \mathrm{SP}(t_i) \neq f(\hat{\theta}) \quad (15)$$

$$\mathrm{AIPI}(\hat{\theta}) = \mathrm{DIPI}(\hat{\theta}) + \mathrm{Const}, \quad \mathrm{Const} = 1/2\,\mathrm{E}\{\|\mathrm{SP}(t_i)\|^2 \mid y(t_j^{i-1})\} \quad (16)$$

where $\mathrm{SP}(t_i)$ is a *slack process* fully independent of $\hat{\theta}$ $[\mathrm{SP}(t_i) \neq f(\hat{\theta})]$ and (16) is the half mean squared Euclidean norm of (15) in the limit as $t_j \to -\infty$.

Remark 3. The solution is given in [6], [7], [8] and thoroughly tested in [9] for identification purposes only. In what follows, we denote the $\mathrm{AIR}(t_i,\hat{\theta})$ by $\varepsilon(t_i,\hat{\theta})$ and use it here for the jointly performed FDD-SIA computational tasks. So from now on, $\varepsilon(t_i,\hat{\theta})$ serves instead of $r(t_i,\hat{\theta})$ in Formulas (8) to (12).

5 Fault Detection and Diagnosis

In accordance with Definition 2 and Remark 3, being the k-th expectation

$$\zeta_j(t_i,\hat{\theta}) = \mathrm{E}\{\left(\frac{\partial\varepsilon(t_i,\hat{\theta})}{\partial\hat{\theta}_j}, \varepsilon(t_i,\hat{\theta})\right[\mid \hat{\theta} = \hat{\theta}_\mu \,\mathrm{or}\, \hat{\theta}_\Upsilon\},\ j = 1,2,\ldots,L \quad (17)$$

of the dot product $(\cdot,\cdot)$ in the immediate vicinity of zero is equipollent to the fact that $\theta^\dagger = \theta_\mu$ or θ_Υ. On this basis we use it as a signature of the fact as it is shown schematically in Fig. 1(b) where ζ is composed of (17), $\zeta \in \mathrm{I\!R}^L$.

In FDD, when we test between (6), the theoretical signature

$$\forall\,\mu \text{ and } j = 1,2,\ldots,L: \quad \zeta_j(t_i,\hat{\theta}_\mu) = 0 \quad (18)$$

is checked by the following approximate decision rule:

$$\forall\,\mu = 1,2,\ldots,\mathcal{M} \text{ and } j = 1,2,\ldots,L: \quad \left|0.5 - \frac{n_j^\mu}{N}\right| \underset{\mathrm{H}_0}{\overset{\mathrm{H}_\mu}{\gtrless}} \gamma_j \quad (19)$$

where n_j^μ is the number of negative outliers of the process $\hat{\zeta}_j(t_i,\hat{\theta}_\mu)$ in the current sample of size N. Here $\hat{\zeta}_j(t_i,\hat{\theta}_\mu)$ is an estimator (for which we recommend the exponential smoothing) of $\zeta_j(t_i,\hat{\theta}_\mu)$ defined for (18) according to (17) for each μ-th *Signature Generator* (SG) within the *Bank of Signature Generators* (BSG), $\mu = 1,2,\ldots,\mathcal{M}$ (Fig. 2(a)). Signature Evaluator and Selector (SES) in this setup operates according to (19), where the sample size N and thresholds γ_j should be determined experimentally to guarantee the maximum decision power with the limited error probability. If several hypotheses among $\mathtt{H}_\mu$, $\mu = 1, 2, \ldots, \mathcal{M}$, have been selected, we discriminate between them by the additional rule: select the $\mathtt{H}_{\mu_*}$ with

$$\mu_* = \arg\min_{\mu} \sum_{j=1}^{L} \left| 0.5 - \frac{n_j^\mu}{N} \right| . \tag{20}$$

If none $\mathtt{H}_\mu$, $\mu = 1, 2, \ldots, \mathcal{M}$, has been selected, we adopt a decision that $\mathtt{H}_0$ is true.

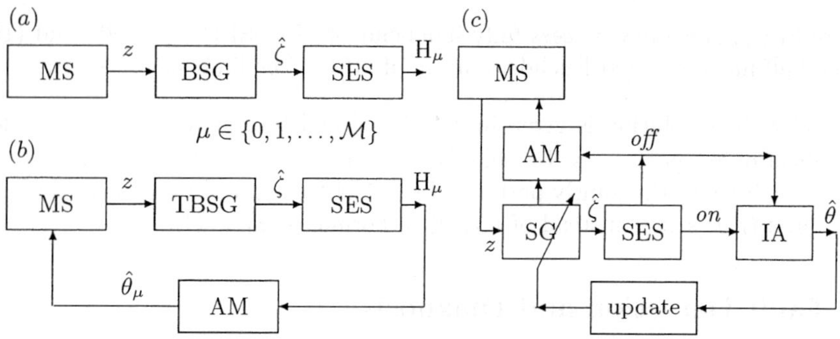

Fig. 2. (*a*) Fault detection and diagnosis (discrimination between $\mathcal{M}+1$ hypotheses). (*b*) Fault diagnosis and system accommodation using TBSG. (*c*) Fault point detection to switch IA and AM. Legend: MS – monitored system, BSG – Bank of Signature Generators, TBSG – Trained BSG, SES – Signature Evaluator and Selector, AM – Accommodation Mechanism, IA – Identification Algorithm

6 Optimal Filter Identification and BSG Training

To reason what change (or what fault) has occurred, one of two strategies may be used: parallel or sequential. The first one is adequate for Case (6), while the latter for Case (7). The parallel strategy implies the existence of the BSG as discussed in Sect. 5. Each SG can be trained beforehand to match the corresponding system mode, i.e. to satisfy (18). Training algorithms for a single (μ-th) SG are standard identification algorithms for $\hat{\theta} = \hat{\theta}_\mu$. For example, one can use the following SLS:

Simplified Least Squares. For $k = 1, 2, \ldots$ and $\Lambda = \mathrm{diag}[\lambda^{(j)}]$, $j = 1, 2, \ldots, L$, with $\Lambda_1 = I$ do (21) and then (22) where k is the step number:

$$\lambda_{k+1}^{(j)} = \lambda_k^{(j)} + \left\| \frac{\partial \varepsilon(t_i, \hat{\theta}_k)}{\partial \hat{\theta}_j} \right\|^2 \tag{21}$$

$$\hat{\theta}_{k+1} = \hat{\theta}_k - \Lambda_{k+1}^{-1} \mathcal{E}\{S^T(t_i, \hat{\theta}_k) \varepsilon(t_i, \hat{\theta}_k)\} \tag{22}$$

where $\mathcal{E}\{\cdot\}$ is a smoothing algorithm. In the case of exponential smoothing we obtain $\hat{\zeta}(t_i, \hat{\theta}) = \mathcal{E}\{\zeta(t_i, \hat{\theta})\}$ from the signature vector $\zeta(t_i, \hat{\theta}) = S^T(t_i, \hat{\theta})\varepsilon(t_i, \hat{\theta})$ by

$$\hat{\zeta}(t_i, \hat{\theta}) = \beta\hat{\zeta}(t_{i-1}, \hat{\theta}) + (1-\beta)\zeta(t_i, \hat{\theta}), \qquad 0 \le \beta < 1 \ . \tag{23}$$

In Case (7) we have to use the sequential strategy of Fig. 2(c) with one SG, one SES and one identification algorithm (IA). In this setup, SG is intended to detect and track all possible abrupt changes as they occur in time in the monitored system. The SG performs this by means of IA, which may be of (21)–(22) type or another. The SES tests between hypotheses (7) so that the rule (19) is implemented with $\mu = 1$. Decision to adopt $\mathtt{H}_1$ starts up the IA; decision to adopt $\mathtt{H}_0$ stops it and starts the accommodation mechanism (AM) in order to bring the feedback of MS and the real situation after change into better agreement and, in doing so, to adapt the system feedback to suit the new (after-change) conditions. Figure 3 shows how it is implemented except AM. Accommodation mechanism is discussed briefly in Sect. 7.

```
begin i := 0;
      for j := 1 to L do
            begin n_j := 0; s_j := off; λ_j := 1 end;
      while i ≤ imax do
            for j := 1 to l do
                  begin i := i + 1;
                        if ζ̂_j(t_i, θ̂) < 0 then n_j := n_j + 1;
                        f_j := 0.5 − n_j/N;
                        if i mod N = 0 then
                              begin n_j := 0;
                                    if abs(f_j) < γ_j then
                                       s_j := off else s_j := on
                              end;
                        if s_j = on then
                              begin
                                    λ^(j) := λ^(j) + || ∂ε(t_i, θ̂)/∂θ̂_j ||^2;
                                    θ̂^(j) := θ̂^(j) − ζ̂_j(t_i, θ̂)/λ^(j)
                              end else
                              if i mod N = 0 then λ^(j) := 1
                  end
end.
```

Fig. 3. Pseudocode sketching the inter-independent parameter identification algorithm

7 Monitored System Accommodation

To accommodate the MS to the newly developed (maybe, faulty) mode, the accommodation mechanism (AM) is expected to be capable to replace parameters $\bar{\Phi}$ and $\bar{K}$ of (3) with the newly identified matrix Φ and optimal SSKF gain K, and $\bar{G}_{\rm FR}$ of (4) with the $G_{\rm FR}$ computed by solving the *Discrete Algebraic Riccati Equation*, DARE, in compliance with the well known LQG control law using matrices of (5).

8 Approach Validation

I. System. Dimensions are $n = 2$, $m = 1$, $q = 1$, $l = 2$. Matrices are as follows:

$$\Phi_\theta = \begin{bmatrix} 0 & 1 \\ \theta_1 & \theta_2 \end{bmatrix}, \qquad \Psi_\theta = \Psi = \begin{bmatrix} 0 \\ 1 \end{bmatrix}, \qquad Q_\theta = Q = \begin{bmatrix} 0 & 0 \\ 0 & 1 \end{bmatrix}$$

$$H = \begin{bmatrix} 1 & 0 \end{bmatrix}, \qquad R_\theta = R = \begin{bmatrix} 0.1 \end{bmatrix}, \qquad \theta^T = \begin{bmatrix} \theta_1 & \theta_2 \end{bmatrix}$$

$$\theta_0^T = \begin{bmatrix} 0.30 & 0.68 \end{bmatrix}, \quad \theta_\Upsilon^T = \begin{bmatrix} 0.40 & 0.10 \end{bmatrix}, \quad \Upsilon^T = \begin{bmatrix} 0.10 & -0.58 \end{bmatrix}$$

$$\bar{\Phi} = \begin{bmatrix} 0 & 1 \\ 0.2 & 0.2 \end{bmatrix}, \qquad \bar{K} = \begin{bmatrix} 0.2 \\ 0.2 \end{bmatrix}, \qquad \bar{G}_{\rm FR} = \begin{bmatrix} 0.5 & 0.5 \end{bmatrix} .$$

II. Residual Generator. According to Remark 3, RG is composed of AE and AIR:

$$g(t_{i+1|i}, \hat{\theta}) = A[g(t_{i|i-1}, \hat{\theta}) + D\eta(t_i, \hat{\theta})] + \Psi u(t_i) \tag{24}$$

$$\eta(t_i, \hat{\theta}) = y(t_i) - Hg(t_{i|i-1}, \hat{\theta}) . \tag{25}$$

For AE (24), (25), we have here $L = 4$ and

$$A = \begin{bmatrix} 0 & 1 \\ a_1 & a_2 \end{bmatrix}, \qquad D = \begin{bmatrix} d_1 \\ d_2 \end{bmatrix}, \qquad \hat{\theta}^T = \begin{bmatrix} a_1 & a_2 & d_1 & d_2 \end{bmatrix} . \tag{26}$$

With reference to Remark 3, we construct AIR as

$$\varepsilon(t_i, \hat{\theta}) = \mathcal{N}(D)\eta(t_{i-s+1}^i, \hat{\theta}), \quad \mathcal{N}(D) = \begin{pmatrix} 1 & 0 & 0 & 0 \\ \ddots & 1 & 0 & 0 \\ d_{n-1} & \ddots & 1 & 0 \\ d_n & d_{n-1} & \ddots & 1 \end{pmatrix} . \tag{27}$$

In the first formula of (27), s is the greatest partial observability index of the system and the second formula holds if MS with $m = 1$ is taken in the standard observable form [7]. Such is here the case and $s = 2$,

$$\eta(t_{i-s+1}^i, \hat{\theta}) = [\eta(t_{i-1}, \hat{\theta}) \quad \eta(t_i, \hat{\theta})]^T . \tag{28}$$

III. Sensitivity Model. To obtain (12) from $\varepsilon(t_i, \hat{\theta})$, (27), in view of Remark 3, we have to generate the sensitivity functions having defined them by

$$\mu^{(j)}(t_{i+1|i}, \hat{\theta}) = \frac{\partial}{\partial \hat{\theta}_j} g(t_{i+1|i}, \hat{\theta}), \quad j = 1, \ldots, L \ . \tag{29}$$

IV. Pattern for Computational Experiments. The composite algorithm we test is outlined by Figs. 2(*c*) and 3. Besides the MS with the above matrices, it involves: RG – (24), (25), (26), (27) and (28); SM – (12) and (29); SG – (17); SES – (19) and (20); and IA – (21), (22) and (23). The algorithm is started up with the non-optimal parameters $\bar{\Phi}$, $\bar{K}$ and $\bar{G}_{\mathrm{FR}}$. The SES is expected to detect this initial non-optimality and to start up the IA precisely in respect to the parameters which require optimisation. The IA is expected to eliminate the initial non-optimality (this is the SG training). After that, at the middle time instant t_{8000}, the MS switches from mode $\mathtt{H}_0$ to $\mathtt{H}_\Upsilon$, cf. (7). The algorithm is expected to detect this mode switch, to identify the new optimal parameters and accommodate the MS feedback to the after-switch conditions.

V. Experimental Results. An example of the results is shown in Figs. 4 and 5. This is one of many results obtained for many different conditions [10]. In these experiments, the following values were chosen: $\gamma_j = 0.05$, $N = 2000$ and $\beta = 0.1$. The estimates formed by the composite algorithm are thus seen to be satisfactory enough. This lends support to the validity of the proposed approach.

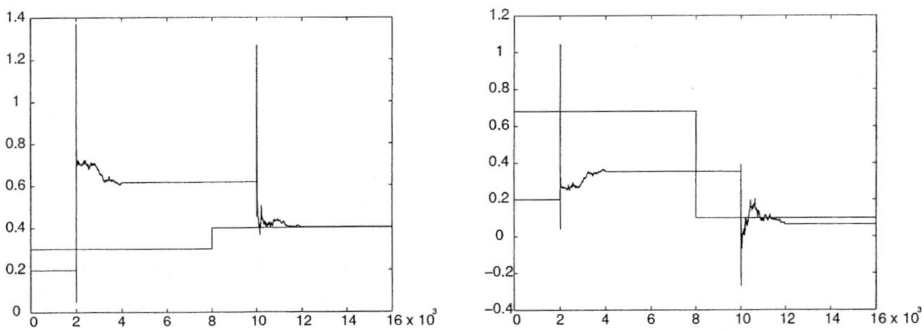

Fig. 4. Tracking the true parameters θ_1 and θ_2 by a_1 (*left*) and a_2 (*right*) of (26)

9 Conclusions and Future Work

We have shown that statistical orthogonality principle and indirect identification performance index provide the basis for the jointly performed computational tasks in the multi-mode stochastic system identification. Further investigations of this approach should be made to study all the aspects, both theoretical and

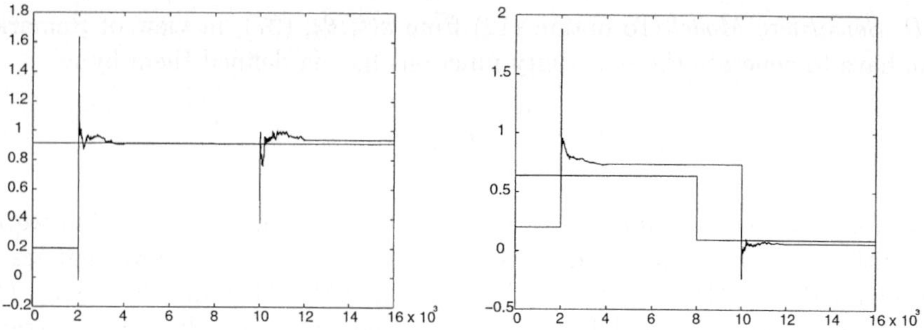

Fig. 5. Tracking the SSKF parameters k_1 and k_2 by d_1 (*left*) and d_2 (*right*) of (26)

applied, for instance – to evaluate it in terms of false alarm and missed detection rates.

References

1. Basseville, M., Nikiforov, I.: Detection of Abrupt Changes: Theory and Applications. Prentice-Hall Inc., New York (1993)
2. Landau, I.D. (ed.): Identification des Systemes. Les Bases de l'Identification des Systemes. Hermes, Paris (2001)
3. Caines, P.: Linear Stochastic Systems. John Willey & Sons, New York Chichester Brisbane Toronto Singapore (1988)
4. Soderstrom, T.: Identification of stochastic linear systems in presence of input noise. Automatica **17** (1981) 713–725
5. Ansay, P., Gevers, M., Wertz, V.: Closed-loop or open-loop models in identification for control? In: Proc. of the European Control Conference (1999) CD-ROM F544
6. Semoushin, I.V.: Adaptive Identification and Fault Detection Methods in Random Signal Processing. Saratov University Publishers, Saratov (1985) [*in Russian*]
7. Semoushin, I.V.: Adaptive control for stochastic linear plants under conditions of uncertainty. Nonlinear Dynamical Systems - Qualitative Analysis and Control: Collected Papers, The Institute for System Analysis of the Russian Academy of Sciences **2** (1994) 104–110
8. Semoushin, I.V., Tsyganova, J.V.: Indirect error control for adaptive filtering. In: Neittaanmaki, P., Tiihonen, T., Tarvainen, P. (eds.): Proc. of the 3rd European Conference on Numerical Mathematics and Advanced Applications. World Scientific, Singapore New Jersey London Hong Kong (2000) 333–340
9. Semoushin, I., Gorokhov, O.: Computational processes in iterative control design. In: Sloot, P.M.A., Kenneth Tan, C.J., Dongarra, J.J., Hoekstra, A.G. (eds.): Computational Science – ICCS 2002. Lecture Notes in Computer Science, Vol. 2329. Springer-Verlag, Berlin Heidelberg New York Barcelona Hong Kong London Milan Paris Tokyo (2002) 186–195
10. Kondratiev, A.E., Fatyanova, O.A.: Start-stop algorithms for adaptive filtering. In: Boulyarski, S.V. (ed.): Young Researchers' Collected Papers of the Ulyanovsk State University. UlSU Publishers, Ulyanovsk (2001) 6–7

Fault Point Detection with the Bank of Competitive Kalman Filters*

Innokenti Semoushin, Julia Tsyganova, and Maria Kulikova

Ulyanovsk State University, 42 L. Tolstoy Str., 432970 Ulyanovsk, Russia
{SemoushinIV, TsyganovaJuV, KulikovaMV}@ulsu.ru
http://staff.ulsu.ru/semousin/

Abstract. The problem of fault point detection in the linear stochastic discrete systems is considered. To solve this problem the algorithm with the finite size of the Bank of competitive Kalman filters is suggested. Theoretical results are confirmed by numerical experiments.

1 Introduction

The fault point detection problem has been discussed in many papers (see, for example [1], [2], [3]). One of the possible solutions to this problem is to use the Bank of competitive Kalman filters and some hypotheses testing algorithm. However, there is the problem of practical implementation of the Bank due to the increase in the number of Kalman filters in direct proportion with the length of testing interval. This causes large computational expenses.

To avoid this drawback, a new algorithm with the Bank of bounded number of Kalman filters is suggested in this paper. The authors multiple numerical experiments indicate the efficiency of this new algorithm for the fault point detection.

2 The SPRT for the Problem of Fault Detection

Consider the problem of detecting whether a stochastic discrete time system has one set of parameters or another.

Let the system be characterized by the following equations:

$$x_{t+1} = \Phi_t x_t + B_t u_t + \Gamma_t w_t \tag{1}$$

$$z_t = H_t x_t + v_t \tag{2}$$

where x_t is the n-dimensional state vector, z_t is the m-dimensional system output, u_t is the control input, and $\{w_0, w_1, \ldots\}$ and $\{v_1, v_2, \ldots\}$ are mutually independent zero-mean Gaussian sequences of independent vectors. Without loss of

* This work was supported in part by the Russian Ministry of Education (grant No. T02-03.2-3427).

P.M.A. Sloot et al. (Eds.): ICCS 2003, LNCS 2658, pp. 417–426, 2003.

generality, their covariances Q and R are assumed to be reduced to identity matrices: $Q = I$ and $R = I$. This can be easily done by normalizing input noise in (1) and measurements in (2). The sequences are considered independent of Gaussian initial x_0 with mean $\bar{x}_0$ and P_0. At any time t, we need to be ascertained, by performing a test on the sequence of measurements $Z_t = \{z_1, z_2, \ldots, z_t\}$, which of the following two hypotheses is true.

Hypothesis $\mathcal{H}_0$: system parameters are Φ_{t0}, B_{t0}, Γ_{t0}, and H_{t0}.
Hypothesis $\mathcal{H}_1$: system parameters are Φ_{t1}, B_{t1}, Γ_{t1}, and H_{t1}.

Consider two Kalman competitive filters designed, correspondingly, under the assumption of hypothesis $\mathcal{H}_0$ or $\mathcal{H}_1$. Let subscript i denote the hypothesis number, i.e. filter number, then the filter equations are:

Time propagation

$$\hat{x}_{ti}^- = \Phi_{ti}\hat{x}_{t-1,i}^+ + B_{ti}u_t, \qquad P_{ti}^- = \Phi_{ti}P_{t-1,i}^+\Phi_{ti}^T + \Gamma_{ti}\Gamma_{ti}^T \ .$$

Vector measurement update

$$\begin{aligned} K_{ti} &= P_{ti}^- H_{ti}^T (H_{ti}P_{ti}^- H_{ti}^T + I)^{-1}, & P_{ti}^+ &= (I - K_{ti}H_{ti})P_{ti}^- \\ \nu_{ti} &= z_t - H_{ti}\hat{x}_{ti}^-, & \hat{x}_{ti}^+ &= \hat{x}_{ti}^- + K_{ti}\nu_{ti} \ . \end{aligned} \tag{3}$$

Each of the following sequences

$$\mathrm{N}_{t0} = \{\nu_{10}, \nu_{20}, \ldots, \nu_{t0}\} \ \text{ and } \ \mathrm{N}_{t1} = \{\nu_{11}, \nu_{21}, \ldots, \nu_{t1}\} \tag{4}$$

consists of mutually independent entries $\nu_{\tau i} = z_\tau - H_{\tau i}\hat{x}_{\tau i}^-$, $\tau = 1, 2, \ldots, t$, subject to the condition that the corresponding hypothesis, $i = 0$ or $i = 1$, is true.

The Wald sequential probability ratio test contains the following recursive algorithm to evaluate the likelihood ratio $\lambda_t = \ln p\{\mathrm{N}_{t1}|\mathcal{H}_1\}/p\{\mathrm{N}_{t0}|\mathcal{H}_0\}$:

$$\lambda_t = \lambda_{t-1} + \mu_t, \quad t \geq 1$$

where

$$2\mu_t = \log\det \Sigma_{t0} - \log\det \Sigma_{t1} + \nu_{t0}^T \Sigma_{t0}^{-1}\nu_{t0} - \nu_{t1}^T \Sigma_{t1}^{-1}\nu_{t1} \ . \tag{5}$$

The mean and variance of sequences (4) for filters $i = 0, 1$ are given by

$$E\{\nu_{ti}\} = 0, \qquad E\{\nu_{ti}\nu_{ti}^T\} = \Sigma_{ti} = H_{ti}^T P_{ti}^- H_{ti} + I \ .$$

The value of λ_t is then tested against two threshold levels A and B (where $A > B$):

$$\begin{cases} \text{if } \lambda_t \leq B, & \text{the test is terminated with the choice of } \mathcal{H}_0. \\ \text{if } \lambda_t \geq A, & \text{the test is terminated with the choice of } \mathcal{H}_1. \\ \text{if } A > \lambda_t > B, & \text{the test is repeated.} \end{cases}$$

The thresholds A and B are chosen in the following way. Let α be the probability of choosing $\mathcal{H}_1$ when $\mathcal{H}_0$ is true, and β be the probability of choosing $\mathcal{H}_0$ when $\mathcal{H}_1$ is true. Then

$$A = \log\frac{1-\beta}{\alpha} \qquad \text{and} \qquad B = \log\frac{\beta}{1-\alpha} \ .$$

The initial value of the test is $\lambda_0 = \log\{P_1/P_0\}$, where P_0 and P_1 are a'priori probabilities of occurrence of hypotheses $\mathcal{H}_0$ and $\mathcal{H}_1$. If $P_0 = P_1 = \frac{1}{2}$ (i.e. hypotheses $\mathcal{H}_0$ and $\mathcal{H}_1$ are equiprobable), then it is clear that the initial value $\lambda_0 = 0$.

The SPRT requires the computation of likelihood ratio function λ_t at each moment t. There was suggested an efficient algorithm to compute a likelihood function for any competitive filter [4]. The basis for this algorithm is the Conventional Kalman Filter with scalar measurement update. The authors obtained two equivalent proofs of the equivalence between two methods of likelihood function evaluation: the vector method with the help of equations (3)–(5) and the scalar method according to the suggested algorithm [4]. The first proof is based on the Information form of Kalman Filter [4], and the second one is algebraic [5]. In the same manner as the Conventional Kalman Filter, one can modify [4] the other data processing algorithms that are described in [6] and based on the covariance factorization: Potter Square Root; Bierman LD-Covariance Factorization; and Carlson Triangular Covariance Square Root.

3 Competitive Fault Point Detection Using a Bank of Kalman Filters

In this section, we solve the problem of fault point detection. The fault point is the time instant at which the parameters of system (1), (2) were changed.

Suppose that hypothesis $\mathcal{H}_0$ describes the nominal mode of system operation, i.e. the fault-free mode.

3.1 The Exact Solution to the Problem of Fault Point Detection

Consider that a ratio test is started at time $t = t_0$, but the point of fault occurrence $t_{01} \in [t_0, t_N]$, (where $t_i \in [t_0, t_N]$ is a testing interval) is unknown. Therefore, in addition to the main hypothesis $\mathcal{H}_0$, $N-1$ alternative hypotheses $\mathcal{H}_j$ are added instead of one alternative hypothesis $\mathcal{H}_1$. Each $\mathcal{H}_j$ means that at the moment t_j, the system parameters are changed from $\{\Phi_{t0}$, B_{t0}, Γ_{t0}, and $H_{t0}\}$ to $\{\Phi_{t1}$, B_{t1}, Γ_{t1}, and $H_{t1}\}$.

Taking into account these prior assumptions, we can write down the expression for the likelihood ratio function:

$$\lambda_t^{j0} = \log \frac{P\{\nu_{j1}^j, \ldots, \nu_{t1}^j | \mathcal{H}_j\}}{P\{\nu_{j0}, \ldots, \nu_{t0} | \mathcal{H}_0\}} = \lambda_{t-1}^{j0} + \mu_t^{j0}, \quad t \geq 1 \tag{6}$$

where

$$2\mu_t^{j0} = \log\det \Sigma_{t0} - \log\det \Sigma_{t1}^j + \nu_{t0}^T \Sigma_{t0}^{-1} \nu_{t0} - \left(\nu_{t1}^j\right)^T \left(\Sigma_{t1}^j\right)^{-1} \nu_{t1}^j \ . \tag{7}$$

The value λ_t^{j0} needs to be calculated at the moment $t = t_0 + j$, where $t \in [t_0, t_N]$.

In Sect. 2, two Kalman filters were required, one based on the assumption of hypothesis $\mathcal{H}_0$ and the other on assumption of $\mathcal{H}_1$. In the present case, we need one Kalman filter per hypothesis, that is, $N+1$ filters.

Let us denote the Kalman filter based on $\mathcal{H}_0$ by $\mathcal{F}_0$ and the Kalman filters based on $\mathcal{H}_j$ by $\mathcal{F}_j$, $j = 1, \ldots, N$. Then in (7), ν_{t1}^j is a residual with the covariance Σ_{t1}^j, obtained from the Kalman filter $\mathcal{F}_j$. So we have the Bank of $N+1$ competitive Kalman filters.

The ratio test for system fault point detection is as follows [3]:

1. If $\forall t < t_N\ \lambda_t^{j0} \le B$, the test is terminated with the choice of $\mathcal{H}_0$ (the system fault point was not detected on the testing interval).
2. If $\exists!\ j : \lambda_t^{j0} \ge A$, the test is terminated with the choice of $\mathcal{H}_j$ (the system fault point was detected at the moment $t_0 + j$).
3. If $\forall j : B < \lambda_t^{j0} \le A$, the test is repeated for $t+1$.
4. If $\exists i \ne j : \lambda_t^{j0} \ge A$ and $\lambda_t^{i0} \ge A$, the test is terminated with the choice of "leading" hypothesis $\mathcal{H}_l$, for which the value of likelihood function is maximum, that is $\lambda_t^{l0} = \max\{\lambda_t^{j0}, \lambda_t^{i0}\}$ (the system fault point was detected at the moment $t_0 + l$).
5. If $\forall i, j\ \lambda_t^{j0} \le B$ and $B < \lambda_t^{i0} < A$, the hypothesis $\mathcal{H}_j$ is excluded from the set of the considered hypothesis and the test is repeated for $t+1$.

(8)

Now we can describe the testing algorithm:

The test is performed on the testing interval $[t_0, t_N]$; from the beginning of the test at each subsequent moment $t = j$, the Kalman filter $\mathcal{F}_j$ based on $\mathcal{H}_j$ is included into the contest of filters (into the Bank of Kalman filters).

Thus, after processing N measurements $N+1$ filters are added to the Bank, moreover each filter $\mathcal{F}_j$, $j = 1, \ldots, N$ matches likelihood against filter $\mathcal{F}_0$.

If all hypotheses $\mathcal{H}_j$, $j = 0, \ldots, N$, are equiprobable, then the initial value of each likelihood ratio function λ_t^{j0} equals zero. When a'priori probabilities of all hypotheses are given, the initial condition for the functioning of filter $\mathcal{F}_j$ is the value $\lambda_0^{j0} = \log\{P_j/P_0\}$, where P_j is an a'priori probability of hypothesis $\mathcal{H}_j$.

So, the value λ_t^{j0} begins to change at Step j of the test at the moment of filter $\mathcal{F}_j$ connection. For all $t < t_0 + j$, $\lambda_t^{j0} = 0$, that corresponds with the condition that the filter is disconnected.

It is clear that according to the suggested algorithm, the number of competitive filters grows at each step of the test. This, in turn, leads to the growth of necessary calculations of likelihood ratio function at each step of the test. The realization of such algorithm in an ideal case will require unlimited computational resources (machine time and an enormous memory size for the data storing).

Therefore in spite of the simplicity of the suggested algorithm and the guaranteed solution to the problem, there is the essential drawback that the algorithm is not practically realizable for the solving of real life problems because of the unbounded growth of computations at each step of the test.

In the next section, we solve the problem of elimination of such obvious drawback of the obtained solution, i.e. the unbounded growth of the number of competitive filters.

3.2 The New Algorithm of Fault Point Detection with the Bank of Finite Number of Filters

From the previous topic, there follows the necessity to find the solution to the problem of Fault Point Detection with the Bank of finite Number of Kalman filters. In other words, the problem to evaluate the necessary number of competitive Kalman filters is posed.

We need to determine the average sample numbers required for decision making between the two hypotheses $\mathcal{H}_1$ and $\mathcal{H}_0$. We first need to know the average values of the increment in the probability ratio λ_t at each stage in the test.

In other words, we need to obtain expressions for

$$\mu_1 = E\left\{\log(\det \Sigma_{t0})^{1/2} - \log(\det \Sigma_{t1})^{1/2} + \frac{1}{2}\{\nu_{t0}^T \Sigma_{t0}^{-1} \nu_{t0} - \nu_{t1}^T \Sigma_{t1}^{-1} \nu_{t1}\} | \mathcal{H}_1\right\}$$

$$\mu_0 = E\left\{\log(\det \Sigma_{t0})^{1/2} - \log(\det \Sigma_{t1})^{1/2} + \frac{1}{2}\{\nu_{t0}^T \Sigma_{t0}^{-1} \nu_{t0} - \nu_{t1}^T \Sigma_{t1}^{-1} \nu_{t1}\} | \mathcal{H}_0\right\}$$

at time $t = n$.

Let the system and filter equations be given by (1), (2), and (3), and all system parameters not depend on time (i. e. $\Phi_{ti} = \Phi_i$, $H_{ti} = H_i$, $B_{ti} = B_i$, $\Gamma_{ti} = \Gamma_i$).

Suppose that in filter equations we use system parameters Φ_1, H_1, B_1, Γ_1 corresponding to the change in the characteristics (i.e. hypothesis $\mathcal{H}_1$) instead of true values of matrix parameters Φ_0, H_0, B_0, Γ_0 (i.e. hypothesis $\mathcal{H}_0$). In this case, in order to determine the values of μ_0 and μ_1, we need to know the actual correlation matrix of residuals $\bar{\Sigma}_{t0} = E\{\nu_{t0}\nu_{t0}^T\}$. In the case of optimal Kalman filter, the residuals $\nu(t)$ have zero mean and covariance matrix Σ_{t0}.

Let us write $\bar{\Sigma}_{t0}$ as

$$\bar{\Sigma}_{t0} = E\{\nu_{t0}\nu_{t0}^T\} = E\{(H_0 x_t - H_1 \hat{x}_t^-)(H_0 x_t^T - H_1 (\hat{x}_t^-)^T)\} + I \ . \qquad (9)$$

After the expansion, we obtain the formula for calculation of correlation matrix of residuals

$$\begin{aligned} \bar{\Sigma}_{t0} = {} & H_0 E\{x_t x_t^T\} H_0^T - H_0 E\{x_t (\hat{x}_t^-)^T\} H_1^T \\ & + H_1 E\{\hat{x}_t^- (\hat{x}_t^-)^T\} H_1^T - H_1 E\{\hat{x}_t^- x_t^T\} H_0^T + I \ . \end{aligned} \qquad (10)$$

For its calculation, we need to first solve the following difference equations:

$$\begin{aligned} E\{x_t x_t^T\} = {} & \Phi_0 E\{x_{t-1} x_{t-1}^T\} \Phi_0^T + \Gamma_0 \Gamma_0^T + B_0 u_{t-1} E\{x_{t-1}^T\} \Phi_0^T \\ & + \Phi_0 E\{x_{t-1}\} u_{t-1}^T B_0^T + B_0 u_{t-1} u_{t-1}^T B_0^T \\ E\{x_t (\hat{x}_t^-)^T\} = {} & \Phi_0 E\{x_{t-1} (\hat{x}_{t-1}^+)^T\} \Phi_1^T + \Phi_0 E\{x_{t-1}\} u_{t-1}^T B_1^T \end{aligned} \qquad (11)$$

$$+B_0 u_{t-1} E\{(\hat{x}_{t-1})^T\}\Phi_1^T + B_0 u_{t-1} u_{t-1}^T B_1^T \tag{12}$$

$$E\{x_t(\hat{x}_t^+)^T\} = E\{x_t(\hat{x}_t^-)^T\}(I - H_1^T K_t) + E\{x_t x_t^T\} H_0^T K_t \tag{13}$$

$$\begin{aligned} E\{\hat{x}_t^-(\hat{x}_t^-)^T\} &= \Phi_1 E\{\hat{x}_{t-1}^+(\hat{x}_{t-1}^+)^T\}\Phi_1^T + \Phi_1 E\{\hat{x}_{t-1}^+\} u_{t-1}^T B_1^T \\ &\quad + B_1 u_{t-1} E\{(\hat{x}_{t-1}^+)^T\}\Phi_1^T + B_1 u_{t-1} u_{t-1}^T B_1^T \end{aligned} \tag{14}$$

$$\begin{aligned} E\{\hat{x}_t^+(\hat{x}_t^+)^T\} &= (I - K_t H_1) E\{\hat{x}_t^-(\hat{x}_t^-)^T\}(I - H_1^T K_t^T) \\ &\quad + (I - K_t H_1) E\{\hat{x}_t^- x_t^T\} H_0^T K_t^T \\ &\quad + K_t H_0 E\{x_t(\hat{x}_t^-)^T\}(I - H_1^T K_1^T) \\ &\quad + K_t H_0 E\{x_t x_t^T\} H_0^T K_t^T + K_t K_t^T \end{aligned} \tag{15}$$

$$E\{\hat{x}_t^+\} = (I - K_t H_1) E\{\hat{x}_t^-\} + K_t H_0 E\{x_t\} \tag{16}$$

$$E\{\hat{x}_t^-\} = \Phi_1 E\{\hat{x}_{t-1}^+\} + B_1 u_{t-1} \tag{17}$$

$$E\{x_t\} = \Phi_0 E\{x_{t-1}\} + B_0 u_{t-1} \tag{18}$$

satisfying the initial conditions

$$\begin{aligned} E\{x_0 x_0^T\} &= E\{x_0(\hat{x}_0^+)^T\} = E\{\hat{x}_0^+(\hat{x}_0^+)^T\} = P_0 + \bar{x}_0 \bar{x}_0^T \\ E\{x_0\} &= E\{\hat{x}_0^+\} = \bar{x}_0 \ . \end{aligned} \tag{19}$$

Similarly, we can obtain the formula for $\bar{\Sigma}_{t1}$ when true values of matrix parameters are Φ_1, H_1, B_1, Γ_1 (this corresponds to hypothesis $\mathcal{H}_1$), but in filter equations parameters Φ_0, H_0, B_0, Γ_0 are used.

Consider the equilibrium solutions

$$\begin{aligned} \Sigma_0 &= \lim_{t\to\infty} \Sigma_{t0}, & \Sigma_{t0}^{-1} &= [\sigma_{ij}^0] \\ \Sigma_1 &= \lim_{t\to\infty} \Sigma_{t1}, & \Sigma_{t1}^{-1} &= [\sigma_{ij}^1] \\ \bar{\Sigma}_0 &= \lim_{t\to\infty} \bar{\Sigma}_{t0} = [\bar{\sigma}_{ij}^0], & \bar{\Sigma}_1 &= \lim_{t\to\infty} \bar{\Sigma}_{t1} = [\bar{\sigma}_{ij}^1] \ . \end{aligned} \tag{20}$$

Then

$$\begin{aligned} 2\mu_1 &= \log\det \Sigma_0 - \log\det \Sigma_1 + \sum_{i,j=1}^{m} (\bar{\sigma}_{ij}^0 \sigma_{ij}^0) - m \\ 2\mu_0 &= \log\det \Sigma_0 - \log\det \Sigma_1 - \sum_{i,j=1}^{m} (\bar{\sigma}_{ij}^1 \sigma_{ij}^1) + m \ . \end{aligned} \tag{21}$$

Now we can find the average sample numbers:
$\bar{N}_B$ is the average number of samples required to reach any threshold from the start of the test, assuming $\mathcal{H}_0$ is true;
$\bar{N}_A$ is the average number of samples required to reach any threshold from the start of the test, assuming $\mathcal{H}_1$ is true.

These sample numbers are given by

$$\begin{aligned} \bar{N}_B &= [\alpha A + (1-\alpha) B]/\mu_0 \\ \bar{N}_A &= [(1-\beta) A + \beta B]/\mu_1 \ . \end{aligned} \tag{22}$$

Let M be the necessary number of samples for decision making in the fault point detection problem. We evaluate M as

$$M = \max(\bar{N}_A, \bar{N}_B) \ . \tag{23}$$

That means that on the testing interval $[t_0, t_N]$ at each time moment only $M+1$ Kalman filters can work simultaneously, and each Kalman filter is based on the corresponding hypothesis.

From the beginning of the test through the first M steps, filters $\mathcal{F}_i$, $i = 1, \ldots, M$, enter the Bank. They match the Filter $\mathcal{F}_0$ corresponding the hypothesis $\mathcal{H}_0$ which means that "the fault point was not detected on the M steps of the test".

The ratio test for system fault point detection is the same as (8).

The only difference is that competitive filters make a queue, moreover each filter works for only a limited number of steps (in our case, M steps). At the $(M+1)$-th step of the test, the filter $\mathcal{F}_i$, which was in the Bank during M steps and has not made the decision (this corresponds to condition $B < \lambda_{1_i 0} < A$), is excluded from the Bank. Instead, the new filter $\mathcal{F}_{M+1}$ is based on the hypothesis $\mathcal{H}_{M+1}$ which means that "the system fault occurred on the $(M+1)$-th step of the test", that is, immediately after moment $t = t_0 + M + 1$.

This algorithm allows us to solve the fault point detection problem with less computational resources consumption.

However, such algorithm is more complex due to the necessity of obtaining the value M. This requires us to solve equations (9)–(19). Moreover, if estimate of M is found incorrectly, and it is less than the real number of samples required for decision making, then according to the suggested algorithm the solution may not be found.

4 Numerical Example

To show how the algorithms of the previous sections may be used to detect the parameter change point for linear dynamical systems, we consider the following example taken from the inertial navigation [7]:

$$x_{t+1} = \begin{bmatrix} 0.75 & -1.74 & -0.3 & 0.0 & -0.15 \\ 0.09 & 0.91 & -0.0005 & 0.0 & -0.008 \\ 0.0 & 0.0 & 0.95 & 0.0 & 0.0 \\ 0.0 & 0.0 & 0.0 & 0.55 & 0.0 \\ 0.0 & 0.0 & 0.0 & 0.0 & 0.905 \end{bmatrix} x_t + \begin{bmatrix} 0.0 & 0.0 & 0.0 \\ 0.0 & 0.0 & 0.0 \\ 24.64 & 0.0 & 0.0 \\ 0.0 & 0.835 & 0.0 \\ 0.0 & 0.0 & 1.83 \end{bmatrix} w_t$$

$$z_t = \begin{bmatrix} 1-e & 0 & 0 & 0 & 1-f \\ 0 & 1-g & 0 & 1-h & 0 \end{bmatrix} x_t + v_t, \qquad e, f, g, h = \{0, 1\}$$

$\{w_t\}$ and $\{v_t\}$ are zero-mean white Gaussian sequences with covariances $Q_t = I_3$, and $R_t = I_2$ (I_n is the n-dimensional identity matrix). These equations describe the damped Shuler loop driven by the exponentially correlated 3-dimensional noise w_t [7].

The program for numerical experiments was written in Pascal.

Let us describe the numerical experiments for the problem of fault point detection. Consider the previous example and two hypotheses: $\mathcal{H}_0$ which means the nominal mode of the system (the system parameters are $\Phi_0 = \Phi(t_i)$, $\Gamma_0 = \Gamma(t_i)$, $B_0 = B(t_i)$, $H_0 = H(t_i)$, where $[e, f, g, h] = [0, 0, 0, 0]$; and $\mathcal{H}_1$ which means the fault mode of the system (the system parameters are $\Phi_1 = \Phi(t_i)$, $\Gamma_1 = \Gamma(t_i)$, $B_1 = B(t_i)$, $H_1 = H(t_i)$, where $\{[e, f, g, h] | e, f, g, h \in \{0, 1\}\} \setminus \{[0, 0, 0, 0]\}$. So, there are 16 types of system faults. All experiments are conducted, provided that the system fault point is random.

Table 1. Detection of the fault point in the system. The type of possible fault is $[0, 0, 1, 0]$. The algorithm with the increasing KFB size.

Experiment number	LF at the moment of detection	The fault point	Detected fault point	Delay in detection	Accepted hypothesis	KFB size
1	16.715688	39	39	0	$\mathcal{H}_1$	44
2	12.159357	21	22	1	$\mathcal{H}_1$	22
3	66.411028	36	36	0	$\mathcal{H}_1$	36
4	12.822501	33	33	0	$\mathcal{H}_1$	35
5	22.600478	11	25	14	$\mathcal{H}_1$	26
6	96.973967	41	41	0	$\mathcal{H}_1$	41
7	67.016929	16	16	0	$\mathcal{H}_1$	16
Mean delay in detection: 2 iterations						

Let us demonstrate the efficiency of fault point detection algorithms considered in Sect. 3.1 and Sect. 3.2.

We conducted a series of experiments for the problem of fault point detection. The conditions of experiments are: the error probabilities $\alpha = 0.00001$, $\beta = 0.00001$, two thresholds $A = 11.512915$ and $B = -11.512915$; the point of possible system fault is unknown and random for each experiment. For all experiments, the system fault really takes place. That means that it is necessary to confirm the hypothesis $\mathcal{H}_1$ and to detect the fault point by using the available measurements. The experiments are conducted for the faults of two types: $[e, f, g, h] = [0, 0, 1, 0]$ and $[e, f, g, h] = [0, 0, 0, 1]$.

The efficiency of the testing algorithm described in Sect. 3.1, can be evaluated according to the data of Table 1. The testing interval here is $[1, 50]$, the number of experiments equals 7, and the series of 500 experiments were also conducted.

According to the experimental data, we conclude that the algorithm with increasing size of the Kalman Filters Bank (KFB) provides a guaranteed solution of the fault point detection problem, but it has one obvious drawback. Since the maximum number of Kalman Filters in the Bank tends to the length of testing interval, the algorithm can not be practically applied to the real life problems.

Now we will try to evaluate the efficiency of the algorithm with the Bank of finite number of competitive Kalman Filters, described in Sect. 3.2.

The results of numerical experiments are shown in Table 2. For the realization of a ratio test before testing, we need to evaluate the necessary parameter M (that is the KFB size) by using the equations (9)–(19). It is clear that this estimate will differ in different types of system faults. Table 2 shows the testing results for the system fault $[0, 0, 1, 0]$ and the testing interval $[1, 100]$. Given the error probabilities α and β and the type of system fault, the maximum number of simultaneously competitive filters needed for the decision making equals 27.

Table 2. Detection of the fault point in the system. The type of possible fault is $[0, 0, 1, 0]$. The algorithm with the finite KFB.

Experiment number	LF at the moment of detection	The fault point	Detected fault point	Delay in detection	Accepted hypothesis	KFB size
1	24.242963	48	49	1	$\mathcal{H}_1$	9
2	39.596861	55	55	0	$\mathcal{H}_1$	24
3	18.171810	5	5	0	$\mathcal{H}_1$	7
4	41.172073	40	40	0	$\mathcal{H}_1$	6
5	19.113414	12	17	5	$\mathcal{H}_1$	27
6	15.477792	31	31	0	$\mathcal{H}_1$	19
7	35.373164	11	11	0	$\mathcal{H}_1$	6
8	41.409267	76	77	1	$\mathcal{H}_1$	12
9	18.612180	19	19	0	$\mathcal{H}_1$	8
Mean delay in detection: 1 iteration						

Table 3. Detection of the fault point in the system. The type of possible fault is $[0, 0, 1, 0]$. The algorithm with the finite KFB.

Experiment number	LF at the moment of detection	The fault point	Detected fault point	Delay in detection	Accepted hypothesis	KFB size
1	24.242963	48	49	1	$\mathcal{H}_1$	9
2	—	55	—	—	—	—
3	18.171810	5	5	0	$\mathcal{H}_1$	7
4	41.172073	40	40	0	$\mathcal{H}_1$	6
5	—	75	—	—	—	—
6	—	67	—	—	—	—
7	35.373164	11	11	0	$\mathcal{H}_1$	6
8	41.409267	76	77	1	$\mathcal{H}_1$	12
9	18.612180	19	19	0	$\mathcal{H}_1$	8

According to the data in Table 2, the algorithm with the finite number of Kalman filters is efficient for solving the fault point detection problem. So, we have practically verified the correctness of the theoretical estimate for the KFB size, because all the experiments required that the real number of competitive filters be less or equal to 27.

The efficiency of such algorithm essentially depends on the correct choice of the KFB size. If parameter M is incorrect, and is less (but not greater) than the real number of required Kalman filters, then the solution may not be found. Such case is confirmed by the data of Table 3. Here the value of M was chosen to be 10, but really it must be 27. According to Table 3, in 50% of cases the fault point was not detected on all testing interval $[1, 100]$ (dash corresponds to these cases).

5 Conclusion

The concept of the Bank of Competitive Kalman Filters is applicable to the problem of fault point detection in stochastic system behavior.

The algorithm with increasing number of Kalman filters provides a guaranteed solution to the fault point detection problem, but it has an essential drawback: it can not be practically realized for the real life problems.

To avoid this drawback, another algorithm with the finite size of Kalman Filters Bank was considered in this paper. The authors suggest a method to find the estimate of required size of Kalman Filters Bank. All theoretical results are confirmed by multiple numerical experiments.

References

1. Newbold, P., M. and Ho Yu-Chi: Detection of Changes in the Characteristics of a Gauss-Markov Process. IEEE Trans. on Aerosp. and Electron. Systems, Vol. **AES-4(5)** (1968) 707–718
2. Hanlon, P., D., Maybeck, P., S.: Equivalent Kalman Filter Bank Structure for Multiple Model Adaptive Estimation (MMAE) and Generalized Likelihood Ratio (GLR) Failure Detection. Proc. of the 36th Conference on Decision & Control: San Diego California USA **5** (1997) 4312–4317
3. Semoushin, I., V.: The Quickest in the Mean Manoeuvre Detection with the Guaranteed Probability Error (Methods). Shipbuilding: Computing Techniques **26** (1990) 3–7 [*In Russian*]
4. Semoushin, I., V., Tsyganova, J., V.: An Efficient Way to Evaluate Likelihood Functions in Terms of Kalman Filter Variables. In: Murgu, A. and Lasker, G., E. (eds.): Adaptive, Cooperative and Competitive Processes in Systems Modelling, Design and Analysis. The International Institute for Advanced Studies in Systems Research & Cybernetics: University of Windsor Windsor Ontario Canada (2001) 67–74
5. Semoushin, I., V., Tsyganova, J., V., Kulikova, M., V.: On Computation of the Likelihood Ratio Function for Gaussian Markov Signals. In: Andreev, A., C. (ed.): Basic problems of mathematics and mechanics: Ulyanovsk State University Ulyanovsk Russia **2(7)** (2000) 93–100 [*In Russian*]
6. Bierman, G., J.: Factorization Methods for Discrete Sequential Estimation. Academic Press, New-York (1977)
7. Stratton, A.: Combination of Inertial Navigation and Radio Systems. In Borisov, N., I. (ed.): Problems of Inertial Navigation, Mashinostroienie, (1961) [*In Russian*]

On Effective Computation of the Logarithm of the Likelihood Ratio Function for Gaussian Signals*

Maria V. Kulikova

Ulyanovsk State University, L. Tolstoy Str. 42, 432970 Ulyanovsk, Russia
KulikovaMV@ulsu.ru

Abstract. In this paper we consider a new way to calculate the logarithm of the Likelihood Ratio Function for Gaussian signals. This approach is based on the standard Kalman filter. Its efficiency is substantiated theoretically, and numerical examples show how such a method works in practice.

1 Introduction and Problem Statement

One of the most important trends of the modern control theory for linear discrete stochastic systems is a fault detection problem. A wide class of the well-known algorithms applied to this situation is based on a concept that is usual in mathematical statistics, namely the logarithm of the Likelihood Ratio Function (Log-LRF) [1]:

$$\text{Log-LRF} \stackrel{\text{def}}{=} \ln \frac{f_{\theta_1}(z)}{f_{\theta_0}(z)} = \ln f_{\theta_1}(z) - \ln f_{\theta_0}(z) \tag{1}$$

where $f_\theta(z)$ is a parametrized probability density and z is a measurement vector. Thus, the efficient evaluation of Log-LRF is very important, with this point of view, for practical use. It follows from (1) that the announced goal can be attained by developing an algorithm for effective calculation of the logarithm of the Likelihood Function (Log-LF) $\ln f_\theta(z)$.

In the paper, we consider the following linear discrete system with the white noise:

$$x_{t+1} = \Phi_t x_t + \Gamma_t w_t, \tag{2a}$$

$$z_t = H_t x_t + v_t \tag{2b}$$

where $x_t \in R^n$, $w_t \in R^q$, $z_t \in R^m$, $\{w_0, w_1, \ldots\}$ and $\{v_1, v_2, \ldots\}$ are mutually independent zero-mean Gaussian sequences of independent vectors. Without loss of generality, their covariances Q and R are assumed to be reduced to identity matrices; i.e., $Q = I$ and $R = I$. The latter can be easily done by normalizing

* This work was supported in part by the Russian Ministry of Education (grant No. T02-03.2-3427).

P.M.A. Sloot et al. (Eds.): ICCS 2003, LNCS 2658, pp. 427–435, 2003.

the input noise in (2a) and the measurements in (2b). We suppose also that the both sequences mentioned above are independent of the Gaussian initial x_0 with a mean $\bar{x}_0$ and a covariance matrix P_0.

It is easy to see that the Likelihood Function (LF) for the t-th measurement z_t in system (2) is given by the formula

$$f(z_t|Z_{t-1}) = \left[(2\pi)^m |\Sigma_t|\right]^{-1/2} \exp\left\{-\frac{1}{2}\left[z_t - H_t\hat{x}_t^-\right]^T \Sigma_t^{-1}\left[z_t - H_t\hat{x}_t^-\right]\right\},$$

where $\Sigma_t \stackrel{\text{def}}{=} H_t P_t^- H_t^T + I$, provided that measurements $Z_{t-1} \stackrel{\text{def}}{=} \{z_1, z_2, \dots z_{t-1}\}$ have been processed. We further denote the updating sequence by $\nu_t \stackrel{\text{def}}{=} z_t - H_t\hat{x}_t^-$. It is characterized by the covariance matrix Σ_t. Then Log-LF has the following form:

$$\ln f(z_t|Z_{t-1}) = -\frac{m}{2}\ln(2\pi) - \frac{1}{2}\ln|\Sigma_t| - \frac{1}{2}\nu_t^T \Sigma_t^{-1}\nu_t. \qquad (3)$$

2 Algorithm of Efficient Evaluation of Log-LRF

When calculating Log-LRF, most of the execution time is spent for finding two of the summands, namely $\ln|\Sigma_t|$ and $\nu_t^T \Sigma_t^{-1} \nu_t$ (see formulas (1) and (3)). With the aim to derive the effective way for evaluating Log-LF, we consider the standard algorithm of Kalman filter [2]. We call such computation *the vector treatment* because the measurement vector z_t at the moment t is being processed as a whole.

Let us now consider another algorithm for computing $\ln|\Sigma_t|$ and $\nu_t^T \Sigma_t^{-1} \nu_t$ which is also formulated on the basis of the Kalman filter. We call the new approach *the scalar treatment* since the measurement vector z_t at the moment t is being processed in a component-wise way. This method is given as follows:

Algorithm 1

I. Set initial values: $\hat{x}_t^0 = \hat{x}_t^-$; $P_t^0 = P_t^-$; $\delta_t^0 = 0$; $\Delta_t^0 = 0$.
II. Compute for $k = 1, 2, \dots, m$:

$$\alpha_k = h_t^{(k)} P_t^{(k-1)} (h_t^{(k)})^T + 1; \qquad (4a)$$

$$K_t^{(k)} = P_t^{(k-1)} (h_t^{(k)})^T / \alpha_k;$$

$$P_t^{(k)} = P_t^{(k-1)} - K_t^{(k)} h_t^{(k)} P_t^{(k-1)};$$

$$\nu_t^{(k)} = z_t^{(k)} - h_t^{(k)} \hat{x}_t^{(k-1)}; \qquad (4b)$$

$$\hat{x}_t^{(k)} = \hat{x}_t^{(k-1)} + K_t^{(k)} \nu_t^{(k)}; \qquad (4c)$$

$$\delta_t^{(k)} = \delta_t^{(k-1)} + \ln\alpha_k; \qquad \Delta_t^{(k)} = \Delta_t^{(k-1)} + (\nu_t^{(k)})^2/\alpha_k.$$

III. Obtain results:

$$\ln|\Sigma_t| = \delta_t^{(m)}; \qquad \nu_t^T \Sigma_t^{-1} \nu_t = \Delta_t^{(m)}; \qquad P_t^+ = P_t^{(m)}.$$

First of all it is necessary to show equivalence of the two methods mentioned above for evaluation of Log-LF: the vector treatment (standard algorithm) and the scalar treatment (according to the proposed algorithm 1). This result is given by

Theorem 1 *Let the* $(1 \times n)$*-matrix* $h_t^{(k)}$ *be the k-th row of* H_t*, and the scalar*

$$z_t^{(k)} = h_t^{(k)} x_t + v_t$$

be the k-th element of z_t *in (2b). Then values of the both summands* $\ln|\Sigma_t|$ *and* $\nu_t^T \Sigma_t^{-1} \nu_t$ *in the right-hand side of formula (3) can be obtained by the scalar treatment (algorithm 1) that is used instead of vector treatment.*

Proof. For convenience, we fix an arbitrary time point t and omit the time from all the formulas. If we now consider the structure of matrix Σ given by the formula $\Sigma \stackrel{\text{def}}{=} H^T P^{(0)} H + I$ then we have

$$\Sigma \stackrel{\text{def}}{=} \begin{bmatrix} h_1^T P^{(0)} h_1 + 1 & h_1^T P^{(0)} h_2 & \cdots & h_1^T P^{(0)} h_m \\ h_2^T P^{(0)} h_1 & h_2^T P^{(0)} h_2 + 1 & \cdots & h_2^T P^{(0)} h_m \\ \vdots & \vdots & \ddots & \vdots \\ h_m^T P^{(0)} h_1 & h_m^T P^{(0)} h_2 & \cdots & h_m^T P^{(0)} h_m + 1 \end{bmatrix}.$$

Further we split the proof of Theorem 1 into two parts.

Part I. Here, we want to prove validity of the formula

$$\ln|\Sigma| = \sum_{k=1}^{m} \ln(\alpha_k) \tag{5}$$

where the magnitudes α_k, $k = 1, \ldots, m$, are calculated by (4a). To complete it, we apply induction with respect to the dimension of matrix Σ as follows:

1. Let $m = 2$. Then we need to verify the equality

$$\ln|\Sigma| = \ln(\alpha_1) + \ln(\alpha_2)$$

or, equivalently,

$$|\Sigma| = \alpha_1 \alpha_2,$$

and α_k, $k = 1, 2$, satisfy formula (4a).

The simple substitution gives

$$\alpha_1 \alpha_2 = \sigma_{11}^{(0)} \left(\sigma_{22}^{(0)} - \frac{1}{\alpha_1} \left[\sigma_{12}^{(0)} \right]^2 \right) = \sigma_{11}^{(0)} \sigma_{22}^{(0)} - \left[\sigma_{12}^{(0)} \right]^2 = |\Sigma|,$$

where $\sigma_{ij}^{(0)}$ denotes the (i,j)-th element of the matrix Σ. Thus, we conclude that Part I of Theorem 1 is valid when $m = 2$.

2. Let now (5) be true for $m = s$. Then the formula

$$|\Sigma| = \prod_{k=1}^{s} \alpha_k \tag{6}$$

with α_k, $k = 1, \dots, s$, calculated by (4a) takes place. Further we want to justify (6) when $m = s + 1$.

To do this, we consider the matrix $\Sigma^{(1)} \stackrel{\text{def}}{=} H^T P^{(1)} H + I$ where $P^{(1)}$ is determined in algorithm 1. The latter yields the formula

$$\Sigma^{(1)} = H^T P^{(0)} H - \frac{1}{\alpha_1} H^T P^{(0)} h_1 h_1^T P^{(0)} H + I$$

or

$$\Sigma^{(1)} = \begin{bmatrix} 1 & 0 & \dots & 0 \\ 0 & \sigma_{22}^{(0)} - \left[\sigma_{12}^{(0)}\right]^2 / \sigma_{11}^{(0)} & \dots & \sigma_{2,s+1}^{(0)} - \sigma_{21}^{(0)} \sigma_{1,s+1}^{(0)} / \sigma_{11}^{(0)} \\ \dots & \dots & & \dots \\ 0 & \sigma_{s+1,2}^{(0)} - \sigma_{s+1,1}^{(0)} \sigma_{12}^{(0)} / \sigma_{11}^{(0)} & \dots & \sigma_{s+1,s+1}^{(0)} - \left[\sigma_{s+1,1}^{(0)}\right]^2 / \sigma_{11}^{(0)} \end{bmatrix}.$$

We see that the matrix $\Sigma^{(1)}$ has been obtained from the matrix Σ by the Gauss method with the pivot $\sigma_{11}^{(0)}$.

From the structure of $\Sigma^{(1)}$ and determinant's properties it follows

$$|\Sigma| = \sigma_{11}^{(0)} |\Sigma^{(1)}|. \tag{7}$$

Let us now denote the minor of matrix $\Sigma^{(1)}$ obtained by omitting the first row and the first column in this matrix as Σ^*. Then the formula

$$\sigma_{ij}^* = \sigma_{i+1,j+1}^{(0)} - \frac{\sigma_{i+1,1}^{(0)} \sigma_{1,j+1}^{(0)}}{\sigma_{11}^{(0)}}, \quad i, j = 1, \dots, s,$$

is valid for each element of the matrix Σ^*.

By the inductive hypothesis, we derive

$$|\Sigma^*| = \prod_{k=1}^{s} \alpha_k^* \tag{8}$$

because the matrix Σ^* is of dimension $s \times s$. Here, the magnitudes α_k^*, $k = 1, \dots, s$, are calculated by (4a) provided that one iteration has been done. Therefore, (7), (8) and the notation introduced above lead to the final result

$$|\Sigma| = \sigma_{11}^{(0)} |\Sigma^{(1)}| = \sigma_{11}^{(0)} |\Sigma^*| = \alpha_1 \prod_{k=1}^{s} \alpha_k^* = \alpha_1 \prod_{k=2}^{s+1} \alpha_k = \prod_{k=1}^{s+1} \alpha_k,$$

where α_k, $k = 1, \dots, s + 1$, satisfy (4a), that completes the proof. Thus, Part I of the theorem has been justified.

Part II. In this Part, it is necessary to prove validity of the formula

$$\nu^T \Sigma^{-1} \nu = \sum_{k=1}^{m} \frac{\nu_k^2}{\alpha_k} \tag{9}$$

where the left-hand side of (9) means the vector treatment, as in equations of the standard Kalman filter, and the right-hand one implies the component-wise treatment by algorithm 1. We now apply induction with respect to the dimension of matrix Σ as well as in Part I.

1. Let $m = 2$. We need to verify the equality

$$\nu^T \Sigma^{-1} \nu = \sum_{k=1}^{2} \frac{\nu_k^2}{\alpha_k}$$

or, equivalently,

$$\frac{\nu_1^2 A_{11} + 2\nu_1 \nu_2 A_{12} + \nu_2^2 A_{22}}{|\Sigma|} = \frac{\nu_1^2}{\alpha_1} + \frac{\nu_2^2}{\alpha_2} \tag{10}$$

where A_{ij} is a relevant cofactor of the matrix Σ. Here we have also used the symmetry of matrix Σ.

Since the left-hand side of (10) means the vector treatment and the right-hand one implies the component-wise treatment, we further denote the i-th component of vector ν in the left-hand side by $(\nu_i)_v$ and the i-th component of vector ν in the right-hand side by $(\nu_i)_{cw}$. Then it is easy to see that

$$(\nu_1)_v = (\nu_1)_{cw} = z^1 - h_1^T x^0,$$

but $(\nu_2)_v \neq (\nu_2)_{cw}$ because

$$\begin{aligned}(\nu_2)_v &= z^2 - h_2^T x^0 \left(\pm h_2^T x^1\right) = z^2 - h_2^T x^1 + h_2^T (x^1 - x^0) \\ &= (\nu_2)_{cw} + h_2^T \left(x^0 + K^1 \nu_1 - x^0\right) = (\nu_2)_{cw} + h_2^T K^1 \nu_1.\end{aligned}$$

Thus, we can transform the vector treatment (left-hand side of formula (10)) to the component-wise treatment by substitution of the above results. We obtain

$$\begin{aligned}&\left\{(\nu_1^2)_v A_{11} + 2(\nu_1)_v (\nu_2)_v A_{12} + (\nu_2^2)_v A_{22}\right\} / |\Sigma| \\ = &\Big\{(\nu_1^2)_{cw}\sigma_{22} - 2(\nu_1)_{cw}(\nu_2)_{cw}\sigma_{12} - 2(\nu_1^2)_{cw}\sigma_{12} h_2^T K^1 + (\nu_2^2)_{cw}\sigma_{11} \\ &+2(\nu_1)_{cw}(\nu_2)_{cw}\sigma_{11} h_2^T K^1 + \sigma_{11}(\nu_1^2)_{cw}\left[h_2^T K^1\right]^2\Big\} / |\Sigma|\end{aligned}$$

where K^1 is calculated by algorithm 1; i.e., $K^1 = P^{(0)} h_1 / \alpha_1$. Now we take into account that from formula (5), proved in Part I of the theorem, it follows $|\Sigma| = \alpha_1 \alpha_2$. Finally, we yield the formula

$$\frac{(\nu_2^2)_{cw}\sigma_{11}}{|\Sigma|} - \frac{2(\nu_1)_{cw}(\nu_2)_{cw}\sigma_{12}}{|\Sigma|} + \frac{2(\nu_1)_{cw}(\nu_2)_{cw}\sigma_{11}\left(h_2^T P^{(0)} h_1\right)}{\alpha_1 |\Sigma|}$$

$$+\frac{(\nu_1^2)_{cw}}{|\Sigma|}\left[\sigma_{22}-\frac{2\sigma_{12}^2}{\sigma_{11}}+\frac{\sigma_{12}^2}{\sigma_{11}}\right]=\frac{(\nu_1^2)_{cw}}{\alpha_1}+\frac{(\nu_2^2)_{cw}}{\alpha_2}$$

which completes the proof of (10) when $m=2$.

2. Let us assume that Theorem 1 is true for $m=s$; i.e., the formula

$$\nu^T\Sigma^{-1}\nu=\sum_{k=1}^{s}\frac{\nu_k^2}{\alpha_k} \tag{11}$$

with α_k, $k=1,\dots,s$, computed by (4a) takes place for any square matrix Σ of dimension s. Our goal now is to check (9) for $m=s+1$.

First of all it is easy to see that

$$\nu^T\Sigma^{-1}\nu=\frac{1}{|\Sigma|}\left[\sum_{j=1}^{s+1}\nu_j\sum_{i=1}^{s+1}\nu_iA_{ij}\right]$$

$$=\frac{1}{|\Sigma|}\left[\sum_{j=2}^{s+1}\nu_j\sum_{i=2}^{s+1}\nu_iA_{ij}+2\sum_{i=2}^{s+1}\nu_i\nu_1A_{i1}+\nu_1^2A_{11}\right]. \tag{12}$$

Here, as above, A_{ij} means the relevant cofactor of the symmetric matrix Σ. Let us apply the first step of Gaussian elimination to Σ and denote the new matrix by $\hat{\Sigma}$. We have to show that formula (11) holds for the matrix $\hat{\Sigma}$.

On the other hand, from formula (12) we obtain

$$\nu^T\hat{\Sigma}^{-1}\nu=\frac{1}{|\hat{\Sigma}|}\left[\sum_{j=2}^{s+1}\nu_j\sum_{i=2}^{s+1}\nu_i\hat{A}_{ij}\right]$$

$$+\frac{1}{|\hat{\Sigma}|}\left[2\sum_{i=2}^{s+1}\nu_i\nu_1\hat{A}_{i1}+\nu_1^2\hat{A}_{11}\right]$$

where $\hat{A}_{ij}$, $i,j=1,\dots,s+1$, are relevant cofactors of the matrix $\hat{\Sigma}$. Having used the structure of matrix $\hat{\Sigma}$ we easily conclude that $\hat{A}_{1j}=0$, $j=2,\dots,s+1$. Thus,

$$\nu^T\hat{\Sigma}^{-1}\nu=\frac{1}{|\hat{\Sigma}|}\left[\sum_{j=2}^{s+1}\nu_j\sum_{i=2}^{s+1}\nu_i\hat{A}_{ij}+\nu_1^2\hat{A}_{11}\right]. \tag{13}$$

Then, taking into account the introduced notation, (6) and the formula

$$\hat{A}_{11}=\prod_{k=2}^{s+1}\alpha_k$$

which follows from the proof of Part I immediately, (13) becomes

$$\nu^T\hat{\Sigma}^{-1}\nu=\frac{\sum_{j=2}^{s+1}\nu_j\sum_{i=2}^{s+1}\nu_i\hat{A}_{ij}}{|\hat{\Sigma}|}+\frac{\nu_1^2}{\alpha_1}. \tag{14}$$

Let now the minor derived by omitting the first row and the first column in the matrix $\hat{\Sigma}$ be denoted as Σ_*. We note that $\hat{A}_{ij} = \sigma_{11} A^*_{i-1,j-1}$, $i,j = 2,\ldots,s+1$, where A^*_{ij} is the relevant cofactor of the matrix Σ_*. It means that the following equality is valid:

$$\sum_{j=2}^{s+1} \nu_j \sum_{i=2}^{s+1} \nu_i \hat{A}_{ij} = \sigma_{11} |\Sigma_*| \nu_*^T \Sigma_*^{-1} \nu_* = |\hat{\Sigma}| \nu_*^T \Sigma_*^{-1} \nu_*.$$

Here, $\nu_* = [\nu_2, \nu_3, \ldots, \nu_{s+1}]^T$.

Having remembered that the inductive hypothesis takes place for the matrix Σ_* of dimension $s \times s$, we come to the formula

$$\nu_*^T \Sigma_*^{-1} \nu_* = \sum_{k=1}^{s} \frac{[\nu_k^*]^2}{\alpha_k^*} \tag{15}$$

where the magnitudes α_k^*, $k = 1,\ldots,s$, are calculated by (4a) provided that one iteration has been done. Therefore, from the notation introduced above and formulas (14), (15) it follows

$$\nu^T \hat{\Sigma}^{-1} \nu = \sum_{k=2}^{s+1} \frac{\nu_k^2}{\alpha_k} + \frac{\nu_1^2}{\alpha_1} = \sum_{k=1}^{s+1} \frac{\nu_k^2}{\alpha_k}$$

where α_k, $k = 1,\ldots,s+1$, satisfy (4a). That proves Part II of the theorem. Thus, Theorem 1 has been completely justified.

Now we show the efficiency of the scalar treatment (Algorithm 1) for evaluation of Log-LF. In this context, we calculate and compare the total number of the arithmetical operations: +, -, /, * needed for the computation of Log-LF as by the vector treatment as by the scalar one (see Table 1).

According to the data of Table 1, we can conclude that the usage of the scalar treatment allows the total number of arithmetical operations to be reduced by the power of 2 with respect to m (i.e., with respect to the dimension of measurement vector z_t). The latter increases the speed of computation of Log-LRF by Algorithm 1 significantly.

3 Numerical Examples

As a test problem, we consider the example from [4]. Let system (2) be defined as follows:

$$x_{t+1} = \begin{bmatrix} 0.75 & -1.74 & -0.3 & 0.0 & -0.15 \\ 0.09 & 0.91 & -0.0005 & 0.0 & -0.008 \\ 0.0 & 0.0 & 0.95 & 0.0 & 0.0 \\ 0.0 & 0.0 & 0.0 & 0.55 & 0.0 \\ 0.0 & 0.0 & 0.0 & 0.0 & 0.905 \end{bmatrix} x_t + \begin{bmatrix} 0.0 & 0.0 & 0.0 \\ 0.0 & 0.0 & 0.0 \\ 24.64 & 0.0 & 0.0 \\ 0.0 & 0.835 & 0.0 \\ 0.0 & 0.0 & 1.83 \end{bmatrix} w_t,$$

Table 1. Total number of arithmetical operations for evaluation of Log-LF with the both treatments (vector and scalar)

Treatment	Total number of operations		
	Additions	Subtractions	Multiplications and Divisions
Vector	$3n^3+n^2(2+q+3m)+n(q^2+2m^2+2m+1)+2m^2+m$	n^2+m	$3n^3+n^2(1+q+3m)+n(q^2+2m^2+2m)+m^3+m^2+2m$
Scalar	$2n^3+n^2(2+q+3m)+n(q^2+3m)+3m$	mn^2+m	$2n^3+n^2(1+q+4m)+n(q^2+4m)+2m$

$$z_t = \begin{bmatrix} 1\,0\,0\,0\,1 \\ 0\,1\,0\,1\,0 \end{bmatrix} x_t + v_t$$

where $\{w_t\}$ and $\{v_t\}$ are zero-mean white noise sequences with covariance matrices $Q_t = I$ and $R_t = I$ (I denotes the identity matrix).

Table 2 shows the results of numerical experiments which confirm Theorem 1 in practice. We see that the differences between the both methods do not exceed the computer accuracy. All software for the numerical tests was written in Pascal.

Table 2. Values of Log-LF computed by the vector and scalar treatments

Discrete Time	Vector Treatment (y_1)	Scalar Treatment (y_2)	$\|y_1 - y_2\|_\infty$
0	-5.42	-5.42	0.0
10000	-4.46	-4.46	4.3E-19
20000	-5.84	-5.84	4.3E-19
30000	-6.35	-6.35	4.3E-19
40000	-4.47	-4.47	4.3E-19
50000	-4.42	-4.42	0.0
60000	-4.51	-4.51	3.8E-19
70000	-4.82	-4.82	0.0
80000	-9.79	-9.79	3.3E-19
90000	-7.83	-7.83	4.3E-19
100000	-8.78	-8.78	0.0

Table 3 shows the execution time (in seconds) required for the vector and scalar treatments in the computation of Log-LF. The advantage of the scalar algorithm will be especially noticeable on systems with large measurement vectors (i.e., when the dimension m is large). According to the data of Table 3, we conclude that the gain in time on the 40000-th iteration is approximately two seconds for our example. (Note that $m = 2$ there, and all the experiments were developed with processor Intel Pentium-II.) It is obvious that increasing of the parameter m results the greater profit of Algorithm 1 in execution time.

Table 3. Execution time (in sec.) for the evaluation of Log-LRF computed by the vector and scalar treatments

Discrete Time	Vector Treatment (t_1)	Scalar Treatment (t_2)	$\|t_1 - t_2\|$
0	0.0	0.0	0.0
10000	4.91	3.65	1.26
20000	9.80	7.87	1.21
30000	13.39	11.93	1.46
40000	17.59	15.69	1.90
50000	22.32	19.59	2.73
60000	26.79	23.37	3.42
70000	31.13	27.37	3.76
80000	35.75	31.28	4.47
90000	40.28	35.50	5.23
100000	44.87	38.97	5.90

Some additional experiments for practical determination of parameters changes in linear dynamic systems were developed for the methods from [1]. The data obtained also confirm the efficiency of the scalar treatment presented in the paper.

4 Conclusion

Thus, we have developed the new effective way for evaluation of Log-LRF for Gaussian signals that is very important for a number of problems in practice. The necessary theory has been given and the efficiency has been investigated on the test examples. We have shown clearly that the vector treatment and the scalar one produce the same result, but the execution time is smaller for the latter method. That is why we recommend this algorithm for practical implementation in real time problems.

Acknowledgment. The author would like to thank Prof. I.V. Semoushin for the problem statement and valuable notes.

References

1. Basseville, M., Nikiforov, I.: Detection of Abrupt Changes: Theory and Applications. Prentice-Hall, Englewood Cliffs, New Jersey, 1993
2. Bierman, G.J.: Factorization methods for discrete sequential estimation. Academic Press, New York, 1977
3. Newbold, P.M., Ho, Yu-Chi.: Detection of changes in the characteristics of a Gauss-Markov process. IEEE Trans. Aerosp. and Electr. Syst. **AES-4** (1968) No. 5, 707–718
4. Semoushin, I.V.: Practical method of fault detection in navigational systems. (*in Russian)* Izvestiya VUZov. Ser. Priborostroenie. (1981) No. 3, 53–56.

Developing a Simulation Tool Box in MATLAB and Using It for Non-linear Adaptive Filtering Investigation*

Oleg Gorokhov and Innokenti Semoushin

Ulyanovsk State University, 42 Leo Tolstoy Str., 432970 Ulyanovsk, Russia
gorohov@icbank.ru SemoushinIV@ulsu.ru http://staff.ulsu.ru/semoushin/

Abstract. In this paper we develop a special purpose tool box for complex computational tasks solution in the area of stochastic adaptive system design. The proposed tool box is used to analyze the influence of different factors on the quality of numerical algorithms.

1 Introduction

Modern system design requires high performance computational modelling tools. MathCAD, MATLAB and Maple are examples. The complexity of problems in the adaptive filtering area often does not allow us to use the standard procedures and tool boxes to analyze the problems even of small dimensions. The reason lies in the amount of time necessary for computing while analyzing the influence of a set of many factors on the algorithm performance. This raises the problem of developing a special tool for complex system investigation.

The purpose of this paper is twofold. The first goal is to develop an efficient tool box for specific problems in the field of adaptive linear or non-linear filtering. The second part of the paper demonstrates the application of the designed tool box and presents the simulations results. The core of the designed tool box is implemented as a dynamic link library, and its interface part is made MATLAB-compatible.

2 Adaptive Non-linear Filtering Problem

Consider the linear stochastic time-invariant discrete-time model

$$\begin{aligned} x_{t+1} &= \Phi x_t + \Gamma w_t \\ z_t &= H x_t + v_t \end{aligned} \tag{1}$$

which is widely used in processing the experimental data for a state of dynamical plants in stochastic environment. Here $x_t \in \mathbb{R}^n$ is the state vector, $z_t \in \mathbb{R}^m$ is the

* This work was supported in part by the Russian Ministry of Education (grant No. T02-03.2-3427).

P.M.A. Sloot et al. (Eds.): ICCS 2003, LNCS 2658, pp. 436–445, 2003.

measurement vector and $\{w_0, w_1, \ldots\}$ and $\{v_1, v_2, \ldots\}$ are zero-mean independent sequences of independent identically distributed random vectors $w_t \in \mathbb{R}^q$ and $v_t \in \mathbb{R}^m$ with covariances Q and R respectively. Measurements are assumed to be incomplete, i.e. $m < n$.

Classical approach to the state estimation for model (1) is that corresponding to the Kalman filtering. However, sometimes it is impossible to use the Kalman filter if the system contains a predesigned part which is nonlinear by its nature. When noise distributions are not Gaussian, optimal filter is not Kalman. In these cases, the whole system should be characterized by nonlinear equations, whose parameters are difficult to optimize. For such situations, an adaptive (or learning) approach seems to be the only possible way. It becomes absolutely necessary if, for a variety of reasons, parameters of model (1) are not precisely known and thus should be identified from experimental data [1], [2].

We assume that the initial time is placed at $-\infty$ and adopt the non-restrictive assumptions that $R > 0$, $Q \geq 0$, and system (1) is stabilizable and completely observable. Additionally, the spectral radius of Φ is assumed to be strictly less than one to have all the processes in (1) wide-sense stationary. In the below experiments, uncertainty inherent to (1) resides in Γ, Q and R only.

In the most case, the estimator $y_t \in \mathbb{R}^n$ is given by a nonlinear functional $\mathcal{O}(\cdot)$ intended to generate a p-step-ahead (predicted) estimate of the state x_{t+p} using all available measurements $z_{t-\tau}, 0 \leq \tau < \infty$. Thus, we have

$$y_t = \mathcal{O}(z_{t-\tau}, 0 \leq \tau < \infty) \tag{2}$$

where each k-th component y_t^k of y_t is desribed by the discrete time Volterra series

$$y_t^k = \sum_{l=1}^{m} \sum_{d=1}^{\infty} \{ \sum_{\tau_1=0}^{\infty} \cdots \sum_{\tau_d=0}^{\infty} h_{kld}(\tau_1, \ldots, \tau_d) z_{t-\tau_1}^{(l)} \cdots z_{t-\tau_d}^{(l)} \}$$

and where $l = 1, 2, \ldots, m$ denotes the component index in the vector of measurements z_t, d is the order of nonlinearity of $h_{kld}(\tau_1, \ldots, \tau_d)$, the d-order kernel of Volterra series. We assume the estimator to be stable so that $\{y_t\}$ is a wide-sense stationary sequence.

Let the system error, original performance index and a problem to be solved are as follows:

$$e_t = x_{t+p} - y_t, \qquad J_o = \mathrm{E}\{\|e_t\|^2\} .$$

Problem 1. Given performance index J_o, formulate the auxiliary performance index J_a depending on some available sequence ε_t so that

$$J_a = \mathrm{E}\{\|\varepsilon_t\|^2\} = J_o + const \tag{3}$$

where *const* does not depend on set of the nonlinear estimator parameters, thus performance indices J_a and J_o have the same optimal points on the set.

We use a new solution to the problem given in [3] with the following notations: $z_{t+1-s}^t \in \mathbb{R}^{sm}$ is a column vector composed of column vectors z_{t+1-s} through z_t, s stands for the observability index, and T is the $n \times n$ observability matrix.

3 Nonlinear Filtering and MATLAB Algorithm Implementation

In this section we present the general computational algorithm for adaptive estimation of parameters. We also demonstrate how the adaptive filtering algorithm is implemented by the designed MATLAB tool box procedures. The MATLAB object `objExperiment` (Fig. 1) includes all information and settings for numerical simulations such as stabilization filter time, signal noise ratio, initial conditions.

```
objExperiment =struct('Tmax', 'TimeStab', 'TimeSwitching', ...
'Alpha', 'Qbefore', 'Qafter', 'Rbefore', 'Rafter', 'PHIbefore',...
'PHIafter', 'P0', 'AdaptiveProcedure');
```

Fig. 1. Tool box experiment object structure

Assuming, for generality sake, that the uncertainty can reside in matrices Φ, Γ and covariances Q and R, we denote the parameterized versions of these matrices as $\Phi_\theta, \Gamma_\theta, Q_\theta$ and R_θ respectively. The system (1) state generation procedure consists of two parts: before change point in the system and after change. The goal of the after-change algorithm is to identify the new parameter values. Tool box function `phiModel` (Fig. 2) demonstrates the top-level programming code without taking into consideration the details hidden in the low-level functions `phiModelInitialization`, `phiGenerateNoise` and others.

The model state estimates are obtained as a result of nonlinear transformation $\mathcal{L}_\theta\{\cdot\}$ of the feedback suboptimal filter estimate

$$\begin{aligned} \tilde{x}(t_{i+1}^-) &= \Phi_{\theta_0}\tilde{x}(t_i^+) \\ \tilde{x}(t_i^+) &= \mathcal{L}_{\theta_0}\{\tilde{x}(t_i^-) + K_{\theta_0}\nu(t_i)\} \\ \nu(t_i) &= z(t_i) - H\tilde{x}(t_i^-)\ . \end{aligned} \quad (4)$$

The gain K_θ is replaced by the result of each iteration (the whole identification process). The initial value for K_{θ_0} is set as some nominal value which is chosen a'priori to satisfy the stability conditions. The corresponding implementation is given at (Fig. 2) where the estimation of the system state vector is performed via Kalman filter before reaching stabilization and when via stabilized filter.

The adaptive model is appended to this system and started with initial state taken from the suboptimal filter (4). It has the following form

$$\begin{aligned} \tilde{g}(t_{i+1}) &= A_\theta\hat{g}(t_i) \\ \hat{g}(t_i) &= \mathcal{L}_\theta\{\tilde{g}(t_i) + D_\theta\eta(t_i)\} \\ \eta(t_i) &= z(t_i) - H_*\tilde{g}(t_i) \end{aligned} \quad (5)$$

```
function [objObject] = phiModel(objExperiment, objObject, t)
    if (t == 0)
        phiModelInitialization(objExperiment, objObject);
    else
        if (t < objExperiment.TimeSwitching)
            objObject = phiBeforeSwitching(objObject,t);
        else objObject = phiAfterSwitching(objObject,t);
        end;
    objObject = phiGenerateNoise(objExperiment, objObject,t);
    objObject = phiGenerateDynamics(objExperiment, objObject,t);
    end;
end;
```

Fig. 2. Tool box model object code generating state of the system at time t

where $A_\theta = T_\theta \Phi_\theta T_\theta^{-1}, H_* = HT_\theta^{-1}$ and T_θ is the observability matrix defined in [3]. Let us assume that the collective parameter θ represents the set of adjustable parameters in the model indexed accordingly (the Kalman gain D_θ, the matrix A_θ and the nonlinear transformation $\mathcal{L}_\theta\{\cdot\}$).

```
function [objFilter] = phiFilter(objExperiment, objObject, t)
    if (t == 0)
        phiFilterInitialization(objExperiment, objObject);
    else
        if (t < objExperiment.TimeStab)
            objFilter = phiKalmanFilter(objObject, objFilter,t);
        end;
    objFilter = phiStabFilter(objFilter,t);
    end;
end;
```

Fig. 3. Tool box filter implementation

Denote the stackable vector of $[\eta(t_{i-s+1}), \ldots, \eta(t_i)]$ as $\mathcal{H}_{i-s+1}^i$ where s is the maximal partial observability index. Then the model error between the adaptive and suboptimal models can be written in the following form

$$\varepsilon(t_i) = \mathcal{N}(D_\theta)\mathcal{H}_{i-s+1}^i \tag{6}$$

where $\mathcal{N}(D)$ is the structure transformation of adaptive model gain D as defined in [3].

```
function [objAdFilter] = phiAdaptiveFilter(objExperiment, objObject, t)
     if (t == objExperiment.timeSwitching)
          objAdFilter = phiAdFilterInitialization(objObject);
     else
          objAdFilter = phiKalmanFilter(objObject, objAdFilter,t);
          objAdFilter = phiSensitivityModel(objAdFilter,t);
               if phiAdaptiveStep(objAdFilter, t)
                    objAdFilter = phiSensitivity(objAdFilter,t);
                    objAdFilter = phiGradient(objAdFilter,t);
                    objAdFilter = phiApproximation(objAdFilter,t);
                    objAdFilter = phiTrialStep(objAdFilter,t);
                    objAdFilter = phiStability(objAdFilter,t);
               end;
     end;
end;
```

Fig. 4. Tool box adaptive filter implementation

The sensitivity model that reflects the influence of the adjustable parameters on the model error (6) and in fact is the partial derivatives of vector $\varepsilon(t_i)$ wrt. vector θ, is defined by three types of recursions according to the placement of adjustable parameter. Let μ denote the sensitivity model state vector. We have

$$\begin{aligned} \tilde{\mu}_j(t_i) &= A_\theta \hat{\mu}_j(t_{i-1}) \\ \hat{\mu}_j(t_i) &= \tfrac{\partial \mathcal{L}_\theta}{\partial x}((I - D_\theta H_*)\tilde{\mu}_j(t_i) + \tfrac{\partial D_\theta}{\partial \theta_j}\eta(t_i)) \end{aligned} \tag{7}$$

where θ_j is a parameter of vector D_θ. The second type of sensitivity equations is defined for parameters θ_j of transition matrix A:

$$\begin{aligned} \tilde{\mu}_j(t_i) &= \tfrac{\partial A_\theta}{\partial \theta_j}\hat{g}(t_{i-1}) + A_\theta \hat{\mu}_j(t_{i-1}) \\ \hat{\mu}_j(t_i) &= \tfrac{\partial \mathcal{L}_\theta}{\partial x}(I - D_\theta H_*)\tilde{\mu}_j(t_i) \ . \end{aligned} \tag{8}$$

The influence of the nonlinear estimator parameters on the error (6) are calculated as follows

$$\begin{aligned} \tilde{\mu}_j(t_i) &= A_\theta \hat{\mu}_j(t_{i-1}) \\ \hat{\mu}_j(t_i) &= \tfrac{\partial \mathcal{L}_\theta}{\partial \theta_j} + \tfrac{\partial \mathcal{L}_\theta}{\partial x}\tilde{\mu}_j(t_i) \ . \end{aligned} \tag{9}$$

All recursions start with initial values $\hat{\mu}_j(t_0) = 0$ for each θ_j. Let vector $\xi_j(t_i)$ be used to denote $-H_*\tilde{\mu}_j(t_i)$. The history for vectors $\xi_j(t_i)$ and $\mathcal{H}^i_{i-s+1}$ should be accumulated during the iterations (4)-(9) till s last values are re-calculated.

The sensitivity matrix $\mathcal{S}(t_i)$ is computed as follows

$$\mathcal{S}(t_i) = \frac{\partial \mathcal{N}(D)}{\partial \theta_j} \mathcal{H}_{i-s+1}^{i} + \mathcal{N}(D) \frac{\partial \mathcal{H}_{i-s+1}^{i}}{\partial \theta_j} \tag{10}$$

where $\frac{\partial \mathcal{H}_{i-s+1}^{i}}{\partial \theta_j}$ is the stackable vector of the s last values $\xi_j(t_k)$. Then the gradient model is defined as the product of transposed sensitivity matrix $\mathcal{S}(t_i)$ and $\varepsilon(t_i)$

$$\begin{aligned} q(t_i) &= \mathcal{S}^T(t_i)\varepsilon(t_i) \\ \hat{q}(t_i) &= \beta\hat{q}(t_{i-1}) + (1-\beta)q(t_i) \end{aligned} \tag{11}$$

where β is the exponential smoothing factor, $0 \le \beta < 1$.

```
objObject = [];
objFilter = [];
objAdFilter = [];
objObject = phiGetProblemDesc(objObject);
while phiIterationsRange(objExperiment) do
      while phiSnrRange(objExperiment) do
            while phiStabilityRange(objExperiment) do
                  for t = 0:objExperiment.Tmax;
                        objObject = phiModel(objObject, t);
                        objFilter = phiFilter(objFilter, t);
                        objAdFilter = phiAdaptiveFilter(objAdFilter, t);
                  end;
            phiSaveResults(objAdFilter);
            end;
      end;
objResults = phiAnalyseResults(objFilter, objAdFilter);
phiMakeGraphs(objResults);
end;
```

Fig. 5. Tool box numerical simulation function

The suboptimal adaptation procedure (SAP) shown here as one of possible variants, is defined for each adjustable parameter θ_j through the recursion

$$\begin{aligned} p_j(t_{i+1}) &= p_j(t_i) + \|\frac{\partial \varepsilon(t_i)}{\partial \theta_j}\|^2 \\ \pi(t_i) &= \hat{\theta}(t_i) - \mathrm{diag}(p_j(t_{i+1}))^{-1}\hat{q}(t_i) \ . \end{aligned} \tag{12}$$

(Here and below $\pi(t_i)$ denotes a trial value for $\hat{\theta}(t_{i+1})$). The stability condition of the linear part of estimator, $\rho\left[(I - DH_*)A\right] < 1$, should be checked for trial estimate $\pi(t_i)$. Also, the constraints of the nonlinear estimator should be satisfied for the next estimate $\hat{\theta}(t_{i+1}) = \pi(t_i)$.

The adaptive filter tool box function is depicted by Fig. 4. The renewal of the adaptive filter parameters with the new calculated values occurs only at each adaptive step, which is obtained by function `phiAdaptiveStep`.

4 Simulations Results

The integral percent error (IPE) will show the algorithm performance. The tool box numerical simulations function (Fig. 5) depends on the possible range of aspects to be evaluated. The extended experiment is planned to analyze the influence of a set of different aspects on the IPE-characteristic of the algorithm, to reveal some effects during the identification process and finally to illustrate the applicability of the nonlinear filtering algorithm and its MATLAB tool box implementation. The set of aspects includes signal-to-noise ratio (SNR defined by $\|Q\|/\|R\|$), stability property of the object, the presence of nonlinear estimator, the type of adaptation procedure (suboptimal, i.e. SAP, optimal, i.e. OAP, or simple stochastic approximation, i.e. SSAP), a number of iterations, and initial values for the estimates. Main user function (Fig. 5) is used to execute the algorithm for the ranges of analyzed aspects (`phiSnrRange`, `phiStabilityRange`), to compare results (`phiAnalyseResults`) and to plot graphs (`phiMakeGraphs`).

```
[RANGES]
     stabInterval=((0.01, 1.000); 0.0; 10.0);
     snrInterval=((0.001, 100.0); 0.0; 10.0);
     snrAdaptiveProcedure=SAP;
     snrvsiterationsIpeLevel=10.0;
     adjMaxAdapationTime=((10000, 10000000); 0.0; 10.0);
[SYSTEM]
     PHI = [-0.8, 0.1];
     Q = 0.04;
     R = 0.06;
     Gd = [0.0; 0.4];
     H = [1.0, 0.0];
```

Fig. 6. Experiment settings

We investigate the properties of the proposed algorithm and define the IPE as follows

$$\rho_{ipe} = \frac{\|\theta^*(t_i)-\theta^{**}\|}{\|\theta^{**}\|} \tag{13}$$

where $\theta^*(t_i)$ is the estimate of parameter θ obtained at time t_i and θ^{**} is the optimal (i.e. true) value of θ.

We consider the following example of the second order model

$$\begin{aligned} x(t_{i+1}) &= \begin{bmatrix} 0 & 1 \\ f_1 & f_2 \end{bmatrix} x(t_i) + \begin{bmatrix} 0 \\ \alpha \end{bmatrix} w_d(t_i) \\ z(t_i) &= Hx(t_i) + v_d(t_i) \end{aligned} \tag{14}$$

with unknown values of the state and measurement noise covariances Q and R and $\alpha = 0.4$, $\beta = 0.0$. The measurement matrix H is $\left[0, 1\right]$ and parameters f_1 and f_2 are known and fixed. Nonlinear estimator has the nonlinear state transformation function which is linear for small state values and constant outside the linear area. The tangent of the linear part θ is unknown and should be identified. In this experiment we simultaneously identify the parameters of adaptive filter and the optimal tangent θ of the nonlinear estimator.

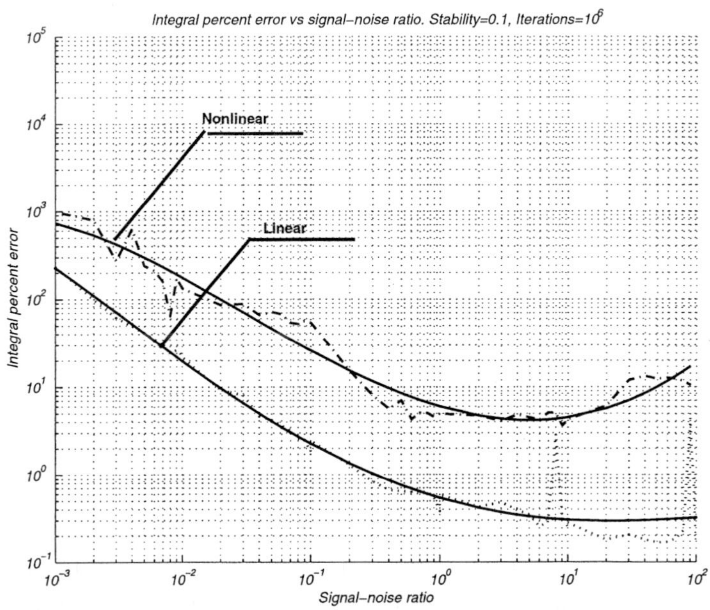

Fig. 7. IPE characteristic for linear and nonlinear problems, stability factor - 0.1

The experiment is a set of algorithm runs with chosen algorithm settings and defined values of modelling aspects. We distinguish the primary and secondary modelling aspects of the experiment that corresponds to an interval of values and single value accordingly. Experimental graphs depict the IPE-characteristic behavior vs the primary aspect interval with the secondary aspects values chosen arbitrarily. The experiment settings for considered example are set in tool box configuration file as shown in Fig. 6.

The primary modelling aspect is SNR, and we successively analyze the SNR-interval $[1.0 \cdot 10^{-3}, \ldots, 1.0 \cdot 10^{2}]$.

Graphs of Fig. 7 and Fig. 8 represent the IPE behavior vs the time scale for nonlinear estimation for different stability factors. In all other cases the quality of the estimates becomes better as the number of iterations grows.

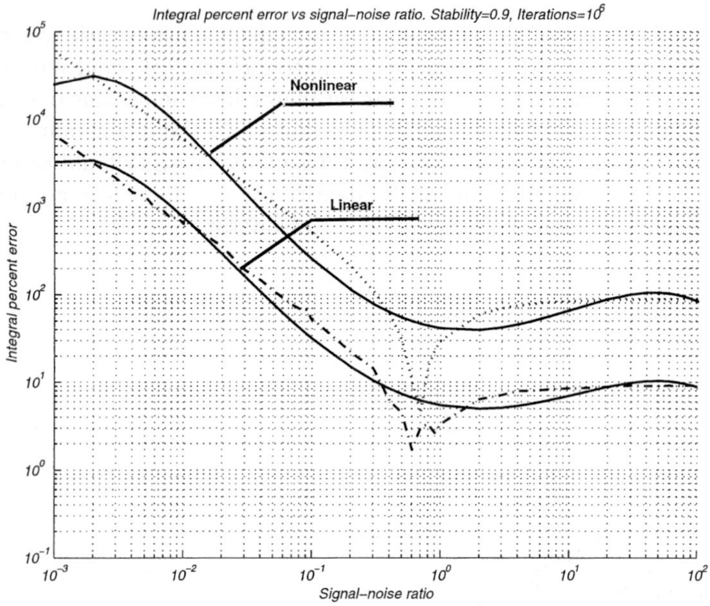

Fig. 8. IPE characteristic for linear and nonlinear problems, stability factor - 0.9

The stability factor depends on location of the eigenvalues of matrix Φ and it occurs that there exists an insensitivity area effect of the considered nonlinearity. This effect entirely defines the shape of the IPE-graph while the SNR interval $[1, \ldots, 10^2]$ does not influence on the quality of estimating process for stability factor 0.9. However, as the number of iterations is increased, the IPE-quality becomes better till the non-improvable level corresponding to the insensitivity area, is attained. This effect can be revealed on Fig. 9.

5 Conclusions

In this paper, we develop the special purpose MATLAB compatible tool box and demonstrate its applicability to computational investigation of complex systems described in terms of high-dimensional vector-matrix stochastic difference equations. This project was motivated by developing novel numerical algorithms for adaptive identification of a non-linear optimal discrete-time steady-state estimator intended to predict the state of the given stochastic system.

This is only one of the possible applications of the designed tool. Another problem investigated with the tool box is iterative control design [4]. The tool

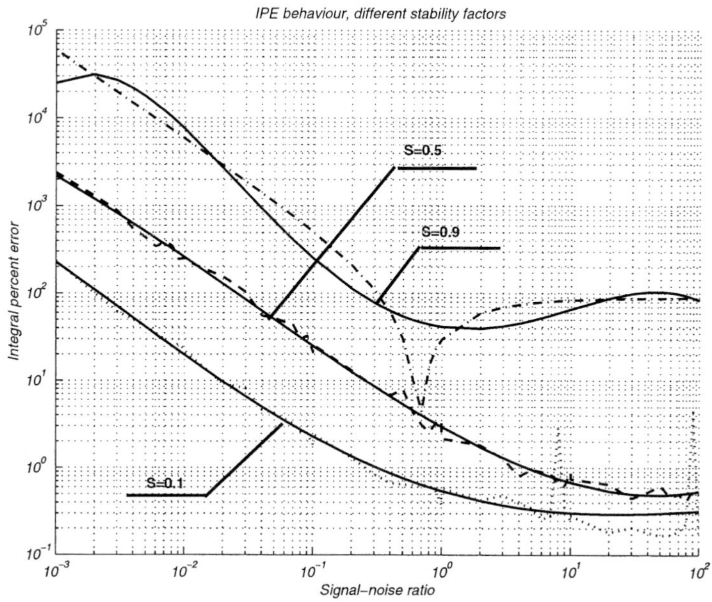

Fig. 9. IPE characteristic for different stability factors

box dramatically extends the capabilities of MATLAB as it makes computational experiments far easier to conduct.

References

1. Landau, I.D. (ed.): Identification des Systemes. Les Bases de l'Identification des Systemes. Hermes, Paris (2001)
2. Caines, P.: Linear Stochastic Systems. John Willey & Sons, New York Chichester Brisbane Toronto Singapore (1988)
3. Semoushin, I.V., Tsyganova, J.V.: Indirect error control for adaptive filtering. In: Neittaanmaki, P., Tiihonen, T., Tarvainen, P. (eds.): Proc. of the 3rd European Conference on Numerical Mathematics and Advanced Applications. World Scientific, Singapore New Jersey London Hong Kong (2000) 333–340
4. Semoushin, I., Gorokhov, O.: Computational processes in iterative control design. In: Sloot, P.M.A., Kenneth Tan, C.J., Dongarra, J.J., Hoekstra, A.G. (eds.): Computational Science – ICCS 2002. Lecture Notes in Computer Science, Vol. 2329. Springer-Verlag, Berlin Heidelberg New York Barcelona Hong Kong London Milan Paris Tokyo (2002) 186–195

Comparative Analysis of Stochastic Identification Methods and Fault Diagnosis for Multi-mode Discrete Systems*

Olga Fatyanova and Alexey Kondratiev

Ulyanovsk State University, 42 Leo Tolstoy Str., 432970 Ulyanovsk, Russia
folga@nm.ru,mouse_fin@mail.ru

Abstract. This paper offers the comparative analysis of stochastic approximation methods applied for multi-mode discrete systems identification from incomplete noisy observations. The main item discussed here concerns the rate of convergence of these methods and several aspects that are considered to affect this property. To corroborate the theoretical arguments, the experimental results are provided.

1 Introduction

A multi-mode system is understood as a system with changeable coefficients that compounds the vector parameter θ. The mode change means that parameter θ switches, say, from value θ_1 to value θ_2, taken from a certain compact set of available parameters Θ. It is assumed that system is stable if and only if the parameter $\theta \in \Theta$. The only data can be used is contained into incomplete noisy observations. So, one can consider two problems to be solved: the problem of mode switching diagnosis, or so-called fault diagnosis, and the problem of identification of the new value θ_2 of parameter θ. We insist on the joint performance of these tasks, using the method of Adaptive Model (AM) and proposing Start-and-Stop Algorithm (SSA), which we. Every time the fault is detected, the SSA launches the identification, which adjusts the AM in order to suit the certain quality requirements. When the identification reaches its goal and demanded agreement with observation is provided, the SSA stops the process. In our research the requirements, mentioned above, are represented in the form of the Auxiliary Functional of Quality (AFQ), compared with certain threshold value. In this article only the identification is discussed, while the description of the SSA can be found in [7].

The outline of the paper is as follows: Section 2 describes the Monitored System (MS) and Kalman filter. In Section 3 we build the AM and the AFQ. Section 4 shows some identification algorithms and certain experimental results are contained in Section 5. Finally, Section 6 concludes the paper.

* This work was supported in part by the Russian Ministry of Education (grant No. T02-03.2-3427).

P.M.A. Sloot et al. (Eds.): ICCS 2003, LNCS 2658, pp. 446–455, 2003.

2 Monitored System

In the frames of this paper we consider MS without control input with an uncertainty parameter θ, $\theta \in \mathbb{R}^N$. First of all, MS includes the object with state equation

$$x(t_{i+1}) = \Phi_\theta x(t_i) + \Gamma w(t_i), \quad i = 0, 1, \dots \tag{1}$$

with $x \in \mathbb{R}^n$, $\Phi_\theta \in \mathbb{R}^{n\times n}$, $\Gamma \in \mathbb{R}^{n\times q}$ and a noise $w(t_i)$ where the random initial $x(t_0)$ at some t_0 has a finite mean $\bar{x}_0$ and a finite covariance $P_0 \geq 0$.

The other part of MS is a sensor with the equation

$$z(t_i) = Hx(t_i) + v(t_i), \quad i = 1, 2, \dots \tag{2}$$

where $H \in \mathbb{R}^{m\times n}$, $\mathrm{rank}(H) < n$ (incomplete observations) and a noise $v(t_i)$. Both $\{w(t_i)\}$ and $\{v(t_i)\}$ are i.i.d. zero mean mutually independent wide-sense stationary sequences whose covariances are $\mathrm{E}\{w(t_i)w(t_i)^T\} = Q_\theta \geq 0$ and $\mathrm{E}\{v(t_i)v(t_i)^T\} = R_\theta > 0$ for all t_i.

The derivation of unbiased estimate $\tilde{x}(t_i)$ for state $x(t_i)$ is considered to be the main problem of the project. Assume that the value θ_0 of parameter θ is known in advance and for certain period it remains permanent. Hence in this situation we have *fixed* (fault-free) system mode specified by a nominal value θ_0 of θ and, correspondingly, the steady-state variant of estimation task. This fact allows us to use the classical approach consisting in the use of Kalman filter to tackle our main problem under these simplified conditions.

The standard Kalman filter equations are as follows.

Initial values:

$$\tilde{P}(t_0^+) = P_0, \quad \tilde{x}(t_0^+) = \bar{x}_0 \tag{3}$$

Step 1: time propagation, $i = 0, 1, \dots$

$$\tilde{x}(t_{i+1}^-) = \Phi_\theta \tilde{x}(t_i^+); \tag{4}$$

$$\tilde{P}(t_{i+1}^-) = \Phi_\theta \tilde{P}(t_i^+)\Phi_\theta^T + \Gamma Q_\theta \Gamma^T. \tag{5}$$

Step 2: measurement update, $i = 1, 2, \dots$

$$K(t_i) = \tilde{P}(t_i^-)H^T[H\tilde{P}(t_i^-)H^T + R_\theta]^{-1}; \tag{6}$$

$$\tilde{x}(t_i^+) = \tilde{x}(t_i^-) + K(t_i)[z(t_i) - H\tilde{x}(t_i^-)]; \tag{7}$$

$$\tilde{P}(t_i^+) = \tilde{P}(t_i^-) - K(t_i)H\tilde{P}(t_i^-). \tag{8}$$

The MS (1), (2) is assumed to be asymptotically stable in all modes of operating referred to by the subscript θ, and so all the processes within the system are wide-sense stationary at every t_i as $t_0 \to -\infty$ (*the main assumption*). The system is designed to hold the *main properties* in all modes:

- ⋆ $(\Phi_\theta, \Gamma Q_\theta^{1/2}\Gamma^T)$ is stabilizable and (Φ_θ, H) is completely observable.

These properties guarantee the existence of the optimal steady-state parameters for a *fixed* (fault-free) system mode

$$\tilde{x}(t_{i+1}^-) = \bar{\Phi}\tilde{x}(t_i^+), \quad \tilde{x}(t_i^+) = \tilde{x}(t_i^-) + \bar{K}[z(t_i) - H\tilde{x}(t_i^-)]$$

with some fixed $\tilde{x}(t_0^+)$, constant transition matrix $\bar{\Phi}$ and constant gain $\bar{K}$ designed to be optimal in a steady state (as $t_0 \to -\infty$).

Unfortunately, because of uncertainty classical theory can not be used as it is. In this work we consider two levels of uncertainty.

Case 1: unknown Q_θ, R_θ. The estimate of Kalman gain K we denote as D. Here vector θ consists of all components of vector K.

Case 2: unknown Q_θ, R_θ, Φ_θ. As in the previous case, we consider D to be the estimate of K, and let A to be the estimate of Φ_θ. Here vector θ consists of all components of vector K and unknown components of matrix Φ_θ.

3 Adaptive Model

The problem concerning the derivation of unbias estimate for Φ_θ in the AM, based on the Kalman filter principle, has a solution only for the case of complete observations [3]. In this work we consider the possibility of evading the strict limitations and accomplishing the task in the case of incomplete observations.

We build AM for MS and Kalman filter with adjustable parameter $\hat{\theta}$ considered to be the estimate of the corresponding MS parameter θ. The property of estimate unbiasedness can be provided only by the optimal choice of the AM structure. In the case of noises with zero mean and gaussian distribution it is the structure of Kalman filter [3]. Let us write the equations of AM:

$$\hat{x}(t_i^-) = A\hat{x}(t_{i-1}^+), \tag{9}$$

$$\hat{z}(t_i) = H\hat{x}(t_i^-), \tag{10}$$

$$r(t_i) = z(t_i) - \hat{z}(t_i) = z(t_i) - H\hat{x}(t_i^-), \tag{11}$$

$$\hat{x}(t_i^+) = \hat{x}(t_i^-) + Dr(t_i). \tag{12}$$

where $\hat{x}(t_i^\pm)$ is the estimate of $x(t_i)$, $r(t_i)$ is a difference (residual) between the observation $z(t_i)$ (immediately obtained from the sensor) and the observation estimate $\hat{z}(t_i)$ (obtained form the AM).

Hence, the AM allows us to derive the unbiased estimate $\hat{x}(t_i^\pm)$ for $x(t_i)$. Let us consider the initial functional of quality to be built on the base of difference $e(t_i) = x(t_i) - \hat{x}(t_i)$. As to [4], error-squared functional of quality is the optimal one for the system with noises having gaussian distribution and zero mean:

$$J_e(t_i) = \frac{1}{2}E[e^T(t_i)e(t_i)], \quad i = 0, 1, \ldots \tag{13}$$

While constructing the AFQ, we must take into account the important condition [3]:

- ⋆ the components of argument $\hat{\theta}^*$ of AFQ minimization must coincide with corresponding actual values of the entire system parameter (unknown coefficients of Φ_θ in the second case of uncertainty) and optimal value of steady-state filter $\bar{K}$.

In practice the value of $x(t_i)$ is unknown, and our goal is to be find it, therefore AFQ should be built up in another way. Usually the difference $r(t_i)$ is used for this purpose. So, we take the AFQ in the following form [2]:

$$J_\varepsilon(t_i) = \frac{1}{2}E[\varepsilon^T(t_i)\varepsilon(t_i)], \quad i = 0, 1, \ldots \tag{14}$$

where $\varepsilon(t_i)$ is the history of residuals $r(t_i)$, transformed in a special manner to provide

$$J_\varepsilon(t_i) = J_e(t_i) + const, \quad i = 0, 1, \ldots \tag{15}$$

Without coming into details of $\varepsilon(t_i)$ construction, we find it necessary to mention here, that the truth of this condition is provided by the known theorem for the general case [1]. We have fixed the system dimension $n = 2$ and checked it for this particular case. While proving the fact the constant mentioned above has been determined: $const = R_\theta$.

As one can see from (15), the argument of the minimization θ^* of J_e coincides with $\hat{\theta}^*$. Hence we will take J_ε as the AFQ, because it is available for practical utilization in the numerical algorithms of identification. Under this point of view the process of identification is evident to turn out the process of optimization or, in other words, of AFQ minimization. The AM is optimal iff (if and only if)

$$\nabla_{\hat{\theta}} J_\varepsilon(t_i) = \varepsilon^T(t_i)S(t_i) = \mathbf{0} \tag{16}$$

where $\nabla_{\hat{\theta}} J_\varepsilon(t_i) \in \mathbb{R}^{1\times N}$, the matrix $S(t_i) = \nabla_{\hat{\theta}}\varepsilon(t_i)$ is the sensibility matrix [7].

In order to write down ε in an explicit form the dimension n is to be fixed, and below we will operate on 2-order systems in standard observable form:

$$x(t_{i+1}) = \begin{bmatrix} 0 & 1 \\ f_1 & f_2 \end{bmatrix} x(t_i) + \begin{bmatrix} 0 \\ \alpha \end{bmatrix} w(t_i), \quad z(t_i) = \begin{bmatrix} 1 & 0 \end{bmatrix} x(t_i) + w(t_i).$$

Since Q_θ, R_θ and Φ_θ are unknown according to list of uncertainty cases, K and Φ_θ are to be estimated as D and A correspondingly, and almost sure convergence is to be demanded:

$$D = \begin{bmatrix} d_1 \\ d_2 \end{bmatrix} \to K = \begin{bmatrix} k_1 \\ k_2 \end{bmatrix}, \quad A = \begin{bmatrix} 0 & 1 \\ a_1 & a_2 \end{bmatrix} \to \Phi_\theta = \begin{bmatrix} 0 & 1 \\ f_1 & f_2 \end{bmatrix}.$$

Now let us write down the equation for ε for $n = 2$:

$$\varepsilon(t_i) = \mathcal{N}(D)\mathcal{H} = \begin{bmatrix} r(t_{i-1}) \\ d_2 r(t_{i-1}) + r(t_i) \end{bmatrix} = \begin{bmatrix} \varepsilon_1(t_i) \\ \varepsilon_2(t_i) \end{bmatrix}.$$

$$\mathcal{N}(D) = \begin{bmatrix} 1 & 0 \\ d_2 & 1 \end{bmatrix}, \quad \mathcal{H} = \begin{bmatrix} r(t_{i-1}) \\ r(t_i) \end{bmatrix}.$$

To obtain the gradient of AFQ (for the first case of uncertainty) the sensibility matrix should be calculated.

$$S_D(t_i)^T = \begin{bmatrix} \frac{\partial \varepsilon_1(t_i)}{\partial d_1} & \frac{\partial \varepsilon_2(t_i)}{\partial d_1} \\ \frac{\partial \varepsilon_1(t_i)}{\partial d_2} & \frac{\partial \varepsilon_2(t_i)}{\partial d_2} \end{bmatrix} = \begin{bmatrix} \frac{\partial r(t_{i-1})}{\partial d_1} & d_2 \frac{\partial r(t_{i-1})}{\partial d_1} + \frac{\partial r(t_i)}{\partial d_1} \\ \frac{\partial r(t_{i-1})}{\partial d_2} & r(t_{i-1}) + d_2 \frac{\partial r(t_{i-1})}{\partial d_2} + \frac{\partial r(t_i)}{\partial d_2} \end{bmatrix}$$

For this purpose we differentiate componentwise the equations of AM by D, taking in account(4), (5), (6), (7), (8).

$$\frac{\partial \hat{x}(t_i^-)}{\partial d_1} = A \frac{\partial \hat{x}(t_{i-1}^+)}{\partial d_1}, \quad \frac{\partial r(t_i)}{\partial d_1} = - \begin{bmatrix} 1 & 0 \end{bmatrix} \frac{\partial \hat{x}(t_i^-)}{\partial d_1},$$
$$\frac{\partial \hat{x}(t_i^+)}{\partial d_1} = \frac{\partial \hat{x}(t_i^-)}{\partial d_1} + \begin{bmatrix} 1 \\ 0 \end{bmatrix} r(t_i) + D \frac{\partial r(t_i)}{\partial d_1}.$$

$$\frac{\partial \hat{x}(t_i^-)}{\partial d_2} = A \frac{\partial \hat{x}(t_{i-1}^+)}{\partial d_2}, \quad \frac{\partial r(t_i)}{\partial d_2} = - \begin{bmatrix} 1 & 0 \end{bmatrix} \frac{\partial \hat{x}(t_i^-)}{\partial d_2},$$
$$\frac{\partial \hat{x}(t_i^+)}{\partial d_2} = \frac{\partial \hat{x}(t_i^-)}{\partial d_2} + \begin{bmatrix} 0 \\ 1 \end{bmatrix} r(t_i) + D \frac{\partial r(t_i)}{\partial d_2}.$$

Hence, $\frac{\partial \hat{x}(t_i^-)}{\partial d_1}$, $\frac{\partial r(t_i)}{\partial d_1}$, $\frac{\partial \hat{x}(t_i^+)}{\partial d_1}$, $\frac{\partial \hat{x}(t_i^-)}{\partial d_2}$, $\frac{\partial r(t_i)}{\partial d_2}$, $\frac{\partial \hat{x}(t_i^+)}{\partial d_2}$ are calculated iteratively and then substituted into the $S_D(t_i)^T$ matrix. The initial values is necessary to set:

$$\frac{\partial \hat{x}(t_0^+)}{\partial d_1} = 0, \quad \frac{\partial \hat{x}(t_0^+)}{\partial d_2} = 0.$$

For second case of uncertainty the sensibility model is designed analogically.

4 Stochastic Approximation

In this paper the process of identification is performed by discrete searchless algorithms, which are also called the methods of stochastic approximation and represented below in the general recurrent form [5]:

$$\hat{\theta}[t_j] = \hat{\theta}[t_{j-1}] - \Lambda[t_j] S^T[t_{j-1}] \varepsilon[t_{j-1}]. \tag{17}$$

The choice of $\Lambda[t_j]$ affects the rate of convergence greatly. The sufficient conditions of algorithm convergence can be found in [5], [6]. The identification method, corresponding diagonal form of $\Lambda[t_j]$ is discussed in 4.3. The case of equal diagonal components $\lambda[t_j]$ see below in 4.1.

4.1 Algorithm 1. Robbins-Monro Algorithm

Let us take $\lambda[t_j] = \frac{1}{j}$. By doing so we get the algorithm representing the stochastic analogue of gradient method. This method is also called the Robbins-Monro multidimensional procedure of stochastic approximation. [6], [7]:

$$\hat{\theta}[t_{j+1}] = \hat{\theta}[t_j] - \lambda[t_{j+1}] S^T[t_j] \varepsilon[t_j], \quad \lambda[t_{j+1}] = \frac{1}{j+1}, \quad j = 1, 2, \ldots \tag{18}$$

4.2 Algorithm 2. Optimal Algorithm

The information about the class of density of distribution the noises belong is of primary importance, because it helps to make an optimal choice of matrix Λ. The algorithm, corresponding to this optimal choice provides the best rate of convergence of $\hat{\theta}[t_j]$ to $\hat{\theta}^*$ among all methods [4]. So, for the case of noises having gaussian distribution with zero mean and unknown covariances least-squares method (LSM) is the optimal one. To derive LSM the general form of identification algorithm is to be written down:

$$\hat{\theta}[t_{j+1}] = \hat{\theta}[t_j] - \Lambda[t_{j+1}]S^T[t_j]\varepsilon[t_j], \quad j = 1, 2, \dots \tag{19}$$

where $\Lambda[t_j] \in \mathbb{R}^{N \times N}$. In LSM $\Lambda[t_j]$ is determined by the formula [4]:

$$\Lambda[t_{j+1}] = \Lambda[t_j] - \Lambda^{-1}[t_j]S^T[t_j](S[t_j]\Lambda[t_j]S^T[t_j])^{-1}S[t_j]\Lambda[t_j]. \tag{20}$$

The equation of approximation written down through the increment is

$$\hat{\theta}[t_{j+1}] = \hat{\theta}[t_j] + \Delta\hat{\theta}[t_j], \quad j = 1, 2, \dots \tag{21}$$

We denote $P[t_j] = \Lambda^{-1}[t_j]$ and from (19), (20), (21) obtain the LSM:

- set the initial value $P[t_1] = I$;
- calculate $P[t_{j+1}] = P[t_j] + S[t_j]^T S[t_j]$;
- find $\Delta\hat{\theta}[t_j]$ from the system of linear equations $P[t_{j+1}]\Delta\hat{\theta}[t_j] = -S^T[t_j]\varepsilon[t_j]$
- calculate new value of parameter $\hat{\theta}[t_{j+1}] = \hat{\theta}[t_j] + \Delta\hat{\theta}[t_j]$

The disadvantage of the LSM is that for each step the solution of the system of linear equations is sought. This operation is quite laborious to perform, so it is worth to modify the method in order to avoid it. Certainly the newly obtained algorithm is not optimal in the strict sense, that is why we call it suboptimal, nevertheless its employment let us to cut down computational costs in comparison with the LSM. In the later section suboptimal method is discussed in details.

4.3 Algorithm 3. Suboptimal Algorithm

So-called suboptimal algorithm is developed on the base of optimal one by transforming the $S^T[t_j]S[t_j]$ matrix to diagonal form. Only diagonal elements of matrix product $S^T[t_j]S[t_j]$ remain unchanged, the rest are substituted for zeros:

$$S^T[t_j]S[t_j] \approx \begin{bmatrix} \|\frac{\partial\varepsilon[t_j]}{\partial\theta_1}\|^2 & \cdots & 0 \\ \cdots & \ddots & \vdots \\ 0 & \cdots & \|\frac{\partial\varepsilon[t_j]}{\partial\theta_N}\|^2 \end{bmatrix} \tag{22}$$

Thus, we have got the suboptimal algorithm:

- set the initial value $P[t_1] = I$;
- calculate diagonal elements of matrix P, $p_{ii}[t_{j+1}] = p_{ii}[t_j] + \|\frac{\partial\varepsilon[t_j]}{\partial\theta_i}\|^2$, for $i = 1, 2, \dots, N$
- calculate new value of parameter $\hat{\theta}_i[t_{j+1}] = \hat{\theta}_i[t_j] - \frac{S^T[t_j]\varepsilon[t_j]}{p_{ii}[t_{j+1}]}$, $i = 1, 2, \dots, N$

4.4 Stability Requirements. Jury's Criteria

From the condition of AM stability $\rho[(I-DH)A] < 1$ the characteristic equation can be obtained [2]:

$$q(\lambda) = b_0\lambda^2 + b_1\lambda + b_2 = 0 \tag{23}$$

where $b_0 = 1$, $b_1 = d_2 - a2$, $b_2 = a_1(d_1 - 1)$. The stability condition according to Jury criteria may be written as [2]:

$$q(1) > 0, \quad (-1)^N q(-1) > 0, \quad b_0 > |b_N|. \tag{24}$$

In particular case of 2-order system these conditions transform to the next ones:

$$1 + b_1 + b_2 > 0, \quad 1_0 - b_1 + b_2 > 0, \quad 1 > |b_2|. \tag{25}$$

Hence, we have got an opportunity to determine the stability of a system at every step of approximation process and to work out the heuristic algorithm, that would be capable of preventing the AM crash.

The main idea of this method is to keep the parameter $\hat{\theta}(t_{j+1})$ within the set of stability Θ. Therefore we have designed the algorithm reasoning from the idea that identification is to be reinitiated, if the AM became unstable, and launched with a new stable value.

- calculate $\hat{\theta}(t_{j+1})$;
- check Jury's criteria (24); if the AM with parameter $\hat{\theta}(t_{j+1})$ is stable, continue identification; else choose some $\hat{\theta}_{new}(t_{j+1}) \in \Theta$.

Now the question is: how to find Θ and choose $\hat{\theta}_{new}$ during the approximation? Let us consider this problem for 2-order system and write down the conditions of stability for d_1 and d_2:

$$1 - \frac{1}{|a_1|} < d_1 < \frac{1}{|a_1|} + 1, \tag{26}$$

$$-(a_1(d_1 - 1) + 1) + a_2 < d_2 < (a_1(d_1 - 1) + 1) + a_2. \tag{27}$$

From (26) it is reasonably to take $d_1 = 1$ and $d_2 = a_2$ from (27). In equivalent manner we take $a_1 = 0$ and $a_2 = d_2$. Thus, it is guaranteed that $\hat{\theta}_{new}(t_{j+1}) \in \Theta$.

5 Experiments

The workability of the proposed ideas is to be approved by the experiments for 2-order system. The subject of our interest is the rate of convergence. We consider several aspects to affect this characteristic of identification: the method used in adaptation process, the case of uncertainty, the level of noise both in object and sensor (Q_θ and R_θ), and the stability of entire system (eigenvalues of matrix Φ_θ from the Juri's criteria point of view).

In real life the parameter θ is beyond our reach, while in the experimental conditions our programm simulates both entire system and AM, and we know θ (we, but not the AM!) and are able to compare it with $\hat{\theta}$. Hence we have "a task with known solution".

Therefore the scheme of all experiments is as follows (unless otherwise arranged): during the first thousand of iteration $\theta = \hat{\theta}$, then at $i = 1000$ we simulate a fault by changing $\hat{\theta}$ and at $i = 1000$ the process of approximation is launched. In this type of experiments the SSA is not used.

Let us consider the first case of uncertainty and take MS with

$$\Phi_\theta = \begin{bmatrix} 0 & 1 \\ 0.30 & 0.67 \end{bmatrix}.$$

which eigenvalues (approximately 0.307 and 0.977) are close to stability limits. The table below contains the convergence time (in iterations) which is averaged over 100 realization of approximation process. The 15-percent ratio error is taken (see fig. 1, fig. 2 and fig. 3).

		Alg 1.		Alg. 2		Alg. 3	
Q	R	d_1	d_2	d_1	d_2	d_1	d_2
1	1	14479	4675	5967	599	6698	963
1	0.1	17766	651	4314	493	4716	1138

Now the MS with eigenvalues which equal approximately -0.358 and 0.558 and, hence, are far from stability limits:

$$\Phi_\theta = \begin{bmatrix} 0 & 1 \\ 0.20 & 0.20 \end{bmatrix}.$$

The average convergence time is represented in the table:

		Alg 1.		Alg. 2		Alg. 3	
Q	R	d_1	d_2	d_1	d_2	d_1	d_2
1	1	38376	39822	13626	18380	11928	21387
1	0.1	16889	8440	4431	8610	5566	8000

In the second case of uncertainty the convergence of A is much slower than the one of D. To explain this fact series of experiments were made. The parameter D was taken coincided with $\bar{K}$ and fixed. The AFQ was considered as a function of a_1 and a_2, and level lines, consisting of the points with coordinates $(a_1, a_2) : J_\varepsilon(a_1, a_2) = const$. These lines look like oblong ellipses. It is typical for ravine surfaces, so one take $f_1 = 0.2$, $f_2 = 0.7$ and in the first realization of approximation process get $a_1[t_{max}] \approx 0.7$ and $a_2[t_{max}] \approx 0.2$, in the second – $a_1[t_{max}] \approx 0.4$ and $a_2[t_{max}] \approx 0.5$, then – $a_1[t_{max}] \approx 0.8$ and $a_2[t_{max}] \approx 0.1$, or even $a_1[t_{max}] \approx -1.3$ and $a_2[t_{max}] \approx 0.4$. The fact that the "correct answer" is also available, is very important, because for all these realizations the error $e(t_{max})$ does not vary greatly, as well as $J_\varepsilon(a_1, a_2)$, so $\hat{x}$ is "good enough". Taking these features in account one may note that $a_1(t_{max}) + a_2(t_{max}) \approx 0.9 = const$, as if points $(a_1[t_{max}], a_2[t_{max}])$ belong to certain line or oblong ellipse.

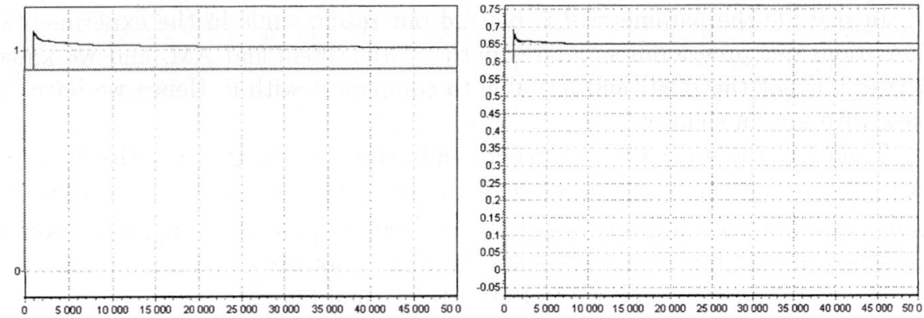

Fig. 1. Tracking the parameters k_1 and k_2 by d_1 (*left*) and d_2 (*right*) for Algorithm 1

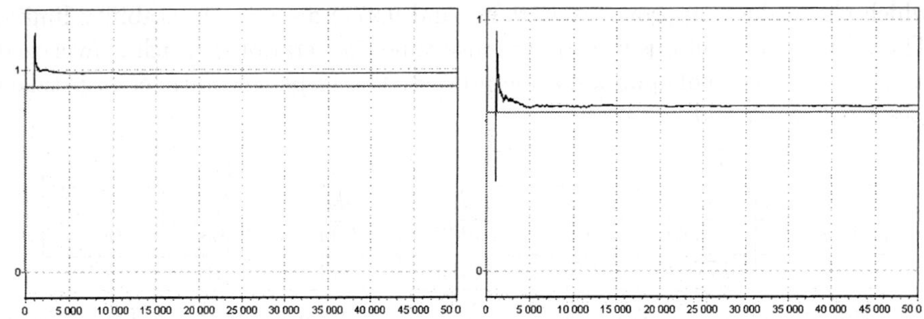

Fig. 2. Tracking the parameters k_1 and k_2 by d_1 (*left*) and d_2 (*right*) for Algorithm 2

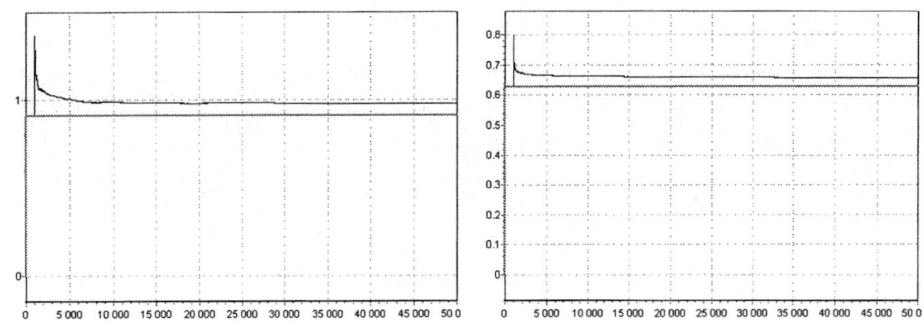

Fig. 3. Tracking the parameters k_1 and k_2 by d_1 (*left*) and d_2 (*right*) for Algorithm 3

5.1 Experimental Conclusions

- The ravine character of AFQ (as a function of a_1 and a_2) degrades seriously the rate of convergence, that does not allow to get correct estimates of Φ_θ for appropriate time. The enlargement of modelling time to 500000 iterations gives nothing, the millions of iterations are demanded. Nevertheless $x(t_i)$ is estimated good enough to say that the main task – to derive $\hat{x}(t_i)$ – is solved.

- The series of experiments approves that LSM has got the best rate of convergence. Suboptimal method considered to be the successful improvement of LSM, because this algorithm provides serious computational shortcut and insignificantly yields to LSM in the rate of convergence.
- The larger R (in comparison with Q) the worse the rate of convergence.
- The rate of convergence is better for MS which eigenvalues are close to the limit of stability. It is easily seen from (1): such MS possess more strongly marked inner dynamics than another ones.

6 Conclusions and Future Work

Basing on the results of the experiments we can say that identification methods can be successfully applied for the solution of the main task, the derivation of unbiased estimate of x. Unfortunately we face the challenge in obtaining correct estimations in the second case of uncertainty, because of the ravine character of AFQ. So one of the directions of the future works is to minimize the influence of this factor to the rate of convergence. Another direction is to study new identification methods such as ones of Newton's type and conjugated gradients algorithm.

References

1. Semoushin, I.V., Tsyganova, J.V.: Indirect error control for adaptive filtering. In: Neittaanmaki, P., Tiihonen, T.,Tarvainen, P. (eds.): Proc. of the 3rd European Conference on Numerical Mathematics and Advanced Applications. World Scientific, Singapore New Jersey London Hong Kong (2000) 333–340
2. Semoushin, I.V.: Adaptive Identification and Fault Detection Methods in Random Signal Processing. Saratov University Publishers, Saratov (1985) [*in Russian*]
3. Semoushin, I.V.: Identification of linear stochastic objects on the base of incomplete noisy observations. Automatics and telemechanics, (1985) 61–71 [*in Russian*]
4. Tsypkin, J.Z.: Optimal identification of dynamic objects. Mensuration control automation, Vol. 3 (1983) 47–59 [*in Russian*]
5. Tsypkin, J.Z.: Adaptation and training in automatic systems. Moscow (1968) [*in Russian*]
6. Vasan M.: Stochastic approximation. Moscow (1972) [*in Russian*]
7. Ponyrko S.A., Semoushin, I.V.: About the choice of start and stop algorithm for minimisation of root-mean-square quality criterion. Autometria, Vol. 2 (1973) 68–74 [*in Russian*]

Computational Expenditure Reduction in Pseudo-Gradient Image Parameter Estimation

Alexandr Tashlinskii

Ulyanovsk State Technical University, 32 Severnyi Venetz Str., 432027 Ulyanovsk
Russia
tag@ulstu.ru

Abstract. An approach enabling to reduce computational expenses at pseudogradient estimation of image parameters based on control of goal function local sample volume is proposed. Local sample volume variation during the process of parameter estimation occurs automatically in correspondence with a preassigned criterion, for instance, sample correlation coefficient. It is shown that for the problems of image mutual spatial deformation parameter estimation, computational expenses can be reduced several times as many.

1 Introduction

For remote exploration of the Earth, in medicine, geology and navigation information extraction systems with embedded spatial apertures of signal sensors are becoming more and more popular. Such systems contain initial data in the form of dynamic arrays with proper feature which is in their space-time correlation and due to this they can be represented in the form of multidimensional images(MI).

When developing algorithmic software of MI processing it is necessary to take into account the dynamics of the scene to be observed, spatial movements of the signal sensors and imperfection of their construction. The influence of the mentioned factors can be described through mathematical models of space-time deformations of multidimensional grids with assigned image. The estimation of the varying parameters of spatial deformations (SD) is required at automated search of a fragment on image, navigational tracking of the mobile object course in the conditions of limited visibility, combination of multidomain images at remote explorations of the Earth, in medical explorations and other problems. A great number of scientific publications are devoted to the aspects of image space-time deformation parameter estimation. In the present work ways of computational expense reduction when using pseudogradient procedures to solve this problem are considered.

Let us assume that the model of MI SD is defined with accuracy of parameters vector $\bar{\alpha}$ and estimation quality criterion $\bar{\alpha}$ is formulated in terms of some functional $J(\bar{\alpha})$ minimization showing expected losses. However it is impossible to find the optimal parameters $\bar{\alpha}^*$ in the appointed sense in view of incompleteness of MI description. In this case the parameters $\bar{\alpha}$ can be estimated on the

P.M.A. Sloot et al. (Eds.): ICCS 2003, LNCS 2658, pp. 456–462, 2003.

basis of MI realization analysis by means of some adaptation procedure which minimizes $J(\bar{\alpha})=J(\bar{\alpha}, Z)$ for a given realization. However it is reasonable to avoid this intermediate stage of exploration and determine $\bar{\alpha}$ directly on values $J(\bar{\alpha})=J(\bar{\alpha}, Z)$ [1]:

$$\bar{\alpha}_t = \bar{\alpha}_{t-1} - \mathbf{\Lambda}_t \, \nabla J(\bar{\alpha}_{t-1}, Z) \tag{1}$$

where $\bar{\alpha}_t$ is next after $\bar{\alpha}_{t-1}$ approximation of minimum point; $\mathbf{\Lambda}_t$ is positively defined matrix determining step value; $\nabla J(\bar{\alpha}_{t-1}, Z)$ is gradient of the functional $J(\bar{\alpha}_{t-1}, Z)$. The application of the procedure (1) in image processing is prevented by the necessity of multiple cumbersome calculations of $\nabla J(\bar{\alpha}_{t-1}, Z)$. It is possible to essentially decrease the volume of calculations if instead of $\nabla J(\bar{\alpha}_{t-1}, Z)$ we use its contraction $\nabla Q_t = \nabla J(\bar{\alpha}_{t-1}, Z_t)$ at some part Z_t of the realization choosing, for example, in the capacity of Z_t a sliding window. The analysis of the approaches [2-5] to the synthesis of large MI SD estimation procedures in real time showed that the algorithms satisfying the requirements of simplicity, fast convergence and capacity to work in various real situations are reasonable to seek in the class of recurrent non-identification adaptive algorithms and the most representative group of these algorithms constitutes pseudogradient algorithms (PGA). The concept of pseudogradient (PG), which was the basis for development of a unified approach to the analysis and synthesis of various procedures of functionals stochastic minimization, was introduced by B.T.Polyak and Y.Z.Tzypkin in work [6] The class of PGA is very wide and includes algorithms of stochastic approximation, random search and many others.

In PGA the following procedure [1] is used

$$\hat{\bar{\alpha}}_t = \hat{\bar{\alpha}}_{t-1} - \mathbf{\Lambda}_t \bar{\beta}_t, \tag{2}$$

where $\bar{\alpha}$ is vector of the parameters to be estimated; $\mathbf{\Lambda}_t$ is positively defined matrix usually called amplification matrix; $\bar{\beta}_t$ is some random direction in the parameter space depending on the values $\hat{\bar{\alpha}}_{t-1}$ and on step number t; $\hat{\bar{\alpha}}_0$ is initial approximation of the parameters vector. The direction $\bar{\beta}_t$ will be PG if the condition $[\nabla J(\hat{\bar{\alpha}}_{t-1})]^T M\{\bar{\beta}_t\} \geq 0$ is satisfied, i.e. if the vector $\bar{\beta}_t$ makes on the average a sharp angle with an exact value of functional gradient. The algorithm [2] will be considered to be pseudogradient, if $\bar{\beta}_t$ is PG at each its step. In this case the steps in (2) are performed on average in the direction of reduction of $J(\hat{\bar{\alpha}})$ and sequence $\hat{\bar{\alpha}}_1, \hat{\bar{\alpha}}_2, \ldots$ will converge to the minimum point $\hat{\alpha}^*$ when satisfying relatively weak conditions [7]).

We should mention that in the algorithm (2) a possibility of calculation of $J(\hat{\bar{\alpha}}_{t-1}, Z_t)$ or $\nabla J(\hat{\bar{\alpha}}_{t-1}, Z_t)$ is not assumed, i.e. $J(\hat{\bar{\alpha}})$ can be non-observable. It is necessary to satisfy only the condition of pseudogradientness. In particular, in the capacity of $J(\bar{\beta}_t)$ we can choose a noisy value of some other functional $J_d(\hat{\alpha})$ which has the same point of minimum $\bar{\alpha}^*$. Admissibility of the dependence $\bar{\beta}_t$ on the preceding values $\hat{\alpha}_t$ enables a possibility of PGA application for MI processing in the order of some sweep. In this case $\bar{\beta}_t$ can depend on the values $\hat{\alpha}_t$ calculated in advance in the preceding rows of the image.

When synthesizing PGA for an assigned goal function (GF) it is necessary to derive some easily calculated PG. Below computational expense reduction ways are considered when designing PG.

2 Pseudogradient Kind Selection

When synthesising PGA (2) the most important moment is in finding the performance index PG $J(\hat{\alpha}, Z)$ or its contraction $J(\bar{\alpha}, Z_t)$. In the work [8] it is shown that if two frames $\mathbf{z}^{(1)} = \left\{ z_{\bar{j}}^{(1)} = x_{\bar{j}} + \theta_{\bar{j}}^{(1)} \right\}$ and $\mathbf{z}^{(2)} = \left\{ z_{\bar{j}}^{(2)} = x\left(\bar{j}, \bar{\alpha}\right) + \theta_{\bar{j}}^{(2)} \right\}$ of MI, defined on grid of samples $\Omega : \{\bar{j} = (j_1, j_2, \ldots, j_n)\}$ constitute additive mixture of informational $\{x_{\bar{j}}\}$ and white $\{\theta_{\bar{j}}\}$ random fields then when minimising GF PG can be obtained at the expense simplification of the gradient

$$\nabla J(\bar{\alpha}, Z) = \sum_{\bar{j},\bar{l}\in\Omega} \frac{\partial x(\bar{j}, \bar{\alpha})}{\partial \bar{\alpha}} V_{Z\bar{j}\bar{l}}^{-1} \left(x_{\bar{l}}(\bar{l}, \bar{\alpha}) - z_{\bar{l}}^{(2)} \right) \tag{3}$$

and when GF maximising is gradient

$$\nabla J(\bar{\alpha}, Z) = - \sum_{\bar{j},\bar{l}\in\Omega} \frac{\partial x(\bar{j}, \bar{\alpha})}{\partial \bar{\alpha}} V_{Z\bar{j}\bar{l}}^{-1} z_{\bar{l}}^{(2)}, \tag{4}$$

where $V_{Z\bar{j}\bar{l}}$ is covariance matrix of the conditional distribution $w\left(\{z_{\bar{j}}^{(2)}\}/\{z_{\bar{j}}^{(1)}, \bar{\alpha}\}\right)$; $x(\bar{j}, \bar{\alpha})$ is deformed frame samples prediction. Reduction of operations number in (3) and (4) can be obtained using the contraction of $\nabla Q_t = \nabla J(\hat{\bar{\alpha}}_{t-1}, Z_t)$, where $Z_t = \left\{ z_{\bar{j},t}^{(2)}, x(\bar{j}, \hat{\bar{\alpha}}_t) \right\}$ local sample of GF Q at t-th iteration; $z_{\bar{j},t}^{(2)} \in \mathbf{z}^{(2)}, \bar{j}_t \in \Omega_t \in \Omega$, and substituting the prediction $x(\bar{j}, \bar{\alpha})$ of deformed frame values by a simpler estimate. This estimate can be obtained, for example, on the basis of interpolation with the estimates $\bar{\alpha}$ used in the capacity of its parameters at next iteration due and obtained at a preceding iteration. Then the local sample of GF at the t-th iteration will be

$$Z_t = \left\{ z_{\bar{j}t}^{(2)}, \tilde{z}_{\bar{j}t}^{(2)} \right\};\ z_{\bar{j}t}^{(2)} \in \mathbf{z}^{(2)}; \tilde{z}_{\bar{j}t}^{(1)} = \tilde{z}^{(1)}(\bar{j}_t, \hat{\alpha}_{t-1}) \in \tilde{\mathbf{z}},$$

where $\tilde{\mathbf{z}}$ is continuous image obtained from $\mathbf{z}^{(2)}$ by means of interpolation; and the relations (3) and (4) for PG will take the form

$$\bar{\beta}_t = \nabla Q_t = \sum_{\bar{j}_t\in\Omega_t} \frac{\partial \tilde{z}^{(1)}(\bar{j}_t, \bar{\alpha})}{\partial \bar{\alpha}} \left(\tilde{z}^{(1)}(\bar{j}_t, \bar{\alpha}) - z_{\bar{j}t}^{(2)} \right) \Big|_{\bar{\alpha}=\hat{\bar{\alpha}}_{t-1}} \tag{5}$$

$$\bar{\beta}_t = -\nabla Q_t = - \sum_{\bar{j}_t\in\Omega_t} \frac{\partial \tilde{z}^{(1)}(\bar{j}_t, \bar{\alpha})}{\partial \bar{\alpha}} z_{\bar{j}t}^{(2)} \Big|_{\bar{\alpha}=\hat{\bar{\alpha}}_{t-1}} \tag{6}$$

We should note that the formula (5) corresponds to the problem of interframe difference mean square minimisation and the formula (6) is the problem of maximization of interframe correlation sample coefficient (ICSC).

In a number of cases in the capacity of GF PG Q it is convenient to choose $\bar{\beta}_t = \bar{\varphi}(\nabla Q_t)$, where $\bar{\varphi}$ is vector function of the same dimensionality as ∇Q. In particular, simple and at the same time fast converging algorithms can be obtained when selecting a sign function $\bar{\beta}_t = sgn(\nabla Q)$ in the capacity of $\bar{\varphi}$.

3 Algorithms with Varying Local Sample Size

In a number of cases the problem of identification (recognition) with decisive rule based on GF values is solved together with the problem of SD estimation. In particular, when searching a fragment position on supporting image, having in relation with the supporting image in addition to SD close to linear brightness distortions, excess of some value of sample correlation coefficient between fragment and supporting image can serve as the criterion of correspondence. In this case for obtaining high confidential probability a large volume μ of GF local sample is required which can not prove itself in the process of convergence of the estimates $\bar{\alpha}$. In such situations it is reasonable to use varying μ, the value of which is regulated automatically during the process of PG algorithm operation and is minimal for some criterion attainment at each iteration. Let us consider an example of such PGA maximizing ICSC and which showed high efficiency when solving the problem of fragment search on a supporting image. Another example is considered in work [9].

Let us choose the value $q_t(Z, \hat{\bar{\alpha}}_{t-1}, \mu_t)$ of GF estimate Q as a magnitude defining μ_t at t-th iteration $q_t(Z, \hat{\bar{\alpha}}_{t-1}, \mu_t)$. Then it is expedient to use (7) in the capacity of GF PG, having complemented it with sign function for attainment of stability of parameters $\bar{\alpha}$ estimates in the condition of noise

$$\operatorname{sgn}\left(\sum_{\bar{j}_t \in \Omega_t} \left(z_+^{(1)}(\bar{j}_t) - z_-^{(1)}(\bar{j}_t)\right)\left(z_{\bar{j}t}^{(2)} - z_m^{(2)}\right) / (\mu \hat{\sigma}_{x1}\hat{\sigma}_{x2})\right),$$

where $z_m^{(2)} = \frac{1}{\mu}\sum_{\bar{j}_t \in \Omega_t} z_{\bar{j}}^{(2)}$ is mean value of the samples $\left\{z_{\bar{j}t}^{(2)}\right\} \in Z_t$; $z_{t(\pm)}^{(1)} = \tilde{z}^{(1)}(\bar{j}_t, \alpha_{1(t-1)}, \ldots, \alpha_{i(t-1)} \pm \Delta_{\alpha_i}, \alpha_{m(t-1)})$; $\Delta_{\alpha i} > 0$ is increment of the parameter α_i, selected for estimation of the derivative in (6); $\hat{\sigma}_{z1}$ and $\hat{\sigma}_{z2}$ are estimates of the variances $\left\{z_{\bar{j}t}^{(2)}\right\}$ and $\left\{\tilde{z}_{\bar{j}t}^{(1)}\right\}$. Correspondingly if the ICSC estimate is less than r_{cr}, then a step in accordance with the PG value is carried out

$$\beta_{i,t}(\mu_t) = \begin{cases} 1, & q_{t+}(\mu_t) > q_{t-}(\mu_t); \\ 0, & q_{t+}(\mu_t) = q_{t-}(\mu_t); \quad i = \overline{1,m}, \\ -1, & q_{t+}(\mu_t) < q_{t-}(\mu_t), \end{cases} \tag{7}$$

where

$$\mu_t = \begin{cases} \mu_{t,k} + 1, & q_{t0}(\mu_{t,k}) \geq r_{cr}; \\ \mu_{t,k}, & q_{t0}(\mu_{t,k}) < r_{cr}, \end{cases} \quad \mu_{t0} = \mu_{min}; \tag{8}$$

$q_{t-}(\mu_t)$, $q_{t+}(\mu_t)$ and $q_{t_0}(\mu_t)$ are GF estimates respectively at $Z_t = \left\{z_{\bar{j}t}^{(2)}, \tilde{z}_-^{(1)}(\bar{j}_t)\right\}$, $Z_t = \left\{z_{\bar{j}t}^{(2)}, \tilde{z}_+^{(1)}(\bar{j}_t)\right\}$ and $Z_t = \left\{z_{\bar{j}t}^{(2)} \tilde{z}^{(1)}(\bar{j}_t, \hat{\bar{\alpha}}_{t-1})\right\}$; $k =$

$\overline{0,(k_{max}-k_{min})}$; r_{cr} is ICSC value at local sample volume μ_{max}, enabling to make a decision about correspondence of a fragment to some domain of a reference image with a pre assigned validity. Thus, at first local sample volume is equal to μ_{min} at each iteration. If the ICSC estimate exceeds r_{cr}, then the value $\mu_{t,k}$ increases until the condition $q_{t_0}(\mu_{t,k}) < r_{cr}$ is satisfied, or $\mu_{t,k}$ attains the value μ_{max}.

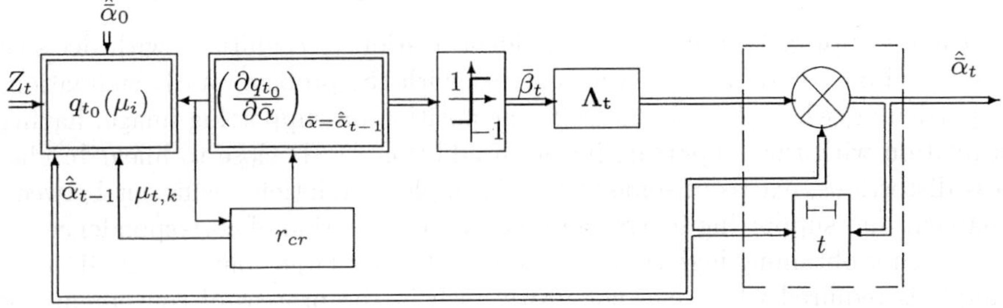

Fig. 1. The algorithm (7) structural chart

The structural chart of the described algorithm is shown on fig.1. It contains functional transformers realizing formation of $q_{t_0}(\mu_t)$ and $(\partial q_{t_0}(\mu_t)/\partial\bar{\alpha})_{\bar{\alpha}=\hat{\bar{\alpha}}_{t-1}}$, relay transformer with $\bar{\beta}_i(\mu_t)$ forming at its output, non-linear transformer $\mathbf{\Lambda}_t$ digrator and block r_{cr} forming the value $\mu_{t,k}$ in accordance with (8). The value $\mu_{t,k}$ increases until the second of the conditions (7) is met and after that a permission for performance of an iteration enters the functional transformer $(\partial q_{t_0}(\mu_t)/\partial\bar{\alpha})_{\bar{\alpha}=\hat{\bar{\alpha}}_{t-1}}$.

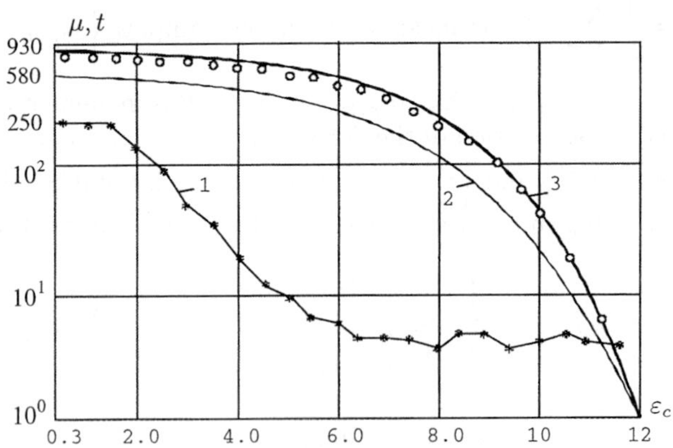

Fig. 2. The plot of connection between local sample size (number of iterations) and ε_c(1 is μ-plotted against ε_c; 2 and 3 are t-plotted against ε_c accordingly at $\mu = var$ and $\mu = const = 250$; $\diamond$ is experimental results at $\mu = var$)

As an example let us present the results obtained when solving the problem of fragment position search on real images of the optical range (with correlation interval approximately equal to 5 steps of sample grid). On fig.2 the dependence of GF local sample size μ on the unbalance ε_c of a fragment centre true position averaged on 800 realizations and its estimations under contrast of fragment scale to a reference image by 25%, slewing angle , slewing angle $\varphi = 30°$ and initial unbalance $\varepsilon_c = 12$ (curve 1, experimental data is asterisks). It is clearly seen that μ increases only at relatively small ε_c. On the same figure dependence of iteration number t on ε_c is shown. The circles correspond to mean iteration numbers averaged on 800 realizations, curve 3 represents the mathematical expectation of iteration number calculated on the basis of the strategy [10]. It is seen that there is good correspondence between experimental and analytical results (circles and curve 3). In this case Gaussian image model with Gaussian correlation function (CF) with signal variance to noise variance ratio equal to 0.01 was used. The algorithm parameters and initial approximations of the parameters vector are consistent with the experimental ones.

Curve 2 is consistent with the iteration number mathematical expectation for the situation of $\mu = const = 250$. The algorithm with $\mu = 250$ obtains $M\{\varepsilon_c\} = 0.3$ in 580 iterations. The algorithm with varying μ affords the same quality of estimation at 930 iterations (curve 2), but computer time expenses prove to be 9,6 times less.

4 Conclusion

The algorithms of multidimensional images sequences spatial deformations parameters estimation developed on the basis of pseudogradient non-identification procedures have high accuracy of estimates to be formed and fast speed of calculation and can be recommended for application in real-time systems. At absence of image brightness distortions it is expedient to choose interframe difference mean square as a goal function and at interframe brightness distortions close to linear-interframe correlation sample coefficient. When calculating the pseudogradient operations number reduction can be obtained by using goal function gradient contraction and substitution of deformed frame value prediction with a simpler estimate through interpolation. Regulation of goal function local sample volume during the algorithm processing enables to essentially reduce computational expenses. In the considered algorithm interframe correlation sample coefficient was used as a goal function. However the presented principles of development of pseudogradient algorithms with regulated volume of local sample can be also applied at other goal functions, in particular, mean square of interframe difference.

References

1. Tzypkin, Y. Z.: Information theory of Identification. Nauka: Fizmatlit, Moskow (1995) [*in Russian*]

2. Tachlinskii, A.G.: Estimation of Image Distortions in Sequence of Frames. Pattern Recognition and Image Analysis, Vol.6(4) (1996) 728–732
3. Tachlinskii, A.G.: Image Sequence Spartial Deformation Parameters Estimation. UlGTU, Ulyanovsk (2000)[*in Russian*]
4. Tashlinskii, A.G.: Pseudogradient estimation of image sequence spatial deformations. Automation, Control and Inrormation Technology: A Publication of The International Association of Science and Technology for Development - IASTED. ACTA Press, Anaheim Calgary Zurich (2002) 382–385
5. Tashlinskii, A.G.: Pseudogradient estimation of image sequence spatial deformations. Naukoemkie Tehnologii, Vol.3(3) (2002) 32–43 [*in Russian*]
6. Polyak, B.T., Tsypkin, Ya.Z.: Pseudogradient Algorithms of Adaptation and Learning. Avtomatika i telemehanika, **6** (1973) 45–68 [*in Russian*]
7. Polyak, B.T., Tsypkin, Ya.Z.: Criterion Algorithms of Stochastic Optimization. Avtomatika i telemehanika, **6** (1984) 95–104 [*in Russian*]
8. Tachlinskii, A.G.: Estimation of Geometric Image Distortions in a Sequence of Frames. Pattern Recognition and Image Analysis, Vol.8(2) (1998) 258–259
9. Tachlinskii, A.G.: Pseudogradient Algorithms for Estimation Spatial Distortions of Image with Varying Volume of the Local Sample of the Goal Function. Pattern Recognition and Image Analysis, Vol.11(1) (2001) 247–250
10. Tachlinskii, A.G., Tikhonov V.O.: A method of multidimentional process parameters pseudogradient measurement error analysis. Izvestija vuzov: Radioelektronika, Vol.44(9) (2001) 75–80 [*in Russian*]

Estimates Conformity Principle in the Problems of Identification

Vladimir Fursov

Image Processing Systems Institute, the Russian Academy of Sciences,
443001, Molodogvardeiskay, 151, Samara, Russia,
fursov@smr.ru

Abstract. A non-traditional approach to solving the identification problems using a small number of observations is developed. The approach is formulated as an estimate conformity principle. The problem of identifying a linear dynamic model is treated within this approach. A significant feature of the above method is the use of a quality criterion not requiring the knowledge of a priori probability noise distributions. The method is well suitable for real-time object identification in adaptive control systems.

1 Introduction

In control systems, the improvement of the system quality is, as a rule, achieved via the adaptation to the varying conditions of operation. The requirement that an adaptive system should have fast response to the changing environment makes us, if possible, use simple linear models and a small number of observations. Because of this, a number of estimation problems solved with the aim of the aircraft adaptation reduce to the following problem. One should find an estimate $\hat{\mathbf{c}}$ of the vector of parameter $\mathbf{c}$, using the $N \times M$ - matrix $\mathbf{X}$ and the $N \times 1$ - vector $\mathbf{y}$ ($N>M$) accessible from the direct observation or generated in one way or another, the matrix and the vector being related via the equation of linear regression

$$\mathbf{Xc} = \mathbf{y} + \boldsymbol{\xi}, \tag{1}$$

where ξ is the $N \times 1$ - vector of the "output" errors caused by the measurement errors and by the limitation imposed on the order of the model. In particular, a nonstationary linear discrete system with one input $u(k)$ and one output $y(k)$ is described by the equation [1]:

$$\mathbf{x}_{k+1} = \mathbf{\Phi}_k \mathbf{x}_k + \mathbf{d}_k u_k, \tag{2}$$

$$y_k = \mathbf{h}^T \mathbf{x}_k + \nu_k, \tag{3}$$

where $\mathbf{x}_k$ is the n-dimensional vector of state, $\mathbf{\Phi}_k$ is the $n \times n$ - matrix of transient state, $\mathbf{d}_k$ and $\mathbf{h}$ are $n \times 1$ - vectors, and ν_k is the random noise of measurements.

P.M.A. Sloot et al. (Eds.): ICCS 2003, LNCS 2658, pp. 463–470, 2003.

We assume that the time of formation of the observation records for identification is small, so that the parameters $\mathbf{\Phi}_k$ and $\mathbf{d}_k$ can be considered as being constant. In that case, for each row of the system (2) after N observations we can put down Eq. (1) in which the $M \times 1$ - vector of estimated parameters will include n elements of the corresponding row of the Φ matrix and one element of the **d** vector, i.e. M=n+1. Solving Eq. (1) n times we can construct estimates for all elements of the Φ matrix and the **d** vector.

From the discrete equations of state, Eqs. (2) and (3), (assuming, as before, that Φ and **d** are constant) by a conventional way [1] we can go over to a single finite-difference equation:

$$y(k+1) = \sum_{i=1}^{n} a_i y(k-i+1) + \sum_{j=1}^{n} b_j u(k-j+1) + \xi(k+1), k = n-1, n-2, \ldots \quad (4)$$

where $\xi(k) = \nu(k) + \sum_{i=1}^{n} a_i \nu(k-i)$ is the equivalent noise of output measurements. In this case, it is quite clear how to form the **X** - matrix and the **y** - vector in Eq. (1) from the input, $u(k)$, and output $y(k)$, measurements. The $M \times 1$ - vector of the sought parameters takes the form $\mathbf{c}^T = [a_1, \ \ldots, a_n, \ b_1, \ \ldots, b_n]$, i.e. at $b_1 \neq 0$, M=2n.

It is common practice to use the statistical theory of estimation for the development of identification algorithms [2]. With this approach, the quality criterion is defined on the basis of a priori information about the distribution of errors. If such an information is lacking a certain probable a priori hypothesis about error distribution is put forth. Then, the procedure constructed on its basis is statistically tested in order to concretize the probability model. Such an approach comes into conflict with the idea of system adaptation. In that case even "good" asymptotic properties of estimates in a particular situation may be misleading.

We develop an algebraic approach to the construction of estimation algorithms. This involves a more realistic formulation of the task as compared with that adopted in the theory of statistical estimation. We discuss the feasibility of solving such problems on the bas*is of the estimates conformity principle*

2 Conformity Principle

The requirement to derive an estimate vector $\hat{\mathbf{c}}$ possibly close to the parameter vector **c** implies that there exists the exact model of the system under estimation corresponding to Eq. (1)

$$\mathbf{y}^* = \mathbf{X}\mathbf{c}, \quad (5)$$

where

$$\mathbf{y}^* = \mathbf{y} - \xi. \quad (6)$$

Obviously, Eq. (5) also holds for all sub-systems of lesser dimensionality constructed from the rows taken in an arbitrary combination. Of course, it appears impossible to attain the exact fulfillment of the equality in (5) for all the rows of the over-defined system because the error vector ξ in the initial set (1) is not known.

However, it is possible to specify a set $\left\{\breve{\xi}_k\right\}$ of so-called correcting vectors $\breve{\xi}_k$ with a view to selecting among them one sufficiently close to ξ. To the set of vectors $\left\{\breve{\xi}_k\right\}$ there will correspond a set of variants of the corrected systems given by

$$\breve{\mathbf{y}}_k = \mathbf{Xc} + \xi - \breve{\xi}_k, \quad k = \overline{1,K}, \tag{7}$$

where $\breve{\mathbf{y}}_k = \mathbf{y} - \breve{\xi}_k$.

For each of the above variants, in its turn, one can specify L variants of the set of equations of lesser dimensionality P: $M < P < N$, with the $P{\times}N$-matrix $\mathbf{X}_l$ and $P{\times}1$ - vectors $\breve{\mathbf{y}}_{k,l}$ and $\xi_l - \breve{\xi}_{k,l}$:

$$\breve{\mathbf{y}}_{k,l} = \mathbf{X}_l\mathbf{c} + \xi_l - \breve{\xi}_{k,l}, \quad k = \overline{1,K}, \quad l = \overline{1,L}. \tag{8}$$

For each (k,l)-variant one can also construct a set of sub-systems of lesser dimensionality, for example, with M rows of the set of equations (8) taken in an arbitrary combination:

$$\breve{\mathbf{y}}_{k,l,q} = \mathbf{X}_{l,q}\mathbf{c} + \xi_{l,q} - \breve{\xi}_{k,l,q}, \quad q = \overline{1,Q}. \tag{9}$$

Here, index q serves to denote the sub-system number and the corresponding-sub-matrix and vectors of the k,l -th variant. If in constructing all sub-systems one takes exactly M rows, then $Q = C_P^M$.

Suppose also that

$$\text{Rank}\, \mathbf{X}_{l,q} = M, \qquad q = \overline{1,Q} \tag{10}$$

for any matrix $\mathbf{X}_{l,q}$. Otherwise, it is always possible to ensure the fulfillment of this requirement by excluding the linearly dependent rows.

The essence of the estimate conformity principle is as follows. For each q-th sub-system of the $q = \overline{1,Q}$ k,l-th variant the estimate vector $\hat{\mathbf{c}}_{k,l,q}$ is calculated. We specify the function $W_l\left[\hat{\mathbf{c}}_{l,q}\right]$ characterizing the mutual closeness of the solutions $\hat{\mathbf{c}}_{k,l,q}$ derived on the k,l-th variant and the selection criterion of the most appropriate variant using the calculated closeness functions $W_{k,l}\left[\hat{\mathbf{c}}_{k,l,q}\right]$, $k = \overline{1,K}, \quad l = \overline{1,L}$. The selection criterion of the point estimate of the sought-for parameter on a set of estimates corresponding to the selected variants and eventually used for constructing the sought-for point estimate $\hat{\mathbf{c}}$ is also specified.

Within the conformity principle, we can specify various functions of mutual closeness, the criteria for selecting the "best" variants based on the mutual closeness functions, and the criteria for constructing the estimate using the selected variants. We consider the situation when the function of mutual closeness of the set of estimates on the variant is given by

$$W_l\left[\hat{\mathbf{c}}_{l,q}\right]=\sum_{i=1}^{M}\sum_{q=1}^{Q}\left(\hat{c}_{q,i}-\bar{c}\right)^2 , \qquad (11)$$

where $\bar{c}_i=\frac{1}{Q}\sum_{q=1}^{Q}\hat{c}_{q,i}$

is the i-th component of the vector $\bar{\mathbf{c}}$ calculated by averaging the estimates $\left\{\hat{\mathbf{c}}_{l,q}\right\}$ derived on the variants for which the values of the mutual closeness function $W_l\left[\hat{\mathbf{c}}_{l,q}\right]$ have turned out to be less than a preset threshold.

Since the function of mutual closeness of estimates, $W\left[\hat{\mathbf{c}}_q\right]$, is not directly related to measuring and estimating errors a natural question arises: whether the application of the estimates conformity principle may guarantee a required accuracy when solving the set (1). To get the answer the qualitative analysis of possible situations has been conducted.

If $\breve{\xi}=\mathbf{Xa}$, where $\mathbf{a}$ is an arbitrary Mx1-vector, which implies that the correcting vector $\breve{\xi}$ is a linear combination of columns of the matrix $\mathbf{X}$, then the component of the estimating error belonging to the null-space $N(\mathbf{X}^T)$ of the matrix $\mathbf{X}^T$ [3] does not depend on the correcting vector and the estimating error will be affected only by the initial data. This means that in this situation the criterion (8) will not respond to variations in the corrections $\breve{\xi}$.

In the general case, when the correcting vector $\breve{\xi}$ also contains the component belonging to $N(\mathbf{X}^T)$ the estimates for different sub-systems will be different. This will result in non-zero values of the criterion (11). For a sequence of the correcting vectors, $\left\{\breve{\xi}_k\right\}$, from $\breve{\xi}_k\to\xi$ it follows that $Q\left(\breve{\xi}_k\right)\to 0$. However, the fact that the criterion (11) converges to zero does not guarantee that the sequence of the correcting vectors will converge to the error vector ξ.

In view of the aforementioned properties we consider one possible strategy for constructing the estimates using the criterion (11). In particular, we discuss a method with which the estimates are being constructed for several variants of less-dimension sub-systems at $\breve{\xi}=0$. For each of them, the criterion (11) is calculated with the use of still less-dimension sub-systems. From the resulting set of sub-system variants and the corresponding estimates we select that for which the criterion value turns out to be the least one. In this paper, we give an example of implementation of the above-described procedure that demonstrates that the Least Square Method's (LSM) estimates can be essentially improved.

3 Analysis of Feasibility of Constructing the Estimates Using Non-statistical Criteria

Since the function of mutual closeness of estimates, $W\lfloor\hat{\mathbf{c}}_q\rfloor$, is not directly related with measurement and estimation errors, a natural question arises if the use of the estimate conformity principle guarantees that the set (1) is solved to the required accuracy. To get the answer we shall conduct a qualitative analysis of possible situations.

Consider the set (8). If Rank(**X**)=M, the vector $\xi-\breve{\xi}$ appearing in the right-hand side of Eq. (8) can be represented as two components:

$$\xi-\breve{\xi}=\mathbf{X}\Delta\mathbf{c}+\mathbf{T}_0\Delta\mathbf{d}\ , \tag{12}$$

где $\Delta\mathbf{c}=\mathbf{F}\boldsymbol{\Lambda}^{-1/2}\mathbf{T}_\lambda^T\left(\xi-\breve{\xi}\right),$ (13)

$$\Delta\mathbf{d}=\mathbf{T}_0^T\left(\xi-\breve{\xi}\right). \tag{14}$$

Here **T** is the $N\times N$ matrix and **F** is the $M\times M$ matrix such that

$$\mathbf{T}^T\mathbf{X}\mathbf{F}=\mathbf{S},\quad \mathbf{X}=\mathbf{T}\mathbf{S}\mathbf{F}^T,$$

$$\mathbf{X}\mathbf{X}^T=\mathbf{T}\mathbf{S}\mathbf{S}^T\mathbf{T}^T=\mathbf{T}\begin{bmatrix}\boldsymbol{\Lambda} & 0\\ 0 & 0\end{bmatrix}\mathbf{T}^T, \tag{15}$$

where **S** is a diagonal $N\times M$-matrix composed of so-called singular numbers, s_i, $i=1,M$, and $\boldsymbol{\Lambda}=diag\left(\lambda_1,\lambda_1,\ldots,\lambda_M\right)$ is a diagonal matrix of eigen-values λ_i, being the corresponding singular numbers s_i squared.

In view of the decomposition in Eqs. (12)-(14), Eq. (8) can be rewritten as follows

$$\breve{\mathbf{y}}=\mathbf{X}\left(\mathbf{c}+\Delta\mathbf{c}\right)+\mathbf{T}_0\Delta\mathbf{d}\,. \tag{16}$$

Premultiply both sides of Eq. (29) by $\mathbf{X}^T$. In view of the property $\mathbf{X}^T\mathbf{T}_0=0$, we have

$$\mathbf{X}^T\breve{\mathbf{y}}=\mathbf{X}^T\mathbf{X}\left(\mathbf{c}+\Delta\mathbf{c}\right) \tag{17}$$

Whence it follows that the root-mean-square estimate calculated on all data,

$$\breve{\mathbf{c}}=\left[\mathbf{X}^T\mathbf{X}\right]^{-1}\mathbf{X}^T\breve{\mathbf{y}}=\mathbf{c}+\Delta\mathbf{c}\,, \tag{18}$$

contains an error (13) that depends only on the components of the difference vector $\xi-\breve{\xi}$, belonging to $R(\mathbf{X})$.

Let us now analyze the solutions to the sub-systems in (9) composed from (8) using corrected data. In view of the decomposition in (12), each of the sub-systems can be represented as

$$\breve{\mathbf{y}}_q=\mathbf{X}_q\left(\mathbf{c}+\Delta\mathbf{c}\right)+\mathbf{T}_{0,q}\Delta\mathbf{d}\,,\ q=\overline{1,\Sigma} \tag{19}$$

where $\mathbf{T}_{0,q}$ is a sub-matrix of the $\mathbf{T}_0$-matrix corresponding to the *q-th* sub-system, and $\Delta\mathbf{c}$, $\Delta\mathbf{d}$ are the same as in (13) and (14). By multiplying from the left both sides of (19) by $\mathbf{X}_q^T$, it can easily be found that as distinct from (18), the LSM estimate calculated on the portion of data (for the q-th sub-system) will contain not only the error vector $\Delta\mathbf{c}$ but also an extra error

$$\Delta\tilde{\mathbf{c}}_q = \left[\mathbf{X}_q^T\mathbf{X}_q\right]^{-1}\mathbf{X}_q^T\mathbf{T}_{0,q}\Delta\mathbf{d} = \left[\mathbf{X}_q^T\mathbf{X}_q\right]^{-1}\mathbf{X}_q^T\mathbf{T}_{0,q}\mathbf{T}_0^T\left(\xi - \breve{\xi}\right). \qquad (20)$$

This is because for sub-systems of dimensionality lesser than *N* the product $\mathbf{X}_q^T\mathbf{T}_{0q}$ is not necessarily equal to zero. Based on the aforementioned relationships, we can formulate the following important properties of the criterion in (11).

If $\breve{\xi} = \xi$, in accordance with (13) and (14), $\Delta\mathbf{c}$ and $\Delta\mathbf{d}$ become zero simultaneously, and as one would expect, the sought-for vector parameter **c** will be determined exactly both from the entire set of data and from data sets of all less-dimension sub-systems. The criterion in (11) that corresponds to the above-described situation takes its minimum value of zero.

If $\breve{\xi} = \mathbf{Xa}$ (where **a** is an arbitrary $M\times 1$-vector, that is the correcting vector $\breve{\xi}$ is a linear combination of the columns of matrix **X**), then from the property $\mathbf{X}^T\mathbf{T}_0 = 0$ and Eq. (14) we have $\Delta\mathbf{d} = \mathbf{T}_0^T\xi$, i.e. $\Delta\mathbf{d}$ does not depend on the correcting vector and the error in (20) will only be determined by the initial data. In this situation the criterion in (11) will not respond to variations in the corrections $\breve{\xi}$.

In the general case, when the correcting vector $\breve{\xi}$ comprises both the components belonging to the space of columns and the null-space, the error $\Delta\mathbf{c}$ appearing in (18) will be determined by the contribution of the component $\mathbf{X}\Delta\mathbf{c}$ in the right-hand side of (12). Whereas the sub-vectors $\mathbf{T}_{0q}\Delta\mathbf{d}$ composed of the null-space component $\mathbf{T}_0\Delta\mathbf{d}$, will determine the set of errors in (20), which cause the estimation difference for different sub-systems. As a result, the value of the criterion in (11) will be non-zero. It is clear that for a sequence of correcting vectors $\left\{\breve{\xi}_k\right\}$ from $\breve{\xi}_k \to \xi$ it follows that $Q\left(\breve{\xi}_k\right) \to 0$. However, the convergence of the criterion (11) to zero does not guarantee that the correcting vector will converge to the error vector.

Based on the above-formulated properties a number of strategies for constructing the estimates using the criterion in (11) have been proposed. In particular, in one method the estimates were constructed for several variants of less-dimension sub-systems at $\breve{\xi} = 0$. For each such a sub-system, the criterion (11) is calculated using still less-dimension sub-systems. From the resulting set of sub-system variants with the corresponding estimates the estimate characterized by the least criterion value is chosen.

4 Example

Below we discuss an example (N=10, M=4) of finding the solution using the criterion (11). The initial set of data was generated using a random number generator. Best conforming estimates were sought using C_{10}^{7}=120 sub-system variants, with 7 observations made for every variant. Then, for every variant, for C_7^4=35 (4×4)-sub-systems a set of LSM-estimates was calculated, for which the value of the criterion (11) was derived. The resulting set of the conformity criterion values was arranged in increasing order. For comparison, the criteria of estimation quality, $\|\Delta c\|$, were arranged in the same order.

Figure 1 shows the values of the conformity criterion. Figure 2 shows the corresponding values of the LMS-estimate quality criterion. The horizontal line shows the quality criterion value for the LSM-estimate calculated on all parameters.

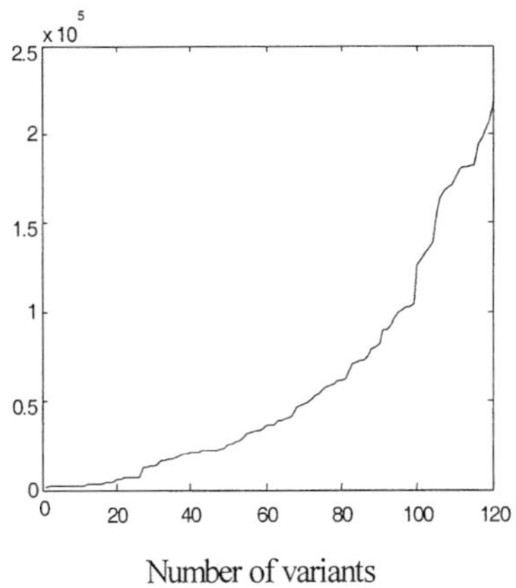

Fig. 1.

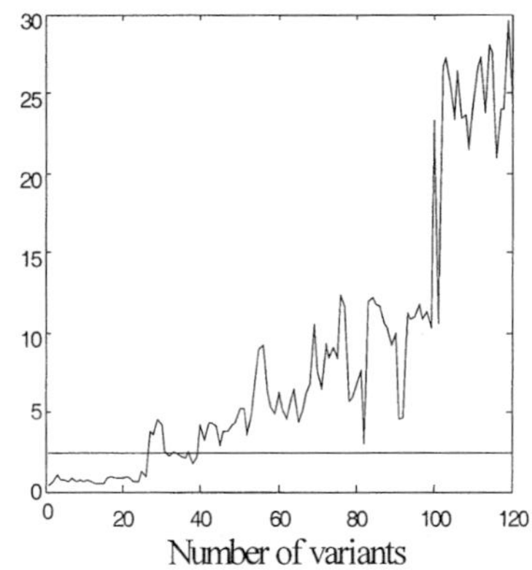

Fig. 2.

5 Conclusions

The aforesaid strategy for seeking the solution is conceptually similar to the ideas by P. Kalman discussed in Ref. [4]. In this case, the criterion (8) ensures the possibility of finding a sub-system with least-noised observations. The suitability of the strategy depends on how well the assumptions of existence of a noise-free sub-system are justified, as well as on the proper choice of dimension of the sub-systems that are assumed to include the above observations.

In order to find a least-noised sub-system it may require to search through a great number of variants. This fact is not astonishing, being the unavoidable fee for a priori information deficiency [5]. It seems reasonable that the application of simple (and, hence, relatively cheap) statistical schemes of data processing becomes feasible as a result of considerable expenses incurred while making a great number of observations at the stage of constructing the system (1).

Acknowledgement. This work was financially supported by the RF Ministry of Education, Samara Region Administration, the American Civilian Research and Development Foundation (CRDF Project SA-014-02) as part of the joint Russian-American program "Basic Research and Higher Education" (BRHE) and by the RFBR (grants N. 00-01-05001, N. 01-01-00097).

References

1. Graupe, D.: Identification of systems. Colorado State University Fort Collins, Robert E. Kriger Publishing Company Huntington, New York, 1978.
2. Ljung, L.: System Identification, Theory for the User. University of Linkoping, Sweden Prentice - Hall, Inc., 1987.
3. Bjorck Ake.: Least Squares Methods, Elsevier Science Publishers B.V. North Holland, 1990.
4. Kalman P.E.: Noised systems identification, Advances of Mathematical Sciences, v. 40, issue 4(244), 1985.
5. Fursov V.: Conformity principle in the problems of evaluating using a small number of observations, Proceedings of the IASTED International Conference Automation, Control, and Information Technology, June 10-13, 2002, Novosibirsk, Russia, p. 279–281.

The ARM-Approach Based Local Modelling of the Gravitational Field

Sultan Valeev and Konstantin Samokhvalov

Ulyanovsk State Technical University, 32 Sev Venets Str., 432027 Ulyanovsk, Russia
{sgv,sam}@ulstu.ru

Abstract. Gravimetrical Measurements revealing particularities of the gravitational field, play an important role in a research after oil thanks to their capacity to reduce the cost of reconnaissance work. Graphic generalization of the field is given by gravimetrical maps of the field anomalies, drawn out with the corresponding software on the basis of one or other version of interpolation formula. Mathematical models of the trend forming regional reference-surface, can extend the capabilities of graphic generalization of measurements. With these models and corresponding contour maps, the local particularities of the gravitational field reveal themselves more clearly.
The use of models in the manner of truncated decompositions in Fourier series on spherical surfaces is offered for practical geophysical tasks. An algorithm of adaptive statistical modelling which allows optimum mathematical models to be built is considered.
In the last section we describe the developed software that allows us to build the statistical potential field models. The corresponding field maps of residual anomalies are described.

1 Introduction

Gravimetrical contour maps (isoanomaly contour maps) which are drawn according to the results of measurements and various reductions are used for geological object identifications which accompany oilfields and gas fields, or just hydrocarbon collectors. This information is used with (or without) some seismic and other types of measurements depending on a number of circumstances.

At present time, gravitational anomalies and, naturally, corresponding isolines on contour maps have mistakes because of incomplete account of global, regional and, in some cases, local components forming the gravitational field of a region. Besides, normal field of the Earth is fixed with sufficiently great mistakes, and mathematical presentation of the field of anomalies in points in which measurements are taken, can contain great casual mistakes because of the model inadequacy.

In practice of geodesic works, local mathematical models are not always used; more often one immediately goes from measurements on points to contouring of anomalies with the help of interpolations in graphic software packages. The usefulness of the local "background" models, averaging the influence of the rock

P.M.A. Sloot et al. (Eds.): ICCS 2003, LNCS 2658, pp. 471–480, 2003.

on "normal" Earth on the area under consideration is obvious. Forming some reference-surface they enable separate local disturbance from geological structures to reveal itself on average background.

So-called surface polynomials (usual two-dimensional algebraic polynomials) are often used to make a description of gravitational field of anomalies in the region. Their factors are defined by a standard method of least squares (MoLS), and dimensions of the model are hard fixed.

It is expected that the approach proposed in this paper will increase the accuracy of potential floor description by the mathematical model and, consequently, the accuracy of identification (in contrast with methods used in practice in Russia and other counties) from 10% to several times as large.

Using ARM-approach, based on adaptive regression modelling [1] the following is planned: 1) determination of global and regional components of the geopotential by a great multifunction array of measurements (overland and satellite data); 2) revision of normal field of the Earth; 3) building of regional and local mathematical models in the manner of optimum decompositions by Fourier series based on the overland gravimetrical measurements.

2 Mathematical Models of Potential Fields

2.1 Global Models of Gravitational Field

Global models of potential fields (gravitational, magnetic and others) are necessary for solution of geophysical and other problems both for the whole Earth, its great regions, and for the building of local models. In the last case global models can be used as reference-models (relationship models).

Decompositions on spherical functions are more often used for the presentation of global models. Harmonic coefficients of those functions are defined by MoLS with lumpy distribution of data [2], [3]. Initial data for processing are the results of the Earth Satellite observing; average anomalies in free air in trapezoids are usually 1^0 x 1^0; average one-degree altimetrical heights of the geoid relative to ocean surface.

Decompositions of anomalies of gravity Δg are researched for the description of the global gravitational field in spherical drawing in the manner of:

$$\Delta g\left(z\right)=\frac{GM}{r^2}\sum_{n=2}^{\infty}\left(n-1\right)\frac{a^n}{r^n}\sum_{m=0}^{n}\left(\Delta\bar{C}_{nm}\cos m\lambda+\Delta\bar{S}_{nm}\sin m\lambda\right)\bar{P}_{nm}\left(\cos\theta\right) \quad (1)$$

where $GM = G\left(M_0 + M_{\text{atm}}\right)$- is geocentric gravitational constant, taking into account the Earth atmosphere, a- big half-axis of the Earth ellipsoid; r,θ=90^0 - φ, λ - are spherical coordinates of the point - accordingly geocentric distance, arctic distance (φ - geocentric width), geographical longitude;$\Delta\bar{C}_{nm}$, $\Delta\bar{S}_{nm}$- are differences of factors of normalizing spherical functions of real and normal fields; $\bar{P}_{nm}\left(\cos\theta\right)$ - are normalized joined Legendre functions of n degree and m order.

Taking into account that model (1) has excluded the planet component, corresponding to normal Earth, it is possible to suppose that they include local and regional components, generated respectively by anomalies of the gravitational field in upper mantle and in the Earths crust. This circumstance ensures a possibility to use with a certain accuracy a mathematical model (1) to determine the identification of geological structures which generate anomalies.

Under standard approach the model (1) is considered to be determined for accepted order of decomposition m, if MoLS - evaluations of amendments to harmonica amplitudes are calculated on the surplus number of measurements of anomalies. Whereupon, we can draw a map of gravitational field (gravimetrical contour map), using expression (1) in the mode of forecast.

On the basis of the theory and computing experiments with great data arrays, we had proved that standard decompositions on spherical functions with MoLS-evaluations of amplitudes of harmonics do not fully correspond to real measurements, on which they are built [1]. The main reason of this fact is the redundancy of mathematical model, containing multiple (up to 60% and more) noise harmonicas. There occurs the reduction of accuracy from 10% up to one order at the study of such models for the forecast (buildings isolines - isoanomalies or isolines in relief description), that causes corresponding displacing of isoanomalies and the loss of accuracy in the identification of geological structures.

We offered to use a statistical approach in the manner of adaptive regression modelling [1] for deciding a problem of adequacy of global models and accordingly gravimetrical contour maps of anomalies of power of gravity. This approach allows to raise the accuracy of identification of structures up to several times already at the stage of elimination of noises. Experiments on its using under comparatively small orders of decomposition are described in [1].

2.2 Local Models of Gravitational Field

Flat functional approximations in the manner of polynomials of low degrees are usually used for local models, but in global model decompositions on spherical functions are researched, using degree n at 360 order. However, spherical functions also can be used for regional and local models as trend functions. It is possible to use noted functional models for the forecast of regular values of anomalies.

The area of the local models application is sufficiently wide. They can assign regional or local, relation surfaces, formed by long geological structures, aside from the description of systematic behavior of a potential field. In the last case residue, for instance$(\Delta g_i - \Delta \hat{g}_i) = e_i$ between observed and computed (forecasted) values of anomalies, can be interpreted as a manifestation of disturbances, generated by changing density of geological objects.

Accuracy of the end result, naturally, depends on the keeping of suggestions, mortgaged in the mathematical device of data processing for getting optimum models, MoLS in particular.

It is possible to offer the model

$$\Delta g\left(\lambda,\theta\right)=\sum_{n=0}^{N}\sum_{m=0}^{n}\left(C_{nm}\cos m\lambda+S_{nm}\sin m\lambda\right)P_{nm}\left(\cos\theta\right), \tag{2}$$

as the local model, describing potential field on a segment of spherical surface. The formula (1) can be offered for the anomaly description.

2.3 Problems of Local Potential Fields Modelling and Ways of Their Solution

Researchers face the problems of non-observance of conditions of MoLS with all resulting consequences, when they're using approximate descriptions and MoLS. MoLS - evaluations are not the best single-line evaluations in practice.

In this paper the following problems are concerned:

1. Algorithm Development of statistical modelling of local potential fields on the basis of adaptive regression modelling (an ARM - approach);

2. The development of ARM - approach software with the use of the decompositions in Fourier series;

3. The research of ARM - approach efficiency on the example of real data array.

3 Methodology and Algorithms of Statistical (Regression) Modelling

3.1 Mathematical Model of Data Processing, to Forecast Potential Field Features

Clearly determined model of observed (measured) phenomena (objects, processes), in mathematical terms is defined as mathematical model of data processing. For instance, the relationship of anomaly of gravity power and measured coordinates of an object in some area can be described as the model

$$MY=\eta\left(X,\beta\right) \tag{3}$$

where Y - is dependent variable (anomalies of gravity power); $X = (x_0, x_1, ..., x_{p-1})$- is matrix of independent variables (measured coordinates, their product or other combinations), which can change in certain areas of space R^0; $\beta = (\beta_0\beta_1...\beta_{p-1})^T$ - is the vector of unknown parameters defined according to the results of experiments; M - is an operator of math mean.

The analysis of standard strategy used in the geophysics for parametric model identification (3), has shown that for essential increasing of accuracy of presentation, it is necessary to solve a number of problems, connected with the use of MoLS.

Standard methodology can be subjected to criticism as methodology, from the point of mathematical statistics and the theory of function reconstruction.

It does not promote getting the whole possible statistical information about the model and its members, and does not help to get the identical structures and parameters according to the results of checking the hypotheses. These statements are considered in detail below.

1. Choice of measures of accuracy for the evaluation of a model quality, its fitness for the whole forecast is limited. Only one measure from the kit of internal measures exists in mathematical statistics - it is a remaining dispersion or its variety. Mixed measure of accuracy used in a number of works in the manner of forecast mistakes expects the observance of the assumption about the absence of systematic mistake in the model that is not always executed. At the same time external measures are not paid attention to, ways of their using are imperfect, and many approaches used are limited.

2. Mathematical model after postulating is considered to be hard given, regardless of its origins: no matter if it is a result of a project or it is approximated. In the last case the members of the model are not analyzed on value in the process of solution, i.e. the procedure of structured identifications is not executed.

3. In papers on modelling in geophysics interest is not shown to the circumstance that evaluations of parameters of the statistical model (3) are the best, i.e. consisted, not-shifted, efficient in the class of single-line not-shifted evaluations only in conditions of observance of a number of suggestions, considered in the following section.

3.2 Methodology of the ARM – Approach

Regression Analysis (RA). *Model of RA.* Data processing Model (1)-(3) can be presented in the manner of [1]

$$Y = X\boldsymbol{\beta} + \boldsymbol{\varepsilon} \ . \tag{4}$$

Usual mathematical presentation of direct and indirect observance resulted in the form (4) when passive experiment in geophysics is used.

RA stipulates the performing of three stages: - model postulating,

- valuation of its parameters,
- solution of the problem of searching its optimum structure.

RA-MoLS suggestions. For correct using of RA it is necessary to observe a number of suggestions. One part of suggestions, given below, is generated by the statistical theory of valuation, in which condition for the sample of data and method of valuation (MoLS) are installed, the other part - is a theory of statistical criteria. Besides, we offer a suggestion (5.1) which concerns the structure identification method taking into consideration the dimensions of the problem.

In the respect of data samples $\{y_i, \ x_{ii}\}$ it is supposed to be sufficiently representative, i.e.

(1.1) - a size of observing is sufficient,

(1.2) - casual selection of observing is ensured,

(1.3) - the set of observing is uniform,

(1.4) - there are no rough misses of inwardly set.

Some suggestions about valued vector parameter β are formulated:

(2.1) - the model (4) is linear on vector β,

(2.2) - no restrictions are imposed on the vector β,

(2.3) - vector β contains the additive constant β_0,

(2.4) - vector elements are calculated with a negligibly small computer inaccuracy.

Let us note suggestions about matrix X.

(3.1) - regressors $x_0, x_1, \ldots x_{p-1}$ (matrix X columns) are linear-independent vectors of matrix X or of matrix: rank X = p,

(3.2) - elements of matrix X are not random quantities.

The main suggestions concerning elements e_i of mistake vector e are:

(4.1) - they are additive casual mistakes,

(4.2) - they are distributed by normal law,

(4.3) - they do not contain systematic displacement,

(4.4) - they have constant dispersion,

(4.5) - they are not correlated and in the suggestion (4.2) are statistically independent.

In conclusion we will point out additional suggestions about vector Y:

(5.1) - the method of searching for optimum model or the identification of optimum kit of regressors $\{x_i : j = \overline{1, p1},\ \ p1 < p\}$ for vector y is exact,

(5.2) - MoLS can be used for each of regressions separately to solve a multi-responsible problem, containing two or more output parameters y_k (k < 2).

Linear normal RA ensures the best (consisted, not-shifted, efficient) evaluations only if all RA conditions are observed. In practice, suggestions (1.1) - (5.2) are broken, so RA evaluations are not optimum. The degree of not-optimality is considered in detail in book [1].

The main stages of RA. There are three main RA stages:

(1) model postulating, (2) valuation of its MoLS parameters, (3) analysis and searching for optimum structure of the model. All these stages are considered in detail in [1].

Adaptive regression modelling. *Consequences of the RA-MoLS suggestions breaking.* Breaking of conditions of RA - MoLS application results in removable, insolvent and inefficient MoLS (for details see book [1]), and evaluations of both the parameters of the model processing β_j $\left(j = \overline{0, p-1}\right)$, and the values of $\hat{Y}$ when using models in the mode of forecast.

Revealing of breaches. To reveal breaking of conditions of RA - MoLS use both corresponding statistics and various graphic procedures can be used [1].

Methodology of the ARM approach. As it was already noted, traditional methodology of the solution of problems concerning data processing is characterized by two moments: 1) the structure of model processing is considered to be given; 2) valuation of model parameters is performed by MoLS. A standard approach for given geophysical problem does not correspond to terms of accuracy. The more so because the simplicity of its use does not compensate forecast losses in characteristics of best linear evaluations from dozens of percent to one order. It is possible to say, that a MoLS computing scheme application (without statistical analysis of the degree of breaking its usage conditions and corresponding adaptation) on the one hand makes the accuracy of the results higher and allows to consider statistically negligible factors. On the other hand (in conditions of some MoLS suggestions breaking) - it results in not-the-best evaluations of parameters and the model's prognoses characteristics become worse.

Certainly, the use of regression analysis is a significant step forward in contrast with MoLS application: 1) model analysis of a some number of criteria (R-, F - criteria) is made; 2) statistical value not only of a model as a whole, but each separate member of the model is analyzed. The last fact allows to develop to some extent the best model structure, using one or another method of structure identification.

Unfortunately, RA does not completely solve the problem of finding the best linear evaluations $\hat{\beta}_j$ and $\hat{Y}$.

Additional stages for standard MoLS methodology (in system ARM - approach developed in [1]) are: 1) evaluation of model adequacy to observation and search for its optimum structure; 2) checking the observance of MoLS suggestions; 3) consequent adaptation of a processing scheme to breaking of MoLS conditions by using a kit of computing procedures (change of postulating models, methods of parametric valuation, structure identification and others. [1]); 4) application of a kit of measures (criterion of models quality, including multi-criteria concept).

3.3 The Algorithm of Building the Local Optimum Model of Field of Anomalies of Gravity Power

At present time, there are at least two rival scenarios of a solution of the geophysical problems of data processing, represented in the manner of models (1) - (4).

Let forecasting model be postulated in the manner of model (4).

One of the perspective scenarios of processing can be described in brief as follows:

1. Single-criteria search in the global criterion of optimum mathematical structure is realized on the base of initial description (4). Such search can be full, if computing possibilities allow, or incomplete. In the last case the method of incomplete change with the restriction can be one of pseudo-bul optimization methods. Casual or systematic mistakes, defined according to the objects being checked, which were not used in processing model building, or values of general F-criterion can serve as global (main) criterion of model quality.

2. The Second and further stages of the structured-parametric identification of optimum model are based on checking of all conditions of RA-MoLS application and consequent adaptation according to the degree of breach importance. In practice the "chains" of adaptation algorithms can be different, and it results in particular requirements to "intellectuality" automated data processing systems.

The Second rival scenario of data processing does not expect the use of global criterion in realization the last, one based on checkpoints is used at the end stage only. The base of this approach is a check of suggestion observance for initial models (4) and its consequent improvement by adaptation to the most serious breaking while the degree of characteristics distortion of best evaluations is decreasing.

3.4 Contouring Algorithm of the Anomalies Field

It is possible to use a method, based on the way from [3] to build isolines in the manner of isoanomalies on the equation (2). Adaptation of this way makes the following algorithm.

Let's calculate two-dimensional curves, defined by optimum mathematical description under initial model (2)

$$\sum_{n=0}^{N}\sum_{m=0}^{n}\left(C_{nm}\cos m\lambda + S_{nm}\sin m\lambda\right)P_{nm}\left(\cos\theta\right) = 0.5k \tag{5}$$

for instance, under k=-4,-3,-2,-1,0,1,2,3,4,5,6, we will get isoanomalies, along which deflections of values δg from 0 will be -2, -1.5,-1 and etc to +3.

We rewrite equation (5) in the manner of

$$f\left(\lambda,\theta,k\right) = 0\ . \tag{6}$$

Let k be fixed, and free spot be chosen as initial (λ_0,θ_0). Then most likely that $f\left(\lambda_0,\theta_0,k\right) \neq 0$. To find the spot (λ_1,θ_1), for which

$$\left|f\left(\lambda_1,\theta_1,k\right)\right| \leq \left|f\left(\lambda_0,\theta_0,k\right)\right| \tag{7}$$

let us enter increments $\delta\lambda$ and $\delta\theta$ and require the equation to be satisfied

$$f\left(\lambda_0+\delta\lambda,\theta_0+\delta\theta,k\right) = 0\ . \tag{8}$$

In the first approximation we get

$$\left(\frac{\partial f}{\partial\lambda}\right)_{\lambda=\lambda_0}\delta\lambda - \left(\frac{\partial f}{\partial\theta}\right)_{\theta=\theta_0}\delta\theta = -f\left(\lambda_0,\theta_0,k\right)\ . \tag{9}$$

Then, solving (9) with equation

$$\left(\frac{\partial f}{\partial\lambda}\right)\delta\lambda - \left(\frac{\partial f}{\partial\theta}\right)\delta\theta = 0 \tag{10}$$

we define

$$\delta\lambda = -\frac{f\left(\frac{\partial f}{\partial\lambda}\right)}{\left(\frac{\partial f}{\partial\lambda}\right)^2 + \left(\frac{\partial f}{\partial\theta}\right)^2} \tag{11}$$

$$\delta\theta = -\frac{f\left(\frac{\partial f}{\partial \theta}\right)}{\left(\frac{\partial f}{\partial \theta}\right)^2 + \left(\frac{\partial f}{\partial \lambda}\right)^2} \ . \tag{12}$$

If these values are added to λ_0 and θ_0, the condition (7) will be satisfied. Values λ_1 and θ_1 which satisfy an equation (6) will be reached under following approximations. When we find one point on the curve (let it be (λ, θ)) we give it an increment along tangent, i.e. the increment must satisfy the condition

$$\left(\frac{\partial f}{\partial \lambda}\right) \delta\lambda + \left(\frac{\partial f}{\partial \theta}\right) \delta\theta = 0 \ . \tag{13}$$

To move forward the inner product of increments $(\delta\lambda, \delta\theta)$ of two consequent points on the curve must be positive. Otherwise, we need to change the signs of both components. Equation (13) gives only the direction, so to move we need to choose a value of the step at each moment.

4 Software

4.1 The First Version of ASSI

The first version of developed automatic system [4] is a specialized system, realizing the strategy of the statistical (regression) modelling [1] for solution of a number of problems of mathematical relief description and gravitational fields of planets. The main purpose of the system is to obtain models of processes or phenomena with their further application for output features (responses) and realization of some functions of management in interactive (display) and packet modes of work. The need for this automatic system is due to great difficulties when performing similar work, requiring both multi-variant calculations, and using different methods of parameters valuation and structural identifications, as well as remainders analysis under chosen scenario of checking the observance of MoLS suggestions.

Module system realizing methods of regression analysis is considered to be base mathematical support. The given modules ensure the solution of normal and surplus (redefined) algebraic equation systems. Both original and modified programs of a well-known package of scientific programs in Fortran are used in this automatic system.

4.2 Modified Version of ASSI

The program package ASSI 1.0 [4] does not use computing facilities very much, the speed of calculations is not high enough because of the use of 16-bit code, so the process of modelling requires a lot of time. Possibilities of the package are limited by modelling when using the spherical functions of order N<40.

New realization differs from initial one by using optimized 32-bit code, and by adding new procedures and modules and new multi-window interface. Methods of mathematical modelling, unceasing and discrete optimization, numeric methods, the theory of probability methods are used in the program. Object-oriented programming methods were used when creating an algorithmic code, which resulted in less complicated package structure, new functions for realization of various methods of calculation have appeared.

In this realization of ASSI 2.0 for the operating system Windows 9x/2000/XP an efficient solution of the problem of relief models adequacy and gravitational fields of planets at the level of measurements accuracy and size of the information used is ensured. Most program code had been developed in Borland C++ Builder 5.0. Also Borland Delphi was used. ASSI 2.0 consists of interface, controlling part and calculation modules. Also a program for making contour maps and a program of sections building are included in the package.

5 Conclusion

The developed mathematical methods and software are used for getting local and regional trend models, as well as drawing maps of iso-anomalies and variations of the anomalies of gravity power for one of the regions at the Volga-river. The results prove the efficiency of the given technique.

The problems to be solved next are:

- adjustment of type models (2) within the framework of ARM-approach (reduction of the effect of interdependence of parameters of Fourier series, adaptation of MoLS use to other breaking of conditions);

- development of other type models, both on spherical and square coordinates (models with generally accepted normalizing harmonic factors in particular) which will provide more exact account in local models of global and regional components;

- expansion of type (2) models application (magnetic and other potential fields);

- practical identification of deposits (territorial expansions of investigated ranges, practical check of accuracy of geological structures localization, consideration of a problem of the depth of their presence bedding);

- improvement of software.

References

1. Valeev, S.G.: Regression Modelling in Observations Treatment. Science, Moscow (1991) [*or* Valeev, S.G.: Regression Modelling in Data Processing. FEN, Kazan (2001)]
2. Torge, W.: Gravimetru. Walter de Gruyter, Berlin New York (1989)
3. Gudas, To.: Decomposition of Relief of the Moon on Spherical Functions / Figure of the Moon and Problems of Moon Topography. Science, Moscow (1968)
4. Valeev, S.G., Dyakov, V.I.: Automatic System for Modelling of the Megarelief and Gravitation Fields of Planets. Izvestiya VUZov. Series: Geodesy and Aerophotography **4-5** (1998) 45–49

Workshop on Numerical Methods for Singular Differential and Differential-Algebraic Equations

The Parameterization Method in Singular Differential-Algebraic Equations*

Vladimir K. Gorbunov and Igor V. Lutoshkin

Ulyanovsk State University, L.Tolstoy street 42,
432970 Ulyanovsk, Russia,
gvk@vens.ru, LutoshkinIV@ulsu.ru

Abstract. The paper is devoted to the circumstantiation of the parameterization method for classical calculus of variation problems corresponding to the non-linear ODEs. The method is based on a finite parameterization of "control" functions (finitely entering in initial system) and on derivation of the problem functional with respect to control parameters. The first and the second derivatives are calculated with the help of adjoint vector and matrix impulses. The problems of arising degeneration of gradients and optimality conditions of the first order are overcome by using the Newton method. Results of the solution to degenerate DAEs, particularly with non-unique solutions, are presented.

1 Introduction

In the paper [3] the parameterization method (PM) of optimal control problems was created. This method is based on sequential development of a simple idea to account that as a rule solutions of real optimal control (OC) or calculus of variation (CV) problems have a rather simple structure and control functions can be well approximated in some parametric function class, for example, by splines with moving knots. The PM consists of a finite-dimensional parameterization of control functions. The parameterized variational problem becomes a finite-dimensional nonlinear programming (NP) problem whereas its functionals are defined on phase trajectories. Their first and second derivatives on the control parameters may be effectively calculated with the help of variational techniques and adjoint variables. By this the problems of optimization and numerical solution of differential equations are separated. Appropriate parameterization allows one to obtain a NP problem of lower dimension than one obtained under traditional finite-difference approximations.

Another advantage of the PM is a possibility to solve OC and CV problems with a continuous admissible control without expansion of the phase space. Continuous control is a natural one in optimization problems of economics dynamic and in variational approach to initial / boundary value problems for differential-algebraic equations (DAEs), especially in cases of their essential degeneration, when the system has differentiation index [8] more then one or it hasn't a finite index [12].

* Supported by the Russian Foundation for Basic Research, project N° 01-01-00731

P.M.A. Sloot et al. (Eds.): ICCS 2003, LNCS 2658, pp. 483–491, 2003.

The paper contains reduction schemes for degenerate DAE to the equivalent basic CV problem. The reduced problems as a rule are the degenerate ones with respect to the first order necessary conditions. Also gradients of reduced functionals degenerate near optimal values of parameters. However, effective calculation of degenerate optimization problems can be obtained by means of the second order numerical methods of the Newton type. This general fact was developed for theoretical investigations of degenerate OC problems by R.Gabasov and F.Kirillova (1971) in [2], where the matrix impulse functions, adjoint to the second variations of the optimal trajectories, were introduced. In [9] some degenerate CV problems were solved thereby using of the second variation of the improved trajectory .

We note here the limited experience of using the second order techniques in variational problems. The PM handles this problem naturally and simply. In difficult unstable problems one can use some regularization of the initial problem [4].

2 Problem Statement

Consider the initial / boundary value problem for DAE

$$\dot{x} = f(x, u(t)), \tag{1}$$

$$h(x(t), u(t)) = 0, \tag{2}$$

with conditions

$$a(x(0), u(0)) = 0, \quad b(x(T), u(T)) = 0. \tag{3}$$

Here $h(x,u) \in R^r$, $a(x,u) \in R^{m_1}$, $b(x,u) \in R^{m_2}$, $m_1 + m_2 = m$, $n \leq m \leq n + r$.

Difficulties of numerical solution to the problem (1)- (3) appear when the Jacobi matrix of the finite subsystem (2)

$$\partial h(x(t), u(t))/\partial u \tag{4}$$

degenerates on the solution $\{x(t), u(t)\}$. There are some regular numerical methods for the initial value problem for the DAE (1) which have a finite index up to 3. However, not every system (1), (2) has a finite index and a verification of this property is rather a complex procedure in the common case. The most complex case occurs when the matrix (4) has a variable range on $[0, T]$ [12].

The problem (1)-(3) is naturally embedded in the class of OC ones with continuous control functions $u(t)$ and can be interpreted as a CV problem. Namely, the problem (1)-(3) is equivalent to the minimization of the functional

$$J[u] = \int_0^T \|h(x(t), u(t))\|_r^2 dt + \|b(x(T), u(T))\|_{m_2}^2, \tag{5}$$

where $\|\cdot\|_k$ is the Euclidean norm in R^k, under conditions

$$\dot{x} = f(x, u(t)), \quad a(x(0), u(0)) = 0. \tag{6}$$

It can be shown [6], such transform leads to the degenerate CV problem (5), (6).

3 Parameterization Method

Introduce a partition of the interval $[0, T]$

$$0 \equiv t_0 \leq t_1 \leq \ldots \leq t_N \equiv T, \tag{7}$$

and define a structure of control functions $u(t)$ on subintervals as

$$u_\mu(t) = u^k_\mu(t; v^k_\mu), \quad t_{k-1} \leq t < t_k, \quad k = 1, \ldots, N, \quad \mu = 1, \ldots, r. \tag{8}$$

Here $v^k_\mu \in R^d$, as well as t_k, are the sought control parameters and $u^k_\mu(t, v^k_\mu)$ are given in analytical form twice-differentiable functions. Further $v^k = (v^k_1, \ldots, v^k_r)$.

Conditions (7), (8) present a class of piece-wise continuous control functions with a given number of discontinuities. A class of continuous admissible functions is described by additional conditions $u^k(t_k; v^k) = u^{k+1}(t_k; v^{k+1})$.

Substituting a control (8) in differential equation (1) we have a solution (its existing is assumed) depended on control parameters

$$x(t) = z(t; v^1, t_1, \ldots, v^{k-1}, t_{k-1}, v^k), \quad t_{k-1} \leq t < t_k. \tag{9}$$

Denote $w^k = (t_k, v^k) \equiv (w^k_{00}, w^k_{11}, \ldots, w^k_{rd})$,

$$\varphi(w^1, \ldots, w^N) = \Phi(z(T; w^1, \ldots, w^{N-1}, v^N)). \tag{10}$$

The problem (5), (6) with the parameterized control (8) is reduced to the finite- dimensional minimization of the function (10) under time-parameter restrictions (7).

Analytical assumptions made above fulfill the conditions of twice-differentiable continuous dependence of the problem (1), (8) solution on parameters, thus

$$\frac{\partial \varphi(w^1, \ldots, w^N)}{\partial w^k_{\mu,\alpha}} = \frac{\partial \Phi(x(T))}{\partial x} \frac{\partial z(T; w^1, \ldots, v^N)}{\partial w^k_{\mu,\alpha}}, \quad \mu = 0, \ldots, r; \quad \alpha = 0, \ldots, d. \tag{11}$$

Denote

$$y^{j\mu\alpha}(t) = \frac{\partial z(t; v^1, \ldots, v^k)}{\partial w^j_{\mu,\alpha}}, \qquad t_{k-1} \leq t \leq t_k, \qquad 1 \leq j \leq k \leq N. \tag{12}$$

These functions are variations of the trajectory (9) on the control parameters. Variations corresponding to parameters v^k are solutions of the DE system

$$\dot{y}^{k\mu\alpha} = \frac{\partial f(x(t), u(t))}{\partial x} y^{k\mu\alpha} + \theta(t_k - t) \frac{\partial f(x(t), u(t))}{\partial u_\mu} \frac{\partial u^k_\mu(t; v^k)}{\partial v^k_{\mu,\alpha}}, \; t_{k-1} \leq t \leq T,$$

with trivial initial conditions $y^{k\mu\alpha}(t_{k-1}) = 0$. Here the Heaviside function $\theta(t) = 0$ for $t \leq 0$ and $\theta(t) = 1$ for $t > 0$. Variations corresponding to the knots t_k, i.e. functions $y^{k00}(t)$, are defined on $[t_k, T]$ by the similar homogeneous DE system with initial condition $y^{k00}(t_k) = f(x(t_k), u^k(t_k; v^k)) - f(x(t_k), u^{k+1}(t_k; v^{k+1}))$. If admissible controls are continuous, then the right hand of the condition equals to zero, correspondingly, $y^{k00}(t) \equiv 0$.

So, one can calculate the derivatives (11) and apply the gradient method for the finite-dimensional minimization problem. However the such scheme is rather complicated because the number of variations equals the number of control parameters. It is essentially simplified via using adjoint functions.

Introduce the Hamilton-Pontrjagin function $H(p, x, u) = \sum_{i=1}^{n} p_i f_i(x, u)$, the vector function $p(t)$ on $[0, T]$ (adjoint one to the first variations (12))

$$\dot{p} = -\frac{\partial H(p, x(t), u(t))}{\partial x}, \quad p(T) = \frac{\partial \Phi(x(T))}{\partial x},$$

and the function $M(t) = H(p(t), x(t), u(t))$.

The first derivatives (11) can be calculated [3] by formulas:

$$\frac{\partial \varphi(w^1, \dots, w^N)}{\partial t_k} = M(t_k - 0) - M(t_k + 0), \qquad 1 \leq k \leq N-1;$$

$$\frac{\partial \varphi}{\partial v^k_{\mu\alpha}} = \int\limits_{t_{k-1}}^{t_k} \frac{\partial H(p(t), x(t), u^k(t; v^k))}{\partial u_\mu} \frac{\partial u^k_\mu(t; v^k)}{\partial v^k_{\mu\alpha}} dt, \ 1 \leq k \leq N.$$

So, in order to calculate the derivatives (11) only one additional Cauchy problem, defined adjoint function $p(t)$, should be solved.

In the considering case control functions $u(t)$ are continuous hence derivatives $\partial\varphi/\partial t_k$, as well as variations $y^{k00}(t)$, are equal to zero.

The second derivation of (11) yields ($0 \leq \mu, \nu \leq r;\ 0 \leq \alpha, \beta \leq d;\ 1 \leq j, k \leq N$)

$$\frac{\partial^2 \varphi(w^1, \dots, w^N)}{\partial w^k_{\nu,\beta} \partial w^j_{\mu,\alpha}} = y^{j\mu\alpha}(T) \frac{\partial^2 \Phi(x(T))}{\partial x^2} y^{k\nu\beta}(T) + \frac{\partial \Phi(x(T))}{\partial x} \frac{\partial^2 z(T; w^1, \dots, v^N)}{\partial w^k_{\nu,\beta} \partial w^j_{\mu,\alpha}}. \quad (13)$$

The second variations $\partial^2 z / \partial w^k_{\nu,\beta} \partial w^j_{\mu,\alpha}$ as well as the first (12) are defined by some Cauchy problems [3,11]. It resolves, in principle, the problem of the second derivatives (13) calculation and using the Newton method. But the calculations also essentially simplified via using appropriate adjoint functions. These functions are the matrix impulses $\Psi(t)$ defined in [2] by the problem

$$\dot{\Psi} = -\left[\frac{\partial f(x(t), u(t))}{\partial x} + \left(\frac{\partial f(x(t), u(t))}{\partial x} \right)^T \right] \Psi - \frac{\partial^2 H(p(t), x(t), u(t))}{\partial x^2},$$

$$\Psi(T) = \frac{\partial^2 \Phi(x(T))}{\partial x^2}.$$

The second derivatives of an admissible trajectory of the system (1) on control parameters in the case of continuous control can be calculated by the next expressions:

$$\frac{\partial^2 \varphi(w^1,\dots,w^N)}{\partial v^j_{\mu,\alpha}\partial v^k_{\nu,\beta}} = \int\limits_{t_{k-1}}^{t_k} \left\{ \frac{\partial u^k_\nu(t;v^k)}{\partial v^k_{\nu,\beta}} \left[\frac{\partial^2 H(p(t),x(t),u(t))}{\partial u_\nu \partial x} + \right.\right.$$

$$\left.\left(\frac{\partial f(x(t),u(t))}{\partial u_\nu}\right)^T \Psi(t)\right] y^{j\mu\alpha}(t) + \delta_{jk}\left[\frac{\partial u^k_\nu(t;v^k)}{\partial v^k_{\nu,\beta}}\frac{\partial u^k_\mu(t;v^k)}{\partial v^k_{\mu,\alpha}}\frac{\partial^2 H(p(t),x(t),u(t))}{\partial u_\nu \partial u_\mu} + \right.$$

$$\frac{\partial u^k_\mu(t;v^k)}{\partial v^k_{\mu,\alpha}}\left[\frac{\partial^2 H(p(t),x(t),u(t))}{\partial u_\mu \partial x} + \left(\frac{\partial f(x(t),u(t))}{\partial u_\mu}\right)^T \Psi(t)\right] y^{k\nu\beta}(t) +$$

$$\left.\left.\delta_{\mu\nu}\frac{\partial^2 u^k_\nu(t;v^k)}{\partial v^k_{\mu,\alpha}\partial v^k_{\nu,\beta}}\frac{\partial H(p(t),x(t),u(t))}{\partial u_\nu}\right]\right\} dt, \quad j \le k;$$

$$\frac{\partial^2 \varphi(w^1,\dots,w^N)}{\partial v^j_{\mu,\alpha}\partial t_k} = \delta_{jk}\frac{\partial u^k_\mu(t_k)}{\partial v^k_{\mu,\alpha}}\frac{\partial H(p(t_k),x(t_k),u^k(t_k;v^k))}{\partial u_\mu}, \quad j \le k;$$

$$\frac{\partial^2 \varphi(w^1,\dots,w^N)}{\partial t_j \partial v^k_{\nu,\beta}} = -\delta_{j(k-1)}\frac{\partial H(p(t_{k-1}),x(t_{k-1}),u^k(t_{k-1};v^k))}{\partial u_\nu}, \quad j < k;$$

$$\frac{\partial^2 \varphi(w^1,\dots,w^N)}{\partial t_j \partial t_k} = \delta_{jk}\left(\frac{\partial H(p(t_k),x(t_k),u^k(t_k;v^k))}{\partial u}\frac{\partial u^k(t_k;v^k)}{\partial t} - \right.$$

$$\left.\frac{\partial H(p(t_k),x(t_k),u^{k+1}(t_k;v^{k+1}))}{\partial u}\frac{\partial u^{k+1}(t_k;v^{k+1})}{\partial t}\right).$$

These formulas are a particular case of more complicate ones for the class of piece-wise continuous admissible control [5,11].

So the calculations of the second derivatives (13) require solution of a few Cauchy problems for the first variations $y(t)$ and the adjoint variables $p(t)$, $\Psi(t)$. It allows ones to apply the Newton method on the base of the penalty function one for the parameterized functional (10) minimization under conditions (7). Note, the variational problem (5), (6), corresponding to the initial problem for DAE (1), can be solved sequentially on sufficiently small intervals $[t_{k-1},\, t_k]$. The control function components $u_i(t)$ can be assigned as square or cubic parabolas. The obtained control will be a polynomial spline.

4 Numerical Example

Example 1. Let's consider on the interval $[0;T]$ a DAE ([10][1])

$$\dot{x}_1 = 10t\exp(5u_2(t)-1)x_2(t), \quad \dot{x}_2 = -2t\ln(u_1(t)),$$
$$u_1(t) = x_1(t)^{1/5}, \quad u_2(t) = \left(x_2^2(t) + u_2^2(t)\right)/2. \tag{E1}$$

[1] On the base of an example from [8].

One of this nonlinear system's solutions is

$$\bar{x}_1(t) = \exp(5\sin(t^2)), \quad \bar{x}_2(t) = \cos(t^2),$$
$$\bar{u}_1(t) = \exp(\sin(t^2)), \quad \bar{u_2}(t) = \sin(t^2) + 1. \qquad (E2)$$

We state here the initial problem corresponding to the solution. Moreover, it is desire to construct numerically all solutions on the interval under given T.

This problem is a base test in some articles of the author [10], but on the interval [1.0708712; 1.423836]. The cause is in degeneration of the Jacobi matrix of the finite subsystem

$$\begin{bmatrix} 1 & 0 \\ 0 & 1 - u_2(t) \end{bmatrix}$$

in points t_k where $u_2(t) = 1$. Such points, corresponding to the solution (E2), are $t_k = \sqrt{k\pi}$ $(k = 0, 1, ...)$.

Solutions of (E1) bifurcate at the points and the system isn't reducible to normal Cauchy form on corresponding intervals, i.e. it isn't a finite index system and it can't be solved by known methods developed for DAE of finite index, particularly, by the methods of [10], assigned to DAE of index one.

The character of degenerating the second finite restriction points at existing two implicit functions $u_2^{\pm}(x_2) = 1 \pm \sqrt{1 - x_2^2}$ at a vicinity of initial conditions $x_2(0) = 1$, $u_2(0) = 1$. So, at the initial point two solution are generated.

To construct numerically two solutions, introduce the functional

$$J_\alpha[u] = \int_0^T \left[(u_2(t) - (x_2^2(t) + u_2^2(t))/2)^2 + \right.$$
$$\left. (x_1(t)^{1/5} - u_1(t))^2 + \alpha(u_2(t) - z(t))^2\right] dt. \qquad (E3)$$

Here $\alpha \geq 0$, $z(t)$ is a trial function presented a priori properties of the solution component $u_2(t)$. Consider the problem to minimize $(E3)$ under restrictions

$$\dot{x}_1 = 10t\exp(5u_2(t) - 1)x_2, \quad \dot{x}_2 = -2t\ln(u_1(t)),$$
$$x_1(0) = x_2(0) = 1. \qquad (E4)$$

This problem is equivalent to the initial one under $\alpha = 0$ and if $\alpha > 0$ it approximates a solution of (E1) which has the component $u_2(t)$ similar to $z(t)$. Having such solution depending on α we use it as initial approximation for a repeating minimization (E3) under $\alpha = 0$.

The problem (E3), (E4) was solved such a way with $T = 0.5$, $z(t) = 1 + \sin(t)$ and $z(t) = 1 - \sin(t)$ consequently on 5 equal intervals $[0; 0.1], \ldots, [0.4; 0.5]$. Every problem was solved in the class of square parabolic "controls" $u_i(t)$. The corresponding Cauchy problems were integrated by the 4-th order Runge-Kutta method with step 0.001.

At the case $z(t) = 1 + \sin(t)$ the obtained approximation (x^+, u^+) of solution $(E2)$ provides $J_0[u^+] = 4.7 \times 10^{-11}$ and maximal coordinate deviations are correspondingly $(4.7 \times 10^{-5}; 8.4 \times 10^{-7}; 3.9 \times 10^{-5}; 2.1 \times 10^{-4})$.

At the case $z(t) = 1 - \sin(t)$ there was obtained another approximation (x^-, u^-) with unknown analytical expression. Here $J_0[u^-] = 1.6 \times 10^{-10}$. The $x_1^-(t)$ is increasing to the value $x_1^-(0.5) = 1.77$ whereas $x_1^+(0.5) = 3.45$. The $u_2^-(t)$ is diminishing to the value $u_2^-(0.5) = 0.81$ whereas $u_2^+(t)$ is increasing to $u_2^+(0.5) = 1.25$.

Example 2. Let's consider singular linear DE system ([1]).

$$A(t)\dot{x} - B(t)x(t) = f(t),$$

where

$$A(t) = \begin{bmatrix} t^2 & 2t^2 & t^2 & t \\ t^2 & t^2 & 2t^2 & t \\ t^2 & t^2 & t^2 & t \\ 0 & 0 & 0 & 0 \end{bmatrix}, \quad B(t) = \begin{bmatrix} 3 & 3t & t & 0 \\ t & 3 & 3t & 0 \\ t & t & 3 & 0 \\ t & t & t & 1 \end{bmatrix},$$

$$f(t) = \begin{bmatrix} 4t^2 - 3t - 3 \\ 4t^2 - 3t - 3 \\ 3t^2 - t - 3 \\ -3t - 1 \end{bmatrix} \exp(t),$$

on the interval $[0; 1]$, with the initial condition $x(0) = (1, 1, 1, 1)^T$, corresponding to the solution

$$x(t) = (1, 1, 1, 1)^T \exp(t).$$

On the $[0; 1]$ the matrix $A(t)$ has a variable range, which equals zero $(t = 0)$ or three $(t > 0)$. Well-known methods are inapplicable for this problem. In the paper [12] the problem of solution existence for such kind degenerate (singular) equations was considered. In the [1] an analogous problem was numerically solved by the special collocation-difference method on the interval $[1; 2]$.

The considered DE problem is linear and the initial problem was solved by the normal spline collocation method in the [7] on the $[0; 1]$. To solve this problem on the $[0; 1]$ by the parameterization method it was reduced to VC problem:

$$\dot{x} = u(t),$$

with initial condition $x(0) = (1,\ 1,\ 1,\ 1)^T$ and minimizing the functional

$$J(u; t_0, T) = \int_{t_0}^{T} \|A(t)u(t) - B(t)x(t) - f(t)\|^2 \, dt,$$

where $t_0 = 0, T = 1$.

Clearly, the functional $J(u; t_0, T)$ is additive, so it can be solved consequently on the partial intervals $[t_k, t_{k+1}]$, which are a partition of the interval $[t_0, T]$. Here the functionals $J(u; t_k, t_{k+1})$ are minimized consequently.

The initial problem was solved consequently on the five equal intervals $[0; 0,2]$, $[0,2; 0,4], \ldots, [0,8; 1,0]$ by two ways. In the first way it was used square parabolas for the approximation of the control $u(t)$:

$$u_i(t) = a_i * t^2 + b_i * t + c_i, \quad i = 1, 2, 3, 4.$$

In the second way exponential functions

$$u_i(t) = a_i * \exp(b_i * t), \quad i = 1, 2, 3, 4,$$

were used.

The Cauchy problems were integrated by the second order Runge-Kutta method. The corresponding NP problems were solved at first by the gradient method, then by the Newton one. The obtained errors (the maximum deviation of components on the integration net) for different steps of integration h are presented in the table 1.

Table 1.

h	0,01	0,005	0,001
way 1	$0,0056$	$0,017$	$0,018$
way 2	4.0×10^{-4}	2.6×10^{-5}	$2,4 \times 10^{-8}$

These results show the importance of using of adequate a priori representation of seeking solution. Besides we should mark paradoxical increase of an error with a diminution of an integration step in the way 1. It can be explained by accumulation of round-off errors, as the usual arithmetic of double precision was used.

5 Conclusion

So, these examples demonstrate a capacity for applying the parameterization method of the second order in the standard form for solving singular DE problems. Success of the PM application to degenerate DA systems can be explained by picking out the regular differential subsystem and fixing the structure of the control solution components on the integration intervals. Moreover, these intervals are more much then a step of a difference scheme that was applied to the regular Cauchy problems.

References

1. Bojarincev, Ju.E., Danilov, V.A., Loginov, A.A., Chistyakov, V.F.: Numeric Methods of Solving Singular Systems. Nauka, Novosibirsk (1989)
2. Gabasov, R., Kirillova, F.: Qualitative theory of optimal processes. Nauka, Moscow (1971) (Russian)
3. Gorbunov, V.K.: The parameterization method for optimal control problems. Computational Mathematics and Mathematical Physics. Vol. 19. **2** (1979)
4. Gorbunov, V.K.: Regularization of degenerate equations and inequalities under explicit data parameterization. J. Inv. Ill-Posed Problems. Vol. 9. **6** (2001) 575–594

5. Gorbunov, V.K., Lutoshkin, I.V.: Second derivatives of parameterized optimal control problems. Proceedings of 11-th Baikal International School-Seminar. Irkutsk **4** (1998) 90–93
6. Gorbunov, V.K., Lutoshkin, I.V.: Development and Experience of Applying the Parameterization Method in Degenerate Problems of Dynamical Optimization. J. of Computer and Systems Sciences International. (2003) (to appear)
7. Gorbunov, V.K., Petrischev, V.V.: Development of the normal collocation spline method for linear differential equations. Comput. Math. and Math. Phys. Vol. 43. (2003) (to appear)
8. Hairer, E., Wanner, G.: Solving Ordinary Differential equations II: Stiff and Differential-algebraic Problems. Springer-Verlag, Berlin (1996)
9. Henrion, R.: La théorie de la variation seconde et ses applications en commande optimale. Bruxelles-Palais des Academies, Bruxelles (1974)
10. Kulikov, G.Yu.: Numerical solution of the Cauchy problem for a system of differential-algebraic equations with the use of implicit Runge-Kutta methods with a untrivial predictor. Comput. Math. and Math. Phys. Vol. 38. **1** (1998)
11. Lutoshkin, I.V.: Using the parameterization method in degenerate problems. Thesis of Ph.D., Phys.-Math. Ulyanovsk (2000) (Russian)
12. März, R., Weinmüller: Solvability of Boundary value problems for systems of singular differential-algebraic equations. SIAM J. Math. Anal. Vol. 24. **1** (1993) 200–215

Development of the Normal Spline Method for Linear Integro-Differential Equations*

Vladimir K. Gorbunov, Vyacheslav V. Petrischev, and Vyacheslav Y. Sviridov

Ulyanovsk State University, L.Tolstoy street 42,
432970 Ulyanovsk, Russia,
gvk@vens.ru, Vyacheslav.Petrischev@transas.com, meria@mv.ru

Abstract. The normal spline method is developed for the initial and boundary-value problems for linear integro-differential equations, probably being unresolved with respect to the derivatives, in Sobolev spaces of the arbitrary smoothness. It allows to solve a high-order systems without the reduction to first-order ones. The solving system can be arbitrary degenerate (with high differentiation index or irreducible to normal form). The method of nonuniform collocation grid creation for stiff problems is offered. Results of numerical solution to test problems are demonstrated.

1 Introduction

The problem under consideration is the system of integro-differential equations

$$A(t)\dot{x}(t) + B(t)x(t) - \int_0^1 K(t,s)x(s)ds = f(t), \quad 0 \le t \le 1, \tag{1}$$

with conditions

$$Cx(0) + Dx(1) = g. \tag{2}$$

Here $x, f, g \in R^n$, $A(t), B(t), K(t,s), C, D$ are square n-order matrices. The function $f(t)$ and the coefficients of matrices are continuous and have so many derivatives as it needs to guarantee appropriate smoothness of the solution $x(t)$ that exists on assumption and belongs to the Hilbert-Sobolev space $W^l_{2,n}$ with norm

$$\|x\|_{l,n} = \left\{ \sum_{i=1}^{n} \left[\sum_{r=0}^{l-1} \left(x_i^{(r)}(0) \right)^2 + \int_0^1 \left(x_i^{(l)}(s) \right)^2 ds \right] \right\}^{1/2}. \tag{3}$$

Here and later $x_i^{(r)}(t)$ are derivatives of the order $r \ge 0$, also $\dot{x}_i = x_i^{(1)}$.

The matrices of the system (1) may be arbitrary degenerate. It covers the problem of solving in universal form to differential-algebraic equations (DAEs) with any differentiation index [1] and singular systems having not the finite index (irreducible to normal Cauchy form) [2].

* Supported by Russian Foundation for Basic Research, project N° 01-01-00731

P.M.A. Sloot et al. (Eds.): ICCS 2003, LNCS 2658, pp. 492–499, 2003.

There exist special methods [1], [3] for solving to degenerate ordinary differential equations (ODEs) working in the case when the matrix rank of the main part of the solving system is constant.

In [4], [5] there was developed a variational normal spline (NS) method for the problem (1),(2). The cases of infinite intervals $[0, \infty)$ and $(-\infty, \infty)$ were also covered. This method is based on the transfer to the finite collocation system of the equalities (1) and the finding an element of the minimal norm. In singular case the problem (1), (2) may have a set of solutions (affine) and the NS method provides then an approximation of the normal solution denoting by x^0 [5].

The theoretical basis of the NS method is the classic functional analysis results: the theorem of embedding of Sobolev spaces in the Chebyshev one [6] and Riesz theorem [7] of canonical representation of linear continuous functional in Hilbert spaces as inner product. Here the coordinate system is not a priori given and is generated by the equation coefficients and the space $W^l_{2,n}$ topology. This representation is based on using some reproducing kernel [8] defined by the norm (3).

One can find the results of the realization of the NS method for ODEs under $l \in \{1, 2\}$ in [4], [5]. The method was realized for arbitrary integer l in [9] where general view of reproducing kernels were created. It allows to solve a high-order systems without their reduction to the first-order ones [10] and to solve problems with a high differentiation index. The NS method also was developed for two-dimensional problems of the computational tomography [11].

The main purpose of this work is to produce a computational scheme of the NS method for the systems of integro-differential equations (1) in $W^l_{2,n}$ with $l \geq 2$. Also the NS method is developed for stiff problems. The using of nonuniform adaptive grids is appropriate in these problems. The different strategies of the collocation grids improvement are based on the theoretical estimates of the NS method precision. The improvement consists in a serial grid modification (addition and/or moving) with the goal to minimize the norm of the current system discrepancy. We present also results of the numerical solving to degenerate and stiff problems.

2 The Normal Spline Method

The NS method is a collocation type one. Introduce some grid

$$0 \leq t_1 < t_2 < \ldots < t_m \leq 1 \tag{4}$$

and consider the collocation system

$$A(t_k)\dot{x}(t_k) + B(t_k)x(t_k) - \int_0^1 K(t_k, s)x(s)ds = f(t_k), \quad k = 1, \ldots, m. \tag{5}$$

For this system we pose a problem of a normal (in norm (3)) solution. In the case of the compatible system (1), (2) the solution exists and is uniquely defined.

This fact is the consequence of the theorem [6] about the embedding of Sobolev space $W^l_{2,n}[0,1]$ to the Chebyshev one $C^{l-1}_n[0,1]$.

Denote the left parts of the system (5), (2) as

$$l_{ik}(x) = \begin{cases} \sum\limits_{j=1}^{n} c_{ij}x_j(0) + d_{ij}x_j(1), \quad k=0; \\ \sum\limits_{j=1}^{n} a_{ij}(t_k)\dot{x_j}(t_k) + b_{ij}(t_k)x_j(t_k) - \int\limits_0^1 k_{ij}(t_k,s)x(s)ds, 1 \le k \le m, \end{cases} \tag{6}$$

where $1 \le i \le n$. Respectively, the system (5), (2) takes the form

$$l_{i0}(x) = g_i, \quad l_{ik}(x) = f_i(t_k), \quad 1 \le i \le n, \quad 1 \le k \le m. \tag{7}$$

The functions (6) may be considered as linear continuous functionals in the vector-functions space $W^l_{2,n}[0,1]$. Respectively, each equation of the system (7) defines a hyperplane and the system solution set (an intersection of the hyperplanes) will be nonempty, convex and closed. The minimal norm element of this intersection is named the normal spline x^m. It is known [4], the sequence of the normal splines x^m converges to a normal solution x^0 in the norm (3) when the maximum step of the grid (4) tends to zero.

According to the Riesz theorem [7] a linear continuous functional may be represented as a inner product corresponding to the norm (3), i.e.

$$l_{ik}(x) = \left\langle h^{\mu(i,k)}, x \right\rangle_{(n)} = \sum_{j=1}^{n} \left\langle h_j^{\mu(i,k)}, x_j \right\rangle_{(1)}. \tag{8}$$

Here $\mu(i,k) = nk + i, 1 \le i \le n, 0 \le k \le m$.

Elements h^m_j of the representation (8) may be found with the help of the reproducing kernel $G(s,t)$ of the scalar function space $W^l_{2,1}[0,1]$. Remind [8], the reproducing kernel is a such function $G(s,t)$ that

1. $G(\cdot,t) \in W^l_{2,1}[0,1]$ for any $t \in [0,1]$,
2. $x_i(t) = \langle G(\cdot,t), x_i\rangle$ for any $x_i(\cdot) \in W^l_{2,1}[0,1]$ and any $t \in [0,1]$.

Correspondingly,

$$h_j^{\mu(i,k)}(s) = \begin{cases} c_{ij}G(s,0) + d_{ij}G(s,1), \quad k=0; \\ a_{ij}(t_k)G_s^{'}(s,t_k) + b_{ij}(t_k)G(s,t_k) - \\ -\int\limits_0^1 k_{ij}(t_k,\tau)G(s,\tau)d\tau, \quad 1 \le k \le m. \end{cases} \tag{9}$$

By virtue of (6) and (8) the system (7) is represented as

$$\langle h^\mu, x\rangle_{(n)} = \bar{f}_\mu, \quad \bar{f}_\mu = \begin{cases} g_i, \quad k=0; \\ f_i(t_k), 1 \le k \le m. \end{cases} \tag{10}$$

According to the generalized Lagrange method the normal solution of the linear equations system (10) can be written in the form

$$x^m(s) = \sum_{\mu=1}^{(m+1)n} u_\mu h^\mu(s). \tag{11}$$

Coefficients u_μ are defined by the system

$$\sum_{\nu=1}^{(m+1)n} g_{\mu\nu} u_\nu = \bar{f}_\mu, \quad 1 \le \mu \le (m+1)n, \tag{12}$$

where $g_{\mu\nu}$ are the coefficients of the Gram matrix of the system $\{h^\mu\}$, i.e.

$$g_{\mu\nu} = \langle h^\mu, h^\nu \rangle_{(n)} = \sum_{i=1}^{n} \langle h_i^\mu, h_i^\nu \rangle_{(1)}. \tag{13}$$

Thus the realization of the NS method with given decomposition of an interval is reduced to the forming of the Gram matrix coefficients according to (13), the solving to the system of linear equation (12) with symmetric, positive defined (as a rule) matrix $\{g_{\mu\nu}\}$ and the creation of the solution $x^m(s)$ at arbitrary point $s \in [0,1]$ with respect to (9), (11).

In [9] it was shown that the kernel G(s,t) for the norm (3) is a Green function of some boundary-value problem and is given by the formula

$$G(s,t) = \begin{cases} \displaystyle\sum_{i=0}^{l-1} \frac{t^i}{i!} \left(\frac{s^i}{i!} - (-1)^{l+i} \frac{s^{2l-i-1}}{(2l-i-1)!} \right), & 0 \le s \le t \le 1; \\ G(t,s), \quad 0 \le t < s \le 1. \end{cases} \tag{14}$$

3 The Adaptive Grid Construction

Define the discrepancy of the equation (1) on the function x^m:

$$\varphi(t) = A(t)\dot{x}^m(t) + B(t)x^m(t) - \int_0^1 K(t,s)x^m(s)ds - f(t). \tag{15}$$

In [5] the evaluation

$$\left\| x^m - x^0 \right\|_{l,n} \le \hat{c} \left\| \varphi \right\|_{(l-1),n} \tag{16}$$

was obtained. The $\hat{c}$ is constant that is depended only on the system (1), (2) coefficients.

The evaluation (16) opens the way to creation of optimal nonuniform grids under given number of nodes m, that is very important for the stiff problem solving. The convergence in norm implies a uniform convergence, therefore in the process of the method realization we should proceed to new grids decreasing the value $\|\varphi\|_{(l-1),n}$.

There are equalities $\varphi(t_k) = 0$ in the collocation nodes hold hence the minimization of the discrepancy norm may be achieved using a simple and reliable heuristic method: to add nodes into subintervals with the greatest intermediate (between nodes) values of $\varphi(t)$. The detailed algorithm of the step-by-step concentration of the grids was offered in [5]. This adaptive grid is named an *adding grid* with two parameters: the number of start nodes and the number of added nodes on each step.

Another strategy of creation of the adaptive grid with fixed nodes number is based on minimization of the value $\|\varphi\|_{(l-1),n}$ as a function of nodes with the constrains (4). Denote

$$\psi_{l-1}(t_1, \ldots, t_m) = \|\varphi(s, t_1, \ldots, t_m)\|^2_{(l-1),n}. \tag{17}$$

Here the index $l-1$ corresponds to the normal spline $x^m \in W^l_{2,n}[0,1]$. The grid (4) providing the minimum of function $\psi_{l-1}(t_1, \ldots, t_m)$ is named *optimal.*

In case, when the system (1),(2) is regular (matrix $A(t)$ is non-degenerate), the smoothness of function ψ_{l-1} is defined by the index of derivative l and the reproducing kernel $G(s,t)$ characteristics. The last has continuous derivatives with order up to $2l-2$. As it is shown in [12] it provides the differentiability of the function ψ_{l-1} with respect to t_k for $l \geq 3$. In this case the appropriate analytic formulas for partial derivatives were obtained. We omit them because of their complexity. In general case one may use some direct method (e.g. Hooke-Jeeves [13]) to minimize the function ψ_{l-1}.

We turn our attention to the solving of an initial problem for the differential equation (1) ($K \equiv 0$), i.e. the condition (2) has the form $x(0) = g$. In this case the most effective NS method realization scheme is the sequential normal spline creation on partial adjoining subintervals with a small number of nodes (up to two). The effectiveness is achieved as a result of reducing of the Gram matrix dimension since the calculation of its elements and solving the linear equation system are the most time-consuming part of the algorithm.

If partial interval is sufficiently small we can use the grid with two nodes and create the spline on two nodes $\{t_{k-1}, t_k\}$ on each step. The initial values on second and following subintervals are defined as final values of the created spline on the previous subinterval.

Define some admissible level ε of local measure of error evaluation $\psi^k_{l-1}(t_{k-1}, t_k)$ defined similar to (17). This value should be estimated by using some quadrature formula.

The simple variant of the algorithm is the next.

1. $k := 1$, $t_0 := 0$.
2. $t_k := 1$.
3. To construct the NS for the Cauchy problem (1) on (t_{k-1}, t_k).
4. If $\psi^k_{l-1} < \varepsilon$ then go to 6.
5. $t_k := (t_k + t_{k-1})/2$, go to 3.

6. If $t_k = 1$, then end else $t_{k-1} := t_k$.
7. $k := k + 1$, go to 2.

As a result we obtain the grid that provides the given level of the local discrepancy norm value.

The similar sequential algorithm can be applied for solution of integro-differential equations of Volterra type. In this case it is necessary to make the next modification of algorithm: on step 3 we construct the normal spline for the equation

$$A(t)\dot{x}(t) + B(t)x(t) - \int\limits_{t_{k-1}}^{t} K(t,s)x(s)ds = f(t) + \int\limits_{0}^{t_{k-1}} K(t,s)x^m(s)ds, \tag{18}$$
$$t_{k-1} \leq t \leq t_k.$$

4 Numerical Implementation

Example 1. The initial problem from [3] for ODE (1) where $K \equiv 0$ and

$$A(t) = \begin{pmatrix} t & 1 \\ 0 & 0 \end{pmatrix}, \quad B(t) = \begin{pmatrix} \gamma & -\delta \\ t & 1 \end{pmatrix}.$$

The system (1) with condition $\gamma + \delta t \neq 1$ has the differentiation index 2 and the unique solution with any right part $f \in C_2^2[0,1]$. Besides, it was shown [3] that classical difference Euler, Runge-Kutta and multistep methods diverge or are not applied for γ, δ such that $|\gamma + \delta t| < 1$. The NS method gives approximate solution in any compatible case.

Table 1 presents results of the initial problem solving for exact desired function

$$x(t) = \begin{pmatrix} e^t sin(t) \\ cos(t) \end{pmatrix}$$

on interval $t \in [0,1]$ with uniform grid of 51 points. Values of the obtained normal splines deviations from exact solution on doubled grid are shown. The cases of mentioned above condition violation is marked by asterix (*). In these cases the non-uniqueness of the solution is possible. The normal spline approximate the normal solution.

One can see that increasing of the spline smoothness leads to the increasing of the approximation accuracy.

Example 2. Consider the second-order initial problem

$$p(t)\ddot{x}(t) + q(t)\dot{x}(t) + r(t)x(t) = f(t), \quad 0 \leq t \leq 1,$$

$$x(0) = g_1, \quad \dot{x}(0) = g_2$$

where $p(t) = sin(10t), q(t) = 10cos(10t), r(s) = -1$. The function $f(t)$ corresponds to the solution $x(t) = te^{-t} + e^{-kt}$, $k > 1$.

Table 1.

$\gamma\backslash\delta$	-1		0		2	
	l=2	l=3	l=2	l=3	l=2	l=3
2	1.36e-1 *	4.66e-04*	1.83e-2	3.43e-05	6.02e-3	1.34e-05
0	1.09e-2	2.74e-05	2.09e-2	4.72e-05	2.06e-2 *	3.50e-05 *
-1	7.79e-3	1.97e-05	1.08e-2	2.73e-05	9.19e-2 *	3.45e-04 *
-2	5.58e-3	1.58e-05	7.31e-3	1.97e-05	1.95e-2	4.59e-05

The coefficients $p(t)$, $q(t)$ have zero values on the interval $[0, 1]$ and the solution is stiff under large k. The variable degeneracy of the coefficients makes difficulties for using known difference methods for stiff and ADEs [1].

We use the transfer to integro-differential equation

$$p(t)\dot{x}(t) + (q(t) - \dot{p}(t))\,x(t) - \int_0^t (\dot{q}(s) - \ddot{p}(s) - r(s))\,x(s)ds =$$

$$= \int_0^t f(s)ds + q(0)x(0) + p(0)\dot{x}(0) - \dot{p}(0)x(0).$$

The effectiveness of this way in stiff cases was shown in [5].

Table 2 presents some results for two variants of the NS method: two-nodes sequential scheme and uniform grid one. The problem was solved in the space W_2^2. Note, that value ε is used in the sequential scheme and it defines the obtained number of nodes.

Table 2.

k	ε	m	sequential	uniform
100	1.00e-3	24	5.64e-02	1.07e+00
-//-	1.00e-5	103	7.69e-03	2.14e-01
1000	1.00e-5	85	1.61e-02	9.30e-01

5 Conclusion

Presented above numerical results also as results of [4], [5], [9], [10], [12] demonstrate the ability of the NS method to obtain appropriate approximation of solutions to arbitrary degenerate problems of linear differential and integro-differential equations. Particularly for equations that can't be reduced to the normal Cauchy form, i.e. equations not having of a differentiation finite index. We know only theoretical works devoted to such kind of ODEs problems [2]. We are inclined to explain the success of the NS method by using in its base of the Sobolev spaces theory that is outstanding achievement of functional analysis.

References

1. Hairer, E., Wanner, G.: Solving Ordinary Differential equations II. Stiff and Differential-algebraic Problems. Springer-Verlag, Berlin (1996)
2. März, R., Weinmüller, E.B.: Solvability of boundary value problems for systems of singular differential-algebraic equations. SIAM J. Math. Anal. Vol. 24. **1** (1993) 200–215
3. Bulatov, M.V., Chistyakov, V.F.: About the numerical method of differantial-algebraic equations solving. Comput. Math. Math. Phys. Vol. 42. **4** (2002) 459–470
4. Gorbunov, V.K.: The method of normal spline-collocation. Comput. Math. Math. Phys. Vol. 29. **2** (1989) 212–224
5. Gorbunov, V.K.: Ekstremalnie zadachi obrabotki rezul'tatov izmerenei. Ilim, Frunze (1990)
6. Sobolev, S.L.: Applications of functional analysis to mathematical physics. Amer. Math. Soc. Providence RI (1963)
7. Balakrishnan, A.: Applied Functional Analysis. Springer-Verlag, New York (1976)
8. Aronszajn, N.: Theory of reproducing kernels. Tranzactions of the AMS **68** (1950) 337–404
9. Gorbunov, V.K., Petrischev, V.V.: Developing of the normal spline-collocation method for linear differential equations. Comput. Math. Math. Phys. (2003) (to appear)
10. Gorobetz, A.S.: Metod normal'nih splainov dlya system ODU vtorogo poryadka i zadach matematicheskoi phiziki. Differencial'nie uravneniya i ih prilogeniya: Sbornik trudov mejdunarodnoy konferencii. Samara (2002) 99–104
11. Kohanovsky, I.I.: Normalnije splaini v vichislitel'noi tomografii. Autometria **2** (1995) 84–89
12. Sviridov, V. Yu.: Optimizaciya setok metoda normal'nih splainov dlya integro-differencial'nih uravnenei. Trudi Srednevolgskogo matematicheskogo obschestva. SVMO, Saransk **3-4** (2002) 236–245
13. Himmelblau, D.M.: Applied nonlinear programming. McGraw-Hill Book Company, Texas (1972)

To Numerical Solution of Singular Perturbed Equations Transformed to the Best Argument*

E.B. Kuznetsov and S.D. Krasnikov

Moscow Aviation Institute, Volokolamskoe shosse 4,
125993, Moscow, Russia
Kuznetsov@mai.ru,
sergeykr@mtu-net.ru

Abstract. We consider the numerical solution of initial value problem for the system of ordinary differential singular perturbed equations. The integral curve of the problem is constructed using method of continuation with respect to a parameter. We can choose the best parameter in any step of integration process. It is found that the best argument of the Cauchy problem is the arc length of the integral curve of the problem. Transformed to the best argument Cauchy problem has a number advantages in comparison with the Cauchy problem over the usual statement [1]. The right - hand side of each transformed equation does not exceed unit. Moreover, the squared norm of the system right - hand sides is always equal to unit. Also the suggested transformation reduces the difficulties that are typical for stiff systems. The efficiency of the approach is shown on test examples.

1 Introduction

Equations with a small parameter by highest derivative are called singular perturbed equations. They form a class of stiff systems convenient for studying the effectiveness of various numerical methods of integration of stiff systems. New advances in asymptotic theory [2], theory of difference schemes [3] and simplicity of qualitative behavior of solution make possible the detailed analyses of such a system.

We consider the simplest singular perturbed equation of form [4]

$$\varepsilon \frac{dy}{dt} = f(y,t), \qquad \varepsilon > 0. \tag{1}$$

Consider the case when a degenerated equation, corresponding to the equation (1),

$$f(x,t) = 0$$

has a unique solution $x = x(t)$ and the value of $\partial f/\partial y$ is negative in the neighborhood of this solution. The latter condition is essential for stability of the solution $x = x(t)$.

* Supported by the Russian Foundation for Basic Research, project N° 01-01-00038

P.M.A. Sloot et al. (Eds.): ICCS 2003, LNCS 2658, pp. 500–506, 2003.

The behavior of the equation (1) solution is as follows. If the value of ε is sufficiently small, the tangents to integral curves are almost parallel to the axis y, even for a small deviation from the function $x(t)$. And the smaller is value of ε, the faster the integral curve and solution $x(t)$ of degenerated equation come close.

The situation can be described in the following way. Two intervals of essentially different behavior take place for any integral curve in the domain considered. The duration of the first of them is much smaller than that of the second one. The first interval, where the desired function varies fast, represents the tending of the integral curve to the curve $x(t)$ and is named the boundary layer. In the second interval the derivatives are essentially smaller and the integral curve practically coincide with the curve $x(t)$. The boundary layer always takes place, except for the case when initial condition is a root of degenerate equation, i.e., when $y_0 = x(t_0)$. The smaller is value of parameter ε , the stronger is the difference of the behavior on the two intervals.

Thus, the solution of the degenerate equation can be used for describing the solution of differential equation (1) outside of the boundary layer. The difficulties of numerical solution of the problems considered result from the fact that the derivative dy/dt increases sharply even for small deviation of initial conditions from the curve $x(t)$ at any of its point.

2 The Best Continuation Parameter

We consider the solution of the system of nonlinear algebraic or trancendental equations

$$F(x) = 0, \quad x \in \mathbb{R}^{n+1}, \quad F : \mathbb{R}^{n+1} \to \mathbb{R}^n. \tag{2}$$

The set of solutions of this equation forms the curve in $\mathbb{R}^{n+1}$.

We will suppose that in equation (2) the unknowns depend of some parameter μ

$$x = x(\mu). \tag{3}$$

Then after differentiating equations (2) with respect to μ we obtain the continuation equations

$$\overline{J}\frac{dx}{d\mu} = 0, \qquad \overline{J} = \overline{J}(x) = \frac{\partial F}{\partial x}. \tag{4}$$

Here $\overline{J}$ coincides with the augmented Jacobi matrix In the vicinity of the point x on the curve of the solutions set we introduce parameter μ that is measured along the axis determined by identify vector $\alpha = (\alpha_1, \dots, \alpha_{n+1})^T \in \mathbb{R}^{n+1}, \quad \alpha\alpha = \alpha_i\alpha_i = 1$. Then in this point

$$d\mu = \alpha dx = \alpha_i dx_i, \quad i = \overline{1, n+1}. \tag{5}$$

Here the summation with respect to repeating indexes in mentioned limits is used.

Choosing the vector α in a different way we can define any continuation parameter.

The continuation equations (4), (5) we write as

$$\begin{bmatrix} \alpha_1 & \alpha_2 & \dots & \alpha_{n+1} \\ F_{1,1} & F_{1,2} & \dots & F_{1,n+1} \\ \vdots & \vdots & \ddots & \vdots \\ F_{n,1} & F_{n,2} & \dots & F_{n,n+1} \end{bmatrix} \begin{bmatrix} x_{1,\mu} \\ x_{2,\mu} \\ \vdots \\ x_{n+1,\mu} \end{bmatrix} = \begin{bmatrix} 1 \\ 0 \\ \vdots \\ 0 \end{bmatrix}. \tag{6}$$

Here $F_{i,j} = \partial F_i / \partial x_j, \quad x_{i,\mu} = dx_i/d\mu$.

We can prove the following assertion [5], [6]

Theorem 1. *For the system of linearized equations (6) to become the best conditioned, it is necessary and sufficient to take the length of the curve of the system solution set as the continuation parameter of the nonlinear system (2)*

3 The Best Argument of the Problem

Consider the Cauchy problem for normal system of ODE's

$$\frac{dy_i}{dt} = f_i(y_1, y_2, \dots, y_n, t), \qquad y_i(t_0) = y_{i0}, \qquad i = \overline{1, n}. \tag{7}$$

Assume that the conditions of existence and the uniqueness theorem are fulfilled for this problem.

Let an integral the problem to be given by the relations

$$\begin{aligned} F_j(y_1, \dots, y_n, t) &= 0, \qquad j = \overline{1, n}, \\ F_j(y_{10}, \dots, y_{n0}, t_0) &= 0, \end{aligned} \tag{8}$$

that define the integral curve of the problem (7) in $(n+1)$ - dimensional Euclidean space $\mathbb{R}^{n+1} : \{y_1, \dots, y_n, t\}$.

The process of finding this curve can be considered as a problem of constructing of the solutions set of the system of equations (8) containing the parameter - argument t for various values of t. We will use parametric continuation method to solve this system. Then the problem (7) can be considered as a Cauchy problem for the continuation equations of the system (8) solution with respect to the parameter t when the system is reduced to the normal form. Hence we can look for the best continuation parameter. We will call it the best argument.

We choose the best argument - parameter locally, i.e.,in a small vicinity of each point of the solution set curve — integral curve of the problem (7).

To solve the problem we assume that y_i and t are such functions of some argument μ that at each point of the integral curve of the problem

$$d\mu = \alpha_i dy_i + \alpha_{n+1} dt, \qquad i = \overline{1, n}. \tag{9}$$

Here $\alpha_k(k = \overline{1,\ n+1})$ are the components of a unit vector $\alpha = (\alpha_1,\ \dots,\ \alpha_{n+1})^T$ determining the direction along which the argument μ is measured. Note that the summation with respect to repeated subscript i is assumed in the expression (9).

Right - hand side of the equality (9) can be regarded as the scalar product of the vector α and the vector - function differential $(dy_1, \dots, dy_n,\ dt)^T$. Assigning various values to the components of the vector α it is possible to consider all possible continuation parameters of the problem (8), i.e., all arguments of the problem (7).

Since the particular form of equation (8) is unknown the change to the argument μ can be implemented immediately for the problem (7). Dividing the equality (9) by $d\mu$ after that, we obtain

$$\begin{aligned} y_{i,\mu} - f_i t_{,\mu} &= 0, \\ \alpha_i y_{i,\mu} + \alpha_{n+1} t_{,\mu} &= 1 \qquad i = \overline{1,\ n}. \end{aligned} \tag{10}$$

If the vector $y = (y_1, \dots, y_n, t)^T \in \mathbb{R}^{n+1}$ is introduced, the system (10)can be written in the matrix form

$$\begin{bmatrix} A \\ \alpha \end{bmatrix} y_{,\mu} = \begin{bmatrix} 0 \\ 1 \end{bmatrix}. \tag{11}$$

Here the matrix A of size $n \times (n+1)$ has the structure

$$A = [Ef],$$

where E is the unit matrix of n – th order and $f = (f_1, \dots, f_n)^T$ is a vector in $\mathbb{R}^n$.

The structure of the system (11) is exactly the same as that of the system (6). Therefore, in according with the Theorem 1, the transition to the normal form of the system (11) is the best conditioned if and only if $\alpha = y_{,\lambda}$, i.e., if the arc length of the curve of the system (8) solution is chosen as the parameter μ. This curve is the integral curve of the problem (7). Thus, the system (11) can be written in the form

$$\begin{aligned} y_{i,\lambda} y_{i,\lambda} + t_{,\lambda}^2 &= 1 \\ y_{i,\lambda} - f_i t_{,\lambda} &= 0, \end{aligned} \tag{12}$$

which can be solved analytically with respect to derivatives. Since the argument does not appear explicitly in the equations we take the initial point of the Cauchy problem (7) as the initial point of λ. Then we arrive to the following form of the Cauchy problem

$$\begin{aligned} \frac{dy_i}{d\lambda} &= \frac{f_i}{\sqrt{1 + f_j f_j}} \quad y_i(0) = y_{i0} \\ \frac{dt}{d\lambda} &= \frac{1}{\sqrt{1 + f_j f_j}} \quad t(0) = t_0 \\ & i,\ j = \overline{1,\ n}. \end{aligned} \tag{13}$$

Below, the argument λ , which provides the best conditioning to the system of equation (11), will be named as the best argument.

Thus the main result has been proved [7].

Theorem 2. *For the Cauchy problem for the normal system of ODE (7) to be transformed to the best argument it is necessary and sufficient that the arc length of the solution curve to be chosen as this argument. In this case the problem (7) is transformed to the problem (13).*

The new formulation of the Cauchy problem (13) has a number advantages in comparison with the Cauchy problem (7). First, the right - hand side of each equation (13) does not exceed unit. Moreover, the squared norm of the system right - hand sides is always equal to unit. This removes many of the problems connected with unlimited growth of the right - hand sides of the system (7), and allows to integrate differential equations which have the limiting points at integral curves where the derivatives become infinite. It becomes possible to solve problem with closed integral curves. Also the suggested transformation reduces the difficulties that are typical for stiff systems.

Example 1. Let consider some problems. The first problem, taken from [3], is nonlinear Edsberg's problem

$$\frac{dy}{dt} = -2ky^2, \qquad y(0) = 10, \tag{14}$$

It has the exact analytical solution

$$y(t) = \frac{10}{1 + 20kt}.$$

After λ - transformation the problem (14) takes the form

$$\frac{dy}{d\lambda} = \frac{-2ky^2}{\sqrt{1 + 4k^2y^4}}, \qquad \frac{dt}{d\lambda} = \frac{1}{\sqrt{1 + 4k^2y^4}},$$

$$y(0) = 10, \qquad t(0) = 0.$$

This problem was integrated by the program PC1 [8] for $k = 10^3$ with accuracy 10^{-10} and initial integration step 0.1 on the interval $t \in [0, 1]$. The execution time for this problem was two times less then for the problem (14).

Example 2. As another example, we consider numerical solution of Cauchy problem for Van der Pol's equation [9]. If a new variable $t = x/\mu$ is introduced in classical Van der Pol's equation $y'' - \mu(1 - y^2)y' + y = 0$, the equation take the form $\varepsilon y'' - (1 - y^2)y' + y = 0$ where $\varepsilon = 1/\mu^2$. And Cauchy problem can be formulated in the form

$$\frac{dy_1}{dt} = y_2, \quad \varepsilon\frac{dy_2}{dt} = (1 - y_1^2)y_2 - y_1,$$

$$y_1(0) = 2, \qquad y_2(0) = -0.66$$

Here the small parameter is $\varepsilon = 10^{-6}$.

This is a well - known benchmark for estimation the efficiency of computational programs intended to solve stiff systems.

The program PC1 was used for solving this problem. The accuracy of computations was checked by comparison with "exact results" obtained in [9] and it did not exceed the value of 10^{-3}. The problem was solved for $t \in [0, 0.01]$. The λ - transformation reduced execution time tenfold. The number of calculations of the right - hand sides was reduced twenty fold. The integration step of the λ - transformed problem was ten times bigger.

Note, that the program PC1 allows to find the solution of λ - transformed Van der Pol problem for $t \in [0, 2]$ whereas the solution the original problem terminated at $t = 0.03$ because of exponent overflow.

Example 3. Consider a initial value problem [3]

$$\varepsilon \ddot{x} + (1+t)x = 2.5(1+t),$$

$$x(0) = -1, \; t \in [0,1], \; \varepsilon = 0.003125.$$

The explicit scheme of Euler method was used for solving system. The program Maple V was used for solving this problem.

Abbreviations.

E – for Euler method results

BP – for results of Euler method for system transformed to the best parameter

h – step size for t or for λ (in case of the best parameter)

Δ – maximal error on $[0, 1]$

Table 1 shows results of numerical experiments.

Table 1.

method	h	Δ	quantity of steps
BP	0.2	0.127	256
BP	0.05	0.029	294
BP	0.01	0.0053	586
BP	0.005	0.0025	935
E	0.005	∞	200
E	0.004	1.950	250
E	0.002	0.584	500
E	0.001	0.239	1000
E	0.0001	0.020	10000

References

1. Shalashilin, V.I., Kuznetsov, E.B.: The method of solution continuation with respect to a parameter and the best parametruzation in applied mathematics and mechanics. Editorial URSS, Moscow (1999) (Russian)
2. Vasil'eva, A.B., Butuzov, V.F.: Asymptotic expansions of solutions of singular – perturbed equations. Nauka, Moscow (1973) (Russian)
3. Doolan, E.P., Miller, J.J.H., Schilders, W.H.A.: Uniform numerical methods for problems with initial and boundary layers. Boole Press, Dublin (1980)
4. Rakitskii, Yu.V., Ustinov, S.M., Chernorutskii, I.G.: Numerical methods for solving stiff systems. Nauka, Moscow (1979) (Russian)
5. Shalashilin, V.I., Kuznetsov, E.B.: The best parameter in the continuation of a solution. Dokladi Academii Nauk of Russia. Vol. 334. **5** (1994) 566–568
6. Kuznetsov, E.B., Shalashilin, V.I.: Cauchy's problem as a problem of continuation with respect to the best parameter. Diff. Urav. V. 30 **6** (1994) 964–971
7. Shalashilin, V.I., Kuznetsov, E.B.: Cauchy problem for nonlinear deformation of systems as a parametric continuation problem. Dokladi Academii Nauk of Russia. Vol. 329. **4** (1993) 426–428
8. Kuznetsov, E.B., Shalashilin, V.I.: Cauchy's problem as a problem of the continuation of a solution with respect to a parameter. Zh. Vychisl. Mat. Mat. Fiz. Vol. 33. **12** (1993) 1792–1805
9. Hairer, E., Wanner, G.: Solving ordinary differential equations II. Stiff and differential – algebraic problems. Springer–Verlag, Berlin (1991)

The Best Parameterization of Initial Value Problem for Mixed Difference-Differential Equation*

A. Kopylov and E. Kuznetsov

Moscow Aviation Institute, Volokolamskoe shosse 4,
125993 Moscow, Russia
akopylov@nes.ru, Kuznetsov@mai.ru

Abstract. We develop an approach to the numerical integration of initial value problem for mixed difference-differential equations that are differential with respect to one argument and difference with respect to others. Preliminary reduction of the problem to a set of Cauchy problems for systems of ordinary differential equations depending on a parameter affords to state it as the problem of the continuing the solution with respect to the best continuation parameter, namely, the integral curve length. This statement has numerous advantages over the usual statement. Namely, the right-hand sides of the transformed system remain bounded even if right-hand sides of the original system become infinite at some points.

1 Introduction

The general type of mixed functional-differential equations (MFDE) [1] is

$$x^{(l+1)}(s,t) = f(t, s, x_{ts}, x'_{ts}, \dots, x_{ts}^{(l)}),$$

Here $a < s < b, -\infty < a < b < \infty, 0 < t < \infty$. The ranges of values of x and f lie in $\mathbb{R}^n$; the i-th order derivative of x_{ts} with respect to t is denoted as $x_{ts}^{(i)}$, $i = 1, \dots, l+1$; $x_{ts}(\vartheta, \xi) = x(t+\vartheta, s+\xi), -g \le \vartheta \le 0$, $|\xi| \le h$; $h, g \ge 0$.

A particular case of general MFDE of the form

$$F(x'(s,t), x'(s+h_1,t), \dots, x'(s+h_k,t), x(s,t), x(s+h_1,t), \dots, x(s+h_k,t)) = 0,$$

where $x : \mathbb{R}^2 \to \mathbb{R}$, $s \in \Omega \subset \mathbb{R}^n$, $h_i \in \mathbb{Z}^n$, $i = 1, \dots, k$, $k \in \mathbb{N}$, is called the mixed difference-differential equation (MDDE).

Despite equations of this type have a lot of different applications in ecology, biology and physics they have been studied systematically not for a long time. The exhaustive review of applications and studies is presented in [2], the general theory of such equations is given in [1]. Paper [3] deals with boundary value problem for MDDEs which appear from variational problems. Paper [4]

* Supported by the Russian Foundation for Basic Research, project N° 01-01-00038

P.M.A. Sloot et al. (Eds.): ICCS 2003, LNCS 2658, pp. 507–515, 2003.

develops an algorithm of numerical integration of boundary value problem for linear MDDE based on finite difference method. This work presents an approach to numerical integration of initial value problem for MDDE transformed to the best argument. The efficiency of the approach is shown on test example.

2 Initial Value Problem for Mixed Difference-Differential Equation

Let consider mixed difference-differential equation

$$R_1(t,s)\frac{\partial}{\partial t}x(t,s) = R_2(t,s)g(x(t,s)) + f(t,s), \qquad ((t,s) \in Q), \tag{1}$$

with boundary condition

$$x(t,s) = 0, \qquad ((t,s) \in (0,T) \times (\mathbb{R}^n \setminus \Omega), \tag{2}$$

and initial condition

$$x(0,s) = \varphi(s), \qquad (s \in \Omega). \tag{3}$$

Here $Q = (0,T) \times \Omega$, $\Omega \subset \mathbb{R}^n$ is a bounded domain with piecewise-smooth boundary, $0 < T < \infty$, $f(t,s) : Q \to \mathbb{R}$ and $g(x) : \mathbb{R} \to \mathbb{R}$;

$$(R_\nu y)(t,s) = \sum_{h \in H} a_\nu^h(t,s)y(t,s+h), \qquad \nu = 1,2,$$

where H is an additive Abelian group of integer vectors, $a_\nu^h(t,s) : Q \to \mathbb{R}$, $\nu = 1,2, \quad h \in H$.

Suppose that $f(t,s)$, $a_\nu^h(t,s)$, $\nu = 1,2$, $h \in H$ are continuous functions with respect to t and piecewise-smooth with respect to $s \in Q$ and $g(x)$ is a continuous function in $\mathbb{R}$.

Let $D \subset \mathbb{R}^{n+1}$ be a bounded domain. Denote by $\tilde{C}^{1,0}(D)$ a set of functions $y(t,s) : D \to \mathbb{R}$ that are continuously differentiable with respect to the first argument for almost all values of s and piecewise-smooth with respect to the second argument. As it was done in [5], introduce operators I_Q, P_Q, R_Q as follows:

$I_Q : \tilde{C}^{1,0}(Q) \to \tilde{C}^{1,0}((0,T) \times \mathbb{R}^n)$ extends a function from the space $\tilde{C}^{1,0}(Q)$ with zero to $((0,T) \times \mathbb{R}^n) \setminus Q$;

$P_Q : \tilde{C}^{1,0}((0,T) \times \mathbb{R}^n) \to \tilde{C}^{1,0}(Q)$ restricts a function from $\tilde{C}^{1,0}((0,T) \times \mathbb{R}^n)$ to Q;

$R_{\nu Q} : \tilde{C}^{1,0}(Q) \to \tilde{C}^{1,0}(Q), \nu = 1,2$ are difference operators given by

$$R_{\nu Q} = P_Q R_\nu I_Q, \qquad (\nu = 1,2).$$

Definition 1. *A function $x(t,s) \in \tilde{C}^{1,0}(Q)$ is the solution to problem (1)–(3), if it satisfies the following equation*

$$R_{1Q}(t,s)\frac{\partial}{\partial t}x(t,s) = R_{2Q}(t,s)g(x(t,s)) + f(t,s), \qquad ((t,s) \in Q), \tag{4}$$

and initial condition

$$x(t,s)=\varphi(s), \qquad ((t,s)\in\{0\}\times\Omega), \tag{5}$$

for almost all $s\in\Omega$.

Note that it is the definition of difference operators $R_{\nu Q}$ that affords to rearrange equation (1) and boundary condition (2) to equation (4).

It has been proved (see [5]) that $\Omega\setminus\left(\bigcup_{h\in H}(\partial\Omega+h)\right)$ breaks into classes of subdomains Ω_{ij} (here and further we use $i=1,\dots,I$ for a class number and $j=1,\dots,J_i$ is a subdomain number in class i). Subdomains $\Omega_{i_1j_1}$ and $\Omega_{i_2j_2}$ belong to the same class iff there is a vector $h\in H$ such that $\Omega_{i_1j_1}=\Omega_{i_2j_2}+h$.

Denote

$$Q_{ij}=(0,T)\times\Omega_{ij}\,.$$

It is evident that

$$\overline{Q}=\bigcup_{i=1}^{I}\bigcup_{j=1}^{J_i}\overline{Q}_{ij}\ \text{and}\ Q_{i_1j_1}\bigcap Q_{i_2j_2}=\emptyset,\ (i_1,j_1)\neq(i_2,j_2)\,. \tag{6}$$

Rearrange problem (4), (5). Let an isomorphism

$$U_i:\tilde{C}^{1,0}\left(\bigcup_{j=1}^{J_i}Q_{ij}\right)\to\tilde{C}^{1,0}_{J_i}(Q_{i1})$$

be given by

$$(U_ix)_j(t,s)=x(t,s+h_{ij}), \qquad ((t,s)\in Q_{i1}),$$

where h_{ij} such that $Q_{ij}=Q_{i1}+h_{ij}$, $\tilde{C}^{1,0}_{J_i}(Q_{i1})=\prod_{1}^{J_i}\tilde{C}^{1,0}(Q_{i1})$ — Cartesian product of J_i spaces $\tilde{C}^{1,0}(Q_{i1})$.

Denote by U_i^{-1} an isomorphism $U_i^{-1}:\tilde{C}^{1,0}_{J_i}(Q_{i1})\to\tilde{C}^{1,0}\left(\bigcup_{j=1}^{J_i}Q_{ij}\right)$ that is reverse to U_i. Using the isomorphism U_i, taking into account condition (6) and permutability of U_i and $\partial/\partial t$, we obtain from equation (4)

$$U_iR_{1Q}U_i^{-1}U_i\frac{\partial x}{\partial t}=U_iR_{2Q}U_i^{-1}U_ig(x)+U_if, \qquad ((t,s)\in Q_{i1})\,. \tag{7}$$

Denote vector-valued function U_ix by x^i, $U_ig(x)$ by $\tilde{g}(x^i)$ and U_if by f^i. Using this notations, we can rewrite (7) as follows

$$U_iR_{1Q}U_i^{-1}\frac{\partial x^i}{\partial t}=U_iR_{2Q}U_i^{-1}\tilde{g}(x^i)+f^i, \qquad ((t,s)\in Q_{i1})\,. \tag{8}$$

The boundary conditions (3) can be rewritten in the same way

$$x^i\big|_{t=0} = \varphi^i(s), \qquad (s \in \Omega_{i\,1}), \tag{9}$$

where by φ^i vector valued function $U_i\varphi$ is denoted.

Denote by R^i_ν, $\nu = 1,2$, $J_i \times J_i$ matrixes with elements $\{R^i_\nu\}_{k\,l}$, $\{R^i_\nu\}_{k\,l}(t,s) = a^h_\nu(t, s + h_{i\,k})$, where $h = h_{i\,l} - h_{i\,k}$, $(t,s) \in Q_{i\,1}$.

Lemma 1. *The operator* $R^i_{\nu Q} = U_i R_{\nu Q} U_i^{-1} : \tilde{C}^{1,0}_{J_i}(Q_{i1}) \to \tilde{C}^{1,0}_{J_i}(Q_{i1})$, $\nu = 1,2$, $i = 1,\dots,I$ *is the operator of multiplication by the matrix* R^i_ν.

The proof follows from lemma 8.6, ch. II, [5].

Theorem 1. *Let* x *be the solution to problem (4), (5), then for* $i = 1,\dots,I$ *vector-valued function* $x^i = U_i x$ *is the solution to the following problem*

$$R^i_1 \frac{\partial x^i}{\partial t} = R^i_2 \tilde{g}(x^i) + f^i, \qquad ((t,s) \in Q_{i1}), \tag{10}$$

$$x^i\big|_{t=0} = \varphi^i(s), \qquad (s \in \Omega_{i1}), \tag{11}$$

and if vector-valued function x^i *is the solution to problem (10), (11), then*

$$x(t,s) = U_i^{-1} x^i, \qquad (t,s) \in \left(\bigcup_{j=1}^{J_i} Q_{i\,j}\right), \qquad i = 1,\dots,I,$$

is the solution to problem (4), (5).

Note that functions R^i_ν, $\tilde{g}$, f^i, φ^i are supposed to provide the existence and uniqueness of the solution to problem (10), (11) that is a continuously differentiable function with respect to t and piecewise continuous function with respect to s, i.e. $x^i \in \tilde{C}^{1,0}_{J_i}(Q_{i1})$.

Proof. To begin the proof recall that the solution to problem (10)-(11) is a function that satisfies equation (10) and boundary condition (11) for almost all $s \in \Omega_{i1}$.

Suppose x is the solution to problem (4), (5). Since equation (7) is equivalent to equation (4) it follows that vector-valued function x^i that is equal to $U_i x$ satisfies equation (8). Taking into account the above notations we obtain that vector-valued function x^i satisfies equation (9). Finally, Lemma implies that x^i satisfies equation (10).

Analogous reasoning implies that if x satisfies initial condition (5) then vector-valued function x^i satisfies initial conditions (11) for almost all s. Thus vector-valued function x^i is the solution to problem (10), (11).

Now, suppose that vector-valued function $x^i(t)$ is the solution to problem (10), (11). It is clear that if $R^i_\nu(t,s)$, $f^i(t,s), \varphi^i(s)$ are defined and continuous for some $s \in \Omega_{i\,j}$ then $x(t,s)$ specified in the statement of the theorem is defined, continuous with respect to s and continuously differentiable with respect to t for the same s. Thus relation (6) implies that $x(t,s)$ has the same properties almost everywhere in Q, i.e. $x \in \tilde{C}^{1,0}(Q)$.

Since x^i, $i = 1, \dots, I$ satisfies equation (10) then due to the Lemma it satisfies equation (8). The latter yield that $x(t,s)$ defined in the statement of the theorem satisfies equation (4) almost for all $s \in \Omega$. The same speculations show that x satisfies initial condition (5) almost for all $s \in \Omega$. Thus, $x(t,s)$ is the solution to problem (4), (5). □

Note to conclude that if $f(t,s)$, $\varphi(s)$, $a_\nu^h(t,s)$, $\nu = 1,2$, $h \in H$, are continuous functions with respect to s for all $s \in \Omega$ may be except for points of the set $\{s \in \partial\Omega_{ij}, i = 1, \dots, I, j = 1, \dots, J_i\}$ then the solution to problem (4), (5) is a continuous function with respect to s for all $s \in \Omega$ may be except for points of the same set. It worth adding that generally continuity of $f(t,s)$, $\varphi(s)$, $a_\nu^h(t,s)$ in Q does not yield continuity with respect to s of the solution to problem (4), (5).

3 The Best Argument of the Problem

To solve the problem of choosing the best argument of problem (4), (5) (the argument of differentiation), we exploit the method of the continuation of a solution. Theorem 1 implies that it is equivalent to seek the best argument of problem (10), (11).

Suppose matrixes R_1^i, $i = 1, \dots, I$, are not singular at $(t,s) \in Q_{i1}$. Hence we can transform system (10), (11) to the normal form considering variable s as a parameter.

$$\frac{dx^i}{dt} = (R_1^i(t,s))^{-1} R_2^i(t,s)\tilde{g}(x^i) + (R_1^i(t,s))^{-1} f^i(t,s), \tag{12}$$

$$x^i(0) = \varphi^i(s), \tag{13}$$
$$(t \in (0,T), s \in \Omega_{i1}).$$

Assume that an integral of problem (12), (13) is given by

$$F(x_1^i(t,s), \dots, x_r^i(t,s), t) = 0, \tag{14}$$

such that

$$F(\varphi_1^i(s), \dots, \varphi_r^i(s), 0) = 0, s \in \Omega_{i1},$$

where x_l^i are components of x^i , $\varphi_l^i - \varphi^i$, $F : \mathbb{R}^{r+1} \to \mathbb{R}^r$ (recall $r = J_i$), defines a smooth integral curve of the problem in $\mathbb{R}^{r+1}$ almost for every $s \in \Omega_{i1}$.

We shall handle a process of this curve construction as a process of solving system (14), where t is the argument of the problem and s is regarded as a parameter, $s \in \Omega_{i1}$. Also we'll investigate this system by the continuation method treating initial value problem (12), (13) as the Cauchy problem for continuation equations for the solution of system (14) with respect to parameter t. Such an approach has been used to solve the Cauchy problem for a system of ordinary differential equation [6] and delayed differential equation [7], [8].

Theorem 2. *To transform the Cauchy problem (12), (13) to the best argument it is necessary and sufficient to choose the arc length λ measured along the integral curve of this problem as a new independent variable for any particular $s \in \Omega_{i1}$. After transformation problem (12), (13) takes the form*

$$\frac{dx^i}{d\lambda} = \pm \frac{G^i(t,s,x^i)}{\sqrt{1+(G^i,G^i)}}, \qquad x^i(0) = \varphi^i(s), \tag{15}$$

$$\frac{dt}{d\lambda} = \pm \frac{1}{\sqrt{1+(G^i,G^i)}}, \qquad t(0) = 0, \tag{16}$$

$$i = 1,\dots,I, \quad t \in (0,T), \quad s \in \Omega_{i1}.$$

Here by $(\cdot,\cdot)$ inner product in $\mathbb{R}^r$ and by $G^i = (G_1,\dots,G_r)$ vector of right-hand side functions of system (12) are denoted. Choice of plus or minus sign defines the direction of moving along the integral curve of the problem.

We call an argument to be the best if it provides the linear system of continuation equations whose matrix is the Jacobi matrix of system (14) with the largest possible condition number at any step of integration process.

Proof. We'll continue a solution to problem (12), (13) locally, in a small neighborhood of any point of the problem integral curve. Suppose that t and all components of x^i are functions of μ such that

$$\Delta\mu = \sum_{l=1}^{r} c_l \Delta x_l^i + c_{r+1}\Delta t. \tag{17}$$

Here Δx_l^i and Δt are increments of corresponding variables, c_l $(l = 1,\dots,r+1)$ are components of the unit vector defining a direction in which μ is measured.

Having differentiated equation (14) with respect to μ as a composite function, we obtain

$$\sum_{l=1}^{r} F_{k,x_l^i} x_{l,\mu}^i + F_{k,t} t_{,\mu} = 0, \qquad k = 1,\dots,r, \tag{18}$$

where the following notations were used

$$x_{l,\mu}^i = \frac{dx_l^i}{d\mu}, \quad t_{,\mu} = \frac{dt}{d\mu}, \quad F_{k,x_l^i} = \frac{\partial F_k}{\partial x_l^i}, \quad F_{k,t} = \frac{\partial F_k}{\partial t}.$$

Assuming nonsingularity of matrix $\left(F_{k,x_l^i}\right)$, we obtain from (18) and (12)

$$\frac{dx_l^i}{dt} = -(F_{k,x_l^i})^{-1}(F_{1,t},\dots,F_{r,t})^T = G_l^i.$$

Hence, equations (18) take the form

$$x_{l,\mu}^i - G_l^i t_{,\mu} = 0, \qquad l = 1,\dots,r. \tag{19}$$

Thus (19) implies that equations (12) multiplied by dt and divided by $d\mu$ can be used as continuation equations.

To obtain $(r+1)$-th continuation equation one has to divide (17) by $\Delta\mu$ and then pass to limit as $\Delta\mu \to 0$.

$$\sum_{l=1}^{r} c_l x^i_{l,\mu} + c_{r+1} t_{,\mu} = 1. \tag{20}$$

Thus (19) and (20) constitute the system of continuation equations.

It has been proved (see [9]) that the transformation of this system to the Cauchy normal form is best conditioned iff $c_l = x^i_{l,\lambda}$, $l = 1, \dots, r$, $c_{r+1} = t_{,\lambda}$. This means that the arc length measured along the integral curve of problem (12), (13) is chosen as μ. Hence, system (19), (20) can be rearranged

$$\begin{cases} x^i_{,\lambda} & - G^i t_{,\lambda} = 0, \\ (x^i_{l,\lambda}, x^i_{l,\lambda}) + & t^2_{,\lambda} \quad = 1. \end{cases} \tag{21}$$

Solving (21) for derivatives and measuring λ from initial point of (12), (13) due to independence of right-hand side functions of the system on λ, we come to the transformed problem that is of the form given by (15), (16). □

Example 1. Let consider the problem

$$x'(t, s-1) + tx'(t, s) + x'(t, s+1) = 1/3x(t, s), \qquad ((t, s) \in (0, T) \times (0, 2)),$$

with boundary condition $x(t, s) = 0, (t, s) \in (0, T) \times (\mathbb{R} \setminus (0, 2))$, and initial condition $x(0, s) = \varphi(s), (s \in (0, 2))$.

In the above notations $I = 1, J_1 = 2$. In this problem difference operator generates decomposition of $Q = (0, T) \times (0, 2)$ into subdomains $Q_{11} = (0, T) \times (0, 1)$ and $Q_{12} = (0, T) \times (1, 2)$. Functions x^i and φ^i correspond to each of these subdomains.

Define matrixes R^1_1, R^1_2

$$R^1_1 = \begin{pmatrix} t & 1 \\ 1 & t \end{pmatrix}, \qquad R^1_2 = \begin{pmatrix} 1/3 & 0 \\ 0 & 1/3 \end{pmatrix}.$$

Calculating a matrix inverse to $(R^1_1)^{-1}$ and denoting x^i by $(x_1, x_2)^T$, we bring the problem to the form of (12), (13).

$$\begin{pmatrix} x'_1 \\ x'_2 \end{pmatrix} = \frac{1}{t^2 - 1} \begin{pmatrix} t & -1 \\ -1 & t \end{pmatrix} \begin{pmatrix} 1/3 & 0 \\ 0 & 1/3 \end{pmatrix} \begin{pmatrix} x_1 \\ x_2 \end{pmatrix}.$$

or equivalently:

$$x'_1 = \frac{1/3tx_1}{t^2 - 1} - \frac{1/3x_2}{t^2 - 1}, \qquad x_1(0) = \varphi_1(s),$$

$$x'_2 = \frac{-1/3x_1}{t^2 - 1} + \frac{1/3tx_2}{t^2 - 1}, \qquad x_2(0) = \varphi_2(s).$$

The solution to this problem

$$x_1 = C_1(t+1)^{1/3} + C_2(t-1)^{1/3},$$
$$x_2 = C_1(t+1)^{1/3} - C_2(t-1)^{1/3},$$

where $C_1 = (\varphi_1(s) + \varphi_2(s))/2$, $C_2 = (\varphi_1(s) - \varphi_2(s))/2$ are obtained from the initial condition. As for the solution to the original problem, it has the form $x(t,s) = x_i(t,s)$, $(t,s) \in Q_{1\,i}$, $i = 1,2$. It is evident that despite the right-hand side functions of system of differential equations are unbounded in a neighbourhood of $t = \pm 1$ the solution to the problem is a continuous function of t.

Having transformed the problem to the best argument, we obtain

$$\frac{dx_1}{d\lambda} = \frac{f_1}{\sqrt{(t^2-1)^2 + f_1^2 + f_2^2}}, \qquad x_1\,|_{\lambda=0} = \varphi_1(s),$$

$$\frac{dx_2}{d\lambda} = \frac{f_2}{\sqrt{(t^2-1)^2 + f_1^2 + f_2^2}}, \qquad x_2\,|_{\lambda=0} = \varphi_2(s),$$

$$\frac{dt}{d\lambda} = \frac{1}{\sqrt{(t^2-1)^2 + f_1^2 + f_2^2}}, \qquad t\,|_{\lambda=0} = 0,$$

where $f_1 = 1/3(tx_1 - x_2)$, $f_2 = -1/3(x_1 - tx_2)$.

Note that right-hand side functions of the transformed system are continuous and uniformly bounded functions of t. It yields that the solution to the problem is a continuously differentiable function of λ.

Numerical tests demonstrate that unboundedness of the right-hand side functions of a system of differential equations may make impossible the implementation of standard numerical integration procedures. Preliminary transformation of a system to the best argument affords to avoid similar difficulties.

References

1. Kamenskii, G.A., Myshkis, A.D.: On the mixed type functional differential equations. Nonlinear Analysis, TMA. Vol. 30. **5** (1997) 2577–2584
2. Kamenskii, G.A.: A review of the theory of mixed functional differential equations. Problems of Nonlinear Analysis in Engineering Systems. Vol. 2. **8** (1998) 1–16
3. Kamenskii, G.A.: Boundary value problems for difference-differential equations arising from variational problems. Nonlinear Analysis, TMA. Vol. 18. **8** (1992) 801–813
4. Kopylov, A.V.: Finite difference method for the numerical solution of boundary value problem for mixed difference-differential equation. Sib. J. Numer. Math. Vol. 3. **4** (2000) 345–355 (Russian)
5. Skubachevskii, A.L.: Elliptic functional differential equations and applications. Birkhauser, Basel Boston Berlin (1997)
6. Kuznetsov, E.B., Shalashilin, V.I.: The Cauchy problem as a problem of continuation with respect to the best parameter. Differ. Eq. Vol. 30. **6** (1994) 893–898
7. Kuznetsov, E.B.: Transformation delayed differential equations to the best argument. Mat. Zametki Vol. 63. **1** (1998) 62–68

8. Kopylov, A.V., Kuznetsov, E.B.: An approach to the numerical integration of the Cauchy problem for Delay Differential Equations. Comput. Math. Math. Phys. Vol. 41. **10** (2001) 1470–1479
9. Kuznetsov, E.B., Shalashilin, V.I.: Method continuation of a solution with respect to the best parameter and the best parameterization in applied mathematics and mechanics. Editorial URSS, Moscow (1999) (Russian)

Numerical Solution of Differential-Algebraic Equations by Block Methods

Michael V. Bulatov

Institute of System Dynamics and Control Theory SB RAS, 134 Lermontov str., 664033 Irkutsk, Russia,
mvbul@icc.ru

Abstract. In this paper some class of nonlinear differential-algebraic equations of high index is considered. For the numerical solution of this problem the family of multistep, multistage difference schemes of high order is proposed. In some cases this difference schemes are Runge-Kutta methods. The estimate of error is found.

In the paper a family of high-order precision difference schemes which are intended for numerical solution of high-index differential algebraic equations (DAEs) is proposed and investigated. The condition of collocation forms the basis for constructing such schemes. The paper is a continuation of the author's works [1], [2], [3], [4].

Consider the following problem

$$f(x^{'}(t), x(t), t) = 0, \ t \in [0, 1], \tag{1}$$

$$x(0) = a. \tag{2}$$

Definition 1 ([5], p. 16). *Let J be an open subinterval of R, D a connected open subset of R^{2n+1}, and f a differentiable function from D to R^n. Then the DAE (1) is solvable on J in D if there is a k-dimensional family of solutions $y(t,c)$ defined on a connected open set $J \times D_1, D_1 \subset R^k$, such that:*

1. $y(t,c)$ is defined on all of J for each $c \in D_1$

2. $(y^{'}(t,c), y(t,c), t) \in D$ for $(t,c) \in J \times D_1$

3. If $z(t)$ is any other solution with $(z^{'}(t), z(t), t) \in D$, then $z(t) = y(t,c)$ for some $c \in D_1$

4. The graph of y as a function of (t,c) is a $k+1$-dimensional manifold.

Let the following mesh be given on the segment $[0;1]$

$$\Delta_h = \{t_i : \ t_i = ih, \ i = 1, ..., M, \ h = 1/M\}.$$

P.M.A. Sloot et al. (Eds.): ICCS 2003, LNCS 2658, pp. 516–522, 2003.

For the purpose of numerical solving of the problem (1)–(2) it is advisable to construct s-stage, m-step difference schemes of the form

$$\begin{cases} f(h^{-1}\sum\limits_{j=0}^{m} k_j^1 x_{i+s-j},\ \sum\limits_{j=0}^{m} l_j^1 x_{i+s-j}, \bar{t}_{i+1}) = 0, \\ f(h^{-1}\sum\limits_{j=0}^{m} k_j^2 x_{i+s-j},\ \sum\limits_{j=0}^{m} l_j^2 x_{i+s-j}, \bar{t}_{i+2}) = 0, \\ \dots \\ f(h^{-1}\sum\limits_{j=0}^{m} k_j^s x_{i+s-j},\ \sum\limits_{j=0}^{m} l_j^s x_{i+s-j}, \bar{t}_{i+s}) = 0. \end{cases} \tag{3}$$

Here $f(h^{-1}\sum\limits_{j=0}^{m} k_j^q x_{i+s-j},\ \sum\limits_{j=0}^{m} l_j^q x_{i+s-j}, \bar{t}_{i+q}) = 0$ is an approximation of the initial problem. We assume that initial values of $x_1, x_2, ..., x_{m-s-1}$ have been computed earlier ($x_0 = a$).

Consider a particular case of schemes (3), say, the interpolation variant. Let an interpolation m-power manifold be passed through the points $x_{i+s}, x_{i+s-1}, ..., x_{i+s-m}$. Then the scheme (3) writes

$$\begin{cases} f(h^{-1}\sum\limits_{j=0}^{m} k_j^1 x_{i+s-j},\ x_{i+1}, t_{i+1}) = 0, \\ f(h^{-1}\sum\limits_{j=0}^{m} k_j^2 x_{i+s-j},\ x_{i+2}, t_{i+2}) = 0, \\ \dots \\ f(h^{-1}\sum\limits_{j=0}^{m} k_j^s x_{i+s-j},\ x_{i+s}, t_{i+s}) = 0. \end{cases} \tag{4}$$

where $h^{-1}\sum\limits_{j=0}^{m} k_j^q x_{i+s-j}$ is an approximation $x'(t_{i+q})$ of the order h^m, $q = 1, ..., s$.

If $s = m, \bar{h} = sh$, and $x_{i+1}, x_{i+2}, ..., x_{i+s-1}$ are considered as intermediate results, then schemes (4) can be interpreted as Runge-Kutta methods with abscissas $c = (1/s, 2/s, ..., 1)$, the weights $b = (1, 0, ..., 0)$ and with the matrix $\mathcal{A}$, determined from the conditions

$$V\mathcal{A}^{\top} = \mathcal{C},$$

where

$$C = \begin{pmatrix} 1/s & 1/(s-1) & \dots 1 \\ 1/(2s^2) & 1/(2(s-1)^2) & \dots 1/2 \\ \cdot & \cdot & \cdots \\ \cdot & \cdot & \cdots \\ \cdot & \cdot & \cdots \\ 1/(s^{s+1}) & 1/(s(s-1)^{s-1}) & \dots 1/s \end{pmatrix},$$

and V is the Vandermonde matrix:

$$V = \begin{pmatrix} 1 & 1 & \dots 1 \\ 1/s & 2/s & \dots 1 \\ \cdot & \cdot & \dots . \\ \cdot & \cdot & \dots . \\ \cdot & \cdot & \dots . \\ (1/s)^{s-1} & (2/s)^{s-1} & \dots 1 \end{pmatrix}.$$

Similar methods for ODEs have probably for the first time been proposed in [6]. In this paper, the abscissas were chosen as follows $c = (0, 1/s, ..., (s-1)/s)$.

Note that presently the theory of methods of difference coefficients for DAEs, which have the index not higher than 3 and the Hessenberg form [5], [7], [8], have been developed rather completely.

Consider a few schemes for the case when $m > s$.

For $s = 1$ we obtain the BDF methods.

For $s = 2, m = 3$ we have

$$\begin{cases} f((2x_{i+2} + 3x_{i+1} - 6x_i + x_{i-1})/6h, x_{i+1}, t_{i+1}) = 0, \\ f((11x_{i+2} - 18x_i + 9x_i - 2x_{i-1})/6h, x_{i+2}, t_{i+2}) = 0, \end{cases}$$

while assuming x_1 to be known and $x_0 = a$.

For $s = 2, m = 4$ we obtain

$$\begin{cases} f((3x_{i+2} + 10x_{i+1} - 18x_i + 6x_{i-1} - x_{i-2})/12h, x_{i+1}, t_{i+1}) = 0, \\ f((25x_{i+2} - 48x_i + 36x_i - 16x_{i-1} + 3x_{i-2})/12h, x_{i+2}, t_{i+2}) = 0, \end{cases}$$

while assuming x_1, x_2 known, $x_0 = a$.

Consider a particular case of problem (1)

$$x(t) + \xi(x'(t), t) = 0, \tag{5}$$

where $\partial\xi(x', t)/\partial x'$ is an upper-triangular matrix with the zero diagonal, furthermore, the number of zero square blocks on the diagonal is r.

It can readily be noted that our system (5) has a unique solution for a sufficiently smooth $\xi(x', t)$. Indeed, when rewriting (5) in the explicit form

$$\begin{cases} x_1 + \xi_1(x_2', x_3', ..., x_r', t) = 0, \\ x_2 + \xi_2(x_3', x_4', ..., x_r', t) = 0, \\ \cdot \quad \cdot \quad \cdot \\ x_r + \xi_r(t) = 0, \end{cases}$$

where $x_1, x_2, ..., x_r$ correspond to zero blocks of the matrix $\partial\xi(x', t)/\partial x'$, we obtain that $x_r = -\xi_r(t), x_{r-1} = -\xi_{r-1}(-\xi_r'(t), t)$ and so on.

Lemma 1 ([4]). *Let $\partial F(x)/\partial x$ be an upper-triangular matrix with the zero diagonal, furthermore, $(\partial F(x)/\partial x)^r = 0$ corresponds to the zero matrix. Hence the system of nonlinear equations*

$$x = F(x), \tag{6}$$

has a unique solution, and the method of simple iteration

$$x^{i+1} = F(x^i), \tag{7}$$

gives a precision solution for the system (6) in r steps for any initial approximation x^0.

Lemma 2. *For the system of nonlinear equations $x = F(x)$, which satisfy the condition of Lemma 1, the Newton's method*

$$x^{i+1} = x^i - (E - \partial F(x)/\partial x|_{x=x^i})^{-1} F(x^i), \tag{8}$$

gives a precision solution of the system (6) in r steps for any initial approximation x^0.

Proving of this result can be conducted likewise in Lemma 2, and so, it is omitted.

Corollary 1. *Let for the system (6) the matrix $\partial F(x)/\partial x$ have the following block form*

$$\partial F(x)/\partial x = \begin{pmatrix} F_{11} & F_{12} & \dots & F_{1k} \\ F_{21} & F_{22} & \dots & F_{2k} \\ . & . & \dots & . \\ F_{k1} & F_{k2} & \dots & F_{kk} \end{pmatrix},$$

where F_{ij} are upper-triangular $(s \times s)$—matrices with the zero diagonal, furthermore, the maximum number of zero blocks on the diagonals F_{ij} is r. Hence this systems has a unique solution. The simple iteration method

$$x^{i+1} = F(x^i)$$

and the Newton's method

$$x^{i+1} = x^i - (E - \partial F(x)/\partial x|_{x=x^i})^{-1} F(x^i),$$

give a precision solution of this system in r steps for any initial approximation x^0.

Now let us turn back to problem (5). For the purpose of numerical solving this problem let us consider the difference schemes (4), which with respect to it have the form

$$\begin{cases} x_{i+1} = \xi(h^{-1} \sum_{j=0}^{m} k_j^1 x_{i+s-j}, t_{i+1}), \\ x_{i+2} = \xi(h^{-1} \sum_{j=0}^{m} k_j^2 x_{i+s-j}, t_{i+2}), \\ . \quad . \quad . \\ x_{i+s} = \xi(h^{-1} \sum_{j=0}^{m} k_j^s x_{i+s-j}, t_{i+s}), \end{cases} \tag{9}$$

where $h^{-1} \sum_{j=0}^{m} k_j^q x_{i+s-j}$ is an approximation of $x'(t_{i+q})$, furthermore,

$$h^{-1}\sum_{j=0}^{m} k_j^q x_{i+s-j} - x^{'}(t_{i+q}) = \sigma_{i+q}/h,$$

$$\sigma_{i+q}/h = h^{m+1}x^{(m+1)}(t_{i+q})/(m+1) + O(h^{m+2}). \tag{10}$$

On account of the corollary of the lemmas, the system (9) has a unique solution, and the method of simple iteration (or the Newton's method) in r steps suggests a precision solution of the given system.

Let us reduce the result concerning the convergence of schemes (9) to the precision solution of problem (5).

Theorem 1. *Let the vector function* $\xi(x^{'},t)$ *in the problem (5) be sufficiently smooth with respect to the set of arguments, and let*

$$\| x_j - x(t_j) \| = O(h^{m+1}),\ j = 0, 1, ..., m - s.$$

Then

$$\| x_i - x(t_i) \| = O(h^{m+2-r}),\ i = m + 1, m + 2, ..., M.$$

Proof. Having substituted the precision value of $x(t)$ into the system (9), we have

$$\begin{cases} x_{i+1} + \varepsilon_{i+1} = \xi(h^{-1}\sum\limits_{j=0}^{m} k_j^1 \varepsilon_{i+s-j} + x^{'}(t_{i+1}) + \sigma_{i+1}/h, t_{i+1}), \\ x_{i+2} + \varepsilon_{i+2} = \xi(h^{-1}\sum\limits_{j=0}^{m} k_j^2 \varepsilon_{i+s-j} + x^{'}(t_{i+2}) + \sigma_{i+2}/h, t_{i+2}), \\ . \quad . \quad . \\ x_{i+s} + \varepsilon_{i+s} = \xi(h^{-1}\sum\limits_{j=0}^{m} k_j^s \varepsilon_{i+s-j} + x^{'}(t_{i+s}) + \sigma_{i+s}/h, t_{i+s}). \end{cases}$$

From the last formula and the initial system (5) we obtain the following result for the solution error:

$$\begin{cases} \varepsilon_{i+1} = A_{i+1}h^{-1}\sum\limits_{j=0}^{m} k_j^1 \varepsilon_{i+s-j} + A_{i+1}\sigma_{i+1}/h, \\ \varepsilon_{i+2} = A_{i+2}h^{-1}\sum\limits_{j=0}^{m} k_j^2 \varepsilon_{i+s-j} + A_{i+2}\sigma_{i+2}/h, \\ . \quad . \quad . \\ \varepsilon_{i+s} = A_{i+s}h^{-1}\sum\limits_{j=0}^{m} k_j^s \varepsilon_{i+s-j} + A_{i+s}\sigma_{i+s}/h, \end{cases} \tag{11}$$

where $A_j = \partial\xi/\partial x^{'}$.

For the purpose of simplicity of our reasoning assume s to be $m-$fold (one can always obtain this by increasing m artificially up to the minimum value m_1, which is multiple to s, and assuming $k_l^p = 0,\ l = m+1, m+2, ..., m_1$).

Introduce the denotations:

$$m_2 = m/s, \quad n_1 = (M - m + s)/s,$$

$$\bar{\varepsilon}_i = (\varepsilon_{m-s+1+is}^\top, \varepsilon_{m-s+2+is}^\top, ..., \varepsilon_{m+is}^\top)^\top,$$

$$\bar{\sigma}_i = ((A_{m-s+1+is}\sigma_{m-s+1+is})^\top, ..., (A_{m+is}\sigma_{m+is})^\top)^\top.$$

On account of these denotations the recurrent relation (11) may be rewritten in the form of a block m_2-diagonal system of linear algebraic equations

$$(\mathcal{N} + \mathcal{E})\Upsilon = \Sigma, \tag{12}$$

where $\mathcal{N} + \mathcal{E} =$

$$\begin{pmatrix} hE + N_{11} & 0 & 0\,0\,0 & . & ..\,0\,0 \\ N_{21} & hE + N_{22} & 0\,0\,0 & . & ..\,0\,0 \\ . & . & ... & . & .... \\ N_{m_2 1} & . & ..\ hE + N_{m_2 m_2} & 0 & ...\ 0 \\ . & . & ... & . & .... \\ 0 & 0 & ... & N_{n_1 n_1 - m_2} & ...\ hE + N_{n_1 n_1} \end{pmatrix},$$

$$\Upsilon = \begin{pmatrix} \bar{\varepsilon}_1 \\ \bar{\varepsilon}_2 \\ . \\ . \\ . \\ \bar{\varepsilon}_{n_1} \end{pmatrix}, \quad \Sigma = -\begin{pmatrix} \bar{\sigma}_1 \\ \bar{\sigma}_2 \\ . \\ . \\ . \\ \bar{\sigma}_{n_1} \end{pmatrix},$$

and

$$N_{ii} = \begin{pmatrix} k_{m-s+1}^1 A_{m-s+is+1} & k_{m-s}^1 A_{m-s+is+1} & ...\ k_0^1 A_{m-s+is+1} \\ k_{m-s+1}^2 A_{m-s+is+2} & k_{m-s}^2 A_{m-s+is+2} & ...\ k_0^2 A_{m-s+is+2} \\ . & . & .... \\ k_{m-s+1}^s A_{m+is} & k_{m-s}^s A_{m+is} & ...\ k_0^s A_{m+is} \end{pmatrix},$$

$$N_{i-j\,i} = \begin{pmatrix} k_{m-s+1}^1 A_{m-s+is+1} & k_{m-s}^1 A_{m-s+is+1} & ...\ k_0^1 A_{m-s+is+1} \\ k_{m-s+1}^2 A_{m-s+is+2} & k_{m-s}^2 A_{m-s+is+2} & ...\ k_0^2 A_{m-s+is+2} \\ . & . & .... \\ k_{m-s+1}^s A_{m+is} & k_{m-s}^s A_{m+is} & ...\ k_0^s A_{m+is} \end{pmatrix}.$$

From the system (12) we have:

$$\| \Upsilon \| = \| (\mathcal{N} + \mathcal{E})^{-1}\Sigma \| . \tag{13}$$

Due to the theorem's condition, A_j are upper-triangular matrices with zero quadratic blocks on the diagonal, whose number is r. It can easily be shown that

$$(\mathcal{N} + \mathcal{E})^{-1} = h^{-1}\mathcal{E} - h^{-2}\mathcal{N} + ... + (-h)^{-r}\mathcal{N}^{r-1}. \tag{14}$$

Each of the addends of the right-hand side of the identity (14) $\mathcal{N}^l$, $l = 1, 2, ...r-1$ is a block matrix, and each of its blocks contains a multiplication of l upper-triangular matrices of the form A_j. Having restored in the memory that

$$\bar{\sigma}_i = ((A_{m-s+1+is}\sigma_{m-s+1+is})^\top, ..., (A_{m+is}\sigma_{m+is})^\top)^\top,$$

we obtain that

$$\mathcal{N}^{r-1}\Sigma = 0.$$

By employing the formula (14), the estimate (10) and the latter identity in (13), we have

$$\| \Upsilon \| = \| (\mathcal{N} + \mathcal{E})^{-1}\Sigma \| = \| h^{-1}\mathcal{E}\Sigma - h^{-2}\mathcal{N}\Sigma + ... + (-h)^{-r}\mathcal{N}^{r-1}\Sigma \| =$$

$$= \| h^{-1}\Sigma - h^{-2}\mathcal{N}\Sigma + ... + (-h)^{1-r}\mathcal{N}^{r-2}\Sigma \| \leq$$

$$\leq K_1 h^{m+1} + K_2 h^m + ... + K_{r-1} h^{m+2-r} = O(h^{m+2-r}).$$

□

In conclusion, we would like to note that multistep, multistage methods of numerical solving of ODEs are presently under intensive development (see, for example [9]). Although their form is different with respect to that of (3).

References

1. Bulatov, M.V., Chistjakov, V.F.: Application of Collocation Methods for Solution of Singular Linear System of Ordinary Differential Equations. Models and Methods of Operation Inverstigation. Nauka, Novosibirsk (1988) 164–170 (Russian)
2. Bulatov, M.V.: On Dependence of Diskretization Step from the Level of Perturbations of Singular Linear Systems for the Ordinary Differential Equations. Ill-posed Problems of Mathematical Physics and Analysis. Krasnoyarsk University, Krasnoyarsk (1988) 211–216 (Russian)
3. Bulatov, M.V.: On Difference Schemes for Differential-Algebraic Systems. Computational Mathematics and Mathematical Physics. Vol. 38. **10** (1998) 1571–1579
4. Bulatov, M.V.: Block-collocation Methods for Solving of Differential-Algebraic Systems. Optimization, Control, Intellect. Vol. 5. **1** (2000) 33–41 (Russian)
5. Brenan, K.E., Campbell, S.L., Petzold, L.R.: Numerical Solution of Initial-Value Problems in Differential-Algebraic Equations. SIAM, Philadelphia (1996)
6. Gudovich, N.N.: On a New Method of Constructing Stable Difference Schemes Assigned Approximation Order for Linear Ordinary Differential Equations. Computational Mathematics and Mathematical Physics. Vol. 15. **4** (1975) 931–945
7. Hairer, E., Wanner, G.: Solving Ordinary Differential Equations II: Stiff and Differential-Algebraic Problems. Springer-Verlag, Berlin (1991)
8. Hairer, E., Lubich, Ch., Roche, M.: The Numerical Solution of Differential-Algebraic Systems by Runge-Kutta Methods. Lecture Notes in Math. Vol. 1409, Springer-Verlag, Berlin (1989)
9. Butcher, J.C., Jackiewicz, Z.: Implementation of Diagonally Implicit Multistage Integration Methods for Ordinary Differential Equations. SIAM Journal of Numerical Analysis. Vol. 34. **6** (1997) 2119–2141

Generalized Jordan Sets in the Theory of Singular Partial Differential-Operator Equations*

Michael V. Falaleev, Olga A. Romanova, and Nicholas A. Sidorov

Irkutsk State University, 664003 Irkutsk, Russia
mihail@ic.isu.ru, olga@baikal.ru, sidorov@home.isu.runnet.ru

Abstract. We apply the generalized Jordan sets techniques to reduce partial differential-operator equations with the Fredholm operator in the main expression to regular problems. In addition this techniques has been exploited to prove a theorem of existence and uniqueness of a singular initial problem, as well as to construct the left and right regularizators of singular operators in Banach spaces and to construct fundamental operators in the theory of generalized solutions of singular equations.

1 Introduction

Let

$$x = (t, x') \text{ be a point in the space } R^{m+1},$$
$$x' = (x_1, \dots, x_m), \quad D = (D_t, D_{x_1}, \dots, D_{x_m}),$$
$$\alpha = (\alpha_0, \dots, \alpha_m), \quad \mid \alpha \mid = \alpha_0 + \alpha_1 + \cdots \alpha_m, \quad \text{where}$$
$$\alpha_i \text{ are integer non-negative indexes, } D^\alpha = \frac{\partial^\alpha}{\partial t^{\alpha_0} \dots \partial x_m^{\alpha_m}}.$$

We also suppose that $B_\alpha : D_\alpha \subset E_1 \to E_2$ are closed linear operators with dense domains in E_1, $x \in \Omega$, where $\Omega \subset R^{m+1}$, $\mid t \mid \leq T$, E_1, E_2 are Banach spaces.

Let us consider the following operator

$$L(D) = \sum_{|\alpha| \leq l} B_\alpha D^\alpha. \tag{1}$$

The operator $\sum_{|\alpha|=l} B_\alpha D^\alpha$ we call the main part of $L(D)$. Due to its theoretical significance and numerous applications (see [1], [2], [8], [9]) the most interesting case is when the operator $L(D)$ contains a Fredholm operator $B_{l0\dots0} \equiv B$ in the higher derivative D_t^l.

If $N(B) \neq \{0\}$, then the operator $L(D)$ is called a singular differential operator. A singular operator of the first order

$$B\frac{d}{dt} + A \tag{2}$$

with the Fredholm operator B when $x = t \in R^1$, has been investigated intensively. The significant results of these investigations were obtained due to new

* Supported in part by INTAS (grant No.2000-15).

P.M.A. Sloot et al. (Eds.): ICCS 2003, LNCS 2658, pp. 523–532, 2003.

research approaches in the semigroup theory (the bibliography and mechanical models are represented in [9]). The investigations similar to our research, considering the case when $x \in R^{m+1}$ are represented in [7], [8]. Below we use some results from [3], [4], [7].

We consider the equation

$$L(D)u = f(x), \tag{3}$$

where $f : \Omega \to E_2$ is an analytical function of x' sufficiently smooth of t. The Cauchy problem for (3), when $E_1 = E_2 = R^n$ and the matrix $B = B_{l0\ldots0}$ is not degenerated, has been thoroughly studied in fundamental papers by I.G. Petrovsky (see [5]). In the case when the operator B is not invertible the theory of initial and boundary value problems for (3) is not developed even for the case of finite dimensions. In general, the standard Cauchy problem with conditions $D_t^i u|_{t=0} = g_i(x'), \quad i = 0, \ldots, l-1$, for (3) has no classical solutions for an arbitrary right part $f(x)$.

The motive of our investigations is the wish to conceive the statement of initial and boundary value problems for the systems of partial differential equations with the Fredholm operator in the main part and also their applications. In this paper we show that we can get a reasonable statement of the initial problems for such systems by decomposing the space E_1 on the direct sum of subspaces in accordance with the Jordan structure ([8], [10]) of the operator coefficients B_α, and imposing conditions on projections of the solution. Here we suppose that B is a closed Fredholm operator, $D(B) \subseteq D(B_\alpha)$ $\forall\alpha$, and among the coefficients B_α there is an operator $A = B_{l_1 0\ldots0}$, $l_1 < l$, with respect to which B has the complete A-Jordan set [10].

In Section 1 the sufficient conditions of existence of the unique solution of equation (3) with the initial conditions

$$D_t^i u|_{t=0} = g_i(x'), \quad i = 0, 1, \ldots, l_1 - 1, \tag{4}$$

$$(I - P) D_t^i u|_{t=0} = g_i(x'), \quad i = l_1, \ldots, l-1, \tag{5}$$

are obtained, where $g_i(x')$ are analytical functions with values in E_1, $Pg_i(x') = 0, \quad i = l_1, \ldots, l-1$. Here P is the projector of E_1 onto corresponding A-root subspace. In Section 2 the left and right regularizators of singular operators in Banach spaces have been constructed. A method of fundamental operators for construction of the solution in the class of Schwarz distributions [6] is considered in Section 3. We hope that these investigations can be useful for considering of new applications [8], [9] of singular differential systems.

2 Selection of Projection Operators and Reduction of the Initial Problem to the Kovalevskaya Form

Suppose the following condition is satisfied:

Condition 1 *[10]. The Fredholm operator B has a complete A-Jordan set $\varphi_i^{(j)}$, B^* has a complete A^*-Jordan set $\psi_i^{(j)}$, $i = \overline{1,n}$, $j = \overline{1,p_i}$, and the systems*

$\gamma_i^{(j)} \equiv A^*\psi_i^{(p_i+1-j)}, z_i^{(j)} \equiv A\varphi_i^{(p_i+1-j)}, i = \overline{1,n}, j = \overline{1,p_i}$, *corresponding to them, are biorthogonal. Here* p_i *are the lengths of the Jordan chains of the operator* B.

Recall that condition 1 is satisfied if the operator $B + \lambda A$ is continuously invertible when $0 <| \lambda |< \epsilon$ [10].

We introduce the projectors

$$P = \sum_{i=1}^{n}\sum_{j=1}^{p_i} < .,\gamma_i^{(j)} > \varphi_i^{(j)} \equiv (< .,\Upsilon > \Phi),$$

$$Q = \sum_{i=1}^{n}\sum_{j=1}^{p_i} < .,\psi_i^{(j)} > z_i^{(j)} \equiv (< .,\Psi > Z),$$

generating the direct decompositions

$$E_1 = E_{1k} \oplus E_{1\infty-k}, E_2 = E_{2k} \oplus E_{2\infty-k},$$

where $k = p_1 + \cdots + p_n$ is a root number. Then any solution of equation (3) can be represented in the form

$$u(x) = \Gamma v(x) + (C(x), \Phi), \tag{6}$$

where $\Gamma = (B + \sum_{i=1}^{n} < ., \gamma_i^{(1)} > z_i^{(1)})^{-1}$ is a bounded operator [10],

$$v \in E_{2\infty-k}, \quad C(x) = (C_{11}(x), \dots, C_{1p_1}(x), \dots, C_{n1}(x), \dots, C_{np_n}(x))^T,$$

$$\Phi = (\varphi_1^{(1)}, \dots, \varphi_1^{(p_1)}, \dots, \varphi_n^{(1)}, \dots, \varphi_n^{(p_n)})^T,$$

where T denotes transposition. The unknown functions $v(x) : \Omega \subset R^{m+1} \to E_{2\infty-k}$ and $C(x) : \Omega \subset R^{m+1} \to R^k$ due to initial conditions (4), (5), satisfy the following conditions

$$D_t^i v|_{t=0} = \begin{cases} B(I-P)g_i(x'), \ i = 0, \dots, l_1 - 1, \\ Bg_i(x'), \ i = l_1, \dots, l-1, \end{cases} \tag{7}$$

$$D_t^i C|_{t=0} = \beta_i(x'), \ i = 0, \dots, l_1 - 1. \tag{8}$$

Here $\beta_i(x')$ are coefficients of projections $Pg_i(x')$, $i = 0, \dots, l_1 - 1$.

Condition 2 *The operator coefficients* B_α *in (3) satisfy at least one of five conditions on* $D(B_\alpha)$*:*

1. $B_\alpha P = QB_\alpha$, *i.e.* B_α (P,Q)*-commute, briefly,* $\alpha \in q_0$;
2. $B_\alpha P = 0$, *briefly* $\alpha \in q_1$;
3. $QB_\alpha = 0$, *briefly* $\alpha \in q_2$;
4. $(I-Q)B_\alpha = 0$, *briefly* $\alpha \in q_3$;
5. $B_\alpha(I-P) = 0$, *briefly* $\alpha \in q_4$;

We introduce the scalar product $(\Phi, C) = \sum_{i=1}^{n} \sum_{j=1}^{p_i} \varphi_i^{(j)} C_{ij}$. Then

$$< B_\alpha(\Phi, C), \Psi >= \mathcal{A}^T{}_\alpha C,$$

where $\Psi = (\psi_1^{(1)}, \ldots, \psi_1^{(p_1)}, \ldots, \psi_n^{(1)}, \ldots, \psi_n^{(p_n)})^T$. Due to condition 1 and lemma 3 [7] $\alpha \in q_0$, if and only if

$$B_\alpha^* \Psi = \mathcal{A}^T{}_\alpha \Upsilon, \ B_\alpha \Phi = \mathcal{A}_\alpha Z.$$

The operators $B \equiv B_{l0\ldots0}$, $A \equiv B_{l_1 0\ldots 0}$ belong to the set q_0. Moreover the matrices of (P, Q)-commutability are symmetrical cell-diagonal matrices:

$$\mathcal{A}_{l0\cdots0} = diag(B_1, \ldots, B_n), \ \mathcal{A}_{l_1 0\cdots 0} = diag(A_1, \ldots, A_n), \tag{9}$$

where

$$B_i = \begin{bmatrix} 0 & 0 & \ldots & 0 \\ 0 & 0 & \ldots & 1 \\ \ldots & \ldots & \ldots & \ldots \\ 0 & 1 & \ldots & 0 \end{bmatrix}, \quad A_i = \begin{bmatrix} 0 & \ldots & 1 \\ \ldots & \ldots & \ldots \\ 1 & \ldots & 0 \end{bmatrix}, \quad i = \overline{1, n},$$

if $p_i \geq 2$ and $B_i = 0, A_i = 1$ if $p_1 = 1$.

$$\mathcal{A}_{l0\cdots0} = 0, \ \mathcal{A}_{l_1 0\cdots 0} = E, \tag{10}$$

if $k = n$.

Note that due to the structure of projectors P, Q the identity $\Gamma Q = P\Gamma$ holds. The spaces E_{2k}, $E_{2\infty-k}$ are invariant subspaces of the operator Γ. Taking into account that operator Γ is a bounded one, $D(B) \subseteq D(B_\alpha)$ and $\overline{D(B_\alpha)} = E_1$, we obtain $B_\alpha \Gamma \in L(E_1 \to E_2)$.

Thus by substituting (6) into (3) and projecting the result onto the subspace $E_{2\infty-k}$, we obtain the equation

$$D_t^l v + (I - Q) \sum{}' B_\alpha \Gamma D^\alpha v = (I - Q)(f - \sum{}'' B_\alpha(D^\alpha C, \Phi)), \tag{11}$$

where

$$\sum{}' = \sum_{|\alpha| \leq l, \ \alpha \in (q_0, q_1, q_2) \setminus (l0\ldots0)}, \quad \sum{}'' = \sum_{|\alpha| \leq l, \alpha \in (q_2, q_4)}$$

with condition (7). Similarly, we project (3), where u is defined by (6), onto the subspace E_{2k} and obtain the system

$$\sum_{|\alpha| \leq l, \alpha \in (q_0, q_3, q_4)} M_\alpha D^\alpha C = b(x, v) \tag{12}$$

with initial condition (8). Thus the initial problem (3), (4), (5) is reduced to problems (11), (7) and (12), (8).

In system (12)

$$M_\alpha = \| < B_\alpha \varphi_l^{(s)}, \psi_i^{(j)} > \|, \quad i,l = 1,\dots n, \; j = 1,\dots,p_i, \quad s = 1,\dots,p_l,$$

are matrices of the dimension $k \times k$, $b(x,v)$ is the vector of projection coefficients

$$Q(f - \sum_{|\alpha| \le l, \alpha \in (q_1,q_3)} B_\alpha \varGamma D^\alpha v).$$

Recall that if $\alpha \in q_0$, then $M_\alpha = \mathcal{A}_\alpha^T$. Thus, for $k = n$ according to (10) it follows that

$$M_{l0\dots0} = 0, \; M_{l_1 0\dots 0} = E,$$

and for $k > n$ the matrices $M_{l0\dots0}$, $M_{l_1 0\dots0}$ are defined from (9).

Theorem 1. *Suppose conditions 1 and 2 are satisfied, the function $f(x)$ is an analytical on x' and sufficiently smooth on t. Suppose*

1. $(q_2, q_4) \subset q_0$ *or* $(q_1, q_3) \subset q_0$;
2. $QB_\alpha P = 0$ *for all* $\alpha \in (q_0, q_3, q_4) \setminus (l0\dots0), (l_1 0\dots 0)$.

Then the problem (3), (4), (5) has a unique classical solution (6).

Proof. Note that for $\alpha \in q_0$, and for any C the equality $(I-Q)B_\alpha(D^\alpha C, \varPhi) = 0$ holds and $QB_\alpha \varGamma v = 0$, where $Qv = 0$. Thus, according to condition 1 of this theorem the right-hand side of (11) is independent on the vector-function $C(x)$, or the right-hand side of (12) is independent on v(x). The equation (11) is solvable with respect to $D_t^l v$, i.e. has the Kovalevskaya form with the bounded operator coefficients. Due to condition 2 of this theorem the system (12) takes the following form

$$M_{l0\dots0} D_t^l C + M_{l_1 0\dots0} D_t^{l_1} C = b(x,v). \tag{13}$$

If $k = n$, then $M_{l0\dots0} = 0$, $M_{l_1 0\dots0} = E$ and system (13) has the order l_1. If $k > n$ then system (13) is split on n independent subsystems:

$$\frac{\partial^{l_1}}{\partial t^{l_1}} C_{ip_i} = b_{ip_i}(x,v), \; \frac{\partial^{l_1}}{\partial t^{l_1}} C_{ip_i - k} + \frac{\partial^l}{\partial t^l} C_{ip_i-k+1} = b_{ip_i-k}(x,v), \tag{14}$$

where $i = \overline{1,n}$, $k = \overline{1, p_i - 1}$. Each subsystem (14) is regular, since it is a recurrent sequence of differential equations of order l_1. Thus systems (11), (12) with boundary conditions (7), (8) have the Kovalevskaya form and therefore have the unique solution. Taking v and C from the regular systems (11), (12) and substituting them into (6), we obtain the needed solution. □

Remark 1. Let the operators B_α in condition 2 depend on x for $\alpha \ne (l0\dots0)$, $(l_1 0\dots0)$. Then the coefficients in the systems (11), (12) also depend on x. If these coefficients are analytical on x' and sufficiently smooth on t, then theorem 1 is valid. Required smoothness on t for these coefficients and $f(x)$ is defined by maximum length of A-Jordan chains of operator B.

3 The Left and Right Regularizators of Singular Operators in Banach Spaces

Let A and B be constant linear operators from E_1 to E_2, where E_1 and E_2 are Banach spaces, $x(t)$ is an abstract function, $t \in R_n$ with the values in $E_1(E_2)$. The set of such functions we denote by $X_t(Y_t)$. We introduce the operator L_t, defined on X_t and Y_t and which is commutable with operators B, A. The examples of such operator L_t are differential and integral operators, difference operators and their combinations. Note that if operators are solved according to higher order derivatives, then they usually generate correct initial and boundary value problems. In other cases, when operators are unsolved according to higher order derivatives, we get the singular problems (see sec.1).

Let us consider the operator $L_tB - A$, which acts from X_t to Y_t, where B, A are closed linear operators from E_1 to E_2 with the dense domains, and $D(B) \subseteq D(A)$. If B is invertable, then the operator $L_tB - A$ can be reduced to regular operator by multiplication on B^{-1}. If B is uninvertable, then $L_tB - A$ is called the singular operator. Let operator B in $L_tB - A$ be Fredholm and $\dim N(B) = n \geq 1$. If $\lambda = 0$ is an isolated singular point of the operator-function $B - \lambda A$, then the operators $L_tB - A, BL_t - A$ admit some regularization. For explicit construction of regularizations we use the Schmidt pseudo resolvent $\Gamma = \hat{B}^{-1}$, where $\hat{B} = B + \sum_{i=1}^{n} <\cdot, A^*\psi_i^{(p_i)}> A\phi_i^{(p_i)}$. On the base of condition 1(sec. 1) and using the equalities $\phi_i^{(j)} = \Gamma A\phi_i^{(j-1)}$, $\psi_i^{(j)} = \Gamma^* A^* \psi_i^{(j-1)}$, $j = 2, \dots, p_i$, $i = 1, \dots, n$ it is easy to check the following equalities

$$(\Gamma - \sum_{i=1}^{n}\sum_{j=1}^{p_i} L_t^j <\cdot, \psi_i^{(p_i+1-j)} > \phi_i)(L_tB - A) = L_t - \Gamma A,$$

$$(L_tB - A)(\Gamma - \sum_{i=1}^{n}\sum_{j=1}^{p_i} L_t^{p_i+1-j} <\cdot, \psi_i > \phi_i^{(j)}) = L_t - A\Gamma.$$

Thus we have the following

Theorem 2. *Suppose condition 1 in section 1 is satisfied . Then*

$$(\Gamma - \sum_{i=1}^{n}\sum_{j=1}^{p_i} L_t^j <\cdot, \psi_i^{(p_i+1-j)} > \phi_i)$$

and

$$\Gamma - \sum_{i=1}^{n}\sum_{j=1}^{p_i} L_t^{p_i+1-j} <\cdot, \psi_i > \phi_i^{(j)}$$

are the left and right regularizators of $L_tB - A$, respectively.

Note that these results are applicable for the investigation of singular differential-operator equations with the Fredholm operator in the main part (see [1]).

4 Fundamental Operator-Functions of Singular Partial Differential and Differential-Difference Operators in Banach Spaces

Since the standard Cauchy problem for equation (3) with the Fredholm operator $B_{l0\ldots0}$ in general has no classical solution, then it will be interesting to extend the notion of solution and to seek generalized solution in a distribution space [6].

The most interesting is the construction of the fundamental operator functions for the singular differential operators in Banach spaces which help to obtain the generalized solutions in closed forms.

Here we construct the fundamental operator functions for the following mappings

$$B\frac{\partial^{2N}u}{\partial x^N \partial y^N} - Au, \quad B\frac{\partial u}{\partial t} - A(u(t,x-\mu) - u(t,x)),$$

where B is a Fredholm operator.

The basic information on generalized functions in Banach spaces, their properties and operations can be found in [3], [8].

Theorem 3. *Suppose that condition 1 is satisfied. Then the mapping* $B\delta'(x)\delta'(y) - A\delta(x)\delta(y)$ *in the space* $K'(E_2)$ *has the fundamental operator function of the form*

$$\mathcal{E}_1(x,y) = \Gamma\mathcal{U}_1(A\Gamma)(x,y)[I-Q]\theta(x,y)$$
$$-\sum_{i=1}^{n}\sum_{k=0}^{p_i-1}\left\{\sum_{j=1}^{p_i-k}\langle\cdot,\psi_i^{(j)}\rangle\varphi_i^{(p_i-k+1-j)}\right\}\delta^{(k)}(x)\cdot\delta^{(k)}(y),$$

where

$$\mathcal{U}_1(A\Gamma)(x,y) = \sum_{i=0}^{\infty}(A\Gamma)^i\cdot\frac{x^i}{i!}\cdot\frac{y^i}{i!}.$$

Proof. In accordance with the definition it is necessary to check up a validity of equality

$$(B\delta'(x)\cdot\delta'(y) - A\delta(x)\cdot\delta(y)) * \mathcal{E}_1(x,y) * u(x,y) = u(x,y)$$

on the basic space $K(E_2^*)$. Let us substitute the expression for $\mathcal{E}_1(x,y)$ into the left-hand side of this equality

$$(B\delta'(x)\cdot\delta'(y) - A\delta(x)\cdot\delta(y)) * \mathcal{E}_1(x,y) * u(x,y)$$
$$= \Bigg(B\Gamma A\Gamma\mathcal{U}_1(A\Gamma)(x,y)[I-Q]\theta(x,y) + B\Gamma[I-Q]\delta(x)\cdot\delta(y)$$
$$-\sum_{i=1}^{n}\sum_{k=1}^{p_i-1}\left\{\sum_{j=1}^{p_i-k}\langle\cdot,\psi_i^{(j)}\rangle\left(B\varphi_i^{(p_i-k+2-j)} - A\varphi_i^{(p_i-k+1-j)}\right)\right\}\delta^{(k)}(x)\cdot\delta^{(k)}(y)$$

$$-A\Gamma\mathcal{U}_1(A\Gamma)(x,y)[I-Q]\theta(x,y)+Q\delta(x)\cdot\delta(y)\Bigg)*u(x,y).$$

Since $B\Gamma = I - \sum_{i=1}^{n}\langle\cdot,\psi_i^{(1)}\rangle z_i$, $z_i = A\varphi_i^{(p_i)}, B\varphi_i^{(j)} = A\varphi_i^{(j-1)}$, then

$$\sum_{i=1}^{n}\langle\cdot,\psi_i^{(1)}\rangle z_i[I-Q]=0,\ \sum_{i=1}^{n}\langle\cdot,\psi_i^{(1)}\rangle z_i A\Gamma\mathcal{U}_1(A\Gamma)(x,y)[I-Q]=0$$

and

$$(B\delta'(x)\cdot\delta'(y)-A\delta(x)\cdot\delta(y))*\mathcal{E}_1(x,y)*u(x,y)=I\delta(x)\cdot\delta(y)*u(x,y)=u(x,y).$$

□

The following theorem can be proved similarly.

Theorem 4. *Suppose condition 1 is satisfied, then the mapping* $B\delta^{(N)}(x)\cdot\delta^{(N)}(y)-A\delta(x)\cdot\delta(y))$ *in the space* $K'(E_2)$ *has the fundamental operator function of the form*

$$\mathcal{E}_N(x,y)=\Gamma\mathcal{U}_N(A\Gamma)(x,y)[I-Q]\theta(x,y)$$

$$-\sum_{i=1}^{n}\sum_{k=0}^{p_i-1}\left\{\sum_{j=1}^{p_i-k}\langle\cdot,\psi_i^{(j)}\rangle\varphi_i^{(p_i-k+1-j)}\right\}\delta^{(k\cdot N)}(x)\cdot\delta^{(k\cdot N)}(y),$$

where

$$\mathcal{U}_N(A\Gamma)(x,y)=\sum_{i=1}^{\infty}(A\Gamma)^{i-1}\cdot\frac{x^{i\cdot N-1}}{(i\cdot N-1)!}\cdot\frac{y^{i\cdot N-1}}{(i\cdot N-1)!}.$$

As a corollary of theorem 3 we obtain

Corollary 1. *Suppose condition 1 is satisfied, the function* $f(x,y)\in C(R_+^2)$ *takes value in* E_2*. Then the boundary value problem*

$$B\frac{\partial^2 u}{\partial x\partial y}=Au+f(x,y),\quad u|_{x=0}=\alpha(y),\ u|_{y=0}=\beta(x),$$

$u(x,y)\in C^2(R_+^2)$, $\alpha(x),\beta(x)\in C^1(R_+^1)$, $\alpha(0)=\beta(0)$, *has a generalized solution of the form*

$$u=\mathcal{E}_1(x,y)*(f(x,y)\theta(x,y)+B\alpha'(y)\delta(x)\cdot\theta(y)$$

$$+B\beta'(x)\theta(x)\cdot\delta(y)+B\alpha(0)\delta(x)\cdot\delta(y)).$$

If additionally the singular components of the generalized solutions are equal to zero then, firstly, generalized solutions coincide with continuous (classical) solutions, and, secondly, we can define a set of the boundary values $\alpha(y)$ and $\beta(x)$ and right sides $f(x,y)$, for which such problems are solvable in the class of functions $C^2(R_+^2)$.

Remark 2. The following boundary value problem can be investigated similarly

$$B\frac{\partial^{2N} u}{\partial x^N \partial y^N} = Au + f(x,y), \frac{\partial^i u}{\partial x^i}|_{x=0} = \alpha_i(y), \frac{\partial^i u}{\partial y^i} u|_{y=0} = \beta_i(x), i = 0, \ldots, N-1.$$

Theorem 5. *Suppose that condition 1 is satisfied with* $p_1 = p_2 = \ldots p_n = 1$. *Then the mapping* $B\delta'(t) \cdot \delta(x) - A\delta(t) \cdot (\delta(x-\mu) - \delta(x))$ *in the space* $K'(E_2)$ *has the fundamental operator function of the form*

$$\mathcal{E}(t,x) = \sum_{k=0}^{\infty} \Gamma e^{-A\Gamma t} \frac{(A\Gamma t)^k}{k!} \theta(t) \cdot \delta(x-\mu) * \left\{ I\delta(t) \cdot \delta(x) + Q\delta'(t) \cdot \sum_{j=0}^{\infty} \delta(x - j\mu) \right\}.$$

Proof. In accordance with the definition it is necessary to check up a validity of equality

$$(B\delta'(t) \cdot \delta(x) - A\delta(t) \cdot (\delta(x-\mu) - \delta(x))) * \mathcal{E}(t,x) * u(t,x) = u(t,x)$$

on the basic space $K(E_2^*)$. Let us substitute the expression for $\mathcal{E}(t,x)$ into the left-hand side of this equality

$$(B\delta'(t) \cdot \delta(x) - A\delta(t) \cdot (\delta(x-\mu) - \delta(x))) * \mathcal{E}(t,x) * u(t,x)$$
$$= \Big[I\delta(t) \cdot \delta(x) + F(t,x) \Big] * u(t,x),$$

where

$$F(t,x) = -\sum_{k=0}^{\infty} Q\left(e^{-t}\frac{t^k}{k!}\theta(t)\right)' \cdot \delta(x-k\mu) + \Big[I\delta(t) \cdot \delta(x)$$
$$- \sum_{k=0}^{\infty} Q\left(e^{-t}\frac{t^k}{k!}\theta(t)\right)' \cdot \delta(x-k\mu)) \Big] * Q\delta'(t) \cdot \sum_{j=0}^{\infty} \delta(x-j\mu) = 0.$$

□

Remark 3. In theorem 4 x can be vector, moreover theorem 5 keeps its validity if $p_i \geq 1,\ i = 1, \ldots, n$, and differential operator can be changed on differential-difference operator

$$B\delta^{(N)}(t) \cdot \delta(x) - A\delta(t) \cdot (\delta(x-\mu) - \delta(x)).$$

Corollary 2. *Suppose condition of theorem 5 is satisfied, the function* $f(t,x) \in BUC(R^1)$ *[4],* $\forall t \geq 0$, *takes value in* E_2. *Then the Cauchy problem for differential-difference equation*

$$B\frac{\partial u}{\partial t} = A(u(t,x-\mu) - u(t,x)) + f(t,x), \quad u|_{t=0} = u_0(x),$$

where $u_0(x) \in BUC(R^1)$, *has a generalized solution of the form*

$$u = \mathcal{E}(t,x) * (f(t,x)\theta(t) + Bu_0(x)\delta(t)).$$

References

1. Chistyakov, V.F.: Algebro-differential equations with finite-dimensional kernel. Nauka, Novosibirsk (1996)
2. Korpusov, M.O., Pletnev,Y.D., Sveshnikov, A.G.: On quasi-steddy process in the conducting medium without dispersion. Comput. Math. and Math. Phys. Vol.40. **8** (2000) 1237–1249
3. Falaleev, M.V.: Fundamental operator functions of singular differential operators in Banach spaces. Sib. Math. J. **41** (2000) 960–973
4. Goldstein, J.A.: Semigroups of Linear Operators and Applications. Oxford University Press, Inc., New York (1985)
5. Petrovsky, I.: Uber das Caushysche problem fur system von partiellen Differentialgleichunden. Math. Sbornic Vol.2. **5** (1937) 815–870
6. Schwartz, L.: Theorie des distributions. I-II Paris (1950–1951)
7. Sidorov, N.A., Blagodatskaya, E.B.: Differential Equations with the Fredholm Operator in the Leading Differential Expression. Soviet Math. Dokl. Vol.44. **1** (1992) 302–305
8. Sidorov N., Loginov B., Sinitsyn A., Falaleev M.: Lyapunov-Schmidt Methods in Nonlinear Analysis and Applications. Book Series: Mathematics and Its Applications: **550** Kluwer Academic Publishers (2002)
9. Sviridyuk, G.A.: On the General Theory of Semigroup of Operators. Uspekhi Mat. Nauk. **49** (1994) 47–74
10. Vainberg, M.M., Trenogin, V.A.: The Theory of Branching of Solutions of Nonlinear Equations. Wolters-Noordhoff, Groningen (1974)

Invariant Manifolds and Grobman-Hartman Theorem for Equations with Degenerate Operator at the Derivative*

Bülent Karasözen[1], Irina Konopleva, and Boris Loginov[2]

[1] Ankara Middle-East Technical University, 06531 Ankara, Turkey,
bulent@metu.edu.tr

[2] Ulyanovsk State Technical University, 432027 Ulyanovsk, Russia,
i.konopleva@ulstu.ru, loginov@ulstu.ru

Abstract. Analog of Grobman-Hartman theorem about stable and unstable manifolds solutions for differential equations in Banach spaces with degenerate Fredholm operator at the derivative are proved. In contrast to usual evolution equation here central manifold arises even in the case of spectrum absence on the imaginary axis. Jordan chains tools and implicit operator theorem are used. The obtained results allow to develop center manifold methods for computation of bifurcation solution asymptotics and their stability investigation.

1 Introduction

Branching theory of solutions of nonlinear equations has various applications in scientific computing [1], [2], [3]. This is one of the areas in applied mathematics intensively developing in last fifty years. The goals of this theory are the qualitative theory of dynamical systems [3], computation of their solutions [4] under absence of conditions guaranteeing the uniqueness of the solutions. The classical Lyapounov-Schmidt method even in contemporary presentation [5] often insufficient for computation of complicated dynamics, like bifurcation to invariant tori. Therefore in the last two decades the center manifold theory [6] and methods are developed. However this theory is completely absent for evolution equations with degenerate operator at the derivative, having numerous applications in filtration theory [7], nonlinear waves theory (the Boussinesq-Love equation) [8] and motion theory of non-Newtonian fluids [9].

The presented work is devoted to invariant manifolds technique and presents the introduction to center manifold methods for evolution equations with Fredholm operator at the derivative. It has found some applications to investigation of the bifurcating solutions stability [10]. The second section of this article contains the necessary tools of generalized Jordan chains, the third, fourth and fifth ones – some aspects of invariant manifolds theory and Grobman-Hartman

* Supported by NATO-TÜBITAK PC program and Russian Foundation for Basic Research (project No: 01-01-0019)

P.M.A. Sloot et al. (Eds.): ICCS 2003, LNCS 2658, pp. 533–541, 2003.

theorem analogs for such equations. Here the nontrivial center manifold arises even at the absence of $\sigma_A(B)$ spectrum on the imaginary axis. It is considered also the simple case of $\sigma_A(B)$ presence on imaginary axis (section 4). For the computation of center manifold in section 3 successive approximation method is suggested.

2 Generalized Jordan Chains Tools

Let E_1 and E_2 be Banach spaces, $A : E_1 \supset D_A \to E_2$, $B : E_1 \supset D_B \to E_2$ be densely defined closed linear Fredholm operators, where $D_B \subset D_A$ and A is subordinated to B (i.e. $\|Ax\| \le \|Bx\| + \|x\|$ on D_B) or $D_A \subset D_B$ and B is subordinated to A (i.e. $\|Bx\| \le \|Ax\| + \|x\|$ on D_A). The differential equation

$$A\frac{dx}{dt} = Bx - R(x), \quad R(0) = 0, \quad R_x(0) = 0 \tag{1}$$

is considered.

It is known [5,10,12] that for the zero-subspaces of the operators A and B $\mathcal{N}(A) = \text{span}\{\phi_1, \dots, \phi_m\}$, $\mathcal{N}(B) = \text{span}\{\varphi_1, \dots, \varphi_n\}$, $\mathcal{N}(A) \cap \mathcal{N}(B) = \{0\}$ and defect-subspaces $\mathcal{N}^*(A) = \text{span}\{\hat{\psi}_1, \dots, \hat{\psi}_m\}$, $\mathcal{N}^*(B) = \text{span}\{\psi_1, \dots, \psi_n\}$, the biorthogonal systems $\{\vartheta_j\}_1^m$, $< \phi_i, \vartheta_j >= \delta_{ij}; \{\zeta_j\}_1^m$, $< \zeta_i, \hat{\psi}_j >= \delta_{ij}$ and $\{\gamma_j\}_1^n, < \varphi_i, \gamma_j >= \delta_{ij}; \{z_j\}_1^n$, $< z_i, \psi_j >= \delta_{ij}$ can be chosen so that the following biorthogonality conditions for the corresponding Jordan chains ($\{\phi_i^{(s)}\}, s = 1, \dots, q_i, \phi_i^{(1)} = \phi_i$, $A\phi_i^{(s)} = B\phi_i^{(s-1)}$, $< \phi_i^{(s)}, \vartheta_j >= 0, s = 2, \dots, q_i, i, j = 1, \dots, m, D_q \equiv \det[< B\phi_i^{(q_i)}, \hat{\psi}_j >] \ne 0; \{\varphi_i^{(s)}\}, s = 1, \dots, p_i, \varphi_i^{(1)} = \varphi_i, \varphi_i^{(s)} = A\varphi_i^{(s-1)}, < \varphi_i^{(s)}, \gamma_j >= 0, s = 2, \cdots, p_i,\ i, j = 1, \dots, n$, $D_p \equiv \det\left[< A\varphi_i^{(p_i)}, \psi_j >\right] \ne 0$; for adjoint operator-functions $A^* - \lambda B^*$ and $B^* - \mu A^*$ Jordan chains $\{\hat{\psi}_j^{(s)}\}, s = 1, \dots, q_i, i = 1, \dots, m$, and $\{\psi_j^{(s)}\}$, $s = 1, \dots, p_j$, $j = 1, \dots, n$ are defined analogously) would be satisfied:

$$\begin{gathered} < \phi_i^{(j)}, \vartheta_k^{(\ell)} >= \delta_{ik}\delta_{j\ell},\ < \zeta_i^{(j)}, \hat{\psi}_k^{(\ell)} >= \delta_{ik}\delta_{j\ell},\ j(\ell) = 1, \dots, q_i(q_k), \\ \vartheta_k^{(\ell)} = B^*\hat{\psi}_k^{(q_k+1-\ell)},\ \zeta_i^{(j)} = B\phi_i^{(q_i+1-j)},\ i, k = 1, \dots, m \end{gathered} \tag{2}$$

$$\begin{gathered} < \varphi_i^{(j)}, \gamma_k^{(\ell)} >= \delta_{ik}\delta_{j\ell},\ < z_i^{(j)}, \psi_k^{(\ell)} >= \delta_{ik}\delta_{j\ell},\ j(\ell) = 1, \dots, p_i(p_k) \\ \gamma_k^{(\ell)} = A^*\psi_k^{(p_k+1-\ell)},\ z_i^{(j)} = A\varphi_i^{(p_i+1-j)},\ i, k = 1, \dots, n \end{gathered} \tag{3}$$

The relations (2), (3) allow to introduce [5,12] the projectors on the root-subspaces $K(A; B) = \text{span}\ \{\phi_i^{(s)}\}$ ($k_A = \sum_{i=1}^{m} q_i = dim\ K(A; B)$ is the root-number for $A - \lambda B$) and $K(B; A) = \text{span}\ \{\varphi_i^{(s)}\}$ ($k_B = \sum_{i=1}^{n} p_i$–the root-number for B-μA):

$$
\begin{aligned}
\mathbf{p} &= \sum_{i=1}^{m}\sum_{j=1}^{q_i} <\cdot,\vartheta_i^{(j)}> \phi_i^{(j)} = <\cdot,\vartheta>\phi : E_1 \to E_1^{k_A} = K(A;B),\\
\mathbf{q} &= \sum_{i=1}^{m}\sum_{j=1}^{q_i} <\cdot,\hat{\psi}_i^{(j)}> \zeta_i^{(j)} = <\cdot,\hat{\psi}>\zeta : E_2 \to E_{2,k_A} = \text{span}\,\{\zeta_i^{(j)}\},\\
\mathbf{P} &= \sum_{i=1}^{n}\sum_{j=1}^{p_i} <\cdot,\gamma_i^{(j)}> \varphi_i^{(j)} = <\cdot,\gamma>\varphi : E_1 \to E_1^{k_B} = K(B;A),\\
\mathbf{Q} &= \sum_{i=1}^{n}\sum_{j=1}^{p_i} <\cdot,\psi_i^{(j)}> z_i^{(j)} = <\cdot,\psi> z : E_2 \to E_{2,k_B} = \text{span}\,\{z_i^{(j)}\}
\end{aligned}
\tag{4}
$$

(where $\phi = (\phi_1^{(1)},\cdots,\phi_1^{(q_1)},\cdots,\phi_m^{(1)},\cdots,\phi_m^{(q_m)})$, the vectors $\vartheta,\hat{\psi},\zeta$ and φ,γ,ψ,z are defined analogously), generating the following direct sums expansions

$$
\begin{aligned}
E_1 &= E_1^{k_A}\dot{+}E_1^{\infty-k_A},\, E_2 = E_{2,k_A}\dot{+}E_{2,\infty-k_A};\\
E_1 &= E_1^{k_B}\dot{+}E_1^{\infty-k_B},\, E_2 = E_{2,k_B}\dot{+}E_{2,\infty-k_B}.
\end{aligned}
\tag{5}
$$

The $\mathbf{p}$, $\mathbf{q}$- and $\mathbf{P}$, $\mathbf{Q}$-intertwining relations are realized

$$
\begin{aligned}
&A\mathbf{p} = \mathbf{q}A \ \text{ on } \ D_A,\ B\mathbf{p} = \mathbf{q}B \text{ on} D_B;\ B\mathbf{P} = \mathbf{Q}B \text{ on } D_B,\ A\mathbf{P} = \mathbf{Q}A \text{ on } D_A,\\
&A\phi = \mathfrak{A}_A\zeta,\ B\phi = \mathfrak{A}_B\zeta,\ B^*\hat{\psi} = \mathfrak{A}_B\vartheta;\ B\varphi = \mathcal{A}_B z,\ A\varphi = \mathcal{A}_A z,\ A^*\psi = \mathcal{A}_A\gamma,
\end{aligned}
\tag{6}
$$

with cell-diagonal matrices $\mathfrak{A}_A = (A_1,\ldots,A_m)$, $\mathfrak{A}_B = (B_1,\ldots,B_m)$; $\mathcal{A}_B = (B^1,\ldots,B^n), \mathcal{A}_A = (A^1,\ldots,A^n)$, where $q_i \times q_i$-cells ($p_i \times p_i$-cells) have the forms

$$
A_i = \begin{pmatrix} 0\,0\,0\ldots 0\,0\\ 0\,0\,0\ldots 0\,1\\ \vdots\ \vdots\ \vdots\ \ddots\ \vdots\ \vdots\\ 0\,0\,1\ldots 0\,0\\ 0\,1\,0\ldots 0\,0 \end{pmatrix}, \qquad
B_i = \begin{pmatrix} 0\,0\,0\ldots 0\,1\\ 0\,0\,0\ldots 1\,0\\ \vdots\ \vdots\ \vdots\ \ddots\ \vdots\ \vdots\\ 0\,1\,0\ldots 0\,0\\ 1\,0\,0\ldots 0\,0 \end{pmatrix}
$$

(B^i is of the type A_i and A^i has the form of B_i). Here

$$
\begin{aligned}
\mathcal{N}(A) \subset E_1^{k_A}, AE_1^{k_A} \subset E_{2,k_A},\ A(E_1^{\infty-k_A}\cap D_A) \subset E_{2,\infty-k_A},\\
\mathcal{N}(B) \subset E_1^{\infty-k_A}, BE_1^{k_A} \subset E_{2,k_A},\ B(E_1^{\infty-k_A}\cap D_B) \subset E_{2,\infty-k_A}
\end{aligned}
\tag{7}
$$

$\overset{\sqcap}{A}= A|_{E_1^{\infty-k_A}\cap D_A}, \overset{\sqcap}{B}= B|_{E_1^{\infty-k_A}\cap D_B}$ and the mappings $B : E_1^{k_A} \to E_{2,k_A}, \overset{\sqcap}{A}$: $E_1^{\infty-k_A}\cap D_A \to E_{2,\infty-k_A}$ are one-to-one. Analogously, the operators B and A act in invariant pairs of the subspaces $E_1^{k_B}$, E_{2,k_B} and $E_1^{\infty-k_B}$, $E_{2,\infty-k_B}$ and also $\overset{\sqcup}{B}= B|_{E_1^{\infty-k_B}\cap D_B} : E_1^{\infty-k_B}\cap D_B \to E_{2,\infty-k_B}$, $A : E_1^{k_B} \to E_{2,k_B}$ are isomorphisms.

3 Analogs of Grobman-Hartman Theorem at $\boldsymbol{\sigma_A^0(B) = \emptyset}$

It is supposed that for the A-spectrum $\sigma_A(B)$ of the operator B $\operatorname{Re}\sigma_A(B) \neq 0$ and the spectral sets $\sigma_A^-(B) = \{\mu \in \sigma_A(B)|\ Re\mu < 0\}$ and $\sigma_A^+(B) = \{\mu \in \sigma_A(B)|\ \operatorname{Re}\mu > 0\}$ be distant from the imaginary axis on some distance $d > 0$.

All solutions of the corresponding to (1) linear Cauchy problem

$$A\frac{dx}{dt} = Bx, \qquad x(0) = x_0 \tag{8}$$

belong to $E_1^{\infty-k_A}$ and (8) is solvable iff $x_0 \in E_1^{\infty-k_A}$. In fact, one sets $x = v+w$, $v(t) = \sum_{i=1}^{m}\sum_{s=1}^{q_i} \xi_{is}(t)\phi_i^{(s)} \in E_1^{k_A}$, $w(t) \in E_1^{\infty-k_A}$, then (8) is splitting into the system

$$\frac{d\xi_{is}(t)}{dt} = \xi_{i,s-1},\ s = 2,\dots,q_i,\ i = 1,\dots,m,\ \xi_{iq_i} = 0;\ \overset{\sqcap}{A}\frac{dw}{dt} = \overset{\sqcap}{B}w. \tag{9}$$

Consequently $\xi_{is}(t) = 0$, solution of (8) takes the form

$$x(t) = \exp(\overset{\sqcap}{A}{}^{-1}\overset{\sqcap}{B}t)x_0, \quad x_0 \in E_1^{\infty-k_A} \tag{10}$$

and $\sigma_A(B) = \sigma(\overset{\sqcap}{A}{}^{-1}\overset{\sqcap}{B})$. Here the function $\exp(\overset{\sqcap}{A}{}^{-1}\overset{\sqcap}{B}\ t)$ has the form of the contour integral $\frac{1}{2\pi i}\int_\gamma (\mu I - \overset{\sqcap}{A}{}^{-1}\overset{\sqcap}{B})^{-1}e^{\mu t}\,dt$ at the assumption about sectorial property [3] of the operator $\overset{\sqcap}{A}{}^{-1}\overset{\sqcap}{B}$ (or, that is the same, about A-sectorial property of the operator B [13]) with some special contour γ belonging to sector $S_{\alpha,\theta}(B)$ in A-resolvent set of the operator B [13].

The more so, this is true when the operator $\overset{\sqcap}{A}{}^{-1}\overset{\sqcap}{B}$ is bounded.

At the generalization of the Grobman-Hartman theorem we will follow to the work [11]. Let us define the spaces D_k, $k = 1, 2$ with graphs norms:

1^0. $D_1 = D_B \subset D_A$ with the norm $\|x\|_1 = \|x\|_{E_1} + \|Bx\|_{E_2}$, $x \in D_1$, if A is subordinated to B,

2^0. $D_2 = D_A \subset D_B$ with the norm $\|x\|_2 = \|x\|_{E_1} + \|Ax\|_{E_2}$, $x \in D_2$, if B is subordinated to A,

and introduce the spaces $X_{k0}, X_{k1}, X_{k2}, Y_{k0}, Y_{k1}, Y_{k2}$ consisting of the bounded uniformly continuous functions $f(t)$ on $[0,\infty)$ with their values correspondingly in D_k, $D_k \cap E_1^{\infty-k_A}$, $E_1^{k_A}, E_2, E_{2,\infty-k_A}$, E_{2,k_A} with supremum norms on the relevant spaces, and the spaces

$$X_{ks}^1 = \{f(t) \in X_{ks} | \dot f(t) \in X_{ks}\}, \|f(t)\|_{X_{ks}^1} = \max\{\|f(t)\|_{X_{ks}}, \|\dot f(t)\|_{X_{ks}}\}.$$

Everywhere below the operator $\overset{\sqcap}{A}{}^{-1}\overset{\sqcap}{B}$ is supposed to be bounded in X_{k1} (for the case k=1 it is evident). Then the operator

$$\mathbf{A}x = A\dot x - Bx \tag{11}$$

acting from X_{k0}^1 to Y_{k0} is linear and continuous with $X_{k2} \subset \mathcal{N}(\mathbf{A})$.

Let be $D_k \supset S_k$ = {initial values of solutions of the equation (8), which are defined and remain in a small neighborhood of zero in D_k for $t \in [0, +\infty)$}

and $U_k =$ {initial values of solutions of (8), which are defined and remain in a small neighborhood of zero in D_k for $t \in (-\infty, 0]$}. From (11) it follows that $S_k \dot{+} U_k = E_1^{\infty - k_A} \cap D_k$. Then the equality $\sigma_A(B) = \sigma(\overset{\sqcap}{A}^{-1}\overset{\sqcap}{B})$ allows to define the projectors $P^- u = \frac{1}{2\pi i}\int_{\gamma_-} (\mu I_{E_1^{\infty-k_A}} - \overset{\sqcap}{A}^{-1}\overset{\sqcap}{B})^{-1} u d\mu$ (γ_- is the contour in $\rho_A(B)$ containing inside itself the points $\mu \in \sigma_A(B)$ with Re $\mu < 0$), and $P^+ = I_{E_1^{\infty-k_A}} - P^-$. Whence $D_k = D_k^- \dot{+} D_k^0 \dot{+} D_k^+$, $D_k^0 = E_1^{k_A}$, $D_k^\pm = P^\pm D_k$. Operator $\mathbf{A}$ is Noetherian [5] with $R(\mathbf{A}) = Y_{k1}$ and

$$\begin{aligned}\mathcal{N}(\mathbf{A}) &= \{f(t) \in X_{k0}^1 | f(t) = \exp(\overset{\sqcap}{A}^{-1}\overset{\sqcap}{B}\, t)P^- f(0) \in D_k^-\} \dot{+} \{f(t) \in D_k^0\} \\ &= \mathcal{N}_1(\mathbf{A}) \dot{+} \mathcal{N}_2(\mathbf{A}) \quad \text{for} \quad t \geq 0 \\ (\mathcal{N}(\mathbf{A}) &= \{f(t) \in X_{k0}^1 | f(t) = \exp(\overset{\sqcap}{A}^{-1}\overset{\sqcap}{B}\, t)P^+ f(0) \in D_k^+\} \dot{+} \{f(t) \in D_k^0\} \\ &\quad \text{for } t \leq 0).\end{aligned}$$

Now setting $x = y + z + v$, $z \in D_k^+$, $v \in D_k^0 = E_1^{k_A}, y \in D_k^-$ one can write the equation (1) in the form ($w = y + z$ in (9))

$$\mathbf{A}z = R(z + y + v) \quad (\mathbf{A}y = R(y + z + v)) \tag{12}$$

and apply the implicit operator theorem to (12) regarding y, v (z, v) as functional parameters (see the relevant theorems 22.1 and 22.2 in [5] for continuous and analytic operator R respectively). It follows that (12) has a sufficiently smooth or analytic (according to the properties of the operator R) solution in some neighborhoods of parameters y, v (z, v) zero values

$$z = z(y + v), \quad z(0) = 0 = Dz(0) \quad (y = y(z + v),\ y(0) = 0 = Dy(0)) \tag{13}$$

Consequently it is true the following Grobman-Hartman theorem [11] analogue asserting that the local solutions behavior for nonlinear equation in hyperbolic equilibrium neighborhood is the same that for its linearization.

Theorem 1. *There exist a neighborhood $\omega^-(\omega^+)$ of zero in $D_k^0 \dot{+} D_k^-$ (in $D_k^0 \dot{+} D_k^+$) and sufficiently smooth mapping $z_R = z_R(\xi, \eta) = z_R(\xi \cdot \phi + \eta) : \omega^- \to D_k^+$, $\eta \in D_k^-$ ($y_R = y_R(\xi, \zeta) = y_R(\xi \cdot \phi + \zeta) : \omega^+ \to D_k^-$, $\zeta \in D_k^+$), such that a) $z_R(0,0) = 0$, $D_\xi z_R(0,0) = 0$, $D_\eta z_R(0,0) = 0$ ($y_R(0,0) = 0$, $D_\xi y_R(0,0) = 0$, $D_\zeta y_R(0,0) = 0$), b) for any solution $x(t)$ of (1) with initial data $x(0) = \xi \cdot \phi + \eta + z_R(\xi \cdot \phi + \eta)$ ($x(0) = \xi \cdot \phi + y_R(\xi \cdot \phi + \zeta) + \zeta$) one has $z(t) = z_R(\xi(t) \cdot \phi + y(t)) \in D_k^+$ for $t \geq 0$ ($y(t) = y_R(\xi(t) \cdot \phi + z(t)) \in D_k^-$ for $t \leq 0$), c) any solution $x(t)$ of (1) with initial data from b) takes the form $x(t) = \xi(t) \cdot \phi + y(t) + z_R(\xi(t) \cdot \phi + y(t))$ ($x(t) = \xi(t) \cdot \phi + y_R(\xi(t) \cdot \phi + z(t)) + z(t)$) and tends to zero when $t \to +\infty (t \to -\infty)$, and belongs, consequently, to local stable manifold $S_k(R)$ (local unstable manifold $U_k(R)$).*

Proof. We give here the proof for the function z_R and local stable manifold $S_k(R)$, the proof of the second part is analogous. Define the projector $\tilde{P}^-$of

X_{k1}^1 onto $\mathcal{N}_1(\mathbf{A})$ by the equality $(\tilde{P}^- f)(t) = \exp(\overset{\sqcap}{A}^{-1}\overset{\sqcap}{B}\, t)P^- f(0)$, $t \geq 0$. If one sets $x(t) = v(t) + y(t) + z(t)$, $v(t) = \mathbf{p}x(t)$, $v(0) = \xi \cdot \phi = \sum_{i=1}^{m}\sum_{s=1}^{q_i} \xi_{is} \cdot \phi_i^{(s)}$, $y(t) = \tilde{P}^- x(t) = \exp(\overset{\sqcap}{A}^{-1}\overset{\sqcap}{B}\, t)\eta$, $\eta = y(0)$, $z(t) = (I_{X_{k1}^1} - \tilde{P}^-)\, x(t)$, then the Lyapounov-Schmidt method (Theorem 27.1 [5] for Noetherian operators with d-characteristic $(n, 0)$ and the indicated above theorems 22.1, 22.2 [5]) implies there is a unique solution of (12) $z = z_R(\xi(t) \cdot \phi + y(t)) \in X_{k1}^1$, such that $x(0) = \xi \cdot \phi + \eta + z_R(\xi \cdot \phi + \eta)$, i.e. the unique solution of (1) $x(t) = v(t) + y(t) + z_R(\xi(t) \cdot \phi + y(t))$, $v(t) = \xi(t) \cdot \phi$, in a sufficiently small semi-neighborhood of $t = 0$, where the function $z_R(\xi, \eta) = z_R(\xi \cdot \phi + \eta)$ is sufficiently smooth by ξ, η, and $z_R(0,0) = 0$, $D_\xi z_R(0,0) = 0$, $D_\eta z_R(0,0) = 0$.

Writing the equation (1) in $\mathbf{p}$, $\mathbf{q}$-projections at the usage of the theorem 1 one can get the system for the determination of $\xi_{is}(t)$ (so-named the resolving system (RS) for the equation (1) [12,10,14]). Here $x(t) = \xi(t) \cdot \phi + w(t)$, where $w(t) = y(t) + z_R(\xi(t) \cdot \phi + y(t))$ for $t \geq 0$ and $w(t) = y_R(\xi(t) \cdot \phi + z(t)) + z(t)$ for $t \leq 0$

$$\overset{\sqcap}{A}\frac{dw}{dt} = \overset{\sqcap}{B}\, w - (I_{D_k} - \mathbf{q})R(\xi \cdot \phi + w) \tag{14}$$

$$\begin{array}{c} 0 = \xi_{iq_i}(t) - < R(\xi(t) \cdot \phi + w), \hat{\psi}_i^{(1)} >, \\ \dot{\xi}_{iq_i}(t) = \xi_{i,q_i-1}(t) - < R(\xi(t) \cdot \phi + w), \hat{\psi}_i^{(2)} >, \\ \cdots\cdots\cdots\cdots\cdots\cdots\cdots\cdots\cdots\cdots \\ \dot{\xi}_{i2}(t) = \xi_{i1}(t) - < R(\xi(t) \cdot \phi + w), \hat{\psi}_i^{(q_i)} >, \\ \xi_{is}(0) = \xi_{is}, \quad s = 1, \ldots, q_i, \quad i = 1, \ldots, m. \end{array} \tag{15}$$

Consequently, the manifold $S_k(R) = \{$ initial values of solutions of the equation (1), which are defined and remain in a small neighborhood of $0 \in D_k$ for $t \in [0, +\infty)\}$ (the manifold $U_k(R) = \{$ initial values of solutions (1), which are defined and remain in a small neighborhood of $0 \in D_k$ for $t \in (-\infty, 0]\}$) has the form $x(0) = \xi \cdot \phi + \eta + z_R(\xi \cdot \phi + \eta)$ $(x(0) = \xi \cdot \phi + y_R(\xi \cdot \phi + \zeta) + \zeta)$, where $\eta \in D_k^-$ $(\zeta \in D_k^+)$ and ξ are small. □

Remark 1. Determined by the function $\xi \cdot \phi + \eta + z_R(\xi \cdot \phi + \eta)$ for $t \geq 0$ $(\xi \cdot \phi + y_R(\xi \cdot \phi + \zeta) + \zeta$ for $t \leq 0)$ invariant manifold $\mathfrak{M}$ can be regarded as center manifold $(\xi \cdot \phi \in D_k^0)$, that is nontrivial for the equation (1) even if $\{\mu \in \sigma_A(B) | Re\ \mu = 0\} = \emptyset$. Here $\{\xi \cdot \phi\}$ can be named as linear center manifold tangent to $\mathfrak{M}$. One can say that $\mathfrak{M}$ has an hyperbolic structure. Thus the RS (15) represents the differential-algebraic system on $\mathfrak{M}$. Of course, if the operator A is invertible, $\mathfrak{M}$ and the system (15) are absent, i.e. in the Grobman-Hartman theorem $z_R = z_R(\eta)$ [11].

Theorem 2. *Let the operators A, B and R in (1) be intertwined by the group G representations L_g (acting in E_1) and K_g (acting in E_2) and the condition I (direct supplements $E_1^{\infty-m}$ to $\mathcal{N}(A)$ and $E_1^{\infty-n}$ to $\mathcal{N}(B)$ are invariant relative to L_g) is satisfied. Then the center manifold $\mathfrak{M}$ is invariant relative to the operators L_g.*

Proof. According to [14] projectors $\mathbf{p}, \mathbf{P}(\mathbf{q}, \mathbf{Q})$ commute with the operators $L_g(K_g)$ and invariant pairs of subspaces reduce the representations $L_g(K_g)$. □

In the article [10] it is proved that the stability (instability) of the trivial solution (even for non-autonomous) equation (1) at sufficiently general conditions is determined by the RS (15) with corollaries for the investigation of the stability (instability) of bifurcating solutions.

It is interesting the case when $\sigma_A^+(B) = \emptyset$. Then $D_k = D_k^- \dot{+} D_k^0$, $x(t) = \xi(t) \cdot \phi + y(t)$ and the center manifold has the form $\xi(t) \cdot \phi + y(\xi(t) \cdot \phi)$. Here the equation (14) gives

$$\overset{\sqcap}{A}\, y'(\xi(t) \cdot \phi)(\tfrac{d\xi}{dt} \cdot \phi) = \overset{\sqcap}{B}\, y(\xi(t) \cdot \phi) + (I - \mathbf{q})R(\xi(t) \cdot \phi + y(\xi(t) \cdot \phi)), \quad y(0) = 0, \;\; y'(0) = 0 \tag{16}$$

In combination with (15) this gives a possibility for the determination of center manifold $w(\xi(t) \cdot \phi) = \xi(t) \cdot \phi + y(\xi(t) \cdot \phi)$ by successive approximations in conditions of sufficiently smooth operator $y(\xi \cdot \phi)$. However on this way essential difficulties arise connected with the fact that the system (15) is differential-algebraic, i.e. the differential equations for the functions $\xi_{i1}(t)$, $i = 1, \ldots, m$, are absent. One can find $y(\xi \cdot \phi)$ iteratively at the differentiation of the first equations (15).

Remark 2. Theorem 1 and all corollaries remain true for the parameter depending equation

$$A\frac{dx}{dt} = Bx - R(x, \lambda), \quad R(0, \lambda) \equiv 0, \quad R_x(0, 0) = 0, \tag{17}$$

($\lambda \in \Lambda$, Λ is some Banach space) in a small neighborhood of $\lambda = 0$, when as earlier Re $\sigma_A(B) \neq 0$, i.e. $\lambda = 0$ is not a bifurcation point. However all functions w, z_R and y_R will depend on small parameter ε.

4 One Case of $\sigma_A^0(B) \neq \emptyset$

Here it is considered the simplest case when $\sigma_A^+(B) = \emptyset$, but $\sigma_A^0(B) = \{\mu \in \sigma_A(B) | Re\ \mu = 0\} \neq \emptyset$ contains some finite number $2n = 2n_1 + \ldots + 2n_\ell$ A-eigenvalues $\pm i\alpha_s$ of multiplicities n_s, $s = 1, \ldots, \ell$, $\alpha_s = \kappa_s \alpha$, $\alpha \neq 0$ with coprime $\kappa_s > 0$ or (and) zero-eigenvalue. Without loss of generality it is supposed that the equation (1) is written in the form of the system

$$\begin{matrix} A_1\dot{x} = B_1 x - f(x, y) \\ A_2\dot{y} = B_2 y - R(x, y), \end{matrix} \quad A = \begin{pmatrix} A_1 & 0 \\ 0 & A_2 \end{pmatrix}, \; B = \begin{pmatrix} B_1 & 0 \\ 0 & B_2 \end{pmatrix} \tag{18}$$

where the linear operators $A_1, B_1 : E_1^{k_{B_1}} \to E_{2,k_{B_1}}$ ($k_{B_1} = 2n_1 p_1 + \ldots + 2n_\ell p_\ell$, p_s are A_1-Jordan chains lengths for $\pm i\alpha_s, s = 1, \ldots, \ell$) act in the invariant pair of finite dimensional subspaces $E_1^{k_{B_1}}, E_{2,k_{B_1}}$ and A_2, B_2 act in the invariant pair of subspaces $E_1^{\infty - k_{B_1}}$, $E_{2,\infty - k_{B_1}}$. Thus, $\sigma_{A_1}(B_1) = \sigma_A^0(B)$ and $\sigma_{A_2}^0(B_2) = \emptyset$.

Here f and R are C^2-functions vanishing together with their first derivatives at the origin.

The main assumption in the simplest case is

$$\mathcal{N}(A_1) = \{0\}, \quad \mathcal{N}(A_2) = \text{span}\,\{\phi_{(2)1}, \ldots, \phi_{(2)m_2}\} \tag{19}$$

Then under section 3 conditions there exists the function $y_R(\xi_2(t)\cdot\phi_{(2)}, x)$ vanishing together with its first derivatives at the origin, such that the second equation (18) is reducing to the system

$$\overset{\sqcap}{A}_2 \frac{dy_R}{dt} = \overset{\sqcap}{B}_2\, y_R - (I - \mathbf{q}_{(2)})R(x, \xi_2(t)\cdot\phi_{(2)} + y_R(\xi_2(t)\cdot\phi_{(2)}, x)) \tag{20}$$

($\mathbf{q}_{(2)} = \sum_{i=1}^{m_2}\sum_{j=1}^{q_{2,i}} < \cdot, \hat{\psi}^{(j)}_{(2),i} > \zeta^{(j)}_{(2)} : E_{2,\infty-k_{A_2}} \to \text{span}\{\zeta^{(j)}_{(2)i}\}$, $\overset{\sqcap}{A}_2, \overset{\sqcap}{B}_2$ act in invariant pair of subspaces $E_1^{\infty-k_{B_1}-k_{A_2}}$, $E_{2,\infty-k_{B_1}-k_{A_2}}$)

$$\begin{array}{c} 0 = \xi_{2iq_{2,i}}(t) - < R(x, \xi_2(t)\cdot\phi_{(2)} + y_R(\xi_2(t)\cdot\phi_{(2)}, x)), \hat{\psi}^{(1)}_{(2),i} >, \\ \dot{\xi}_{2iq_{2,i}}(t) = \xi_{2i,q_{2,i}-1}(t) - < R(x, \xi_2(t)\cdot\phi_{(2)} + y_R(\xi_2(t)\cdot\phi_{(2)}, x)), \hat{\psi}^{(2)}_{(2),i} > \\ \cdots\cdots\cdots\cdots\cdots\cdots\cdots\cdots\cdots\cdots\cdots\cdots\cdots\cdots\cdots \\ \dot{\xi}_{2i2}(t) = \xi_{2i1}(t) - < R(x, \xi_2(t)\cdot\phi_{(2)} + y_R(\xi_2(t)\cdot\phi_{(2)}, x)), \hat{\psi}^{(q_{2,i})}_{(2),i} >, \\ \xi_{2i\sigma}(0) = \xi_{2i\sigma},\ \sigma = 1, \ldots, q_{2,i},\ \ i = 1, \ldots, m_2. \end{array} \tag{21}$$

If the system (18) is equipped with initial values $x(0)$, $y(0)$, then they must satisfy the equality

$$y(0) = \xi_2\cdot\phi_{(2)} + y_R(\xi_2\cdot\phi_{(2)}, x(0)). \tag{22}$$

Now one has to solve the problem

$$A_1\dot{x} = B_1 x - f(x, \xi_2(t)\cdot\phi_{(2)} + y_R(\xi_2(t)\cdot\phi_{(2)}, x)) \tag{23}$$

at the initial data $x(0)$ satisfying (22).

Thus one has two systems (21) and (23) on the center manifold $y = y_R(\xi_2(t)\cdot\phi_{(2)}, x)$.

5 Grobman-Hartman Theorem Analog for Maps

According to section 3 the equation (14) can be written in the form

$$\frac{dw}{dt} = \overset{\sqcap}{A}^{-1}\overset{\sqcap}{B}\, w - \overset{\sqcap}{A}^{-1} (I_{D_k} - \mathbf{q})R(\xi\cdot\phi + w) \tag{24}$$

in the space X^1_{k1}. Then the assumption about the boundedness of the operator $\overset{\sqcap}{A}^{-1}\overset{\sqcap}{B}$ in X_{k1} allows to prove Grobman-Hartman theorem for maps [15]. In fact, then for small ξ there exists the resolving operator $U_\xi(t,\cdot) : X_{k1} \to X^1_{k1}, w_0 \mapsto w(t)$ for the problem (24) with the initial value $w(0) = w_0$ (at $\xi = 0, U_0(t)$ is linear). Thus the following assertion is true:

Theorem 3. *For small ξ at $\sigma_A^0(B) = \emptyset$ and operator $\overset{\sqcap}{A}^{-1}\overset{\sqcap}{B}$ boundedness assumption there exits the resolving operator $U_\xi(t, w_0)$ and a homeomorphism $\Phi_\xi : X_{k1}^1 \to X_{k1}^1, ||\xi|| \ll 1$, such that for $t \in R$ and $w_0 \in X_{k1}$ the following relation*

$$U_0(t)\Phi_\xi(w_0) = \Phi_\xi(U_\xi(t, w_0)) = \Phi_\xi(w(t)) \tag{25}$$

is true, where the function $w(t)$ and the initial values w_0, ξ_0 satisfy the initial value problem for differential-algebraic system (15).

Remark 3. The results of this article remains true for the more general operators subordinateness (A is subordinate to B if on D_B $\|Ax\| \leq \|Bx\| + \alpha\|x\|$, $\alpha \geq 0$).

The authors are thankful to Prof. V.S. Mokeychev (Kazan' State University) for this remark on improvement of our article.

References

1. Gurel, O. (Ed.): Bifurcation Theory and its Applications in Scientific Disciplines. Annals of the New York Academy of Sci. Vol. 316. (1979)
2. Joseph, D.J.: Stability of Fluid Motions. Springer Verlag (1976)
3. Henry, D.: Geometric Theory of Semilinear Parabolic Equations. Lect. Notes in Math, Vol. 840. Springer Verlag (1981)
4. Govaerts, W.J.F.: Numerical Methods for Bifurcations of Dynamical Equilibria. SIAM, Philadelfia (2000)
5. Vainberg, M., Trenogin, V.: Branching Theory of Solutions of Nonlinear Equations. Wolters Noordorf, Leyden (1974)
6. Carr, J.: Applications of Centre Manifold Theory. Appl. Math. Sci. Vol. 35. Springer Verlag (1981)
7. Barenblatt, G.I., Zheltov, Yu.P., Kochina, I.N.: On the principal conceptions of the filtration theory in jointing media. Appl. Math. Mech. Vol. 24. **5** (1960) 58–73
8. Whitham, G.B.: Linear and Nonlinear Waves. Wiley-Int. Publ. (1974)
9. Oskolkov, A.P.: Initial-boundary value problems for equations of Kelvin-Foight and Oldroidt fluids. Proc. Steklov Math. Inst. AN SSSR. Vol. 179 (1988) 126–164
10. Loginov B., Rousak, Ju.: Generalized Jordan structure in the problem the stability of bifurcating solutions. Nonlinear Anal. TMA. Vol. 17. **3** (1991) 219–232
11. Hale, J.: Introduction to dynamic bifurcation. Bifurcation Theory and Appl. Lect. Notes in Math. Vol. 1057. Springer Verlag (1984) 106–151
12. Loginov, B.: Branching equation in the root subspace. Nonlinear Anal. TMA. Vol. 32. **3** (1998) 439–448
13. Sviridiuk, G.: Phase spaces of semilinear Sobolev-type equations with relatively strong sectorial operator. Algebra and Anal. Vol. 6. **5** (1994) 252–272
14. Loginov, B., Konopleva, I.: Symmetry of resolving systems for differential equations with Fredholm operator at the derivative. Proc. of Int. Conf. MOGRAN-2000 Ufa. USATU (2000) 116–119
15. Volevich, L., Shirikyan, A.: Local dynamics for high-order semilinear hyperbolic equations. Izvestya RAN. Vol. 64. **3** (2000) 439–485

Poster Papers

Modeling of the Potential Energy Surface of Regrouping Reaction in Collinear Three-Atom Collision System Using Nonlinear Optimization

A. S. Gevorkyan[1], A. V. Ghulyan[2], and A.R. Barseghyan[2]

[1]Institute of Informatics And Automation Problems NAS of Armenia
[2]State Engineering University of Armenia
g_ashot@sci.am, garthur@arm.hpl.com,
ast_prog@yahoo.com

Abstract. A two-dimensional analytical model with a set of adjusting parameters [1] is proposed for an interaction potential of three-atom collinear system. Different numerical methods for obtaining the optimal set of parameters are considered in the present work. These parameters are used to approximate a two-dimensional numerical array (as obtained from another quantum-chemical *ab initio* calculations for reaction surfaces) with model potential to given accuracy. Appropriate methods of numerical simulation have been analyzed and the optimal one was selected. Based on this model an algorithm for numerical solution of the problem of finding the adjusting parameters has been realized using successive iterations. It was implemented for two specific cases discussed below.

1 Formulation of the Problem

The absence of a universal analytic representation of the interaction potential in the three-atom system impedes the development of a sufficiently universal package of programs for numerical calculation of bimolecular chemical reactions. Note that the numerical solutions are also highly important for deeper understanding of the fundamentals of quantum mechanics, i.e., the relation of classical nonintegrability to the quantum chaos [1-3]. A modeled potential of two-dimensional surface for reactions (in curvilinear coordinates) of the following type:

$$U(u,v)=U_1(u)\left[A(u)+(C(u)+G(u)v+a(u)v^2)\exp(-2\alpha(u)v)-B(u)\exp(-\beta(u)v)\right], \quad (1)$$

was proposed in [1] after analyzing the geometrical and topological features of different reaction surfaces, where the functions $U_1(u)$, A(u), C(u), G(u), a(u), $\alpha(u)$, B(u) and $\beta(u)$ were

$$F(u)=C_{i\_}+\frac{C_{i+}-C_{i\_}}{1+\exp(-2\gamma_{C_i}u)}+\frac{C_{i0}\gamma_{C_i}^{\ 2}}{\left[\exp(\gamma_{C_i}u)+\exp(-\gamma_{C_i}u\right]^2}, \qquad i=1,..,8, \quad (2)$$

and $C_{i\_},C_{i+},C_{i0},\gamma_{C_i}$ were some adjusting parameters.

P.M.A. Sloot et al. (Eds.): ICCS 2003, LNCS 2658, pp. 545–554, 2003.

The present paper aims at the determination of an optimal algorithm and development on its basis of a program for computing the adjusting parameters of potential (1) by means of comparative analysis of numerical arrays of reaction surfaces of bimolecular chemical reactions with the potential (1).

2 Numerical Methods of the Problem Solution

The expressions (1)-(2) are seen to comprise 32 adjusting parameters. To find these, below we analyze and apply different methods for solution of systems of nonlinear equations. The most frequently used of these are the relaxation method, the Newton method, and the modified Newton method [4,5]. All the mentioned methods are based on the procedure for inverse Jacobian matrix computation with respect to the adjusting parameters. Note that the Newton method is one of the most efficient methods for minimization of convex functions in problems of unconstrained optimization.

The Newton method is based on the substitution of a function with the first two terms of the Taylor series expansion and following minimization of the obtained quadratic form;

$$f(x+\delta) \approx f(x) + f'(x)\delta + \frac{f''(x)}{2}\delta^2 + \ldots \tag{3}$$

For sufficiently small δ and well-behaved functions, we may assume that $f(x+\delta)=0$ and obtain from expression (3)

$$\delta = -\frac{f(x)}{f'(x)} \tag{4}$$

However, all these methods are efficient, i.e., readily converging, in case of successful selection of the initial approximation that is located close to the root of equation. It follows from model (1) that although the given function can have numerous extremums, they may prove local, and this does not provide global convergence, i.e., does not give reliable results. Even cursory analysis shows that these methods are inapt for solving the tasks.

The methods considered in [6] next to the above ones, belong to the class of the so-called “global methods”, which provide global convergence almost from any initial point. Considered here were the Newton method with backtracking, as well as the Broyden method (quasi-Newton method). One of the distinctive features of the Broyden method is that for evaluation of Jacobian the final differences are used, which is inexpedient here because the Jacobian may be evaluated analytically. For that reason, in what follows the method of Newton with backtracking will be analyzed in detail.

Here follows a brief description of the method in question for a specific task. We shall try now to solve a system of transcendental equations with 32 unknowns (the adjusting parameters)

$$F\mathbf{x}=0 \tag{5}$$

where $\mathbf{x}$ is the vector of adjusting parameters.

The Newton step is computed from

$$\mathbf{x}_{new} = \mathbf{x}_{old} + \delta\mathbf{x} \tag{6}$$

where

$$\delta\mathbf{x} = -\mathbf{J}^{-1} \cdot \mathbf{F} \tag{7}$$

where $\mathbf{J}^{-1}$ is Jakobian.
The problem is minimizing the following expression

$$\mathbf{f} = \frac{1}{2}(\mathbf{F} \cdot \mathbf{F}) \tag{8}$$

The searching strategy is simple enough; at first the full Newton step is executed, if after that the expression (8) is minimized, then the procedure of minimization is continued in the same direction, otherwise the traversal along the Newton direction is reversed.
The backtracking procedure is as follows;
Expression (6) can be written as

$$\mathbf{x}_{new} = \mathbf{x}_{old} + \lambda \cdot \mathbf{p} \tag{9}$$

where $\mathbf{p} = \delta\mathbf{x}$, $0 < \lambda \leq 1$.
The execution of Newton step is suggested in [6] when the following condition is observed:

$$f(\mathbf{x}_{\mathbf{new}}) \leq f(\mathbf{x}_{\mathbf{old}}) + \alpha \cdot \nabla\mathbf{f} \cdot (\mathbf{x}_{\mathbf{new}} - \mathbf{x}_{\mathbf{old}}) \tag{10}$$

where $0 < \alpha < 1$.
The searching strategy in the backtracking algorithm implies:
The determination of the following function

$$g(\lambda) = f(\mathbf{x}_{\mathbf{old}} + \lambda \cdot \mathbf{p}), \tag{11}$$

so as

$$g'(\lambda) = \nabla\mathbf{f} \cdot \mathbf{p}. \tag{12}$$

After execution of the first Newton step, an analysis of minimization procedure of expression (10) is carried out. If it was not minimized, then the function $g(\lambda)$ is calculated from the following equation

$$g(\lambda) = [g(1) - g(0) - g'(0)]\lambda^2 + g'(0)\lambda + g(0). \tag{13}$$

The solution of equation (13) regarding λ is

$$\lambda = -\frac{g'(0)}{2[g(1) - g(0) - g'(0)]} \tag{14}$$

At successive iterations g is modeled to the cubic form that may be written as

$$g(\lambda) = a\lambda^3 + b\lambda^2 + g'(0) + g(0). \tag{15}$$

To find correct results for λ, it is necessary to solve two equations with respect to *a* and *b*;

$$\begin{bmatrix} a \\ b \end{bmatrix} = \frac{1}{\lambda_1 - \lambda_2} \begin{bmatrix} 1/\lambda_1^{\ 2} & -1/\lambda_2^{\ 2} \\ -\lambda_2/\lambda_1^{\ 2} & \lambda_1/\lambda_2^{\ 2} \end{bmatrix} \bullet \begin{bmatrix} g(\lambda_1) - g'(0)\lambda_1 - g(0) \\ g(\lambda_2) - g'(0)\lambda_2 - g(0) \end{bmatrix} \tag{16}$$

The minimum for the given function is reached in the vicinity of the value

$$\lambda = \frac{-b + \sqrt{b^2 - 3ag'(0)}}{3a}, \tag{17}$$

So, one is made certain that this method may be applied for calculation of the adjusting parameters. The next is the class of methods of variable metrics [6]. Now we shall describe these methods in more detail. Note, that if function (1) is minimized along some direction **u,** then its gradient is to be normal to the vector **u,** otherwise it will imply that a nonzero derivative along the direction **u** is found.

Suppose that $\mathbf{x_0}$ is the initial approximation. Hence, a desired function can be approximated by means of decomposition

$$f(\mathbf{x}) = f(\mathbf{x}_0) + \sum_i \frac{\partial f}{\partial x_i} x_i + \frac{1}{2} \sum_i \frac{\partial^2 f}{\partial x_i \partial x_j} x_i x_j + ... \tag{18}$$

$$\approx \mathbf{c} - \mathbf{b} \bullet \mathbf{x} + \frac{1}{2} \mathbf{x} \bullet \mathbf{A} \bullet \mathbf{x}$$

where $\mathbf{c} = f(\mathbf{x}_0)$, $\mathbf{b} = -\nabla \mathbf{f}|_{\mathbf{x}_0}$, $|\mathbf{A}|_{ij} = \dfrac{\partial^2 \mathbf{f}}{\partial x_i \partial x_j}$ is the Gessian matrix.

It is easy to calculate $\nabla \mathbf{f}$ from (18)

$$\nabla \mathbf{f} = \mathbf{A} \cdot \mathbf{x} - \mathbf{b}. \tag{19}$$

The change in the direction of gradient is calculated as follows:

$$\delta(\nabla \mathbf{f}) = \mathbf{A} \cdot \delta \mathbf{x} \tag{20}$$

If the direction of motion is along **u**, in case of which the function is minimized and it is required to continue the motion along **v** , then the condition ought to be fulfilled

$$\mathbf{0} = \mathbf{u} \cdot \delta(\nabla \mathbf{f}) = \mathbf{u} \cdot \mathbf{A} \cdot \mathbf{v}. \tag{21}$$

The main difference between the methods of conjugate gradients and of the variable metrics consists in the fact that in the latter case the intermediate results, which have been accumulated at previous stages, are stored and updated . The main feature of methods of variable metrics is an iterative approximation of the inverse matrix $\mathbf{A}^{-1}$ with given accuracy. The methods of the last class under study refer to methods based on the model-trust region approach. Here the least-square method, and in particular, the method of Levenberg-Marquardt will be considered. Let us determine the function

$$\chi^2 \approx \gamma - \mathbf{d} \cdot \mathbf{a} + \frac{1}{2} \mathbf{a} \cdot \mathbf{D} \cdot \mathbf{a}, \tag{22}$$

where **d** - is the vector of unknown quantities (the adjusting parameters M) and **D** is the matrix of $M \times M$ dimensions:

$$\mathbf{a}_{\mathbf{min}} = \mathbf{a}_{\mathbf{cur}} + \mathbf{D}^{-1}\left[\nabla\chi^2\left(\mathbf{a}_{\mathbf{cur}}\right)\right]. \tag{23}$$

The next approximation can be determined in the following way;

$$\mathbf{a}_{\mathbf{next}} = \mathbf{a}_{\mathbf{cur}} - constant \cdot \nabla\chi^2\left(\mathbf{a}_{\mathbf{cur}}\right) \tag{24}$$

Here the *constant* was taken to be rather small not to infringe the descending trend. In expressions (23) and (24) it is necessary to compute the Gessian of function χ^2 for all unknowns. The main difference from the previous class of methods is that the direct calculation of the Gessian by means of minimization methods is impossible. The minimization methods permit the iterative evaluation of a function and of its gradients. The second essential difference from all previous methods is that all the above methods are based on the linear search or linear minimization. The methods in issue are based on nonlinear procedures with the use of least-squares formalism.
The calculation procedure of the gradient and Gessian in the Levenberg-Marquardt method is described in [6].
The model under consideration can be represented as

$$U = U\left(\mathbf{u}, \mathbf{v}, \mathbf{C}\right) \tag{25}$$

where **C** is the vector of adjusting parameters having the dimension M. The function χ^2 can be determined in the following way

$$\chi^2 = \sum_{i=1}^{N} \frac{U\left(u_i, v_i\right) - U\left(u_i, v_i, \mathbf{C}\right)}{\sigma_i^2} \tag{26}$$

The gradient of function χ^2 is

$$\frac{\partial\chi^2}{\partial C_k} = -2\sum_{i=1}^{N} \frac{\left[U_l - U\left(u_l, v_l, \mathbf{C}\right)\right]}{\sigma_l^2} \cdot \frac{\partial U\left(u_l, v_l, \mathbf{C}\right)}{\partial C_k} \tag{27}$$

The second derivative is computed as follows:

$$\frac{\partial^2\chi^2}{\partial C_k \partial C_l} = 2\sum_{i=1}^{N} \frac{1}{\sigma_i^2}\left[\frac{\partial U\left(u_i, v_i, \mathbf{C}\right)}{\partial C_k} * \frac{\partial U\left(u_i, v_i, \mathbf{C}\right)}{\partial C_l} - \right. \tag{28}$$

$$\left. -\left[Ui - U\left(u_i, v_i, C\right)\right]\frac{\partial^2 U\left(u_i, v_i, \mathbf{C}\right)}{\partial C_l \partial C_k}\right]$$

Here we introduce some designations

$$\beta_k = \frac{1}{2}\frac{\partial\chi^2}{\partial C_k} \qquad \alpha_{kl} = \frac{1}{2}\frac{\partial^2\chi^2}{\partial C_k \partial C_l} \tag{29}$$

Using expressions (23) and (29) we obtain the relation

$$\sum_{l=1}^{M} \alpha_{kl} \delta C_l = \beta_k \tag{30}$$

Taking into account (24) a system of linear equations is obtained

$$\delta_{a_l} = constant \cdot \beta_l . \tag{31}$$

The second partial derivatives are computed using the formula

$$\alpha_{kl} = \sum_{i=1}^{N} \frac{1}{\sigma_i^2} \left[\frac{\partial U(u_i, v_i, \mathbf{C})}{\partial C_k} \frac{\partial U(u_i, v_i, \mathbf{C})}{\partial C_l} \right] \tag{32}$$

So, here all necessary formulas for closed computation of adjusting parameters are presented. All these methods are implemented in the system of NIQC. The adjusting parameters of energy surfaces for a number of chemical reactions have been computed by means of the Newton method with backtracking and the Levenberg-Marquardt method.

3 Analysis of Results and Visualization of Data

The text below witnesses the effectiveness of the numerical method used. The numerical array, by means of which the energy surface is built (Fig.1), may be generated by substitution of a selected set of adjusting parameters to formula (1).

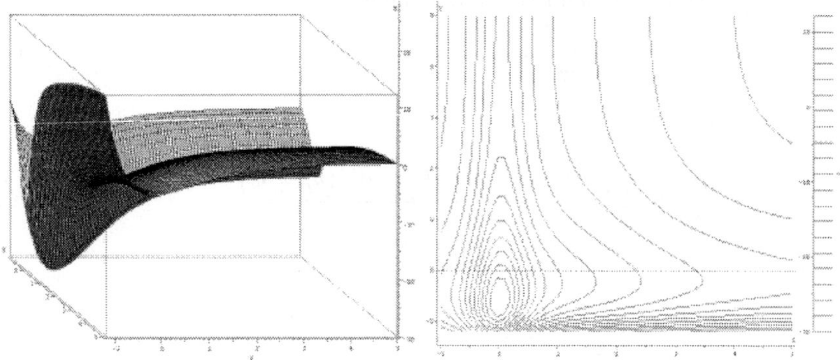

a) b)

Fig. 1. Three-dimensional and two-dimensional representations of the surface by substitution of adjusting parameters in the model (1).

The three-dimensional surface of interaction potential obtained after the substitution of a selected set of adjusting parameters in interaction potential (1) is shown in Fig.1a. The two-dimensional projection of interaction potential (the isoline of potential energy), in the plane $\{u, v\}$, is shown in Fig.1b.

On the basis of this array the program in the system NIQC has been implemented.

For computation of adjusting parameters the following scheme is proposed:

1. By means of an appropriate test function the numerical data array is generated into the file. The file contains the values of appropriate coordinates u, v potential computed by the model (1) or another model.
2. Then the first approximation for 32 adjusting parameters is introduced.
3. At this stage the array written in file at the step 1 is read onto respective arrays (u, v and $U(u,v)$).
4. The invocation of minimization procedure realized on the basis of the Levenberg-Marquardt method is made.
5. The values of the first approximation for adjusting parameters and an appropriate array with 32 zeros and units, by means of which the flag is set to allow or prohibit the changes of parameter with the corresponding index, are fed to the input of above function. The quadratic divergences are set to the value of unit.
6. Then the procedure with appropriate parameters is called in loop. The expression (26) is the minimization function. The preset minimum value gives the convergence criterion.
7. At the output of function the array of adjusting parameters providing the preset minimum value is obtained and saved in the file.
8. At the final stage, the substitution of appropriate adjusting parameters is performed and the difference between the initial values of function and those computed according to the above procedure is calculated for corresponding u and v. All the results are saves in the file.

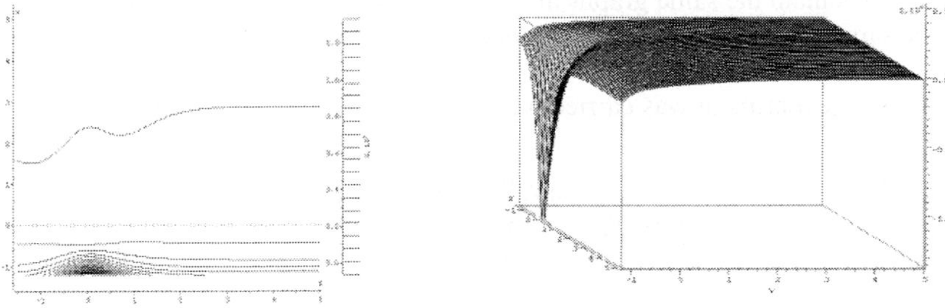

a) b)

Fig. 2. In 2a) and 2b) the two-dimensional surface for initial approximation and its differential graph obtained by computing the difference between the values of initial test function and those computed by model (1) based on the first approximation are shown respectively.

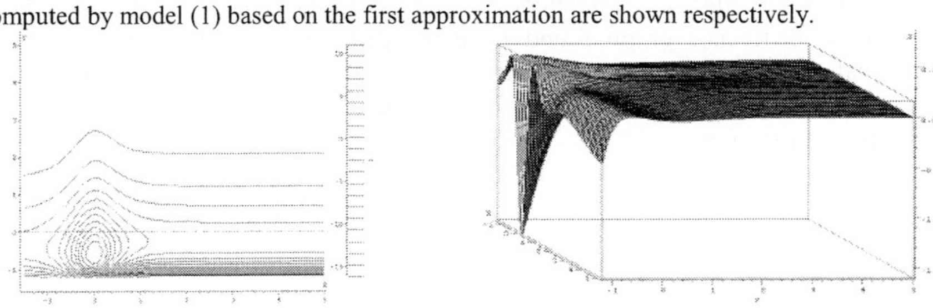

a) b)

Fig. 3. In 3a) and 3b) the two-dimensional surface for initial approximation and its differential graph obtained by computing the difference between the values of initial test function and those computed by model (1) based on the first approximation are shown respectively.

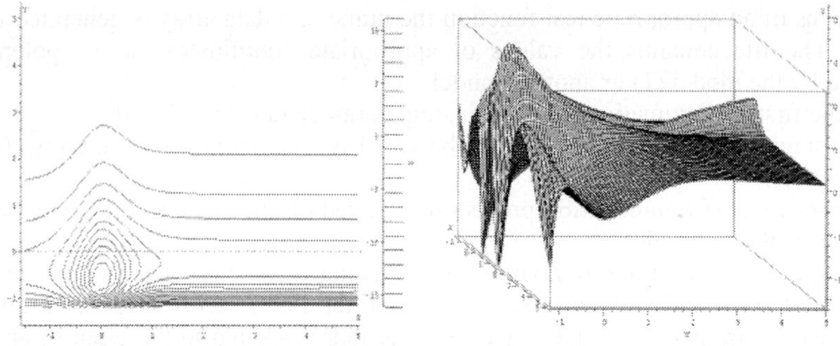

a) b)

Fig. 4. In 4a) and 4b) the two-dimensional surface for initial approximation and its differential graph obtained by computing the difference between the values of initial test function and those computed by model (1) based on the first approximation are shown respectively.

In Fig.5 the minimization process of expression (26) is illustrated.
The minimization process is illustrated by the graphs. In Figs. 2a) and 2b) the two-dimensional surface for initial approximation and its differential graph obtained by computing the difference between the values of initial test function and those computed by model (1) based on the first approximation are shown respectively. Figs. 3a) and 3b) contain the same graphs after 10 iterations, and, finally, Figs. 4a) and 4b) show the same graphs after the convergence of the process.

The following experiment was carried out with test function in the form

$$U(u,v)=\frac{\sin\sqrt{(u^2+v^2)}}{\sqrt{(u^2+v^2)}}. \tag{33}$$

For appropriate definition ranges of u and v the following three-dimensional 6a) and two-dimensional 6b) surfaces were obtained. Then, appropriate adjusting parameters of model (1) for the surface given by expression (33) were obtained. On the basis of these adjusting parameters and model (1) the numerical array has been generated, its difference graph found from the difference between the initial test function value and model (1) based on the initial approximation (Fig.6c). In Fig. 6d) the minimization process of expression (26) is shown, where the expression (33) is used as the test function.

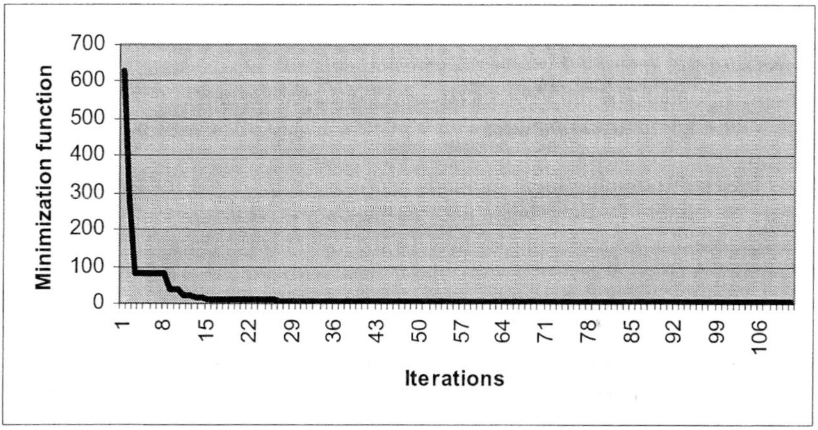

Fig. 5. The minimization function according to expression (26) versus the number of iterations.

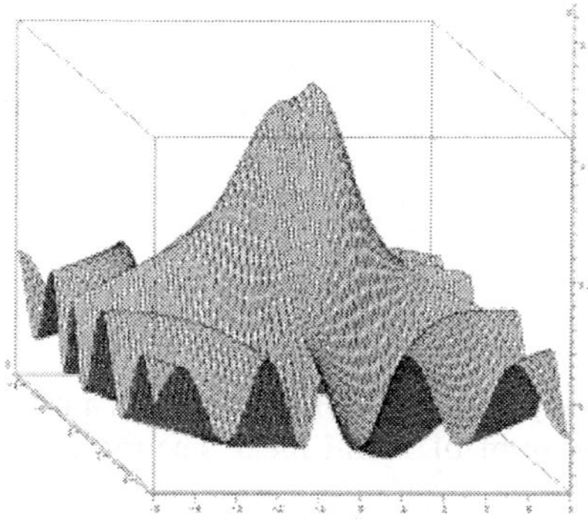

a)

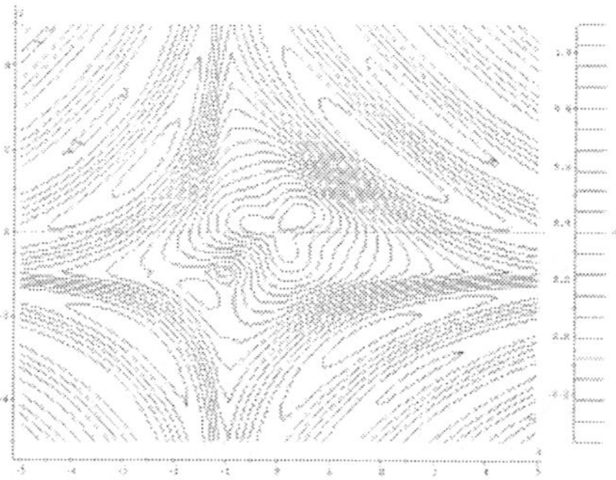

b)

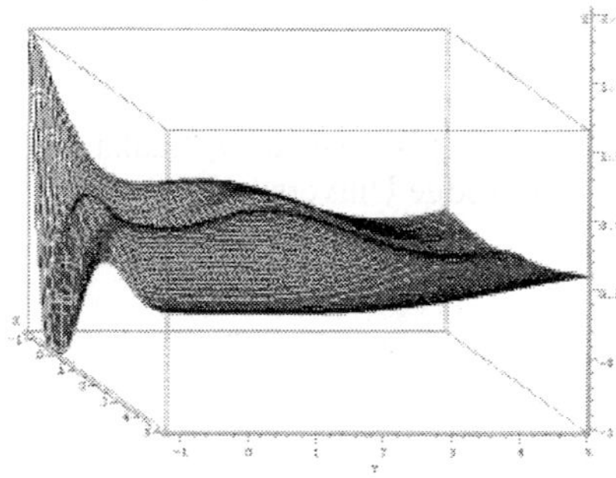

c)

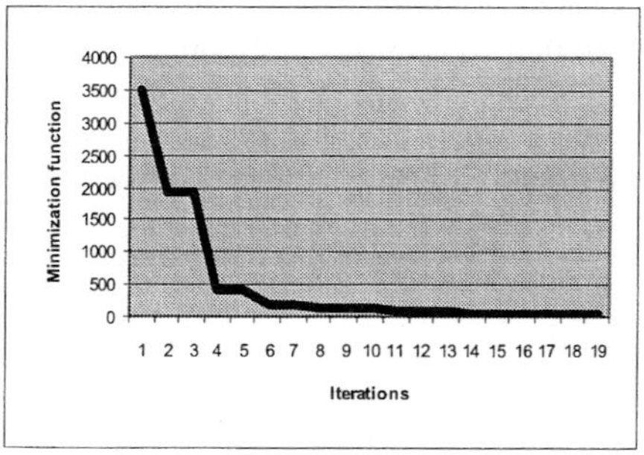

d)

Fig. 6. For appropriate definition ranges of u and v the following three-dimensional 6a) and two-dimensional 6b) surfaces were obtained. Then, appropriate adjusting parameters of model (1) for the surface given by expression (33) were obtained. On the basis of these adjusting parameters and model (1) the numerical array has been generated, its difference graph found from the difference between the initial test function value and model (1) based on the initial approximation (Fig.6c). In Fig. 6d) the minimization process of expression (26) is shown, where the expression (33) is used as the test function.

4 Conclusion

The analysis of the analytical model (1) shows that:
1. It is proved that the energy surfaces of bimolecular reactions with different topological and geometrical features can be described by means of potential (1) with sets of adjusting parameters.
2. Some classes of computational methods and algorithms, used for finding adjusting parameters of model (1), have been analyzed in details. It is shown that the method most acceptable for solution of the task is the Levenberg-Marquardt method. The results obtained (Fig. 5 and Fig. 6d) show that with the help of this method the expression (26) is minimized rather quickly and a good approximation of function (1) is provided.
So, a numerical experiment proves the efficiency of the description of the two-dimensional reaction surface of various bimolecular chemical reactions by means of analytical potential (1).

Acknowledgement.
A. Gevorkyan thanks ISTC Grant A-823 for support and also gratefully acknowledges Prof. Yu. Shoukourian and Dr. V. Sahakyan for their kindness to this work.

References

1. Gevorkyan A.S. Dissertation of the Dr.Sci., 2000 St.Petersburg(S.Un.St-P.)
2. Bogdanov A.V., Gevorkyan A. S. "Three-body multichannel scattering as a model of irreversible quantum mechanics" Symposium of Non-Linear Theory and Its applications. Hilton Hawaiian Village, 1997.
3. A.V.Bogdanov, A.S.Gevorkyan, A.G.Grigoryan Internal Time Peculiarities as a Cause of Bifurcations Arising in Classical Trajectory Problem and Quantum Chaos Creation in Three-body System, AMS/IP Studies in Advanced Mathematics, V.13. p. 69-80, (1999).
4. Samarskiy A.A., "Numerical methods", 1997., Moscow, Nauka
5. Bakhvalov N.S., "Introduction to the numerical methods". Moscow, Nauka 1987.
6. William H. Press "Numerical Recipes in C" Cambridge University Press 1992.

Workflow for Simulators Based on Finite Element Method

Felix C.G. Santos, Mardoqueu Vieira, and Maria Lencastre

Federal University of Pernambuco - Department of Mechanical Engineering
Rua Acadêmico Hélio Ramos, s/n - Recife - PE 50740-530 - Brazil
fcgs@demec.ufpe.br, {msv,mlpm}@cin.ufpe.br

Abstract. Current workflow systems usually do not provide adequate support for workflow modeling. Real life work processes can be much richer in variations and more dynamic than a typical workflow model is capable of expressing; this means that the users need to be able to adjust workloads and modify workflow models on the fly [10]. Plexus, a system for the development of simulators [5,6], is a typical case where such a difficulty arises. Simulators provide an economical means of understanding and evaluating the performance of abstract and real-world systems; their design and implementation is almost as complex as the systems being simulated, to be efficient they must be adaptable to an ever-increasing system complexity. The use of workflow technology helps the development of more flexible and versatile strategies. This paper proposes a workflow management framework, called GIG, for controlling the simulator workflows in the Plexus context.

1 Introduction

Due to tremendous ongoing activity in the fields of application of the Finite Element Method (FEM), there is a need for tools, which could help the development of simulators with a high reusability degree in both the academic and industrial worlds. Nowadays simulation systems supporting coupled multi-physics phenomena can be important predictive tools in many industrial activities. However, the need for suitable numerical tools, which could more appropriately simulate a large amount of coupled phenomena, and the need for computational environments, which could help the building of those tools, are still a reality. The Finite Element Method is a way of implementing a mathematical theory of physical behavior. Simulations using FEM can become very complex, particularly when the designer wants to guarantee high levels of abstraction and reuse of the developed solutions. Those requirements comprise the main strategies in saving the production costs of high quality simulation software.

This work was done as a part of the activities of Plexus, a project for the development of a computational environment that helps the design and implementation of simulation software for coupled phenomena, through flexible and friendly tools, based on the FEM [3,4,5,6]. By simulator we mean a computational system aimed at obtaining approximate solutions to systems of coupled

P.M.A. Sloot et al. (Eds.): ICCS 2003, LNCS 2658, pp. 555–564, 2003.

partial differential equations, together with a set of restrictions (differential-algebraic relationships involving one or more vector fields). In this paper we analyse the importance of workflows in Plexus.

This work was devised from the experience obtained during the implementation of several simulators in the FEM context. Researchers of the Mechanical Engineering Department - UFPE wanted to organize their code in a way that was easier to adapt to new strategies and also to allow process reuse. So they designed and implemented the GIG, a generic interface graph, which provides a process to achieve the mentioned advantages. This workflow has been used in the development of FEM-systems and a variety of other numerical methods. Due to their good performance in different case studies we decided to standardize the proposed solution, extending it with some other workflow concepts, and also evaluating the final result.

The paper motivation can be summarize by the need for: (i) easiness of translating from the natural language representation of the processes into a computer representation; (ii) simplicity of use (iii) versatility in the implementation of solution processes in the FEM Domain (iv) reduction of the possibility of errors in the coupling of processes (v) support of adaptability at run time.

The paper is organized in the following way: section 2 presents some background about workflows; section 3 details Plexus System; section 4 describes the proposed workflow. Section 5 presents some conclusions.

2 Workflow Technology

Workflow technology and process support lies at the centre of modern information systems architectures [1]. Workflow can be defined as representing the operational aspect of a work procedure: (i) the structure of tasks and the application and humans that perform them; (ii) the order of task invocation; (iii) task synchronization and the information flow to support the tasks and; (iv) the tracking and reporting mechanisms that measure and control the tasks.

One of the chief tasks of workflow management system is to separate process logic from task logic, which is embedded in individual user applications. This separation allows workflow users to modify one without affecting the other, that is, the two can be independently modified and the same logic can be reused in different processes. This promotes software reuse and the integration of heterogeneous software applications [1].

The workflow management system is a system that completely defines, manages and executes "workflows" through the execution of software whose order of execution is driven by a computer representation of the workflow logic. These management systems automate the process logic, while humans and software applications perform workflow tasks, implementing the task logic. The majority of workflow systems share a small set of common features. Some common features of workflow systems are: (i) flow independence, (ii) domain independence, (iii) monitoring and history and (iv) manual interventions.

Considering the levels defined in [12], a Workflow Model system may be characterized as providing support for: (i) Built functions, concerned with defining, and possibly modeling, the workflow process and its constituent activities; (ii) Run-time control functions, concerned with managing the workflow processes in an operational environment and sequencing the various activities to be handled in each process; (iii) Runtime interactions with human users and IT application tools for processing the various activity steps.

3 Plexus System

Plexus is a system whose objective is to reduce the complexity and cost involved in the development and implementation of a simulation system, providing a more flexible environment with efficient techniques for coupled phenomena simulation. Despite of being a specific simulation environment, which uses FEM in coupled phenomena solution, the Plexus system can be used in several types of applications, allowing at first hand the modelling of different classes of simulators, considering a set of pre-defined features, for example, a solution scheme involving time stepping and adaptation.

The Plexus system is divided into 4 subsystems, representing the main processes, Fig. 1: (i) Administration/ System Loading, which supports the system management and the loading of general system data and metadata; (ii) Preprocessing, where the user inputs problem data, and where dynamic structures for a simulation are built; (iii) Simulation Processing, where data are processed to obtain the solution and where the verification occurs; (iv) Post-processing, where the solution is processed in order to obtain the quantities of interest for the user and for the needed visualization. This component also deals with system validation.

The system manages great volumes of data, previously built components, phenomena, phenomena coupling, algorithms components, definition of persistent data and simulation knowledge reuse. To give support to the high level of abstraction, flexibility, reusability, and data security available in the Plexus environment, there is a Database Management System (DBMS), which maintains the general abstract data related to the context, the algorithms that take part in different simulation strategies, the simulation problem's data and also the simulation's intermediary data and results.

The Plexus simulation implementation is represented with the use of algorithm skeletons and also with other predefined object-oriented structures, like computational phenomena, exploring the FEM polymorphism. A computational phenomenon (for example a computational representation of heat transfer) is described by its vector field and weak forms defined in its geometric entity together with boundary conditions information, which is also implemented as fictitious phenomena defined on the respective geometric entity of the boundary of its domain. It has also Math Methods that implement: Mesh generation, Integration Rules, Shape Functions, etc [3,4,5,6].

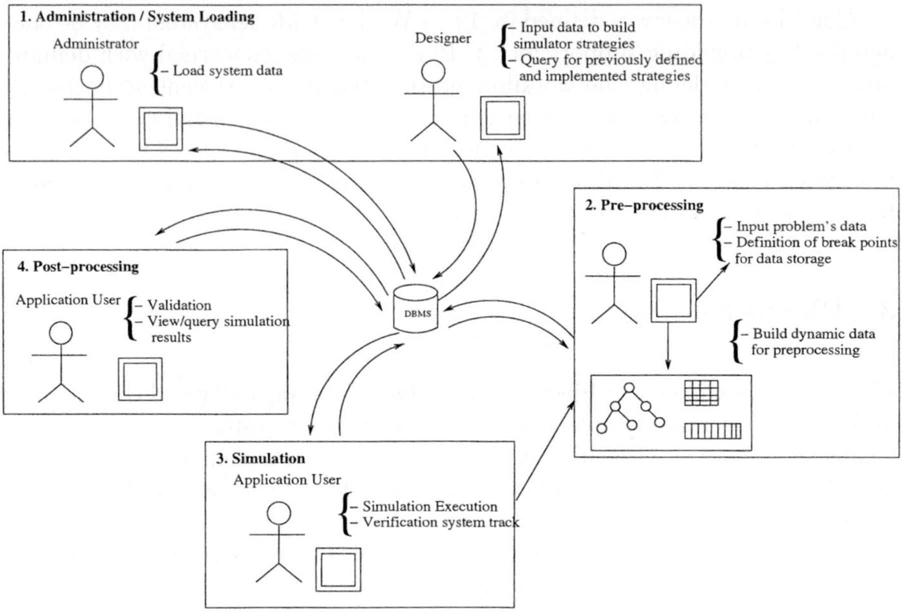

Fig. 1. Plexus Overview

The FEM-Simulator Skeleton pattern [3] suggests a FEM-Simulator algorithm organization within 4 levels of computational demands: Global Skeleton, Block Skeletons, Group Skeleton and the Phenomenon level. These levels were defined due to the high number of repeated (similar) structures and the degree of reusability of the involved algorithms, see subsection 4.1.

4 FEM Workflow

This section shows that the implementation complexity for coupled phenomena simulation can be greatly reduced by the use of predefined object oriented structures, due to FEM polymorphism, and also by the use of workflow and dataflow management. It proposes a generic framework for Plexus Workflow called GIG. By frameworks we mean, reusable semi-complete applications that can be specialized to produce custom applications [9].

GIG already attends, or can be easily adapted to, satisfy the following Plexus application requirements: easiness of translating from the natural language representation of the processes into a computer (executable) representation; simplicity of use; versatility and flexibility in the implementation of solution processes; reduction of the possibility of errors in the coupling of processes; need for support of adaptability at run time, due to dynamic change of business rules, and need of dynamic data creation. GIG framework follows the object-oriented style (modeling and programming). For purposes of simplicity of use and easy correctness verification, the GIG is restricted to be a direct acyclic graph (DAG).

We assume that the simulator building and assembling will be based on a variable designer data model, which describes: the initial scenery, algorithm skeletons and auxiliary numerical methods, phenomena, geometry and so on. The initial scenery defines the class of problems that the simulator will be able to tackle in a broad sense. The simulator model is able of considering the use of many procedures (for instance: Time Loop; Adaptation Iteration; Time Step Estimation; Solution of Algebraic Systems; Error Estimation; etc), which may be either present or not, depending on the configuration of the initial scenery. The implementation of those procedures is done through the algorithm skeletons and auxiliary numerical methods, which define a procedure within the 4 levels of computational demand. The other data model abstractions (like computational phenomena, geometry and so on) are used to describe the problem domain. Some processes used during the simulation are encapsulated in those data models; in this work we will not consider such level of detail.

In what follows we will describe the FEM process and the GIG solution.

4.1 FEM Simulator Process

The FEM simulator process can be basically defined by 4 levels of computation demands (skeletons and methods). From the first level down to the last one, the procedures get more and more specific, making it possible to separate the most reusable components. Thus, in each level, the so-called skeletons represent the activity flows. The skeletons are those parts of that solution process which can be replaced, making it possible to build different solution strategies. Each skeleton is able of articulating skeletons in the immediate lower level. When they are assembled together, they represent the entire solution process.

We detail, in what follows, each level:

- Global Skeleton is the first level of computation and represents the global algorithm skeleton (the core of the simulator). It is unique for each simulator, but may be replaceable, producing another simulator. The global algorithm skeleton articulates the procedures involving all blocks.
- Block Skeletons articulate the Groups of Phenomena in the execution of tasks demanded by the Global Skeleton. Each block has a set of skeletons (Block Skeletons), which satisfies the demands from the Global Skeleton by decoding them into demands for the groups in a previously defined order. A simulator may have a Block Skeleton changed without needing to change its Global Skeleton. Nevertheless, a well-designed Block Algorithm Skeleton is also very reusable and it is not supposed to be substituted even in the case of very severe changes in the solution algorithm in the level of the Group of phenomena.
- Group Skeletons articulate the Phenomena in the execution of tasks demanded by the Block Skeletons. A Group is provided with a set of Group Skeletons, which represent very specific procedures and may not be very reusable. Its purpose is to encapsulate the parts from the solution scheme, which are specific of the particular solution method being used for a group

of phenomena. Usually, the more reusable parts of the solution scheme are best located either in a Block Skeleton or in the Global Skeleton.

– Phenomenon Procedures represent the lowest level of all procedures in the simulation and are specific of all possible contributions its Phenomenon can provide to any solution scheme. Starting from the computation of the Global Skeleton and going through the two other levels of articulation, what remains to be defined are the contributions of each phenomenon to its Group solution scheme in a uniform parameterised way. The phenomena classes will be composed of phenomenon data and a group of numerical methods (Math-Methods), which are replaceable.

4.2 GIG Solution

The GIG solution propouses starting from an algorithm in natural language. The procedure is first divided into different algorithm nodes and is organized in the form of a graph. For the whole algorithm there is a data repository, which contains the existing data decomposed into different algorithm data classes.

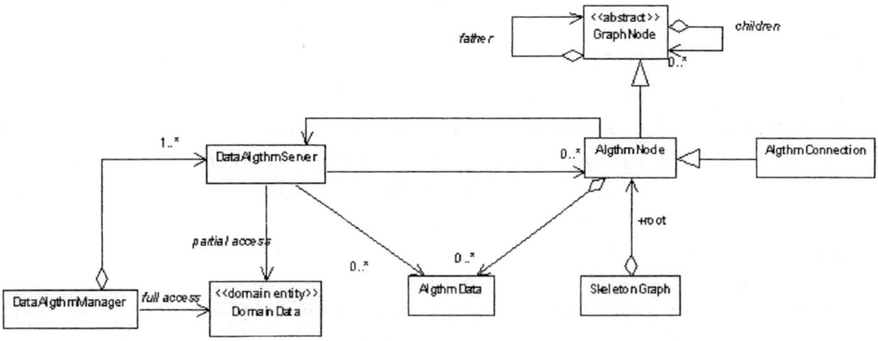

Fig. 2. GIG UML Class Diagram

The GIG framework structure is show in the Fig. 2. As we can realize, the GIG is composed of the following participants:

DataAlgthmManager its function is to control the whole workflow, managing its lifecycle. This involves build, reprogramming the application workflow at runtime.

DomainData represents the whole set of objects, which represent the problem domain.

GraphNode represents the relationship of the workflow nodes. It is an abstract class that implements low level operations to visit the graph.

SkeletonGraph it has the root of a workflow graph (skeleton). It manages the building and the modification of its skeleton. It is responsible to execute the workflow when requested to.

AlgthmNode represents the procedure (algorithm) of each workflow node. It is used as a base class for all algorithm classes of the application. We can say that it is a white box [9] of the GIG framework.

AlgthmData represents a data type to be used by an AlgthmNode during workflow execution. It is used as a base class for all algorithms data classes of the application.

DataAlgthmServer it provides a service that relates AlgthmNode with AlgthmData to be used in the building and modification of a workflow graph.

AlgthmConnection represents an algorithm that was not expanded yet, that is, it references an algorithm that was not incorporated yet to the workflow. So when it is executed it fetches the algorithm and replaces itself with the fetched algorithm.

4.3 Workflow Building and Execution

The workflow building and execution process is decribed in the UML sequence diagram of the Fig. 3. Initially the application send a request to *DataAlgthmManager* for building a workflow starting with an identification of the desired driver component. *DataAlgthmManager* forwards this request to *DataAlgthmServer*, which is the provider of all data and algorithm components. Then, *DataAlgthmServer* creates an object of the class *SkeletonGraph* and the *AlgthmNode* correspondent to the driver component. Next, *DataAlgthmServer* asks the *SkeletonGraph* to build the graph which in turn asks the Root (the driver) to recursively build the entire graph. The driver, then, asks *DataAlgthmServer* for its *AlgthmData* and its children *AlgthmNodes* objects. After that it asks each one of its children *AlgthmNodes* to build the graph recursively. This process goes on until all nodes of the workflow are created and assembled.

When it is the time for the application to execute the workflow, it asks SkeletonGraph to execute it. Then the SkeletonGraph asks its Root (driver) to execute its algorithm. Next, the root starts executing its procedure, which involves the execution of its children, and the execution of the entire workflow unfolds.

4.4 Considerations

Plexus deals with a fixed domain, where high levels of FEM abstraction were modeled [3,4,5,6,]. For the whole solution, hierarchical levels of processes were defined, each one with several possibilities of algorithms. Plexus suggests an architecture of component-based systems, which are significantly more powerful than that of traditional monolithic integrated solutions because they are easier to understand, adapt, reuse, customize and extend. Plexus clearly identifies levels of its architecture where the workflows are to be dynamically defined, built and controlled.

A Workflow Model system may be characterized as providing support for: built-functions, run-time control functions and runtime interactions (see section 2). We have separated the run-time control functions in the GIG solution; the other workflow model functions were just left to Plexus.

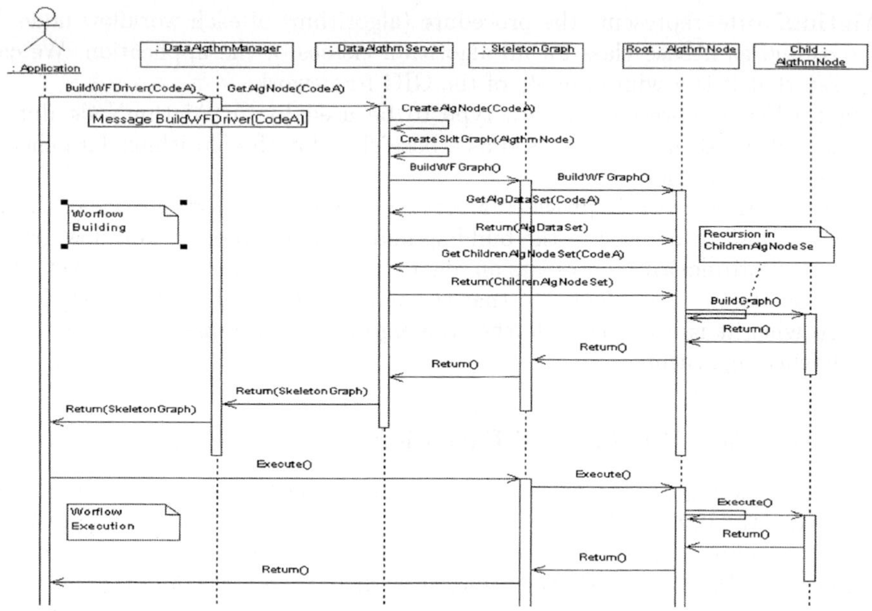

Fig. 3. Workflow building and execution

We can identify many advantages that GIG framework brings to FEM simulators development, like:

- Simplicity and easier production of algorithms from natural language: Plexus main expected users are scientists and engineers who develop or use FEM simulators, and already deal with development of FEM codes in some level. This users usually program in procedural style. The requirements of simplicity, easier production of algorithms from natural language into a computer (executable) representation and easier integration with object-oriented application, are very relevant. The GIG solution tries to attend this requirement. It allows an easier organization in a graph level, allowing the distribution of code in a very flexible way, not compelling a rigid division of code, like in [1]. The skeletons levels of computation (global skeleton, blocks and groups skeleton algorithms) are easily implemented in the GIG proposal.
- Restriction of Flow Granularity: An important consequence of the scale difference between micro workflow and macro workflow concerns users and activities [1]. Macro workflow targets process designers who typically are non-technical users; workflow activities involve applications and humans. In contrast, micro workflow targets people who build applications. At a smaller scale, micro workflow involves the objects that make up object-oriented applications. Plexus needs a mixture of small-scale and large-scale workflow, once its designers are also the programmers. The scales go from largest down to smallest, when the levels of computation go from the highest down to the

lowest. GIG allows a flexible representation for a mixture of scales, since it does not restrict the levels of programming into which the code is defined.
- Solution independence, easiness to change: Due to the frequently changing of numerical methods (for achieving the more suitable ones), FEM simulators usually suffer adaptations, forcing heavy reprogramming. The solution independence is important in order to allow the designer to specify the system features and strategies, supporting high level of control. The support for system reconfiguration and versatility is a desired feature. Flow independence guarantees more flexibility and reuse. In GIG, it's easy to change parts of the GIG graph, maintaining the desired ones intact. Also the levels of computation allow the tracking of the right process substitution.
- Providing Dynamic Procedural Programming and OO abstractions: Plexus maintains two paradigms: object oriented abstractions and the procedural programming (so common in scientific algorithms). The abstractions found in the FEM domain yield powerful and reusable systems through the established concepts of simulator, computational method, algorithm skeletons, and the defined levels of computation. On the other hand, the workflow perspective solves problems related to dynamic programming, which is very important in the production of FEM simulators by Plexus. So workflows following an object-orientation style can be of great help.

5 Conclusion

In this work we analysed the GIG solution, identifying some aspects of the solution domain that lead to more simplified workflow systems considering also some object-oriented aspects. As any workflow, GIG separates process logic (simulator definition and configuration) from task logic (problem strategies), which is embedded in individual user applications, allowing the two to be independently modified and the same logic to be reused in different cases.The best of GIG is that it is an object-oriented framework, which provides a powerful workflow control with simplicity of specification, programming and use. However, GIG, is not yet implementing monitoring, history and manual intervention and explicit parallelism control.

The Plexus system clearly identifies levels of its architecture where the workflows are to be dynamically defined, built and controlled. It is worthwhile observing that the proposed solution makes it easier and flexible the simulator definition and implementation, with a smaller overhead than other workflow management systems.

The Plexus system, focus on the development of simulators for chemo-thermo-mechanical interactions, which occur inside a given system and between such a system and its surrounding environment. However, the proposed GIG framework can be applied to the development of general-purpose systems, despite of being adequate for the underlining Plexus context.

References

1. Dragos-Anton Manolescu, *Micro-Workflow: A Workflow Architecture Supporting Compositional Object-Oriented Software Development*, Ph.D, Department of Computer Science University of Illinois at Urbana-Champaign, 2001.
2. Fiorini S. Leite J.P. and Lucena C.J., *Process Reuse Architecture*, CaiSE 2001, LNCS pp 284–298.
3. Lencastre M., Santos F. Rodrigues I., *FEM Simulator based on Skeletons for Coupled Phenomena (FEM – Simulator Skeleton)*, The Second Latin American Conference on Pattern Languages of Programming SugarloafPLoP'2002 Conference, Itaipava, Rio Janeiro, Brasil.
4. Lencastre M., Santos F., Rodrigues I. *Data and Process management in a FEM Simulation Environment for Coupled Multi-Physics Phenomena*, Fifth International Symposium On Computer Methods In Biomechanics And Biomedical Engeneering; 2001 Rome- Italy.
5. Lencastre M., Santos F., *FEM Simulation Environment for Coupled Multi-physics Phenomena.* Simulation and Planning In High Autonomy Systems – AIS2002; Theme: Towards Component-Based Modeling and Simulation. Lisboa, Portugal, 2002.
6. Santos F., Lencastre M., Araújo J., *A Process Model for FEM Simulation Support Development*, SCSC 2002 – Summer Computer Simulation Conference, US Grant Hotel- San Diego, California, 2002.
7. Yoder J., Balaguer F. and Johnson R., *Architecture and Design of Adaptative Object-Models*, 2000.
8. Roberts D., Johnson R., *Evolving Frameworks: A Pattern Language for Developing Object Oriented Frameworks*, University of Illions.
9. Fayad M., Douglas S., Johnson R., *Building Application Frameworks: Object- Oriented Foundations of Framework Design*, Wiley Computer Publishing, 1999.
10. Kwan M.M., Balasubramanian P.R., Dynamic Workflow Management: A Framework for Modeling Workflows,System Sciences, 1997, Proceedings of the Thirtieth Hwaii International Conference on , Volume: 4 , 7–10 Jan 1997 Page(s): 367–376 vol.4.
11. Workflow Management Coalition: The Workflow Reference Model, Workflow Management Coalition Specification, – Winchester, Hampshire – UK, 95.

Parallel Implementation of the DSMC Method Coupled with a Continuum Solution: Simulation of a Lubrication Problem in Magnetic Disc Storage

Sergey Denisikhin[1], Vladimir Memnonov[2], and Svetlana Zhuravleva[3]

[1] Baltic State University, St.Petersburg, Russia
s_d_d@aport.ru
[2] St.Petersburg State University, St.Petersburg, Russia
pokusa@star.math.spbu.ru
[3] St.Petersburg State University, St.Petersburg, Russia
svig@vega.math.spbu.ru

Abstract. We present some Monte Carlo simulation results for a two-dimensional unsteady problem of gas flow in an extremely narrow channel with an inclined upper wall and moving lower one. This is a model of gas film lubrication in modern magnetic disk storage, which is now under development. In order to account for non-uniform inner and outer structure of the boundary incoming and outgoing flows the DSMC simulations in the channel as well as in adjacent to inlet and outlet regions of linear size 20• are coupled at their boundaries to continuum finite element solution for the outer flow around the magnetic head. The coupling of DSMC simulation with this continuum solution is realized as step by step approximation. It is found that the pressure before and after the channel differs markedly from the atmospheric one because of the slowing down of the flow by magnetic head. Thus previously much used atmospheric pressure boundary conditions are unreal. Space and time distributions of different flow parameters are given.

1 Introduction

Flows with some of its linear dimension R having the order of several nanometers, are now often found in high technological processes and thus are of growing importance. The mean free path (mfp) λ for NTP in air is equal to 62 nm so the Knudsen number $Kn=\lambda/R$ may be close to unity. They are in transitional regime and it is just the case for the gap flow in the modern Winchester-type disk drives [1]-[3]. Thus the current trend toward nanoscale technology in the slider air bearing causes the previous design calculations based on continuum Reynolds equation or on the simulation of the whole slider region with surrounding external flow by solving the Navier-Stokes equations [4] to be changed by some microscopic solution at least in the part of the computational region. In this occasion the direct simulation Monte Carlo (DSMC) method introduced by Bird [5] is particularly suitable and was employed in [2], [3], [6] for a stationary flow problem by calculating some important parameters of a slider air bearing.

P.M.A. Sloot et al. (Eds.): ICCS 2003, LNCS 2658, pp. 565–574, 2003.

In this method molecular collisions and their free motion are directly simulated, but at the same time it is introduced a stochastic process which approximately replaces comprehensive description of particle dynamics, especially for molecular collisions.

Under fulfillment of certain restrictions on the mean numbers of molecules in the computational cells and the value of the time step this process can be approximated by a Poisson process [7]. But the same restrictions demand changing Bird's selection procedure for calculating of molecular collisions to a selection without replacement back already collided in this time step molecules. The new selection procedure was implemented with the help of a special additional array and the displacement in it of the collided particles to the right and reducing the number of molecules during selection only to those on the left, not yet collided. Thus removing an odious phenomenon in the Bird's algorithm namely potential repeated collision of the pair just collided molecules which is possible because of too strong violation of the microscopic dynamics produced by that selection with replacement.

For performance evaluation of a Winchester-type disk drive under the transient conditions one needs to know unsteady characteristics of the flow. In this paper we present some results from Monte Carlo simulations of two-dimensional unsteady problem of gas flow in an extremely narrow channel with an inclined upper wall and moving lower one. This is a model of gas film lubrication, which occurs in modern magnetic disk storage, now being under development. Due attention is given to non-uniform inner and outer structure of the boundary incoming and outgoing flows. This is accomplished by simultaneous continuum finite element solution for the outer flow around the magnetic head. By parallel implementation of our program on the clusters of St. Petersburg University with special load balance technique against computer breakdown to secure simultaneous run of it on several clusters we were able to obtain enough independent realizations for an ensemble average of the simulation results.

2 Statement of the Problem

It is considered a two-dimensional simulation problem for flow around the magnetic head and inside the channel, which is formed by the latter with the hard disc moving with constant velocity U_w, see Fig.1. In this channel and in the additional regions of 20λ height and long before and after the head a numerical solution was obtained with the help of DSMC method which was coupled along the boundaries at 0C and CD on the left and GF, FE on the right with a continuous finite element solution of Navier-Stokes equations. The latter produces boundary conditions for DSMC simulations by forming fluxes, which are calculated from mean velocities and its derivatives on these boundaries. Molecular interactions are taken to be the hard sphere one and their interaction with the wall was diffuse reflection. The square cells in

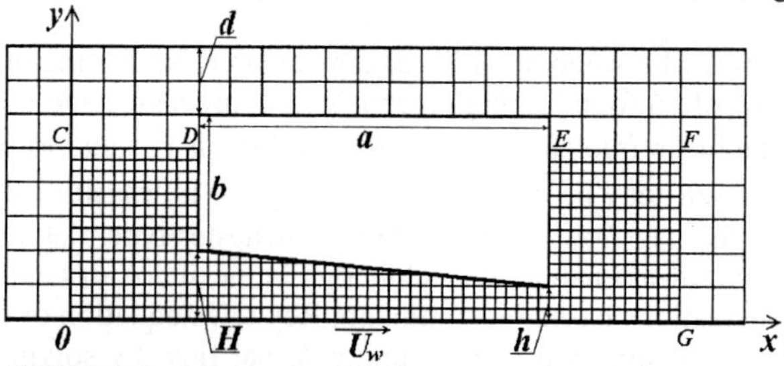

FIGURE 1. Computational region.

DSMS computational region had the linear size of approximately $\lambda/3$ and time step Δt was less than one third of the molecular mean free time (mft) $\Delta t = 4\cdot10^{-11}$. The height of the inlet and outlet in the channel was equal to 2λ and λ accordingly. The initial rotation of the lower wall which models the magnetic media of the disk drive for small time interval can be approximated as being set moving instantly with linear speed U_w from the rest in a direction parallel to itself and thereafter maintained in uniform motion. The speed U_w was taken to be 75m/s, which is representative for the devices of 10^4 rpm of the media with the 3.5in diameter. The flow itself is formed because the air molecules after diffuse reflection from this wall are carried along this motion.

The coupling procedure was realized as a step by step approximation. First in the whole computational region Navier-Stokes equations were solved by finite element method with diminishing cell size in these additional regions and the channel. As the boundary conditions for this solution is quite far from the head at $x=X_b$ and $x=-X_b$ asymptotic solution of the Rayleigh problem obtained by Cercignani and et al. [8] was used.

$$U_{as} = U_w \cdot \left(1 - erf(q) - s_1 g^{1/2} e^{-q^2} + s_2 g^{1/2} e^{-s_3 \cdot \frac{y}{\lambda}} \right), \tag{1}$$

where $erf(q)$ is the probability integral [5], $q = 0.77y \cdot (\lambda \cdot t \cdot V_T)^{-1/2}$, $g = \lambda(\pi t V_T)^{-1/2}$, $s_1=1.24$, $s_2=0.29$, $s_3=7.86$ and t is the time in seconds.

Though employment of continuum equations near inlet, outlet and inside of the channel is inadequate one can expect that on the boundaries 0CDEFG which are 20λ apart from them the influence of these small bad parts will be insignificant. So, at the next step within this part the DSMC method was employed with incoming molecular fluxes calculated from the stored mean velocities and their derivatives at these boundaries obtained at the previous step. Details of this procedure are outlined in the next section.

3 Simulations of the Incoming Fluxes

The linear size of the cells in finite element solution was four times and its time step was 25 times larger than in DSMC simulation, so that an interpolation of the stored macroscopic parameters was used. Molecular properties in a Navier-Stokes solution are represented through the Chapmen-Enskog velocity distribution function $f(u,v,w)$. It contains spatial derivatives of mean velocities and temperature. As in our problem there is no boundary temperature differences in the data and all mean velocities are relatively small in comparison with the sound velocity a_s (a_s=320m/s) we have neglected the temperature derivatives in $f(u,v,w)$. Simulations confirmed this assumption and the corresponding evidence will be presented later on in Sect. 4.

But now with decart thermal velocity components $u=(u'-u_{av})/V_T$, $v=(v'-v_{av})/V_T$, $w=w'/V_T$ being related to the most probable velocity V_T , $V_T^2=2kT/m$, where m is the molecular mass and k – Boltzmann's constant, it can be represented in the form:

$$f(u,v,w)= f_0(u,v,w)\left(1-\alpha_1 u^2-\alpha_2 v^2+\alpha_3 w^2-\alpha uv\right), \tag{2}$$

where all the coefficients in the bracket are less then 1, and $f_0(u,v,w)$ is the Maxwell velocity distribution function,

$$f_0(u,v,w)=\pi^{-3/2}\exp\left[-\left(u^2+v^2+w^2\right)\right]$$

And coefficients α_i are given by the formulas:

$$\alpha_1=c_0\cdot\frac{1}{3}\left(2\frac{\partial u_{av}}{\partial x}-\frac{\partial v_{av}}{\partial y}\right);\ \alpha_2=c_0\cdot\frac{1}{3}\left(2\frac{\partial v_{av}}{\partial y}-\frac{\partial u_{av}}{\partial x}\right);\ c_0\equiv\frac{5}{4}\cdot\frac{\lambda\sqrt{\pi}}{V_T}$$

$$\alpha_3=c_0\cdot\frac{1}{3}\left(\frac{\partial u_{av}}{\partial x}+\frac{\partial v_{av}}{\partial y}\right);\quad \alpha=c_0\cdot\left(\frac{\partial u_{av}}{\partial y}+\frac{\partial v_{av}}{\partial x}\right);$$

It is seen that coefficients α_i are connected by relation $\alpha_1+\alpha_2=\alpha_3$. The value of the dimensional constant c_0 is only $0.35\cdot10^{-9}$s. From finite element solution mean velocities and their derivatives on the boundary 0CD and EFG (cl. Fig.1) at different time moments are stored in special file and then they are red during DSMC simulations and used for simulating boundary incoming fluxes inside the DSMC computational region. For the latter procedure the magnitude of these coefficients α_i. is very important. On Fig.2 a) their absolute values *Ab* for some particular time moment are represented. It is seen that on the vertical boundaries 0C and FG coefficient α, which is shown by dashed line, exceeds strongly the others. Large cells of the finite element solution numbered by N_c with the help of the letters 0, C, D, E, F, G are made to correspond with the boundary shown on the Fig.1. Magnitudes of all coefficients change little for other time moments so that their value relation is maintained. In order to see more details of the coefficients α_i they are shown on the Fig.2 b) on a larger scale. The largest combination of derivatives in our problem for α amounts to or less than $10^8 s^{-1}$. As the value of the dimensional constant c_0 is only $0.35\cdot10^{-9}$seconds, the squared value α^2 reaches correspondingly utmost 10^{-3} and in what follows it will be neglected in comparison with 1. The same is true for the other coefficients α_i. Thus in the simulation of incoming fluxes one can neglect the coefficient squared in comparison with 1.

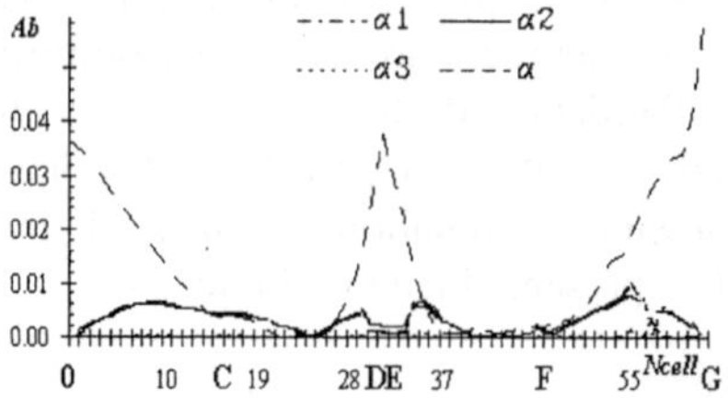

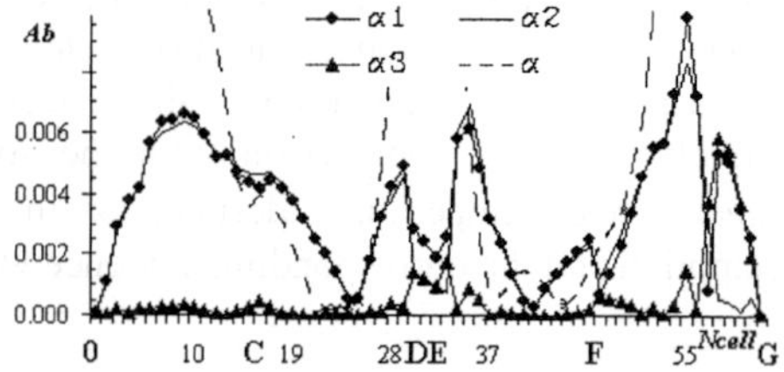

Fig. 2. Absolute values *Ab* of the coefficients α_i

Simulation of the decart velocity component for the incoming molecules from the multidimensional distribution can be proceeded by forming consequently conditional densities. As an example let us take velocity simulation of the molecules incoming through the boundary CD (cl. Fig.1). Consider simulation Z-component that is normal to the plane in Fig.1. Going over to one-dimensional density *P(w)* according to expression (3).

$$P(w) = \int_{-\infty}^{\infty} du \int_{-\infty}^{\infty} dv f = \frac{e^{-w^2}}{\sqrt{\pi}}(1 - \frac{\alpha_1 + \alpha_2}{2} + \alpha_3 w^2) \tag{3}$$

Note that additional integration over w in this formula results in the expression $1-(\alpha_1 + \alpha_2)/2+\alpha_3/2$ which is equal exactly to 1 because of aforementioned relation between the coefficients. Now the expression (3) with adopted precision could be rewritten as

$$P(w) = \frac{1 - \alpha_3/2}{\sqrt{\pi}} \cdot e^{-w^2(1-\alpha_3)}, \tag{4}$$

Now with the help of, for instance, Box-Muller procedure we have a simulated value of velocity component w_1.

$$w_1 = \frac{\sqrt{-\ln(\gamma_1)}}{1-\alpha_3/2} \cdot \cos(2\pi\gamma_2), \tag{5}$$

where γ1 and γ2 are consecutive uniform random numbers. Consider the two-dimensional density $P_2(v,w)$.

$$P_2(v,w) = \int_{-\infty}^{\infty} du f = \frac{e^{-(v^2+w^2)}}{\pi} \cdot (1 - \frac{\alpha_1}{2} - \alpha_2 v^2 + \alpha_3 w^2), \tag{6}$$

By making use of the already simulated value w_1, we form the conditional density $P_2(v|w_1)=P_2(v,w_1)/P(w_1)$

$$P_2(v \mid w_1) = \frac{e^{-v^2}}{\sqrt{\pi}} \cdot \frac{1 - \alpha_1/2 - \alpha_2 v^2 + \alpha_3 w_1^2}{1 - (\alpha_1 + \alpha_2)/2 + \alpha_3 w_1^2}, \tag{7}$$

which with adopted accuracy could be written through the expression

$$P_2(v \mid w_1) = \frac{e^{-v^2(1+\alpha_2)}}{\sqrt{\pi}} \cdot (1 + \alpha_2/2), \tag{8}$$

We are considering the molecules entering from above through the line CD (cl. Fig.1) so the possible values for v fall in the interval $-\infty<v<-v_{av}/V_T$ and for those values of u one has $-\infty<u<\infty$. Yet in order a molecule could enter this way in a particular cell with the boundary surface ΔS during time Δt it has had to be before inside of an adjacent cylinder of the volume $\Delta S|v\ V_T + v_{av}|\Delta t$. Thus the final expression for the v-simulating density will be as follows

$$P(v \mid w_1) = C \cdot (-v - V_{av}/V_T) \cdot P_2(v \mid w_1), \tag{9}$$

where in fact by the adopted precision there is no dependence on w_1 according to (8) and C is a norm constant not important for simple acceptance-rejection technique [5] utilized for obtaining the simulated value v_1.

Now by using (2) with adopted precision the conditional density

$$P_3(u|v_1, w_1) = f(u, v_1, w_1)/P(w_1) \cdot P_2(v_1/w_1),$$

where v_1 is a value of v velocity component already simulated by above mentioned technique, could be written in the form

$$P_3(u \mid v_1, w_1) = \frac{e^{-(1+\alpha_1)\cdot(u+\alpha \cdot v_1/2)^2}}{\sqrt{\pi} \cdot (1-\alpha_1/2)} \cdot (1 + O(\frac{\alpha^2}{4} \cdot v_1^2)), \tag{10}$$

Now rejecting automatically in the code very rarely realized values v_1 greater than 6, which means excluding the molecules with dimensional velocity component $v_d > 6 * V_T$ and taking into account that the maximum value of α is only 0.035 one can with accuracy one part to 100 neglect the small term in the bracket of (10).

And finally simulating from the pure exponent term we obtain [5]

$$u_1 = \frac{\sqrt{-\ln \gamma_1}}{(1+\alpha_1/2)} \cdot \cos(2\pi \cdot \gamma_2) - \frac{\alpha}{2} v_1$$

with γ_1 and γ_2 being two next random numbers.

4 Simulation of the Temperature

In order to confirm the possibility of using the expression (2) for the Chapman-Enskog velocity distribution function $f(u,v,w)$ for our case we performed simulations to evaluate the temperature changes. Determine kinetic temperature for X-velocity component by $T_x = m(<u^2> - <u>^2)/k$, where $<u^2>$ and $<u>$ are simulated mean values. Determining in the same way T_y and T_z as well as the general temperature T through $T = (T_x + T_y + T_z)/3$ we have simulated this temperature T. On the Fig.3 are shown the results of this simulation.

On the left a) of magnetic head (cl. Fig.1) it is depicted for different times: by solid line for $t_1 = 10^{-8}$s, dashed for $t_2 = 10^{-7}$s., and chain-dotted line for $t_3 = 5 \cdot 10^{-7}$s. And on the right b), it is represented for region after the outlet with the same designation of the curves. In both cases the distance is given in mfp units. It is clearly seen that except for some statistical scattering the space temperature changes are not noticeable so that in general Chapmen-Enscog expression the space temperature derivatives could be really neglected.

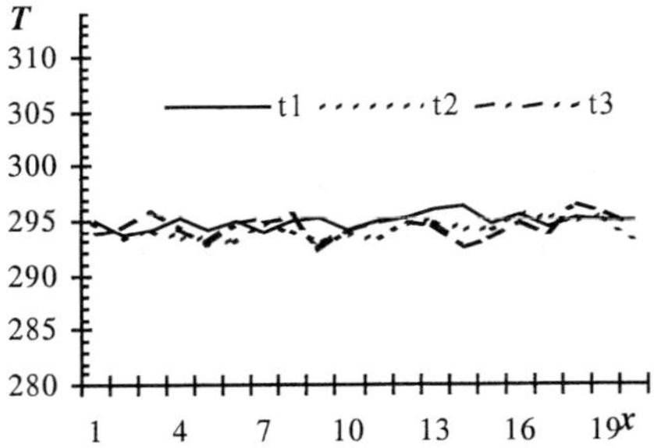

Fig. 3a). Before the inlet

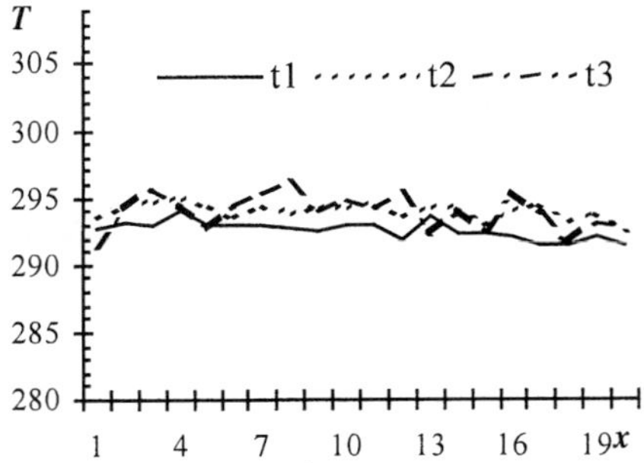

Fig. 3b) After the outlet

5 Numerical Evaluation of the Present Coupling Scheme

The step by step approximation used in the present paper for coupling of two solutions was evaluated by comparison of successive approximations for mean velocity components u_{av} and v_{av} as well as for density n_{av} at the boundary 0CDEFG (cl. Fig.1) inevitably at particular time moment. But as a matter of fact for the other times the whole picture is roughly the same. The relative differences of zero approximation as obtained from the finite element solution and the first one after DSMC simulations $Ru=(u^0_{av}-u^1_{av})/u^1_{av}$ and in the same way for Rv and Rn are shown on the Fig.4. The relative differences for X-velocity component Ru as shown on the left in Fig.4 oscillate near a zero level with peak-to-peak values mostly less than 10% but somewhat grow between $26<N_c<50$ where the component u_{av} is itself small so that statistical scattering becomes larger being apparently the origin of it.

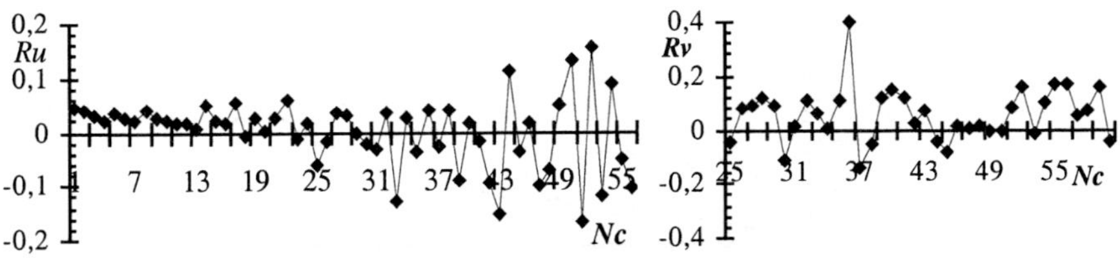

Fig. 4. Relative differences Ru and Rv, composed from successive approximations.

The relative differences for Y-velocity component $Rv_{av}=(v^0_{av}-v^1_{av})/v^1_{av}$ are shown in Fig.4 on the right. Their chaotic oscillations near zero level with peak values amount to 15%. As the v_{av} magnitude is mostly much less than that of u_{av}, it can be explained again by statistical scattering. The relative density differences $Rn_{av}=(n^0_{av}-n^1_{av})/n^1_{av}$ are shown on Fig.5. It is clearly seen that they vary near the 1%-2% level of precision with peak-to-peak values of about 3%.

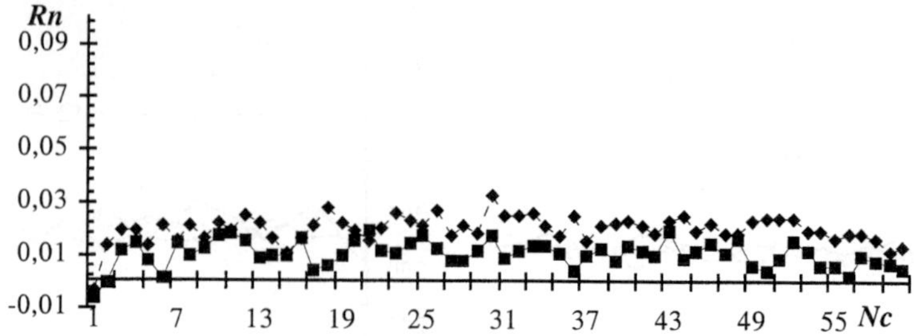

Fig. 5. Relative density differences *Rn* of successive approximations.

Thus one can say that except the random statistical scattering inherent to DSMC method the mean level of these relative differences does not exceed 15%. So that convergence of this step-by-step coupling approximation is satisfactory.

6 Numerical Simulation Results

Simulations for our DSMC parallel code were performed on the St. Petersburg university clusters. The largest of them contains 40 processors Intel Pentium III 933 MHz. Parallelization of the code was proceeded through independent realizations when each processor produced the whole solution, the results being gathered then on some leading one for averaging and outputting. This scheme allowed us sometimes use an additional cluster located far away but connected with one of the ours through the Internet channel into a metacomputer with good efficiency, for details see [9].

All previous researchers [2], [3], [6] have used the assumption that before the inlet and after outlet of the channel unperturbed atmospheric pressure could be used as the boundary condition. Thus they entirely neglected the slowing down of the flow by magnetic head. We are now in a position to check it.

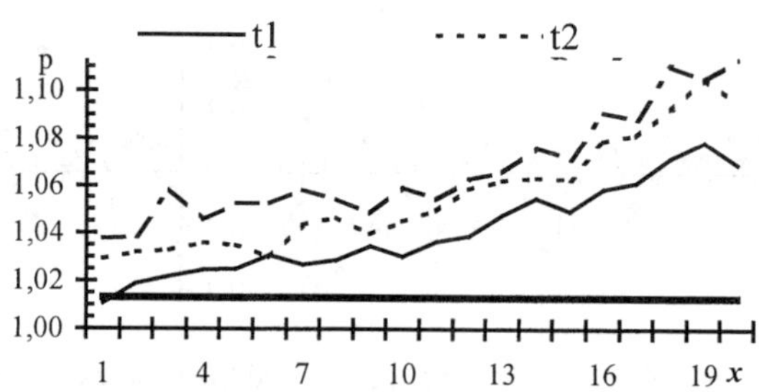

Fig. 6. Increase of the pressure *p* before the channel inlet for different times.

For this purpose we simulated pressure before the channel and represented it for the height y=λ/3 on the Fig.6 with the time symbols for the curves as in the foregoing and by the horizontal bold solid line the atmospheric pressure. It is clearly seen that the slowing down of the flow is present by a 8% increase in pressure before the channel inlet.Thus one may conclude that the pressure boundary conditions used in previous papers at least for $U_w \geq 75$m/s are not fulfilled. Both of the pressures before and after the magnetic head are not atmospheric and are quite different near the inlet and outlet.

On the Fig.7 the pressure along the channel is shown for different times with the curves: solid – for $t_1=5\cdot10^{-9}$s, pointed – for $t_2=3\cdot10^{-8}$s and dashed – for $t_3=6\cdot10^{-8}$s. On the Fig.8 streamwise velocity u_{av} is presented for different height values.

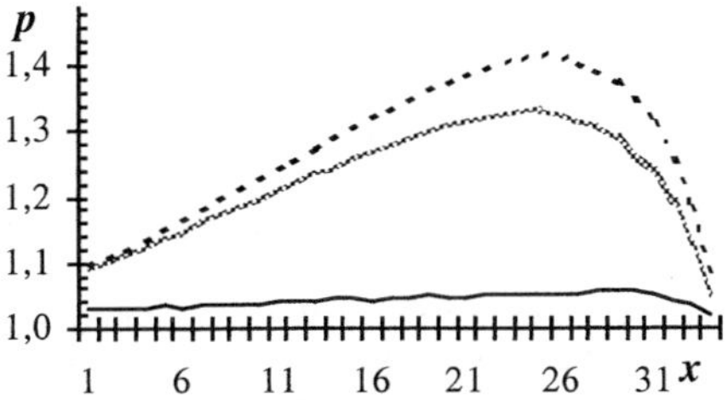

Fig. 7. Pressure along the channel for different times.

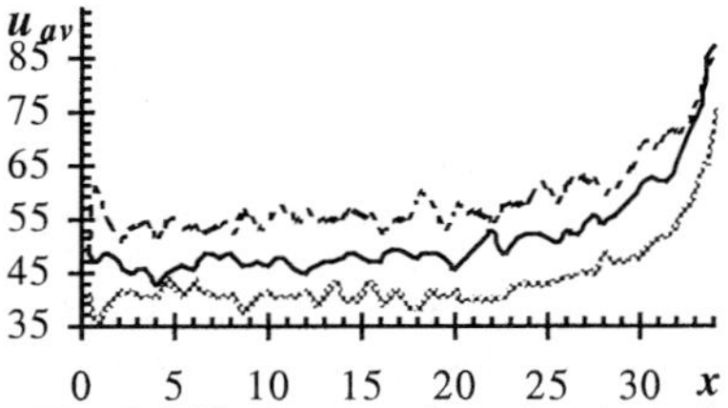

Fig. 8. Velocity u_{av} along the channel: dashed - y=λ/3, solid - y=2λ/3, pointed - y=λ.

Near the end of the channel all of them are growing thus forming back rapid decrease of the pressure maximum in Fig.7. Finally on the Fig.9 it is shown the time development of the whole pressure on magnetic head.

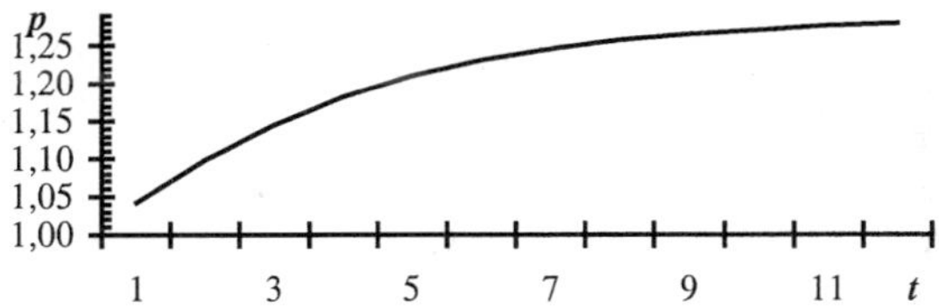

Fig. 9. General pressure on the upper wall, time t in units of $5\cdot10^{-9}$s .

Conclusions

Based on the coupling of the DSMC method with a continuum solution at the outer regions it is found considerable deviations from the atmospheric pressure before and after the channel. Thus quality of the pressure boundary conditions much used in the previous papers on this topic seems to be doubtful. Space and time distributions of different parameters of the flow are given.

Acknowledgments. This work was partially supported by the RFBR grant N01-01-00315.

References

1. Cowburn, R.P. The attractions of magnetism for nanoscale data storage. Phil. Trans. R. Soc. Lond. A358 (2000)281–301
2. Igarashi, S. Analysis of Gas Film Lubrication Using Monte Carlo Direct Simulation Method. Progr. in Astron. and-Astron. 159 (1992) 303–310
3. Alexander, F.J., Garcia, A.L., and Alder, B. Direct Simulation Monte Carlo for Thin-Film Bearings. Phys. Fluids. 6 (1994) 3854–3860
4. Henshaw, W.D., Reyna, L.G., and Zufria, J. Compressible Navior-Stokes Computations Air-Bearings. J.Tribology, 18 (1991) 73–79
5. Bird, G.A.: Molecular Gas Dynamics and the Direct Simulation of the Gas Flows. Claren. Press, Oxford, (1994)
6. Wu, J.-S., and Tseng, K.-C. Analysis of Internal Microscale Gas Flows with Pressure Boundaries Using the DSMC Method. In Bartel T.J., and Gallis M.A. (eds.): Rarefied Gas Dynamics. Proc. 22d. Int.Symp. Sydney 2000, AIP Conference Proceedings 585, American Institute of Physics, N.Y. (2001) 486–493
7. Memnonov, V.P. Direct Simulation Monte Carlo Method: the Other Way of Reunion for Decoupled Processes (in Russian). Math. modeling.11 (1999) 77–82
8. Cercignani C., and Sernagiotto F., "Rayleigh's Problem at Low Mach Numbers according to Kinetic Theory," in The Fourth Int. Symp. on RGD, edited by J.H.d Leeuw, Acad. Pr., N.Y., 1965 v.1, pp.332–353.
9. Galyuk, Y. P., Memnonov, V.P., Zhuravleva, S.E., and Zolotarev, V.I., "Grid Technology with Dynamic Load Balancing for Monte Carlo Simulations," in Applied Parallel Computing, edited by J. Fagerholm, J. Haataja, J. Jarvinen, M. Lyly, P. Raback, V. Savolainen, Lecture Notes in Computer Science, Vol. 2367. Springer-Verlag, Berlin Heidelberg New York, 2002, pp.515–520.

Markowitz-Type Heuristics for Computing Jacobian Matrices Efficiently

Andreas Albrecht, Peter Gottschling[1], and Uwe Naumann[2]

[1] Department of Computer Science, University of Hertfordshire
{A.Albrecht, P.1.Gottschling}@herts.ac.uk
[2] Mathematics and Computer Science Division, Argonne National Laboratory
naumann@mcs.anl.gov

Abstract. We consider the problem of accumulating the Jacobian matrix of a nonlinear vector function by using a minimal number of arithmetic operations. Two new Markowitz-type heuristics are proposed for vertex elimination in linearized computational graphs, and their superiority over existing approaches is shown by several tests. Similar ideas are applied to derive new heuristics for edge elimination techniques. The well known superiority of edge over vertex elimination can be observed only partially for the heuristics discussed in this paper. Nevertheless, significant improvements can be achieved by the new heuristics both in terms of the quality of the results and their robustness with respect to different tiebreaking criteria.

1 Introduction

Consider the nonlinear vector function $F : I\!R^2 \rightarrow I\!R^2$ that is given by the following sequence of scalar assignments:

$$v_1 = v_{-1}v_0;\ v_2 = \sin(v_1);\ v_3 = v_1 v_2;\ v_4 = \cos(v_3);\ v_5 = \exp(v_3).$$

For simplicity, all variables carry a unique index. There are $n = 2$ independent, $p = 3$ intermediate, and $m = 2$ dependent variables. Mathematical functions that are implemented as computer programs written in an imperative programming language such as C or Fortran can always be decomposed to meet this requirement. The structure of such computations can be visualized by a directed acyclic graph (dag) $G = (V, E)$ as shown in Figure 1. If one assumes that jointly continuous local partial derivatives

$$c_{j,i} \equiv \frac{\partial}{\partial v_i}\varphi_j(v_k)_{k\prec j}, \quad j = 1, \ldots, p+m \quad ,$$

of the elemental functions (e.g., $*, \sin, \cos, \exp$) with respect to its arguments exist in some neighborhood of the current point, the corresponding numerical values, computed as

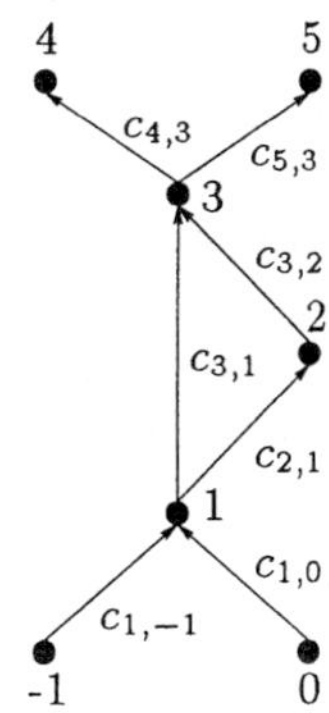

Fig. 1. F

$$c_{1,-1} = v_0;\ c_{1,0} = v_{-1};\ c_{2,1} = \cos(v_1);\ c_{3,1} = v_2;\ c_{3,2} = v_1$$

P.M.A. Sloot et al. (Eds.): ICCS 2003, LNCS 2658, pp. 575–584, 2003.

$$c_{4,3} = -\sin(v_3); \; c_{5,3} = v_5 \quad ,$$

can be attached to the edges in the dag. The notation $k \prec j$ is used to indicate that v_k is an argument of φ_j. Thus one gets the linearized computational graph (or c-graph) $G = (V, E)$ of F as displayed in Figure 1. Its vertices $V = X \cup Z \cup Y$, where $X = \{1-n, \ldots, 0\}$, $Z = \{1, \ldots, p\}$, and $Y = \{p+1, \ldots, p+m\}$, are numbered consistently with respect to dependence, that is, $i \prec^+ j \Rightarrow i < j$. Here, $\prec^+$ denotes the transitive closure of the dependence relation $\prec$. The vertices in X (Z; Y) are referred to as minimal or independent (intermediate; maximal or dependent) vertices.

The objective is to transform the program that implements F into one that computes the Jacobian matrix (or Jacobian)

$$F' = F'(\mathbf{x}_0) = \left(\frac{\partial y_i}{\partial x_j}(\mathbf{x}_0) \right)_{i=1,\ldots,m,\; j=1,\ldots,n}$$

of F with respect to the n inputs for a given argument $\mathbf{x}_0$ such that a minimal number of scalar fused multiply-add floating-point operations (**fmas**) are performed. Once numerical values have been computed for all local partial derivatives, scalar floating-point multiplications and additions are the only arithmetic operations required to accumulate F'. The accumulation of F' can be regarded as an elimination procedure in the c-graph G of F, as introduced in [1]. The original c-graph is transformed into a subgraph of the directed complete bipartite graph $K_{n,m}$ such that the labels on the remaining edges are exactly the nonzero elements of the Jacobian. The result of this transformation applied to the c-graph from Figure 1 is displayed in Figure 2.

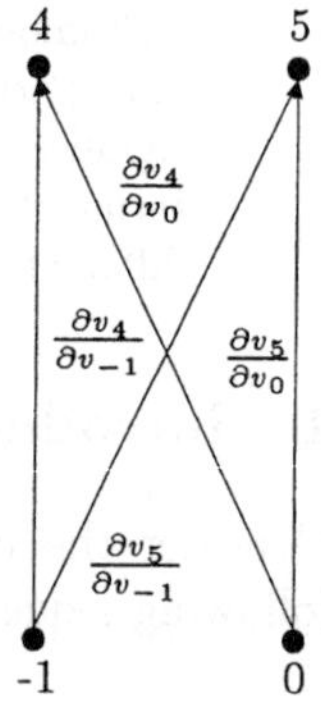

Fig. 2. F'

In this paper we develop heuristics for eliminating vertices and edges such that the overall cost of computing F' is minimized. A detailed discussion of the corresponding theory can be found in [2]. Here, we introduce only a minimal subset of the framework, in order to focus on new heuristics for vertex and edge elimination in c-graphs. The structure of the paper is as follows. In Section 2 we introduce vertex and edge elimination in c-graphs. Various well known and new Markowitz-type heuristics for both vertex and edge elimination are presented in Section 3. Their properties and performance are discussed in Section 4, and numerical results are presented. Conclusions are drawn in Section 5.

2 Jacobians by Vertex and Edge Elimination

The origins of both vertex and edge elimination in c-graphs are in *automatic differentiation* (AD) [3–6]. This technique modifies the semantics of numerical programs such that derivatives of the underlying vector function can be computed efficiently with machine accuracy. In contrast to divided difference

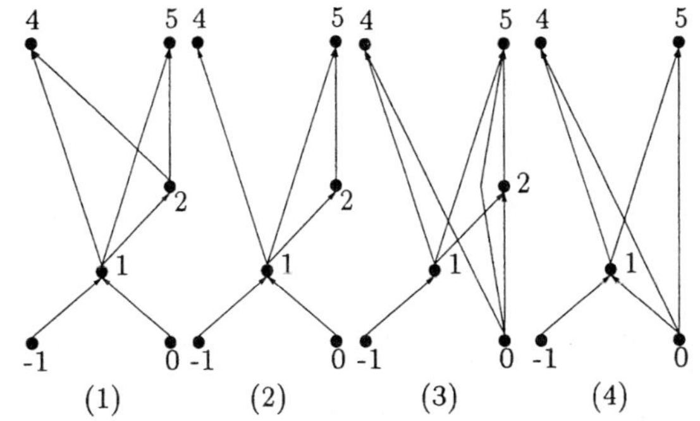

Fig. 3. Vertex and edge elimination

approximations, AD exploits the chain rule to compute Jacobian times vector products $\dot{\mathbf{y}} = F'\dot{\mathbf{x}}$ in forward mode and transposed Jacobian times vector products $\bar{\mathbf{x}} = (F')^T\bar{\mathbf{y}}$ in reverse mode (see [6, Chapter 3] for details). In particular, the Jacobian itself can be obtained at a cost of $n|E|$ in forward mode and $m|E|$ in reverse mode by letting the vectors $\dot{\mathbf{x}}$ and $\bar{\mathbf{y}}$ range over the Cartesian basis vectors in $\mathbb{R}^n$ and $\mathbb{R}^m$, respectively. The number of edges in the c-graph of $G = (V, E)$ is denoted by $|E|$.

Alternatively, F' can be computed by eliminating all intermediate vertices or edges in G as follows. When eliminating an intermediate vertex j, new edges are introduced connecting the predecessors of j with its successors. A new edge (i, k) is labeled with the product of the labels of (j, k) and (i, j). Parallel edges are merged, and the corresponding edge labels are added. Finally, j is removed together with its incident edges. The elimination of vertex 3 from the c-graph shown in Figure 1 leads to graph (1) in Figure 3. For example, the new label of $(1, 4) \in E$ is equal to $c_{4,1} = c_{4,3}c_{3,1}$. The elimination of vertex 2 in graph (3) leads to graph (4) and, for example, $c_{5,0} = c_{5,0} + c_{5,2}c_{2,0}$. The correctness of the vertex elimination rule is shown in [1]. The number of **fmas** involved in the elimination of j is referred to as the *Markowitz degree* of j, and it is equal to $\mu_j = |\{i : i \prec j\}||\{k : j \prec k\}|$.

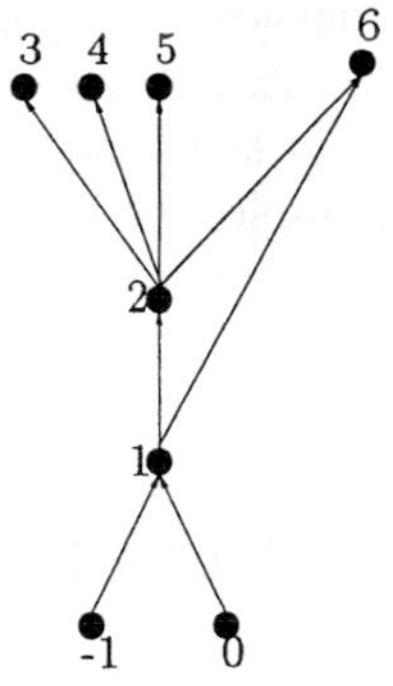

Fig. 4. Lion

The elimination of a vertex j is equivalent to the simultaneous *front elimination* of all edges leading into it. Similarly, the elimination of j is equivalent to the simultaneous *back elimination* of all edges emanating from it. An edge (i, j) is front eliminated by connecting i with all successors of j. For all successors k of j, the new edges (i, k) are labeled with $c_{k,i} = c_{k,j}c_{j,i}$. If (i, k) existed before, then $c_{k,i} = c_{k,i} + c_{k,j}c_{j,i}$. The edge (i, j) is removed after this. The number of **fmas** required to front eliminate (i, j) is equal to $|\{k : j \prec k\}|$. The front elimination of $(0, 1)$ transforms graph (2) into graph (3) in Figure 3. Analogously, an edge (j, k)

is back eliminated by connecting all predecessors of j with k. For all predecessors i of j the new edge (i,k) is labeled with $c_{k,i} = c_{k,j}c_{j,i}$. Again, the label becomes $c_{k,i} = c_{k,i} + c_{k,j}c_{j,i}$ if (i,k) existed before. Finally, (j,k) is removed. The number of **fmas** required to back eliminate (j,k) is equal to $|\{i : i \prec j\}|$. In Figure 3, graph (2) can be obtained from graph (1) by back elimination of $(2,4)$. Newly generated edges are referred to as fill-in. Absorption takes place whenever two parallel edges are merged.

The set of all valid vertex eliminations is contained within the set of all valid edge eliminations. Hence, the optimal vertex elimination sequence is contained within the set of all edge elimination sequences, and thus the optimal edge elimination sequence performs at most the same number of arithmetic operations as the optimal vertex elimination sequence. The *lion graph* [7] displayed in Figure 4 represents one example where the optimal edge elimination sequence involves fewer operations than does the optimal vertex elimination sequence. Both elimination sequences require $4+8=12$ **fmas**. The back elimination of $(2,6)$ followed by the elimination of 1 and 2 reduces this number by one. Refer to [2] for a more detailed investigation of this *vertex-edge discrepancy*. It is the motivation for introducing edge elimination in addition to the conceptually much easier vertex elimination method.

We assume the chain rule to be associative over the floating point numbers. In other words, the numerical values of the entries in F' do not depend on the order in which the intermediate vertices or edges are eliminated. However, the computational cost varies. In Figure 1, for example, the vertex elimination sequence $[1,2,3]$ performs $4+2+4=10$ **fmas**. The reader may wish to verify that $[2,1,3]$ takes only $1+2+4=7$ **fmas**. The cost of computing F' by using either the forward or the reverse mode of AD is $2 \cdot 7 = 14$. Even on this small example, the operations count can be reduced by a factor of two. This is the primary motivation for investigating heuristics for vertex and edge elimination in Section 3.

3 Heuristics

Both the vertex and edge elimination problems in c-graphs are conjectured to be NP-complete [8, 1, 7]. No polynomial algorithm is known for solving them exactly. The problem of minimizing the fill-in under vertex elimination was shown to be NP-complete by Herley [9] in an unpublished adaption of a note by Gilbert [10] on a result by Rose and Tarjan [11] about vertex elimination techniques for solving sparse linear systems. So far, it remains unclear whether the same is true for edge and *face* [2] elimination.

In the following we write $H(G) = i$ whenever the application of a heuristic H to a c-graph G gives the vertex i as a result. An analogous notation is used for edge elimination heuristics. All heuristics H are defined such that $|H(G)| = 1$, that is, the result of applying H to G should contain a single vertex or edge.

3.1 Forward Vertex and Edge Elimination

The forward vertex elimination mode FM_v eliminates the intermediate vertices in ascending order with respect to their indices, that is, $\mathrm{FM}_v(G) = j \Leftrightarrow \forall i \in V: j \le i$. The same idea can be applied to edge elimination. For reasons of consistency, we require the forward edge elimination mode FM_e to have the same computational cost as FM_v. This can be achieved by pure back edge elimination sequences in lexicographical order or by pure front edge elimination sequences in switched lexicographical order. For example, the latter can be written as $\mathrm{FM}_e(G) = (i,j)_f$ if

$$(i,j) \in E \cap (V \times Z) \quad \wedge \quad \forall (k,l) \in E \cap (V \times Z): j \le l \ \vee \ j = l \ \wedge \ i \le k.$$

The fact that an edge (i,j) is front (back) eliminated is denoted by $(i,j)_f$ $((i,j)_b)$. Note that the set of edges that can be front eliminated is restricted to those having an intermediate vertex as target.

3.2 Reverse Vertex and Edge Elimination

In reverse vertex elimination mode the intermediate vertices are eliminated in descending order starting with p, that is, $\mathrm{RM}_v(G) = j \Leftrightarrow \forall i \in V: j \ge i$. The extension to edge elimination sequences is similar to FM, for example, $\mathrm{RM}_e(G) = (i,j)_f$ if

$$(i,j) \in E \cap (Z \times V) \quad \wedge \quad \forall (k,l) \in E \cap (Z \times V): j \ge l \ \vee \ j = l \wedge i \ge k.$$

To ensure uniqueness of the result, one must combine all heuristics with a tiebreaking criterion or even a hierarchy of tiebreakers. We use either FM or RM (for vertices or edges, depending on the context) as the "bottom line," since we always have $|\mathrm{FM}(G)| = 1$ as well as $|\mathrm{RM}(G)| = 1$.

3.3 Lowest Markowitz

The lowest Markowitz (LM) degree-first heuristic was introduced in [1]. It was motivated by a similar idea from the theory of direct methods for solving sparse linear systems.

Heuristic 1 *$LM_v(G) = j$ if*

$$\forall i \in V \ : \ \mu_j \le \mu_i \quad \wedge \quad \forall i \in V', V' = \{k \in V: \mu_j = \mu_k\}: j = RM_v(G') \quad ,$$

where G' is the subgraph of G that is induced by V'.

$G' = (V', E')$ is said to be induced by a vertex set $V' \subseteq V$ if $E' = \{(i,j) : i \in V' \wedge j \in V'\}$. In general, we always have the choice to use either FM_v or RM_v as the ultimate tiebreaker to make the result of a heuristic unique. This choice, however, may affect the performance of the heuristic significantly.

As for FM_v and RM_v, the computational effort of LM_v is constant per vertex and per iteration. The number of iterations is equal to the number of intermediate vertices p, and the vertices taken into account are the intermediate vertices that have not been eliminated yet. Hence, the maximal overall effort is $p(p+1)/2 \in \mathcal{O}(|V|^2)$.

The Markowitz degree of $(i,j) \in E$ is defined as $\min(|\{k\colon j \prec k\}|, |\{k\colon k \prec i\}|)$. The edge (i,j) is front eliminated if $|\{k\colon j \prec k\}| \leq |\{k\colon k \prec i\}|$. Otherwise, it is back eliminated. The lowest-Markowitz heuristic for edge elimination LM_e is defined analogous to Heuristic 1. Its complexity, however, also depends on the fill-in that is generated. Although, the termination of edge elimination was shown in [7], it remains unclear whether the cost of edge elimination sequences is still polynomial in the worst case. This question is the subject of ongoing research.

3.4 Lowest Relative Markowitz

The lowest relative Markowitz (LRM) degree-first heuristic is an extension of LM. It was introduced in [12]. Let $\iota_j = |\{k\colon k \in X \wedge k \prec^+ j\}|$, $\delta_j = |\{k\colon k \in Y \wedge j \prec^+ k\}|$, and $\hat{\mu}_j = \mu_j - \iota_j \cdot \delta_j$. The relative Markowitz degree is defined as the difference between the Markowitz degree and the *dependence degree* $\iota_j \cdot \delta_j$ of the vertex j. The idea is to maximize the dependence degree while, at the same time, minimizing μ_j. This heuristic is usually combined with LM_v as tiebreaker.

Heuristic 2 *$LRM_v(G) = j$ if*

$$\forall i \in V : \hat{\mu}_j \leq \hat{\mu}_i \;\wedge\; \forall i \in V', V' = \{k \in V\colon \hat{\mu}_j = \hat{\mu}_k\}\colon j = LM_v(G') \quad ,$$

where G' is the subgraph of G that is induced by V'.

See [12] for further details.

To formulate LRM for edge elimination, we set

$$\hat{\mu}((i,j)) = \min(|\{k : j \prec k\}| - \delta_j, |\{k : k \prec i\}| - \iota_i).$$

As for LM_e, an edge (i,j) is front eliminated if $|\{k : j \prec k\}| - \delta_j \leq |\{k : k \prec i\}| - \iota_i$ and else back eliminated. Thus, LRM_e can be derived immediately from Heuristic 2.

3.5 Maximal Overall Markowitz Degree Reduction

The maximal overall markowitz degree reduction (MOMR) heuristic represents the global pendant to LM. It considers the effect of the elimination of a vertex on the Markowitz degrees of its neighbors instead of its own Markowitz degree. Therefore we define the overall Markowitz degree of a c-graph $G = (V,E)$ as $M = M(G) = \sum_{i \in V} \mu_i$. The overall Markowitz degree reduction of a vertex $i \in V$ is defined as $\mu_j^- = M(G) - M(G-j)$, where $G-j$ is the c-graph after the elimination of j.

Heuristic 3 *$MOMR_v(G) = j$ if*

$$\forall i \in V : \mu_i^- \leq \mu_j^- \ \wedge \ \forall i \in V', V' = \{k \in V : \mu_j^- = \mu_k^-\} : j = LM_v(G') \quad ,$$

where G' is the subgraph of G that is induced by V'.

Initially, MOMR_v attempts to reduce the Markowitz degrees of its neighbors as much as possible, thus making use of absorption while trying to avoid excessive fill-in. Only as a second step does it take the vertex's own Markowitz degree into account. MOMR_v reduces the cost of eliminating the remaining vertices as much as possible.

To calculate the Markowitz degree reduction of k caused by the elimination of some $j \prec k$, one must compute the difference of their respective predecessor sets. Similarly, the successor sets must be considered if $k \prec j$. Hence, MOMR_v involves at most $\mathcal{O}(a^2\,|V|^2)$ operations, where a is the average number of predecessors (or successors) per vertex over all elimination steps.

The formulation of MOMR for edge elimination is straightforward by comparing the overall Markowitz degree of G before and after the front and back elimination of an edge.

3.6 Lowest Markowitz Minimal Damage

This lowest Markowitz minimal damage (LMMD) heuristic combines LM and MOMR. The effect of MOMR is twofold. On one hand, it enforces the elimination of vertices that reduce the Markowitz degrees of their neighbors maximally (or at least do not increase them too much). On the other hand, it implicitly prefers vertices with high Markowitz degrees, and hence incurs a high elimination cost.

The idea behind LMMD is to look for vertices with low Markowitz degrees that cause a minimal increase of the Markowitz degree on other vertices (minimal damage). If the overall Markowitz degree without considering $j \in V$ of a c-graph $G = (V, E)$ is defined as $\hat{M}_j(G) = \sum_{i \neq j} \mu_i$, then the damage caused by the elimination of j is set to $d_j = M(G - j) - \hat{M}_j(G)$. In order to increase the flexibility of LMMD, the damage is scaled with a weight w.

Heuristic 4 *$LMMD_v(w, G) = j$ if $\forall i \in V \ \mu_j + wd_j \leq \mu_i + wd_i$ and $\forall i \in V', V' = \{k \in V : \mu_j + wd_j = \mu_k + wd_k\} : j = RM_v(V')$.*

W.l.o.g., we choose RM_v as a tiebreaker. Other criteria can be used as well; however, especially LM_v as tiebreaker would be somewhat redundant because of the already present implicit orientation toward the lowest Markowitz degree. Choosing a small factor w focuses on the Markowitz degree of the vertex and reduces the value of the damage to a tiebreaker. Large values, on the other hand, emphasize the Markowitz degree reduction of the neighboring vertices and degrade the current degree to a secondary criterion. The computational effort is identical to that of MOMR_v.

Again, the formulation of LMMD_e is analogous to Heuristic 4, when considering the damage caused by front and back edge elimination, respectively.

Table 1. Elimination Costs for Graphs from Structured Random C-Graph Generator

n	p	m	FM	RM	LM1	LM2	LRM1	LRM2	MOMR1	MOMR2	LMMD
3	64	2	108	87	89	77	93	86	83	81	77
			108	87	104	87	106	86	83	81	78
3	206	3	638	418	374	316	392	389	337	342	312
			638	418	553	351	448	417	333	336	334
10	236	10	1216	609	480	425	520	499	434	441	417
			1216	609	666	462	507	467	448	454	447
10	243	3	1567	435	468	369	448	430	411	406	367
			1567	435	837	392	813	418	428	398	380
10	245	1	1605	349	454	326	461	430	356	357	325
			1605	349	801	349	614	352	339	339	332
3	236	10	718	649	463	420	526	519	435	462	415
			718	649	646	460	669	488	440	459	453
1	236	10	337	651	423	373	444	432	342	394	356
			337	651	337	404	742	426	334	368	352

4 Numerical Results

The generation of c-graphs from real-world application programs requires a fully functional compiler front-end to transform the program into an abstract syntax tree. We are using the tool EliAD [13] to perform this task. The heuristics were applied to the accumulation of the Jacobian of Roe's numerical flux [14], coating thickness standardization, combustion of propane, full formulation, human heart dipole, and flow in a channel. All but the Roe flux are part of the MINPACK-2 test problem collection [15]. The computational graphs were generated by unrolling all loops for a given input.

Unfortunately, the structure of the resulting graphs is not very complicated, and it shows a strong regularity. The LM heuristic works well in all these cases. Although, this feature is likely to be exhibited by many real-world simulation codes, the corresponding graphs are not very useful for studying the effect of heuristics on the number of `fmas` that are required to accumulate the corresponding Jacobian. Therefore, we have developed a random c-graph generator [16]. It allows us to check the behavior of various heuristics on a large number of c-graphs of varying sizes.

In Table 1 we have listed the results obtained by applying the heuristics discussed in this paper to a number of randomly generated c-graphs. For a given graph, the first row shows the results of the heuristics for vertex elimination, and the second row shows the values obtained by the corresponding heuristic for edge elimination. We observe that MOMR and LMMD show the most consistent behavior, while LMMD delivers the best results in most cases. RM outperforms FM if there are fewer outputs than inputs. FM delivers better results if the number of outputs exceeds the number of inputs. If there are as many outputs as there are inputs, then RM seems to perform better. A likely reason is that for the intermediate vertices the number of successors is arbitrary, whereas the number of predecessors is limited by two. LM appears to be a reasonable compromise for a large number of c-graphs. Typically, it performs almost as well as the best heuristic.

An important goal of the research that led to this paper was the investigation of Markowitz degree-based heuristics for edge elimination. We expected our results to be improved by simply reformulating the ideas behind the vertex elimination heuristics in the context of edge elimination. The opposite appears to be the case. Apart from the last example, none of the edge elimination heuristics delivers better results than the best vertex elimination heuristic. Apparently, the heuristics are influenced negatively by the considerably widened search space. We conclude that the implicit locality of eliminating all edges incident to a vertex is usually more beneficial than the higher degree of freedom when considering the heuristics that are discussed here.

5 Conclusion

Both MOMR and LMMD represent good choices for Markowitz-based vertex elimination heuristics. On a representative test set (a subset of which was presented here) they exhibit a consistent behavior in terms of the quality of the elimination sequences that were generated. In particular, they exhibit an increased robustness with respect to the choice of different tiebreaking criteria. Similar ideas were applied to edge elimination with no noticeable improvement in the cost of the elimination sequences generated. We conclude that different criteria must be developed in order to exploit the power of edge elimination for accumulating Jacobians. How edge elimination sequences can reduce the cost of vertex elimination sequences is the subject of ongoing research.

Similar to the approach taken in [17], we are implementing logarithmic simulated annealing algorithms [18] for edge elimination. We hope to observe discrepancies between the optimal vertex and edge elimination sequences for real-world applications. From the structure of such problems we expect to learn more about suitable criteria for edge elimination heuristics. Furthermore, a detailed investigation of the energy landscape of the various combinatorial optimization problems arising in Jacobian computation will allow us to gain insight into potential improvements. Finally, we conclude that, in view of our promising results, robust heuristics for accumulating Jacobians efficiently should become a key feature of software tools for automatic differentiation.

Acknowledgments

This research is supported by the UK's Engineering and Physical Sciences Research Council under grant GR/R/38101/01.

Naumann is supported by the Mathematical, Information, and Computational Sciences Division subprogram of the Office of Advanced Scientific Computing Research, U.S. Department of Energy, under Contract W-31-109-ENG-38.

References

1. Griewank, A., Reese, S.: On the calculation of Jacobian matrices by the Markovitz rule. In: [5]. (1991) 126–135
2. Naumann, U.: Optimal accumulation of Jacobians by elimination methods on the dual computational graph. Mathematical Programming (2002) to appear.
3. Berz, M., Bischof, C., Corliss, G., Griewank, A., eds.: Computational Differentiation: Techniques, Applications, and Tools, Philadelphia, SIAM (1996)
4. Corliss, G., Faure, C., Griewank, A., Hascoet, L., Naumann, U., eds.: Automatic Differentiation of Algorithms – from Simulation to Optimization, New York, Springer (2002)
5. Corliss, G., Griewank, A., eds.: Automatic Differentiation: Theory, Implementation, and Application, Philadelphia, SIAM (1991)
6. Griewank, A.: Evaluating Derivatives. Principles and Techniques of Algorithmic Differentiation. Number 19 in Frontiers in Applied Mathematics. SIAM, Philadelphia (2000)
7. Naumann, U.: Efficient Calculation of Jacobian Matrices by Optimized Application of the Chain Rule to Computational Graphs. PhD thesis, Technical University Dresden (1999)
8. Bischof, C., Haghighat, M.: Hierarchical approaches to automatic differentiation. In: [3]. (1996) 82–94
9. Herley, K.: A note on the NP-completeness of optimum Jacobian accumulation by vertex elimination. Presentation at: Theory Institute on Combinatorial Challenges in Computational Differentiation (1993)
10. Gilbert, J.: A note on the NP-completeness of vertex elimination on directed graphs. J. Alg. Disc. Meth. **1** (1980) 292–294
11. Rose, D., Tarjan, R.: Algorithmic aspects of vertex elimination on directed graphs. J. Appl. Math. **34** (1978) 176–197
12. Naumann, U.: An enhanced Markowitz rule for accumulating Jacobians efficiently. In Mikula, K., ed.: ALGORITHMY'2000 Conference on Scientific Computing, Slovak University of Technology, Bratislava, Slovakia (2000) 320–329
13. Tadjouddine, M., Forth, S., Pryce, J., Reid, J.: Performance issues for vertex elimination methods in computing Jacobians using Automatic Differentiation. In: Proceedings of the ICCS 2000 Conference. Volume 2330 of Springer LNCS. (2002) 1077–1086
14. Roe, P.: Approximating Riemann solvers, parameter vectors, and difference schemes. J. Comp. Physics (1981) 357–372
15. Averik, B., Carter, R., Moré, J.: The MINPACK-2 test problem collection (preliminary version). Technical Report 150, Mathematical and Computer Science Division, Argonne National Laboratory (1991)
16. Albrecht, A., Gottschling, P., Naumann, U.: Logarithmic simulated annealing for optimal derivative code. Technical Report 372, University of Hertfordshire (2002)
17. Naumann, U., Gottschling, P.: Prospects for simulated annealing in automatic differentiation. In Steinhöfel, K., ed.: SAGA 2002 - Stochastic Algorithms, Foundations and Applications. Volume 2264 of LNCS., Springer, Berlin (2001) 131–144
18. Albrecht, A., Wong, C.: On logarithmic simulated annealing. In van Leeuwen, J., Watanabe, O., Hagiya, M., Mosses, P., eds.: Proc. IFIP International Conference on Theoretical Computer Science. LNCS, Springer (2000)

Propagation of the Hydraulic Head in an Elastic Pipeline

Blanka Filipová, Pavel Nevřiva, Štěpán Ožana

VSB – Technical University of Ostrava, 17. listopadu 15, Czech republic
{blanka.filipova, pavel.nevriva, stepan.ozana}@vsb.cz

Abstract. This paper deals with the measurement and simulation of the dynamic processes on the systems with distributed parameters. As an example of such a system the elastic pipeline transporting the elastic fluid has been chosen. The effort was mainly aimed at the measurement and modeling of prime hydraulic head in the pipeline which arises after sudden closing the valve. The problem described here can be considered as signals and systems analysis because we actually look for an output signal of the system that transforms the input signal. Finally the measured and simulated data are compared and their correspondence was evaluted.

1 Physical Model of the System

Physical model of the system is shown on Fig. 1. It composes of the pipeline, pressure sensors and valve. The measured part of the pipeline is the part of pipeline between pump and valve. Measured quantities were the pressures in points P1 and P2. After closing the valve the moving water column starts the dynamic process.

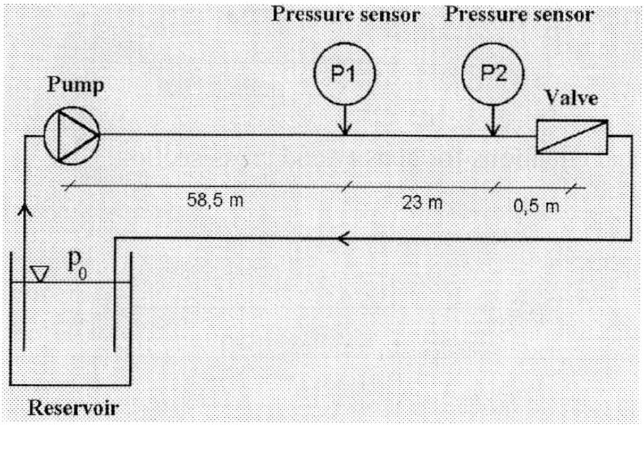

Fig. 1. Schema of the laboratory system.

P.M.A. Sloot et al. (Eds.): ICCS 2003, LNCS 2658, pp. 585–592, 2003.

2 Setting up the Mathematical Model

The model below describes flowing of the elastic fluid in the segment of elastic pipeline. The original mathematical model is described by three nonlinear partial differential equations of the first order. The fundamental of the model consists of Newton equation (1), Equation of continuity (2) and Bernoulli equation (3). The model is extended by equations (4) and (5) with respect to the elasticity of pipeline and transferred fluid.

- Newton equation

$$\frac{\partial p}{\partial x} + \rho\, v \frac{\partial v}{\partial x} + \rho \frac{\partial v}{\partial t} + \rho\, g \frac{\partial z}{\partial x} + \frac{\lambda\, \rho\, v\, |v|}{2\, dn} = 0 \tag{1}$$

- Equation of continuity

$$\begin{aligned} &\rho\, S \frac{\partial v}{\partial x} + \rho\, v \left(\frac{\partial S}{\partial p}\frac{\partial p}{\partial x} + \frac{\partial S}{\partial T}\frac{\partial T}{\partial x} + \frac{\partial S}{\partial x}\right) + S\, v \left(\frac{\partial \rho}{\partial p}\frac{\partial p}{\partial x} + \frac{\partial \rho}{\partial T}\frac{\partial T}{\partial x}\right) + \\ &+ \rho \left(\frac{\partial S}{\partial p}\frac{\partial p}{\partial t} + \frac{\partial S}{\partial T}\frac{\partial T}{\partial t}\right) + S \left(\frac{\partial \rho}{\partial p}\frac{\partial p}{\partial t} + \frac{\partial \rho}{\partial T}\frac{\partial T}{\partial t}\right) = 0 \end{aligned} \tag{2}$$

- Bernoulli equation

$$\frac{\partial}{\partial t}\left(\rho \left(c\, T + \frac{v^2}{2}\right)\right) + \frac{\partial}{\partial x}\left(\rho\, v \left(c\, T + \frac{v^2}{2}\right)\right) + \frac{\partial (p\, v)}{\partial x} + \frac{\partial (\rho\, v\, g\, z)}{\partial x} - \frac{\gamma\, (T_{ok} - T)}{S} = 0 \tag{3}$$

- Additional equations for density and cross-section:

$$\rho = \frac{\rho_0}{\exp\left(\frac{1}{K}(p_0 - p)\right) \exp(\beta\, (T - T_0))} \tag{4}$$

$$S = S_0 \exp\left(\frac{dn}{E\, d}(p - p_0)\right) \exp(2\, \alpha\, (T - T_0)) \tag{5}$$

where

c	=	internal energy of liquid
d	=	wall thickness of the pipeline
dn	=	internal diameter of the pipeline
E	=	modulus of elasticity
g	=	acceleration of gravity
K	=	liquid elasticity bulk modulus
l	=	length of pipeline
p	=	pressure of liquid
p_0	=	relative pressure
S	=	cross sectional area of the pipeline
S_0	=	relative cross sectional area of the pipeline
t	=	time
T	=	temperature of liquid
T_{ok}	=	ambient temperature
T_0	=	relative temperature
v	=	flow velocity of liquid
x	=	coordinate along pipeline axis
z	=	elevation of the pipeline
α	=	bulk expansivity of pipeline
β	=	bulk expansivity of liquid
γ	=	heat transfer coefficient
λ	=	friction factor of liquid
ρ	=	density of liquid
ρ_0	=	relative density of liquid

The original mathematical model was adapted for the laboratory system. The temperature processes were omitted, they would be significant mainly in pipeline with a gas or with a steam.

The mathematical model was therefore modified as follows:

$$\frac{\partial p}{\partial x} + \rho\, v \frac{\partial v}{\partial x} + \rho \frac{\partial v}{\partial t} + \rho\, g \frac{\partial z}{\partial x} + \frac{\lambda\, \rho\, v\, |v|}{2\, dn} = 0 \qquad \textbf{(6)}$$

$$\rho\, S \frac{\partial v}{\partial x} + \rho\, v \left(\frac{\partial S}{\partial p}\frac{\partial p}{\partial x} + \frac{\partial S}{\partial x}\right) + S\, v \left(\frac{\partial \rho}{\partial p}\frac{\partial p}{\partial x}\right) + \rho \left(\frac{\partial S}{\partial p}\frac{\partial p}{\partial t}\right) + S \left(\frac{\partial \rho}{\partial p}\frac{\partial p}{\partial t}\right) = 0 \qquad \textbf{(7)}$$

Appropriate equations for density and cross-section were modified, too:

$$\rho = \frac{\rho_0}{\exp\left(\frac{1}{K}(p_0 - p)\right)} \tag{8}$$

$$S = S_0 \exp\left(\frac{dn}{E\,d}(p - p_0)\right) \tag{9}$$

In order to find a unique solution of set of equations (6)-(9), the boundary and initial conditions must be added.

These are as the follows:

The boundary conditions:

- The velocity in the end of the pipeline – in the point where the valve is placed. After closing valve the velocity of the water in this point is v = 0.
- The pressure in the beginning of the pipeline – in the point where the pump is placed. This pressure is described by load characteristic of the pump, see Fig. 2.

The initial conditions:

- The distribution of the pressure within the whole length of the pipeline, this was linearly interpolated of the measured pressures P1, P2. The distribution of the pressure within the whole length of the pipeline is shown on Fig. 3.
- The distribution of the velocity within the whole length of the pipeline. The velocity in t = 0 was constant and equals to $v = 1.66\ \mathrm{m.s^{-1}}$.
- The distribution of the elevation within the whole length of the pipeline. The derivative of the elevation equals to zero, because the pipeline was rolled up horizontally.
- The distribution of the cross-section within the whole length of the pipeline. The cross-section in time t = 0 is constant, dn = 0, 032 m.

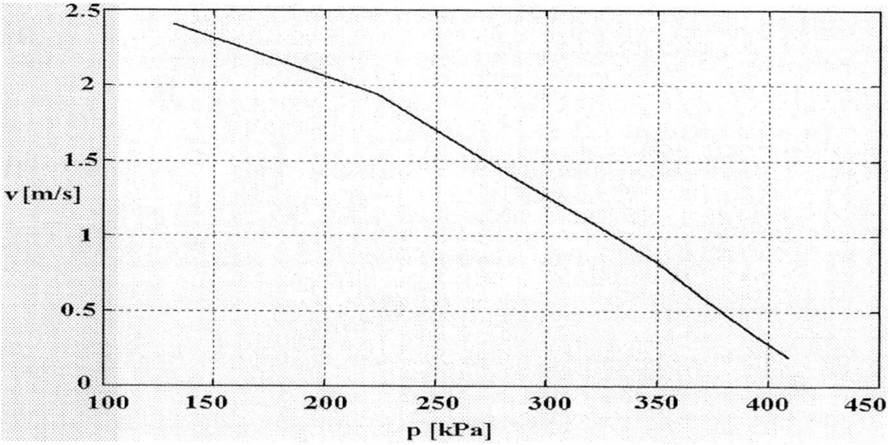

Fig. 2. The load characteristic of the pump

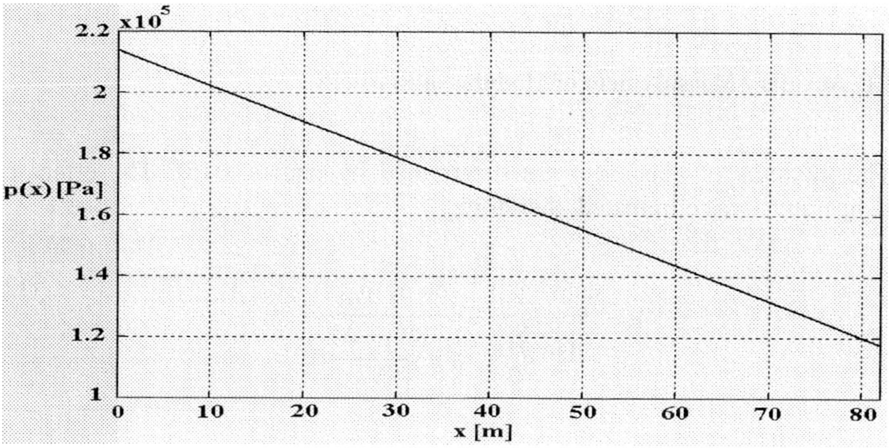

Fig. 3. The distribution of the pressure within the whole length of the pipeline.

3 Simulation of the Model

The system of the equations (6) - (9) was solved using the numerical methods, particularly the method of final differences and method of Adams – Bashforth. Simulation was performed using MATLAB.

Here only the first step of setting up a Matlab code will be shown. It is the idea to express the terms $\frac{\partial v}{\partial t}$ and $\frac{\partial p}{\partial t}$ from equations (6) and (7).

$$\frac{\partial v}{\partial t} = -\frac{\lambda\, v\, |v|}{2\, dn} - \frac{\partial z}{\partial x} - v\frac{\partial v}{\partial x} - \frac{1}{\rho}\frac{\partial p}{\partial x} \tag{10}$$

$$\frac{\partial p}{\partial t} = \frac{\rho\, S\frac{\partial v}{\partial x} + \rho\, v\,(\frac{\partial S}{\partial p}\frac{\partial p}{\partial x} + \frac{\partial S}{\partial x}) + S\, v\,(\frac{\partial \rho}{\partial p}\frac{\partial p}{\partial x})}{\rho\frac{\partial S}{\partial p} + S\frac{\partial \rho}{\partial p}} \tag{11}$$

Terms are integrated using Adams-Bashforth method of the 4th order.

4 Measurement on the System

4.1 Auxiliary Measurement of Modulus of Elasticity

The modulus of elasticity was determined by equation (5). The effect of temperature was eliminated, therefore

$$E = \frac{dn(p - p_0)}{d\ln\frac{S}{S_0}} = \frac{dn(p - p_0)}{d\ln\frac{V_0 + \Delta V}{V_0}} \tag{12}$$

where $V_0 = 3\pi dn^2$ is initial volume at initial pressure. Measuring procedure: there is the controlled source of pressure at the beginning of the pipeline. At the opposite end there is an outlet valve and pressure sensor. The pipeline of the length of 3 meters was pressured and the differences ΔV of its volume were measured.

Two sets of measurements were performed:

- The pipeline was always pressured from initial pressure p_0 to maximal value of 560 kPa.
- The pipeline was progressively pressured from initial pressure p_0 to maximal value in 10 steps of about 56 kPa.

The modulus of elasticity was determined in range of $E = 0.99 \cdot 10^7 \div 1.45 \cdot 10^7$ Pa.

4.2 Measurement of Hydraulic Head and Its Comparison with Simulated Data

The measuring of the hydraulic head is performed on the measuring system from Fig. 1. At first the valve is open and water flows in pipeline. Then the valve is suddenly closed and the hydraulic head arises. The propagation of the hydraulic heads in places P1 and P2 and its comparison with simulated data is presented on Fig. 4 and Fig. 5.

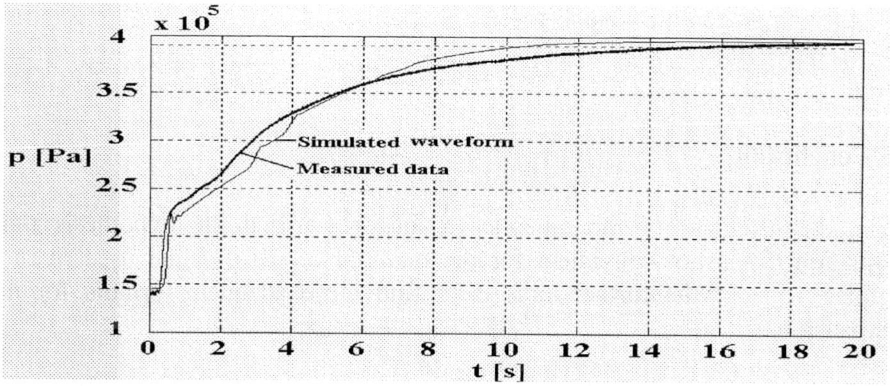

Fig. 4. Comparison of the measured and simulated data in point P1.

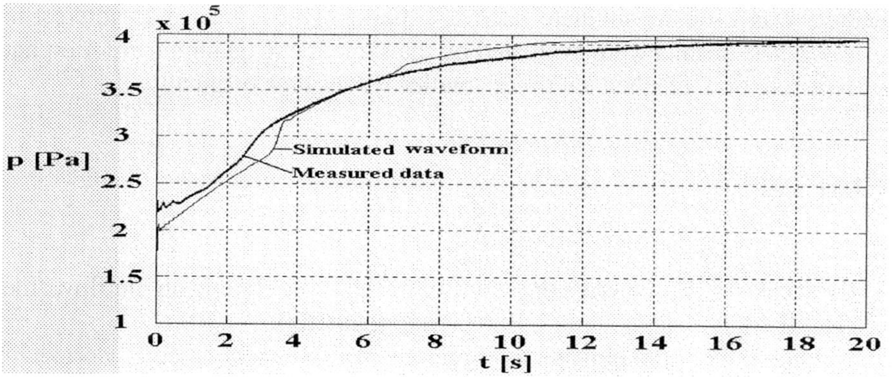

Fig. 5. Comparison of the measured and simulated data in point P2.

4.3 The Evaluation of the Mathematical Model

For the evaluation of the mathematical model the mean quadratic criterion was used. The following loss function is minimized:

$$J = \frac{1}{n} \sum_{i=1}^{n} | (y_{i,ref} - y_i) |^2 \qquad (13)$$

where y_i are values of quantity which are refereed to reference values $y_{i,ref}$. For measuring point P1 the computed value is $J = 9, 9 \cdot 10^7$, for measuring point P2 the computed value equals to
$J = 1, 068 \cdot 10^8$.

5 Conclusion

The measurement of the pipeline expansion verified the hypothesis that equation (9) is not convinient for the material of the pipeline used in this project, one of the further aims is to find more adequate formula for its expansion.

The simulation proved that the model is very sensitive to the modulus of elasticity of the used material which has some non-linearities and other dynamic qualities. Therefore the subsequent aim of the poject is to explore and include these dynamic qualities to the current model. The heuristic method was used to simulate such a situation and it leaded to interesting results. These results were evaluated by the same mean quadratic criterion and the loss function reached ten times smaller values than originally.

References

1. Demčáková, B., Nevřiva, P.: Control of the Flow Rate in an Outflow line. 15th IFAC World Congress. Barcelona, Spain 2002.
2. Filipová, B.: Modelování a regulace rázu pružného média v pružném potrubí. Dissertation. Ostrava, 2002.
3. Ožana, Š.: Měření dynamických jevů na soustavách tvořených pružnými potrubími. Thesis. Ostrava, 2002.

Numerical Revelation of the Molecular Structure for Reaction Effective Stimulator or Inhibitor by the Method of Hamiltonean Systematization of Chemical Reaction System Kinetic Models

Levon A. Tavadyan, Gagik A. Martoyan, Seyran H. Minasyan

Institute of Chemical Physics, National Academy of Sciences, Reublic of Armenia, 5/2 Sevak Street, Yerevan 375014, Armenia. Fax: (3742) 28-17-42, E-mail: tavadyan@ichph.sci.am

1 Introduction

The fundamental question of theoretical chemistry is to find the relationship between the electronic structure of reacting particles and their reactivity. In the case of reactions proceeding through complex multi-step mechanisms the problem to determine specie reactivity's is of difficult solution. This report presents a new numerical method, which enables to reveal the structure of reaction optimal (effective) stimulator (catalyst, promoter, *etc.*) and inhibitor based upon the kinetic model available for the multi-step chemical reaction.

1.1 Theoretical Fundamentals

Calculus of variations using the Pontjagin principle of maximum forms the basis for the non-empirical solution of the problem concerning the revelation of molecular structure an optimal stimulator is to be of, in the case of a complex (multistep) reaction [1,2].

In order to solve the mentioned above problem we have suggested for the characteristics of reaction component molecular structure to be presented as control parameters. The solution infers to step by stage execution of the following steps:

a) Based upon chemical reaction kinetic model, selection of the aimed control character $-F$: the functional, characterizing the chosen indicator of reaction stimulator or inhibitor reactivity (for example, the rate of purposeful product formation, the rate of initial species consumption, *etc.*);

$$I(t)=\int_0^t F(t)dt = extremum \tag{1}$$

b) Presentation of rate constants for individual steps in the reaction system kinetic model with participation of reaction stimulator and intermediate products from its conversion as a function of parameters, which characterize the molecular structure of reaction stimulator or inhibitor

$$k_j = \varphi_j(D) \tag{2}$$

where D is a numerical parameter characterizing the molecular structure of reaction stimulator and inhibitor (e.g. bond energy, ionization potential, steric parameters, *etc*);

P.M.A. Sloot et al. (Eds.): ICCS 2003, LNCS 2658, pp. 593–599, 2003.

c) setting up kinetic equations and the respective Hamiltonian H with selecting the control parameters;

$$dc_i / dt = f_i(k, c, D) \qquad i = 1, 2, \ldots m, \tag{3}$$

$$H = -F + \sum_{i=1}^{m} \psi_i f_i(k, c, D) \tag{4}$$

where $c(t)$ is the m-vector for the concentration of c_i components, $c(t_0) = c^0$; k - the n-vector of rate constants; ψ_i – the function conjugate to the concentrations c_i, D – p -vector of parameters for a reaction stimulator or inhibitor molecular structure.

The control parameters of D are assumed to be varied in a range of

$$D_l^{min} \le D_l \le D_l^{max}, \quad l=1,2,\ldots,p \tag{5}$$

d) setting up a system of differential equations for conjugate functions $\psi_i(t)$ (value of components)

$$\frac{d\psi_i}{dt} = -\frac{\partial H}{\partial c_i}, \qquad i = 1, 2, \ldots, m \tag{6}$$

e)Determination of conjugate function values $\psi_i(t_0)$ for the initial moment of time [3].

f) Finding of the optimum via parameters of reaction stimulator molecular structure. In respect to the principle of maximum the conditions for the optimum are:

$$\sup H(\psi^*, c^*, D^*) = 0 \tag{7}$$

The solution of the system of kinetic equations (3) and that of differential ones (6) with simultaneous observance of extremum conditions (7) correspond to optimal value for D^*. At this D^* is chosen from values of D^{min}, D^{max} and D^{cl}, where D^{cl} is the value of D, corresponding to condition $\partial H / \partial D = 0$.

The time-constancy condition for molecular structure parameters of reaction stimulator and inhibitor: $D = const$, significantly simplifies the problem. The value of D found for the initial moment of time corresponds to parameter values characterizing the molecular structure of reaction optimal stimulator for the given conditions of that reaction proceeding.

g) Determination of chemical reaction effective stimulator molecular structure using the calculated values of D^*.

h) Numerical ranging of stages and components in accordance with their value contributions and significance to have the dominating chemical stages and components be revealed, which in the result, determine the molecular structure of the most effective (optimal) stimulator or inhibitor for chemical reaction under the given conditions [3]. (It makes the obtained results evident from the chemical point of view).

2 An Illustrative Example. Computational Determination of the Molecular Structure for Reaction Effective Antioxidant in the Inhibited Reaction of the Ethylbenzene Liquid Phase Oxidation

For the example presented, the solution of the stated problem infers to the determination of molecular structure of the effective inhibitor from the class of p-substituted phenols (InH) for ethylbenzene (RH) liquid phase oxidation.

Table 1. Kinetic model of ethylbenzene (RH) oxidation reaction inhibited by *p*-substituted phenols (InH).

№	Reactions	Rate constant, k, lgk= $\varphi(D_{OH})$ /60°C
1.	$2RH + O_2 \rightarrow 2R^{\cdot} + H_2O$	9,26 x 10^{-13}
2.	$\dot{R} + O_2 \rightarrow R\dot{O}_2$	8,75 x 10^{8}
3.	$R\dot{O}_2 + RH \rightarrow RO_{2H} + \dot{R}$	2,74
4.	$RO^{\cdot} + RH \rightarrow ROH + \dot{R}$	2,32 x 10^{6}
5.	$OH + RH \rightarrow H_2O + \dot{R}$	10^{9}
6.	$RO_2H \rightarrow RO^{\cdot} + \dot{O}H$	10^{-9}
7.	$RO_2H \rightarrow R(-H)O + H_2O$	3 x 10^{-9}
8.	$R\dot{O} + RO_2H \rightarrow ROH + R\dot{O}_2$	4,9 x 10^{8}
9.	$R\dot{O}_2 + R\dot{O}_2 \rightarrow 2R\dot{O} + O_2$	5,5 x 10^{6}
10.	$R\dot{O}_2 + R\dot{O}_2 \rightarrow ROH + R(-H)O + O_2$	10^{7}
11.	$R\dot{O}_2 + InH \rightarrow ROH + \dot{I}n$	lgk_{11}=34,24-0,082 D_{OH}
12.	$R\dot{O} + InH \rightarrow \dot{I}n + ROH$	lgk_{12}=17,5-0,025 D_{OH}
13.	$RO_2H + InH \rightarrow \dot{I}n + R\dot{O} + H_2O$	lgk_{13}=50,1-0,157 D_{OH}
14.	$\dot{I}n + R\dot{O}_2 \rightarrow In(-H)O + ROH$	7 x 10^{8}
15.	$\dot{I}n + \dot{I}n \rightarrow InH + In(-H)$	3,5 x 10^{8}
16.	$\dot{I}n + RH \rightarrow InH + \dot{R}$	lgk_{16}=-25,2+0,066 D_{OH}
17.	$\dot{I}n + RO_2H \rightarrow InH + RO_2$	lgk_{17}=-22,24+0,072 D_{OH}

Note: rate constants are given in units of M, sec.; the value for phenolic OH bond energy in the molecule of *p*-substituted phenol (D_{OH}) - kJ/mole. The bond energy is changed in a range of 355-382,5 kJ/mole. The temperature of the reaction equals to 60°C. The structural formula of the substituted phenol is: (R - substituting group)

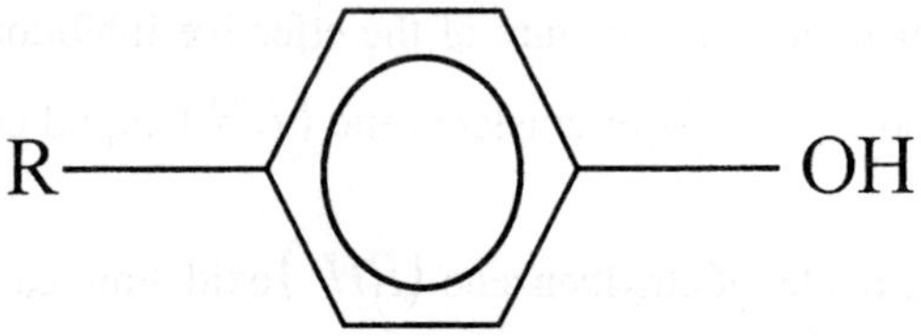

The solution of the problem is performed using the reaction scheme presented in Table 1[4]. Authors of the present report experimentally approve the choice of this kinetic model.

For the reaction model given in Table 1, the value of D_{OH} relating to OH bond energy of *p*-substituted phenol is presented as the control parameter. Following this purpose there are given the correlation equations in Table 1, for the reaction steps (11)-(13), (16), and (17). These equations describe the dependence of constants for stages proceeding with participation of p -substituted phenol and phenoxyl radical on OH bond energy in p -substituted phenol.

The purpose functional was chosen taking into account that the effective inhibitor is to be considered the one that the most retards the chemical reaction, decreasing its overall rate (r). In respect with this statement the following purpose functional is to be chosen for this problem:

$$I(t) = \int_0^t r dt = \min \tag{8}$$

where $r = dc_{RH}/dt$; c_{RH} is the concentration of ethylbenzene.

Then, performing the following operations, in accordance with procedures (method) described in items (b)-(g), the Hamiltonian values are accounted depending on D_{OH} for various initial concentrations of the inhibitor.

For a certain $[InH]_{to}$ the value of optimal D^*_{OH} and respectively, the molecular structure of the effective antioxidant-inhibitor for oxidation reaction of ethylbenzene correspond to the minimal value of Hamiltonian.

Numerical accounts have been performed using computer program VALKIN that we have developed. The algorithm of the program VALKIN is developed on the basis of Hamiltonian systematization for chemical reaction system mathematical models [1,3]. In this program the differential equations are solved by the program ROW-4A [5]. Simulations were carried out using personal computer. Results from simulations may be reduced to graphics, diagrams and other visual convenient forms.

The following result is obtained in the result of simulations:

In a wide range of inhibitor initial concentration (10^{-4} - 10^{-2} M) D^*_{OH} equaled to its minimal possible value (355 kJ/mole) and the following molecular structure of p-substituted phenol corresponds to it

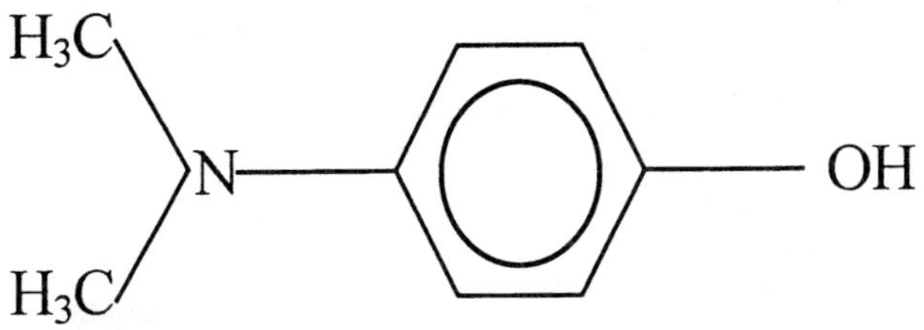

As it was mentioned above, the useful information contains the significance of value contribution individual steps (11-13, 16, 17) with participation of effective inhibitor and phenoxile radical, the speed constants of which are sensitive to the index of reaction ability of D_{OH} . Be reminded that the value contribution (h_j) stage characterizing its kinetic significance is calculated with joint solution of equation (3) and (6) of Hamilton systematization [1,3]

$$h_j(t) = v_j(t) G_j(t), \qquad G_j(t) = \frac{\partial F\left[v_1(t),...v_n(t)\right]}{\partial v_j}\Big|_{v_j=v_j(t_0)}, \quad (9)$$

where $v_j(t)$ is the speed of j -th stage, n -the number in kinetic model of the chemical reaction. In the case being examined $F \equiv r$.
According to the findings acquired from the Table 2, much more ponderable contributions have the step (11) and the step (17) which is opposite to (11).
It's obvious that in the process of inhibited oxidation the balance is carried out namely by means of these reaction steps;

$$RO_2^{\bullet} + InH \underset{k_{17}}{\overset{k_{11}}{\rightleftarrows}} RO_2H + In^{\bullet} \qquad (v_{11} \approx v_{17})$$

Now it's quite evident that the effective inhibitor with minimal significance of Don ensures the highest displacement of the given balance in the law: from the carrier of chain peroxide radical to the side of the phenoxide radical formation.

Table 2. The acquired value step contributions dependinf on D_{OH} for the liquid-phase auto-oxidation of ethil-benzene are inhibited by effective anti-oxidant of p – dimethilaminophenol. The temperature is equal to 60^0 C. The level of antioxidant conversion is 70%. The initial antioxidant and hydroperoxide ethyl-benzene concentrations are 10^{-3}, 10^{-5} M, respectively.

№	Reactions	$\bar{h}_j$
11.	$R\dot{O}_2 + InH \rightarrow ROH + \dot{I}n$ 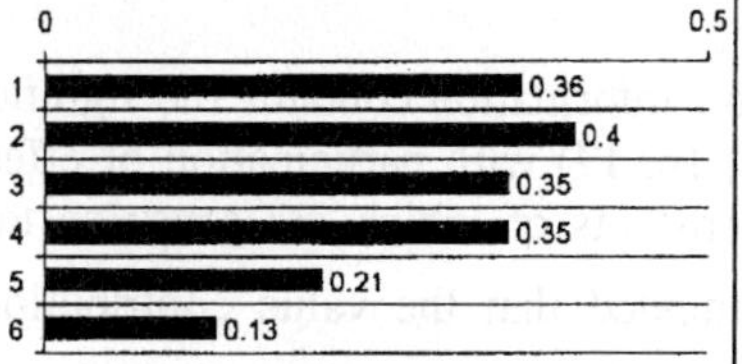	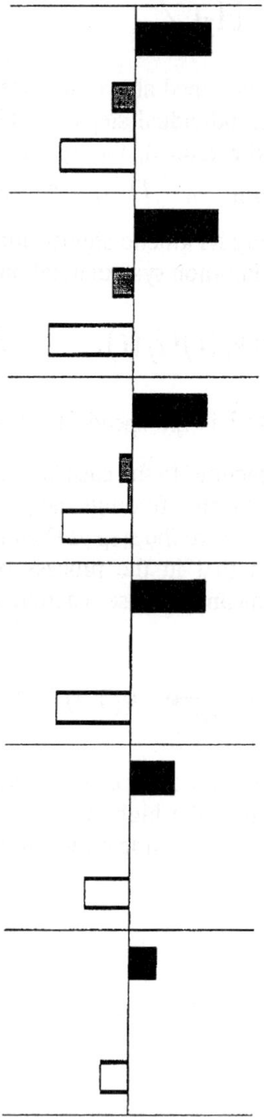
12.	$R\dot{O} + InH \rightarrow \dot{I}n + ROH$ 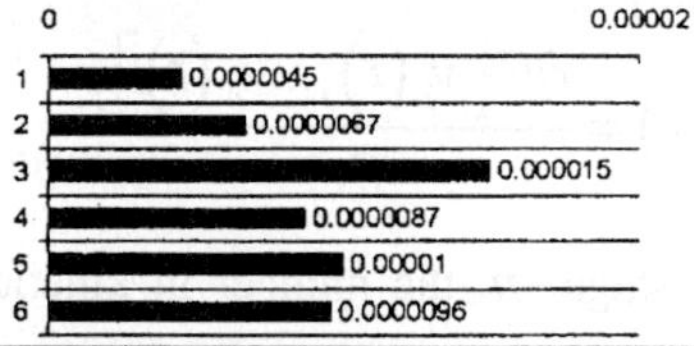	
13.	$RO_2H + InH \rightarrow \dot{I}n + R\dot{O} + H_2O$ 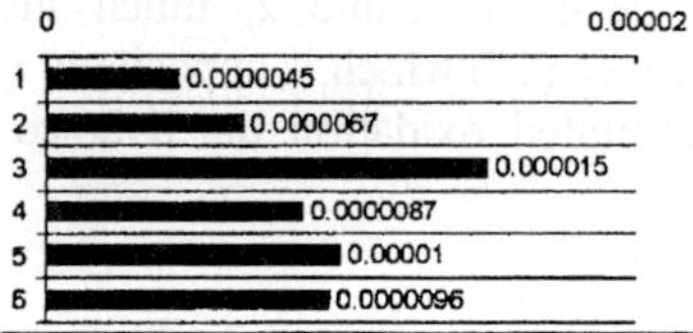	
16.	$\dot{I}n + RH \rightarrow InH + \dot{R}$ 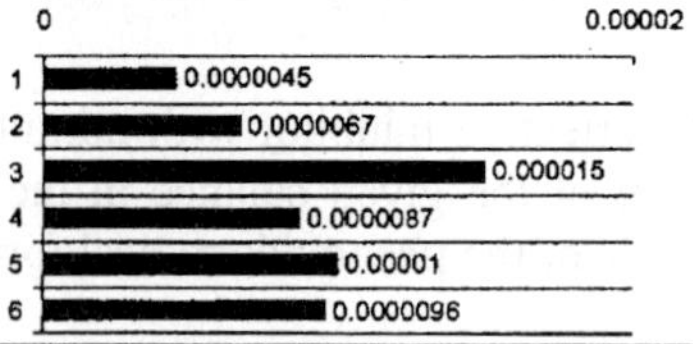	
17.	$\dot{I}n + RO_2H \rightarrow InH + R\dot{O}_2$ 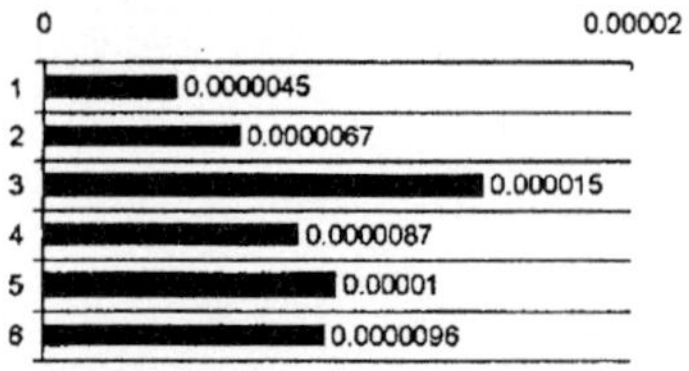	

where $\overline{h}_j$ is the radical value contribution of j -th stage

$$\overline{h}_j = h_j \left(\sum_{j=1}^{17} h_j^2 \right)^{-1/2}$$

3 Conclusions

Based upon the presented research the following conclusions are made:
Hamiltonian systematization using the Pontjagin principle of maximum is an operative method to be applied for multi-stage chemical reaction system mathematical model to have the molecular structure of chemical reaction optimal stimulator be numerically revealed.
The definition of kinetic significance of the stage by means of value contributions made it possible to define stages including inhibitor and its intermediate-phenoxile radical which determine the efficacy of the action of the initial inhibitor.
Computer program VALKIN worked out on the basis of value analysis for kinetic models of complex chemical reaction systems is an effective program to be used for the computational solution of analogous problems in chemistry and relating disciplines.

References

1. Tavadyan L.A., Martoyan G.A. Chem. Phys. Rep. 1994, v.13, p.793.
2. Pontrjagin L.S., Boltyanskii V.G., Gankrelidze R.D., Mishenko E.F. Mathematical theory of Optimal Processes, Moscow, Fizmatgiz, 1961 (Rus.).
3. Tavadyan L. A., Martoyan G. A., Chem. Physics, 2000, V20, No 2, p. 26 (Rus.).
4. Denisov E. T.,Azatyan V. V.: The Inhibition of Chain Reactions. Chernogolovka (1996)
5. Goffwald B.A. Simulation. 1981, v.33, p.169.

Numerical Revelation of Kinetic Significance for Steps and Species in Complex Chemical Reaction Mechanisms by Hamiltonian Systematization Method

Gagik A. Martoyan and Levon A. Tavadyan

Institute of Chemical Physics, National Academy of Sciences, Reublic of Armenia, 5/2 Sevak Street, Yerevan 375014, Fax: (3742) 28-17-42, tavadyan@ichph.sci.am

Abstract. The method of numerical calculation of kinetic significance (value magnitude) of the individual steps is suggested by means of Hamiltonian systematization multystep chemical reaction model. The given computation method is desplayed on the Chapman analyse model for ozone kinetics,as well as a tools of reduction model of formaldehyde oxidation reaction in the presence of carbon (II) oxide.

1 Introduction

The key question for mathematical simulation of complex (multistep) chemical reaction systems is to reveal the kinetic significance of steps and species in these reactions. This is connected with the solution of the following important tasks.

- to reveal the chemical basis of the multistep reaction - the role of molecular chemical conversion of species in the reaction system;
- to reduce the kinetic model of chemical reaction systems with revelation of the basic mechanism;
- to find ways for oriented effecting on multistep chemical reactions and their optimal control;
- to plan the experiment.

The objective of the present report is the development of a new universal method to have the steps and the species included in the chemical reaction kinetic model be numerically estimated.

2 Results

2.1 Characteristic of the Role of Individual Reaction Steps and Species via Value Magnitudes

A uniform chemical system at isothermal conditions may be presented in space as a set of usual differential kinetic equations:

$$\frac{dc_i}{dt} = f_i\left(k, c\right) \tag{1}$$

P.M.A. Sloot et al. (Eds.): ICCS 2003, LNCS 2658, pp. 600–609, 2003.

where $c(t)$ is the m-vector of species concentrations of $c(t_0)=c^0$, and k - the n - vector of rate constants.

Vector f involves a linear combination of step rates v_j which are presented in a form

$$v_j = k_j \prod_{l=1}^{m} c_e^{a_{je}} \tag{2}$$

where $a_{je} \geq 0$ and $\sum_{l=1}^{m} a_{je} \leq 3$.

The kinetic significance of reaction steps and species included in the reaction mechanism was determined using the approach of value magnitudes we have introduced before [1-3].

The value of G_j step is determined as the relation of response by a chosen characteristic magnitude of the reaction system, at any t moment of time, to perturbation of j reaction step rate at the t_0 moment of time:

$$G_j(t) = \frac{\partial F\left[v_1(t),...v_n(t)\right]}{\partial v_j}\bigg|_{v_j=v_j(t_0)},\quad j=1,2,..n, \tag{3}$$

where $F(t)$ is the dynamic parameter of integral behavior; this parameter characterizes the chosen output value of reaction, e.g. the concentration of species, the yield of purposeful products, the reaction retardation period, etc.

Analogous, the value of a reaction system species is determined as the relation of output value (F) response, at a t moment of time, to perturbation of species concentration change rate at the t_0 moment of time:

$$\psi_i(t) = \frac{\partial F\left[f_1(t),...f_m(t)\right]}{\partial f_i}\bigg|_{f_i=f_i(t_0)},\qquad i=1,2,..,m \tag{4}$$

where $f_i(t) = dc_i(t)/dt$ is the rate of concentration change for an i-th component. It follows from (3) and (4) that the value of a i-th step equals to the difference of value between the species formed and participating in this reaction step:

$$G_j(t) = \sum_{i=1}^{m} \psi_i(t) \frac{\partial f_i}{\partial v_j} = \sum_p a_{jp}\psi_{jp}(t) - \sum_l a_{jl}\psi_{jl}(t) \tag{5}$$

where a_{jp}, a_{jl} are stoichiometric coefficients of the j-th elementary step, "p", "l" - indexes relating, respectively, to products and to initial substances that step. Value contributions by h_j reaction steps and b_i species are determined taking into account the reaction steps rates and the rates of species accumulation:

$$h_j(t) = v_j(t)G_j(t) \quad \text{and} \quad b_i(t) = f_i(t)\psi_i(t) \tag{6}$$

2.2 Account of Value Parameters

The approach of Hamiltonian systematization for dynamic systems was used to calculate the value parameters.
Let the chosen property or the purpose functional be represented in an integral form:

$$I(t) = \int_{t_0}^{t} F(t)\,dt \tag{7}$$

For example, if it is aimed to reveal the role of reaction individual steps and species in the dynamics of change in concentration of the i-th species or several species, then the functionals, respectively, will be:

$$I_i(t) = c_i(t) = \int_{t_0}^{t} f_i(t)\,dt \qquad I(t) = \sum_{i=1}^{r} c_i(t) = \sum_{i=1}^{r}\int_{t_0}^{t} f_i(t)\,dt \tag{8}$$

r - is the number of chosen species in reaction.
Rewriting equation (4) in an integral form

$$F = \sum_{i=1}^{m} \psi_i f_i + const, \tag{9}$$

and denoting

$$H = -F + \sum_{i=1}^{m} \psi_i f_i, \quad H = const, \tag{10}$$

correspondingly

$$\frac{dH}{dt} = \sum_i \left(\frac{\partial H}{\partial c_i}\frac{dc_i}{dt} + \frac{dH}{\partial \psi_i}\frac{\partial \psi_i}{dt} \right) = 0 \tag{11}$$

As in regard to (9) the system of equations might be represented in a form

$$\frac{dc_i}{dt} = \frac{\partial H}{\partial \psi_i} = f_i \qquad i = 1,2,...m \tag{12}$$

then proceeding from (11)

$$\frac{dH}{dt} = \sum_i \left(\frac{\partial H}{\partial c_i} + \frac{d\psi_i}{dt} \right) \frac{dc_i}{dt} = 0, \tag{13}$$

Proceeding from (13) for an arbitrary $dc_i(t)/dt$ it follows that

$$\frac{d\psi_i}{dt} = -\frac{\partial H}{\partial c_i} \tag{14}$$

Thus it follows from equations (10), (12) and (14) that H is the Hamiltonian of the system. Values of components ψ_i and then, in compliance with (5) and (6) the value contributions of species and reaction steps are accounted by solving the system of conjugated equations (14) jointly with the kinetic equation (12).It must be mentioned that value magnitudes acquired by the present method are easy interpretable. In order to explain it from the point of view of physical meaning, the equation (10) is to be represented as

$$F = -H + \sum_{j=1}^{n} h_j = -H + \sum_{i=1}^{m} b_i \tag{15}$$

It follows from equation (15) that value contributions of h_j steps and those of the b_i species characterize their relative contribution into the chosen output parameter F for a multystep reaction, with an accuracy up to a constant.

Initial values of $\psi_i(t_0)$ required for the simulation of value parameters are determined by (4).

Thus, the procedure of accounting the value parameters, which forms the basis of the computer program VALKIN, is inferred to the following steps:

a) setting up a system of kinetic equations (1);
b) formal representation of the purpose functional for a multistep reaction (equation (7)) and the respective Hamiltonian of the system (equation (10));
c) setting up a system of differential equations for $\psi_i(t_0)$ (values) - conjugated concentrations of species, in accordance with equation (14);
d) estimation of conjugated function $\psi_i(t_0)$ values for the zero moment of time, proceeding from the purpose functional;

e) computation of $\psi_i(t)$ dynamics by simultaneous solving the system of kinetic equations (1) and the system differential equations for conjugated functions (14);
f) computation of dynamics for steps values, value contributions of individual reaction steps and species, according to equations (5) and (6), respectively.

3 Illustrative Examples

3.1 Ozone Kinetics

Chapman mechanism relating to the kinetics of atmosphere ozone conversion presented in Table 1 has been used as an illustrative example. This choice is conditioned by the fact that this example is assumed to be considered as a test model for new methods; other different methods of model analysis, in particular the method of sensitivity analysis were tested considering this example [4,6].
Mathematical program VALKIN that we have developed was used for numerical computation of reaction species concentration time profiles as well as reaction steps relative contributions. In this program the systems of differential equations are solved by the program ROW-4A [7].
The initial concentrations expressed in terms of number of particles to centimeter in cube, are the following: $[O]_0 = 10^6$, $[O_3]_0 = 10^{12}$ and $[O_2]_0 = 3,7 \cdot 10^6$.
At the first step one has to set up a system of kinetic equations (1) in accordance with the model represented in table 1.
For the revelation of stage contributions the purpose functional (8) is to be represented as:

$$I(t) = [O_3]_t = \int_{t_0}^{t} f_{O_3} dt \qquad (16)$$

Here f_{O_3} - is the rate of change in ozone concentration.
The respective Hamiltonian, in compliance with (10), will be reduced to the following form

$$H = -f_{O_3} + \psi_O f_O + \psi_{O_3} f_{O_3} + \psi_{O_2} f_{O_2} \qquad (17)$$

Then taking into account this form of Hamiltonian (17) a system of conjugated equation ψ_i (14) is to be set up.

The dynamics of $\psi_O, \psi_{O_3}, \psi_{O_2}$ magnitudes was determined solving the system of differential equations (1) and (14). In accordance with the purpose functional (16) the initial values for ψ_O, ψ_{O_3} and ψ_{O_2} equal respectively t_0 : 0, 1, 0. Based on values obtained for conjugate functions, according to equations (5) and (6), time dependencies of reaction (see Table 1) contributions were computed.

$$h_1 = \left(\psi_{O_3} - \psi_O - \psi_{O_2}\right) v_1, \qquad h_2 = (2\psi_{O_2} - \psi_O - \psi_{O_3}) v_2$$

$$h_3 = \left(2\psi_O - \psi_{O_2}\right) v_3, \qquad h_4 = \left(\psi_O + \psi_{O_2} - \psi_{O_3}\right) v_4$$

The results from value approach sufficiently enough coincide with the ones obtained by T. Hwang [4] by the analysis of ozone concentration sensitivity to variations of step rate constants. Though in general, in case of more complex reaction mechanisms, data from value approach could differ as well. This assertion is associated with the fact that in the case of reaction system response to step rate variation the role of steps is revealed to a more extent than when only the rate constant of that step is varied.

Table 1. Model of ozone kinetics. Time dependencies of steps relative value contributions ($\overline{h}_j$)

Reactions	Rate constant, (mole,sec,dm^3)	N	Value contribution of steps, $\overline{h}_j$
$O_3 \rightarrow O + O_2$	2.50x10^{-4}	1	
$O_2 \rightarrow 2O$	5.00x10^{-11}	2	
$O + O_3 \rightarrow 2O_2$	4.66x10^{-16}	3	
$O + O_2 \rightarrow O_3$	1.63x10^{-16}	4	

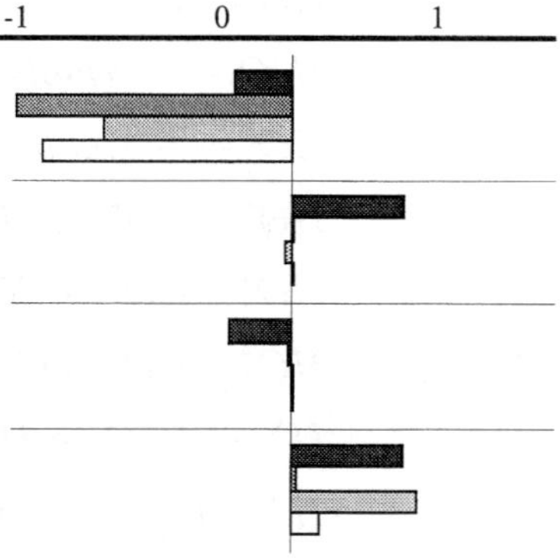

Time, sec □ - $2 \cdot 10^{-2}$ ▨ - $3 \cdot 10^{6}$ ▨- 10^{7} ■ $5 \cdot 10^{7}$

3.2 The Reduction of the Reaction Mechanism of Gas Phase Chain Reaction of Phormaldehyde Oxidation in the Presence of Carbon (II) Oxide

The conscious choise of this object is conditianed by the fact that it turned out to be very useful approbation of reaction kinetic models reduction methods based on the method of sensitivity analysis [5,8,9].

Both the dominant and non-essential steps of reaction mechanism, shown in Table 3, were defined in work [5], on the bases of the isolated massive of the given species concentration sensitivity in accordance with the reaction step rate constant speech $5 \cdot 10^{-3}$sec. for reaction time. Further, the use of a more systematic analyze method of the given sensitivity of concentration and of the accumulation rate of the species to the variations of step rate constant (the method of principal component analysis) succeeded in ranging the roles of reaction mechanism individual step of phormaldehyde oxidation in the presence of carbon(II) oxide. In these works the minimal mechanism reaction consists of 13 steps, at the average with an accuracy up to 2% giving the same results (the reaction time is $5 \cdot 10^{-3}$ sec.) as the initial kinetic model consisted of 25 steps (see Table 2).

Table 2. The Reaction Mechanism of Gas Phase Chain Reaction of Phormaldehyde Oxidation in the Presence of Carbon(II) Oxide

№	Reactions	Rate Constant
1.	$HC^{\bullet}O + O_2 \rightarrow HO_2^{\bullet} + CO$	$1{\bullet}10^{-13}$
2.	$HO_2^{\bullet} + CH_2O \rightarrow H_2O_2 + HC^{\bullet}O$	$5.7{\bullet}10^{-14}$
3.	$H_2O_2 + M \rightarrow 2HO^{\bullet} + M$	$6.66{\bullet}10^{-18}$
4.	$HO^{\bullet} + CH_2O \rightarrow H_2O + HC^{\bullet}O$	$1.6{\bullet}10^{-10}$
5.	$HO^{\bullet} + H_2O_2 \rightarrow H_2O + HO_2^{\bullet}$	$5.1{\bullet}10^{-12}$
6.	$H_2O_2 \longrightarrow molec.products$	$1.05{\bullet}10^{+2}$
7.	$HO_2^{\bullet} \longrightarrow molec.products$	10.5
8.	$HO_2^{\bullet} + HO_2^{\bullet} \rightarrow H_2O_2 + O_2$	$3{\bullet}10^{-12}$
9.	$HO^{\bullet} + CO \rightarrow CO_2 + H^{\bullet}$	$3.3{\bullet}10^{-13}$
10.	$HO_2^{\bullet} + CO \rightarrow CO_2 + HO^{\bullet}$	$1.2{\bullet}10^{-15}$
11.	$H^{\bullet} + CH_2O \rightarrow H_2 + HC^{\bullet}O$	$2.7{\bullet}10^{-12}$
12.	$H^{\bullet} + O_2 \rightarrow HO^{\bullet} + {}^{\bullet}O^{\bullet}$	$5.51{\bullet}10^{-14}$
13.	$H^{\bullet} + O_2 + M \rightarrow HO_2^{\bullet} + M$	$1{\bullet}10^{-32}$
14.	$HO_2^{\bullet} + M \rightarrow H^{\bullet} + O_2 + M$	$4.7{\bullet}10^{-19}$
15.	${}^{\bullet}O^{\bullet} + H_2 \rightarrow HO^{\bullet} + H^{\bullet}$	$3.02{\bullet}10^{-13}$
16.	${}^{\bullet}O^{\bullet} + CH_2O \rightarrow HO^{\bullet} + HC^{\bullet}O$	$1{\bullet}10^{-10}$
17.	$H^{\bullet} + H_2O_2 \rightarrow HO_2^{\bullet} + H_2$	$1.3{\bullet}10^{-12}$
18.	$H^{\bullet} + H_2O_2 \rightarrow H_2O + HO^{\bullet}$	$5.9{\bullet}10^{-12}$
19.	${}^{\bullet}O^{\bullet} + H_2O_2 \rightarrow HO^{\bullet} + HO_2^{\bullet}$	$1{\bullet}10^{-13}$
20.	$HC^{\bullet}O \rightarrow H^{\bullet} + CO$	$4.\ 6{\bullet}10^{-12}$
21.	$HO^{\bullet} + H_2 \rightarrow H_2O + H^{\bullet}$	$1{\bullet}10^{-11}$
22.	$CH_2O + O_2 \rightarrow HC^{\bullet}O + HO_2^{\bullet}$	$2.9{\bullet}10^{-20}$
23.	$H^{\bullet} + HO_2^{\bullet} \rightarrow 2HO^{\bullet}$	$5{\bullet}10^{-12}$
24.	$H^{\bullet} + HO_2^{\bullet} \rightarrow H_2O + {}^{\bullet}O^{\bullet}$	$5{\bullet}10^{-11}$
25.	$H^{\bullet} + HO_2^{\bullet} \rightarrow H_2 + O_2$	$4.5{\bullet}10^{-11}$

Comments: The significance of the rate constant of the reaction is reduced to units: the number of the particles, cm, second, T=679 °C. The initial concentrations are

$[CH_2O]_0 = 6.77 \cdot 10^{16}$, $[O_2]_0 = 1.27 \cdot 10^{18}$, $[CO]_0 = 2.83 \cdot 10^{18}$ and $[M]_0 = 7.09 \cdot 10^{18} cm^{-3}$. Now, let the stages of value analysis of kinetic model of formaldehyde oxidation in the presence of carbon (II) oxide be represented in a scheme. To begin with, let's single out the purpose functional characterizing the concentration change of the initial things: - formaldehyde, carbon oxide (II) and oxygen

$$I(t) = \Delta[CH_2O]_t + \Delta[CO]_t + \Delta[O_2]_t = -\int_0^t [f_{CH_2O} + f_{co} + f_{O_2}]dt \quad (18)$$

The system of the differential equation (12) and (13) is solved under the following initial significances of ψ_i. From the purpose functional (18) we have, that $\psi_{CH_2O}(0) = \psi_{CO}(0) = \psi_{O_2}(0) = -1$.

For other species ψ_i of the reaction during the initial moment of time the significance is equal to zero. Table 3 shows the calculated value contributions of the reaction steps, where the absolute significance is reduced gradually from 10^{-5} sec to $3 \cdot 10^{-3}$ sec. (conversion CH_2O 1.23%).

Table 3. The Comparison of the Step Numbers Being Arranged in the Reduction of their Kinetic Significance, Acquired by Different Methods, as well the Base (Minimal) Mechanism of the Reaction

Analyses Methods, Base model	Step Numbers															
Value Analysis	1	2	10	4	3	9	13	12	11	22	16	8	14	6	25	18
Sensitivity Analysis[3]	10	22	3	2	9	4	8	12	11	1	-	-	-	-	-	-
Principal Component Analysis[7]	22	10	4	9	2	16	12	11	1	13	14	3	6	7	24	8
Base Minimal Model of the Reaction [7,8]	1	2	3	4	6	8	9	10	11	12	13	16	22	-	-	-

Analyzing the findings from Table 3, it's obvious that the value contribution of the absolute magnitude the first 14 steps include the base mechanism of the reaction. The relative contributions of the reaction, not being included in the reaction mechanism, are only a few under the reaction time considered. It is worth mentioning that the individual step (14), being out of the base mechanism [7,8] occupies only the 13th

place in the isolated range. The base mechanism , consisted of 14 steps , including the step (14), describes the time profile of hydrogen atom concentration , which is deflected from the calculation of the initial model by 4%, and the isolated base model of the reaction in [7,8] consisted of 13 steps is deflected by 2,7%. However, for all species and especially for oxygen atoms, during the reaction time (from 10^{-5} sec. to $3 \cdot 10^{-3}$ sec.) this model is much less deflected from the calculation made by the use of initial models (25 steps). To compare, we can pay attention on the fact, that in (17) the first 16 steps of the analogous rate were taken into consideration for the isolated base mechanism of the reaction. The individual step 8, being included in the base mechanism, according to the calculation, has little significance and occupies the 16th place in the range. Whereas, the step 14, being isolated as ponderable enough for the reaction, isn't included in the base mechanism.

Now let's focus our attention on the role of the decrease order of the step. The decrease order of kinetic significance of the reaction steps, isolated by value methods, generally coincides with literature data, being defined by sensitivity analyze method [5,8,9]. Some of the observed differences apparently are connected with the difference of the value characteristics and parameters of sensitivity, which define the kinetic significance of the step [1-3]. Here the essential role can take the very circumstance, which by the value analysis of kinetic model reaction is distinctly selected as an index of the reaction quality: a purpose functional on which wholly depends the kinetic significance of the reaction step.

So, the examples of kinetic model of formaldehyde oxidation in the presence of carbon (II) oxide makes it quite possible to be completely convinced that the value analysis is a reliable mathematical tools of reduction mechanisms which reveals the base kinetic model of chemical conversion.

Thus we have worked out a new numerical method and a relevant computer program VALKIN for the analysis of chemical reaction models, which widen the scope of means for specialists in the field of chemical kinetics and engineering.

References

1. Tavadyan, L. A.: Selective Inhibition and Initiation of Multicenter Chain Reactions. Arm. Chem. Journ. **2** (1987) 81 – 92 (in Rus.)
2. Tavadyan, L. A., Martoyan, G. A.: Value Principle of Studing the Kinetics of Complex Chemical Reactions. Chem.Phys. Rep. **13** (1994) 793–797
3. Martoyan, G. A., Tavadyan, L. A.: Value Method of Revealing of Steps' Kinetic Significance in the Models of Chemical Reactions. Chem. Physics. **20:2** (2001) 25–33 (in Rus.)
4. Hwang, T.: Sensitivity Analysis in Chemical Kinetics by the Method of polinomal Approximations. Int. J. Chem. Kinet. **15** (1983) 959–987
5. Gougherty, E. P., Hwang, J.-T., Rabitz, H.G.:Furter Developments and Applications of the Green's Function Method of Sensitivity Analysis in Chemical Kinetics. Chem. Phys., **71:4** (1979) 1974–1808.
6. Saltelli, A., Chan, K., Scott, E.M. (eds): Sensitivity Analysis. John Wiley Sons, New York (2000)

7. Gottwald, B. A.: Kiss- A digital Simulation System for Coupled Chemical Reactions. Simulation. **33** (1981) 169–173.
8. Vaida, S., Valko, P., Turanyi, T.:Principal Component Analysis of Kinetic Models. Int. J. Chem. Kinet. **17** (1985) 55–81
9. Turanyi T., Berces T., Vaida S.: Reaction Rate Analysis of Complex Kinetic Systems. Int. J. Chem. Kinet. **21** (1989) 83–89

Optimization of Computations in Global Geopotential Field Applications

J.A.R. Blais[1] and D.A. Provins[2]

[1,2] Department of Geomatics Engineering
[1] Pacific Institute for the Mathematical Sciences
University of Calgary, Calgary, AB, T2N 1N4, Canada
blais@ucalgary.ca and provinsd@telusplanet.net

Abstract. Most boundary value problems of the geopotential field have integral and series solutions in terms of Green's convolution kernels. These solutions are advantageously evaluated using fast Spherical Harmonic Transforms (SHTs) for regular arrays of simulated or observed global data. However, the computational complexity and numerical conditioning of SHTs for relatively dense data are quite challenging and recent algorithmic developments warrant further investigations for geodetic and geophysical applications.

Global multiresolution applications for scalar, vector and tensor fields on the Earth and its neighborhood require spherical harmonic analysis and synthesis using convolution filters with data decimation and dilation. For global spherical grid applications, efficient and reliable SHTs are needed just as Fast Fourier Transforms (FFTs) are used in regional planar applications.

With the availability of enormous quantities of space, surface and subsurface data, extensive data structuring and management are unavoidable for most array computations. Different methodologies imply very different strategies and conflicting claims often appear in the literature. Discussions of the implicit and other assumptions with simulated results would undoubtedly help to clarify the situation and help decide on appropriate data structuring strategies for different computational applications.

1 Introduction

For every linear boundary value problem of the Earth's geopotential, it is possible to define a source or a Green's function. If this Green's function can be formulated explicitly, then the boundary value problem (BVP) is solved formally in terms of integral or series forms. In general, Green's function for a linear partial differential operator BVP is the solution for a Dirac delta impulse and homogeneous boundary conditions. Corresponding to the nonhomogeneous Dirichlet, Newmann and Robin BVPs of the Laplace operator, Green's functions for the sphere are well known with the solutions in integral and series forms readily available. Explicitly, the BVP solutions are expressed as convolutions of Green's functions and normal derivatives thereof with discrete measurements on or near the surface of the Earth.

Convolution operations are fundamental in linear filtering, solving BVPs using Green's functions and numerous other applications. In planar contexts, Fourier transform techniques are well established with FFTs, as the computational complexity

P.M.A. Sloot et al. (Eds.): ICCS 2003, LNCS 2658, pp. 610–618, 2003.

is generally reduced from $O(N^2)$ to O(NlogN). In spherical contexts such as in geopotential applications, the situation is somewhat different because of the lack of some efficient SHTs for analysis, synthesis and convolution operations. Furthermore, as spherical displacements (or equivalently, spatial rotations) are generally non commutative, convolutions on the sphere are quite different from convolutions in the plane but can advantageously be evaluated using SHTs.

From a sampling perspective, it is often desirable to sample a band-limited function in such a way that the original function can be exactly recovered from the samples. In the case of functions on the real line, Shannon's sampling theorem states that a function whose Fourier transform has bounded support may be recovered from its samples provided that these are chosen uniformly at a rate of at least twice the bounding frequency. An important consideration is that the Fourier transform and hence the frequency spectrum can be computed from the discrete samples for analysis purposes.

For geodetic and geophysical applications, it is often desirable to sample a band-limited function on the sphere so that the SHT can be efficiently evaluated as weighted sums of the samples. The investigation of numerical integration formulas has a long history and most relevant quadrature results on the sphere have drawbacks (see e.g. [16]). In the following, the primary objectives are for multiresolution applications using equiangular and Gaussian data grids for convolution and decimation/dilation operations in the study of geopotential BVPs (see e.g. [3] for more details). Brief comments about the asymptotic behavior of the Legendre functions of high degrees and orders are included.

Data structuring is becoming an important consideration with very large global datasets. Recently, Bond et al. [4] proposed a hierarchical structure known as an igloo pixelization for spherical harmonic transforms of the Cosmic Microwave Background (CMB) datasets. Another data structure known as HEALPix has been proposed by Wandelt et al. [22] for the same purposes. These data structuring strategies are quite different from the usual global approaches in geodesy and geophysics (e.g., [18]). Optimality in data structuring obviously depends on the intended applications.

2 Sperical Harmonic Analysis and Synthesis

The orthogonal or Fourier expansion of a function $f(\theta, \lambda)$ on the sphere $\mathbf{S}^2$ is given by

$$f(\theta,\lambda)=\sum_{n=0}^{\infty}\sum_{|m|\le n} f_{n,m}\, Y_n^m(\theta,\lambda)$$

using polar angles θ and λ, where the basis functions $Y_n^m(\theta,\lambda)$ are called the spherical harmonics as $\Delta_{S^2} Y_n^m(\theta,\lambda)=0$, for all $|m| \le n$ and n = 0, 1, 2, ... This is an orthogonal decomposition in the Hilbert space $\boldsymbol{L}^2(\mathbf{S}^2)$ of functions square integrable with respect to the standard rotation invariant measure $d\sigma = \sin\theta \, d\theta \, d\lambda$ on $\mathbf{S}^2$. In particular, the Fourier or spherical harmonic coefficients appearing in the preceding expansion are obtained as inner products

$$f_{n,m} = \int_{S^2} f(\theta,\lambda)\, \overline{Y}_n^m(\theta,\lambda)\, d\sigma$$
$$= \sqrt{\frac{(2n+1)(n-m)!}{4\pi\,(n+m)!}} \int_{S^2} f(\theta,\lambda)\, P_n^m(\cos\theta)\, e^{-im\lambda}\, d\sigma$$

where the $P_{nm}(\cos\theta) = (-1)^m P_n^m(\cos\theta)$ are the associated Legendre functions, with the overbar denoting the complex conjugate (e.g., [12]). In most practical applications, the functions $f(\theta,\lambda)$ are band-limited in the sense that only a finite number of those coefficients are nonzero, i.e. $f_{n,m} \equiv 0$ for all $n \geq N$.

The usual geodetic spherical harmonic formulation is slightly different with

$$f(\theta,\lambda) = \sum_{n=0}^{\infty} \sum_{m=0}^{n} [\overline{C}_{nm} \cos m\lambda + \overline{S}_{nm} \sin m\lambda]\, \overline{P}_{nm}(\cos\theta)$$

where

$$\begin{Bmatrix} \overline{C}_{nm} \\ \overline{S}_{nm} \end{Bmatrix} = \frac{1}{4\pi} \int_{S^2} f(\theta,\lambda) \begin{Bmatrix} \cos m\lambda \\ \sin m\lambda \end{Bmatrix} \overline{P}_{nm}(\cos\theta)\, d\sigma$$

and

$$\overline{P}_{nm}(\cos\theta) = \sqrt{\frac{2(2n+1)(n-m)!}{(n+m)!}}\, P_{nm}(\cos\theta) \qquad \text{for } m = 1, \ldots, n$$

$$\overline{P}_n(\cos\theta) = \sqrt{2n+1}\, P_n(\cos\theta) \qquad \text{for } n = 0, 1, \ldots$$

in which the overbars refer to the usual geodetic normalization.

Geopotential models are fundamental in geodesy and geophysics. Figure 1 shows the geopotential model GPM98B of degree and order 1800 evaluated using software and coefficient dataset from Wenzel [23]. The latter geopotential model has been derived from the EGM96 model [10] through differential corrections, with integral formulas from [15, 25]. From investigations of the numerical accuracy of the fully normalized Legendre functions, accuracy of the available software was limited to the degree and order indicated [24]. Denoting the above normalizing factor for $P_{nm}(\cos\theta)$ by G_{nm}, their magnitudes

$$G_{1800,1} \approx 0.047, \quad \ldots.., \quad G_{1800,1800} \approx 1.68 \times 10^{-5619}$$

confirm that special care is required in the computations for meaningful numerical results.

For very high degree and order computations, asymptotic analyses of the behavior of the associated Legendre functions are required [21]. Considering the functions

$$\psi_n^m(\theta) = \sqrt{\sin\theta}\, P_n^m(\cos\theta)$$

it can be shown that these satisfy the following simple Schrödinger differential equation

$$\frac{d^2}{d\theta^2}\psi_n^m(\theta) = -\left[(n+1/2)^2 - (m^2 - 1/4)\csc^2\theta\right]\psi_n^m(\theta)$$

which has the asymptotic solution behavior by the Wentzel-Kramers-Brillouin (WKB) method

$$\psi_n^m(\theta) \approx \exp\left\{ i \int^{\theta} \left[(n+1/2)^2 - (m^2 - 1/4)\csc^2 t \right]^{1/2} dt \right\}$$

when the square root is real in the integrand. The behavior of $\psi_n^m(\theta)$ is oscillatory with very large magnitudes near 0 and π (see Figure 2 for plots of $P_{60}^{20}(\cos\theta)$ and $\psi_{60}^{20}(\theta)$). Mohlenkamp [12] has reformulated the spherical analysis and synthesis in terms of these functions with advantageous computational results.

3 Discretization and Numerical Analysis

Colombo [5] has discretized the preceding geodetic formulation with $\theta_j = j\pi/N$, j = 1, 2, …, N and $\lambda_k = k\pi/N$, k = 1, 2, …, 2N. This is only an approximate quadrature approach that uses summations to estimate the numerical integrals for the frequency spectrum and the computational effort required is $O(N^4)$, without counting the computations involved in the evaluation of the Legendre functions. The longitude computations can easily be carried out using FFTs and the corresponding computational effort is reduced to $O(N^3 \log N)$.

Legendre quadrature is well known to provide an exact representation of polynomials of degrees up to 2N – 1 using only N data values at the zeros of the Legendre polynomials. Driscoll and Healy [7] have exploited these quadrature ideas in an exact algorithm for a reversible SHT using the following N^2 data grid for degree and order N:

For synthesis purposes, given a band-limited square integrable function $f(\theta, \lambda)$, with co-latitude θ and longitude λ, given normalized spherical harmonic coefficients a_{nm} and b_{nm} for $m \le n$, n = 0, 1, 2, … , N-1,

$$\begin{aligned} f(\theta,\lambda) &= \sum_{n=0}^{N-1}\sum_{m=0}^{n} (a_{nm}\cos m\lambda + b_{nm}\sin m\lambda)\, P_{nm}(\cos\theta) \\ &= \sum_{m=0}^{N-1}\sum_{n=m}^{N-1} (a_{nm}\cos m\lambda + b_{nm}\sin m\lambda)\, P_{nm}(\cos\theta) \\ &= \sum_{m=0}^{N-1}\left\{ \left(\sum_{n=m}^{N-1} a_{nm}P_{nm}(\cos\theta)\right)\cos m\lambda + \left(\sum_{n=m}^{N-1} b_{nm}P_{nm}(\cos\theta)\right)\sin m\lambda \right\} \end{aligned}$$

and defining

$$A_m(\theta) = \sum_{n=m}^{N-1} a_{nm}P_{nm}(\cos\theta)$$

and

$$B_m(\theta) = \sum_{n=m}^{N-1} b_{nm}P_{nm}(\cos\theta)$$

one has

$$
\begin{aligned}
f(\theta,\lambda) &= \sum_{m=0}^{N-1}\{A_m(\theta)\cos m\lambda + B_m(\theta)\sin m\lambda\} \\
&= \frac{1}{2}\sum_{m=0}^{N-1}\{[A_m(\theta)+iB_m(\theta)]e^{-im\lambda} + [A_m(\theta)-iB_m(\theta)]e^{im\lambda}\} \\
&= \frac{1}{2}\{DFT[A_m(\theta)+iB_m(\theta)] + \overline{DFT[A_m(\theta)+iB_m(\theta)]}\} \\
&= \mathrm{Re}\, DFT\,[A_m(\theta)+iB_m(\theta)]
\end{aligned}
$$

assuming discrete longitudes $\lambda_k = 2\pi k/N$, k=0, 1, 2, …, N-1, with the Discrete Fourier Transform (DFT), for discrete co-latitudes θ to be determined. Writing $C_m(\theta) = A_m(\theta) + iB_m(\theta)$ and $c_{nm}(\theta) = a_{nm}(\theta) + ib_{nm}(\theta)$, one then has

$$f(\theta,\lambda_k) = \mathrm{Re}\, DFT[C_m(\theta)]$$

where

$$C_m(\theta) = \sum_{n=m}^{N-1}(a_{nm} + ib_{nm})P_{nm}(\cos\theta) = \sum_{n=m}^{N-1} c_{nm}P_{nm}(\cos\theta)$$

that is,

$$
\begin{aligned}
C_0(\theta) &= c_{00}P_{00}(\cos\theta) + c_{10}P_{10}(\cos\theta) + c_{20}P_{20}(\cos\theta) + \ldots + c_{N-1,0}P_{N-1,0}(\cos\theta) \\
C_1(\theta) &= c_{11}P_{11}(\cos\theta) + c_{21}P_{21}(\cos\theta) + c_{31}P_{31}(\cos\theta) + \ldots + c_{N-1,1}P_{N-1,1}(\cos\theta) \\
C_2(\theta) &= c_{22}P_{22}(\cos\theta) + c_{32}P_{32}(\cos\theta) + c_{42}P_{42}(\cos\theta) + \ldots + c_{N-1,2}P_{N-1,2}(\cos\theta) \\
&\ldots \\
C_{N-1}(\theta) &= c_{N-1,N-1}P_{N-1,N-1}(\cos\theta)
\end{aligned}
$$

for discrete co-latitudes θ to be determined.

For analysis, given a band-limited square integrable function $f(\theta, \lambda)$, with co-latitude θ and longitude λ, with observations at $\lambda_k = 2\pi k/N$ and $\theta_j = (j + ½)\pi/N$, k, j = 0, 1, 2, …, N-1, assuming N to be a power of 2, using Driscoll and Healy's formulation [7], the normalized spherical harmonic coefficients a_{nm} and b_{nm} for $m \le n$, n = 0, 1, 2, … , N-1, can be evaluated as follows:

$$\begin{Bmatrix} a_{nm} \\ b_{nm} \end{Bmatrix} = \frac{2\pi(-1)^m}{N}\sum_{j=0}^{N-1}\sum_{k=0}^{N-1} d_j f(\theta_j,\lambda_k)\begin{Bmatrix} \cos m\lambda_k \\ \sin m\lambda_k \end{Bmatrix} P_{nm}(\cos\theta_j)$$

or

$$
\begin{aligned}
c_{nm} &= \frac{2\pi(-1)^m}{N}\sum_{j=0}^{N-1}\sum_{k=0}^{N-1} d_j f(\theta_j,\lambda_k)e^{+im\lambda_k}P_{nm}(\cos\theta_j) \\
&= \frac{2\pi}{N}\sum_{j=0}^{N-1}\sum_{k=0}^{N-1}(-1)^m d_j f(\theta_j,\lambda_k)e^{+im\lambda_k}P_{nm}(\cos\theta_j) \\
&= \frac{2\pi}{N}\sum_{j=0}^{N-1}(-1)^m d_j P_{nm}(\cos\theta_j)\, IDFT_k[f(\theta_j,\lambda_k)]
\end{aligned}
$$

with the Inverse Discrete Fourier Transform (IDFT), where the quadrature weights

$$d_j = \frac{2\sqrt{2}}{N}\sin[(j+½)\pi/N]\sum_{h=0}^{N/2-1}\frac{1}{2h+1}\sin[(2h+1)(j+½)\pi/N]$$

assuming the usual normalization of the Legendre functions. In practice, DFTs and IDFTs would of course be replaced by FFTs and IFFTs for efficiency. The offsetting of the co-latitudes θ_j is to avoid the unacceptable situations at the poles for geocomputations. Other formulations such as the SpherePack software [1] require that both poles be included with an equiangular grid or a Gaussian grid. Mohlenkamp [12] uses a Gaussian grid for better accuracy.

Other authors have used "igloo-like" and "equal-area, iso-latitude" constructions to facilitate spherical harmonic transforms and other operations on the sphere (see, e.g. [13, 6, 9]). In particular, the latter construction has been used for fast Haar wavelet transforms.

Linear convolutions are among the most fundamental operations in digital signal processing. They correspond to the filtering operation of a (data) sequence by another sequence that characterizes the filter. Considering square integrable functions over the sphere, i.e. $\mathbf{L}^2(S^2)$, a convolution algebra of spherical harmonic polynomials can be obtained, but as translations in the plane correspond to rotations on the sphere which do not always commute, the corresponding convolution algebra is non commutative or non Abelian. Using the SHTs for convolution equation $Z(P) = X(P)*Y(P)$ for points P on the sphere $\mathbf{S}^2$,

$$Z_{n,m} = \frac{4\pi}{2n+1} X_n\, Y_{n,m} \neq \frac{4\pi}{2n+1} Y_n\, X_{n,m}$$

with the inequality due to the non-commutativity of displacements on the sphere, where as above, n denotes the degree and m, the order in the SHTs. Examples of convolutions on the sphere and in neighboring space are given in [3], mainly for global multiresolution applications.

4 Data Stucturing Considerations

Enormous quantities of space, surface and subsurface data imply extensive data structuring and management requirements for most array computations. For some years, various methods have been proposed for data analysis, storage and visualization [e.g. 2, 11, 8, 26]. More recently, attempts have been made to address the data management and processing problems. The Hierarchical Data Format (HDF) file organization has recently been promoted for many processing tasks and discipline areas [20]. HDF's limitations have largely been overcome in Version 5 by employing parallelism in data storage and computation [14].

The use of spherical quadtrees has been proposed by NASA to solve storage, multiple resolution analysis and to improve interactive browsing [17]. For the analysis of the Cosmic Microwave Background (CMB), both the igloo [6] and HEALPix [9] approaches have been proposed. The latter is said to meet the needs for a hierarchical database structure, an equal area pixelization and an isolatitude distribution of area elements. A software package that implements this discretization and provides the means of performing fast spherical harmonic transforms is available. The igloo approach gives an exact azimuthal symmetry at each latitude and hence exact spherical harmonic transforms are possible.

5 Concluding Remarks

Spherical computations involve analysis and synthesis using SHTs for convolutions in linear filtering, boundary value problems and numerous other applications. With discrete data on regular spherical grids, different quadrature formulations exist in the literature and different conventions make the intercomparative analyses quite challenging. Different strategies imply different grids such as equiangular, Gaussian, etc. and the normalization used in the resulting frequency spectra is often different.

Considerable work has been done on solving the computational complexities, and enhancing the speed of calculation of spherical harmonics. The approach of Driscoll and Healy [7] is exact for exact arithmetic, while the earlier approach by Swarztrauber [19] which was a quadrature method as well, also provided an exact analysis for a continuous function expressible in terms of a discrete surface harmonic basis. The latter method was quite accurate, with cited errors of the order of 10^{-13} for a 2.5 degree grid. Both methods provide recovery of the original signal provided it is not aliased in the sampling process.

With respect to gridding, both Gaussian quadrature and equiangular grids have been used to obtain accurate results. In particular, the SpherePack code, which in part is based on Swarztrauber's work, provides both grids, and the mechanism to move between the two. More recently, Mohlenkamp [12] has developed the means to expand a function on a Gaussian grid in $O(N^2 \log^2 N)$ time, although the speed is said to depend in part on the use of pre-computed and compressed representations of the associated Legendre function bases. More optimization research is clearly warranted for fast SHTs to be comparable to two-dimensional FFTs.

Acknowledgements.

The authors would like to acknowledge the sponsorship of the Natural Science and Engineering Research Council in the form of a Research Grant to the first author on Computational Tools for the Geosciences.

References

1. Adams, J.C. and P.N. Swarztrauber. SPHEREPACK 2.0: A Model Development Facility, 1997. http://www.scd.ucar.edu/softlib/SPHERE.html
2. Augenbaum, J.M. and C.S. Peskin. On the Construction of the Voronoi Mesh on a Sphere. Journal of Computational Physics, 59, pp. 177-192, 1985.
3. Blais, J.A.R. and D.A. Provins. Spherical Harmonic Analysis and Synthesis for Global Multiresolution Applications. Journal of Geodesy, vol.76, no.1, pp.29-35, 2002.
4. Bond, J.R., R.G. Crittenden, A.H. Jaffe and L. Knox. Computing Challenges of the Cosmic Microwave Background. Computing in Science and Engineering, March-April, pp.21-35, 1999.
5. Colombo, O. Numerical Methods for Harmonic Analysis on the Sphere. Ohio State University, Report no. 310, 1981.
6. Crittenden, R.G. and N.G. Turok. Exactly Azimuthal Pixelizations of the Sky, 1998. http://xxx.lanl.gov/list/astro-ph/9806[374] (I.e. article 374 in the 9806 directory).

7. Driscoll, J.R. and D.M. Healy, Jr. Computing Fourier Transforms and Convolutions on the 2-Sphere. Advances in Applied Mathematics, 15, pp. 202-250, 1994.
8. Gold, C.M. Problems with Handling Spatial Data – The Voronoi's Approach. CISM Journal ACSGC, vol. 45, no. 1, pp. 65-80, 1991.
9. Gorski, K.M., E. Hivon and B.D. Wandelt. Analysis Issues for Large CMB Data Sets. Proceedings: Evolution of Large Scale Structure, Garching, 1998.
10. Lemoine, F.G., D.E. Smith, L. Kunz, R. Smith, E.C. Pavlis, N.K. Pavlis, S.M. Klosko, D.S. Chinn, M.H. Torrence, R.G. Williamson, C.M. Fox, K.E. Rachlin, Y.M. Wang, S.C. Kenyon, R. Salman, R. Trimmer, R.H. Rapp and R.S. Nerem. The Development of the NASA GSFC and NIMA Joint Geopotential Model. International Symposium Gravity, Geoid and Marine Geodesy, Tokyo, International Association of Geodesy Symposia, Vol. 117, pp. 461-469, Springer-Verlag, 1996.
11. Lukatela, H. Hipparchus Geopositioning Model: An Overview. Proceedings of AUTO-CARTO 8, 1987. http://www.geodyssey.com.
12. Mohlenkamp, M.J. A Fast Transform for Spherical Harmonics. The Journal of Fourier Analysis and Applications, vol. 5, nos. 2/3, pp. 159-184, 1999.
13. Muciaccia P.F., P. Natoli and N. Vittorio. Fast Spherical Harmonic Analysis: A Quick Algorithm for Generating and/or Inverting Full-Sky, High Resolution Cosmic Microwave Background Anisotropy Maps. The Astrophysical Journal, 488, pp. L63-L66, 1997.
14. NCSA Introduction to HDF5 Release 1.0, National Center for Supercomputing Applications, 1999.
15. Paul, R.H. Recurrence Relations for the Integrals of Associated Legendre Functions. Bulletin Géodésique, Vol. 52, pp. 177-190, 1978.
16. Ricardi, L.J. and M.L. Burrows. A Recurrence Technique for Expanding a Function in Spherical Harmonics. IEEE Transactions on Computers, June, pp. 583-585, 1972.
17. Short, N.M., Jr., R.F. Cromp, W.J. Campbell, J.C. Tilton, J.L. LeMoigne, G. Fekete, N.S. Netanyahu and G. Sylvain. AI Challenges within NASA's Mission to Planet Earth. Workshop Notes for the 1994 National Conf. on Artificial Intelligence, C.L. Mason (Ed.).
18. Sneeuw, N. Global Spherical Harmonic Analysis by Least-Squares and Numerical Quadrature Methods in Historical Perspective. Geophys. J. Int. (1994) **118**, 707-716.
19. Swarztrauber, P.N. On the Spectral Approximation of Discrete Scalar and Vector Functions on the Sphere. SIAM Journal of Num. Analysis, Vol.16, No. 6, pp. 934-949, 1979.
20. Tan, C.J., J.A.R. Blais and D.A. Provins. Large Imagery Data Structuring Using Hierarchical Data Format for Parallel Computing and Visualization. High Performance Computing Systems and Applications, edited by A. Pollard, D. J.K. Mewhort and D.F. Weaver, Kluwer Academic Publishers, Chapter 39, pp. 371-386, 2000.
21. Varshalovich, D.A., A.N. Moskalev and V.K. Khersonskij. Quantum Theory of Angular Momentum. World Scientific Publishing, Singapore, 1988.
22. Wandelt, B.D., E. Hivon and K.M. Gorski. Topological Analysis of High-Resolution CMB Maps. Theoretical Astrophysics Centre, Copenhagen, Denmark, 1998, 4 pages.
23. Wenzel, G. Ultra High Degree Geopotential Model GPM3E97A to Degree and Order 1800 Tailored to Europe. Proceedings of the Second Continental Workshop on the Geoid in Europe, Budapest, 1998.
24. Wenzel, G. Ultra High Degree Geopotential Models GPM98A, B and C to Degree 1800. Preprint, Bulletin of International Geoid Service, Milan, 1998.
25. Wenzel, G. Hochaufloesende Kugelfunktionsmodelle fuer das Gravitationspotential der Erde. Wissenschaftliche Arbeiten der Fachrichtung Vermessungswesen der Universitat Hannover, Nr. 135, Hannover, 1985.
26. Zheng, C., J. Nie and J.A.R. Blais. Applicability of the Hipparchus Software in Geoscience Information Systems. Proceedings of the Canadian 1994 GIS Conference in Ottawa, pp. 434-442.

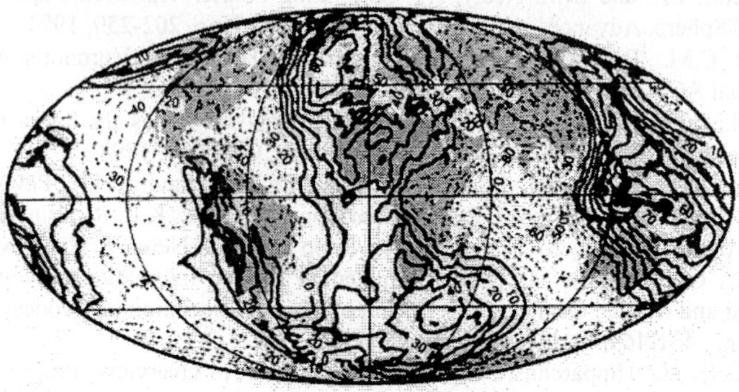

Fig. 1. GPM 98 B Undulation Map of Degree and Order 1800.

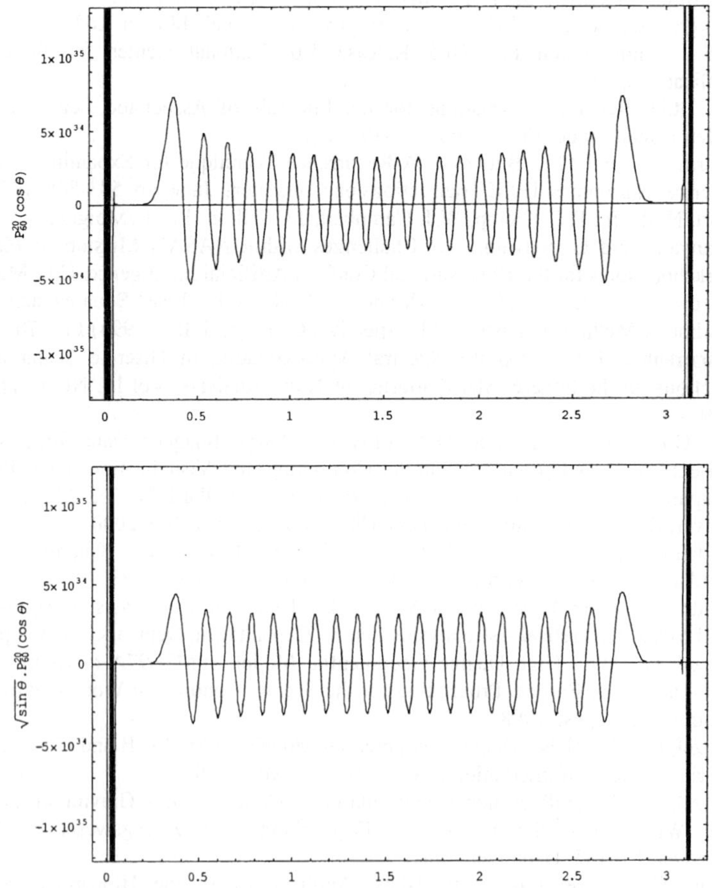

Fig. 2. Plots of $P_{60}^{20}(\cos\theta)$ and $\psi_{60}^{20}(\theta) \equiv \sqrt{\sin\theta}\, P_{60}^{20}(\cos\theta)$, $0 \leq \theta \leq \pi$.

Data Assimilation for 2-D Advection-Dispersion Equations

Sergey Kivva

Institute of Mathematical Machines and System Problems, National Academy of Sciences of Ukraine, 42, Glushkova pr., 03187, Kiev, Ukraine
slk@env.immsp.kiev.ua

Abstract. Data assimilation based on a variational principle for parameter estimation of 2-D advection-dispersion equations is considered. It is assumed that a priori estimations for the model parameters and the initial condition are available. Improvement of parameters for the radionuclide transport model is reduced to an optimization problem with quadratic cost functional. The cost functional comprises the measurement model and model errors, the initial condition uncertainty and parameter penalties. The existence of a unique solution for the state equations and the adjoint system is proved. Differential properties of the cost functional are investigated and a necessary condition for cost functional minimum is derived. Restrictions on the cost functional weights, which guarantee the existence of a unique stationary point, are derived.

1 Introduction

In this paper we concentrate on parameter estimation for the advection-dispersion equations, which, for example, can be used to describe radionuclide transport by surface water flow. The radionuclide transport in the aqueous phase and on suspended sediments by surface water flow is governed by the equations with the sink-source term describing physical-chemical interactions and erosion-deposition exchange processes [1-2]

$$\frac{\partial(hc_l)}{\partial t}+\frac{\partial}{\partial x_i}(v_i hc_l)=\frac{\partial}{\partial x_i}\left(he_i\frac{\partial c_l}{\partial x_i}\right)-\lambda hc_l-\alpha_s hS\left(k_d^s\rho^{-1}c_l-c_s\right)-$$

$$-(1-\phi)z_*\alpha_b\left(k_d^b\rho_b\rho^{-1}c_l-c_b\right) \tag{1}$$

$$\frac{\partial(hSc_s)}{\partial t}+\frac{\partial}{\partial x_i}(v_i hSc_s)=\frac{\partial}{\partial x_i}\left(he_i\frac{\partial Sc_s}{\partial x_i}\right)-\lambda hSc_s+\alpha_s hS\left(k_d^s\rho^{-1}c_l-c_s\right)+$$

$$+\rho_b^{-1}q_b c_b-q_s c_s \tag{2}$$

where $h(x,t)$ is the surface water depth; $v_i(x,t)$ are the surface water velocity components; $\phi(x)$ is the porosity of upper soil layer; ρ is the density of water; $\rho_b(x)$ is the density of soil matrix; $S(x,t)$ is the suspended sediment concentration; $c_l(x,t)$ is the

P.M.A. Sloot et al. (Eds.): ICCS 2003, LNCS 2658, pp. 619–628, 2003.

volumetric radionuclide activity in aqueous phase; $c_s(x,t)$ is the radionuclide activity on suspended sediment; $c_b(x,t)$ is the volumetric radionuclide activity in upper soil layer; $z_*(x,t)$ is the thickness of active upper soil layer; λ is the radionuclide decay constant; $q_s(x,t)$ and $q_b(x,t)$ are the deposition and erosion rates; $e_i(x,t)$ – the diffusion coefficients; k_d^s, k_d^b and a_s, a_b are the partition coefficients and the exchange rates, respectively; i=1,2.

Contamination of the active upper soil layer is described by the equations

$$\frac{\partial}{\partial t}\left((1-\phi)\,z_* c_b\right)=\frac{\partial}{\partial x_i}\left((1-\phi)\,z_* e_i^b\,\frac{\partial c_b}{\partial x_i}\right)+$$

$$+(1-\phi)\left[\alpha_b z_*\left(k_d^b\rho_b\rho^{-1}c_l-c_b\right)-\lambda z_* c_b\right]-\rho_b^{-1}q_b c_b+q_s c_s \qquad (3)$$

Values of the partition coefficients $k_d^s(x)$, $k_d^b(x)$ and exchange rates $\alpha_s(x)$, $\alpha_b(x)$ are determined by the soil physical-chemical properties and forms of radionuclide contamination. These coefficients characterize the interchange between the various radionuclide forms. So a forecast quality of radionuclide contamination of surface reservoirs depends on accuracy determination of their values. Experience of the Chernobyl accident has shown that exact values of the partition coefficient and exchange rates are unknown just after accident and at the best their estimations are only available.

Nowadays data assimilation methods used for parameter estimation is based on estimation theory or control theory [3]. Estimation theory usually involves a sequential estimations of both model and data errors by using a statistical approach. Control theory methods are generally based on variational principles.

In this paper, we consider variational assimilation approach [4] for parameter estimation of the model of radionuclide transport by surface water flow. In this case, parameter estimation problem reduces to minimization of a quadratic cost functional on a set of admissible values of the model parameters in neighborhood of their a priori estimations. On the set of admissible values of the model parameters the differential properties of the cost functional are investigated, a necessary condition for a minimum of the cost functional is derived, and a sufficient condition for existence of a unique stationary point is obtained.

2 Problem Statement

Rewrite the model equations (1)-(3) in the vector form

$$\frac{\partial}{\partial t}(A_0 c)+\frac{\partial}{\partial x_i}(A_i c)=\frac{\partial}{\partial x_i}\left(D_i\frac{\partial c}{\partial x_i}\right)-G(\mu,\nu)c+r,\quad (x,t)\in Q=\Omega\times(0,T) \qquad (4)$$

where $c(x,t)=(c_1,c_2,c_3)^T=(c_l,Sc_s,c_b)^T$ is the state vector; $r(x,t)=(r_1,r_2,r_3)^T$ is the function of the model error; $\mu=(\mu_1,\mu_2)^T=\left(\alpha_s k_d^s/\rho,\alpha_b k_d^b\rho_b/\rho\right)^T$ and $\nu=$

$(v_1, v_2)^T = (\alpha_s, \alpha_b)^T$ is the parameter vector; Ω is a connected bounded set in E^2 with twice boundedly differentiable boundary $\partial\Omega$; E^n is the n-dimensional Euclidean space. The function of model errors $r(x,t)$ we shall consider as an auxiliary variable, which is needed to minimize.

Matrices A_0, A_i, D_i are diagonal matrices

$$A_0 = diag\left\{a_0^{11}, a_0^{22}, a_0^{33}\right\} = diag\{h, h, z\};$$

$$A_i = diag\left\{a_i^{11}, a_i^{22}, a_i^{33}\right\} = diag\{v_i h, v_i h, 0\};$$

$$D_i = diag\left\{d_i^{11}, d_i^{22}, d_i^{33}\right\} = diag\left\{he_i, he_i, ze_i^b\right\}.$$

Matrix $G(\mu, \nu)$ is represented as

$$G(\mu,\nu) = \begin{pmatrix} \lambda h + \tilde{\mu}_1 hS + \tilde{\mu}_2 z & -\tilde{\nu}_1 h & -\tilde{\nu}_2 z \\ -\tilde{\mu}_1 hS & \lambda h + \tilde{\nu}_1 h + q_1 & -q_2 \\ -\tilde{\mu}_2 z & -q_1 & \lambda z + \tilde{\nu}_2 z + q_2 \end{pmatrix},$$

where $z = (1-\phi) z_*$; $q_1 = q_s / S$; $q_2 = q_b / \rho_b$; $\mu_i, \nu_i \in E^m$ and $\tilde{\mu}_i = \mu_i^T \psi(x)$, $\tilde{\nu}_i = \nu_i^T \psi(x)$; the function $\psi(x) = (\psi_1, \ldots, \psi_m)^T$ satisfies to relations

$$\psi_k(x) = \begin{cases} 1, & x \in \Omega_k \\ 0, & x \notin \Omega_k \end{cases}; \quad \Omega = \bigcup_{k=1}^{m} \Omega_k; \qquad \Omega_i \bigcap \Omega_j = \varnothing, \quad i \neq j.$$

Let elements of the matrices A_0, A_i, D_i and G are bounded measurable functions on Q. The matrices A_0, A_i and D_i are differentiable with respect to t and x and the following inequalities are valid

$$\upsilon \le d_i^{kk} \le \pi, \qquad \upsilon_1 \le a_0^{kk} \le \pi_1, \quad \upsilon, \upsilon_1 > 0, \quad k = \overline{1,3}; \quad i = \overline{1,2}$$

$$\operatorname{vrai} \max_{(x,t) \in Q} \left(\left\| \frac{\partial A_0}{\partial t} \right\|_E^2, \sum_{i=1}^{2} \left\| \frac{\partial D_i}{\partial t} \right\|_E^2, \sum_{i=1}^{2} \left\| \frac{\partial A_i}{\partial t} \right\|_E^2 \right) \le \pi_2$$

$$\operatorname{vrai} \max_{(x,t) \in Q} \left(\sum_{i=1}^{2} \left\| \frac{\partial A_0}{\partial x_i} \right\|_E^2, \sum_{i=1}^{2} \left\| \frac{\partial D_i}{\partial x_k} \right\|_E^2, \sum_{i=1}^{2} \left\| \frac{\partial A_i}{\partial x_i} \right\|_E^2 \right) \le \pi_3 \tag{5}$$

$$\underset{\substack{(x,t)\in Q\\ u\in U_{ad}}}{\text{vrai max}} \left(\sum_{i=1}^{2} \|A_i\|_E^2,\ \|G\|_E \right) \le \pi_4 ,$$

where $\|A\|_E^2 = \sum_{i,j=1}^{n} \left(a^{ij}\right)^2$ is the Euclidean norm of a $n\times n$-matrix $A = \left\{a^{ij}\right\}_j^i$.

For the equation (4), let's consider the first boundary value problem with homogeneous boundary condition

$$c(x,0) = c^0(x), \qquad x\in\Omega \tag{6}$$

$$c(x,t) = 0 , \qquad (x,t)\in\Gamma = \partial\Omega\times[0,T] \tag{7}$$

The measurement model is described by equation

$$c^d(x) = H(x)c(x,T) + \vartheta(x) \tag{8}$$

where $c^d(x)$ is the observation data; ϑ is the measurement error; $H(x)$ is the observation matrix, elements of which are bounded and measurable on Ω and have bounded derivatives with respect to x.

We shall assume that exact values of the model parameters and the initial condition are unknown and their a priori estimations are only available

$$\mu_i = \mu_i^a + \vartheta_i^\mu, \quad \nu_i = \nu_i^a + \vartheta_i^\nu, \quad c^0(x) = c_a^0(x) + \vartheta_c^0(x),$$

where μ_i^a and ν_i^a are the given a priori estimations of the model parameters; c_a^0 is the known a priori estimation of the initial state; ϑ_i^μ and ϑ_i^ν are the errors of the model parameters; ϑ_c^0 is the error of the initial state.

Denote by U the space $E^{2\times m}\times E^{2\times m}\times L_2(Q)\times \dot{W}_2^1(\Omega)$ with elements $u=(u_1, u_2, u_3, u_4)^T=(\mu,\nu,r,c^0)^T$. Introduce in the space U the following scalar product

$$\langle u,w\rangle_U = \langle u_1,w_1\rangle_{E^{2\times m}} + \langle u_2,w_2\rangle_{E^{2\times m}} + \langle u_3,w_3\rangle_{L_2(Q)} + \langle u_4,w_4\rangle_{W_2^1(\Omega)}$$

for $\forall u,w\in U$, where $\langle\xi,\eta\rangle_{E^{n\times k}} = \sum_{i=1}^{n}\sum_{j=1}^{k}\xi_{ij}\eta_{ij}$ is the scalar product in $E^{n\times k}$ (further we shall omit dimension of the Euclidean space in scalar product when it can be done without ambiguity); $L_2(Q)$ is the Hilbert space of measurable functions on Q with finite scalar product

$$\langle f,g\rangle_{L_2(Q)} = \int_0^T\int_\Omega \langle f,g\rangle_E\, dxdt .$$

$\dot{W}_2^1(\Omega)$ is the subspace of the Hilbert space $W_2^1(\Omega)$, dense set in which is set of all infinitely differentiable finite functions on Ω. The scalar product in $W_2^1(\Omega)$ is defined by

$$\langle f,g\rangle_{W_2^1(\Omega)} = \int_\Omega \left\{ \langle f,g\rangle_E + \sum_{i=1}^{2} \left\langle f_{x_i}, g_{x_i} \right\rangle_E \right\} dx .$$

Let U_{ad} is a closed convex bounded set in U, defining admissible values of the model parameters (4), the initial state (6) and the model error. Further, the set U_{ad} is called as set of admissible model parameters. Then, we shall consider the problem of improvement of the model parameters $u=(\mu,\nu,r,c^0)^T$ for the equation (4) by using the observations (8) as a problem of minimization the following cost functional on U_{ad}

$$\begin{aligned} \Im(u,c(x,t;u)) = &\int_\Omega \left\{ c^d(x) - H(x)c(x,T;u) \right\}^T R_0 \left\{ c^d(x) - H(x)c(x,T;u) \right\} dx + \\ &+\int_Q r^T R_1 r dx dt + \int_\Omega \left(c^0(x) - c_a^0(x) \right)^T R_2 \left(c^0(x) - c_a^0(x) \right) dx + \\ &+\sum_{i=1}^{2} \left\{ \left(\mu_i - \mu_i^a\right)^T R_{\mu_i} \left(\mu_i - \mu_i^a\right) + \left(\nu_i - \nu_i^a\right)^T R_{\nu_i} \left(\nu_i - \nu_i^a\right) \right\}, \end{aligned} \qquad (9)$$

where $R_0(x)$, $R_1(x,t)$, $R_2(x)$, R_{μ_i}, R_{ν_i} are a positive-definite symmetric weight matrices. We shall assume that elements of the matrices $R_0(x)$, $R_2(x)$, $R_1(x,t)$ are a bounded measurable functions on Ω and Q; the functions $c^d(x), c_a^0(x) \in \dot{W}_2^1(\Omega)$. Moreover, the matrix $R_0(x)$ have bounded derivatives with respect to x on Ω. We can consider the weight matrices as parameters of data assimilation and their choice depends on users.

Using the method of the Lagrange multipliers, we get the following adjoint system

$$A_0 \frac{\partial c^*}{\partial t} + A_i \frac{\partial c^*}{\partial x_i} + \frac{\partial}{\partial x_i}\left(D_i \frac{\partial c^*}{\partial x_i} \right) - G^T(\mu,\nu) c^* = 0 \qquad (10)$$

$$c^*(x,T) = A_0^{-1} R_0 H \left(c^d - Hc \right), \qquad x \in \Omega \qquad (11)$$

$$c^*(x,t) = 0, \qquad (x,t) \in \Gamma \qquad (12)$$

3 Existence of Solution of the State and Adjoint Problems

Consider a system of equations

$$B_0(x,t)\frac{\partial w}{\partial t}-\frac{\partial}{\partial x_i}\left(B_{ij}(x,t)\frac{\partial w}{\partial x_j}+B_i(x,t)w\right)+C_i(x,t)\frac{\partial w}{\partial x_i}+C(x,t)w=-f(x,t) \tag{13}$$

where $w(x,t)$ and $f(x,t)$ are a m-dimensional vector-functions; B_0, B_{ij}, B_i, C_i and C are $m\times m$-matrices (B_0, B_{ij} are the diagonal matrices and $B_{ij}=B_{ji}$) with elements $b_0^{kk}(x,t)$, $b_{ij}^{kk}(x,t)$, $b_i^{kl}(x,t)$, $c_i^{kl}(x,t)$ and $c^{kl}(x,t)$, respectively; $x=(x_1,x_2)$.

We shall assume that the following relationships are valid

$$\sigma\zeta^T\zeta\le\sum_{i,j=1}^{2}b_{ij}^{kk}\zeta_i\zeta_j\le\gamma\zeta^T\zeta\,,\qquad \sigma>0,\quad k=\overline{1,m}$$

$$\sigma_1\xi^T\xi\le\xi^T B_0\xi\le\gamma_1\xi^T\xi\,,\qquad \sigma_1>0 \tag{14}$$

for any real $\xi=(\xi_1,\ldots,\xi_m)^T$ and $\zeta=(\zeta_1,\zeta_2)^T$. The matrices B_0, B_i, B_{ij} are differentiable with respect to t and x and

$$\underset{(x,t)\in Q}{\text{vrai max}}\left(\left\|\frac{\partial B_0}{\partial t}\right\|_E^2,\sum_{i,j}\left\|\frac{\partial B_{ij}}{\partial t}\right\|_E^2,\sum_i\left\|\frac{\partial B_i}{\partial t}\right\|_E^2\right)\le\gamma_2$$

$$\underset{(x,t)\in Q}{\text{vrai max}}\left(\sum_i\left\|\frac{\partial B_0}{\partial x_i}\right\|_E^2,\sum_{i,j}\left\|\frac{\partial B_{ij}}{\partial x_k}\right\|_E^2,\sum_i\left\|\frac{\partial B_i}{\partial x_i}\right\|_E^2\right)\le\gamma_3 \tag{15}$$

$$\underset{(x,t)\in Q}{\text{vrai max}}\left(\sum_{i=1}^{2}\left\|B_i\right\|_E^2,\ \sum_{i=1}^{2}\left\|C_i\right\|_E^2,\ \left\|C\right\|_E\right)\le\gamma_4\,.$$

Consider the first initial-boundary value problem for the equation (13) with the following initial and boundary conditions

$$w(x,0)=\phi_0(x)\,,\qquad x\in\Omega\,; \tag{16}$$

$$w(x,t)=0\,,\qquad (x,t)\in\Gamma\,. \tag{17}$$

Definition. The function $w(x,t)\in W_{2,0}^{2,1}(Q)$ is called a generalized solution of (13), (16)-(17) if for all t from $[0,T]$ and any vector-function $\eta(x,t)\in L_2(Q)$ the following identity is fulfilled

$$\int_0^t\int_\Omega\eta^T\left\{B_0w_t-\left(B_{ij}w_{x_j}+B_iw\right)_{x_i}+\left(C_iw_{x_i}+Cw+f\right)\right\}dxdt=0 \tag{18}$$

and

$$\int_{\Omega} [w(x,\Delta t)-\phi_0(x)]^2 \, dx \to 0 \qquad \text{as } \Delta t \to +0 . \tag{19}$$

$W_{2,0}^{2,1}(Q)$ is the Hilbert space of functions with zero-value at Γ, and the scalar product is defined by

$$(w,v)_{W_2^{2,1}(Q)} = \int_Q \sum_{k=1}^{m} \left[w_k v_k + \frac{\partial w_k}{\partial t}\frac{\partial v_k}{\partial t} + \sum_{i=1}^{2} \frac{\partial w_k}{\partial x_i}\frac{\partial v_k}{\partial x_i} + \sum_{i,j=1}^{2} \frac{\partial^2 w_k}{\partial x_i \partial x_j}\frac{\partial^2 v_k}{\partial x_i \partial x_j} \right] dxdt$$

Theorem 1. *Let the coefficients and free terms in the equation* (13) *satisfy to restrictions* (14)-(15). *Ω is a open connected bounded domain in E^2 with boundary $\partial\Omega$, having second bounded derivatives.*

Then, for an arbitrary function $\phi_0(x)$ from $\dot{W}_2^1(\Omega)$ and for an arbitrary function $f(x,t) \in L_2(Q)$, there is a unique solution of the problem (13), (16)-(17) *in $W_{2,0}^{2,1}(Q)$ such that*

$$\|w(x,t)\|_{W_2^{2,1}(Q)}^2 \le M\left(\|\phi_0(x)\|_{W_2^1(\Omega)}^2 + \|f\|_{L_2(Q)}^2 \right)$$

and for all $t \in [0,T]$

$$\|w(x,t)\|_{L_2(\Omega)}^2 \le \sigma_1^{-1} \exp\{\gamma_5 t/\sigma_1\} \left(\gamma_1 \|\phi_0(x)\|_{L_2(\Omega)}^2 + \|f\|_{L_2(Q)}^2 \right),$$

where the constant *M depends only on σ, σ_1, γ, γ_1, γ_2, γ_3, γ_4, T, $\partial\Omega$ and $\gamma_5 = 1+\gamma_2+2\gamma_4+2\gamma_4\sigma^{-1}$.*

The solvability of the initial-boundary value problem (4), (6)-(7) can be proved by using the Galerkin method in similar way as in [5].

Corollary. *Let the coefficients of the equation* (4) *satisfy the restrictions* (5) *on U_{ad}. The matrices $H(x)$, $R_0(x)$ are bounded and measurable on Ω and have bounded derivatives with respect to x, and $c^d(x) \in \dot{W}_2^1(\Omega)$.*

Then, for each $u = (\mu, \nu, r, c^0)$ from U_{ad}, the problems (4), (6)-(7) *and* (10)-(12) *admit a unique solution in $W_{2,0}^{2,1}(Q)$.*

4 Existence of Cost Functional Minimum

Theorem 2. *Let state vector $c(x,t;u)$ is governed by the first initial-boundary value problem* (4), (6)-(7), *the coefficients of which satisfy to the restrictions* (5). *Let the set of admissible model parameters U_{ad} is a closed convex bounded set in U. Cost functional is defined by the relationship* (9).

Then $U^* = \left\{ u = \left(\mu, \nu, r, c^0\right)^T \in U_{ad} : \Im\left(u, c(x,t;u)\right) = \inf_{v \in U_{ad}} \Im\left(v, c(x,t;v)\right) \right\}$ *is non-empty and weakly compact set. Any minimizing sequence* $\{u^k\} \in U_{ad}$ *converges weakly to* U^*.

Proof. Let $\{u^k\} \in U_{ad}$ is a minimizing sequence, i.e.

$$\Im\left(u^k, c(x,t;u^k)\right) \to \inf_{u \in U_{ad}} \Im\left(u, c(x,t;u)\right) \quad \text{as } k \to \infty. \tag{20}$$

According to the Theorem 1, the problem (4), (6)-(7) admits a unique solution on U_{ad} such that $\|c(x,t;u)\|_{W_2^{2,1}(Q)} \le M_1$, where the constant M_1 does not depend on *u*.

Due to boundedness of the solutions $c(x,t;u)$ on U_{ad} and boundedness of the set U_{ad} there is a subsequence $\{u^N\}$ of the sequence $\{u^k\}$ that

- $c(x,t;u^N) \to \tilde{c}(x,t)$ weakly in $W_{2,0}^{2,1}(Q)$,
- $\mu^N \to \tilde{\mu}$ and $\nu^N \to \tilde{\nu}$ in $E^{2\times m}$,
- $c^{0N} \to \tilde{c}^0$ weakly in $\dot{W}_2^1(\Omega)$,
- $r^N \to \tilde{r}$ weakly in $L_2(Q)$.

Moreover, from the imbedding theorems [6, pp.35-36], we have that c^{0N} converges to $\tilde{c}^0$ in $L_2(\Omega)$ and $\left\|c(x,t;u^N) - \tilde{c}(x,t)\right\|_{L_2(\Omega)} \to 0$ as $N \to \infty$ uniformly with regard to $t \in [0,T]$. Since U_{ad} is a closed convex set, the set U_{ad} is weakly closed and $\tilde{u} = \left(\tilde{\mu}, \tilde{\nu}, \tilde{r}, \tilde{c}^0\right)^T \in U_{ad}$.

It is easy to show that $\tilde{c}(x,t)$ coincides with $c(x,t;\tilde{u})$. To show this, it is sufficiently in (18)-(19) written for the $c(x,t;u^N)$ to pass to the limit as $N \to \infty$. Hence, the function $\tilde{c}(x,t)$ is a solution of the problem (4), (6)-(7) for $u = \tilde{u}$.

The functional $\Im\left(u, c(x,t;u)\right)$ is a convex continuous functional of the arguments, so it is weakly lower semicontinuous and

$$\underline{\lim}_{N \to \infty} \Im\left(u^N, c(x,t;u^N)\right) \ge \Im\left(\tilde{u}, \tilde{c}(x,t)\right)$$

Therefore from (20) follows

$$\Im\left(\tilde{u}, \tilde{c}(x,t)\right) \le \inf_{u \in U_{ad}} \Im\left(u, c(x,t;u)\right).$$

Thus $\Im\left(\tilde{u}, \tilde{c}(x,t)\right) = \inf_{u \in U_{ad}} \Im\left(u, c(x,t;u)\right)$ and $\tilde{u} \in U^*$.

It remains for us to show that the set U^* is weakly compact. Let's take an arbitrary sequence $\{u^k\} \in U^*$. Then $\{u^k\} \in U_{ad}$ and there is a subsequence $\{u^N\}$ which converges weakly to some point $\tilde{u} \in U_{ad}$. But $\Im\left(u^N, c(x,t;u^N)\right) = \inf_{u \in U_{ad}} \Im\left(u, c(x,t;u)\right)$,

N=1,2,..., so $\{u^N\}$ is a minimizing sequence. As shown above, the $\{u^N\}$ converges weakly to U^*, therefore $\tilde{u} \in U^*$. Hence, the set U^* is weakly compact that completes the proof.

Let's remark that the set of solutions of the problem (4), (6)-(7) is non-convex, therefore a minimum of the cost functional can be non-unique.

Theorem 3. *Let the assumptions of Theorem 2 are fulfilled.*

Then the cost functional (9) *is continuously differentiated on* U_{ad} *and its gradient* $\Im'(u) = \left(\frac{\partial \Im}{\partial \mu}, \frac{\partial \Im}{\partial \nu}, \frac{\partial \Im}{\partial r}, \frac{\partial \Im}{\partial c^0} \right)^T$ *is represented by*

$$\frac{1}{2}\frac{\partial \Im}{\partial \mu_i}(u, c(x,t;u)) = R_{\mu_i}\left(\mu_i - \mu_i^a\right) + \chi^{\mu_i},$$

$$\frac{1}{2}\frac{\partial \Im}{\partial \nu_i}(u, c(x,t;u)) = R_{\nu_i}\left(\nu_i - \nu_i^a\right) + \chi^{\nu_i}, \tag{21}$$

$$\frac{1}{2}\frac{\partial \Im}{\partial r}(u, c(x,t;u)) = R_1 r(x,t) - c^*(x,t;u),$$

$$\frac{1}{2}\frac{\partial \Im}{\partial c^0}(u, c(x,t;u)) = R_2\left[c^0(x) - c_a^0(x)\right] - A_0 c^*(x,0;u),$$

where $c^*(x,t;u)$ *is the solution of the adjoint problem* (10)-(12), $\chi^{\mu_i}, \chi^{\nu_i} \in E^m$ *and*

$$\chi_k^{\mu_i}(u) = \int_0^T \int_{\Omega_k} c^{*^T}(x,t;u) \frac{\partial G}{\partial \mu_i} c(x,t;u)\,dx\,dt,$$

$$\chi_k^{\nu_i}(u) = \int_0^T \int_{\Omega_k} c^{*^T}(x,t;u) \frac{\partial G}{\partial \nu_i} c(x,t;u)\,dx\,dt. \tag{22}$$

For any point $u^* \in U^*$, *it is necessary the following inequality is fulfilled*

$$\left\langle \Im'\left(u^*\right), u - u^* \right\rangle_U \geq 0 \qquad \text{for all } u \in U_{ad}.$$

5 Sufficient Condition of Existence of Stationary Points

In the previous section has been shown that the set of minimum of the cost functional (9) on U_{ad} is non-empty. But the point of cost functional minimum can be both interior, and boundary point of U_{ad}. It is clear that an interior point of cost functional minimum is the stationary point.

Theorem 4. *Let state function* $c(x,t;u)$ *is governed by the first initial-boundary value problem* (4), (6)-(7), *the coefficients of which satisfy to restrictions* (5). *Let the set of admissible model parameters* U_{ad} *is a convex closed bounded set and the following inequalities are valid for all* $u = \left(\mu, \nu, r, c^0\right)^T \in U_{ad}$

$$\left\|\mu_i - \mu_i^a\right\|_{E^m} \le \delta_{\mu_i}, \qquad \left\|\nu_i - \nu_i^a\right\|_{E^m} \le \delta_{\nu_i},$$

$$\left\|r\right\|_{L_2(Q)} \le \delta_r, \qquad \left\|c^0 - c_a^0\right\|_{W_2^1(\Omega)} \le \delta_c.$$

Then it is sufficient for existence of a unique stationary point on U_{ad} that the weight functions R_0, R_1, R_2, R_{μ_i}, R_{ν_i} satisfied to the following inequalities for all $u, w \in U_{ad}$ and $q \in (0,1)$

$$\left\|R_1^{-1} c^*(x,t;u)\right\|_{L_2(Q)} \le \delta_r, \qquad \left\|R_2^{-1} A_0 c^*(x,0;u)\right\|_{W_2^1(\Omega)} \le \delta_c,$$

$$\left\|R_{\mu_i}^{-1} \chi^{\mu_i}(u)\right\|_{E^m} \le \delta_{\mu_i}, \qquad \left\|R_{\nu_i}^{-1} \chi^{\nu_i}(u)\right\|_{E^m} \le \delta_{\nu_i},$$

$$\left\|R_1^{-1}\left[c^*(x,t;u) - c^*(x,t;w)\right]\right\|_{L_2(Q)}^2 + \left\|R_2^{-1} A_0\left[c^*(x,0;u) - c^*(x,0;w)\right]\right\|_{W_2^1(\Omega)}^2 +$$

$$+\left\|R_{\mu_i}^{-1}\left(\chi^{\mu_i}(u) - \chi^{\mu_i}(w)\right)\right\|_{E^m}^2 + \left\|R_{\nu_i}^{-1}\left(\chi^{\nu_i}(u) - \chi^{\nu_i}(w)\right)\right\|_{E^m}^2 \le q^2 \left\|u - w\right\|_U^2,$$

where $c^*(x,t;u)$ is the solution of the adjoint problem (10)-(12), *and $\chi^{\mu_i}(u), \chi^{\nu_i}(u) \in E^m$ are defined by* (22).

Really, we can rewrite the stationarity conditions (21) as the following operator equation $u = F(u)$. It is obviously that under assumptions of Theorem 4, the operator F is contracting and mapping the set U_{ad} into itself.

References

1. Onishi, Y., Serne, J., Arnold, E., Cowan, C., Thompson, F.: Critical review: radionuclide transport, sediment transport, water quality, mathematical modeling and radionuclide adsorption/desorption mechanism. NUREG/CR-1322. Pacific Northwest Laboratory, Richland (1981)
2. Zheleznyak, M., Demchenko, R., Khursin, S., Kuzmenko, Yu., Tkalich, P., Vitjuk, N.: Mathematical modeling of radionuclide dispersion in the Pripyat-Dnieper aquatic system after the Chernobyl accident. The Science of the Total Environment, Vol. 112. (1992) 89–114
3. Robinson, A.R., Lermusiaux, P.F.J., Sloan III, N.Q.: Data Assimilation. The sea, Vol.10. John Wiley & Sons Inc. (1998) 541–593
4. Penenko, V.V.: Some aspects of mathematical modeling using the models together with observational data. Bull. Nov. Comp. Center, Num. Model. in Atmosph., Ocean and Environment Studies, Vol. 4. Computer Center, Novosibirsk (1996) 31–52
5. Ladyzhenskaya, O.A., Solonnikov, V.A., Uraltseva, N.N.: The linear and quasilinear equations of parabolic type. Nauka, Moscow (1967) (In Russian)
6. Ladyzhenskaya, O.A.: A mixed problem for hyperbolic equations. Gostekhizdat, Moscow (1953) (In Russian)

Mathematical Modelling the Ethnic System

Victor Korobitsin and Julia Frolova

Omsk State University, Computer Science Division,
55A, Mira pr., 644077 Omsk, Russia
{korobits, frolova}@univer.omsk.su
http://www.univer.omsk.su/socsys

Abstract. The main of presented research is a demonstration of creating the society model on ethnic solidarity level. This model describes the behavior of ethnic system. The ethnic system includes a few ethnoses and provides their interactions. The interactions transmit by ethnic fields. The model is described by system of parabolic differential equations. The software TERRI is used for the forecast of arising the ethnic conflicts. Based on simulation result the researcher can compute the direction of ethnic field distribution and the most probable points of skirmish between ethnoses.

1 Introduction

The modelling of biosphere (ecological) processes gave rise to the research of society development. These models were destined for solving the problem of global change the ecological situation. Now the problem of interethnic conflicts is growing in society. It forces the international organizations to find the way for its adjustment. The modelling of global ethnic processes will allows to evaluate the world ethnic situation.

The aim of this research is the construction of mathematical model of ethnic field. The model is described by the system of parabolic equations. It is the tools for research the evolution of interactive ethnic systems under landscape influence.

2 Ethnic Solidarity Level

On the ethnosphere level the traditions play the special role in the society. The people get the behavior stereotypes from them. Thereby the general function of this level is the sample maintenance. The individuals strive for conservation of culture as a collection of history experience.

The *ethnos* is a people group, formed on basis of the original behavior stereotype. It exists as a energy system, opposing itself to other like groups. Thereby people are divided on own and alien man. The main ethnos attribute is a behavior stereotype. It is a complex of behavior standards of ethnos members. The collection of behavior stereotypes is defined by ethnic tradition differed the ethnos from biological population.

P.M.A. Sloot et al. (Eds.): ICCS 2003, LNCS 2658, pp. 629–635, 2003.

The *passio energy* is an excess of biochemical energy of living substance. It suppresses the self-preservation instinct of man and defines the ability to goal-directed ultratension. The ethnic field is formed by the passio energy. It provides the interaction of ethnos members and regulates the joint goal-directed activity of their. Each ethnos forms the unique field and each ethnos member responds to this field. The behavior stereotypes, landscape, and culture values of ethnos characterize the field influence.

The primary motive for arising the ethnic conflicts is a skirmish of two not solidary ethnoses. The skirmish is an effect of distribution of some ethnic field on the territory of another ethnos. There are the territories occupied by the people of different ethnic systems. Such territory is a border or buffer zone placed between two ethnoses. The ethnic conflicts mostly arise on these zones. Therefore the actual problem is to discover the buffer zones and to forecast the ethnic conflicts. For this problem decision, we propose to use the methods of mathematical modelling. The model of level is created on the basis of Lev N. Gumilev's theory of ethnogenesis [1].

3 Mathematical Model of Ethnic Field

The ethnic system includes a few ethnoses and provides their interactions. The interactions is transmited by ethnic fields. This field is distributed on the landscape as hot gas in the space. We constructed the model of ethnic field from this analogy.

Consider the interaction of k ethnoses in the field $G \subset R^2$ with boundary Γ. Let the passio energy of i ethnos (U_i) satisfies the energy conservation law in any given area. Define the passio energy density u_i by

$$U_i(t) = \iint\limits_G u_i(x, y, t)dxdy.$$

The ethnos state is defined by the passio tension. This characteristic is the ratio of passio energy volume to ethnos population quantity. The function $E(x, y, t)$ passio tension of ethnic field is constructed on base of the measurement strategy of history events frequency.

Interrelate the passio tension and density $u(x, y, t)$ of ethnic field energy by

$$u(x, y, t) = k_S q(x, y, t)E(x, y, t),$$

where $q(x, y, t)$ is the density of field receptivity by ethnos members, k_S is the coefficient. The function $q(x, y, t)$ is defined by the relation

$$Q(t) = \sum_j Q_j(t) = \iint\limits_G q(x, y, t)dxdy,$$

where, the function $Q_j(t)$ describes the degree of receptivity and goal-directed use of passio energy by j^{th} ethnos member. The summation is made on all ethnos members fallen in G area.

Construct the integral balance equation describing change to density of ethnic field energy $u_i(x,y,t)$ of ith ethnos $(i = 1, \dots, k)$, k is amount of ethnoses.

$$U_i(t_2) - U_i(t_1) = \int_{t_1}^{t_2} \Big[R_i(t) + P_i(t) + T_i^+(t) + T_i^-(t) + K_i(t) \Big] dt, \qquad (1)$$

where $U_i(t) = \iint_G u_i(x,y,t)dxdy$. The flows of passio energy are described by following expressions:

- R_i is the passio energy inflowing in G through boundary Γ,

$$R_i(t) = \oint_\Gamma \varepsilon_i(x,y,t)\frac{\partial u_i}{\partial n}(x,y,t)d\gamma,$$

  the coefficient $\varepsilon_i(x,y,t)$ characterizes the velocity of passio energy distribution.
- $P_i(t)$ is the passio energy inflowing in G under the influence of directional moving energy through boundary Γ,

$$P_i(t) = \oint_\Gamma -(\boldsymbol{a}_i, \boldsymbol{n})u_i(x,y,t)d\gamma,$$

  the vector field $\boldsymbol{a}_i$ gives the direction of energy moving, the vector $\boldsymbol{n}$ is exterior normal to boundary section $d\gamma$. Let rot $\boldsymbol{a}_i = 0$ then the scalar function φ_i exists and $\boldsymbol{a}_i = -\text{grad}\ \varphi_i(x,y,t)$.
- $T_i^+(t)$ is inflow of passio energy under the induction process in G,

$$T_i^+(t) = \iint_G \beta_i^+(x,y,t)u_i(x,y,t)dxdy,$$

  the coefficient $\beta_i^+(x,y,t)$ is the velocity of induction process.
- $T_i^-(t)$ is outflow of passio energy to life support of ethnos members and landscape maintenance,

$$T_i^-(t) = \iint_G -\beta_i^-(x,y,t)u_i(x,y,t)dxdy,$$

  the coefficient $\beta_i^-(x,y,t)$ is the velocity of passio energy losses.
- $K_i(t)$ is outflow of passio energy under the skirmish of two ethoses,

$$K_i(t) = \iint_G -\Big(\sum_{j=1}^{k} \gamma_{ij}(x,y,t)u_j(x,y,t)\Big)u_i(x,y,t)dxdy,$$

  where u_j is the density of passio energy of hostile ethnos, the coefficient $\gamma_{ij}(x,y,t)$ is the velocity of energy losses under the rivalry i^{th} and j^{th} ethnoses. The ratio $\gamma_{ii}u_i^2$ describes the internal conflicts in ethnos.

The system of integral equations (1) is equivalent to the system of parabolic differential equations (add see [2])

$$\frac{\partial u_i}{\partial t} = \frac{\partial}{\partial x}\left(\frac{\partial \varphi_i}{\partial x}u_i + \varepsilon_i \frac{\partial u_i}{\partial x}\right) + \frac{\partial}{\partial y}\left(\frac{\partial \varphi_i}{\partial y}u_i + \varepsilon_i \frac{\partial u_i}{\partial y}\right) + \left(\beta_i^+ - \beta_i^- - \sum_{j=1}^{k} \gamma_{ij} u_j\right) u_i, \tag{2}$$

Define the initial and edge conditions for the system of parabolic equations by

$$\begin{array}{ll} u_i(x,y,0) = u_i^0(x,y), & (x,y) \in G, \\ \frac{\partial u_i}{\partial n}(x,y,t) = 0, & (x,y) \in \Gamma. \end{array} \tag{3}$$

Define the functions as follows:

– moving the passio energy

$$\varphi_i(x,y) = \frac{\lambda_i}{2\mu_i} e^{-\mu_i((x_i^0 - x)^2 + (y_i^0 - y)^2)}, \quad \lambda_i > 0,\ \mu_i > 0,\ (x_i^0, y_i^0) \in G,$$

– the passio energy distribution

$$\varepsilon_i(x,y) = I_G(\xi_{\varepsilon_i} \circ l_\omega)(x,y), \quad l_\omega : \omega \to L, \quad \xi_{\varepsilon_i} : L \to R^+,$$

where ω is the discrete grid on G area, I_G is the interpolation operator of discrete functions on ω to continuous functions on G, L is a set of landscape types,

– outflow of the passio energy

$$\beta_i^-(x,y) = I_G(\xi_{\beta i} \circ l_\omega)(x,y), \quad \xi_{\beta i} : L \to R^+,$$

– inflow of the passio energy

$$\beta_i^+(t) = \max\{0, \beta_i^0 - \beta_i^1 \cdot (t - T_0^i)\}, \quad \beta_i^0, \beta_i^1 \in R^+, \quad T_0^i \geq 0,$$

– the passio energy losses $\gamma_{ij} \in R^+$.

The system of parabolic differential equations (2) with the initial and edge conditions (3) is a mathematical model of ethnic field interactions.

Given model is a way for formalization of Lev N. Gumilev's theory. The model accentuates the energy and geographical aspects of theory and gives the clear formal description of internal processes.

4 Simulation Tools TERRI for Modeling the Ethnic Fields

The simulation tools TERRI is created for modeling of ethnosphere level. The tools realize the method for solving the system of parabolic differential equations that described the model of ethnosphere. The modeling result is demonstrated on the computer display as a dynamic map of ethnic fields.

The initial data for modeling are the number of ethnoses k, map of landscapes, rates of changing the passio energy (functions ε_i, φ_i, β_i, γ_{ij}), initial distribution of passio energy density u_i^0.

Consider the simulation result of ethnosphere on real example. The aim of simulation was to define the landscape dependence of division of territory between ethnoses. The dependence is discovered on real geographical features of Europe, North Africa, and Middle East. Examine the interaction of three ethnic systems: West European, East Slavonic, Asia Minor. Each ethnos was described by the set of features (the function in the system (2)).

After run the modeling software TERRI, the map of landscapes is appeared on the display. On this map the different landscapes are marked by various colors. The ethnos is born in some point on the map. So the ethnic field is got the initial pulse. According to dynamic rule (2) the field is distributing on the landscape. The ethnic field is marked by color area on the display. Each ethnos has own color: first ethnos – blue, second – red, third – green. Given picture is demonstrated the distribution of ethnoses on the landscape. The value of passio energy density is shown by the brightness of color. The three stages of ethnos dynamics is shown on figure 1.

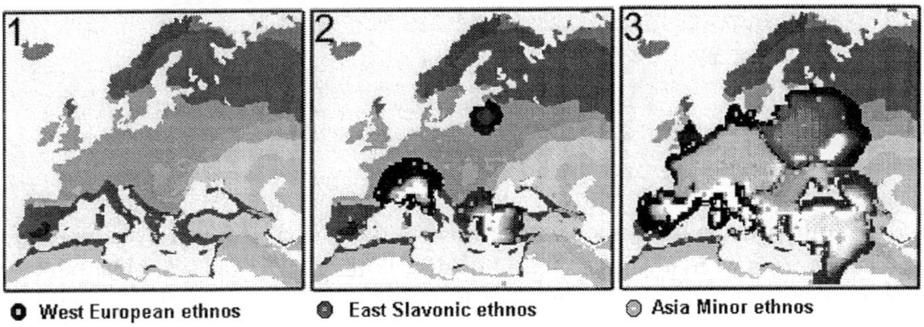

Fig. 1. Distribution of ethnic fields

Initially the born ethnoses is developed on the isolation with each other. In time they come into collision observed by the ethnic field crossing. Under conflicts the passio energy of hostile ethnoses is loss. Since there are not solidary ethnoses then all they can not coexist on common territory. We can observe two way of conflict adjustment. Either the most powered ethnos forces out the feeble one or the equal-powered ethnoses separate the landscape. The buffer zone is formed between them.

The software TERRI allows doing a lot of tests with model. We fixed the part of initial parameters but were changed other parameters in various tests. We were getting the various pictures of ethnic dynamics. For analyzing the model behavior we was collecting the data of ethnos field distribution. The statistical analysis is demonstrated the dependence of ethnic field distribution on the landscapes.

The simulation result is the statistical distribution of super-ethnoses on the landscape (figure 2).

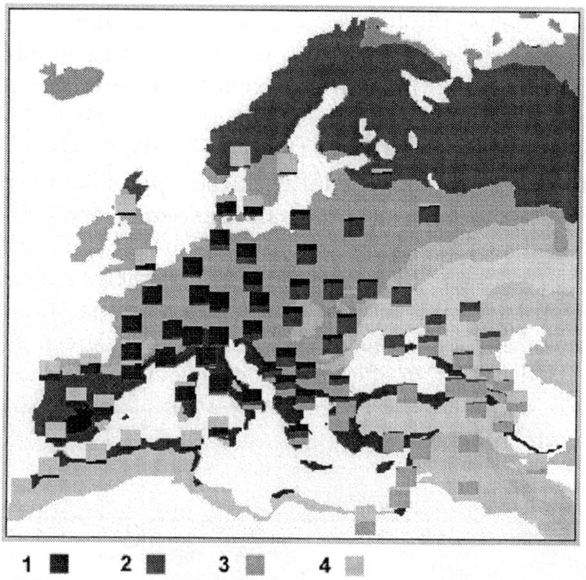

Fig. 2. Statistics of Ethnos Distribution

The comparative analysis of experimental data with the facts was made for the confirmation of hypothesis on ethnos distribution. The facts are the percentage composition of population by church in the cityes of region. These data was given from electronic library of Utrecht University, The Netherlands. The result of comparative analysis is shown on table 1.

The cityes on table is sorted in ascending order of value Δ. It is a deviation of experimental data from the facts, defined by

$$\Delta = \frac{1}{2}\sum_{i=1}^{4} |a_i - b_i|,$$

where a_i is the facts, b_i is the experimental data, i is super-ethnos number.

The analysis of computer simulation results allows doing the following conclusions:

- the distribution of territories between ethnoses really depends on landscape;
- the obtained statistical data demonstrates the correlation of settling the ethnos on landscapes;
- the size of buffer zone is depended on the hostility of neighbor ethnoses.

Table 1. Experimental Data vs the Facts. **Church**: I - Roman Catholic, II - Orthodox, III - Islam, IV - other; **Super-Ethnos**: 1 - West-European, 2 - East-Slavonic, 3 - Asia Minor, 4 - other.

City	I	II	III	IV	1	2	3	4	Δ
Paris	89	4	3	4	87.2	8.0	0.0	4.8	4.8
Andorra	86	0	0	14	76.8	5.8	0.0	17.4	9.2
Zagreb	77	11	0	12	68.8	29.8	1.4	0.0	20.2
Rome	83	0	0	17	78.4	14.0	7.4	0.2	21.4
Buchares	6	80	0	14	22.8	71.6	5.6	0.0	22.4
Minsk	8	60	0	32	15.6	82.2	1.0	1.2	30.8
Vienna	85	0	0	15	64.8	35.2	0.0	0.0	35.2
Saraevo	15	31	40	14	49.2	32.2	18.6	0.0	35.4
Sophia	1	87	8	4	35.0	49.8	15.2	0.0	41.2
Prague	50	2	0	48	66.2	33.8	0.0	0.0	48.0
Berlin	37	0	2	61	66.6	33.4	0.0	0.0	63.0
Athens	0	97	1	2	33.2	33.0	31.6	2.2	64.0
Budapest	68	0	0	32	35.6	64.0	0.4	0.0	64.4
Tbilisi	0	75	11	14	0.0	10.4	75.6	14.0	64.6
Copenhagen	1	0	0	99	60.0	11.8	0.0	28.2	70.8
Jerusalem	0	3	15	82	0.4	0.0	86.4	13.2	71.8
Yerevan	0	100	0	0	0.0	6.8	71.0	22.2	93.2

5 Conclusion

We constructed the mathematical model of ethnic system. On results of presented research we can make up the following conclusions:

- this model is the tools for investigation in global development society area. Based on simulation result the researcher will have got the numerical evaluation of historical hypothesis on ethnosphere evolution;
- the software TERRI is used for the forecast of arising the ethnic conflicts. In that case, it is necessary to keep track of the passio energy pulse. Then we can compute the direction of ethnic field distribution and the most probable points of skirmish between ethnoses;
- one of the ways for ethnic conflict prevention is to fix the territory for certain ethnos. The landscape features characterized for this ethnos define these territories. So the separation of influence area of ethnos on territories is realized.

References

1. Gumilev, L.N.: Ethnogenesis and Biosphere of Earth. Tanais DI-DIK, Moscow (1994)
2. Guts, A.K., Korobitsin, V.V., Laptev, A.A., Pautova, L.A., Frolova, J.V.: Mathematical Models of Social Systems. Omsk State University Press, Omsk (2000)

Simulations for Thermal Analysis of MOSFET IPM Using IMS Substrate

Malgorzata Langer[1], Zbigniew Lisik[1], Ewa Raj[1], Nam Kyun Kim[2], and Jan Szmidt[3]

[1] Institute of Electronics, Technical University of Lodz, Poland
90-924 Lodz, Stefanowskiego 18/22, Poland
{malanger, lisikzby, ewaraj}@ck-sg.p.lodz.pl
[2] Korea Electrotechnology Research Institute, P.O. BOX 20, Chang Won, 641-600 Korea,
nkkim@keri.re.kr
[3] Institute of Microelectronics & Optoelectronics, Warsaw University of Technology, Poland
00-662 Warsaw, Koszykowa 75, Poland
j.szmidt@elka.pw.edu.pl

Abstract. The project was focused on the thermal aspects of the new IPM module design. The investigations were aimed to estimate the influence of the particular design of the MOSFET transistor location versus the steady-state thermal feature of the considered IPM and to determinate its thermal impedance as well as to evaluate its equivalent RC network (ladder) model. It required working out the 3-D thermal model of the considered system that was used in numerical simulations and next, tested with commercial software, ANSYS 5.7 based on Finite Element Method (FEM).

1 Introduction

In any electronics equipment, the temperature rise above the permissible level results in its worse work or even in its destruction. Therefore, the proper thermal design, known as the thermal management is more and more important part of any design process in electronics industry. One can find that effective draining the heat off the source to ambient becomes the problem of the greatest importance, as one must keep the safe work temperature [1,2]. According to the general assumptions concerning the MOSFET IPM under consideration, the analysed system covered 50x40 mm^2 IMS (Insulated Metal Substrate) plate as the substrate on which two MOSFET Toshiba transistors 2SK2866 in TO-220AB packages were to be placed symmetrically and fastened by soldering technique.
The investigations were aimed to evaluate the influence of the particular design of the transistor location versus the steady-state thermal feature of the considered IPM.

2 Thermal Model of IPM

The Fig. 1 introduces two possible designs of the module with two considered layouts for the transistor packages. The both designs use the IMS substrate that is introduced in Fig. 2. section. The preliminary simulations proved that the layout introduced in Fig. 1 a) should be considered as the optimal one, so the further simulations were led for this layout. As the structure is fully symmetrical, only one of its halves is taken to further investigations.

P.M.A. Sloot et al. (Eds.): ICCS 2003, LNCS 2658, pp. 636–643, 2003.

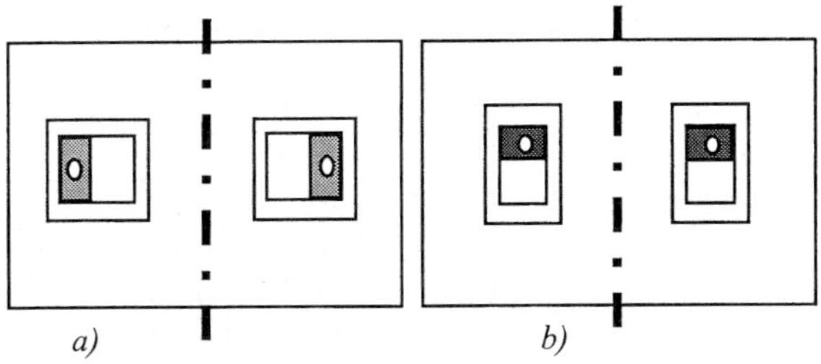

Fig.1. Two possible locations of the transistors on the IMS plate with symmetry axis as the dot line (copper foil is marked as an additional rectangular surrounding the transistor)

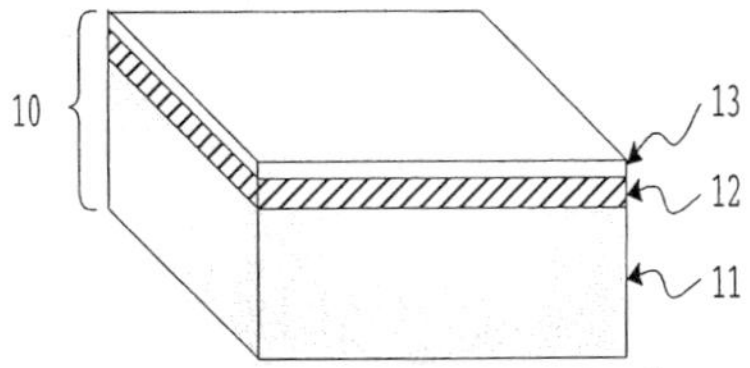

Fig.2. General view of IMS plate presenting the construction layers: 10 – the whole structure (2.22mm); 11 – aluminium baseplate (2.0mm); 12 – polymer insulating (0.12mm); 13 – copper foil for circuit pattern (0.10mm)

The final view of the 3-D thermal model of IPM used in the simulations is shown in Fig.3. It covers one half of the IPM with the transistor fastened by a homogeneous solder layer to the IMS plate. In the transistor thermal model three elements are distinguished only. There are the bottom copper layer (lead frame) soldered to the copper foil of IMS, the silicon chip placed on the lead frame and the epoxy layer covering the silicon chip.

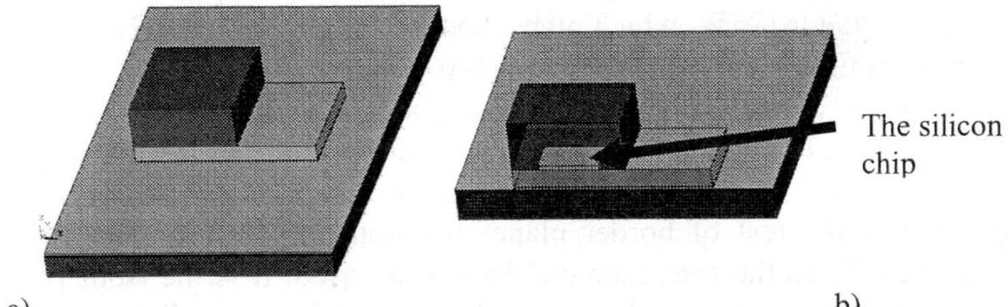

Fig.3. 3-D view of the IPM thermal model: (a) top view; (b) the cross section

The Table 1 collects dimensions of the elements for the model from Fig.3. The copper foil length and width are not determined since they can vary depending on the circuit pattern and they are one of the parameters changed during the investigations. The physical parameters of the materials the parts are made from are collected in the Table.2. It has been assumed that the thermal conductivity of silicon can vary according to the actual temperature in the following way:

$$\lambda = a_t T^{-b} \quad (1)$$

where: a_t=4350, b=1.4, and T is the temperature [K].

Table 1. The dimensions of the elements of thermal model

Material	length [mm]	width [mm]	thickness [mm]	Comments
Al base plate (11)	50	40	2.0	IMS plate
Insulate layer (12)	50	40	0.12	IMS plate
Cu foil (13)	var	var	0.10	IMS plate
Solder	10	10	0.10	
Lead frame	10	10	0.7	Transistor
Silicon chip	4.8	4.8	0.30	Transistor
Epoxy				Transistor

Table 2. The physical parameters of materials

Material	ρ[kg/m^3]	c_p[J/kg K]	λ [W/m K]
Aluminium	2700	900	205
Insulator in IMS	2300	800	1.5
Copper	8960	385	398
Solder	9290	167	48
Epoxy	1270	1050	1.7
Silicon	2330	167	var

In the model, it has been assumed that the silicon chip is the only heat source in which the heat is dissipated homogeneously. The magnitude of heat dissipation has been fixed on the base of Toshiba transistor 2SK2866a data sheet [3]. Taking into account its catalogue maximum ratings, so called the P_D drain power dissipation has been chosen as equal to125W, which gives the heat generation of 18.1e^9 W/m^3 when taking into account the volume of the silicon chip.
The boundary conditions for the border planes have been chosen according to their role in the heat exchange in the real IPM construction. For the plane that corresponds to the symmetry axis the adiabatic boundary condition is the only one that can be applied. For the rest of border planes the boundary conditions using the ambient temperature T_a as the reference one have been used. It is the isothermal one for the bottom of IMS substrate aluminium layer, which is usually kept at the ambient temperature and the convection one for the other surfaces. In the simulation the convection coefficient of 7 W/m^{2}K that is typical for free convection inside a well-ventilated case and the ambient temperature equalled to 25^{0}C has been used.
The full geometry of the considered structure was introduced to Preprocessor of Ansys 5.7 software [4] and all the material properties as well. The manual meshing was chosen to optimize the mesh according to the areas of the greatest interest. Fig. 3 shows the introduced model, and Fig. 4 the optimized mesh.

3 The Simulation Results

We aimed to find the critical value for the Cu foil surface for the heat disposal, and to obtain the results we considered five different dimensions (see Table 1) outward to the TO-220AB package; i.e. 0.5 mm, 1 mm, 2 mm, 4 mm, whole surface. The maximum temperature (always on the top part of the silicon chip) cannot exceed 125^0C as it is the upper edge of the permitted temperature range that allows the devices to operate safely.

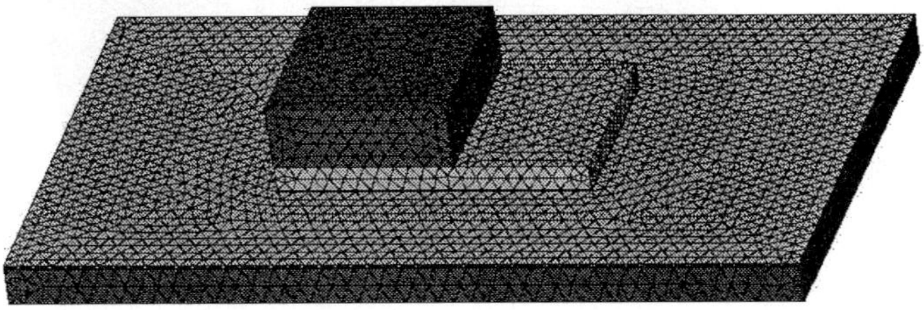

Fig. 4. Meshing. The figure shows the model for Cu on the whole surface

The Fig. 5 introduces the example of maximum temperature distribution for the structure (a – end view; b – section), where the Cu layer exceeds the surface of TO-220AB by 4 mm. The case when Cu covers the whole surface was checked to make no difference comparing to 4 mm. The figures 6 and 7 show calculated curves of the thermal resistance and the maximum temperature versus the copper width. Checking the temperature distribution, one sees that for the copper width less than 4 mm the possible temperature is exceeded, even for these ideal conditions of the heat removal. Only for 4 mm width the maximum temperature in the silicon chip gains the value of 124^0C and assures the safe operation of the electronic devices. The temperature does not lower when one enlarges the copper surface more than 4 mm. So the heat conditions are stable then.

To estimate the heat dissipation in the whole structure we prepared the heat vector diagram (Figs. 8 and 9). The Figure 8 introduces the heat flow for the case when copper covers 1 mm path beyond the transistor package surface, and the Figure 9 the one when the Cu path width amounts to 4 mm. The further widening of copper did not give any changes in the heat flow distribution.

4 Conclusions

The simulations have been led with the assumption that the temperature at the bottom of the whole package equals to an ambient one (25^0C), i.e. the cooling method is very efficient. Also the assumption of the convection coefficient at the value of 7 W/m^2K can be valid for efficient ventilating, only. That is why, the results should be considered as made in good cooling conditions. If any real conditions are worse, the maximum power range should not be obtained if the silicon is to be kept in a safe temperature. The maximum temperature value for silicon (125^0C) should be met then, without any exceptions. The simulations showed that the surface covered by Cu is very important here. If the circuit design does not allow to cover the surface with Cu in a sufficient way, the heat cannot be removed when the maximum power is applied. The width of 4 mm has resulted from our simulations.

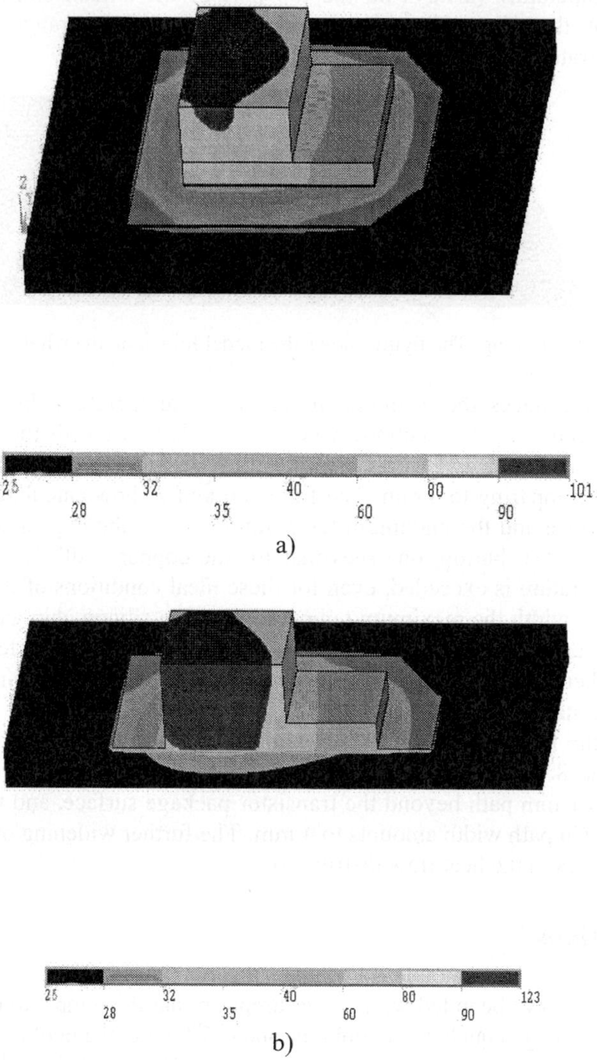

Fig. 5. The temperature [°C] distribution for Cu width 4.0 mm; a) end view, b) section

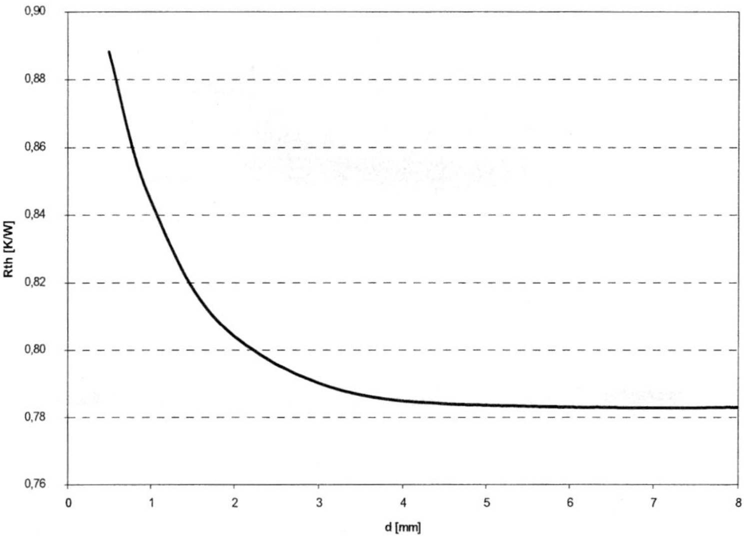

Fig. 6. Thermal resistance versus the growing width of copper

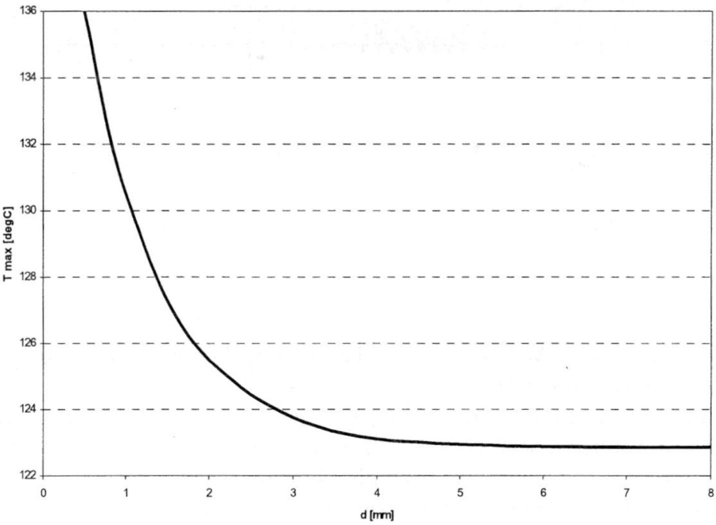

Fig. 7. Maximum temperature (upper surface of silicon chip) versus the growing width of copper

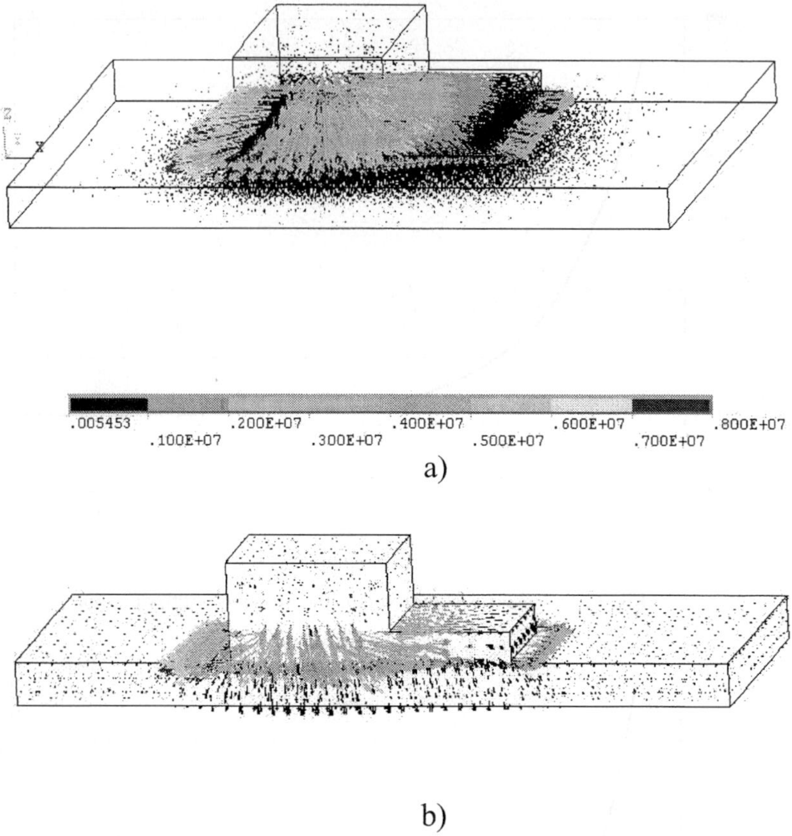

Fig. 8. Heat dissipation for Cu path width of 1 mm; a) end view, b) section

References

1. Kim S.J., Lee S.W.:Air Cooling Technology For Electronic Equipment; CRS Press, Boca Raton (1996)
2. Kakac S., Yucu H., Hijikata K., Cooling of Electronic System; NATO ASI Series: Applied Sciencies, vol. 258, Kluwer Academic Publishers, (1994)
3. Toshiba ed., Catalogue – pages with 2SK2866, (1998)
4. ANSYS, User's Manual for Revision 5.7; SASI, (2000)

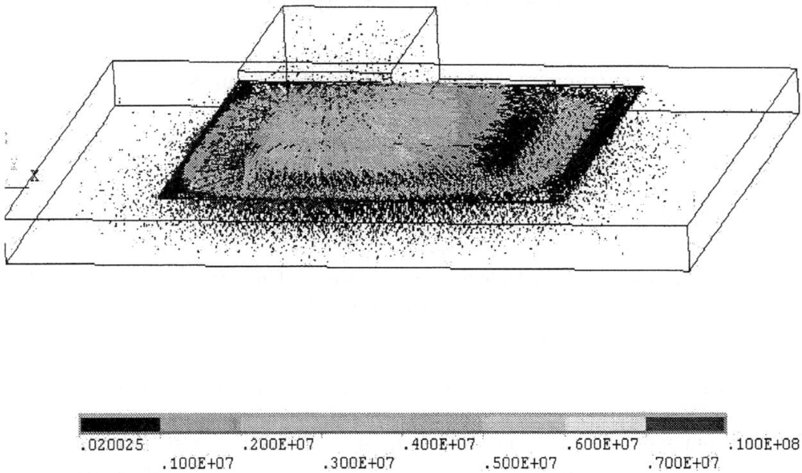

a)

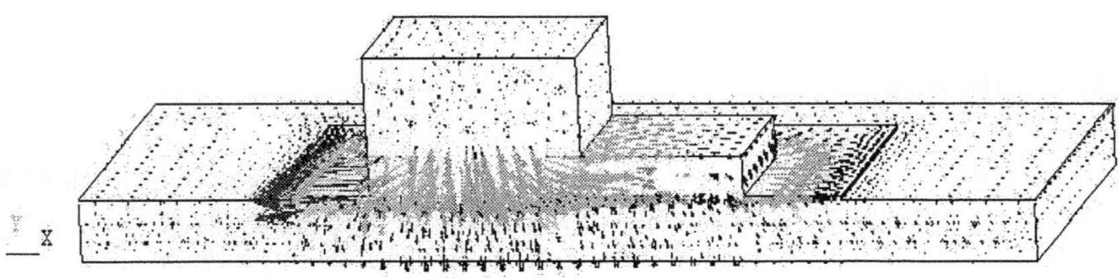

b)

Fig. 9. Heat dissipation for Cu path width of 4 mm; a) end view, b) section

Dynamic Simulations for Thermal Analysis of MOSFET IPM on IMS Substrate

Malgorzata Langer[1], Zbigniew Lisik[1], Ewa Raj[1], Nam Kyun Kim[2], and Jan Szmidt[3]

[1] Institute of Electronics, Technical University of Lodz, Poland
90-924 Lodz, Stefanowskiego 18/22, Poland
{malanger, lisikzby, ewaraj}@ck-sg.p.lodz.pl
[2] Korea Electrotechnology Research Institute, P.O. BOX 20, Chang Won, 641-600 Korea,
nkkim@keri.re.kr
[3] Institute of Microelectronics & Optoelectronics, Warsaw University of Technology, Poland
00-662 Warsaw, Koszykowa 75, Poland
j.szmidt@elka.pw.edu.pl

Abstract. The authors introduce some further research on the problem depicted in [1]. The transient model has been worked out to check how rapidly the temperature of the chip changes when the applied power changes abruptly This means searching for the answer: how the thermal resistance and the maximum temperature vary versus time. The simulations required using the 3-D thermal model of considered system that was used in numerical simulations and next, tested with commercial software, ANSYS 5.7 based on Finite Element Method (FEM) and additional calculations.

1 Introduction

The temperature rise above the permissible level results in much worse work of the device or even in its destruction. Enlarging frequency and the requested speed of switching make the conditions more difficult yet. The work introduces our results of dynamic simulations performed for the system: 2 MOSFETs on IPM, described in [1]. The 3-D thermal model of IPM used in the simulations is discussed there. In the model, it has been assumed that the silicon chip is the only heat source in which the heat is dissipated homogeneously. The magnitude of heat dissipation has been fixed on the base of Toshiba transistor 2SK2866a data sheet [2]. Taking into account its catalogue maximum ratings, so called the PD drain power dissipation has been chosen as equal to125W, which gives the heat generation of 18.1e9 W/m3 when taking into account the volume of the silicon chip.

2 The Thermal Model of IPM

The thermal model used for transient simulations is similar to the one described in [1]. It covers one half of the examined structure, introduced in Fig. 1 and its basic parameters are collected in Table 1. The structure above was the basis to work out the procedure in Ansys software that made possible the dynamic simulations of the heat flow inside the considered IPM module, for the step heat load function.

P.M.A. Sloot et al. (Eds.): ICCS 2003, LNCS 2658, pp. 644–649, 2003.

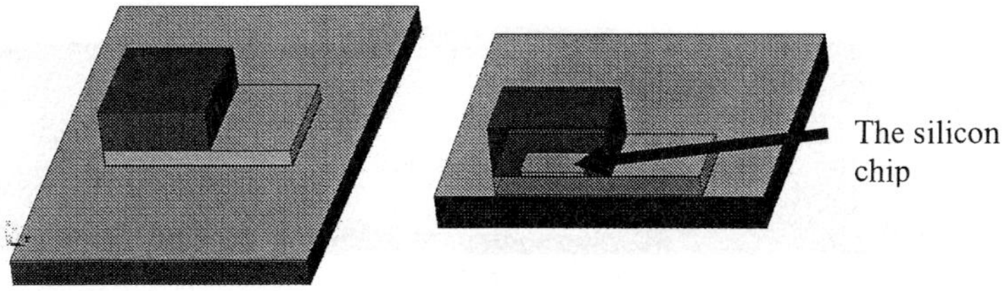

a) b)

Fig.1. 3-D view of the IPM thermal model: (a) top view; (b) the cross section

Table 1. The dimensions of the elements of thermal model

Material	length [mm]	width [mm]	thickness [mm]	Comments
Al base plate (11)	50	40	2.0	IMS plate
Insulate layer (12)	50	40	0.12	IMS plate
Cu foil (13)	var	var	0.10	IMS plate
Solder	10	10	0.10	
Lead frame	10	10	0.7	Transistor
Silicon chip	4.8	4.8	0.30	Transistor
Epoxy				Transistor

3 The Dynamic Simulations

The simulation has been performed for the step function of power dissipation. Its magnitude has been assumed as equaled to the value of 18.1 e9 W/m^3, what meets 125W heat source in the silicon chip. The Figures 2-5 compose an example of one typical series of the temperature development to show the changes during heating since the step start till the thermal equilibrium state is reached. We have kept the same scale of temperature for the diagrams to make possible easy comparing.

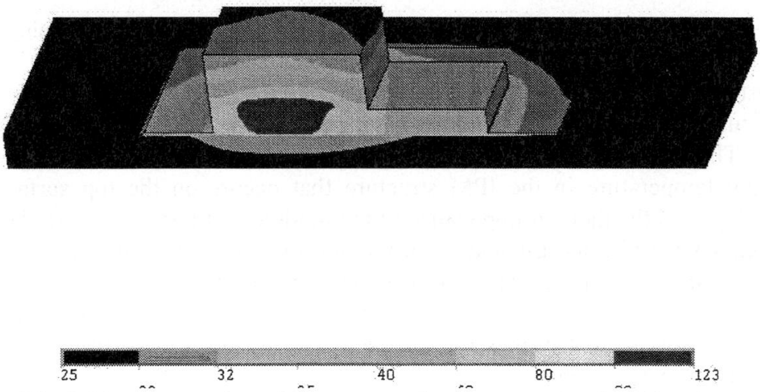

Fig. 2. The cross-section of the structure. Temperature after 1 s of heating

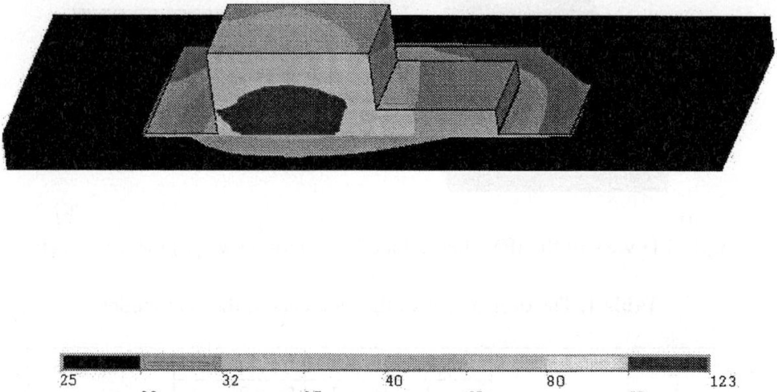

Fig. 3. The cross-section of the structure. Temperature after 4.6 s of heating

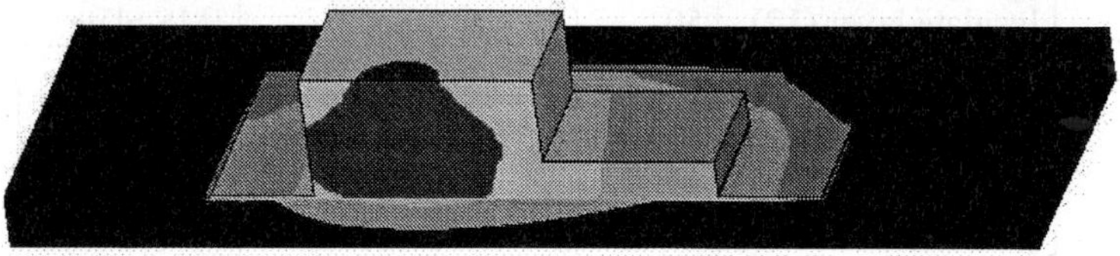

Fig. 4. The cross-section of the structure. Temperature after 10 s of heating

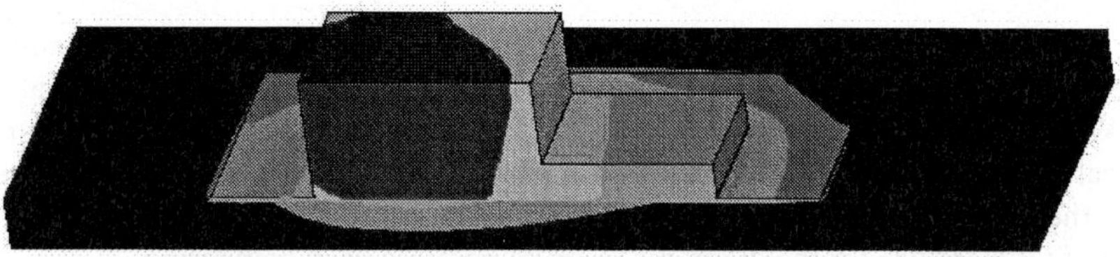

Fig. 5. The cross-section of the structure. Temperature after 100 s of heating

The last figure (after 100 s of heating) gives the steady-state conditions. The temperature does not change any more, having achieved its maximum of 123^0C.

The thermal maps from figs. 2-5 permit to estimate the changes of the maximum temperature in the IPM structure that occurs on the top surface of the silicon chip, and the thermal impedance of the module that corresponds to them. Such dependencies have been estimated for different dimensions of Cu foil. The fig. 6 gives the maximum temperature, which is captured on the hottest place, and the fig. 7 thermal impedance curves for the whole structure. One can see that the curves for Cu width 4.0mm coincide the ones for Cu on whole the surface, so the conditions are steady-state with this value.

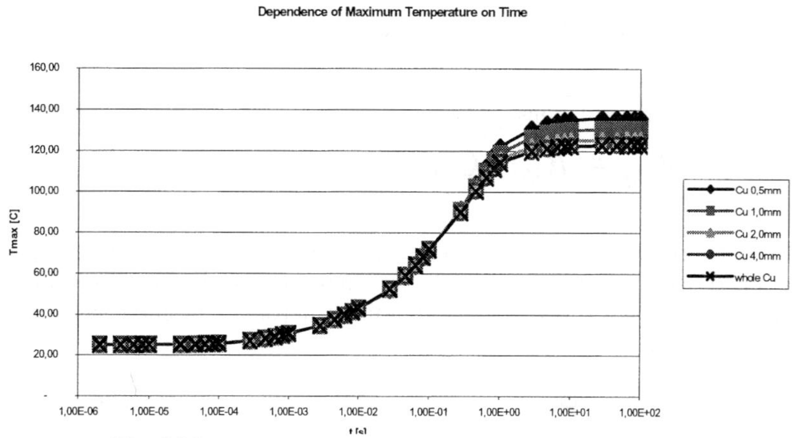

Fig. 6. Maximum temperature curves for different Cu path width.

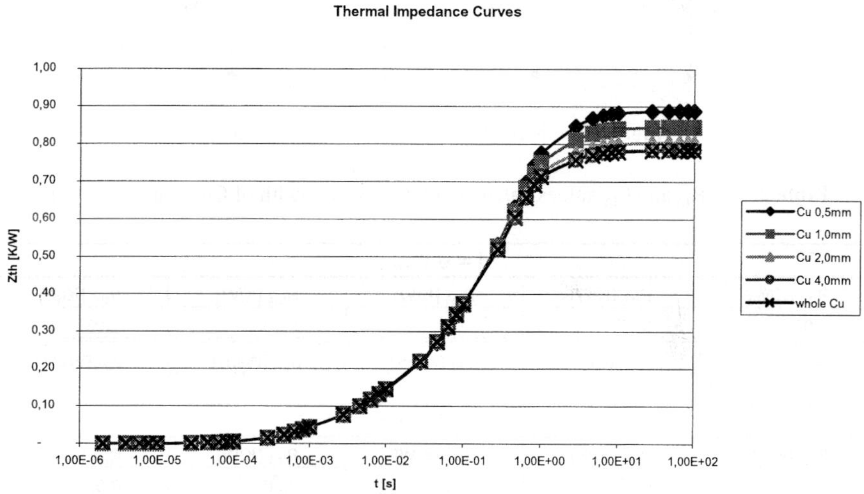

Fig. 7. Thermal impedance curves for different Cu path width

4 The Estimation of the RC Ladder

The simulations for the input step function were run to calculate the heat properties for the considered package. The final numbers have been evaluated for dynamic thermal models, covering thermal resistances (R_{th}) and capacitances (C_{th}). Fig. 8 shows the scheme of such a RC ladder, and Table 2 introduces the values, calculated for enlarging Cu width values. With them one can calculate the thermal conditions for any boundary conditions and schemes, where the tested package is applied.
The values are calculated versus time (t) according to the formula (1) [4]:

$$Z_{th}(t) = \sum_{i=1}^{n} R_{thi}\left(1 - \exp\left(-\frac{t}{\tau_{thi}}\right)\right) \quad (1)$$

where n=5 for our simulations, and

$$\tau_{thi} = R_{thi} \cdot C_{thi} \quad (2)$$

The RC network (ladder) model is five-stages (Fig. 8), what allows us to exam which part of the structure changes and which one can be even neglected. The results proved that the two end stages (IMS) give the significant impact to the curve shape, i.e. to the thermal capacitance.

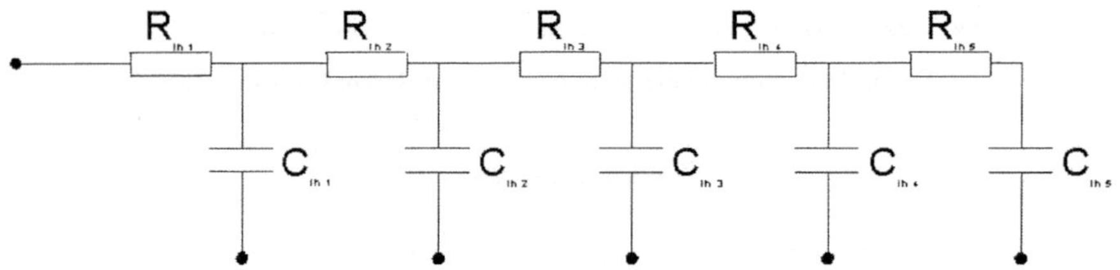

Fig. 8. The thermal circuit for considered heat flow

Table 2. The R_{th} and C_{th} values calculated for growing width of Cu (marked as 'd')

d = 0.5mm				
R_{th1} [K/W]	R_{th2} [K/W]	R_{th3} [K/W]	R_{th4} [K/W]	R_{th5} [K/W]
0.03	0.07	0.17	0.53	0.08
C_{th1} [Ws/K]	C_{th2} [Ws/K]	C_{th3} [Ws/K]	C_{th4} [Ws/K]	C_{th5} [Ws/K]
0.03	0.06	0.24	0.81	41.33
d = 1.0mm				
R_{th1} [K/W]	R_{th2} [K/W]	R_{th3} [K/W]	R_{th4} [K/W]	R_{th5} [K/W]
0.03	0.07	0.18	0.50	0.07
C_{th1} [Ws/K]	C_{th2} [Ws/K]	C_{th3} [Ws/K]	C_{th4} [Ws/K]	C_{th5} [Ws/K]
0.02	0.06	0.23	0.83	51.62
d = 2.0mm				
R_{th1} [K/W]	R_{th2} [K/W]	R_{th3} [K/W]	R_{th4} [K/W]	R_{th5} [K/W]
0.02	0.07	0.19	0.47	0.06
C_{th1} [Ws/K]	C_{th2} [Ws/K]	C_{th3} [Ws/K]	C_{th4} [Ws/K]	C_{th5} [Ws/K]
0.02	0.05	0.22	0.84	63.39
d = 4.0mm				
R_{th1} [K/W]	R_{th2} [K/W]	R_{th3} [K/W]	R_{th4} [K/W]	R_{th5} [K/W]
0.02	0.08	0.19	0.45	0.05
C_{th1} [Ws/K]	C_{th2} [Ws/K]	C_{th3} [Ws/K]	C_{th4} [Ws/K]	C_{th5} [Ws/K]
0.02	0.05	0.22	0.85	72.00

Whole Cu				
R_{th1} [K/W]	R_{th2} [K/W]	R_{th3} [K/W]	R_{th4} [K/W]	R_{th5} [K/W]
0.02	0.07	0.19	0.45	0.05
C_{th1} [Ws/K]	C_{th2} [Ws/K]	C_{th3} [Ws/K]	C_{th4} [Ws/K]	C_{th5} [Ws/K]
0.02	0.05	0.22	0.85	74.08

5 Conclusions

The 3D simulations presented here allow for the sight into the thermal processes taking part in the considered IPM module. They make possible estimating the maximum temperature inside the examined structure that can be used for the thermal impedance evaluation. The obtained curve has been applied to determine the lumped thermal model of the module in the form of the RC ladder.

References

1. Langer M., Lisik Z., Raj E., Kim N.K., Simulations For Thermal Analyze of MOSFET IPM Using IMS Substrate, ICCS2003
2. Toshiba ed., Catalogue – pages with 2SK2866, (1998)
3. ANSYS, User's Manual for Revision 5.7; SASI, (2000)
4. Lisik Z., [in Polish], Modeling of Thermal, Electrical & Electrothermal Properties of Semiconductor Power Devices, ZN Techn. Univ. of Lodz, Poland, (1992)

Correlation between Mutation Pressure, Selection Pressure, and Occurrence of Amino Acids

Aleksandra Nowicka[1], Paweł Mackiewicz[1], Małgorzata Dudkiewicz[1], Dorota Mackiewicz[1], Maria Kowalczuk[1], Stanisław Cebrat[1], and Mirosław R. Dudek[2]*

[1] Department of Genetics, Institute of Microbiology, University of Wroclaw, ul. Przybyszewskiego 63/77, PL-54148 Wroclaw, Poland
{nowicka, pamac, malgosia, dorota, kowal, cebrat}@microb.uni.wroc.pl
http://smORFland.microb.uni.wroc.pl
[2] Institute of Physics, University of Zielona Góra, ul. A. Szafrana 4a, PL-65516 Zielona Góra, Poland
mdudek@proton.if.uz.zgora.pl

Abstract. With the help of the empirical mutation table for nucleotides in the *Borrelia burgdorferi* genome we have performed Monte Carlo simulation of the pure mutation pressure experienced by the genes of the genome. We have examined the divergence of the mutated genes from the ancestral ones and we have constructed MPM1 matrix (Mutation Probability Matrix) of the substitution rates between amino acids of the diverging genes. The results have been compared to mutation data matrix PAM1 PET91 representing mutation and selection data of 16130 homologous genes od different organisms. We have found that the effective survival time of amino acids in organisms follows a power law with respect to frequency of their occurrence in genes. This makes possible to find the effect of the pure mutational pressure and the selection on the amino acid composition of genes. The results are universal in the sense that the survival time of amino acids calculated from the higher order PAMk matrices ($k > 1$) follows the same power law as in the case of PAM1 matrices.

1 Introduction

Determining the evolutionary distances between two protein sequences requires the knowledge of the substitution rates of amino-acids. It is generally accepted that the more substitutions are necessary to change one sequence into another, the more unrelated they are and the larger their distance to the common ancestor. The most widely used method for the calculation of distances between sequences is based on the mutation data matrix, M_{ij}, published by Dayhoff et al. [1], where i, j represent amino acids, and an element M_{ij} of the matrix gives

* corresponding author

P.M.A. Sloot et al. (Eds.): ICCS 2003, LNCS 2658, pp. 650–657, 2003.

the probability that the amino acid in column j will be replaced by the amino acid in row i after a given evolutionary time interval. The interval corresponding to 1 percent of substitutions between two compared sequences is called one PAM (Percent of Accepted Mutations), and the corresponding matrix is denoted as PAM1 matrix. There is assumed a Markov model of sequence evolution and a simple power M^k of the PAM1 matrix (multiplied by itself k times) denotes a matrix, PAMk, that gives the amino acid substitution probability after k PAMs. Today, a much more accurate PAM matrix is available, generated from 16130 protein sequences, published by Jones et al. [2]. The large number of compared genes guarantees that the matrix has negligible statistical errors and it can be considered to be the reference matrix during the calculations of the phylogenetic distances. The matrix is also known as PET91 matrix.

Recently, by comparing intergenic sequences being remnants of coding sequences with homologous sequences of genes, we have constructed an empirical table of the nucleotide substitution rates in the case of the leading DNA strand of the *B. burgdorferi* genome [3],[4],[5]. We have found that substitution rates, which determine the evolutionary turnover time of a given kind of nucleotide in third codon positions of coding sequences, are highly correlated with the frequency of the occurrence of that nucleotide in the sequences. There is a compositional bias produced by replication process, introducing long-range correlation among nucleotides in the third positions in codons, which is very similar to the bias seen in the intergenic sequences [6].

We have used the empirical table of nucleotide substitution rates to simulate mutational pressure on the genes lying on the leading DNA strand of the *B.burgdorferi* genome and we have constructed MPM1 matrix (Mutation Probability Matrix) for amino acid substitutions in the evolving genes. Thus the resulting table represents the percent of amino acid substitutions introduced by mutational pressure and not by selection. Next, we compared the survival times of the amino acids in the case without any selection with the effective survival times of the amino acids, counted with the help of the PAM1 PET91 matrix.

2 Mutation Table for Nucleotides

DNA sequence of the *B.burgdorferi* genome was downloaded from the website *www.ncbi.nlm.nih.gov.* The empirical mutation table for nucleotides in third positions in codons, which we used in the paper, is the following [3],[4]:

$$M = \begin{pmatrix} 1 - uW_A & u\,W_{AT} & u\,W_{AG} & u\,W_{AC} \\ u\,W_{TA} & 1 - uW_T & u\,W_{TG} & u\,W_{TC} \\ u\,W_{GA} & u\,W_{GT} & 1 - uW_G & u\,W_{GC} \\ u\,W_{CA} & u\,W_{CT} & u\,W_{CG} & 1 - uW_C \end{pmatrix} \tag{1}$$

where [1]

$$\begin{array}{llll} W_{GA} = 0.0667 & W_{GT} = 0.0347 & W_{GC} = 0.0470 & W_{AG} = 0.1637 \\ W_{AT} = 0.0655 & W_{AC} = 0.0702 & W_{TG} = 0.1157 & W_{TA} = 0.1027 \\ W_{TC} = 0.2613 & W_{CG} = 0.0147 & W_{CA} = 0.0228 & W_{CT} = 0.0350 \end{array} \tag{2}$$

[1] The transpose matrix convention has been chosen in [3].

and the elements of the matrix give the probability that nucleotide in column j will mutate to the nucleotide in row i during one replication cycle. The symbols W_{ij} represent relative substitution probability of nucleotide j by nucleotide i, and u represents mutation rate. The symbols W_j in the diagonal represent relative substitution probability of nucleotide j:

$$W_j = \sum_{i \neq j} W_{ij}, \tag{3}$$

and

$$W_A + W_T + W_G + W_C = 1. \tag{4}$$

The expression for the mean survival time of the nucleotide j depends on W_j as follows (derivation can be found in [3])

$$\tau_j = -\frac{1}{ln(1 - u\ W_j)} \approx \frac{1}{u\ W_j}. \tag{5}$$

The above approximated formula is true for small values of the mutation rate u.

In papers [3],[4],[5], we concluded that in a natural genome the frequency of occurrence f_j of the nucleotides, in the third position in codons, is linearly related to the respective mean survival time τ_j,

$$f_j = m_0\ \tau_j + c_0, \tag{6}$$

with the same coefficients, m_0 and c_0, for each nucleotide. The Kimura's neutral theory [7] of evolution assumes the constancy of the evolution rate, where the mutations are random events, much the same as the random decay events of the radioactive decay. However, the linear law in (6) is not contrary to the Kimura's theory. Still, the mutations represent random decay events but they are correlated with the DNA composition.

3 Mutational Pressure MPM1 Matrix Construction for Amino Acids

In order to compare an effect of the pure mutational pressure and the selection pressure on amino acid composition of genes we used the results of Monte Carlo simulation of the mutational pressure applied on 564 genes from the leading DNA strand of the *B. burdorferi* genome. This enabled us to calculate amino acid substitution rates which, next we could compare with the ones originating from the PAM1 PET91 substitution rates matrix [2]. The way, the experimental PET91 matrix has been constructed, determined our simulation algorithm, which consisted of the following steps:

(i) for each gene, considered to be an ancestral one, make two copies of the gene at $t = 0$,
(ii) increase time step t ($t = t + 1$) and with frequency u mutate nucleotides of the two gene copies with the probability distribution defined by the elements of the mutation matrix in (1),

(iii) goto (ii) unless the number of amino acid substitutions between the homologous protein sequences reaches 1%,

The applied value of the mutation pressure was $u = 0.01$. The steps (i)–(iii) have been repeated 10^5 times in order to calculate the averaged values of the substitution rates between the homologous protein sequences. These values we used to construct a mutation probability matrix MPM1 according to the procedure of Dayhoff et al.[1] and Jones et al. [2]. The resulting mutation table, with substitution probabilities M_{ij}, the amino acid mutability m_j, and the fraction f_j of amino-acid in the compared sequences have been presented, respectively, in Table 1 and Table 2.

The elements M_{ij} of the MPM1 matrix in Table 1 have been scaled with the parameter λ, which related them to the evolutionary distance of one percent of substitutions and it is equal to 0.00009731 in our simulations.

4 Discussion of Results

The major qualitative difference between the MPM matrix introduced in the previous section and the PAM1 PET91 matrix published in the paper by Jones et al.[2] is that the first one is a result of pure mutational pressure whereas the second one is a result of both mutational and selection pressures. Thus, we have two evolutionary mechanisms responsible for the resulting PAM matrices.

With the help of formula (5) (extended to amino acids) we have calculated effective survival times of amino acids in the case of the MPM1 matrix (Table 1) and the mutational/selectional PAM1 PET91 matrix ([2]). The value of the parameter λ is a counterpart of u in (5). In Fig.1, we presented the relation between the calculated survival time o amino acids and their fractions in the *B. burgdorferi* proteins, in the pairs of diverged genes, in a log-log scale. One can observe that the data are highly correlated and in both cases the dependence of the mean survival time of amino acid on the fraction of the amino acid represents a power law:

$$\tau_j \sim F_j^{\alpha} \tag{7}$$

with a negative value of $\alpha \approx -1.3$ in the case of selection and a positive value of $\alpha \approx 0.2$ in the case of mutation pressure on the leading DNA strand of the *B. burgdorferi* genome. The value of α for the analogous mutational PAM1 matrix calculated in the case of the lagging DNA strand of the *B. burgdorferi* genome is about twice as small. It is worth to underline that the slopes α are the same for the matrices PAMk with high values of k, and thus, they are universal with respect to evolution.

In Fig.1 we may observe a kind of evolutional scissors acting on amino acids. Once the less frequent amino acids, like W (tryptophane), C (cysteine), have much shorter turn over time compared with other amino acids (as can be seen from the lower line) the selection pressure (upper line) counteracts with the effect. On the other hand, the most mutable amino acids, like L (leucine) or I (isoleucine) , which are very frequent in genes, seem to be much weakly influenced by selection.

Table 1. Simulated Mutation Probability Matrix for an evolutionary distance of 1 PAM (splitted into two parts). Values of the matrix elements are scaled by a factor of 10^5 and rounded to an integer. The symbols in the first row and the first column represent amino-acids and numbers following colons -number of codons representing a given amino-acid in the universal genetic code.

	A:4	R:6	N:2	D:2	C:2	Q:2	E:2	G:4	H:2	I:3
A:4	99027	0	0	60	1	0	44	56	0	0
R:6	0	98784	1	0	168	91	1	151	123	39
N:2	0	1	98925	255	3	1	1	1	217	126
D:2	77	0	220	98935	2	0	232	157	129	0
C:2	0	33	0	0	97443	0	0	64	1	0
Q:2	0	51	0	0	0	99243	32	0	350	0
E:2	63	1	1	258	0	99	99132	173	1	0
G:4	69	227	0	151	483	0	149	99089	0	0
H:2	0	38	37	25	1	194	0	0	98313	0
I:3	1	103	180	1	2	0	0	1	1	99025
L:6	1	61	0	0	3	130	0	0	322	116
K:2	0	325	295	1	0	184	276	1	1	95
M:1	0	47	0	0	0	0	0	0	0	106
F:2	1	0	0	0	467	0	0	0	2	128
P:4	42	35	0	0	1	54	0	0	126	0
S:6	166	219	140	1	717	0	0	118	1	61
T:4	225	31	46	0	1	0	0	0	0	127
W:1	0	43	0	0	127	0	0	24	0	0
Y:2	0	0	153	150	580	1	0	0	414	0
V:4	329	1	1	162	1	0	132	166	0	175

	L:6	K:2	M:1	F:2	P:4	S:6	T:4	W:1	Y:2	V:4
A:4	0	0	0	0	79	94	304	0	0	230
R:6	21	129	93	0	54	101	35	334	0	0
N:2	0	213	1	0	0	118	93	0	241	1
D:2	0	1	0	0	0	0	0	0	203	145
C:2	0	0	0	50	0	66	0	199	100	0
Q:2	25	41	0	0	47	0	0	0	0	0
E:2	0	191	1	0	0	0	0	1	1	131
G:4	0	0	0	0	0	82	0	282	0	143
H:2	35	0	0	0	60	0	0	0	110	0
I:3	106	98	557	178	1	74	366	0	1	261
L:6	99267	0	228	412	342	94	1	304	1	124
K:2	0	99245	231	0	0	1	97	1	1	1
M:1	40	46	98686	0	0	0	38	1	0	38
F:2	268	0	1	98928	1	153	1	1	221	123
P:4	77	0	0	0	98948	107	77	0	0	0
S:6	71	0	1	177	359	98951	261	47	85	1
T:4	0	35	69	0	108	109	98724	0	0	1
W:1	13	0	0	0	0	3	0	98827	0	0
Y:2	0	0	0	137	0	46	0	1	99033	0
V:4	76	0	133	115	0	1	2	1	1	98802

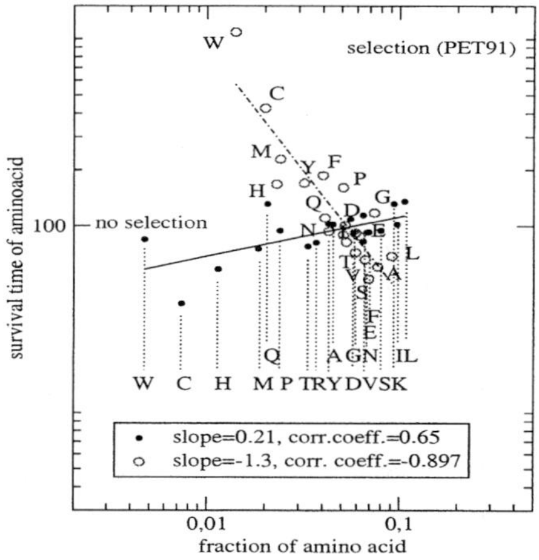

Fig. 1. Relation between survival time of amino-acids and their fractions in compared pairs of diverged homologous genes in the case with selection (PET91) and in the case without selection (simulated mutational pressure of the *B. burgdorferi* genome).

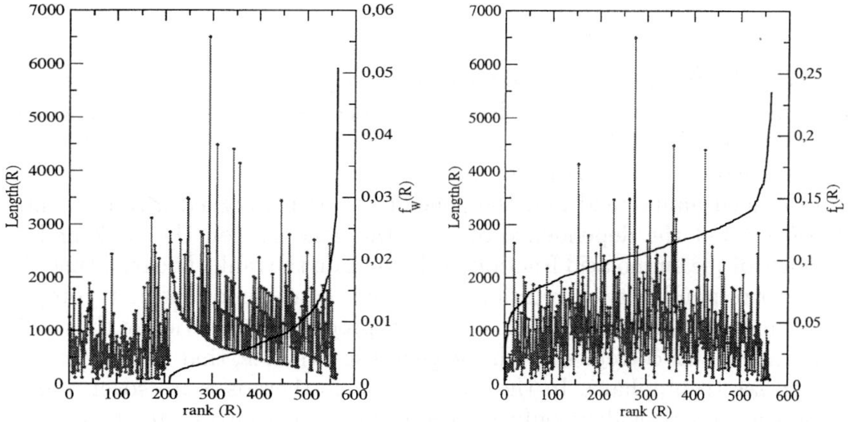

Fig. 2. Gene length versus gene rank, where each gene has assigned a rank with respect to fraction of tryptophane (left graph) and with respect to leucine (right graph). In each graph there are two plots: the dots represent gene's size vs. rank, whereas the second plot represent the fraction vs. rank.

Table 2. Relative mutabilities and fractions of 20 amino acids in the compared simulated sequences. We used the convention that mutabilities are relative to alanine and it is arbitrarily assigned a value of 100.

amino acid	relative mutability (m_j)	fraction (f_j)
A	100.00	0.0449
R	126.09	0.0369
N	110.42	0.0671
D	109.29	0.0579
C	262.91	0.0073
Q	77.67	0.0206
E	89.88	0.0644
G	92.21	0.0582
H	173.18	0.0114
I	100.12	0.0964
L	75.23	0.1060
K	77.51	0.0930
M	135.01	0.0184
F	110.12	0.0690
P	108.02	0.0238
S	107.78	0.0796
T	131.05	0.0333
W	127.44	0.0047
Y	99.25	0.0427
V	123.56	0.0645

In genes the fraction of the amino acids most protected by selection strongly depends on gene size, it is diminishing when gene's size is increasing. The effect weakens if we go into right the direction of the evolutionary scissors in Fig.1. To show this, we ordered all 564 genes under consideration with respect to fraction of an examined amino acid and the genes have been assigned a rank number. Next, we plotted the dependence of both the gene size on the rank and the dependence of the amino acid fraction in the gene on the rank. The resulting plots in Figs.2 correspond to two evolutionary extreme cases, representing tryptophan and leucine. It is evident that in the case of tryptophan the fraction of that amino acid in genes is anti-correlated with the gene's size (notice, that about 1/3 genes do not posses tryptophan). In the case of leucine there is a crossover and the effect of selection is evident only for the genes which have more than 10% of that amino acid. When the fraction of leucine is less than 10% there is even a reverse effect, i.e., the increasing fraction is correlated with the increasing gene's size. If we look at the evolutionary scissors in Fig.1, we can see that in the case of leucine the survival time, which originates from pure mutational pressure is longer than its selectional counterpart. Recently, there has appeared a paper by Xia and Li [8] discussing which amino acid properties (like polarity, isoelectric point, volume etc.) affect protein evolution. Thus, there is a possibility to relate these properties to our discussion of the evolutionary scissors.

5 Conclusions

With the help of computer simulations of the mutational pressure experienced by genes in the *B.burgdorferi* genome we have shown that the amino acids which experience the highest selectional pressure have the shortest turn over time with respect to mutation pressure. The fraction of these amino acids in genes depends on the gene's size. Much different is the selectional role of the amino acids, like leucine, from the right hand side of the selectional scissors. Although they have long turn over time with respect to the mutational pressure, their fraction cannot be too high. This could be considered as an effect of optimisation of the genetic information on coding with processes of mutagenesis and phenotype selection for protein functions.

Acknowledgemets. The work was supported by the grant number 1016/S/IMi/02.

References

1. Dayhoff, M.O., Schwartz, R.M., and Orcutt, B.C.: A Model of Evolutionary Change in Proteins. In: Atlas of Protein Sequence and Structure , Vol. 5 Suppl. 3 (1978) 345–352
2. Jones, D.T., Taylor, W.R., and Thornton, J.M.: The rapid generation of mutation data matrices from protein sequences. In: CABIOS, Vol. 8 no. 3, (1992) 275–282
3. Kowalczuk, M., Mackiewicz, P., Szczepanik, D., Nowicka, A., Dudkiewicz, M., Dudek, M.R., and Cebrat, S.: Multiple base substitution corrections in DNA sequence evolution. Int. J. Mod. Phys. C **12** (2001) 1043–1053
4. Mackiewicz, P., Kowalczuk, M., Mackiewicz, D., Nowicka, A., Dudkiewicz, M., Łaszkiewicz, A., Dudek, M.R., Cebrat, S.: Replication associated mutational pressure generating long-range correlation in DNA. Physica A **314** (2002) 646–654
5. Kowalczuk, M., Mackiewicz, P., Mackiewicz, D., Nowicka, A., Dudkiewicz, M., Dudek, M.R., and Cebrat, S: High correlation between the turnover of nucleotides under mutational pressure and the DNA composition. BMC Evolutionary Biology **1** (2001) (1):13
6. Cebrat, S., Dudek, M.R., Gierlik, A., Kowalczuk, M., Mackiewicz, P.: Effect of replication on the third base of codons. Physica A **265** (1999) 78–84
7. Kimura, M.: The Neutral Theory of Molecular Evolution, Cambridge University Press, Cambridge (1983)
8. Xia, X., Li, W-H.: What Amino Acid Properties Affect Protein Evolution. J. Mol. Evol. **47** (1998) 557–564

Introducing CEES: Complex Economic Environments Simulator*

Ismael Rodríguez and Manuel Núñez

Dept. Sistemas Informáticos y Programación.
Universidad Complutense de Madrid, E-28040 Madrid. Spain.
e-mail:{isrodrig,mn}@sip.ucm.es

Abstract. The combination of powerful computers and complex simulators allows us to accurately simulate the behavior of real system. This is specially important in the case of systems that cannot be manipulated, as the economic behavior of a society. CEES is a program that allows a user to simulate the behavior, both in the short and long-term, of an economy. Besides, users of the system can be part of the simulation by taking charge either of a company or of the State. Let us remark that, in contrast with most economic models, CEES takes into account a huge amount of variables (more than one thousand) in order to compute the current *state* of the economy.

Keywords: Complex Systems; Simulators; Macroeconomics;

1 Introduction

Traditionally, the empirical study of new political, demographic, or social systems has been confronted with the difficulty of managing, with enough precision, environments containing a huge amount of relevant variables. Thus, the definition of new models has been sometimes lessened because it was difficult to verify their validity. The main problem consisted, on the one hand, in the impossibility of manipulating the real environment and, on the other hand, in the disability to generate realistic artificial environments. Hence, these fields have been favored due to the possibilities offered by new powerful computers. Thus, the validity of a new theory can be contrasted by simulating the environment and by comparing the obtained and expected results. Afterwards, by using different test values, one can estimate the behavior of the real system in conditions that could never be applied to the real environment. Examples of such systems can be found in [11, 4]. Actually, together with verbal argumentation and mathematics, computer simulation is becoming the third *symbol system* available to social scientists [8]. A good candidate to be *simulated* is the economic behavior of a society. In fact, there are already good simulators and algorithms taking into account partial views of the economy (e.g. [3] at the microeconomic level).

In this paper we present CEES (acronym for Complex Economic Environments Simulator). The economic models underlying CEES are much more complex than most macroeconomics models in use. So, our system is able to explain

* Work partially supported by the CICYT project TIC2000-0701-C02-01.

P.M.A. Sloot et al. (Eds.): ICCS 2003, LNCS 2658, pp. 658–667, 2003.

how the economy of a society works in a more adequate way than models found in the literature usually focussing on one particular characteristic. For instance, we consider several productive sectors simultaneously. A new feature that we have considered specially important is that we split the primary sector into two sectors: *Alimentation* and *raw materials*. In this case, the latter must be understood in a broad system as it contains those industries needed to assure the productive cycle of companies. That is, in addition to proper raw materials we consider sectors as energy, industrial machinery, industrial construction, etc. Let us remark that some of these industries are usually included in the secondary sector. As a measure of its complexity, CEES controls more than one thousand *heterogenous* different variables in order to compute the current state of the economy, while most theoretical economic models control, in the best cases, a couple of tens.

We would like to emphasize two main characteristics of our system. First, our design methodology can be described as a *symbiosis* between the implementation of the simulator and the learning of advanced macroeconomic concepts. Initially, we identified the set of basic operators (at the human level) producing the economic process (e.g. purchasing, selling, work, demand of goods, etc). Next, we identified the main results giving raise to the correct interrelation among these concepts (e.g. business cycles and their control, state regulations by means of taxes and unemployment benefits, monopolies and oligopolies, etc). Afterwards, we looked for economic explanations to the basic forces producing the high level behavior of the system (e.g. if there is an excess of supply then the entrepreneur will decrease the prices). We generated preliminary models and studied their validity, so that we could incrementally improve the models until they reached the desired behavior.

It is worth to point out how we were dealing with *unexpected* behaviors. By default, we did not consider that they were due to conceptual errors. Actually, it is a fact that in any society almost *anything* can happen if the specific conditions appear simultaneously. Hence, we tried to take advantage of these behaviors instead of discarding them. For example, during the development of the simulator we found out the existence of a very rare situation where a depression was accompanied by an inflationary spiral (called *stagflation*). Then, we gathered documentation in the specific literature about this strange behavior in real economies. We found that one of its possible causes is that, due to some factors, companies are forced to increase salaries though their profit is negative. Afterwards, we compared this situation with standard behaviors, where salaries use to decrease in depressions. We identified the main factors provoking this rare behavior in our model (e.g., excessive influence of the *still* low unemployment in the beginning of the depression), and we corrected them. Nevertheless, instead of managing the rules of our model so that the rare situation was not possible, we *generalized* the model so that *both* situations were possible. In the new model, the behavior depended on whether the exceptional factors provoking the situation were present or not. This management of *strange* situations favored the increase *by-demand* of the set of relevant factors, as well as the necessary learning by the authors of advanced concepts in the area. Thus, we may properly claim that the simulator was teaching to the authors and that the design process was guided by the simulator itself.

The second main characteristic of CEES is that the system is a complete and complex simulator that adequately reflects the majority of relevant macroeconomic aspects. In CEES, the high level behavior (at the macroeconomic level) appears as a consequence of the interrelation between the different economic agents (providers of goods and providers of work) when they try to increase their own profit (microeconomic level). Actually, traditional economic theory was always separating macroeconomics from microeconomics. However, it is nowadays recognized that macroeconomic phenomena cannot be correctly explained without taking into account the microeconomic side (see e.g. [10]). In a wider sense, we assume the Axelrod simplicity principle (*KISS, Keep It Simple, Stupid*), which states that very simple agents can generate complex emerging structures [1].

It is worth to point out that CEES is more than a simulator where a user passively observes the behavior of the economy. On the contrary, it is an *interactive* system that allows a user to play the role of the most important economic institutions (i.e. the State or a private company), so that the effect of different decisions on the global economy can be checked. Thus, adequate interfaces are provided where the relevant parameters can be modified. For instance, if the user plays the role of an entrepreneur then he can regulate the demanded amount of both work and raw materials, the price of his products, the salaries of the workers, etc. Besides, as State, the player regulates parameters as the progressiveness of income taxes, the indirect taxes rate, and unemployment benefits. Actually, CEES allows various users to play the role of different institutions in the same simulation (e.g., one user is the State, and several users represent private companies), so that the decisions of each player affect the others. Finally, CEES provides several *views* of the economic behavior, by means of graphical representations, according to the different parameters to be considered (e.g. prices, employment levels, stocks, etc).

The rest of the paper is structured as follows. In Section 2, we describe the main features of the algorithm underlying the behavior of CEES. We will concentrate on how the next state of an economy is computed from the current state. In Section 3 we give some implementation details as well as examples of the behavior of our system. Finally, in Section 4 we present our conclusions and some lines for future work.

2 The Algorithm Controlling the Behavior of CEES

In this section we describe the algorithm governing the behavior of our simulator. Given the current state of the economy, the simulator gives the state of the economy in the next instant. So, time is split into *turns* (e.g. weeks, months, years). An iterated execution of the algorithm gives us the state of the economy in any arbitrary future time. As we said before, we slightly depart from the usual conception that economy consists of three sectors. We will suppose four productive sectors: *Alimentation*, *industry of goods*, *services*, and *raw materials*. We denote the first three productive sectors by *sectors of goods*. The latter sector must be understood, in a broad sense, as those materials that are needed by companies to perform the productive process. For instance, it includes industries as energy, mining, heavy industry, construction, etc. As we said in the introduction, CEES allows some of the *players* in the economy to be real persons. By *Society* we denote all of the companies which do not belong to any of the human participants. Human participants will be simply called *Participants*.

The algorithm includes 9 phases split into 20 steps. Next we briefly describe the purpose of each of these phases as well as the different steps included in each of them. We will write phase names in boldface characters while the different steps will be given in italics. Due to space limitations we will not give the mathematical formulation of our algorithm. However we present in Phase 1, as an example, the definition of demand functions for specific products.

Phase 1. Computing the aggregate demand of goods. First, we have to consider the willingness of each individual to save money. Let us note that if the savings of an individual cannot afford his minimal necessities then he will save nothing. So, the rate of money each individual saves depends on the capacity of his wealth (salary and savings) to afford his basic necessities. In the case of the demand of goods by entrepreneurs (as they are also individuals), we must take into account that some money will be taken away from the savings to afford the company costs in next steps. So, it will be necessary to estimate the maintenance costs of their companies. As the quality of their products strongly influences the production cost (more quality requires better supply and more man-power), the selection of future quality will be done in this step.

Next, we will get the way all of the purchasers spend the money they had decided not to save. Let us remark that if the purchasing power of an individual is low then his expenditure rate in alimentation is higher (as well as he will spend less in services). On the contrary, an individual with a high purchasing power will have opposite preferences. In all cases, preferences will be strongly influenced by both the price and the quality of the products. However, quality increases production cost. Thus, it influences the selling price. Essentially, the influence of price on the demand by wealthy people will be less than the corresponding to poor people. As a result, low quality will maximize demand of poor people, while it will be the opposite for well-off people.

As an example of the mathematical basis underlying CEES we present the rate of expenditure by an individual in each of the available products. From the total amount that an individual decides to spend, $ProductDemand_i$ indicates the percentage to be spent in the product i. Each product i is associated with a *consumption* sector (alimentation, industry, and services, respectively) by means of a function $s : \mathbb{N} \longrightarrow \{1,2,3\}$. In addition, $\pi : \{1,2,3\} \longrightarrow \mathbb{R}$ computes the *priority* that the individual assigns to consumption in each of the sectors. Finally, b and c are constant and different for each individual. They indicate the minimal quality threshold for the product and the additional influence of price in demand, respectively.

$$aux_i = \frac{PurchasingPower^{\pi(s(i))} \cdot \left(b - \frac{1}{Quality_i}\right)}{Price_i^{\frac{1}{PurchasingPower}+c}} \qquad ProductDemand_i = \frac{aux_i}{\sum_j aux_j}$$

The terms aux_i compute the *rough* percentage of money to be spent in each product. These terms are afterwards *normalized* so that $ProductDemand_i$ gives the real percentage. Let us remark that the demand increases with quality and diminishes with price. Besides, the influence of price in demand is bigger if the purchasing power is low.

Phase 2. Computing the amount of goods that are sold. In this phase we distinguish four steps. First, we need to compute the *difference between demand and supply of goods.* As demand could be greater than supply, it could be impossible for purchasers to buy all the goods they wanted, so they could save some additional money. In this case, some of this saved wealth could be spent in other goods that were not initially required. This allows some expensive and bad products to be sold after cheap and good products are exhausted. Next, *products will be sold under stock restrictions.* In this case, savings of entrepreneurs will be increased and savings of purchasers (entrepreneurs included) will be decreased according to the purchased products. The stock will be recalculated according to the goods actually sold. In addition, indirect taxes corresponding to these sales will be collected by the State.

Once sales are computed, entrepreneurs will decide whether they *modify the prices of their products* for the next period. In order to do so, they will take into account the demand/supply ratio in the current period. Let us note that new prices will be strongly influenced by the degree of monopolism. Besides, they will also depend on the type of good and on its quality (see e.g. [2] for an explanation on sticky prices). However, only prices of the Society will be automatically computed because Participants make these decisions voluntarily by using the interface provided by the system. The last step in this phase is used to compute the *expectation of future profit rates for each sector of goods using the previously computed prices.* For each company, this rate will be given by the ratio between income and total costs (including taxes).

Phase 3. Computing the amount of wealth devoted to new investments. In this phase we distinguish three steps. First, we have to compute the set of *individuals who desire (an can) invest in each sector.* People in general (i.e. Society entrepreneurs, employees of both the Society and the Participants, and unemployed people) may decide either to invest in a new company or to enlarge an existing one. Participants are not included in this step as they perform their investments from the application interface. The amount of new investment in each company will depend on the ratio between demand an supply in the previous period, on the expectations of profit, and on the current wealth of investors. The higher are those values, the higher amount of money will be invested. So, a sector with excessive production but good profit may receive additional new investment. Finally, in order to consider an investment feasible, it will be required that the invested money surpasses a minimal threshold. Each sector has a specific threshold of minimal supply that has to be purchased to make the investment work. For instance, opening a bakery is cheaper than opening a car factory.

Let us remark that there is not a Stock Exchange market in CEES (we plan to include it in a future release). So, we suppose that companies (or part of them) cannot be sold/bought.

Next, we have to compute the *new investment and add it to the total investment*. The aggregate amount of investment, as calculated in the previous step for each individual, could be either excessive or insufficient. For instance, if somebody decides to open a bakery in its street and notices that 10 other people had the same idea, he could reconsider his decision. So, an optimum of investment will be calculated and compared with the amount of desired investment. Individual investments will be reduced if they were excessive and increased if they were insufficient. In the latter case, this will be done only if it is possible according to the savings of investors.

Finally, we have to *remove those entrepreneurs who give up due to a lack of profit*. The number of entrepreneurs leaving their sectors will be proportional to the amount of money that should be taken away from the corresponding sector. A sector could also be reduced if savings of entrepreneurs are not enough to afford the production costs. In this case, the reduction will ensure that the new size of companies can be afforded by entrepreneurs. Let us remark that size reductions may happen even if the sector has positive profits.

Phase 4. Computing the demand of employment by companies. According to the old companies that stay and the newly created companies, an amount of man-power will be demanded. Each company will require a rate of specialists (i.e. people with specific knowledge on the specific area) and a rate of non-specialists. The man-power required by companies exponentially rises as the desired quality of the final products increases.

Phase 5. Computing the demand of raw materials by companies. Each productive sector has specific necessities of supply to keep the production working each turn. In case of new companies or companies enlarged during this period, we have to add the materials required to create/increase them. Companies will have different preferences with respect to the different available materials. While quality increases the preference, price decreases it. Besides, price will have a lower influence in companies which desire to produce high quality products.

Phase 6. Computing the maintenance/creation of companies. In this phase we consider three different steps. First, according to the demanded man-power, *workers try to find a job as a wage-earner*. Jobs are sorted from the highest to the lowest salary. Then, we take the *best* job and we look for candidates. Specialized jobs require workers with the corresponding knowledge, while everybody is a good candidate for a non-specialized job. Next, we calculate the proportion between suitable candidates and the number of required jobs. The corresponding rate of candidates will take the job. Then, we take the next best paid job and repeat the process. Let us remark that not all of the suitable candidates who are still unemployed will accept the best available job. The decision will be influenced by the relation (given by the utility function of each individual) among the salary of the job, the cost of living, and unemployment benefits. However, current workers lessen the effect of the previous factors in order to continue in the same position. Let us note that a current employee loses his job only if the sector is reduced and he is fired, if he gives up voluntarily to look for another job, or if he thinks that it is better for him to be unemployed.

Next, we have to *provide raw materials to companies.* The total amount of demanded raw materials is added. This quantity is compared with the total stock of raw materials. If demand is bigger than supply then the rate between demanded and available supply is calculated, and companies are supplied in this same rate. Additionally, stocks of some raw materials providers could be exhausted, and companies trying to buy them will save the money corresponding to the amount they could not buy. This money will be available to be spent in supply stocks of other companies. This fact allows again, as it was the case of usual goods, that bad and expensive raw materials enter the market right after cheap and good ones are sold out.

Finally, we have to compute the *amount of man-power and raw materials demanded by companies which has been actually filled.* If the demand was not satisfied, it will be calculated which one of these resources was obtained in the lower proportion. This resource will mark the bottleneck of the productive process: The amounts of the other resources surpassing this rate will actually be spared (but entrepreneurs must still pay for them). Let us remark that new companies which did not surpassed the minimal threshold of needed supply to be *created* will close down.

Phase 7. Computing the amount of raw materials that are bought by companies. Here, we consider three steps. First, for each productive sector, we subtract the demand of raw materials from the corresponding supply. Then we perform the *selling of raw materials.* We reduce the savings of entrepreneurs according to the materials that they bought, and we add this money to the savings of the corresponding owners of raw materials companies. Let us remark that companies that are unproductive have also to pay the quantity of materials that they were ordering. Then, the new stocks will be computed. In addition, the State collects the indirect taxes associated with the selling of these products. Finally, raw materials companies decide whether they *increase/decrease their prices according to the demand/supply relation.* The process is again similar to that in the second phase.

Phase 8. Computing the wages of workers. In this phase we consider three different steps. First, we compute *the difference between demand and supply of employment.* Next, by *paying salaries*, we have that the wealth of entrepreneurs is reduced according to the hired man-power. Simultaneously, savings of employees increase according to their salaries. Let us remark that unproductive companies have to pay the man-power they hired. The State will pay the corresponding unemployment benefits. In addition, both wage-earners and unemployed people will pay their income taxes. Finally, *new salaries will be computed according to the the demand/supply relation.* Once again, Participants may voluntarily modify these values. Let us remark that in a labor sector presenting unemployment, salaries must not necessarily decrease. If unemployment is small then salaries could even rise, because finding employees will be still hard. Otherwise, salaries would decrease continuously as every society has structural unemployment.

Phase 9. Computing the aggregate production. According to the real availability of resources in companies (man-power and raw materials) the final production is computed and added to the corresponding stocks. New theoretical productivity is calculated according both to the suitable or excessive exploitation of natural resources and to technological research. Besides, the limit to consider that the natural resources are excessively exploited is increased according to technological research. Taking into account the proportion of hired employees over the standard amount of needed employees, the quality of the acquired raw materials, and the quality of the machinery (the *raw materials* from which the company was created), the quality of the final products is computed. A part of the stored stock is spoiled each turn as well as a part of the machinery. By doing so we can simulate the effect of obsolete and old machinery.

3 Implementation Details and Experiments

CEES was implemented by using Borland C++ Builder™ and the code is around 11000 lines. The structure of the economy model guided the decomposition of the design in classes (State of the Economy, Productive sector, Labor sector, Entrepreneur, Worker, etc.), the needed operations (buy, sell, consume, invest, look for a job, migrate of sector, etc.), and the needed attributes (salary, savings, amount of people, etc.).

Let us remark that, depending on the *length* of the turns (i.e., the length of the period of time they represent - days, months, years, etc.), some of the operations described in the previous section are appropriate in all turns but others are not. For instance, while it is clear that all of them are suitable in the case of years, it is in general false that taxes are paid every month or salaries are paid every week. So, depending on the length of turns, some operations will be disabled in most of them, which will reduce their computational cost. Obviously, as the length is reduced, more turns are needed to reach each future time, which dramatically increases the time required to compute the simulation, but the economic precision of the model improves. For example, it allows us to consider the effect of temporal singularities (e.g., paydays or months in which taxes are paid) in real economies.

Next we briefly comment on some of the experiments that we have performed with CEES. First, as we expected, the sector of raw materials was guided by the derivative of the other sectors. In Figure 1 (left) the evolution of the aggregated industry and raw materials sectors is depicted. In order to reduce the required space, we have deleted the rest of the CEES graphical interface. As the given initial conditions were not natural enough, according to the model, there is a first stage of stabilization where the Economy tries to fit properly into the model. Afterwards, the stable cyclic behavior begins. As we can see, the raw materials sector (down) follows almost the derivative of the industry sector (up). The

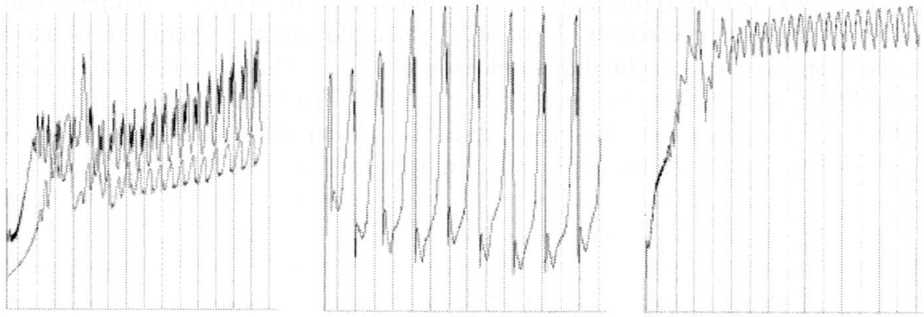

Fig. 1. Industry and Raw Materials sectors (left), completely unrestricted economy (center), and lowly restricted economy (right)

reason is that a great part of the demand of the sector of raw materials comes from the creation of new companies.

We were also trying to determine conditions to soften the cyclic behavior of the economy. In fact, a very reduced form of state can help in this task. In Figure 1 (center) we show a simulation of a completely free market, while in the right hand side we present the behavior of an economy where a simple state (with no public companies) collects some taxes that are used to pay low unemployment benefits. Besides, the cyclic behavior can also be affected if we consider that a huge company appears in the economy. For instance, if we play the role of one of such companies, it is even possible to stop a expansion or a depression as a result of the investments of our company.

Finally, we studied the influence of different company tactics in an economy. In particular, we analyzed the effects of *price dumping*. This technique consists in selling products under the production costs until the competence disappears. Once the market is controlled, the company can monopolistically act (with the consequent increase of prices). As we expected, it worked only after a minimal threshold of portion of the market was initially owned and a great stored capital was available by the company (in order to confront the initial loses).

We would like to remark that as CEES is endowed with the needed complexity to simulate most of the relevant aspects of the economy, it allows to perform experiments it was not specifically devoted to when it was designed. So, we expect that many other interesting situations will be confronted in the future.

4 Conclusions and Future Work

In this paper we have presented CEES. The interest of CEES consists of two main points. First, its success to simulate complex economical behaviors using basic low level economical concepts. Second, its designing methodology where the learning in the specific area, the development of the model, and the treatment of unexpected behaviors have been jointly and coordinately dealt with.

As future work we would like to migrate the architecture of CEES to a more standardized technology of economic simulation (e.g. Swarm [6]). It would allow us to easily compare simulation results and to share components in order to study design alternatives. Besides, we plan to include some of the main economic factors which actually could not be included in the current version. This is the case of the loan system, a stock exchange, or international commerce. In addition, we plan to apply some of the economical mechanisms underlying CEES to the systems presented in [7, 5]. In these works, simple microeconomic concepts are applied to concurrent and e-commerce systems, respectively. We are confident that relevant *high level* emerging properties can be found in the distributions of resources using (a subset of) CEES. So, basic rules governing the exchange of resources in the previously mentioned systems could be improved so that better distributions are achieved. Finally, we are also working on improving our performance by parallelizing our application using the language Eden [9].

Acknowledgements. We would like to thank Julián Gutiérrez, María Teresa Gutiérrez, and Jesús Nieves, who collaborated in the implementation of the first version of CEES.

References

1. R. Axelrod. Advancing the art of simulation in the social sciences. In *Simulating Social Phenomena, Lecture Notes in Economics and Mathematical Systems 456*, pages 21–40. Springer, 1997.
2. A.S. Blinder. On sticky prices: academic theories meet the real world. In G. Mankiw, editor, *Monetary Policy*, pages 155–182. University of Chicago Press, 1994.
3. J.Q. Cheng and M.P. Wellman. The WALRAS algorithm: A convergent distributed implementation of general equilibrium outcomes. *Computational Economics*, 12:1–24, 1998.
4. J. Frolova and V. Korobitsin. Simulation of gender artificial society: Multi-agent models of subject-object interactions. In *ICCS 2002, LNCS 2329*, pages 226–235. Springer, 2002.
5. N. López, M. Núñez, I. Rodríguez, and F. Rubio. A multi-agent system for e-barter including transaction and shipping costs. In *Symposium on Applied Computing, SAC 2003.* ACM Press, 2003. 8 pages. In press.
6. F. Luna and B. Stefansson. *Economic Simulations in Swarm: Agent-Based Modelling and Object Oriented Programming.* Kluwer Academic Publishers, 2000.
7. M. Núñez and I. Rodríguez. PAMR: A process algebra for the management of resources in concurrent systems. In *FORTE 2001*, pages 169–185. Kluwer Academic Publishers, 2001.
8. T. Ostrom. Computer simulation: the third symbol system. *Journal of Experimental Social Psychology*, 24:381–392, 1998.
9. R. Peña and F. Rubio. Parallel Functional Programming at Two Levels of Abstraction. In *PPDP'01*, pages 187–198. ACM Press, September 2001.
10. J.E. Stiglitz. *Principles of Macroeconomics.* W.W. Norton & Company, Inc, 1993.
11. R. Suppi, P. Munt, and E. Luque. Using PDES to simulate individual-oriented models in ecology: A case study. In *ICCS 2002, LNCS 2329*, pages 107–116. Springer, 2002.

Structure of Bilayer Membranes of Gemini Surfactants with Rigid and Flexible Spacers from MD Simulations

Dmitry Yakovlev[1] and Edo S. Boek[2]

[1] St.Petersburg State University, Department of Chemistry, 26 Universitetsky pr., 198504 St.Petersburg, Russia
[2] Schlumberger Cambridge Research, High Cross, Madingley Road, Cambridge CB3 0EL, United Kingdom,
boek@cambridge.oilfield.slb.com

Abstract. Molecular Dynamics simulations were performed for 9×9 bilayers formed by gemini surfactants $p\text{-}[C_{19}H_{39}N^{+}(CH_3)_2CH_2]_2C_6H_4 \cdot 2Cl^{-}$ or $[C_{19}H_{39}N^{+}(CH_3)_2CH_2]_2CH(OH) \cdot 2Cl^{-}$ with rigid hydrophobic or flexible hydrophilic types of spacer respectively. The structure of the bilayers is rather different depending on the type of spacer regarding the effect of $NaCl$ salt addition. It is shown that the structure of the bilayer strongly depends on the interaction between the surfactant head groups and the counter ions.

1 Introduction

An experimental study of gemini surfactants with different types of spacer has been reported in a number of papers [1,2,3]. So-called gemini or bis-surfactants are formed by two novel surface active molecules connected via a flexible or rigid linkage between the hydrophilic head groups. These systems are of interest due to a number of unusual properties. One of these is that the CMC value increases when the hydrocarbon chain of the hydrophobic tail reaches some value. This observation is in contradiction with the well-known fact that the CMC value - in general - monotonically decreases with tail lengthening. It is worth to mention that some novel surfactants with long hydrophobic tail display similar behaviour. To understand the physical backgroud of these unusual micellar properties, Molecular Dynamics (MD) simulations were used.

Molecular Dynamics study is a method to observe the microscopic structure of aggregates and to extract a number of macroscopic parameters to be used in thermodynamic models of micellization. Spontaneous formation of spherical or rodlike micelles (depending on surfactant concentration) from a random initial configuration has been reported in a number of papers [4,5,6,7]. Self-organization was observed for systems where the concentration of surfactants exceeds the CMC value. MD simulation with random starting configuration allows us to estimate the CMC value of a surfactant with a given molecular structure. However, this method cannot be applied to systems with a CMC value below some

P.M.A. Sloot et al. (Eds.): ICCS 2003, LNCS 2658, pp. 668–677, 2003.

limit due to computational expenses. For example, in the case of a CMC value of 10^{-3}M, the number of water molecules per surfactant is greater than 55000; for a spherical micelle with average aggregation number about 60, the size of the system is then too large to observe aggregation in a standard MD simulation starting from a random initial configuration. The reasonable way of MD simulations of such systems is to construct a bicontinuous bilayer surrounded by water [8].

In this work, MD simulations have been performed to compare the properties of bilayers formed by gemini surfactants (see Fig. 1) with rigid and flexible spacers. The length of the hydrocarbon chains in both cases was taken as 19 carbon atoms. To observe the effect of electrolyte addition on the bilayer structure, the simulations were carried out also for systems with 3 percent of $NaCl$ salt. Overall four systems have been considered. The first one includes gemini surfactant with rigid hydrophobic spacer $[(C_{19}H_{39}N^{+}(CH_3)_2CH_2)_2Ar] \cdot Cl_2^-$. We will label this system as "Gemini-Ar". The second system is the same but with 3 percent of added $NaCl$; the label is "Gemini-Ar-NaCl". The last two systems include a gemini surfactant with flexible hydrophilic spacer $[(C_{19}H_{39}N^{+}(CH_3)_2CH_2)_2CH(OH)] \cdot Cl_2^-$. We will refer to these as "Gemini-OH" and "Gemini-OH-NaCl" depending on salt addition.

2 MD Simulation Details

The simulations were carried out using GROMACS [9] software running in parallel [10] under MPI environment on an SGI Origin 2000 supercomputer equipped with 32 CPUs. Actually not more than 12 processors were utilized due to communication overhead. The GROMACS united atom force field was used. Equilibrium molecular geometry and atomic charge distributions were taken from semiempirical calculations using the MNDO hamiltonian model. The quantum chemical calculations were performed using the GAMESS [11] program package. The charges of united atom groups (CH_n) were calculated as the sum of atomic charges over all atoms of given functional group. The major part of the total charge (+2) is found to be distributed over the surfactant head group and the tail hydrocarbon group directly connected to nitrogen. The MOPAC atomic charges on united atoms are given in Fig. 1.

The basic cell has orthogonal geometry with periodic boundary conditions applied in all dimensions. The bilayer was positioned in the middle of the box perpendicular to z axis. The rest of the basic cell was filled with water (SPC model) and ions. To build a bilayer with a density close to the density of liquid hydrocarbons, a special technique was developed. The problem is that - due to the equilibrium molecular geometry of the surfactants under investigation - the constructed bilayer has many cavities between the tails, which are filled with water and ions. The idea is to generate a large number of conformations by running MD simulation of a single surfactant molecule and than find a geometry that provides a suitable density. A number of criteria was used to select the best configuration from formally acceptable ones. The resulting bilayers were

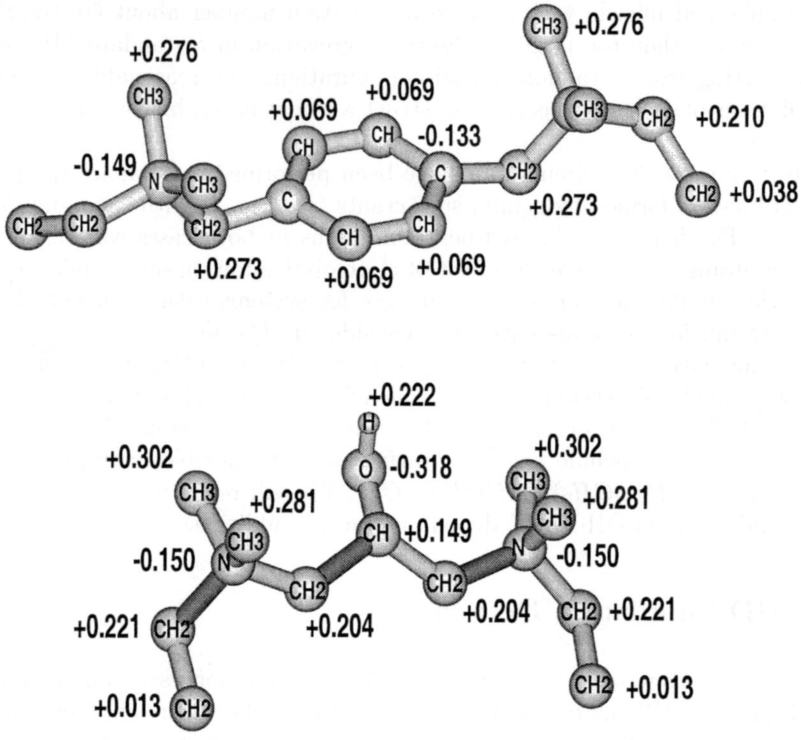

Fig. 1. United atom model and charges of the head groups of gemini surfactants with rigid hydrophobic (top) and flexible hydrophilic (bottom) spacers. Only two carbon atoms of tails are shown (the rest of the tails is cut).

formed by 162 surfactant molecules (9×9) and hardly contain any water inside the hydrophobic core. A summary of all systems studied is presented in Table 1.

The MD simulations were performed at a constant temperature of 300 K and a constant pressure of 1 bar using the Berendsen external bath coupling [12]. To take into account the effect of water penetration inside the bilayer, the basic cell was scaled isotropically in x and y directions but differently in z direction (so-called "semiisotropic" coupling). The bonds were constrained using the SHAKE algorithm and it's special implementation for rigid water molecules (SETTLE) [13]. The leap-frog method with time step 2 fs was used to integrate the equations of motion. A cut-off distance of 1 nm was applied to all intermolecular interactions. Due to slow penetration of water into the surfactant bilayer, the geometry of basic cell changes. To speed up equilibration, the simulation was split up in two parts: first, long-range electrostatic interactions were cut at

Table 1. Details of the molecular dynamics simulations. #Surf and #W are numbers of surfactant and water molecules, #Cl and #Na are numbers of chloride and sodium ions and #Total is total number of atoms in the basic box. t_{PME} is time of simulation with PME electrostatic summation and t_{prod} is time used for average production, *ps*.

System	#Surf	#W	#Cl	#Na	#Total	t_{PME}	t_{prod}
Gemini-Ar	162	6163	324		27237	1160	260
Gemini-Ar-NaCl	162	6047	392	58	27005	1060	310
Gemini-OH	162	6841	324		28785	1060	310
Gemini-OH-NaCl	162	6709	390	66	28521	1000	500

a distance of 1.0 nm during 1 ns. This was followed by more than 1 ns simulation with PME (Particle Mesh Ewald) summation of coulomb interactions [14]. Switching on the PME summation results in a rapid decrease of the cell size in z direction. Equilibrium was defined as the point where the cell geometry becomes stable.

3 Analysis of MD Data

The aim of our MD simulation is to clarify the dependence of bilayer structure on the chemical nature of the surfactant and salt addition. The only difference between the surfactants under investigation is the structure of the spacer: rigid hydrophobic in case of Gemini-Ar and flexible hydrophilic in case of Gemini-OH. As will be shown below, the nature of the spacer has a dramatic effect on the behaviour of bilayer.

The equilibrium size of the basic box in x, y and z dimensions provides us with information about the average surface area a per surfactant (see Table 2). Note that we neglect the curvature of the bilayer when calculating the average surface area as $x \cdot y$ divided by number of surfactant in one layer. It is surprising that the surface area per molecule is smaller for Gemini-Ar than for Gemini-OH: obviously the excluded volume of the $CH_2C_6H_4CH_2$ spacer (and therefore the size of the head group) is larger than for $CH_2CH(OH)CH_2$ (see Fig. 1). Furthermore, in the case of adding 3 weight per cent $NaCl$ to the Gemini-Ar system results in an expansion of the cell in z but squeezing in x and y dimensions. Hence the area per surfactant decreases from 77.7 Å^2 to 69.9 Å^2 (see Table 2). The effect of the salt on the geometry of basic box in case of Gemini-OH is quite opposite: the area a increases from 97.7 Å^2 up to 106.4 Å^2. These effects can be related with a partial binding of the chloride ions, screening electrostatic repulsion between the charged head groups.

3.1 Density Profiles

The number density profiles along the z axis of the systems studied are presented in Fig. 2. These plots demonstrate significant penetration of solution into the

Table 2. Sizes of the basic box in x, y and z dimentions (nm), density ρ (g/cm^3), surface area per one surfactant a (Å^2), thickness of the bilayer H (nm) and measure of water penetration into bilayer core S_{WT}. See text for details.

System	x	y	z	ρ	a	H	S_{WT}
Gemini-Ar	7.43	8.47	6.50	0.975	77.7	3.1	770
Gemini-Ar-NaCl	7.05	8.03	7.21	0.984	69.9	3.5	860
Gemini-OH	8.27	9.57	5.23	0.982	97.7	2.4	670
Gemini-OH-NaCl	8.63	9.99	4.79	0.991	106.4	2.2	560

hydrocarbon core of the membrane so that the head groups are surrounded by water. Note that the bilayer formed by Gemini-Ar surfactant has a more diffuse boundary with bulk water than the bilayer formed by Gemini-OH. A sharp water-membrane boundary is typical for ionic surfactants with a negatively charged head group [8]. In the case of gemini surfactants, the boundary is not well-defined, so one needs a way to measure the thickness of the bilayer. It is suitable to determine the thickness of the membrane as the distance between the points where the density profiles of water and hydrocarbon groups cross. In this way the thickness will depend on water penetration into bilayer. The degree of water penetration into the membrane for different systems can be estimated as an overlap integral S_{WT} between the number density profiles of water n_W and hydrocarbon groups of the tail n_T:

$$S_{WT} = \int_0^Z n_W(z) n_T(z) dz \tag{1}$$

The overlap integral S_{WT} as well as the thickness H of the membrane are presented in Table 2. The thickness of the bilayer formed by Gemini-OH surfactant is significantly smaller than for Gemini-Ar systems and decreases when adding salt while the thickness of Gemini-Ar bilayer increases. The dependence of the S_{WT} parameter on addition of salt is the same. The density profiles of head groups, chloride and sodium ions (see Fig. 2, bottom) also show that Gemini-OH systems have a more compact structure of the water-bilayer boundary. Note that the head group and chloride ion profiles are very close to each other. This means that the best part of the counter ions is bound to the head groups.

3.2 Radial Distribution Functions

The analysis of radial distribution functions (RDF) provides us with more detailed information about the structure of the bilayer. As mentioned above, changing from a rigid hydrophobic to a flexible hydrophilic spacer causes a dramatic distinction in membrane behaviour. The bilayer properties strongly depend on the chemistry of the head group but not on the hydrophobic tail. Interaction of charged heads with ions also plays a key role. Keeping these in mind, we will

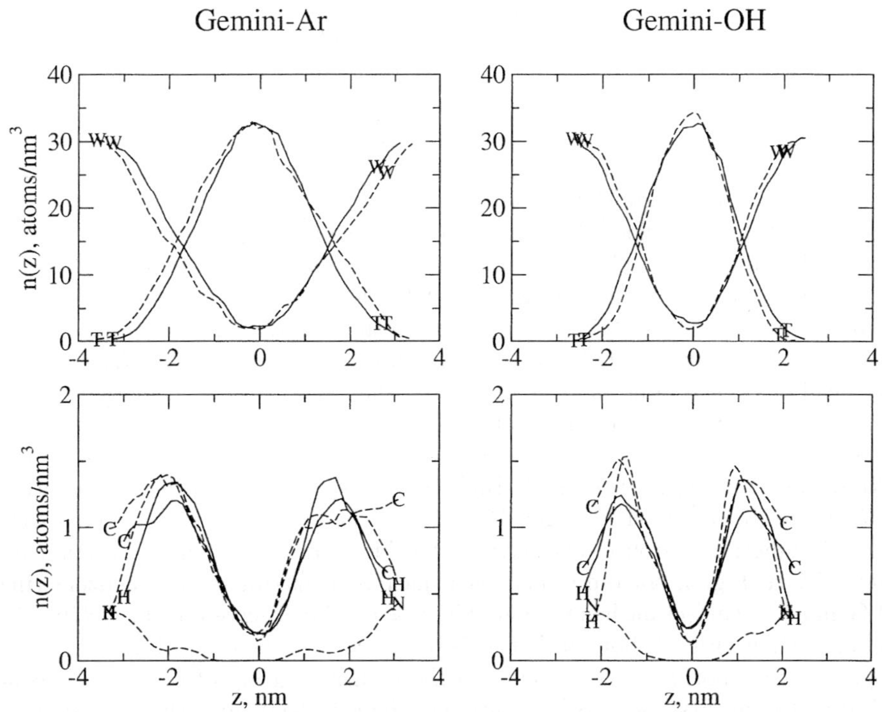

Fig. 2. Number-density profiles across bilayer surface of Gemini-Ar, Gemini-Ar-NaCl (left side) and Gemini-OH, Gemini-OH-NaCl (right side) systems. Systems without salt addition are solid, with salt are dashed. Symbols W, T, H, C and N mark the density profiles of water, hydrocarbon groups of tails, head groups (nitrogen atoms), chloride and sodium ions respectively. Origin of the z axis is shifted to the middle of the basic box for each system. Note that some profiles have different value at left and right edges due to smoothing.

focus our investigation on the set of RDFs related to head-head, head-ions and head-water interaction. The aim is to clarify the origin of the different salt effect on bilayer structure.

The nitrogen-nitrogen radial distribution functions (see Fig. 3, left top) have a well-defined peak that corresponds to two nitrogen atoms of the same surfactant molecule. For Gemini-OH systems this peak is at r=0.46 nm while a second maximum at r=0.86 nm corresponds with non-bonded nitrogen atoms. In the case of the Gemini-Ar system, these maxima are very close to each other and form one asymmetric peak. The position and magnitude of the peaks of nitrogen-nitrogen RDFs are almost independent of salt addition (see Table 3). The number of neighbors N_c in the first coordination shell was calculated as integral over the RDF up to the first minimum. The nitrogen-nitrogen coordination number for Gemini-Ar is about 10 but only 6.5 in case of Gemini-OH system. This means that the bilayer formed by Gemini-Ar surfactants has a diffuse boundary with water because the coordination number is too high for a planar structure. The analysis shows that the positions of the extrema and values of surface area per surfactant for the Gemini-Ar bilayer correspond to hexagonal packing, while the Gemini-OH membrane more likely has an ortorhombic packing. Salt addition causes N_c to increase from 9.8 up to 10.2 for Gemini-Ar but to decrease from 6.5 to 6.2 for the Gemini-OH system.

The nitrogen-chloride RDFs demonstrate a strong coupling between chloride ions and head groups. The position of the first maximum remains the same for all systems but the magnitude is subject to change (see Fig. 3, top right and Table 3). Nitrogen has 2.0-2.2 chloride ions in the closest shell for Gemini-Ar and 2.3-2.7 for the Gemini-OH bilayer. Taking into account that each CH_3 group of the head has exactly one chloride in the first shell and using molecular geometry, one can conclude that the anions around nitrogen are the ones bound with positively charged CH_3 groups connected to nitrogen. The second maximum of the CH_3(head)-chloride RDF (Fig. 3, Table 3) corresponds to chloride of the other CH_3 group. In case of the Gemini-OH bilayer, chloride ions are also bound with the OH functional group. The oxygen-chloride radial distribution function for Gemini-OH (Fig. 3, bottom right) has a sharp first maximum. The benzene ring of Gemini-Ar surfactant has no bound ions as can be observed in the CH(head)-chloride RDF (Fig. 3, bottom left) where CH are the benzene ring atoms (see Fig. 1). The chloride ion connected to the hydroxyl group of Gemini-OH contributes to the first peak of the nitrogen-chloride RDF and the second peak of CH_3(head)-chloride and increases the corresponding coordination numbers. Salt addition always decreases the number of chloride ions in coordination shells because part of chlorides become involved in the formation of ionic pairs with sodium (see Table 3). This is confirmed by the observation that the number of chloride ions in the first coordination shell of sodium (1.3) is greater than one. The positions of the extrema of CH_3(head)-oxygen(water) RDFs are the same for all systems. The number of water molecules in the first hydration shell of the CH_3 group is 0.5 only. The reason is that the head groups are surrounded mainly by chloride ions, not by water.

4 Results and Discussion

The data presented above demonstrates many distinctions in the structure of bilayers formed by gemini surfactants with different types of spacer. The surfactant with a flexible hydrophilic spacer gives us a thin membrane with a well-defined surface. Changing to a hydrophobic rigid spacer leads to a bilayer with a diffuse boundary. The most interesting question is why salt addition affects the membrane properties in an opposite way. The analysis of radial distribution functions allows us to observe that the interaction between head group and chloride ions controls the behaviour of the bilayer. Packing, bilayer thickness, area per surfactant, water penetration are different for Gemini-Ar and Gemini-OH systems but these properties are secondary. Comparing radial distribution functions for

Table 3. Positions of maxima r_{max}, minima r_{min} (nm) and coordination numbers N_c for nitrogen-nitrogen, nitrogen-chloride, CH_3(head)-chloride, CH_3(head)-oxygen(water), CH(head)-chlorine (for Gemini-Ar and Gemini-Ar-NaCl) and oxygen(head)-chloride (for Gemini-OH and Gemini-OH-NaCl systems) radial distribution functions (see Fig. 3). The value of $g(r)$ is separated by a slash. In case of well separated first and second maxima, two coordination numbers are given (second coordination number is calculated over both maxima).

System	$r_{max}/g(r_{max})$	$r_{min}/g(r_{min})$	N_c
	Nitrogen-Nitrogen		
Gemini-Ar	0.80/3.9 0.85/1.8 1.50/1.1	1.24/1.0	9.8
Gemini-Ar-NaCl	0.80/3.9 0.85/1.8 1.53/1.1	1.28/1.0	10.2
Gemini-OH	0.49/12.8 0.86/1.9 1.24/1.1	1.05/1.1	1.2 6.5
Gemini-OH-NaCl	0.49/12.8 0.86/1.9 1.25/1.1	1.08/1.0	1.2 6.2
	Nitrogen-Chloride		
Gemini-Ar	0.49/4.0 0.75/1.4	0.65/1.5	2.2
Gemini-Ar-NaCl	0.49/3.5 0.75/1.4	0.65/1.5	2.0
Gemini-OH	0.49/4.7 0.70/1.7	0.65/1.8	2.7
Gemini-OH-NaCl	0.49/4.1 0.70/1.5	0.65/1.5	2.3
	CH_3(head)-Chloride		
Gemini-Ar	0.39/4.1 0.59/2.2	0.51/1.7	1.0 3.2
Gemini-Ar-NaCl	0.39/3.8 0.59/2.0	0.51/1.6	1.0 3.0
Gemini-OH	0.39/4.4 0.59/2.9	0.51/1.9	1.1 3.7
Gemini-OH-NaCl	0.39/3.9 0.59/2.5	0.51/1.7	1.0 3.3
	CH_3(head)-Oxygen(water)		
Gemini-Ar	0.34/2.2 0.55/1.3	0.47/1.1	0.5 2.0
Gemini-Ar-NaCl	0.34/2.2 0.55/1.3	0.47/1.0	0.4 2.0
Gemini-OH	0.34/2.1 0.55/1.4	0.47/1.1	0.4 2.0
Gemini-OH-NaCl	0.34/2.1 0.55/1.4	0.47/1.1	0.5 2.1
	CH(head)-Chloride		
Gemini-Ar	0.43/2.0 0.65/2.1	0.53/1.7	0.7 6.4
Gemini-Ar-NaCl	0.43/1.9 0.65/2.0	0.53/1.5	0.6 6.0
	Oxygen(head)-Chloride		
Gemini-OH	0.34/3.8 0.56/2.7	0.42/1.5	0.4 6.5
Gemini-OH-NaCl	0.34/3.3 0.56/2.4	0.42/1.2	0.4 6.0

surfactants with various types of spacer, one can state that the difference is in the interaction of the head group with chloride ions. The heads of Gemini-OH surfactants coordinate a little bit more chloride ions in the closest shell than Gemini-Ar. The equilibrium thickness of the surfactant membrane is found to be smaller for Gemini-OH than for Gemini-Ar, because the former has a more flexible molecular structure.

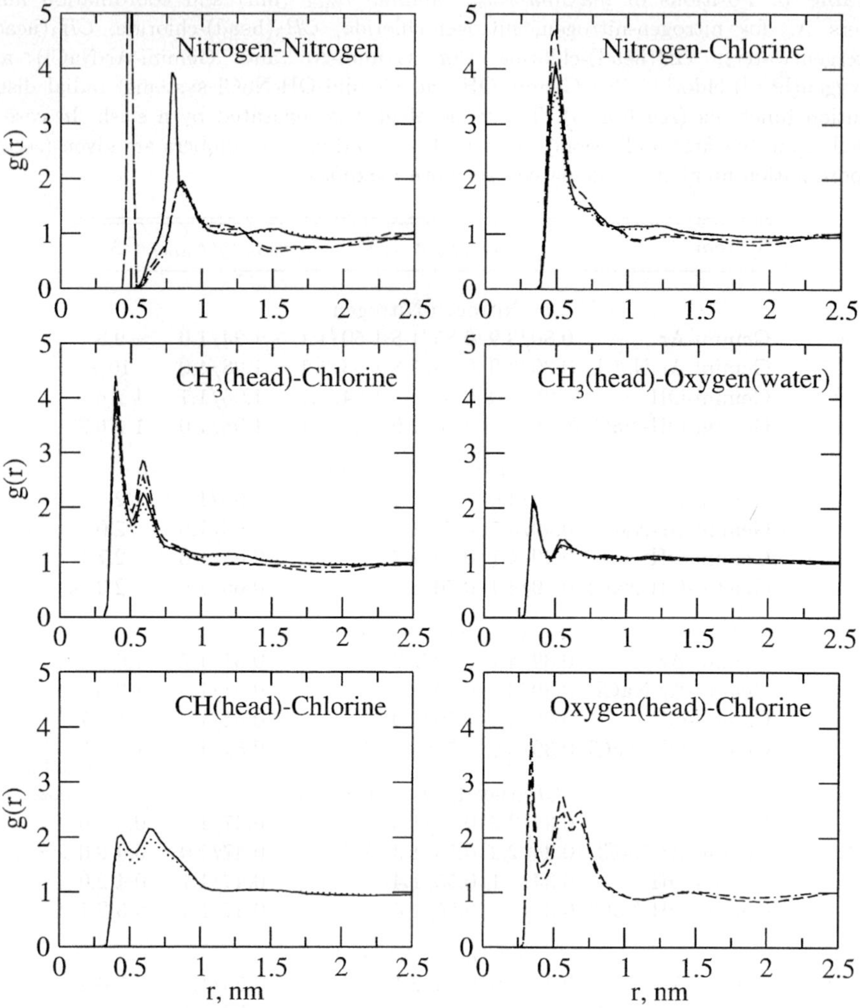

Fig. 3. Nitrogen-nitrogen, nitrogen-chloride, CH_3(head)-chloride, CH_3(head)-oxygen(water), CH(head)-chloride and oxygen(head)-chloride radial distribution functions of Gemini-Ar (solid), Gemini-Ar-NaCl (dotted), Gemini-OH (dashed) and Gemini-OH-NaCl (dash-dotted) systems.

References

1. Li D. Song and Milton J. Rosen. Surface properties, micellization, and premicellar aggregation of gemini surfactants with rigid and flexible spacers. *Langmuir*, 12(5):1149–1153, 1996.

2. Alexander A. Yaroslavov, Oleg Yu. Udalykh, Nickolay S. Melik-Nubarov, Victor A. Kabanov, Yuri A. Ermakov, Vladimir A. Azov, and Fredric M. Menger. Conventional and gemini surfactants embedded with bilayer membranes: contrasting behavior. *Chemistry - A European Journal*, 7(22):4835–4843, 2001.
3. Fredric M. Menger and Jason S. Keiper. Gemini surfactants. *Angewandte Chemie Int. Ed.*, 39(11):1906–1920, 2000.
4. B. Smit, A. G. Schlijper, L. A. M. Rupert, and N. M. van Os. Effect of chain length of surfactants on the interfacial tension: molecular dynamics simulations and experiments. *J. Phys. Chem.*, 94(18):6933–6935, 1990.
5. B. Smit, P. A. J. Hilbers, K. Esselink, L. A. M. Rupert, N. M. van Os, and A. G. Schlijper. Structure of a water/oil interface in the presence of micelles: a computer simulation study. *J. Phys. Chem.*, 95(16):6361–6368, 1991.
6. Rüdiger Goetz, Gerhard Gompper, and Reinhard Lipowsky. Mobility and elastisity of self-assembled membranes. *Phys. Rev. Lett.*, 82(1):221–224, 1999.
7. S. J. Marrink, D. P. Tielenan, and A. E. Mark. Molecular dynamics simulation of the kinetics of spontaneous micelle formation. *J. Phys. Chem. B*, 104(51):12165–12173, 2000.
8. E. S. Boek, A. Jusufi, H. Löwen, and G. C. Maitland. Molecular design of responsive fluids: Md studies of viscoelastic surfactant solutions. *J.Phys.:Condens.Matter*, 14:9413–9430, 2002.
9. E. Lindahl, B. Hess, and D. van der Spoel. Gromacs 3.0: A package for molecular simulation and trajectory analysis. *J. Mol. Mod.*, 7:306–317, 2001.
10. H. J. C. Berendsen, D. van der Spoel, and R. van Drunen. Gromacs: A message-passing parallel molecular dynamics implementation. *Comp. Phys. Comm.*, 91:43–56, 1995.
11. M. W. Schmidt, K. K. Baldridge, J. A. Boatz, S. T. Elbert, M. S. Gordon, J. H. Jensen, S. Koseki, N. Matsunaga, K. A. Nguyen, S. J. Su, T. L. Windus, M. Dupuis, and J. A. Montgomery. Gamess program package. *J. Comput. Chem.*, 14:1347–1363, 1993.
12. H. J. C. Berendsen, J. P. M. Postma, A. DiNola, and J. R. Haak. Molecular dynamics with coupling to an external bath. *J. Chem. Phys.*, 81:3684–3690, 1984.
13. S. Miyamoto and P. A. Kollman. Settle: An analytical version of the shake and rattle algorithms for rigid water models. *J. Comp. Chem.*, 13:952–962, 1992.
14. U. Essman, L. Perela, M. L. Berkowitz, T. Darden, H. Lee, and L. G. Pedersen. A smooth particle mesh ewald method. *J. Chem. Phys.*, 103:8577–8592, 1995.

Algorithms for All-Pairs Reliable Quickest Paths *

Young-Cheol Bang[1†], Nageswara S.V. Rao[2], and S. Radhakrishnan[3]

[1] Department of Computer Engineering, Korea Polytechnic University
Kyunggi-Do, Korea
ybang@kpu.ac.kr

[2] Computer Science and Mathematics Division, Oak Ridge National Laboratory
Oak Ridge, Tennessee 37831-6364
raons@ornl.gov

[3] School of Computer Science, University of Oklahoma
Norman, Oklahoma 73019
Sridhar@cs.ou.edu

Abstract. We consider the reliable transmission of messages via quickest paths in a network with bandwidth, delay and reliability parameters specified for each link. For a message of size σ, we present algorithms to compute all-pairs quickest most-reliable and most-reliable quickest paths each with time complexity $O(n^2m)$, where n and m are the number of nodes and links of the network, respectively.

1 Introduction

We consider a computer network represented by a graph $G = (V, E)$ with n nodes and m links. Each link $l = (i, j)\in E$ has a *bandwidth* $B(l)\geq 0$, *delay* $D(l)\geq 0$, and *reliability* $0\leq \pi(l) \leq 1$, which is the probability of l being fault free. A message of σ units can be sent along the link l in $T(l) = \sigma / B(l) + D(l)$ time with reliability $\pi(l)$ as in [13].

Consider a *path* P from i_0 to i_k given by $(i_0, i_1), (i_1, i_2), \ldots, (i_{k-1}, i_k)$, where $(i_j, i_{j+1}) \in E$, for $j = 0, 1, \ldots(k-1)$, and $i_0, i_1, \ldots, i_k$ are distinct. The *delay* of path P is $D(P) = \sum_{j=0}^{k-1} D(l_j)$, where $l_j = (i_j, i_{j+1})$. The *bandwidth* of P is $B(P) = \min_{j=0}^{k-1} B(l_j)$. The *reliability* of P is $R(P) = \prod_{j=0}^{k-1} \pi(i_j, i_{j+1})$. The end-to-end delay of P in transmitting a message of size σ is $T(P) = \sigma / B(P) + D(P)$ with reliability $R(P)$.

* Research of Rao is sponsored by Defense Advanced Research Projects Agency under MIPR No. K153, and by Engineering Research Program and High-Performance Networking Program of Office of Science, U. S. Department of Energy under Contract No. DE-AC05-00OR22725 with UT-Battelle, LLC. Research of Bang is sponsored by Electronics and Telecommunications of Research Institute under Contract No. 1010-2002-0057.

† Corresponding Author

P.M.A. Sloot et al. (Eds.): ICCS 2003, LNCS 2658, pp. 678–684, 2003.

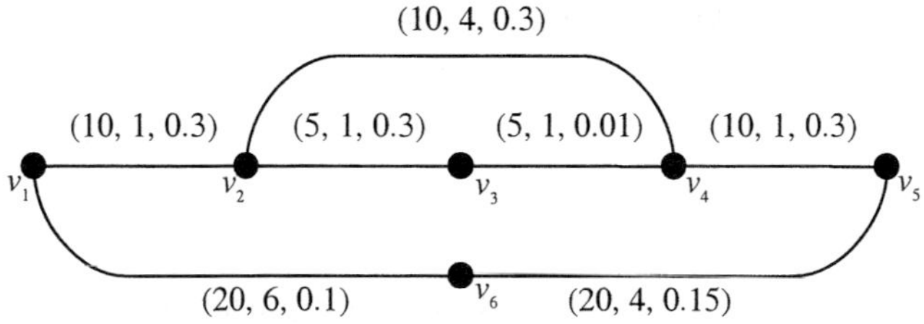

	B(.)	D(.)	R(.)	T(.)
$P_1 : \{v_2, v_2, v_3, v_4, v_5\}$	$B(P_1) = 5$	$D(P_1) = 4$	$R(P_1)= 0.00027$	$T(P_1) = \sigma /5 + 4$
$P_2 : \{v_1, v_2, v_4, v_5\}$	$B(P_2) = 10$	$D(P_2) = 8$	$R(P_2) = 0.027$	$T(P_2) = \sigma /10 + 8$
$P_3 : \{v_1, v_6, v_5\}$	$B(P_3) = 20$	$D(P_3)= 10$	$R(P_3) = 0.015$	$T(P_3)= \sigma /20 + 10$

Fig. 1. Example network

The path P from s to d is the *most-reliable* (MR) if $R(P)$ is the maximum among all paths from s to d. The path P is the quickest for message size σ if $T(P)$ is the minimum among all paths from s to d. The path P is the *quickest most-reliable* (QMR) if it is the quickest for σ among all MR paths from s to d. The P is the *most-reliable quickest* (MRQ) if it has highest reliability among all quickest paths from s to d for σ. For the network in Figure 1, for $s = v_1$ and $d = v_5$, P_2 is the MR path with $R(P_2) = 0.027$, and it is also the QMR path for any σ. For $\sigma < 40$, P_1 is the quickest path, and for $\sigma >$ 40, and P_3 is the quickest path. All P_1, P_2, and P_3 are quickest paths for $\sigma = 40$. Then P_1 is the MRQ path for $\sigma < 40$, and P_3 is the MRQ path for $\sigma > 40$, and P_2 is the MRQ path for $\sigma = 40$.

The classical quickest path problem was extensively studied [7, 4, 12, 2], and has received increased attention recently due to its applicability to computer networks [9, 13]. In real-life computer networks, communication links may fail, and hence it is important to determine quickest paths that are reliable as well. Recently, such reliability aspects have been studied by Xue [13], and $O(rm + rn \log n)$ time algorithms were proposed for computing both QMR and MRQ paths from s to d for message size σ, where r is the number of distinct link bandwidths.

In this paper, we consider the all-pairs versions of computing QMR and MRQ paths considered in Xue [13]. The all-pairs version of the classical quickest path problem was solved in [3, 6] with time complexity of $O(n^2m)$. By applying the algorithms of Xue [13] for each $s \in V$, we can compute QMR and MRQ paths between all pairs s and d with time complexity $O(nrm + rn^2 \log n)$; since $r \leq m$, we have the complexity $O(nm^2 + n^2m \log n)$. In this paper, we present $O(n^2m)$ time algorithms to compute all-pairs QMR and MRQ paths, which match the best-known complexity for the all-pairs classical quickest path problem.

2 All-Pairs Quickest Most Reliable Paths

For any pair $u, v \in V$, a MR path is computed using the All-pairs Shortest Path (ASP) algorithm [1] with weight $\pi'(l) = \log(1 / \pi(l))$, for $l \in E$, where the weight of a path is the sum of the weights of its links. We compute $[\Phi[i, j]_{i, j \in V}] = [\Phi[i, j]]$ using ASP, where $\Phi[u, v]$ is the shortest weight of a path from i to j under the weight π'. There can be more than one MR paths from which the QMR path may be chosen. To account for such MR paths, we identify each link (i, j) that is on some most MR path from u to v, by checking the condition $\Phi[u, i] + \pi'(l) + \Phi[j, v] = \Phi[u, v]$ at an appropriate step in our algorithm.

The rest of the algorithm is a modification of the All-pairs Quickest Path (AQP) algorithm of Chen and Hung [3] which ensures that only edges on appropriate MR paths are considered in computing QMR paths. We use three arrays, denoted by $d[u, v]$, $b[u, v]$ and $t[u, v]$ for $u, v \in V$, to represent the delay, bandwidth and end-to-end delay of the quickest path from u to v, respectively, at any iteration of the algorithm. The edges of G are considered in the non-increasing order in lines 5-13; top operation in line 6 returns the top element and removes it from the heap. In each iteration, the chosen edge (i, j) is checked if it is on some MR path from u to v in line 8. If yes, the algorithm identical to that in [3]: it checks if the path via (i, j) is quicker than the quickest path computed so far, and replaces that latter if so. If not, this edge (i, j) is not considered further.

Algorithm AQMR (G, D, B, π, σ)

1. compute $[\Phi[i, j]]$ using ASP (G, π')
2. **for** pair $u, v \in V$ **do**
3. $d[u, v] = \infty$; $b[u, v] = 0$; $t[u, v] = \infty$;
4. arc_heap $\leftarrow$ top-heavy heap of all $l \in E$ according to bandwidth;
5. **while** arc_heap $\neq \emptyset$ **do**
6. (i, j) = top(arc_heap); let $l = (i, j)$;
7.   **for** each pair $u, v \in V$ **do**
8.   **if** ($\Phi[u, i] + \pi'(l) + \Phi[j, v] = \Phi[u, v]$) **then**
9. $D_1 = d[u, i] + D(l) + d[j, v]$
10.   **if** $D_1 < d[u, v]$ **then**
11. $d[u, v] = D_1$
12.   **if** $\sigma / B(l) + d[u, v] < t[u, v]$ **then**
13. $b[u, v] = B(l)$; $t[u, v] = \sigma / B(l) + D_1$;

Lines 2-11 of algorithm AQMR are identical to that of [3] except for the condition in line 8 which ensures that only the edges on MR paths are considered in computing the quickest paths, and hence its correctness directly follows. All pairs shortest path algorithm has the complexity of $O(n^3)$ in line 1. The complexity of lines 2-13 is $O(n^2 m)$. Thus, the all-pairs QMR paths for message size σ are computed by algorithm AQRM with time complexity $O(n^2 m)$ and space complexity $O(n^2)$.

3 All-Pairs Most-Reliable Quickest Paths

To compute an MRQ path from s to d, we have to "account" for all quickest paths from s to d. Note that AQP returns a quickest path from s to d, which may not be a MRQ path, and hence a simple condition similar to line 8 of AQMR does not work. In particular, it is not sufficient to check if an edge l is on a quickest path P_1 with bandwidth $B(l)$; in fact, l can be on a quickest path with any $b = B(P_1) \leq B(l)$. In our algorithm, we compute the largest of such b and place l at an appropriate step in the computation, which is an iterative process similar to AQMR. Let $t[u, v]$ represent the end-to-end delay of quickest path from u to v for σ .

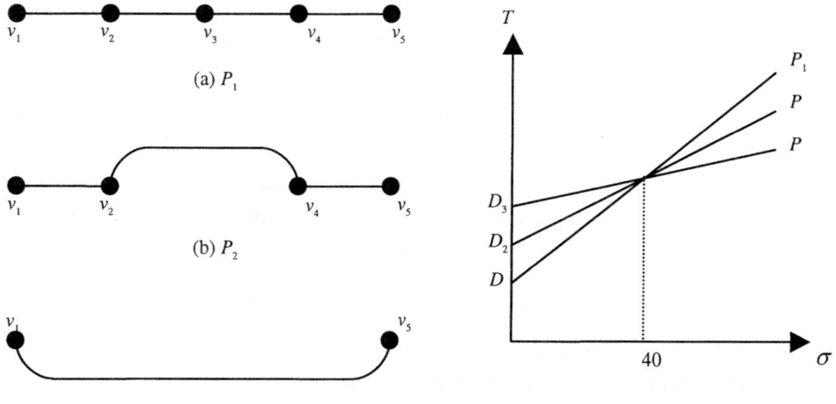

(d) Delay profiles of path P_1, P_2, and P_3

Fig. 2. (a), (b), and (c) represent paths P_1, P_2, and P_3 for the network in Fig.1. (d) shows the plot representation of P_1, P_2, and P_3.

Let $b_1 < b_2 < \ldots < b_r$ be the distinct values of $B(l)$, $l \in E$. Let G_b be the subnetwork with all edges of G whose bandwidth is greater than or equal to b. To compute MRQ paths, we first compute all-pairs quickest paths in G using AQP with the following enhancement. For each bandwidth value b_k and pair u, $v \in V$, we store a matrix $[\, d_{b_k}[u,v]\,]$ where $d_{b_k}[u,v]$ is the delay of the shortest path from u to v in G_{b_k} . These matrices can be computed during the execution of AQP. Let $\Theta(u,v,\sigma)$ be the set of bandwidths of all quickest paths from u to v for given σ . As shown in Fig.2, we have $\Theta(v_1,v_5,20) = \{5\}$, $\Theta(v_1,v_5,40) = \{5, 10, 20\}$, and $\Theta(v_1,v_5,60) = \{20\}$ for the example of Fig.1. The following is an important property of $\Theta(u,v,\sigma)$, which is due to the fact that a quickest path from u to v is a shortest path in G_{b_i} for some b_i.

Lemma 3.1 $\Theta(u,v,\sigma) \neq \emptyset$ if and only if there is a shortest path from u to v in G_b for some $b \in \Theta(u,v,\sigma)$.

In AMRQ, we organize the sets $\Theta(u,v,\sigma)$'s as stacks with bandwidths decreasing top to bottom. We use AQP [3] to compute $[t[u, v]]$ and $[d_b[u, v]]$ in line 1. In line 2, we compute all $\Theta(u,v,\sigma)$'s for each pair $u, v \in V$ with time complexity $O(m)$: for each b_k we simply check for the condition $t[u, v] = \sigma / b_k + d_{B_k}[u, v]$. There are $O(n^2 m)$ iterations in the rest of the algorithm, where edges are considered in decreasing order of bandwidth with which they participate in quickest paths (if at all). In each iteration, we consider the current link bandwidth $B(l)$, and pair $u, v \in V$. Lines 9-10 compute the maximum bandwidth with which the edge l is used in a quickest path from u to v. The reliability of new path via l from u to v is then computed and the existing value is replaced appropriately in lines 11-12. Consider that as a result of while loop in lines 9-10, the retrieved bandwidth $b_{[u, v]}$ is strictly smaller than $B(l)$ if $b_{[u, v]}$ corresponds to link l_1, no more pop operations on $\Theta(u,v,\sigma)$ will performed until all edges with bandwidths in the range $[B(l_1), B(l)]$ have been retrieved from the heap and processed. For each pair $u, v \in V$, this algorithm can be viewed in terms of alternating subsequences of top operations on arc_heap and pop operations on stack $\Theta(u,v,\sigma)$ with no backtracking involved. In actual execution, however, all these subsequences corresponding to various $u - v$ pairs are intermingled among themselves as well as subsequences of top operations.

Algorithm AMRQ (G, D, B, π, σ)

1. compute $[t[u, v]]$ and $[d_b[u, v]]$ using AQP (G, B, D, σ)
2. compute stack $\Theta(u,v,\sigma)$ for each pair $u, v \in V$
3. **for** each pair $u, v \in V$ **do**
4. $\quad b_{[u, v]} = \text{top}(\Theta(u,v,\sigma))$;
5. arc_heap = top_heavy heap of all edges of G according to the bandwidth
6. **while** not arc_heap $\neq \varnothing$ **do**
7. $\quad (i, j) = \text{top(arc\_heap)}$; let $l = (i, j)$;
8. $\quad$ **for** each pair $u, v \in V$ **do**
9. $\quad\quad$ **while** $(B(l) < b_{[u, v]})$ and $(\Theta(u,v,\sigma) \neq \varnothing)$ **do**
10. $\quad\quad\quad b_{[u, v]} = \text{pop}(\Theta(u,v,\sigma))$;
11. $\quad\quad$ **if** $(B(l) \geq b_{[u, v]})$ and $(d_{b_{[u,v]}}[u,v] = d_{b_{[u,v]}}[u,i] + \text{D(i, j)} + d_{b_{[u,v]}}[j,v])$ **then**
12. $\quad\quad\quad \Phi[u,v] \leftarrow min\{\Phi[u,v], \Phi[u,i] + \pi'(\text{i, j}) + \Phi[j,v]\}$;

Consider the correctness of the algorithm. If an edge l is on a quickest path from u to v, there is a corresponding $b \in \Theta(u,v,\sigma)$ by lemma 3.1. Consider a MRQ path P_1 and let l_{p_1} be the link with the lowest bandwidth, which implies $B(l_{p_1}) \in \Theta(u,v,\sigma)$. All other links l of P_1 are retrieved in line 7 before l_{p_1}: each $B(l)$ is checked with $b' \geq b$ or higher in line 11, and hence is accounted for in computing the shortest paths in G_b. Hence, by end of iteration in which l_{p_1} is examined, the reliability of P_1 is computed in line 12, since all its edges would have satisfied the condition in line 11 and hence accounted for in the reliability computation.

The complexity of lines 1 and 2 is $O(n^2m)$. For each pair u, v, each edge is considered at most one time in lines 7 – 12, and hence the time and space complexities of AMRQ are both $O(n^2m)$.

4 Conclusion

We presented algorithms to compute most-reliable quickest and quickest most-reliable paths between all pairs of nodes in a network. These algorithms match the best known computational complexity for the classical all-pairs quickest path problem, namely without the reliability considerations. It would be interesting to obtain all-pairs algorithms for other variations of the quickest path problem such as general bandwidth constraints [11], random queuing errors [8], dynamic bandwidth constraints [5] and various other routing mechanisms [10]. Another future direction is the on-line computation of quickest paths so that path's quality can be traded-off for computational speed.

References

1. A. V. Aho, J. E. Hopcroft, and J. D. Ullman, *The Design and Analysis of Computer Algorithms*, Addison Wesley Pub., Reading, MA, 1974.
2. Y. C. Bang, S. Radhakrishnan, N. S. V. Rao, and S. G. Batsell, On update algorithms for quickest paths, *Computer Communications*, vol 23, pp. 1064–1068, 2000.
3. G. H. Chen and Y. C. Hung, On the quickest path problem, *Information Processing Letters*, vol 46, pp. 125–128, 1993.
4. Y. L. Chen and Y. H. Chin, The quickest path problem, *Computers and Operations Research*, vol. 17, no. 2, pp. 53–161, 1990.
5. W. C. Grimmell and N. S. V. Rao, On source-based route computation for quickest paths under dynamic bandwidth constraints, *Journal of Foundations of Computer Science*, 2002, in press.
6. D. T. Lee and E. Papadopoulou, The all-pairs quickest path problem, *Information Processing Letters*, vol. 45, pp. 261–267, 1993.
7. J. F. Mollenauer, On the fastest routes for convoy-type traffic in flowrate-constrained networks, *Transportation Science*, vol. 10, pp. 113–124, 1976.
8. N. S. V. Rao, LetLets for end-to-end delay minimization in distributed computing over Internet using two-paths, *International Journal of High Performance Computing Applications*, vol. 16, no. 3, 2002.
9. N. S. V. Rao and S. G. Batsell. Algorithm for minimum end-to-end delay paths, *IEEE Communications Letters*, vol. 1, no. 5, pp.152–154, 1997.
10. N. S. V. Rao, W. C. Grimmell, S. Radhakrishnan, and Y. C. Bang, Quickest paths for different network router mechanisms, Technical Report ORNL/TM-2000/208, Oak Ridge National Laboratory, Oak Ridge, TN, 2000.

11. N. S. V. Rao and N. Manickam, General quickest paths and path-tables, *Computer Systems: Science and Engineering*, vol. 17, no. 4/5, pp. 235–239, 2002.
12. J. B. Rosen, S. Z. Sun, and G. L. Xue, Algorithms for the quickest path problem and the enumeration of quickest paths, *Computers and Operations Research*, vol. 18, no. 6, pp. 579–584, 1991.
13. G. Xue, End-to-end data paths: Quickest or most reliable?, *IEEE Communications Letters*, vol. 2, no. 6, pp. 156–158, 1998.

The Unified Design Methodology for Application Based on XML Schema

Yoon Bang Sung [1], Mun-Young Choi [2], and Kyung-Soo Joo [3]

[1]Hanseo University Computer Center,
Hanseo University Computer Center, SeoSan, ChungNam, Korea, 356-706
sybang@hanseo.ac.kr.
[2]Dept. of Computer Science, Graduate School Soonchunhyang Uni.
SoonChunHyang Uni., P.O Box 97, Asan, ChungNam, Korea, 336-745
E-mail : griffin@hyejeon.ac.kr.
[3]Dept.of Computer Science and Engineering, College of Engineering SoonChunHyang Uni.,
P.O Box 97, Asan, ChungNam, Korea, 336-745
gsoojoo@sch.ac.kr

Abstract. In this paper, we introduce a XML modeling methodology to design W3C XML schema using UML and we propose a unified design methodology for relational database schema to store XML data efficiently in relational databases. In the second place, in order to verify objectivity of the unified design methodology. Using Ronald Bourret's method, first we introduce the method of the transformation from XML schema to the object model and second we translate the object model into the relational database schema. Therefore we show the mutual consistency between those consequences, and so can verify the unified design methodology, we proposed in this paper.

1 Introduction

XML is a markup language for documents that includes the structured information[1]. W3C XML Schema can define more diverse data types than XML DTD, and it has a convenient advantage in using as a diverse application by using the powerful expression. The unified design modeling describes 3WC XML Schema using UML Class and store interchanged data by the modeling in RDB. However, until recently, there has been no effective means to verify objectivity of the unified design modeling. In this study, it describes a unified design methodology in Chapter 2, verification of unified design methodology in Chapter 3, and the conclusion in the last Chapter 4.

2 Unified Design Methodology Using UML

The sequential diagram takes on the role to clarify the flow of events from the application program and may convert the use case that is in the form of writing into the picture expressed in massage. In particular, it helps to make easy works when

P.M.A. Sloot et al. (Eds.): ICCS 2003, LNCS 2658, pp. 685–694, 2003.

making class diagram for us. Then it can be converted to XML modeling and data modeling as shown in the Fig.1 by the class diagram.

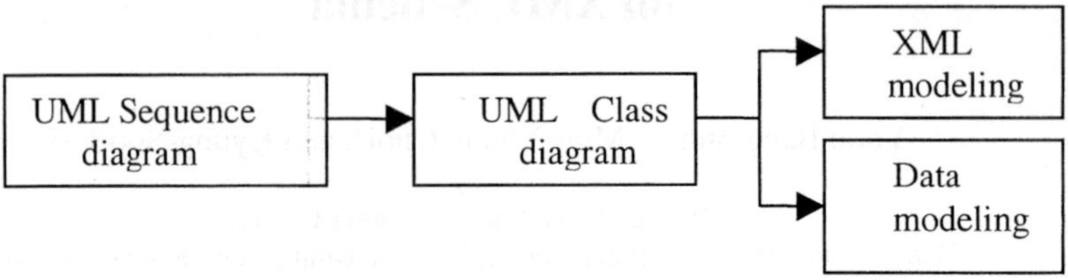

Fig. 1. XML modeling and Data modeling using UML

On Fig.2, the Order and Distributor are aggregate relationships in the relativity. However, It do not illustrate the process, only the information items which participate in the process and the relationships or dependencies between those objects[3].

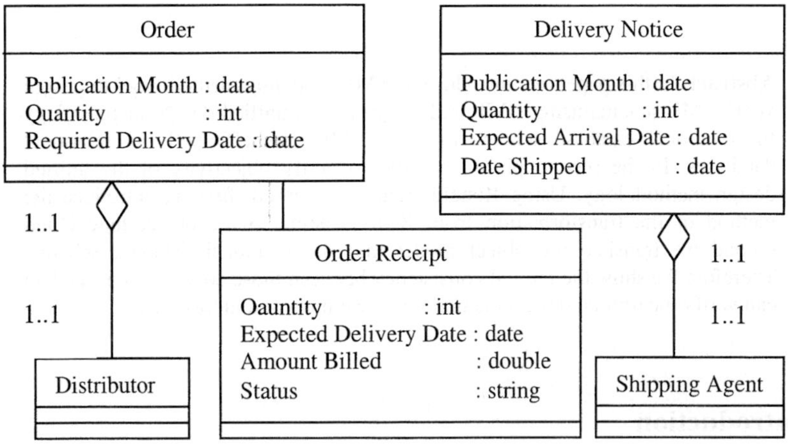

Fig. 2. Class diagram

2.1 XML Modeling

The following methods are used to formulate W3C XML Schema from the UML class proposed in the Fig.2[3].

1 Do not allow empty elements or attributes. Instead, ensure that they either have some value or are not present. If an element or attribute is mandatory(such an *minOccurs="1"* or *use="required"*), allowing an empty string as the value is questionable. If they are "*allowed*" to be empty, then it is better to set *minOccurs="0"* or *use="optional"*, and take their absence to mean "*no value*" ; if they do exist, force them to have some content.

2 Since whitespace can increase the size of an instance document significantly, it may be necessary for an application receiving an XML instance document to confirm the file size is under a certain limit.
3 Create limited versions of the base types *string*, *decimal*, and *intege*r in order to restrict character set and field size.
4 Make use of the *unique* element to assert uniqueness of element or attribute values.
5 Use *key / keyref* elements to associate two locations in a document. Any path in the document can be specified as a named *key*, such that all paths that are a *keyref* to that named *key* must contain a value specified as *key*.
6 If the schema is in danger of becoming large, consider moving sections into other files using the group reference method. The principle of abstraction applies equally here. These secondary schema files do not need to have even one global element.
7 Wherever we change namespaces in an instance document, use an *any* element and explicitly specify the namespace and number of child elements permitted at that location, or include an element *ref=". . ."* element to explicitly name the element in the new namespace.
8 If an element or attribute does not exist, does the application take this to mean there is a default value ? If so, this can explicitly be stated by using the *default* attribute on the element or attribute declaration.
9 Include the *blockDefault="#all"* attribute on the schema element. This will prevent derived types being allowed to substitute in an instance document. This is a security issue, and depends on the application's mechanism for schema validation.
10 Include the *finalDefault="#all"* attribure on the schema element. This will prevent any new types deriving from those declared in this schema. This is a much stricter constraint than *blockDefault.*

2.2 Example of XML Modeling

On Fig.2, the 'Order' object and the 'Distributor' object are aggregate relationships in the relativity and its meaning is that the 'Order' object has to have the 'Distributor', and accordingly, the 'Distributor' side is shown to have the multiplicity value of 1..1. On the other hand, the multiplicity value of the 'Order' object being 1..1 means that the 'Distributor' can exist depending on the orders. In order to make XML modeling, modeling with XML Schema by applying the XML modeling method of 1 and 4for XML Schema for Schema is the same as shown in Fig.3.

```
<element name="Distributor">
 <complexType>
   <attribute name="Distributor" type="ID" use="required"/>
 </complexType>
</element>
```

Fig. 3a. XML Schema of Distributor Object(Continute in the next page)

```
<element name="Order">
 <complexType>
  <sequence>
   <element name="Publication Month" type="date"
         minOccurs="1" maxOccurs="1"/>
   <element name="Quantity type="int" minOccurs="1"
         maxOccurs="1"/>
   <element name="Required Delivery Date" type="date"
         minOccurs="1" maxOccurs="1"/>
   <element REF="Distributor"
  </sequence>
    </complexType>
</element>
```

Fig. 3b. XML Schema of Order object(Continue in before page)

On Fig.3, the 'Order' object has 3 child objects and 1 reference object. Also, the 'Distributor' object that refers from the 'Order' object has no child object.

2.3 Data Modeling

The conversion methods of class diagram of Fig.2 into RDB Schema is as follows[8].

1 UML class becomes table.
2 UML attribute in class becomes column in table.
3 UML attribute type in class becomes column type in table through type transformation table.
4 IF nullable UML attribute tag, attribute has NULL constraint ; otherwise, NOTNULL constraint.
5 IF UML attribute has initializer, add DEFAULT clause to column.
6 For classes with no generalization(root or independent) and implicit identity, create integer primary key; for oid, addoid tagged columns to PRIMARY KEY constraint; ignore composite aggregation and association classes.
7 For subclasses, add the key of each parent class to the PRIMARY KEY constraint and to a FOREIGN KEY constraint.
8 For association classes, add primary key from each role-playing table to PRIMARY KEY constraint and FOREIGN KEY constraint.
9 IF alternate oid = <n> tag, add columns to UNIQUE constraint.
10 Add CHECK for each explicit constraint.
11 Create FOREIGN KEY columns in referencing table for each 0..1, 1..1 role in association.
12 Create PRIMARY KEY for composite aggregation with FOREIGN KEY to aggregating table(with CASCADE option), add additional column for PRIMARY KEY.
13 Optimize binary association classes by moving into to-many side table where appropriate.
14 Create tables for many-to-many, ternary associations with no association classes.
15 Create PRIMARY KEY, FOREIGN KEY constraints from keys of role-playing tables in many-to-many, ternary associations.

2.4 Example Conversion of RDB Schema

(1) The 'Distributor' object by Fig.2, It is converted as in Fig.4 with the 'Distributor' table that stores the object type attribute of 'DistributorID' following the characteristics of RDB conversion method 1, 2 and 6.

```
SQL> CREATE  TABLE Distributor(
  DistributorID INTEGER PRIMARY KEY
  )
```

Fig. 4. Distributor Table

(2) The 'Order' object by Fig.2, It stores the 'Order' object type attribute following the characteristics of RDB conversion method number 1, 2 and 6, and the object type of 'DistributorID' is applied with 'Distributor' defined under Fig.4 by conversion method number 11.

```
SQL> CREATE TABLE Order (
    OrderID       INTEGER    PRIMARY KEY
    Publication Month  date
    Quantity          int
    Required Delivery   date
    DistributorID  INTEGER  REFERENCE  Distributor
    ExpectedDeliveryDateID  INTEGER  REFERENCE
                            OrderReceipt
CONSTRAINT  Order_PK  PRIMARY  KEY
        (OrderID, DistributorID, Expected DeliveryDateID)
)
```

Fig. 5. Order Table

3 Verification of the Unified Design Methodology Using UML

In this section, we apply the existing Ronald Bourret's method in order to verify objectivity of the unified design methodology. It transforms XML schema into the object model and transform the object model into relational database schema. Therefore we show the mutual consistency between them and relational database chema according to relational data modeling using UML, and verified objectivity of design methodology.

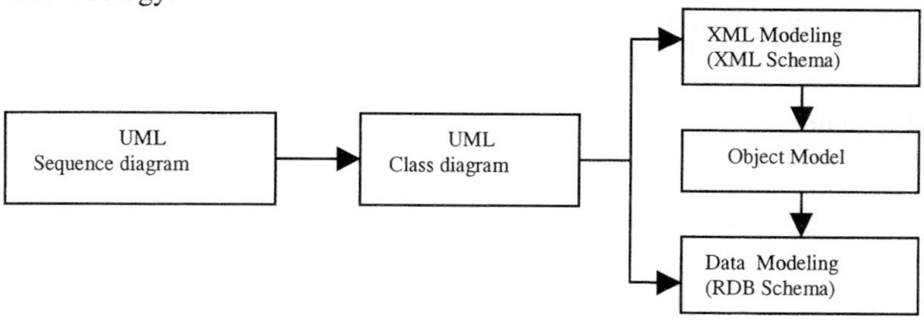

Fig. 6. Verification of design methodology

Fig.6 describes two different processes that transform XML Schema through the object model into the relational database schema. The first part of the process, transform XML Schema generally known as "XML data binding" to the object model. The second, transform object model known as "object-relational" mapping to the relational database schema[2].

3.1 Cyclic Graph Expression of XML Schema

The transformation XML Schema to object model is best understood by viewing an instance of the schema data model as a directed, possibly cyclic graph consisting only of schema, attribute, complex element type, simple element type, attribute group, model group. From this graph, the attribute and the simple element type nodes are transformed into the scalar types. the edges pointing from these nodes are transformed into the properties in these calsses[3]. Fig.7 is an example of XML Schema that transforms Fig.2 into XML modeling.

```
<element name="Order">
 <complexType>
  <sequence>
   <element name="Publication Month" type="date"
        minOccurs="1" maxOccurs="1"/>
   <element name="Quantity type="int" minOccurs="1" maxOccurs="1"/>
   <element name="Required Delivery Date" type="date"
        minOccurs="1" maxOccurs="1"/>
   <element REF="Distributor"/>
  </sequence>
 </complexType>
</element>
```

```
<element name="Distributor">
 <complexType>
   <attribute name="Distributor" type="ID" use="required"/>
 </complexType>
</element
```

Fig. 7. XML Schema of Distributor(Continue in before page)

3.2 Element

(<!ELEMENT>) is the fundamental component in XML DTD and it defines Tag used in XML document. Element in XML Schema is one of the basic components. We use to assign the items that support all each of property in <element>, such as an element name, type of Tag content, minimun/maximun number of occurrence of Tag. The element type is classified in two type. Simple element type has no children element and complex element type has children element and relationship between elements. Specially, it is useful to define complex type. Simple element type node is generally transformed into scalar data type. Complex element type node is transformed into class. The edges pointing element type node are transformed into property. Fig.8 shows the example that transformed Fig.7 into cyclic graph.

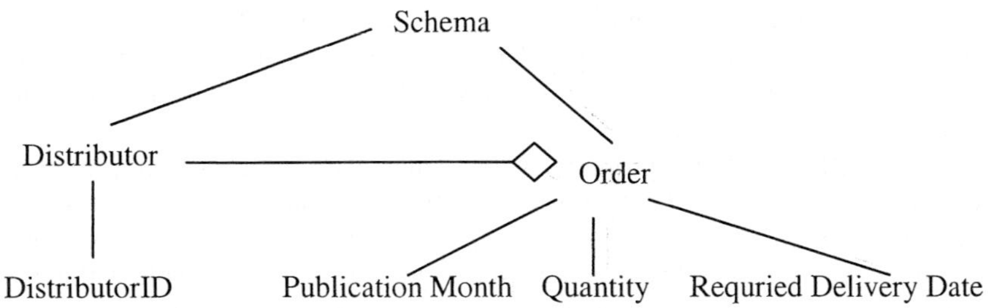

Fig. 8. Cyclic graph of schema component

3.3 Attribute

Every element can have an attribute and the attribute has an attribute name and type. The attribute is included in element. The attribute node is transformed into the scalar type and the edges pointing at the attribute node are transformed into their properties. The distributor element node has a property of DistributorID in Fig.9 and becomes a property of Order element node. The Distributor node pointing at Order node with edge(—◇) in Fig.8 represented in the object model to mean aggregation between two nodes.

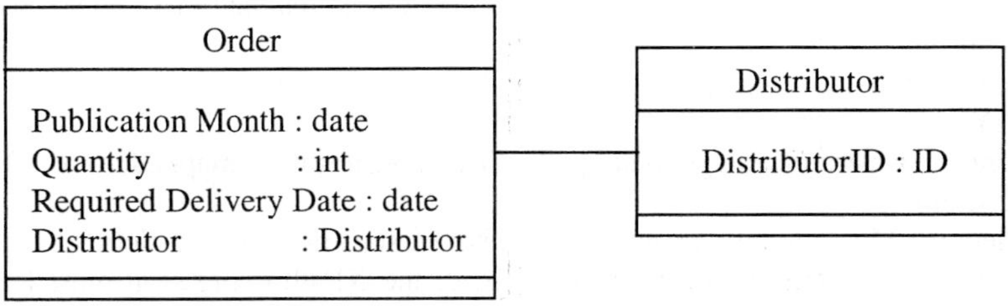

Fig. 9. Order and Distributor object

3.4 Transforming Object Model into RDB

We transformed XML Schema Example of Fig.7 in this section 3.1 into the object model. The following method describes how transformed object model are transformed into the relational database schema.

1 Class : Table. this is known as a class table. an object is represented by a row in a class table.
2 Inheritance : The superclass and subclass are mapped to separate tables with a unique Key/foreign key joining them. The unique key is in the superclass table. An object is represented by a row in each table.
3 Single-valued property with scalar data type : Column in class table. This is known as a property column. The data type determines the set of possible data types of the column. A property is represented by a value in a property column. The property table is joined to the class table with a unique key/foreign key relationship and the unique key is in the class table.
4 Multi-valued(collection) property with scalar data type : Multiple property columns in class table. Each position in the collection is mapped to a specific property column. Property column in a property table. There is one row for each value in the collection. The property table is joined to the class table with a unique key/foreign key relationship and the unique key is in the class table.
5 Single-valued property with union data type : There is one table for each data type in the union and the unique key is in the class table.
6 Single-valued property with class data type : The class containing the property is called the parent class and the class corresponding to the property is called the child class. Unique key/foreign key relationship between the tables of the two classes. The unique key is in the table of the parent class.
7 Multi-valued property with class data type : Unique key/foreign key relationship between the tables of the two classes. The unique key is in the table of the parent class. There is one row in the table of the child class for each property value.
8 Identity constraint : See sections 3.5.

3.5 Transforming Identity Constraint into RDB

1 Identity constraint category(key) : The constraint is mapped to a PRIMARY KEY constraint.
2 Identity constraint category(keyref) : The constraint is mapped to a FOREIGN KEY.
3 Identity constraint category(unique) : The constraint is mapped to a UNIQUE constraint.
4 Selector : Must identity a single table. That is, starting from the element type node on which the identity constraint is defined, the XPath expression must lead to a complex element type node.
5 Field : Must identity one or more columns in the table in 4. That is, starting from the node identified in 4, each XPath expression must lead to an edge leading from the node in 4 to simple element type node or an attribute node.
6 Reference key : The selector and fields properties of the referenced key identity constraint are used to determine the table and key columns referenced in a FOREIGN KEY constraint.

3.6 Example Conversion of RDB Schema

(1) The 'Order' object by Fig.9 is converted as in Fig.10 with conversion method that uses RDB conversion method number 1, 4 and RDB identifies constraint number 1 ~ 6. It shows the mutual consistency of Fig. 5 and Fig. 10.

```
SQL> CREATE TABLE Order (
    OrderID       INTEGER    PRIMARY KEY
    Publication Month  date
    Quantity         int
    Required Delivery   date
    DistributorID  INTEGER  REFERENCE  Distributor
    ExpectedDeliveryDateID  INTEGER  REFERENCE  OrderReceipt
CONSTRAINT  Order_PK  PRIMARY  KEY
        (OrderID, DistributorID, Expected DeliveryDateID)
```

Fig. 10. Order table

(2) The distributor' object by Fig.9 is converted as in Fig.11 with conversion method that uses RDB conversion method number 1, 3. It shows the mutual consistency of Fig.4 and Fig.11.

```
SQL> CREATE   TABLE  Distributor(
      DistributorID    INTEGER PRIMARY KEY
      )
```

Fig. 11. Distributor table

4 Conclusion

In order to store and manage the information mutual exchanged systematically and stably by using XML, there are a variety of studies on XML application and database linkage have been performed on the basis of relational database to this point. However, for there is a limit not to define a variety of data types of DTD in the XML application, there is a difficulty in smoothly linking it to the data base. Furthermore, there is an inherent limit in storing the XML data that has a variety of level structure into the relational database, the aggregation of 2-dimensional table.

In this paper, we proposed modeling methodology to store in relational database for structured information that transformed hierarchical architecture into 2-dimension information using conversion rule. We applied guideline that designed XML modeling and Relational data modeling using UML Class. However, it needs to verify objectivity to repose trust on modeling. In order to verify objectivity, first we transformed XML Schema into relational database and second we transform UML Class into relational database and third we showed the mutual consistency comparing with each other's consequence. For this goal first we represented XML Schema on cyclic graph and transformed edge with node into the object model and transformed object the model into the relational database. We used existing Ronald Bourret's method.

Acknowledgement. This work was supported by the University IT Research Supporting Program under the Ministry of Information Communication of Korea.

References

[1] What is XML ?, http://www.xml.com/pub/a/98/10/guide1.html#AEN58.
[2] Mapping W3C Schemas to Object Schemas to Relational Schemas, http://www.rpbourret.com/xml/SchemaMap.htm.
[3] Duckett Jon, Ozu Nik, Williams Kevin, Mohr Stephen, Cagle Jurt, Griffin Oliver, Norton Francis, Stokes-Rees Ian, and Tennison Jeni. Professional Xml Schemas , Wrox Pr Inc, 2001.
[4] Florescu, D., Kossmann, D.: Storing and Querying XML Data using an RDBMS. Data Engineering 22:3 (1999), 27–34.
[5] XML Modeling, http://www.xmlmodeling.com
[6] Modeling XML vocabularies with UML, http://www.xml.com/pub/a/2001/09/19/uml.html, 09/19/2001.
[7] Bang Sung-yoon, Joo Kyung-soo, "XML Application Design Methodology using Model of UML Class", Korear of Institute of CALS/EC, The Journal of Korean Institute of CALS/EC, Vol.7, No.1, pp.154–166, 2002.
[8] Bang Sung-yoon, Joo Kyung-soo, "A Unified Design Methodology using UML for XML Application based on Database", Korean Society for Information Management, Journal of the Korean Society for Information Management Vol.19, No.2, pp.50–67, 2002.

Automatic Recognition of Alzheimer's Disease Using Genetic Algorithms and Neural Network[1]

Sunyoung Cho[1], Boyeon Kim[2], Eunhea Park[3], Yunseok Chang[4], Jongwoo Kim[5], Kyungchun Chung[6], Weiwan Whang[5], and Hyuntaek Kim[3]

[1] Basic Science Research Institute, Chungbuk National University, Chungju, Korea, sycho@chungbuk.ac.kr

[2] Department of Electrical & Computer Engineering, Kangwon National University, Chuncheon, Korea

[3] Department of Psychology, Korea University, Seoul, Korea

[4] Department of Computer Engineering, Daejin University, Pocheon, Korea

[5] Department of Oriental Neuropsychiatry, Kyunghee University, Seoul, Korea

[6] Department of Neurology, Kyunghee University, Seoul, Korea

Abstract. We propose an Alzheimer's disease (AD) recognition method combined the genetic algorithms (GA) and the artificial neural network (ANN). Spontaneous EEG and auditory ERP data recorded from a single site in 16 early AD patients and 16 age-matched normal subjects were used. We made a feature pool including 88 spectral, 28 statistical and 2 nonlinear characteristics of EEG and 10 features of ERP. The combined GA/ANN was applied to find the dominant features automatically from the feature pool, and the selected features were used as a network input. The recognition rate of the ANN fed by this input was 81.9% for the untrained data set. These results lead to the conclusion that the combined GA/ANN approach may be useful for an early detection of the AD. This approach could be extended to a reliable classification system using EEG recording that can discriminate between groups.

1 Introduction

A number of quantitative EEG analysis have been used to detect the brain's functional changes in the Alzheimer's disease (AD). Investigators have extracted specific quantitative features from the EEG, which would be characteristics for each stage of this disease. Various spectral and nonlinear analyses were employed and some progress has been established [1-2].

To the spectral nature of the EEG changes in the AD, there is a general agreement that the earliest changes are an increase in theta and a decrease in beta mainly over

1. This work was supported by Korean Ministry of Health and Welfare, 00-PJ9-PG1-CO05-0002 and a result of research activities of Advanced Biometric Research Center (ABRC) supported by KOSEF.

P.M.A. Sloot et al. (Eds.): ICCS 2003, LNCS 2658, pp. 695–702, 2003.

parieto-occipital area, followed by a decrease of alpha activity [3-4]. Delta activity increases later in the course of disease [5]. Claus et al.(1998) reported that a slowing spectral EEG could predict the rate of subsequent cognitive and functional decline in the AD, using multiple linear regression analysis [6]. Intra- and inter-hemispheric EEG coherence, which is considered to be a measure of the cortical synchronization and possibly to reflect a functional status of the intracortical communication, was significantly lower in alpha and beta frequency in AD patients [7].

Recent progress in the theory of the nonlinear dynamics has provided new methods for the study of the time-series physiological data. The nonlinear analysis of the EEG data could be a useful tool to differentiate normal and pathologic brain state. Several studies of the EEG in AD patients estimated the correlation dimension (D2) and Lyapunov exponent (L1) [8-9]. They showed significantly lower values of D2 and L1 in AD than age-matched normal subjects, reflecting less complex signal dynamics.

Another useful quantitative electrophysiological assessment for the monitoring of the cortical function is the event-related potential (ERP). Since Goodin et al.(1978) demonstrated the prolonged latency in the P3 component with aging [10], many researchers have studied the ERP components in AD patients but this is still a matter of debate, and the diagnostic sensitivity and specialty of the ERP remain yet to be confirmed [11-12].

In this study, we propose an automatic AD recognition method combined the genetic algorithms (GA) and the artificial neural network (ANN), using the spontaneous EEG and auditory ERP recorded from a single site. The EEG and ERP were analyzed to compute their spectral features as well as statistical and nonlinear features, to make a feature pool. The combined GA/ANN approach was applied to select the dominant features that are most efficient to classify two groups. The selected features were used as a neural network input for training and testing the network.

2 Method

We adopted the artificial neural network as a usual classifier to discriminate the AD patients from the normal subjects, using the computed EEG and ERP features. Applying the ANN as an effective classifier, we have to find the optimum and minimum features as a network input. To solve this problem we used the genetic algorithm to find the dominant input features from a feature pool.

2.1 The Feature Pool

With the electrophysiological data of the AD patients and normal subjects, we made a feature pool that represents their data. The spontaneous EEG data were divided into 30s segments and each segment was analyzed to compute their spectral, statistical and nonlinear characteristics, to generate 118 features. The ERP data to target tone with an averaging epoch 1s including 100ms of the prestimulus baseline, were analyzed to generate 10 features that would describe the characteristic of the patterns.

The final feature pool includes as follow;

- 88 power spectral measurements: for example, the maximum power, the frequency at the maximum power, the accumulated and relative power, the mean and variance of the power in δ, θ, α, β, γ band separately
- 28 statistic measurements: for example, the average amplitude, the range between the maximum and minimum amplitude, the ratio between the maximum and mean amplitude, the variance
- 2 chaotic features: the central tendency, the box-counting dimension
- 10 ERP features: for example, the latency and amplitude of the largest peak, the left second peak and right second peak in 300-700ms post-stimulus, the difference of amplitude and latency

2.2 Design of the Chromosomes and the Fitness Function

In the genetic algorithms, the concepts of chromosome are used to encode and manipulate the solution [13]. Each chromosome defines an individual of a population. We set a chromosome as a string consisted of 35 constants, that are representing the feature number in the feature pool and that will be used as a network input.

With this input of a chromosome after learning the ANNs to every training segment, the fitness function gives back a value for the chromosome, which is measuring the performance on the solution. The fitness value of a chromosome was defined as the inverse of the sum of mean square errors of the ANNs, by equation 1 where N is the number of ANNs, m is the number of output nodes of ANNi , do_j is the desired output of output node j and no is the network output of output node j.

$$Fittness = 1/ \sum_{i=1}^{N} mean(\sum_{j=1}^{m} (do_j - no_j)^2) \tag{1}$$

2.3 The Genetic Operation

To create a population for a new generation, three basic genetic operations were used: crossover, mutation, and reproduction. One or two chromosomes in a formal generation were selected by the roulette wheel method as the parent chromosomes, with a probability based on its fitness value. In the crossover operation, two offspring chromosomes were produced from two parents by the one-point crossover. In the mutation, only one terminal of a parent chosen randomly would be mutated to generate a new chromosome. The reproduction operation copied a parent chromosome to the next generation.

2.4 The Combined GA/ANN Approach

1. Generate an initial population of the first generation with random proportions of the terminals.
2. Repeat the following steps until the terminating criterion has been satisfied. The evolution would be terminated, once a fitness value reach to 10,000.
   - Evaluate each chromosome in the population. After training the ANN using the features in the chromosome as a network input, it would be assigned a fitness value for each chromosome.
   - Create a population for the next generation by the genetic operations. These operations are applied to the chromosomes in a formal generation with the probabilities based on their fitness.
3. Choose a chromosome that has the maximum fitness value in each generation. Using these chromosomes chosen from several generations, we selected a dominant feature set. We made a histogram showing the number of the selection by these chromosomes for each feature, as shown in the figure 1, which would represent the significance of the feature to fit the solution. We selected the 35 dominant features in order of their significances.
4. Train and test the ANN with these dominant features as a network input.

Table 1 summarized the control parameters related to the execution of our combined GA/ANN approach.

Table 1. Control parameters of GA/ANN approach.

GA	the number of chromosomes in a generation	200
	the maximum number of generations	200
	crossover rate	0.95
	mutation rate	0.05
	reproduction rate	0.001
ANN	ANN model	multi-layered perceptron
	ANN learning rule	backpropagation
	the number of input node	35
	the number of output node	1
	the number of hidden layer	1
	the number of hidden node	13
	learning rate	0.1
	the maximum number of learning iteration	2000

3 Experiments and Results

3.1 Data Acquisition

Subjects. Two groups of the early AD patients and the age-matched normal subjects were studied. Sixteen AD patients were recruited from the oriental neuropsychiatric and neurological sections of Kyunghee University Hospital, aged between 61-82 (72 ± 6.4, mean ± SD). The patients with probable AD were diagnosed using the K-DRS (Korean-dementia rating scale) criteria [14] and their MMSE scores ranged from 15 to 27 with an average score of 19.5. The K-DRS consists of five subcategories including attention, initiation & preservation, construction, computation, and memory. Other medical conditions that are known to produce dementia were excluded following neurological and neuroimaging studies. None of the patients have been previously diagnosed with the psychiatric disorders, such as a depression, attention deficit, or schizophrenia, nor have they any history of significant head trauma, intracranial mass lesion, or any other neurological condition that could be associated with cognitive decline. Sixteen volunteers were the normal subjects aged between 61-78 (70 ± 5.3). They were carefully screened to eliminate individuals with the medical and neuropsychiatric disorder.

EEG recording. The EEG was recorded from an Ag-AgCl electrode placed at P4 based on the 10-20 system, referenced to linked earlobes with recording the EOG. The impedance of the electrodes was kept below 5 kΩ. The EEG was filtered (bandpass 0.5-35 Hz), amplified, digitized (250Hz) and stored on a hard disk for the off-line analysis. Spontaneous EEG with the eyes open was recorded for about 5min, of which artifact-free segments were selected for the analysis. The EEG data was divided into 30s segments, and in each segment 118 features were computed.

Auditory ERP procedure. Event-related potentials were acquired during the auditory oddball task. The stimuli consisted of series of sine-wave tones of 1 kHz and 1.5 kHz, each lasting 300ms. The two tones occurred respectively in 75% and 25% of the trials, in a random sequence. A total of 100 stimuli, including frequent 1 kHz tone and infrequent 1.5 kHz tone were delivered with an inter-stimulus interval of 0.8-1.2s. The subjects were instructed to count internally the number of the 1.5 kHz 'target' tones. The ERP data of the target tone that is averaged with an epoch of 1s including 100ms of the prestimulus baseline, were analyzed to generate 10 features to be included in the feature pool.

3.2 Selection of Dominant Features and Performance of Neural Network

We used 137 EEG segments from 11 AD patients and 10 normal subjects as a training data set. After training, the combined GA/ANN approach found the 35 dominant features including 24 spectral, 8 statistical, 1 nonlinear and 2 ERP features. Figure 1 shows a histogram for selecting the dominant features. It indicates the number of the selection by the 17 chromosomes that have the highest fitness value each in the last 17 generations. We selected the 35 dominant features in order of their heights, which were marked by the rectangular boxes in the figure.

The selected dominant features were applied as a network input to train the ANN again with 137 training EEG segments. After training, the weight values of the ANN were determined to able to test the performance of the network. The 72 EEG segments for the test were from 5 AD patients and 6 normal subjects. Table 2 reports the performance of the network for these untrained data set. For the EEG of the AD patients, the ANN recognized 22 segments out of the 30 test segments. And for the normal EEG, 37 segments out of the 42 test segments were recognized correctly. The 5 segments the network fail to recognize were all from one normal subject.

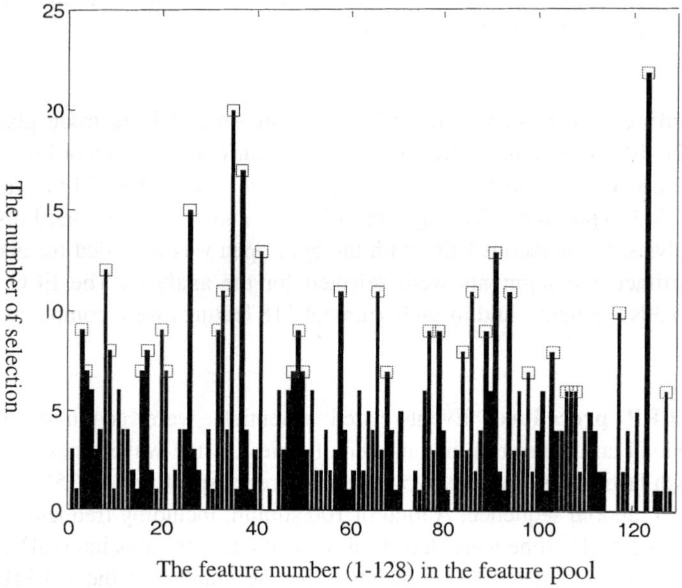

Fig. 1. The histogram for selecting the dominant features. The y axis indicates the number of the selection by the 17 chromosomes that have the highest fitness value each in last 17 generations. The rectangular boxes marked the selected dominant features (1-88: spectral features, 89-116: statistical features, 117-118: nonlinear features, 119-128: ERP features)

Table 2. Network performance for untrained segments

Target / NN output	Segments of AD patients	Segments of normal controls	Performance
AD	22	5	22/30 = 73%
Normal	8	37	37/42 = 88%
Total	30	42	59/72 = 81.9%

4 Discussion

The main goal for the clinical research in the AD is enhancement of the diagnostic accuracy and an earlier diagnosis. It is crucially important for the proper medical treatment and slowing down of the illness progress. We propose a reliable method to recognize the AD, only using one site of EEG recording. Single channel recordings of EEG are extremely easy and convenient to perform, even on the AD patients. If this simple tool has had enough accuracy to differentiate the AD patients from the normal adults, it would be highly helpful to diagnose the disease and to reduce costs. With a single channel EEG, our combined GA/ANN approach could find the dominant feature set and show good performance in determining the AD or normal EEG.

Our network was able to recognize successfully the EEG of normal subjects except with one subject. Even though her DRS and MMSE scores are in the normal range, the EEG of this subject could be deviant. In case of AD patients, 22 of the 30 segments were recognized correctly, so that the global recognition rate of the network was 73%. Note that it was only the case using the features from the 30s segments of the EEG data, and not with the every whole EEG data of each subject. With each subject, the whole EEG data consisted of 4-8 segments. The network failed to recognize only one or two segments with each AD patient. The remaining segments, that are more numerous, were recognized correctly as AD. Therefore, there should be no flaw whether a subject has AD or not. The spontaneous EEG of AD patients may vary a lot. In fact, the standard deviations in each segment (the 92nd statistic feature of the feature pool) of the AD group were significantly higher than those of the normal group ($F(1,206)=8.162$, $p<.005$). This feature was included in the dominant feature set selected by the GA.

We selected the dominant features by the genetic algorithms and used them as an optimum input of the neural network. We think this procedure enhanced the efficiency of the network. The redundant input from the raw EEG data or the manipulated data could make it rather worse. It was also useful to include the nonlinear characteristics and the ERP features in the feature pool, verified by the experimental results.

It seems reasonable to conclude that a single channel EEG data might be enough to recognize the AD using our combined GA/ANN approach. The suggested approach could be extended to a reliable classification system using EEG recording that can discriminate between groups.

References

1. Stam, C.J., Jelles, B., Achtereekte. H.A.M., Van Birgelen, J.H., Slaets, J.P.J.: Diagnostic Usefulness of Linear and Nonlinear Quantitative EEG Analysis in Alzheimer's Disease. Clin Electroencephalogr, Vol. 27.2. (1996) 69–77
2. Huang,C., Wahlund, L.-O., Dierks, T., Julin P., Winblad, B., Jelic, V.: Discrimination of Alzheimer's Disease and Mild Cognitive Impairment by Equivalent EEG Sources: A Cross-Sectional and Longitudinal Study. Clinical Neurophysiology. Vol. 111. (2000) 1961–1967
3. Prinz, P.N., Vitiello, M.V.: Dominant Occipital (Alpha) Rhythm Frequency in Early Stage Alzheimer's Disease and Depression. Electroencephalogr Clin Neurophysiol, Vol. 73. (1989) 427–432
4. Jelic, V., Shigeta, M., Julin, P., Almkvist, O., Winblad, B., Wahlund, Lo.: Quantitative Electroencephalography Power and Coherence in Alzheimer's Disease and Mild Cognitive Impairment. Dementia, Vol. 7. (1996) 314–323
5. Elmstaêhl, S., Roseân, I., Gullberg, B.: Quantitative EEG in Elderly Patients with Alzheimer's Disease and Healthy Controls. Dementia, Vol. 5. (1994) 119–124
6. Claus, J., Kwa, V., Teunisse, S., Walstra, G., Van Gool, W., Koelman, J., Bour L., Ongerboer De Visser, B.: Slowing on Quantitative Spectral EEG is a Marker for Rate of Subsequent Cognitive and Functional Decline in Early Alzheimer Disease. Alzheimer Dis Assoc Disord, Vol. 12.3. (1998) 167–174
7. Wada, Y., Y. Nanbu, Y. Koshino, N. Yamaguchi, T. Hashi-Moto.: Reduced Interhemispheric EEG Coherence in Alzheimer'S Disease: Analysis During Rest And Photic Stimulation. Alzheimer Dis. Assoc. Disord, Vol. 12. (1998) 175–181
8. Pritchard, W.S., Duke, D.W., Coburn, K.L.: Altered EEG Dynamical Responsivity Associated with Normal Aging and Probable Alzheimer's Disease. Dimentia, Vol. 2. (1991) 102–105
9. Jeong, J., Kim, S.Y., Han S.H., Nonlinear Analysis of Chaotic Dynamics Underlying EEGs in Patients with Alzheimer'S Disease. Electro Encephalogr Clin Neurophysiol, Vol. 106. (1998) 220–228
10. Goodin, D.S., Squires, K.C., Henderson, B., Starr, A.: Age-Related Variations in Evoked Potentials to Auditory Stimuli in Normal Human Subjects. Electroenceph clin Neurophysiol, Vol. 44. (1978) 447–458
11. Patterson, J.V., Michalewski, H.J., Starr A.: Latency Variability of the Compo-Nents of Auditory Event-Related Potentials of Infrequent Stimuli in Aging.Alzheimer-Type Dementia, and Depression. Electroenceph Clin Neuro-Physiol, Vol. 71. (1988) 450–460
12. Ray, P.G., Meador, K.J., Loring, D.W., Murro, A.M., Buccafusco, J.J., Yang, X.H., Zamrini, E.Y., Thompson, W.O., Thompson, E.E.: Effects of Scopolamine on Visual Evoked Potentials in Aging and Dementia. Electroenceph. Clin. Neurophysiol, Vol. 80. (1991) 347–351.
13. Davis, L.: Handbook of Genetic Algorithms. Van Nostrand Reinhold (1991)
14. Choi, J.: Expert Manual in Korean Dementia Rating Scale. (1998)

Traffic Characterization of the Web Server Attacks of Worm Viruses

Kihun Chong, Ha Yoon Song, and Sam H. Noh

School of Information and Computer Engineering, Hongik University
{khchong, song, noh}@cs.hongik.ac.kr

Abstract. With the explosive popularity of the Internet, the number of accessible web servers has proliferated as well. Subsequently, malicious attacks on these servers via viruses have become more prevalent. Due to the self-propagation and self-duplication nature of these viruses, such attacks can congest the network quickly, aggravating the already limited bandwidth available and curtail service provided by the server, eventually leading to denial of all services. The IIS, in particular, has been gravely affected by such Denial of Service (DoS) attacks. Hence, various methods to prevent such attacks from affecting the network and server have been researched and proposed. In this paper, we analyze the characteristics of worm virus attack traffics, by extracting and analyzing virus attack logs. With the use of various statistical methods, we show that worm attack patterns show self-similarity with Hurst parameter H. Our purpose is to use this characteristic in annulling the negative effects of worm attacks.

1 Introduction

The number of the web servers has proliferated with the explosive popularity of Internet, especially with the advent of the World Wide Web (WWW) and its various services. Simultaneously, computer virus attacks on web servers over the Internet have also increased. Due to viruses' charateristic to self-propagate, the number of attacking packets of viruses tend to increase exponentially, eventually leading to network congestion. Web servers, the IIS in particular, are usual victims of such Denial of Service (DoS) attacks. It is imperative to recognize the pattern of such virus attacks in order to take measure to defend from such attacks.

It has been well known that network traffic shows self-similarity. It is also well known that the WWW traffic shows self-similar charateristics[1]. However, there have been little research on self-similar characteristics of traffic generated by viruses that attack web servers. So, we analyzed the viruses' requests that attack web servers and measured whether the distribution of those requests seems to be self-similar. The requests of viruses are a subset of the total requests to web servers and the distribution of the total requests for web servers are self-similar. Hence, checking to see whether the distribution of the requests of viruses that attack web servers are self-similar is an interesting endeavor, as if it is so, this characteristic could be used to defend against such attacks. We analyzed

P.M.A. Sloot et al. (Eds.): ICCS 2003, LNCS 2658, pp. 703–712, 2003.

and measured the requests of the worms which attack the IIS using the logs made by the Apache web server. We were ablt to verify that the distribution of worms' requests are self-similar with self-similarity parameter H, that is, the Hurst parameter,.

The rest of this paper is organized as follows. The next section reviews related works. Section 3 introduces self-similarity and summarizes statistical tests for self-similarity. Section 4 describes worm viruses' properties and attacking pattern. We analyze the real data of logs of attacking web server in Section 5 and conclude with directions for future research in Section 6.

2 Related Works

There are a considerable number of statistical studies about traffics of network and WWW [1][2]. Early studies are about the self-similarity of network traffics, e.g. LAN, WAN, etc. [2][3], and later studies are about the self-similarity of WWW traffics [1], video stream traffics [4], wavelets [5], etc. For viruses, vaccines are made in general by analyzing occurrence day, properties, and factors of danger, and a real state of worms, recently run actively, is researched about the ways of attacks and recovery [6][7]. Therefore, the work presented here is different from existing studies that serves as a means to cure the viruses in that the study of traffic patterns of worm's attacking web servers is presented as a startng point to safeguard servers from work attacks. There are two basic related topics that relate to our research: fractal and self-similarity.

2.1 Fractal

Fractal is something that has characteristics of *self-similarity* and *recursiveness* [8]. Fractal's characteristic can be derived from a Fractal diagram. A Fractal diagram is similar to its sub-diagram and its area is finite but its circumference is infinite.

2.2 Self-Similarity

Self-similarity follows the self-similarity and recursiveness properties of a Fractal. A self-similar object is also similar to its subset from a statistical view. It manages a time process or a time series with time parameter t as shown in figure 1. Figure 1 shows a sort of network traffic and it is similar to a Fractal diagram with the change of the scale of time unit. This characteristic is called Self-similarity.

3 The Mathematics of Self-Similarity

We will summarize the mathematical and statistical techniques regarding self-similarity that will be used to analyze the traffic patterns of worm attacks.

3.1 Mathematical Foundation of Self-Similarity

Stationarity and second-order self-similarity are theoretical basis underlining self-similarity. This subsection describes the mathematical and statistical foundations of *stationarity* and the *derivation of self-similarity* [9].

Stationarity and Second-order Stationarity. We denote the time series as follows.

$X(t)$ is time series, where $t \in N$ and N is a natural number.

We denote the time series with a cumulative parameter of time t as $Y(t)$. We will then reserve $X(t)$ to be the increment process corresponding to $Y(t)$, that is,

$$X(t) = Y(t) - Y(t-1) . \tag{1}$$

For traffic modeling purposes, $X(t)$ has to be *stationary*, which means that the behavior and structure of $X(t)$ are the same as one that is shifted in some time. $X(t)$ is strictly stationary if $(X(t_1), X(t_2), \cdots, X(t_n))$ and $(X(t_{1+1}), X(t_{2+1}), \cdots, X(t_{n+1}))$ has the same joint distribution for all $n, t_n, k \in Z$. Hence, when we denote X_k as the k-shifted time process or time series, X and X_k are said to be equivalent in the sense of finite-dimensional distributions, and we denote it as follows.

$$X =_d X_i . \tag{2}$$

Stationarity is too strict to implement in real processes, henceforth a weaker form of stationarity is used in general and is called *second-order stationarity.* Second-order stationarity is weaker than stationarity, and it requires the *autocovariance function* as follows.

$$\gamma(r,s) = E\left[(X(r) - \mu)(X(s) - \mu)\right] . \tag{3}$$

When $\gamma(r, s)$ has to satisfy the *translation invariance* as follows.

$$\gamma(r,s) = \gamma(r+k, s+k) \qquad (\text{for all } r, s, k \in Z) . \tag{4}$$

Suppose that there are r, s which are finite, and we set $\mu = E[X(t)]$, $\rho^2 = E[(X(t) - \mu)^2)]$ for all $t \in Z$. And also assume $\mu = 0$, then $\gamma(r, s) = \gamma(r - s, 0)$ by stationarity, and it is denoted the autocovariance by $\gamma(k)$.

We define $X^{(m)}$, the aggregated process of X at aggregation level m, to formulate scale invariance as follows.

$$X^{(m)}(i) = \frac{1}{m} \sum_{t=m(i-1)+1}^{m \times i} X(t) . \tag{5}$$

That is, $X(t)$ is partitioned into nonoverlapping blocks of size m, their values are averaged and i is used to index these blocks. We denote $\gamma^{(m)}(k)$ to be the autocovariance function of $X^{(m)}$. Under the supposition of this second-order stationarity, we can define the second-order self-similarity.

Second-order Self-similarity. $X(t)$ is *second-order self-similar* with Hurst parameter H if

$$\gamma(k) = \frac{\rho^2}{2}((k+1)^{2H} - 2k^{2H} + (k-1)^{2H}) . \tag{6}$$

where $\gamma(k)$ is the autocovariance function of $X(t)$, ρ^2 is the variance of $X(t)$ and H is *Hurst parameter.* And $X(t)$ is *asymptotically second-order self-similar* when $X(t)$ satisfies the following.

$$\lim_{m\to\infty} \gamma^{(m)}(k) = \frac{\rho^2}{2}((k+1)^{2H} - 2k^{2H} + (k-1)^{2H}) . \tag{7}$$

Distributional Self-similarity. Consider the cumulative process $Y(t)$ mentioned in section 3.1. $Y(t)$ is *self-similar with self-similarity parameter H* $(0 < H < 1)$, that is, the Hurst parameter, denoted H-ss, if

$$Y(t) =_d a^{-H} Y(at) \qquad (^{\forall}a,\, t, \quad a > 0,\, t \geq 0) . \tag{8}$$

That is $Y(t)$, and its time scaled and normalized version have the same distribution. Suppose $Y(t)$ has these finite variances as follows.

$$\begin{aligned} &E\left[Y(t)\right] = 0, \\ &E\left[Y^2(t)\right] = \rho^2|t|^{2H}, \\ &\gamma(k) = \frac{\rho^2}{2}\left(|t|^{2H} - |t-s|^{2H} + |s|^{2H}\right) \end{aligned}$$

Thus, $Y(t)$ can be defined as follows.

$$Y(t) =_d t^H Y(1) \qquad (Y(t) =_d a^{-H} Y(at), \quad t = 1, a = t) . \tag{9}$$

And also suppose $X(t)$, we can show $X^{(m)}$ as follows.

$$\begin{aligned} X^{(m)} &= \frac{1}{m}\sum_{i=1}^{m} X(t) \\ &= m^{-1}(Y(m) - Y(0)) \\ &=_d m^{-1} m^H (Y(1) - Y(0)) \\ &= m^{H-1} X . \end{aligned}$$

Therefore, if $Y(t)$ shows H-self-similarity and has *stationarity increments,* denoted H-sssi, then its increment process $X^{(m)}$ satisfies the following.

$$X^{(m)} =_d m^{H-1} X \tag{10}$$

And $X(t)$ is *exactly self-similar* or *asymptotically self-similar* if

$$\lim_{m\to\infty} =_d m^{H-1} X \tag{11}$$

for all m, $m \geq 0$. In this paper, we say that X is self-similar with parameter H when X satisfies the following.

$$X_t =_d m^{-H} \sum_{i=(t-1)m+1}^{m} X_i \qquad \text{(for all } m \in \mathrm{N}) \tag{12}$$

And also, a time series X is self-similar with parameter H if it and $X^{(m)}$ have the same *autocorrelation function*, $\gamma(k) = \frac{E[(X_i-\mu)(X_{i+k}-\mu)]}{\rho^2}$. Thus, parameter H has to satisfy the following.

$$\frac{1}{2} < H < 1 \tag{13}$$

3.2 Statistical Tests for Self-Similarity

This section presents statistical methodologies for self-similarity. These statistical methodologies are variance-time plot, R/S plot, periodogram, etc. [1].

Variance-time plot. $\mathtt{var}(X^{(m)})$, the variance of $X^{(m)}$ for large number m, is plotted against m on a log-log plot. Suppose β is an approximate slope of the plotted distribution, and H satisfies $H = 1-\beta/2$. Then $X^{(m)}$ is self-similar with parameter H if $\frac{1}{2} < H < 1$.

R/S plot. The rescaled version $X^{(m)}$ by R/S is plotted against n on a log-log plot. R and S are denoted as follows.

$$\begin{aligned} S(n) &= \sqrt{\mathtt{var}(X^{(m)})}, \\ R(n) &= [\mathtt{max}(0, W_1, W_2, \cdots, W_n) - \mathtt{min}(0, W_1, W_2, \cdots, W_n)] \\ &\quad \left(W_k = (X_1 + X_2 + \cdots + X_k) - k\overline{X}(n), \qquad k = 1, 2, \cdots, n\right) \end{aligned}$$

Thus, H is the estimated slope of the graph on a log-log plot, and $X^{(m)}$ is self-similar with parameter H if $\frac{1}{2} < H < 1$.

Periodogram. The periodogram is plotted on a log-log plot. A periodogram can be formulated as follows.

$$I(\lambda) = \frac{1}{2\pi N} \left| \sum_{j=1}^{N} X_j e^{ij\lambda} \right|^2 \tag{14}$$

where λ is a period and N is the number of series. Suppose β is an approximate slope of the plotted distribution, and H satisfies $H = 1-\beta/2$. Then, $X^{(m)}$ is self-similar with parameter H if $\frac{1}{2} < H < 1$.

4 Worm Attacks against the Web Server over the Internet

Worm attacks against general servers as well as web servers have increased tremendously. In particular, the Internet Information Server (IIS) [10] is the most popular target of these web server attacks. IIS is a product of Microsoft and runs on the Windows operating system. Hackers frequently choose the IIS as their target due to its weak security. Though IIS provides security measures as fast as possible, new attacks continue to appear as rapidly as these new measures are provided.

A typical attack method of a virus is to infect the system and attack the file system. However, in recent years, the attacks are inflicted so that the infected system also becomes an attacker inflicting harm to neighboring systems or thereby congesting the network itself. Virus attacks on web servers also take this form, infecting all of clients connected to the infected web server or attacking the web server by a Denial of Service (DoS) attack. For example, Code Red and Nimda Worm(W32/Nimda worm) are a couple of the more well known notorious ones. Particularly, the Nimda worm, which appeared later than Code Red, acts intelligently by searching the back door, which was made by the Code Red virus. The attack patterns of the Nimda worm are described as follows.

```
(1)  GET /scripts/root.exe?/c+dir
(2)  GET /MSADC/root.exe?/c+dir
(3)  GET /c/winnt/system32/cmd.exe?/c+dir
(4)  GET /d/winnt/system32/cmd.exe?/c+dir
(5)  GET /scripts/..%5c../winnt/system32/cmd.exe?/c+dir
     ...
(6)  GET /scripts/..%2f../winnt/system32/cmd.exe?/c+dir
```

The attack logs between line (1) and line (4) are attempts to search the back door made by the Code Red virus and the other logs are to attack the weakness of the directory traversal.

5 Characterization of Worm Attack Traffics

We used log data that was made by the Apache web server of the Department of Computer Engineering, Hongik Univ. This section describes information of log data and the web server system, and explains the results of the data analysis.

5.1 Collection of Worm Attack Traffic Logs

The almost attacks against web servers are concentrated onto the MS Windows and IIS, and cannot attack UNIX systems and the Apache web server [11]. The failed worm virus attacks are written on the log files of the Apache web server, thus we could collect a large amount of logs. The size of raw log data is 343MB and the size of attack data is 13MB. The web server system runs on the Apache version 1.3.9 on the SUN SPARC station-10 using UNIX System V release 4.0 (SunOS 5.5). The duration for gathering the log data is from 10/1/2001 00:00 to 2/28/2002 24:00.

5.2 Self-Similarity of Worm Attack Traffics

Figure 1 shows the distribution of the log data at time unit 1, 10, 100, and 1000 of worm attack traffic patterns. The x-axis is for time and y-axis is for the number of attacks at time x. We can see that the distributions of each time unit are all similar in Figure 1. It seems to shows that the distribution pattern of web server attacks is self-similar.

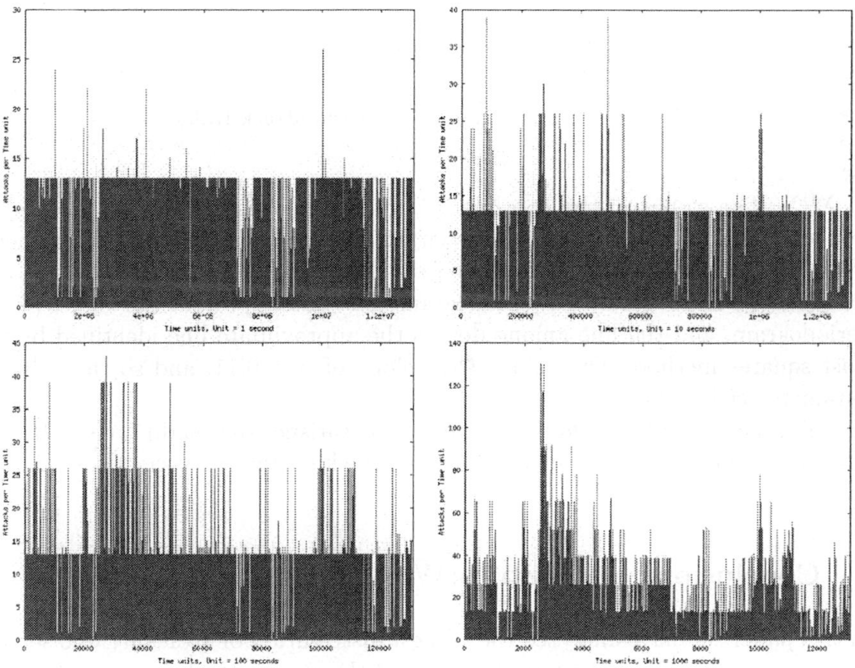

Fig. 1. The measured frequency of worm attacks over time domain.

Figure 2 shows a variance-time plot. The dotted line is the value of the variance and the solid line is the approximate line calculated by the method of least squares. We found the value of β on a log-log plot to be 0.65 and the value of parameter H to be 0.67.

Figure 3 shows a R/S plot. In Figure 3, concentrated plotting of dots express the value of r/s, and the value of n on the x-axis is the same as the value of m

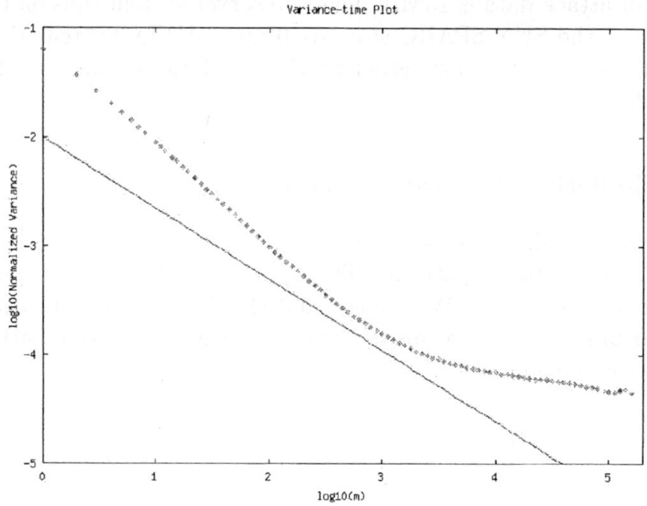

Fig. 2. Variance-time plot of worm attack traffic.

at $X^{(m)}$. The slopes of the two lines above and under the dots are 0.5 and 1, respectively, showing that H is between 0.5 and 1. The real value of parameter H is 0.57.

Finally, Figure 4 shows a periodogram. The dots represent the values of the periodogram, and the line among dots is the approximate line identified by the least squares method. The value of the slope of β is 0.11, and so the value of parameter H is 0.94.

All three methods shown that parameter H satisfies the condition $\frac{1}{2} < H < 1$. Therefore we may safely conclude that the attack pattern of worm viruses is self-similar with parameter H.

6 Conclusions and Future Works

In this paper, we have analyzed the request distribution of attacking web servers by worm viruses. From the fact that most of the worm attacks are concentrated on the IIS of Windows platforms, we could collect logs of worm attacks on the Apache web servers on UNIX systems. These logs were analyzed to obtain the characteristics of the attacks. We used statistical methods, e.g. variance-time plot, R/S plot and periodogram to test for self-similarity. We found the request distribution of worm attacks is self-similar with parameter H. Therefore we can determine worm attacks and defend against such attacks by using the value of parameter H.

For the future works, we plan to consider the possibility of detecting worm attacks in real time by using parameter H, thereby blocking or taking appropriate action to such attacks as it happens.

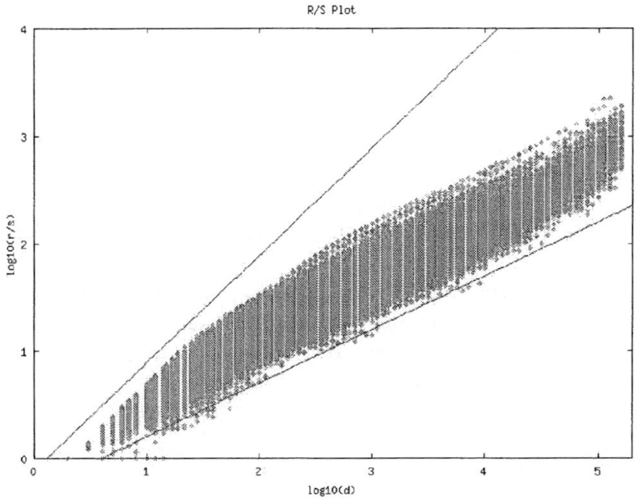

Fig. 3. R/S plot of worm attack traffic.

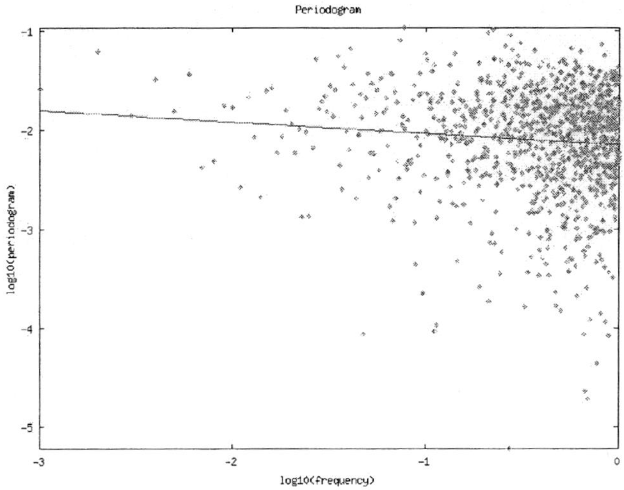

Fig. 4. Periodogram of worm attack traffic.

References

1. Mark E. Crovella and Azer Bestavros. Self-similarity in World Wide Web traffic: evidence and possible causes. *IEEE/ACM Transactions on Networking*, 5(6):835–846, December 1997.

2. Walter Willinger, Murad S. Taqqu, Robert Sherman, and Daniel V. Wilson. Self-similarity through high-variability: statistical analysis of Ethernet LAN traffic at the source level. *IEEE/ACM Transactions on Networking*, 5(1):71–86, February 1997.
3. Boris Tsybakov and Nicoals D. Georganas. On self-similar traffic in ATM queues: definitions, overflow probability bound, and cell delay distribution. *IEEE/ACM Transactions on Networking*, 5(3):397–409, June 1997.
4. Mark W. Garrett and Walter Willinger. Analysis, modeling and generation of self-similar vbr video traffic. In *Proceedings of ACM SIGCOMM 94*, pages 269–280, 1994.
5. Patrice Abry and Darryl Veitch. Wavelet analysis of long-range-dependent traffic. *IEEE Trans. Information Theory*, 44(1):2–15, January 1998.
6. Korea Information Security Agency. `http://www.kisa.or.kr`.
7. Ahnlab, Inc. `http://www.ahnlab.com`.
8. Benoit B. Mandelbrot. *The Fractal Geometry of Nature.* W.H. Freeman & Company, New York, 1977.
9. Peter J. Brockwell and R. A. Davis. *Time Series: Theory and Methods (Springer Series in Statistics).* Springer Verlag, New York, February 1991.
10. Microsoft Corporation. Internet Information Services Features. `http://www.microsoft.com/windows2000/server/evaluation/features/web.asp`.
11. Apache Software Foundation. The Apache Web Server. `http://www.apache.org`.

An Object-Oriented Software Platform for Examination of Algorithms for Image Processing and Compression

Bogusław Cyganek and Jan Borgosz

University of Mining and Metallurgy
Department of Electronics
Al. Mickiewicza 30, 30-059 Kraków, Poland
{cyganek, borgosz}@uci.agh.edu.pl

Abstract. This paper presents a design and implementation of the innovative software development system for image processing and compression. This platform allows to make 3D image processing as well as fundamental operations on images. One of the most important features of this system, that we will focus on in this paper, is its ability to evaluate performance of different stereo matching methods. This is accomplished by a special software module that verifies machine computed disparities with values provided by a person. Additionally, due to OOP technology, the software constitutes an open architecture that allows for easy addition of new components for image processing and compression. The presented platform was verified experimentally and also compared with existing commercial packages.

1 Introduction

This paper reports the work on development of the software platform for the 2D and 3D image processing and compression. Part of the functionality of the presented software supports fundamental image operations, like arithmetic operators, filtration, etc., as well sophisticated stereo processing, scale space based methods and neural networks [1][2]. The described platform was designed and implemented by means of object-oriented programming techniques as well as with multiple design patterns [6]. Thanks to this feature, almost all system components can be modified and changed independently of each other.

The next very important feature of the presented system poses the module for qualitative measure of performance of a stereo image algorithm. This feature is accomplished by a verification of machine computed disparities with values provided by a person thus allowing for quality assessment of a given stereo method.

In this paper we present a description of those system modules that deal with stereo image processing and stereo quality verification, focusing on their functionality rather than programming details. We start from the description of the class hierarchy and end with the explanation of the user interface and the disparity verification module.

P.M.A. Sloot et al. (Eds.): ICCS 2003, LNCS 2658, pp. 713–720, 2003.

2 Description of the Software Platform

2.1 Representation of Images

A flexible data structure for image representation is crucial for efficient image processing. There must be a trade off between different input formats of images and their internal representation. Further, considering allowable size of images and time complexity of algorithms, it has been chosen that images will be represented internally as square matrices, programmatically denoted by the base template class `TImageFor<T>` [11], where `T` stands for a given data type chosen for representation of a single pixel. The class notation used hereafter complies with the UML notation [5]. Fig. 1 depicts the class hierarchy for the internal digital image representation.

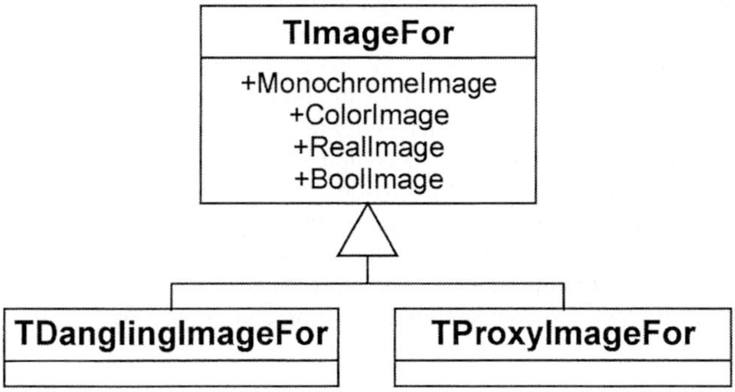

Fig. 1. The template class hierarchy for representation of digital images

The four instances of the base class in Fig. 1 are as follows: `MonochromeImage` for monochrome pixels of one-byte depth, `ColorImage` for color pixels of three-bytes depth, `RealImage` for pixels in floating point format suitable for precise computations such as convolution, and `BoolImage` for binary images.

There are two derivatives of the base class: `TDanglingImageFor` and `TProxyImageFor`. The former is used when at each pixel it is necessary to store a different length list of values rather than a single value. The latter approximates image-in-image concept that enables operations on a part of a given image as if it were a separate image by itself – however, there is only one underlying data storage. This feature creates a data abstraction that facilitates a uniform treatment of all images despite of their real representation.

2.2 Image Processing Modules

The image processing module, represented by the class hierarchy in Fig. 2, is the most important part of the evaluation platform. Due to its uniform interface, each algorithm can be used basically in the same manner, at the same time hiding all implementation details. Such an approach contributes to the platform independence and allow for easy addition of new classes implementing another image processing algorithms, such as stereo methods.

The class hierarchy interface is defined in the base class `TTwoMonochromeImageProcessor` that exposes the virtual method:

```
virtual MonochromeImage * operator() (
                 const MonochromeImage & leftImage,
                 const MonochromeImage & rightImage );
```

It is implemented in the form of the binary call operator, that takes two references to images as its input, and outputs a pointer to the outcome image [11].

The auxiliary three classes: `TAddGenerator`, `TSubtractGenerator`, and `TXORGenerator` are used simply to add, subtract or exclusively-or the two input images.

The derived class named `T_Disparity_Generator` builds a foundation for all point-oriented stereo processing algorithms [3]. All its derivatives are specializations for point-oriented stereo algorithms. These are: `T_SAD_Metric_Matching` that uses the *SAD* measure, `T_SSD_Metric_Matching` with *SSD* measure, `T_Census_Metric_Matching` that relies on *Census* measure, `T_Rank_Metric_Matching` with *Rank* measure [3]. At the same time `T_Census_Metric_Matching` is a base for all neural oriented stereo methods, such as `TNeuralCensus` that implements the Hopfield neural network and its derivatives `TCastingNetNeuralCensus` and `TCastingNetNeuralCensus` that implements a modified versions of the Hopfield neural network [4].

There are also two classes for the tensor based stereo methods [1]. The first one, named `TTensorStereo`, implements the point-oriented version of this method. The second, `Tensor_Based_Algorithm`, implements the disparity-oriented version of the tensor stereo method.

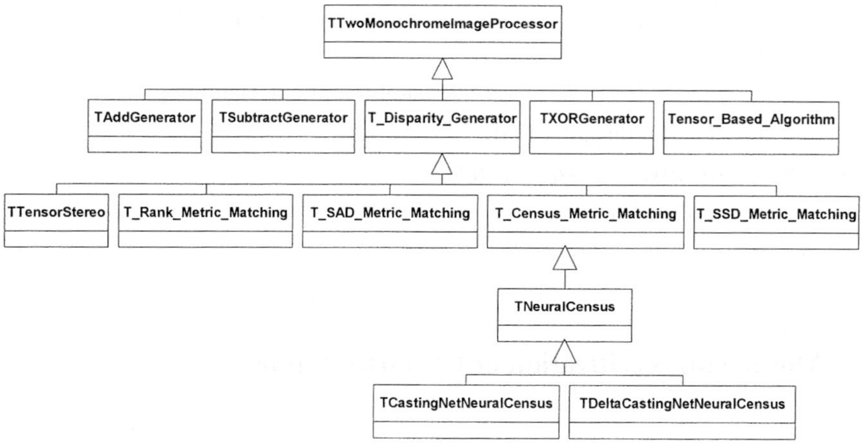

Fig. 2. The main class hierarchy for image processing modules

There are many additional classes that support the aforementioned hierarchies. All classes are well documented, so there is no trouble with the software improvements. Additionally due to the object-oriented programming style, changes in one module do not affect other software parts. This feature protects user against software bugs, and at the same time increases development speed of new components.

2.3 User Interface and Its Functionality

Fig. 3 depicts user interface of the presented system. Visible are different kinds of windows, such as: input stereo pair, disparity map, binary representation of an image, etc. User interface was implemented with the Microsoft's Visual C++ 6.0 and the Microsoft Foundation Classes. Exemplary output data from the disparity verification module are presented in Fig. 4.

The following list presents functional specification of the stereo processing related part of the software:

- loading single and stereo images in many recognizable formats,
- preprocessing on the loaded images: filtration with median, gauss, binominal and user mask, extraction of edges
- histogram generation,
- level adjusting based on histogram, brightness adjustment,
- operations like add, subtract and XOR on the loaded images,
- image value thresholding,
- correction picture geometry,
- test image generation for the stereo processing algorithms
- standard disparity calculation algorithms (*SAD*, *SSD*, *SCP*, *SSD-N*, *SCP-N*, *ZSSD-N*, *CoVar*, *Census*, *Rank*, *Tensor, etc.*),
- neural network based disparity calculator (Hopfield, Hamming),
- advanced tensor disparity finder,
- stereo map cross-checking module
- verification module,
- injection of different types of noise.

3 Module for Verification of Disparity Values

In practice, quality estimation of a computed disparity map is a very difficult task, because there is no objective measure and all known metrics require a true disparity map which is usually known only for synthetic stereo images. Many researchers tried to solve this problem (e.g. Ishikawa and Geiger [8], Leclerc, Luong and Fua [9]), but it still seems to be open.

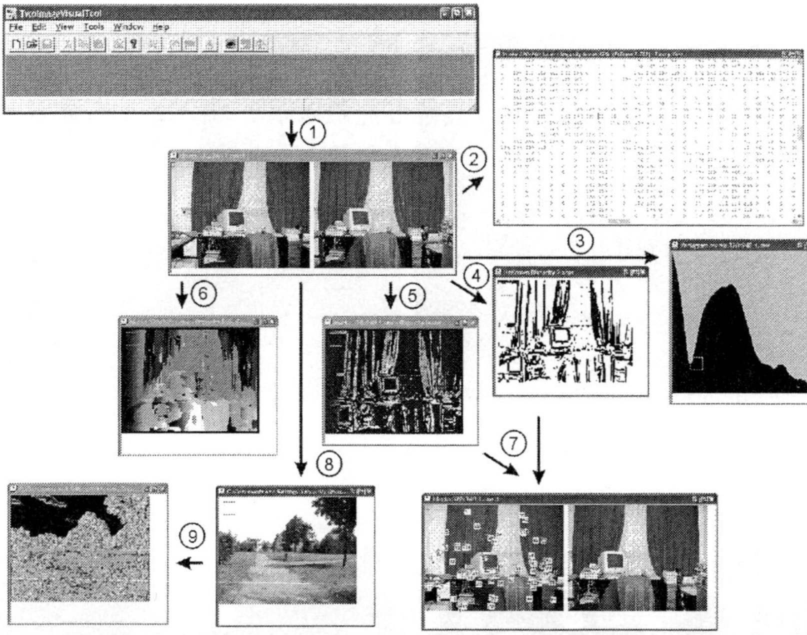

Fig. 3. Example of the view windows of the user interface: 1) loaded stereo-pair window, 2) binary view of the selected part of the image, 3) generation of the stereo–variogram, 4) tensor processing (local structures), 5) stereo tensor processing (disparity for the places with structure), 6) stereo – processing with *SAD*, 7) disparity quality verification, 8) a bitmap loaded independently and filtered with 3x3 median filter, 9) tensor angular component for the loaded bitmap

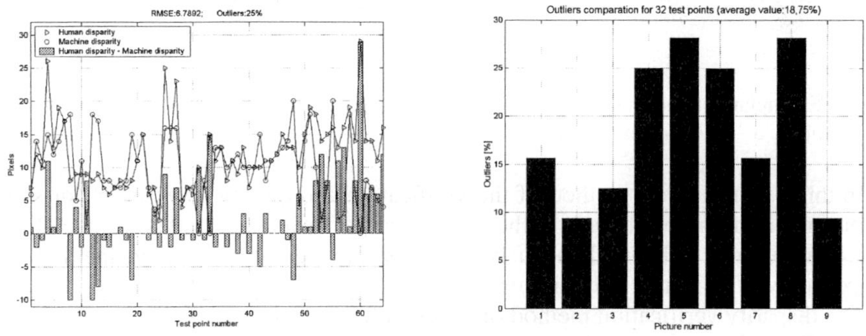

a b

Fig. 4. Results of the disparity quality verification module: a – a bar chart with marked differences between machine and human generated disparities for test points, b – a bar chart with stereo matching outliers ratio for tested stereo-pairs

Fig. 5. Exemplary stereo images with marked test points and epipolar line (green color) for a chosen pixel: a – „*Phone*", b – „*Punto*"

In this paper the new method of the verification and quality assessment for stereo algorithms is proposed, based on the comparison of matching results produced by a stereo algorithm and simultaneously by a testing person – similar method was proposed by Wei, Brauer and Hirzinger [12].

The disparity verification method operates as follows: test points are generated in a Monte Carlo fashion but only at image regions that are highly textured. Those areas are computed by means of the structural tensor operator [1], which is quite different and much faster than the method proposed by Wei, Brauer and Hirzinger [12]. Then disparity values are found at the generated test points. This is done independently by a testes stereo algorithm and a testing person. Both disparity sets are then compared and the final quality measure of the tested algorithm is computed. The method operates under tacit assumption that human generated disparity values outperform in quality those generated by a machine. This is true at least in an aspect of evident false matches, frequently encountered in machine stereo processing.

The presented software exposes the following interface for disparity verification:

- generation of chosen number of test points,
- epipolar line drawing for each test point,
- guided selection of a point lying on an epipolar line,
- calculation of error after testing
- export of results in text and graphical formats.

Exemplary windows involved in a disparity verification process can be seen in Fig. 5.

4 Comparison with Other Image Processing Platforms

The short comparison of the presented and other software platforms for the image processing was summarized in Table 1. The two flexible and user modifiable platforms were considered: Intel® Software Library [7] and Matlab® Image Processing Toolbox [10].

Table 1. Comparison of Image Processing Platforms

Feature	Intel® Library	Matlab®	This Platform
OS platform	Windows	Windows, Unix	Windows
User interface	Left for user development	Left for user development	Prepared, easily modifiable
Programming language	C++	Matlab script	C++
Programming style	Object	Classic/Object	Object
Library contents	From basic to advanced algorithms	Basic operations	From basic to advanced algorithms, focused on the stereo images processing
Speed	High	Low	High
Test application development cost	Medium	Low	Medium
Advanced application development cost	Medium	High	Low
Platform documentation	Rich	Rich	Rich

5 Conclusions

This paper presents structure and experimental results of the software development platform and its stereo processing capabilities. The described solution of the evaluation platform proves to be very useful and easily extensible. At the other side, execution times have been verified to be shorter than times of corresponding algorithms which run with commercially available scientific packages.

Simultaneously, a researcher has a freedom of easy addition of almost any new ideas, such as for example new neural network configurations.
The It is strongly recommended to well design all modules prior to the programming. Especially important is a choice of the interface methods chosen for each class. Also, a usage of the design patterns can greatly help and clarify the design stage.
The presented software framework was extended recently to the other image processing functions, such as image compression. In this new form it is used also by students of our faculty for implementation and testing of their image processing ideas.

Acknowledgments. This paper and its publication were sponsored by the Polish State Committee for Scientific Research (KBN) grant number 4 T11D 005 23.

References

1. Cyganek, B.: Stereo Matching with Tensor Representation of Local Neighborhoods. Ph.D. thesis (in Polish). University of Mining and Metallurgy, Kraków, Poland (2001)
2. Cyganek, B.: Neural Networks Application to the Correlation-Based Stereo-Images Matching, Proceedings of the EANN '99, Warsaw, Poland (1999) 17–22
3. Cyganek, B., Borgosz, J.: A Comparative Study of Performance and Implementation of Some Area-Based Stereo Algorithms, CAIP'2001, Warsaw, Poland (2001) 709–716
4. Cyganek, B., Korohoda, P.: Improved Neural Network for Fast Extraction of Information from Stero-Images, Fourth Conference Neural Networks, Zakopane, Poland (1999)
5. Douglass, B.P.: Doing Hard Time. Developing Real-Time Systems with UML, Objects, Frameworks, and Patterns, Addison-Wesley (1999)
6. Gamma, E., Helm, R., Johnson, R., Vlissides, J.: Design Patterns. Elements of Reusable Object-Oriented Software, Addison-Wesley (1995)
7. Intel: Documentation for the Intel Image Processing Library, Intel (2000)
8. Ishikawa, H., Geiger, D.: Occlusions, Discontinuities, and Epipolar Lines in Stereo. ECCV' 98, 5th European Conference on Computer Vision, Vol. 1, June (1998)
9. Leclerc, Y.G., Luong, Q.-T., Fua, P.: Self-Consistency: A Novel Approach to Characterizing the Accuracy and Reliability of Point Correspondence Algorithms. Artificial Intelligence Center, SRI International (1999)
10. Matlab: Image Processing ToolBox Documnetation, MathWorks (2002)
11. Stroustrup, B.: The C++ Programming Language. 3rd edition, Addison-Wesley (1998)
12. Wei, G-Q., Brauer, W., Hirzinger, G.: Intensity- and Gradient-Based Stereo Matching Using Hierarchical Gaussian Basis Functions. IEEE Transactions on Pattern Analysis and Machine Intelligence, Vol. 20, No. 11, November (1998)

Combined Detector of Locally-Oriented Structures and Corners in Images Based on a Scale-Space Tensor Representation of Local Neighborhoods of Pixels

Bogusław Cyganek

University of Mining and Metallurgy
Department of Electronics
Al. Mickiewicza 30, 30-059 Kraków, Poland
cyganek@uci.agh.edu.pl

Abstract. Detection of low-level image features such as edges or corners has been an essential task of image processing for many years. Similarly, detectors of such image features constitute basic building blocks of almost every image processing system. However, today's growing amount of vision applications requires at least twofold research directions: search for detectors that work better than the other, at least for a chosen group of images of interest, and – at the other hand – search for new image features, such as textons or oriented structures of local neighborhoods of pixels. In this paper we present a new approach to the old problem of corner detection, as well as detection of areas in images that can be characterized by the same angular orientation. Both detecting techniques are based on a scale-space tensor representation of local structures, and present computationally attractive image feature detectors.

1 Introduction

The paper addresses the problem of detection of locally-oriented structures and corners in images or sequences of images. Although, the problem of feature detection has been studied for many years and by many researchers [5][4][9][7][8][15][16], there is still growing demand on computationally efficient algorithms. One of the measures of a detector efficiency is its time-memory complexity, but also or even more frequently, its robustness in respect to different scenes for real images. The latter can be stated in terms of a repeatability and detection quality in respect to the amount of thresholds that need to be defined for a given detector. Very profound overview of over six corner detectors, as well as advanced methods of their examination, was presented by Schmid and Mohr [11].

The combined method for detection of corners and locally-oriented structures, presented in this paper, belongs to a group of signal based methods. For corner detection it can be conceived as a modified version of the Harris detector [5], mostly due to the precise computation of directional derivatives. Simultaneously, this method allows for better localization of image features by restricting a detection area to contours found in images. The method can be easily applied not only for detection of

P.M.A. Sloot et al. (Eds.): ICCS 2003, LNCS 2658, pp. 721–730, 2003.

corners, but also for detection of common pixel clusters that are characterized by the same local orientation.

The presented theoretical foundations, as well as practical implementation issues, came from the research in the field of stereo processing methods and a quest for an image feature detector that lets avoid false matches during stereo processing [3]. However, the presented concepts are by no means limited only to the stereo processing. They can be used in almost any image processing system, such as for example image and video indexing which consists of retrieval of interested points that are further used in a database search. Comparative video indexing method, based on a multiresolution contrast energy, presents paper by Bres and Jolion [2].

2 Tensor Detector of Local Structures in Pixel Neighborhoods

Let us analyze an image with local neighborhood U defined around a point x_0 (Fig. 1) where each point has been additionally endowed with a directional vector, e.g. an intensity gradient.

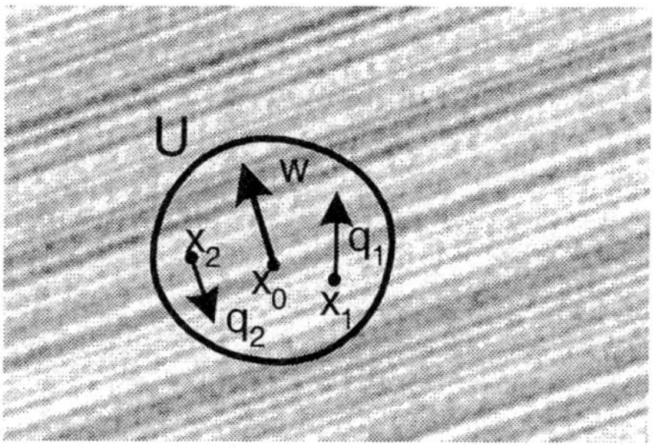

Fig. 1. Local neighborhood of a pixel x_0 with shown local gradient vectors q_i and structural vector w

The goal now is to find such a vector w that in a uniform way represents all other directional vectors q_i from $U(x_0)$. In order to compare vectors we use their inner product.

Additionally we assume the following:

1. Direction of w is invariant under rotation of π radians;
2. Angle and module of w follow signal changes in an image;
3. There is an additional measure of coherency of the local structure.

Thus, the vector w at a point x_0 is an estimator of an average orientation in a neighborhood $U(x_0)$ that maximizes the following formula [6]:

$$Q = \int_{U(x_0)} \left(q^T(\vec{x})w(\vec{x}_0)\right)^2 d\vec{x}. \tag{1}$$

The square of the inner product in (1) fulfills the invariant assumption on rotation of π radians. Otherwise parallel and anti-parallel configurations of vectors would cancel out.

Let us introduce a symmetric tensor **T**, defined as follows:

$$\mathbf{T}(\overrightarrow{x_0}) = \int_U \vec{\mathbf{q}}(\vec{x})\vec{\mathbf{q}}^T(\vec{x})d\vec{x}, \tag{2}$$

where $q(x)\ q^T(x)$ stands for an outer product of vectors and U is a local neighborhood of pixels around the point x_0. Components of **T** can be described by the following formula:

$$T_{ij} = \int_U q_i(\vec{x})q_j(\vec{x})d\vec{x}. \tag{3}$$

Taking into account (1) to (3), the problem of finding structural vector w is reduced to the solution of the maximization problem which is as follows [6]:

$$\max_w(Q) = \max_w(\vec{w}^T \mathbf{T}\vec{w}). \tag{4}$$

Expression (4) is fulfilled if w is an eigenvector corresponding to the maximum eigenvalue of **T**. Thus, the problem of finding structural vector w reduces to the analysis of eigenvalues of **T**. It can be solved analytically, as follows [3]:

$$\lambda_{1,2} = \frac{1}{2}\left(\left(T_{xx} + T_{yy}\right) \pm \sqrt{\left(T_{xx} - T_{yy}\right)^2 + 4T_{xy}{}^2}\right), \tag{5}$$

whereas trace Tr(**T**) of **T,** which will be helpful further on during the classification of image structures, can be expressed as:

$$\lambda_1 + \lambda_2 = T_{xx} + T_{yy} = \mathrm{Tr}(\mathbf{T}). \tag{6}$$

Based on (5), (6), and basic algebra we can try to classify local structures in images based on eigenvalues and trace of **T**. Results summarizes the Table 1.

Table 1. Classification of local structures based on eigenvalues and rank of **T**

Rank of **T**	Eigenvalues	Type of a local structure in an image
0	$\lambda_1 = \lambda_2 = 0$	Constant intensity value.
1	$\lambda_1 > 0,\ \ \lambda_2 = 0$	Ideal local orientation. The eigenvector, corresponding to the eigenvalue different from zero, points in a direction of maximum changes of intensity.
2	$\lambda_1 > 0,\ \ \lambda_2 > 0$	Both eignevalues greater than zero mean changes in all directions of a local neighborhood. The structure vector can be found if one eigenvalue is dominating.
2	$\lambda_1 = \lambda_2 > 0$	An isotropic gray value structure. Intensity changes equally in all directions.

Eigenvectors of **T** are formed as a columns of an adjoint matrix $[\mathbf{T}\text{-}\lambda\mathbf{1}]_{ad}$, as follows:

$$[\mathbf{a} \quad \mathbf{b}] = \begin{bmatrix} a_1 & b_1 \\ a_2 & b_2 \end{bmatrix} = [\mathbf{T} - \lambda_i \mathbf{1}_2]_{ad} = \begin{bmatrix} T_{yy} - \lambda_i & -T_{xy} \\ -T_{xy} & T_{xx} - \lambda_i \end{bmatrix}, \tag{7}$$

where *a* and *b* are linearly dependent eigenvectors, λ_i denotes an eigenvalue of **T**. Taking into an account (5) we obtain:

$$\begin{bmatrix} a_1 & b_1 \\ a_2 & b_2 \end{bmatrix} = \begin{bmatrix} \dfrac{(T_{yy} - T_{xx}) \pm \sqrt{(T_{xx} - T_{yy})^2 + 4T_{xy}^2}}{2} & -T_{xy} \\ -T_{xy} & \dfrac{-(T_{yy} - T_{xx}) \pm \sqrt{(T_{xx} - T_{yy})^2 + 4T_{xy}^2}}{2} \end{bmatrix} \tag{8}$$

A proper sign of the square root in (8) is chosen based on an eigenvalue.

In order to fulfill the rotation invariant assumption, an angle of the searched structural vector must change as a doubled angle (denoted further on as ξ) of an eigenvector corresponding to the maximum eigenvalue. Thus, from the simple tangent relationship

$$\mathrm{tg}(2\xi) = \frac{2\,\mathrm{tg}(\xi)}{1 - \mathrm{tg}^2(\xi)} \quad , \quad \mathrm{tg}(\xi) \neq 1, \tag{9}$$

and from (8) we obtain easily:

$$\begin{aligned} \mathrm{tg}(2\xi) &= \frac{-4T_{xy}\left[(T_{yy} - T_{xx}) \pm \sqrt{(T_{xx} - T_{yy})^2 + 4T_{xy}^2}\right]}{(T_{yy} - T_{xx})^2 \pm 2(T_{yy} - T_{xx})\sqrt{(T_{xx} - T_{yy})^2 + 4T_{xy}^2} + (T_{yy} - T_{xx})^2 + 4T_{xy}^2 - 4T_{xy}^2} = \\ &= \frac{-4T_{xy}\left[(T_{yy} - T_{xx}) \pm \sqrt{(T_{xx} - T_{yy})^2 + 4T_{xy}^2}\right]}{2(T_{yy} - T_{xx})\left[(T_{yy} - T_{xx}) \pm \sqrt{(T_{xx} - T_{yy})^2 + 4T_{xy}^2}\right]} = \frac{2T_{xy}}{T_{xx} - T_{yy}} \quad , \quad T_{xx} \neq T_{yy}. \end{aligned} \tag{10}$$

Because of the orthogonality of eigenvectors, result (10) does not depend on the choice of a particular eigenvalue in (5).

Defining an angle $\theta = 2\xi$ of a structural vector *w*

$$\mathrm{tg}(\theta) = \mathrm{tg}(2\xi) = \frac{2T_{xy}}{T_{xx} - T_{yy}} = \frac{w_2}{w_1} \quad , \quad T_{xx} \neq T_{yy}, \tag{11}$$

we can finally discover components of *w* which can be described as:

$$\mathbf{w} = \begin{bmatrix} w_1 \\ w_2 \end{bmatrix} = \begin{bmatrix} T_{xx} - T_{yy} \\ 2T_{xy} \end{bmatrix}. \tag{12}$$

The structural vector (12) can now be extended by a third component which is equal to the trace of the tensor **T**:

$$\mathbf{w}' = \begin{bmatrix} \mathrm{Tr}(\mathbf{T}) \\ \mathbf{w} \end{bmatrix} = \begin{bmatrix} T_{xx} + T_{yy} \\ T_{xx} - T_{yy} \\ 2T_{xy} \end{bmatrix}. \tag{13}$$

Thanks to this new component in (13) we can distinguish the case $\lambda_1 = \lambda_2 = 0$ (constant intensity) from the case $\lambda_1 = \lambda_2 > 0$ (ideal isotropy).

For low-level features detection, the modified version of the structural vector w' takes the form of the vector s with components defined as follows:

$$\mathbf{s} = \begin{bmatrix} T_{xx} + T_{yy} \\ \angle \mathbf{w} \\ c \end{bmatrix}, \text{ where } \angle w = \begin{cases} \arctan\left(\dfrac{2T_{xy}}{T_{xx} - T_{yy}} \right) & , \quad T_{xx} \neq T_{yy} \\ \dfrac{\pi}{2} & , \quad T_{xx} = T_{yy} \ \wedge \ T_{xy} \geq 0 \\ -\dfrac{\pi}{2} & , \quad T_{xx} = T_{yy} \ \wedge \ T_{xy} < 0 \end{cases}. \tag{14}$$

The first component in (14) is not changed from the corresponding component of w'. The third component c of (14) constitutes a convenient coherency measure [10]:

$$c = \begin{cases} \left(\dfrac{\lambda_1 - \lambda_2}{\lambda_1 + \lambda_2} \right)^2 = \dfrac{\|\mathbf{w}\|^2}{(\mathrm{Tr}(\mathbf{T}))^2} & , \quad \mathrm{Tr}(\mathbf{T}) \neq 0 \\ 0 & , \quad \mathrm{Tr}(\mathbf{T}) = 0 \end{cases}. \tag{15}$$

Coefficient c takes on 0 for ideal isotropic areas or structures with constant intensity value, and up to 1 for ideally directional structure.

2.1 Determining Structural Tensor for Different Scales

Introducing some window function $h(x)$ into (3) the structural tensor components can be expressed as follows [6]:

$$T_{ij} = \int_{-\infty}^{+\infty} h(\vec{x}_0 - \vec{x}) \frac{\partial I(\vec{x})}{\partial x_i} \frac{\partial I(\vec{x})}{\partial x_j} d\vec{x}. \tag{16}$$

Formula (16) describes a convolution of a window function and product of directional gradients. Thus, for digital realizations (16) reduces to the following computation:

$$\hat{T}_{ij} = F(R_i R_j), \tag{17}$$

where $\hat{T}_{ij}$ stands for discrete values of the structural tensor **T**, F is a smoothing operator in a local neighborhood of pixels, R_k is a discrete gradient operator in *k-th* direction. The required quality of the detectors is achieved in practice by application of proper discrete gradient masks R_k in (17) – the most isotropic filter the better results. We found that the best results were obtained with matched filters such as the one proposed by Simoncelli [12], with the smoothing prefilter p_k and differentiating filter d_k masks in a form as follows:

$$\begin{aligned} p_5 &= \begin{bmatrix} 0.035697 & 0.248874 & 0.430855 & 0.248874 & 0.035697 \end{bmatrix} \\ d_5 &= \begin{bmatrix} 0.107663 & -0.282671 & 0 & 0.282671 & -0.107663 \end{bmatrix} \end{aligned}. \tag{18}$$

The formulas (3) and its implementation (17) can be extended to comprise information on different scales [13]. In this method it is accomplished by applying F in a form of Gaussian smoothing kernels of different size – an idea proposed by Weickert [14][1]. Based on (17) we obtain the general expression for the components of the structural tensor **T** at different scales:

$$\hat{T}_{ij}(\rho) = F_\rho(R_i R_j), \tag{19}$$

where ρ denotes a scale and F_ρ is a Gaussian kernel.

3 Detector of Locally-Oriented Structures in Images

The equations (14), (15), and (19) can be directly used for detection of locally-oriented structures in the sense of the tensor detector of local structures in pixel neighborhoods. For a searched pixel blobs Ω that angular orientation is in a given range s_ϑ, based on (14), we propose the following rule for their detection:

$$\Omega = \left\{ (x_i, y_i) : \ \mathbf{s}(x_i, y_i) = \begin{bmatrix} s_1 \\ s_2 \\ s_3 \end{bmatrix} \ \wedge \ \begin{bmatrix} s_1 > 0 \\ |s_2 - s_\vartheta| < \tau_\vartheta \\ s_3 > \tau_c \end{bmatrix} \right\}. \tag{20}$$

where τ_ϑ is a threshold for detection accuracy, while τ_c is a threshold for coherency. In our implementation, components of the structural tensor are computed in accordance with formula (19) and with differentiating filters (18).

4 Corner Detector

For detection of corners we compute eigenvalues (5) of the structural tensor which components for different scales are computed in accordance with the formula (19) and filters (18). Thus we obtain the following rule for detection of corners:

$$\mathrm{K} = \{(x_i, y_i): \ \lambda_1(x_i, y_i) \geq \lambda_2(x_i, y_i) \geq \tau_K\}. \tag{21}$$

where τ_K is a threshold for eigenvalues.

To avoid a cumbersome selection of a threshold parameter for eigenvalues, a special *priority queue* was developed – Fig. 2.

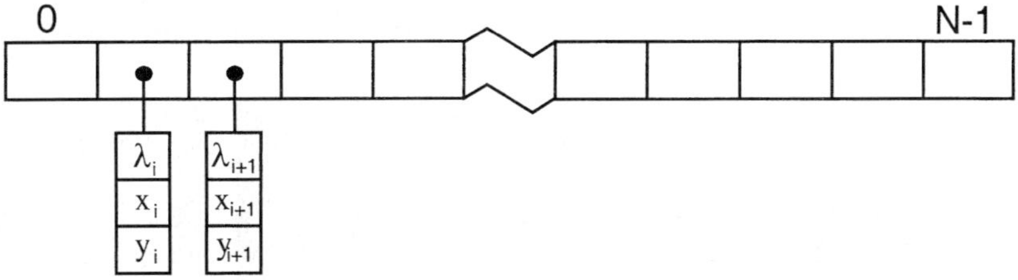

Fig. 2. Priority queue for selection of corners that comply with an assumed quality

For all elements of the queue it holds that $\lambda_i \leq \lambda_{i+1}$. If a new corner point is found then it can be inserted into the proper cell of the queue depending on its lowest eigenvalue. All the time the queue fulfills the constraint $\lambda_i \leq \lambda_{i+1}$. If a new element is inserted, then the element at index 0 is removed. This voting scheme allows for automatic selection of the best *N* interest points.

With this data structure it is also possible to impose additional constraints, e.g. on a minimal allowable distance among adjacent corner-points. This way we can search for more "distributed" corner positions in an image. For example we can search for interest points that are at least two pixels apart.

5 Experimental Results

To test the presented concept we used many real and artificial images, four of them are shown in Fig. 3. Presented tests in this paper are comparative and do not check for repeatability of the detectors (that is the issue of a further research).

a b c d

Fig. 3. Test images: real images (a, b), artificial images (c, d)

5.1 Experimental Results of Locally-Oriented Structures

Tests for locally-oriented structures for experimental images in Fig. 3 and exemplary orientations are presented in Fig. 4.

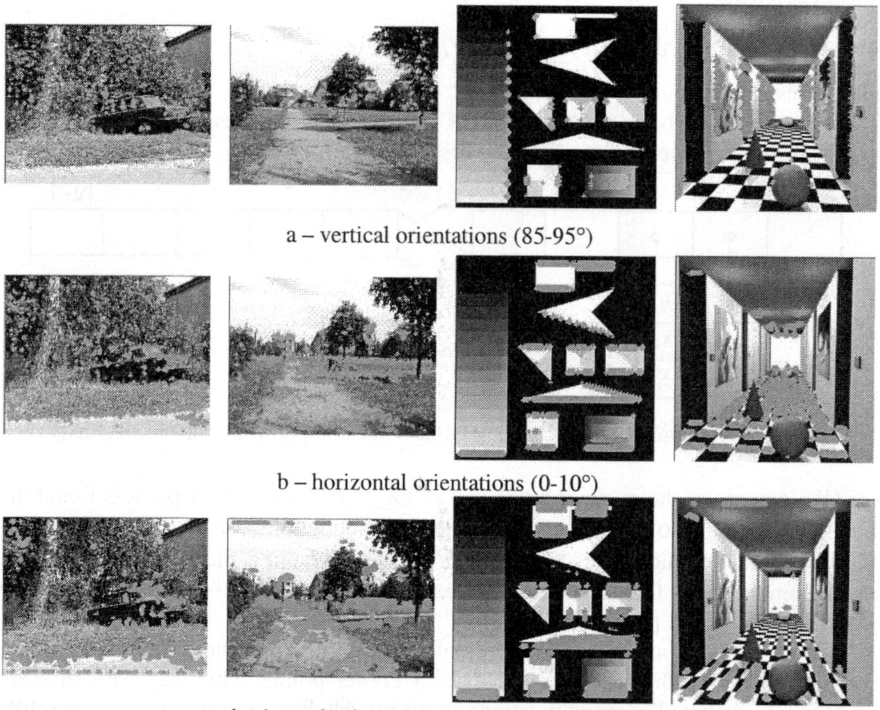

a – vertical orientations (85-95°)

b – horizontal orientations (0-10°)

c – horizontal orientations (0-10°) at a coarse scale

Fig. 4. Results of the tensor detector of locally-oriented structures in test images. Found interest points are marked with a green color

The first row of Fig. 4 shows interest points that angular orientation of their local structures is in the range 85-95°. Most of the points have been detected correctly, although for gradually changing intensity in artificial images not all points are detected. This is due to flawed discrete gradient computation in areas of uniformly changing intensity values. Fig. 4b depicts interest points with a horizontal angular orientation (0-10°), while Fig. 4c shows the same computations but for different scale (Gaussian mask 13×13). In all experiments the threshold τ_ϑ was set to 0.05 and τ_c to 0.

5.2 Experimental Results of Corner Detection

Fig. 5 contains experimental results of the corner detector with different working parameters for test images in Fig. 3.

a – max. 200 corners, self distance of detected points not limited

b – max. 200 corners, self distance set to be at least 1 pixel

c – the same as (b) but at a coarse scale

Fig. 5. Results of the corner detector for test images. Found interest points are marked with a green color

Fig. 5a depicts test images with marked corner points without any constraint on their mutual position. No threshold value was needed as well, because of the voting technique (Fig. 2). The length of the priority queue was: N=200.

An additional constraint on pixel positions was set for experiments presented in Fig. 5b. In this case we search for the best interest points, but also it is guaranteed that the found points are at least 1 pixel from each other. Fig. 5c depicts results obtained with the same settings but at a different scale (Gaussian filter 9×9).

6 Conclusions

In this paper we have presented a novel concept of detection of interest points based on the tensor representation of local structures in images. Two types of detectors was considered: the detector of angular orientations and the corner detector. Theoretical foundations of both have been shown, as well as their practical implementations. The advantages of the detectors can be stated as follows:

- Combined detector of corners and oriented-structures.
- Precise localization due to matched directional filters.
- Scale space operations.
- No threshold for corner detection due to the voting mechanism.

- Tensor method can be easily extended to a *third* dimension and then directly applied to examination of interest points, e.g. in video sequences.

At the other hand, the known drawback are as follows:

- Gradient computation cannot work properly for uniformly changed intensities.
- The presented detector has not been yet tested in regard to its repeatability in a sense stated by Schmid and Mohr [11].

Numerous experiments showed great usefulness of the presented techniques for low-level image processing. The proposed corner detector was designed especially for stereo matching. It was successively used then for computation of the epipolar geometry for real images.

References

1. Aubert, G., Kornprobst, P.: Mathematical Problems in Image Processing. Applied Mathematical Sciences Vol. 147, Springer (2002)
2. Bres, S., Jolion, J-M.: Detection of Interest Points for Image Indexation. Visual Information and Information Systems, D.P. Huijsmans et A.W. Smeulders editors, LNCS 1614, Springer, Proceeding of the third International Conference on Visual Information Systems, Amsterdam, The Netherlands (1999) 427–434
3. Cyganek, B.: Novel Stereo Matching Method That Employs Tensor Representation of Local Neighborhood in Images, Machine Graphics & Vision, Special Issue on Stereogrammetry, Vol.10, No.3 (2001) 289–316
4. Deriche, R., Giraudon, G.: A Computational Approach for Corner and Vertex Detection. Int. Journal of Computer Vision, 10 (2) (1993) 101–124
5. Harris, C., Stephens, M.: A combined corner and edge detector, Proc. of 4th Alvey Vision Conf. (1988) 147–151
6. Hauβecker, H., Jähne, B.: A Tensor Approach for Local Structure Analysis in Multi-Dimensional Images. Interdisciplinary Center for Scientific Computing, University of Heidelberg (1998)
7. Ji, Q., Haralick, R., M.: Corner Detection with Covariance Propagation. Technical Report, Intelligent Systems Laboratory (1997)
8. Lindeberg, T.: Edge detection and ridge detection with automatic scale selection. Computational Vision and Active Perception Laboratory. Technical report ISRN KTH/NA/P-96/06-SE (1996)
9. Smith, S.M., Brady, J.M.: SUSAN – A New Approach to Low Level Image Processing. Int. Journal of Computer Vision, 23 (1) (1997) 45–78
10. Jähne, B.: Digital Image Processing. 4th edition, Springer-Verlag, (1997)
11. Schmid, C., Mohr, R.: Comparing and Evaluating Interest Points. International Conference on Computer Vision, Bombay (1998)
12. Simoncelli, E.,P.: Design of Multi-Dimensional Derivative Filters. IEEE International Conference on Image Processing (1994)
13. Sporring, J., Nielsen, M., Florack, L., Johansen, P.: Gaussian Scale-Space Theory. Kluwer Academic Publishers (1997)
14. Weickert, J. : Anisotropic Diffusion in Image Processing. Teubner-Verlag (1998)
15. Würtz, R., Lourens, T.: Corner detection in color images by multiscale combination of end-stopped cortical cells. LNCS 1327, Proceedings of ICANN (1997) 901–906
16. Zheng, Z., Wang, H., Teoh, E.,K.: Analysis of gray level corner detection, Pattern Recognition Letters, 20 (1999) 149–162

Telecommunication Jitter Measurement in the Transmultipexer Systems Method Based on the Wavelet Pattern Correlation

Jan Borgosz and Bogusław Cyganek

Electronic Engineering and Computer Science Department,
Academy of Mining and Metallurgy, Mickiewicza 30, 30-059 Kraków, Poland
{borgosz, cyganek}@agh.edu.pl

Abstract. The paper presents a new method of the measure of the jitter in telecommunication. Jitter is the phenomenon which is especially harmful to the transmultiplexer systems, so its exact measure is very important to the system engineer. Described scheme is based on the recognition and correlation of the patterns generated by means of the wavelet transform. This research is a continuation of the work on development of the wavelet based jitter measurement methods that do not use the reference clock. Presented algorithm allows to estimate the jitter parameters in relation to the jitter wavelet base functions. Practical computer algorithm is also presented. Additionally, such topics like the wavelet type, its order, as well as calibrating methods are also discussed. This paper extends previous research and gives not only a proposal but also implementation details.

1 Introduction

A jitter is an unwanted, spurious transmission clock phase modulation that originates from the physics of semiconductor device [6][11].

Estimating the jitter of a transmission clock is an important problem in telecommunication measurements, especially for the transmultiplexer systems. Each transmultiplexer is time to a frequency-domain multiplexer and may be treated as a discrete multitone modulator (DMT) [1]. It is common knowledge that DMT blocks are fundamental for the Digital Subscriber Line (DSL) technology so presented solution seems to be interesting for DSL engineers and designers.

The classic approach to the jitter measurement analysis usually consists of processing steps that use a reference clock [6][11][12]. The most troublesome part of the measurement process is to correlate slopes of the reference and received clocks. The purpose of this paper is to present a totally different wavelet based approach to jitter measurement analysis as compared to the aforementioned methods. Presented approach extends the research work which was presented during the ICWAA 2001 conference [5].

P.M.A. Sloot et al. (Eds.): ICCS 2003, LNCS 2658, pp. 731–740, 2003.

Jitter test equipment is used with an Equipment Under Test (EUT). Generated test signals are transmitted over telecom line into the Equipment Under Test. The EUT retransmits received data to the meter over a telecom line. Test equipment processes all received information and calculates results [3][3].

In the classic approach, the implementation of the jitter meter requires on the reference clock.

2 Mathematical Representation of Jitter

A jitter as an unwanted, spurious transmission clock phase modulation. This phenomenon can be modeled using a modulation scheme, which allows us to describe it with the multitone technique [3][10]. A case of single tone modulation can be described as follows:

$$y(t) = A \cdot \cos(2 \cdot \pi \cdot f_{clock} \cdot t + \varphi_m(t)) \,, \tag{1}$$

where y is the jittered clock signal with amplitude A [V] and base frequency f_{clock} [Hz]. The phase modulating function $\varphi_m(t)$ can be described as follows:

$$\varphi_m(t) = 2 \cdot \pi \cdot k \, sin \cdot (2 \cdot \pi \cdot f_{jitt} \cdot t) \tag{2}$$

where f_{jitt} is jitter frequency [Hz], $k \geq 0 \wedge k \in R$ the jitter amplitude in telecommunication UI units (UI means *Unit Interval* which is equal to one cycle of transmission clock). All calculations will be presented for the sine function which easily can be changed to other wave shapes. Note that the sine waves after the comparator will be square wave with 50% duty cycle – ideal clock signal [7][11][12].

A single tone modulated signal can be written as follows:

$$y_{PM}(t) = A \cdot cos(\Omega \cdot t + \Delta\Theta_{PM} \cdot sin \varpi \cdot t) \,. \tag{3}$$

As shown in [10] this representation can be replaced by a more appropriate form that makes use of the Bessel function:

$$y_{PM}(t) = A \cdot \sum_{n=-\infty}^{\infty} J_n(\Delta\Theta_{PM}) \cdot cos(\Omega + n \cdot \varpi) \cdot t \,, \tag{4}$$

where J_n is an n-th order Bessel function of the first kind. In this case equations (1) and (2) can be rewritten as follows:

$$y(t) = A \cdot \sum_{n=-\infty}^{\infty} J_n(2 \cdot \pi \cdot k) \cdot \cos(\Omega + n \cdot \varpi_j) \cdot t \,, \tag{5}$$

where $\Omega = 2 \cdot \pi \cdot f_{clock}$ and $\varpi_j = 2 \cdot \pi \cdot f_{jitt}$.

3 Preprocessing of the Jittered Signal

A jittered sine signal can be integrated by a circuit with a much higher cut-off frequency than the maximum frequency in the input signal. In this case, the integrator is a kind of an accumulator. A jittered signal given by (5) after the integration is equal to:

$$y_{INT}(t) = \frac{A}{\Delta T} \cdot \int_{t}^{t+\Delta T} \sum_{n=-\infty}^{\infty} J_n(2 \cdot \pi \cdot k) \cdot \cos(\Omega + n \cdot \varpi_j) \cdot t\,, \qquad (6)$$

where ΔT is the integration period. Because a direct analysis of (6) is somewhat cumbersome, therefore some numerical computations were performed and are presented in Fig. 1a, 1b. Numerical computations were performed for square waves as well – sine waves after the comparator are presented in Fig. 1c, 1d. Note, that there is no need for the reference clock for detection of jitter. This is a confirmation of the presented here calculations [5].

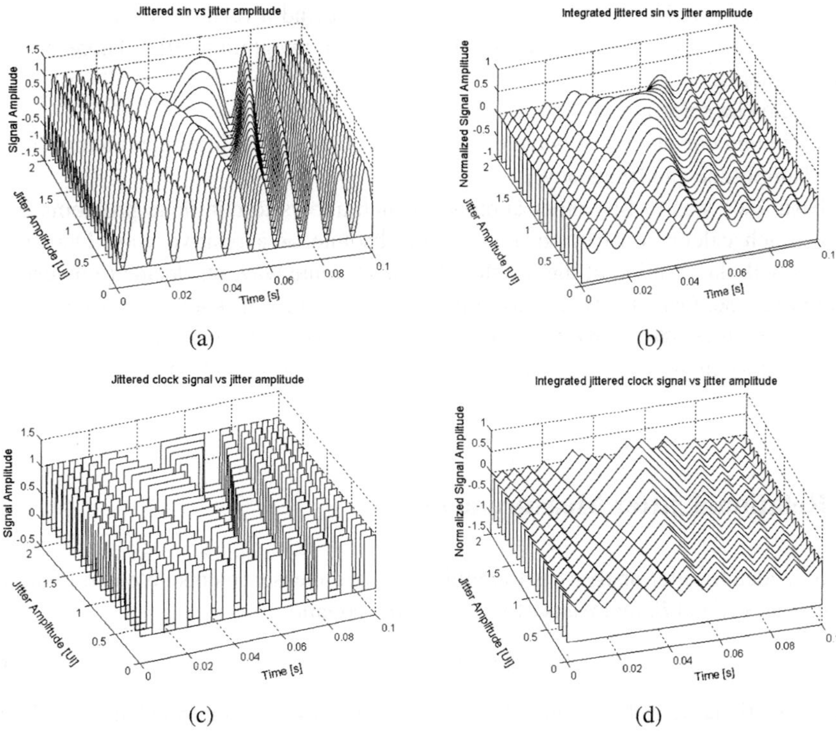

Fig. 1. a) Set of jittered sine waves with jitter frequency f_{jitt}=10Hz, carrier frequency f_n = 100 Hz for different jitter amplitudes, b) the same sine waves after integration, c) set of jittered clock signals with jitter frequency f_{jitt}=10Hz, carrier frequency f_n = 100 Hz for different jitter amplitudes, d) the same clock signals after integration.

4 Wavelet Transform Applied to Jitter Analysis

High quality jitter measurements involve measurements of an amplitude and frequency of signals, as well as their change in time. In authors opinion, the best tool for such an analysis is the Continuous Wavelet Transform (CWT) [1][8] which is represented by the following equation:

$$C(s,p)=\int_{-\infty}^{\infty} f(t)\cdot\Psi(s,p,t)dt\,, \tag{7}$$

where $\Psi(s,p,t)$ is a mother wavelet, s the scale and p the position. Inserting (6) into (7) provides us to the following formula:

$$C(s,p)=\int_{-\infty}^{\infty}\left(\left(\frac{A}{\Delta T}\cdot\int_{t}^{t+\Delta T}\left(\sum_{n=-\infty}^{\infty}J_n(2\cdot\pi\cdot k)\cdot\cos(\Omega+n\cdot\varpi_j)\cdot t\right)dt\right)\cdot\Psi(s,p,t)\right)dt \tag{8}$$

As can be observed in (8), there is no easy way to find a relationship between jitter parameters and *CWT* coefficients for a signal after integration [9]. There are two basic problems:

1) wavelet selection
2) finding correlation of wavelet coefficients $C(s,p)$ with jitter parameters (i.e. its amplitude and frequency).

The basic idea was to find other approach which allows to eliminate problems listed above. Each calculation can be done in an absolute or a relative way. Due to the troubles with the solution of the problem by the absolute way, we decide to investigate the relative one. Our idea was to use jittered signal with well known jitter parameters – called the "jitter base". We try to find relation between any jittered signal and jitter base and find unknown jitter parameters from the relation which has been found. Of course linear one would be the best.

5 Relational Jitter Measurement

The new idea to jitter analysis is to employ relative measurements. We can relate the two matrices A and B, by means of the following equation:

$$A\cdot r=B\,. \tag{9}$$

In real situations we rather attempt to solve for r in a sense of a minimization problem:

$$\min_r\,(B-A\cdot r)\,. \tag{10}$$

For the purpose of this paper we solve (10) by means of the least squares method. We introduce a notation presented below:

$$r = LSQR(A, B), \tag{11}$$

which means that r will be found with the least squares method.

In all of the following cases the matrix (vector) A will be called "the base" and matrix (vector) B will be called "the test". So in all measurements relation r will be searched which allows to transform "base" space to the "test" space. A selection of spaces will be discussed in the next sections. We have checked all possibilities of the spaces generation (results are presented below), to show that our choice is not accidental.

5.1 Relation between Integrated Jittered Signals

First attempt was to check relation between non transformed signals in a time domain that are in the form of integrals (6). Due to this supposition, A will be a base vector generated as an integrated jittered signal with known jitter parameters, and B will be a test vector, for which relation r will be searched. Numerically calculated exemplary relation r for the varying test vector is presented in Fig.2a. It can be seen that there is no easy relation between r and changing jitter parameters, thus it is not clear how the two spaces can be related.

5.2 Relation between CWT of the Jittered Signals

Due to a fail of the first method, the next attempt was made based on the observation that if jitter parameters are changed, then jittered signal "changes scale". It can be observed also that increasing jitter amplitude results in something like a local stretching of the signal. A similar conclusion may be drawn for situation with a changing jitter frequency. In this case A will be a base matrix generated as an CWT coefficients of the jittered signal with known jitter parameters, and B will be a test matrix with CWT coefficients of the jitter function with the other parameters, for which relation r will be searched. Numerically calculated exemplary relation r for the varying test vector parameters is presented in Fig.2b. Results are not very promising, like in the previous case.

5.3 Relation between CWT of the Integrated Jittered Signals

In this case the measure space was selected as the CWT of the integrated jittered signal. It can be observed that in this case, increasing jitter amplitude results in something like changing parameters of the standing wave. Similar conclusion may be done for changing jitter frequency. In this case A will be a base matrix generated as an CWT coefficients of the integrated jittered signal with known jitter parameters and B will be a test matrix with CWT coefficients of the integrated jitter function with the other parameters for which relation r will be searched. Numerically calculated exemplary relation *r* for the varying test vector parameters is presented in Fig.2c. This way obtained results are very interesting. Some linear relation may be observed between the jitter parameters and the CWT jitter base. Detailed discussion of this phenomenon will be presented in the next section.

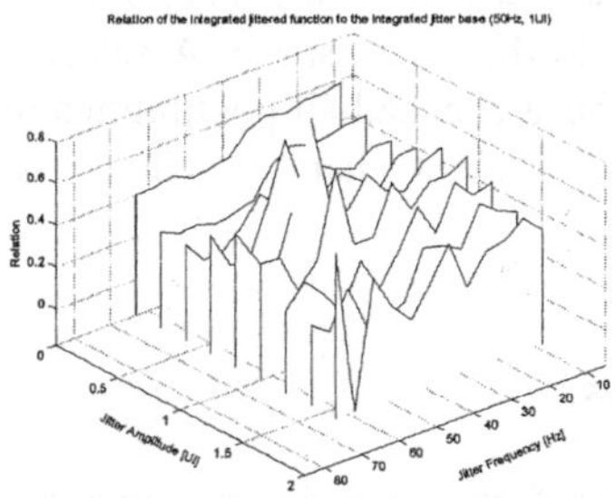

(a)

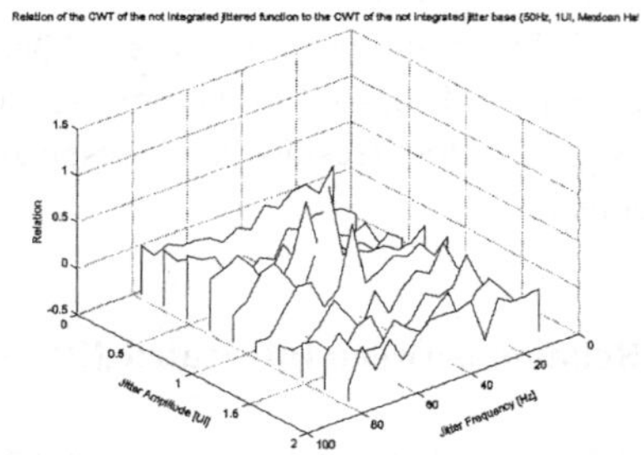

(b)

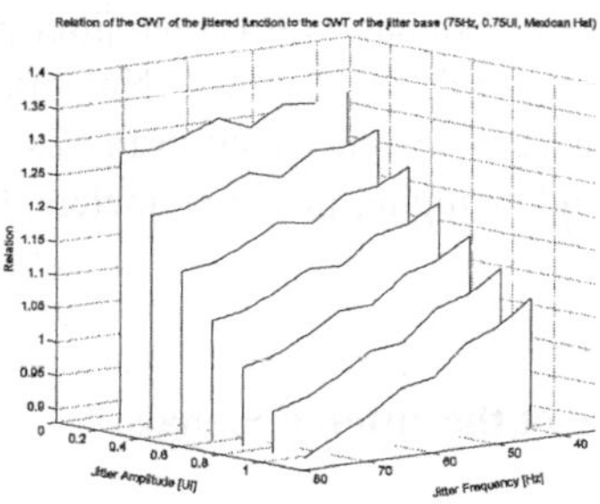

(c)

Fig. 2. a) Relation r for the time domain jitter signals, b) relation r for the varying jitter parameters and for the CWT jitter space, c) Relation r in respect to the varying jitter parameters for the CWT of the integrated jitter signal – Mexican Hat wavelet (75Hz, 0.75UI)

6 Relational Jitter Measurement with Wavelet Base of the Integrated Signal

6.1 Selection of the Wavelet Type

Due to discussion presented in previous sections and experiments which were described in the ICWAA 2001 paper [5], bases with CWT of the integrated jitter signal with different wavelet types were checked.

The following wavelets were tested:

- *Mexican Hat*,
- *Morlet*,
- *Coiflets 2-5*,
- *Biorthogonal 2.6 2.8 4.4 5.5 6.8*,

because their shapes appeared to be the most appropriate to analyze the signals shown in Fig.4b and Fig.4c (see section 4.1.). Exemplary results are presented only for the *Biorthogonal 4.4, Morlet* and *Mexican Hat* – see Fig.3.

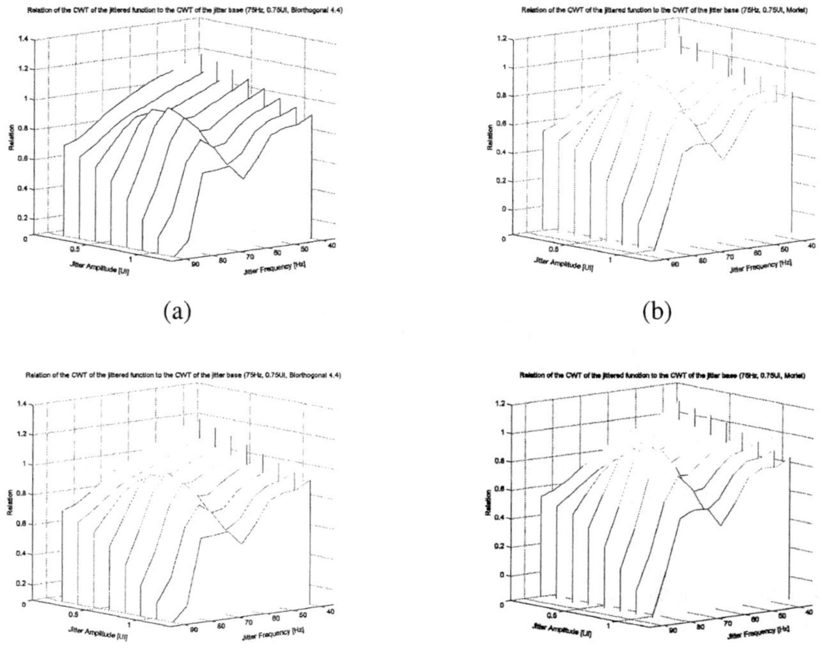

Fig. 3. a) Relation r for the varying jitter parameters and for the CWT of the integrated jitter space – Biorthogonal 4.4 wavelet, b) relation r for the varying jitter parameters and for the CWT of the integrated jitter space – Morlet wavelet, c) relation r for the varying jitter parameters and for the CWT of the integrated jitter signal – Mexican Hat wavelet (100Hz, 0.25UI), d) Relation r for the varying jitter parameters and for the CWT of the integrated jitter signal – Mexican Hat wavelet (50Hz, 1UI)

Only the Mexican Hat wavelet gives almost linear relation, when in other cases parabolic or other relations were obtained. Thus, there is a first important remark that a wavelet type used in base function is significant and decides about type of relation.

Presented results show that the *Mexican Hat* wavelet gives the best results, due to nearly linear characteristic of *r*. Set of the numerical tests were done which confirmed our observations.

Figure 4 depicts relation *r* in the 2D-style plot. It may be observed that relation *r* may be described by the classic linear equation ($y=ax+b$), where parameters *a* and *b* depend on jitter frequency – this can be written as follows:

$$r = a(f_{jitt}) \cdot A_{jitt} + b(f_{jitt}) . \tag{12}$$

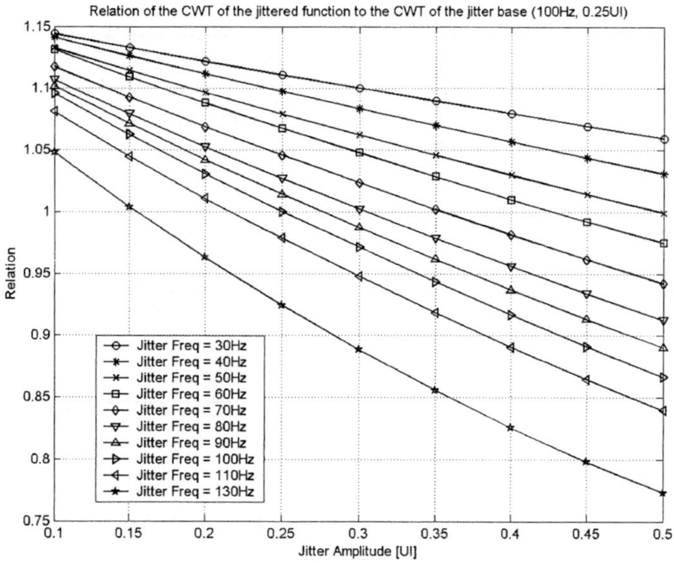

Fig. 4. Relation *r* for the varying jitter parameters and for the CWT of the integrated jitter signal – Mexican Hat wavelet

6.2 Relation between CWT of the Integrated Jittered Signals – Analytic Analysis

Results obtained by means of the numerical analysis may be verified by the analytic relations. Let us assume that:

$$C(s,p) = \int_{-\infty}^{\infty} f(t) \cdot \Psi(s,p,t) dt = CWT(f(t),\Psi) . \tag{13}$$

The relation r may be rewritten as follows:

$$r = LSQR\left(CWT(y_{INT}(t), \Psi), CWT(y_{BASE\_INT}(t), \Psi)\right), \tag{14}$$

where:

$$y_{INT}(t) = \frac{A}{\Delta T} \cdot \int_{t}^{t+\Delta T} \sum_{n=-\infty}^{\infty} J_n(2 \cdot \pi \cdot k) \cdot \cos(\Omega + n \cdot \varpi_j) \cdot t\,, \tag{15}$$

$$y_{BASE\_INT}(t) = \frac{A_{BASE}}{\Delta T} \cdot \int_{t}^{t+\Delta T} \sum_{n=-\infty}^{\infty} J_n(2 \cdot \pi \cdot k_{BASE}) \cdot \cos(\Omega + n \cdot \varpi_{jBASE}) \cdot t\,. \tag{16}$$

Thus, we can rewrite (16) as follows:

$$r = \frac{A}{A_{BASE}} \cdot LSQR\left(\begin{array}{l} CWT\left(\int_{t}^{t+\Delta T} \sum_{n=-\infty}^{\infty} J_n(2 \cdot \pi \cdot k) \cdot \cos(\Omega + n \cdot \varpi_j) \cdot t, \Psi \right), \\ CWT\left(\int_{t}^{t+\Delta T} \sum_{n=-\infty}^{\infty} J_n(2 \cdot \pi \cdot k_{BASE}) \cdot \cos(\Omega + n \cdot \varpi_{jBASE}) \right. \end{array} \right. \tag{17}$$

7 Conclusions

In this paper we conclude that a jitter measurement may be realized as calculation of the relation between *CWT* coefficients of integrated jittered signals and *CWT* coefficients of a base jitter signal.

This method provides a novel way of jitter measurement without usage of the reference clock. An introduction to the analytic description of the method was presented. Different types of wavelets were analyzed in a context of their application to the presented method. However, only wavelets with the Mexican Hat mother functions produce sought linear results.

There are still subjects that need further investigation, however. These are:

- detailed analytic solution of the problem,
- analytic solution including wavelet type,
- detailed analysis of the range of jitter parameters change.

Presented solution due to its innovatively may be very interesting for the modern telecommunication business associated with DSL and transmultiplexer techniques, which has to look for the new techniques which allow to use cheap hardware resources (like twisted-pair telephone lines) to deliver modern multimedia to the client.

Acknowledgments. This paper is a result of research work registered in The State Committee for Scientific Research of Poland (KBN) number 4 T11D 005 23 and its publication is sponsored by means of the KBN founds.

References

1. Akansu A., Duhamel P., Lin X., Courville M.: Orthogonal transmultiplexers in communication: A review. IEEE Trans. SP, vol. 46 (1998) 979–995
2. Białasiewicz J.: Wavelets and Approximations (in Polish). WNT (2000)
3. Borgosz J., Cyganek B., Korohoda P.: Jitter Telecom Test Equipment Based On FPGA Structure, International Conference on Signals and Electronic Systems ICSES'2001, Łódź, Poland (2001) 397 – 402
4. Borgosz J., Cyganek B.: Digital Implementation of Jitter Generator for Telecom Test Equipment, Digital Signal Processing for Multimedia Communications and Services, EURASIP Conference: ECMCS 2001, Budapest, Hungary (2001)
5. Borgosz J., Cyganek B.: A Proposal of Jitter Analysis Based on a Wavelet Transform, The Second International Conference on Wavelet Analysis and Its Applications ICWAA01 Hong Kong Baptist University, Hong Kong (2001) 257–268
6. Feher and Engineers of Hewlett-Packard: Telecommunication Measurements Analysis and Instrumentation, Hewlett-Packard (1991)
7. Glover I. A., Grant P.M.: Digital Communications. Prentience Hall (1991)
8. Prasad L., Iyengar S.S.: Wavelet Analysis. CRC Press (1997)
9. Starck J.-L., Murtagh F., Bijaoui A.: Image Processing and Data Analysis. The Multiscale Approach. Cambridge University Press, (2000)
10. Szabatin J.: Foundations of Signals Theory (in Polish). WKŁ (2000)
11. Trischitta P.R., Varma E.L.: Jitter in Digital Transmission System. Artech House Publishers (1989)
12. Takasaki Y., Personick S.D.: Digital Transmission Design and Jitter Analysis. Artech House Publishers (1991)

Self-Organizing Compact Modeling Methodology for High-Speed Passive Electrical Interconnection Structures

Tom Dhaene

University of Antwerp, Middelheimlaan 1, 2020 Antwerpen, Belgium
tom.dhaene@ua.ac.be
Agilent Technologies, EEsof Comms EDA, Lammerstraat 20, 9000 Gent, Belgium
tom_dhaene@agilent.com

Abstract. New modeling technology is developed that allows engineers to define the frequency range, layout parameters, material properties and desired accuracy for automatic generation of simulation models of general passive electrical structures. It combines electromagnetic (EM) accuracy of parameterized passive models with the simulation speed of analytical models. The adaptive algorithm doesn't require any a priori knowledge of the dynamics of the system to select an appropriate sample distribution and an appropriate model complexity. With this technology, designers no longer must put up with legacy modeling techniques or invest resources in examining new ones.

1 Introduction

Component and circuit models are a cornerstone of EDA (Electronic Design Automation) technology. With wireless and wireline designs constantly increasing in complexity and operating at higher frequencies, design engineers push the limits of their EDA tool's passive analytical models. Often, these passive models are used outside their operational range, causing the EDA tool to return inaccurate simulation results. The inconsistencies of legacy modeling techniques from the 1970s and 1980s hinder the accuracy of these models when applied to different processes and frequencies. Exceeding a model's frequency limit causes errors due to the model's failure to account for higher-order propagation modes. Limitations of the equivalent circuit model, such as frequency independent inductive or capacitive elements, also lead to simulation errors. Since most EDA tools do not proactively report such errors, they propagate through the design flow and may not be discovered until a prototype fails to perform as expected. To avoid errors and inconsistencies, full-wave EM simulation is required to fully characterize the structure and produce an accurate S-parameter model of the discontinuity that is then used by the circuit simulator.

Developing new models is not a trivial task! To model a single parameter over a range of values, several sample points are required. Since the model can be a function of many layout parameters (line width, length, metal thickness, dielectric constant, substrate thickness, loss tangent, etc.) there is an exponential growth in the number of samples as the number of layout parameters increases. Also, developing a new model usually requires a highly skilled person working for an extended period — several weeks or even months—to build, test and produce the desired analytical model. If the

P.M.A. Sloot et al. (Eds.): ICCS 2003, LNCS 2658, pp. 741–747, 2003.

requirement is for a complete library of models, the total effort is multiplied by the number of models sought. This task needs to be weighed against measurement-based or EM-based modeling on a case-by-case basis.

Some common approaches to modeling issues have limiting factors [1]-[2]. Methods using pre-calculations of equivalent circuits, using a variety of look-up tables, fitting equations and interpolation techniques can have a limited number of samples and have insufficient interpolation methods. One clear example where the dependability of these techniques comes into question is with high-Q resonant circuits such as those used in narrow band filters. Using discrete data grids and interpolation techniques with such circuits might cause the generated model to suffer from either "oversampling" or "undersampling." With oversampling, too many data samples are collected and model generation is inefficient; on the other hand, with undersampling, too few data samples are collected and the model is not completely defined.

As an alternative to building classic analytical models, engineers can utilize a full-wave EM modeling tool to fully characterize a given passive component. This method permits accurate characterization of the actual passive structure to be used, accounting for higher-order mode propagation, dispersion and other parasitic effects. However, the calculation time required for full-wave EM simulation of a given component makes real-time circuit tuning impossible.

A new efficient adaptive sampling and modeling technique addresses this model accuracy dilemma. The '*Multidimensional Adaptive Parameter Sampling*' algorithm (MAPS) selects a limited set of data samples in consecutive iterations, and interpolates all S-parameter data using rational and multinomial fitting models. This algorithm allows important details to be modeled by automatically sampling the response of the structure more densely where the S-parameters are changing more rapidly. The goal is minimizing the total number of samples needed, while maximizing the information provided by each new sample. The modeling technique combines the speed and flexibility of analytical models, and accuracy and generality of full-wave EM simulation in one compact parameterized passive model [3]-[4].

2 Adaptive Modeling and Sampling Technique

The MAPS technique builds a global fitting model of the chosen parameters, handling frequency and geometrical dependencies separately. Multidimensional polynomial (or multinomial) fitting techniques are used to model the geometrical dependencies, while rational fitting techniques [5]-[6] are used to handle frequency dependencies. The modeling process does not require any *a priori* knowledge of the circuit under study. Different adaptive algorithms are combined to efficiently generate a parameterized fitting model that meets the predefined accuracy. This includes the adaptive selection of an optimal number of data samples along the frequency axis and in the geometrical parameter space, and adaptive selection of the optimal order of the multinomial-fitting model.

The number of data points is selected to avoid oversampling and undersampling. The process of selecting data points and building models in an adaptive way is called *reflective exploration* [7]. Reflective exploration is useful when the process that provides the data is very costly, which is the case for full-wave EM simulators. Reflective exploration requires *reflective functions* that are used to select new data points. For example, the difference between two fitting models can be used as a reflective function. Also, some physical rules, such as a passivity-check, can be used as a reflective function. The modeling process starts with an initial set of data points. New data points are selected near the maximum of the reflective function until the desired accuracy is reached.

The model complexity is automatically adapted to avoid overmodeling (overshoot or ringing) and undermodeling, and the model covers the whole parameter and frequency space and can easily be used for optimization purposes.

The MAPS modeling technique follows four steps to adaptively build a model.

- *Step One:* The frequency response of the circuit is calculated at a number of discrete sample points (using the Agilent Momentum full-wave EM simulator [8]). The Adaptive Frequency Sampling (AFS) algorithm [6] selects a set of frequencies and builds a rational model for the S-parameters over the desired frequency range (Figure 1).
- *Step Two:* A multinomial is fitted to the S-parameter data at multiple discrete frequencies (Figure 2).
- *Step Three:* This model is written as a weighted sum of orthonormal multinomials. The multinomials only depend on the layout parameters. The weighting coefficients preceding the orthonormal multinomials in the sum are only frequency dependent (Figure 3).
- *Step Four:* Using the AFS models built in step one, the coefficients can be calculated over the whole frequency range (Figure 4). These coefficients, together with the orthonormal multinomials, are stored in a database for use during extraction afterwards.

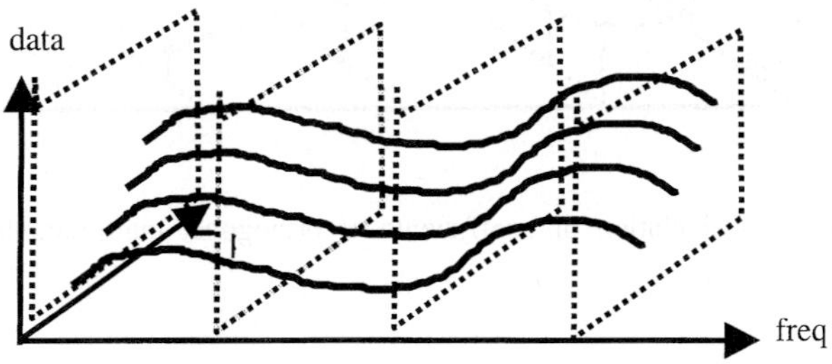

Fig. 1. *Step One:* AFS rational models over the desired frequency range, derived from full-wave EM simulation.

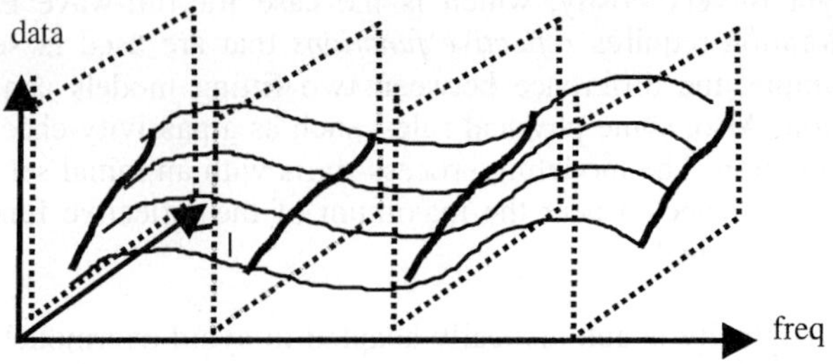

Fig. 2. *Step Two:* Multinomial models are created at discrete frequencies.

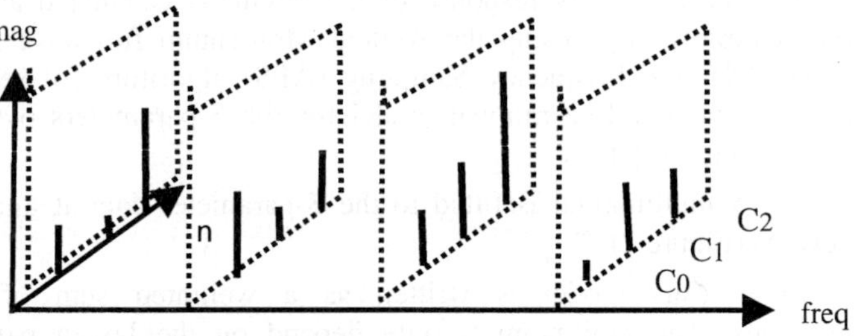

Fig. 3. *Step Three:* Creation of the coefficients of orthogonal multinomials at discrete frequencies.

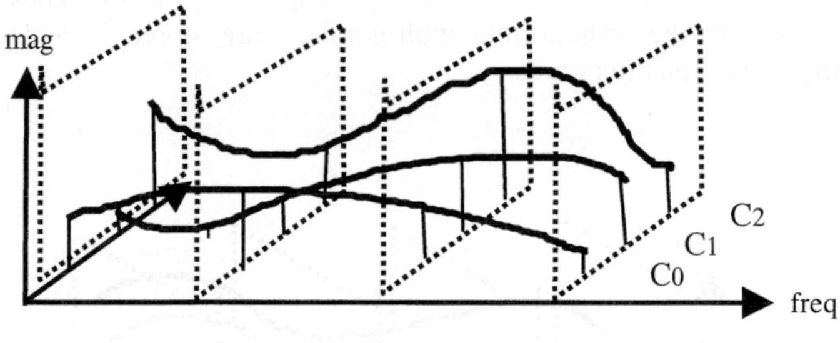

Fig. 4. *Step Four:* Calculation of coefficients of orthogonal multinomials over the entire frequency range.

3 Example

The automated modeling technique was used to generate analytical circuit models for *all* sub-parts (*transmission line*, *open end*, *slot coupler*, *step in width*, *corner-fed patch*) of a *slot-coupled microstrip-fed patch antenna* structure (figure 5). This modeling step is a one-time, up-front time investment. A double sided duroid substrate was used (thickness = 31 mil & 15 mil, $\varepsilon_r = 2.33$, tg $\delta = 0.0012$).

First, parameterized circuit models were built for *all* substructures of the circuit. For example, the *corner-fed patch* (figure 6) circuit model was built over the following parameter range (table 1):

Table 1. Parameter ranges of corner-fed patch

variable	min	max
L_patch	320 mil	400 mil
W_feed	5 mil	30 mil
f	5 GHz	15 GHz

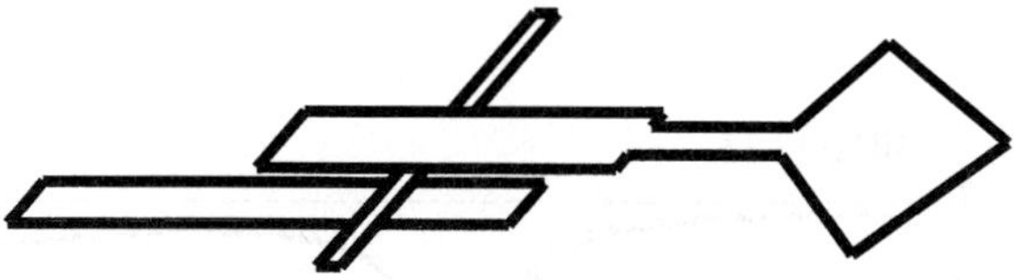

Fig. 5. Slot-coupled microstrip-fed patch antenna structure

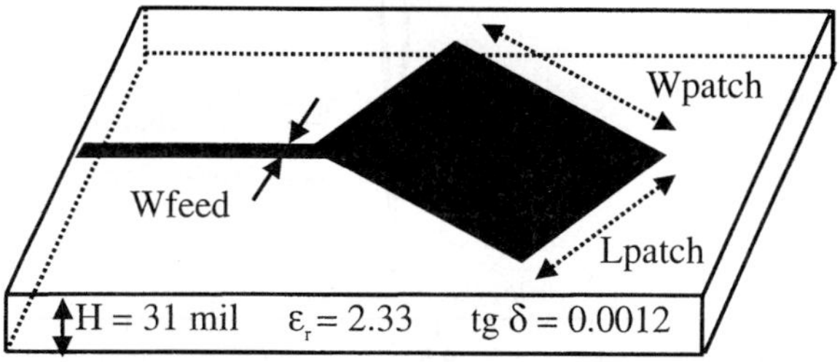

Fig. 6. Layout of *corner-fed patch*

The automated modeling tool (ADS Model Composer) selected 25 data points (= discrete layouts) in an adaptive way, and grouped all S-parameter data all in one global, compact, analytical model. ADS Momentum was used as planar EM simulator [8]. The desired accuracy level was set to -55 dB. In figure 7, the reflection coefficient S_{11} of the *corner-fed patch* is shown as a function of frequency and width.

Then, the parameterized circuit models were used to simulate the overall antenna structure (figure 5). Figure 8 shows S_{11} simulated with Momentum, and with the new analytical circuit models for all sub-components (divide and conquer approach). Both results correspond very well. However, the simulations based on the circuit models easily allow optimization and tuning, and took only a fraction of the time of the full-wave simulation (2 seconds compared to 96 minutes on a 450 MHz Pentium II computer).

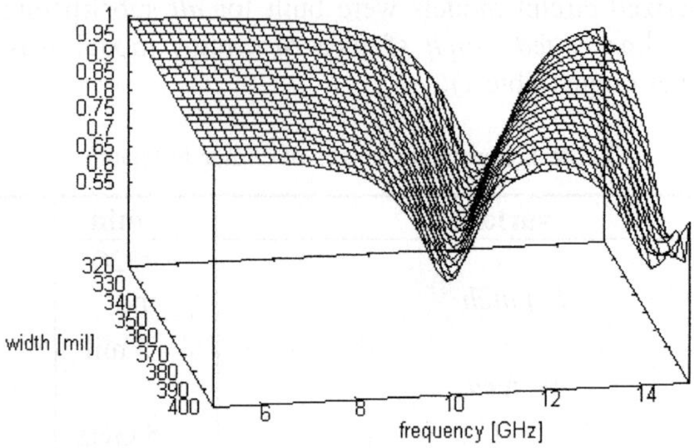

Fig. 7. Reflection coefficient S_{11} of corner-fed patch (W_feed = 8 mil)

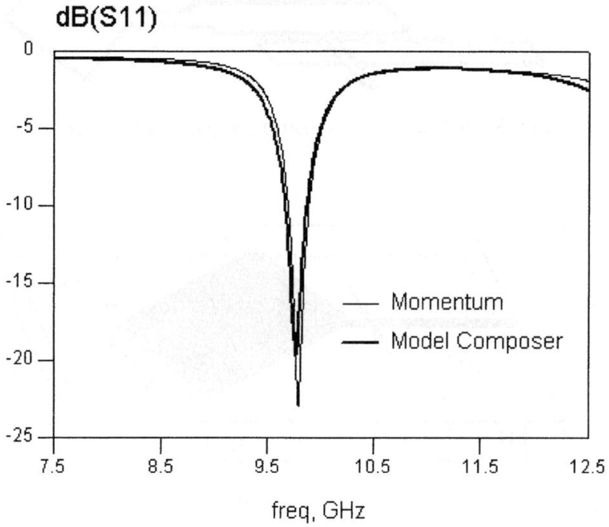

Fig. 8. Reflection coefficient S_{11} of slot-coupled microstrip-fed patch antenna

4 Conclusions

An advanced modeling technique was presented for building parameterized models for general passive microwave and RF structures. The models are based on full-wave EM simulations, and have a user-defined accuracy. Once generated, the analytical models can be grouped in a library, and incorporated in an EDA tool where they can be used for simulation, design and optimization purposes. A patch antenna example was given to illustrate the technique. The results based on the parameterized models correspond very well with the global full-wave simulations. However, the time required for a simulation using the compact analytical circuit models was only a fraction of the time required for a global full-wave simulation.

References

1. Chaki S., Aono S., Andoh N., Sasaki Y., Tanino N., Ishihara O.: Experimental Study on Spiral Inductors, Proceedings of the IEEE Symposium on Microwave Theory and Techniques, (1995) 753–756.
2. Liang J-F., Zaki K. A.: CAD of Microwave Junctions by Polynomial Curve Fitting, Proceedings of the IEEE Symposium on Microwave Theory and Techniques, (1993) 451–454.
3. De Geest J., Dhaene T., Fache N., De Zutter D.: Adaptive CAD-Model Building Algorithm for General Planar Microwave Structures, IEEE Transactions on Microwave Theory and Techniques, vol. 47, no. 9, (1999) 1801–1809.
4. Dhaene T., De Geest J., De Zutter D.: EM-based Multidimensional Parameterized Modeling of General Passive Planar Components, Proceedings of the IEEE Symposium on Microwave Theory and Techniques, Vol. 3, (2001) 1745–1748.
5. Dhaene, T., Ureel, J., Fache, N., De Zutter, D.: Adaptive Frequency Sampling Algorithm for Fast and Accurate S-parameter Modeling of General Planar Structures, Proceedings of the IEEE Symposium on Microwave Theory and Techniques, (1995) 1427–1430.
6. Dhaene, T.: Automated Fitting and Rational Modeling Algorithm for EM-Based S-Parameter, Proceedings of the Applied Parallel Computing Conference, LNCS 2367, (2002), pp. 99–105.
7. U. Beyer and F. Smieja, Data Exploration with Reflective Adaptive Models, Computational Statistics and Data Analysis, vol. 22, (1996) 193–211.
8. Momentum software, Agilent EEsof Comms EDA, Agilent Technologies, Santa Rosa, CA.

Specification and Automated Recognition of Algorithmic Concepts with ALCOR *

Beniamino Di Martino[1] and Anna Bonifacio[2]

[1]Dipartimento di Ingegneria dell' Informazione -
Second University of Naples - Italy
[2] Covansys s.p.a - Rome - Italy
beniamino.dimartino@unina.it

Abstract. Techniques for automatic program recognition, at the algorithmic level, could be of high interest for the area of Software Maintenance, in particular for knowledge based reengineering, because the selection of suitable restructuring strategies is mainly driven by algorithmic features of the code. In this paper a formalism for the specification of algorithmic concepts, based on an automated *hierarchical concept parsing* recognition technique, is presented. Based on this technique is the design and development of ALCOR, a production rule based system for automatic recognition of algorithmic concepts within programs, aimed at support of knowledge based reengineering for high performance.

1 Introduction

The automatization of program comprehension techniques, even if limited to the algorithmic level, could be of high interest for the area of Software Maintenance, in particular for knowledge based reengineering to improve program performance.

We have designed and developed an original technique for automatic algorithmic concept recognition, based on production rule-based hierarchical concept parsing.

Based on this technique is the development of ALCOR [2], a tool for automatic recognition of algorithmic concepts within programs, aimed at support of program restructuring and porting for high performance [3]. First order logic programming, and Prolog in particular, has been utilized to perform the hierarchical concept parsing, thus taking advantage of Prolog's deductive inference rule engine. The input code is C, and the recognition of algorithmic concept instances within it is performed without any supervision from the user. The code recognition is nevertheless partial (only the concept instances specified in

* This work has been supported by the CNR - Consiglio Nazionale delle Ricerche, Italy (Agenzia 2000 Project *ALCOR* - n. CNRG00A41A), and by the Campania Reg. government (Projects "L41/96 Smart ISDN" and "Regional Competence Center on Information and Communication Technologies").

P.M.A. Sloot et al. (Eds.): ICCS 2003, LNCS 2658, pp. 748–757, 2003.

the Alcor's inferential engine are recognized) and limited to the functional (algorithmic) level: concepts related to the application domain level of the source code aren't taken into consideration, in order to design a completely automatic procedure. These limitations are nevertheless irrelevant, due to the purpose of recognition: to drive parallelization strategies, the algorithmic level comprehension is sufficient, and it can be applied on partial portions of the code.

In this paper we describe a formalism for the specification of algorithmic concepts recognition, based on higher order attributed grammars.

The paper proceeds as follows: an overview of the developed technique is presented in sec. 2; in sec. 3 we describe a formalism for the specification of algorithmic concepts recognition, based on higher order attributed grammars. Sec. 4 provides with examples of specification of algorithmic concepts.

2 The Recognition Strategy

The recognition strategy is based on hierarchical parsing of algorithmic concepts. The recognition process is represented as a hierarchical abstraction process, starting from an intermediate representation of the code at the structural level in which *base concepts* are recognized; these become components of *structured concepts* in a recursive way. Such a hierarchical abstraction process can be modeled as a hierarchical parsing, driven by *concept recognition rules*, which acts on a description of *concept instances* recognized within the code.

The concept recognition rules are the production rules of the parsing: they describe the set of characteristics that allow for the identification of an algorithmic concept instance within the code.

The characteristics identifying an algorithmic concept can be informally defined as the way some abstract entities (the subconcepts), which represents a set of statements and variables linked by a functionality, are related and organized within a specific abstract control structure. By "abstract control structure" we mean structural relationships, such as control flow, data flow, control and data dependence, and calling relationships.

More specifically, each recognition rule specifies the related concept in a recursive way, by means of:

- a compositional hierarchy, recursively specified through the set of subconcepts directly composing the concept, and their compositional hierarchies;
- a set of conditions and constraints, to be fulfilled by the composing subconcepts, and relationships among them (which could involve subconcepts at different levels in the compositional hierarchy, thus not only the direct subconcepts).

In section 3 a formalism for the specification of the recognition rules is presented.

The properties and relationships which characterize the composing concepts have been chosen in such a way to privilege the structural characteristics respect to the syntactic ones. We have decided to give the data and control dependence relationships a peculiar role: they become the characteristics that specify the *abstract control structure* among concepts. For this purpose, they undergo an abstraction process during recognition, such as the concept abstraction process. This abstraction has been represented by the introduction of the notions of *abstract control and data dependence* among concept instances. The set of abstract data and control dependence relationships is produced within the context of the concept parsing process, and is explicitly represented within the program representation at the algorithmic level.

The direction of the concept parsing has been chosen to be *top-down* (descendent parsing). This choice is motivated by the particular task of the recognition facilities in the framework of the parallelization process. Since we are interested in finding instances of parallelizable algorithmic patterns in the code, an algorithmic recognition of the whole code is not mandatory: thus a top-down parsing (demand-driven), which leads to partial code recognition, is suitable, and allows for a much deeper pruning of the search space associated with the hierarchical parsing than the bottom-up approach.

The base concepts, starting points of the hierarchical abstraction process, are chosen among the elements of the intermediate code representation at the structural level. The code representation at the structural level (*basic representation*) is thus a key feature that affects the effectiveness and generality of the recognition procedure; we have chosen the *Program Dependence Graph* [4] representation, slightly augmented with syntactical information (e.g. tree-like structures representing expressions for each statement node) and control and data dependence information (edges augmented e.g. with control branch and data dependence level, type, dependence variable). Two main features make this representation suitable for our approach: (1) the structural information (data and control dependence), on which the recognition process relies, is explicitly represented; (2) it's an inherently delocalized code representation, and this plays an important role in solving the problem of concept delocalization. An overall *Abstract Program Representation* is generated during the recognition process. It has the structure of a *Hierarchical PDG* (HPDG), reflecting the hierarchical strategy of the recognition process.

3 A Formalism for the Specification of Algorithmic Concepts Recognition

Attributed grammars [8] have been selected as formalism for the specification of the recognition rules of the hierarchical concept parsing. Its effectiveness to specify relationships among structured informations, well exploited for the specification and structural analysis of programming languages, makes this formalism suitable for the program analysis at the algorithmic level too.

$CG = (G, A, R, C)$ is thus our *Concept Grammar*, with $G = (T, N, P, Z)$ its associated context-free grammar.

In the following the several components of the grammar are described, together with their relationships with the recognition process.

The set of *terminal symbols* of the grammar, T, represents the *base concepts*. These are terminals of the hierarchical abstraction process: they are thus elements of the program representation at the structural level. We give the role of base concepts to the elements of the structural representation which represent the executable program statements. The set of grammar terminals is thus, for Fortran 77: $T = \{\texttt{do}, \texttt{assign}, \texttt{if}\}$

The set of *nonterminal symbols* of the grammar, N, represents the algorithmic concepts recognized by the concept parsing.

The set of *start symbols* of the grammar, Z, represents the subset of algorithmic concepts, named PAPs (Parallelizable Algorithmic Patterns), which are associated with a specific set of parallelization strategies.

The set of production rules of the grammar, P, specifies the composition of the concepts represented by the lhs non-terminal symbols, and the relationships and constraints to be fulfilled by the instances of their subconcepts, represented by the rhs symbols.

The syntax of a production rule is as follows:

Rule =
  rule *concept* →
    composition
      { *subconcept* }
    condition
      [**local** LocalAttributes]
      { Condition }
    attribution
      { AttributionRule }

LocalAttributes =
  attribute : Type { *attribute* : Type }

concept ∈ N
subconcept ∈ $N \cup T$
attribute ∈ A
Condition ∈ C
AttributionRule ∈ R

A production rule specifies:

- a set {*subconcept*} of (terminal and nonterminal) symbols which represent the set of subconcepts composing the concept represented by the lhs symbol *concept*;
- the set {Condition} of the production's conditions; it represents the set of relationships and constraints the subconcept instances of the set {*subconcept*} have to fulfill in order for an instance of `concept` to be recognized;
- the set {AttributionRule} of the production's attribution rules. These assign values to the attributes of the recognized concept, utilizing the values of at-

tributes of the composing subconcepts (in this way we restrict the attributes of the grammar CG to be only synthesized ones).

Optionally, local attributes can be defined (the definition is marked by the keyword **local**).

The set of grammar *conditions*, C, is composed of predicates, or predicative functions, defined as follows:

Conditions =
  | Condition & Conditions
  | Condition '|' Conditions
  | ¬ Conditions
  | **if** Conditions **then** Conditions [**else** Conditions]

Condition =
  | [(] Conditions [)]
  | *condition* ([ParameterList])
  | ['{'] ParameterList ['}'] := *condition* ([ParameterList])
  | *attribute* = *attribute*
  | *attribute* ≠ *attribute*
  | *attribute* ∈ *attributelist*
  | ∀ Condition Condition
  | ∃ Condition s.t. Condition

ParameterList =
  Parameters { , Parameters }

Parameters =
  | [*attribute* :] Type
  | ([*attribute* :] Type '|' [*attribute* :] Type)

Conditions have grammar attributes, of different types, as parameters; in the case of predicative functions, they return a set of values (to be assigned to attributes of the corresponding type) if the condition on the input parameters is verified, and an undefined value otherwise. The conditions present in a grammar rule represent the constraints the corresponding concept has to fulfill, or the relationships among its composing subconcepts. The conditions represent constraints imposed to the parsing process, because if one of them is not satisfied, the current application of the corresponding rule fails, and the corresponding concept instance is not recognized. Conditions can be composed, by means of the usual logical connectives, the universal and existential quantifiers can be utilized, and alternatives can be tested by means of the conditional connective **if**. Conditions can applied to local attributes, and to attributes of symbols at the right hand side of the production. This constraint, together with the presence of synthesized attributes only, ensures that all the attributes which could be needed by a condition assume defined values when the rule is fired. This allows for the conditions' verification to be performed during the parsing process.

The set of grammar *attributes*, A, is composed of the attributes associated to the grammar symbols. As already mentioned, the attributes associated to a concept add an auxiliary semantical information to the semantical meaning represented by the concept; in addition, they permit to verify constraints and properties of the concept, and relationships among concepts; finally, they characterize the particular instances of the concept: two occurrences of a grammar symbol, presenting different values for at least one attribute, represent two different instances of the same concept.

4 Examples of Algorithmic Concepts' Specifications

We now highlight the flexibility and expressivity power of the formalism for the specification of the hierarchy, the constraints and the relationships among concepts through an example.

The algorithmic concept we consider is the *matrix-matrix product*, and the concept *vector-matrix product*, which is the subconcept of the previous. Due to space restrictions, we leave unexplained, even though we mention, all the other hierarchically composing subconcepts. The generic *vector-matrix product* operation is usually defined (in a form which includes application to multidimensional arrays) as:

$$VMP^{hkl} = [\cdots, DP^{hk}_m, \cdots] = \left[\cdots, \sum_i e(A(\cdots, \overset{h}{i}, \cdots) \times B(\cdots, \overset{k}{i}, \cdots, \overset{l}{m}, \cdots)), \cdots\right] \quad (1)$$

where DP^{hk} is the generic *dot product* operation, where $e(x)$ is a linear expression with respect to x, whose coefficient is invariant with respect to the i index of the sum. The vector-matrix operation involves the h-th dimension od the A array (the "vector") and the k-th and l-th dimensions of the array B (the "matrix"). The result is a monodimensional array, which can be assigned to a column of a multidimensional array.

The recognition rule for the `matrix_vector_product` is presented in figure 1, with related attributes presented in table 1. We illustrate the 1 in the following.

Its main component is the `dot_product` concept. The other components are two instances of the `scan` concept. The first scan instance (`scan[1]`) scans a dimension of the array, which records the result: this is specified by verifying that the scan statement is an assignment (to elements of the scanned array), sink of a dependence chain whose source is the assignment statement which records the result of the dot product. The second scan instance (`scan[2]`) scans the l-th dimension (cfr. eq: 1) of one of the arrays involved in the dot product. It has to be of course different from the dimension (k-th in 1) involved in the dot product. The two scans must share their `count_loop` subconcepts, in order to scan the two dimensions of the two arrays at the same time. These dimensions have to be completely scanned, in sequence (for both the scan concepts, the *range* attribute

Table 1. Attributes of the `matrix_vector_product` concept.

Attributes	kind
`in`	: instance
`hier`	: hierarchy
`vector_struct`	: struct of `ident` : identifyer `inst` : expression `dot_prod_subscr_exp` : expression `dot_prod_subscr_pos` : integer `dot_prod_index` : identifyer endstruct
`matrix_struct`	: struct of `ident` : identifyer `inst` : expression `dot_prod_subscr_exp` : expression `dot_prod_subscr_pos` : integer `dot_prod_index` : identifyer `matr_vec_subscr_exp` : expression `matr_vec_subscr_pos` : integer `matr_vec_index` : identifyer endstruct
`res_vector_struct`	: struct of `ident` : identifyer `inst` : expression `matr_vec_subscr_exp` : expression `matr_vec_subscr_pos` : integer `matr_vec_index` : identifyer endstruct

must be `whole` and *step* must be `sweep`). Finally, the `dot_product` instance must be control dependent from the `scan[1]` instance.

The *matrix-matrix product* operation is defined in terms of the matrix-vector product, as:

$$MMP^{hskl} = [\cdots, VMP^{hkl}_{m}, \cdots] = \begin{bmatrix} \cdots & \cdots & \cdots \\ \cdots & DP^{hk}_{mn} & \cdots \\ \cdots & \cdots & \cdots \end{bmatrix} =$$

$$\begin{bmatrix} \cdots & \cdots & \cdots \\ \cdots \sum_i e(A(\cdots, \overset{h}{i}, \cdots, \overset{s}{m}, \cdots) \times B(\cdots, \overset{k}{i}, \cdots, \overset{l}{n}, \cdots)) & & \cdots \\ \cdots & \cdots & \cdots \end{bmatrix} \quad (2)$$

```
rule matrix_vector_product →
    composition
        scan[1]
        dot_product
        scan[2]
    condition
        local countLoopList1, countLoopList2 : [hierarchy]
              TERM[1], TERM[2] : instance
        control_dep(dot_product,scan[1],true)
        scan[1].hier = -(-,countLoopList1,-(TERM[1],-))
        TERM[1] = assign
        subexp_in_exp(scan[1].array_scan.array_inst,TERM[1].lhsExp)
        scan[2].hier = -(-,countLoopList2,-)
        countLoopList1 = countLoopList2
        scan[1].array_scan.scan_index = scan[2].array_scan.scan_index
        scan[1].range = scan[2].range = whole
        scan[1].stride = scan[2].stride = sweep
        ((scan[2].array_scan.array_inst = dot_product.array1_struct.inst &
          scan[2].array_scan.subscr_pos ≠ dot_product.array1_struct.subscr_pos)
         |
         (scan[2].array_scan.array_inst = dot_product.array2_struct.inst &
          scan[2].array_scan.subscr_pos ≠ dot_product.array2_struct.subscr_pos)
        )
        TERM[2] = last(dot_product.accum_struct.stmList)
        dep_chain(TERM[2],TERM[1])
```

Fig. 1. Recognition rule of the `matrix_vector_product` concept.

where VMP^{hkl} is the matrix-vector product, defined in 1, and DP^{hk}_{mn} is the dot product operation. The operation involves the h-th and s-th dimensions of the A array, and the k-th and l-th dimensions of the B array. The result is a bidimensional array, which can be assigned to a page of a multidimensional array.

The recognition rule for the `matrix_matrix_product` is presented in figure 2. We illustrate the 2 in the following.

Its main component is the `matrix_vector_product` concept. The other components are two instances of the `scan` concept. The first scan instance (`scan[1]`) scans a dimension of the array, which records the result of the matrix-vector product. The second scan instance (`scan[2]`) scans the s-th dimension (cfr. eq: 2) of the "vector" array of the matrix vector product. It has to be of course different from the dimension (h-th in 2) involved in the matrix-vector product. The two scans must share their `count_loop` subconcepts, in order to scan the two dimensions of the two arrays at the same time. These dimensions have to be completely scanned, in sequence (for both the scan concepts, the *range* attribute must be `whole` and *step* must be `sweep`). Finally, the `matrix_product` instance must be control dependent from the `scan[1]` instance.

```
rule matrix_matrix_product →
   composition
      scan[1]
      matrix_vector_product
      scan[2]
   condition
      local countLoopList1, countLoopList2 : [hierarchy]
      control_dep(matrix_vector_product,scan[1],true)
      scan[1].array_scan.array_inst =
         matrix_vector_product.res_vector_struct.inst
      scan[1].array_scan.subscr_pos ≠
         matrix_vector_product.res_vector_struct.matr_vec_subscr_pos
      scan[1].hier = -(-,countLoopList1,)
      scan[2].hier = -(-,countLoopList2,-)
      countLoopList1 = countLoopList2
      scan[1].array_scan.scan_index = scan[2].array_scan.scan_index
      scan[1].range = scan[2].range = whole
      scan[1].stride = scan[2].stride = sweep
      scan[2].array_scan.array_inst =
         matrix_vector_product.vector_struct.inst
      scan[2].array_scan.subscr_pos ≠
         matrix_vector_product.vector_struct.dot_prod_subscr_pos
```

Fig. 2. Recognition rule for the `matrix_matrix_product` concept.

5 Conclusion

In this paper a production-rule based hierarchical concept parsing recognition technique, and a formalism for the specification of algorithmic concepts have been presented.

The main contributions of the work presented can be summarized in the following: – definition of a formalism for the specification of algorithmic concepts, based on Higher Order Attributed Grammars, suitable for expressing in a flexible but exact way the compositional hierarchy and relationships among them, at any level within the hierarchy; – systematic utilization of structural properties (control, data dependence structure) more than syntactical one, in the definition and characterization of the algorithmic concepts; the utilization of powerful techniques for the analysis at the structural level (such as array dependences), and their abstraction; – utilization of symbolic analysis techniques for expressions, to face the syntactic variation problems not solvable through the algorithmic characterization at the structural level; – development of a technique for automated *hierarchical concept parsing*, which implements the mechanism of hierarchical abstraction defined by the grammar, utilizing the first order logic programming (Prolog); – representation of the concept instances recognition within the code, and of their hierarchical structure, through an *Abstract Hierarchical Program De-*

pendence Graph; – focus on algorithms and algorithmic variations within code developed for scientific computing and High Performance Computing.

References

1. T.J. Biggerstaff, "The Concept Assignment Problem in Program Understanding", Procs. *IEEE Working Conf. on Reverse Engineering*, May 21-23, Baltimore, Maryland, USA, 1993.
2. B. Di Martino, "ALCOR - an ALgorithmic COncept Recognition tool to support High Level Parallel Program Development", in: J. Fagerholm, J. Haataja, J. Jrvinen, M. Lyly, P. Rback, V. Savolainen (Eds.): *Applied Parallel Computing. Advanced Scientific Computing*, Lecture Notes in Computer Science, n. 2367, pp. 150-159, Springer-Verlag, 2002.
3. A. Bonifacio, B. Di Martino "Algorithmic Concept Recognition support for Skeleton Based Parallel Programming", Proc. of *Int. Workshop on High-Level Parallel Programming Models and Supportive Environments (HIPS'03)* - Nice (FR), 22-26/4 2003, Apr. 2003. IEEE CS Press.
4. J. Ferrante, K.J. Ottenstein and J.D. Warren, "The Program Dependence Graph and its use in Optimization", *ACM Trans. Programming Languages and Systems*, 9(3), pp. 319-349, June 1987.
5. J. Grosh and H. Emmelmann, "A Tool Box for Compiler Construction", *Lecture Notes in Computer Science*, Springer - Verlag, n. 477, pp. 106-116, Oct. 1990.
6. "Puma - A Generator for the Transformation of Attributed Trees", Compiler Generation Report n. 26, GMD Karlsruhe, July 1991.
7. M.T. Harandi and J.Q. Ning, "Knowledge-Based Program Analysis", *IEEE Software*, pp. 74-81, Jan. 1990.
8. D. E. Knuth, "Semantics of context-free languages", *Math. Syst. Theory*, 2(2) pp. 127-145, 1968.
9. W. Pugh, "A practical algorithm for Exact Array Dependence Analysis", *Communications of ACM*, 8(35), pp. 102-115, Aug. 1992.
10. H. Vogt, S. Swiestra and M. Kuiper, "Higher Order Attribute Grammars", Proc. of *ACM SIGPLAN Conference on Programming Language Design and Implementation*, pp. 131-145, June 1989.
11. L.M. Wills, "Automated Program Recognition: a Feasibility Demonstration", *Artificial Intelligence*, 45, 1990.

Modelling of Complex Systems Given as a Mealy Machine with Linear Decision Diagrams

P. Dziurzanski

Faculty of Computer Science and Information Systems, Technical University of Szczecin, Zolnierska 49, 71-210 Szczecin, POLAND
pdziurzanski@wi.ps.pl

Abstract. We propose a novel approach to transform an arbitrary Mealy machine, which is a kind of Finite State Machine (FSM) with outputs, into so-called Linear Binary Moment Diagrams (LBMDs). We stress the attractive features of this approach, and demonstrate the results of experiments on benchmarks in comparison with a state-of-the art Reduced Ordered Binary Decision Diagrams technique.

1 Introduction

Many practical problems in a field of sequential system optimization are described by finite state system over a sequence of state transitions. As classical Finite State Machine (FSM) offers only two outputs, namely 'accepted' and 'unaccepted', modifications of FSM with allows multiple output values are often used instead. The most popular FSMs of that type are Mealy and Moore machines. In this paper, we focus on Mealy machine, but it does not restrict applications of the approach, as each Moore Machine can be transformed into Mealy ones and vice-versa [6].

Although classic algorithms construct an explicit representation [3], these techniques are practical only for small-size problems. In order to overcome the problem, symbolic state graph methods were developed, in which the next state behavior is represented with a Boolean function [2]. Then, a Reduced Ordered Binary Decision Diagram (ROBDD) can be constructed [8]. It is beneficial if all the future operations are carried out only on that diagram.

Although successful for many class of switching functions, the above mentioned method has a number of drawbacks. Firstly, the number of nodes in ROBDD depends on variable ordering. Moreover, for a majority of functions the number of nodes remains exponential. As a result, a construction of ROBDD for characteristic functions is often unsuccessful due to the ROBDD size explosion.

In this paper we deal with word-level Binary Moment Diagrams (BMDs), introduced by Bryant et. al. in [1]. Since BMDs also depend on variable ordering in general, we take advantage of the only class of Decision Diagrams which are insensitive to the variable ordering, known as Linear Decision Diagrams (LDDs). Besides, as our experiments proved, for popular benchmark sets we can build diagrams with lower number of nodes than the typical ROBDD.

P.M.A. Sloot et al. (Eds.): ICCS 2003, LNCS 2658, pp. 758–765, 2003.

2 Principles

2.1 Characteristic Functions of Mealy Machines

Mealy machine is a 6-tuple $(S, I, O, \delta, \lambda, s_0)$, where S is the set of states, I is the set of input values, O is the set of output values, $\delta : S \times I \to O$ is the next state function, $\lambda : S \times I \to S$ is the next state function, and $s_0 \in S$ is the initial function.

In order to synthesis of FSM, we need to assign binary codes to the states. This process is called *state assignment*. After assigning, we can build a truth table, which input set consists of primary inputs x and current states s, whereas an output set consists of next states $\overset{+}{s}$ and primary outputs y. After minimizing, truth table is often referred as *cube table*.

As in mentioned above definition of Mealy machine there are two functions, from which one determines the next state and the other output values, it is convenient to build two characteristic functions for describing a behavior of the automata.

For more details about Mealy machines see, for example, [5,6].

The characteristic function χ_δ which describes the next state of Mealy machine is the function $f : S \times I \times S \to \mathbf{B}$, $f(x, i, y) = 1$ iff $y = \delta(x, i)$, where $i \in I$, $x, y \in S$, and $\mathbf{B} = \{0, 1\}$.

The characteristic function χ_λ which describes outputs of Mealy machine is the function $f : S \times I \to \mathbf{B}$, $f(x, i, z) = 1$ iff $z = \lambda(x, i)$, where $i \in I$, $x \in S$, $z \in O$, and $\mathbf{B} = \{0, 1\}$.

Example 1 *Let us consider a State Transition Graph presented in Fig. 1a. After assigning st0 with '0' and st1 with '1', we obtain a cube table presented in Fig. 1b. A characteristic function χ_δ has inputs x_1, x_2, s_1, $\overset{+}{s}_1$, whereas χ_λ has inputs x_1, x_2, s_1, y. Consequently, the characteristic functions in a SOP forms equal*

$$\chi_\delta = x_1 x_2 \overline{s}_1 \overline{s}_1^+ \vee \overline{x}_2 \overline{s}_1 \overline{s}_1^+ \vee \overline{x}_1 x_2 \overline{s}_1 s_1^+ \vee \overline{x}_1 s_1 s_1^+ \vee x_1 x_2 s_1 \overset{+}{s}_1,$$
$$\chi_\lambda = x_1 x_2 \overline{s}_1 \overline{y} + \vee \overline{x}_2 \overline{s}_1 \overline{y} \vee \overline{x}_1 s_1 y \vee x_1 x_2 s_1 \overline{y}.$$

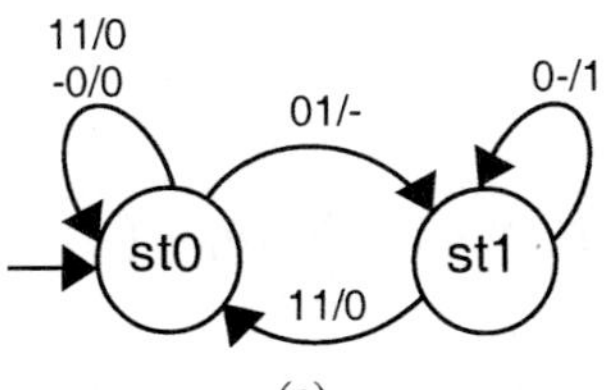

(a)

x_1	x_2	s_1	$\overset{+}{s}_1$	y
1	1	0	0	0
-	0	0	0	0
0	1	0	1	-
0	-	1	1	1
1	1	1	0	0

(b)

Fig. 1. An example of a State Transition Graph (a) and its cube table (b)

In our previous work, described in [4], we propose a usage of one characteristic functions, which incorporated both of the next state and outputs ones. Our previous approach makes the process of simulation, i.e. evaluating the outputs and next state performance, much more complex and time-consuming, what makes it inefficient in case of complex systems.

2.2 Linear Arithmetic Expressions

Having a cube table of both of characteristic functions as an input, we can derive a minimal SOP. The SOP consisting of p products is a two-level representation. In the first and the second level p ANDs and one OR functions are implemented, respectively. Each of these functions can be extended to a multi-output function generating a Linear Arithmetic Expressions (LAR) (1), and, therefore, by a Linear Binary Moment Diagram (LBMD).

An arbitrary function can be described by a corresponding arithmetical expression (AR), for example $\overline{x} = 1 - x$, $x_1 \vee x_2 = x_1 + x_2 - x_1 x_2$. In particular cases only these arithmetical expressions have a linear structure, including terms with no more than one variable. LAR for the n-variable logic function can be generally described by

$$d_0 + d_1 x_1 + \ldots + d_n x_n, \tag{1}$$

where d_i are arbitrary integers. Thus the term linearization means a transformation from nonlinear AR into corresponding LAR.

Taking into account that in a case of a word-level expression the value of a given Boolean function are on a specific bit, it is necessary to use a masking operator, which allow us to select an arbitrary bit. The masking operator selecting the i-th bit from an m-output function f is denoted by $\Xi^i_m\{f\}$.

The method of describing an n-variable elementary function (such as AND and OR) by a LAR is presented in [7]. Now we shortly recall this result.

Theorem 1 *(re-formulated from [7]). The n-variable elementary functions can be represented as follows:*

(i) n-variable AND function $AND = \bigwedge_{i=1}^{n} x_i^{\sigma_i}$ by the LAR

$$LAR_{AND} = 2^{j-1} - n + \sum_{i=1}^{n} (\sigma_i + (-1)^{\sigma_i} x_i), \tag{2}$$

generated by an m-output function, in which the function f is in the most left position indicated by the masking operator $\Xi^m_m\{LAR_{AND}\}$;

(ii) n-variable OR function $OR = \bigvee_{i=1}^{n} x_i^{\sigma_i}$ by the LAR

$$LAR_{OR} = 2^{j-1} - 1 + \sum_{i=1}^{n} (\sigma_i + (-1)^{\sigma_i} x_i) \tag{3}$$

generated by an m-output function, in which the function f is the most significant bit, that is indicated by the masking operator $\Xi^m_m\{LAR_{OR}\}$, where

$$j = \lceil \log_2 n \rceil + 1, \tag{4}$$

and

$$x_i^{\sigma_i} = \begin{cases} x_i \text{ if } \sigma_i = 0, \\ \overline{x}_i \text{ if } \sigma_i = 1. \end{cases} \tag{5}$$

The proof of the Theorem is given in [7].

2.3 Linear Decision Diagrams

Our goal is to represent a given sequential digital circuit by a set of two LBMDs. In Decision Diagrams a function is expanded with respect to one splitting variable x. As a result of this expansion there are two subfunctions. For a switching function f, f_x and $f_{\overline{x}}$ denote the positive and negative cofactor of f with respect to variable x, correspondingly. The subfunction f_x is obtained by substituting a constant 1 to x, whereas $f_{\overline{x}}$ is obtained by substituting a constant 0 to x.

In our method, we use masking operators and decision diagrams with arithmetic positive Davio (pD_A) decomposition $f = f_{\overline{x}} + x(f_x - f_{\overline{x}})$. The method of deriving this equation is presented, for example, in [1].

If arithmetic positive Davio is applied for decomposition of an LAR, it splits the function into two subfunctions $f_{\overline{x}}$ and $(f_x - f_{\overline{x}})$. It is noticeable, that both of them does not depend on the splitting variable x and the subfunction $(f_x - f_{\overline{x}})$ is always constant and equal to the coefficient of the variable x in LAR.

3 Strategy

Lemma 2 *An arbitrary characteristic function can be represented by a double LBMD.*

The proof is based on the following two statements:

- Let us have a characteristic function given by a cube table. Each cube in the cube table corresponding to t-th, $t \in (1, \ldots, p)$, product $y_t = x_1^{\alpha_1} \wedge \ldots \wedge x_l^{\alpha_l}$, $x_i^{\alpha_i} = \overline{x}_i$ when $\alpha_i = 0$ and $x_i^{\alpha_i} = x_i$ when $\alpha_1 = 1$, $l \in (1, \ldots, n)$, can be represented by an l-node LBMD A^X, and p products $y_1, \ldots, y_p$ with no more than l literals each, can be represented by the same LBMD,
- Any set of cubes in the cube table can be treated as a SOP $f = y_1 \vee \ldots \vee y_p$ of p product terms can be represented by an LBMD B^Y, and an r-output function with no more than p products each can be represented by the same LBMD.

Let us consider a characteristic functions. Their outputs describe a SOP with p products with no more than n literals. We extend each product y_t (AND function) to an n_t-output function in order to generate $LAR(y_t)$. The LARs for other products y_t form the weighed arithmetic sum:

$$A^X = 2^{J_1} LAR(y_1) + \ldots + 2^{J_p} LAR(y_p), \tag{6}$$

and, substituting the $LAR(y_t)$ (1) into the above equation yields the LAR

$$A^X = TW_0 + TW_1 x_1 + \ldots + TW_n x_n, \tag{7}$$

where the coefficients TW_i, TW_0 are formed as below

$$\begin{aligned} TW_i &= a_{i,1} 2^{J_1} + \ldots + a_{i,p} 2^{J_p}, \\ TW_0 &= a_{0,1} 2^{J_1} + \ldots + a_{0,p} 2^{J_p}, \end{aligned} \tag{8}$$

for $a_{i,t} \in \{0, +1, -1\}$ and positive integer numbers $a_{0,t}$;

$$\begin{aligned} J_1 &= j_0 = 0, \\ J_t &= j_0 + \ldots + j_{t-1}, \end{aligned} \tag{9}$$

and (equation (4))

$$j_t = \lceil \log_2 n_t \rceil + 1,$$

n_t is the number of outputs in the function obtained by extending of y_t. So, AND functions of the first level of the SOP are described by the LAR (7), and, so, by an LBMD A^X.

Then, characteristic functions χ_δ and χ_λ form a SOP $f = y_1 \vee \ldots \vee y_p$, i.e. a p-input OR function. The output f can be extended to a multi-output function, in order to generate $LAR(f)$. They are combined to the LAR:

$$B^Y = LAR(f),$$

that is equal to

$$B^Y = BW_0 + BW_1 y_1 + \ldots + BW_p y_p, \tag{10}$$

where the coefficients JB_j are formed using equations (9) and

$$\begin{aligned} BW_t &= b_{t,1} 2^{JB_1} + \ldots + b_{t,r} 2^{JB_r} \\ BW_0 &= b_{0,1} 2^{JB_1} + \ldots + b_{0,p} 2^{JB_r}. \end{aligned} \tag{11}$$

So, all OR functions (the second level of the SOPs) are described by (10), and so we can generate the LBMD B^Y.

Example 2 *(Continuation of Example 1) Since both of the characteristic functions χ_δ and χ_λ of the function f include 9 products, there is a need to extend these products y_t to an n_t-output function in order to generate $LAR(y_t)$, where $t = 1, \ldots, 9$.*

As the $t = 1$st product $y_1 = x_1 x_2 \overline{s}_1 \overline{s}_1^+$ includes $n_1 = 4$ variables, parameter $j_1 = 3$ (equation (4)). Accordingly to equations (9) and (6), $J_1 = 0$ and thus weight of $LAR(y_1)$ equals $2^{J_1} = 1$. Taking into account equation (2), $LAR(y_1) = 2 + x_1 + x_2 - s_1 - s_1^+$.

The remaining LARs of the A^X tree and their masking operators are presented in Table 1. As a result, LAR for all the cubes is equal to

$$\begin{aligned} A^X &= 2^0(2 + x_1 + x_2 - s_1 - s_1^+) + 2^3(4 - x_2 - s_1 - s_1^+) \\ &+ 2^6(2 - x_1 + x_2 - s_1 + s_1^+) + 2^9(2 - x_1 + s_1 + s_1^+) \\ &+ 2^{12}(x_1 + x_2 + s_1 + s_1^+) + 2^{15}(2 + x_1 + x_2 - s_1 - y) \\ &+ 2^{18}(4 - x_2 - s_1 - y) + 2^{21}(2 - x_1 + s_1 + y) \\ &+ 2^{24}(1 + x_1 + x_2 + s_1 - y). \end{aligned}$$

In the first level, there are nine products y_t. The characteristic functions χ_δ and χ_λ include y_1 to y_5 and y_6 to y_9, respectively. Consequently, in the second level the characteristic function χ_δ can be described as $\chi_\delta = y_1 \vee y_2 \vee \ldots \vee y_5$ and similarly χ_λ as $\chi_\lambda = y_6 \vee y_7 \vee \ldots \vee y_9$.

Since χ_δ includes $n_1 = 5$ literals, $j_1 = 4$ (equations (4) and (3)), the second LAR is equal to

$$B^Y = 2^0(7 + y_1 + y_2 + y_3 + y_4 + y_5) + 2^4(7 + y_6 + y_7 + y_8 + y_9).$$

Table 1. LARs and their parameters built from the cube table of the function f (Example 1)

t	y_t	j_t	J_t	$LAR(y_t)$	Masking operator
1	$x_1 x_2 \overline{s}_1 \overline{s}_1^+$	3	0	$2 + x_1 + x_2 - s_1 - s_1^+$	$\Xi_{27}^{3}\{A^X\}$
2	$\overline{x}_2 \overline{s}_1 \overline{s}_1^+$	3	3	$4 - x_2 - s_1 - s_1^+$	$\Xi_{27}^{6}\{A^X\}$
3	$\overline{x}_1 x_2 \overline{s}_1 s_1^+$	3	6	$2 - x_1 + x_2 - s_1 + s_1^+$	$\Xi_{27}^{9}\{A^X\}$
4	$\overline{x}_1 s_1 s_1^+$	3	9	$2 - x_1 + s_1 + s_1^+$	$\Xi_{27}^{12}\{A^X\}$
5	$x_1 x_2 s_1 s_1^+$	3	12	$x_1 + x_2 + s_1 + s_1^+$	$\Xi_{27}^{15}\{A^X\}$
6	$x_1 x_2 \overline{s}_1 \overline{y}$	3	15	$2 + x_1 + x_2 - s_1 - y$	$\Xi_{27}^{18}\{A^X\}$
7	$\overline{x}_2 \overline{s}_1 \overline{y}$	3	18	$4 - x_2 - s_1 - y$	$\Xi_{27}^{21}\{A^X\}$
8	$\overline{x}_1 s_1 y$	3	21	$2 - x_1 + s_1 + y$	$\Xi_{27}^{24}\{A^X\}$
9	$x_1 x_2 s_1 \overline{y}$	3	24	$1 + x_1 + x_2 + s_1 - y$	$\Xi_{27}^{27}\{A^X\}$

Finally, characteristics functions can be extracted with the masking operators $\chi_{delta} = \Xi_7^4\{B^Y\}$ and $\chi_{lambda} = \Xi_7^7\{B^Y\}$.

So, both of LBMDs A^X and B^Y representing the function f include 14 non-terminal nodes and 16 terminal nodes. These LBMDs are depicted in Fig. 2.

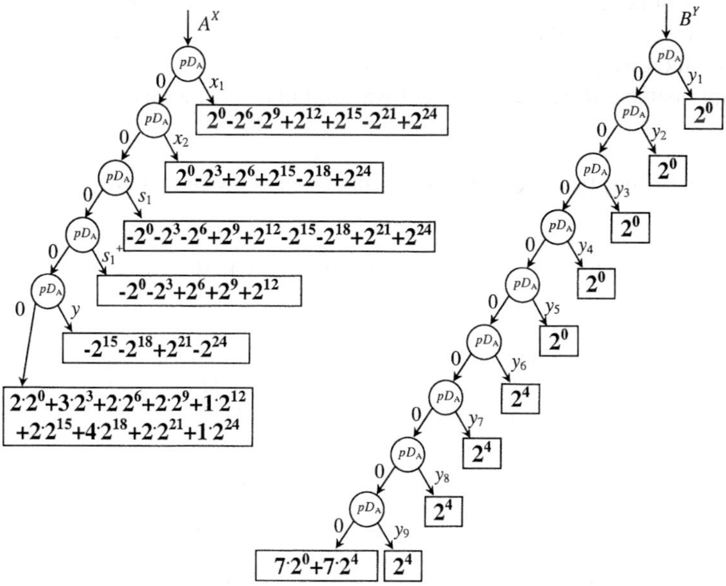

Fig. 2. Double LBMD of the function f (Example 2)

4 Experiments

The software implementation of the described above method were done in C++. In order to test the approach with standard benchmark library, we use FSMs given in KISS format from the LGSynth93 set. Using SIS [9], we performed a state assignment and conversion into 2-level SOP format. Each SOP, treated as the cube table of characteristic function, was then used for building a typical ROBDD using CUDD from VIS 1.3 [10] and LBMD, as proposed in this paper. Since variable ordering influences on the size of ROBDD, we used a well-known dynamic reordering method (sifting) to obtain ROBDDs of a reasonable size.

In Table 2 we give the number of nodes to represent some benchmarks by ROBDDs and LBMDs. The first two columns contain the FSM name and the number of primary inputs/outputs (**I/O**) and states (**S**). Next two columns show the number of nodes for ROBDD technique, and next columns contain the number of nodes for our LBMDs accordingly to the method described in this paper. The last two columns also contain the number of nodes for ROBBs and LBMDs, but these figures are taken from our previous work [4], where the two characteristic functions χ_δ and χ_λ were united into one.

In this experiment we use JEDI from SIS to perform state assignment, which makes an assignment in natural binary code. This assignment is useful in a case of implementation into CPLD structures. From the table it follows, that the proposed approach reduces the number of nodes by 96 per cent in comparison with traditional ROBDD technique: a total number of nodes in ROBDD equals 38246, whereas in LBMDS it is equal to 1485. Moreover, a separation of two characteristic functions χ_δ and χ_λ even slightly improved the result which we obtained in our previous work. However, benefits of this separation are mainly visible during the task of software simulation of a Mealy machine.

Table 2. Comparison of number of nodes of ROBDDs and LBMDs for LGSynth93 benchmarks

		Nodes in		Nodes in ([4])	
TEST	**I/O/S**	**ROBDD**	**LBMD**	**ROBDD**	**LBMD**
s1488	8/19/ 48	4101	164	3902	189
s1494	8/19/ 48	3907	171	3704	184
s208	11/2/ 18	603	49	551	55
s27	4/1/ 6	135	26	113	25
s298	3/6/ 218	11795	430	11713	417
s386	7/7/ 13	678	61	545	57
s1	18/19/ 20	3321	142	3181	107
s832	18/19/ 25	3031	128	2895	152
s820	18/19/ 25	5008	138	5260	152
s420	19/2/ 18	640	55	587	61
s510	19/7/ 47	5027	121	4855	104
Total:		38246	1485	37306	1503

5 Conclusion

We showed that characteristic table of an arbitrary Mealy machine can be transformed to a double Linear Binary Moment Diagrams. In comparison with popular method of representing characteristic function by Reduced Ordered Decision Diagrams, we obtain diagrams with much less nodes (in our experiments we obtained a reduction equal to 96%).

Acknowledgment. The author would like to thank Prof. V.P. Shmerko and Prof. S.N. Yanushkevich for helping and making valuable comments.

References

1. R. Bryant, Y. Chen, Verification of Arithmetic Functions Using Binary Moment Diagrams, *Proc. Design Automation Conf.*, 1995, pp. 535–541
2. J.R. Burch, E.M. Clarke, K. McMillan, Symbolic Model Checking, *Fifth Annual IEEE Symposium on Logic in Computer Science*, 1990, pp. 428–439
3. E.M. Clarke, E.A. Emerson, A.P. Sistla, Automatic verification of finite-state concurrent systems using temporal logic specifications, *ACM Transactions on Programming Languages 8,2*, 1986, pp. 244–263.
4. P. Dziurzanski , Representation of Finite State Machines with Linear Decision Diagrams, *Int. Conference on Signals and Electronic Systems (ICSES'2002)*, Swierardow Zdroj, Poland, 2002, pp. 161–168.
5. G.D. Hachtel, F. Somenzi, Ligic Synthesis and Verification Algorithms, *Kluwer Academic Publisher*, 1996
6. J. E. Hopcroft, J. D. Ullman, *Introduction to Automata Theory, Languages, and Computation*, Addison-Wesley Publishing Company, 1979
7. V. Malyugin, Realization of Boolean Function's Corteges by Means of Linear Arithmetical Polynomial, *Automation and Remote Control (USA)*, vol.45, no. 2, Pt. 1, 1984, pp. 239-245
8. R. Bryant, Graph - based algorithm for Boolean function manipulation, *IEEE Trans. on Computers*, 1986, vol. C-35, no. 8, pp. 667-691
9. SIS and Release 1.2, Logic synthesis benchmarks, *UC Berkeley Soft. Distr*, http:/www.cbl.ncsu.edu/www/, 1994.
10. VIS: A system for Verification and Synthesis, *Proceedings of the 8th Internationa conference on Computer-Aided Verification, Springer Lecture Notes in Computer Science, # 1102, Edited by R. Alur and T. Henziger, NJ, July 1996*, pp. 428-432

Empirical Evaluation of the Difficulty of Finding a Good Value of k for the Nearest Neighbor

Francisco J. Ferrer–Troyano, Jesús S. Aguilar–Ruiz, and José C. Riquelme

Department of Computer Science, University of Sevilla
Avenida Reina Mercedes s/n, 41012 Sevilla, Spain
{ferrer,aguilar,riquelme}@lsi.us.es

Abstract. As an analysis of the classification accuracy bound for the Nearest Neighbor technique, in this work we have studied if it is possible to find a *good value* of the parameter k for each example according to their attribute values. Or at least, if there is a pattern for the parameter k in the original search space. We have carried out different approaches based on the Nearest Neighbor technique and calculated the prediction accuracy for a group of databases from the UCI repository. Based on the experimental results of our study, we can state that, in general, it is not possible to know a priori a specific value of k to correctly classify an unseen example.
Keywords: Nearest Neighbor, Local Adaptive Nearest Neighbor.

1 Introduction

In Supervised Learning, systems based on examples (CBR, Case Based Reasoning) have been object of study and improvement from their introduction at the end of the fifties. These algorithms extract knowledge through inductive processes from the partial descriptions given by the initial set of examples or instances. Machine learning process is usually accomplished in two functionally different phases. In the first phase of Training a model of the hyperspace is created by the labelled examples. In the second phase of Classification the new examples are labelled based on the constructed model. In the Nearest Neighbor algorithm (from here on *NN*) the training examples are the model itself. *NN* assigns to each new query the label of its nearest neighbor among those that are remembered from the phase of Training. In order to improve the accuracy with noise present in data, the *k-NN* algorithm introduces a parameter k so that for each new example q to be classified the classes of the k nearest neighbors of q are considered: q will be labelled with the majority class or, in case of tie, it is randomly broken. Another alternative consists in assigning that class whose average distance is the smallest one or introducing a heuristically obtained threshold $k_1 < k$ so that the assigned class will be that with a number of associated examples greater than this threshold [10]. Extending the classification criterion, the k-NN_{wv} algorithms (Nearest Neighbor Weighted Voted) assign weights to the prediction made by each example. These weights can be inversely proportional to the distance with respect to the example to be classified [4, 6]. Therefore, the

P.M.A. Sloot et al. (Eds.): ICCS 2003, LNCS 2658, pp. 766–773, 2003.

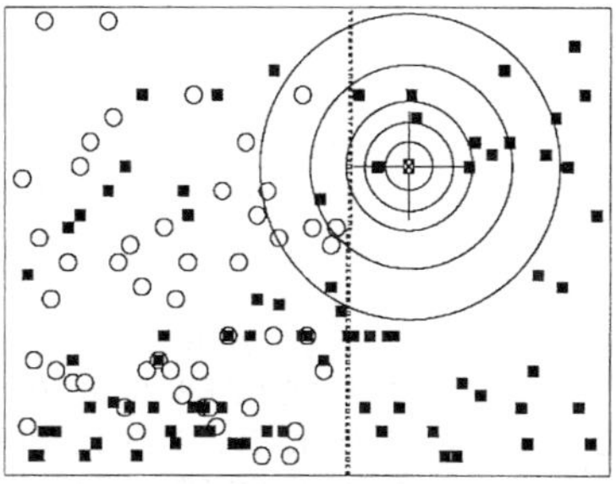

Fig. 1. Horse Colic database. If the new example to be classified is a central point, the k value slightly determines the assigned label.

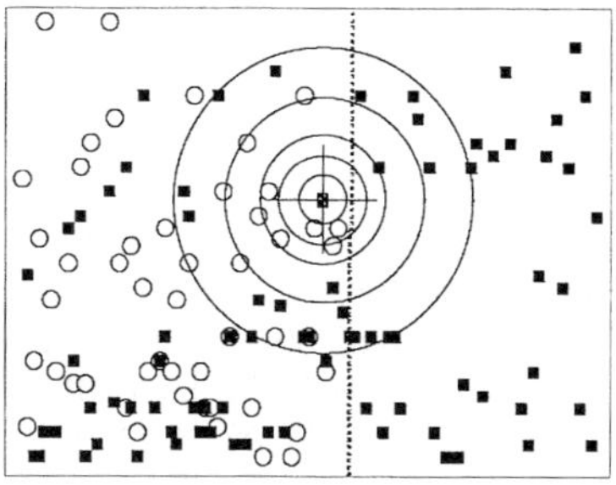

Fig. 2. Horse Colic database. If the new query is a border point, the k value can be decisive in the classification.

k number of examples observed and the metric used to classify a test example are decisive parameters. Usually k is heuristically determined by the user or by means of cross-validation [9]. The usual metrics of these algorithms are the Euclidean distance for continuous attributes and the Overlap distance for nominal attributes (both metrics were used in our experiments).

In the last years have appeared interesting approaches that test new metrics [13] or new data representations [2] to improve accuracy and computational complexity. Nevertheless, in spite of having a wide and diverse field of application, to determine with certainty when *k-NN* obtains higher accuracy than *NN* [1] and viceversa [8] is still an open problem. In [5] it was proven that when the distance among examples with the same class is smaller than the distance among examples of different class, the probability of error for *NN* and *k-NN* tends to 0 and $\frac{1}{2}$, respectively. But, not always this distribution for input data appears, reason why *k-NN* and $k\text{-}NN_{wv}$ can improve the results given by *NN* with noise present in the data. In [12] a study of the different situations in which *k-NN* improves the results of *NN* is exposed, and four classifiers are proposed (Locally Adaptive Nearest Neighbor, $localKNN_{ks}$) where for each new example q to be classified the parameter k takes a value k_q which is near to the values that classified the M (an extra parameter) nearest neighbors e_q of q.

In this work we intend to study the limits that the *k-NN* algorithm presents even when the value of k is not fixed but variable for each example. When an example as the Figure 1 illustrates is interior to a region of examples with its same label (it is a central point), the assigned label will depend little on the value of k. However, with an example near the decision boundaries (a border point, see Figure 2) the choice such parameter can be decisive. In the following section we explain several results obtained after applying the standard and weighted k–Nearest Neighbor algorithm (from now *k-NN* and $k\text{-}NN_{wv}$) with databases from the UCI repository [3]. In principle it seems logical to consider that the classification accuracy can improve when the k value is adjusted locally. In a previous work [7] we introduced a local nearest neighbor classifier which evaluates

Table 1. Percentage of examples that are correctly classified by kNN and kNN_{wv} where k is an odd number belongs to the interval [1,11].

	k-NN						*k-NN$_{wv}$*					
DB	k=1	k=3	k=5	k=7	k=9	k=11	k=1	k=3	k=5	k=7	k=9	k=11
An	92.53	88.97	87.63	83.96	85.30	85.52	92.53	91.2	91.87	91.42	91.42	92.09
Aud	74.33	66.81	63.71	59.73	58.40	60.61	74.33	76.1	73.45	72.56	69.02	69.46
Aut	75.60	66.82	60.97	57.07	57.07	57.56	75.6	77.07	78.04	75.6	73.17	71.7
BS	79.03	79.84	80.32	86.4	88.96	88.80	79.03	79.84	80.32	87.03	89.6	89.28
BC	70.27	69.58	72.02	74.12	74.12	74.82	70.27	68.53	71.32	72.72	74.12	74.47
CHD	74.58	81.84	81.51	82.17	81.51	81.18	74.58	80.19	80.85	82.17	81.51	81.51
CR	81.88	85.94	86.52	86.52	86.81	86.37	81.88	84.63	85.21	85.65	85.94	85.94
GC	72.60	73.00	73.30	72.89	72.89	73.20	72.6	73.0	73.0	73.1	73.1	73.7
Gl	70.09	68.22	64.01	61.21	58.87	57.47	70.09	71.49	72.89	70.56	68.69	68.22
HS	75.55	79.25	80.00	81.11	80.37	81.48	75.55	78.88	80.0	81.85	80.74	81.11
He	80.64	82.58	83.87	83.87	84.51	84.51	80.64	82.58	82.58	83.22	82.58	81.93
HC	68.47	69.29	69.02	70.38	69.83	69.02	68.47	70.65	71.46	73.64	73.09	70.92
Io	86.89	86.03	85.47	84.04	84.33	84.04	86.89	86.03	85.75	84.04	84.33	84.04
Ir	95.33	95.33	95.33	96.66	95.33	94.66	95.33	95.33	95.33	96.0	94.66	94.66
PD	70.57	74.08	74.08	75.26	73.82	73.43	70.57	73.95	73.69	74.86	73.95	73.56
PT	34.21	29.20	33.03	35.69	34.21	35.39	34.21	27.13	30.38	31.56	31.85	32.15
Son	87.50	83.65	82.21	80.28	75.96	72.59	87.5	83.65	82.69	82.69	81.25	77.4
Soy	91.80	91.80	91.06	90.48	90.19	89.31	91.8	91.8	91.94	91.21	91.36	91.06
Ve	69.85	68.43	67.73	68.91	67.49	66.90	69.85	71.04	71.63	71.74	69.85	70.21
Vot	91.03	91.49	92.64	93.56	93.10	93.56	91.03	91.26	91.95	92.87	92.87	93.1
Vow	99.39	97.97	94.24	89.89	83.53	42.72	99.39	98.08	97.27	96.06	94.94	94.74
Wi	95.50	96.62	96.06	96.06	96.06	95.50	95.5	96.62	96.06	96.62	96.62	96.06
WBC	95.27	96.56	96.99	96.85	96.85	96.70	95.27	96.56	96.99	96.85	96.7	96.7
Zoo	96.03	92.07	93.06	91.08	89.10	89.10	96.03	92.07	95.04	95.04	93.06	92.07
Av.	80.37	79.81	79.36	79.09	78.27	76.43	80.37	80.74	81.24	81.63	81.02	80.67

several k values to decide the label for a new query. But the results obtained in the really interesting domains were very similar to the results given by *k-NN*, so the added computational complexity (the calculation of this local k value) can not be worth. In the next empirical analysis we show that local classifiers based on geometric proximity present a limit for classifying very near the Nearest Neighbor prediction ability, and we try to find somehow an empirical measure for that limit.

2 Empirical Evaluation

In first place we obtained the error rates by leave-one-out validation increasing the value of k. The chosen limits for the maximum values of k were three: 11, 31 and 51. The results obtained for the odd numbers in the interval [1,11] applying *k-NN* and *k-NN$_{wv}$* are showed in Table 1. Table 2 shows the average values for k in [1,11], [1,31] and k in [1,51]. Observing both Tables we can state that

Table 2. Average percentage of examples that are classified by *k-NN* and *k-NN*$_{wv}$ with an odd value of k in the intervals [1,11], [1,31] and [1,51].

	k-NN			*k-NN*$_{wv}$		
DB	k in [1,11]	k in [1,31]	k in [1,51]	k in [1,11]	k in [1,31]	k in [1,51]
Anneal−	87.32	84.51	83.18	91.75	91.34	90.77
Audiology−	63.93	58.51	54.11	72.49	70.46	67.75
Autos−	62.52	56.28	53.45	75.20	73.32	71.76
Balance-Scale+	83.89	87.56	88.06	84.18	87.72	88.44
Breast-Cancer	72.49	73.68	73.60	71.91	73.51	73.66
Cleveland-HD+	80.47	81.47	82.07	80.14	81.82	82.40
Credit-Rating+	85.67	85.90	86.18	84.87	85.74	86.34
German-Credit	72.98	72.93	72.78	73.08	73.38	73.39
Glass−	63.31	60.57	59.99	70.32	67.31	65.79
Heart-Statlog+	79.62	81.20	82.16	79.69	81.06	81.92
Hepatitis	83.33	82.86	82.03	82.25	82.54	82.03
Horse-Colic−	69.33	67.45	66.96	71.37	69.98	69.34
Ionosphere−	85.13	83.26	80.11	85.18	83.60	81.10
Iris	95.44	95.58	95.25	95.22	95.41	95.56
Pima-Diabetes+	73.54	74.20	74.58	73.43	74.51	74.73
Primary-Tumor+	33.62	37.62	38.88	31.21	33.99	34.78
Sonar−	80.36	73.58	72.61	82.53	75.72	74.90
Soybean−	90.77	86.91	80.23	91.53	90.60	88.39
Vehicle−	68.22	67.20	65.97	70.72	69.49	68.36
Vote	92.56	92.39	91.94	92.18	92.35	91.90
Vowel−	84.62	34.40	21.17	96.75	95.63	95.63
Wine	95.97	96.34	96.45	96.25	96.52	96.65
Wisconsin-BC	96.54	96.54	96.35	96.51	96.55	96.38
Zoo−	91.74	86.26	80.69	93.89	92.69	90.74
Averages	78.89	75.72	74.12	80.94	80.63	80.11

the performance of both algorithms is very similar, although there is a slight tendency in favor of *k-NN*$_{wv}$ with regard to the obtained accuracy. From Tables 1 and 2 we can also observe that some databases have a high difficulty to be classified, for instance, *Primary-Tumor* or *Glass*. Aiming for obtaining a priori the best value of k for each example, we wonder: "what would the gain be if it is possible to find such a value for k?" , i.e., "how many examples are correctly classified for some value of k?". In other words: "how many examples are not correctly classified for any value of k by the Nearest Neighbor algorithm?".

That is, there is not a value of k for which most of the k nearest neighbors of an example has the same label as such an example. This value provides an interesting rate because it gives an error bound for the Nearest Neighbor algorithm, and generally, for any classification method based on geometric proximity.

To answer this question, we measured for each example all those values of k (among 1 and 51) that classified it correctly. If there was not value of k which classified a certain example correctly, this example was indicated as non-classifiable.

Table 3. Percentage of examples that are not able to be correctly classified by *k-NN* and $k\text{-}NN_{wv}$ for any k in the intervals [1,11], [1,31] and [1,51].

	k-NN			$k\text{-}NN_{wv}$		
DB	k in [1,11]	k in [1,31]	k in [1,51]	k in [1,11]	k in [1,31]	k in [1,51]
Anneal	2.78	2.56	2.45	4.90	4.56	4.34
Audiology	17.69	15.92	15.92	17.69	16.37	15.92
Autos	12.68	9.27	9.27	17.07	14.14	13.65
Balance-Scale	8.32	8.16	8.16	8.32	8.16	8.16
Breast-Cancer	14.68	12.93	12.58	17.83	16.78	16.43
Cleveland-HD	11.22	9.90	8.58	12.87	11.55	10.89
Credit-Rating	9.13	7.82	7.68	11.15	9.71	8.55
German-Credit	13.0	10.50	10.39	13.40	11.29	10.90
Glass	19.15	13.55	13.08	19.15	16.35	15.42
Heart-Statlog	9.63	9.26	8.89	10.0	9.26	8.89
Hepatitis	9.68	7.74	7.10	12.90	10.96	9.03
Horse-Colic	17.11	14.94	14.13	17.39	15.21	14.13
Ionosphere	8.55	6.84	6.84	8.55	7.69	7.69
Iris	3.33	2.67	2.0	4.0	3.33	2.0
Pima-Diabetes	14.19	11.19	10.28	14.71	11.97	11.19
Primary-Tumor	48.37	42.18	40.70	57.52	53.39	51.91
Sonar	6.25	4.33	4.33	6.25	4.33	4.33
Soybean	5.12	4.10	4.10	5.42	4.25	4.25
Vehicle	14.53	11.58	10.04	15.24	12.41	11.46
Vote	4.14	4.14	4.14	4.83	4.83	4.83
Vowel	0.50	0.50	0.50	0.50	0.50	0.50
Wine	1.68	1.12	1.12	1.68	1.12	1.12
Wisconsin-BC	2.0	1.72	1.72	2.0	1.72	1.72
Zoo	2.97	1.98	1.98	2.97	1.98	1.98
Averages	10.67	8.95	8.58	11.93	10.49	9.97

Table 3 indicates the percentage of non-classifiable points for both techniques according to the fixed limits. Let's observe *Horse-Colic*. The 17.11% of examples does not correctly classify with any value of k in [1,11], the 14.94% of examples does not correctly classify with any k in [1,31] and the 14.14% is not correctly classified with k in [1,51]. Thus, we can state that there is not significant difference among the limits 31 and 51, as well as between *k-NN* and $k\text{-}NN_{wv}$.

From Table 3 a maximum bound of the classification ability of *k-NN* can be obtained, still knowing a priori the value of k. That is, although *k-NN* could adapt locally so that for each example to be classified, according to the values of its attributes, we calculated the *best* k, the error rates given in Table 3 can not be avoided. However, there are some databases in which the improvement in the accuracy can be worth the computational effort (the calculation of that local k). So, taking again *Horse-Colic*, we can observe in Tables 1 and 3 that we would have an error rate of 14.14% instead of 30.77%, i.e., an improvement of around

Table 4. Percentage of examples that have at least a number *cvk* of common values of k which classify it correctly and classify its nearest neighbor correctly by means of *k-NN*, when $k \in [1, 51]$ and $cvk \in \{1, 3, 5, 7, 9, 11, 31, 51\}$.

DB/*cvk*	1	3	5	7	9	11	31	51
Anneal	93.65	91.09	88.86	86.41	85.30	84.18	78.06	63.91
Audiology	78.31	68.58	62.83	61.94	61.06	59.73	39.38	23.89
Autos	79.51	70.24	63.90	60.0	57.56	57.07	35.60	15.12
Balance-scale	84.0	82.24	81.92	81.44	81.12	81.12	78.88	53.12
Breast-cancer	76.92	68.53	66.78	66.43	66.08	65.73	61.18	29.37
Cleveland-HD	79.86	76.56	75.24	74.25	73.26	71.94	68.31	53.46
Credit-rating	84.20	82.6	81.44	80.28	80.28	80.0	75.79	60.43
German-credit	77.10	69.19	66.0	64.50	63.70	62.70	56.49	32.80
Glass	71.49	64.95	61.68	57.0	55.14	54.20	46.72	31.30
Heart-statlog	80.37	73.33	73.33	72.59	71.11	71.11	69.62	55.18
Hepatitis	85.16	82.58	81.93	77.41	76.77	76.77	69.67	59.35
Horse-colic	70.65	63.58	58.96	56.79	55.97	54.89	51.63	31.52
Ionosphere	89.17	85.18	83.19	82.90	82.05	81.48	69.80	61.53
Iris	96.0	96.0	95.33	94.66	94.66	94.66	94.66	86.66
Pima-diabetes	75.52	70.44	68.75	67.57	66.53	66.01	57.94	38.28
Primary-tumor	36.87	34.21	31.26	30.97	30.08	29.79	23.0	9.44
Sonar	91.34	87.01	83.17	79.32	74.03	69.23	61.05	39.42
Soybean	92.38	91.36	90.04	88.72	87.84	87.70	73.20	54.02
Vehicle	76.71	69.26	65.13	62.17	60.28	59.33	50.35	33.68
Vote	92.41	90.80	90.11	90.11	89.19	89.19	88.04	84.13
Vowel	99.49	96.66	90.70	85.65	77.07	35.75	0	0
Wine	98.87	98.31	97.75	97.75	96.62	96.62	94.94	87.07
Wisconsin-BC	95.56	94.27	94.13	94.13	94.13	94.13	93.13	89.98
Zoo	96.03	94.05	91.08	89.10	86.13	85.14	78.21	60.39
Averages	83.39	79.20	76.81	75.08	73.58	71.18	63.15	48.08

50%. In general, logically, the highest increment is given for those databases that we point out previously as difficult to be classified by means of *k-NN*.

Related to our initial objective that was to find a relationship among the attributes values of any example and some *correct value* of *k* to classify it correctly, we chose two databases with a significant gain, *Glass* and *Horse-Colic*. Next, for each domain we built a new database, where the label of each example was replaced by the minimum value of *k* for which such example was correctly classified. Then different approaches were attempted to predict the parameter *k*: lineal and quadratic regression through traditional statistical methods, *C4.5* (where the leaves in the decision tree obtained are possible values of *k*, and the Nearest Neighbor algorithm itself in a similar form to Locally Adaptive NN methods [11]. None of these techniques was able to improve the average error rate obtained by the standard *k-NN* by applying ten-folds cross-validation. Note that it is not necessary applying again the Nearest Neighbor algorithm to validate this last prediction approach because for each point we had calculated if a certain k value gave a correct classification.

In a second approach we consider that maybe the problem could be in the choice of the minimum k as the label of the database, due to the possible relationship between the original space of attributes and the k value could be formed by a set of different k values. These new values might not necessarily coincide with the smallest k. In order to solve this problem and to obtain more exact information, a second database was built. In this new database, the label of each example was replaced with a set of values that classified correctly such example. Given the special features of these data sets (multi-labelled), we carried out different approaches through regressions (lineal, quadratic, and quotient of polynomials). In this manner, the adjustment was correct if for each point the value obtained by means of regression was some of the *k-labels* associated the point. However, like the previous case, this type of regressions presented some results that did not improve the basic Nearest Neighbor algorithm.

Finally, in order to measure what extent reaches the relationship between the k obtained for each example and the region of the space where this example is located, it was calculated the number of common k values that classified an example and its nearest neighbor. The results are shown in Table 4. In this Table we can observe that, again for *Horse-Colic*, only the 70.65% of examples have at least a k value shared with its nearest neighbor. In addition, this percentage decreases quickly when increasing the requirement that the number of shared k values must be higher. This means that if we tried to predict the k that classifies a point correctly according to the k that classified its nearest neighbors correctly, we would have a minimum error rate of 29.35%, that is, the points for which its nearest neighbor have not any value that classifies it correctly. It is important to notice that the values in Table 4 provide a superior bound of the probability to *guess* the parameter k in function of the nearest neighbor of an example, but it does not mean that this probability will be reached easily. In fact, we can observe that for the databases that we have denominated difficult, with 3 or 4 common values, the percentage is so low that it seems complicated to determine the correct k by means of the Nearest Neighbor.

3 Conclusions and Future Directions

A priori, we could consider that the value of k to classify a new query through the Nearest Neighbor must depend on the space region in which the such example is located. Thus, when the point is central, the value of k can be low, and when it is near to the decision boundaries, such value must be higher. Nevertheless, after our study, we can conclude that it is not possible to determine with certainty the relationship between the attribute values for a particular point and the values of k that classifies it correctly through *k-NN*. To reach this observation, different tests have been carried out on databases from the UCI repository trying to establish which are the accuracy that *k-NN* gives as classification method. In this sense, we infer that to find a space distribution of the values of k in the attributes space is not an easy task. As sample, it is enough with verifying that the percentage of common values of k between a point and its nearest neighbor falls quickly and, therefore, the disposition in concrete regions of the values of k for a later correct estimation of the same one does not seem feasible. At least, by applying the traditional tools as regression, 1-NN or C4.5.

For future works we are trying to predict the correct value of k for each point from the original search space using genetic programming, which provides a capacity for obtaining regressions that are not bound by previous conditions. Another possible approach that we are studying is to consider the value of k based on the *enemy* instead of the nearest neighbor, since this can provide us a measurement of the proximity from a point to the decision bound of the region in which it is.

Acknowledgment.

The research was supported by the Spanish Research Agency CICYT under grant TIC2001-1143-C03-02.

References

1. D. W. Aha, D. Kibler, and M. K. Albert. Instance-based learning algorithms. *Machine Learning*, 6:37–66, 1991.
2. S. Arya, D. M. Mount, N. S. Netanyahu, R. Silverman, and A. Wu. An optimal algorithm for nearest neighbor searching. In *Proceedings of 5th ACM SIAM Symposium on discrete Algorithms*, pages 573–582, 1994.
3. C. Blake and E. K. Merz. Uci repository of machine learning databases, 1998.
4. S. Cost and S. Salzberg. A weighted nearest neighbor algorithm for learning with symbolic features. *Machine Learning*, 10:57–78, 1993.
5. T. M. Cover and P. E. Hart. Nearest neighbor pattern classification. *IEEE Transactions on Information Theory*, IT-13(1):21–27, 1967.
6. S.A. Dudani. The distance-weighted k-nearest-neighbor rule. *IEEE Transactions on Systems, Man and Cybernetics*, SMC-6, 4:325–327, 1975.
7. F. J. Ferrer, J. S. Aguilar, and J. C. Riquelme. Nonparametric nearest neighbor with local adaptation. In *Proceedings of the* 10th *Portuguese Conference on Artificial Intelligence*, Porto, Portugal, December 2001.
8. R. C. Holte. Very simple classification rules perform well on most commonly used datasets. *Machine learning*, 11:63–91, 1993.
9. M. Stone. Cross-validatory choice and assessment of statistical predictions. *Journal of the Royal Statistical Society B*, 36:111–147, 1974.
10. I. Tomek. An experiment with the edited nearest-neighbor rule. *IEEE Transactions on Systems, Man and Cybernetics*, 6(6):448–452, June 1976.
11. D. Wettschereck and T. G. Dietterich. An experimental comparison of nearest neighbor and nearest hyperrectangle algorithms. *Machine Learning*, 19(1):5–28, 1995.
12. D. Wettschereck and T.G. Dietterich. Locally adaptive nearest neighbor algorithms. *Advances in Neural Information Processing Systems*, (6):184–191, 1994.
13. D. R. Wilson and T. R. Martinez. Improved heterogeneous distance functions. *Journal of Artificiall Intelligence Research*, 6(1):1–34, 1997.

Replicated Ambient Petri Nets

David de Frutos Escrig and Olga Marroquín Alonso

Departamento de Sistemas Informáticos y Programación
Universidad Complutense de Madrid, E-28040 Madrid, Spain
{defrutos,alonso}@sip.ucm.es

Abstract. Recently we have introduced Ambient Petri nets, as a multilevel extension of the Elementary Object Systems, that can be used to model the concept of nested ambients from the Ambient Calculus. Both mobile computing and mobile computation are supported by that calculus, and then by means of our Ambient Petri nets we get a way to introduce in the world of Petri nets these important features of nowadays computing. Nevertheless, our basic proposal does not yet provide the suitable background for the modeling of replication, one of the basic operators from the original calculus, by means of which infinite processes are introduced and treated in a very simple way. In this paper we enrich our framework by introducing that operator. We obtain a simple and nice model in which the basic nets are still static and finite, since the dynamics of the systems can be covered by the adequate notion of marking, where all the copies generated by the application of the replication operator will live together, without interfering in an inadequate way.

1 Introduction

Internet provides a computational infrastructure that spans the planet. Using it we get a nice support for both *mobile computing* and *mobile computation*. Mobile computing refers to virtual mobility (mobile software), while mobile computation refers to physical mobility (mobile hardware). Both kinds of mobility are elegantly modelled by the *Ambient Calculus* [3, 4].

The Ambient Calculus allows the movement of self-contained nested environments that include data and live computation. Those computational ambients, which are defined as bounded places where computation proceeds, have a *name*, a collection of *local processes*, and a collection of *subambients*. Ambients can move in and out of other ambients by means of *capabilities*, that are associated with ambient names. There exist three kinds of capabilities: *in* for entering an ambient, *out* for exiting an ambient, and *open* for opening up an ambient.

We are interested in translating this framework into the world of Petri nets. R.Valk has already studied his *Elementary Object Systems* (EOS) [9–11], that are composed of a system net where several object nets move. These object nets behave like ordinary tokens of the system net, that is, they lie in places and are moved by transitions, but also they may change their internal state (their marking), either when executing their own internal transitions or as a result of an interaction with a system transaction.

P.M.A. Sloot et al. (Eds.): ICCS 2003, LNCS 2658, pp. 774–783, 2003.

Elementary Object Systems provide two-level systems, but to cope with the features of the Ambient Calculus, we need arbitrary nesting. Then we had to unify both frameworks by defining *ambient Petri nets* [5], which allow the arbitrary nesting of object nets in order to model the arbitrary nesting of ambients.

Although the Ambient Calculus is a relatively simple model, we have found several technical (but interesting) difficulties related with the covering of different features of the calculus. Therefore, in order to give a clear and well motivated presentation, we have decided to introduce the model in an incremental way, by means of a series of papers, such that each one of them will focus on some of those features. So, in our first paper we have just considered the mobility primitives of the calculus and the parallel operator, which are enough to get a first notion of ambient net.

In this paper we enrich its definition in order to introduce the replication operator from the Ambient Calculus, $!P$, which generates an unbounded number of parallel copies of P. Besides, we will shortly explain how the restriction operation, $(\nu n)P$, which is used to introduce new names and limit their scope, interacts with that new operator.

Our final goal is to use ambient Petri nets to provide a denotational semantics for the Ambient Calculus. The way we follow, in order to encompass an algebraic formalism together with another Petri net based one, is that of the *Petri Box Calculus* (PBC) [1, 2, 6], which has been proved to be a suitable framework for these purposes. Therefore, our new model can be also interpreted as an extension of PBC that includes ambients [5].

In fact, we have a large experience in the development of PBC extensions. So, in [8] we have defined an elaborate timed extension, while in [7] we have presented a stochastic extension. By means of this new mobile version of PBC we try to introduce the ideas from the Ambient Calculus, which allows to model mobile systems in a simple but formally supported way. We hope that the developers of mobile systems who are familiar with Petri nets, will find in our formal model a tool to define those rather complicated systems, providing a formal basis for proving that the modelled systems fulfill their specifications.

1.1 Background

In this section we give a short overview of our basic model of ambient Petri nets.

A simple *ambient Petri net* is a finite collection of named Petri nets $A = \{n_1 : A_1, n_2 : A_2, \ldots, n_k : A_k\}$ for which we introduce a *location pair*, $\langle loc, open \rangle$, that defines both the current location of each component net, $A_i = (P_i, T_i, W_i)$ with $i \geq 2$, and its (open or closed) state. Intuitively, nets $\{A_2, \ldots, A_k\}$ can be seen as net tokens that move along the full set of places of A, thus representing the movement of ambients. As a consequence, it is possible to find in the places of an ambient Petri net both ordinary and high-level tokens. By unfolding the latter we obtain the collection of nested marked ordinary nets that constitute the ambient net, in which we say that n_1 is the *root net*.

In order to adequately support the mobility of ambients due to the execution of capabilities, each ambient Petri net has two kinds of transitions. Besides the ordinary ones there is a set of high-level transitions, that we call *ambient transitions*, $Amb(A) = \{In\ n_i, Out\ n_i, Open\ n_i \mid i \in \{2, \ldots, k\}\}$. Those ambient transitions are used for controlling the movement of the object tokens in A. Their firing is synchronized with the firing of the transitions in the component nets labelled by elements in $\mathcal{C} = \{in\ n_i, out\ n_i, open\ n_i \mid i \in \{2, \ldots, k\}\}$, thus modifying both the location of the modified component and the internal state of the triggering low-level token transition which represents the capability to move that component.

In the Ambient Calculus, the entry capability can be used by a process $in\ m.P$, to instruct the surrounding ambient to enter into a sibling ambient named m, as stated by the reduction rule $n[in\ m.P|Q]|m[R] \rightarrow m[n[P|Q]|R]$. In the case of ambient Petri nets we say that two net tokens are siblings if they are located in places of the same component. The firing of $(In\ n_i, in\ n_i)$ pairs will move a component net into another, but since we need the particular place where the token will be allocated, we will provide together with the capability the name of that place, thus having some $(in\, n_i, p_i)$ as the label of the corresponding transition.

Concerning the exit transitions, modelled in the Ambient Calculus by the capabilities $out\, m$, they produce the exit of the surrounding ambient of $out\, m.P$ from its containing ambient m, such as the corresponding reduction rule shows: $m[n[out\, m.P|Q]|R] \rightarrow m[R]|n[P|Q]$. Then, in the ambient Petri net model whenever a net token n_j located at n_i may fire an $out\ n_i$ transition, we can fire the ambient transition $Out(n_i, n_j)$ by moving n_j into the location of n_i.

Finally, ambients can be open (and destroyed) by using an opening capability, $open\, m$. Thus, $open\, m.P$ provides a way of dissolving the boundary of an ambient named m located at the same level that this process, according to the rule $open\ m.P|m[Q] \rightarrow P|Q$. The firing of pairs of the form $(Open\ n_i, open\ n_i)$ has the same effect in an ambient Petri net that the one described. Nevertheless, since we are interested in an static description of ambient nets, we do not remove the open net, but just attach to it a label that describes its (open) state. In this way, its contents will be treated in the future as parts of the containing component.

The execution of ordinary transitions, which are not labelled by capabilities, follows the firing rule for the ordinary Petri nets: The tokens in the preset places of the involved transition are consumed, and instead new tokens are added into the postset places.

As a consequence of mixing both ordinary and ambient transitions, we obtained a new class of Petri nets suitable for modeling mobile agents based systems. Nevertheless, this new framework does not support any mechanism of replication, which is widely used in order to represent replication of services. With this purpose, the Ambient Calculus provides expressions of the form $!P$ that represent an unbounded number of parallel replicas of P, and whose behaviour is reflected by the structural congruence relation between $!P$ and $P|!P$: $!P \equiv P|!P$.

The extension of the ambient nets with such a mechanism results in the definition of *Replicated Ambient nets*, which are described in the following section.

2 Formal Definitions

In order to define replicated ambient Petri nets we extend the ambient Petri nets in [5] to include the elements that will make possible the translation of the replication operator from the Ambient Calculus. We will motivate each necessary extension by means of a simple example. In each case we will study a term from our calculus, which combines operators from both the Ambient Calculus and PBC.

This calculus mixes together capabilities and ordinary actions that belong to a countable alphabet of labels $\mathcal{A}$, and provides some operators to combine processes in a compositional way. Amongst those operators, the new language includes the sequential composition (_; _), the parallel composition (_|_) and the synchronization (_ sy a), all inherited from PBC, together with the replication operator (!_).

The sequential composition is a generalized form of the prefix operator. In the parallel composition $P|Q$ the involved actions are independently executed by P and Q, either sequentially or concurrently, but without any communication between them. Therefore, if $P \xrightarrow{\Gamma} P'$ and $Q \xrightarrow{\Delta} Q'$ then $P|Q \xrightarrow{\Gamma+\Delta} P'|Q'$, where $\Gamma, \Delta \in \mathcal{M}(\mathcal{A})$. Finally, in order to support synchronization, we assume the existence of a bijection $\widehat{\ }: \mathcal{A} \longrightarrow \mathcal{A}$, called *conjugation*, by means of which we associate to each label $a \in \mathcal{A}$ the corresponding $\hat{a} \in \mathcal{A}$. This function must satisfy that $\forall a \in \mathcal{A} \quad \hat{a} \neq a \wedge \hat{\hat{a}} = a$. Then synchronization, which is never forced, is modeled by means of pairs of conjugated actions in such a way that if $P \mathsf{sy}\, a \xrightarrow{\{\alpha+a\}+\{\beta+\hat{a}\}+\Gamma} P' \mathsf{sy}\, a$ then $P \mathsf{sy}\, a \xrightarrow{\{\alpha+\beta\}+\Gamma} P' \mathsf{sy}\, a$. By applying this operator over terms of the form $P = P_1|P_2$ we obtain the usual communication mechanism between processes, although, as stated above, it is not mandatory, since $P \mathsf{sy}\, a$ can still mimic the behaviour of P: Whenever $P \xrightarrow{\Gamma} P'$ we have also $P \mathsf{sy}\, a \xrightarrow{\Gamma} P' \mathsf{sy}\, a$.

Example 1. Let us consider the term $!a$. In order to represent the creation of a new clone of the body of the replication, as stated by the expansion $!a \equiv a|!a$, we introduce a τ-transition connected to the entry place of the process (Figure 1(a)). In this way, whenever a new copy of a is needed, the τ-transition is fired generating two new tokens: One of them will occupy the precondition of a, which allows to initiate a new execution of that body expression, while the other one stays in the precondition of the τ-transition awaiting for a new replication of the process. In our graphical representations we will omit τ-transitions by replacing them with arcs that directly connect the involved places. As a consequence, for the net in Figure 1(a) we would get that in Figure 1(b).

By applying this definition we would lose safeness of the markings, since by firing the initial τ-transition one can create as many tokens as desired in the entry places of the net corresponding to the body of the operator. Nevertheless, this is necessary in order to represent in the same net the parallel execution of an unbounded number of replicas of the net. However, in some cases it is important to separate the execution of those copies, to adequately translate the semantics of the replication operator.

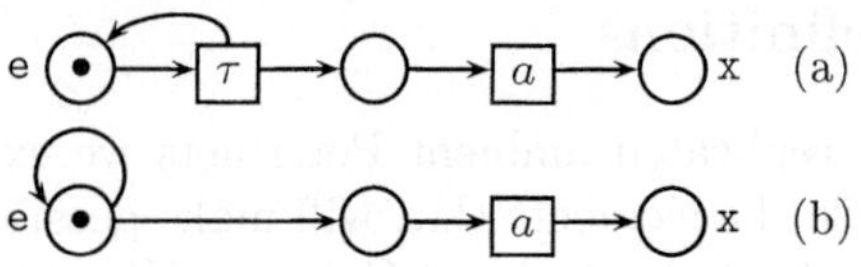

Fig.1. (a) Representing replication (b) Compact representation (omiting τ's)

Example 2. Let us consider the term $!((a|\hat{a})\mathtt{sy}\,a)$. If we apply the simple translation from the previous example we would get the net in Figure 2(a), once we just remove all the tokens in its places. Then, if we expand twice the replication operator we would get the expression $((a|\hat{a})\mathtt{sy}\,a)|((a|\hat{a})\mathtt{sy}\,a)|!((a|\hat{a})\mathtt{sy}\,a)$, which in the net will produce a couple of tokens in each of the entry places of the encoding of the term $(a|\hat{a})\mathtt{sy}\,a$. How do we represent the joint firing of a from the first copy and $\hat{a}$ from the second?. If the textual expression performs both actions we obtain a term in which we cannot fire the synchronizing transition without a new expansion of the replication operator:

$$\begin{aligned}((a|\hat{a})\mathtt{sy}\,a)|((a|\hat{a})\mathtt{sy}\,a)|!((a|\hat{a})\mathtt{sy}\,a) &\xrightarrow{a} (\hat{a}\mathtt{sy}\,a)|((a|\hat{a})\mathtt{sy}\,a)|!((a|\hat{a})\mathtt{sy}\,a)\\ &\xrightarrow{\hat{a}} (\hat{a}\mathtt{sy}\,a)|(a\mathtt{sy}\,a)|!((a|\hat{a})\mathtt{sy}\,a)\end{aligned}$$

Instead, in the net representation we can use the available tokens from the two replicas to fire the synchronizing transition. We conclude that it is necessary to personalize the tokens from each activation of a replication, to prevent from consuming tokens of several copies in the firing of a transition, that should be restricted to single copies of the body. This is done by labelling tokens in the scope of a replication operator with a natural number that identifies the serial number to which it corresponds. So we get the representation in Figure 2(b).

But in order to cover the case of nested replicated operators we still need another generalization.

Example 3. Let us consider the term $!(a;!b)$. Each activation of the main replication operator generates a term $a;!b$, where still a collection of different copies of b would be generated. This behaviour is reflected by the following computation, in which we can execute b actions belonging to different replicas of the term $a;!b$:

$$\begin{aligned}!(a;!b) \equiv (a;!b)|(a;!b)|!(a;!b) &\xrightarrow{a} !b|(a;!b)|!(a;!b) \xrightarrow{a} !b|!b|!(a;|b)\\ \equiv (b|b|!b)|(b|!b)|!(a;!b) &\xrightarrow{b} (b|b|!b)|!b|!(a;!b)\end{aligned}$$

The solution to link each replica of b with the corresponding clone of $a;!b$, that is, to individualize the involved tokens in a simple and satisfactory way, is to

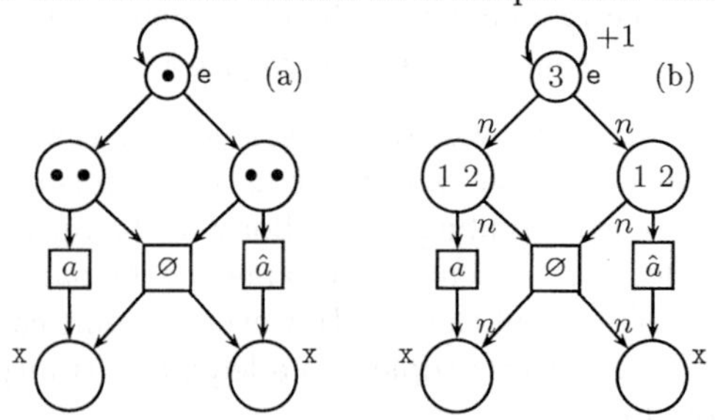

Fig.2. (a) Encoding with plain tokens (b) Encoding with individualized tokens

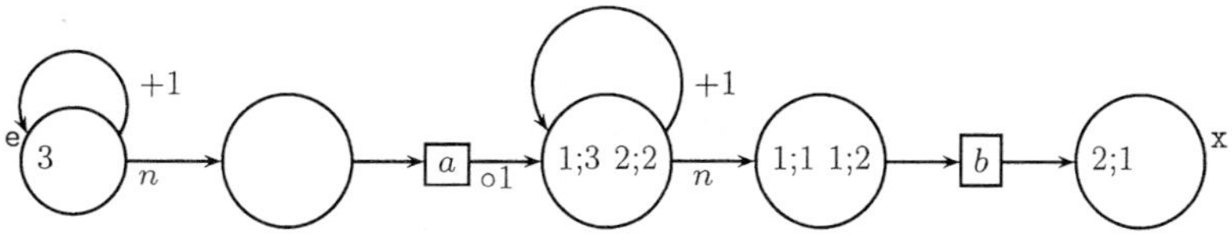

Fig.3. Encoding with sequences of natural numbers

label the tokens with a sequence of natural numbers, with an element for each nested replicated operator. Thus, for the given computation we would get the marked net in Figure 3.

For any place in a subnet which is not in the scope of any replication operator, the tokens in it would be labelled with the empty sequence ϵ.

With all this in mind we can give the definitions for replicated ambient Petri nets. Following the ideas from the Petri Box Calculus we distinguish static (unmarked) nets and dynamic (marked) ones. The first constitute the fixed architecture of the corresponding system, which remains the same along its execution. Therefore, static replicated ambient Petri nets are defined as in the simple non-replicated case (see [5]), although now component nets can contain τ-transitions which will be used to generate the copies of the replicated subnets of the system. Besides, we need some annotations in the arcs to cope with the adequate labelling of the tokens involved in the firing of some transitions. Therefore, we start with a given set of ambient names $\mathcal{N}$, and then we have:

Definition 4 (Component net). *A* ***component net*** *is an ordinary Petri net* $N = (P, T, W)$, *where:*

- P *is the set of places, which is partitioned into three subsets:* E *of* ***entry places****,* X *of* ***exit places*** *and* I *of* ***internal places****,*
- T *is the set of transitions, disjoint from* P*, that contains two subsets:* $\mathcal{I}nt(T)$*, composed of* ***internal transitions****, and* $\mathcal{A}mb(T)$*, constituted by* ***ambient transitions*** *which are labelled by pairs of the form* (cap, n) *with* $cap \in \{in, out, open\}$ *and* $n \in \mathcal{N}$*,*
- $W \subseteq (P \times T) \cup (T \times P)$ *is the set of connection arcs. Some of its elements could have an annotation in the set* $L \cup \{+1, \circ 1\}$*, where* L *is a set of variables to denote labels of tokens,* $+1$ *represents the addition of one unit to the last term of the label of the involved token, and* $\circ 1$ *represents the concatenation with a new* 1*.*

Example 5. The (single) component net for the term $!(a; !b)$ would be the net in Figure 3, if we just remove all the tokens from its places.

Definition 6 (Static replicated ambient Petri net). *A* ***static replicated ambient Petri net*** *is a finite collection of named component nets* $A = \{n_1 : A_1, n_2 : A_2, \ldots, n_k : A_k\}$ *where* $k \geq 1$, $n_1 = root$ *and* $\forall i, j \in \{1, \ldots, k\}\ n_i \in \mathcal{N} \wedge (i \neq j \Rightarrow n_i \neq n_j)$*.*

In the following, we will call to A_1 the root net of A and $\mathcal{N}(A) = \{n_2, \ldots, n_k\}$ will denote the ambient nets that can be used in the system. For each one of them we have the corresponding component which defines the behaviour of each copy of this ambient. In ordinary ambient nets we cannot replicate any term, which is captured by the fact that each component can be only performed once, but now we can begin the execution of a new copy of any ambient whenever a τ-transition is fired, reflecting the replication of the corresponding subnet. As a consequence, in a dynamic replicated ambient Petri net we may have several copies of some ambients. Each one of those replicas can be identified by the set of tokens that are involved in its execution.

Definition 7 (Ambient copies). *Given a replicated ambient net A, a set of **ambient copies** for it is defined by means of a function $Copies : \mathcal{N}(A) \rightarrow \mathcal{P}_f(\mathbb{N}^*)$ which verifies that $\forall c, c' \in Copies(n)\ |c| = |c'|$.*

As stated above, the values in $Copies(n)$ denote the set of labelled tokens that have been used to fire a new copy of the ambient. Since the replicated ambient will be situated in a fixed place of the system, then all the tokens that can visit that place will have the same length, this is why we impose that all the copies of each ambient will be labelled by strings of the same length. More exactly, any replica of an ambient n will be denoted by a pair composed of the name of this ambient and the label c of the token that is involved in its activation.

The behaviour of the copies of an ambient is not interconnected, so the location of each ambient copy of $n \in \mathcal{N}(A)$ must be defined independently by means of the location pair corresponding to A. Due to the arbitrary nesting of component nets, an ambient copy of $n \in \mathcal{N}(A)$, $\langle n, c\rangle$, could be located in a place p' of an ambient copy of $n' \in \mathcal{N}(A)$, $\langle n', c'\rangle$, which is denoted by $loc(\langle n, c\rangle) = (\langle n', c'\rangle, p')$, and because of the performance of the *open* capability, the boundary of any ambient replica could be broken, which is denoted by $open(\langle n, c\rangle) = true$.

Definition 8 (Location pair for ambient copies). *Given a static replicated ambient Petri net A and a set of ambient copies for it defined by the function Copies, we define a **location pair** for them as a pair of partial functions $\langle loc, open\rangle$ with $loc : \mathcal{N}(A) \times (\mathbb{N}^* \cup \{0\}) \rightarrow ((\mathcal{N}(A) \cup root) \times (\mathbb{N}^* \cup \{0\})) \times P$ verifying*

- $loc(\langle n, c\rangle) \downarrow \Longleftrightarrow (c \in Copies(n) \vee c = 0)$,
- $(c \in \mathbb{N}^* \wedge loc(\langle n, c\rangle) = (\langle n', c'\rangle, p')) \Longrightarrow (n' = root \wedge c' = \epsilon) \vee c' \in Copies(n')$,
- $loc(\langle n, 0\rangle) = (\langle n', c'\rangle, p') \Longrightarrow (n' = root \wedge c' = \epsilon) \vee (n' \neq root \wedge c' = 0)$,

and $open : \mathcal{N}(A) \times \mathbb{N}^* \rightarrow Bool$, with $open(\langle n, c\rangle) \downarrow \Longleftrightarrow c \in Copies(n)$.

In this way, if n is the name of a component net, then we have two kinds of associated ambients: $\langle n, 0\rangle$ represent the original ambient from which we can generate copies and it will be located in some place of the system. Whenever a token labelled by c arrives to that place, we are able to generate the corresponding copy $\langle n, c\rangle$, whose execution starts. The original ambient token will remain at the same place, ready to generate new copies when new generating tokens will arrive to the involved place. These generated replicas will be the active net tokens of the system. In particular, they can be moved by means of the corresponding ambient transition. Instead, the original ambients are static: Neither they are executed nor moved along the system.

Definition 9 (Located replicated ambient Petri net). *A **located replicated ambient Petri net** is a static replicated ambient net for which we have defined the location pair corresponding to its full set of ambient copies.*

Definition 10 (Dynamic replicated ambient Petri net). *A **dynamic replicated ambient Petri net** is a located replicated ambient net for which we have defined an ordinary marking $M : P \longrightarrow \mathcal{M}_f(\mathbb{N}^*)$, where P is the full set of places of the ambient net, that is, $P = \bigcup_{i=1}^{k} P_i$ with $A_i = (P_i, T_i, W_i)$.*

All the markings of the different copies of each replicated ambient are put together. This is not a problem since the ordinary tokens for the execution of a copy $\langle n_i, c_i \rangle$ will be those tokens in the places of P_i labelled by sequences extending the sequence c_i.

Markings of replicated ambient Petri nets consist of two components: the function *Copies*, which defines the set of ambient tokens in the net, and M, which defines the ordinary marking for each ambient token.

Definition 11 (Initial marking). *The **initial marking** of a located replicated ambient Petri net A, $\langle Copies_{init}, M_{init} \rangle(A)$, is that with $Copies_{init}(n_i) = \varnothing$ $\forall n_i \in \mathcal{N}(A)$ and where only the entry places of A_1 are marked, that is, $\forall p \in E(A_1)$ $M_{init}(p) = \{\epsilon\}$ and $\forall p \in P \backslash E(A_1)$ $M_{init}(p) = \varnothing$.*

Definition 12 (Activation rule). *Whenever we have a marking $\langle Copies, M \rangle$ of a dynamic replicated ambient net such as for some $p_j \in P$ and $c \in \mathbb{N}^*$ verifies that $M(p_j)(c) > 0$, and there exists some n_i such that $loc(\langle n_i, 0 \rangle) = (\langle n_j, 0 \rangle, p_j)$, we can fire an internal activation transition which consumes an ordinary token in p_j labelled by c, producing a new copy of n_i (then we have $Copies'(n_i) = Copies(n_i) \cup \{c\}$), whose entry places will be also marked by tokens labelled with the same sequence c.*

Dynamic ambient tokens will move along the system by means of the firing of those transitions labelled by ambient operations. As an example of the different firing rules for these high-level transitions associated to those transitions expressing capabilities, next we give the rule for the entry operation.

Definition 13 (Entry operation). *Let $c_j \in Copies(n_j)$ be such that under the ordinary marking for $\langle n_j, c_j \rangle$ we can fire a transition $t_j \in T_j$ labelled by $(in\ n_i, p_i)$. If $loc(\langle n_j, c_j \rangle) = (\langle n_k, c_k \rangle, p_k)$ and there exists some $c_i \in Copies(n_i)$ with $loc(\langle n_i, c_i \rangle) = (\langle n_k, c_k \rangle, p'_k)$ then we can fire the high-level transition associated to t_j, getting as reached state of A that defined by the following changes in M:*

- *The local marking of A_j (more exactly, that of its c_j-copy) changes as the ordinary firing rule for t_j says.*
- *The location of the replica $\langle n_j, c_j \rangle$ changes, getting $loc(\langle n_j, c_j \rangle) = (\langle n_i, c_i \rangle, p_i)$.*

Note that the different replicas of an ambient n_i could move in different directions, and therefore they could stay at different locations of the same system. Nevertheless, the original copies (denoted by $\langle n, 0 \rangle$) , which are static by definition, will remain at their original location without ever moving.

3 Replication and the Restriction Operation

The restriction operator νn is introduced in the Ambient Calculus as a mechanism for defining internal names which cannot be used in the outer environment, neither by chance nor if we have known those names in an inappropriate way. The only way to get internal names is by a willingly communication of the internal process, which would produce the extrusion of the restriction operator. In this way a simple but powerful security mechanism is formalized.

For instance, in the term $m[in\ n]|(\nu n)n[a]$ we use the same ambient name n inside m, but from there we cannot reach the so named ambient. In fact, restricted names can be renamed without any change in the semantics of the system, since the Ambient Calculus identifies processes up to renaming of bound names. Therefore, process $m[in\ n]|(\nu n)n[a]$ is identical to $m[in\ n]|(\nu p)p[a]$.

Restriction can be treated by means of simple renamings if we consider systems without recursion and replication. But things are rather more complicated if, as in this paper, the replication operator is allowed.

For instance, if we consider the process $!(\nu n)(n[a]|m[in\ n])$ we have different internal ambient names in each of the copies of the system, due to the fact that replication creates new names ($!(\nu n)P \not\equiv (\nu n)!P$). As a consequence, each single copy should use each concrete name, although the behaviour of all the replicas is the same, once we take into account the corresponding renaming. Then it is possible to preserve the static definition of systems, including a fixed collection of ambient names and net components. In order to support the dynamic creation of new names, we use the fact that they can be seen as (different) copies of the original ones.

We have found that the structured labels of the tokens provide a nice way to support the sets of restricted names. So we only have to change the set $\mathbb{N}^*$ into $(\mathcal{P}(\mathcal{N}) \times \mathbb{N})^* \times \mathcal{P}(\mathcal{N})$, where the indexes associated to applications of the replication operator are intercalated with the sets of ambient names which are composed of the names that are restricted before and after the application of each replication operator.

For instance, for the process $(\nu n)!P$ we would get labels such as $\{n\}1\varnothing$, while $!(\nu n)P$ would produce instead $\varnothing 1\{n\}$. Then, if a copy of the ambient n is activated with these tokens, we would obtain a renamed copy whose name is n_c where c is the prefix of the place of the sequence where n is restricted. So in the examples above we would get n_ϵ and n_1. The new name would be taken into account to avoid unsuitable accesses to the restricted name. More on the subject in the next forthcoming paper, completely devoted to the study of the restriction operator.

4 Conclusions and Future Work

In this paper we have extended the basic model of ambient Petri nets in order to support the replication operator from the Ambient Calculus. We have seen that even if replication produced the dynamic creation of nets, which would represent the copies of the replicated ambients, we can still base our definitions on a static system describing the set of basic components of its architecture. Net tokens in the markings of these systems represent the dynamic activation of ambients.

We have seen that although we had to afford some technical difficulties a nice solution was still possible. Besides, this solution only has to be slightly modified to cope with the interrelation between the replication and the restriction operators.

In our opinion it is very important to give the designers who are familiar with Petri nets a (relatively) simple extension by means of which they will be able to develop mobile systems, whose correctness could be formally proved.

We are currently working on the remaining features of the Ambient Calculus and PBC in order to get the full framework in which we are interested.

Acknowledgements.
Research supported by CICYT project Desarrollo Formal de Sistemas Basados en Agentes Móviles (TIC 2000-0701-C02-01)

References

1. E. Best, R. Devillers and J. Hall. *The Petri Box Calculus: A New Causal Algebra with Multi-label Communication.* Advances in Petri Nets 1992, LNCS vol.609, pp.21-69. Springer-Verlag, 1992.
2. E. Best, R. Devillers and M. Koutny. *Petri Net Algebra.* EATCS Monographs on Theoretical Computer Science Series. Springer-Verlag, 2001.
3. L. Cardelli. *Abstractions for Mobile Computation.* Secure Internet Programming: Security Issues for Mobile and Distributed Objects, LNCS vol.1603, pp.51-94. Springer-Verlag, 1999.
4. L. Cardelli. *Mobility and Security.* Proceedings of the NATO Advanced Study Institute on Foundations of Secure Computation, pp.3-37. IOS Press, 2000.
5. D. Frutos Escrig and O. Marroquín Alonso. *Ambient Petri Nets.* Submitted for publication.
6. M. Koutny and E. Best. *Operational and Denotational Semantics for the Box Algebra.* Theoretical Computer Science 211, pp.1-83, 1999.
7. H. Maciá, V. Valero and D. Frutos Escrig. *sPBC: A Markovian Extension of Finite Petri Box Calculus.* Petri Nets and Performance Models PNPM'01, pp.207-216. IEEE Computer Society, 2001.
8. O. Marroquín Alonso and D. Frutos Escrig. *Extending the Petri Box Calculus with Time.* Applications and Theory of Petri Nets 2001, LNCS vol.2075, pp.303-322. Springer-Verlag, 2001.
9. R. Valk. *Petri Nets as Token Objects: An Introduction to Elementary Object Nets.* Applications and Theory of Petri Nets 1998, LNCS vol.1420, pp.1-25. Springer-Verlag, 1998.
10. R. Valk. *Relating Different Semantics for Objects Petri Nets, Formal Proofs and Examples.* Technical Report FBI-HH-B-226, pp.1-50. University of Hamburg, Department for Computer Science, 2000.
11. R. Valk. *Concurrency in Communicating Object Petri Nets.* Concurrent Object-Oriented Programming and Petri Nets, Advances in Petri Nets, LNCS vol.2001, pp.164-195. Springer-Verlag, 2001.

Neural Networks for Event Detection from Time Series: A BP Algorithm Approach

Dayong Gao[1], Y. Kinouchi[1], and K. Ito[2]

[1] Faculty of Engineering
[2] Faculty of the Integrated Arts and Sciences
The University of Tokushima, Japan

Abstract. In this paper, a relatively new event detection method using neural networks is developed for financial time series. Such method can capture homeostatic dynamics of the system under the influence of exogenous event. The results show that financial time series include both predictable deterministic and unpredictable random components. Neural networks can identify the properties of homeostatic dynamics and model the dynamic relation between endogenous and exogenous variables in financial time series input-output system. We also investigate the impact of the number of model inputs and the number of hidden layer neurons on forecasting. ...

1 Introduction

Neural networks have been applied in time series to improve multivariate prediction ability [1–3]. Neural networks map an input space (present and past values of the time series) onto an output space (future value). Future values of variables often depend on their past values, the past values of other correlated variables, and other uncertain factors. For example, the future price of a stock depends on political and international events as well as various economic indictors.

Most studies on time series analysis using neural networks address the use of neural networks as a forecasting tool, but limited research focuses on the application of neural networks as a data mining technique to extract event patterns from time series. These forecasts rely on historical trends or past experiences to predict the future. The measurement of past trends may be interrupted by changes in policy or data availability. When the past cannot be used to predict the future due to major unexpected shifts in economic, political, or social conditions, the future can be more uncertain due to the occurrence of major events. Evaluating the impact of these events (exogenous variables) on stock markets are interesting to economists, because the markets are affected by less measurable but important forces in economic analysis and financial prediction.

In this paper, a systematic approach for time series using neural networks is developed to generate an event sequence and to model the dynamic relation between stock market index (endogenous variables) and macroeconomic environment (exogenous variables). Because time series related to the input-output

P.M.A. Sloot et al. (Eds.): ICCS 2003, LNCS 2658, pp. 784–793, 2003.

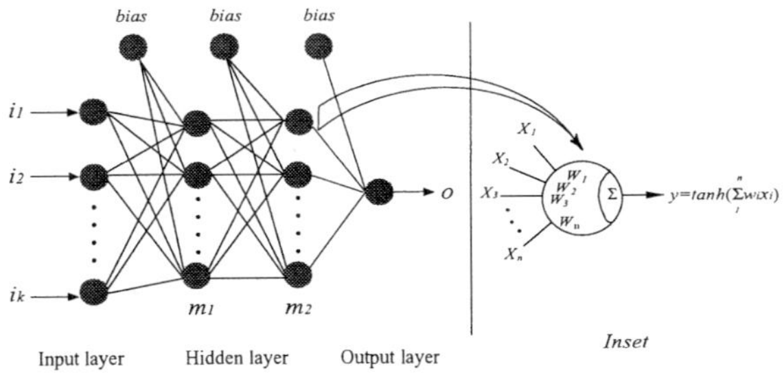

Fig. 1. Neural Network Configuration and Nonlinear Neuron Model (*Inset*). The *input*(x) – *output*(y) relation of nonlinear neurons is given by a conventional sigmoid function such as $y = tanh(x)$. The *input*(x) is expressed by a form $\Sigma(w_i x_i)$, where w_i is a weight for an input component x_i.

system of financial markets may include non-linearity, back propagation (BP) neural networks are used to clarify basic predictive properties. This model can acquire automatically the dynamics included in the data by use of an error back propagation (BP) algorithm [4].

2 Methodological Consideration

2.1 Identification by Neural Networks

Identification by neural networks for time series data $x_1, x_2, \ldots, x_k, \ldots, x_N$ requires predicting or estimating a datum x_k based on preceding observed n data $x_{k-n}, \ldots, x_{k-2}, x_{k-1}$, *i. e.*, prediction by a nth order MA model, where $x_1, x_2, \ldots, x_k, \ldots, x_N$ are discrete time sequences. In order to implement the prediction, a multiplayer neural network model as shown in *Fig.* 1 is used. The network has one input layer $i_1 = x_{k-n}$, … $i_n = x_{k-1}$ and one output layer x_k.

Each weight w_i is adjusted to coincide with output o_k and target x_k by a gradient method. For error function F defined by

$$F = \sum_{k \in S_t} \varepsilon_k^2, \qquad \varepsilon_k = o_k - x_k \tag{1}$$

where S_t is a set of training patterns, a correction δw_i of w_i to minimize error function is given by

$$\delta w_i = -\eta \frac{\partial F}{\partial w_i} \tag{2}$$

where η is the learning rate parameter, a small positive constant. To calculate δw_i for neurons in hidden layers, an error back propagation (BP) algorithm is used to train the network [4]. After training, the network is tested by the use of the rest patterns to verify whether the network acquires source dynamics, *i. e.*, generalization. If the network output for each test pattern is almost close to

Table 1. Data Acquisition and Normalized Results. N is the maximum number of trading days for TOPIX during 1995–2000; m is the normalized factor. Min, Mean, Max and MaxDev are the minimum, mean, maximum and maximum deviation of acquired and normalized data respectively.

Data Acquisition				
N	Min	Mean	Max	MaxDev
1481	980.00	1415.89	1754.78	435.89
Normalized Results				
m	Min	Mean	Max	MaxDev
500	−0.87178	−4.2E-06	0.67778	0.87178

the respective target, the network is considered to have acquired the underlying dynamics from the training patterns.

2.2 Data Acquisition

The most popular treatment of input data is to feed neural networks with either the data at each observation, or the data from several successive observations. In this paper, we consider daily data of TOPIX (Tokyo Stock Price Index) from Japanese stock market in the period 1995–2000.

Succeeding data of TOPIX, denoted by t_k, are converted to the deviation from their mean value and scaled by a factor m to normalize their values in the range of -1 to 1 by equation:

$$x_k = \frac{t_k - \bar{t}}{m}, \qquad k = 1 \sim N \tag{3}$$

where k is the number of trading days, corresponding to 6 years from 1995 to 2000 and N is the maximum number. $\bar{t}$ is the average of t_k.

In Table 1, historical data of TOPIX are divided into training phases and test phases, with each phase containing half of the collected data. The data of the earlier period (1995-1997) are used for training; the data of the latter period (1998-2000) are used for test.

Real world data contain varying degree of noise. To demonstrate our method and to compare with real data from stock markets, two types of the data except TOPIX are generated as follows: 1) random data with uniform distribution between -1 to 1, and 2) the data from a sinusoidal wave, *i. e.*, deterministic data, with unit amplitude at an interval of T/60.1 (T: period). These two types of generated data represent a sequence of true random numbers where no model has a chance and simple noise-free data, respectively.

2.3 Model Identification

Various measurement methods such as approximate entropy, auto-correlation, average mutual information, and number of false nearest neighbors can be used to evaluate time series data [5]. These methods can quantify data regularity and randomness if there are large amounts of the data. However, stock market data are relatively limited and this study uses even more limited data containing various market environments at fixed periods. If the data periods are too long, market environments change much; financial patterns and the neural network models also change much.

The following "R-method" of quantifying data regularity and randomness can be applied to a relatively small amounts of the data. This R-method can also be used to measure multi-valued property of the data and determine neural network configuration. To calculate values of R, let x denote the input vector (i_1, i_2, !D i_{N-n}) and y denote the output o. The data produce input vectors x_1, x_2,!D x_{N-n} and corresponding outputs y_1, y_2, !D y_{N-n}. If x_i and x_j are similar, then y_i and y_j are also similar for deterministic data, but y_i and y_j may be quite different for random data and multi-valued function. The change of $|y_j - y_i|$ for y_i corresponds to all x_j in the neighborhood of each x_i as a measurement of data randomness. Values of R are defined by

$$R = \frac{1}{\sum_i |\Delta_i|} \sum_{i=1}^{N-n} \sum_{j \in \Delta_i} f(z_{ji}) \tag{4}$$

where

$$\begin{aligned} \Delta_i &= \left\{ j \middle| |x_i - x_j| / \sqrt{n} x_{rms} \le \beta, j \neq i \right\}, \quad if\ x_j\ exists \\ &= \left\{ j \middle| \min |x_i - x_j| / \sqrt{n} x_{rms}, j \neq i \right\}, \quad otherwise \end{aligned} \tag{5}$$

Here, $|\Delta_i|$ represents the number of x_j in the neighborhood Δ_i. The x_{rms} is a root square value for the data x_k ($Eq.1$), and the y_{rms} is a root mean square value for the data y_i. The $\sqrt{n}x_{rms}$ is the mean length of the vector x_i taken in the n dimensional space. The z_{ji} is given by

$$z_{ji} = \left(|y_j - y_i| \right) / \left(y_{rms} / \sqrt{n} x_{rms} \right) \tag{6}$$

The function $f(z_{ji})$ is a logistic function as following:

$$\begin{aligned} f(z_{ji}) &= 0, \quad if |x_j - x_i|) / \sqrt{n} x_{rms} \le \beta\ and\ |y_j - y_i|) / y_{rms} \le \beta \\ &= 1 - (1 + \alpha z_{ji}) exp(-\alpha z_{ji}), \quad otherwise \end{aligned} \tag{7}$$

where α and β are positive constants. If the data x_1, x_2,!D x_N are obtained by sampling a deterministic single-valued time function, using $Eq.7$, $f(z_{ji}) = 0$, because $y_i \approx y_j$ for all x_j existing in the neighborhood Δ_i of any x_i. Values of R are zero when the data number N is relatively large. If the data come from a random function, $f(z_{ji}) = 1$ because y_j is independent of y_i, therefore R=1 for the relatively large data numbers. R for any data will have a value between zero and one, indicating the degree of randomness in the data. High values of R

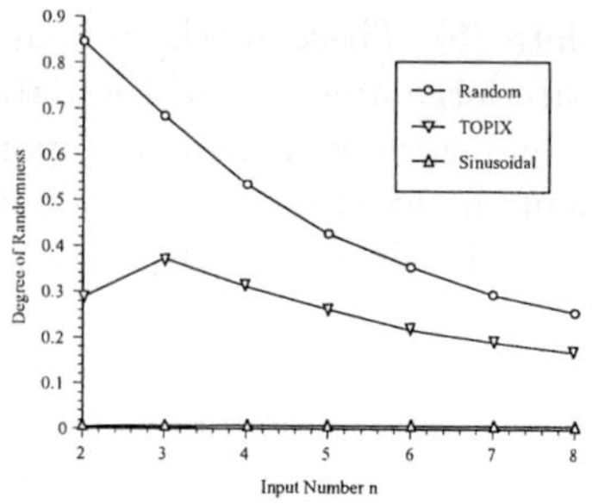

Fig. 2. Optimum Neural Networks. Values of R for TOPIX start low at n=2, but increase at n=3, and then start to decrease from n=4. This process characterizes the deterministic nature of the data.

indicate high randomness and high statistical dependence between the data; low values of R indicate more recognizable patterns and high statistical independence of the data.

3 Empirical Results

3.1 Model Development

Network size is determined by the input number n, the number of hidden layers (1 or 2) and the number of neurons (m_1 and m_2) in the hidden layers. If different network sizes have similar values for the error function, the smallest network size is optimum [6]. In this paper, we use R-method (see Sect. 2.3) to characterize the deterministic and random properties of the data in relation to the input data so as to determine optimum neural networks.

Samples from TOPIX are used to measure the degree of randomness. Values of R are determined for the input sizes n = 2, 3, 4, 5, 6, 7, 8. Each sample is grouped into an overlapping cluster with 3 years of the data, enough to characterize the dynamic properties of the system. 6 month graduations are used to characterize moving group samples MGS(1), MGS(2), !D MGS(7) such as

$$MGS_k : x_k, x_{k+1}, \ldots, x_{k+n}, k = 1 \sim 7 \tag{8}$$

where n corresponds to 3 years of the data.

After calculating values of R for both TOPIX and two types of generated data (see Sect. 2.2), mean values of R corresponding to moving group samples for TOPIX are used to evaluate deterministic and random properties of the data. Values of R for two other types of generated data are calculated as comparison. Here, the data number for these two types of the data is 750 because training data number for the stock market indexes is about 750. Values of R for random data are mean values of five generated data sets. For values of R, we choose &A = 0.4 and &B= 0.01.

Table 2. Forecasting Performance for $NN341$ (iteration=3000). &G learning rate parameter; RMS: root mean square error.

Phase	Data Period	&G	RMS
Training	1995–1997	0.005	0.001926
Test	1998–2000	0.005	0.036991

Fig. 2 shows that the degree of randomness for TOPIX lays between values of R for random data and sinusoidal data. Therefore, the data from TOPIX might have both deterministic and random properties.

Values of R can also be used to determine the minimum input number of dynamic variables needed to train the networks. At input number n=3, there is the maximum statistical independence between distance observations of data series, so input number n=3 maximize the network ability to learn from the data and should be the optimum number of nodes in the input layer. To verify the optimum input number, RMS error corresponding to various input number of nodes is calculated. It is found that neural networks corresponding to input number n=3 give minimum RMS error.

To finally determine the optimum sizes of neural networks, depending on the determined input number, the networks with 1 hidden layer and 2 hidden layers are used for training and test by changing neuron sizes in the hidden layer(s). It is found that the networks with 1 hidden layer and 4 neurons can make RMS error converge to lower values. Optimum network size NN341 shows the number of neurons in each layer for TOPIX. Therefore, the neural network forecasters are constructed based on determined input pattern and the number of neurons in the hidden layer(s).

3.2 Estimation of Determinstic Component

To verify whether the optimum neural network acquires underlying dynamics of the system from the data, both training RMS error and test RMS error should be estimated.

All the neurons in the multiplayer perceptron should desirably learn at the same rate. To a good degree of approximation, the misadjustment of back propagation (BP) algorithm is directly proportional to the learning rate parameter [7]. Therefore, if the learning rate parameter is increased so as to accelerate the learning process, then the misadjustment is increased. Careful attention has therefore to be given to the choice of the learning rate parameter in the design of the network in order to produce a satisfactory overall performance.

Table 2 shows the simulation results for $NN341$. During convergence, weight updates not only include the change dictated by the learning rate, but also include a portion of the last weight change. The training process will terminate when the RMS error is less than 0.001 or the learning times are over 3,000 iterations. Since the RMS errors for both training and test have small values,

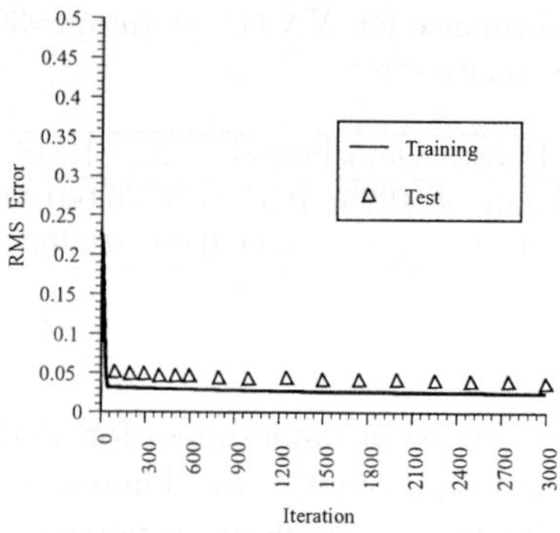

Fig. 3. Convergence Process of $NN341$. The solid line shows the training process. The dotted line describes corresponding test process.

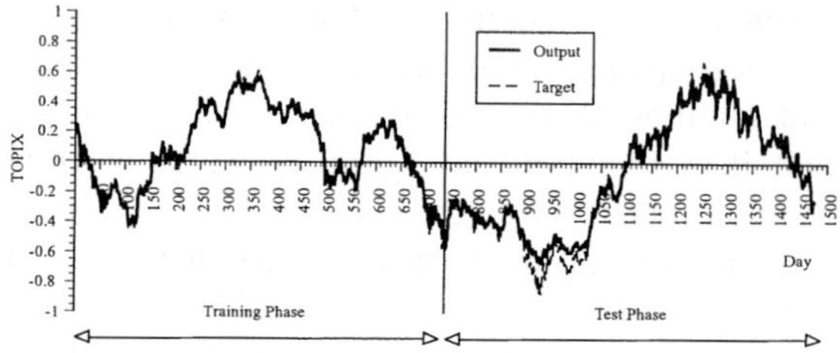

Fig. 4. Comparison of Neural Network Outputs (predicated values) and Target Data (x_k) after training of 3000 iterations.

smaller than the fluctuation of the data, the network obtains good generalization and acquires the source dynamics.

Fig. 3 shows that before training, synaptic connections w_{ij} have random volumes and the network yields large RMS error. After training of 3000 iterations, where one iteration corresponds to one round of training using all training data, the average RMS error becomes as small as 0.019626. RMS error in test phases is somewhat higher than that in training phases. However, RMS errors in both phases show a similar trend, so neural networks acquire underlying dynamics of the system from the data.

It is found from *Fig.* 4 that the network outputs follow the data change in both early training and later test phases. Clearly, the network has acquired the regularity in both phases.

3.3 Estimation of Random Component

To evaluate the predictability of neural networks and to analyze the random components of the data, training errors (the residuals between the outputs of neural networks and corresponding target data) are used as input data, and neural networks with the same optimum sizes are retrained.

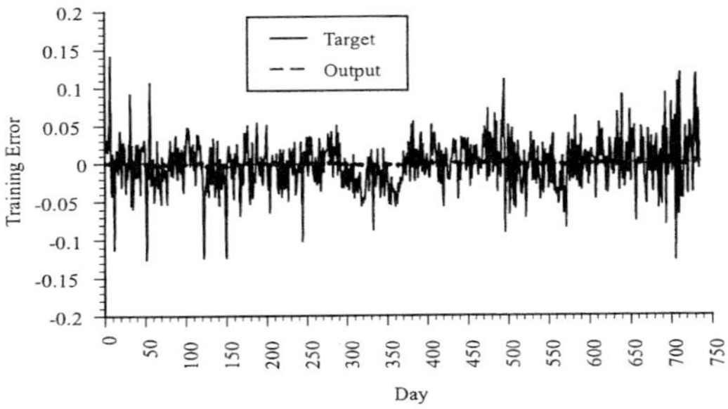

Fig. 5. Prediction by $NN341$ Using Training Error as Input Data. When the Training errors are used as the input data, the outputs of neural networks are close to zero.

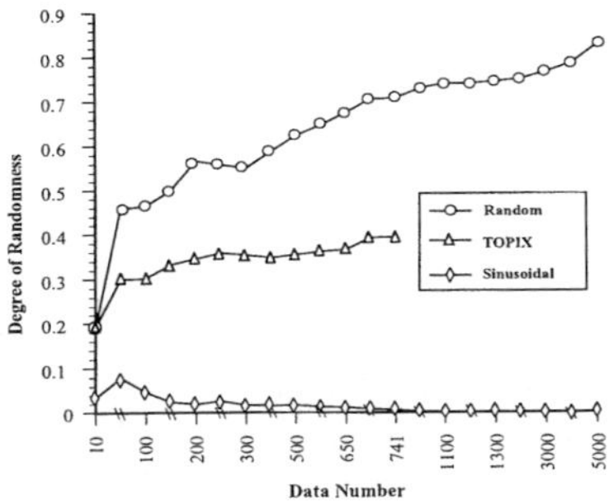

Fig. 6. Degree of Randomness for Random Data, TOPIX, and Sinusoidal Data. Maximum number of TOPIX is 741. Maximum number of random and sinusoidal data is 5000. Values of R for random data are the mean of ten data series. The x-axis graduation varies.

It is found from *Fig.* 5 that the outputs of neural networks are close to zero. This may mean that training errors are white noise, so we can deduce that TOPIX data are composed of a regular (deterministic) part and a random part.

To verify this further, both TOPIX and two types of generated data (see Sect. 2.2) are trained similarly using neural networks. After training, the network outputs for random data are close to zero and the network outputs for deterministic data close to the target data. The network outputs for TOPIX fall between outputs for random data and outputs for deterministic data.

It is found from *Fig.* 6 that values of R gradually approach one (1) for random data and zero (0) for sinusoidal data as the data number increases. Values of R for random data reduce with decreasing the data number because randomness may be dificult to find using relatively small amounts of the data. As in the case of neural network outputs, values of R for TOPIX are between values of R for random data and sinusoidal data. This evidence gives a convincing picture that TOPIX are considered to include both deterministic and random components. Calculating values of R gives us important insights into regularity and noise, showing how neural networks can forecast performance in financial markets.

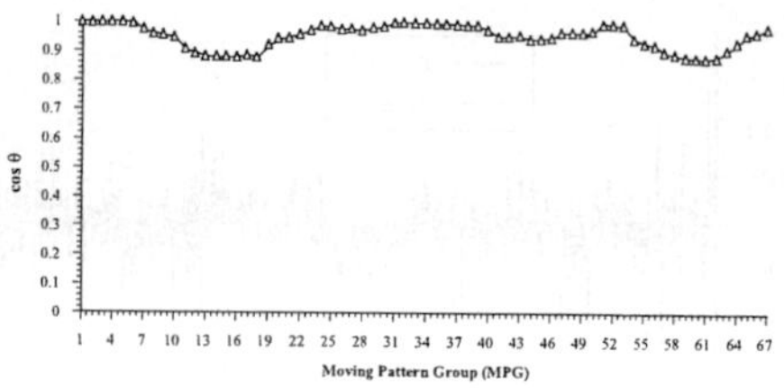

Fig. 7. Change of Internal Representation. Input data are half-year TOPIX daily data, and 1-month shift. MPG_1 = the period from Jan. 1995 to Jun. 1995, !D, MPG_{67} = the period form Jul. 2000 to Dec. 2000.

3.4 Network Adaptability and Environment Influence

Since the neural networks acquire dynamic properties from the data, stock market mechanism can be understood well by examining how external forces influence the network adaptability. The properties of trained networks are represented by weights w_i between neurons, a kind of internal representation of the networks. The optimum network $NN341$ has 21 weights denoted by vector w. The dynamic properties of the system vary with time, and this adaptive change may be evaluated by the change of weight vector w. The networks are trained successively by moving pattern groups MPG_1, MPG_2, ..., MPG_{67} such as

$$MPG_k : P_k, P_{k+1}, \ldots, P_{k+n}, k = 1 \sim 67 \tag{9}$$

where n corresponds to 6 months of the data. Each group has 6 months of the data, enough to acquire dynamics of the system. Convergence of training by using each pattern group produces weight vectors, w_1, w_2, ..., w_{67} . Dynamics of the system is characterized by the matrix of transition probabilities between network groups. The adaptive change of acquired dynamics may be evaluated by the direction change of weight vectors from w_1. The direction change is measured by cos &H, that is

$$\cos\theta_k = \frac{w_k \cdot w_1}{|w_k| \cdot |w_1|}, k = 1 \sim 67 \tag{10}$$

where &His an angle between two weight vectors.

Inspection of the change of $\cos\theta_k$ in *Fig.* 7 provides insight into how environment influences the network properties. Plotting the weight vectors as a function of time reveals that the environment influences during the corresponding periods. Since $\cos\theta_k$ is more than about 0.6 as a whole, the properties of the network is considered to be maintained almost the same. Generally, $\cos\theta_k$ is kept relatively constant for a period of time, but there are gradual changes in some periods. It is therefore considered that homeostatic dynamics of the system may be adjusted occasionally due to environment changes and then stay in steady state for some periods. At some points, $\cos\theta_k$ changes, indicating that the networks obtain different weight vectors, such as in MPG_6, corresponding to June 1995–Nov. 1995; MPG_{40}, corresponding to Apr. 1998–Sep. 1998; and

MPG_{53}, corresponding to May 1999–Oct. 1999, the networks obtain slightly different weight vectors.

4 Conclusion

This paper demonstrated the usefulness of neural networks as an event detection technique for financial time series. Empirical results show that dynamic properties of the regularity components were captured by neural network model. The results can be attributed to the high ability of the network to capture nonlinear market patterns. Calculation of data randomness shows that randomness combines with determinism to create statistical order. Prediction using training errors as inputs of networks and calculating degree of randomness prove that financial time series include both predictable components and unpredictable components. The property that weights in neural networks remain relatively stable in some periods characterizes the structure of financial markets. Such property shows that stock market has memory, maybe because investors often make market decisions based on past situations. Evaluating the change of weights in neural networks shows that homeostatic dynamics change occasionally, perhaps because of adaptation to environmental factors, which may induce the fluctuations. Neural networks can acquire exogenous regularities and endogenous dynamics.

In our research, we have systematically examined effects of various factors such as the number of model inputs, the number of hidden layer neurons and influence of the event. The question that still remains is under what conditions are they better than tranditional statistical methods. Future experiments could focus on comparative study of neural network approach with other statistical time series approach to understand the nature of financial markets well.

References

1. Apostolos-Paul N. Refenes, A. Neil Burgess, and Yves Bentz: Neural Networks in Financial Engiineering: A Study in Methodology. IEEE Transactions on Neural Networks, **8-6** (1997) 1222–1267.
2. Weigend, A., Huberman, B. and Rumelhart, D.: Predicting the future: a connectionist approach. International Journal of Neural Systems, **1-3** (1990) 193–209.
3. Jingtao Yao, Chew Lim Tan and Hean-lee Poh.: Neural network for technical analysis: a study on KLCI. International Journal of Theoretical and Applied Finance, **2-2** (1999) 221–241.
4. Rumelhart, D. E., Hinton, G. E., Williams, R. J.: Learning internal representation by error propagation. In Rumelhart, D. E., McClelland, J. L., and the PDP research group(Eds), parallel distributed processing, MIT press, Cambridge, MA (1986) 318–362.
5. Perambur S. Neelakanta: Information-Theoretic Aspects of Neural Networks. CRC Press LLC, (1999).
6. Matsuba, I., Masui, H., Hebishima, S.: Optimizing multiplayer neural networks using fractal dimensions of time series data. Proc IEEE Trans Ind Elec, **44** (1997) 707–716.
7. Simon Haykin: Neural Network: A Comprehensive Foundation. Macmillan College Publishing Company (1994).

Data Loss Reparation Due to Indeterminate Fine-Grained Parallel Computation

Ekaterina O. Gorbunova[1], Yuri V. Kondratenko[2], and Michael G. Sadovsky[3]

[1] Institute of Computational Modelling of SB RAS, Krasnoyarsk, Russia, 660036, gkat@icm.krasn.ru
[2] Krasnoyarsk State University, Krasnoyarsk, Russia, 660042
[3] Institute of Biophysics of SB RAS, Krasnoyarsk, Russia, 660036, uvenal@ktk.ru

Abstract. The new method of a gap recovery in symbol sequences is presented. A covering is combined from the suitable reasonably short strings of the parts of a sequence available for an observation. Two criteria are introduced to choose the best covering. It must yield the maximum of entropy of a frequency dictionary developed over the sequence obtained due to the recovery, if an overlapping combined from the copies of strings from the available parts of the sequence exists. The second criterion identifies the best covering in case when one has to use any string to cover the gap; here the best covering must yield the minimum of specific entropy of the frequency dictionary developed over the available parts of the sequence against that one developed over the entire sequence obtained due to the recovery. Kirdin kinetic machine that is the ideal fine-grained structureless computer has been used to resolve the problem of the reconstruction of a gap in symbol sequence.

1 Introduction

A reconstruction of the lost data seems to be an acute problem both for fundamental and applied sciences. The results of the reconstruction depend significantly on the method to do it and on the information providing the basis for that latter. The up-to-date methods of the reconstruction of lost data implement various additional information concerning the data, or the knowledge concerning the properties of these data. We shall consider the problem of data loss reconstruction using the most strict definitions allowing to avoid such discrepancies.

Finite symbol sequences will be considered as a data source. The absence of a part of sequence would be considered as a data loss. One should distinguish the situation when the length of a gap is known from the situation where this length is unknown. Similarly, the alphabet could be either known, or unknown to a researcher; everywhere below we assume that both the alphabet and the length of a gap are known. Moreover, the finiteness of the alphabet will be assumed.

Simply speaking, the basic principle of the gap fulfillment force to do that in a manner that the sequence resulted from the fulfillment looks mostly similar to

P.M.A. Sloot et al. (Eds.): ICCS 2003, LNCS 2658, pp. 794–801, 2003.

the originally available parts of that former. Additionally, the fulfillment must bring the minimum external, additional information concerning the sequence. The principle has two forms:

- a maximum of entropy of the augmented dictionary resulted due to the reconstruction of a sequence, and
- a minimum of specific entropy of the reference frequency dictionary against the augmented one. Let now turn onto the strict formulations and exact statements.

This paper is devoted to the consideration of some preliminary results in data loss reparation obtained due to the first approach (i.e. maximization of the augmented frequency dictioanry).

2 Criteria for a Gap Fulfillment

Consider a sequence of the length N:

$$N = N_1 + N_2 + L.$$

Here N_1 and N_2 are the lengths of available parts of the sequence, and L is the length of the fragment that must be reconstructed. The parts of the sequence known to a researcher are considered to be continuous. A word of the length q is a continuous subsequence (string) of that length. A list of all the words occurred at the available parts of a sequence accompanied with their frequency is called the reference frequency dictionary W_q (of the thickness q); f_ω is the frequency of word ω. The frequency dictionary developed over the entire sequence obtained due to the fulfillment of a gap is called augmented frequency dictionary $\overline{W}_q$. Obviously, the constraint

$$1 \leq q \leq min\{N_1, N_2\}$$

holds true. A word of the length t, $0 \leq t \leq q-1$ located at the right border (at the left border, respectively) of fragment available to a researcher is called the left basement (the right basement, respectively). To develop a gap recovery means to figure out a string

$$\omega_1,\ \omega_2,\ \omega_3,\ \ldots,\ \omega_{L+2t-q},\ \omega_{L+2t-q+1} \tag{1}$$

of the length $L + 2t$, where ω_i is a word, and

$$\omega_j = i_1\overline{\omega}, \quad \omega_{j+1} = \overline{\omega}i_q.$$

Here $\omega_1 = l_t\alpha$ and $\omega_{L+2t-q+1} = \alpha r_t$ are the first and the last words bearing the basement (the left one, and the right one, respectively), and $\overline{\omega}$ is a word of the length $q-1$. To fulfill a gap, one shall use the words available at the frequency dictionary of thickness q. There are two options here. The first one is to use the words from the reference dictionary only, and the second one implies a fulfillment with any possible words of the given length, since no fulfillment exists combined from the words from the reference dictionary. To choose the best fulfillment, we propose the extremal principle, for each case.

Consider the first case. If the unique fulfillment with the words from the reference dictionary exists, then the problem is resolved. If a fulfillment is ambiguous, then one must choose the string among all possible entities that yields the maximum of entropy of the augmented frequency dictionary:

$$S = -\sum_{\omega} \tilde{f}_{\omega} \ln \tilde{f}_{\omega}, \tag{2}$$

here $\tilde{f}_{\omega}$ is the frequency of word ω observed at the text obtained due to the gap recovery, i.e. the frequency of word ω at the augmented dictionary $\overline{W}_q$.

Consider now the case, when the fulfillment with the words from the reference dictionary does not exist. It must be done then with any possible words. Here the fulfillment must be chosen yielding the minimum of specific entropy of the reference frequency dictionary W_q against the augmented one $\overline{W}_q$:

$$\overline{S} = \sum_{\omega} f_{\omega} \ln \frac{f_{\omega}}{\tilde{f}_{\omega}}. \tag{3}$$

f_{ω} is the frequency of word ω observed at the reference dictionary, and $\tilde{f}_{\omega}$ is the frequency obtained over the recovered sequence; obviously, $f_{\omega'} = 0$ for some ω', while $\tilde{f}_{\omega'} > 0$ for them. This second criterion is universal, since it is applicable always. The applicability of the first criterion is not guaranteed *á priori*.

The existence of the recovery with the words from the reference dictionary is not guaranteed; moreover, a complete searching for all strings of the length $L + 2t$ composed from the words from the reference dictionary of the thickness q seems to be the only way to do that. An implementation of highly paralleled structureless computation devices is the most efficient way to make up such fulfillment. Kirdin kinetic machine that is the ideal highly paralleled grain computation device, pretends to be the most efficient entity to resolve the problem of data loss reparation [1,2,3].

3 Kirdin Kinetic Machine in the Problem of Gap Reparation

To begin with, let's describe the principle of execution of Kirdin kinetic machine (KKM) in detail. KKM is an ideal fine-grained parallel computer, similar to Turing machine from the point of view of the abstraction [1,2,3]. KKM is algorithmically universal [3,4], i.e. any algorithm could be implemented in the terms of KKM. Besides, it realizes the fine-grained parallelism. Let Ω be an alphabet of symbols, and Ω^* is the set of all finite strings (or words) in that alphabet. An ensemble M of the words from that alphabet is a processed entity; that latter is identified with the function F_M on a finite support from Ω^*. The function takes value in positive integers: $F_M\colon \Omega^* \mapsto N \cup \{0\}$. The value of $F_M(\omega)$ is interpreted as the number of copies of a word ω at the ensemble M.

Properly, KKM operation consists of an assembly of elementary events that take place in parallel and non-deterministic way. An elementary event $S\colon M \mapsto M'$ consists in a removal of the ensemble K^- from the ensemble M, and the addition of the ensemble K^+ to the ensemble M, so that $F_{M'} = F_M - F_{K^-} + F_{K^+}$. The removal is possible, if $F_{K^-}(\omega) \leq F_M(\omega)$ for all the words from the ensemble M. The ensembles K^+ and K^- are defined unambiguously by the commands making a programme. Three types of commands are possible:

1. ***Disintegration***. $uvw \longrightarrow uf + gw$, where u, w are the arbitrary words from Ω^*, while v, f, g are fixed words from Ω^*.
2. ***Synthesis***. $uk + qw \longrightarrow usw$, where u, w are the arbitrary words from Ω^*, while k, q, s are fixed words from Ω^*.
3. ***Mutation***. $uvw \longrightarrow usv$, where u, w are the arbitrary words from Ω^*, while v, s are the fixed ones from Ω^*.

Informally, KKM looks like a chemical reactor. One has a tank where the words are suspended. Then one provides the suspension with catalysts (that are the commands), and the words interact. The interaction of a word with catalyst may result in a disintegration. A couple of words interacting due to the proper command may yield a new word. Finally, a new word may appear due to the interaction of some word with the command resulting in a substitution of a substring inside a word.

Consider now the method to fulfill a gap in a sequence with a help of KKM in more detail. To do that, we shall provide the programme of a dictionary development, and the programme of the gap reparation, itself. Let a frequency dictionary W_q be developed for the text T. The KKM programme yielding the dictionary consists of a single command:

$$uf^1v^{q-1}g^1w \longrightarrow uf^1v^{q-1} + v^{q-1}g^1w,$$

where the ensemble M contains a single word T. Upon a completion of KKM processing, the ensemble M contains all the words of the length q occurred at the original text, with respect to the number of their copies.

The KKM programme covering a gap in a sequence looks as following:

$$\begin{array}{lcl} \alpha_l + \alpha_l v^{q-t} & \longrightarrow & \alpha_l v^{q-t}\star, \\ v^{q-t}\alpha_r + \alpha_r & \longrightarrow & \star v^{q-t}\alpha_r. \end{array} \tag{4}$$

The symbol $\langle\star\rangle$ falls beyond the alphabet Ω and marks the word that has passed the initialization successfully. Metasymbol $\langle\star\rangle$ identifies the words of the length q which start from the left basement (and end with the right basement, respectively). The commands producing a growth of infill itself are the following:

$$\begin{array}{l} uv^{q-1}\star + v^{q-1}v^1 \longrightarrow uv^{q-1}v^1\star, \\ v^1v^{q-1} + \star v^{q-1}w \longrightarrow \star v^1v^{q-1}w, \end{array} \tag{5}$$

and the following is the command that glues two strings into a covering:

$$u\star + \star v \longrightarrow uv. \tag{6}$$

An original ensemble of this programme consists of several copies of the basements (these are the left (α_l) and the right (α_r) ones), and several copies of the dictionaries obtained due to the execution of the previous programme. KKM runs non-determinally and in parallel way. Nevertheless, the programme is implemented so that initially the ensemble bears no words which could be processed by the commands (5, 6). Thus, a programme recovering a gap in the sequence could be considered to consist of three stages.

The first stage — "Basement initialization" (see Eq.(4)) consists in the connection of words from the dictionary W_q to the left (α_l) and the right (α_r) basement.

The second stage — "Growth" (see Eq.(5)) consists in the connection of words from the dictionary W_q to the primers obtained due to the initialization. It should be said that the number of words marked with $\langle\star\rangle$ depends on the structure of the fragments available for an observation. It might be that the reference dictionary W_q would bear no word which can provide the initialization. Sufficiently long execution of the command (5) yields the occurrence of the words of $\alpha_l u\star$ and $\star v\alpha_r$ types.

The third stage — "Glue". Finally, one must glue up the words of $\alpha_l u\star$ and $\star v\alpha_r$ types due to the last command (6). The strings obtained due to that command make the final ensemble for the given programme, since no one command is applicable to them. Now one can choose the strings of the length $L + 2t$ from the ensemble and select that one satisfying a criterion.

4 Imitator of Kirdin Kinetic Machine for the Gap Recovery Problem

To solve the problem of a gap recovery, we have implemented the consequent imitator of the algorithm for KKM. A number of copies of the left and right basements, as well as a number of copies of the dictionary are the input data for the imitator. To improve the imitator processing, we have modified that latter.

- The primers grow upright, only. As they reach the proper length of $L+2t-q$, a word containing the right basement should be connected to a string.
- To eliminate the inefficient steps, the original frequency dictionary has been modified. The function F_M was replaced with the function $\widetilde{F}_M$: $\Omega^* \mapsto N \cup \{0\}$, $\widetilde{F}_M = F_M(uv^{q-1}v^1\star) + F_M(v^1v^{q-1})$. This modification separates the words which can interact with a primer, due to the commands (5).
- The gap was splitted into several segments by check-points. Some strings resulted as a continuation of primer fail to grow up to the length $L + 2t$. Such truncated strings were twice eliminated during a course of a fulfillment development. Simultaneously, the population of strings that had reached the check-point successfully, were multiplied in number with specific factor.

5 Results

The primary goal of the computation experiments presented here was to learn whether the developed methodology would be fruitful for a gap recovery, and the programme implementing that former would be efficient enough for PC. A detail study of the properties of the recoveries themselves obtained at these experiments falls beyond the scope of this paper. One should bear in mind, that the length of the gap used in the computational experiments is too great to observe any statistically valid data concerning the relative number of the recoveries in comparison to the total number of the possible strings of such length.

To carry out the computational experiments on the gap recovery, we have used the sequence of a complete genome of avian CELO adenovirus, from the four–letter alphabet A, C, G, T. The genome is 43804 symbols in length, its accession number is U46933 in EMBL–bank. Artificial gaps of various length were created within this sequence. The length of these gaps are indicated at the Table 1. All the gaps were located approximately at the center of the sequence. The number of copies of primers was equal to 200, for each experiment. The recovery was executed for the family of the dictionaries of the thickness varied from 2 to 8. The length of a basement was equal to $q - 1$; a continuation of a primer was developed over a dictionary of the thickness $q - 1$, as well.

Table 1. Entropy values observed for various thickness q of dictionaries used to cover a gap; L is the length of the gap; S_1 is the entropy of the frequency dictionary obtained over the original sequence; S_2 is the entropy of the reference frequency dictionary; S_3 is the entropy of the augmented dictionary.

		$L = 2040$		$L = 6120$		$L = 10200$	
q	S_1	S_2	S_3	S_2	S_3	S_2	S_3
2	2.762391	2.762560	2.763216	2.764007	2.764846	2.766264	2.767316
3	4.139174	4.139590	4.140491	4.141834	4.143873	4.145174	4.146399
4	5.510682	5.511079	5.512071	5.513930	5.515805	5.517836	5.518700
5	6.874128	6.874118	6.874410	6.876863	6.876947	6.880571	6.879550
6	8.207368	8.205078	8.202776	8.203433	8.198796	8.202798	8.194168
7	9.408337	9.396622	9.388944	9.376704	9.357722	9.349918	9.318023
8	10.201937	10.171654		10.110157		10.035009	

In our computation experiments, the multiplication factor 2 was used, for the population of strings that had reached the check-points. Here all the coverings have been obtained over the reference dictionary. The computation experiments show that the methodology presented above is quite efficient for reparation of gaps in symbol sequences. The data presented here are obtained over the experiment with a single genetic text. Hence, one is not able to distinguish the effects resulted from the entity under consideration from those ones peculiar to the method itself.

The conditions of our computation experiments allowed 4800 strings which might be the coverings, for each frequency dictionary thickness (see Table 1). It is evident, some of them fail to be complement to the right basement. The number of the available coverings goes down almost exponentially, as the thickness of the dictionary grows up. The reference dictionary of the thickness 8 failed to produce a recovery. Meanwhile, this result may follow from the low number of copies of basement used to carry out the experiment.

A correspondence between the number of the recoveries obtained by KKM and the entire possible number of them was evaluated due to the following computation experiment. The original sequence has been replaced with the surrogate one that is a realization of a random process with the same probability distribution of the isolated symbols, as at the original entity. Totally, three realizations have been studied. Also, the number of primers in various series of the experiment increased from 750 up to 45000; all the other parameters (such as the gap length, its location inside the sequence, multiplication factor etc.) were the same. The number of proper fulfillments grows up, as the number of primers increases. This fact allows to assume the further growth of number of proper fulfillments following the growth of the number of primers. Thus, one hardly could expect that the best recovery has been obtained, in our computation experiments.

6 Discussion

The efficiency of KKM implementation to the problem of the lost data reconstruction is evident. Nevertheless, still there are some problems to be discussed. **The problem of a test object** is the basic one. A random non-correlated sequence is the standard test object. There could be other types of test objects. For example, a one-dimensional fractal with a relevant number of elementary cells could be a good test object. Probably, the symbol sequences generated with clear and unambiguous rules could be the test objects; one can consider expansion of Lioville numbers, or transcendental numbers.

The existence of the exact recovery is another problem. It might be observed for some specific sequences, and peculiar lengths of words used to make up a recovery. Anyway, the answer on this question will provide researchers with important knowledge concerning the basic properties of the method presented here. Since the exact recovery is impossible, in general, then one needs to compare the versions of the recovery obtained due to the method presented above, and the original sequence (see, e.g., [6]). The comparison of an original entity and that one obtained with the method presented above allows to clarify some peculiarities of the methodology.

A study of relationship between the entities obtained under the different extremal principles is of great importance. The question arises whether these two principles may produce the same covering or not, and if not, is a researcher able to evaluate the difference between them. That question is still waiting for the student.

A relation between the dictionary used to develop a covering and that one to evaluate the best entity among several is another substantial subject to be considered. Everywhere above we have searched for the best covering through the calculation of the entropy for the frequency dictionary of the same thickness, as that one used to develop the covering. There is no constraint to use the dictionaries of different thickness to develop a string covering a gap, and to find the best one [5]. We believe, this matter requires special, careful and comprehensive investigation; further discussion of that subject falls beyond the scope of our paper.

Finally, a number of questions concerning the improvement and/or modification of the imitator of KKM are to be studied carefully. For example, a development of self-training algorithms for a choice of the points of multiplication of primers that are selected from the entire pool of growing strings, as well, as the factor of that multiplication is the important question. We used the version with two point of replications; the factor of this replication was 2, 3 and 4. This choice was the matter of experience. A development of a proper modification of a frequency dictionary can also contribute significantly the progress in Kirdin kinetic machine applications.

Acknowlegdement. We are grateful to Prof. Alexander N.Gorban for his permanent and encouraging support of our efforts and this work. This was partially granted by Krasnoyarsk Regional Science Foundation (grant 11F0114C).

References

1. Kirdin A.N. Ideal ensemble model of parallel computations. In: "Neural informatics and its applications". Krasnoyarsk, KGTU, 1997. p.101.
2. Gorban A.N., Gorbunova K.O., Wunsch D.C. Liquid Brain: Kinetic Model of Structureless Parallelism. // Advances in Modelling & Analisis. AMSE, v.**5**, No.5, 2000.
3. Gorban A.N., Gorbunova K.O., Wunsch D.C. Liquid Brain: The Proof of Algorithmic Universality of Quasichemical Model of Fine-Grained Parallelism. Neural Network World, **4**/2001, p.391–412.
4. Katya O. Gorbunova. Kinetic Model of Parallel Data Processing // (Lecture notes in computer science; Vol. 1662) Parallel computing technologies: 5th International Conference; Rroc./ PaCT-99, St.Petersburg, Russia, September 6–10, 1999. Victor Malyshkin (ed.). Springer, 1999, P.55–59.
5. Sadovsky M.G. Information capacity of symbol sequences / Open Systems & Information Dynamincs, 2002, **v.9**, pp. 231–247.
6. Sadovsky M.G. Comparison of symbol sequences: no editing, no alignment / Open Systems & Information Dynamincs, 2002, **v.8**, pp. 123–132.

Measurement Models for Survivability and Competitiveness of Very Large E-marketplace

Jingzhi Guo and Chengzheng Sun

School of Computing and Information Technology, Griffith University
Nathan, QLD 4111, Australian
{J.Guo, C.Sun}@cit.gu.edu.au

Abstract. E-Marketplace is a man-made profit center with emergent customer requirements. This emergence nature poses many difficulties for e-marketplace to satisfy its customers and thus affects its survivability and competitiveness. To challenge this issue, this paper has introduced a novel emergent e-marketplace model, and analyzed and quantified the emergent properties of e-marketplace from simple ESMs to very large complex electronic intermediary EIMs. The measurement models developed in this paper are very useful for producing dynamic interaction records appropriate for further simulation model of e-marketplaces.

1 Introduction

Business-to-business electronic marketplace (e-marketplace) is man-made, which involves costs [8] such as setup cost, maintenance fees, content creation fees, telecommunication link fees and software development fees [10], and needs revenues such as service fees [7] to offset the costs and earn profits for survivals and increasing market competition ability. To achieve this enterprise goal, the best practice is to satisfy customer requirements to deliver the needed services [1], [3], [5], [11].

However, an e-marketplace is an emergent organization, a very large distributed Internet application that supports the mediation of *sellers and buyers* (simply called as "customers") to finish business deals, in which customers' requirements are in continuous change [2], [6]. This changing nature results in difficulties in capturing customer requirements to design an e-marketplace system [4].

To challenge the above issue, an e-marketplace must be continuously analyzed, measured and dynamically negotiated with customers about emergent requirements. System structure must be flexible to adapt to the incomplete and ambiguous specifications for continuous redevelopment of emergent e-marketplace [12]. Nevertheless, despite the widespread adoption and active roles that e-marketplaces have played, to our best knowledge there is no direct work published on the formal analysis of the emergent e-marketplace requirements for survivability and competitiveness. It is not clear what indicators should be adopted to measure survivability and competitiveness and what properties should be considered to design and simulate an emergent e-marketplace.

P.M.A. Sloot et al. (Eds.): ICCS 2003, LNCS 2658, pp. 802–811, 2003.

The purpose of this paper, therefore, is to undertake a formal analysis of the requirements for emergent e-marketplaces to quantify the measurements of what are the key factors to the survivability and competitiveness of an emergent e-marketplace.

In the rest of this paper, Section 2 will introduce a novel but generic model of emergent e-marketplace. In Section 3, we will develop the key survivability measurements of a single seller/buyer owned e-marketplace. Section 4 analyzes the key competitiveness measurements of an electronic intermediary e-marketplace. Finally in Section 5, a conclusion will be given and future work will be mentioned.

2 Emergent E-marketplace

An emergent e-marketplace is a shared reality in which actors continuously interact with each other to achieve a common goal, lowering business cost and increase revenue but for their own [12]. This shared reality forms the current abstract shape of an e-marketplace and therefore exposes a set of fixed requirements to support the existence of the e-marketplace. These requirements are the "*persistent properties* (PP)" to satisfy the existence of an e-marketplace. In contrast, the emergent requirements of actors make the e-marketplace in constant negotiation and in continuous change by referencing the original e-marketplace identity. We call these emergent requirements as "*emergent properties* (EP)", which are the forces to make an e-marketplace fluid.

2.1 Persistent Properties

Persistent properties of an e-marketplace can be derived by the analysis of the commonalities and differences between traditional market and e-marketplace. Traditional market is defined as "an arrangement by which buyers and sellers of a commodity interact to determine its price and quantity" [9]. The resemblance between an e-marketplace and a traditional market is that both have products, sellers, buyers, and an arrangement for trading interactions. The difference is the means of how to arrange interaction for sellers and buyers. In traditional market, sellers and buyers interact by face-to-face meeting, verbal or written descriptions of their products to deliver trading information. In e-marketplace, product information is delivered and shared between sellers and buyers in electronic representations via network (generally Internet). Sharing electronic product representations is fundamental to transform a traditional market to an e-marketplace and is the primary task and condition to build an e-marketplace. Although transactional (such as negotiation and payment) and institutional functions (such as legal systems) are often in e-marketplace definition domain [1], they are not the primary task to make e-marketplace alive. So only sellers S, buyers B, products P, electronic product representations R, and public accessible network N are considered as persistent properties of an e-marketplace. These persistent properties exist in the lifetime of an emergent e-marketplace.

The persistent properties are footstones of an e-marketplace. Applying for them, we define an e-marketplace as an emergent two-stage functional business organization on Internet. Its first stage functions are about *e-matching* by forming an e-marketplace. Its second stage functions are about *e-trading* by realizing transactional and institutional functions. E-matching is fundamental to e-trading and has two major tasks: *e-representation* defined as electronically representing real-world products into machine-readable and multi-firm sharable data, and *e-mapping* defined as logically mapping sellers' and buyers' electronic product representations into e-tradable forms for trade interactions. In this paper we focus our analysis on the e-matching stage and assume that the profitability of e-trading stage is proportional to the efforts of e-matching stage. So the survivability and competitiveness of an e-marketplace is thus dependent on how the first stage tasks are fulfilled.

2.2 Persistent Relationships

Persistent properties are not enough to understand how an e-marketplace operates. Only relationships between persistent properties, *persistent relationships* (PR), reflect the operation of an e-marketplace.

PR 1: Sellers *S* owns a set of products $SP \subseteq P$, expressed as *S Own SP*.

PR 2: Buyers *B* owns a set of products $BP \subseteq P$, expressed as *B Own BP*.

PR 3: A set of sellers' products $SEP \subseteq SP$ is electronically represented as *SER*, expressed as *SEP RepAs SER*.

PR 4: A set of buyers' products $BEP \subseteq BP$ is electronically represented as *BER*, expressed as *BEP RepAs BER*.

PR 5: A set of sellers' product representations *SER* is displayed on *N*, expressed as *SER DispOn N*

PR 6: A set of buyers' product representations *BER* is displayed on *N*, expressed as *BER DispOn N*.

These six relationships are persistent during the lifetime of an e-marketplace. The sum of them defines an e-marketplace.

Definition 1 (*Emergent E-marketplace* "*EEM*"): An emergent e-marketplace *EEM* exists if and only if there exist persistent relationships *PR1* and *PR2* and *PR3* and *PR4* and *PR5* and *PR6*.

2.3 Emergent Properties and Relationships

Definition 1 provides a static framework and existent conditions of an EEM. To study how an EEM is emergent, we assume that there is a mechanism to make a market arrangement for sellers *S* and buyers *B* to interact [9], which is called as *interaction mechanism I* and releases a dynamic force, originated from sellers *S* or buyers *B*, acted on network *N* to have *S* and *B* interacting each other. The effects that *I* generates are the normalized emergent requirements (emergent properties *EP*) of customers. The fundamental relationships between *EP*s (emergent relationship *ER*) are:

- When an *EP* generates a set of positive effects, a set of persistent properties (*PP*s) change themselves in quantity with favorable outcomes for an EEM and further generate a new set of *EP*s to require EEM to adapt to them.

- When an *EP* generates a set of negative effects, a set of *PP*s change themselves in quantity with unfavorable outcomes for an EEM and further generate a new set of *EP*s to require the EEM to improve them.
- If the effects that a set of *EP*s generates lead to any *PP* disappeared in an EEM such that the quantity of the *PP* goes to zero, then the EEM is dead.

2.4 EEM Operational Model

To describe the interactions between sellers and buyers and to quantify the emergent requirements of an EEM, we define the following interactive relationships:
Definition 2 (*Logical Intersection* "$\mathfrak{I}$"): Given three sets of emergent properties X, Y and I_{xy}, then $I_{xy} = X \mathfrak{I} Y$, called X logically intersected with Y at I_{xy} if and only if (1) X is originated from S, Y is originated from B and I_{xy} is originated from I, and (2) I_{xy} logically agrees with both at least zero number of X and at least zero number of Y.
Definition 3 (*Interaction Model*): An EEM is said to be interactive if and only if (1) there exists an I, and (2) a set ER is generated based on $I_{xy} = X \mathfrak{I} Y$ and $ER \neq \varnothing$.

Emergent relationships are generated in the process of continuous interactions between sellers and buyers. The matched results by interaction mechanism of their emergent requirements become the emergent properties and they further require an EEM to react.
Corollary 1: If I is missing in an EEM, then an EEM generates no *EP* and is static as Definition 1, then it only provides the e-representation service such as a plain online product catalogue displaying product information.

Corollary 1 shows an interactive EEM is conditional to the existence of I. Actually different I qualities lead to different types of EEMs:

Table 1. Types of Electronic Marketplaces

No	EEM Type	EEM Owner
1	One seller, multiple buyers	Seller
2	One buyer, multiple sellers	Buyer
3	Multiple buyers, multiple sellers	Third-party market provider

Type 1 and type 2 are most common. Each company website can be regarded as an EEM as it at least provides e-representation service if they do not provide interaction mechanism. Type 3 is the advanced form of type 1 and 2. It aggregates multiple sellers and buyers and runs by a separate firm to serve as an e-marketplace via interaction mechanism.

3 Survivability Analysis

In this section, we will first study the simplest e-marketplace, type 1 and 2 of EEMs and refer them as "ESM". To unveil how they survive, we begin by investigating how the emergent properties generated by a seller's or buyer's interaction mechanism I affect their survivability.

3.1 Tradability Analysis

Given an emergent ESM, then its *I* generates an emergent property – tradability. This part describes how it is generated and affects the ESM's market efficiency.

Definition 4 (Total Marketing Size "*TMS*"): ESM total marketing size *TMS* is defined as the sum of *SER* and *BER* defined in PR3 and PR4, or expressed as *(SER + BER).*

TMS reflects an ESM's total market scale at a certain time.

Definition 5 (*Tradable Product Representations* "*T*"): Given *SER* and *BER* defined in PR 3 and PR 4, a set of product representations *T* is called "tradable" if and only if $T = SER \mathfrak{J} BER$.

T determines the current exchangeable products between sellers and buyers. It reflects how many electronically represented products are actually tradable.

Definition 6 (*Tradability* "$\mathfrak{I}$"): Given *T* and *TMS* defined in definition 4 and 5, the tradability $\mathfrak{I}$ in an ESM is measured as:

$$\mathfrak{I} = T \,/\, (TMS - T) \tag{1}$$

Tradability measures the capability of an EEM to link sellers and buyers together for trading. It tells how efficient an ESM to satisfy customers' emergent requirements to match the product of sellers or buyers. The ESM's market efficiency measured by $\mathfrak{I}$ is dependent on the capability of interaction mechanism *I* either provided by sellers or buyers that assists the conversion and mapping of product representations.

3.2 Profitability Analysis

A fully efficient ESM may not be necessarily profitable in an emergent e-business environment. For example, the increase of tradability may be increased by decreasing both *T* and *TMS* and decrease total profit. As a profit center, an important emergent property – profitability realized at the seller/buyer's interaction mechanism must be introduced to measure and control cost and revenue. To analyze, we assume:

Assumption 1: A piece of electronic product representation $R_i \subseteq TMS$ defined in Definition 4 has an average cost *AC*.

Assumption 2: A piece of electronic tradable product representation T_i defined in definition 5 generates an average revenue *AR*.

Immediately, we have two corollaries:

Corollary 2: If given total marketing size *TMS* defined in definition 4, then according to Assumption 1, an ESM's total cost $TC = TMS*AC$.

Corollary 3: If given a set of tradable product representations *T* defined in definition 5, then according to assumption 2, an ESM's total revenue $TR = T*AR$.

These assumptions and corollaries enable us to define profitability.

Definition 7 (Profitability "$\mathcal{P}$"): Given *TC* and *TR* derived from Corollary 2 and 3, the profitability of an ESM is measured as:

$$\mathcal{P} = (TR - TC) \,/\, TR \tag{2}$$

Corollary 4: If profitability $\mathcal{P} \leq 0$, then an ESM is in difficulty to survive. If profitability $\mathcal{P} > 0$, then an ESM is in a favorable survival position. If $\mathcal{P}$ goes to 1, then the cost of an ESM goes to 0 and the ESM tends to be frictionless like the traditional market assuming that market is efficient and no informational cost.

Obviously, the profitability of ESM is positively proportional to the tradable product representations *T* and negatively proportional to the total market size *TMS*. However high profitability may not contribute to improving tradability if both *TMS* and *T* remain constant but the unit revenue of *T* is increased when a seller or buyer adopts a radical profit strategy or the ESM tends to be frictionless.

3.3 Survivability Analysis

The analysis of Section 3.1 and 3.2 shows the tradability and profitability may independently increase or decrease. It suggests that we must further develop survivability to include the joint effects of the above two emergent properties.

Definition 8 (*Survivability* "$\mathfrak{S}$"): At a given time if profitability is $\mathcal{P}$ and tradability is $\mathfrak{T}$, then the survivability is.

$$\mathfrak{S} = \mathcal{P} * \mathfrak{T} \qquad \textbf{(3)}$$

Survivability reflects at a given time the quality of an ESM in terms of e-market efficiency and profits. To further analyze the dynamic effects of tradability and profitability, we gather a time series of survivability data to reflect the changes of an ESM's survivability.

Definition 9 (*Market Growth Rate* "$\mathcal{G}$"): If from time t_a to t_b, survivability is from $\mathfrak{S}_a$ to $\mathfrak{S}_b$, then the market growth rate is:.

$$\mathcal{G} = (\mathfrak{S}_b - \mathfrak{S}_a) / (t_b - t_a) \qquad \textbf{(4)}$$

Market growth rate reflects the ability of an ESM to cope with the long-run emergent ESM requirements.

Corollary 5: If $\mathcal{G}$ is positive, an ESM is in a favorable survival condition. If $\mathcal{G}$ is negative, an ESM is in an unfavorable survival condition.

In summary, an ESM's survivability depends on the emergent requirements of tradability and profitability. However, it is worth to note that the live of an ESM may be independent of survivability in that a seller/buyer may only want to benefit from the *e-representation* to display products as promotional purpose.

4 Competition Analysis

Emergence theory assumes an organization is in the state of being continual process, never arriving but always in transition [12]. Coincidently, economic theories prove that changes of market power between sellers and buyers change the competition power and market type [9]. Often competition power can be obtained by reconstructing market types by alliance. Applying these theories, this section will analyze the competitiveness of type 3 e-marketplace evolved from a set of ESMs.

4.1 Electronic Intermediary Marketplace

Drives of E-intermediary Marketplace. The ESM described in Section 3 is a typical EEM under perfect competition where all sellers and buyers are responsible for their own e-matching service and fully compete with each other. It is often presented as a website such as a standalone electronic catalogue or a corporate website with embedded catalogue [10] such as GE Polymerland[1]. It is comparatively small and has no enough power to exert influences on selling or buying. Its customers are limited to the related partners in a narrow industry or random product seekers. From the viewpoint of all *TMS*s of ESMs, an individual ESM is insignificant in market share because its product information is difficult to reach to and share by all customers in e-marketplaces. To attract more customers, one choice is to aggregate multiple ESMs into an electronic intermediary marketplace (EIM) representing a superset of some ESMs, in which a larger set of sellers and/or buyers map their electronic product representations via brokering, mediation, federation and mapping. A practical example is CommerceOne[2] who provide third-party B2B marketplaces. This choice is a natural evolution of the ESM development. The benefits of adopting an EIM are:

– Aggregating a set of individual ESMs to increase competition power.
– Reaching some consensus to adopt a unified product representation standard for all its sellers and buyers to increase tradability. The proper standards enable an EIM to compete with those un-joined sellers, buyers and other EIMs.
– Integrating the increased number of de facto product standards, international standards and ad-hoc electronic product representations.

EIM Persistent Properties and Relationships. As a superset of some ESMs, an EIM has its special persistent properties that are: an *EIM provider* who represents a certain set of sellers and buyers, and a set of *mediating product representations* (*IER*) that mediates sellers' and buyers' product representations *SER* and *BER* defined in PR3 and PR4. So the new persistent relationships in EIM are:

PR 7: Some sellers' *SER* is mediated via a set of mediating product representations *IER* provided by an EIM provider, expressed as *SER MedBy IER*.

PR 8: Some buyers' *BER* is mediated via a set of mediating product representations *IER* provided by an EIM provider, expressed as *BER MedBy IER*.

PR 9: A set of product representations *IER* is owned by an EIM provider, expressed *EIM Own IER*.

PR 10: A set of product presentations *IER* is displayed on network *N*, expressed as *IER DispOn N*.

Definition 10 (*Electronic Intermediary Marketplace* "EIM"): An electronic intermediary marketplace EIM exists if and only if there exist persistent relationships PR1, PR2, PR3, PR4, PR7, PR8, PR9 and PR10.

Unlike an ESM, an EIM acts as an independent interaction mechanism. The relationships existed in an EIM reflect the existence conditions of an EIM.

---

[1] http://www.gepolymerland.com

[2] http://www.commerceone.com

4.2 EIM Mediability

Similar to tradability in an ESM, the emergent property, mediating ability, for sellers' and buyers' product representations is important to an EIM's survival and should be accurately defined.

Definition 11 (*Target Mediating Volume* "TMV"): Given *SER* and *BER* defined in PR 3 and PR4, a set of product representations *TMV* is called "target mediating volume" if and only if there exist two sets of product representations *MSR* and *MBR* such that $MSR \subseteq SER$ and $MBR \subseteq BER$, and $TMV = MSR + MBR$.

MSR and *MBR* are sellers' and buyers' electronic product representations ready to be mediated in an EIM. The sum of them is the target volume for an EIM to mediate. It must be noted that the quantity of *SER* and/or *BER* in an EIM may have increased compared with that in an ESM where SER/*BER* is from only one seller/buyer if this ESM is a seller's or buyer's EEM.

Definition 12 (*Mediable Representations* "*M*"): Given *MSR* and *MBR* defined in Definition 11, *IER* defined in PR 9, a set of product representations *M* is said to be "mediable" if $M = MSR \,\mathcal{I}\, IER \cap IER \,\mathcal{I}\, MBR$.

The volume of mediable product representations indicates how many e-mapping services an EIM can provide to satisfy its sellers and/or buyers. It reflects the fulfillment of the target market.

Definition 13 (*Mediability* "$\mathcal{M}$"): Given *TMV* and *M* defined in definition 11 and definition 12, the mediability of $\mathcal{M}$ is measured as:

$$\mathcal{M} = \mathrm{M} / (TMV - M) \quad \textbf{(5)}$$

Mediability reflects an EIM's market efficiency about its target market. An EIM's profitability, survivability and market growth rate are similar to ESM. The indicators can be directly given as:

Definition 14 (*EIM Profitability* "$\mathcal{IP}$", *EIM Survivability* "$\mathcal{IS}$" and *EIM Market Growth Rate* "$\mathcal{IG}$"): If given *TMC=TMV*AC* and *TMR=M*AR* derived from Assumption 1, 2 and definition 11 and 12, then:

$$\mathcal{IP} = (TMR - TMC) / TMR \quad \textbf{(6)}$$

$$\mathcal{IS} = \mathcal{IP} * \mathcal{M} \quad \textbf{(7)}$$

$$\mathcal{IG} = (\mathcal{IS}_b - \mathcal{IS}_a) / (t_b - t_a) \quad \textbf{(8)}$$

The above indicators jointly describe the long-run survivability of an EIM.

4.3. EIM Competition Power

To completely reflect the competitiveness of an EIM in an emergent EEM, we should compare with other EIMs. We compare survivability between EIMs to determine an EIM's competition power.

Definition 15 (*EIM Competition Power* "$\mathcal{CP}$"): Given a set of $EIM = {}_d\{0 \le i,j \le n \mid EIM_{i,j}\}$ defined in Definition 10, and a set of *EIM* survivability $\mathcal{SS} = {}_d\{0 \le i,j \le n \mid \mathcal{SS}_{i,j}\}$ defined in Definition 14(2), then:

EIM_i's relative competition power against EIM_j is:

$$\mathcal{CP}_{i,j} = (\mathcal{SS}_i - \mathcal{SS}_j) / \mathcal{SS}_j \quad \textbf{(9)}$$

EIMi's absolute competition power against the whole set of EIM is:

$$\mathcal{CP}_i = (\mathcal{SS}_i - \mathcal{SS}) / \mathcal{SS} \quad \textbf{(10)}$$

Relative competitive power compares an EIM with another EIM. Absolute competitive power compares an EIM with the whole intermediary provider market. A positive figure means a strong market competition power to attract sellers and buyers to join in.

Corollary 6: For any two EIMi and EIMj, if any $\mathcal{CP}$i,j is approximate to 0, the overall EIM marketplace goes to perfect competition.

Corollary 7: If there exist one EIM_i such that $\mathcal{CP}_{i,j}$ is always approximate to 1 against EIM_j, EIM_i goes to be monopolistic in the overall EIM marketplace.

Corollary 7 and 8 measures the market power of an EIM over the sellers, buyers and other EIMs, which changes e-marketplace types.

4.4 Competitive Analysis

The above sections have described the changes of an EIM's competition power. But how to increase EIM competition power is still not clear. This section, we will analyze the changing relationships between emergent properties to find out the solution. As Definition 14, EIM profit is determined by the factors of $TMR_{EIM} = M*AR_{EIM}$ and $TMC_{EIM}=TMV*(AC_{EIM}+AC_{SB})$. So an EIM's total profit is $TMR_{EIM} - TMC_{EIM}$ and the target is to either increase M and AR_{EIM}, or decrease $AC_{EIM} + AC_{SB}$. For the total customers of the EIM, if, for each mediable product representation, it can gain AR_{SB}, then its profit from using the EIM is $M*AR_{SB} - TMV*AC_{SB}$. Obviously, AR_{EIM} partially and proportionally determines AC_{SB} in that part of AC_{SB} is the AR_{EIM}. For example, EIM increases service fees to increase AR_{EIM} but it also increases the cost of customers' AC_{SB}. Another part of AC_{SB} is the sinking cost of sell customers who electronically represent their products by themselves. Therefore the limit for an EIM to increase AR_{EIM} is that the decreasing total profit of customers is still positive and the conditions that other EIMs have not provided customers better profit rates. Otherwise customers will either exit from the EIM or turn to other EIMs. Another method to increase profit is to decrease *TMV*, but it is not under an EIM's control, customers may increase product representation volume hence increase their e-marketplace sinking cost but still try to find a better EIM to absorb it. Further method is to decrease AC_{EIM} but it may hinder an EIM to provide better e-mapping service due to the less input of the development fees. The obvious better solution is to increase *M*, which has no limit and has the greatest potential. To increase volume *M* is to map more sellers' and buyers' electronic product representations.

5 Conclusion

By introducing a novel emergent e-marketplace model, this paper has formally analyzed the emergent properties of simple ESMs and large and complex EIMs. Concomitantly, it has quantified the key measurements of survivability and competitiveness of a very large e-marketplace. As an analysis result, it points out that the key to adapt to the customers' emergent requirements is to increase *mediability*. Satisfying this requirement can keep an EIM in a favorable market position with high survivability and strong competition power and obtain a win-win outcome. Measurement models developed in this paper have many implications in which the most direct is to apply these models to simulation models of very large e-marketplaces. The changing values of emergent properties over time are appropriate interaction records for simulation. The future work includes to explore the appropriate electronic representation of products and to further develop measurement model for e-trading of the second stage e-marketplace.

References

1. Bakos, Y., The Emerging Role of Electronic Marketplaces on the Internet, *Communications of the ACM*, Vol. 41/No. 8, August 1998, 35-42.
2. Damsgaad, J., and Truex, D., Binary Trading Relations and the Limits of EDI Standards: The Procrustean Bed of Standards, *European Journal of Information Systems*, 2000, 9(3) 173-188.
3. Hoffner, Y., Facciorusso, C., Field, S., and Schade, A., Distribution Issues in the Design and Implementation of a Virtual Market Place, *Computer Networks*, 32 (2000) 717-730.
4. Maidantchik, C., Montoni, M., and Santos, G., Learning Organizational Knowledge: An Evolutionary Proposal for Requirements Enginnering, in Proceeding of *ACM SEKE'02*, July 15-19, 2002, Ischia, Italy, 151-157.
5. Malone, T. W., Yates, J., and Benjamin, R., Electronic Markets And Electronic Hierarchies, *Communications of the ACM*, Vol. 30/No. 6, June 1987, 484-497
6. Ngwenyama, O. K., Groupware, social action and organizational emergence: on the process dynamics of computer mediated distributed work, *Accounting, Management and Information Technology,* 8 (1998) 127-146.
7. Phillips, C. and Meeker, M., Collaborative Commerce, The B2B Internet Report, *Morgan Stanley Dean Witter*, April 2000.
8. Rose, M., Implications of Costly Information, *The Library of Economics and Liberty*, April 22, 2002, http://www. econlib.org/library/Columns/Teachers/information.html.
9. Samuelson, P.A., Nordhaus, W. D., Richardson, S., Scott, G., and Wallace, R., Economics – Volume 1 Microeconomics, Third Australian Edition, *McGraw-Hill Book Company* Sydney, 1992.
10. Segev, A., Wan, D., and Beam, C., Electronic Catalogs: a Technology Overview and Survey Results, in proceedings of *ACM CIKM'95*, Baltimore, MD, USA, 1995, 11-18.
11. Strader, T., and Shaw, M., Characteristics of Electronic Markets, *Decision Support Systems*, 21 (1997) 185-198.
12. Truex, D. P., Baskerville, R., and Klein, H., Growing Systems in Emergent Organizations, *Communications of the ACM*, Vol. 42/No. 8, August 1999, 117-123.

Embedded Fuzzy Control System: Application to an Electromechanical System

R.E. Haber [1,2], J.R. Alique[1], A. Alique[1], and J.E. Jiménez[1]

[1] Instituto de Automática Industrial (CSIC)
Nº-III km. 22800, La Poveda. 28500. Madrid. SPAIN.
rhaber@iai.csic.es

[2] Escuela Técnica Superior
Ciudad Universitaria de Cantoblanco
Ctra. de Colmenar Viejo, km. 15. 28049 - SPAIN
Rodolfo.Haber@ii.uam.es

Abstract. Nowadays with open computerized numerical controls internal control signals can be gathered and mathematically processed by means of integrated applications. Working with a commercial open computerized numerical control, a fuzzy control system has been designed, implemented and embedded that can provide an additional optimization function for cutting speed. The results show that, at least in rough milling operations, internal signals can double as an intelligent, sensorless control system. The integration process, design steps and results of applying an embedded fuzzy control system are shown through the example of real machining operations.

1 Introduction

One of the main activities the manufacturing industry has to deal with is machining, a process that includes operations that range from rough milling to finishing. There is a number of angles from which to view the optimization of the machining process, angles where minimum production cost, maximum productivity and maximum profit are significant factors [1].

There are also various different ways of implementing machining process optimization. The implementation on which we will focus here attains optimal goals via automatic control of the machining process. The spectrum of "conventional" methods (so named to distinguish them from intelligent methods) available for designing control systems is enormous. In the incessant pursuit of better performance, newer approaches have been also tested such as model reference adaptive control (MRAC) [2], which incorporates an on-line estimation scheme to tune controller parameters for time-varying process dynamics. Another recently applied method is robust control based on the Quantitative Feedback Theory (QFT) [3]. The results of these tests, however, have not lived up to expectations, because all these approaches have the indispensable design requisite of an accurate (traditional) process model,

P.M.A. Sloot et al. (Eds.): ICCS 2003, LNCS 2658, pp. 812–821, 2003.

e.g., differential equations, transfer functions and state equations. Unfortunately, accurate models of this sort cannot yet be attained for the machining process.

The complexity and uncertainty of processes like the machining process are what make what is known as *intelligent systems technology* a feasible option to classical control strategies. Indeed, Artificial Intelligence techniques have received considerable interest from scientific community and have been applied to machining [4]-[5]. Two interesting approaches are the neural network and expert rule based on adaptive control constraints (ACC) [6] and the evolutionary algorithms based on adaptive control optimization (ACO) approach [7].

However, the main disadvantage related with the above-mentioned approaches is that neural network and evolutionary algorithm based computation requires time, and therefore it limits the performance of the intelligent control system. Nowadays, open computerized numerical control is powerful to build up an intelligent system. However, there are constraints for real-time signal processing and to implement complex control algorithms.

On the other hand, the majority of the work in machining optimization is devoted to the issue of adaptive techniques. Adaptive controllers are highly expensive considering the time requirements because they estimate parameters on-line and adjust the controller gains accordingly. Likewise, these systems must be carefully tuned and exhibit complex, and sometimes undesirable behavior.

In order to improve machining efficiency, the current study focuses on the design and implementation of an intelligent controller in an open computerized numerical control system. From all available techniques, Fuzzy Logic (FL) is selected because it has proven useful in control and industrial engineering as a highly practical optimizing tool. To the best of our knowledge, the main advantage of the present approach includes: (i) embedded fuzzy controller in an open computerized numerical control to deal with a real life industrial process, (ii) a simple computational procedure to fulfill the time requirements, and (iii) without restrictions concerning sensor cost (sensorless application), wiring and synchronization. The results with fuzzy control strategy through actual industrial tests show a higher machining efficiency.

This paper is organized as follows. In Section 2 we present a brief study of the machining process, explaining why it is considered a complex process and setting up the milling process as a case study. In Section 3 we design the fuzzy controller to optimize the milling process. Next, in Section 4 we describe how the fuzzy controller can be embedded in open computerized numerical control, and we discuss the key design and programming stages. In Section 5 we share the experimental results and explore some comparative studies. Finally we give some concluding remarks.

2 The Machining Process

This section introduces the milling process, which is characterised by both nonlinear characteristics and varying process parameters. The task of controlling this complex plant is performed with the help of a fuzzy controller that will be introduced in Section 3. A panoramic view of a typical machining center is given in Figure 1.

Fig. 1. Overall view of a typical machining center

The machining process, also known as the metal removal process, is widely used in manufacturing. One of the most complex of the four operations is the milling process. One important study of cutting-process monitoring and control by means of current measurement shows the feasibility of motor-current measurement for adaptive control [8]. The direct relationship between the current consumed and the cutting force is fair enough for a real industrial implementation of the control system to be made on the basis of a main spindle's drive current [9].

Nowadays the development of open control systems in the NCK offers more facilities for using digital drive signals without the need to install more sensors. Indeed, cutting processes show significant effects on drive signals such as actual drive current and drive torque. Digital drive signals have many limitations for process monitoring alone, because of the ratio between process-related components of the signal and non-process-related disturbances. However, main-spindle drive current can be used to optimize cutting speed. Signal behavior is complex because of specific design considerations such as star-triangle switching in drive configuration. Nevertheless, correcting the current offsets during the signal-processing stage (before entering the fuzzy algorithm) solves the problem.

In terms of control system design, the most important aspects are the variables, parameters and typical measures of performance that we use to characterize the system. After a preliminary study we selected: the spatial position of the cutting tool, considering the Cartesian coordinate axes (*x, y, z*) [*mm*], spindle speed (*s*) [*rpm*], relative feed speed between tool and worktable (*f*, feedrate) [*mm/min*], cutting power invested in removing metal chips from the workpiece (P_c)[*kW*], current consumed in the main spindle during the removal of metal chips (I_s) [A], radial depth of cut (*a*, cutting depth) [*mm*] and cutting-tool diameter (*d*) [*mm*].

In order to evaluate system performance, we need to select certain suitable performance indices. The milling process basically consists of two operations, rough milling and finishing. The differences in these operations' objectives are what will decide which performance index is useful in each operation. The quality and geometric profile of the cutting surface is paramount in finishing operations, whereas the quantity of metal removed from the workpiece is the main issue in rough milling operations. This work dealt essentially with rough milling, so the main index was the metal-removal rate (MRR).

3 Fuzzy Logic Controller for Machining Processes

The fuzzy controller follows rather classic lines. The controller's actual core performs on-line actions to control the feedrate. The three basic tasks known as fuzzification, decision-making and defuzzification were implemented on-line. The main design steps for this controller and their corresponding implementation were as follows:

1) Defining input and output variables, fuzzy partitioning and building the membership functions.

The input variables included in the error vector $\boldsymbol{e}$ are the current-consumed error (ΔI_S in ampere) and the change in current-consumed error ($\Delta^2 I_S$ in ampere). The manipulated (action) variable we selected is the feed-rate increment (Δf in percentage of the initial value programmed into the computerized numerical control), whereas the spindle speed is considered constant and preset by the operator.

$$e^T = \left[KE \cdot \Delta I_S \quad KCE \cdot \Delta^2 I_S\right]; u = \left[GC \cdot \Delta f\right]. \tag{1}$$

where KE, KCE and GC are scaling factors for inputs (error and change in error) and output (change in feed rate), respectively.

The fuzzy partition of universes of discourse and the creation of the rule base were drawn from the criteria of skilled operators, although we did have to apply a "cut and trial" procedure as well. Figure 2 shows the fuzzy partition thus obtained.

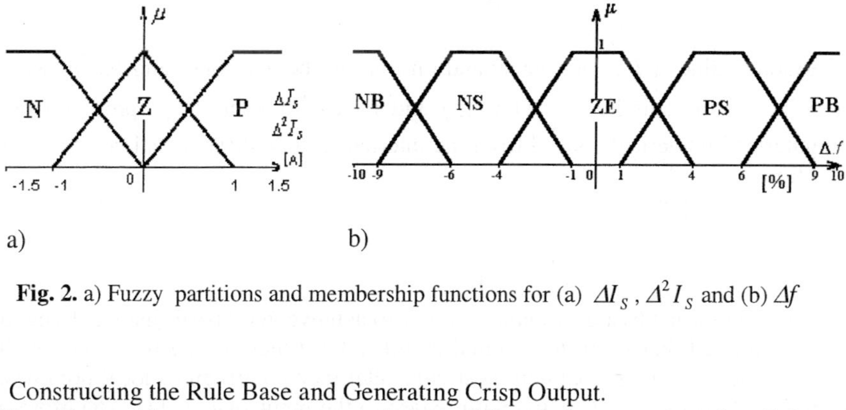

a) b)

Fig. 2. a) Fuzzy partitions and membership functions for (a) ΔI_S, $\Delta^2 I_S$ and (b) Δf

2) Constructing the Rule Base and Generating Crisp Output.

We will consider a set of rules consisting of linguistic statements that link each antecedent with its respective consequent, having, for instance, the following syntax:

IF ΔI_S is positive AND $\Delta^2 I_S$ is positive THEN Δf is Positive Big

The controller output is inferred by means of the compositional rule. The Sup-Product compositional operator was selected for the compositional rule of inference.

$$\mu(\Delta I_S, \Delta^2 I_S, \Delta f) = \overset{m\times n}{\underset{i=1}{S_1}}\left[T_2\left[\mu_{\Delta I_{S\,i}}(\Delta I_S), \mu_{\Delta^2 I_{S\,i}}(\Delta^2 I_S), \mu_{\Delta f_i}(\Delta f)\right]\right]. \tag{2}$$

where T_2 represents the algebraic product operation and S_1 represents the union operation (max), $m \times n = 9$ rules. The crisp controller output, which is used to change the machine-table feedrate, is obtained by defuzzification employing the center of area (COA) method [10] defined as

$$\Delta f^* = \frac{\sum_i \mu_R(\Delta fi) \cdot \Delta fi}{\sum_i \mu_R(\Delta fi)} . \tag{3}$$

where Δf^* is the crisp value of Δfi for a given crisp input (ΔI_{S_i} , $\Delta^2 I_{S_i}$) .

The crisp control action (generated at each sampling instant) defines the final actions that will be applied to the set points. The strategy used to compute f determines what type of fuzzy regulator is to be used. In this case it can be a PI- FLC

$$f^*(k) = f(k-1) + \Delta f(k) . \tag{4}$$

or a PD- FLC

$$f^*(k) = f_0 + \Delta f(k) . \tag{5}$$

Feedrate values (f) were generated on-line by the controller and fed in with the set point for the spindle current (I_{Sr}) and measured value (I_S) provided by the computerized numerical control from the internal spindle-drive signal, as is explained in section 4.

The static input-output mapping can be represented by the nonlinear control surface shown in Figure 3, considering that $\Delta I_S \in [-1.5, 1.5]$, $\Delta^2 I_S \in [-1.5, 1.5]$ and $\Delta f \in [-10, 10]$.

These nonlinearities are essential in order to achieve good performance. Recently it has been proved that when trapezoidal membership functions are used, the resulting system is the sum of a global nonlinear controller (static part) and a local nonlinear PI controller (dynamically changing with regard to the input space) [11]. Therefore, it is expected that this kind of membership functions can be relevant for dealing with nonlinear process behavior.

4 Open Computerized Numerical Control

The demand for open control systems is on the rise as the result of the current need for enhanced control functionality in industry. To date, only a small group of machine-tool manufacturers has accepted adaptive controllers. One main reason for their reluctance is the need for an accurate model for calculating controller parameters. Additionally, adaptive controllers cannot be applied reliably to various combinations of machining processes, materials and tools without major changes being required in the control algorithm and its parameters. Some European computerized numerical control manufacturers do not include any adaptive functions in their products [12], while others offer an "AC control" whose performance is very restricted [13]. The adaptive module is rarely used in a flexible machining environment containing many types and combinations of materials and tools.

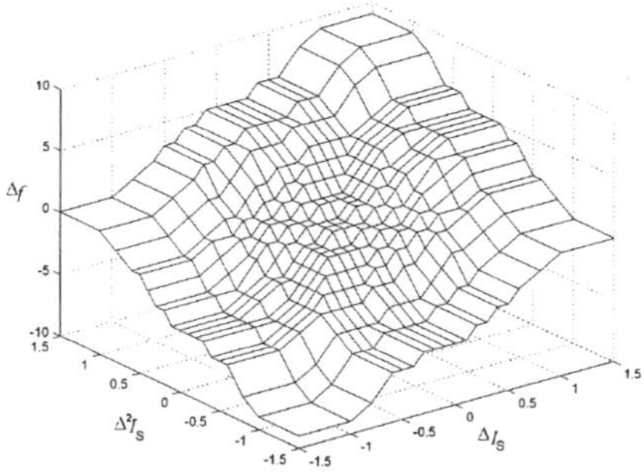

Fig. 3. Fuzzy control surface

At present two levels of opening are available from machine-tool and computerized numerical control manufacturers. The degree varies, however, depending on part involved. There are two categories, depending on how important it is from the control-system viewpoint: opening of Man/Machine Communication (MMC) and opening of the Numerical Control Kernel (NCK). The first category is the one responsible for interaction with the user (e.g., office applications can be developed using the DDE Dynamic Data Exchange protocol). Use of a bus (e.g., Multi-Point Interface bus) enables communication with low-level data. The more important level of opening from the control-system viewpoint, can be found in the NCK, where real-time critical tasks (e.g., axes control) are scheduled and performed.

4.1 Embedding the Fuzzy Controller

This section explains how the fuzzy controller designed in Section 3 is embedded in the open CNC. Nowadays the integration of any control system in the CNC is a complex task that requires the use of various software utilities, technologies and development tools. Three classical technologies can be used: a software-developing tool (C++, Visual Basic), an open- and real-time computerized numerical control and communications technology.

The general outline of the control system is depicted in Figure 6. First the fuzzy controller was programmed in C/C++ and compiled, and as a result a dynamic link library (DLL) was generated. Other tools were used as well. A Sun Workstation, the UNIX operating system and C++ were used to program the NCK. A PC, the WINDOWS 9X operating system and Visual Basic were used for programming the MMC. Inter-module communications between the MMC and the NCK was established through DDE. Finally, and for the sake of simplicity, the user interface was programmed in Visual Basic.

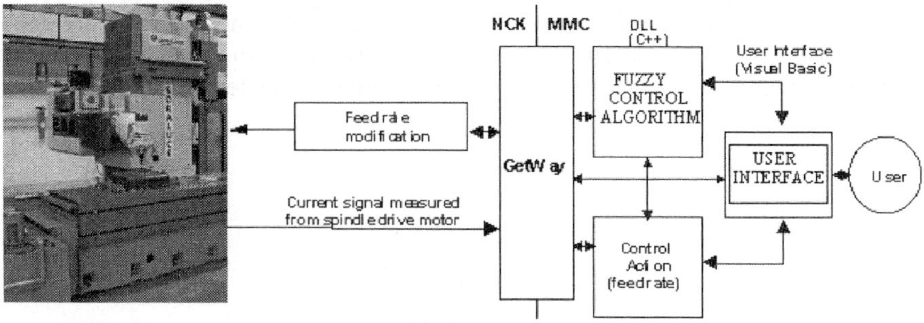

Fig. 4. Diagram of the control system.

The application was developed on the basis of a Sinumerik 840D CNC. The process of integrating the software application into the NCK involved a series of steps. The development was run at a workstation, including edition, compiling and code linking. After that, the file was transferred to the PC, where OEM software ran debugging routines. Finally, the code was copied on a PCMCIA card and inserted into the computerized numerical control [13].

An experimental control internal data-acquisition system was developed and used to record the control internal spindle-drive signal. The system enables a selected drive signal to be recorded and provides stored data on the hard drive of the SINUMERIK 840D user interface PC (MMC). Internal data-acquisition software was developed to obtain control information. The maximum sampling frequency could not be any longer than 500 Hz, defined by the servosystems' control cycle. Therefore the acquisition-signal software works at the same time for which servosystems are configured in the computerized numerical control (i.e., 2 ms.). The software consists of a data-acquisition module in the NCK that records the selected data into an internal buffer and an MMC background task that receives the completed measurement and stores it in the hard drive of the user-interface PC. Data transfer is performed by splitting the recorded data into a number of fragments, due to limitations in the Sinumerik 840D's file systems. The signals to be recorded (in our case the current consumed at the spindle) are configured using specially added machine data. Recording can be started and stopped either manually through an MMC application or under the numerical control program.

5 Industrial Tests. Evaluation

Tests were carried out in the SL400 SORALUCE machining center, which was equipped with a SIEMENS open- Sinumerik 840D. The SL400 model SORAMILL milling machine. It possesses high-precision slideways that allow all three axes to reach a speed of up to 15 m/min. The machine has high-rigidity, high-precision features, and the workpiece has no influence whatsoever on the moving part.

The cutting tool was a Sandvik D38 R218.19 - 2538.35 - 37.070HA (020/943) face mill 25 mm in diameter with four inserts of the SEKN 12 04 AZ (SEK 43A, GC-A p25) type. The workpiece material used for testing was F114 quality steel. The maximum depth of cut was 20 mm, the nominal spindle speed was s_0=850 rpm, and the nominal feedrate, f_0=300mm/min. The actual dimensions of the profile were 334x486 (mm). The profile is depicted in Figure 5b.

a)

b)

Fig. 5. a) Tool used for industrial tests, b) workpiece for industrial tests

We used control internal spindle-drive signal. The processing-signal analysis was done on-line. The signal processing was centered in the correction of current offsets. The current offsets were corrected on the basis of the model that relates spindle speed vs. current. This model was created using linear regression. Moreover, further filtering was performed using the mean value of the signal in the sampling period.

The set point was estimated according to constraint given by the power available at the spindle motor, material and the tool characteristics. In order to compare the performance of the PI and PD- FLC, a reference value of 18A was set during the experiments. In the second study, the set point used was 22A in order to make an appropriate comparison with a commercial product. The results of applying two fuzzy controllers (PI and PD- FLC) to mechanize an irregular profile (see Fig. 5b) are depicted in Figure 6a. The negative effect of integral components is reflected in the transient response.

An illustrative example was inserted to uphold the validity of the theoretical approaches examined in this paper. The second study compared an embedded PD-FLC, the OPTIMIL [14] and the CNC working alone. The results are shown in Figure 6b. The reduction in machining time was very similar for the OPTIMIL and the embedded PD-FLC (10%), yet the overshoot was bigger for the embedded PD-FLC (23.7%). From the technological viewpoint, this overshoot is allowable for rough milling operations. It is important to note that in some approaches the goal of the controller design is to limit the percent overshoot to 20% [3].

Regarding OPTIMIL, when used as external equipment, requires synchronization steps and a tedious calibration procedure including a learning phase. On the other hand, the embedded PD-FLC does not require any calibration procedure or external sensors, and the application can be supplied with a computerized numerical control (CNC) alone, and no other external equipment.

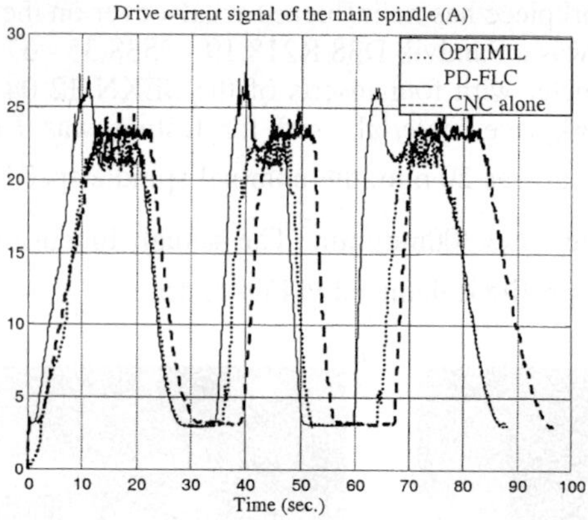

Fig. 6. Time responses of the control systems

Conclusion

The results of transferring technology to a machine-tool manufacturer through cooperation with a technological center show the effectiveness of a fuzzy control system for dealing with the nonlinear behavior of the machining process. The embedded fuzzy controller is able to work using only internal CNC signals. Moreover, it can run in parallel with other applications without any synchronization problems. However, future work is necessary to refine the fuzzy controller's performance and so to improve the transient response.

All this work, from hypothesis to implementation and experimentation, including design and analysis, is done following the classical patterns. Our focus at all times lay on the practical implementation, so that the results we have presented in this work are for a real life industrial plant. The results show that internal CNC signals can double as an intelligent, sensorless control system. Actual industrial tests show a higher machining efficiency: in-process time is reduced by 10% and total estimated savings provided by installing the system are about 78%.

References

1. Koren Y.: Control of machine tools. Journal of Manufacturing Science and Engineering 119 (1997) 749–755
2. Lauderbaugh L.K., Ulsoy A. G.: Model reference adaptive force control in milling. ASME Journal Engineering of Industry 111 (1989) 13–21
3. Rober S.J., Shin Y.C., Nwokah O.D.I.: A digital robust controller for cutting force control in the end milling process. Journal of Dynamic Systems, Measurement and Control 119 (1997) 146–152
4. Haber R.E., Peres C.R., Alique A., Ros S., Alique J.R.: Towards intelligent machining: hierarchical fuzzy control for end milling process. IEEE Transactions on Control Systems Technology 6 (2) (1998) 188–199

5. Lin L.-C, Lee G.-Y.: Hierarchical fuzzy control for C-axis of CNC turning centers using genetic algorithms. Journal of Intelligent and Robotic Systems: Theory and Applications 25(3) (1999) 255–275
6. Liu Y., Zhuo L., Wang C.: Intelligent adaptive control in milling process. International Journal of Computer Integrated Manufacturing 12(5) (1999) 453–460
7. Liu Y., Wang C.: Neural networks based adaptive control and optimization in milling process. International Journal of Advanced Manufacturing Technology 15 (1999) 791–795
8. Altintas Y.: Prediction of cutting forces and tool breakage in milling from feed drive current measurements. ASME Journal of Engineering of Industry 11 (1992) 386–392
9. Kim T., Kim P.: Adaptive cutting force control for a machining centre by using indirect cutting force measurements. International Journal of Machine Tool Manufacturing 36(8) (1996) 925–937
10. Yager R., Filev D.: Essentials of Fuzzy Modeling and Control, (New York, John Wiley&Sons) pp. 313–354 (1994)
11. Ying, H.: Analytical structure of a typical fuzzy controllers employing trapezoidal input fuzzy sets and nonlinear control rules. Information Science 116 (1999) 177–203
12. Fagor Automation: Needs of New Numerical Controls (in Spanish). Producción Mecánica 3 (1999) 74–83
13. Siemens AG: Sinumerik 840D, OEM-package NCK, software release 4, user's manual, Erlangen 1999.
14. Nosrat A.: Adaptive controls lead to peak efficiency, American Machinist (1997) 78-80

A Dynamically Grouped Multi-multicast Stream Scheduling Strategy for Video-on-Demand Systems*

Dafu Deng, Hai Jin, and Zongfen Han

Huazhong University of Science and Technology, Wuhan, 430074, China
hjin@hust.edu.cn

Abstract. Network bandwidth is the key performance bottleneck for a video-on-demand (VoD) server. It controls the number of clients the server can support simultaneously. Previous works have shown some strategies, such as the batching strategy and the stream merging strategy, that use one multicast stream to serve different clients requesting the same video object at the same time. They improve the performance of server bandwidth effectively. But the batching strategy results in long start-up latency and the traditional stream merging strategy also wastes lots of server bandwidth. In this paper, we propose a *dynamically grouped multi-multicast* stream scheduling strategy, called DGMM, and analyze its performance in two factors: the start-up latency and the average bandwidth consumption.

1 Introduction

Video-on-Demand (VoD) is a service that allows clients to communicate with a video server selecting and viewing a video of their choice simultaneously. A few companies, such as IBM, Microsoft and Apple, successfully ran VoD systems serving a lot of clients with high cost. With broadband network technology, the client-end bandwidth can achieve almost 100Mbits/s. It can support up to 60 MPEG-I streams (approximately 1.5Mbits/s per MPEG-1 stream). Therefore, server I/O bandwidth and network bandwidth are the major limiting factors in the widespread usage of VoD systems. They control the number of streams concurrently sent by a video server.

A conventional video server simply schedules one requested stream for each client, while most clients often request the same hot video at the same time [5][6]. This phenomenon results in that conventional VoD servers send some streams of same video data at the same time. It wastes a mass of server bandwidth. Still conventional systems do not scale well, and better solutions are necessary.

The batching strategy [1][5][6][9] presents one such solution. It uses a single multicast stream of data to serve clients that request the same hot video during the same short time interval. In the batching strategy, a time threshold must be set firstly. Video servers just schedule the multicast stream at the end of each time threshold. In order to obtaining a good efficiency in usage of bandwidth, value of this time threshold must be at least 7 minutes [5]. Therefore, the expected start-up latency is approximate 3.5 minutes. This long delay increases clients reneging rate and decreases the popularization of VoD systems.

* This paper is supported by Wuhan Key Hi-Tech Project under grant 20011001001.

P.M.A. Sloot et al. (Eds.): ICCS 2003, LNCS 2658, pp. 822–831, 2003.

Stream merging strategies [2][3][4][7][8][10][11][12] present a good idea to solve long start-up latency problem. In stream merging strategy, video server immediately schedules one or two kinds of streams to serve each client request: the complete multicast stream and the patching unicast stream. For the first arrived request, the video server schedules a complete multicast stream to transmit the integrated video object immediately. For later requests with the same object, the video server firstly notifies them to join the multicast group for receiving the remainder of scheduled complete multicast stream, and then, schedules a patching multicast stream to transmit the lost video data. Thus, later starters must concurrently receive two streams and merging them into an integrated video stream. This strategy guarantees the zero start-up latency and high server bandwidth efficiency when client request rates are low and moderate. However, the server bandwidth efficiency is decreased dramatically at high request rates due to mass retransmission of same video data.

This paper contributes in developing a novel stream scheduling strategy, called *Dynamically Grouped Multi-Multicast*, which can significantly improving server bandwidth efficiency over a wide range of client request rates. The subsequent sections are organized as follows. In section 2, we describe the basic idea of dynamically grouped multi-multicast stream strategy, and in section 3, introduce an example algorithm. Section 4 presents performance evaluation based on experimental research. Finally, section 5 contains our concluding remarks and future works.

2 Dynamically Grouped Multi-multicast Scheduling Strategy

2.1 Preliminaries

Consider that clients requesting different hot video objects are served independently. Clients request the same hot video object will be discussed in the following sections. Time is divided into fixed size intervals *T*. The time slot in which the first client request arrived is labeled t_0. The following time slots are labeled t_1, …, t_n $(0<n<+\infty)$. The hot video with L minutes is divided into $\lceil L/T \rceil$ fixed size segments, denoted as V_0, V_1, …, $V_{\lceil L/T \rceil - 1}$. Other conceptions are defined as follows.

Clients requests: For requests arriving at the same time slot, video servers can batch them together and serve them all by a multicast stream. Therefore, those requests are considered as one request. We use C_i to represent the request arriving at the time slot t_i.

Requests group: Given a clients' requests sequence C_0, …, C_i, …, C_j, …, C_n $(0<i<j<n<+\infty)$, the sequence is not continuous because requests may not arrive in some time slots. We sequentially divide those requests into several flexible sized requests groups G_0, G_1,…, G_k $(0<k<+\infty)$, while G_k is called the k-th requests group.

Complete multicast stream (CS): For requests group G_i, in order that all clients can receive all the video segments, video servers must schedule a multicast stream, which transmits all the segments. This multicast stream is defined to be the complete multicast stream. We use CS_i to represent the complete multicast stream in requests group G_i, $N_{cs}(i)$ represents the serial number of the time slot at the end of which CS_i is scheduled, and $\ell(CS_i)$ represents the number of video segments transmitted on CS_i.

Patching multicast stream (PS): In a requests group G_i, once CS_i has been initialized, clients with requests arrival time later than time $t_{Ncs(i)}$ may miss some video segments that have been transmitted. Therefore, video server must initialize several streams to patch the lost video segments to clients. We use patching multicast streams to define those streams and use PS_k ($0<k\leq+\infty$) to present the patching multicast stream initialized at the end of time slot t_k. $\ell(PS_k)$ is used to represent the number of video segments transmitted on PS_k.

Segment transmission time (STT): Assuming that video server use CBR (*constant bit rate*) transmission mode, any segment must be transmitted completely in a time slot. Therefore, if a video segment is transmitted on a stream, the transmission must begin at a fixed time point, called *segment transmission time* (STT). We use $\Gamma(m,n)$ to represent the segment transmission time of segment V_m transmitted on stream PS_n. The following formula can be deduced: $\Gamma(m,n)=t_n+m*T$.

2.2 Basic Idea of DGMM

DGMM scheduling strategy determines how to divide the requests sequence into different requests groups and how to schedule streams for intra-group requests and inter-group requests so that the bandwidth efficiency can be improved significantly. Rules of DGMM strategy are described as follows.

Grouping rule: A video server dynamically maintains several request groups G_0, $G_1,\ldots,G_n$. When the first client request arrived, the video server creates requests group G_0 and puts the request C_0 into G_0. For other request C_i, if $t_i<t_{cs}+(\lceil L/T\rceil-1)*T$, where t_{cs} is the first request arrival time slot of the latest created requests group G_n, the video server puts the request into G_n. Otherwise, it creates a new requests group G_{n+1} and puts the request into it. When all requests of the requests group G_i have been served, G_i is deleted. For example, as shown in Figure 1, requests C_0, C_1, C_2, C_3, C_4, C_5, C_6, C_7 are put into requests group G_0 because their requests arrival time slots are less than $t_0+(\lceil L/T\rceil-1)*T$. Request C_{14} is grouped into G_1 because time slot t_{14} is later than $t_0+(\lceil L/T\rceil-1)*T$.

Inter-group rule: According to the above grouping rule, there is no video segments can be shared among inter-group requests. Therefore, the video server handles each grouped requests independently.

Intra-group rule: The objective of each client request in a requests group G_i is to receive all the $\lceil L/T\rceil$ parts of transmission and view them without any interruption. For the first client request of requests group G_i, the video server schedules a complete multicast stream CS_i at the end of the first request arrival time slot. The corresponding clients join the multicast group of CS_i to receive and then playback all segments of requested videos. For any other request C_k ($N_{cs}(i)<k< N_{cs}(i)+\lceil L/T\rceil-1$), as video segments $V_0, \ldots, V_{k-Ncs(i)-1}$ have been transmitted, the video server firstly notifies the corresponding clients to join the multicast group of CS_i to receive and buffer video segments $V_{k-Ncs(i)}, \ldots, V_{\lceil L/T\rceil-1}$. Then, for each segment of $V_0, \ldots, V_{k-Ncs(i)-1}$, the video server searches all the existed patching multicast streams $PS_{Ncs(i)+1}, \ldots, PS_{k-1}$.

If there are no patching multicast streams existed, it schedules a patching multicast stream PS_k to transmit video segments $V_0, \ldots, V_{k-Ncs(i)-1}$ and notifies the corresponding clients to receive and playback those video segments. If video segment V_j ($0\leq j\leq k-$

$N_{cs}(i)-1$) is transmitted on a existed patching multicast stream PS_m ($N_{cs}(i)<m<k-N_{cs}(i)$) and the transmit time $\Gamma(j, m) \geq t_k$, the video server notifies those clients to join the multicast group of PS_m to receive segment V_j. At last, for video segments not transmitted on existed streams or the segment transmission time of them is less than t_k, the video server schedules a patching multicast stream PS_k at the end of time slot t_k and transmits them at the corresponding segment transmission time. The corresponding clients C_k must also join the multicast group of PS_k to receive those segments.

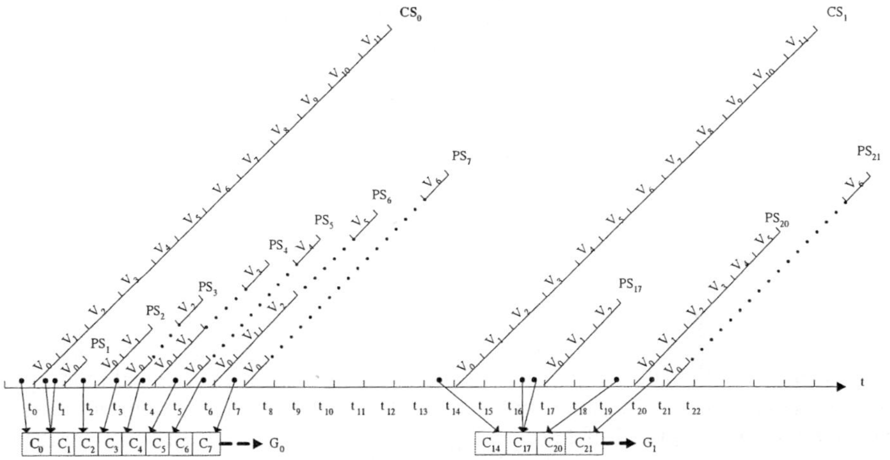

Fig. 1. A scheduling example of DGMM

Figure 1 shows a scheduling example of DGMM. The solid lines describe multicast streams, while the dotted lines indicate the skipped video segments. Because it is similar between requests group G_0 and requests group G_1, we just discuss the scheduling for group G_1. CS_1 is the complete scheduled multicast stream and the client request C_{14} joins the multicast group of CS_1 to receive and playback all segments. Because segments V_0, V_1, V_2 have been transmitted completely when request C_{17} arrives, the video server schedules PS_{17} to transmit those missed segments. The client C_{17} join the multicast group of CS_1 to receive and buffer segments V_3, …, V_{11} and join the multicast group of PS_{17} to receive and playback segments V_0, V_1, V_2.

The similar case is happened for request C_{20}. For serving the request C_{21}, the video server firstly notifies the corresponding clients to join the multicast group of CS_1 to receive and buffer segments V_7, V_8, V_9, V_{10}, V_{11}. Then, it searches the existed patching multicast streams P_{17}, P_{18} and finds that segments V_1, V_2, V_3, V_4, V_5 can be received on PS_{18} by client C_{21} (i.e. $\Gamma(1, 18)$, …, $\Gamma(5, 18) \geq t_{21}$). Therefore, the server notifies C_{21} to join the multicast group of PS_{18} to receive those segments. At last, the server schedules a patching multicast stream P_{21} to transmit segments V_0, V_6 at the corresponding segment transmission time. The client C_{21} also needs to join the multicast group of PS_{21} to receive segments V_0, V_6.

3 An Example Algorithm of DGMM

A video server is usually comprised of three parts--scheduler, data server, and video storage. A scheduler is a controller of the video server. It is responsibility for exchanging control messages with clients on a control channel and for scheduling a data server to transmit video segments. Functions of a data server are reading video segments from a video storage and allocating data channels to transmit video segments. When a client request arrives, the scheduler determines data channels a client should join and sends a message list M_l to notify the information of those channels. The information of each data channel is embedded in one element of M_l. We use a triple M_i(a, A, S_{Id}) to describe the i-th element of M_l, where *a* is the multicast IP address of that channel, *A* is a bit array indicating which video segment is transmitted on that channel, and S_{Id} is the exclusive identity of that channel. Here we discuss an example algorithm for the client, the scheduler, and the data server.

3.1 Client Algorithm

Each client contains a set-top box, a disk, and a display monitor. Clients communicate with a video server by the set-top box. There is a controlling thread, a network listening thread, a task thread, a display thread, and a buffer management thread running in the set-top box. The controlling thread requests its desired video, waits to receive message list M_l comprised of M_0(a, A, S_{Id}), …, M_i(a, A, S_{Id}). For each message, the controlling thread notifies the network listening thread to listen in the corresponding data channel. When a video segment arrives on a multicast stream, the network listening thread notifies the task thread to run the corresponding task to receive them. Each task is responsible for receiving video segments on one data channel, and sending them to the display thread or save them to disk. Figure 2(a) shows the client data receiving algorithm in detail.

3.2 Server Scheduler Algorithm

The key function of a scheduler is to determine in which data channel a client can receive valid video segments and how many video segments should be transmitted on that channel. In order to implementing this function, a stream information list is built for all data channels allocated to serve a group requests. Each unit of the stream information list describes the information of one data channel in a triple S_i(t, a, A), where *t* is the initiating time of a data channel, *a* is the multicast IP address of the data channel, and *A* is an array with each element the serial number of a video segment transmitted on data channel. Furthermore, a message list M_l comprised of M_0(a, A, S_{Id}), …, M_i(a, A, S_{Id}) is built to record the information of data channels allocated for one request. The video server uses this message list to notify that client to receive video segments on the corresponding channels.

Figure 2(b) illustrates the server scheduler algorithm. The scheduling thread is driven by two kinds of events: the request arrival event and the time out event. The request arrival event is started when a request arrives, while the time out event is started at the end of each time slot. Each event includes a parameter to indicate the requested video object so that the scheduler can find out appropriate scheduling parameters.

```
Client Main Control Thread:
1. Send a request to video server;
2. Wait for messages M0(a,A,S1d),...,Mi(a,A,S1d);
3. For each message Mi(a,A,S1d),inform network listening thread to listen
   in a data channel Sid and then initiate a task to receive data from it;

Client Task Thread:
1.Wait for listened data channels to be notified that some media data has
  arrived;
2.According Sid of notified data channel, select the corresponding task
  to run on it;
3.If the value returned by current task is -1
4.   delete the current task;
5.go to 1;

Client Task:
1.while(1)
2. Receive arrived media data on the corresponding data channel;
3. If the received media data is not NULL
4.    If the received media data must be displayed now,
5.       put it to the display buffer which is shared by the display thread;
6.    else
         write media data to disk;
7. If the received media data is the last media data on the data channel.
8.    return -1;
9. else
      inform network listening thread to listen to the data channel again;
10.   return 0;
```

(a)Client receiving algorithm

```
Notations:
tcur:        the current time;
Si(t,a,A):  the ith stream information unit;
Data Server Thread:
1   m = the length of array A in Si(t,a,A);
2   n = 0;
3   for(j=n;j<m;j++)
4     if(t of Si(t,a,A)+Aj*T<=tcur)
         read the media unit MAj from video storage and send it to client.
5       n++;
     else
6       sleep(t of Si(t,a,A)+Aj*T-tcur);
```

(c)Data server algorithm

```
Notations:
L:   Length of the requested video;
T:   Time interval;
Tc:  The last complete multicast stream initial time.
Tp:  The last patching multicast stream initial time.
tr : The request arrival time.
Q:   The queue for batching clients that are not served immediately.

Scheduler Thread Algorithm:

1  If a request arrival event start
   {
2      Finds out corresponding parameters Q,Tc, and L based on
requested video object.
3      if ((tr>=(Tc+L-T))||(tr==t0))
       {
4          Schedules a complete multicast stream to transmit the
requested video.
5          Updates Tc (Tc=tr).
6          Initializes a new timer and set time out interval to T;
       }
       else
7          Put the arrived request into Q;
   }
8  Else if a time out event start
   {
9      Finds out Q and Tp based on the request video object parameter
included in the
              event.
10     if (Q is not null)
       {
11         Searches the existed stream information list to find out
which video
           segments will be sent on other existed patching streams.
12         Builds the message information list Mi and sends it to all
clients batched
              in Q.
13         Schedules a patching multicast stream to transmit the lost
video segments.
14         Update Tp (Tp = tc).
15         Delete all requests batched in Q;
       }
16     Update timer to the end time of next time slot.
   }
17 wait for next event start.
```

(b) Server scheduler algorithm

Fig. 2. An example algorithm of DGMM.

3.3 Data Server Algorithm

Each data server thread is responsible for serving a multicast stream. The function of a data server thread is reading media data from a video storage, and then sending them to clients on a data channel according to the corresponding stream information unit $S_i(t, a, A)$. On a special data channel, a video segment must be transmitted at a fixed time point, and the multicast stream may be not continual. Therefore, we must interrupt the data server thread and wake up it at the appropriate time point. Figure 2(c) shows the data server algorithm.

4 Performance Evaluations

In this section we will give out experimental results with comparison with the FCFS batching strategy and the stream merging strategy.

4.1 Experimental Parameters

We consider two factors for each video: its length and popularity. The data from *Internet Movie Database* (http://www.imdb.com) has shown a normal distribution with a mean of 102 minutes and a standard deviation of 16 minutes. In order to reserve some client bandwidth used to obtain other services (i.e FTP, WWW, Email), we assume that the maximum bandwidth for a client is 30 concurrent MPEG-1 streams (near 45 Mb/s), and a video is split into several parts (usually 2 parts). The length of each part is less than 60 minutes, and each part is considered as an integrated video. Clients who request a video longer than 60 minutes must send request more than once. The time interval *T* for each request is 1 minute. Therefore, a video is divided into 45~60 fixed size segments.

The popularity of each video was modeled using a Zipf-like distribution with parameter 0.271. This is the mostly used distribution for VoD studies. In our experiment, the number of videos stored on a video storage is 100. Client requests are generated using a Poisson arrival process with an interval time of $1/\lambda$, for varying λ values, between 100 to 1400 arrivals per hour. Clients simply selected a video, waited for their request to be served, and then watched the video until it was completed.

Video server and network are main components in VoD systems. For a video server, we utilize *Darwin-streaming server* (http://www.apple.com), which can support 600-800 MPEG-1 unicast streams simultaneously. We also use 1000M Ethernet to simulate the broadband network. Because the bandwidth for each NIC is 100Mb/s. Figure 3 shows the main experimental environment parameters.

Video length (minutes) L	45~60
Number of videos Nv	100
Video Format	MPEG-I
Clients' maximum bandwidth (Mbits/s)	100
Server's maximum bandwidth(Mbits/s)	1000
Clients arrival rate λ(/hour)	100~1400

Fig. 3. Experiment Parameters

4.2 Results

For a video server, there are two most important performance factors: start-up latency, which is the amount of time clients must wait to watch the chosen videos; and average bandwidth consumption, which indicates the bandwidth efficiency of a video server.

4.2.1 Start-up Latency

Figure 4 displays the system average start-up latency at client request rates 100, 500, 800, 1400 per hour, and shows how these latencies are affected by the number of served clients. We choose 100 to be the low request arrival rate, 500 and 800 to be the middle request arrival rate, and 1400 to be the high request arrival rate. As one can see, the system average start-up latency is changed very small under different request arrival rate. It always remains on 25±5 seconds when there are enough clients' requests arrived. This latency can be accepted by almost all clients.

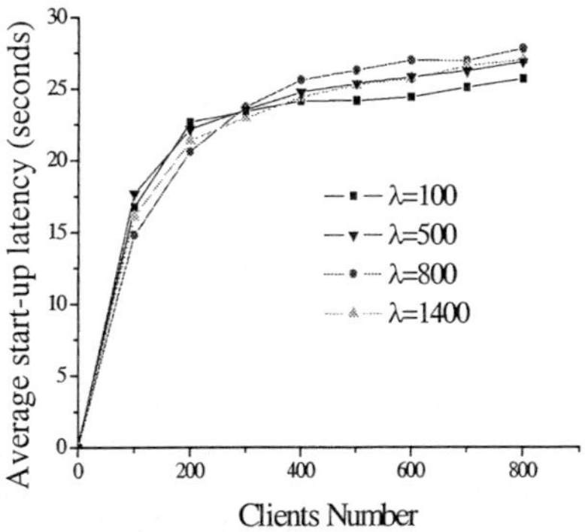

Fig. 4. Request arrival rate vs. average the start-up latency.

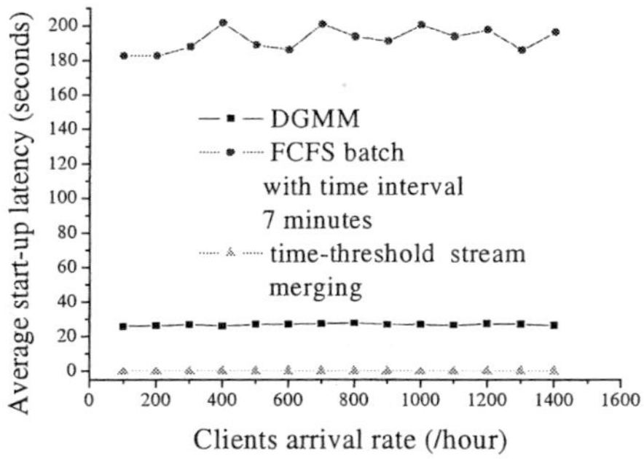

Fig. 5. A start-up latency comparison among FCFS batching strategy with the time interval 7 minutes, the time-threshold stream merging strategy with parameter W=20 minutes and unconstrained clients buffer, and the DGMM strategy with time slot =1 minute

Figure 5 displays the start-up latency comparison among FCFS batching with time interval T=7 minutes, the stream merging strategy with parameter W=20 minutes and unconstrained client disk space and our DGMM. We choose 7 minutes because paper [5] have presented that FCFS batching could obtain a good tradeoff between start-up latency and bandwidth efficiency at this time interval. The same as to the stream merging strategy, paper [8] have proposed a time-threshold stream merging strategy and showed that the total performance was better than other traditional stream merging strategies when the time-threshold *W* was 20 minutes and the client storage disk was unconstrained. As one can see, DGMM strategy outperforms the FCFS strategy and just is little poorer than the time-threshold stream merging strategy in the aspect of the system average start-up latency. The reason of little poor performance compared with the time-threshold stream merging strategy is that DGMM batches clients' requests arrived at the same time slot. This will increase the bandwidth efficiency.

4.2.2 Bandwidth Consumption

Figure 6 shows how request arrival rate affects the average server bandwidth consumption. We can find out that the average server bandwidth consumption is increased in some degree with the increasing of request arrival rate. The reason is that server will support more clients at a period of time. As showed on Figure 7, when request arrival rate is less than 150 requests per hour, the bandwidth consumption of three kinds of scheduling strategies are held in same level. But by the increasing of request arrival rate, the increasing degree of DGMM is distinctly less than FCFS batching and the time-threshold stream merging strategy. When request arrival rate is 500, the average bandwidth consumption of DGMM is approximate 170Mb/s.

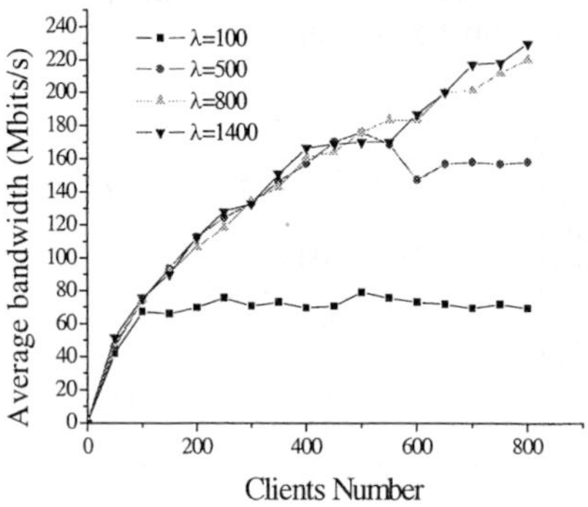

Fig. 6. Request arrival rate vs. average Bandwidth.

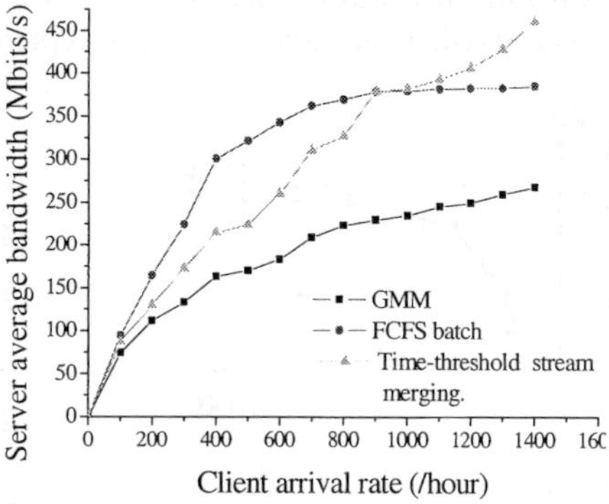

Fig. 7. An average bandwidth comparison among the FCFS batching strategy with time interval 7 minutes, the time-threshold stream merging strategy with parameter W=20 minutes and unconstrained clients buffer, and the DGMM strategy with time slot =1 minute

At the same request arrival rate, the average bandwidth consumption of the FCFS batching is approximate 320Mb/s and that of the time-threshold stream merging strategy is approximate 220Mb/s. At middle request arrival rate, DGMM can save approximate 45% and 23% bandwidth consumption compared with FCFS batching and time-threshold stream merging strategy, respectively. When request arrival rate is higher than 1000 requests per hour, the bandwidth performance of the time-threshold stream merging strategy is decreased dramatically. It is worse than FCFS batching. In any case, they are all worse than DGMM. (FCFS: approximate 380Mb/s; Time-threshold stream merging: approximate 400Mb/s; DGMM: 240Mb/s; at request rate: 1100 request per hour). Therefore, DGMM outperforms the FCFS batching strategy and the stream merging strategy at the aspect of bandwidth performance.

5 Conclusions and Future Works

This paper presents a novel stream scheduling strategy that significantly reduces the demand on the server network-I/O bandwidth. Unlike existing batching strategy and stream merging strategy, DGMM dynamically groups the clients' requests according to the request arrival time and schedules two kinds of multicast stream: continuous completely multicast stream and no-continuous patching multicast stream. This guarantees that no redundant video data are transmitted at the same time and the transmitting video data are shared among grouped clients. Furthermore, we give out an example algorithm and study the performance of DGMM from start-up latency and bandwidth consumption.

Time interval *T* is an important issue in DGMM. It determines the number of streams that can be concurrently received by a client. If *T* is too small, the number of concurrently received streams may be increased dramatically. Therefore, the bandwidth of clients may be exhausted. But if *T* is too large, the start-up latency of clients may be too long to be endured. In this paper, we simply choose 1 minute to be the value of *T* because most clients can accept the start-up latency less than 1 minute and current broadband network technology improves the bandwidth of client to 100Mb/s. In the near future, we will study the effects of time interval *T* and propose an optimal time interval *T*.

References

[1] C. C. Aggarwal, J. L. Wolf, and P. S. Yu, "The Maximum Factor Queue Length Batching Scheme for Video-on-Demand Systems", *IEEE Transactions on Computers*, Vol.50, No.2, pp.97–110, Feb. 2001.

[2] S. W. Cater and D. D. E. Long, "Improving video-on-demand server efficiency through stream tapping", *Proc. of ICCCN*, Las Vegas, NV, Sept. 1997, pp.200–207.

[3] S. W. Cater and D. D. E. Long, "Improving bandwidth efficiency on video-on-demand servers", *Computer Networks,* Vol.30, No.1-2, pp.99–111, Jan. 1999.

[4] S.-H. G. Chan and E. Chang, "Providing scalable on-demand interactive video services by means of client buffering", *Proc. IEEE ICC*, Helsinki, Finland, June 2001.

[5] J.-K. Chen and J.-L. C. Wu, "Heuristic batching policies for video-on-demand services", *Computer Communications*, Vol.22, pp.1198–1205, 1999.

[6] A. Dan, D. Sitaram, and P. Shahabuddin, "Dynamic batching policies for an on-demand video server", *Multimedia Systems*, Vol.4, pp.112–121, June 1996.

[7] D. Eager and J. Z. M. Vernon, "Bandwidth Skimming: A Technique for Cost-Effective Video-on-Demand", *Proc. Multimedia Computing and Networking 2000*, San Jose, CA, January 2000.

[8] L. Gao and D. Towsely, "Threshold-based multicast for continuous media delivery", *IEEE Transactions on Multimedia*, Vol.3, No.4, pp.405–414, December 2001.

[9] S.-H. G. Chan and F. Tobagi, "Tradeoff between System Profit and User Delay/Loss in Providing Near Video-on-Demand Service", *IEEE Transactions on Circuits and Systems for Video Technology*, Vol.11, No.8, Aug. 2001.

[10] K. A. Hua, Y. Cai, and S. Sheu, "Patching: A multicast technique for true video-on-demand services", *Proc. of ACM Multimedia*, Sept. 1998.

[11] H. Shachnai and P. Yu, "Exploring Wait Tolerance in Effective Batching for Video-on-Demand Scheduling", *Multimedia Systems*, Vol.6, No.6, pp.382–394, 1998.

[12] S. Viswanathan and T. I. Metropolitan, "Video-on-demand service using pyramid broadcasting", *Multimedia Systems*, Vol.4, No.4, pp.197–208, Aug. 1996.

Multilevel System as Multigraph*

Waldemar Korczyński[1], José de Jesús Cruz Guzmán[2], and
Zbigniew Oziewicz[2]**

[1] University of Arts and Science, ulica Wesoła 52 PL - 25353 Kielce, Poland
korwald@wsu.kielce.pl
[2] Universidad Nacional Autónoma de México,
Facultad de Estudios Superiores Cuautitlán, Apartado Postal 25, 54700 Cuautitlán
Izcalli, Estado de México {cruz,oziewicz}@servidor.unam.mx
oziewicz@ift.wni.wroc.pl

Abstract. Graph based models of hierarchical systems are usually seen as "graphs equipped with some refinements", understood as the homomorphisms or (bi)simulations. In such a model it is not possible to consider phenomena happened on different levels of the system. We propose a new formalism of multi-graphs allowing to see a hierarchical system similar as a formula of second order logic, i.e. to consider all levels "at the same time". The concurrency in hierarchical system is modelled in terms of multi-graphs.

1 Introduction

Hierarchical systems are usually described by refinement of some of its elements namely modules.(W. Korczyński 2000, [6]) This way of thinking about hierarchical systems as a sum of its modules leads to consideration of any part of the system on another abstraction level which make impossible to consider at the same time properties of parts of systems being of different levels of abstraction (that could be comparable by means of the hierarchical order).

In the paper we propose a way of seeing hierarchical systems, which allow to consider the elements of all levels of abstractions in exactly the same way. The ideology can be seen as a generalization of the notions of the graph ($\equiv$ one-graph). Instead of considering two levels of abstraction, namely vertices and edges, we can treat a hierarchical system as a sequence of elements, called in these context cells, connected by operations having exactly the same properties as the well known operations of source and target in a one-graph. The structure thus obtained is called multi-graph or more precisely an n-graph. The aim of the paper is to describe the construction of n-graph based on some typical presentations of graphs.

* Supported by el Consejo Nacional de Ciencia y Tecnología (CONACyT) de México, proyecto # U41214-F, and by UNAM, DGAPA, Programa de Apoyo a Proyectos de Investigacion e Innovacion Tecnologica, proyecto # IN 105402.

** Zbigniew Oziewicz is a member of Sistema Nacional de Investigadores, México, No de expediente 15337.

P.M.A. Sloot et al. (Eds.): ICCS 2003, LNCS 2658, pp. 832–840, 2003.

The paper is organized as follows: Section 1 are presented the principal aims of these work. Section 2 recall some well known definitions of graphs. Section 3 are proposed some interpretations on n-graphs as models of hierarchical systems. Section 4 a extension of n-graphs into n-category as been described. Section 5 category structure is presented for refining concurrency of processes in multi level systems. Section 6 some historical remarks are presented. The paper is written in the language of category theory. We do not assume the reader is familiar with the category theory because the paper is in essence self-container; we explain all the notions at the very beginning level.

2 Some Presentations of Graphs

A directed graph can be seen as a sequence $s, t : V \longrightarrow E$ being the source and the target,

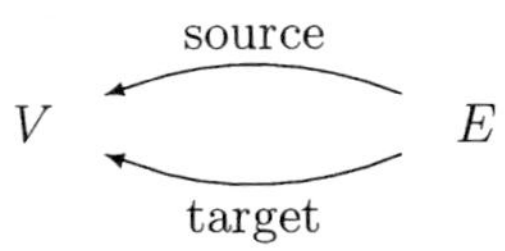

Fig. 1. A directed graph.

The elements of the set V are called vertices of the graph and are illustrated as the zero-dimensional points. The elements of E are called arrows or edges of the graph and are illustrated as one-dimensional directed segments of some lines. However more adequate terminology is 0-cell $\in V$ and 1-cell $\in E$ (A. Burroni 1981, [3]). A graph (Figure 1) is said to be a 1-graph. In the theory of automata there is also alternative terminology: a state or 0-transition $\in V$ and 1-transition $\in E$. (A. Obtułowicz 2001, [12])

In some situations it is suitable to consider a graph consisting of 'arrows' only (a 'globular' graph).

Definition 1 (One-sorted graph). *Let $id, s, t : X \longrightarrow X$ be an alphabet of a monoid, w be the homogeneous word in this alphabet and $|w| \in \mathbb{N}$ be a length of w. A graph (a 1-graph) is defined by (globular) relations*

$$\forall\, |w| = 1, \quad s \cdot w = w, \quad \& \quad t \cdot w = w.$$

This presentation of a graph was originated from Hasse & Michler (M. Hasse & Michler L. 1966, [5]). We call it one sorted presentation of a graph, or a globular graph.

Let $X = E \cup V$. One can extend domain of s and t on Figure 1, from E to X by $s|V = \text{id}$ and $t|V = id$. Then $s, t : X \longrightarrow X$ is a one sorted graph.

Definition 2 (Reflexive graph). *A graph (V, E, s, t) is said to be reflexive if there is an injective section $i : V \longrightarrow E$, $s \circ i = t \circ i = id_V$. A reflexive graph is an algebra (V, E, s, t, i).*

Another presentation of graphs defines them as (indexed) families of sets.

Definition 3 (Graph as a family of sets). *A graph is a map* $G : V \times V \longrightarrow 2^E$. *The elements of* V *are called the vertices and those of the disjoint union*

$$arrows(G) = \bigsqcup_{(u,v)\in V\times V} G(u,v)$$

the arrows of G. *For any* $u, v \in V$ *the elements of the set* $G(u,v)$ *are called arrows from* u *to* v *and for any* $\alpha \in G(u,v)$ *the vertex* u *is called the source and* v *the target of* α.

The sets of the family $(G(x,y))_{x,y\in V}$ *are pointwise disjoint and we use the standard set-theoretical union instead of the direct sum of sets.*

Example 1. Consider the following graph with two vertices: In the following picture (figure 2) the sets $G(x,y)$ are illustrated. We have here $G(u,u) = \{\delta\}$, $G(u,v) = \{\alpha,\beta,\gamma\}$, $G(v,u) = \{\varepsilon,\varphi\}$, $G(v,v) = \emptyset$. The set of vertices of this graph is $V = \{u,v\}$ and the set of arrows

$$\bigsqcup_{(x,y)\in V\times V} G(x,y) = G(u,u) + G(u,v) + G(v,u) + G(v,v) =$$
$$\{\delta\} + \{\alpha,\beta,\gamma\} + \{\varepsilon,\varphi\} + \emptyset = \{\alpha,\beta,\gamma,\delta,\varepsilon,\varphi\}.$$

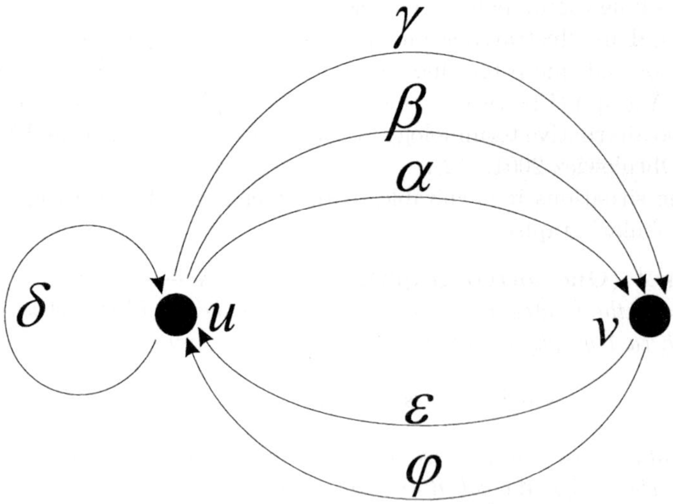

Fig. 2. A graph pictured "traditionally"

The above definitions allow to see a graph as a presheave. The corresponding categories are closed under the formations of product, homomorphic image and taking subalgebras.

Graphs are tools for describing various phenomena. Let us list some of them:

- the arrows of a graph are processes and vertices are situations being their initializations (births) and deads.
- schemes of connections, schemes of electrical or telephonic networks, flow-charts (flow diagrams in computer sciences) or (dependence) relations in sets of institutions or persons.
- schemes of dynamical systems. The vertices G_0 of a graph $G \equiv G_0 \leftleftarrows G_1$ are resources, and the arrows G_1 are interpreted as activities (events) consuming these resources.

3 Multi-graphs

Graphs are a good description tool for dynamical systems with processes running on one level only. The systems, one meets in many branches of every day life, do not have this property. Typical examples are systems of management where one has to consider the control on the fundamental (lowest) level, the control of processes running on this fundamental level, the control of control processes and so on.

Example 2 (Medical). An example in medicine: process running in a biological cell, processes running in an organ, in a group of organs and at the level of the whole organism. A biological cell $c \in G_1$ in an organ O : the situation $sc \in G_0$ is its birth and $tc \in G_0$ its death. The process $\alpha \in G_2$ in the organ O is an arrow of a graph having situations being cells,

$$G_1 \ni c_1 \xrightarrow{\alpha} c_2 \in G_1.$$

In a new 1-graph $G_1 \leftleftarrows G_2$ vertices G_1 are arrows of a 1-graph $G_0 \leftleftarrows G_1$ illustrating processes in the biological cells,

$$\text{vertices}(G_0) = V_0 \overset{d_1^0}{\underset{d_0^0}{\leftleftarrows}} A_0 = arrows(G_0) = vertices(G_1) \overset{d_1^1}{\underset{d_0^1}{\leftleftarrows}} A_2 = arrows(G_1)$$

Continuing this reasoning we obtain a sequence of directed 1-graphs $G_i \leftleftarrows G_{i+1}$, $0 \geq i \leq n-1$, called a directed n-graph, describing different levels of an organism.

We formalize the above considerations as follows.

Definition 4 (n-graph). *A directed n-graph G is a sequence of families of i-cells $\{G_i\}$, $0 \leq i \leq n$, such that $G_i \leftleftarrows G_{i+1}$ is a directed 1-graph with a source map $s_i : G_{i+1} \longrightarrow G_i$ and target map $t_i : G_{i+1} \longrightarrow G_i$, (Z. Oziewicz 2003, W. Lawvere 1989, [13], [8])*

$$V_0 \overset{d_1^0}{\underset{d_0^0}{\leftleftarrows}} A_0 = V_1 \overset{d_1^1}{\underset{d_0^1}{\leftleftarrows}} A_1 = V_2 \overset{d_1^2}{\underset{d_0^2}{\leftleftarrows}} A_2, ..., V_n \overset{d_1^n}{\underset{d_0^n}{\leftleftarrows}} A_n.$$

The corresponding definition in the case of reflexive graphs can be the following one.

Definition 5 (n-graph from reflexive graphs). *A directed n-graph is said to be reflexive if there are sections* $\{i_i : G_i \longrightarrow G_{i+1}\}$, *satisfying the conditions* $s_i \circ i_i = id_{G_i} = t_i \circ i_i$.

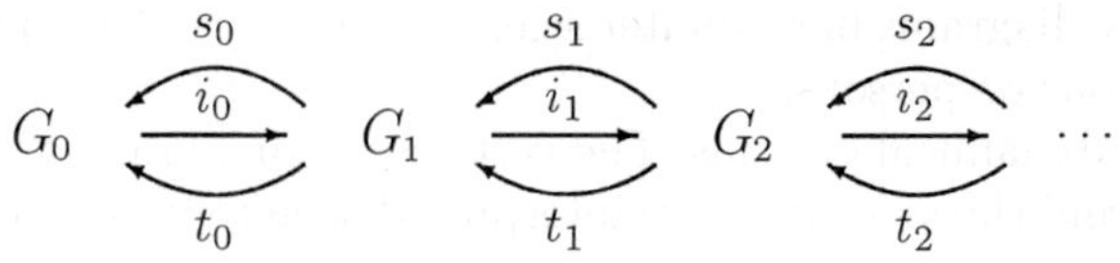

Homomorphisms of n-graphs are defined componentwise.

Definition 6. *A homomorphism* $f \in \hom(G, H)$ *of an n-graph G into an n-graph H is a sequence* $f \equiv \{f_i : G_i \longrightarrow H_i\}$,

$$\begin{array}{ccc} G_i & \overset{\longleftarrow}{\longleftarrow} & G_{i+1} \\ {\scriptstyle f_i}\downarrow & & \downarrow{\scriptstyle f_{i+1}} \\ H_i & \overset{\longleftarrow}{\longleftarrow} & H_{i+1} \end{array}$$

such that for any $i \in \mathbb{N}$, (f_i, f_{i+1}) *is a homomorphism of 1-graphs.*

Let us note a difference between the approach used in the above example and the standard modular modelling of multilevel systems. Seeing a system as an n-graph we can consider it on all levels *at the same time* (A. Obtułowicz 2001, [11], [12]). This point of view is similar to that known from the high order logic when one has to consider variables being individuals, sets of individuals, sets of sets of individuals, etc. Seeing a system as an object consisting of some other objects called in this context modulus we are always at the same level. Refining of a model doesn't lead to a new situation; we are still at the same level because one hasn't to consider any relation between objects of different levels. The last is characteristic for n-graphs and higher order logic. Let us consider an analogous construction for graphs seeing as families of sets. Having a graph one can consider the family $\Gamma(G)$, of all graphs with vertices being arrows in G, i.e. graphs of the form

$$(G(\alpha, \beta))_{(\alpha,\beta) \in arrows(G) \times arrows(G)}$$

This construction can be repeated several times and one can for a given graph G_0 obtain a sequence of families of graphs

$$G_0, \Gamma(G_0), \Gamma^2(G_0) = \Gamma(\Gamma(G_0)), ..., \Gamma^n(G_0)$$

where $\Gamma(\Gamma(G))$ is the family of all graphs obtained by the operation Γ from the elements of the family $\Gamma(G_0)$, $\Gamma(\Gamma(\Gamma(G_0)))$ denotes the family of graphs obtained in the same way from the elements of the family $\Gamma(\Gamma(G_0))$ and so on.

Fact 1. For any sequence

$$G = G_0, G_1, G_2, \ldots, G_n$$

of graphs such that for $i < n$ it holds $G_i \in \Gamma^i(G_0)$ we have arrows(G_i) =vertices(G_{i+1}).

Definition 7. *By an n-graph we mean any sequence*

$$G_0, G_{1,} G_2, \ldots, G_n$$

of graphs satisfying the condition $G_{i+1} \in \Gamma^i(G_i)$ for any i. The elements of the sets

$$V_0, V_1, V_2, \ldots, V_n, E_n(= V_{n+1})$$

will be called cells, more precisely 0-cells (those from V_0), 1-cells (those belonging to V_2), ..., n-cells (the elements of V_{n+1}).

Remark 1 (Normal graph). For a reflexive graph G we have

$$\text{vertices}(G) \subseteq \text{arrows}(G)$$

In other words in the above definition we make no constraints on the function $G : V \times V \to Set$. Particularly we do not exclude the case when for some arrow α and vertex u of a graph $G \in \Gamma^i(G_0)$ the set $G(\alpha, u)$ is not empty. n-graphs in which for any arrow α and vertex u it holds $G(\alpha, u) = \emptyset$ for any $i \leq n$ and $G \in \Gamma^i(G_0)$ will be called normal.

Example 3. In the following picture a 2-graph is illustrated. In this case G_0 is the graph pictured in Figure 3. The operation Γ adds 4 sets of arrows: $G(u, \delta)$, $G(u, \alpha)$, $G(\alpha, \varepsilon)$, $G(\varphi, \gamma)$. In the picture only nonempty sets of arrows has been illustrated. So the considered graph is not normal.

4 Multi-category

An n-graph is a carrier for an n-category by adding of compositions (W. Marcinek & Oziewicz Z. 2001, [10]) as a partial functions $\circ_i$, of the form

$$\circ_i : arrows(G_i) \times arrows(G_i) \to arrows(G_i)$$

being the set theoretical union of total functions of the form

$$G_i(u, v) \times G_i(v, w) \to G_i(u, w)$$

for $u, v, w \in vertices(G_i)$.(that means $u, v, w \in G_{i-1}$. Such a composition is assumed to be associative and having two sided identities id_u and id_v such that for any arrow $\alpha \in G_i(u, v)$ it holds

$$id_u \circ_i \alpha = \alpha \circ_i id_v = \alpha$$

In other words we make any of graphs

$$G_0, G_{1,} G_2, \ldots, G_n$$

a multiplicative graph with associativity and having two sided identities multiplication, i.e. a category. For any $i \leq n$ the category thus obtained will be denoted by $C(G_i)$.

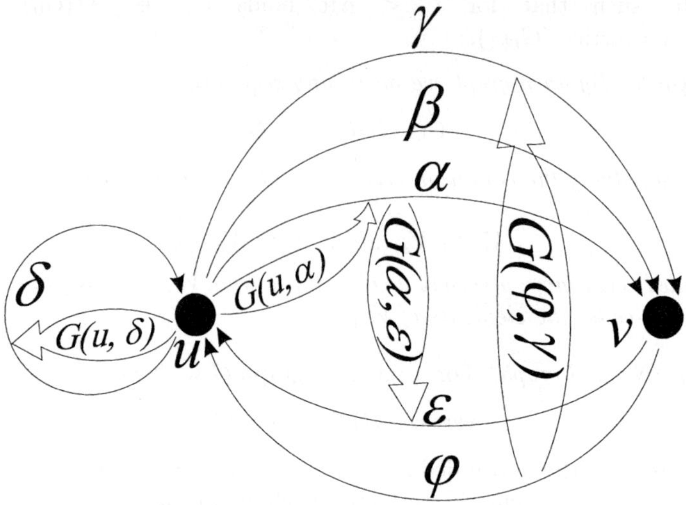

Fig. 3. A 2-graph on the graph picture

Definition 8 (n-category (W. Marcinek & Oziewicz Z. 2001, [10])). *An n-graph G*

$$V_0 \leftleftarrows E_0 = V_1 \leftleftarrows E_1 = V_2 \leftleftarrows E_2 = V \leftleftarrows, ..., V_n \leftleftarrows E_n$$

is said to be an n-category we mean any sequence

$$C(V_0 \leftleftarrows E_0), C(V_1 \leftleftarrows E_1), ..., C(V_n \leftleftarrows E_n).$$

n-categories and some of them applications have been described in Hasse & Michler (M. Hasse & Michler L. 1966, [5])

5 Refining and Concurrency

Refining is a method for system describing and analysis. An advantage of this way of treating complicated composed systems is that we are always on the same level of description. We use in any moment the first order logic. Sometimes this approach to systems is not suitable. Typical example of such situations can be found while considering concurrency of processes running in different levels of a hierarchical system. This kind of concurrency can not be described by the modular hierarchical modelling point. One of the examples of such systems can be the biological system mentioned early. Other example can be found in hierarchical management system. Let us consider such a system consisting of two levels. The first one, let us call it production, can be seen as a really production system in a factory, in an office or in economical organization. The second level manage the processes in the first one. So the way of working of such a system can be seen as follows. Firstly, an object, let us call it m, performs a planning action A. As a result one obtains a project, say p, which will be realized by an object w on the level 1. Now, one can consider a concurrent activities $A\prime$ and p. Next another project, say $p\prime$ being the result of the activity $A\prime$, can be realized concurrently with another designing activity $A\prime\prime$, and so on. The generalization of this concurrency to multilevel systems is evident.

This kind of concurrency is different from the concurrency from Petri nets where one considers concurrency in the set of elements of the same level (B. Baumgarten 1990, M. Leszak and H. Eggert 1988, M. van Sinderen, L. Ferreira, C. A. Vissers and J. Katoen 1995, [1], [9], [14]). The category structure (given by the multiplication of arrows)allows for considering a hierarchy of sequential systems. In this case the only concurrency is the concurrency described above. But one can also consider another kind of concurrency - the concurrency of the same level of the system. This concurrency can be modelled by introducing in the categories corresponding to the levels of the system a new operations expressing the parallel composition of the elements (arrows and vertices) of the system. It can be done in the same way as it has be done in Korczyński 2001 (W. Korczyński 2001, [7]). This leads to the notion of (partial) monoidal n-category. We do not consider here these methods.

6 Historical Remark

Some notions used in the paper are not very popular and frequently used. Let us recall the origin of them. The one sorted presentation of graphs has been introduced by Maria Hasse & L. Michler in 1966 (M. Haese & Michler L. 1966, [5]). One can say it belongs to the language of the so called "French school of category theory" originated by Schuetzenberger and Ehresmann (Ehresmann 1950, [4]) Such an approach allows to consider categories as a kind of "partial semigroups" and leads to interesting applications of(semi)group theory seeing as a (meta)language for such branches of mathematics as for example topology. Reflexive graphs have been considered firstly by Burroni (A. Burroni 1981 [3]) Gray and Lavwere (F. W. Lawvere 1989, [8]). The idea was similar to that of introducing one-sorted graphs; one wants to consider categories as a kind of of monoids. The embedding of vertices into arrows shows how an "identity" can be introduced into the classical graphs. This identity assigns to every vertex of a graph a "loop" being an arrow having its beginning and end - point in these vertex. Burroni (A. Burroni 1981, [3]) also introduces n-graphs as a tool for describing of some notions in "pure mathematics", more precisely in category theory seeing as a kind of mathematical language. Another interesting papers exploring n - graphs and very interesting applications of n-graphs to some problems in computer sciences have been described by Obtułowicz (A. Obtułowicz 2001, [11], [12]).

References

1. Baumgarten B., Petri-Netze, Grundlagen und Anwerdungen, Wissenschaftesverlag, Mannheim, 1990
2. Brown R. and P. J. Higgins, On the algebra of cubes, Journal of Pure and Applied Algebra **21** (3) (1981) 233–260
3. Burroni Albert, Algèbres graphiques, Cahier (1981)
4. Ehresmann, Les Connections infinitesimales dans un espace fibre differentiable, Coloque de Topologie, Bruxelles (1950).
5. Hasse Maria and L. Michler, Theorie der Kategorien, VEB Deutscher Verlag der Wissenschaften, Berlin 1966
6. Korczyński Waldemar , Motody Sieciowe - Elementy Filozofii Podejścia, Kielce 2000
7. Korczyński Waldemar, On some presentations of graphs and hypergraphs, Instytut Matematyki PAN, preprint 2001
8. F. William Lawvere, Display of Graphics and their applications, as exemplified by 2-categories and the Hegelian "taco", State University New York at Buffalo, typescript, April 1989.
9. Leszak M., Eggert H., Petri Netz Methoden und Werkezeuge, Springer Informatik Fachberichte, 197, Berlin, 1988
10. Marcinek Władysław and Zbigniew Oziewicz, Miscellanea Algebraicae (Kielce), Rok 5, No 2/2001, ISBN 83-87798-23-1; math.CT/0104136
11. Obtułowicz Adam, On n-dimensional graphs and n-graphs, Miscellanea Algebraicae (Kielce), Rok 5, No 1/2001, pp. 87–96, ISBN 83-87798-22-3
12. Obtułowicz Adam, A note on the connections of automata and weak double categories, Miscellanea Algebraicae (Kielce), Rok 5, No 1/2001, pp. 75–86, ISBN 83-87798-22-3
13. Oziewicz Zbigniew, Operad of graphs, convolution and quasi-Hopf algebra, Contemporary Mathematics 2003, in press
14. Marten van Sinderen, Luís Ferreira Pires, Chris A. Vissers, Joost-Pieter Katoen, A design model for open distributed processing systems, Computer Networks and ISDN Systems, **27** (1995) 1263–1285

Fast Exponentiaion over GF(2^m) Based on Cellular Automata

Kyo-Min Ku [1], Kyeoung-Ju Ha [2], and Kee-Young Yoo [3]

[1] Mobilab Co.Ltd, 952-3, 4F Plus Bldg., DongChun-Dong, Buk-Gu, Daegu, Korea, 702-250
kmku@nate.com

[2] KyungSan University, 75 San, JumChon-Dong, Kyungsan, KyungPook, Korea, 712-715
kjha@kyungsan.ac.kr

[3] KyungPook National University, 1370 Sankyuk-Dong, Puk-Gu, Daegu, Korea 702-701
yook@knu.ac.kr

Abstract. In this paper, we present a new exponentiation architecture and multiplier/squarer which are the basic operations for exponentiation on GF(2^m). The proposed multiplier/squarer is used as kernal architecture of exponentiation and processes the modular multiplication and squaring at the same time for effective exponentiation on GF(2^m) using a cellular automata. Proposed architecture can be used efficiently for the design of the modular exponentiation on the finite field in most public key crypto systems such as Diffie-Hellman key exchange, ElGamal, etc. Also, the cellular automata architecture is simple, regular, modular, cascadable and therefore, can be utilized efficiently for the implementation of VLSI.

1 Introduction

For the past 30 years, studies on finite fields have been conducted in many areas, including crypto systems[1], and most public key crypto systems, such as Diffie-Hellman key exchange and ElGamal, are based on modular exponentiation computations in a finite field[2][3]. Such modular exponentiation uses a modular multiplier as the basic structure for its implementation. The Elliptic Curve Cryptosystem is also based on constant multiplication[4]. Examples of the algorithms used to implement multipliers include the LSB-first multiplication algorithm[5], MSB-first multiplication algorithm[6], and Montgomery algorithm[7]. Previous research and development on modular multiplication is as follows: First, for a one-dimensional systolic array, in the case of an LSB-first algorithm, the modular multiplication is performed within $3m$ clock cycles using m cells[5]. While in the case of an MSB-first algorithm, the modular multiplication can be performed within $3m$ clock cycles using m cells[6]. With an LFSR structure, the modular multiplication can be performed within $2m$ clock cycles using m cells[10], the modular multiplication can be performed within m clock cycles using m cells and the modular squaring can be performed within m clock cycles using m cells[11]. The structures proposed in [5, 6, 10, 12] are simple modular multipliers.

P.M.A. Sloot et al. (Eds.): ICCS 2003, LNCS 2658, pp. 841–850, 2003.

However, when computing exponentiation, such structures must be repeated twice for modular multiplication and squaring. In case of [11], the structures of multiplication and squaring must be used together to simultaneously perform the modular multiplication and squaring.

The purpose of the current paper is to reduce the time and the space, and to investigate and develop a simple, regular, modular, and cascadable architecture for the VLSI implementation of exponentiation in GF(2^m) based on cellular automata, which is the basic computation in any public key crypto system.

Accordingly, this paper proposes a fast exponentiation architecture over GF(2^m) based on a cellular automata. The proposed architecture uses the basic architecture that can simultaneously perform multiplication and squaring in m clock cycles using $3m$-1 AND gates, $3m$-1 XOR gates, and $4m$-1 registers. Based on the properties of LSB-first multiplication, the parts of modular multiplication and squaring that can be performed in common are identified, then the remainder is processed in parallel. As a result, the multiplication and squaring can be performed much more efficiently as regards time and space compared to repeating the structure as proposed in [5, 6, 10] and can be performed much more efficiently as regards space compared to repeating the structure as proposed in [11,14]. Furthermore, the performance of the exponentiation is much more efficient than that of the [16] as regards time and space.

The remainder of the paper is as follows: Chapter 2 gives an overview of the concept of cellular automata, while Chapter 3 reviews the general exponentiation algorithm in GF(2^m). Chapter 4 introduces the structure of the proposed multiplier/squarer for efficient exponentiation using a cellular automata. Chapter 5 gives the exponentiation architecture over GF(2^m). Finally, Chapter 6 offers some conclusions.

2 Cellular Automata(CA)

Cellular automata consist of numbers of interconnected cells arranged spatially in a regular manner[8][9]. A cell of the CA has the "0" state or "1" state at a certain time. The next state of a cell depends on the present states of 'k' of its neighbors, for a k-neighborhood CA. The neighbor in CA means the cell which affect the state of cell update. Fig.1 is the example of the two-way CA.

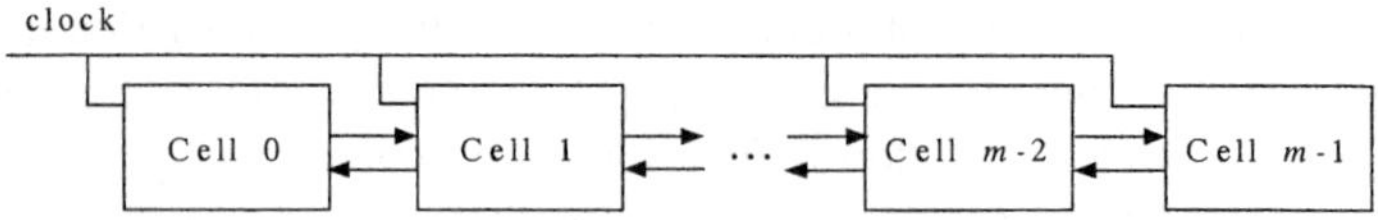

Fig. 1. m-cell two-way CA

The m-cell of the two-way CA is operated at the same time by a clock. The state of a cell at time t is determined by the states of the neighbors at time t-1. In this paper, we assume that the leftmost cell and rightmost cell are adjacent.

We use the modified CA which is the one-way CA(OCA) to solve the problem. Fig.2 shows the OCA structure. It is the same as two-way CA in Fig.1 except that the data stream is one-way.

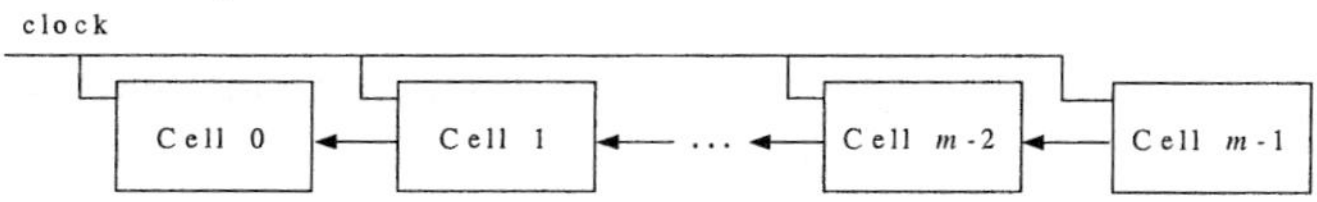

Fig. 2. m-cell OCA

The characteristic matrix shows the entire rules of the CA. An example of the characteristic matrix with 4-cell of which state is renewed by the state of its right adjacent cell is as follows:

$$D = \begin{bmatrix} 0 & 1 & 0 & 0 \\ 0 & 0 & 1 & 0 \\ 0 & 0 & 0 & 1 \\ 1 & 0 & 0 & 0 \end{bmatrix}$$

In the above example, it is shown that element "1" of the matrix on the *i*th line of the *j*th row shows that the *i*th cell is dependent on the neighbor of the *j*th cell.

3 General Algorithm for the Exponentiation over GF(2^m)

In this chapter, general algorithms for obtaining $M(x)^E$ mod $P(x)$ on GF(2^m) are illustrated[12].

Polynomial $P(x)$ of arbitrary degree with coefficients from GF(2) is called an irreducible polynomial if $P(x)$ is not divisible by any polynomial over GF(2) of degree greater than 0 but less than the degree of $P(x)$ [1]. Let $P(x)=x^m+p_{m-1}x^{m-1}+ \ldots +p_1x^1+p_0$ be an irreducible polynomial over GF(2) and α be a root of $P(x)$.

Let's suppose that $A(x)$ and $B(x)$ are the elements on GF(2^m). Then two polynomials $A(x)$, $B(x)$ are as follows:

$$A(x)=a_{m-1}x^{m-1}+ \ldots +a_1x^1+a_0 \tag{1}$$
$$B(x)=b_{m-1}x^{m-1}+ \ldots +b_1x^1+b_0 \tag{2}$$

Firstly, computation of $M(x)^E$ mod $P(x)$ is divided into the LSB-first method and MSB-first method according to the method of processing of the exponent E, $[e_{m-1}, e_{m-2}, \ldots, e_1, e_0]$, where the method of computation is as follows:

LSB-first exponentiation: Compute in the order of from e_0 to e_{m-1}

$$M(x)^E = M(x)^{e_0}(M(x)^2)^{e_1}(M(x)^4)^{e_2} \ldots (M(x)^{2^{m-1}})^{e_{m-1}}$$

MSB-first exponentiation: Compute in the order of from e_{m-1} to e_0

$$M(x)^E = (..((M(x)^{e_{m-1}})^2 M(x)^{e_{m-2}})^2 \ldots M(x)^{e_1})^2 M(x)^{e_0}$$

In this chapter, general algorithm for the method of LSB-first exponentiation is reviewed [13], the structure of a multiplier using the cellular automata that can perform its computation efficiently is proposed in the next chapter.

Algorithm 1 : LSB-first Exponentiation Algorithm[13]
Input : $A(x)$, E, $P(x)$
Output : $C(x)=A(x)^E \bmod P(x)$
STEP 1 : $C(x)=\alpha^0 \cdot T(x)=A(x)$
STEP 2 : for i=0 to m-1
STEP 3 : if e_i ==1 $C(x)= T(x)C(x) \bmod P(x)$
STEP 4 : $T(x)= T(x)T(x) \bmod P(x)$

The general method for implementation of Algorithm 1 is to design an exponentiation architecture by using two multipliers, or to use one multiplier and one squarer. However, in the next chapter, an efficient multiplier/squarer is designed by identifying the commonly computed part of two operations (modular multiplication and squaring) perform it at the same time, and processing the remaining operation in parallel. Proposed structure can perform exponentiation efficiently.

4 Multiplier/Squarer Using the Cellular Automata

In this chapter, we show the architecture that simultaneously process the modular multiplication and squaring over GF(2^m) in m clock cycles using a cellular automata[15].

According to Alg. 1 which is proposed in Chapter 3, in order to compute the LSB-first exponentiation, it is necessary to compute $M(x)= T(x)C(x)$ which is the modular multiplication used in Step 4 and $S(x)= T(x)T(x)$which is squaring used in Step 3.

These two computations can be expressed in a recurrence form again[15]. First, the recurrence form of the modular multiplication is as follows:

$$M^{(i)}(x)=M^{(i-1)}(x)+c_{i-1}T^{(i-1)}(x),\ T^{(i)}(x)=T^{(i-1)}(x)x \bmod P(x),\ \text{for } 1\le i\le m \qquad (3)$$

where $T^{(0)}(x)=T(x)$, $M^{(0)}(x)=0$, $M^{(i)}(x)=c_0T(x)+c_1[T(x)\,x \bmod P(x)] + c_2[T(x)\,x^2 \bmod P(x)] +\cdots+ c_{i-1}[T(x)x^{i-1} \bmod P(x)]$. For $i=m$, $M^{(m)}(x)=M(x)=T(x)C(x) \bmod P(x)$. In equation 3, two equations can be performed in parallel. And second, squaring can be also converted into the LSB-first recurrence form similar to equation 3 as follows:

$$S^{(i)}(x)=S^{(i-1)}(x)+t_{i-1}T^{(i-1)}(x),\ T^{(i)}(x)=T^{(i-1)}(x)x \bmod P(x),\ \text{for } 1\le i\le m \qquad (4)$$

where $T^{(0)}(x)=T(x)$, $S^{(0)}(x)=0$, $S^{(i)}(x)=t_0T(x)+t_1[T(x)\,x \bmod P(x)] + t_2[T(x)\,x^2 \bmod P(x)] + \ldots+ t_{i-1}[T(x)\,x^{i-1} \bmod P(x)]$. For $i=m$, $S^{(m)}(x)=S(x)=T(x)T(x) \bmod P(x)$. In equation 4, two equations can be performed in parallel.

The following bit-wise LSB-first algorithm both modular multiplication and squaring simultaneously can be derived from the above equation 3 and 4 :

Algorithm 2: MS(*C*(*x*), *T*(*x*), *P*(*x*))
Multiplication and Squaring Algorithm

Input : $C(x)$, $T(x)$, $P(x)$

Output : $M(x)=C(x)T(x)\ mod\ P(x)$, $S(x)=T(x)T(x)\ mod\ P(x)$

Step1 : $M^{(0)}(x)=0$, $T^{(0)}(x)=T(x)$, $S^{(0)}(x)=0$

Step2 : **for** i=1 **to** m

Step3 : $T^{(i)}(x)=T^{(i-1)}(x)x \bmod P(x)$

Step4 : $M^{(i)}(x)=M^{(i-1)}(x)+c_{m-i-1}T^{(i-1)}(x)$, $S^{(i)}(x)=S^{(i-1)}(x)+t_{m-i-1}T^{(i-1)}(x)$

Therefore, a structure in which the exponentiation can be computed efficiently on GF(2^m) by performing the modular multiplication and squaring simultaneously in the same amount of time as that of modular multiplication by computing $T^{(i)}(x)=T^{(i-1)}(x)x$ mod $P(x)$ for $1\le i\le m$, which is the common part in equations 3 and 4, only once without duplicate computation, using the result, and obtaining the remaining part of equations 3 and 4 in parallel was proposed in [15].

The characteristic matrix of the CA to operate the step 3 fo Algorithm 2 is as follows:

$$D=\begin{bmatrix}0&1&0&0\\0&0&1&0\\0&0&0&1\\1&0&0&0\end{bmatrix}$$

And the CA structure is in Fig.3.

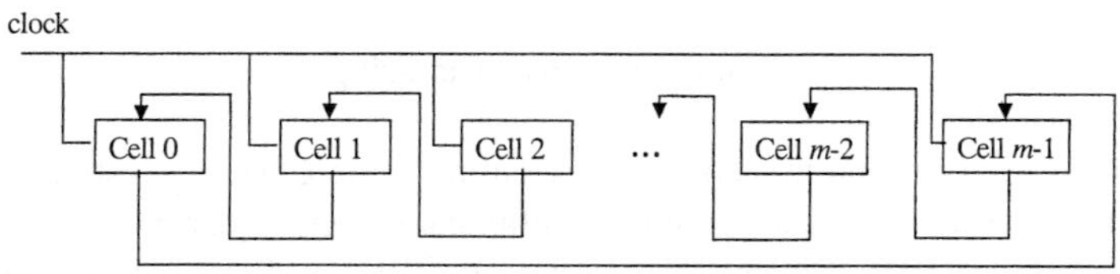

Fig. 3. OCA having the characteristic matrix *D*

The Fig.4 shows the entire suructure of common operation of multiplication and squaring. It accomplish the step 3 of Algorithm 2.

Now, how to perform the modular multiplication and squaring simultaneously is described. In order to perform step 4 of Algorithm 2, $M^{(i)}(x)=M^{(i-1)}(x)+c_{m-i-1}T^{(i-1)}(x)$ for $1\le i\le m$, the $c_{m-i-1}T^{(i-1)}(x)$ operation is reviewed firstly. For $c_{m-i-1}T^{(i-1)}(x)$ operation with

1≤*i*≤*m*, the *m* bits obtained as a result of CA of Fig. 4 and c_{i-1} are inputted into *m* AND gates at i^{th} clock. The next result and $M^{(i-1)}(x)$ are subject to XOR, and the result is

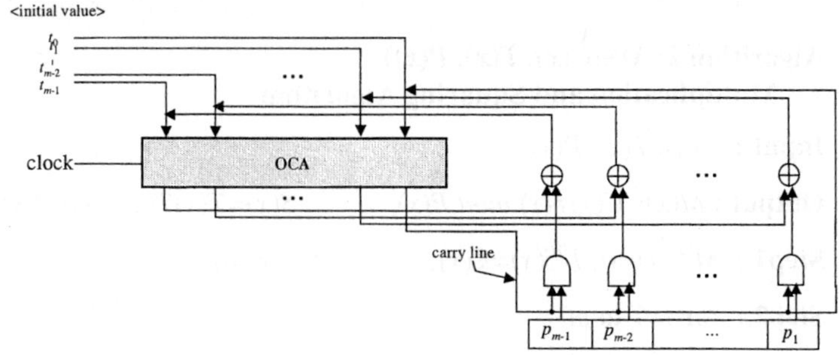

Fig. 4. Structure of $T^{(i-1)}(x)x$ mod $P(x)$ operation

stored at $M^{(i-1)}(x)$ again. The value of each $M^{(0)}(x)$ register at the beginning is initialized to be 0. The structure for that is as shown in Fig. 5, which shows the computation in the *i*th clock for 1≤*i*≤*m*:

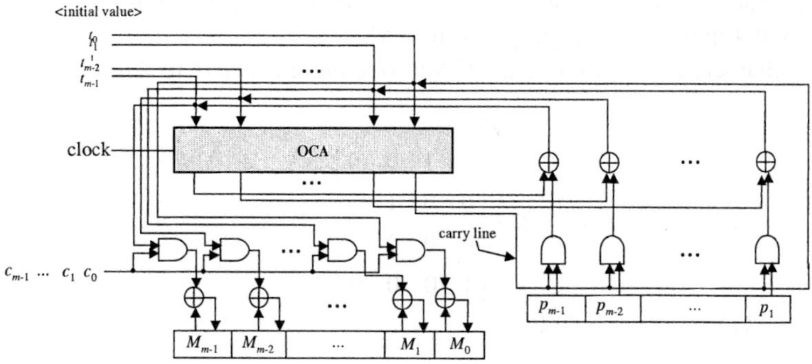

Fig. 5. Structure of modular multiplication

For squaring operation in step 4 of Algorithm 2, the $C(x)$, is substituted with $T(x)$. So the structure in which the modular multiplication and squaring can be performed simultaneously using CA, is shown in Fig. 6(at i^{th} clock). Each initial value is as follows:

- Initial values of OCA : $T(x) = t_{m-1} \ldots t_2 t_1 t_0$
- Initial values of the *P* register : $P(x) = p_{m-1} \cdots p_2 p_1 p_0$
- Initial values of the *S*, *M* register : all 0

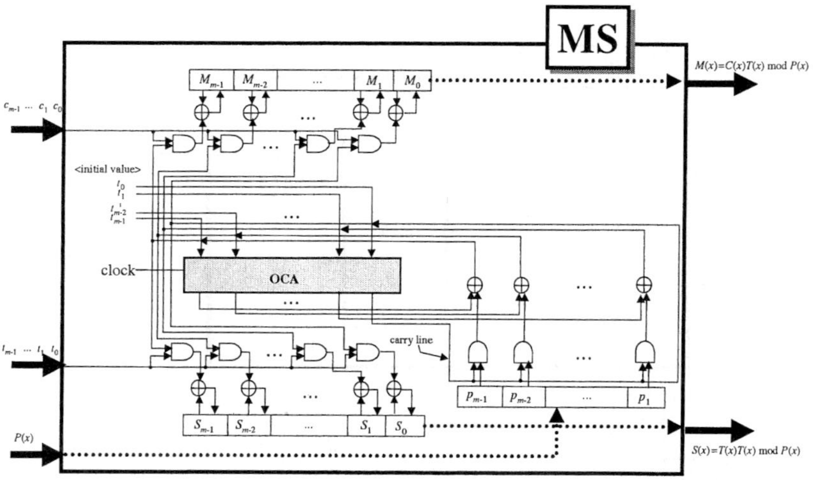

Fig. 6. Structure of simultaneously performing modular multiplication and squaring using CA

5 Exponentiation Architecture Using Multiplier/Squarer over GF(2^m)

Ordinary LSB-first exponentiation algorithm in [12] can be slightly changed to Algorithm 3. Exponentiation algorithm which uses a new multiplication/squaring algorithm, MS algorithms in chapter 4, as a sub function for exponentiation algorithm is as follows:

Algorithm 3 : EXP($A(x)$, E, $p(x)$)
Exponentiation Algorithm using MS Algorithm.
Input : $A(x)$, E, $P(x)$
Output : $C(x)=A(x)^E \bmod P(x)$
STEP 1 : $C(x)=\alpha^0$, $T(x)=A(x)$
STEP 2 : for i=0 to m-1
STEP 3 : if e_i ==1 $(C^{(i+1)}(x), T^{(i+1)}(x))$=MS$(C^{(i)}(x), T^{(i)}(x))$
else $C^{(i+1)}(x)=C^{(i)}(x)$, $T^{(i+1)}(x)$= MS$(C^{(i)}(x), T^{(i)}(x)$.

In step 3, MS algorithm is called with two parameters $C^{(i)}(x)$ and $T^{(i)}(x)$, and it returns two computation results of multiplication, $T^{(i)}(x)C^{(i)}(x) \bmod P(x)$ and squaring $T^{(i)}(x)T^{(i)}(x) \bmod p(x)$. Square result is stored to $T^{(i+1)}(x)$. But, the variable $C^{(i+1)}(x)$ stores a value depends on the value e_i.
If e_i has value 1 then returned multiplication result is passed to $C^{(i+1)}(x)$, but the other case, $C^{(i)}(x)$ is passed to $C^{(i+1)}(x)$.

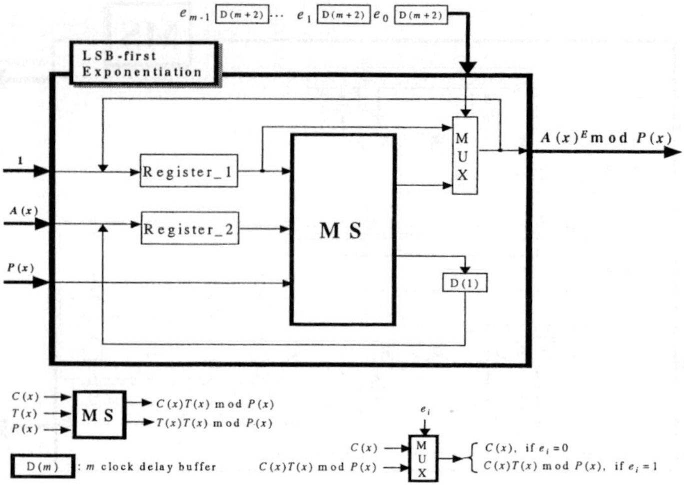

Fig. 7. A structure for performing an exponentiation using MS structure

Fig 7. represents a structure for performing an exponentiation using a new MS structure in Fig.6, which operates Algorithm 3. In Fig.7, the architecture includes a MS structure as its kernel architecture. This MS structure for multiplier/squarer is described in Fig. 6. While the multiplier/squarer computes the multiplication operation, it also processed the squaring operation concurrently forming the square term. The MUX depicted in Fig. 7 decides the multiplication result depends on the exponent.

6 Analysis and Conclusion

In this paper, we propsed a new fast exponentiation architecture over GF(2^m) using a cellular automata. The performance of the proposed archtecture is compared with that of previous study. Table 1 shows the result.

In conclusion, the proposed structure in this paper is much more efficient than systolic structure in view of the space and time. And we generally consider construction simplicity, defined by the number of transistors needed for its construction and the time needed for the signal change to propagate through the gate[17]. So the comparison of area-time product[17] for LSB first exponentiation is shown in Table 2. In Table 2, the proposed architecture in this paper is more efficient in view of the space and time than that of [16]. Even if when m=512, the proposed architecture get 24% speed up. In view of the space, the area complexity of the proposed architecture is O(m^2), but architecture based on systolic array is O(m^3).

Architecture proposed in this paper can be efficiently compute the exponentiation over GF(2^m) in public key cryptosystems.

Table 1. Comparison of performance

Structure	Systolic array [16]	Proposed paper	
Operation	exponentiation	exponentiation	
NO. of basic components	2(m-1) multipliers	1 MS*	
NO. of AND gates	$4m^2(m-1)$	$3m-1$	
NO. of XOR gates	$4m^2(m-1)$	$3m-1$	
NO. of one bit latches	$14m^2(m-1)$	m^2+2m+1	
NO.of MUXes	m	*1*	
NO.of registers	0	m bit 5	m-1 bit: 1
Execution time (clock cycles)	$2m^2+m$	m^2+3m	

* MS : Structure of simultaneously performing modular multiplication and squaring

Table 2. Comparison of Area-Time Product for LSB first exponentiation

Structure	Systolic array [16]	Proposed paper
AREA	$4m^2(m-1)A_{2AND}$ $+4m^2(m-1)A_{2XOR}$ $+mA_{2MUX}$ $+(14m^3-14m^2)\ A_{1LATCH}$ $= (192m^3-192\ m^2+20m\ )\phi$	$(3m-1)A_{2AND}$ $+(3m-1)A_{2XOR}$ $+A_{2MUX}$ $+(6m-1)A_{1FF}$ $+\ (m^2+2m+1)\ A_{1LATCH}$ $= (8m^2+184m-10)\ \phi$
TIME	$(2m^2+m)(2T_{2AND}+4T_{2XOR})$ $= (13.6\ m^2+6.8\ m)\Delta$	$(m^2+3m)(T_{2AND}+T_{2XOR}+\ T_{1FF})=$ $(10.4\ m^2+31.2\ m)\ \Delta$
AREA × TIME	$(2611.2m^5-1305.6m^4-1033.6m^3+136m^2)\ \phi\Delta$	$(83.2m^4+2163.2m^3+5636.8m^2-312m)\ \phi\Delta$

Acknowledgements. This research was supported by Kyungpook National University Research Team Fund, 2002.

References

[1] R.J. McEliece, *Finite Fields for Computer Scientists and Engineerings*, New York:Kluwer Academic, 1987

[2] W. Diffie and M.E. Hellman, "New directions in cryptography," *IEEE Trans. on Info. Theory*, VOL. 22, pp.644–654, Nov. 1976.

[3] T. ElGamal. "A public key cryptosystem and a signature scheme based on discrete logarithms," *IEEE Trans. on Info. Theory*, VOL. 31(4). pp. 469–472, July 1985.

[4] A.J. Menezes, *Elliptic Curve Public Key Cryptosystems*, Kluwer Academic Publishers,1993.

[5] C.–S. YEH, IRVING S. REED, T.K. TRUONG, "Systolic Multipliers for Finite Fields GF(2^m)," *IEEE TRANSACTIONS ON COMPUTERS*, VOL. C-33, NO. 4, pp. 357–360, April 1984.

[6] C.L. Wang, J.L. Lin, "Systolic Array Implementation of Multipliers for Finite Fields GF(2^m)," *IEEE TRANSACTIONS ON CIRCUITS AND SYSTEMS*, VOL. 38, NO. 7, pp. 796–800,July 1991.

[7] P.L. Montgomery, "Modular multiplication without trial division," *Mathematics of Computation*, 44(170):519–521, April, 1985.

[8] M. Delorme, J. Mazoyer, *Cellular Automata*, KLUWER ACADEMIC PUBLISHERS 1999.

[9] STEPHEN WOLFRAM, *Cellular Automata and Complexity*, Addison-Wesly Publishing Company, 1994.

[10] ELWYN R. BERLEKAMP, "Bit-Serial Reed-Solomon Encoders," *IEEE TRANSACTIONS ON INFORMATION THEORY*, VOL. IT-28, NO. 6, pp. 869–874, November, 1982.

[11] P.P. Choudhury, R. Barua, " Cellular Automata Based VLSI Architecture for Computing Multiplication And Inverse In GF(2^m)," *IEEE Proceeding of the 7th International Conference on VLSI Design*, pp.279–282, January 1994.

[12] Knuth, *THE ART OF COMPUTER PROGRAMMING*, VOL. 2/Seminumerical Algorithms, ADDISON-WESLEY, 1969.

[13] P.A. Scott, S.J. Simmons, S.E. Tavares, and L.E. Peppard, "Architectures for Exponentiation in GF(2m)", *IEEE J. Selected Areas in Comm.*, VOL.6., NO. 3, pp. 578–586, April, 1988.

[14] C.Parr, "Fast Arithmetic for Public-Key Algorithms in Galois Fields with Composite Exponents," *IEEE TRANSACTIONS ON COMPUTERS*, VOL.48, NO. 10, pp. 1025-1034, October, 1999.

[15] Kyo-Min Ku, Kyeoung-Ju Ha, Hyun-Sung Kim, Kee-Young Yoo, "New Parallel Architecture for Modular Multiplication and Squaring Based on Cellular Automata.", *PARA 2002, LNCS 2367*, pp.359–369, 2002

[16] C.L. Wang, "Bit-Level Systolic Array for Fast Exponentiation in Gf(2^m)", *IEEE TRANSACTIONS ON COMPUTERS,* VOL. 43, NO. 7, pp. 838–841, July, 1994.

[17] D. D. Gajski, *Principles of Digital Design*, Prentice-Hall International, Inc., 1997.

Interacting Automata for Modelling Distributed Systems*

Irina A. Lomazova

Program Systems Institute of the Russian Academy of Science
Pereslavl-Zalessky, 152020, Russia
irina@univ.botik.ru

Abstract. A *community of interacting automata* is a set of nondeterministic finite automata which can execute actions autonomously, synchronize with each other, or generate new members of the community. Thus interacting automata allow to model distributed systems with unlimited number of interacting agents. We show that the formalism of interacting automata has a clear semantics and nice semantic properties, leading to decidability of some important behavioral properties of modelled systems.

1 Introduction

Finite automata are widely used in modelling systems, especially when handling control aspects is important. The benefits of finite automata are in particular connected with clear mathematical semantics, systems visualization and decidability of many crucial behavioral properties.

The up-to-date needs for modelling concurrent and distributed systems with a complex structure lead to different extensions of the finite automata formalism, such as hierarchical, interacting, dynamic automata [1,5,9,10], which allow to model hierarchy, concurrency and communication among concurrent components.

Here we introduce the *interacting automata*, which together with concurrency and communication allow to model dynamic activation of new agents. There is a rather large variety of formalisms for representing distributed systems in the modern theoretical computer science. Modelling systems with a dynamically changing number of agents requires extending standard formalisms to capture a possibility of changing a system structure (see e.g. [6] for an extension of Petri nets for modelling multi-agent systems).

Interacting automata represent in some sense the most simple model of a system consisting of agents, which can do autonomous actions, interact with each other, be generated by other agents and die. Interacting agents form an unstructured community with no hierarchical structure, all agents have thus equal

* This work was partly supported by the Presidium of the Russian Academy of Science, program "Intellectual computer systems", project 2.3 – "Instrumental software for dynamic intellectual systems" and INTAS-RFBR (Grant 01-01-04003).

P.M.A. Sloot et al. (Eds.): ICCS 2003, LNCS 2658, pp. 851–860, 2003.

rights to do actions autonomously or synchronize. Agents are represented by finite automata. Initially there is one automata which can generate new members of community, which in turn can also generate new members. An agent automata dies, when it comes to its terminal state. Concurrent behavior is represented by interleaving semantics. A community comes to its final state, when it is empty, i.e. all its members have died.

The paper is organized as follows. In section 2 we give the definitions of interacting automata community. In section 3 we study languages, accepted by interacting automata. It is shown, that interacting automata accept some context-free nonregular and even some noncontext-free languages. In section 4 it is shown, that interacting automata can be considered as well-structured transition systems, and this leads to decidability of some behavioral properties for them. In section 5 interacting automata are compared with Petri nets. It is proved, that interacting automata can be simulated by Petri nets. As the consequence of this result we obtain decidability of the Reachability problem for interacting automata. Also it implies, that there exist a context-free language, which is not accepted by interacting automata. Together with the results of section 3 this means, that the class of interacting automata languages is incomparable with the class of context-free languages.

2 Interacting Automata

We start by recalling that a (nondetermitistic) finite automaton is defined as a tuple $A = (Q, U, \theta, q_0, F)$, where

- Q is a finite set of inner states;
- F — a finite set of terminal states, $F \cap Q = \emptyset$;
- U — a finite alphabet of action names;
- $\theta : Q \times U \to 2^{(Q \cup F)}$ — a transition map;
- $q_0 \in Q$ — an initial state.

Here the set F of terminal states is not included into the set Q of inner states (as it is usually done in the standard definition of a finite automaton) by reasons of convenience. In our definition the transition map can't be applied to terminal states, so after coming to any terminal state an automaton stops. It can be easily shown, that our definition is equivalent to the standard one concerning expressibility, accepted languages and other behavioral properties of automata. We do not consider here the questions of automata size and minimization.

A finite automaton can be considered as a finite oriented graph with a set $Q \cup F$ of nodes and with arcs labelled by symbols from U. As usual, we suppose U contains a special symbol τ for the silent (invisible) action. In the graph representation inner states are designated by circles. The set Q of inner states contains a designated initial state q_0. Terminal nodes are designated by squares and do not have outgoing arcs.

A community of interacting automata will be a multiset of finite automata with actions of three types: usual autonomous actions, actions of synchronization, and actions generating a finite automaton — a new member of the community.

Recall, that a *multiset* m over a set S is a mapping $m : S \to \mathbb{N}$, where $\mathbb{N}$ is the set of natural numbers, i.e. a multiset may contain several copies of the same element. A multiset m is *finite* iff the set $\{s \in S \mid m(s) > 0\}$ is finite. For two multisets m, m' we write $m \subseteq m'$ iff $\forall s \in S : m(s) \leq m'(s)$ (the inclusion relation). The sum of two multisets m and m' is defined as usual: $\forall s \in S : (m+m')(s) = m(s)+m'(s)$. We define also $(m-m')(s) = m(s)-m'(s)$, if $m(s) > m'(s)$, and $(m-m')(s) = 0$, otherwise.

Definition 2.1. *Let L be a finite set of synchronizing labels, s.t. for each label $l \in L$ the adjacent label $\bar{l} \in L$ is defined and $\bar{\bar{l}} = l$. Let U be a finite set of action names, $\mathcal{A} = \{\alpha_1, \ldots, \alpha_k\}$ — a set of automata names.* A community of interacting automata *is defined as a collection $\Sigma = (A_1, \ldots, A_k)$ $(k \geq 1)$ of finite nondeterministic automata, where for $i = 1, \ldots, k$ an automaton $A_i = (\alpha_i, Q_i, (U \cup L \cup \mathcal{A}), \theta_i, q_0^i, F_i)$ is specified by the following components:*

- α_i *— a name of the automaton A_i;*
- Q_i *— a finite set of inner states of the automaton A_i;*
- $(U \cup L \cup \mathcal{A})$ *— a (common for all automata) set of names for autonomous (U), synchronized (L) and generating $(\mathcal{A})$ actions;*
- $\theta_i : Q_i \times (U \cup L \cup \mathcal{A}) \to 2^{(Q_i \cup F_i)}$ *— a transition map for A_i;*
- $q_0^i \in Q_i$ *— an initial state for A_i, and*
- F_i *— a set of terminal states for A_i.*

Among this collection a distinguished parent automaton *is chosen, we suppose it to be A_1.*

Now we define the behavior of an automata community $\Sigma = (A_1, \ldots, A_k)$.

A *configuration of an automaton A_i* is represented by a pair (α_i, q), where α_i is the name, $q \in Q_i \cup F_i$ — a state of A_i. An automaton configuration (α_i, q) is called *terminal*, iff $q \in F_i$ — a terminal state.

A *configuration of an automata community* Σ is a multiset K over the set of all nonterminal configurations of automata $A_1, \ldots, A_k$.

A *step* for an automata community Σ is a triple (K, u, K'), where K, K' — configurations for Σ, $u \in (U \cup \mathcal{A} \cup \{\tau\})$ — an action name. We define three kinds of steps.

- **An autonomous step.** Let the configuration K include an automaton configuration (α_i, q) and for some $u \in U$ we have $q' \in \theta_i(q, u)$, i.e. an autonomous action u transforms a state q of the automaton A_i into the state q'. Then the configuration K of Σ is transformed to K' (write $K \xrightarrow{u} K'$), where $K' = K - (\alpha_i, q) + (\alpha_i, q')$ provided $q' \in Q_i$ and $K' = K - (\alpha_i, q)$, if $q' \in F_i$. Thus terminal automaton configurations are deleted from the community.
- **A synchronization step.** Let K include two automata configurations (α_i, q), (α_j, p) and for some label $l \in L$, $q' \in \theta_i(q, l)$, $p' \in \theta_j(p, \bar{l})$. Then K can be transformed to K' (write $K \xrightarrow{\tau} K'$), where K' is obtained by deleting all terminal automaton configurations from $K - (\alpha_i, q) - (\alpha_j, p) + (\alpha_i, q') + (\alpha_j, p')$.

- **A generating step.** Let K include an automaton configuration (α_i, q) and for some automaton name $\alpha_j \in \mathcal{A}$ we have $q' \in \theta_i(q, \alpha_j)$. Then K can be transformed to K' (write $K \overset{\alpha_j}{\to} K'$), where K' is obtained by deleting a terminal automaton configuration (if any) from $K-(\alpha_i, q)+(\alpha_i, q')+(\alpha_j, q_0^j)$.

We write $K \to K'$, if there exists a step from K to K'.

An automata community behavior is defined in terms of *consecutive runs* (or just *runs*). A run R is a finite or infinite sequence of automata community configurations $K_0, K_1, \ldots$, such that

1. The *initial configuration* K_0 consists of the only automaton configuration (α_1, q_0^1), i.e. the parent automaton in its initial state;
2. $K_i \to K_{i+1}$ for all $i = 0, 1, \ldots$;
3. R is finite iff it can't be continued.

Thus the last configuration in a finite run may be either empty (a terminating run), or a nonempty configuration in which no action can occur (a deadlock). The *behavior* of an automata community Σ is defined as the set of all its runs.

3 Languages of Interacting Automata

Figure 1 shows an example of very simple automata community Σ_1 without synchronization; Σ_1 consists of the only automaton A, which can generate its own copies.

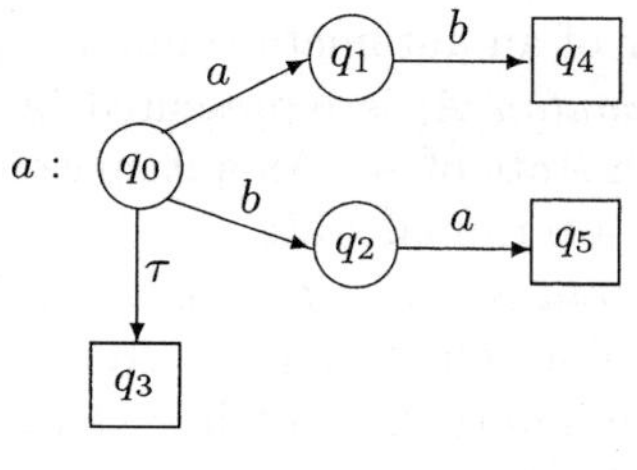

Fig. 1. The automata community Σ_1.

Here a is the name of the automaton A, $A = (a, Q, (U \cup \mathcal{A}), \theta, q_0, F)$, where $Q = \{q_0, q_1, q_2\}$ is the set of inner states of A with the initial state q_0, $U = \{b, \tau\}$ — the set of autonomous actions, the set of synchronization labels is empty, $\mathcal{A} = \{a\}$ — the set of automata names, θ — the transition map, shown by arcs in the graph representation of A, $F = \{q_3, q_4, q_5\}$ is the set of terminal states.

The automaton A can

- either do a silent action and terminate,
- or can generate its own copy, after that at some further moment do the action b and terminate,

– or first do the action b, then at some further moment generate its own copy and also terminate.

The visible actions of the automata community, generated by the automaton A, are actions a and b. During the life cycle of this automata community the unlimited number of copies of A can appear.

Definition 3.1. *Let Σ be an automata community. The language $\mathcal{L}(\Sigma)$ accepted by Σ is the set of all finite words over the alphabet $(U \cup \mathcal{A}) \setminus \{\tau\}$, s.t. for each word $w \in \mathcal{L}(\Sigma)$ there exists a terminating run R for Σ with a sequence w of visible actions.*

It's easy to see, that for an automata community Σ_1, represented in Figure 1, $\mathcal{L}(\Sigma_1) = \{w \mid w \in \{a, b\}^* \text{ and } n_a(w) = n_b(w)\}$, where $n_a(w)$ and $n_b(w)$ designate the number of symbols 'a' (correspondingly, symbols 'b') in the word w.

Since the language $\mathcal{L}(\Sigma_1)$ can't be accepted by any finite nondeterministic automaton, the immediate consequence of this example is the following

Proposition 3.2. *The formalism of automata communities is strictly more expressive than nondeterministic finite automata.*

It is well-known, that the language $\mathcal{L}(\Sigma_1)$ is a context-free language, so it can be accepted by a push-down automaton. Further in Section 5 we show that there are contest-free languages, that are not accepted by any automata community.

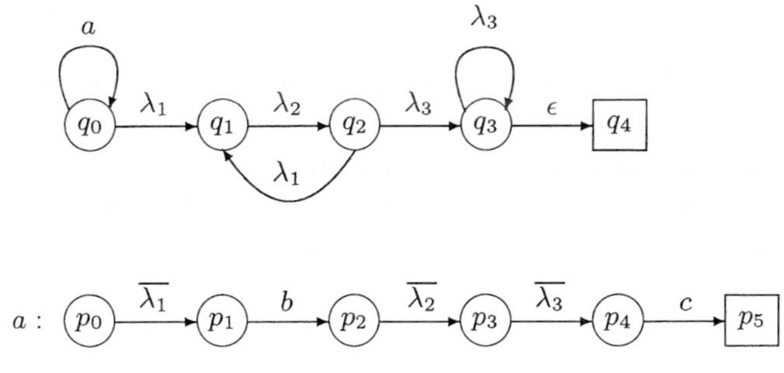

Fig. 2. The automata community Σ_2.

However, interacting automata can accept also some languages, which are not context-free. Figure 2 shows the automata community Σ_2, which accepts the language $\{a^n b^n c^n \mid n \geq 1\}$. It is well-known, that this language is not context-free and can't be accepted by any push-down automaton.

4 Interacting Automata as Well-Structured Transition Systems

Interacting automata may generate new members of their community, and hence they are infinite state-space systems. In this section we show, that interacting automata satisfy a nice monotonicity property: a bigger community can produce all actions that are possible for a smaller one. Moreover, we prove that interacting automata can be considered as a Well-Structured Transition System (WSTS) [2]. The theory of WSTS helps to obtain some decidability results for interacting automata.

Recall that a transition system is a pair $\mathcal{S} = \langle S, \rightarrow \rangle$, where S is an abstract set of states (or configurations) and $\rightarrow \subseteq S \times S$ is any transition relation. For a transition system $\mathcal{S} = \langle S, \rightarrow \rangle$ we write $Succ(s)$ for the set $\{s' \in S \mid s \rightarrow s'\}$ of immediate successors of s. $\mathcal{S}$ is finitely branching if all $Succ(s)$ are finite.

Definition 4.1. *A* quasi-ordering *is any reflexive and transitive relation* $\leq$ *(over some set* X*). A* well-quasi-ordering *(a wqo) is any quasi-ordering* $\leq$ *such that, for any infinite sequence* $x_0, x_1, x_2, \ldots$*, in* X*, there exist indexes* $i < j$ *with* $x_i \leq x_j$*.*

Note, that if $\leq$ is a wqo, then any infinite sequence contains an infinite increasing subsequence: $x_{i_0} \leq x_{i_1} \leq x_{i_2} \ldots$.

Definition 4.2. *A* well-structured transition system *(a WSTS) is a transition system* $\Sigma = \langle S, \rightarrow, \leq \rangle$ *equipped with an ordering* $\leq$ *between states such that* $\leq$ *is a wqo, and* $\leq$ *is "compatible" with* $\rightarrow$*, where "compatible" means that for all* $s_1 \leq t_1$*, and transition* $s_1 \rightarrow s_2$*, there exists a transition* $t_1 \rightarrow t_2$*, such that* $s_2 \leq t_2$*.*

Now we define a wqo on configurations of an automata community and show that together with transformation relation $K \rightarrow K'$ on configurations it forms WSTS.

Theorem 4.3. *Let* Σ *be an automata community,* $\mathcal{K}$ *— the set of all configurations for* Σ*. Then*

- *the relation* $\subseteq$ *of multiset inclusion on* $\mathcal{K}$ *is a wqo,*
- *if* $K_1 \subseteq K_2$ *are configurations for* Σ *and* $K_1 \rightarrow K_1'$*, then there exists a configuration* K_2'*, s.t.* $K_2 \rightarrow K_2'$ *and* $K_1' \subseteq K_2'$*.*

Thus every automata community is a WSTS w.r.t. the multiset inclusion relation.

The following diagram illustrates the statement of the theorem:

$$\begin{array}{ccc} K_1 & \subseteq & K_2 \\ {\scriptstyle u}\big\downarrow & & \big\downarrow{\scriptstyle u} \\ K_1' & \subseteq & K_2' \end{array}$$

Proof sketch: To prove that the multiset inclusion on $\mathcal{K}$ is a wqo note, that for a given automata community the number of different automaton configurations is finite. So, configurations of an automata community are multisets over a finite set of automaton configurations and can be encoded by vectors of some fixed size n with nonnegative integers as elements. The multiset inclusion then corresponds to the component-wise comparison of integer vector elements, which is a wqo.

Compatibility is proved straightforward from the definition of an automata community behavior by the analysis of all three kinds of steps. □

We say, that a configuration K accepts an action word ω, if there exists a sequence of steps $K \to K_1 \to \dots$ with the sequence ω of visible actions.

Corollary 4.4. *Let K, K' be two configurations for automata community Σ, such that $K \subseteq K'$. Then K accepts an action word ω implies K' accepts ω.*

Though a transition system with an infinite state-space has an infinite reachability tree, for each WSTS with an effective ordering relation and computable transition relation a finite coverability tree can be effectively constructed [3]. For interacting automata it can be done as follows.

We start from the root node labelled by the initial configuration K_0. It is easy to see, that the number of configurations K, s.t. $K_0 \to K$, is finite. For each such K we construct a node, labelled by K and an arc from K_0 to K.

Then we continue this process taking each K instead of K_0 in turn, i.e. we start to construct the reachability tree. Each branch of this tree is continued until one of the following situations occurs:

1) we reach the terminal empty configuration, which is a leaf, or
2) we reach a deadlock configuration, which is also a leaf, or
3) we reach a configuration K, s.t. the branch leading to K contains a node, labelled by a configuration K' and $K' \subseteq K$, and then we also do not continue further this branch and declare the node K to be a leaf.

Since the multiset inclusion relation is a wqo, all branches in a coverability tree are finite. Since automata communities are finitely branching, for every automata community its coverability tree is finite. A coverability tree can be used for checking some behavioral properties of a modelled system.

A system terminates if there exists no infinite run (*Termination Problem*). The *Control-State Maintainability Problem* is to decide, given an initial state K and a finite set $\{K_1, K_2, \dots, K_m\}$ of configurations, whether there exists a computation starting from K with all its inner configurations covering (not less than w.r.t. the multiset inclusion) one of the K_i's. The dual problem, called the *Inevitability Problem*, is to decide whether all computations starting from K eventually visit a state not covering one of the K_i's. Thus for example, for automata communities we can ask whether there is a run with all configurations containing at least one copy of a given automaton (control-state maintainability), or whether all copies of a given automaton, which can evolve during a life-cycle of an automata community, will eventually come to its terminating state and disappear (inevitability).

Theorem 4.5. *Termination, the control-state maintainability problem and the inevitability problem (w.r.t. $\subseteq$) are decidable for automata communities.*

Proof sketch: An automata community doesn't have infinite runs iff all leaves in its coverability tree are marked by either empty configurations or deadlocks.

For a given automata community a computation starting from K with all its inner configurations not less than one of $\{K_1, K_2, \dots, K_m\}$ exists iff in the coverability tree with the configuration K in the root there exists a path from the root to a leaf with all inner configurations not less than one of the K_i's. □

5 Interacting Automata and Petri Nets

In this section we compare automata communities with Petri nets — a popular formalism for modelling and analysis of parallel and distributed systems [8].

Definition 5.1. *A Petri net is a finite directed graph with two types of nodes, referred to as places and transitions. An arc in a Petri net goes either from a place to transition, or from a transition to a place.*

For a transition t every place p with an arc going from p to t is called an input place, and a place q with an arc going from t to q is called an output place.

A marking M in a Petri net is a mapping from the set P of places to non-negative integers. We say, that there are k tokens in the place p, if $M(p) = k$. A marked Petri net is a Petri net together with its initial marking.

We now define the behavior of a Petri net. A transition t is enabled in a marking M iff each input place of t contains at M at least one token. An enabled transition t may fire by removing one token from every input (for t) place and by adding one token to every its output place.

A Petri net run is defined as a sequence of its firings starting from the initial marking.

Transitions in a Petri net may be labelled by names of actions to make a behavior of a Petri net observable. Some of transitions may be invisible (labelled by a silent action τ).

We show, that the behavior of every automata community can be simulated by a labelled Petri net. Let $\Sigma = (A_1, \ldots, A_k)$ be an automata community with inner states $Q_1, \ldots, Q_k$, terminal states $F_1, \ldots, F_k$ and transition maps $\theta_1, \ldots, \theta_k$ correspondingly. Without loss of generality we can suppose, that all sets of inner and terminal states here are pairwise disjoint. A Petri net $PN(\Sigma)$ simulating Σ will be constructed as follows:

1. For each inner state $q \in Q_1 \cup \ldots \cup Q_k$ there will be a place $\tilde{q}$ in $PN(\Sigma)$. A configuration K for Σ will be represented by a marking $\tilde{K}$, so that n copies of an automaton configuration (α, q) will be encoded by n tokens residing in the place $\tilde{q}$.
2. For each arc (q, q'), labelled by an internal action u, in an automaton A_i of the community Σ we construct a transition t in $PN(\Sigma)$, also labelled by u, with the input place $\tilde{q}$ and the output place $\tilde{q'}$. If q' is a terminal state in A_i, then the transition t doesn't have output places.
3. For each arc (q, q'), labelled by the name α_i of an automaton A_i, in some (other or the same) automaton we construct a transition t, labelled by α_i, with the input place $\tilde{q}$ and two output places: the place corresponding to the initial state q_i^0 of A_i and the place $\tilde{q'}$. If the state q' is terminal in its automaton, then the transition t has only one output place $\tilde{q_i^0}$.
4. For each two arcs (q, q'), (p, p') labelled by mutually adjacent synchronization labels (these two arcs may be either in two different automata of the community, or in one automaton) we construct a transition t, labelled by τ with two input places $\tilde{q}$, $\tilde{p}$ and two output places $\tilde{q'}$, $\tilde{p'}$. As before, if q', or p', or both are terminal states in their automata, then the corresponding output places for the transition t are omitted.

Figure 3 shows a Petri net, simulating the automata community Σ_1, shown in Figure 1.

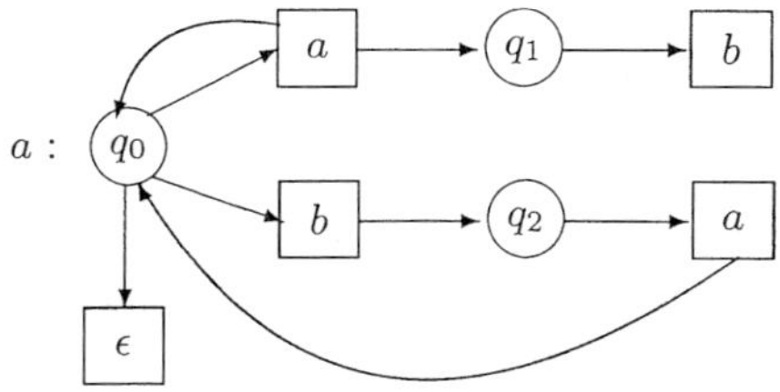

Fig. 3. A Petri net, simulating Σ_1.

The observable behavior of a Petri net can be described in terms of Petri net languages [4].

Definition 5.2. *Let PN be a labelled Petri net with transition labels from X. Let $\mathcal{R}_\emptyset$ be the set of all runs for PN, leading to the empty marking. A terminal language of PN is a set of all words (over X) labelling runs from $\mathcal{R}_\emptyset$.*

Note, that in the definition of a terminal Petri net language an arbitrary fixed marking can be used as the target marking instead of the empty one. These two definitions are equivalent. The following theorem states, that for every automata community Σ there exists a Petri net $PN(\Sigma)$, simulating the behavior of Σ.

Theorem 5.3. *For every automata community Σ there exists a Petri net $PN(\Sigma)$, such that the terminal language of $PN(\Sigma)$ coincides with the language $\mathcal{L}(\Sigma)$, accepted by the automata community Σ.*

Proof sketch: It can be shown, that for each run of Σ there exists a corresponding run of $PN(\Sigma)$ and vice versa. The proof can be done by induction on the length of the run by the analysis of all cases of state transitions. □

In the theory of Petri net languages it is known, that context-free languages are incomparable with Petri net languages [4]. More precisely, there exist context-free languages, which are not terminal Petri net languages. An example of such a language is the language $\{ww^R \mid w \in \{a,b\}^*\}$, where w^R denotes a word w in the inverse order, i.e. written from the right to the left. Then from theorem 5.3 we get the following

Proposition 5.4. *There exist context-free languages, which are not accepted by any automata community.*

The Petri net $PN(\Sigma)$ simulates runs of Σ not only as sequences of actions, but also as sequences of reachable states. These means, that the Reachability Problem for automata communities can be reduced to the Reachability Problem for Petri nets. Recall, that the Reachability Problem is to decide, given an initial state S and a finite state S', whether there exists a run, leading from S to S'.

For Petri nets reachability is decidable [7], hence we obtain

Proposition 5.5. *Reachability Problem is decidable for interacting automata communities.*

References

1. R. Alur, S. Kannan, and M. Yannakakis. Communicating Hierarchical Automata. In *ICALP'99*, volume 1644 of *Lecture Notes in Computer Science*, pages 169–178. Springer, 1999.
2. A. Finkel. Reduction and covering of infinite reachability trees. *Information and Computation*, 89(2):144–179, 1990.
3. A. Finkel, and Ph. Schnoebelen. Well-structured transition systems everywhere! *Theoretical Computer Science*, 256:1–2, pages 63–92, 2001.
4. M. Jantzen. Language Theory of Petri Nets. In *Petri nets: Central Models and their Properties*, volume 254 of *Lecture Notes in Computer Science*, pages 397–412. Springer, 1987.
5. R. Lanotte, A. M. Schettini, A. Peron, S. Tini. Dynamic Hierarchical Machines. In *Proc. of the Concurrency Specification and Programming (CS&P'2000) Workshop. 7–9 October 2002. Vol.2*, Informatik-Bericht Nr.161, Humboldt-Universität zu Berlin, pages 205–216, 2002.
6. I. A. Lomazova. Nested Petri nets — a Formalism for Specification and Verification of Multi-Agent Distributed Systems. *Fundamenta Informaticae*, 43(1–4): 195–214, 2000.
7. E. W. Mayr. An Algorithm for the General Petri Net Reachability Problem. *SIAM Journal on Computing*, 13:441-460, 1984.
8. W. Reisig. *Petri Nets. An Introduction*, volume 4 of *EATCS Monographs on Theoretical Computer Science.* Springer, 1985.
9. B.A. Trakhtenbrot. Automata, Ciruits, and Hybrids: Facets of Continious Time. volume 2076 of *Lecture Notes in Computer Science*, pages 4–23. Springer, 2001.
10. M. Yannakakis. Hierarchical State Machines. In *IFIP Theoretical Computer Science'2000*, volume 1875 of *Lecture Notes in Computer Science*, pages 315–330. Springer, 2000.

The Reachability Problem in a Concave Region: A New Context

Ali Mohades[1], Mohammad Ebrahim Shiri[1], and Mohammadreza Razzazi[2]

[1] Faculty of Math. and Computer Sc. AmirKabir University of Tech.
424 Hafez Ave. Tehran, Iran.
[2] Faculty of Computer Eng. at AmirKabir University of Tech.
424 Hafez Ave. Tehran, Iran.
{mohades, shiri, razzazi}@aut.ac.ir

Abstract. In this paper we present an algorithm that moves a chain confined in a T-shaped rectilinear region from an initial configuration to a final configuration where the end point of chain reaches a given point p. This work is an extension of the our previous results in concave region. In our algorithm links my cross over one another and none of end points of the link chain are fixed. It is shown that the algorithm takes a quadratic time and works when a certain condition is satisfied.

1 Introduction

The problem related to the movement of n-link chain in two dimensions is considered. The n-link chain consists of n links, which are line segments of arbitrary length. The links are joined end-to-end by freely rotating joints. One problem of interest in this area is the reachability problem. In this problem, we decide whether an n-link chain can be moved from one position in a bounded region to another position in order to reach a point in the region.

The movement of an n-link chain was first studied by Hopcroft *et al.* [1]. An algorithm running in O(n) time was then given for an n-link chain confined in a circular region of the plane. In [2] they showed that reachability decision problems are NP-hard if the arm constrained by arbitrary polygonal walls. The case when the arm is confined in a square has been discussed in [3].

All the results mentioned above are on the analysis of a given arm in a convex region. In [5] and[6] we extended the reachability problem for some special concave regions. We proved that under some conditions an n-linked chain that is located in one part of a T-shaped concave region could reach all points of the region. The present paper extends our previous works. We show that under some additional conditions we can improve our previous algorithm for a given n-linked chain without considering its placement in the region. The rest of the paper is organized as follows. Section 2 introduces the technical definitions and the problem statement. Section 3 presents our extended algorithm.

P.M.A. Sloot et al. (Eds.): ICCS 2003, LNCS 2658, pp. 861–868, 2003.

2 Preliminaries and Previous Work

Assume $\Gamma[0,1,...n]$ is an n-link chain and l_i is the length of the i-th link L_i with endpoints A_{i-1} and A_i and $|\Gamma| = max_{1 \leq i \leq n} l_i$. We say that an n-link chain Γ is *bounded by* b if $|\Gamma| < b$, i.e no link has length greater than or equal to b.

A simple motion is the one in which only a few joints are moved simultaneously (at most four angles), or the chain is translated or rotated as a rigid object.

We say that Γ is in *Rim Normal Form* (denoted RNF), if all its joints lie on ∂S.

We say that Γ is in *Ordered Normal Form* (denoted ONF), if Γ is in RNF and moving from A_0 toward A_n along Γ is always either clockwise or counterclockwise.

Consider two line segments γ_1 and γ_2 which intersect at q and $\angle\gamma_1\gamma_2$ is in $[\pi, 2\pi]$. Let ρ be the line segment which starts at q and divides the angle $\angle\gamma_1\gamma_2$ into two angles $\angle\gamma_1\rho$ and $\angle\rho\gamma_2$ in such a way that $\angle\gamma_1\rho$ is in $[\pi/2, \pi]$. Initial configuration of $\Gamma[1,2,3]$ is defined as follows: A_1 placed at point p on line segment γ_1, A_2 at q and A_3 at point r on line segment γ_2. By this assumption we can define our movement in a concave region as follows.

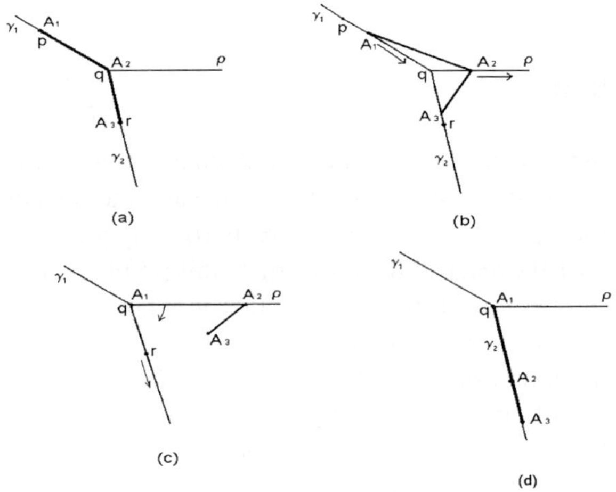

Fig. 1. *(a): Initial configuration of Γ, (b): middle-joint-up motion, (c): front-link-forward motion, (d): final configuration of Γ.*

Definition 1. The *mot(A_1,A_2,A_3,ρ)* movement changes the initial configuration of $\Gamma[1,2,3]$ to a final configuration by which Γ lies on γ_2. This is done by two consecutive motions (Fig 1):

– *Middle-joint-up(A_1,A_2,A_3,ρ)*: moves A_2 along ρ away from q until A_1 reaches q. During the movement A_1 remains on γ_1.
– *Front-link-forward(A_1,A_2,A_3,ρ)*: fixes A_1 at q then by turning counterclockwise A_2A_3 about A_2 and turning clockwise Γ about A_1 brings down A_3 on γ_2 (if not already there). To straighten Γ, it moves A_3 along γ_2 away from q.

It is possible to show that the mot(A_1,A_2,A_3,ρ) movement can be done in finite number of simple motions. This will be done by showing how each of middle-joint-up motion and front-link-forward motion can be done in finite number of simple motions (see [5] and[6]).

A T-shaped rectilinear S with boundary ∂S can be considered as the union of two rectangles S_1 and S_2, with boundaries ∂S_1 and ∂S_2, and with sides s_1, s_2 and s_3, s_4 respectively, attached together via the side ρ (Fig 2).

Suppose $p \in S_2$ and $\Gamma[0,1,...n]$ confined within S_1 and $|\Gamma| < \mathrm{Min}\{\frac{\sqrt{2}}{2}s_1, |\rho|\}$, then A_n can reach the point p using the following algorithm.

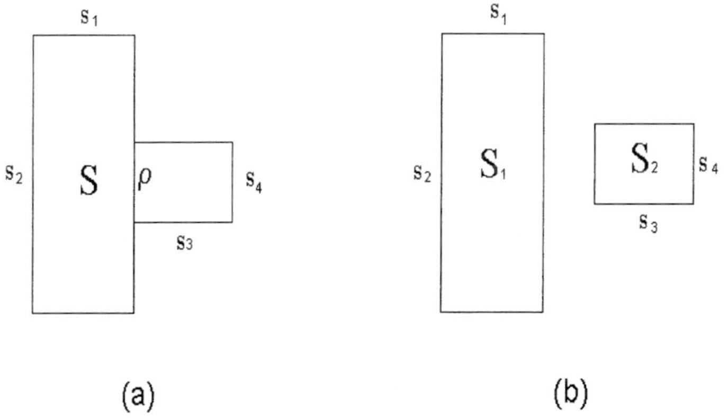

Fig. 2. *(a):A T-shaped rectilinear, (b):T-shaped rectilinear can be considered as the union of two rectangles with sides $s_1, s_2 and s_3$, s_4.*

Algorithm 1: Move all joints of Γ to ∂S_1, then move link $A_{n-1}A_n$ inside S_2 and bring joints A_{n-2}, A_{n-3},...inside S_2 consecutively, using mot(A_{i-1},A_i,A_{i+1},ρ) until p is reached by A_n.

Sketch of proof: We introduce an algorithm to bring A_n to the point p using O(n^2) simple motions, in the worst case. The algorithm consist of two cases:

$$d(p,v_1) \geq |A_{n-1}A_n| \text{ and } d(p,v_1) < |A_{n-1}A_n|.$$

In case 1, i.e $d(p,v_1) \geq |A_{n-1}A_n|$, A_n reaches p in three steps. In the first step we place all joints of Γ on ∂S_1 and reach v_1 by A_n. It takes O(n).

In the second step we move link $A_{n-1}A_n$ inside S_2, then define k_0 = min {k $|d(p,v_1) \geq \sum_{i=k+1}^{n} l_i$} and use mot($A_{i-1}, A_i, A_{i+1}, \rho$), $k_0 < i \leq n-1$ to bring joints $A_{n-2}, A_{n-3},...,A_{k_0}$ inside S_2 consecutively.

Finally p is reached during mot($A_{k_0-1}, A_{k_0}, A_{k_0+1}, \rho$). This step takes O($n^2$) in the worst case.

In the third step, if k=0, we bring the whole Γ inside S_2, and then reach the point p (Fig 3-a).

In case 2, i.e $d(p,v_1) < |A_{n-1}A_n|$, p is close to v_1 and the technique of case 1 is not applicable. In this case we present three subcases (Fig 3-b, c and d). It has been shown that each case takes O(n) time to reach p. For more details see [5, 6].

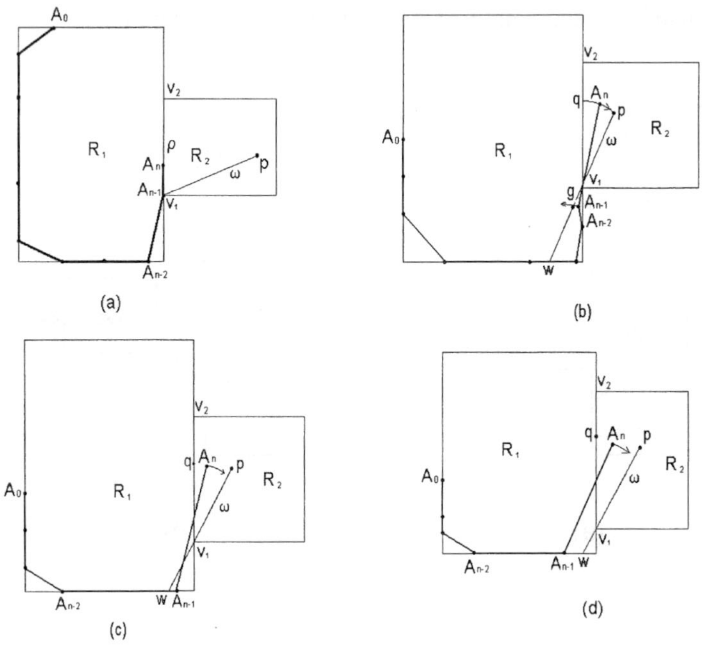

Fig. 3. *(a):* $d(p,v_1) \geq |A_{n-1}A_n|$, *(b):* A_{n-1} *belongs to the same edge as* v_1, *(c):* A_n *and* A_{n-1} *are in both sides of* ω, *(d):* A_n *and* A_{n-1} *are in the same side of* ω.

3 Main Result

The main result is; Moving the chain in a T-shaped region from a given configuration to a final configuration in which A_n reaches p.

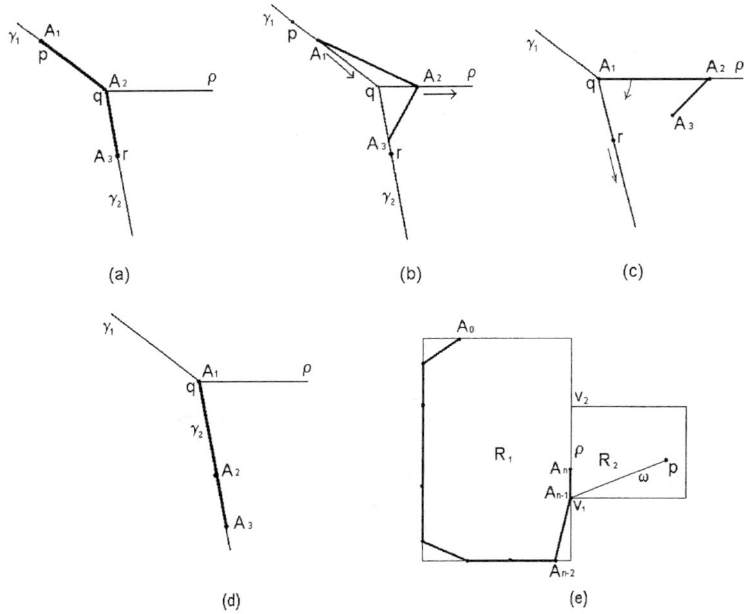

Fig. 4. *q is the farthest distance of A_i from $\partial S_1 \backslash \rho$*

Algorithm 2: Assume Γ is an n-link chain with $|\Gamma| < min\{s_1, s_2, s_3, s_4, \frac{|\rho|}{2}\}$ confined within T-shaped rectilinear S and p a point in S, then p is reachable by the end point of Γ.

We will present a few results that will be used in the explenation of the algorithm.

Assume $A_{i+1} \in S_2$, $\Psi = \partial S_1 \setminus \rho$,

$$q_i = min\{\text{Farthest distance of } A_i \text{ from } \psi, |A_i v_i|, |A_i v_2|\} \text{ (refer to Fig 4)},$$

and i is the lowest-indexed joint such that $A_i \in S_1 \backslash \rho$ and $A_{i+1} \in S_2$, and assume j is a index such that,

$$\Sigma_{m=j+1}^{i} l_m < q_i \leq \Sigma_{m=j}^{i} l_m,$$

(if j=i define $\Sigma_{m=j}^{i} l_m = \Sigma_{m=j+1}^{i} l_m = 0$).

Lemma 1. Assume $|\Gamma| < min\{s_1, s_2\}$, $A_0 \in S_1$, i is lowest-indexed joint such that $A_i \in S_1 \backslash \rho$ and $A_{i+1} \in S_2$ (at least i=0). For some j, as defined before, it is possible to bring $\Gamma[0,..j]$ to RNF or straighten $\Gamma[0,..i]$ with O(n) simple motion.

Proof: Bring the joints of $\Gamma[0,...j]$ one by one to ∂S_1 in decreasing order of their indices or straighten $\Gamma[0,..i]$ in two steps. In the first step A_j is brought into ∂S_1 or $\Gamma[0,..i]$ is straightened. In the second step $\Gamma[0,..j]$ is brought to ∂S_1.

Step 1: Let A_{i-1} and A_{i-2} move, turning A_{i-1} about A_i, keeping A_i and $\Gamma[0,...i-3]$ fixed (if i=2, let A_1 and A_0 move, if i=1, let A_0 move and if i=0 there is nothing to do) so that A_{i-1} gets close to ∂S_1. One of the following events occurs:

A_{i-1} hits ∂S_1 (Fig 5-a), A_{i-2} hits ∂S_1 (Fig 5-b), or the joint angle A_{i-2} straightens (Fig 5-c).

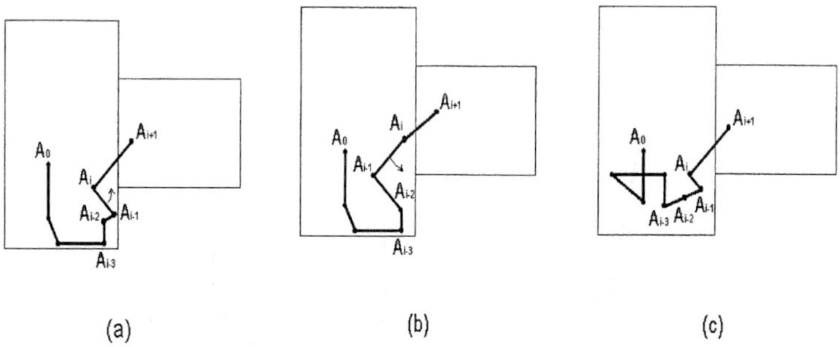

Fig. 5. *(a): A_{i-1} hits ∂S_1, (b): A_{i-2} hits ∂S_1, (c): angle A_{i-2} straightens.*

In cases A_{i-1} or A_{i-2} hits ∂S_1, move to the step 2, in case the joint angle A_{i-2} straightens begin again with L_{i-1} replaced by line segment $A_{i-1}A_{i-3}$ and A_{i-2} by A_{i-3}. Unless for some $k < i-3$, A_k hits ∂S_1, the angles will be successfully straighten, after which repeatedly apply the same procedure to successively lower-indexed joint angles. If all joint angles become straight, the chain $\Gamma[0,...i-1]$ is simply turned to the direction of the A_i-to-A_{i-1} until $\Gamma[0,..i]$ straightens or A_0 hits ∂S_1. This step takes O(n) simple motion at most.

Step 2: Consider for some $k <= j$, A_k lies on ∂S_1, first bring the joints of $\Gamma[0,...k]$ one by one to ∂S_1 in decreasing order of their index joints, then bring the joints between A_k and A_j (if $k \neq j$) to ∂S_1 (see [4]). This step also takes O(n) simple motion.

Lemma 2. Assume $|\Gamma| < min\{s_1, s_2, s_3, s_4, \frac{|\rho|}{2}\}$, $A_0 \in S_1$ and i is lowest-indexed joint such that $A_i \in S_1 \backslash \rho$ and $A_{i+1} \in S_2$ (at least i=0) and assume for some $j <= i$, $\Gamma[j,...i]$ is straight and $A_j \in \partial S_1$, if $j \neq 0$. It is possible to bring A_{i+1} to S_2.

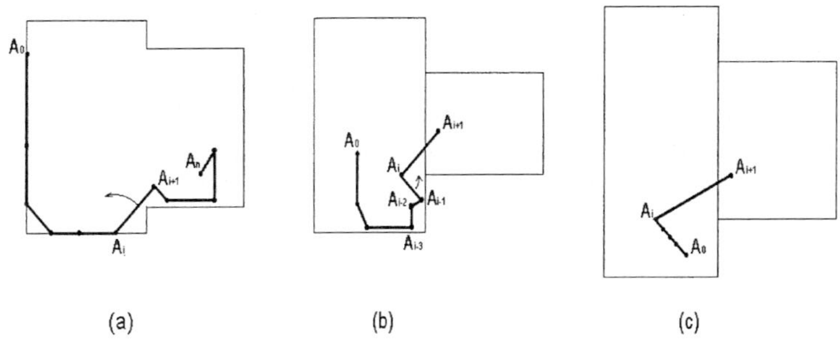

Fig. 6. *(a): turn A_{i+1} about A_i, (b): move A_{i+1} along λ toward the initial place of A_i, (c): A_i is simply rotated about A_0*

Proof: W.l.g. assume v_1 is farthest point of ρ from p. There are tree cases to consider as follows:

In case $j = i$, i.e. $A_i \in \partial S_1$, turn A_{i+1} about A_i until A_{i+1} exit S_2 and enter S_1. By assumption ($|\Gamma| < \frac{|\rho|}{2}$) during rotation A_{i+1} does not hit ∂S_2 (Fig 6-a).

In case $j < i$, i.e. $A_i \notin \partial S_1$, let λ be the straight line trough A_iA_{i+1}. MoveA_{i+1} along λ toward the initial place of A_i letting joint A_i move, keeping the joints between A_i and A_j straight, and fixing A_j. If A_{i+1} exits S_2, we are done. If A_i hits ∂S_1, do the same as the case j=i. (Fig 6-b).

In case j=0, $\Gamma[0,...i]$ is straight and $A_0 \notin \partial S_1$, A_i is simply rotated about A_0. All these cases takes a finite number of simple motion.

The algorithm consists of four consecutive steps as follows.

Step 1: Fix A_i and depending on distance of A_i from Ψ, for some j (as defined), bring $\Gamma[0,...j]$ to ∂S_1 or straighten $\Gamma[0,...i]$, using lemma 1.

Step 2: Move all chains of Γ to S_1 by repeatedly applying step.1 and lemma 2 successively.

Step 3: Move Γ to ONF in S_1 using Kantabutra Algorithm (see [4]).

Step 4: Move chain to a configuration in which A_n reaches p using algorithm 1. All these steps take $O(n^2)$ simple motions together at most.

References

1. J. Hopcroft, D. Joseph, and S. Whitesides. *Movement problems for 2-dimensional linkages*, SIAM J. Compt., 13: pp. 610-629, 1984.

2. J. Hopcroft, D. Joseph, and S. Whitesides. *On the movement of robot arms in 2-dimensional bounded regions*, SIAM J. Compt., 14: pp. 315-333, 1985.
3. V. Kantabutra. *Motions of a short-linked robot arm in a square.* Discrete and Compt. Geom., 7:pp. 69-76, 1992.
4. V. Kantabutra. *Reaching a point with an unanchored robot arm in a square.* International jou. of comp. geo. & app., 7(6):pp. 539-549, 1997.
5. A. Mohades and M. Razazi, *Reachability On a Region by two Attached Squares*, ICCS'01, USA, Lecture Notes in Computer Sc., Vol. 2120, 2001.
6. A. Mohades and M. Razazi, *Reachability Problem in T-shaped Rectilinear Polygon*, Proc. 2th Fun With Algorithms, Italy, pp.199-213, 2001. 2001.

Generalized Coordinates for Cellular Automata Grids

Lev Naumov

Saint-Peterburg State Institute of Fine Mechanics and Optics, Computer Science Department, 197101 Sablinskaya st. 14, Saint-Peterburg, Russia
levnaumov@mail.ru

Abstract. After some cellular automata basics here is stated an approach of universal data organization for different automata grids. It allows to identify any cell with only one non-negative integer index even for multidimensional grids

1 Introduction

Cellular automata are simple models, which are used for studying complex systems behavior in different fields of science. They found applications in physics, mathematics, computer sciences, chemistry, psychology, meteorology, social sciences and others.

These automata are discrete dynamic systems, which work can be completely described with the terms of local interactions [1, 2]. One of the most important causes of interest to cellular systems is that they form the common paradigm of parallel computations as Turing machines do for the consecutive computations [3].

2 Cellular Automata Basics

Here we will use the following theoretical formalism of cellular automata: cellular automaton A is a set of four objects $A = <G, Z, N, f>$, where

- G – set of cells, automatons workspace and data storage. It is named as "grid";
- Z – set of possible cells states;
- N – description of cells neighborhood. Neighborhood is a set of cells which have an influence on the currently considered one;
- f – next-state function. It can operate so $Z^{|N|+1} \to Z$ (if cells state on the current step has an influence on its state in the next moment, so it can be said that such automaton have cells "with memory") or so $Z^{|N|} \to Z$ (only cells neighbors are significant, then cells can be named "memoryless") [4].

The difference between these two variants of next-state functions operation isn't important on practice. It is just theoretical. For real tasks next-state function has been generally defined as a computable program on a programming language (may be, a

P.M.A. Sloot et al. (Eds.): ICCS 2003, LNCS 2658, pp. 869–878, 2003.

specialized language for cellular automata next-state function definition) [1], instead of a mathematical (or logical) expression. It can be also named as "rules" of cellular automaton.

The set Z is to be finite. Although, for example, in physics task solving it can represent continuum, an interval of possible physical magnitude values, but on practice, any floating-point variable has a finite set of values.

Grid G is an array of cells, each of them can contain a value (or "be in state") from a set Z. Grids can be one-dimensional, two-dimensional or multidimensional, it depends on task. However here we will view two-dimensional grids, composed of regular polygons. There are only three such grids: a grid of triangles (Fig. 1), a grid of squares (Fig. 2) and a grid of hexagons (Fig. 3).

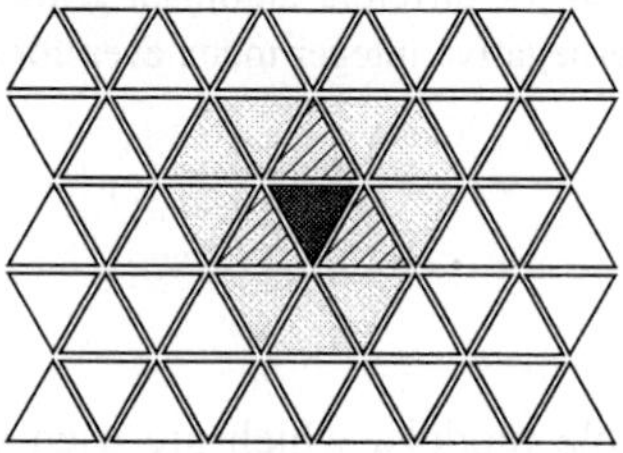

Fig. 1. Grid of triangles and cells nearest neighbors on it

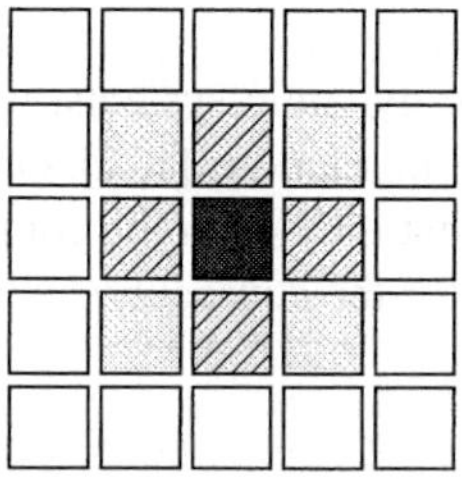

Fig. 2. Grid of squares and cells nearest neighbors on it

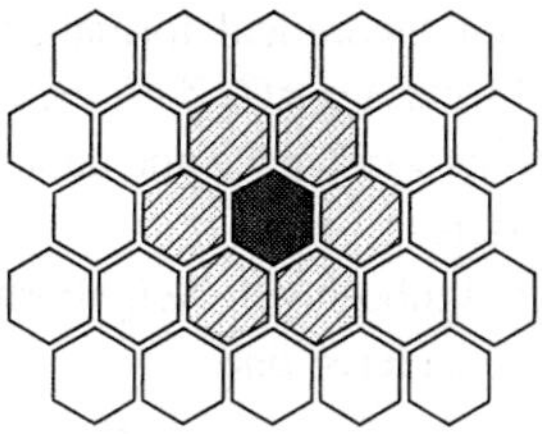

Fig. 3. Grid of hexagons and cells nearest neighbors on it

Cells neighborhood N is to be common for all of grids cells. So, after application of a metrics for a grid, this set can be determined as a collection of displacements relative to the currently considered cell. As usual neighborhood of a cell is a subset of its nearest cells. On Fig. 1-3 they are shown for all three considered grids. The involved cell is dark, its nearest neighbors are darken a little. The cells, that have a common edge with the involved one can be named as a "main neighbors" (they are showed with the hatching). The set of actual neighbors of the cell a, which can be found according to N, we will denote as $N(a)$.

Cellular automata basic properties can be formulated as following:

- *Laws are local.* All cells neighbors (elements of neighborhood) are to be on a finite distance from it;
- *System is similar for all the cells.* Neighborhood and laws are always the same. There are no two cells on the grid, which can be distinguishable by the landscape;
- *All cells get their new values simultaneously, at the end of the timestep*, after all new values were calculated for all grids cells.

In previous narration we several times mentioned that there is a use in metrics for a grid of cellular automata. Let us assume the nearest neighbors of the cell to be the cells "of first ring" of the involved one. For the current cell, the cells of the i-th ring are the cells of the first ring of all cells, which are the member of (i–1)-th ring, excluding cells of (i–1)-th and (i–2)-th rings. Formally, if $R(a, i)$ is a set of cells of i-th ring of cell a, then we can write

$$R(a,i) = \{b | \exists c : b \in R(c,1), c \in R(a,i-1), \\ b \notin R(a,i-1), b \notin R(a,i-2)\} \quad (1)$$

On Fig. 4-6 cells of different rings are showed with the different hatching or color.

It is useful to assume $\{a\} = R(a, 0)$. The important property of ring concept is that if a is in $R(b, i)$ then b is in $R(a, i)$.

So now we can denote a distance function [5] between cells a and b, $D(a, b)$. For example, it could be done as follows:

$$D(a,b) = i : a \in R(b,i) \quad (2)$$

The notion of ring may be generalized for multi-dimensional grids. And the definition of distance function, given by formula (2), would remain the same.

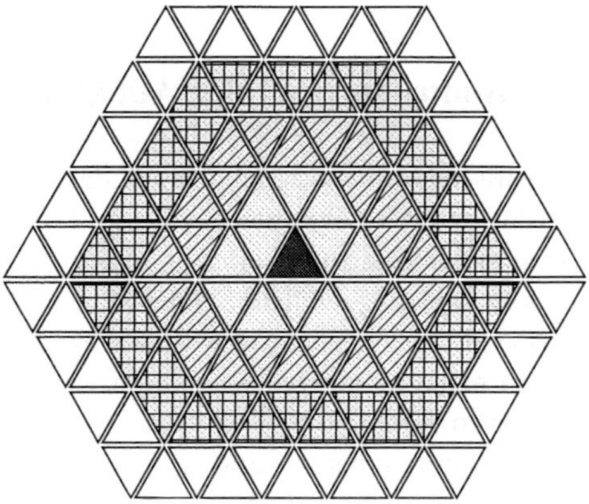

Fig. 4. First four rings for the cell of the grid of triangles

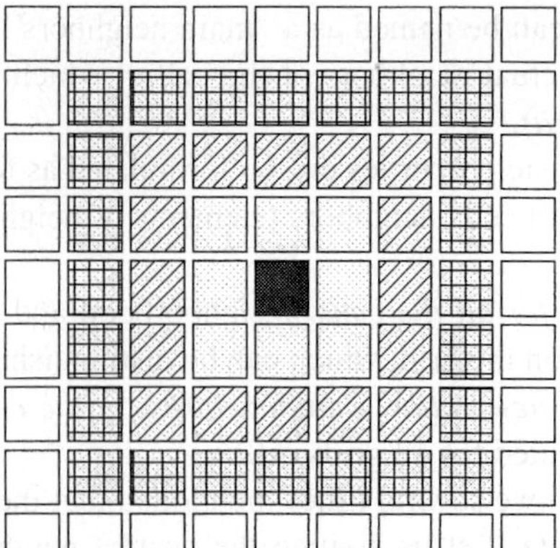

Fig. 5. First four rings for the cell of the grid of squares

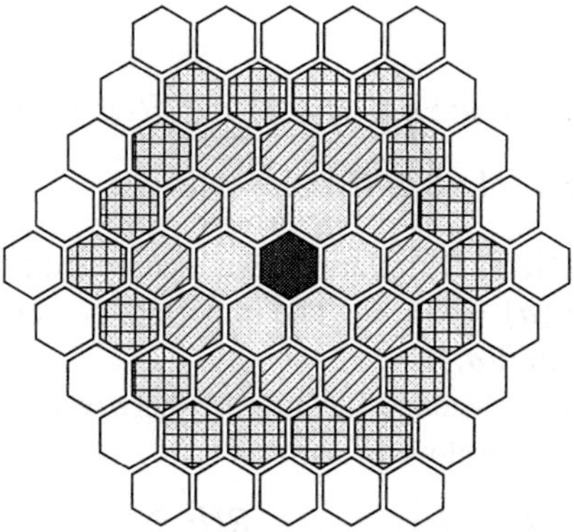

Fig. 6. First four rings for the cell of the grid of hexagons

3 Generalized Coordinates for Cellular Automata Grids

When the task is to deal with the n-dimensional array of data with the help of cellular automaton, the obvious way is to use the n-dimensional grid. Each cell can be defined with the help of n-dimensional coordinate (Cartesian in most cases). Usage of generalized coordinates is the way to manage with one-dimensional array always [6].

The idea is to enumerate all cells of the grid with, for example, non-negative integer numbers, which we will name as generalized coordinates. It must be done without any blanks. Each number is to have one and only one corresponding cell. Last sentence is correct with the assumption that the grid is infinite. In fact, of course, generalized coordinates have to be bounded above. But the bound can move aside if needed and new cells can be appended to the end of the array.

The n-dimensional arrays are very suitable for quick foundation of the nearest neighbors. If it would be shown that there is still an ability to find cells nearest neighbors, after applying generalized coordinates, it would mean that they can be used for cellular automata tasks solving. In case when N contains not nearest neighbors, they would be applicable too. Then cells could be found as nearest neighbors of nearest neighbors and so on.

The ways of associating cells with generalized coordinates can be different. The main aim is to introduce them in the way, which allows to get cells neighbors as fast as possible. Lets look at two examples of such ways. First – method of the spiral coordinates introduction for the grid of hexagons. Second is applicable for the grid of triangles, but it is based on the first approach. Other methods of general coordinates introduction and their introduction for other grids remain out of this paper [6].

3.1 Spiral Generalized Coordinates by the Example of the Grid of Hexagons

The rule of enumeration of cells is very simple: chose any zero-cell and enumerate in each its ring, for example, clockwise. Possible result is shown on Fig. 7.

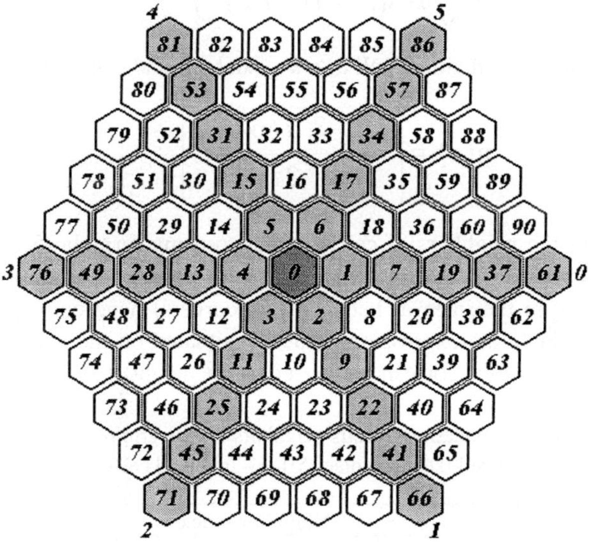

Fig. 7. Spiral generalized coordinates for grid of hexagons. Rings corner cells and zero-cell are darken here

Definition of a cell, which gets zero coordinate is not important. The system is invariant relative to its position.

The task is to find cells nearest neighbors now. Lets enumerate first rings cells with indexes from 0 to 5 as it is shown of Fig. 8.

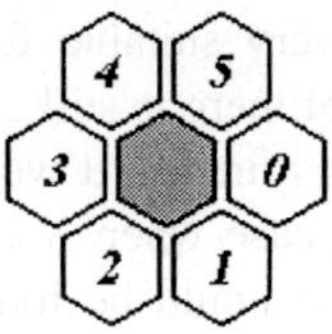

Fig. 8. Cells neighbors indexes for the grid of hexagons

Let $N_i(a)$ be the i-th neighbor of the cell a (i is from 0 to 5). We may find just four of six neighbors, because of the formulae (3), but we will show how to find all of neighbors without recursion.

$$N_0(a) = N_5(N_1(a)) = N_1(N_5(a)) \qquad (3)$$
$$N_3(a) = N_4(N_2(a)) = N_2(N_4(a))$$

On Fig. 7 cells that are the corners of the rings were darked. Each ring has 6 corners, which are also enumerated from 0 to 5 as shown on Fig. 7. Let $C_{i,j}$ be the coordinate of j-th corner of the i-th ring (j is from 0 to 5). It could be shown [6] that

$$C_{i,j} = 3(i^2 - i) + 1 + i \cdot j \qquad (4)$$

The number of ring containing cell a, $n(a)$, can be found using formula (5)

$$n(a) = \left\lceil \frac{\sqrt{1 + 8\lceil a/6 \rceil} - 1}{2} \right\rceil \qquad (5)$$

Calculation of the neighbors of cell x comes to using formula (5). So you can determine the position of the cell inside the ring. All possible 13 positions (it could be any corner cell or in the interval between two corners) have corresponding rows in table 1. Then evaluate the expression, which is situated on the necessary row of the table, in the column, which corresponds to a definite neighbor index.

Table 1. Expressions for foundation of cells neighbors for the grid of hexagons (n – the number of ring, which contains the cell (see formulae (5)), x – involved cell)

Position	Neighbor index					
	0	1	2	3	4	5
$C_{n,0}$	$C_{n+1,0}$	$C_{n+1,0}+1$	$x+1$	$C_{n-1,0}$	$C_{n-1,0}-1$	$C_{n+2,0}-1$
$(C_{n,0};\ C_{n,1})$	$C_{n+1,0}-C_{n,0}+x$	$C_{n+1,0}-C_{n,0}+x+1$	$x+1$	$C_{n-1,0}-C_{n,0}+x$	$C_{n-1,0}-C_{n,0}+x-1$	$x-1$
$C_{n,1}$	$C_{n+1,1}-1$	$C_{n+1,1}$	$C_{n+1,1}+1$	$x+1$	$C_{n-1,1}$	$x-1$

$(C_{n,1};$ $C_{n,2})$	$x-1$	$C_{n+1,1}-C_{n,1}$ $+x$	$C_{n+1,1}-C_{n,1}$ $+x+1$	$x+1$	$C_{n-1,1}-C_{n,1}$ $+x$	$C_{n-1,1}-C_{n,1}$ $+x-1$
$C_{n,2}$	$x-1$	$C_{n+1,2}-1$	$C_{n+1,2}$	$C_{n+1,2}+1$	$x+1$	$C_{n-1,2}$
$(C_{n,2};$ $C_{n,3})$	$C_{n-1,2}-C_{n,2}$ $+x-1$	$x-1$	$C_{n+1,2}-C_{n,2}$ $+x$	$C_{n+1,2}-C_{n,2}$ $+x+1$	$x+1$	$C_{n-1,2}-C_{n,2}$ $+x$
$C_{n,3}$	$C_{n-1,3}$	$x-1$	$C_{n+1,3}-1$	$C_{n+1,3}$	$C_{n+1,3}+1$	$x+1$
$(C_{n,3};$ $_{n,4})$	$C_{n-1,3}-C_{n,3}$ $+x$	$C_{n-1,3}-C_{n,3}$ $+x-1$	$x-1$	$C_{n+1,3}-C_{n,3}$ $+x$	$C_{n+1,3}-C_{n,3}$ $+x+1$	$x+1$
$C_{n,4}$	$x+1$	$C_{n-1,4}$	$x-1$	$C_{n+1,4}-1$	$C_{n+1,4}$	$C_{n+1,4}+1$
$(C_{n,4};$ $C_{n,5})$	$x+1$	$C_{n-1,4}-C_{n,4}$ $+x$	$C_{n-1,4}-C_{n,4}$ $+x-1$	$x-1$	$C_{n+1,4}-C_{n,4}$ $+x$	$C_{n+1,4}-C_{n,4}$ $+x+1$
$C_{n,5}$	$C_{n+1,5}+1$	$x+1$	$C_{n-1,5}$	$x-1$	$C_{n+1,5}-1$	$C_{n+1,5}$
$(C_{n,5};$ $C_{n+1,0}$ $-1)$	$C_{n+1,5}-C_{n,5}$ $+x+1$	$x+1$	$C_{n-1,5}-C_{n,5}$ $+x$	$C_{n-1,5}-C_{n,5}$ $+x-1$	$x-1$	$C_{n+1,5}-C_{n,5}$ $+x$
$C_{n+1,0}$ -1	$C_{n+2,0}-1$	$C_{n,0}$	$C_{n-1,0}$	$C_{n,0}-1$	$x-1$	$C_{n+2,0}-2$

In [6] formalism of spiral generalized coordinates is applied also to grid of triangles and squares.

3.2 Example of Generalized Coordinates for Grid of Triangles Based on Spiral Generalized Coordinates for Grid of Hexagons

Another example of introduction of generalized coordinates will be made for the grid of triangles. It is based on the previous method.

Every hexagon can be presented as if it consists of the six triangles how it is shown on Fig. 9.

Fig. 9. Hexagon consists of six triangles

Fig. 10. Generalized coordinates for grid of triangles based on spiral generalized coordinates for grid of hexagons (small numbers – triangles coordinates, big numbers – hexagons coordinates)

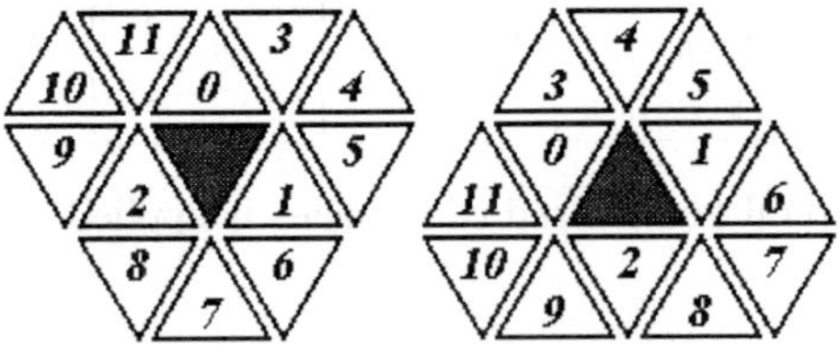

Fig. 11. Cells neighbors enumeration for the grid of triangles

Generalized coordinates may be introduced in this way: after applying of generalized coordinates for such hexagons, we can get each triangles coordinate as a coordinate of hexagon, multiplied by 6 with the added index of triangle inside the hexagon (these indexes are presented on Fig. 9). The result is shown on Fig. 10 (note that grid of hexagons was turned there in comparison with Fig. 7). Neighbor cells can be easily computed for this method. Lets enumerate cells neighbors as showed on Fig. 11 (there are to cells orientation variants, so we need two numerations).

In this case the algorithm of computing of neighbor cells for cell x is following: let h equals residue of division of x by 6 and H equals integer part of division of x by 6. So h would be triangle index inside the hexagon and H would store hexagons generalized coordinate. In table 2 we adduced the accordance between h and the neighbor indexes (Fig. 11). Each cell can contain one or two numbers, separated with hyphen. If there is only one number then it means that desired cell has the specified index inside the same hexagon H. If there are two number then the first is neighbor index of hexagon (Fig. 8) to which neighbor cell belongs, and second is a triangle index in the founded hexagon. So all numbers are from 0 to 5.

Table 2. Indexes for foundation of cells neighbors for the grid of triangles

H	Neighbor index											
	0	1	2	3	4	5	6	7	8	9	10	11
0	5-3	1	5	5-2	0-5	0-4	2	3	4	4-2	4-1	5-4
1	0-4	2	0	0-3	1-0	1-5	3	4	5	5-3	5-2	0-5
2	1	1-5	3	0-4	0-3	1-0	1-4	2-1	2-0	4	5	0
3	2	2-0	4	1-5	1-4	2-1	2-5	3-2	3-1	5	0	1
4	5	3	3-1	0	1	2	2-0	2-5	3-2	3-0	4-3	4-2
5	0	4	4-2	1	2	3	3-1	3-0	4-3	4-2	5-4	5-3

We used rather successful neighbors numeration (Fig. 11). The success is that there is no need in two separate tables of different cells orientations.

Tests shows [6], that this way of introducing of generalized coordinates for the grid of triangles allows to calculate the nearest neighbor cells several times faster than a spiral variant for the triangle grid [6]. It is caused by complexity of a ring for the grid of triangles and by recursion which was used for spiral coordinates.

4 Resume

Usage of generalized coordinates can be rather useful and gives several opportunities:

1. Generalized coordinates provide a universal way of data storage for different grids. The only thing need to be changed for transformation from one cellular automaton to another one – the table of expressions of the cells nearest neighbors;

2. Grid may be easily enlarged if it is necessary. Appending cells to the end of the chain is much more easer than, for example, reallocation of multidimensional block structures. It can found application for the tasks like diffusion processes studying, when data propagate in all directions;

3. Serial data is easier to serialize and store;

4. Independency from the zero-cells position gives the opportunity to move it as it would be more useful. May be – to the center of activity, to operate with the smaller

numbers. May be – far from the activity, so the majority of calculations would be on the long edges of the rings, so some neighbors could be found by addition or subtraction of unity to cells coordinates (it is said for the spiral method);

5. Generalized coordinated is just a concept, so it could be adopted for the definite task, for example, according to the internal symmetry of system.

Cellular automata are the models of parallel computations with infinite extent of parallelism. It could be achieved with the help of infinitive amount of Turing machines. Using the generalized coordinates the system with infinite extent of parallelism can be emulated with the help of finite quantity of Turing machines. A single machine used to work with data storage (the tape) and finite quantity of machines is used for neighbors calculation and synchronization. Their exact amount depends on realization (on functions programmed in machines for neighbor cells calculations).

Main disadvantage of the offered approach is that calculations of neighbors are, obviously, slower than, for example, for the Cartesian case.

For investigations of cellular automata a project CAME&L was started in Saint-Petersburg State Institute of Fine Mechanics and Optics. CAME&L is Cellular Automata Modeling Environment & Library. It is a Windows-based software package, that is desired to be simple, extensible workspace for complicated cellular calculations. Each automaton is projected to be represented by a trinity of components:

- grid – that implements visualization of grid and cells navigation;
- metrics – that provides the relationship of neighborhood and, optionally, it can allow to calculate distance between cells (may be useful for rules writing);
- data – that maintains data storage and some aspects of data visualization.

This union is to cooperate with two other components:

- rules – defines the rules of automaton and controls the iterations;
- analyzer – allows to keep an eye on definite properties of automaton, draw diagrams, draw up the reports and all of this kind.

A part of CAME&L, named CADLib (Cellular Automata Developing LIBrary) is designing to present an easy-to-use and rich set of instrument for developers and researchers. Its base is a class hierarchy for further enlarging by the users. CAME&L is planned to be provided with the set of examples of different classes and some components.

References

1. Toffoli, T., Margolus, N.: Cellular Automata Machines. Mir (World), Moscow (1991)
2. Sarkar, P.: A Brief History of Cellular Automata // ACM Computing Surveys, Vol. 32. N1. (2000)
3. von Neumann, J.: Theory of Self-Reproducing Automata. University of Illinois Press (1966)
4. Naumov, L., Shalyto, A.: Cellular Automata: Realization and Experiments. In preparation
5. Korn, G., Korn, T.: Mathematical Handbook for Scientists and Engineers. McGraw-Hill Book Company, London (1968)
6. Naumov, L.: Applying and Using of Generalized Coordinates in Cellular Automata with the Grid of Triangles, Squares and Hexagons. In preparation

On Generating Random Network Structures: Trees*

Alexey S. Rodionov[1] and Hyunseung Choo[2]

[1] Institute of Computational Mathematics and Mathematical Geophysics
Siberian Division of the Russian Academy of Science
Novosibirsk, RUSSIA +383-2-396211
alrod@rav.sscc.ru

[2] School of Information and Communication Engineering
Sungkyunkwan University
440-746, Suwon, KOREA +82-31-290-7145
choo@ece.skku.ac.kr

Abstract. Random trees (RTs) are widely used for testing various algorithms on tree-type networks and also for generating connected graphs similar to real nets. While random topologies based on RTs are generally accepted as network models, the task of their generation is almost unexplored. In this paper we discuss the set of basic algorithms for generating random trees. The fast algorithms with proven properties are presented for generating random trees under conditions for given restrictions, such as a limited node degree, fixed node degrees, and different probabilities of edge existence. Generating random graphs similar to physical networks are underway.

1 Introduction

In this paper we present the set of basic algorithms for generating random trees (RTs). RTs are used for testing various algorithms on tree-structural networks and as a base for further algorithms for generating connected graphs. Random graphs are widely used as a model of communication networks, usually for the testing of control algorithms [1,2,3]. Using real network structures for an algorithm testing is impossible with a simple reason: to achieve reliable results the large number of different structures with given properties is necessary while actual data for real network structures are usually inaccessible and the number of such real structures is small. While random trees and graphs are widely used as a network model, the task of their generation is almost unexplored. It seems that most authors who use random trees and graphs as a model consider the task as an obvious one. Still there are a lot of problems in the task of random graphs generation, especially when complex limitations are loaded on the graph structure. The usual requirements are to guarantee the uniformity on the given

* This paper was partially supported by BK21 program. Dr. H.Choo is the correspondent author

P.M.A. Sloot et al. (Eds.): ICCS 2003, LNCS 2658, pp. 879–887, 2003.

space of graphs (to the enumeration) and attainability of all graphs from the given space. Hereafter we discuss some base approaches for the generation of RTs and present some effective algorithms. The rest of the paper is organized as follows. Section 2 is devoted to the base approaches, notations and concepts. In Section 3 we discuss about the generation algorithms of RTs for equal probabilities of edge existence. Section 4 presents the algorithm for generating RTs with different probabilities of edges existence. Section 5 is a brief conclusion.

2 Preliminaries

The variable types are clear from the context usually, or are stated in comments. The following procedures are used in the algorithms descriptions:

- `INC(n)`, `DEC(n)` – increment and decrement operators;
- `FindPlace(x,V,OK)` – integer function, returns the index of the first element in vector `V` equal to `x`. `OK` is TRUE if such element is found and FALSE in the opposite case;
- `remove(x,V)` – removes element `x` from vector `V` (the last element replace `x`). If `x` is not belongs to `V` then no action is done;
- `rand()` – random number uniformly distributed on $(0, 1)$;
- `random(n,m)` – integer random number uniformly distributed on $[n, n+1, \ldots, m]$, $n \leq m$;
- `random(P,n,m)` – integer random number distributed on $[n, n+1, \ldots, m]$, $n \leq m$ with probabilities $p_i = P(x_i = n + i - 1)$, $i = 1, 2, \ldots, m - n + 1$;
- `Call Alg(Input,Output)` – Call the algorithm or procedure `Alg` with input `Input` and output `Output`.

Let us denote arbitrary non-oriented graph with N nodes and M edges and without multiple edges as $G(N, M)$. Several approaches can be used for the random graphs generation. First and most inefficient approach (but still widely used) is the *trial method.* A random graph with given number of nodes (and possibly edges) is generated and then its properties are checked to answer the given limitations. As the number of possible $G(N, N-1)$ graphs (with all possible enumerations of nodes) is

$$S = \binom{\frac{N(N-1)}{2}}{N-1} \frac{N(N-1)}{2}! = \frac{\left[\frac{N(N-1)}{2}!\right]^2}{(N-1)!\left[\frac{N(N-1)}{2} - N + 1\right]!}, \tag{1}$$

while the total number of all N-node trees (with all possible enumerations of nodes also) is only N^{N-1}, the trial method is obviously ineffective for RT generation.

Next basic approach is the "sequential growth": each new element is added to the generated graph without violating the limitation and with guaranteeing its properties. Once more there are two possibilities: either to use trial or "smart" method on each step. We use the sampling without repetition for trial method.

As a "smart" method we use the *method of admissible choice (MAC)*. This is the method in which on each step the choice is done only from the set of elements that 1) keeps the graph in a given class and 2) does not allow to break any limitation. Let us denote:

- X_i – set of nodes in the generated graph on the i-th step;
- E_i – set of edges in the generated graph on the i-th step;
- A_i – set of edges that are allowed to be added to the generated graph on the i-th step;
- In_i – set of edges that are to be added to A_i before the $(i+1)$-st step;
- Ex_i – set of edges that are to be excluded from A_i before the $(i+1)$-st step;
- v_i – i-th node (vertex) of a graph;
- e_{ij} – edge that connects v_i and v_j;
- $X(G), E(G)$ – sets of nodes and edges of the graph G;
- $C(N)$ – the complete graph with N nodes.

Thus, $A_{i+1} = A_i \setminus Ex_i \cup In_i$. Some limitations can lead to the empty A_{i+1}. In this case the rollback is needed by one step. If e_{st} is the last edge chosen before the deadlock then it transfers from A_i to Ex_i.

The simple example is to generate random trees in the case of equal probability of an edge existence. If the trial method is used, then after each new edge selection the test is done if addition of this edge makes a cycle or multi-edge in a tree. If yes then the edge is rejected and choice repeats. If MAC is used then first we choose uniformly any node as a root, let it be v_f. Thus

$$A_1 = \{e_{f\,1}, e_{f\,2}, \ldots, e_{f\,f-1}, e_{f\,f+1}, \ldots, e_{f\,N}\}. \quad (2)$$

If on the i-th step the edge e_{st} is selected (t-th node is new) then

$$Ex_i = \{e_{ct} | v_c \in X_i\}, \quad In_i = \{e_{td} | v_d \in \bar{X}_i\}. \quad (3)$$

3 Algorithms

First we formally describe the algorithm for generating random trees sketched above and then discuss about more complex cases of random trees with limitations.

3.1 The Algorithm for Generating Random N-Vertex Trees

For generating random N-vertex trees (the isomorphism is possible) the following fast algorithm is offered:

Fast Random Tree Generating Algorithm.[1]
Input: N is the number of nodes.
Output: $OU[1..N-1]$ and $IO[1..N-1]$ are the vectors of numbers starting and ending nodes for the resulting tree edges (i-th edge connects node OU_i with IO_i).

[1] Some algorithms discussed in this paper including this one were first programmed for the application package Graph-ES, see [4,5,6]

```
for i:=1 to N do
   I[i]:=i;                     \\ Initial vector of node numbers
endfor;
l:=random(1,N);                 \\ Choose the root
K:=I[N]; I[N]:=I[l]; I[l]:=K;\\ Replacement with the last node
for m:=1 to N-1 do
   t:=random(1,m);              \\ New node
   s:=random(m+1,N);            \\ Node of "growth"
   OU[m]:=I[t]; IO[m]:=I[s]; \\ Edge registration
   K:=I[t]; I[t]:=I[N-m];       \\ Replacement: from N-m+1 to N -
   I[N-m]:=K;                   \\ nodes that are in a tree already
endfor.
```

Remark: It is clear that this algorithm needs only $2N-1$ calls to the random number generator and uses the minimum memory ($N+4$ integers for the case of equal probabilities and $N+1$ reals additionally in the case of different probabilities of edge existence).

The following theorem is proved for this algorithm.

Theorem 1. *Fast Random Tree Generating Algorithm creates any of N-vertex tree with equal probability* $1/N^{N-1}$.

Proof is rather obvious but tiresome. At first the equal probability of entire results is proved and then it is proved that the number of different results is equal to N^{N-1}. As it is the well-known number for different N-node trees (all renumbering included), the attainability is proved also. For the complete proof, refer to [7].

3.2 Generating Trees with the Restriction on a Maximum Node Degree

Let us have the limitation on the maximum degree of nodes in a tree: $deg(v_i) \leq Deg$. This leads to the new definition of Ex_i in comparison with the previous case:

$$Ex_i = \begin{cases} \{e_{ct}|v_c \in X_i\} & \text{if } deg(v_s) \leq Deg, \\ \{e_{ct}|v_c \in X_i\} \bigcup \{e_{sd}|v_d \in \bar{X}_i\} & \text{otherwise.} \end{cases} \tag{4}$$

The addition of new edges is similar to the case of the unconditional random tree generation, but the set of nodes that *are in the tree already* is divided into two parts: those with maximum degree (N_m) and those with degrees under it ($N_<$). New edge must be chosen to connect some free node with the one in $N_<$. The algorithm looks as follows:

Algorithm for Generating Trees with Limited Node Degrees (GTD).
Input: N is the number of nodes; Deg is the possible maximum degree of nodes.
Output: $OU[1..N-1]$ and $IO[1..N-1]$ are the vectors of starting and ending nodes for the resulting tree edges, respectively (i-th edge connect node OU_i with IO_i), $Degrees(1 \times N)$ is the vector for the degrees of nodes.

```
for i:=1 to N do
  I[i]:=i; Degree[i]:=0;    \\Initial vectors
endfor;
l:=random(1,N);             \\ Choose the root
R:=I[N];I[N]:=I[l];I[l]:=R;\\ Replacement with the last element
m:=1;                  \\ current numbers of nodes in a tree and
k:=0;                  \\ nodes restricted for further choice
while m<N do                \\ Main cycle
  s:=random(N-m+1,N-k);     \\ For admissible node in a tree
  t:=random(1,N-m);         \\ For new node
  IO[m]:=I[t]; OU[m]:=I[s];\\ Registration
  INC(Degree[I[t]]); INC(Degree[I[s]]);
 \\ Now let's Check for the maximum possible degree
 \\ If YES then restrict the node for future choice
  if Degree[I[s]]=Deg then
    I[s]:=I[N-k]; INC(k);
  endif;
  R:=I[t];      \\ Rearrange the working vector: from N-k+1 to
  I[t]:=I[N-m]; \\ N - nodes restricted for further selection
  I[N-m]:=R;
  INC(m);                    \\ Working step is done
endwhile.
```

Remark: There cannot be a deadlock in this algorithm: as $2 \leq Deg \leq N-1$, the subset $X_i^* = \{v_j | v_j \in X_i \& deg(v_j) < Deg\}$ is always non-empty (at least the last node added to the generated tree always has degree one). And A_{i+1} includes all the edges that connects these nodes with free nodes.

From this we can simply construct the algorithm for generating unbalanced binary tree (B-tree) with the number of nodes $N > 4$. Note that node degree in B-tree is limited to 3. So we can use the following generating scheme:

Algorithm for Generating B-trees.
Input: N is the number of nodes.
Output: $OU[1..N-1]$ and $IO[1..N-1]$ are the vectors of numbers starting and ending nodes of the resulting tree edges (i-th edge connect node OU_i with IO_i), *Root* is the number of root node.

```
N1:=random(1,N+1); N2:=N-N1+1;
Call GTD(N1,3,OU1,IO1,Degree1);
Call GTD(N2,3,OU2,IO2,Degree2);
Using vectors Degree1 and Degree2 the nodes s and t are chosen
from nodes with degree 1 in both trees, respectively;
Renumbering of nodes: the node s becomes node number N1 in
first tree, the node t becomes node number N1 in second tree.
Root:=N1;                         \\ Root is assigned
for i:=1 to N1-1 do
```

```
   IO[i]:=IO1[i];OU[i]:=OU1[i]; \\ First tree is the left part
endfor;
for i:=N1 to N1+N2-1 do        \\ Second tree is the right part
   IO[i]:=IO2[i]+N1-1; OU[i]:=OU2[i]+N1-1;
endfor.
```

Remark: If we assume the possibility of one-side B-tree then $N2$ can be 0. In this case we simply generate one tree with restriction $Deg = 3$ and then choose a root from all nodes with a degree of 1 or 2.

3.3 Generating Trees with Given Node Degrees

Now let us have the strict values for $deg(v_i)$ in the resulting tree. This is more complicated case, because there can be a deadlock in direct use of the basic algorithm. In fact, it is enough that $\sum_j deg(v_j|v_j \in X_i) = 2(i-1)$. For example, if some v_s and v_t both have degree one where v_s is selected as a root on the preliminary step, and e_{st} is the first choice, then we have the deadlock on the very first step of the algorithm. To avoid the deadlock we employ the following rule: each time when we have $\left(\sum_j deg(v_j|v_j \in X_i) = 2(i-1) - 1\right) \&(i < N-2)$ we add all edges that connects nodes from X_i with free nodes with prescribed degree one to the Ex_i.

Let us prove the correctness of this rule based on the proof by contradiction. Assume that $\sum_j deg(v_j|v_j \in X_i) = 2(i-1) - 1$. The deadlock will take place if and only if next node has degree one. Assume that all free nodes (their number is $N-i$) have the prescribed degree one. So the total sum of degrees by all nodes is $2(i-1)-1+N-i = N+i-3$ while for a tree it must be $2(N-1)$. We have a contradiction, so our rule is proved to be correct.

4 Generating Trees with Different Probabilities of Edge Existence

This case differs from the previous one only in the procedure of a new edge choice. In the case of equal probability the set A_i is virtual and choice is made between nodes, and now we need to have it in some form that allows us to make choice of new edge as a numbered pair of nodes. Note that in general case the sum of probabilities of the edge existence is not equal to 1 as these events are not alternative. Thus for the choice of *one* edge from the set we use normalized probabilities $p_i^* = p_i / \sum_{j=1}^{n} p_j$.

Let us denote $\frac{N(N-1)}{2}$ as n. In hereafter we use the presentation of undirected graph with N nodes by the up-diagonal part of its adjacency matrix VV unfolded into a vector of the length n. We denote this vector as $D(G)$ or simply D.

Obviously there is one-to-one correspondence between any element VV_{ij} and some element D_l. Let us derive a formula for l through i, j and N.

Each element VV_{ij} has $i-1$ rows above it and k-th row contains $N-k$ elements that belong to the up-diagonal part. The i-th row contains $j-i-1$ elements to the left of VV_{ij} which belong to this up-diagonal part. Thus

$$l = [(N-1)+(N-2)+\ldots+(N-i+1)]+j-i = \frac{(2N-i)(i-1)}{2}+j-i. \quad (5)$$

To obtain i and j from the values of l and N the following primitive algorithm is provided.

```
procedure RowCol(N,l : integer; var Row,Col: integer);
var m : integer;
begin i:=1;
  while (i<N) and (m>0) do
    m:=m-N+i; INC(i);
  endwhile;
  Row:=i-1;
  Col:=N+m;
end RowCol;
```

Note that if we don't need to save the computer memory it is simpler to use the array of pairs (i, j) instead of transformation described above. Now we go on to the algorithms.

As any node inevitably belongs to the spanning tree, we can choose any one as a root on the preliminary step with equal probability. Then we construct the initial list (vector) of possible edges as list of edges incident to the root.

It's simple to obtain the maximum possible length L_m of the list of possible edges: recollect that with each new i-th edge chosen we eliminate from it $i-1$ edges that connected new node with the nodes from X_i and include $N-i$ edges that connect new node with the nodes from $\bar{X}_i$. Thus, after completion the i-th step we have

$$L = (N-1)-1+(N-2)-2+(N-3)-3+\ldots-(i-1)+(N-i) = (N-i)i, \quad (6)$$

and $\lceil L_m = N^2/4 \rceil$. Any additional limitation will only decrease this number.

The pseudo code of the algorithm looks as follows.

Algorithm for Generating Tree with Different Probabilities of Edge Existence and Limited Maximum Node Degree.

Input: $N > 1$ is the number of nodes, $Prob[1..n]$ is the vector of probabilities of edges existence (this vector corresponds to the $\|\varepsilon_{ij} = P(e_{ij}\text{exists})\|$), Deg is the possible maximum node degree.

Output: $OU[1..N-1]$ and $IO[1..N-1]$ are the vectors of the resulting tree edges starting and ending nodes numbers (i-th edge connect node OU_i with IO_i), $Degrees(1 \times N)$ is the vector of the nodes degrees.

Working arrays and variables: $X[1..N-1]$ is the vector for numbers of nodes that are in a generated tree already, $W[1..K]$, $P[1..K]$ – vectors of numbers of edges admissible for choice on a step and their normalized probabilities, respectively. M is current number of nodes that are in a tree already and K is a number of edges that are admissible for choice on a current step.

```
1.  n:=N*(N-1)/2;                  \\ Whole number of possible edges
    \\ Initialization of the vector for edge numbers
2.  for i:=1 to n do I[i]:=i; endfor;
    \\ Initialization of the vector for node degrees
3.  for i:=1 to N do Degree[i]:=0; endfor;
4.  l:=random(P,1,n); X[1]:=l;  \\ Root is selected
5.  S:=0.0; j:=0; M:=1;  \\ M - current number of selected nodes,
    K:=N-1;              \\ K - current number of free nodes
    \\ Now let's obtain the vector for admissible edges
    \\ and Sum of probabilities for future normalization
6.  for v:=1 to l-1,l+1 to N do
7.    INC(j); W[j]:=I[InUpDiag(j,l)];
8.    S:=S+Prob[InUpDiag(N,v,l)];
9.  endfor;
10. for i:=1 to N-1 do             \\ Main cycle
      \\ Probabilities normalization
11.   for v:=1 to K do P[v]:=P[v]/S; endfor;
12.   j:=W[random(P,1,m)];INC(M);OldS:=S; \\ New edge is selected
      \\ Let's obtain the numbers of adjusted nodes and increment
      \\ thier degrees
13.   Call RowCol(N,j,s,t); INC(Degree[s]); INC(Degree[t]);
      \\ Now let us define which node is new one
14.   w:=FindPlace(s,X,OK);
15.   if OK then NewNode:=t else NewNode:=s; s:=t; endif;
      \\ s is the "old" node in the selected edge
16.   X[M]:=NewNode;                \\ Registration of a new node
17.   if Degree[s]=Deg then         \\ Check for maximum degree
18.     for t:=1 to N do
    \\ All free edges connected with a node with maximum possible
    \\ degree must be removed from the set of admissible
19.       w:=FindPlace(t,X,OK);
20.       if not OK then
            q:=InUpDiag(N,t,s); S:=S-P[q]; remove(q,W); DEC(K);
21.   endif; endfor; endif;
22. \\ Now let us exclude all edge that connects new node with
    \\ old ones from the set of admissible and include edges that
    \\ connect new node with free ones into it
23.   SS:=0.0;
24.   for v:=1 to NewNode-1,NewNode+1 to N do
25.     t:=InUpDiag(X[v],NewNode); w:=FindPlace(t,W,OK);
26.     if OK then S:=S-P[w]; remove(t,W); DEC(K);
27.     else SS:=SS+Prob[t]; INC(K); W[K]:=t;
```

```
28.    endif endfor;
     \\ Let us recalculate a sum for normalization
29.    S:=SS+S*OldS;
30. endfor.
```

The procedure InUpDiag (line 8) is for expression (5).

5 Conclusion

We have presented general methods and computer algorithms for generating random trees with many essential properties. It is shown that the use of trial method is mostly ineffective. Meanwhile the time complexity of the proposed algorithms based on the consequential tree growth are proven to be effective in terms of the number of operations and the memory use. Random trees are widely used in different applications by themselves, and furthermore can be efficiently employed for generating various connected graphs. As graphs used in applications have different limitations, their spanning trees must satisfy them too. Our basic algorithms allow those. In the future we present algorithms for generating connected graphs and graphs "similar to real nets." The last task assumes definitions of some special limitations that are typical for such graphs.

References

1. M. Doar, Multicast in the ATM environment. PhD thesis, Cambridge Univ., Computer Lab., September 1993.
2. M. Doar, "A Better Mode for Generating Test Networks," *Proc. Global Telecommunication Conf. GLOBECOM'96*, pp. 86–93, 1996.
3. Chai-Keong Toh, "Performance Evaluation of Crossover Switch Discovery Algorithms for Wireless ATM LANs," *Proc. of INFOCOM'96*, pp. 1380–1387, 1993.
4. A.S. Rodionov, L.N. Postnikova, and O.K. Rodionova, "Program Complex for Graph Generating in the GRAPH-ES Package," *Materials of the Conf. "Computers and System Analysis," Novosibirsk, 1982*, pp. 13–15., 1982. (in Russian)
5. A.S. Rodionov, "Graph Generating in the GRAPH-ES Package," *Materials of the Conf. "Methods and Programs for Solving Optimization Problems on Graphs and Nets," Novosibirsk, 1982, Part 1*, pp. 170–171., 1982. (in Russian)
6. M.I. Netchepurenko, V.K. Popkov, S.M. Mainagashev, etc, *Algorithms and Programs for Solving Optimization Problems on Graphs and Networks.* Novosibirsk, Nauka P.A., 1990, 515 p. (in Russian).
7. G. Omarova, "Program Complex for Graph Generating," *Proc. of Young Scientists Annual Conf., Institute of Computational Mathematics and Mathematical Geophysics of Siberian Division of the Russian Academy of Science, Novosibirsk*, pp. 174–181, 1998. (in Russian)

Sensitivities for a Single Drop Simulation*

Christian H. Bischof, H. Martin Bücker, Arno Rasch, and Emil Slusanschi

Institute for Scientific Computing, Aachen University, D–52056 Aachen, Germany
{bischof, buecker, rasch, slusanschi}@sc.rwth-aachen.de
http://www.sc.rwth-aachen.de

Abstract. In process engineering, a single drop is investigated to better understand its physical and chemical behavior. Laboratory experiments using the Nuclear Magnetic Resonance (NMR) technology are prepared by numerical simulations aiming at finding a suitable geometry of the measuring cell. In the underlying numerical optimization problem, derivatives of the flow field around a single drop with respect to geometric parameters are needed. Rather than using numerical differentiation based on divided differencing, a technique called automatic differentiation is used to compute truncation-error free derivative values. It is shown that automatic differentiation is comparable to numerical differentiation in terms of CPU time but eliminates potential problems in accuracy encountered when using numerical differentiation.

1 Introduction

In an interdisciplinary project, process engineers and computer scientists are aiming at properly designing the geometry of the measuring cell by numerical simulations. In the actual laboratory experiments, one is trying to fix a single drop at a certain position in a liquid in order to analyze fluid-fluid interactions by using the Nuclear Magnetic Resonance (NMR) technology. The goal is to avoid spatial movement of the single drop enabling appropriate NMR measurements. From a conceptual point of view the numerical simulation is minimizing a suitable objective function representing the position of a single drop in the flow field with respect to parameters characterizing the geometry of the underlying computational domain. A crucial ingredient of most numerical optimization algorithms is the derivative of the objective function with respect to the free variables. In the present situation, the objective function is given by a large-scale computational fluid dynamics code that, given the geometric parameters of the computational domain, computes the flow field by solving the Navier-Stokes equations and some additional post-processing of the flow field which we neglect throughout this note for the sake of simplicity.

One option to evaluate the derivatives of the flow field with respect to geometric parameters consists of a numerical approach based on divided differencing.

* This research is partially supported by the Deutsche Forschungsgemeinschaft (DFG) within SFB 540 "Model-based experimental analysis of kinetic phenomena in fluid multi-phase reactive systems," Aachen University, Germany.

P.M.A. Sloot et al. (Eds.): ICCS 2003, LNCS 2658, pp. 888–896, 2003.

The advantage of divided differencing is its simplicity: approximations to derivatives of certain accuracy are available by evaluating the function whose derivatives are sought in a black box fashion at suitably perturbed input arguments. However, there is always truncation error and, when the input arguments differ only slightly, the resulting cancellation error may also dramatically influence the accuracy of the derivative approximations.

In this note, we apply a truncation error-free technique called automatic differentiation to obtain derivatives of the flow field with respect to geometric parameters. In automatic differentiation (AD), a given computer program to evaluate some function is transformed into another program capable of evaluating the original function along with its derivatives. More precisely, we apply the AD system ADIFOR [2] to a general purpose finite element package SEPRAN [14] written in Fortran 77. In doing so, we extend previous work [3] where the flow solver module of SEPRAN was transformed by ADIFOR to evaluate derivatives of the flow field with respect to flow parameters such as the Reynolds number or the angle of attack. Now, we additionally apply AD to SEPRAN's grid generator enabling derivatives with respect to geometric parameters. A similar approach in the area of aerodynamics is studied in [6] where AD is applied to the WTCO wing grid generator and the TLNS3D Navier–Stokes flow solver, both Fortran 77 programs. Details on developing an AD-enhanced version of TLNS3D are given in [10,9]. Another application of AD applied to a grid generator called CSCMDO which is written in C is described in [4]. Aerodynamic sensitivities with respect to geometric parameters are also reported in [8].

The structure of this note is as follows. In Sec. 2, the basic idea behind automatic differentiation is briefly summarized and issues of applying AD to SEPRAN are sketched. The simulation of a single drop using SEPRAN and the corresponding derivatives obtained from AD are given in Sec. 3.

2 Automatic Differentiation and the SEPRAN Package

When a function is given in the form of a computer code, AD may be used to generate another computer program capable of evaluating the function *and* its derivatives. The basic idea of AD is to treat a computer program as the composition of a large sequence of elementary functions and intrinsics whose derivatives are known. Then, these derivatives are accumulated using the chain rule. When using an AD tool a user has to specify the output variables whose derivatives are sought, called *dependent* variables, and the input variables with respect to which one is differentiating, called *independent* variables. A variable is called *active* if its value depends on the value of an independent variable and it influences the value of a dependent variable. More details on AD are given in [11, 13,7,12,1]. The AD technology is not only applicable to small computer codes but also scales to large programs of several hundred thousand lines of code.

In this note, we apply the AD tool ADIFOR [2] to the general purpose finite element package SEPRAN [14], developed at "Ingenieursbureau SEPRA" and Delft University of Technology. The SEPRAN package solves the continuity and Navier-Stokes equations. SEPRAN is employed in a wide variety of engineering applications [5,15,16,18,17,19,20] including laminar or turbulent flow of

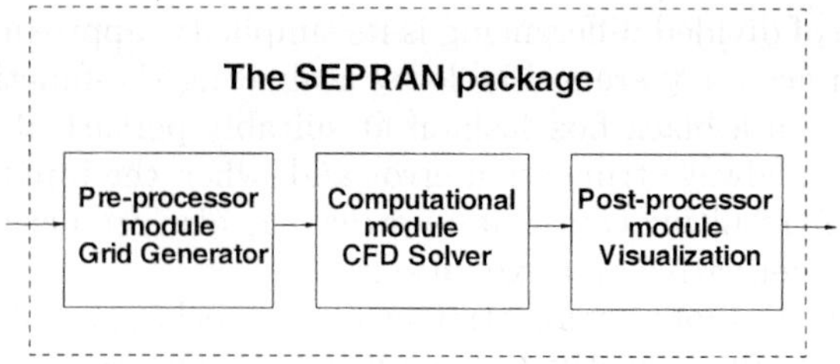

Fig. 1. The structure of the finite element package SEPRAN

incompressible liquids. The SEPRAN package consists of about 800,000 lines of Fortran 77 in several modules, such as the pre-processing (grid generation), computational and post-processing modules schematically depicted in Fig. 1. In a previous work [3], we successfully applied AD to the computational module where approximately 400,000 lines were automatically transformed. In this note, we not only apply AD to the CFD solver but also to the grid generation part consisting of 130,000 additional lines of code in order to get derivatives for the single drop problem to be described in Sec. 3.

In order to be able to use the ADIFOR tool to obtain the augmented version of the code, SEPRAN.AD, a preprocessing step has to be taken. This comprises of massaging the SEPRAN source code so that it adheres to the Fortran77 language standard. It is worth noticing that SEPRAN is generally a "clean" code, which requires little changes by hand in the original code in order to be able to apply ADIFOR to it. However, some problems, like type mismatches and calling of some subroutines with different number of arguments were encountered. Usually, normal compilers do not detect such problems because they are examining only one Fortran file at a time, whereas ADIFOR does a global data flow and dependency analysis and therefore is able to detect these inconsistencies. More details are given in [3]. A problem that was a bit harder to detect was encountered in the grid generation part of the SEPRAN code, and it has to do with a certain way of referencing memory in Fortran through the use of "read" and "write" instructions. When the ADIFOR tool encounters an active variable in a statement that does a "read" or a "write", it generates a warning message. In this particular case, a value was written to memory using "write" and was later used by calling a corresponding "read." Therefore, any AD tool is unable to recognize whether or not this value has changed. In order to maintain the "logic link" between the dependent and independent variables, a common block was inserted instead of the mentioned use of the "read/write" instructions. In this way the augmented code SEPRAN.AD, consisting of the grid generation and computational part totaling about 703,000 lines of code, produced correct results as presented in the following section. Note that the code massaging part, which requires some manual work is done only once in the developing cycle of the augmented version of SEPRAN.

3 Sensitivities of a Single Drop

The study of kinetic phenomena in a single drop is the topic of an interdisciplinary project at Aachen University within the Collaborative Research Center SFB 540. To this end a two-dimensional numerical simulation of a measuring cell is implemented using SEPRAN. This simulation is stationary and rotationally symmetrical under the assumption of Newtonian and incompressible fluids with a non-deformable drop. The goal of the project is to enable the measurements of the velocity and material properties of a single drop using the NMR technology. The design of the measuring cell has to be optimized so that the drop remains stationary for a certain period of time during which the measurements are taking place. The idea is to maintain the drop stable using a counter current flow. The drop is initiated at the bottom of the measuring cell with a special pump and it keeps on rising, until its velocity and that of the counter current flow are the same, and in this way the drop reaches its desired stable position and the NMR measurements can then take place. After the measurements, the counter current flow is stopped and the drop rises until it reaches a collector where it is absorbed. Given the velocity of the inflow v_0, the density ρ, and viscosity μ of the two phases (drop and counter current), the SEPRAN package is able to compute the velocity and pressure fields at any point in the measuring cell. The design of this measuring cell depends on the following geometric parameters schematically depicted in Fig. 2. The first parameter, α, is the angle between the vertical axis and the wall of the measuring equipment. The second parameter, r_1, denotes the radius of the inlet and the third parameter, r_2, specifies the smallest possible radius of the measuring cell, representing the distance between the symmetric axis and the wall.

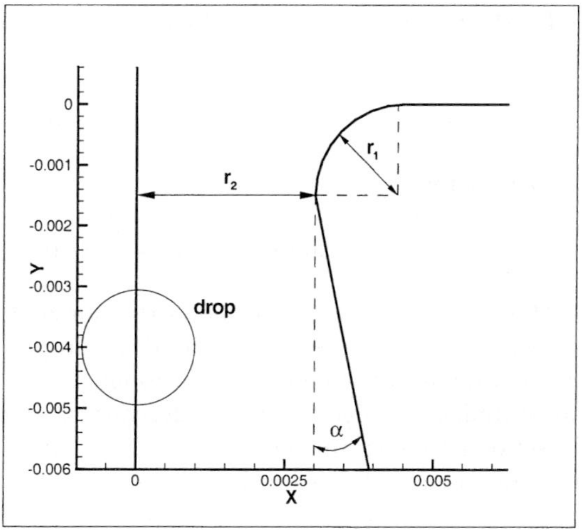

Fig. 2. Geometry of the measuring cell

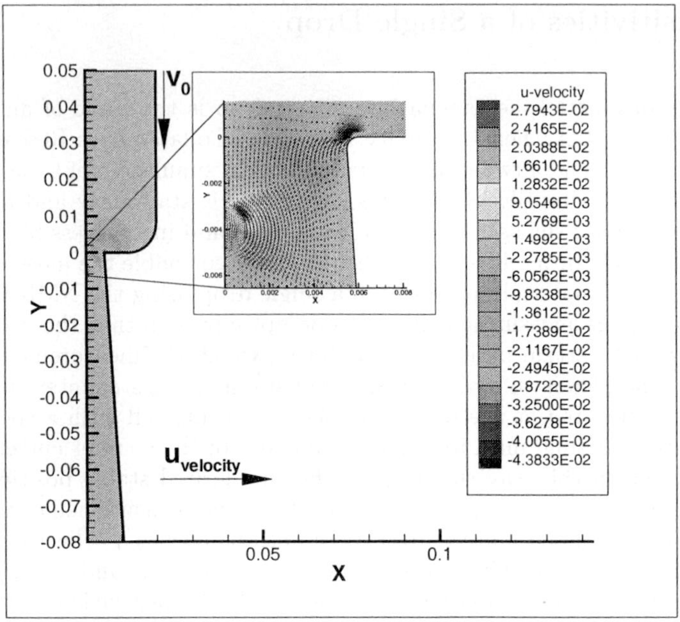

Fig. 3. Original function: horizontal component u of the velocity

From an abstract point of view, the computer program implements a function $f(v_0, \rho, \mu, \alpha, r_1, r_2)$ representing the flow field. For the sake of simplicity, we will use f to denote the velocity of the fluid. Note that the actual simulation also computes the pressure field for which a sensitivity analysis can be carried out similar to the one for the velocity given in the following discussion. The velocity field computed by SEPRAN for the flow parameters

$$v_0 = -4.2 \cdot 10^{-3}, \quad \rho = 1.0 \cdot 10^3, \quad \mu = 1.0 \cdot 10^{-3},$$

and the geometric parameters

$$\alpha = 3.6411, \quad r_1 = 5.0 \cdot 10^{-4}, \quad r_2 = 5.5 \cdot 10^{-3} \tag{1}$$

is given in Fig. 3. More precisely, the figure shows the horizontal component, u, of the velocity field with a zoom in the region where the drop is kept stable. The vectors of the flow field are also presented in this zoom.

In a numerical optimization algorithm, the derivatives of the velocity field with respect to the set of design parameters (1) are needed. We use automatic differentiation for the computation of these derivatives, $\partial f/\partial\alpha$, $\partial f/\partial r_1$ and $\partial f/\partial r_2$. As an illustrating example, the results presented in Fig. 4 show the derivatives of the horizontal component of the velocity with respect to the angle α. The figure demonstrates that the largest increase of this velocity component with respect to changes in α occurs in the vicinity of the drop.

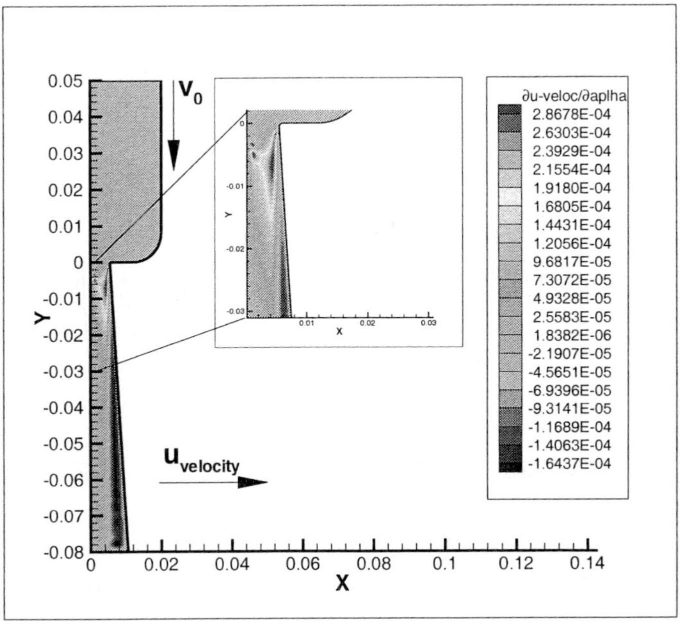

Fig. 4. AD-generated derivative $\partial u/\partial\alpha$ field

To illustrate the advantage of the AD approach in terms of accuracy, we present a comparison with the traditional numerical approach of computing derivatives making use of divided differences. We will use DD as a shorthand notation for first-order forward divided differences, defined by

$$\frac{\partial f(v_0,\rho,\mu,\alpha,r_1,r_2)}{\partial\alpha} \approx \frac{f(v_0,\rho,\mu,\alpha+h,r_1,r_2)-f(v_0,\rho,\mu,\alpha,r_1,r_2)}{h}, \tag{2}$$

where h is a suitable chosen step size for the approximation. Equation (2) can also be written for the other geometric parameters, r_1 and r_2. DD is easy to implement in a black-box manner. On the other hand though, the accuracy of the approximations provided depends very much on choosing a suitable step size h. The problem with DD is that, in general, there is no a priori knowledge of a suitable step size which may also be hard to determine. Furthermore, there are cancellation and truncation errors that the user has to contend with, all contributing to unavoidable errors on the end result. When employing automatic differentiation, however, there is no truncation error to contend with, the results are accurate up to machine precision and there is no need to experiment with a step size.

For the derivative $\partial u/\partial\alpha$, it turns out that it is possible to find a suitable step size $h = 10^{-3}$ where the infinity norm of the difference between the DD and AD-generated derivatives is $1.1\cdot10^{-7}$. However, when considering the derivative $\partial u/\partial r_2$, it is no longer possible to find an accurate DD-approximation. After

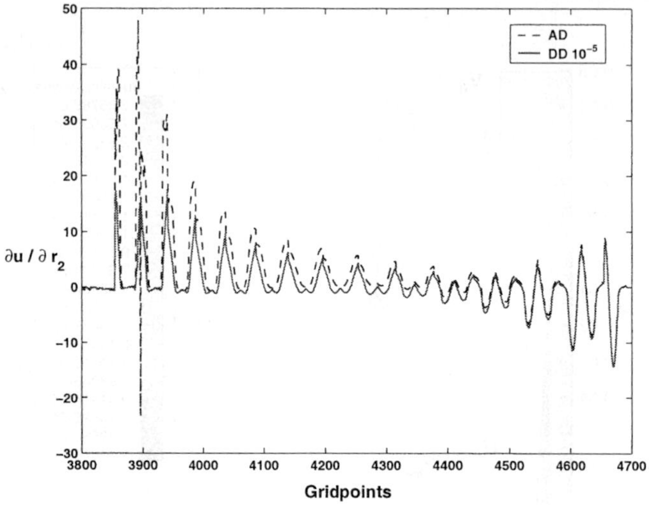

Fig. 5. AD-DD comparison of the derivative of the $\partial u/\partial r_2$

identifying a suitable step size of $h = 10^{-5}$, the corresponding infinity norm is 38.94 representing a large difference between the values computed by AD and DD. An explanation is given by Fig. 5 showing the derivative $\partial u/\partial r_2$ versus a subset of selected grid points. The grid points shown in this figure are numbered from 3800 to 4700 and are located in the inlet region near the drop. At certain grid points, the divided difference approximations are not able to give us the exact derivatives computed by AD. A similar analysis holds for the derivative $\partial u/\partial r_1$ where the infinity norm of the difference between AD and DD is 39.32.

Therefore we conclude that using a numerical approach to calculating the required sensitivities in order to optimize the geometry of the measuring cell is not appropriate, as these approximations may be inaccurate. In contrast, automatic differentiation provides reliable derivative values. Moreover, considering that an optimization must be conducted with respect to a series of geometry parameters (α, r_1, r_2) it is not advantageous to have different suitable step sizes, namely 10^{-3} and 10^{-5}, for different derivatives.

When it comes to analyzing the performance of the AD augmented code compared to the original simulation, we observe an increase in execution time by a factor of 4.15 and an increase in memory usage by a factor of 3.41. The simulations were carried out on a 1733 MHz Pentium 4 Linux machine with 2 GB of main memory. The number of independent variables in this simulation is three, the angle α, and the two radii r_1 and r_2. Therefore the expected increase in memory requirement is within normal limits. Concerning the execution times, the approximations obtained by forward divided differences for three partial derivatives require at least four evaluations of the original function, provided a "suitable" step size is already known. When employing AD, on the other hand, the expected increase in execution time is proportional to the number of independent variables specified. Therefore, the observed ratio of 4.15 in the increase in execution time, can be considered very good, as the results are reliable and precise up to machine precision.

4 Conclusions

Laboratory experiments for the analysis of fluid-fluid interactions are prepared by numerical simulations. In particular, the geometry of the underlying measuring cell is designed in order to keep a constant position of a single drop in a liquid. The SEPRAN package is used to simulate the flow field for given geometric parameters. The ADIFOR system is applied to obtain derivatives of the flow field with respect to geometric parameters. This approach using automatic differentiation is compared to a numerical approach based on divided differencing. The major advantage of automatic differentiation in this application is the availability of derivative values without truncation error. Moreover, time complexity is comparable to that of a first-order forward divided differencing approach.

Currently, we are trying to determine a suitable objective function for the future optimization of the design of the measuring cell. However, from the point of view of algorithmic ingredients, the major computational challenge is to efficiently and accurately evaluate the derivatives of the flow field with respect to the geometric parameters as described in this work. An implementation of the objective function will consist of only a few lines of code compared to the complete SEPRAN package and its remaining differentiation will easily be accomplished by the application of an AD tool.

Acknowledgments. The authors would like to thank the collaborators at the Institute of Thermal Unit Operations, Aachen University. In particular, Edwin Groß-Hardt deserves special recognition for making available the grid and problem specifications of the underlying problem used in the SEPRAN simulation. We would also like to thank Jakob Risch for his notable contribution during the initial phase of applying ADIFOR to SEPRAN. This research is partially supported by the Deutsche Forschungsgemeinschaft (DFG) within SFB 540 "Model-based experimental analysis of kinetic phenomena in fluid multi-phase reactive systems," Aachen University, Germany.

References

[1] M. Berz, C. Bischof, G. Corliss, and A. Griewank. *Computational Differentiation: Techniques, Applications, and Tools.* SIAM, Philadelphia, 1996.

[2] C. Bischof, A. Carle, P. Khademi, and A. Mauer. ADIFOR 2.0: Automatic differentiation of Fortran 77 programs. *IEEE Computational Science & Engineering*, 3(3):18–32, 1996.

[3] C. H. Bischof, H. M. Bücker, B. Lang, A. Rasch, and J. W. Risch. On the Use of a Differentiated Finite Element Package for Sensitivity Analysis. In V. N. Alexandrov, J. J. Dongarra, B. A. Juliano, R. S. Renner, and C. J. K. Tan, editors, *Computational Science – ICCS 2001, Proceedings of the International Conference on Computational Science, San Francisco, USA, May 28–30, 2001. Part I*, volume 2073 of *Lecture Notes in Computer Science*, pages 795–801, Berlin, 2001. Springer.

[4] C. H. Bischof, A. Mauer, W. T. Jones, and J. Samareh. Experiences with automatic differentiation applied to a volume grid generation code. *Journal of Aircraft*, 35(4):569–573, 1998.
[5] E. G. T. Bosch and C. J. M. Lasance. High accuracy thermal interface resistance measurement using a transient method. *Electronics Cooling Magazine*, 6(3), 2000.
[6] A. Carle, L. L. Green, C. H. Bischof, and P. A. Newman. Applications of automatic differentiation in CFD. In *Proceedings of the 25th AIAA Fluid Dynamics Conference, Colorado Springs, CO, USA, June 20–23, 1994*, AIAA Paper 94–2197, 1994.
[7] G. Corliss, C. Faure, A. Griewank, L. Hascoët, and U. Naumann, editors. *Automatic Differentiation of Algorithms: From Simulation to Optimization.* Springer, New York, 2002.
[8] S. A. Forth and T. P. Evans. Aerofoil optimisation via AD of a multigrid cell-vertex Euler flow solver. In Corliss et al. [7], chapter 17, pages 153–160.
[9] L. Green, P. Newman, and K. Haigler. Sensitivity derivatives for advanced CFD algorithm and viscous modeling parameters via automatic differentiation. In *Proceedings of the 11th AIAA Computational Fluid Dynamics Conference, Orlando, FL, USA, July 6–9, 1993*, AIAA Paper 93–3321, 1993.
[10] L. L. Green, P. A. Newman, and K. J. Haigler. Sensitivity derivatives for advanced CFD algorithm and viscous modeling parameters via automatic differentiation. *Journal of Computational Physics*, 125(2):313–324, 1996.
[11] A. Griewank. *Evaluating Derivatives: Principles and Techniques of Algorithmic Differentiation.* SIAM, Philadelphia, 2000.
[12] A. Griewank and G. Corliss. *Automatic Differentiation of Algorithms.* SIAM, Philadelphia, 1991.
[13] L. B. Rall. *Automatic Differentiation: Techniques and Applications*, volume 120 of *Lecture Notes in Computer Science.* Springer-Verlag, Berlin, 1981.
[14] G. Segal. *SEPRAN Users Manual.* Ingenieursbureau Sepra, Leidschendam, NL, 1993.
[15] G. Segal, C. Vuik, and F. Vermolen. A conserving discretization for the free boundary in a two-dimensional Stefan problem. *Journal of Computational Physics*, 141(1):1–21, 1998.
[16] A. P. van den Berg, P. E. van Keken, and D. A. Yuen. The effects of a composite non-Newtonian and Newtonian rheology on mantle convection. *Geophys. J. Int.*, 115:62–78, 1993.
[17] P. E. van Keken, C. J. Spiers, A. P. van den Berg, and E. J. Muyzert. The effective viscosity of rocksalt: implementation of steady-state creep laws in numerical models of salt diapirism. *Tectonophysics*, 225:457–476, 1993.
[18] P. E. van Keken, D. A. Yuen, and L. R. Petzold. DASPK: a new high order and adaptive time-integration technique with applications to mantle convection with strongly temperature- and pressure-dependent rheology. *Geophysical & Astrophysical Fluid Dynamics*, 80:57–74, 1995.
[19] N. J. Vlaar, P. E. van Keken, and A. P. van den Berg. Cooling of the Earth in the Archaean: consequences of pressure-release melting in a hot mantle. *Earth Plan. Sci. Lett.*, 121:1–18, 1994.
[20] C. Vuik, A. Segal, and F. J. Vermolen. A conserving discretization for a Stefan problem with an interface reaction at the free boundary. *Computing and Visualization in Science*, 3(1/2):109–114, 2000.

A Simple Model of Drive with Friction for Control System Simulation*

Adam Woźniak

Institute of Control and Computation Engineering
Warsaw University of Technology,
ul. Nowowiejska 15/19, 00-665 Warszawa, Poland
wozniak@ia.pw.edu.pl

Abstract. This paper proposes a simple discrete model of drive with friction. The nonlinear part of the model is based on the observation that friction is maximum just before two objects begin to move, and decreases when the objects are in motion with constant relative velocity. Although it falls in the realm of classical Static+Coulomb+Viscous models of friction it is a dynamical one. Linear part is based on the typical linearized equations of a DC motor. The obtained model equations can be easily modeled, e.g., in SIMULINK. The complete model has four parameters and simple identification procedure is presented. An example of model parameter identification is provided.

1 Introduction

Friction is always present in operation of machines. Its reduction is one of the ways to minimize its undesirable influence on the behavior of the system. Designers of precision servo-drives go pains to reduce the hindering effects of this phenomenon. However problems caused by friction remain and it is necessary to contend with them when one designs precision control system. The main undesirable effects are stick-slip, limit cycling ("hunting") and "hangoff". Stick-slip occurs, when the ramp input has a relatively low velocity and a motion is a series of starts and stops. The latter effects affect steady-state behavior: steady-state error hunts about zero or is not reduced to zero (effect similar to dead-zone). For the control engineer the challenge is not the reducing friction, but the reduction of its influence on motion of the controlled system. To find the remedies, a simple, easy to identify and valid model of friction is necessary.

2 Loaded Drive Model

The literature presents two distinct approaches to building a mathematical model of friction.

Within the first one, attempts are made to derive the complete physical model. The starting point of considerations is based, for example, on micro-

*This research was supported by KBN Grant No. 7T11A02220.

P.M.A. Sloot et al. (Eds.): ICCS 2003, LNCS 2658, pp. 897–906, 2003.

scopic observation that the source of frictional force (or torque) between moving surfaces can be described as rubbing elastic bristles. The deflection of the bristles produces the friction. Less or more complicated models based on the behavior of the bristles are presented, e.g., [1], [2] and [3]. The main drawback of such models is their dependence on many parameters, which are usually difficult to identify. So the various identification procedures are presented in the literature, e.g., [4], [5], [6].

The second, more classical, approach [7], [8], is based on macroscopic observation that friction between two objects is at maximum just before they begin to move, and decreases when the objects are in motion with constant relative (angular) velocity. Another observation is that friction increases with velocity. These lead to the three-term equation of friction torque T_f (we are interested in rotational motion further on)

$$T_f = T_s + T_c + T_v. \tag{1}$$

The first term, T_s, is a friction at rest (static friction), the second, T_c, is the constant friction during motion – called Coulomb or kinetic friction, and the last component, T_v, is viscous friction – proportional to the velocity. In the sequel we will follow this classical approach assuming, like the early investigators, that friction is a function of velocity, acceleration and applied torque. This means that friction is inherent in our dynamical model of moving mechanism as opposed to prevailing approach where friction is the memoryless element added to the dynamical model of machine.

The satisfactory model of Coulomb friction is represented by nonlinear discontinuous function

$$T_c(\omega) = C_o \mathrm{sgn}(\omega), \tag{2}$$

where ω is the angular velocity and C_o is the Coulomb friction coefficient (for simplicity we assume a symmetric model). Also, we assume that viscous friction is proportional to the velocity

$$T_v(\omega) = B\omega, \tag{3}$$

where B is the viscous friction coefficient.

The most difficult to model is the static friction torque, which exists only when the velocity is near zero or the object is at rest ($\omega = 0$) but tends towards motion (external torque is not equal to zero). When the motion begins, the static friction vanishes, but at rest it may have any value between $T_{sm} \leqslant -C_o$ and $T_{sM} \geqslant C_o$. This observation leads to the equation:

$$T_s(\omega) = 0, \qquad \omega \neq 0. \tag{4}$$

The expression for static friction torque when $\omega = 0$, can be obtained by using d'Alembert's principle of conservation, i.e., by equating torques applied to the rotating body

$$T_e(t) - J\frac{d\omega(t)}{dt} = T_s(t) + C_o \mathrm{sgn}(\omega(t)) + B\omega(t), \tag{5}$$

where T_e is an external (input) torque and J is the equivalent moment of inertia referred to the axis of the rotating body. In general the external torque can be described as a sum of net torque causing acceleration, T_a, and disturbance torque (e.g., due to gravitational forces and/or interactions), T_d,

$$T_e = T_a + T_d. \tag{6}$$

The above leads, for the instant t_0, when the velocity equals zero, $\omega(t_0) = 0$, to the equations:

- for $\dfrac{d\omega(t_{0-})}{dt} \neq 0$

$$T_s(t_0) = \begin{cases} T_e(t_0) & \text{when } \left|T_e(t_0) - J\,\dfrac{d\omega(t_{0-})}{dt}\right| < C_o \\ C_o \lim\limits_{t \to t_{0-}} \operatorname{sgn}(\omega(t)) & \text{otherwise} \end{cases} \tag{7}$$

- for $\dfrac{d\omega(t_{0-})}{dt} = 0$

$$T_s(t_0) = \begin{cases} T_e(t_0) & \text{when } |T_e(t_0)| < C_s \\ C_o \lim\limits_{t \to t_{0-}} \operatorname{sgn}(T_e(t_0)) & \text{otherwise} \end{cases} \tag{8}$$

where C_s is the static friction coefficient, $C_s \geqslant C_o$ (we have assumed symmetry again).

The structure of the model (2) – (8) is shown in fig. 1. It is worth noting that while our model falls in the realm of classical SCV (Static+Coulomb+Viscous) models of friction, it is a dynamical one.

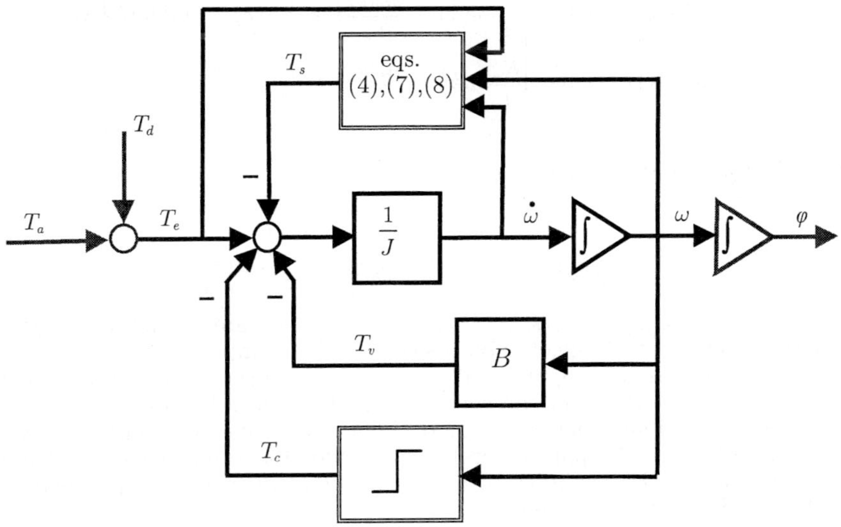

Fig. 1. Structure of the model

The computer simulation models have, in fact, discrete nature. So we transform the above continuous description to a discrete one, having in mind the application, i.e., position servomechanisms. Therefore we limit our consideration to situation when armature controlled permanent magnet DC motor is the source of motion of the machine. Basing on classical equations of DC motor, neglecting armature inductance, assuming non-flexible load and zero order hold on input [9] we adopt, for the situation with a short sampling period T_p, the following discrete dynamical model of DC drive with friction, where linear equations of the loaded motor are:

$$\begin{aligned} \varphi(k+1) &= \varphi(k) + T_p\omega(k) \\ \omega(k+1) &= h\omega(k) + K_v(1-h)v(k) \\ v(k) &= u(k) - f(k) \\ h &= \exp(-\frac{T_p}{T}), \end{aligned} \tag{9}$$

where φ is the angular position, ω – angular velocity, K_v – velocity gain of the motor, T – resultant motor time constant taking into account all moments of inertia referred to the motor shaft and viscous friction, v – net input causing acceleration, u – control fed by voltage source, f – drag, i.e., input from Coulomb and static friction components.

Coulomb and static friction can be expressed:

- for $|\omega(k)| \geqslant \alpha$ as:

$$f(k) = c_o \operatorname{sgn}(\omega(k)) \tag{10}$$

- for $|\omega(k)| < \alpha$ and $|\Delta\omega(k)| \geqslant \beta$ as:

$$f(k) = \begin{cases} u(k) & \text{when } \left|\frac{K_v}{T}u(k) - \frac{\Delta\omega(k)}{T_p}\right| \geqslant \frac{K_v}{T}c_o \text{ and } |u(k)| < c_o \\ c_o \operatorname{sgn}(u(k)) & \text{when } \left|\frac{K_v}{T}u(k) - \frac{\Delta\omega(k)}{T_p}\right| \geqslant \frac{K_v}{T}c_o \text{ and } |u(k)| \geqslant c_o \\ c_o \operatorname{sgn}(\omega(k)) & \text{when } \left|\frac{K_v}{T}u(k) - \frac{\Delta\omega(k)}{T_p}\right| \geqslant \frac{K_v}{T}c_o \end{cases} \tag{11}$$

- for $|\omega(k)| < \alpha$ and $|\Delta\omega(k)| < \beta$ as:

$$f(k) = \begin{cases} u(k) & \text{when } |u(k)| < c_s \\ c_o \operatorname{sgn}(u(k)) & \text{otherwise} \end{cases} \tag{12}$$

where $\Delta\omega(k) = \omega(k) - \omega(k-1)$, c_o and c_s, $c_o \leqslant c_s$, are Coulomb and static friction coefficients, respectively, scaled in units of input (control).

The constants α and β, used above, form an approximate zero crossing detector for the velocity and the acceleration. The suitable value for β is near 0.005 [rad/sec] and suitable value for α should be adjusted experimentally (it depends on other model parameters) to preclude chattering behavior of modeled static friction. For validity of model in our identification experiments, α near 0.09 [rad/sec] was sufficient.

In this section we derived simple, discrete model of loaded drive with friction: equations (9) – (12); now we present method of its identification.

3 Identification Procedure

The proposed drive model has four parameters: T, K_v, c_o and c_s. Resultant motor time constant, T, is relatively easy to identify. Velocity gain, K_v, and Coulomb friction coefficient, c_o, must be identified together because Coulomb friction decreases gain. The simplest method for determining these parameters is transient analysis based on step response and known motor specification [10]. The problem of determining the static friction coefficient, c_s, is harder and the method proposed for identification will be presented in the sequel.

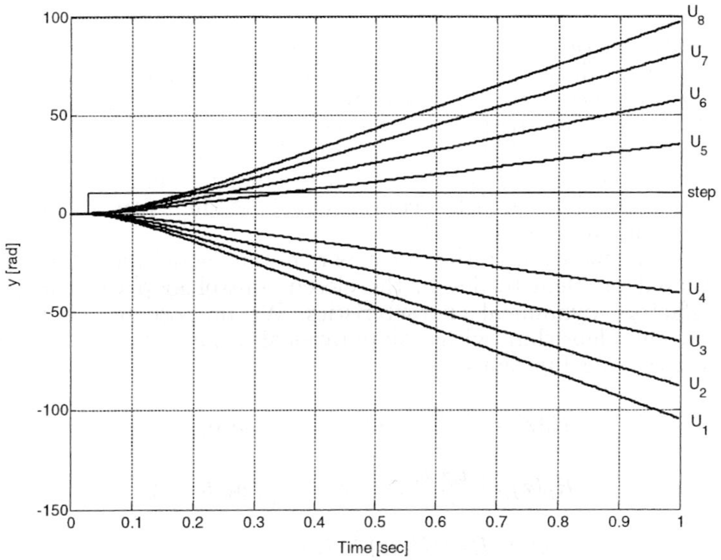

Fig. 2. Step response for identified motor.

A simple experiment that produce the transient response is to excite the loaded motor by a step input with known amplitude and to measure the resulting output. Because the velocity response is always recorded with high level of noise (cf. fig. 3) we base our identification on position response. Typical step response data obtained experimentally from the loaded motor appears on fig. 2 where U_i, $i = 1,...,N$, is the amplitude of step measured in input units of controlled voltage source. We know that linear approximation of our system response is a combination of transients of integrator and first order system (cf. equations (9)). Therefore, for each response, using paper and pencil calculations or least-squares method of parameters determination [11], it is easy to obtain initial estimates of parameters K_v and T. Next, using a computer for plotting, we tune the parameters to achieve the best fit. This iterative selection gives better estimators k_{ei} and T_{ei}, $i = 1,...,N$, of the parameters. As the final estimate of the time constant T we choose the mean value of obtained estimators

$$T = \frac{1}{N}\sum_{i=1}^{N} T_{ei} , \tag{13}$$

where N is the number of excitations.

As we remember estimates of K_v and c_o must be determined together. We exploit our assumption that Coulomb friction coefficient, c_o, does not depend on velocity, therefore estimators k_{ei} obtained above, velocity gain K_v and Coulomb coefficient c_o should be coupled by equations

$$K_v(U_i - c_o \operatorname{sgn}(U_i)) = k_{ei}U_i, \quad i = 1,...,N. \tag{14}$$

Hence we can obtain final estimated values of parameters K_v and c_o, using least-squares approach with differences

$$\epsilon_i = k_{ei}U_i - K_v(U_i - c_o \operatorname{sgn}(U_i)), \quad i = 1,...,N, \tag{15}$$

treated as errors.

The static friction causes hunting, so this phenomenon can be used as a vehicle for identification of static friction coefficient. It is well known that oscillations may occur in position control system with integral action. When in such a system, due to friction, the error is not equal to zero, the control increases with time until the motion starts and subsequently the control rapidly decreases, because Coulomb friction is smaller than static friction, $c_o < c_s$. Hence the maximal value of the control equals to the static friction coefficient c_s (scaled in the units of control).

At the moment we have the linear part of our model identified, so it is not a complicated problem to design the closed control loop with integral action and satisfactory dynamical characteristics. We propose to use the PI+Lead Compensation algorithm with computational time delay described by the following z-transfer functions:

$$\begin{aligned} R_I(z) &= \frac{(1+\gamma)z-(1-\gamma)}{z-1}, \qquad \gamma > 0, \\ R_D(z) &= \frac{b_0 z - b_1}{z-a}, \qquad a,\, b_0,\, b_1 > 0 \\ u(z) &= R_I(z) R_D(z) \frac{1}{z} e(z) \end{aligned} \tag{16}$$

where $e = \varphi - r$ is control error and r – the reference input. Parameters γ, a, b_0, b_1 can be easily chosen basing on classical frequency-response method using the w-transform [9], [12], assuming linear model of motor with load with identified parameters K_v and T:

$$\begin{aligned} \varphi(z) &= HG(z)u(z) \\ HG(z) &= K_v \frac{[T_p - T(1-h)]z + [T(1-h) - hT_p]}{(z-1)(z-h)} \\ h &= \exp(-\frac{T_p}{T}). \end{aligned} \tag{17}$$

The last step of identification procedure is model validation, that is investigation if it can be accepted, given its possible use. A natural test is to simulate the model from measured input and then compare obtained output with the measured one. The presented model (equations (9) – (12)) is so simple, that its implementation in SIMULINK do not cause difficulties and validation is a not complicated task.

4 Example of Model Identification

The prototype RNT robot manipulator was designed at Warsaw University of Technology [13]. It has six degrees of freedom and a DC motor actuates each axis. The robot is intended to use as a milling device, so precision positional control of each axis is needed. In the sequel we present the identification of the first axis drive model.

According to our procedure, we started with transient analysis. Positional step response of the drive was presented for eight step amplitudes on fig. 2. Using least-squares method and then tuning obtained values of parameters, from the data presented in fig. 2, the estimators presented in Table 1 have been obtained.

Table 1. Parameter estimators

U_i	-150	-130	-100	-70	70	100	130	150
T_{ei} [sec]	0.0591	0.0624	0.0590	0.0604	0.0579	0.0762	0.0842	0.0877
k_{ei} [rad/sec]	0.7542	0.7348	0.7048	0.6342	0.6254	0.6925	0.7295	0.7481

Basing on these data final estimated values of time constant, velocity gain and Coulomb friction coefficient were calculated as:

$$T = 0.0684 \text{ [sec]}, \quad K_v = 0.8546 \text{ [rad/sec]}, \quad c_o = 18. \tag{18}$$

The received part of the model fits well the set of data (fig. 3), where ω_{data} is the registered velocity and ω_{model} is the simulated velocity.

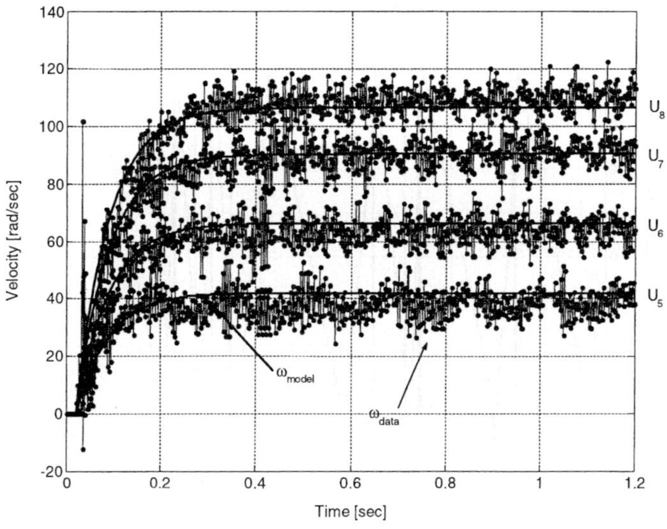

Fig. 3. Recorded velocity and simulated velocity comparison.

To obtain oscillations at steady state we excited closed loop system with reference input signal having rectangular profile of velocity. Raw measured profile of control when system was in hunting mode is presented in fig. 4. The mean value of control maxima we adopt as static friction coefficient, i.e., for the case

$$c_s = 21. \tag{19}$$

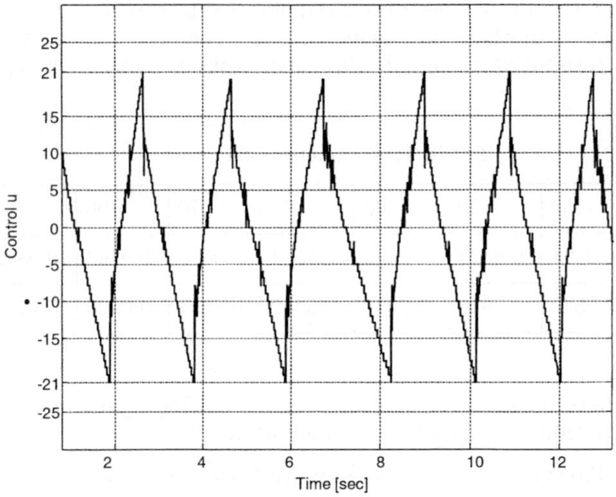

Fig. 4. Measured profile of control in hunting mode.

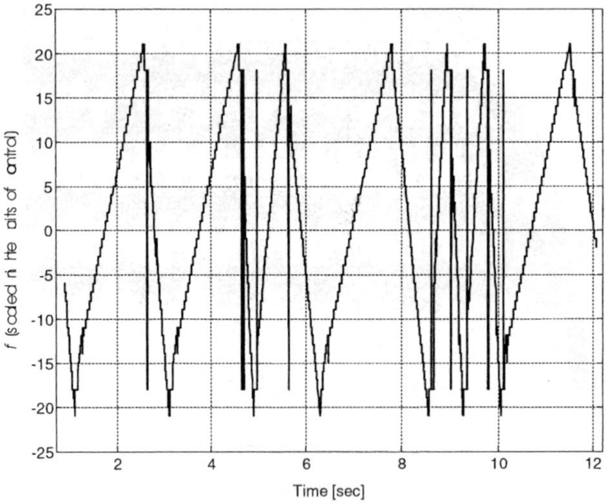

Fig. 5. Modeled static and Coulomb friction f.

The fig. 4 on the previous page shows the randomness of friction phenomena although fluctuations are not large.

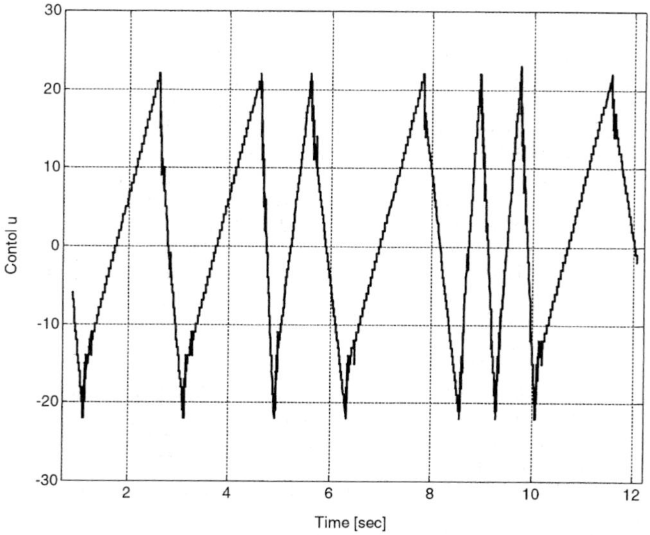

Fig. 6. Simulated control with the same reference input as used when Fig. 4 was obtained.

The described model of loaded motor with identified parameters together with controller was implemented in SIMULINK. The exemplary results of simulation appear in fig. 5 and fig. 6, where simulated friction expressed in units of control and control itself are presented. Comparing fig. 4 and fig. 6 we can conclude that difference is small enough and our model describes influence of friction with desired accuracy. The irregularity of simulated control compared with almost periodical measured control is due to the necessity of use approximate zero crossing detector in simulation.

5 Conclusions

This paper concentrated on the modeling of friction in servo-drives. A four parameter discrete dynamic model of moving mechanism (motor and load) with friction has been presented. Those parameters are relatively easy to identify and suitable identification procedure was described. So the creation of a simulation model for the determination/examination of advanced control algorithm is enabled. The proposed procedure was illustrated on practical example of identification of drive model for one of RNT robot manipulator axis. The other five drive models were identified using this approach also. As presented, validation of derived and identified simulation model comes off well. However the author realizes preliminary nature of presented results, and necessity of described model and identification procedure further examinations.

References

1. Armstrong-Helouvry, B. et al.: A Survey of Models, Analysis Tools and Compensation Methods for the Control of Machines with Friction. Automatica, **30**(7) (1994) 1083 – 1138
2. Canudas de Wit, C., Olsson, H., Åstrom, K.J., Lischinsky, P.: A New Model for Control of Systems with Friction. IEEE Trans. on AC, **40**(3) (1995) 419 – 425
3. Haessig Jr., D.A., Friedland, B.: On the Modeling and Simulation of Friction. J. Dyn. Sys. Meas. and Control (Trans. ASME). **113**(3) (1991) 354 – 362
4. Elhami, M.R., Brookfield, D.J.: Sequential Identification of Coulomb and Viscous Friction in Robot Drives. Automatica **33**(3) (1997) 393 – 401
5. Canudas de Wit, C., Carillo, C.J.: A Modified EW-RLS Algorithm for Systems with Bounded Disturbances. Automatica, **26**(3) (1990) 599 – 606
6. Tzes, A. et al.: Genetic-based Fuzzy Clustering for DC-motor Friction Identification and Compensation. IEEE Trans. on Control Systems Technology **6**(4) (1998) 462 – 472
7. Tou, J., Schultheiss, P.M.: Static and Sliding Friction in Feedback System. J. Appl. Phys. **24**(9) (1953) 1210 – 1217
8. Popov, E.P., Paltov, I.P.: Approximate Methods for Analysis of Nonlinear Automatic Systems (in Russian). Gos. Izd. FizMatLit, Moskva (1960) (English translation: AFSC Wright-Patterson AFB, O. Report FTD-TT-62-910, (1962))
9. Franklin, G.F., Powell, J.D., Workman, M.L.: Digital Control Systems. Addison-Wesley, Reading, Ma. (1994)
10. Ljung, L., Glad, T.: Modeling of Dynamic System. Prentice Hall, Englewood Cliffs, N.J. (1994)
11. Franklin, G.F., Powell, J.D., Emami-Naeini, A.: Feedback Control of Dynamic Systems. Addison-Wesley, Reading, Ma. (1994)
12. Chen, Y,: Implementation of Lag-lead Compensator for Robots. Proc. of 27th IEEE Conf. on Decision and Control (1988) 174 – 179
13. Bidziński, J. et al.: A Manipulator with an Arm of Serial Parallel Structure. Arch. Mech. Eng. **39**(1-2) (1992) 65 – 78

Research on Fuzzy Inventory Control under Supply Chain Management Environment

Guangyu Xiong and Hannu Koivisto

Automation and Control Institute,
Tampere University of Technology, P.O. Box 692,
FIN-33720 Tampere. Finland
xgy@ac.tut.fi

Abstract. The inventory control has been paid great attention for a long time because of its important in the cost control. This paper presents an approach to use fuzzy control system applied in controlling inventory under supply chain management. A fuzzy control system based on the (s, S) framework is given, and the fuzzy controller in inventory control makes it easier to design the system. Two inventory control approach are applied in two cases study, which are single stage inventory and multiple stage inventory in supply chain. For each case, the inventory control approaches are used based on the give data. The cases are given to illustrate the design procedure of the fuzzy controller and its effectiveness. Some conclusions are drawn out based on the simulation.

1 Introduction

The inventory control has been paid great attention for a long time because of its important in the cost control since Ford Harris' famous EOQ model was first given in 1913[1]. One of the objectives is to minimize the total expected inventory costs per unit time while satisfying the customer demand on time. Under the influence of the supply chain management, traditional inventory control theories and methods are no longer adapted to the new environment. So it will have great practical implication to find new methods for inventory control.

In this paper, we analyse and discuss an approach to use fuzzy control applied in inventory control under supply chain management based two cases, which are the single stage inventory and the multiple stage inventory in supply chain, where the demand process is uniform distribution. Firstly, the classic inventory control methodologies is discussed, where the classic inventory model cost equation is given. Meanwhile, a fuzzy control system based on the (s, S) framework is introduced. The core advantages of a fuzzy controller are robustness under uncertainty, expert experiments considered, and inaccurate information. Therefore, the fuzzy controller in inventory control makes it easier to design the system. Next, we apply two inventory control approach to two cases, the fuzzy control system will be compared with the classical inventory control system for each case. Then, both systems were simulated over a

P.M.A. Sloot et al. (Eds.): ICCS 2003, LNCS 2658, pp. 907–916, 2003.

span of the demand periods for each case using Matlab. The design procedure and fuzzy control effectiveness are illustrated. We compared the fuzzy control system with the classical control system for the different aspect. Finally, some conclusions are drawn out based on the simulation.

2 The Inventory Control Approaches

Inventory is considered as an accumulation of materials or product that will be used to satisfy future demand for materials or product.

Inventory requires the policy to mange the inventory, which may involve the quantitative problems dealing with controlling material are [6]:

- The amount of raw material ordered,
- Time to order raw material,
- Right method used,
- A minimum cost.

Obvious the cost involves the related to the previous actions, i.e. three factors contribute to inventory costs:

- Carrying Cost
  Cost of storage and space,
  Costs incurred from taxes and insurance,
  Cost of obsolescence.
- Shortage Cost
  Cost of lost sales and good will.
- Ordering Cost
  Cost of preparing, placing orders,
  Cost of shipping and handling.

2.1 The Classic Inventory Control Approach

As mentioned above, the choice of the policy will be done with the objective of the minimum costs. An inventory policy is the review and ordering discipline used to control inventory.

The most common policies are: periodic review, which need to review a new order of the amount specified by the order quantity at equal interval of time [2]. Inventory levels are observed at equal intervals of time T. Policy defined by:

$$Q_i = 0 \qquad \text{if } I_i > s \,. \tag{1}$$

Or

$$Q_i = R - I_i \qquad \text{if } I_i \le s \,. \tag{2}$$

Where:
R = target inventory level when order placed
s = reorder level

Ii = inventory level at end of period i
Qi = order size at period i (R-Ii)

Where no order is placed when the inventory level is greater than the reorder level; and an order quantity of (R – Ii) is placed when inventory level is at or below the reorder level.

Three parameters define this policy: R, r, and T. The optimal values for R, r, and T are determined to minimize total inventory costs. We can calculate the general inventory quantity for periodic review as the following equation:

$$\text{Invtoday}(t+1) = \text{Invtoday}(t) - \text{Demand}(t) + \text{Order}(t)\,. \tag{3}$$

Where:
Invtoday (t+1)___Next inventory quantity;
Invtoday (t)___Current inventory quantity;
Order (t)___ Current order quantity.

We can summarize the total cost of a general inventory model for periodic review as a function of its principal component in the following manner [2]:

(Total inventory cost) = (purchasing cost) + (setup cost) + (holding cost) + (shortage cost)

In practice, however, that an inventory model need not include all four types of costs, either because some of the costs are negligible or will render the total cost function too complex for mathematical analysis, so we can delete a cost component only if its effect on the total cost model is negligible. For this paper, we just calculate the total inventory costs using the following cost equation:

$$C_T = C_h + C_o + C_s\,. \tag{4}$$

Where:
C_T___Total inventory cost
C_h___Total holding cost
C_o___Total ordering cost
C_s___Total shortage cost

2.2 Fuzzy Model

As stated previously, a classical (s, S) policy is: at the moment when the inventory drops below level s, order units up to level S [1,4]. This implies there is always a fixed number Q=S-s units ordered. But for desecrate case, if the inventory level is lower than s, place an order. It's possible that the inventory level drops to zero before the new order arrives. We present an improvement approach inventory control for the supply chain using fuzzy control system. With this approach, several aspects of the system were handled in the same manner as in the crisp runs. The number of orders, number of stockouts, and average inventory were all calculated in the same way. The fuzzification occurred when an order was being placed.

When the inventory level was dropped below a point (inventory level), the inventory level and the current period's demand were given membership function values. The membership values were based on the logic described below. To maintain flexibility in the model, all the parameters indicated below are in terms of the model's inputs. This allows the model to adapt to different cases. The core advantages of a fuzzy controller are robustness under uncertainty, expert experiments considered, and inaccurate information.

Fuzzification. We classify the demand, inventory level and order into the three sorts: low, medium and high. The corresponding the membership function is shown in fig. 1.

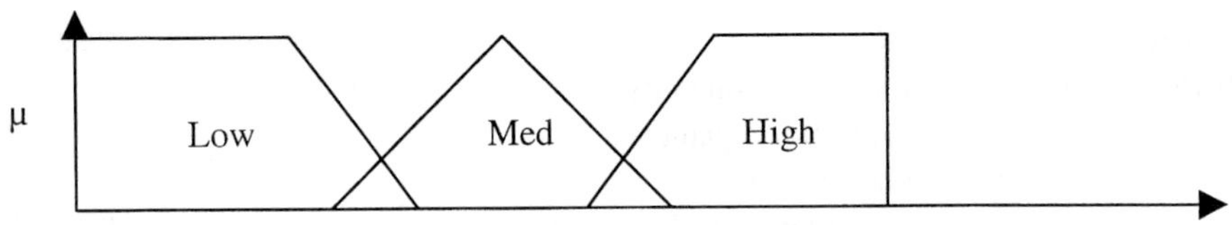

Fig. 1. Membership function

The values on the x-axis represent the different values for different variables. The scalar factor could be changed easily. Varying the value of this scaling unite can be tuning the membership function to make the performance better.

Fuzzy Control Rules. The fuzzy control rules are based on the experiences on inventory ordering policy. The relationship between demand, current inventory level and the amount to be ordered is summarized in Table 1. This table should be read "If demand is High, and the inventory level is Medium, then order High amount of items."

Fuzzy Operators. Selection of the intersection and union operators was constrained by the need to reduce the frequency of ordering. If the operators were compensatory, that could cause the resulting purchase amounts to be further from the extremes, eventually causing an increase in the frequency of orders. This would increase the cost of the system. Another factor of interest was computing cost; complex operators drastically increase the number of computations necessary to run the system. For these reasons, the maximum and minimum operations were selected as the union and intersection operators.

Table 1. Demand, Inventory Level and Order Amount Relations

	Demand		
Inventory Level	Low	Medium	High
Low	High	Med	High
Medium	Med	High	High
High	Low	Low	High

3 Case Study

To compare the performance and implementation of the fuzzy and classical inventory control systems. We apply two inventory control approach to two cases, which are the single stage inventory and the multiple stage inventory in supply chain (Figure 2). Our supply chain looks like the following [6]:

We are a distributor for a certain product. The demand for our product comes from a single source, a retailer (for case 1) or multiple sources, three retailers (for case 2). We get supplied the product from a manufacturer, whom we order from whenever our inventory is at a certain level. For the multiple stage inventory problem, the manufacturer in turn gets supplied from a vendor.

We make the assumptions for the cases:

- The demand has a uniform distribution. Demand is a function of t, which is Demand (t).
- Average daily demand (D) from past data.
- The manufacturer (for case 1) or the vendor (for case 2) is a reliable supplier.
- About delay time: For the singer stage inventory, the delay time for us to receive an order after the order had been placed with the manufacturer is some days (delay). For the multiple stage inventory, the delay time to receive an order after the order had been placed is some days (delay).
- Initialize inventory: For the singer stage inventory, to cover the first delay days, initialize inventory by: inv = D * delay. For the multiple stage inventory, initialize inventory based on D, delay, and number of stages from reliable source.

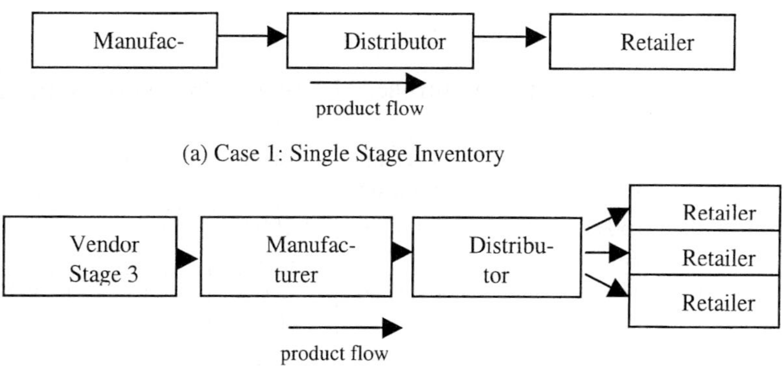

(a) Case 1: Single Stage Inventory

(b) Case 2: Multiple Stage Inventory

Fig. 2. Inventory under supply chain management environment

4 Simulation

Because the order is placed in the beginning of the period, we don't know the exact demand, we create the demand based on its distribution functions. In Matlab, we use rand ( ) to create random values for demand at every period and store the values in an array. The simulation could be based on the cost evaluation equation (4) used the following data [4]:

$$C_T=C_h+C_o+C_s=h\times Q+K\times N+p\times \text{(numbers of Stockouts)}\ . \tag{5}$$

Where:
N= number of times of ordering
Q= average inventory quantity
K= setup cost for placing an order K=12000$
h = holding cost per unit inventory per unit time h= 0.3$/unit/period
p = shortage or stockout cost p=5$/ unit stockout

Table 2. Simulation for singer stage inventory

Series	Fuzzy Inventory Controller			Classical Inventory controller		
	Total cost	Order times	Average inventory	Total cost	Order times	Average inventory
1	8.0640e +005	67	8.0029e+ 003	2.2462e+ 006	187	7.4120e+ 003
2	8.9036e +005	74	7.8758e+ 003	2.4742e+ 006	196	7.2187e+ 003
3	8.5436e +005	71	7.85940e+ 003	2.4622e+ 006	205	7.2713e+ 003

Table 3. Simulation for multiple stage inventory

Series	Fuzzy Inventory Controller			Classical Inventory controller		
	Total cost	Order times	Average inventory	Total cost	Order times	Average inventory
1	9.5160e +006	793	59.6206	1.1904e +007	994	42.6793
2	9.5040e +006	792	61.8706	1.0560e +007	880	42.4915
3	9.4800e +006	790	65.3649	1.1928e +007	994	42.6873

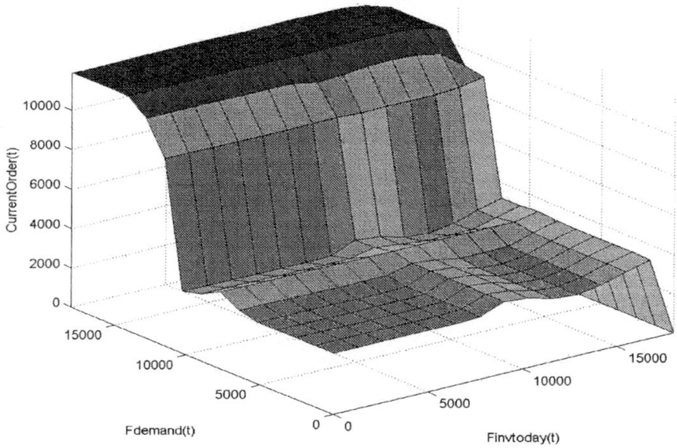

Fig. 3. Control surface for the singer stage inventory

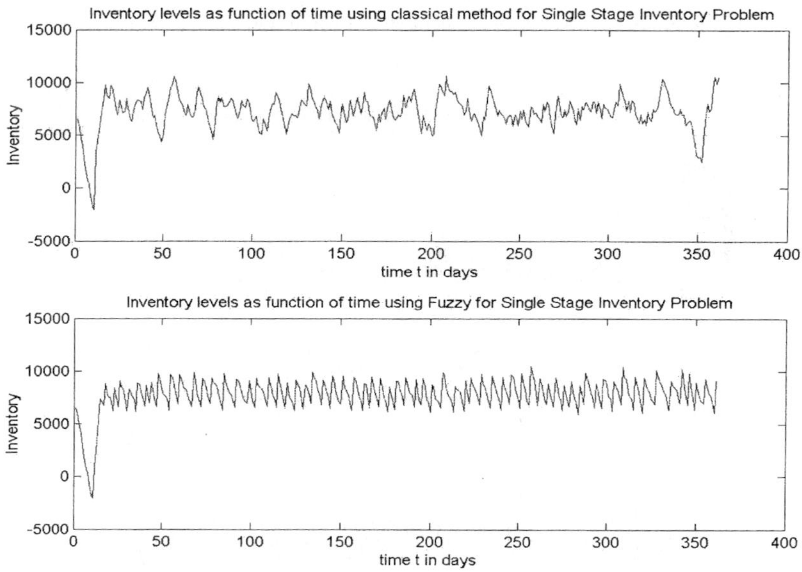

Fig. 4. Case 1: Inventory curve for using fuzzy and classical approach

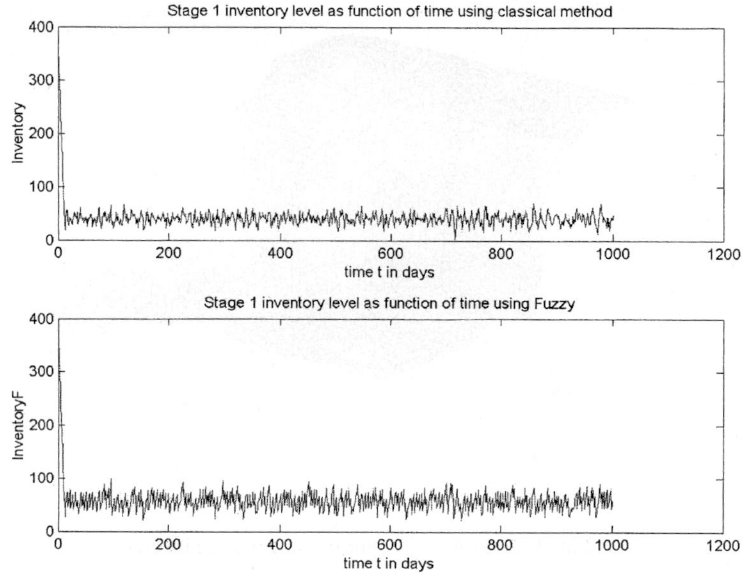

Fig. 5. Case 2: Inventory curve of stage1 using fuzzy and classical approach

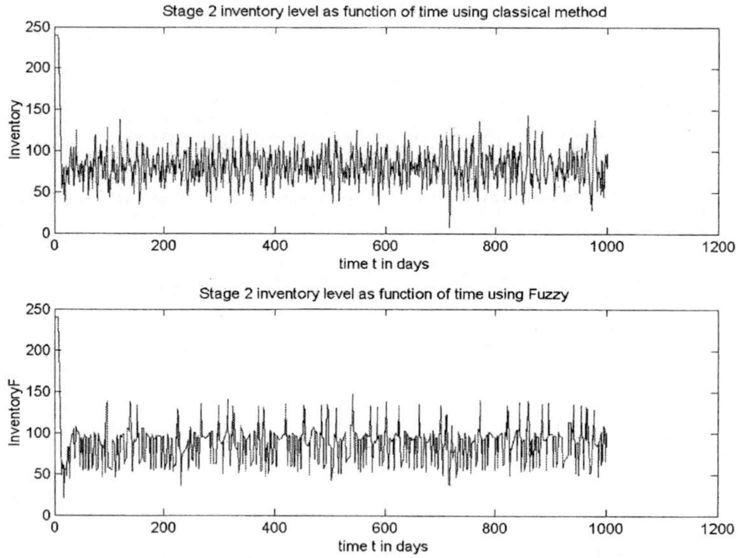

Fig. 6. Case 2: Inventory curve of stage2 using fuzzy and classical approach

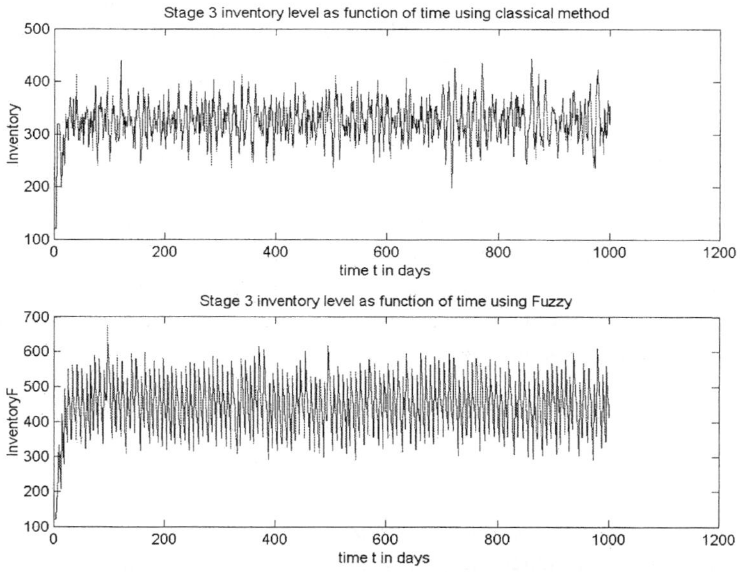

Fig. 7. Case 2: Inventory curve of stage 3 using fuzzy and classical approach

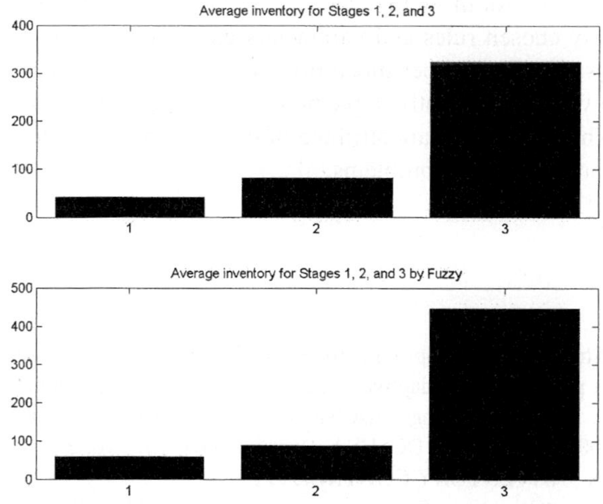

Fig. 8. Case 2: Average inventory using fuzzy and classical approach

To compare the fuzzy and classical inventory control systems based the above cases. This was done by comparing the total cost, excluding ordering cost, of each system using identical demand streams for some periods for two cases. Based on the uniform demand distribution, we ran simulation of 360 periods to evaluate the performance for the two cases. We started with the control rules as the Table 1 and fuzzy operator as above. See Figure 4 -8, Tables 2 and 3 for explicit results.

Table 2 and Table 3 compare the simulation between the fuzzy inventory system and the classical inventory system. Clearly, for the fuzzy inventory system, but its total cost is much lower because its number of times of order is much less than using the classical inventory system even though its average inventory is higher than the classical approach. Also its inventory level is much stable than using the classical inventory system. The simulation has demonstrated that the fuzzy logical approach can successfully tune a fuzzy controller using tuning membership function. The fuzzy controller can cope with the imprecise information, its matrix can be improved by the experts based on the experiences.

5 Conclusion

In this paper we have provided a simple fuzzy logic approach for inventory control for the supply chain environment subject to uniform distribution demand. Our approach explicitly accounts for the differences from the classical approach in design. Based on our simulation studies we see that the approach yields benefits for different levels of variability in the in a supply chain environment, and fuzzy logical shows a powerful tool for inventory control in supply chain. However, fuzzy rules must be chosen carefully. Improperly chosen rules and parameters can result in worse performance than the classical systems. The proper-tuned fuzzy control system is superior to its classical control system when intelligently implemented. Therefore, the fuzzy logic approach is fairly robust. This is a significant attribute of our approach as it increases the viability of its application to inventory problems.

References

1. Shervais, S. Shannon, T.T.: Improving theoretically-optimal and quasi-optimal inventory and transportation policies using adaptive critic based approximate dynamic programming. Neural Networks, 2001. Proceedings. IJCNN '01. International Joint Conference on, Volume: 15–19 July 2001, Washington, DC, USA. Page(s): 1008–1013 vol.2
2. J.B.G.FRENK.: 'INVENTORY CONTROL'. (1999),
3. Hamdy A.Taha.: 'Operations Research, An introduction', Macmillan Publishing Company, 15th Edition. New York
4. Barbara Ballard, Andrew Zozom Jr.: A Comparison of Inventory Systems: (Q, r) vs. Fuzzy Control. OR591: Fuzzy Optimization and Decision Making. (7, May.1996).
5. J E Beasley.: http://www.ms.ic.ac.uk/jeb/or/invent.htmlOR-Notes.
6. http://www.uh.edu/~djwells/guidelines/guidelines.html
7. Guruprasad Pundoor.: Supply Chain Simulation Models for Evaluating the Impact of Rescheuling Frequencies. MASTER'S THESIS, the University of Maryland, Sep, 2002.

Independent Zone Setup Scheme for Re-configurable Wireless Network

Jae-Pil Yoo[1], Kee-cheon Kim[1], and SunYoung Han[1]

School of Computer Science & Engineering. Konkuk University, Seoul, Korea,
+82-2-450-3518

Abstract. ZRP (Zone Routing Protocol), is especially considered to be suitable for moves of dynamic nodes and scalability of RWN. In configuring the ZRP zone, if we remove redundant function of a node, we can get performance improvement in routing. This improved performance of routing is usually from configuring a ZRP zone in which a node doest not have functioning zone head, relaying (internal) node, and peripheral node simultaneously. In such cases, internal routing information size based on link state can be significantly reduced. Hop count is also reduced when routing is reduced and routing loop prevention steps are taken. In this paper[1] we introduce a zone setup scheme based on ZRP in which a node does not have overlapping function as head and relay, and border. Also we introduce a routing scheme for proposed new zone setup scheme.

1 Introduction

RWN (Re-configurable Wireless network) is a scalable network in which a number of nodes can moves around dynamically. A node itself has a sending or receiving wireless interface and with these interfaces nodes can communication each other. When they communicate each other, each node needs to relay some traffic for others without relying on fixed transmission infrastructure. We assume that almost all the nodes in RWN have limited wireless transmission radius and power supply.

Routing protocol that reflects features mentioned above is classified into two types: proactive scheme and reactive scheme. Proactive is a scheme that all the nodes constantly and periodically exchange network path information prior to sending packets such as OSPF [1]. On the other hand, reactive scheme sends path setup signal only when it is needed. It is also called 'on-demand' scheme [2][3]. But it needs packet buffering. Since all the nodes using proactive scheme already know their destination path information before they send data traffic, they are able to send data traffic without delay or buffering. In spite of its merits, the use of proactive scheme heavily depends on the number of nodes in RWN. Therefore, it has a scalability problem. In order to overcome this problem, various clustering (grouping) scheme were introduced [4][5][6][7][8]. Clustering scheme is

[1] This work was supported by grant No. R01-2001-000-00349-0(2002) from the Basic Re-search Program of the Korea Science Engineering Foundation.

P.M.A. Sloot et al. (Eds.): ICCS 2003, LNCS 2658, pp. 917–924, 2003.

to grouping local nodes into a cluster and then separates inter-cluster routing information and intra-cluster routing information for routing efficiency. One of above scheme is ZRP, ZRP uses hybrid scheme; proactive method for intra-zone routing such as link-state and on-demand based reactive method for inter-zone routing for dynamic RWN

2 ZRP and Independent Zone Setup Scheme for RWN

ZRP is a hybrid of reactive and proactive scheme [9][10]. A zone is established as setting the zone center node as the central point of zone and the zone center node has n-hop radius within the zone, in which nodes make maximum n-hop path. Intra-zone routing protocol such as IARP monitors intra-zone routing information and keeps routing table constantly. Inter-zone routing protocol such as IERP finds paths using so-called border-casting. If the zone center node named zone head needs to send some packets to the destination, it sends query packets to its peripheral nodes using border-casting, which groups the same zone. At the same time, one of peripheral nodes that receives query packet also functions as a center node of a certain zone. This node checks whether the destination node resides in its zone or not. If the destination exists, that node sends reply packet to the source node in the reverse order. If not, it repeats border-cast to its peripheral nodes. When a query packet travels around zone's center nodes and peripheral nodes, it adds node's unique ID to its packet header. Finally, when a source node receives a reply packet, it sets reverse path of the reply packet as a routing path. In case of ZRP routing, all the nodes have their zones and also belong to other zones. Each node has functions of zone head, relaying (internal) nodes, and peripheral nodes simultaneously. The number of zones matches exactly with the number of nodes. It means that the number of different intra-zone routing information is set on each zone. Figure 1 Depicts ZRP zone topology. Solid line circles (cells) are needed to route from the source 'node S' to the destination 'node D'. Dotted line circles (cells) also represent each zone, but it does not participate in routing from 'node S' to 'node D'

However, if we possibly prevent a zone area from overlapping each other, and we only permit overlapping peripheral nodes, each node in RWN should function as a center node, internal node, or peripheral node. A node will function as a center node, peripheral node or a relay node when setting up a zone. This removal of overlapping zones results in simplification of node function and routing efficiency and it also reduce the traffic that is used to prevent the routing loop. In case of the existing ZRP, each node belongs to many different zones depending on n-hop radius. It means that if a node belongs to ten different zones, it must monitor different ten routing information. But if we keep zones not to be overlapped, a node does not have to function as a center, internal, and peripheral node at the same time, all the nodes in a zone only need to keep its internal routing information.

At this time, the only problem is to configure the initial zone that does not overlap and keeps the zone without overlaps constantly. If we just adopt

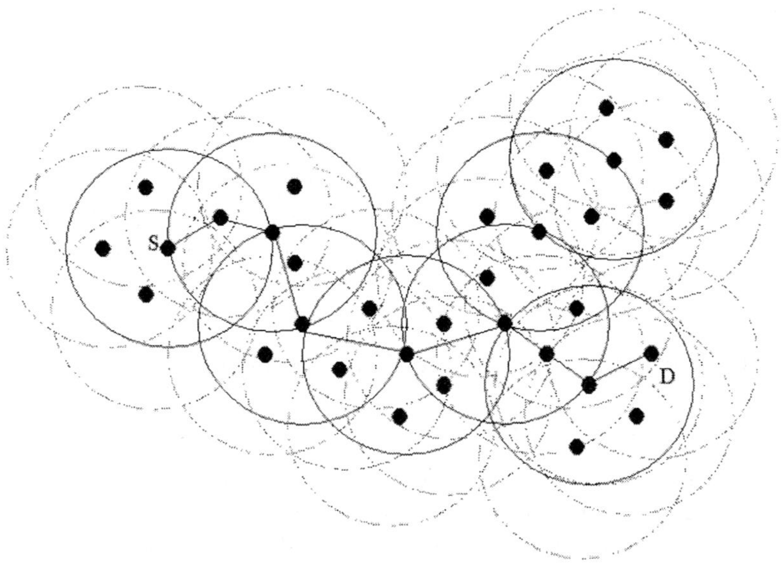

Fig. 1. Zone Topology of ZRP and an example of its IERP operation

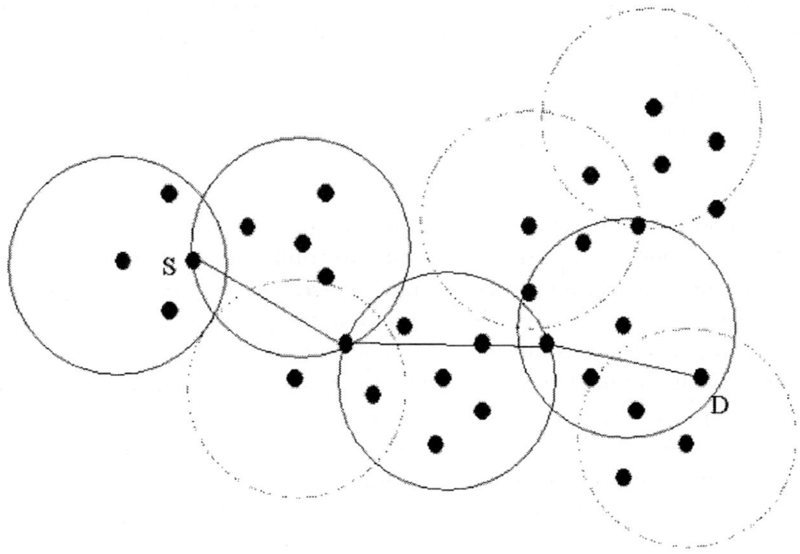

Fig. 2. Newly configured Independent Zone Topology and an example of its IERP operation

previous research result for setting up a simple cluster or hierarchical cluster [5][6], it is not easy to reflect zone radius of ZRP. It results in imbalance of each zone. After all, we need new algorithm for a zone configuration that does not overlap that named independent zone. We also need a scheme that maintains overlap-free zones even when the node moves and zones with modified routing schemes. Figure 2 depicts a newly configured independent zone topology and an example of its IERP operation. Solid line circles (cells) are needed to route from the source 'node S' to the destination 'node D'. Dotted line circles (cells) also mean each zone, but it does not participate in routing from 'node S' to 'node D'. The details of its operation will be explained in the coming sections of this paper.

3 Algorithms for Independent Zone Scheme

3.1 Independent Zone Configuration

Algorithm as follows, tries to make a zone without overlaps during the initial zone configuration allowing as many nodes as possible in a zone radius. Once a zone is configured, it starts making another zone and shares a peripheral node that belongs to each zone to make a connection from one to others.

a) Each node finds out the number of the neighbor node contained within suitable signal strength. Such signal strength also becomes an n-hop radius of independent zone.
b) Each node exchanges neighbor counts with its neighbors. A node that has maximum neighbor node count becomes the center node of an independent zone. It informs their neighbors that it is the center node of the zone for them to join a specific zone. It implicitly tells that all the nodes once joined to a specific zone are not to respond to a zone configuration request except for the peripheral nodes. (Center node for zone X: cX, internal node for zone X: iX, peripheral node for zone X: pX)
c) This step finds candidate zones. A center node for zone X, cX, selects one of its peripheral nodes, pX, and delegates pX to configure the zone by sending request message. pX temporarily becomes a center node, cA, of the virtual zone A. And then, cA (=pX) tries to find out the peripheral node p'A that has the largest neighbor count in its 'virtual zone A'. Undoubtedly, neighbor count means the number of nodes that does not configure the zone at that time. To find out a node that has the largest number of neighbors, we need to just apply step a). If the nodes are uniformly distributed, the peripheral node p'A with the largest neighbor count would be located in a straight line from cX via pX, that is to say, p'A will be the farthest peripheral node belongs to the virtual zone A from cX. This scheme makes each zone not to be overlapped each other and allows a zone to have the maximum number of nodes as possible. Fig.3 depicts this step.
d) p'A with the largest neighbor count configures a new zone Y. At the same time, p'A becomes a center node cY of zone Y. Step b) can be applied in

configuring the zone. At this time, pX naturally belong to the zone Y. so pX also becomes pXY, it then acts as a peripheral node of zone X and zone Y at the same time.

e) Repeat step c) and d) to extend the zone coverage. If a node heard requesting messages to configure a zone that already belongs to a certain zone, it just discards the requesting message.

f) During the step 1 through 5, if some nodes could not configure a zone for the critical time value, they just make zones with them. Also, step b) can be applied in configuring the zones. However, a configured zone needs to have at least on minimum peripheral node sharing with other zones to constitute a routing path.

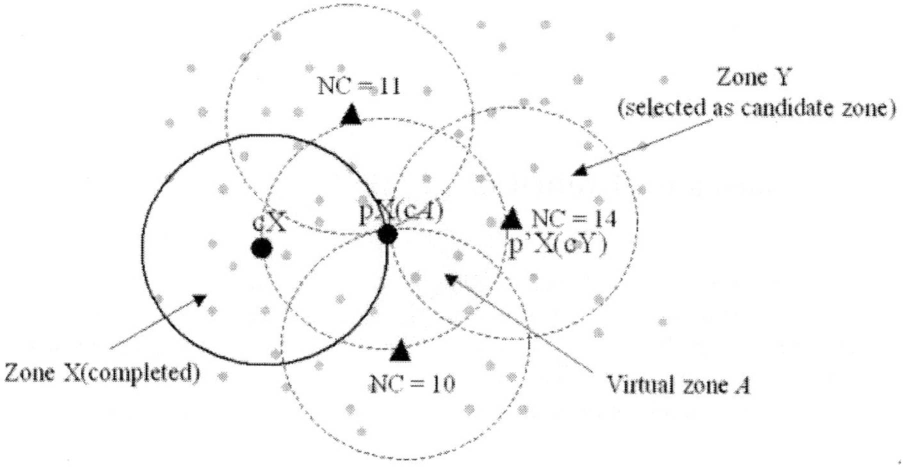

Fig. 3. Finding next candidate zone

3.2 Independent Zone Maintenance

a) Movement of a node within a zone does not affect the external routings. It only affects the internal routing information. So it easily fixes up the internal routing information changes

b) If a peripheral node of a zone moves to another location, and if the peripheral node just breaks routing path between two zones, one of the center node of two zones needs to strengthen the signal power to find out the substitute peripheral node to connect the two broken zones.

c) If the center node of a zone moves to another location, and if the nodes within the zone cannot communicate each other, the zone should be divided into two zones to restore the connection. Some time later, if some nodes reach a location that connects the split zones into one, they can be merged.

3.3 Independent Zone Routing

A source node s of a zone X, sX, needs to follow the following steps to find a routable path to the destination node d of zone Z, dZ.

a) sX tries to find out dZ in its routing table belongs to zone X. if dZ is found in the internal routing table of sX, sX can send packets to dZ directly .
b) If sX could not find dZ, sX border-casts query message to all the peripheral nodes pX in the zone X.
c) One of the peripheral node pX in zone X may also belongs to other zones and also has many routing tables depends on the number of zones it belongs to. So, pX that receives a query message tries to find out the destination node dZ in its routing table except for the zone from which query message arrived.
d) If dZ is not found in each routing table on the peripheral node, pX border-casts the query message to the rest of the peripheral nodes except for the zone from which query message came.
e) Repeats c), d) until source node sX finds the destination node dZ.

4 Performance Evaluation

The results of our simulation are presented in the following graphs. We use the assumption that a network of 200 nodes moves in an area of 1000x1000m Squares. In this model every node can select a random location and moves there at a uniformly chosen speed of between 0 and 1m/s. We ignored border-casting delay and IARP retrieving delay setting as 0ms respectively. A node can send query packets by maximum of ten times in one second.

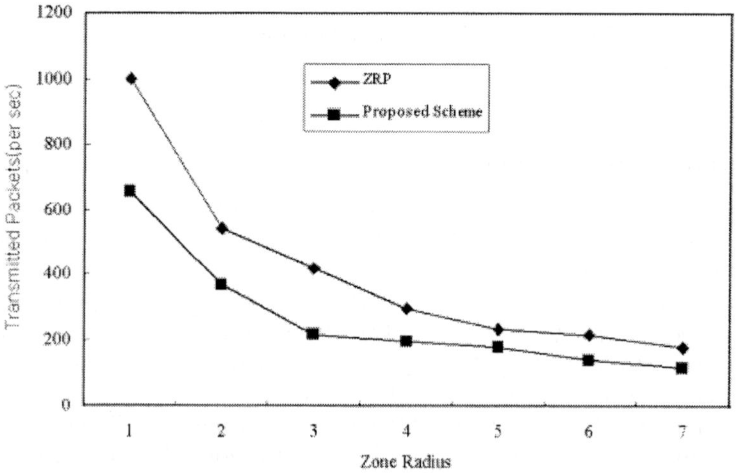

Fig. 4. Traffic per Route Query after initial setup

Figure 4 illustrates the transmitted packets of basic ZRP and the proposed scheme supporting the independent zone with IERP operation. The number of transmitted packets depends on the zone radius. But, in most cases, the proposed scheme shows less IERP operation packets in a network. IERP Operation is mostly depends on the number of zones. So it is natural that the proposed scheme is more efficient than the basic ZRP, because the proposed scheme has by far less number of zones than that of the basic ZRP. Besides, the proposed scheme just needs nearly half the number of zones than the basic ZRP with IERP operation.

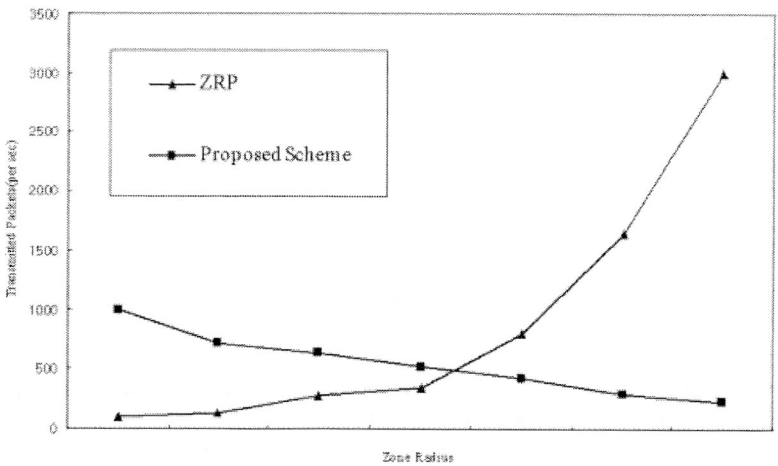

Fig. 5. Control Traffic variation due to mobile

Figure 5 illustrates the transmitted packets of basic ZRP and the proposed scheme supporting the independent zone with re-configuration due to the movement of a node. Basic ZRP shows when a zone radius grows larger, much more control traffic flows on the network. In case of the proposed scheme, when its zone radius is small, it uses much traffic to re-configure the zone than the basic ZRP. However, as the zone radius grows larger, it uses less traffic to re-configure the zone in comparison with the basic ZRP. We expect that intra-zone movement and inter-zone movement ratio causes this result. With the smaller zone radius, almost all the movement of node are considered to be the inter-zone movement. So it needs to re-configure the zone frequently. But the larger zone radius means more inter-zone movement. So it will have less chance to re-configure the zones.

5 Conclusion

In this paper, we proposed a new independent zone configuration scheme for ZRP in which a node does not have multiple functions as head, internal,

and peripheral node. We also introduced a modified routing scheme for the newly established zone scheme. This scheme tries to make a zone without overlaps during the initial zone configuration and allows as many nodes as possible in a zone radius. This new independent zone configuration scheme has the following advantages as shown in the performance evaluation. Internal routing information size based on the link state can be significantly reduced. A hop count is reduced and the routing loop prevention steps are possible with less cost. Out simulation results tells us that our independent zone scheme is more suitable for less dynamic large RWN.
Zone re-configuration algorithm of the proposed scheme is more complex than ZRP since It needs to keep the zone not to be overlapped all the time. This produces negative performance especially for dynamically moving node in narrow radius zone. Since the performance of RWN is affected by such parameters as node movement and zone radius, as a future research, we will research on the dynamically adaptive RWN to such parameters to enhance the performance.

References

1. J. Moy., OSPF version 2. RFC 2178, Internet Engineering Task Force, July 1997.
2. C.E. Perkins and E.M. Royer., Ad-Hoc On-Demand Distance Vector Routing. In Proceedings of the Second Annual IEEE Workshop on Wireless and Mobile Computing Systems and Applications, February 1999.
3. J.J.Garcia-Luna-Aceves and M.Spohn, Source-Tree Routing in Wireless Networks, In Proc, IEEE ICNP 99, 7th Intl, Conf. On Network Protocols, Toronto, Canada, Oct 1999.
4. D.J. Baker and A. Ephremides., A Distributed Algorithm for Organizing Mobile Radio Telecommunication Networks. In Proceedings of the Second International Conference on Distributed Computer Systems, April 1981.
5. C.-C. Chiang., Routing in Clustered Multi-hop, Mobile Wireless Networks.In Proceeding of ICOIN Nobemver,1996.
6. M. Gerla and J.T.-C. Tsai, Multi-cluster, Mobile, Multimedia Radio Network, Wireless Networks, October 1995.
7. R. Krishnan, R. Ramanathan, and M. Steenstrup., Optimization Algorithms for Large Self-Structuring Networks. In Proceedings of IEEE INFOCOM '99, March 1999.
8. J. Zavgren., NTDR Mobility Management Protocols and Procedures. In Proceedings of the IEEE Military Communications Conference, November 1997.
9. Z.J. Hass and M.R. Pearlman., The Performance of Query Control Schemes for the Zone Routing Protocol. In Proceedings of SIGCOMM '98, September 1998.
10. M.R. Pearlman and Z.J. Haas., Determining the Optimal Configuration of the Zone Routing Protocol. IEEE Journal on Selected Areas of Communications, August 1999.

Finding Synchronization-Free Parallelism for Non-uniform Loops

Volodymyr Beletskyy

Faculty of Computer Science, Technical University of Szczecin, Zolnierska 49 st., 71-210 Szczecin, Poland,
vbeletskyy@wi.ps.pl

Abstract. A technique, permitting us to find synchronization-free parallelism in non-uniform loops, is presented. It is based on finding affine space partition mappings. The main advantage of this technique is that it allows us to form constraints for finding mappings directly in a linear form while known techniques result in building non-linear constraints which should next be linearized. After finding affine space partition mappings, well-known code generation approaches can be applied to expose loop parallelism. The technique is illustrated with two examples.

1 Introduction

A lot of transformations have been developed to expose parallelism in loops, minimize synchronization, and improve memory locality in the past [1],[3],[4],[6],[7],[8],[9],[10],[11],[12],[14],[15],[19]. However, there are the following questions. Which of these methods permit us to find synchronization-free parallelism and what is their complexity for non-uniform loops.

According to a study by Sass and Mutka[18], a majority of the loops in scientific code are imperfectly nested, and a majority of the performance-increasing techniques developed in the past assume that loops are perfectly nested, that is, imperfectly nested loops deserve more attention from the research community.

This paper presents a technique permitting us to find synchronization-free parallelism in non-uniform loops. We refer to a particular execution of a statement for a certain iteration of the loops, which surround this statement, as an operation. The operations of a loop are divided into partitions, such that dependent operations are placed in the same partition. A partitioning is described by an affine mapping for each loop statement.

An *m*-dimensional affine partition mapping for statement *s* in a loop is an *m*-dimensional affine expression $\overline{\phi}_s = C_s\overline{i} + \overline{c_s}$, which maps an instance of statement *s*, indexed by its iteration vector $\overline{i}$, to an *m*-dimensional vector. Given affine mappings, well-known techniques for generating parallel code can be applied, for example, [2],[4],[5],[17].

P.M.A. Sloot et al. (Eds.): ICCS 2003, LNCS 2658, pp. 925–934, 2003.

2 Dependence Analysis

Our algorithm is based on the dependence analysis proposed by Pugh and Wonnacott [16]. That analysis permits us to extract exact dependence information for any single structured procedure in which the expressions in the subscripts, loop bounds, and conditionals are affine functions of the loop indices and loop-independent variables, and the loop steps are known constants. Dependences are presented with dependence relations. A dependence relation is a mapping from one iteration space to another, and is represented by a set of linear constraints on variables that stand for the values of the loop indices at the source and destination of the dependence and the values of the symbolic constants. A dependence relation is a tuple relation. An integer *k*-tuple is a point in Z^k. A tuple relation is a mapping from tuples to tuples.

The basic merits of the dependence analysis proposed by Pugh and Wonnacott are as follows: i) it is exact; ii) it is valid for both perfectly and imperfectly nested loops; iii) it permits value-based dependences to be calculated.

A dependence between operations $\overline{I}$ and $\overline{J}$, which are the source and destination of the dependence, respectively, is value-based if: $\overline{I}$ is executed before $\overline{J}$; $\overline{I}$ and $\overline{J}$ refer to a memory location *M*, and at least one of these references is a write; the memory location *M* is not written between operation $\overline{I}$ and operation $\overline{J}$.

The dependence analysis by Pugh and Wonnacott is implemented in Petit, a research tool for doing dependence analysis and program transformations. To carry out dependence analysis manually, the Omega calculator can be applied [13].

An affine loop nest is non-uniform if it originates non-uniform dependence relations represented by an affine function *f* that expresses the dependence sources $\overline{I}$ in terms of the dependence destinations $\overline{J}$ ($\overline{I}=f(\overline{J})$) or vice versa.

An algorithm proposed in this paper is applicable for those loops that meet the restrictions of the dependence analysis proposed by Pugh and Wonnacott [16].

3 Space-Partition Constraints

Our approach is applicable to the following imperfectly nested loop considered in [19]

$$
\begin{array}{ll}
 & \text{do } x_1 = L_1, U_1 \\
S_{1a}: & \quad H_{1a}(x_1) \\
 & \quad \text{do } x_2 = L_2, U_2 \\
S_{2a}: & \quad\quad H_{2a}(x_1, x_2) \\
 & \quad\quad \ldots \\
 & \quad\quad \text{do } x_n = L_n, U_n \\
S_n: & \quad\quad\quad H_n(x_1, \ldots, x_n) \\
 & \quad\quad\quad \ldots
\end{array}
\qquad (1)
$$

S_{2b}: $H_{2b}(x_1, x_2)$

S_{1b}: $H_{1b}(x_1)$,

where the loop bounds of x_k are affine constraints over surrounding loop variables $x_1, \ldots, x_{k-1}$ and some symbolic integer constants, or formally:

$$L_k = \max(l_{k,1}, l_{k,2}, \ldots)$$
$$U_k = \max(u_{k,1}, u_{k,2}, \ldots),$$

where

$$l_{k,p} = \lceil (l_{k,p}^0 + l_{k,p}^1 x_1 + \ldots + l_{k,p}^{k-1} x_{k-1}) / l_{k,p}^k \rceil$$

$$u_{k,p} = \lfloor (u_{k,p}^0 + u_{k,p}^1 x_1 + \ldots + u_{k,p}^{k-1} x_{k-1}) / u_{k,p}^k \rfloor$$

and all $l_{k,p}$ and $u_{k,p}$ are integer constants, except possibly for $l_{k,p}^0$ and $u_{k,p}^0$, which may be symbolic constraints but must still be loop invariants in the loop nest. The ceiling and floor functions are introduced to convert rationals to integers. In general, a lower(upper) loop bound is a maximum(minimum) of affine constraints with rational coefficients. This ensures that the space defined by any set of loops in the loop nest is a convex polyhedron.

In this section, we consider the following task. Given a set of dependences originated by a loop and presented with dependence relations

$$D = \{\overline{I}_j -> \overline{J}_j, j = 1,2,\ldots,q\},$$

find m_s-dimensional affine space partition mappings $\overline{\phi}_s = C_s \overline{i} + \overline{c}_s$ for each *s=1a, 2a, ..., n,...,2b, 1b,* such that $\overline{\phi}_{si}(\overline{I}_j) = \overline{\phi}_{sk}(\overline{J}_j) = \overline{P}_m$, where *si, sk* are the statements which instances originate the source and destination of the dependence $\overline{I}_j -> \overline{J}_j$, C_s is a matrix of dimensions m_s x *n,* $\overline{c}_s$ is an m_s-dimensional vector representing a constant term, $\overline{P}_m$ is a vector representing the identifier of a processor to execute the source and destination of the dependence $\overline{I}_j -> \overline{J}_j$.

Let $\overline{I}_j, \overline{J}_j$ be represented in the following form

$$\overline{I}_j = A_{1j} * \overline{i1}_j + \overline{B1}_j, \quad \overline{J}_j = A_{2j} * \overline{i2}_j + \overline{B2}_j,$$

where $\overline{I}_j, \overline{J}_j$ are $m1_j$ and $m2_j$-dimensional vectors, respectively, $m1_j <= n$, $m2_j <= n$, *n* is the number of the loop nests, $\overline{B1}_j$, $\overline{B2}_j$ are $m1_j$ and $m2_j$-dimensional vectors, respectively, $\overline{i1}_j$ and $\overline{i2}_j$ are *n*-dimensional vectors, A_{1j} and A_{2j} are matrices of dimensions $m1_j$ x *n* and $m2_j$ x *n,* respectively.

Let us write matrices A_{1j} and A_{2j} in the following form

$$A_{1j} = [\overline{A}_{1j}^{1} \ \overline{A}_{1j}^{2} \ \cdots \overline{A}_{1j}^{n}], \quad A_{2j} = [\overline{A}_{2j}^{1} \ \overline{A}_{2j}^{2} \ \cdots \overline{A}_{2j}^{n}],$$

where $\overline{A}_{1j}^{i}$ and $\overline{A}_{2j}^{i}$, $i=1,2,\ldots,n$ represent the columns of A_{1j} and A_{2j}, respectively.

If a dependence $\overline{I}_j -> \overline{J}_j$, $j \in [1, \ q]$ is the self dependence, that is, it is originated with the same statement s, we seek an affine space partition mapping $\overline{\phi}_s = C_s \overline{i} + \overline{c}_s$ such that the following condition is satisfied

$$C_s \overline{I}_j + \overline{c}_s = C_s \overline{J}_j + \overline{c}_s$$

or

$$C_s \overline{I}_j - C_s \overline{J}_j = 0,$$

which means that the same processor executes operations $\overline{I}_j$ and $\overline{J}_j$.

Let us rewrite the equation above as follows

$$C_s(\overline{A}_{1j}^{1} i1_j^1 + \overline{A}_{1j}^{2} i1_j^2 + \ldots + \overline{A}_{1j}^{n} i1_j^n + \overline{B1}_j) -$$

$$- C_s(\overline{A}_{2j}^{1} i2_j^1 + \overline{A}_{2j}^{2} i2_j^2 + \ldots + \overline{A}_{2j}^{n} i2_j^n + \overline{B2}_j) = 0,$$

where $i1_j^1, i1_j^2, \ldots, i1_j^n$ and $i2_j^1, i2_j^2, \ldots, i2_j^n$ are the coordinates of $\overline{i1}_j$ and $\overline{i2}_j$, respectively, and transform it to the form

$$\langle \overline{C}_s, \overline{A}_{1j}^{1} \rangle i1_j^1 + \langle \overline{C}_s, \overline{A}_{1j}^{2} \rangle i1_j^2 + \ldots + \langle \overline{C}_s, \overline{A}_{1j}^{n} \rangle i1_j^n + \langle \overline{C}_s, \overline{B1}_j \rangle -$$

$$- \langle \overline{C}_s, \overline{A}_{2j}^{1} \rangle i2_j^1 - \langle \overline{C}_s, \overline{A}_{2j}^{2} \rangle i2_j^2 - \ldots - \langle \overline{C}_s, \overline{A}_{2j}^{n} \rangle i2_j^n - \langle \overline{C}_s, \overline{B2}_j \rangle = 0, \quad (2)$$

where $\langle \overline{x}, \overline{y} \rangle$ denotes the inner product of vectors $\overline{x}$ and $\overline{y}$, $\overline{C}_s$ represents an arbitrary row of C_s.

If a dependence $\overline{I}_j -> \overline{J}_j$, $j \in [1, \ q]$ is originated with two different statements $s1$ and $s2$, we seek two affine space partition mappings $\overline{\phi}_{s1} = \overline{C}_{s1} \overline{i} + \overline{c}_{s1}$ and $\overline{\phi}_{s2} = \overline{C}_{s2} \overline{i} + \overline{c}_{s2}$ such that the following condition is satisfied

$$C_{s1} \overline{I}_j + \overline{c}_{s1} = C_{s2} \overline{J}_j + \overline{c}_{s2}.$$

Let us rewrite the above condition as follows

$$C_{s1}(\overline{A}_{1j}^{1} i1_j^1 + \overline{A}_{1j}^{2} i1_j^2 + \ldots + \overline{A}_{1j}^{n} i1_j^n + \overline{B1}_j) + \overline{c}_{s1} =$$

$$C_{s2}(\overline{A}_{2j}^{1} i2_j^1 + \overline{A}_{2j}^{2} i2_j^2 + \ldots + \overline{A}_{2j}^{n} i2_j^n + \overline{B2}_j) + \overline{c}_{s2},$$

and transform it to the form

$$\langle \overline{C}_{s1}, \overline{A}^{1}_{1j} \rangle i1^{1}_{j} + \langle \overline{C}_{s1}, \overline{A}^{2}_{1j} \rangle i1^{2}_{j} + \ldots + \langle \overline{C}_{s1}, \overline{A}^{n}_{1j} \rangle i1^{n}_{j} + \langle \overline{C}_{s1}, \overline{B1}_{j} \rangle + c_{s1} - \tag{3}$$

$$- \langle \overline{C}_{s2}, \overline{A}^{1}_{2j} \rangle i2^{1}_{j} - \langle \overline{C}_{s2}, \overline{A}^{2}_{2j} \rangle i2^{2}_{j} - \ldots - \langle \overline{C}_{s2}, \overline{A}^{n}_{2j} \rangle i2^{n}_{j} - \langle \overline{C}_{s2}, \overline{B2}_{j} \rangle - c_{s2} = 0,$$

where $\overline{C}_{s1}, \overline{C}_{s2}$ represent an arbitrary row of C_{s1}, C_{s2}, respectively, c_{s1}, c_{s2} are unknown constant terms which are dependent on $\overline{C}_{s1}, \overline{C}_{s2}$.

Let us introduce an r_j-dimensional vector $\bar{i}_j$ which consists of all uncommon coordinates(having different names) of $\overline{I}_j, \overline{J}_j$ and the coordinates of this vector be $i^1_j, i^2_j, \ldots, i^{r_j}_j$, $r_j <= m1_j + m2_j$. Rewrite equations (2) and (3) in the following form

$$\sum_{k=1}^{r_j} D^k_j \; i^k_j + d_j = 0, \tag{4}$$

where D^k_j and d_j are formed as follows:

i) for self dependences, it is the sum of all those $\langle \overline{C}_{s1}, \overline{A}^{m}_{1j} \rangle$ and $-\langle \overline{C}_{s1}, \overline{A}^{p}_{2j} \rangle$ for which the following condition holds $i1^m_j = i2^p_j = i^k_j$; $d_j = \langle \overline{C}_{s1}, \overline{B1}_j \rangle - \langle \overline{C}_{s1}, \overline{B2}_j \rangle$;

ii) for dependences originated with two different statements, it is the sum of all those $\langle \overline{C}_{s1}, \overline{A}^{m}_{1j} \rangle$ and $-\langle \overline{C}_{s2}, \overline{A}^{p}_{2j} \rangle$ for which the following condition holds $i1^m_j = i2^p_j = i^k_j$; $d_j = \langle \overline{C}_{s1}, \overline{B1}_j \rangle - \langle \overline{C}_{s2}, \overline{B2}_j \rangle + c_{s1} - c_{s2}$.

Algorithm. Find affine space partition mappings for a loop originating the dependences defined by set *D*.

1. From each dependence $\overline{I}_j -> \overline{J}_j$, $j = 1,2,\ldots,q$, build the constraint in the form of (4).
2. Construct a system of linear equations of the form

   $$D^k_j = 0,$$
   $$d_j = 0,\; j = 1,2,\ldots,q,\; k = 1,2,\ldots,r_j$$

   which we rewrite as

   $$A\bar{x} = 0,$$

   where $\bar{x}$ is a vector representing all the unknown coordinates of $\overline{C}_s$ and constant terms c_s of the affine space partition mappings, *s=1a, 2a, ..., n,..., 2b, 1b*.

The remaining steps are the same as in the algorithm proposed in [15], namely

3. Eliminate all the unknowns c_s from $A\bar{x}=0$ with the Gaussian Elimination algorithm. Let the reduced system be $A'\bar{x}'=0$, where $\bar{x}'$ represents the unknown coordinates of $\overline{C}_s$.
4. Find the solution to $A'\bar{x}'=0$ as a set of basis vectors spanning the null space of A'.
5. Find one row of the desired affine partition mapping from each basic vector found in step 4. The coordinates of $\overline{C}_s$ are formed directly by the basic vector; the constant terms c_s are found from the coordinates of $\overline{C}_s$ using $A\bar{x}=0$.

After finding mappings, well-known techniques for generating parallel code can be applied, for example, [2],[4],[5],[17] and they are out of the scope of this paper.

4 Examples

Let us illustrate the technique presented by means of the two following examples.

Example 1:

```
for (i = 1; i <= n; i++)
   for (j = 1; j<= n; j++)
      for (k = 1; k <= n; k++)
            s1: a(j)=b(i);
```

For this loop, the dependences found with Petit are as follows

output *s1: a(j)* → *s1: a(j)*
{[i,j,k] → [i,j,k'] : 1 <= k < k' <= n && 1 <= i <= n && 1 <= j <= n},

output *s1: a(j)* → *s1: a(j)*
{[i,j,k] → [i',j,k'] : 1 <= i < i' <= n&&1 <= j <= n&&1 <= k<=n&&1<= k' <= n}.

We seek a mapping of the form $\bar{\phi}_{s1}=[C_{11}\ \ C_{12}\ \ C_{13}]\bar{i}$. According to our approach, we first form the following constraint

$$C_{11}*i+C_{12}*j+C_{13}*k=C_{11}*i+C_{12}*j+C_{13}*k'$$

$$C_{11}*i+C_{12}*j+C_{13}*k=C_{11}*i'+C_{12}*j+C_{13}*k',$$

which we simplify to the following form

$$C_{13}*(k-k')=0,$$
$$C_{11}(i-i')+C_{13}(k-k')=0.$$

The resulting constraint is as follows

$$C_{11} = 0$$
$$C_{13} = 0 .$$

The linearly independent solution to this system is

$$C_{11} = 0, C_{12} = 1, C_{13} = 0 .$$

Applying the Omega code generator (free available at ftp://ftp.cs.umd.edu/pub/omega) for the transformation of the source loop by means of the space partition mapping $C_1 = [0\ 1\ 0]$, we have got the following parallel code

```
parfor(p = 1; p <= n; p++)
  for(t1 = 1; t1 <= n; t1++)
    for(t2 = 1; t2 <= n; t2++)
      s1: a(p) = b(t1);
```

where `for` and `parfor` denote serial and parallel loops, respectively. The outer loop gives space partitioning while the inner loops define the statement instances executed serially by a given processor *p*.

Consider the following imperfectly nested loop.
Example 2:

```
for (i = 1; i <= n; i++){
   for (j = 1; j <= n; j++){
      for (k = 1; k <= n; k++){
         s1: c(i,j,k)=a(N-j,k);
      }
      s2: a(N-j+1,i)= b(j,k);
   }
}
```

This loop originates the following dependences found with Petit

anti *s1: a(N-j,k)* → *s2: a(N-j+1,i)*

{[i,j,i] → [i,j+1] : 1 <= i <= N && 1 <= j < N,},

anti *s1: a(N-j,k)* → *s2: a(N-j+1,i)*

{[i,j,k] → [k,j+1] : 1 <= i < k <= N && 1 <= j < N},

flow *s2: a(N-j+1,i)*→ *s1: a(N-j,k)*

{[i,j] → [i',j-1,i] : 1 <= i < i' <= N && 2 <= j <= N}.

We seek mappings of the form $\bar{\phi}_{s1} = [C_{11} \quad C_{12} \quad C_{13}]\bar{i} + c_1$ and $\bar{\phi}_{s2} = [C_{21} \quad C_{22}]\bar{i} + c_2$. Firstly, we form the following constraint

$$C_{11}*i + C_{12}*j + C_{13}*i + c_1 = C_{21}*i + C_{22}*(j+1) + c_2$$

$$C_{11}*i + C_{12}*j + C_{13}*k + c_1 = C_{21}*k + C_{22}*(j+1) + c_2$$

$$C_{21}*i + C_{22}*j + c_2 = C_{11}*i' + C_{12}*(j-1) + C_{13}*i + c_1$$

and next transform it to the form

$$(C_{11} - C_{21} + C_{13})*i + (C_{12} - C_{22})*j + c_1 - c_2 - C_{22} = 0$$

$$C_{11}*i + (C_{12} - C_{22})*j + (C_{13} - C_{21})*k + c_1 - c_2 - C_{22} = 0$$

$$(C_{21} - C_{13})*i + (C_{22} - C_{12})*j - C_{11}*i' + c_2 - c_1 + C_{12} = 0.$$

On the basis of the equations above, we construct the following constraint

$$\begin{aligned} &C_{11} - C_{21} + C_{13} = 0 \\ &C_{12} - C_{22} = 0 \\ &c_1 - c_2 - C_{22} = 0 \\ &C_{11} = 0 \\ &C_{13} - C_{21} = 0 \\ &C_{21} - C_{13} = 0 \\ &c_2 - c_1 + C_{12} = 0. \end{aligned} \tag{5}$$

Eliminating c_1, c_2, we get

$$\begin{aligned} &C_{11} - C_{21} + C_{13} = 0 \\ &C_{12} - C_{22} = 0 \\ &C_{12} - C_{22} = 0 \\ &C_{11} = 0 \\ &C_{13} - C_{21} = 0 \\ &C_{21} - C_{13} = 0. \end{aligned}$$

The linearly independent solution to the system above is

$$C_{11} = 0, C_{12} = 1, C_{13} = 1, C_{21} = 1, C_{22} = 1.$$

From system (5) we find that $c_1 = 1$, $c_2 = 0$.

Applying the Omega code generator for the transformation of the source loop by means of the space partition mappings found, we have got the following parallel code

```
parfor(p=2; p<= 2*N+1; p++) {
  for(t1=1; t1<= N; t1++) {
    for(t2=max(-N+p-1,1); t2<=min(-t1+p-1,N); t2++) {
```

```
      s1(t1,t2,p-t2-1);
    }
    if (t1 >= 2 && t1 <= p-1 && t1 >= -N+p) {
      s1(t1,p-t1,t1-1);
    }
    if (t1 >= 2 && t1 <= p-1 && t1 >= -N+p) {
      s2(t1,p-t1);
    }
    for(t2 = max(-t1+p+1,1); t2 <= min(p-2,N); t2++) {
      s1(t1,t2,p-t2-1);
    }
    if (p <= N+1 && t1 <= 1) {
      s2(1,p-1);
    }
  }
}
```

where `for` and `parfor` denote serial and parallel loops, respectively; *s1* and *s2* are the statements of the source loop.

The outer loop gives space partitioning while the inner loops define the statement instances which should be executed serially by a given processor *p*.

5 Related Work and Conclusion

Unimodular loop transformations[3],[19], permitting the outer loop to be parallelized, find synchronization-free partitions. But unimodular transformations do not allow such transformations as loop fission, fusion, scaling, reindexing, or reordering.

Techniques presented in [1],[11] enable finding synchronization-free partitioning only for perfectly nested loops, supposing statements within each loop iteration are indivisible.

The affine partitioning framework, considered in many papers, for example, [8],[9],[10],[15], unifies a large number of previously proposed loop transformations. It is the most powerful framework for the loop parallelization today allowing us to parallelize loops with both uniform and non-uniform dependences.

Work [15] is most closely related to ours. In contrast to that work and other known approaches, our technique permits us to form constraints for finding affine space partition mappings for non-uniform loops directly in a linear form without the necessity of applying the Farkas lemma to linearize the constraint, and hence it is less time-consuming than that of work [15] and other known approaches.

In the future research, we plan to extend our technique to find affine time partition mappings for the non-uniform loops which do not allow synchronization-free parallelization.

References

[1] Amarasinghe, S.P., Lam, M.S.: Communication optimization and code generation for distributed memory machines. In: Proceedings of the SIGPLAN'93 (1993) 126–138

[2] Ancourt, C., Irigoin, F.: Scanning polyhedra with do loops. In: Proceedings of the Third ACM/SIGPLAN Symposium on Principles and Practice of Parallel Programming, ACM Press (1991) 39–50

[3] Banerjee, U.: Unimodular transformations of double loops. In: Proceedings of the Third Workshop on Languages and Compilers for Parallel Computing (1990) 192–219

[4] Boulet, P., Darte, A., Silber, G.A., Vivien, F.: Loop paralellelization algorithms: from parallelism extraction to code generation. Technical report (1997)

[5] Collard, J.F., Feautrier, P., Risset, T.: Construction of do loops from systems of afffne constraints. Technical Report 93–15, LIP, Lyon (1993)

[6] Darte, A., Risset, T., Robert, Y.: Loop nest scheduling and transformations. In Dongarra, J., Tourancheau, B., eds.: Enviroments and tools for parallel science computing. North Holland (1993)

[7] Darte, A., Silber, G., Vivien, F.: Combining retiming and scheduling techniques for loop parallelization and loop tiling. Technical Report 96–34, Laboratoire de l'Informatique du Parallelisme (1996)

[8] Feautrier, P.: Some efficient solutions to the affine scheduling problem, part i, one dimensional time. International Journal of Parallel Programming 21 (1992) 313–348

[9] Feautrier, P.: Some efficient solutions to the affine scheduling problem, part ii, multidimensional time. International Journal of Parallel Programming 21 (1992) 389–420

[10] Feautrier, P.: Toward automatic distribution. Journal of Parallel Processing Letters 4 (1994) 233–244

[11] Huang, C., Sadayappan, P.: Communication-free hyperplane partitioning of nested loops. Journal of Parallel and Distributed Computing 19 (1993) 90–102

[12] Kelly, W., Pugh, W.: A framework for unifying reordering transformations. Technical Report CS-TR-2995.1, University of Maryland (1993)

[13] Kelly, W., Maslov, V., Pugh, W., Rosser, E., Shpeisman, T., Wonnacott, D.: The omega library interface guide. Technical Report CS-TR-3445, University of Maryland (1995)

[14] Lim, W., Lam, M.S.: Communication-free parallelization via affine transformations. In: Proceedings of the Seventh Workshop on Languages and Compilers for Parallel Computing (1994) 92–106

[15] Lim, W., Lam, M.S.: Maximizing parallelism and minimizing synchronization with affine transforms. In: Conference Record of the 24th ACM SIGPLAN-SIGACT Symposium on Principles of Programming Languages (1997)

[16] Pugh, W., D.Wonnacott: An exact method for analysis of value-based array data dependences. In: Workshop on Languages and Compilers for Parallel Computing (1993)

[17] Quillere, F., Rajopadhye, S., Wilde, D.: Generation of efficient nested loops from polyhedra. International Journal of Parallel Programming 28 (2000)

[18] Sass, R., Mutka, M.W.: Enabling Unimodular transformations. In: Proceedings of Supercomputing'94 (1994) 753–762

[19] Wolf, M.E.: Improving locality and parallelism in nested loops. Ph.D. Dissertation CSL-TR-92-538, Stanford University, Dept. Computer Science (1992)

A Transformation to Provide Deadlock–Free Programs*

Pablo Boronat and Vicente Cholvi

Dept. LSI, Universidad Jaume I, Campus Riu Sec s/n,
12071Castellón, SPAIN
{boronat, vcholvi}@lsi.uji.es

Abstract. A commonly employed technique to control access to shared resources in concurrent programs consists of using critical sections. Unfortunately, it is well known that programs using several critical sections may suffer from deadlocks.
In this paper we introduce a new approach for ensuring deadlock–freedom in a transparent manner from the programmer's point of view. Such an approach consists of obtaining a deadlock–free "version" of the original program by using some program transformations. We formally prove the correctness of those transformations and we analyze their applicability.

1 Introduction

In the development of concurrent programs, considerable effort has been devoted to study the resource allocation problem. While local resources are accessed only by one process, shared resources can be accessed by many. If two or more processes simultaneously use the same resource, they can leave it in an incoherent state. Therefore, some mechanism is necessary to ensure that at most one process is accessing a shared resource at a time.

Maybe the most widely used mechanism for such a task consists of using *critical sections*. Roughly speaking, a critical section can be seen as a section of code where processes have exclusive access to the resources allocated within.

In order to access a critical section, a process must acquire and release it by executing especial code. This code forces shared resources to be "sequentially" accessed. Consequently shared resources are left in coherent states. Unfortunately, it is widely-known that programs which use several critical sections may suffer from deadlocks (i.e., the program stands in an infinite wait).

In order to avoid deadlocks, three approaches have been traditionally used. The first approach consists of detecting when a deadlock occurs and then recovering to a safe state (i.e., a previous state where the program is not deadlocked) [10,7]. The second approach prevents deadlock states in execution time using knowledge about the current state [3], in particular the state of the shared resources and the pending demands. Finally, the third approach consists of performing a "good" program design so that no deadlock states are reached. This last approach, since it is done prior to the execution of the program, does not require the evolution of its execution to be monitored [2,11,5,14,8].

* This work is partially supported by the CICYT under grant TEL99-0582.

P.M.A. Sloot et al. (Eds.): ICCS 2003, LNCS 2658, pp. 935–944, 2003.

Our Work

In this paper, we focus on a potential approach for ensuring deadlock–freedom (due to critical section access) in a transparent manner from the programmer's point of view (i.e., it does not force programmers to pay attention to deadlocks when designing the program). That technique consists of using a number of operations to access some (artificially introduced) critical sections so as to guarantee that deadlock states will not be reached. Furthermore, that technique does not require monitoring program's executions.

In particular, we use a program transformation which introduces some new operations in the original program. As a result, it generates a deadlock–free "version" of the original program which maintains its behavior. In turn, it may result in a decrease of the program's concurrency. However, we think that there are many situations where programmers, in order to guarantee deadlock–freedom, may accept a concurrency reduction in their program (which, on the other hand, may be necessary if one wants to guarantee deadlock-freedom). We have used a well–known problem (that of the *dining philosophers*) to show how our approach behaves (see Section 5).

It must be noted that the only way to transform an existing program without additional information (neither from programmers nor from runtime values) consists of making use of *static analysis*. However, static analysis has several limitations which come from its lack of runtime data and from its exponential complexity. Consequently, the theoretical cost of our transformation are, in the worst case, non-tractable. We have tested the programs in the SPLASH II benchmark suite [13] using a "brute force" static analysis algorithm (see Section 2) and two of the applications were not analyzable due to lack of memory (on a personal computer Intel Pentium III with 1 Gigabyte of main memory).

Related Work

One of the first works dealing with static analysis of deadlocks is that of Taylor [15] where the author focus on blocking communications (*rendezvous*) in Ada programs.

Following [15], Masticola [11] presents some algorithms to detect deadlock–freedom in polynomial time. The work is also applied to Ada programs with blocking communications, but it is shown how these techniques can be applied to programs using binary semaphores to access shared resources. However, polynomial cost is attained by relaxing the knowledge of state reachability, and consequently it provides pessimistic solutions. Furthermore, neither [15] nor [11] provide any hint about how to ensure deadlock–freedom in the case where programs may become deadlocked. Static analysis of Java programs, following previous cited works, can be found in [6,12] (our work could benefit of the analysis presented in these references).

Maybe the only work (to the best of our knowledge) that focus on ensuring deadlock–freedom by using a prevention transformations is that of Taubin et al. [14]. By using their approach, the authors apply transformations to Petri net models to reduce the cost of deadlock states detection. Unfortunately, the overall cost remains exponential and it may occur that they are unable to provide any deadlock-free solution, even a pessimistic one.

In any case, we want to strength the goal of our work. The purpose is to point out the chance for automatic deadlock prevention techniques. The work is not specialized on static analysis of concurrent programs (it could complement those approaches) nor it takes into account all possible sources of deadlocks (i.e. accesses to replicated resources or cycles in data/control dependence as is the case of blocking communications).

The rest of the paper is organized as follows. In Section 2 we provide some definitions and we characterize when a program is deadlock–free. In Section 3, we introduce a transformation for safely removing one "source of deadlocks". Based on this transformation, in Section 4 we propose a program transformation algorithm which provides a deadlock–free version of a given program. An example of how our approach behaves is presented in Section 5. Finally, in Section 6 we present some concluding remarks.

2 Preliminaries

In this paper we use a simplified version of concurrent programs in which there are only operations to access critical sections: a $lock(cs)$ operation is used to acquire the critical section cs and an $unlock(cs)$ operation is used to release it. The use of these synchronization operations are common in the most popular languages used in concurrent programming, such as Java, Ada, C, etc. Operations are issued so that operations from the same process are executed sequentially. We use the relationship $\prec$ to characterize the order in which operations from the same process are intended to be executed. We also assume that a process can not request any critical section which it already locks, nor it can release any critical section which it does not lock.

Even though the model of programs seems quite restrictive, the conclusions drawn about deadlock prevention methods can also be used to reason about *structured programs* (i.e., programs that, besides sequential constructs, allow loops and conditionals). A program with conditional constructs is equivalent to a number of sequential programs covering all the possible paths that the different execution flows may follow. Such a number of sequential programs, even though it may be high, will be finite. Thus, it is enough to verify that every one of those programs is deadlock–free to ensure that the program (with conditional constructs) is deadlock–free. With regard to looping constructs, the situation is simpler if each acquired critical section within a given iteration is released in the same iteration[1]. That is because, for each one of the computations, the states corresponding to the process that performs the loop will be repeated in each one of the iterations, which allows us to treat loops as sequential constructs. A pre–processor (using a "brute force" algorithm) to obtain, given a real structured program, a set of programs such as those we consider, can be found in [1].

[1] A very reasonable assumption since, otherwise, it may be difficult to ensure that no critical section will be acquired/released more than once without it having been released/acquired meanwhile.

For our work, we will denote as op_{match} the unlock operation intended to release the critical section acquired by op (obviously op denotes a lock operation from the same process that op_{match}). Furthermore, we say an operation $op = lock(cs)$ is a *wrapper* of another operation op', denoted $op \in wrappers(op')$, if $op \prec op' \prec op_{match}$.

$$\begin{array}{lllll} P_1: & op_1 = lock(cs_1) & op_2 = lock(cs_2) & op_3 = unlock(cs_1) & op_4 = unlock(cs_2) \\ P_2: & op_5 = lock(cs_2) & op_6 = lock(cs_1) & op_7 = unlock(cs_2) & op_8 = unlock(cs_1) \end{array}$$

Fig. 1. A two–process program with 8 operations and 2 critical sections.

In Fig. 1, operations op_3 and op_4 are the respective matching operations to op_1 and op_2. Also, op_1 is a wrapper of op_2, which is in turn a wrapper of op_3.

Now, we introduce the "source of deadlock" (SoD) concept which is the key definition for characterizing deadlock-free programs. However, first we will introduce the "contemporariness" concept (on which the SoD definition is based). Roughly speaking, a set of operations is contemporary if it is possible an execution with a state in which all operations in the set are "active" at the same time.

Definition 1. *A set of operations OP of a program P is* contemporary *if there is at least one possible execution of P in which each operation in OP is the next operation to be executed in its respective process.*

It can be readily seen that operations op_2 and op_6 in Fig. 1 are contemporary. It can also be seen that operations op_3 and op_6 are not contemporary since both are in different accesses to critical section cs_2.

Definition 2. *Two disjoint sets of operations $OP = \{op_i\}_{i=1,\ldots,n,\ n>1}$ and $OP' = \{op'_i\}_{i=1,\ldots,n,\ n>1}$ form a* source of deadlocks (SoD) *if OP' is contemporary and $\forall op'_i = lock(cs)$ $(\exists op_i \in wrappers(op'_i)\ :\ op_{(i \bmod n)+1} = lock(cs))$.*

Note that this definition involves the existence of a set of critical sections $\{cs_i\}_{i=1,\ldots,n,\ n>1}$ in which, for each section, there is a lock in both sets, OP and OP'. Otherwise, the set OP' would not be contemporary. The sets $OP = \{op_1,\ op_5\}$ and $OP' = \{op_2,\ op_6\}$ in Fig. 1 form a SoD.

In [4], it is shown that programs are deadlock-free if and only if they do not have any SoD[2].

Theorem 1. *A program is deadlock–free iff it does not contain any SoD.*

[2] There, the concept of *stopper* is used. This is equivalent to the definition of SoD given here.

3 A Transformation for Safely Removing One SoD

In this section it is proposed a program transformation which eliminates a SoD. Our approach consists of breaking, for a given SoD $\{OP, OP'\}$ in the original program, the contemporariness of operations in OP'. For such a task, we only have to ensure that at least a pair of operations in this set become non–contemporary. We achieve that by adding two pairs of operations that access a new (artificially introduced) critical section and that are wrappers of two different operations in OP'.

Transformation 1 *Given a program P with a SoD $d \equiv (OP,\ OP')$, we define the* non–contemporary transformation $NonContemporary(d,\ P)$ *as one which adds two accesses to a new defined critical section where each access contains just one (different) operation in OP'.*

We will call op_{prev} and op'_{prev} the lock operations introduced in the transformation $NonContemporary()$.

However, this transformation does not guarantee that new SoD's will not be induced. An example can be seen in Fig. 2 (b).

$$\begin{array}{l} P_1\text{: } lock(cs')\ \underline{lock(cs)}\ unlock(cs')\ \underline{lock(cs')}\ \ldots \\ P_2\text{: } \underline{lock(cs')}\ \underline{lock(cs)}\ \ldots \end{array}$$

(a)

$$\begin{array}{l} P_1\text{: } lock(cs')\ lock(cs)\ unlock(cs')\ \underline{lock(cs_{prev})}\ \underline{lock(cs')}\ unlock(cs_{prev})\ \ldots \\ P_2\text{: } \underline{lock(cs')}\ \underline{lock(cs_{prev})}\ lock(cs)\ unlock(cs_{prev})\ \ldots \end{array}$$

(b)

$$\begin{array}{l} P_1\text{: } lock(cs_{prev})\ lock(cs')\ lock(cs)\ unlock(cs')\ lock(cs')\ unlock(cs_{prev})\ \ldots \\ P_2\text{: } lock(cs_{prev})\ lock(cs')\ lock(cs)\ unlock(cs_{prev})\ \ldots \end{array}$$

(c)

Fig. 2. In Fig. (a), the underlined operations form a SoD d. After applying transformation $NonContemporary(d,\ P)$ (Fig. (b)), the former SoD no longer exists. However, the underlined operations form a new one. By also applying Lemma 1 (Fig. (c)), we ensure that new Sod's are not induced.

As the following lemma shows, if operations op_{prev} and op'_{prev} have not any wrapper, then no SoD is generated in the transformed program.

Lemma 1. *Let P be a program with a SoD $d \equiv (OP, OP')$ and let us apply $NonContemporary(d,\ P)$. If op_{prev} and op'_{prev} do not have any wrappers then no SoD is induced in the transformed program.*

Proof: By contradiction. Assume that program P' is the result of applying $NonContemporary(d,\ P)$. Assume also that op_{prev} and op'_{prev} do not have wrappers and that a new SoD $(OP'',\ OP''')$ has been generated.

Step 1: op_{prev} or op'_{prev} are included in $OP'' \cup OP'''$.
Proof: Let us assume that neither op_{prev} nor op'_{prev} are included in $OP'' \cup OP'''$. Thus:

1. All operations in $OP'' \cup OP'''$ are in P.
   Proof: Trivial since neither op_{prev} nor op'_{prev} are included in $OP'' \cup OP'''$.
2. The operations in OP''' are contemporary in P.
   Proof: As $(OP'',\ OP''')$ is a SoD in P', then OP''' is contemporary in P'. This set would also be contemporary in P since the $NonContemporary()$ only introduces new restrictions on the possibility of being contemporary.

Therefore, $(OP'',\ OP''')$ is a SoD in P, contradicting the assumption that it is a new SoD generated in the transformation.

Step 2: Either op_{prev} or op'_{prev} is included in OP'' and the other is included in OP'''.
Proof: Immediate given the definition of SoD and the fact that they are the only locks on the critical section defined in the transformation.

By Steps 1 and 2 we have that either op_{prev} or op'_{prev} is in OP'''. Furthermore, from the SoD's definition, all operations in OP''' have a wrapper. This contradicts the assumption that neither op_{prev} nor op'_{prev} have wrappers. Therefore, (OP'', OP''') is not a new SoD.

□

Fig. 2 (c) shows an example of how this result is applied. It must be noted that this is a simple example; in a general case only a part of a program is executed sequentially.

Remark. Whereas there can be several approaches for providing a program transformation capable to eliminate a given SoD, it has to be taken into account that such a transformation must ensure that new behaviors will not be introduced into the resulting program. Therefore, we can not use any technique based on making internal changes (such as moving an operation from one location to another) since they may not guarantee that the behavior of the program will be affected. Note that our approach only reduces the set of potential executions.

4 A Transformation for Making Programs Deadlock–Free

In this section we introduce a transformation which provides deadlock-free versions of concurrent programs in a transparent manner. Such a transformation, denoted $DirectTransform(P)$, is obtained by using $NonContemporary(d,\ P)$ and applying Lemma 1 directly. It uses the following functions:

- $GetSetofSoD(P)$: returns the SoDs of program P.
- $MoveBack(op, P)$: swaps the operation op with its immediately preceding one in the same process.

```
DirectTransform(P)::
  D ← GetSetofSoD(P)
  for each d ∈ D do
    NonContemporary(d, P)
    while wrappers(op_prev) ≠ ∅ do MoveBack(op_prev, P)
    while wrappers(op'_prev) ≠ ∅ do MoveBack(op'_prev, P)
  return P
```

Fig. 3. Algorithm of $DirectTransform(P)$.

Fig. 3 shows the algorithm of this transformation. The next theorem proves that the program resulting from applying this algorithm is deadlock-free.

Theorem 2. *Given a program P, the program resulting from $DirectTransform(P)$ is deadlock-free.*

Proof:

Step 1: In each iteration of the **for** loop, a SoD is removed without generating a new one.
Proof: Immediate by Lemma 1. (Note that the number of iterations in the **while** loops is finite since, in the worst case, op_{prev} and op'_{prev} operations would be placed in the first position of the respective process).

Step 2: The algorithm ends.
Proof: Immediate, since the number of SoDs of a program with a finite number of operations is also finite.

By Steps 1 and 2, the transformed program is free of SoDs and by Theorem 1, it is also deadlock-free.

□

Regarding the complexity of $DirectTransform()$, it must be taken into account that function $GetSetofSoD()$ has non-polynomial cost. That is because it is hindered by the well–known state explosion problem: the number of states in a concurrent system tends to increase exponentially with the number of processes [15]. However, this problem can be solved by not checking the contemporariness of OP' set (obviously, obtaining an approximate result[3]). In that case, the cost of function $GetSetofSoD()$ is to $O(n + p^2c^4 + p^3)$, being n the number

[3] The function, in this case, returns all SoDs but can also return some false SoDs.

of operations of the program, p the number of processes and c the number of critical sections used.

Whereas here we have introduced a transformation that directly applies Lemma 1, other transformation can also be introduced. For instance, it can be defined a new transformation that, whenever a SoD is eliminated, the operations op_{prev} and op'_{prev} are selectively moved back, applying Lemma 1 only in the worst case. This transformation would be more costly to apply (to eliminate a SoD it must be checked that new SoD's are not induced) but the resulting program may have a lower reduction of its concurrency.

5 An Example

As it has been said previously, a drawback of applying the above mentioned transformations is that they restrict the way in which processes can execute their operations. Therefore, programs may suffer a loss on the number of states the program can reach (from now called *potential concurrency*), which may become in a loss on the program's performance. In this section, we use the widely know *dining philosophers* problem to show how our proposed transformations affect the concurrency.

For our comparison we will take, first, the "classical" solution. By using this solution, each philosopher, say philosopher i, locks sections cs_i and $cs_{(i \bmod n)+1})$ in order, except for one of the philosophers that operates in the opposite way. It is known that this solution guarantees that philosophers will not get deadlocked.

On the other hand, we also use $DirectTransform()$ to obtain a deadlock–free version for the philosophers problem. As Fig. 4 shows, the version obtained by using $DirectTransform()$ has not a significant concurrency loss with respect to the classical algorithm. However, whereas the classical solution burdens programmers with a new task (i.e., finding that solution), by using one of our transformation the algorithm is obtained in a transparent manner from the programmer's point of view.

6 Conclusion and Future Work

In this paper we have introduced a method that provide deadlock-free versions of concurrent programs in a transparent manner from the programmer's point of view. Even though the cost of the transformation is exponential, it can be drastically reduced using a pessimistic approach.

There are some other issues, which we are currently working in, that need further research. The decrease in concurrency that $DirectTrasform()$ provides depends on which operations are chosen to be synchronized in $NonContemporary()$. So, it will be worthwhile to try different combinations. Also, there are tools as [5,6] and [9] that can be used to check different program properties as deadlock–freedom. So, it will be worth to study how integrate our prevention techniques in such programming environments.

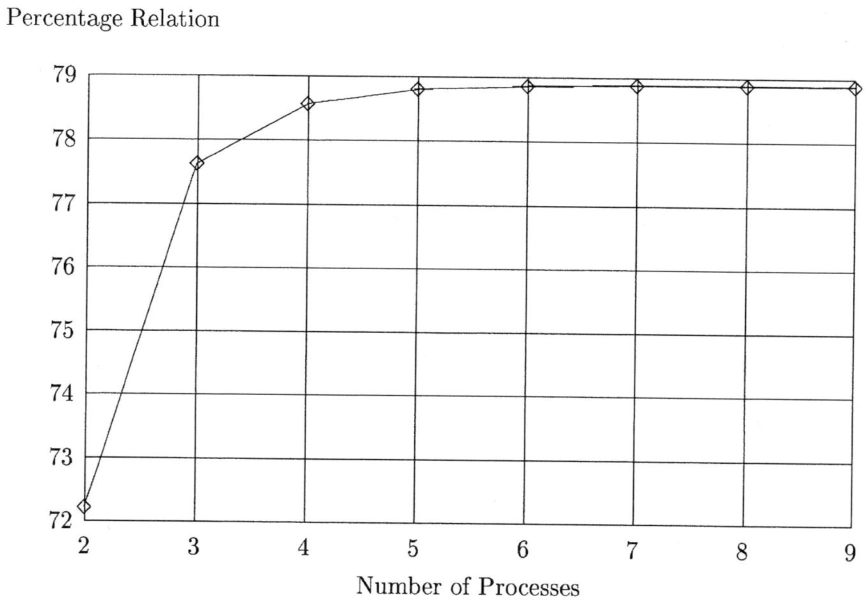

Fig. 4. Relationship between the potential concurrency provided by using *DirectTransform*() and the "classical" solution in the *dining philosophers* problem. Only one iteration for each philosopher has been considered.

References

1. L. Almajano. Desarrollo de un preprocesador para el an lisis est tico de programas concurrentes, 2001. Proyecto final de carrera para la obtenci n del t tulo de Ingeniero en Inform tica (Universidad Jaume I).
2. Ö. Babaoglu, E. Fromentin, and M. Raynal. A unified framework for the specification and run-time detection of dynamic properties in distributed computations. Technical Report UBLCS-95-3, Department of Computer Science, University of Bologna, February 1995.
3. F. Belik. An efficient deadlock avoidance technique. *IEEE Transactions on Computers*, 39(7):882–888, July 1990.
4. V. Cholvi and P. Boronat. A minimal property for characterizing deadlock-free programs. *Information Processing Letters*, (77):283–290, 2001.
5. J.C. Corbett. Constructing compact models of concurrent java programs. In Michal Young, editor, *ISSTA 98: Proceedings of the ACM SIGSOFT International Symposium on Software Testing and Analysis*, pages 1–10, 1998.
6. J.C. Corbett, M.B. Dwyer, J. Hatcliff, S. Laubach, C.S. Pasareanu, Robby, and H. Zheng. Bandera: extracting finite-state models from java source code. In *International Conference on Software Engineering*, pages 439–448, 2000.
7. J.R. Gonz lez de Mend vil, J.M. Bernab u, A. Demaille, and J.R. Garitagoitia. A distributed deadlock resolution algorithm for the and request model. *IEEE Transactions on Parallel and Distributed Systems*, 1999.

8. C. Flanagan and M. Abadi. Types for safe locking. In *ESOP 99: European Symposium on Programming, LNCS 1576*, pages 91–108, 1999.
9. Gerard J. Holzmann. The model checker SPIN. *IEEE Transactions on Software Engineering*, 23(5):279–295, May 1997.
10. T.F. Leibfried. A deadlock detection and recovery algorithm using the formalism of a directed graph matrix. *Communications of the ACM*, 23(2):45–55, April 1989.
11. S.P. Masticola. *Static detection of deadlocks in polynomial time.* PhD thesis, The State University of New Jersey, May 1993.
12. G. Naumovich, G.S. Avrunin, and L.A. Clarke. A conservative data flow algorithm for detecting all pairs of statements that may happen in parallel. In *ACM SIGSOFT Software Engineering Notes, Proceedings of the 7th European Engineering*, volume 24, 1999.
13. J. Singh, W. Weber, and A. Gupta. SPLASH: Stanford parallel applications for shared memory. *Computer Architecture News*, 20(1):5–44, March 1992.
14. A. Taubin, A. Kondratyev, and M. Kishinevsky. Deadlock Prevention Using Petri Nets and their Unfoldings. Technical Report 97-2-004, Department of Computer Hardware, The University of Aizu, March 1997.
15. R.N. Taylor. A general-purpose algorithm for analyzing concurrent programs. *Communications of the ACM*, 26:362–376, 1983.

Building the Program Parallelization System Based on a Very Wide Spectrum Program Transformation System

Alexander Alexeyevich Bukatov

Rostov State University, Computer Center, 200/1, build. 2, Stachki Ave.,
344090 Rostov-on-Don, Russia
baa@rsu.ru
http://uginfo.rsu.ru

Abstract. An approach to build an automated Program Parallelization System based on a multipurpose very wide spectrum program transformation system is discussed. The main advantage of the proposed approach is a nonprocedural way in which the parallelizing transformations are described in the Program Transformation System. It allows doing any required customization of the Program Parallelization System for a new parallel computer platform or new parallelization techniques. General organization of the proposed Multipurpose Program Transformation System is discussed, some examples of nonprocedural representations of parallelizing transformations are analyzed and comparison with related works is provided.

1 Introduction

Automated parallelization tools provide the means for porting legacy sequential programs to parallel computer platforms and using sequential programming languages to develop new programs for parallel computers. The program parallelization techniques were being widely investigated in the last decades. For some representative classes of parallel computers with shared memory, vector and SMP architecture computers included, effective parallelization techniques had been developed. Basic vendors of parallel computers with such architectures have provided the parallelizing compilers for these types of computers. Nevertheless, for the parallel computer architectures with distributed memory (DM) such as MP, cluster or NUMA architectures more or less effective parallelization techniques are still poorly developed and numerous investigations in this direction are being leaded. Some new techniques of program parallelization for DM computers are being introduced from time to time. Therefore there is a need to change each time the implementation of experimental program parallelization tools that have been developed for DM computers to incorporate the new techniques. Building an automated program parallelization tools based on nonprocedural description of parallelizing transformations may dramatically simplify the evolutionary development of such tools.

The other reason for building the program parallelization tools around the nonprocedural description or parallelizing transformations arises from the practical need how

P.M.A. Sloot et al. (Eds.): ICCS 2003, LNCS 2658, pp. 945–954, 2003.

to "teach" the existing parallelization tools to apply new parallelizing transformation developed by skilled users of such tools. The application domain specific parallelization methods that can be developed by skilled application programmers for parallelizing some high-level multi-component program structures is of special interest.

This paper discusses the way of building an extendable Automated Program Parallelization System (APPS) based on a very wide spectrum Multipurpose Program Transformation System (MPTS) [1,2]. The next section of the paper discusses the main features of the MPTS and of underlying program transformation language. In Section 3 the general organization of the MPTS and APPS is observed. In Section 4 two examples of parallelizing program transformations are discussed. In Conclusion general assessment of this paper and comparison of discussed results with related investigations are provided.

2 Main Features of the Multipurpose Program Transformation System and Language

The Multipurpose Program Transformation System (MPTS) like most others program transformation systems [3,4] is based on a nonprocedural representation of program transformations in the form of Schematic Transformation Rules.

The basic notion used in the area of schematic program transformation is a program scheme [3]. A *program scheme* is a representation of some class of similar programs or similar program structures (statements). It originates from the concrete program or from the program structure by parameterization: some components of the program (of the program structure) are replaced by the parameters named.

Schematic transformation rule (STR) is the mapping (probably a partial one) from the space of program schemes to the same space. Usually [3,4] this mapping is recorded in the following form:

input-pattern & enabling-condition ⇒ output-pattern

where the *input-pattern* is the scheme of the transformed structure and the *output-pattern* is the scheme of transformation result. The *enabling-condition,* which usually depends on the *input-pattern* parameters, restricts the STR applicability.

The STRs can be applied to any program for its transformation. The STR application includes the following stages. First, all the structures (statements) of the transformed program matching the input pattern are found. Simultaneously for each matching the values of input pattern schema parameters are determined (unified). Then for each matching the enabling condition is evaluated. If for some source structure matched to the input pattern this value is 'TRUE', then this source structure is replaced by the structure produced from the output pattern by substituting the unified parameter values.

A single STR or some set of STRs can be applied once or repeatedly "by exhaustion", i.e. until no matches of its input pattern are found. The order of the STRs application can be specified in a *program transformation scenario* expressed in an appropriate program transformation language. This provides the means to represent both the STRs and the program transformation scenarios.

The STR form discussed above is used in all known program transformation systems (PTS) [3,4]. It provides the way for nonprocedural representation of rather simple program transformations and cannot cover all the needs for implementation of the well-known program parallelization and vectorization techniques (see [5-7]). Local transformations can only be represented in the STR form. But not all program parallelization transformations are the local ones. The typical example of a non-local transformation is the loop restructuring because usually the statements that declare and initialize the loop variables are involved in this transformation. Moreover not all the local program transformations can be expressed in a natural way in the STR form. For example, to restructure some loop statement to a parallelized program structure some Diophantine equations should be solved. It's obvious enough that the algorithm used for solution of equations cannot be expressed efficiently in the form of the program rewriting rules.

So the expressive power of the traditional schematic program transformation languages and systems is not good enough to be used for the implementation of program parallelization systems. The MPTS suggested in [1,2] extends the traditional program transformation languages in two main directions. First, sophisticated compound multi-component patterns are introduced. And second, some functional capabilities are introduced into a program transformation language. Additionally, some interactive features are introduced to the MPTS.

Let's consider the Multipurpose Program Transformation Language (MPTL) and System (MPTS). The core construction of the MPTL is the declaration of the Program Transformation Rule (PTR) that is of the following form:

rule *rule-name* (*rule-parameters-list*) [**var** *variable-declarations-list* ;]
  compound-input-pattern
  && enabling-condition =>
  compound-output-pattern
end

Here and everywhere in this paper the names of non-terminal constructions are printed in the *italic* font and the square brackets are used to enclose optional constructions.

In the variable declarations list rule parameters as well as input and output pattern parameters are declared. The declaration of each parameter includes the name of this parameter syntax type (the name of the corresponding non-terminal symbol of the transformed program grammar). It means that only the string derivable from the defined non-terminal symbol can be matched to this parameter. It provides the basis for the efficient implementation of the matching process.

Now let's consider the most important parts of the PTR declaration, the compound input and output patterns. Each of these patterns can be composed of several pattern or procedural terms. The definition of compound input pattern, for example, has the following syntax:

compound-input-pattern ::= *pop-term* { & *pop-term* }
pop-term ::= *pattern-term* | *procedural- term*

Here and further in the paper the '::=' symbol is used to separate the left and right parts of the syntax rule; the figure brackets '{' and '}' are used to enclose a construction that can be repeated 0 or more times; the '|' symbol separates alternatives of the syntax rule and the ' ' symbol ends the syntax rule.

The first term of the input pattern is usually the *pattern-term*. It provides the schema of the root (or, in other words, the 'anchor') component of a compound program structure. Each of the subsequent pattern terms is as a rule connected with one of the previous terms by a semantic relation or by common parameters. The procedural terms are used to transform their input parameters to the output ones when the algorithm of the required transformation is rather sophisticated. The simplified syntax of the pattern term has the following form:

pattern-term ::= *identifying-expression* : *schema*
identifying-expression ::=:<*s-type-name*>{. *relation-name*} [(*alias-name*)] |
alias-name{. *relation-name*} [(*alias-name*)] |
$*parameter-name*{. *relation-name*}[(*alias-name*)

The form of identifying expressions is mostly based on a semantic model of the transformed program that is used in the MPTS. According to this model the transformed program is treated as a hierarchical structure of the syntax objects connected by the binary semantic relations such as "declared in". The hierarchical links are also considered as the named binary relations. The whole program structure is considered to be a directed graph whose nodes are the syntax objects and edges are the named relations.

The simplest case of an identifying expression is the syntax type name (*s-type-name*). In this case all the syntax objects of *s-type-name* type are compared with the *schema* during the matching process. The fact that the syntax type of the matched object is known before matching simplifies the matching process dramatically and provides the means for its efficient implementation. A more complex form of identifying expressions is based on the principles of a navigational query language of special object-oriented database [8,9]. In this case the identifying expression has the form of the dot separated chain of relation names that defines some path in the graph program structure leading from the named object to the object being identified.

The alias name optionally provided in the identifying expression is used to refer to the matched object from others pattern terms. In the PTR output pattern the three following additional forms of identifying expression of the pattern terms can also be used: *before*(*alias-name*), *instead-of*(*alias-name*) and *after*(*alias-name*). Note that the simple *alias-name* form is equal to *instead-of*(*alias-name*) when used in the out pattern.

The *procedural-pattern* is the way to capture procedural transformation steps into schematic transformation process. The simplest form of the procedural term is an external procedure call. This procedure can be implemented in some traditional programming language (the interface to C language is supported in the MPTS).

The last PTR component, which has not been discussed yet, is the enabling condition. It is the logical expression that depends on the input pattern parameters. This expression can include some external predicates (i.e. functions that return Boolean value) as operands. One of such predicates predefined in the MPTS is used to introduce interactivity to a transformation process. It asks a programmer a parameterized question and transforms his 'yes' or 'no' reply to the Boolean 'TRUE' or 'FALSE' values respectively that are returned to the transformation engine. A more complex form of an enabling condition is beyond of the scope of this paper.

The MPTL features discussed in this section can sufficiently widen the area of the program transformation approach applicability, so that MPTL can be used to represent rather complex parallelizing transformations. Almost all of the discussed extensions of traditional program transformation languages have been introduced by us. Let us note, that in works of V.L. Winter, dealing with HATS system [10,11], the pattern schemas and parameters are also marked by syntax types of corresponding syntax objects. This, as it has been mentioned above, provides the means for efficient implementation of the matching process. Winter has also tried to overcome the gap between the local and global transformations by introducing special wildcard symbol '*', which can be matched with an arbitrary string. But this attempt is not considered to be successful one because a wildcard symbol cannot be implemented efficiently and it does not really provide means to represent complex semantic structures composed of distributed components.

3 General Organization of the Multipurpose Program Transformation and Automated Program Parallelization Systems

The transformed programs are represented in the MPTS in the internal form and are stored in the object database (ODB) [8-9]. The primary form of the internal representation of the transformed programs is the Abstract Syntax Tree (AST) extended by additional semantic links (relations), which connect the AST nodes. On the bases of the extended AST some additional secondary representations such as the Control Flow Graph (CFG) and the Data Dependency Graph (DDG) are produced. These secondary representations are used to implement efficiently different predicates intended for application in PTR enabling conditions.

The MPTS is composed of the tools (subsystems), which provide the following main functions:

- conversion of source programs to their internal presentation stored in the ODB and vice versa;
- support the PTRs and PTSs (program transformation scenarios) creation an modification (the aggregate of PTRs and PTSs play the role of the APPS's knowledge base);
- application of the PTRs and PTSs to the transformed program to produce the transformation result.

The tools, converting source programs to their internal presentation and vice versa (below, the *program conversation tools*) can be customized to the syntax of programming languages that are used to represent source sequential programs and target parallel programs, to be created as the result of the parallelization. In fact a combined language, containing constructions of both sequential and parallel language is constructed and its syntax in the form of an L-attributed grammar [12] is defined. Then the definition of this grammar ***G*** is entered into the MPTS. The syntax rules of the grammar ***G*** are compiled to C-procedures, which make the transformation of the source program to an attributed AST and vice versa. On the AST basis the required secondary presentations are produced such as CFG and DDG. Let us note that all the PTRs when applied to the transformed programs actually operate on the AST level that is very close to the source program level. Secondary representations are reconstructed by the MPTS after each AST transformation. The techniques of such reconstruction are discussed in the paper [13] written under our guidance.

Tools intended for the PTRs and PTSs creation and modification (below, *the PTR creation tools*) do compile the transformation programs expressed in the form of PTSs and PTRs to C-procedures that perform the corresponding transformation. Such approach provides the simple way to implement the MPTL functional (procedural) capabilities including procedural terms of the PTR patterns and "embedded" predicates used in enabling conditions. The PTR creation tools also provide the check of PTR structural correctness. Each pattern term is checked for consistency with the ***G*** grammar.

The *transformation engine* provides the PTSs and/or PTRs (MPTL programs) application to the transformed program. It is the main part of the MPTS. A MPTL program can be applied to the entire program, which is to be transformed, or to some identified part of this program. Both the interactive interface and the application programmer interface (API) to the transformation engine are provided by the MPTS. The API interface is provided through the *apply* function. Note, if the transformation engine API is provided, any traditional programming language can be used as means to code program transformation scenarios.

The automated program parallelization system (APPS) is built on the MPTS basis in the following way. The set of "predefined" predicates and functions (procedures) intended for application in the PTR's enabling conditions and procedural pattern terms is implemented in the C programming language. This set forms the procedural part of the APPS knowledge base. The nonprocedural part of the APPS knowledge base is represented in the PTRs form. This approach was used in paper [14] to implement classical parallelization techniques of R. Allen and K. Kennedy [5] in the form of the APPS knowledge base.

4 Examples of the Nonprocedural Representation of Parallelizing Transformations

Two examples are provided in this section. The first example is simplified to be expressed mainly with the MPTL features discussed in section 2. The second example is rather complicated and demonstrates some additional features of the MPTL.

Let's consider the following program structure that is coded in Pascal like language as a first example:

s: **real**; . . . a: **real array of** [1..n]; . . .
s := 0; . . . **for** I := 1 **to** n **do** s := s + a[i];

The loop, contained in this structure, can be replaced by two consecutive loops. The first one can be run in parallel where as the second one is similar to the source loop but involves n/2 (if n is even) iterations. The following PTR defines the transformation discussed.

```
rule sum_loop_tr :
var <name> $i, $a, $s, $b, $type; <int_const> $n, $n1;
<loop_stmt> (par_loop): for $i=1 to $n do
                        $s := $s + $a[$i];  &
<assign_stmt> (ss):     $s:=0; &
$a.declared_in (adcl):  $a: array [1..$n] of $type;
&& precedes(ss,loop_tr) &
unchanged_between($s,ss,loop_tr) & even($n)
=>
$n1 = int_const:$n/2  &  $b = new_name &
after(adcl): $b: array [1..$n/2] of $type; &
instead_of(par_loop): for $i=1 to $n1 do
                        $b[$i] := $a[i] + $a[$n + $i];
        for $i=1 to $n1 do $s := $s + $b[$i];
end
```

In this example the input pattern of the `sum_loop_tr` PTR consists of three pattern terms. The identifying expression of the second term defines the path from the syntax construction matched with \$a parameter to the structure that should be matched with the second pattern term. The output pattern consists of two procedural terms and two pattern terms. The enabling condition is the conjunction of the three predicates.

Of course the `sum_loop_tr` PTR can be generalized. For example, the '+' operation in the loop statement as well as the '0' value in the assignment statement should be parameterized for any operation that is commutative and associative and for the 'NULL' value of this operation group. Moreover, the syntax of the \$s + \$a[\$i] expression should also be parameterized to be matched with expressions like MIN(\$s,\$a[\$i]). The MPTL language provides the mean of the required expression parameterization not discussed above. Pattern terms can be written in the attributed AST notation eliminating the difference between the infix and functional forms of the operation syntax.

As the second example we consider the PTR that implements the loop unrolling (see [5], for instance).

```
rule loop_unroll:
var <name> $i, $n, $j;  <stmt> $loop_body;
<loop_stmt>(L): for $i=1 to $n do  $loop_body
&& few_iterations(L) & no_jumps_in(L)
=>
instead_of(L):
forall $j in (1:$n)
   $loop_body_j = apply(($i => $j), $loop_body)
end
```

In this example, which in fact is also simplified, the loop unrolling transformation is represented.

Let's discuss two additional constructions of the MPTL that are used in the above example. The first one is the **forall-in** construction. This construction is used to produce sequentially the indexed samples of the embedded statement. In the MPTL there are some other constructions, which provide the means to handle and to explore the sets of the schemas. The second construction provides the way of the recursive call of the transformation engine. The `apply` function is called recursively in this example to apply the embedded anonymous transformation rule to the loop body.

There are some other features of the MPTL and MPTS that are not discussed in this paper. But the features discussed do demonstrate the expressive power of the MPTL, the potential efficiency of the MPTS implementation, and the applicability of the discussed tools for building the program transformation systems.

5 Conclusion

The main aim of this paper is to develop an approach to implement extendable program parallelization tools. This approach should provide the means for evolutionary extension of such tools to incorporate innovations of the underlying parallelization techniques. According to this aim the main achievements discussed in the paper are the following:

First, the principle has been suggested to build a program parallelization system on a basis of a schematic transformation system that allows representing parallelizing transformations in a nonprocedural form that is clearer for understanding and easier for modification than a procedural implementation of these transformations. As we know there are no other investigations that suggest any approach to the nonprocedural implementation of the parallelizing transformation and, partially, an approach based on a program transformation system.

Second, development of a schematic program transformation approach has been performed with the aim of increasing the expressive power of a program transformation language, which would be suitable for clear representation of complicated parallelizing transformations. The MPTL program transformation language providing new expressive features has been developed. The prototype implementation of the MPTS program transformation system based on the MPTL has been performed. The MPTS and MPTL do develop ideas of the traditional program transformation systems [3-4] in two main directions. The transformations of the compound multi-component program structures (the semantic structures composed of distributed components) can be defined in MPTL and applied by MPTS. And the procedural means (procedural terms and predicates) can be used where required. There are no other program transformation systems that provide the facilities mentioned above. The HATS system [11,12] provides only some means to express non-local transformations (see the end of Section 2). But these means are restricted and are not powerful enough to represent complicated transformations. And there is no way to implement the wildcard feature efficiently.

And third, the MPTL application to represent of complex parallelizing transformations has been shown in the Section 4 and in the paper [14] written under our guidance.

Note that MPTS based PPS is, in fact, the meta-system that makes it possible to implement different parallelization techniques. In paper [14], the parallelizing transformations of R. Allen and K. Kennedy are represented in a non-procedural form. Other classical parallelization techniques [6,7], different extensions of these techniques (see, [15]) as well as parallelization techniques based on completely different approaches to parallelization can also be represented in this form. At present we are planning to implement the V-Ray parallelization technology [16] based on the MPTS.

Since we implement in nonprocedural form the parallelization techniques developed by other researchers the parallelizing power of our implementation is mainly determined by the underlying parallelization techniques. That is why we do not provide any test data based on PTS running on the Livermore Loops benchmark. By the similar reason no comparison of the PPS performance time parameters with other parallelization systems has been made.

References

1. Bukatov A.A. Development of Tools Intended for Creation of the Source Program Transformation Systems. Information Technologies, N 2, 1999, 22–25 (published in Russian).
2. Bukatov A. A. Development of Tools for Nonprocedural Implementation of Program Parallelizing Transformations. Proceedings of the Russian Conference "Fundamental and Application Aspects of Development of the Large Distributed Program Complexes", Abrau-Durso, MSU Publishing, 1998, 109–116 (published in Russian).
3. Partch H., Steinbruggen R. Program Transformation Systems. ACM Computing Surveys, 1983, v. 15, N 3, 199–236.
4. Visser E. A Survey of Rewriting Strategies in Program Transformation Systems. Electronic Notes in Theoretical Computer Science, v. 57 (2001), 35 p. http://www.elsevier.nl/locate/entcs/volume57.html
5. Allen R., Kennedy K. Automatic Translation of FORTRAN Programs to Vector Form, 1987.
6. Bannerjee U. Dependence Analysis for Supercomputing. Kluver Academic Publishers, New York, 1988.
7. Wolfe M.J. Optimizing Supercompilers for Supercomputers. MIT Press, Cambridge, Mass, 1889.
8. Bukatov A.A., Zastavnoy D.A. High-level navigational language for querying complex data objects and its applications to CASE systems // Proc. of the 3rd Joint Conference on Knowledge-Based Software Engineering, Smolenice, Slovakia, 1998, pp. 103–107.
9. Zastavnoy D.A., Bukatov A.A. Representation of Complex Structures Extracted from Object Databases, and Access to their Components // In: Hruska T., Hasimoto M. (Eds) Knowledge-Based Software Engineering, Amsterdam: IOS Press, 2000, 93–100.
10. Winter V.L. An Overview of HATS: A Language Independent High Assurance Transformation System. Proc. of the IEEE Symposium on Application-Specific Systems and Software Engineering Technology (ASSET), March 24–27, 1999.

11. Winter V.L. Program Transformation in HATS. Proceedings of the Software Transformation Systems Workshop, May 17, 1999.
12. Lewis P.M., Rosenkranz D.J., Stearns R.E. Attributed Translations. Journal of Computer and System Sciences, v. 9, N 3, 1974, 279–307.
13. Lugovoy V.V. Development of the Internal Representation Modification Techniques in the Program Parallelization CASE-system. Proceedings of the Conference "High Performance Computing and its Applications", Chernogolovka, Russia, 2000, 133–136 (published in Russian).
14. Zhegulo O. Representation of Knowledge on Program Parallelization Techniques in the Expert System Supporting Program Parallelization. Artificial Intelligence, the Journal of National Academy of Sciences of Ukraine, Institute of Artificial Intelligence, No 2001'3, Donetsk, 2001, 323–330 (published in Russian).
15. Pinter S.S., Pinter R.Y., Program Optimization and Parallelization Using Idioms. ACM Transactions on Programming Languages and Systems, 1994, vol. 16, N 3, 305–327.
16. Voevodin V.V., Voevodin Vl.V. V-Ray technology: a new approach to the old problem: Optimization of the TRFD Perfect Club Benchmark to CRAY Y-MP and CRAY T3D Supercomputers. Proceedings of the High Performance Computing Symposium'95, Phoenix, Arizona, USA, 1995, 380–385.

Performance Evaluation of the Striped Checkpointing Algorithm on the Distributed RAID for Cluster Computer

Yun Seok Chang[1], Sun Young Cho[2], and Bo Yeon Kim[3]

[1] Department of Computer Engineering, Daejin University, Pocheon, Korea
[2] Basic Science Research Institute, Chungbuk University, Chungju, Korea
[3] BK21 Department of Electrical and Computer Engineering, Kangwon National University Chuncheon, Korea

Abstract. The distributed RAID for serverless cluster computer is used to save the checkpoint files periodically according to the checkpointing algorithm for rollback recovery. Striped checkpointing algorithm newly proposed in this paper can adopt the merits of the coordinated and the staggered checkpointing algorithms. Coordinating enables parallel I/O on distributed disks and staggering avoids network bottleneck in distributed disk I/O operations. With a fixed cluster size, we reveal the tradeoffs between these two speedup techniques. The striped checkpointing approach allows dynamical reconfiguration to minimize checkpointing overhead among concurrent software processes. We demonstrate how to reduce the overhead by striping and staggering dynamically. For communication-intensive computational programs, this new scheme can significantly reduce the checkpointing overhead. Linpack HPC Benchmark results prove the benefits of trading between stripe parallelism and distributed staggering. These results are useful to design efficient checkpointing algorithm for fast rollback recovery from any single node failure in a cluster computer.

1 Introduction

In the typical cluster computer system, all concurrent processes on different nodes should communicate with each other through message passing. Checkpointing algorithm can maintain the global consistency among the nodes by saving the checkpoint data to the stable storage. Several checkpointing algorithms have been studied on the cluster computer. Coordinated checkpointing[1] takes checkpoints from each node simultaneously at every time interval to yield a consistent recovery globally in case of any failure. The drawback of the coordinated checkpointing is in the loss of freeze time for collecting checkpoints from nodes at the same time and heavy network created by synchronized checkpointing. Simultaneous writing of checkpoint files into local disks may cause a heavy network contention problem and disk IO bottleneck. To solve the disk contention problem, diskless checkpointing algorithm was introduced[6]. But the fault coverage is rather limited in a diskless checkpointing algorithm. So,

P.M.A. Sloot et al. (Eds.): ICCS 2003, LNCS 2658, pp. 955–962, 2003.

staggered checkpointing algorithm was introduced to write checkpoints onto a central stable storage[8]. It can greatly reduce the storage contention to some extent. But the staggered checkpointing algorithm has to pay additional overhead to achieve correct rollback recovery to avoid inconsistent states among cluster nodes. Therefore, if the size of the cluster increases, message logging overhead also increases. In a serverless cluster system, the distributed RAID system is suitable for saving distributed checkpoint files simultaneously in disks attached to cluster nodes[4]. Based on the distributed RAID, we propose striped checkpointing scheme. This scheme has two advantages. First, simultaneous distribution of the checkpoint files greatly alleviates the network IO contention problem by taking advantage of parallel IO. Second, rollback recovery latency is reduced on a distributed RAID system. So, mean time to repair of cluster is reduced and high cluster availability can be achieved. To take advantage of parallelism of the distributed RAID, the staggered checkpointing algorithm can be combined with striped checkpointing algorithm. Only a subset of disks constitutes a stripe group to take the checkpoint simultaneously. After one stripe group finishes checkpointing, another stripe group starts checkpointing in a staggered manner. This striped checkpointing algorithm has two benefits. First, when the IO contention is more serious a problem, parallel IO of striped group helps lessening the problem. Second, if network contention dominates, staggering can reduce it. To evaluate the performance of the striped checkpointing scheme, we carried out benchmark tests and compared with other checkpointing algorithm.

2 Distributed RAID and Single IO Space

To build a distributed RAID, all disks embedded in the cluster must establish the single IO space(SIOS). The SIOS should have high scalability, availability, and compatibility with IO centric cluster application. These should also imply a total transparency to the users, who can utilize all disks without knowing the physical locations of the data blocks. On NFS of centralized cluster file server, all checkpoint files are saved on a network file system. On SIOS, checkpoint files spread over the distributed disks. At the striped checkpointing scheme, checkpoint files are saved over the distributed RAID system. Distributed RAID is crucial to building scalable cluster of computers[5], and SIOS can be implemented by using cooperative device drivers working at the system kernel[3]. These drivers work together to establish the SIOS across all physical distributed disks. Once the SIOS is established, all disks will be used collectively as a single global virtual disk. Then cluster can be built serverless and offers remote disk access directly at the kernel level. Parallel IO is made possible on any subset of local disks, since all distributed disks form SIOS. So, no heavy cross-node or process system calls are needed to perform remote file access.

3 Striped Checkpointing Scheme

The original concept of staggered checkpointing allows only one process to store the checkpoint at a time. Figure 1 shows the timing diagram of staggered checkpoint scheme. Although this scheme introduces non-blocking, lessen IO and network contention, it suffers from message logging[2] overhead and many control messages to avoid the inconsistency problem among nodes. Therefore, staggered checkpointing scheme increases the checkpoint overhead dramatically according to the cluster size and causes the traffic for stable storage to save message log information. Simultaneous writing of multiple processes in coordinated checkpointing also causes the network contention and IO bottleneck problem as the cluster size is increased. Parallel IO capacity of a distributed RAID is applied to

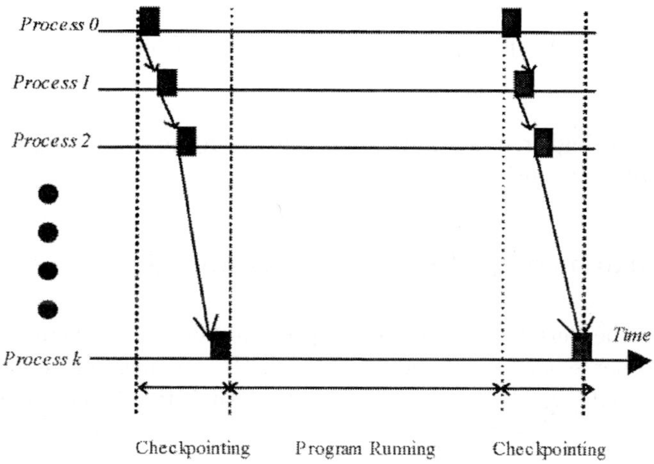

Fig. 1. Staggered checkpointing scheme

achieve fast checkpointing in a cluster system. The striped checkpointing scheme saves the checkpoint files over distributed disks that forming a stripe group. A stripe group is defined as a subset of disks that can be accessible in parallel. Only the disks in the same stripe group can initiate checkpointing simultaneously. To alleviate the network contention, staggered writing method is combined with the striped groups. Figure 2 shows this concept clearly. On the striped checkpointing scheme, each stripe group takes its checkpoints one after another. The stripe group leads to parallelism and avoids IO contention of the storage. This scheme can enhance both network utilization and IO performance. There exists a tradeoff between stripe parallelism and staggering depth. One can reconfigure the relative stripe size and staggering depth to achieve maximum benefits from it on a particular application. Higher parallelism can lead higher aggregate disk

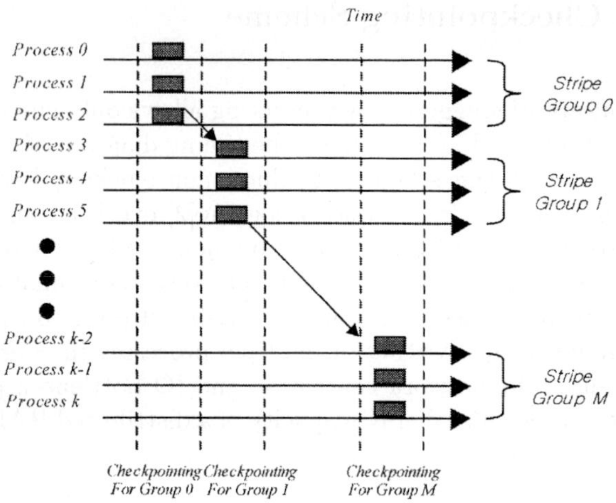

Fig. 2. Striped checkpointing scheme

bandwidth and higher staggering depth can lessen the contention problem with the least cost of inconsistency.

4 Performance Evaluation Environment

To evaluate the performance of the striped checkpointing scheme and to compare with other schemes, we choose Linpack HPC Benchmak to carry out the massive computational cluster experiments with MPI(Message Passing Interface). The size of checkpoint data is used to be proportional to the problem size of the computational program. The library function program Libckpt[7] is inserted in Linpack HPC computational program to implement a full-state, non-incremental checkpoint with the least system load in the Linux cluster environment. The performance is directly indicated by the checkpoint overhead and IO throughput. Our cluster platform is configured as 16 system nodes connected through the 100Mbps dedicated ethernet switch. Each node consists of Pentium-II processor, 128MB main memory, a 4GB local disk in single partition with Redhat Linux 6.2. Two storage schemes are used for saving checkpoints. On NFS, checkpoints of all nodes are saved only one disk on a dedicated node. On SIOS, 8 disks among the 16 disks are participated in the SIOS scheme. In our experiments, three checkpointing schemes, coordinated checkpoint, staggered checkpoint and striped checkpoint, are conducted on benchmark program. The coordinated checkpoint scheme forces all processes to take checkpoints at the same time. Although the cluster size would not be enough to show the drawback of the coordinated checkpointing scheme, we show the performance just to compare it's relative performance with staggered checkpointing scheme. The staggered checkpoint scheme allows only one node takes checkpoint at a time.

The striped checkpoint scheme takes three stripe configurations: 2, 4, and 8 disks in a striped group. The computational problem size of Linpack HPC Benchmark is varied from 1000 to 5000 to drive various workloads in our cluster system like light, medium and heavy workload reaches from 2.3 to 12.1Gflops during experiments.

5 Result and Discussion

Through the performance evaluation in various configuration, we had lots of impressive results. Figure 3 and 4 show the checkpointing overheads vs. problem size for all checkpointing schemes on NFS and 8 disks SIOS configuration. Checkpointing time represents total IO time including checkpoint file saving time, message logging time, and other communication overheads. Figure 5 and 6 also show the IO throughput for all checkpointing algorithm on NFS and SIOS configuration. IO throughput represents data transfer rate from nodes to NFS file server or distributed RAID system for checkpointing files. The results show that coordinated checkpointing scheme has the best performance in checkpointing time and IO throughput because of the small cluster size and parallel IO operation of the distributed RAID at first. But coordinated checkpointing will cause serious network contention when the cluster size in increases enough to over the network capacity. So, it makes sense to compare striped checkpointing scheme with staggered checkpointing scheme under our cluster platform.

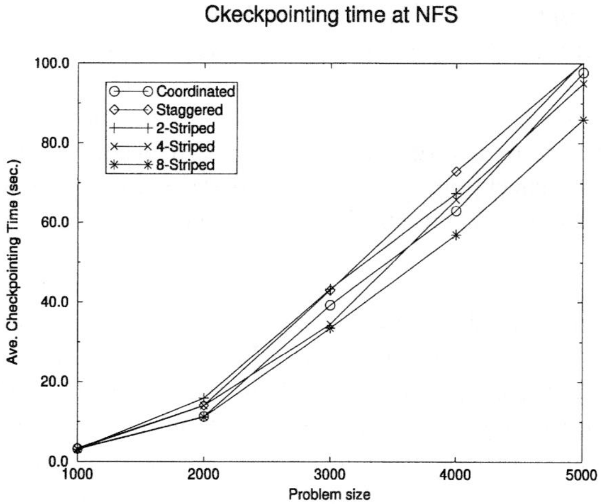

Fig. 3. Checkpointing overhead on NFS

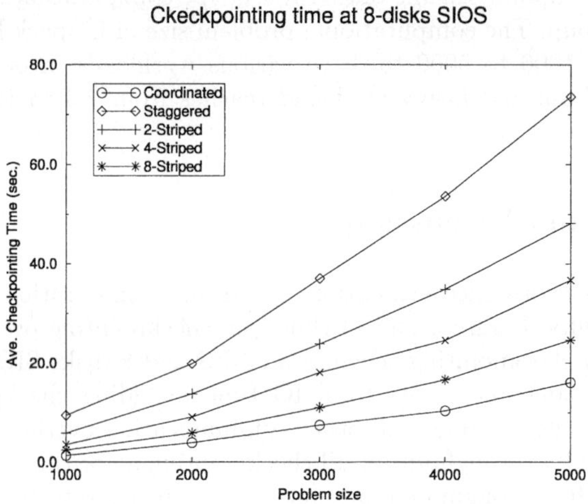

Fig. 4. Checkpointing overhead on SIOS

On NFS, all checkpointing schemes have not so much difference among their checkpointing time and throughput. It shows that IO contention on the cluster file server dominates over network contention. Therefore, checkpointing schemes do not make much sense on the checkpointing performance on NFS. Results on SIOS show the effect of the distributed RAID system results in its parallel IO operation. On SIOS, striped checkpointing scheme has better performance than staggered checkpointing scheme. Moreover, larger stripe group size leads to better performance until parallel IO bandwidth of a stripe group does not exceed total network capacity. On our cluster platform, 8-striped checkpointing scheme shows over 4 times better checkpointing time and over 3 times IO throughput than staggered checkpointing scheme. In the Linpack HPC Benchmark, large problem size generates a large checkpoint file. It results in performance gap between staggered checkpointing scheme and striped checkpointing scheme according to the problem size in checkpointing time.

The IO throughput of the striped checkpointing schemes are slightly down at the high problem size because of the IO bottleneck at distributed RAID system and still better performance than staggered checkpointing scheme. These results prove that maximum size of stripe group within the network bandwidth can produce the best performance on the distributed RAID system with SIOS although the cluster size becomes very large enough to stall network when coordinated checkpointing scheme is applied. This means that striped checkpointing scheme needs some negotiation between network bandwidth and size of stripe group. Therefore, a cluster system designed with ate size of stripe group will pay the least overhead for checkpointing operation under

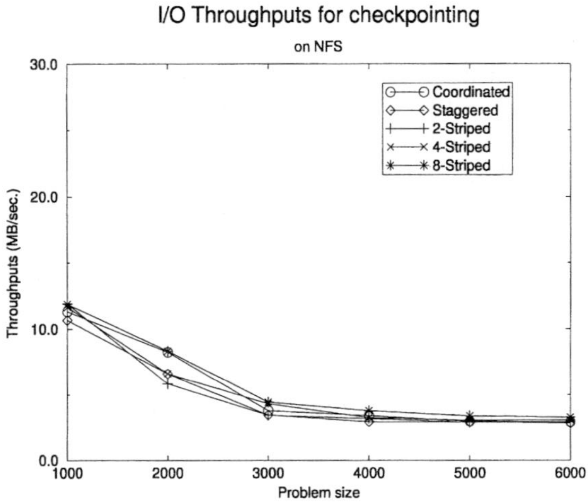

Fig. 5. Throughput on NFS

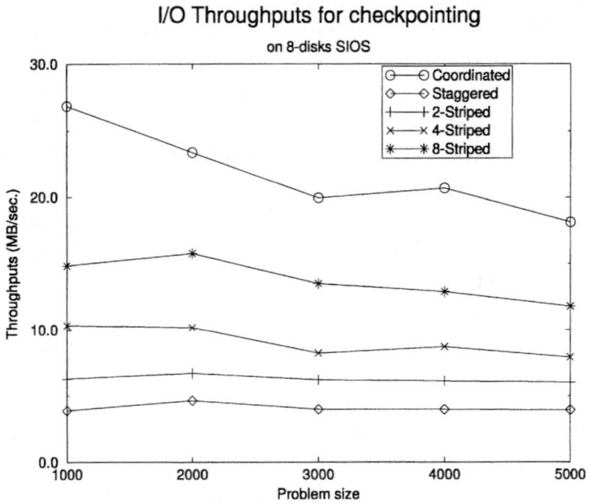

Fig. 6. Throughput on SIOS

the striped checkpointing scheme. If the cluster size is increased, we can easily guess the checkpointing overhead and the throughput would show something different result. We are going to deal the correlation between cluster size and rollback recovery performance with heavier computational problem at the further research.

6 Conclusion

The striped checkpointing scheme has been shown effective to perform checkpointing on cluster with SIOS. Striping exploits parallelism across multiple disks. Staggering alleviates the IO bottleneck on each disk. Therefore, striped checkpointing scheme can show better performance than other checkpointing schemes when the cluster size becomes large enough to over the IO and network bandwidth limitation. The tradeoff between striping and staggering should be applied dynamically for different applications. The tradeoff results can be applied to yield the striped checkpointing configuration with the lowest cost for a given application. Higher stripe parallelism can lead to higher aggregate disk bandwidth and higher staggering depth copes better with the network contention problem. Therefore, the stripe size suitable on a particular application should be established to get the best checkpointing performance on the distributed RAID system.

References

1. Cao, G., Singhal, M. : On Coordinated Checkpointing in Distributed Systems. IEEE Transactions on Parallel and Distributed Systems. **9(12)** (1998)
2. Elnozahy, E., Zwaenepoel, W. : On the Use and Implementation of Message Logging. Proceedings of 24th International Symposium on Fault-Tolerant Computing. (1994)
3. Hwang, K., Jin, H., Ho, R. : RAID-x: A New Distributed Disk Array for I/O-Centric Cluster Computing. Proceedings of 9th High-Performance Distributed Computing Symposium. (2000)
4. Hwang, K., Jin, H., Ho, R., Ro, W. : Reliable Cluster Computing with a New Checkpointing RAID-x Architecture. Proceedings of 9-th Workshop on Heterogeneous Computing. (2000)
5. Hwang, K., Jin, H., Chow, E., Wang, C., Xu, Z. : Designing SSI Clusters with Hierarchical Checkpointing and Single IO Space. IEEE Concurrency Magazine. (1999)
6. Plank, J., Li, K., Puening, M. : Diskless Checkpointing. IEEE Transactions on parallel and Distributed Systems. (1998)
7. Plank, J., Beck, M., Kingsley, G., Li, K. : Libckpt: Transparent Checkpointing Under UNIX. Proceedings of USENIX Winter 1995 Technical Conference. (1995)
8. Vaidya, N. : Staggered Consistent Checkpointing. IEEE Transactions on Parallel and Distributed Systems. **10(7)** (1999)

An Evaluation of Globus and Legion Software Environments

M.A.R. Dantas[1], J.N.C. Allemand[2], and L.B.C. Passos[2]

[1] Department of Informatics and Statistics (INE),
Federal University of Santa Catarina (UFSC)
88040-900 - Florianopolis - Brazil
{mardantas@computer.org}

[2] Department of Computer Science (CIC)
University of Brasília (UnB)
70919-970, Brasília, Brazil

Abstract. In this article we present a case study comparison of the implementation characteristics of two software environments that are well known in grid computing configurations. We evaluate the performance of these environments during the execution of distributed parallel MPI tasks. Therefore, first we consider some concepts of the grid paradigm and then we present a comparison between the two software environments. Our case study is based on the *Globus* and *Legion* environments, because these two research projects are in more developed stage when compared to others research initiatives. Our experimental results indicate that the grid computing approach can be interesting to execute distributed parallel MPI applications with a performance improvement.

1 Introduction

Advances in the communication and computer technologies are providing to an applications programmer access to a large quantity of computer resources distributed in wide area networks. As [1, 6] mentioned, it is possible to image that many users´ problem can be solved without the requirement of a local supercomputer.

In this context, we base our goal to build a *grid computing environment* which could represent a global infrastructure of hardware and software. This environment can provide users with a reliable access and low cost to share distributed resources. This concern is expressed by others research groups [2, 21, 10, 3, 12]. We can consider a *grid computing environment* as an infrastructure because in many aspects this paradigm represents the gathering action of resources, information and people which are widely distributed. As a result, it is necessary the use of a complex software environment to manage this configuration and also provides a useful computer power to application programmers.

P.M.A. Sloot et al. (Eds.): ICCS 2003, LNCS 2658, pp. 963–970, 2003.

Many research initiatives [4, 5, 11] are in advanced stage, providing several software tools, which can represent the answer for a grid configuration. In this paper, we present a case study performance evaluation considering the *Globus* and *Legion* software environment to execute distributed parallel MPI tasks. Therefore, in section 2 we describe some characteristics of these two software packages. The research grid environment used in this paper is introduced in section 3. Implementation aspects of the two software environments and our experimental results, executing distributed parallel MPI applications, are presented in section 4. Finally, in section 5 we present our conclusions and future work.

2 Software Characteristics

In section we address a brief overview of the *Globus* and *Legion* environments. Our target is to present some important components and their functions in these powerful grid tools.

2.1 Globus

The *Globus* software environment [15, 17] is a project developed by *Argonne National Laboratory* (ANL) and *University of Southern California.* In this paper we use the version 1.1.4 of the Globus software package because this release provides support to MPI applications. The Globus environment is composed by a set of components implementing basic services to resource allocation, communication, security, process management and access to remote data [17, 20].

The resource allocation component of the Globus environment (GRAM - *Globus Resource Allocation Manager*) is the element that acts as an interface between global and local services. Application programmers use the GRAM element, through the *gatekeeper* software portion. This element is responsible for the user authentication and association with a local computer account. The mechanism to identify users of the *grid* is based on a file called *map-file*. In this file exists information about authorized users of the *grid configuration.* Any requirement for resource should be translated to the *Resource Specification Language (RSL).*

Communication in the Globus environment is performed using a communication library called *Nexus* [18, 19]. This component defines low a level API to support high level programming paradigms. Examples of high level programming paradigms supported are message passing, remote procedure call and remote I/O procedures. The information about the system and the *grid configuration* are management by a component called *Metacomputing Directory Service (MDS)* [9, 16,23].

An important aspect of the Globus software environment is the security. This software tool employs the certificate approach, which is carried by a CA (Certificate Authority) using the protocol *Secure Socket Layer* (SSL) [13, 14].

2.2 Legion

The *Legion* software environment is a system object oriented which is being developed since 1993 at University of Virginia. This environment has an architecture concept of grid computing providing a unique virtual machine for users´ applications. The approach of the *Legion* is to have some important concepts of a grid configuration (e.g. scalability, easy to program, fault tolerance and security) transparent to final users [7].

In the *Legion*, every entity such as processing power, RAM memory and storage capacity is represented as objects. Objects communicate with each other using services calls to a remote mechanism [7, 22]. The security component of the *Legion*, as the others elements of this software, is based on an object. The application programmer specifies the security related to an object, where it is defined which type of mechanism is allowed [22]. In addition, the *Legion* provides some extra basic mechanism to ensure more security. The *May I* method is an example. Every class should define the method *May I*, which checks for a called object the related allowed access.

The traditional system file is emulated in the *Legion* environment through the combination of persistent objects with the global information of object identification. This approach simplifies the manipulation of files to application programmers. In addition, it is allow to users to add fault tolerance characteristics to applications using rollback and recovery mechanisms [7].

3 The Grid Environment

After providing some characteristics of the Globus and Legion software tools, in this section we present our grid configuration environment. It is important to mention that all the machines were in the same laboratory. However, using a *Ethernet Layer 3 Switch* we were able to have the abstraction of a WAN (Wide Area Network) inside this box. In other words, this equipment could prove the abstraction of a distributed resource environment for our experiments.

Table 1. The grid configuration environment.

Computer Name	**AIX 1**	**AIX 2**	**AIX 3**	**AIX 4**
Operating System	AIX 4.3	AIX 4.3	AIX 4.3	AIX 4.3
Processor	PowerPC_ 604 233 MHz	PowerPC_ 604 233 MHz	PowerPC_ 604 233 MHz	PowerPC_ 604 233 MHz
Memory RAM	256 MB	128 MB	128 MB	512 MB
Hard disk	Two disks of 9 GB	Two disks of 4 GB	Two disks of 4 GB and one 2 GB disk	Two disks of 4 GB and one 2 GB disk
Software	*Legion*	*Globus*	*Globus*	Legion

Our configuration environment was formed by four IBM RS/6000 workstations connected using an *Ethernet Layer 3 Switch.* The decision to use the RS/6000 was based in the fact that we are planning to connect an IBM supercomputer SP4 to the grid configuration. In addition, it is interesting to mention that we have used before some IBM-PC machines running Linux. We have previously information about some problems with the AIX operating system. Therefore we decide to learn about the software environments in an open source operating system.

Table I shows the configuration environment of our *grid.* The environment could be considered heterogeneous once some elements, as memory and disk capacity, were different.

4 Evaluation of the Environments

After understanding and using both software environments, with a special attention to the concepts of wide distributed resources and passing through some problems related to AIX operating system, we were able to configure the *grid* environment for experiments. In this section, we are going to present comments about the use of two software environments and our results executing distributed parallel MPI tasks.

4.1 General Aspects of the Environments

The *Legion* software provides a homogeneous view of the *grid* to the application programmer. The environment uses its own tools to create the homogeneity. The procedure to install the software does not represent any problem, because the application programmer needs only to uncompress binary files and execute some script files. However, for the AIX environment it is necessary more information then those available from the software documents. We fixed some problems using our background on AIX and exchanging several e-mails with other AIX systems managers. The *Legion* concept of file system represents an advantage of the environment. The *Legion* file system presents a unique identifier for each object. This approach provides application programmers the facility to access files widely distributed only using their names. In other words, the users only use the name of the file, which can be storage in a local or remote machine. On the other hand, we have verified some problems with the package. As a first problem, we can mention the necessary installation of the entire environment when the *bootstrap host* has a power failure. The *bootstrap host* is responsible for the domain control. Another drawback of the environment is the low communication rate between objects. The paradigm of the *Legion* is to be a *framework environment,* where users can develop their own tools, such as security and fault tolerance facilities. This freedom can represent some flexibility to any developers group. However, it does not allow the use external tools.

The *Globus* approach allows users to use existing system available tools and have a uniform interface to the gird environment. Interesting features of the *Globus* environment are related to the security and to the autonomy of the configuration. The system has an infrastructure based on X509 certificate [13, 14] and the use the mutual autentification. On the other hand, one drawback of the software is the scalability, which can be understood as the capability to add new resources and new sites. When considering new facilities application programmers are required to have account into all new hosts.

4.2 Experimental Results

The next stage of our research work was the implementation of a distributed parallel application. We decided to use a parallel matrix multiplication using the MPI message-passing library. The *Globus* environment has a tool called MPICH-G2 to execute the application. On the other hand, the *Legion* has an internal tool called *legion_mpi_run* for executing MPI applications. As we mentioned before, because of the *Legion* framework, it was required some fixes on our MPI original code using the *Legion libraries*. Our results represent the average elapsed-time of twenty executions.

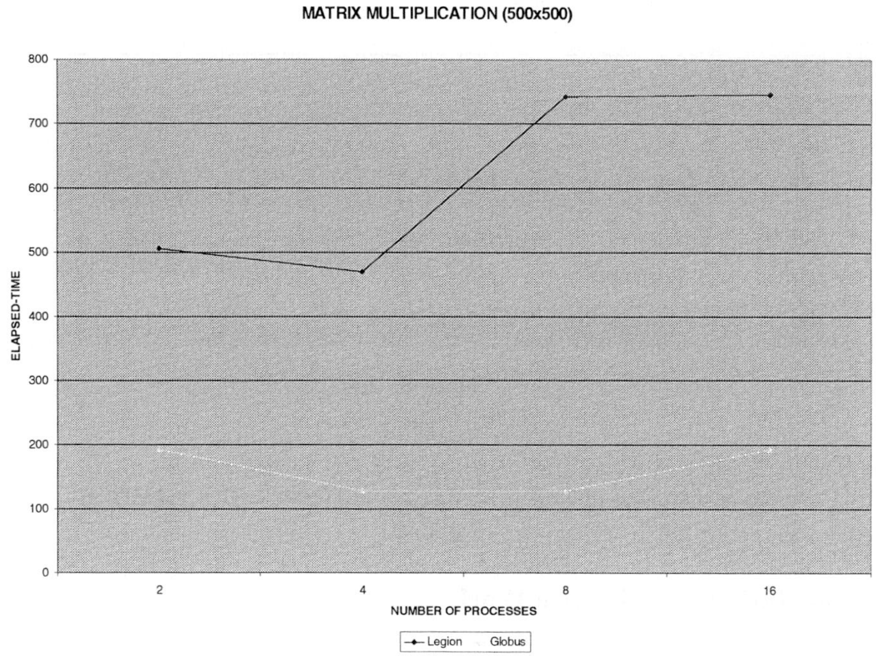

Fig. 1. Execution of the parallel MPI code on Globus and Legion Environments

Figure 1 shows the comparison to execute a 500 x 500 matrix multiplication MPI parallel code in the Globus and Legion environments. As we have mentioned before, the low communication rate between objects is responsible for the low performance of the Legion environment.

Our second experiment was configured using the native *MPICH* and comparing with the two software environments. The MPICH package is an implementation of MPI library and we considered the version mpich-1.2.2.3. Figure 2 presents results of parallel matrix multiplication of Globus, Legion and MPICH. This graphic shows that the Globus scheduling mechanism can represent an interesting feature.

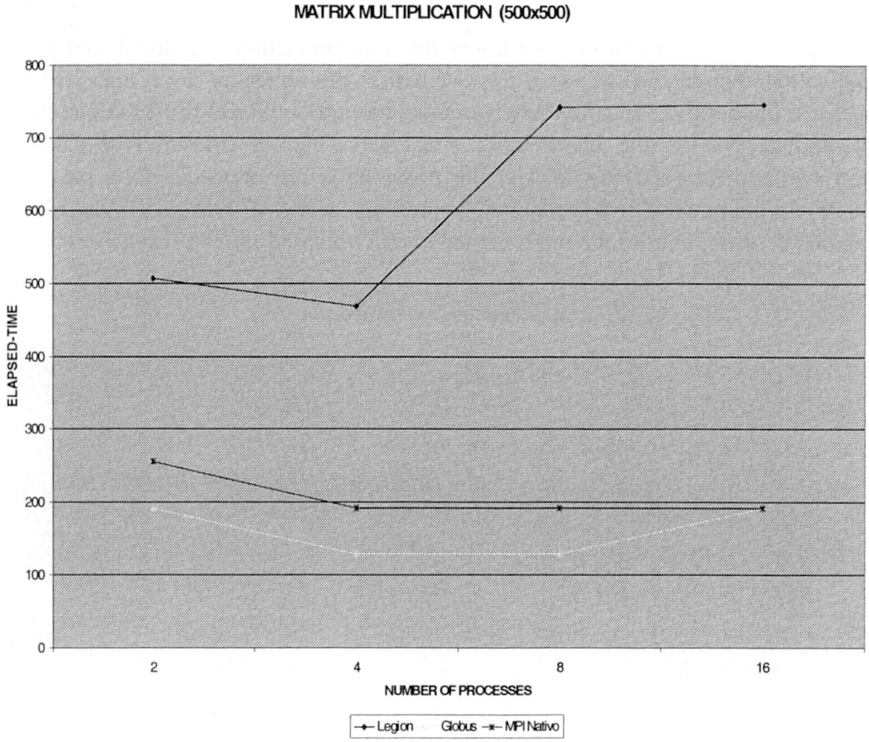

Fig. 2. Execution of the parallel MPI code on Globus, Legion and MPICH

5 Conclusions and Future Work

In this research work we have presented the implementation characteristics of the *Globus* and *Legion* software environments. In addition, we have evaluated these two environments when executing a distributed parallel MPI application.

Globus and *Legion* are important tools to configure grid configurations. The *Globus* environment has presented a more robust features, because the software includes security and fault monitoring mechanisms together with many others services. On the other hand, because it is an object-oriented package the *Legion* environment is more efficient to present the grid abstraction. This software is a *framework* and it is not a finished tool. However, we believe that *Legion* can address those users that are expecting a grid configuration that can be customized for their organisations.

One direction as a future work for this research is to understand how these two grid environments can perform under heterogeneous operating systems (e.g. Linux, Irix, HP-UX and Solaris). As we notice in this work, several problems can appear to an application programmer when using these packages. Another future path is to develop a friendly interface (e.g. via *browser)* to present the grid environment to users. As we have presented in [8], users can be awarded of the available resources in the grid environment. This approach should be easy as an ordinary interface used on an operating system (e.g. Linux or Windows) to display resources available in the local computer.

Acknowledgements. The first author was funding by the Brazilian National Research Council (CNPq). The Grid environment used in our experiments comes from collaboration between UnB-CiC/TECSOFT-DF.

References

1. Baker, M., *Cluster Computing Trends*, Physics Seminar, Liverpool University, http://www.dcs.port.ac.uk/~mab/Talks/Liverpool00/, May 2000.
2. Baker, M., *Technologies for MultiCluster/Grid Computer*, Cluster 2001, Newport Beach, http://www.dcs.port.ac.uk/~mab/Tutorials/ Los Angeles, EUA, 2001.
3. Bester, J., Foster, I., Kesselman, C., Tedesco, J., and Tuecke, S., *Gass: A Data Movement and Access Service for Wide Area Computing System,* In 7 (1997), ftp://ftp.globus.org/pub/globus/papers/gass.pdf.
4. Brunett, S., Davis, D., Gottschalk, T., Messina, P., and Kesselman, C., *Implementing Distributed Synthetic Forces Simulations in Metacomputing Environments,* In Proceedings of the Heterogeneous Computing Workshop (Mar. 1998), ftp://ftp.globus.org/pub/globus/papers/sf-express.pdf.
5. Brunett, S., Czajkowski, K., Foster, I., Fitzgerald, S., Johnson, A., Kesselman, C., Leigh, J., and Tuecke, S., *Application Experiences with the Globus Toolkit.,* In Proc. 7th IEEE Symp. on High Performance Distributed Computing (July 1998), IEEE Computer Society Press, ftp://ftp.globus.org/pub/globus/papers/globus-apps.pdf.
6. Buyya, R. e Baker, M., *Emerging Technologies for MultiCluster/Grid Computing,* www.cacr.caltech.edu/cluster2001/program/abstracts/buyya.html, 2001.
7. Czajkowski, K., Foster, I., Karonis, N., Kesselman, C., Martin, S., Smith, W., and Tuecke, S, *A Resource Management Architecture for Metacomputing Systems.,* In The 4th Workshop on Job Scheduling Strategies for Parallel Processing (Mar. 1998), IEEE-P, pp. 4-18, ftp://ftp.globus.org/pub/globus/papers/gram97.pdf.

8. Dantas, M.A.R,J.G.C Lopes, T.G. Ramos, *An Enhanced Scheduling Approach in a Distributed Parallel Environment using Mobile Agents, In Proc. 16th Annual International Symposium on High-Performance Computing and Systems,* Moncton, Canada, 2002, IEEE Computer Society Press.
9. Fitzgerald, S., Foster, I., Kesselman, C., von Laszewski, G., Smith, W., and Tuecke, S., *A Directory Service for Configuring High-performance Distributed Computations,* In Proc. 6th IEEE Symp. on High Performance Distributed Computing (1997), IEEE Computer Society Press, pp. 365-375, *ftp://ftp.globus.org/pub/globus/papers/hpdc97-mds.pdf.*
10. Foster, I., Kesselman, C. e Tuecke, S. *The Anatomy of the Grid : Enabling Scalable Virtual Organizations,* www.globus.org/research/papers/anatomy.pdf, 2001.
11. Foster, I., *Grid Technologies & Applications: Architecture & Achievements*, Computing in High Energy and Nuclear Physics 2001 - CHEP'01, Pequim, China, September 2001, http://www.ihep.ac.cn/~chep01/paper/10-047.pdf.
12. Foster, I. e Kesselman, C.;*The Grid: BluePrint for a new computing infrastructure*, Morgan Kaufmann, 1999.
13. Foster, I., Kesselman, C., and Tsudick, S. T.G., *A Security Architecture for Computational Grids,* In Proc. of the 5th ACM Conference on Computer and Communication Security (Nov. 1998), ACM Press, ftp://ftp.globus.org/pub/globus/papers/security.pdf
14. Foster, I., Karonis, N. T., Kesselman, C., and Tuecke, S.., *Managing Security in High-performance Distributed Computations,* Cluster Computing 1, 1 (1998), 95-107, *ftp://ftp.globus.org/pub/globus/papers/cc-security.pdf.*
15. Foster, I., and Kesselman, C., *The Globus Project: A Progress Report,* In Proceedings of the Heterogeneous Computing Workshop (Mar. 1998), ftp://ftp.globus.org/pub/globus/papers/globus-hcw98.pdf
16. Foster, I., and von Laszewski, G., *Usage of LDAP in Globus*, TR, ANL, 1997, ftp://ftp.globus.org/pub/globus/papers/ldap_in_globus.pdf.
17. Foster, I., and Kesselman, C., *Globus: A Metacomputing Infrastructure Toolkit,* International Journal of Supercomputer Applications 11, 2 (1997), 115-128, ftp://ftp.globus.org/pub/globus/papers/globus.pdf
18. Foster, I., and Tuecke, S. , *Nexus: Runtime Support for Task-parallel Programming Languages,* ftp://ftp.globus.org/pub/globus/papers/nexus_paper_ps.pdf , TR, ANL, 1994.
19. Foster, I., Kesselman, C., and Tuecke, S., *The Nexus Task-parallel Runtime System,* In Proc. 1st Intl Workshop on Parallel Processing. Tata McGraw Hill, 1994, pp. 457-462, ftp://ftp.globus.org/pub/globus/papers/india_paper_ps.pdf.
20. The Globus Project, *Globus Toolkit 1.1.3 Sytem Administration Guide*, University of Shouthern California, http://www.globus.org, December 2000.
21. Grimshaw, A., Ferrari, A., Knabe, F., Humprey, M.; *Legion: An Operating System for Wide-Area Computing*, 1999.
22. University of Virginia, *Legion 1.8 System Administrator Manual*, http://legion.virginia.edu, 2001.
23. Stelling, P., Foster, I., Kesselman, C., Lee, C., and von Laszewski, G. , *A Fault Detection Service for Wide Area Distributed Computations,* In Proc. 7th IEEE Symp. on High Performance Distributed Computing (July 1998), IEEE Computer Society Press, ftp://ftp.globus.org/pub/globus/papers/hbm.pdf.

An Agent Model for Managing Distributed Software Resources in Grid Environment

Jingbo Ding and Weiqin Tong

School of Computer Engineering and Science,
Shanghai University, Shanghai 200072, China
dingjingbo@etang.com

Abstract. Grid technologies enable large-scale sharing of many types of resources, among which software resources are a vital part. Abundant software resources exist in geographically distributed hosts in grid environment. The characteristic "distributed" makes the discovery, characterization, and monitoring of them challenging due to the considerable diversity, large numbers, dynamic behavior, and geographical distribution of the software entities. Thus, we put forward here an agent model, which is built on grid technologies, to manage the software resources in grid environment. The agent can store, retrieve and manage the information of software resources through a resource directory. Our model also provides a visualizing uniform interface by which user can easily find software resources in grid and utilize them. To sum up, this agent is a middleware between end user and software resources.

1 Introduction

Grid computing technologies enable widespread sharing and coordinated use of networked resources, and make remote utilization of distributed software resources possible [4, 5, 10, 15, 1]. But users often feel inconvenient when using these distributed resources in grid environment. For this reason, it is necessary to design an agent model, which is designated to support the initial discovery and ongoing utilization of the existence and characteristics of software resources.

The goal of our model is to provide a software resource pool, which can collect the information of software resources, and to provide an interface and a mechanism by which end users can discover and then utilize appropriate software entities interested.

The "distributed" characteristic of software resources in grid environment can be viewed as Fig. 1. All the software entities are geographically distributed and owned by Internet hosts. As Fig. 2 shows, we can regard them together as a *resource tree*. A virtual grid root is defined at the top of tree for convenience of undertaking. Software entities belong to hosts, which are a part of domain or connected to Internet directly.

Our agent model provides a software resources pool, which can be seen as a *logical information tree*. Fig. 3 describes an example. In this tree, all the software resources are managed centrally and classified in order to provide a uniform view to users. This mechanism makes managing software resources in grid computation environment more efficient and provides an encapsulated, uniform interface for user. For example, users can use software resources in grid without knowing what they are

P.M.A. Sloot et al. (Eds.): ICCS 2003, LNCS 2658, pp. 971–980, 2003.

and where they exist. By using this agent, user can find adequate information of software resources conveniently.

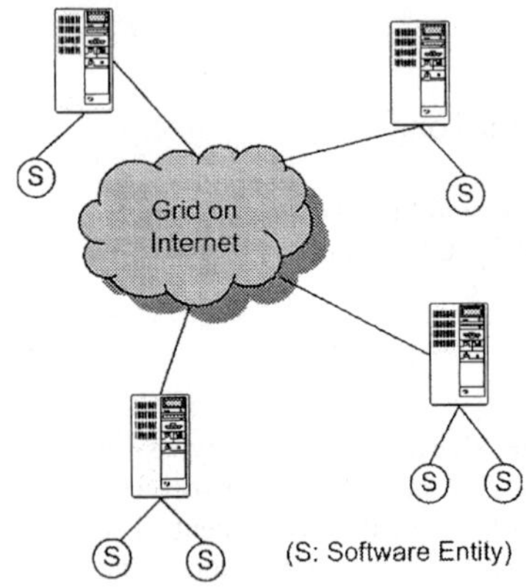

Fig. 1. Distributed software resources in grid environment

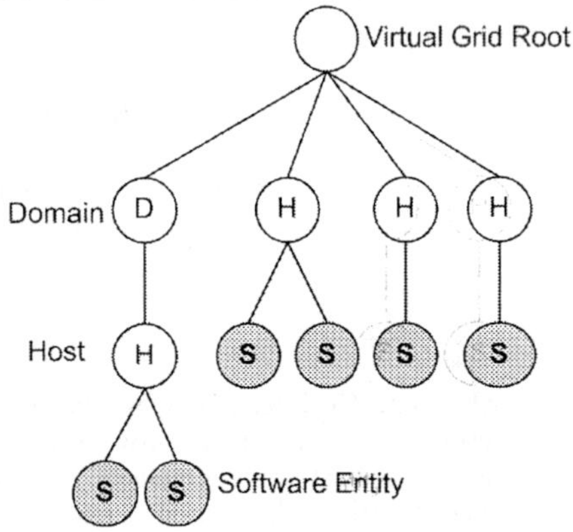

Fig. 2. A sample of resource tree

Let us compare the two trees. The former "resource tree" reflects the original characteristic of software resources – distributed; while the later "logical information tree", which satisfies users' needs, does conceal this characteristic. The main goal of our agent is to provide a logical view of software resources to end-users.

In a word, our agent for managing distributed software resources is a middleware between users and software resources that can: 1) for software resources, construct a centralized resource pool to store information of software resources and some maintenance tools to manage it; 2) for users, provide a visualizing uniform interface by with users can easily find and utilize them in grid environment.

The architecture of our model consists of four basic elements:

- **Resource directory.** This is a database that contains the information of software resources. In our model, we use LDAP (Lightweight Directory Access Protocol) [6, 2, 3] as its based protocol.
- **Registry tool for software resources.** Software resources should be initially registered to a centralized resource directory, which stores the information of software resources, to enable themselves to be discovered. Registry tool provides operation interface by which software resources can be registered in the directory. But software resources in grid environment are highly dynamic. To adapt for this characteristic, the agent uses a corresponding mechanism that updates the resource directory when some changes happen in software resources.
- **Discovery tool for users.** The main function of this tool is to accept querying request from users, search information in directory, and reply results back. Another purpose of this tool is to make users able to maintain the resource directory, such as add, delete, and modify information in it.
- **Web interface.** To provide a platform-irrespective operation and a friendly interface, web interface adopts XML technologies [12] to produce uniform

data format. An interpreter converts the data between web interface and agent kernel commands, such as "register" and "query".

The rest of this paper is structured as follows. In the next section, we briefly review the background of grid computation and some related work. In subsequent sections we first outline our architecture and then describe our agent model in detail: data model, registry tool, and discovering process. In Sect. 7 and Sect. 8, we present our trace of OGSA proposal and give a famous related work. We summarize the paper and conclude with a discussion of some future directions in Sect. 9.

2 Background

2.1 Grid and Globus Toolkit

A computational grid is a hardware and software infrastructure that provides access to high-end computational capabilities. Grid computing is concerned with coordinated resource sharing and problem solving in dynamic, multi-institutional virtual organizations [5, 10, 15]. The key concept is the ability to negotiate resource-sharing arrangements among a set of participating parties (providers and consumers) and then to use the resulting resource pool for some purpose. The Globus Project provides software tools that make it easier to build computational grids and grid-based applications. These tools are collectively called the Globus Toolkit, which provides three elements (Resource Management, Information Services, and Data Management) necessary for computing in a grid environment [9, 11].

The relationship between our agent model and Globus Toolkit is as following. Our agent model is based on Globus Toolkit. Parts of our agent work through the API of Globus Toolkit.

2.2 Directory, Directory Service, and Grid Information Services

Directory. Directories are used to store and retrieve information [3, 6]. Thus, directories are similar to databases.

Directory Service. A Directory Service provides a directory that can be accessed via a network protocol. Often, directory services include mechanisms for replication and data distribution. An example of directory service is Domain Name System (DNS), which resolves the names of computers to appropriate addresses [2, 3].

LDAP. The abbreviation LDAP stands for Lightweight Directory Access Protocol. LDAP defines a standard directory protocol that includes the following features [3, 6]:

- A network protocol for accessing information in the directory.
- An information model defining the form and character of the information.
- A namespace defining how information is referenced and organized.
- An emerging distributed operation defining how data may be distributed and referenced (LDAP version 3 [6]).
- An extensible protocol.
- An extensible information model.

Information Services are utilities that provide information about grid resources, including the Monitoring and Discovery Service (MDS), which provides a configurable information provider component called a Grid Resource Information Service (GRIS) and a configurable aggregate directory component called a Grid Index Information Service (GIIS) [1, 7, 8, 9]. The MDS uses the LDAP as a uniform interface to such information. A variety of software components, ranging from commercial servers to open source code, can be used to implement the basic features of MDS: generating, publishing, storing, searching, querying, and displaying the data.

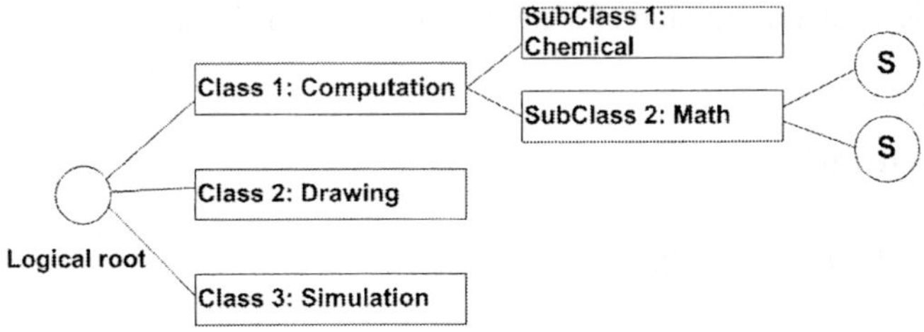

Fig. 3. A sample of logical view of resource tree

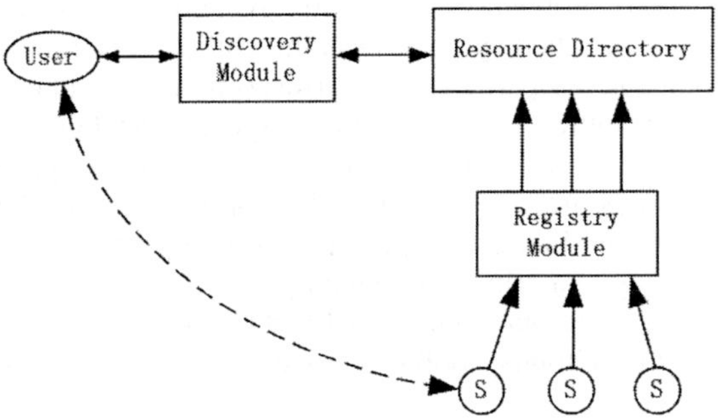

Fig. 4. Architecture overview

3 Architecture Overview

As described in Sect. 2, in Grid Information Services, with using the GRid Information Protocol (GRIP), users can query aggregate directory services to discover relevant entities, and/or query information providers to obtain information about individual entities. And description services are normally hosted by a grid entity directly, or by a front-end gateway serving the entity [1]. Now that software resource is a kind of resources, and it can be specified based on its information, we put forward such a motivation that utilizing the Grid Information Services to make software resources sharing possible.

The architecture of our agent model, which is described as Fig. 4, comprises three main parts – specialized aggregate resource directory, registry module, and discovery module. Software entities put their description information into resource directory through registry module. So directory stores gathered information of them. Using discovery module, users can query aggregate resource directory to discover relevant software entities and obtain information about individual entities.

Grid Information Service defines two basic protocols, the GRid Information Protocol (GRIP) and the GRid Registration Protocol (GRRP), to build a directory service [1, 9]. GRIP is used to access information about entities, while GRRP is used to notify aggregate directory services of the availability of this information [1]. We utilize these two protocols to realize our model, as we will describe in detail below.

Registry and discovery module are defined as services. In the process of registry, information of resources is collected in directory by registry module based on GRRP. Users use GRIP to query resources entities through discovery module. That is to say, registry service is used to register information of software resources, while discovery service is utilized to look up appropriate software entities for users.

4 Data Model

4.1 Utilizing Data Model of LDAP

In LDAP, the information model and namespace are based on entries. An entry is used to store attributes. An attribute has an associated type and can have one or more values. Each entry in the namespace has a distinguished name that allows easy identification [3, 6].

Entries are organized to form a hierarchical, tree-structured name space called a directory information tree (DIT). The distinguished name for an entry is constructed by specifying the entries on the path from the DIT root to the entry being named [2, 3, 6].

The data model in our agent is organized by logical information tree of software resources (See Fig. 3). Each object is stored in the directory as an item. The attribute softwareclass indicates the class to which the resource belongs. Hence, the agent can classify software resources by their attributes. On the other hand, the data structure of our model accords with the LDAP data model. So we can use LDAP to communicate between registry tool and resource directory.

An example is given to illustrate data model of our agent model, where a distinguished name could be as following:

< sn = software1,
sc = math,
sc = computation,
hn = node5.shugrid.shu.edu.cn,
ou = CS,
o = Shu,
c = CN >

Like MDS, the components of the distinguished name are listed in little endian order, with the component corresponding to the root of the DIT [2]. Within a distinguished name, abbreviated attributes are typically used. In this example, the names of the distinguishing attributes are: software name (SN), software class (SC), host name (HN), organizational unit (OU), organization (O), and country (C). Thus, a country entry is at the root of the DIT, while host entries are located beneath the organizational unit level of the DIT. So the software entities belonging to specified host can be located distinctively.

4.2 Specification of Software Resources Information

Each object of software entities is stored in resource directory as an item. To search software resources and acquire adequate information, the attributes of each item should at least include: software name, class to which resource belongs, owner, host name and IP address of the host in which resource exists, local directory in the remote host, execution file name, software description and other information necessarily. A specialized MDS object class "GridSoftware" is created to contain these attributes. Thus, a software entity can be represented by data model of agent. For the example above, the directory should include an item such as:

dn: <sn=equation, sc=math, sc=computation, hn=node1.shugrid2.shu.edu.cn, ou=shugrid2, o =Shu, c=CN>
objectclass: GridSoftware
owner: <cn=Person1, ou=CS, o=Shu, c=CN>
softwareclass: math
softwareclass: computation
hostname: node5.shugrid.shu.edu.cn
IP: 202.120.113.199
localdir: /usr/local/mathtool
execution: math1
description: Description text of this entity

To make process of searching entities more efficient, an additional database is used to store information of software resources separately from DIT. It is used for speeding up searching process. Each item in this database is corresponding to the same entity in DIT. When users look for some entity in resource pool, additional database is searched first, and then the agent searches relevant entries in DIT.

4.3 Information Directory

Information Directory is a vital part of our agent model. A vital one of the three elements of our agent model (See Sect. 1) is information directory, which is used to store information of software resources. We now describe the realization of information directory.

Since we use LDAP as our based protocol, let us first review DIT used by MDS. As introduced in Sect. 2 and Sect. 4.1, the information model of LDAP are based on entries, which are organized to form a DIT. A subset of DIT defined in MDS, which is showed in Fig. 5, is typical for LDAP directory structures, looking similar to the organization used for multinational corporations [2].

As we described in Sect. 3.1, our motivation of designing this agent model is to utilize Grid Information Services in grid technologies to manage distributed software resources. In our agent, we merge the information of software resources into DIT. An additional quick database is used to speed up searching process. It stores information of software resources in DIT in particular. Each item in this database is corresponding to the same entity in DIT. When users look for some entity in resource pool, additional database is searched first, and then the agent searches relevant entries in

DIT. The situation that information merged into DIT and the relationship between DIT and its additional quick database is showed in Fig. 6.

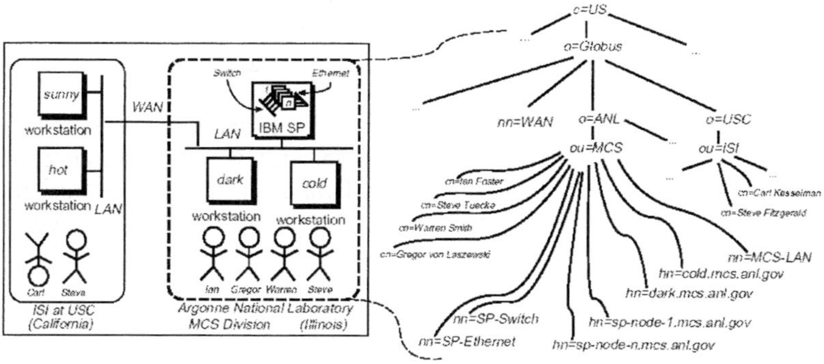

Fig. 5. A subset of DIT defined by MDS, showing the organizational nodes for Globus, ANL, and USC; the organizational units ISI and MCS, and people, hosts, and networks [2].

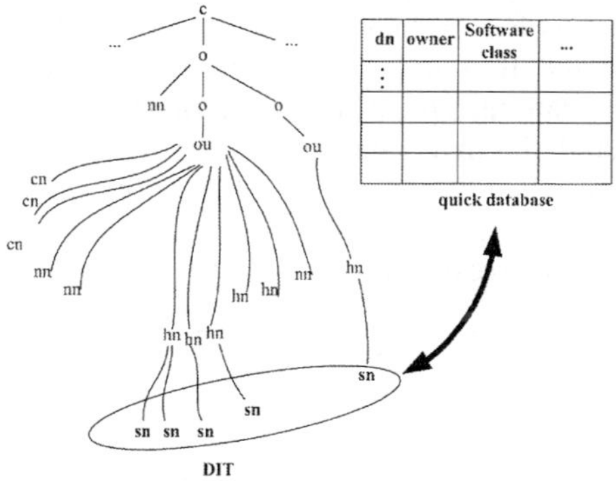

Fig. 6. Using quick database in the information directory to speed up searching process.

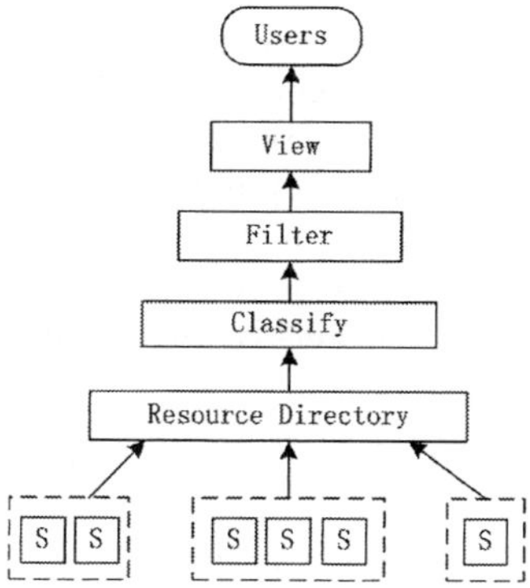

Fig. 7. Layers to provide logical view to users

4.4 Logical View

In the agent model, software resources are arranged as a logical information tree from users' point of view. As showed in Fig. 7, in our agent architecture, there are three layers on top of resource directory to provide logical view to users.

Classify layer. Items in directory are classified by their attributes in this layer.

Filter Layer. To give exact information in which users are interested, filter layer is used above the classified layer. Data are filtered according to users' demand in filter layer.

View layer. This layer gives the final logical view to users by some arrangement and process of data from filtered layer.

5 Registry of Software Resources

To initialize the directory, each software entity should be registered as an item in a centralized resource directory that is managed by agent. As is shown in Fig. 8, the registry tool in agent, which is based on Globus API, acts as a LDAP client to store information in the directory.

The realization of register module in our agent is based on the API of MDS (Metacomputing Directory Service). It communicates with the resource directory using LDAP protocol. Users initially register software entities using registry interface. Each entity is processed as an entry, with its information stored in resource directory according to the agent data model (See Sect. 4).

6 User Discovering Resources

An XML-based web interface is used to make users' operation easy. Using this interface, Users can query software resources by specifying their demand such as software name or a given software class. As Fig. 9 shows, between web interface and searching engine is a *command interpreter*. Its function is to convert the data from web page to searching engine, which can recognize and then make appropriate searching policy regarding users' demand. To make sure that every software entity replied to users can be used immediately, when finding results in directory, the agent tests the availability of entities among them, then just returns the available results to users. Unavailable ones are marked and will be tested later.

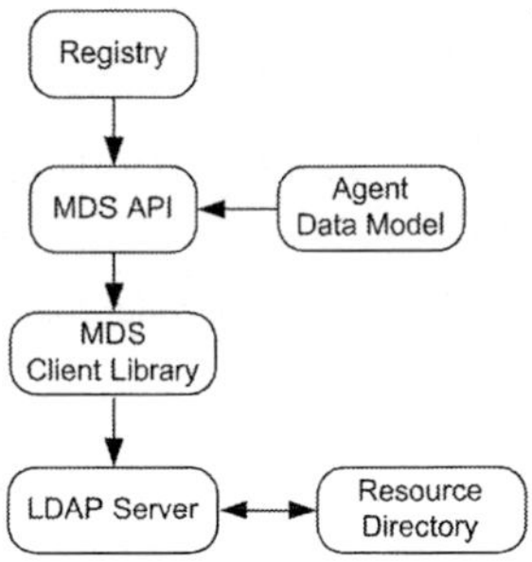

Fig. 8. Implementation architecture of registry tool

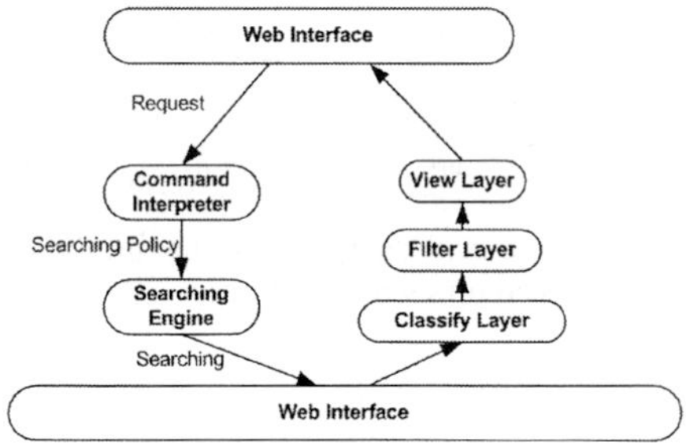

Fig. 9. A *command interpreter* converts data from web interface to searching engine

7 Trace of OGSA Proposal

Building on both Grid and Web services technologies, the Open Grid Services Architecture (OGSA) defines mechanisms for creating, managing, and exchanging information among entities called Grid services. Succinctly, a Grid service is a Web service that conforms to a set of conventions (interfaces and behaviors) that define how a client interacts with a Grid service [9].

We have implemented our agent model based on Globus Toolkit 2.0. To take advantage of OGSA, we are now trying to build a service to make distributed software resources sharing. The result is to provide software libraries, which is transparent to users, in grid environment.

8 Related Work

A famous related work is Universal Description, Discovery and Integration (UDDI) protocol. It is one of the major building blocks required for successful Web services. UDDI creates a standard interoperable platform that enables companies and applications to quickly, easily, and dynamically find and use Web services over the Internet. UDDI also allows operational registries to be maintained for different purposes in different contexts. UDDI is a cross-industry effort driven by major platform and software providers, as well as marketplace operators and e-business leaders within the OASIS standards consortium [18].

Because of some similarity between UDDI and our agent model, it's necessary to present the difference of them. In brief, the main function of UDDI is to provide a "meta service" for locating web services by enabling robust queries against rich metadata. The purpose of our agent model is to manage a kind of specific resource – software resource in grid environment.

9 Conclusions and Future Work

We have presented an agent model for sharing distributed software resources in grid computations. It not only simplifies the operation of user, but also provides a uniform interface to users. User can search the resources he needs conveniently and efficiently and get some information about them. In the future, we expect to use the following technologies:

Automatically identify installed software. To avoid registering manually, we would have agents scan file systems to automatically identify installed software.

Multiple agent servers (or backup servers). To realize fault-tolerant operation for user, we can build multiple servers to store information cooperatively. The redundant directories will take over the job when original directories fail due to some faults such as network partition.

Monitoring tool and allocation mechanism. Resources in grid environment can be redundant. If some kind of software resource exists in more than one host, agent should allocate an appropriate one to user. The resources state such as using rate can be obtained through monitoring tool. Then agent may generate an optimized strategy. The benefit of such mechanism is balancing the load of hosts.

References

1. K. Czajkowski, S. Fitzgerald, I. Foster, C. Kesselman: Grid Information Services for Distributed Resources Sharing. Proc. 10th IEEE International Symposium on High-Performance Distributed Computing (HPDC-10), IEEE Press (2001)
2. S. Fitzgerald, I. Foster, C. Kesselman, G.V. Laszewski, W. Smith, S. Tuecke: A Directory Service for Configuring High-Performance Distributed Computations. Proc. 6th IEEE Symp. on High Performance Distributed Computing
3. G.V. Laszewski, I. Foster: Usage of LDAP in Globus. http://www.globus.org (2002)
4. I. Foster, C. Kesselman, S. Tuecke: The Anatomy of the Grid Enabling Scalable Virtual Organizations. Intl J. Supercomputer Applications (2001)
5. I. Foster, C. Kesselman, The grid: Blueprint for a Future Computing Infrastructure, Morgan-Kaufmann (1999)
6. OpenLDAP 2.1 Administrator's Guide. http://www.openldap.org (2002)
7. MDS 2.2 User's Guide. http://www.globus.org
8. MDS 2.2: Creating a Hierarchical GIIS. http://www.globus.org
9. S. Tuecke, K. Czajkowski, I. Foster, J. Frey, S. Graham, C. Kesselman: Grid Service Specification Draft 4 (10/4/2002). http://www.gridforum.org/ogsi-wg
10. I. Foster, C. Kesselman: Computational Grids. http://www.globus.org
11. Globus: A Metacomputing Infrastructure Toolkit. http://www.globus.org
12. Extensible Markup Language (XML) 1.0 (Second edition) W3C Recommendation. http://www.w3.org/TR/REC-xml
13. M. Livny: Matchmaking: Distributed Resource Management for High Throughput Computing. Proc. 7th IEEE Symp. on High Performance Distributed Computing (1998)
14. J. Rosenberg, H. Schulzrinne, B. Suter: Wide Area Network Service Location. IETF Internet Draft (1997)
15. S. Gullapalli, K. Czajkowski, C. Kesselman, S. Fitzgerald: The Grid Notification Framework. Grid Forum Working Draft GWD-GIS-019. http://www.gridforum.org
16. T.A. Howes, M. Smith: A Scalable Deployable Directory Service Framework for the Internet. Technical report, Center for Information Technology Integration, University of Michigan (1995)
17. R. L. Ribler, J. S. Vetter, H. Simitci, D. A. Reed: Autopilot: Adaptive Control of Distributed Applications. Proc. 7th IEEE Symp. on High Performance Distributed Computing (1998)
18. http://www.uddi.org

Parallel DNA Sequence Alignment Using a DSM System in a Cluster of Workstations

Renata Cristina Faray Melo, Maria Emília Telles Walter, Alba Cristina Magalhaes Alves de Melo, and Rodolfo B. Batista

Department of Computer Science, Campus Universitario - Asa Norte, Caixa Postal 4466, University of Brasilia, Brasilia – DF, CEP 70910-900, Brazil
{renata, mia, albamm, rodolfo}@cic.unb.br

Abstract. Distributed Shared Memory systems allow the use of the shared memory programming paradigm in distributed architectures where no physically shared memory exist. Scope consistent software DSMs provide a relaxed memory model that reduces the coherence overhead by ensuring consistency only at synchronisation operations, on a per-lock basis. Much of the work in DSM systems is validated by benchmarks and there are only a few examples of real parallel applications running on DSM systems. Sequence comparison is a basic operation in DNA sequencing projects, and most of sequence comparison methods used are based on heuristics, that are faster but do not produce optimal alignments. Recently, many organisms had their DNA entirely sequenced, and this reality presents the need for comparing long DNA sequences, which is a challenging task due to its high demands for computational power and memory. In this article, we present and evaluate a parallelisation strategy for implementing a sequence alignment algorithm for long sequences. This strategy was implemented in JIAJIA, a scope consistent software DSM system. Our results on an eight-machine cluster presented good speedups, showing that our parallelisation strategy and programming support were appropriate.

1 Introduction

In order to make shared memory programming possible in distributed architectures, a shared memory abstraction must be created. This abstraction is called Distributed Shared Memory (DSM). The first DSM systems tried to give parallel programmers the same guarantees they had when programming uniprocessors. It has been observed that providing such a strong memory consistency model creates a huge coherence overhead, slowing down the parallel application and bringing frequently the system into a thrashing state[13]. To alleviate this problem, researchers have proposed to relax some consistency conditions, thus creating new shared memory behaviours that are different from the traditional uniprocessor one.

In the shared memory programming paradigm, synchronisation operations must be used every time processes want to restrict the order in which memory operations should be performed. Using this fact, hybrid Memory Consistency Models guarantee that processors only have a consistent view of the shared memory at synchronisation

P.M.A. Sloot et al. (Eds.): ICCS 2003, LNCS 2658, pp. 981–990, 2003.

time [13]. This allows a great overlapping of basic read and write memory accesses that can potentially lead to considerable performance gains. By now, the most popular hybrid memory consistency models for DSM systems are Release Consistency (RC) [3] and Scope Consistency (ScC)[7].

JIAJIA is a scope consistent software DSM system proposed by [5] that implements consistency on a per-lock basis. When a lock is released, modifications made inside the critical section are sent to the home node, a node that keeps the up-to-date version of the data. When a lock is acquired, a message is sent to the acquirer process containing the identification of the data that are cached at the acquirer node that are no longer valid. These data are, then, invalidated and the next access will cause a fault and the up-to-date data will be fetched from the home node. On a synchronisation barrier, however, consistency is globally maintained by JIAJIA and all processes are guaranteed to see all past modifications to the shared data [5].

In DNA sequencing projects, researchers want to compare two sequences to find similar portions of them, that is, they want to search similarities between two substrings of the sequences, and obtain good local sequence alignments. In practice, two families of tools for searching similarities between two sequences are widely used - BLAST [1] and FASTA, both based on heuristics. To obtain optimal local alignments, the most commonly used method is based on the Smith-Waterman algorithm [17], based on dynamic programming, with quadratic time and space complexity.

Many works are known that aim to efficiently implement the Smith-Waterman algorithm for long sequences of DNA. Specifically, parallel implementations were proposed using MPI [12] or specific hardware [4]. As far as we know, this is the first attempt to use a scope consistent DSM system to solve this kind of problem.

In this article, we present and evaluate a parallelisation strategy for implementing the Smith-Waterman algorithm. A DSM system was used since the shared memory programming model is often considered easier than the message passing counterpart. As the method proposed by [17] calculates each matrix element $A_{i,j}$ by analysing the elements $A_{i-1,j-1}$, $A_{i-1,j}$ and $A_{i,j-1}$, we used the "wavefront method" [14]. In this method, the parallelism is small at the beginning of the calculation, increases to a maximum across the matrix diagonal and then decreases again. The work was assigned to each processor in a column basis with a two-way lazy synchronisation protocol. The heuristic proposed by [12] was used to reduce the space complexity to O(n).

The results obtained in an eight-machine cluster with large sequence sizes show good speedups when compared with the sequential algorithm. For instance, to align two 400KB sequences, a speedup of 4.58 was obtained, reducing the execution time from more than 2 days to 10 hours.

The rest of this paper is organized as follows. Section 2 briefly describes the sequence alignment problem and the serial algorithm to solve it. In section 3, DSM systems and the JIAJIA software DSM are presented. Section 4 describes our parallel algorithm. Some experimental results are presented and discussed in section 5. Finally, section 6 concludes the paper and presents future work.

2 Smith-Waterman's Algorithm for Local Sequence Alignment

To compare two sequences, we need to find the best alignment between them, which is to place one sequence above the other making clear the correspondence between similar characters or substrings from the sequences [15]. We define *alignment* as the insertion of spaces in arbitrary locations along the sequences so that they finish with the same size.

Given an alignment between two sequences *s* and *t*, an score is associated for them as follows. For each column, we associate *+1* if the two characters are identical, *-1* if the characters are different and *–2* if one of them is a space. The *score* is the sum of the values computed for each column. The maximal score is the *similarity* between the two sequences, denoted by *sim(s,t)*. In general, there are many alignments with maximal score. Figure 1 illustrates this strategy.

G	A	-	C	G	G	A	T	T	A	G	
G	A	T	C	G	G	A	A	T	A	G	
+1	+1	-2	+1	+1	+1	+1	-1	+1	+1	+1	= 6

Fig. 1. Alignment of the sequences s= GACGGATTAG and t=GATCGGAATAG, with the score for each column. There are nine columns with identical characters, one column with distinct character and one column with a space, giving a total score 6 = 9*(+1)+1*(-1) + 1*(-2)

Smith-Waterman [17] proposed an algorithm based on dynamic programming. As input, it receives two sequences *s*, with $|s|=m$, and *t*, with $|t|=n$. There are $m+1$ possible prefixes for *s* and $n+1$ prefixes for *t*, including the empty string. An array $(m+1)x(n+1)$ is built, where the (i,j) entry contains the value of the similarity between two prefixes of *s* and *t*, $sim(s[1..i],t[1..j])$.

$$sim(s[1..i],t[1..j]) = \max\begin{cases} sim(s[1..i],t[1..j-1])-2 \\ sim(s[1..i-1],t[1..j-1])+p(i,j) \\ sim(s[1..i-1],t[1..j])-2 \\ 0. \end{cases} \qquad \text{Equation 1}$$

Figure 2 shows the similarity array between *s=AAGC* and *t=AGC*. The first row and column are initialised with zeros and the other entries are computed using equation 1. In this equation, $p(i,j) = +1$ if $s[i]=t[j]$ and -1 if $s[i]\neq t[j]$. Note that if we denote the array by *a*, the value of $a[i,j]$ is exactly $sim(s[1..i],t[1..j])$.

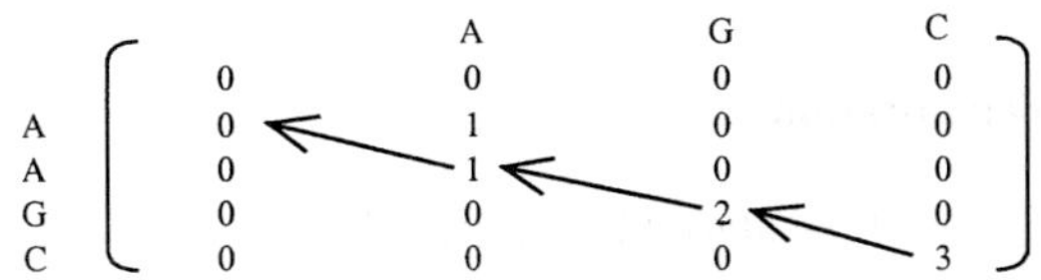

Fig. 2. Array to compute the similarity between the sequences s=AAGC and t=AGC.

We have to compute the array *a* row by row, left to right on each row, or column by column, top to bottom, on each column. Finally arrows are drawn to indicate where the maximum value comes from, according to equation 1. Figure 3 presents the basic dynamic programming algorithm for filling the array *a*. Notice that the score value of the best alignment is in *a[m,n]*.

```
Algorithm Similarity
Input: sequences s and t
Output: similarity between s and t

m ← |s|
n ← |t|
For i ← 0 to m do
a[i,0] ← i x g
For j ← 0 to n do
a[0,j] ← j x g
For i ← 1 to m do
For j ← 1 to n do
a[i,j] ← max(a[i-1,j]-2, a[i-1, j-1]±1, a[i,j-1]-2, 0)

Return a[m, n]
```

Fig. 3. Basic dynamic programming algorithm to build a similarity array a[m][n].

An optimal alignment between two sequences can be obtained as follows. We begin in a maximal value in array *a*, and follow the arrow going out from this entry until we reach another entry with no arrow going out, or until we reach an entry with value *0*. Each arrow used gives us one column of the alignment. If we consider an arrow leaving entry (i,j) and if this arrow is horizontal, it corresponds to a column with a space in *s* matched with *t[j]*, if it is vertical it corresponds to *s[i]* matched with a space in *t* and a diagonal arrow means *s[i]* matched with *t[j]*. An optimal alignment is constructed from right to left if we have the array computed by the basic algorithm. It is not necessary to implement the arrows explicitly, a test can be used to choose the next entry to visit. The detailed explanation of this algorithm can be found in [15].

Many optimal alignments may exist for two sequences because many arrows can leave an entry. In general, the algorithms for giving optimal alignments return just one of them, giving preference to the vertical edge, to the diagonal and at last to the horizontal edge.

The time and space complexity of this algorithm is $O(m\ n)$, and if both sequences have approximately the same length, *n*, we get $O(n^2)$.

2.1 Sequential Implementation

We implemented a variant of the algorithm described in Section 2 that uses two linear arrays [15]. The bi-dimensional array could not be used since, for large sequences, the memory overhead would be prohibitive. The idea behind this algorithm is that it is possible to simulate the filling of the bi-dimensional array just using two rows in memory, since, to compute entry *a[i,j]* we just need the values of *a[i-1,j]*, *a[i-1,j-1]* and *a[i,j-1]*. So, the space complexity of this version is linear, *O(n)*. The time complexity remains $O(n^2)$.

The algorithm works with two sequences *s* and *t* with length $|t|$. First, one of the arrays is initialised with zeros. Then, each entry of the second array is obtained from the first one with the algorithm described in Section 2, but using a single character of *s* on each step.

We denote *a[i,j]=sim(s[1..i,1..j])* as *current score*. Besides this value, each entry contains: *initial* and *final alignment coordinates*, *maximal and minimal score*, *gaps, matches and mismatches counters* and a *flag* showing if the alignment is a candidate to be an optimal alignment. These information allow us to keep a candidate optimal alignment with a score greater than a certain value. When computing the *a[i,j]* entry, all the information of *a[i-1,j]*, *a[i-1,j-1]* or *a[i,j-1]* is passed to the current entry.

To obtain the above values for each entry, we used some heuristics proposed by [12]. The *minimal* and *maximal scores* are updated accordingly to the *current score*. The *initial coordinates* are updated if the *flag* is *0* and if the value of the *maximal score* is greater than or equal to the *minimal score* plus a *parameter* indicated by the user, where this *parameter* indicates a minimum value for opening this alignment as a candidate to an optimal alignment. If it is the case, the *flag* is updated to *1*, and the *initial coordinates* change to the current position of the array. The *final coordinates* are updated if the *flag* is *1* and if the value of the *current score* is less than or equal to the *maximal score* minus a *parameter*, where the *parameter* indicates a value for closing an alignment. In this case, this alignment is closed and passed to a queue *alignments* of the reached optimal alignments and the *flag* is set to *0*.

The *gaps, matches* and *mismatches counters* are employed when the *current score* of the entry being computed comes from more than one previous entry. In this case, they are used to define which alignment will be passed to this entry. We use an expression *(2*matches counter + 2*mismatches counter + gaps counter)* to decide which entry to use. In this heuristic [12], gaps are penalized and matches and mismatches are rewarded. The greater value will be considered as the origin of the current entry. These counters are not reset when the alignments are closed, because the algorithm works with long sequences, and the scores of candidate alignments can begin with good values, turn down to bad values and turn again to good values.

If these values are still the same, the heuristic adopted is different from the one described in Section 2. Our preference will be to the horizontal, to the vertical and at last to the diagonal arrow, in this order. This is a trial to keep together the gaps along the candidate alignment [12]. At the end of the algorithm, the coordinates of the best alignments are kept on the queue *alignments*. This queue is sorted and the repeated alignments are removed. The best alignments are then reported to the user.

3 Distributed Shared Memory Systems

Distributed Shared Memory has received a lot of attention in the last few years since it offers the shared memory programming paradigm in a distributed or parallel environment where no physically shared memory exists.

One way to conceive a DSM system is by the Shared Virtual Memory (SVM) approach [11]. SVM implements a single paged, virtual address space over a network of computers. It works basically as a virtual memory system. Local references are executed exclusively by hardware. When a non resident page is accessed, a page fault is generated and the SVM system is contacted. Instead of fetching the page from disk, as do the traditional virtual memory systems, the SVM system fetches the page from a remote node and restarts the instruction that caused the trap.

Relaxed memory models aim to reduce the DSM coherence overhead by allowing replicas of the same data to have, for some period of time, different values [13]. By doing this, relaxed models no longer guarantee strong consistency at all times, thus providing a programming model that is complex since, at some instants, the programmer must be conscious of replication.

Hybrid memory models are a class of relaxed memory models that postpone the propagation of shared data modifications until the next synchronisation point [13]. These models are quite successful in the sense that they permit a great overlapping of basic memory operations while still providing a reasonable programming model. Release Consistency (RC) [3] and Scope Consistency (ScC) [7] are the most popular memory models for software DSM systems.

The goal of Scope Consistency (ScC) [7] is to take advantage of the association between synchronisation variables and ordinary shared variables they protect. In Scope Consistency, executions are divided into consistency scopes that are defined on a per lock basis. Only synchronisation and data accesses that are related to the same synchronisation variable are ordered. The association between shared data and the synchronisation variable (lock) that guards them is implicit and depends on program order. Additionally, a global synchronisation point can be defined by synchronisation barriers [7]. JIAJIA [5] and Brazos [16] are examples of scope consistent software DSMs.

JIAJIA is a software DSM system proposed by [5] which implements the Scope Consistency memory model. In JIAJIA, the shared memory is distributed among the nodes in a NUMA-architecture basis. Each shared page has a home node. A page is always present in its home node and it is also copied to remote nodes on an access fault. There is a fixed number of remote pages that can be placed at the memory of a remote node. When this part of memory is full, a replacement algorithm is executed.

In order to implement Scope Consistency, JIAJIA statically assigns each lock to a lock manager. The functions that implement lock acquire, lock release and synchronisation barrier in JIAJIA are jia_lock, jia_unlock and jia_barrier, respectively [6].

Additionally, JIAJIA provides condition variables that are accessed by jia_setcv and jia_waitcv, to signal and wait on conditions, respectively. The programming style provided is SPMD (Single Program Multiple Data) and each node is distinguished from the others by a global variable jiapid [6].

4 Parallel Algorithm to Compare DNA Sequences

The access pattern presented by the algorithm described in section 2 presents a non-uniform amount of parallelism and has been extensively studied in the parallel programming literature [14]. The parallelisation strategy that is traditionally used in this kind of problem is known as the "wave-front method" since the calculations that can be done in parallel evolve as waves on diagonals.

Figure 4 illustrates the wave-front method. At the beginning of the computation, only one node can compute value a[1,1]. After that, values a[2,1] and a[1,2] can be computed in parallel, then, a[3,1], a[2,2] and a[1,3] can be computed independently, and so on. The maximum parallelism is attained at the main matrix anti-diagonal and then decreases again.

$$\begin{bmatrix} 0 & 0 & 0 & 0 & 0 & 0 \\ 0 & a_{1,1} & a_{1,2} & a_{1,3} & a_{1,4} & a_{1,5} \\ 0 & a_{2,1} & a_{2,2} & a_{2,3} & a_{2,4} & a_{2,5} \\ 0 & a_{3,1} & a_{3,2} & a_{3,3} & a_{3,4} & a_{3,5} \\ 0 & a_{4,1} & a_{4,2} & a_{4,3} & a_{4,4} & a_{4,5} \\ 0 & a_{5,1} & a_{5,2} & a_{5,3} & a_{5,4} & a_{5,5} \end{bmatrix}$$

Fig. 4. The wave-front method to exploit the parallelism presented by the algorithm.

We propose a parallel version of the algorithm presented in section 2.1 and, thus, only two rows are used. Each processor p acts on two rows, a writing row and a reading row. Work is assigned in a column basis, i.e., each processor calculates only a set of columns on the same row, as shown in figure 5. Synchronisation is achieved by locks and condition variables provided by JIAJIA (section 3). Barriers are only used at the beginning and at the end of computation.

In figure 5, processor 0 starts computing and, when value $a_{1,3}$ is calculated, it writes this value at the shared memory and signals processor 1, that is waiting on jia_waitcv. At this moment, processor 1 reads the value from shared memory, signals processor 0, and starts calculating from $a_{1,4}$. Processor 0 proceeds calculating elements $a_{2,1}$ to $a_{2,3}$ When this new block is finished, processor 0 issues a jia_waitcv to guarantee that the preceeding value was already read by processor 1. The same protocol is executed by every processor i and processor i+1.

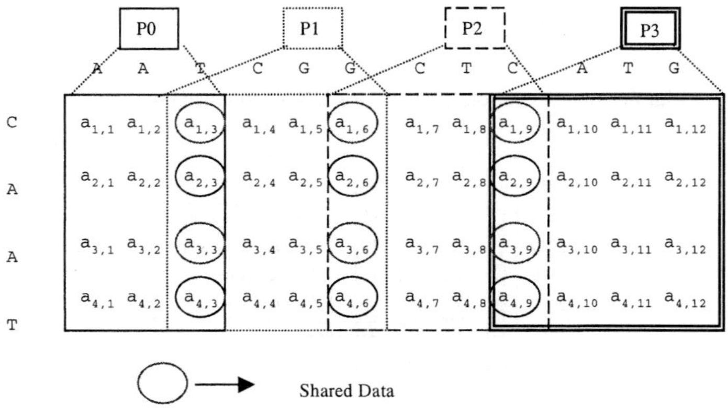

Fig. 5. Work assignment in the parallel algorithm. Each processor p is assigned N/P rows, where P is the total number of processors and N is the length of the sequence.

5 Experimental Results

The proposed parallel algorithm was implemented in C, using the software DSM JIAJIA v.2.1.

To evaluate the gains of our strategy, we ran our experiments on a dedicated cluster of 8 Pentium II 350 MHz, with 160 MB RAM connected by a 100Mbps Ethernet switch. The JIAJIA software DSM system ran on top of Debian Linux 2.1.

Our results were obtained with real DNA sequences obtained from the site *www.ncbi.nlm.nih.gov/PMGifs/Genomes*. Five sequence sizes were considered (15KB, 50KB, 80KB, 150KB and 400KB). Execution times and speedups for these sequences, with 1,2,4 and 8 processors are shown in Table 1 and illustrated in figure 6. Speedups were calculated considering the total execution time and thus include times for initialisation and collecting results.

Table 1. Total execution times (seconds) and speedups for 5 sequence comparisons

Size	Serial Exec	2 proc Exec /Speedup	4 proc Exec /Speedup	8 proc Exec /Speedup
15K x 15K	296	283.18/1.04	202.18/1.46	181.29/1.63
50K x 50K	3461	2884.15/1.20	1669.53/2.07	1107.02/3.13
80K x 80K	7967	6094.19/1.31	3370.40/2.46	2162.82/3.68
150K x 150K	24107	19522.95/1.23	10377.89/2.32	5991.79/4.02
400K x 400K	175295	141840.98/1.23	72770.99/2.41	38206.84/4.58

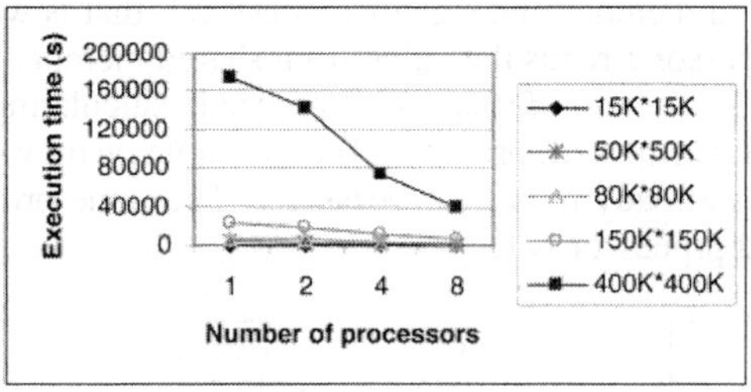

Fig. 6. Total execution times (s) for DNA sequence comparison

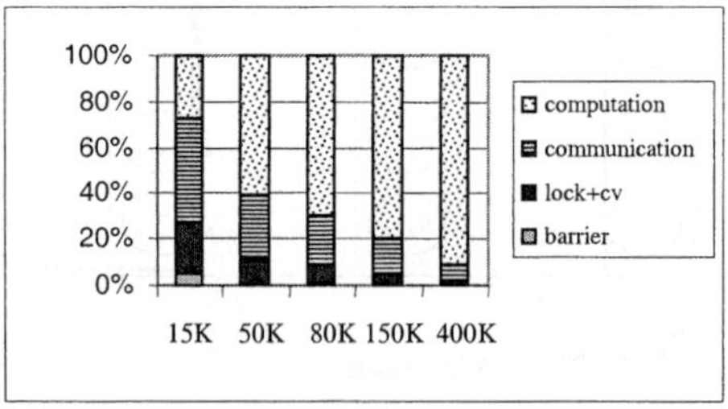

Fig. 7. Execution time breakdown for 5 sequence sizes, containing the relative time spent in computation, communication, lock and condition variable and barrier.

As can be seen in table 1 and figure 6, for small sequence sizes, e.g. 15K, very bad speedups are obtained since the parallel part is not long enough to surpass the amount of synchronisation inherent to the algorithm. As long as sequence sizes increase, better speedups are obtained since more work can be done in parallel. This effect can be better noticed in figure 7, which presents a breakdown of the execution time of each sequence comparison.

Martins et al. [12] presented a parallel version of the Smith-Waterman[1] algorithm using MPI that ran on a Beowulf system with 64 nodes each containing 2 processors. Speedups attained considering the total execution time were very close to ours, e.g., for 800Kx500K sequence alignment, a speedup of 16.1 were obtained for 32 processors and we obtained a speedup of 4.58 with 8 processors for 400K x 400K sequences. Besides that, our solution is cheaper and the DSM programming model is easier for this kind of problem.

6 Conclusions and Future Work

In this paper, we proposed and evaluated one parallel algorithm to solve the DNA local sequence alignment problem. A DSM system was chosen since, for this kind of problem, DSM offers an easier programming model than its message passing counterpart. The *wavefront* method was used and work was assigned in a column basis. Synchronisation was achieved with locks and condition variables and barriers.

The results obtained for large sequences in an eight machine cluster present good speedups that are improved as long as the sequence lengths increase. In order to compare two sequences of approximately 400KB, we obtained a 4.58 speedup on the total execution time, reducing execution time from 2 days to 10 hours. This shows that that our parallelisation strategy and the DSM programming support were appropriate to our problem.

As future work, we intend to port the algorithm implemented in MPI proposed in [12] to solve the same problem to our cluster and compare its results with ours. Also, we intend to propose and evaluate a variant of our approach, which will use variable block size to take advantage of the non-uniform type of parallelism presented by the *wavefront* approach.

References

1. S. F. Altschul et al. *Gapped BLAST and PSI-BLAST: a new generation of protein database search programs.* Nucleic Acids Research, v. 25, n. 17, p. 3389–3402, 1997.
2. J. Carter. *Efficient Distributed Shared Memory Based on Multi-Protocol Release Consistency.* PhD dissertation, Rice University, 1993.
3. K. Gharachorloo, D. Lenoski, J. Laudon, P. Gibbons, A. Gupta, J. Hennessy. *Memory Consistency and Event Ordering in Scalable Shared-Memory Multiprocessors.* Proc. Int. Symp. On Computer Architecture, May, 1990, p15–24.
4. L. Grate, M. Diekhans, D. Dahle, R. Hughey, *Sequence Analysis With the Kestrel SIMD Parallel Processor.* 1998.
5. W. Hu., W. Shi., Z. Tang. *JIAJIA: An SVM System Based on A New Cache Coherence Protocol.* In Proc. of HPCN'99, LNCS 1593, pp. 463–472, Springer-Verlag, April, 1999.

6. W.Hu, W.Shi. *JIAJIA User´s Manual*. Technical report, Chinese Academy of Sciences, 1999.
7. Iftode L., Singh J., Li K. *Scope Consistency: Bridging the Gap Between Release Consistency and Entry Consistency*. Proc. Of the 8th ACM SPAA'96, June, 1996, pages 277–287.
8. Keleher, P., Cox, A., Dwarkakas, S., Zwaenenpoel, W. *TreadMarks: Distributed Shared Memory on Standard Workstations and Operating Systems*. Proc. of USENIX, 1994, p.115–132.
9. Lamport L. *How to Make a Multiprocessor Computer that Correctly Executes Multiprocess Programs*. IEEE Transactions on Computers, 1979, 690–691.
10. D. Lenosky et al. *The DASH Prototype: Logic Overhead and Performance. IEEE Transactions on Parallel and Distributed Systems*, January, 1993.
11. K. Li. *Shared Virtual Memory on Loosely Coupled Architectures*. PhD Dissertation, Yale University, 1986.
12. W. S. Martins, J. B. Del Cuvillo, F. J. Useche, K. B. Theobald, G. R. Gao. *A Multithread Parallel Implementation of a Dynamic Programming Algorithm for Sequence Comparison*. Proceedings of the Symposium on Computer Architecture and High Performance Computing, 2001, Pirenopolis, Brazil, p.1–8.
13. Mosberger D. *Memory Consistency Models*. Operating Systems Review, p. 18-26, 1993.
14. G. Pfister,. *In Search of Clusters – The Coming Battle for Lowly Parallel Computing*. 1995.
15. J. C. Setubal, J. Meidanis, *Introduction to Computational Molecular Biology*. Pacific Grove, CA, United States: Brooks/Cole Publishing Company, 1997.
16. *E. Speight, J. Bennet. Brazos*: a Third Generation DSM System. Proc. Of the USENIX/WindowsNT Workshop, p.95-106, August, 1997.
17. T. F. Smith, M. S. Waterman. *Identification of common molecular sub-sequences*. Journal of Molecular Biology, 147 (1) 195-197–1981.

CSA&S/PV: Parallel Framework for Complex Systems Simulation

Ewa Niewiadomska-Szynkiewicz[1,2] and Maciej Żmuda[2]

[1] Research and Academic Computer Network (NASK), Wąwozowa 18, 02-796 Warsaw, Poland
e-n-s@ia.pw.edu.pl
http://www.nask.pl

[2] Institute of Control and Computation Engineering, Warsaw University of Technology, Nowowiejska 15/19, 00-665 Warsaw, Poland
mzmuda@elka.pw.edu.pl
http://www.ia.pw.edu.pl

Abstract. The paper[1] presents an integrated environment CSA&S/PV (*Complex Systems Analysis & Simulation - Parallel Version*), which can be used for design and simulation of large scale systems such as data networks, complex control systems and many others operating in inherently parallel environments. CSA&S/PV provides framework for simulation experiments carried out on parallel computers. It was used to investigate several real-life problems. Two examples are provided to illustrate the operation of the presented software tool: a hierarchical optimization algorithm for flow control in communication and computer networks and a hierarchical control structure for flood control in water networks.

1 Introduction

It is natural to model complex systems as a set of calculation processes, which can then be handled by distributed machines or processors. Recently parallel and distributed simulation has been an active research area, [3,4]. Distributed simulation allows to reduce the computation time of the simulation program, to execute large programs, which cannot be executed on single machine, to reflect better the structure of the physical system, which usually consists of several components.

In order to perform simulation experiments efficiently it is required to have good software tool. Numerous systems have been engineered to aid programmers. In most cases they are specified to certain solving problems [1,8,7,9,10]. Since parallel and distributed simulation is fast becoming the dominant form of model execution the focus is on experiments carried on parallel and distributed hardware platforms. High Level Architecture (HLA) standard for distributed

[1] This work was supported by Research and Academic Computer Network (NASK) and Polish Committee for Scientific Research under grant 7 T11A 022 20.

P.M.A. Sloot et al. (Eds.): ICCS 2003, LNCS 2658, pp. 991–1001, 2003.

discrete-event simulation was defined by the United States Department of Defense. During last years numerous integrated environments for parallel and distributed processing were developed. These software tools apply different techniques for synchronization and memory management, and focus on different aspects of distributed implementation. Many of them are built in Java, [5]. SimJava was among the first publically released simulators written in Java. This paper deals with the description of an integrated framework for parallel simulation CSA&S/PV *Complex Systems Analysis & Simulation - Parallel Version.*

2 Description of CSA&S/PV

CSA&S/PV presented in preliminary sequential version, in [6], is the software environment for different types of real systems simulation. The main idea of its development was to minimize user's effort during design and simulation of complex systems. CSA&S/PV provides a framework, which allows to perform simulations on parallel computers. It offers the graphical environment (shell) for supporting the considered case study implementation and a library of functions providing communication between the user's applications (LPs) and the system interface. CSA&S/PV manages calculations and communication between running processes, provides tools for on-line monitoring of the computed results. The asynchronous version of simulation is applied, [4]. In asynchronous simu-

receiver address	flags	time stamp	transmission delay	data

Fig. 1. Contents of messages from LPs simulating the considered physical system graph)

lation shared data object, the global clock and global event list, are discarded. Each logical process maintains its own local clock (*LVT* - Local Virtual Time). The local time of different processes may advance asynchronously. Events arriving at the local input message queue of a logical process are executed according to the local clock and schedule scheme. LPs can operate in two modes:

time-driven: The increment of LVT of each logical process is fixed and defined by the user. LP is executed every defined time step (repetition time), which means that LVT changes at regular intervals. We assume that for different LPs different repetition times may be introduced.

event-driven: Logical processes are executed after event occurrence. LVTs change at irregular intervals. A conservative scheme similar to the CMB algorithm, described in [3] is used for synchronization. Each event is executed only if it is certain that no event with an earlier time-stamp can arrive. At current time t each logical process LPi computes the minimum time $LVT_i = \min_{j \in N(i)}(t_{ij} + \tau_{ij})$, where t_{ij} is the time-stamp of the last message received from LPj process, $N(i)$ is a set of processes transmitting data to LPi and τ_{ij} is a transmission delay from node j to i (transfer cost). Next, every LPi simulates all events with time-stamps smaller than the LVT_i. The processes exchange messages as presented in Fig. 1. When the execution of the considered event begins, LPi sends to all their neighbours null messages with the time-stamp $LVT_i + \Delta T_i$ (where ΔT_i denotes the pending events time), which is the earliest possible time of next message. Null messages are used to announce absence of messages with new data.

Both types of LPs can be executed during one simulation.

2.1 System Structure

CSA&S/PV is composed of five components (see Fig. 2): 1) *shell* - the Graphical User Interface (GUI), responsible for user-system interaction, 2) *calculation module (manager)* - the system kernel that manages calculations and communication between running processes, 3) *communication library* - the library of functions that provides communication between the graphical shell and the system kernel, 4) *user library* - the library of functions that provides interface between the user application and "manager" (system kernel), 5) *user applications* - simulators of the physical subsystems (developed by the user).

The user's task is to implement the subsystems' simulators corresponding to the nodes of the considered graph and responsible for adequate physical systems simulation. These modules may be written in Java or C, C++. As it was mentioned above the CSA&S/PV package supplies the library of functions, which provides the interface between these programs and the system. This allows the user to focus on the numerical part of the program only. In addition if the functions unique to the operating system are not used by the user, applications can be moved as necessary between different computing platforms.

In general each user's application consists of six functions: *csasInit*, the task of which is to prepare the environment for future calculations and to calculate the initial conditions, *csasExecuteArgs*, which gathers data for calculations from CSA&S/PV, *csasExecute*, which provides calculations (the main part of the user's application), *csasExecuteResults*, which sends the results of calculations to CSA&S/PV, *csasStore*, the task of which is to store all current calculation results after system termination (simulation can be continued), *csasEnd*, an additional function for removing all data structures dynamically allocated during program operation.

Each node independently executes its program and communicates each other and with system kernel using *user library* functions. The system kernel manages only communication between the user interface (*shell*) and calculation processes corresponding to the nodes.

2.2 Implementation Details

The CSA&S/PV system has been implemented in Java, so it may operate under MS-Windows, Windows-NT and Unix operating systems. All calculation

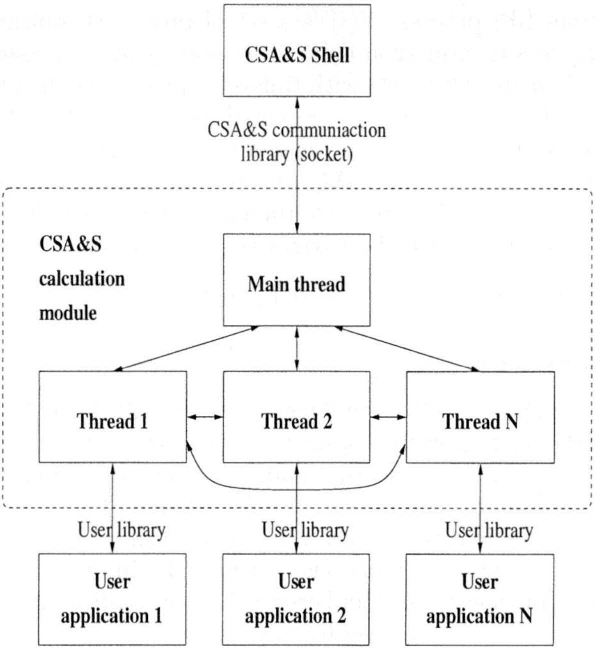

Fig. 2. The architecture of the CSA&S/PV system

processes communicate with each other via shared memory. The mechanism for parallel implementation is based on threads (see Fig. 2). Each calculation process (user's application) runs a thread. Because the software system is heterogeneous (*shell* and *manager* may operate under different operating systems) CSA&S/PV establishes a two-way socket connection between the user interface and the system kernel. The additional library of functions that provide communication between these two processes was developed.

2.3 Simulation under CSA&S/PV

During simulation experiment performed under CSA&S/PV one can distinguish two main stages: preparatory stage and experimental stage. At the *preparatory stage* the model and the properties of the system to be simulated are investigated. The calculation process is partitioned into several subsystems (subtasks) with respect to functionality and data requirements. The directed graph of the considered system $G = (N, L)$ is created and entered into CSA&S/PV using graph editor. The set of nodes is equal N with node i representing the i-th subsystem ($i = 1, \ldots, N$). The set of links is equal L. The presence of an arc $(i, j) \in L$ indicates the possibility that the i-th subsystem influences the j-th subsystem. Each node of the graph represents the program (LP) executing the tasks of the simulated physical subsystem. As it was mentioned above all these programs have to be prepared by the user and they must be ready to run. As far as the CSA&S/PV system is concerned, the goal of each node application program is to gather data from the connected nodes and to generate another data for the other nodes. Within the next step the user is asked to provide some information related to the nodes of the considered graph. The information includes: the name of the calculation program corresponding to each node, the repetition time period (if necessary), the decision delay - time required to execute the events in the physical application. Next the user is asked to provide some information related to all inputs of these nodes: the name of each input, the transmission delay related to data transmission to a particular location. The currently considered graph of simulated system may be saved into the disc file. In this way the application can be used in many future simulations.

The *experimental stage* begins when all decisions regarding the simulated system are made. The simulation time horizon is defined and the experiment starts. The adequate programs corresponding to the nodes of the system graph are executed. The results of the calculations are displayed. The user employs monitoring and analysis of the current situation. All results may be recorded into the disc file during the experiment. The simulation may be interrupted and restarted. There is a possibility to extend the simulation horizon if it is desired.

3 Practical Examples

The CSA&S/PV software package was used to perform simulations of several large scale systems. In this paper the case studies connected with application of hierarchical technics to optimization and control are described. The results of analysis of pricing algorithms for data networks and systems for operational flood control in multiple reservoir system obtained under CSA&S/PV are presented.

3.1 Optimization Network Flow Control

The first considered case study was related to the optimization approach to flow control in computer or communication networks. The asynchronous link algorithm for pricing of network services, based on the Price Method was implemented and tested. The detailed description of this method together with the discussion of its convergence one can find in [2].

Consider a network consisting of a set $L = \{1, \ldots, L_n\}$ of unidirectional links of capacities c_l, $l \in L$ and a set $S = \{1, \ldots, S_m\}$ of traffic sources. Each source is defined by a fourtuple $(L(s), U_s(x_s), x_{s_{min}}, x_{s_{max}})$, where x_s denotes transmission rate, $U_s(x_s)$ source utility function defined over interval $X_s = [x_{s_{min}}, x_{s_{max}}] \subseteq R_+$, $x_{s_{min}}$, $x_{s_{max}}$ respectively, minimum and maximum transmission rates. For each link l let $S(l)$ be the set of sources that use the link l, so $l \in L(s)$ if and only if $s \in S(l)$. The objective is to maximize the aggregate source utility over their transmission rates, so the flow optimization problem can be formulated as follows:

$$\max_{x_s \in X_s} \sum_s U_s(x_s), \qquad \sum_{s \in S(l)} x_s \leq c_l, \qquad l \in L_n \tag{1}$$

If the feasible set is nonempty and the performance function is strictly concave then the unique maximizing solution, $\hat{x}$ exists (see [2]).

IP networks consist of many subsystems, i.e. sources, routers, etc. They are constrained by the common resources - network capacity. The usage of methods with decomposition and coordination seems to be natural for such systems control. The optimization problem (1) can be solved by the Price Method (dual method using price coordination) in the parallel or distributed environment. Define the Lagrange function of (1)

$$\begin{aligned} L(x,\lambda) &= \sum_s U_s(x_s) - \sum_l \lambda_l \left(\sum_{s\in S(l)} x_s - c_l\right) = \\ &= \sum_s \left(U_s(x_s) - x_s \sum_{l\in L(s)} \lambda_l\right) + \sum_l \lambda_l c_l \end{aligned} \tag{2}$$

where $\lambda_l \geq 0$, i.e. the Lagrange multipliers associated with capacity constraints denote the link prices.

We can formulate the local (source) and coordinator level optimization problems:

LPs : $s = 1,\ldots,S_m$, for given λ_l find maximum w.r.t. x_s of the local performance index

$$\max_{x_s\in X_s} \left[L_s(x_s,\lambda) = U_s(x_s) - x_s \sum_{l\in L(s)} \lambda_l \right] \tag{3}$$

CP : for the results of LPs find minimum w.r.t. λ_l of the coordinator performance index

$$\min_{\lambda_l\geq 0,\, l=1,\ldots,L_n} \left[\varphi(\lambda) = \sum_s L_s(\hat{x}_s,\lambda^s) + \sum_l \lambda_l c_l \right] \tag{4}$$

where $\lambda^s = \sum_{l\in L(s)} \lambda_l$.

Synchronous and asynchronous distributed algorithms for prices computing were proposed by Low and Lapsley in [2]. They are the descent algorithms for the dual function minimization, with the price projection on $R_+^{S_m}$. In the synchronous version the l-th link price at the iteration instant $k+1$ is calculated as follows:

$$\lambda_l(k+1) = \left[\lambda_l(k) - \gamma \frac{\partial \varphi(\lambda(k))}{\partial \lambda_l}\right]_+ = \left[\lambda_l(k) + \gamma \left(\sum_{s\in S(l)} \hat{x}_s(k) - c_l\right)\right]_+ \tag{5}$$

where $[y]_+ = \max(y,0)$ and γ is sufficiently small step size.

Thus, in the (5) approach all sources receive, at a given time instant k, prices $\lambda_l(k)$, compute respective source prices $\lambda^s(k)$ and calculate optimal source rates $\hat{x}_s(k)$ solving LPs problems. The obtained values of the source rates $\hat{x}_s(k)$ are then send to the links and the new link prices $\lambda_l(k+1)$ are computed according to (5).

In the case of asynchronous approach both sources and link algorithms use weighted averages of the past values of the link prices and the locally optimal source rates. So, the l-th link price at the iteration instant $k+1$ is

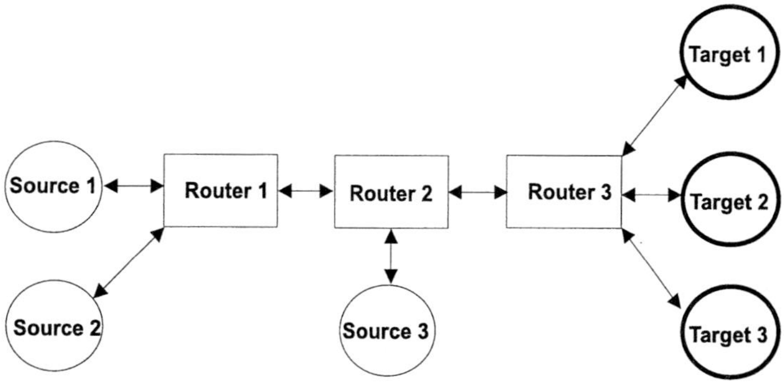

Fig. 3. Considered IP network under CSA&S/PV system

calculated according to (5) assuming $\hat{x}_s(k) = \sum_{k'=k-k_0}^{k} a_{ls}(k',k)x_s(k')$ with $\sum_{k'=k-k_0}^{k} a_{ls}(k',k) = 1$, for all k, l and $s \in S(l)$; k_0 denotes the length of past window taken into account. Respectively, the s-th source rate at time $k+1$ is calculated solving LPs, assuming $\lambda^s(k) = \sum_{l \in L(s)} \sum_{k'=k-k_0}^{k} b_{ls}(k',k)\lambda_l(k')$ with $\sum_{k'=k-k_0}^{k} b_{ls}(k',k) = 1$, for all k, s and $l \in L(s)$.
The described above asynchronous pricing algorithm was applied for flow control in experimental computer network, as presented in Fig. 3. It consists of nine nodes: three sources, three routers, three destination nodes and eight bidirectional links (the algorithm requires communication between sources and routers). The maximal capacity of the links *Router1-Router2* and *Router2-Router3* was equal 290. The capacity of other links was unlimited. The network was implemented in CSA&S/PV system using nine calculation processes. The processes exchanged messages as presented in Fig. 1, containing adequate data: link prices - messages from routers and source rates - messages from sources. All calculation processes corresponding to the nodes in Fig. 3 could communicate and update their controls asynchronously at different time instants, with different frequencies, and after transmission delays. The utility functions U_s of the sources were set to $\alpha_s log(1+x_s)$, with $\alpha_s = 10^4$ for all sources. Only the last received rate $x_s(\tau)$, for $\tau \in k-k_0,\ldots,k$ was used to estimate the locally optimal source rates and the link prices, i.e. a_{ls} and a_{ls} were set to 1 for $k' = k$ and 0 for $k' < k$. Each source transmitted data for a total of 120000 time units; *source 1* started transmission at time 0, *source 2* at time 40000, *source 3* at time 80000. The whole horizon of simulation was equal 240000 time units. The goal was to test the convergence of the algorithm w.r.t. value of the step size in (5) and transmission delays in the network. Several experiments were performed taking into account different values of step size $\gamma = \{1e{-}3,\ 1e{-}4,\ 1e{-}5\}$, and different transmission delays $\tau_D = \{1,\ 10,\ 100\}$ time units. It was assumed that the transmission was delayed for all links. The results are presented in table 1 and figures 4, 5. Table 1 contains percentage of rejected data packets w.r.t. all packets passed during the experiment. Figures 4 and 5 show respectively the destination receive rates

Table 1. Percentage of rejected packages with respect to different values of γ and τ_D

	$\tau_D = 1$	$\tau_D = 10$	$\tau_D = 100$
γ = 1e-3	14.59	15.62	98.41
γ = 1e-4	63.54	63.55	65.51
γ = 1e-5	92.22	92.23	92.28

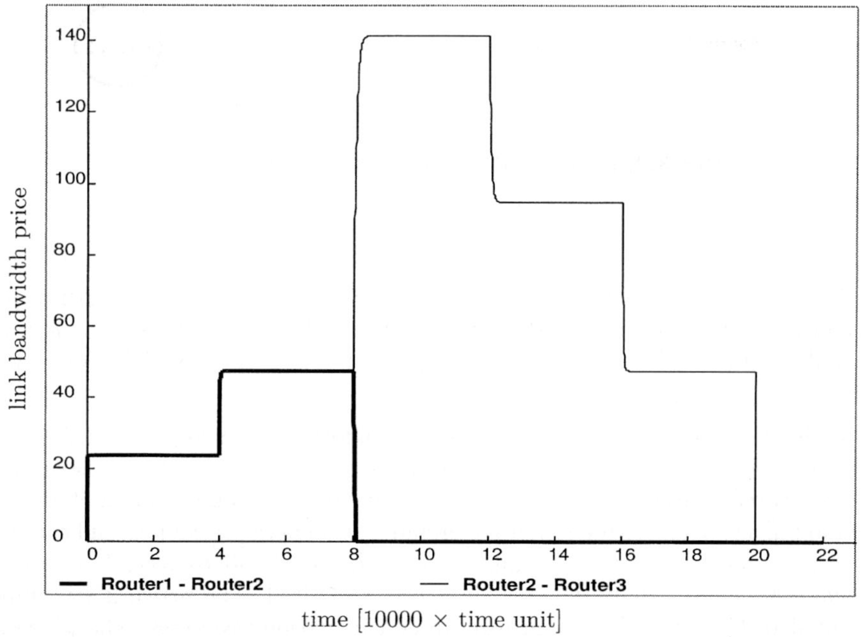

Fig. 4. The link prices for $\gamma = 1e - 3$ and $\tau_D = 1$

and the link prices. We can observe that the source rates adjusted dynamically as new sources started or stopped transmitting. As expected, the number of rejected packages increased for the longer transmission delay. For $\tau_D = 100$ and step size $\gamma = 1e - 3$ the algorithm was not convergent to the global optimum (see Table 1). After decreasing the step size better solution was achieved. On the other hand the decreasing γ took longer for the algorithm (5) to arrive at the proper price values. In the case of very small value ($\gamma = 1e - 5$) the algorithm seemed to track the optimum but the solution was not reached. The presented results show that the considered pricing algorithm for flow control is very sensitive to the value of the step size in optimization proces. The estimation of the proper γ may involve many problems especially in the case of huge network traffic.

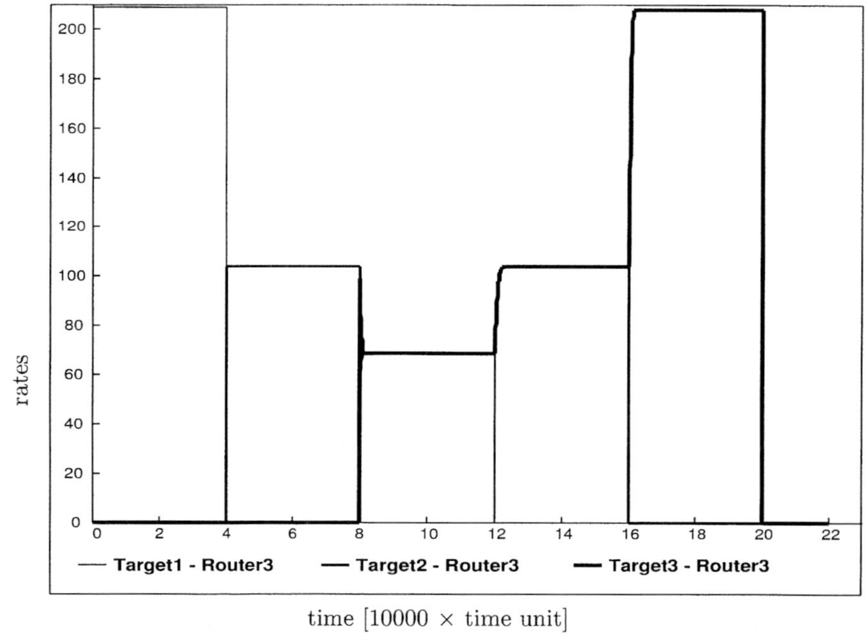

Fig. 5. The source rates for $\gamma = 1e-3$ and $\tau_D = 1$

3.2 Flood Control in Multireservoir System

The second case study, implemented under CSA&S/PV was related to flood control in the Upper Vistula river-basin system in the Southern part of Poland. Three retention reservoirs, located on Sola, Raba, Dunajec rivers were considered. The optimal release problem was defined as the problem of minimization the flood damages related to the peak flows at the measurement points in the whole river system. The hierarchical control mechanism (HDM) for reservoirs management, capable of satisfying the global objectives was investigated (see Fig. 6). This mechanism is based on the use of the repetitive optimization of the outflow trajectories, using the predicted inflows - forecasts (see [7] for details). It incorporates two decision levels each (see Fig. 6): the upper level with the control center (coordinator) and the local level formed by the operators of the reservoirs. The local decision rules are designed in such way that a central authority, the coordinator, may adjust them in the process of periodic coordination so as to achieve the coordination of reservoirs in minimizing the global damages. Hence, the decision problem of the i-th local reservoir operator ($i = 1, 2, 3$) at time t_l is as follows:

$$\min_{u_i} \left[q_i(u_i(.), a_i) = \max_{t \in [t_l, t_f]} (u_i(t) \cdot \alpha_i(t)) \right] \tag{6}$$

where $[t_l, t_f]$ denotes local level optimization horizon, q_i local cost function, a_i parameters specified by the coordinator. The vector a_i of coordinating param-

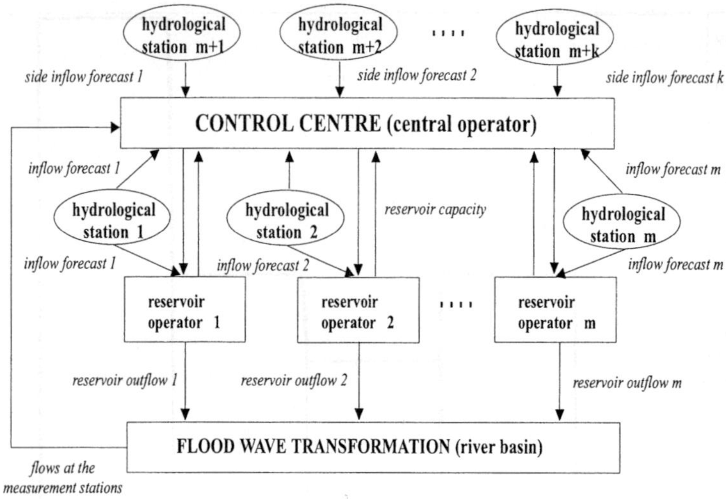

Fig. 6. Flood control in Vistula reservoir system

eters for the i-th reservoir is related to the weighting function $\alpha_i(.)$ defined as follows: $\alpha_i(t) = 1 + (c_i - 1) \cdot 1(t - T_i^\star)$, i.e., $\alpha_i(t) = 1$ for $t \in [t_l, T_i^\star)$ and $\alpha_i(t) = c$ for $t \in [T_i^\star, t_f]$.

The goal of the control center is to calculate the optimal values of parameters a, such that minimize the damages in the whole river basin.

$$\min_{a \in A} J(Q_{[t_c,t_f]}), \quad Q(t) = F\left(Q(t_c),\, \hat{u}_{[t_c,t]}(a),\, \overline{d}\,^{t_c}_{[t_c,t]}\right) \tag{7}$$

where $[t_c, t_f]$ denotes control center optimization horizon ($t_c \leq t_l$), $Q(t)$ vector of flows at the measurement points, $Q(t_c)$ vector of real flows measured at time t_c, $\overline{d}\,^{t_c}$ vector of forecasts of all the inflows calculated at time t_c, $\hat{u}$ vector of optimal outflows from the reservoirs (associated with vector of parameters a), $J(Q_{[t_c,t_f]})$ a performance (loss) function. Every iteration of the optimization process the value of $J(.)$ is computed based on simulation of the lower decision level (reservoirs' operators) and flow transformation in the whole river basin. Presented control structure was implemented under CSA&S/PV. The whole system was decomposed into several subsystems (processes) associated with the nodes in Fig. 6: control center, (coordination parameters calculation), reservoirs' operators (releases calculation), hydrological stations (inflow forecasts computing), rivers (flow transformation). Simulations were performed for the set of historical data. The results obtained for presented control system were compared with an independent reservoir management. The robustness of hierarchical control mechanism with respect to decision and transmission delays was tested too.

4 Conclusion

Simulation tools play an important role in the computer-aided analysis and design of complex control systems. CSA&S/PV is such a tool written in Java programming language on different platforms. The presented applications demonstrate the effectiveness and efficiency of the CSA&S/PV system. In both case studies CSA&S/PV was used as an environment for testing hierarchical techniques for control and optimization. It should be pointed that the considered applications could be solved in sequential manner but in both cases parallel computations were natural. In general the CSA&S/PV framework is suitable to solve many other small and large scale problems based on computer analysis and simulation. The package may be easily extended by "toolbox" of software modules, which are specific to a chosen case study.

References

1. Di, Z., Mouftah, H.T., QUIPS-II: a Simulation Tool for the Design and Performance Evaluation of Diffserv-based Networks, Computer Communications, 25, pp.1125–1131, 2002.
2. Low, S., Lapsey, D.E., Optimization Flow Control I: Basic Algorithm and Convergence, IEEE/ACM Transactions on Networking, 7(6), 1999.
3. Misra J., Distributed Discrete-Event Simulation, Computing Surveys, Vol. 18, No. 1, 1986.
4. Nicol D.M., Fujimoto R., Parllel Simulation Today, Annals of Opertions Research, Vol. 53, pp. 249–285, 1994
5. Nicol D.M., Johnson, M., Yoshimura, A., Goldsby, M., Performance Modeling of the IDES Framework, Proc. of the Workshop on Parallel and Distributed Simulation, pp.38–45, Lockenhaus, Austria, 1997.
6. Niewiadomska-Szynkiewicz E., Pooenik P., Bolek P., Malinowski K., Software Environment for Complex Systems Analysis and Simulation, Preprints IFAC/IFORS/IMACS Symposium "Large Scale Systems: Theory and Applications", London, UK, 1995
7. Niewiadomska-Szynkiewicz, E., Software Environment for Simulation of Flood Control in Multiple-Reservoir Systems, *Proc. International Conference "Advances in Hydro-Science and Engineering" ICHE2002*, Warsaw, Poland, 2002.
8. ns-2 (network simulator). *http://www.isi.edu/nsnam/ns/ns-documentation.html.*
9. Omnet (network simulator). *http://www.hit.bme.hu/phd/vargaa/omnetpp.htm.*
10. Szymański, B., Liu, Y., Sastry, A., Madnani, K., Real-Time On-Line Network Simulation, Proc. 5th IEEE International Workshop on Distributed Simulation and Real-Time Applications DS-RT 2001, pp.22–29, Los Alamitos, CA, 2001.

A Parallel Framework for Computational Science

Fernando Rubio and Ismael Rodríguez*

Departamento de Sistemas Informáticos y Programación
Universidad Complutense, 28040 – Madrid, Spain
{fernando,isrodrig}@sip.ucm.es

Abstract. Parallel languages based on skeletons allow the programmer to abstract from implementation details, reducing the development time of the parallelizations of large applications. Unfortunately, these languages use to restrict the set of parallel patterns that can be used. The parallel functional language Eden extends the lazy functional language Haskell with expressions to define and instantiate process systems. These extensions also make possible to easily define skeletons as higher-order functions. By doing so, skeletons can be both defined and used in the same language, using a high level of abstraction. Due to these facts, the advantages of skeleton-based languages are kept in Eden, while we do not inherit the restrictions they have, as the set of skeletons can grow as needed. Moreover, in our approach the sequential code of the programs can be written in any language supporting a COM interface.

Keywords: Parallel computing, skeletons.

1 Introduction

Due to the size of the applications in computational science, it is particularly important to be able to take advantage of parallel architectures to reduce the computation time. Unfortunately, conventional parallel programming languages use to require too much programming effort to correctly implement the parallel versions of the applications. Moreover, usually the parallelization of these programs heavily depend on the underlying architecture. Thus, porting a program to a different machine is not a trivial task at all.

Fortunately, during the last years several parallel languages (see e.g. [8,1,7] based on skeletons have been developed. A skeleton [2] is a *parallel problem solving scheme* applicable to certain families of problems. For example, the divide&conquer family can be abstracted in a single skeleton. The specific functions to be performed in the nodes of the process topology are abstracted as parameters. Thus, for instance, to parallelize a mergesort it is enough to specify it in terms of the divide&conquer method, while the actual parallel implementation will be delegated to the underlying skeleton.

The main advantages of using skeletons are two. Firstly, the parallelization effort is reduced, as predefined skeletons can be used; secondly, the probabilities of errors are reduced a lot, as the programmer does not need to handle all the gory details of the

* Work partially supported by the Spanish CICYT projects AMEVA and MASTER, and by the Spanish-British Acción Integrada HB1999-0102.

P.M.A. Sloot et al. (Eds.): ICCS 2003, LNCS 2658, pp. 1002–1011, 2003.

parallelization. However, skeleton-based languages use to restrict the set of skeletons available, so that all the programs must fit somehow a structure representable with those available skeletons. Fortunately, the language we present in this paper allows the user both to use predefined skeletons, and to define new ones. Let us remark that being able to extend the set of skeletons is a very important issue, as programmers can add new skeletons specific of their particular working areas.

In this paper we show how the Eden language can be used to define skeletons dealing with any process topology. These skeletons will not only be simple *schemes* that the programmer can *follow*: They will be actual programs parameterized by the code that need to be executed in each processor. Moreover, the programmer will be able to slightly modify the skeletons in case he want to include any characteristic particular to his working area. Thus, for each area of computational science, it could be possible to adjust the skeletons to the corresponding peculiarities.

Let us remark that, as Eden is a functional language, it is needed some knowledge of the functional paradigm to understand how to *define* new topologies. However, it is not necessary this knowledge to *use* the topologies, and it is even possible to *modify* them with only some knowledge.

The rest of the paper is structured as follows. In the next section we introduce the basic features of our language, while in Section 3 we present how we can use our language to develop generic topologies. In Section 4 we show the actual speedups obtained with one application running in a Beowulf architecture. Finally, in Section 5 we present our conclusions.

2 The Eden Language

Eden [9,6] extends the (lazy evaluation) functional language Haskell [11] by adding syntactic constructs to explicitly define processes. A new expression of the form `process x -> e` is added to define a *process abstraction* having variable `x` as input and expression `e` as output. Process abstractions can be compared to functions, the main difference being that the former, when instantiated, are executed in parallel.

Process abstractions are not actual processes. In order to really create a process, a *process instantiation* is required. This is achieved by using the predefined infix operator `#`. Given a process abstraction and an input parameter, it creates a new process, and it returns the output of the process. Each time an expression `e1 # e2` is evaluated, the instantiating process will be responsible for evaluating and sending `e2`, while a new process is created to evaluate the application `(e1 e2)`.

Once a process is running, only fully evaluated data objects are communicated. The only exceptions are lists, which are transmitted in a *stream*-like fashion, i.e. element by element. Each list element is first evaluated to full normal form and then transmitted. Concurrent threads trying to access not yet available input are temporarily suspended. This is the only way in which Eden processes synchronize. Notice that process creation is explicit, but process communication (and synchronization) is completely implicit.

In addition to the previous constructions, a process may also generate a new *dynamic channel* and send a message containing its name to another process. The receiving process may then either use the received channel name to return some information to the sender process (*receive and use*), or pass the channel name further on to another process (*receive and pass*). Both possibilities exclude each other, to guarantee that two processes cannot send values through the same channel.

Eden introduces a new expression `new` (*ch_name*, *chan*) *e* which declares a new channel name *ch_name* as reference to the new input channel *chan*. The scope of both is the body expression *e*. The name should be sent to another process to establish the communication. A process receiving a channel name *ch_name*, and wanting to reply through it, uses an expression *ch_name* `!*` e_1 `par` e_2 . Before e_2 is evaluated, a new concurrent thread for the evaluation of e_1 is generated, whose result is transmitted via the dynamic channel. The result of the overall expression is e_2, while the communication through the dynamic channel is a side effect.

Let us remark that it is trivial to extend the previous constructions to provide functions that deal with list of dynamic channels. For instance, the following function creates a list of n dynamic channels, where `[]` denotes an empty list, while `x:xs` is a list with `x` as head, and with `xs` as tail:

```
generateChannels 0 = []
generateChannels n = new (cn,c) ((cn,c) : generateChannels (n-1))
```

while the next function sends a list of values through their dynamic channels, and returns the evaluation of its second argument:

```
sendValues [] e            = e
sendValues ((v,ch):more) = ch !* v par sendValues more e
```

In most situations —in particular in all the topologies presented in this paper— by using only process abstractions and instantiations it is possible to create the same topologies that could be created by using dynamic channels. However, dynamic channels can be used to *optimize* the efficiency of the implementations. In this sense, this feature can be seen as an optimization using a low-level construct provided by the language rather than as a radically new concept.

Let us remark that, in contrast to most parallel functional languages, Eden includes high-level constructions both for developing *reactive* applications and for dynamically establishing direct connections between any pair of processes. This allows handling *low-level* parallel features that cannot be used in conventional functional languages. Thus, Eden provides an intermediate point between very high-level parallel functional languages (whose performance use to be poor), and classical parallel languages (which do not allow using high-level constructions). We do not claim that Eden can obtain optimal speedups, but it can obtain quite *acceptable* speedups with small programming effort.

Eden's compiler[1] has been developed by extending the most efficient Haskell compiler (GHC [4,10]). An important feature of Eden's compiler is that it reuses GHC's capabilities to interact with other programming languages. Initially, GHC only allowed to include C code, but currently it also provides a COM interface. Thus, the sequential

[1] The compiler can be freely downloaded from `http://www.mathematik.uni-marburg.de/inf/eden`

parts of our programs could be written in any language supporting COM interfaces. So, Eden can be used as a coordination language, while the computation language can be, for instance, C.

In order to easily port the compiler to different architectures,[2] Eden's runtime system has been designed to work on top of a message passing library. In the current compiler, the user can choose between PVM [3] and MPI [13].

3 Defining Topologies

In this section we present how processors topologies can be expressed in Eden. For the sake of clarity, we start presenting the most simple skeleton (`map`), then we introduce an example of how to define a simple topology of processors (a ring), which can be easily extended to deal with other typical topologies as a grid or a torus. After this introductory example, we present how to define a general topology dealing with any graph of connections amongst processors.

3.1 Defining Simple Skeletons

Let us remark that process abstractions in Eden are not just annotations but first class values which can be manipulated by the programmer (i.e. passed as parameters, stored in data structures, and so on). This facilitates the definition of skeletons as higher order functions. The most classical and simple skeleton is `map`. Given a list of inputs `xs`, and a function `f` to be applied to each of them, the Haskell specification is as follows:

```
map f xs  =  [f x | x <- xs]
```

This can be trivially parallelized in Eden using a different process for each task:

```
map_par f xs =  [pf # x | x <- xs]
                where pf = process x -> f x
```

The process abstraction `pf` wraps the function application `(f x)`. It determines that the input parameter `x` as well as the result value will be transmitted on channels.

Let us remark that developing skeletons in Eden is relatively easy. Due to the lack of space, we only show the simplest example, but many other skeletons have already been implemented, and in most of the cases their source code fit in half a page. Details about their implementation can be found in [12].

3.2 Ring Topology

A ring is a well-known topology where each process receives values from its left neighbor and sends values to its right one. Additionally, the first and last processes are also

[2] Currently, we have tested it on Beowulf clusters of up to 64 processors running Linux, on clusters of workstations running Solaris, and on a shared memory UltraSparc machine running also Solaris.

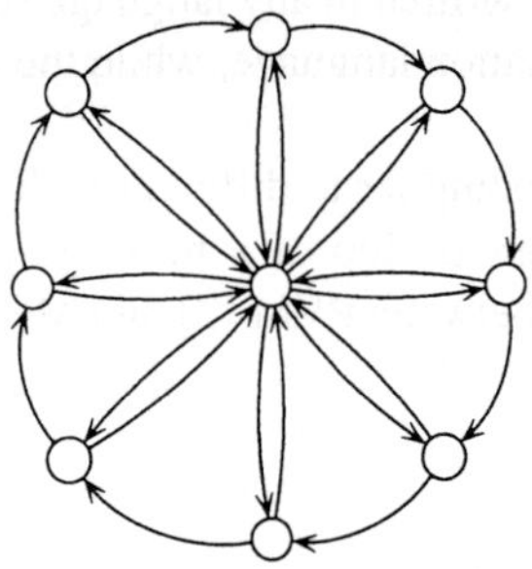

Fig. 1. A ring topology

considered to be neighbours, to form a real ring. In addition to that, all the processes can communicate with the main one — See Figure 1. This topology is appropriate for uniform granularity algorithms in which the workers at the nodes perform successive rounds. Before the first round, the main process sends the initial data to the workers. After that, at each round, every worker computes, receives messages from its left neighbour, and then send messages to its right neighbour. Eden's implementation uses lists instead of synchronization barriers to simulate rounds.

In order to create the topology, the skeleton receives two parameters: the worker function `f` that each of the processes will perform, and the initial list of `inputs` that will be provided initially to the processes. Let us remark that the length of such list will be the same as the number of processes in the ring. Let us also remark that the function `f` receives an initial datum from the parent and a list of data from the left neighbour, and it produces a list of data for its neighbour and a final result for its parent. In the following piece of code (that includes the whole skeleton), the `ring` function creates the desired topology by properly connecting the inputs and outputs of the different `pring` processes. As we want processes to receive values from its previous process, it is only necessary to shift the outputs of the list of processes before using them as inputs of the same list. Each `pring` receives an input from the parent, and one from its left sibling, and produces an output to the parent and another one to its right sibling:

```
ring f inputs = outsToParent   where
    outs    =  [(pring f) # outA' | outA' <- outs']
    (outsToParent,outsA) = unzip outs
    outsA' = last outsA : init outsA
    outs'  = zip inputs outsA'

pring f = process (fromParent, inA) -> out
    where out = f (fromParent, inA)
```

The previous definition can be optimized by using the lower level constructions of the language, that is, the *dynamic channels*. Fortunately, it is not neccessary to repeat the design, as we can take advantage of the methodology defined in [12] to automatically

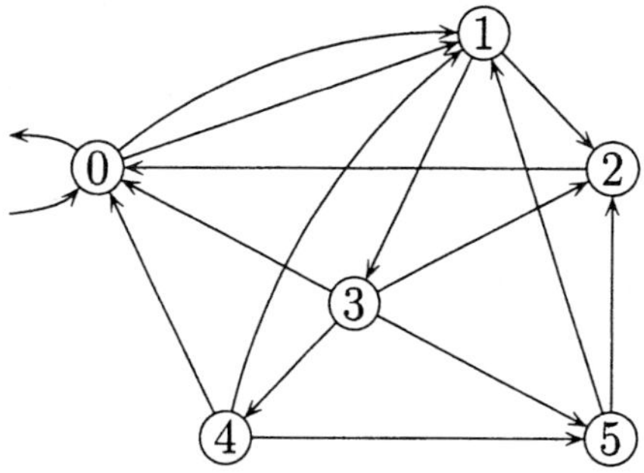

Fig. 2. A graph topology

transform high level definitions into lower level ones. The main idea of such transformation (not shown due to lack of space) is that a new dynamic channel is created by each process desiring to receive a value, and the name of such channels are to be sent to the producers of the values. Thus, by applying the transformation, now each `pring` receives an input from the parent, and a channel name to be used to send values to its sibling, and produces an output to the parent and a channel name to be used to receive inputs from its sibling, as shown below.

```
pring f = process (fromParent,outChanA) -> out
   where out = new (inChanA, inA)
                   let (toParent,outA) = f (fromParent,inA)
                   in outChanA !* outA par (toParent,inChanA)
```

Let us remark that it is trivial to extend the ring skeleton to deal with two-dimensions. In fact, in [12] it can be found the definition of several topologies, like a grid or a torus.

3.3 A General Graph

In order to proof the expressive power of our approach, we will now describe the most general topology: a graph of processes. Let us remark that conventional skeleton-based languages do not provide such a topology, even though it is becoming more and more important. Traditionally, parallel computers only used specialized topologies, and a general graph was not very unuseful. However, nowadays due to the wide implantation of the Internet, parallel computation is not restricted to specialized expensive architectures: several standard computers connected to the web can cooperate in the computation of a single problem. So, the actual topology they are using is a general graph. Thus, a parallel language should facilitate the use of such topologies.

As in the previous examples, in order to define a graph, we need to take care of two tasks: Defining the behaviour of the processes, and defining the connection topology. In this case, we will directly present the optimized implementation using dynamic channels, which has been obtained taking as basis the corresponding definition using only process abstractions and instantiations, as in the ring case.

Each process will be identified by a unique number, and its behaviour will be parameterized by the function to be computed, and by the list of identities of the processes that have direct connections with it. The source code is the following, where the process uses `generateChannels` to create as many dynamic channels as input connections are needed, while `sendValues` is used to actually send the output values:

```
pabs i f senders = process chansOut -> sends
  where channelsIn = generateChannels (length senders)
        (namesIn,valuesIn) = unzip channels
        valuesOut = f valuesIn
        sends = sendValues (zip ChansOut valuesOut)
                           (zip namesIn senders)
```

Let us remark that there must be an initial process in the graph, as shown in Figure 2. The difference with the other processes is that it will receive the inputs of the problem, and it will return the outputs. That is, it will be the interface of the topology with the outside world. The definition of this initial process is done in the same way as a normal one, but receiving an extra input `ins` and producing an extra output value `outs`:

```
pabs0 f senders = process (chansOut,ins) -> (sends,outs)
  where channelsIn = generateChannels (length senders)
        (namesIn,valuesIn) = unzip channels
        (outs:valuesOut) = f (ins:valuesIn)
        sends = sendValues (zip ChansOut valuesOut)
                           (zip namesIn senders)
```

Once the basic processes have been defined, we only need to properly connect them. In order to do that, we need as parameters the list of functions `fs` to be computed by each process, the list of connections `c` amongst processes, and the inputs `ins` of the initial process. In order to access to the function that process i should perform, it will be enough to use the predefined operator `!!`, that extracts the i-th element from a list. The connection topology only need to create $n - 1$ *normal* processes, and one extra *initial* process. After doing that, function `reorganize` uses the information encoded in the list of connections in order to establish them in the right way:

```
graph fs c ins = outs       where
  sendss = [(pabs i (fs!!i) (senders i)) # (receivers!!i)
              | i <- [1..n-1]]
  (sends0,outs) = (pabs0 (fs!!0) (senders 0)) # (receivers!!0,ins)
  senders i = map fst (filter ((== i).snd) c)
  receivers = reorganize (sends0:sendss)
  reorganize xss =  toList2 . sort2 . concat
  n = length fs
```

Let us remark that, even though it could seem that the previous program is complex to understand, it is not really important to understand the details of it. The important thing is that it can be written in a compact way and, most importantly, it can be used without understanding it: It will only be necessary to pass as parameters the list of connections, and the list of behaviours.

Let us also remark that it is completely trivial to nest several subgraphs. That is, in case several graphs have been defined to solve a set of problems, it is straighforward to create a new graph connecting all the subgraphs. In order to do that, it is enough to consider each subgraph as a function of the list `fs`. When doing that, each subgraph will be a node of the overall graph, obtaining the desired topology. The reason why this can be done so easily is that Eden processes are first-class citizens in a higher-order language. Thus, they can be used as parameters of other functions or processes.

4 Measuring an Application: Interactions amongst Particles

In this section we present the actual results obtained for one example using the previous topologies. For the sake of clarity, we have chosen a not too complex example, using a simple method to compute the forces amongst a set of particles, but we have already used our approach to parallelize other complex examples, like a simulator of the evolution of the economy of a country. Unfortunately, this simulator is too complex for explaining it briefly, but the speedups we have obtained with it are similar to those described in the following example.

The experiments have been performed in a 64-processor Beowulf cluster at the University of St. Andrews. Nodes are 450MHz Pentium II running Linux RedHat 5.2, with 348MB of DRAM and connected through a CISCO 2984G full duplex 100Mb/s fast Ethernet switch, being the latency $\delta = 142\mu s$. So, it is a low cost environment with high latencies. Eden RTS was running on top of PVM 3.4.2. Due to administrative reasons, it has not been possible to use all the processors in the tests.

Let us assume that we want to determine the force undergone by each particle in a set of n atoms. The total force vector f_i acting on each atom x_i, is

$$f_i = \sum_{j=1}^{n} F(x_i, x_j)$$

where $F(x_i, x_j)$ denotes the attraction or repulsion between atoms x_i and x_j.

This constitutes an example of pairwise interactions. For a parallel algorithm, we may consider n independent tasks, each devoted to compute the total force acting on a single atom. Thus, task i is given the datum x_i and computes $\{F(x_i, x_j) \mid i \neq j\}$. It is however inconceivable to have a separate process for each task when dealing with a large set of particles, as it is usually the case. Therefore, we distribute the atoms in as many subsets as the number of processors available. We use a ring structure, so that all the data can flow around. In the first iteration, each process will compute the forces between the local particles assigned to it. Then, in each iteration it will receive a new set of particles, and it will compute the forces between its own particles and the new ones, adding the forces to the ones already computed in the previous iterations.

In the next program, function `force` solve the problem by creating a ring of processes. The list of particles `xs` is split into `np` parts, being `np` the number of processors of the ring. Afterwards, the forces are computed by using the sequential function `force'`:

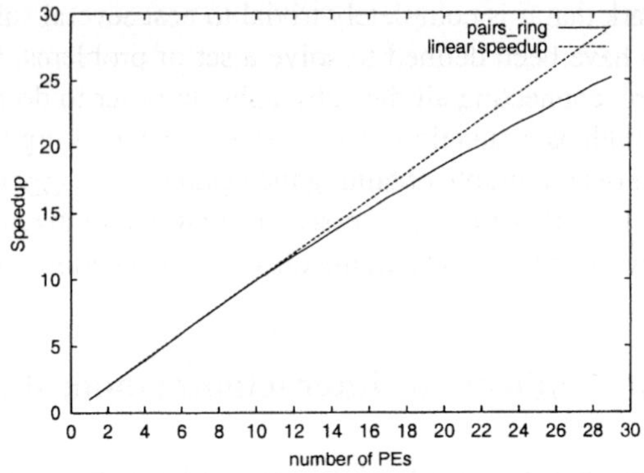

Fig. 3. Speedups of pair interactions

```
force xs = ring (force' np) (splitIntoN np xs)
force' np (local,ins) =  (total,outs)
  where outs            = take (np - 1) (local : ins)
        total           = foldl1' f forcess
        f acums news = zipWith addForces acums news
        forcess         = [map (faux ats) local | ats <- (local:ins)]
        faux xs y       = sumForces (map (forcebetween y) xs)
        sumForces l     = foldl' addForces nullvector l
```

Let us remark that it has not been necessary to deal with processes in the definition of the example. Let us also remark that the definition of `force'` could have been done in other programming language, as C. In fact, the efficiency could be improved not only by using a more efficient language, but also by using more efficient algorithms. The important point is that the main difficulty will be finding the right sequential solution, while parallelizing it will not require much effort.

Figure 3 shows the speedups obtained using 7000 particles, the sequential execution time being 194.86 seconds. As we pointed out when presenting the language, the speedups obtained are *acceptable*. In fact, in this particular example the speedups are quite good, although in other examples with less inherent parallelism the speedups are not so high. Anyway, our results are always *competitive* with those obtained by using C+MPI: Although slightly slower when running, our applications are developed much faster.

5 Conclusions

In this paper we have presented a framework that facilitates the task of parallelizing large applications. By using our language, the programmer can concentrate on the development of the sequential applications, while the parallelization effort will be minimized. The main advantage of our language is the combination of high-level facilities (that enable fast development) and lower-level constructions (that improve the efficiency).

The experiments we have performed so far are encouraging. As it can be seen in [5], our efficiency is always comparable to that obtained by using C+MPI: Although our results are always slightly worse, the programming effort needed is much smaller. It is important to remark that in [5] we have already compared the efficiency of Eden and other two parallel languages: GpH [14] and PMLS [7]. The first one represents the state of the art in parallel functional programming, while the second one is a good representative of the skeletons community. The results obtained were encouraging: Even though we are still working on optimizing our Eden compiler, the runtimes obtained were better (on average) than those obtained with the other (more mature) languages.

References

1. G. H. Botorog. *High-Level Parallel Programming and the Efficient Implementation of Numerical Algorithms*. PhD thesis, RWTH-Aachen, January 1998.
2. M. Cole. *Algorithmic Skeletons: Structure Management of Parallel Computations*. MIT Press, 1989. Research Monographs in Parallel and Distributed Computing.
3. A. Geist, Ad. Beguelin, J. Dongarra, and W. Jiang. *PVM: Parallel Virtual Machine*. MIT Press, 1994.
4. S. L. Peyton Jones, C. V. Hall, K. Hammond, and W. Partain. The Glasgow Haskell Compiler: a Technical Overview. Department of Computer Science, University of Glasgow, December 1992.
5. H. W. Loidl, F. Rubio, N. Scaife, K. Hammond, S. Horiguchi, U. Klusik, R. Loogen, G. J. Michaelson, R. Peña, Á. J. Rebón Portillo, S. Priebe, and P. W. Trinder. Comparing Parallel Functional Languages: Programming and Performance. *Higher-Order and Symbolic Computation*, 2003. To appear.
6. R. Loogen, Y. Ortega-Mallén, R. Peña, S. Priebe, and F. Rubio. Parallelism Abstractions in Eden. In F. A. Rabhi and S. Gorlatch, editors, *Patterns and Skeletons for Parallel and Distributed Computing*, pages 95–128. Springer-Verlag, 2002.
7. G. Michaelson, N. Scaife, P. Bristow, and P. King. Nested Algorithmic Skeletons from Higher Order Functions. *Journal of Parallel Algorithms and Applications*, 16:181–206, August 2001.
8. S. Pelagatti. *Structured Development of Parallel Programs*. Taylor and Francis, 1998.
9. R. Peña and F. Rubio. Parallel Functional Programming at Two Levels of Abstraction. In *PPDP'01*, pages 187–198. ACM Press, September 2001.
10. S. L. Peyton Jones. Compiling Haskell by Program Transformation: A Report from the Trenches. In *ESOP'96 — European Symposium on Programming*, volume 1058 of *LNCS*, pages 18–44, Linköping, Sweden, April 22–24, 1996. Springer-Verlag.
11. S. L. Peyton Jones and J. Hughes. Report on the Programming Language Haskell 98. Technical report, February 1999. `http://www.haskell.org`.
12. F. Rubio. *Programación Funcional Paralela Eficiente en Eden*. PhD thesis, Dpto. Sistemas Informáticos y Programación. Universidad Complutense de Madrid (Spain), November 2002. In Spanish.
13. M. Snir, S. W. Otto, S. Huss-Lederman, D. W. Walker, and J. Dongarra. *MPI: The Complete Reference*. MIT Press, Cambridge, MA, USA, 1996.
14. P. W. Trinder, K. Hammond, J. S. Mattson Jr., A. S. Partridge, and S. L. Peyton Jones. GUM: a Portable Parallel Implementation of Haskell. In *Programming Language Design and Implementation (PLDI'96)*, pages 79–88. ACM Press, 1996.

Application Controlled IPC Synchrony – An Event Driven Multithreaded Approach

Susmit Bagchi and Mads Nygaard

Department of Computer and Information Science,
Norwegian University of Science and Technology (NTNU),
N-7491, Trondheim, Norway
{susmit, mads.nygaard}@idi.ntnu.no

Abstract. Interprocess communication (IPC) is an important phenomenon in distributed computing and operating systems. Microkernels of modern operating systems use synchronous IPC semantics for every individual process. On the other hand, a process may exploit non-blocking IPC semantics. In either case, the controlling mechanism belies in the hand of the underlying operating system. IPC monitors open up for misinterpretation of IPC timeout events due to thread unavailability in dynamic multithreaded systems. In this paper we propose a software architecture applicable to distributed systems, which confers the decision on IPC semantics during execution to the processes so that they can admix blocking and non-blocking semantics in a flexible way, case by case, as needed. Moreover, the concept of thread pool is introduced to eliminate the possibility of misinterpretation of IPC timeout events by monitors. Worker threads in a thread pool are effectively scheduled to minimize the waste of processing time and dynamic thread overhead. Event driven and multithreaded system models are diagonally opposite during execution. However, our architecture utilizes the benefits of an event driven model with that of a multithreaded model in a fruitful manner to exploit concurrency and protection. The software implementation of our proposed architecture is made as a middleware extension on the communication subsystem of Windows operating systems.

1 Introduction

Interprocess communication (IPC) is one of the most common concepts used in operating systems (OS) and distributed computing systems. It deals with how multiple processes can communicate among each other. In a distributed system, cooperating processes communicate by sending messages. High performance communication is a very critical facility in the distributed computing systems [3]. On the other hand, IPC monitoring enables the examination of any IPC between the source and destination machines [4]. Monitoring of IPC is useful for debugging, logging and security purposes. However, the performance of IPC and monitors is heavily dependent on the IPC semantics and protocols used to implement them [3]. Unfortunately, the IPC mechanism has been considered an expensive operation in the recent past [5]. However, "extensibility" has become an important phenomenon in OS research [5],

P.M.A. Sloot et al. (Eds.): ICCS 2003, LNCS 2658, pp. 1012–1021, 2003.

and new IPC mechanisms along with monitors are being implemented as extensions of the basic kernel. There exists two main semantics in implementing IPC; i.e. synchronous and asynchronous. Such synchrony features are solely controlled by the underlying operating system [2]. In modern microkernels, synchronous IPC semantics is used [4]. This indicates that the communicating process gets blocked until the message is delivered to the destination or an error occurs. This method often leads to an unnecessary waste of execution time. In addition, every IPC message is stamped with a source specified timeout value. Unfortunately, a monitor may disrupt the interpretation of such timeout values if it does not have a thread ready to send or receive the message [4]. Eventually this leads to a loss of transparency of the IPC mechanism.

In this paper, we investigate a different scheme for IPC semantics and its realization, which encompasses a new control mechanism of IPC synchrony and concurrency. In effectively protected environments, we can provide flexible message synchronization relying on decisions taken by processes engaged in IPC. We provide a flexible and reliable mechanism so that individual processes may decide whether they should block for a message delivery and reception, or continue while processing the IPC as a background task. In addition, IPC concurrency is achieved in a better way using the event driven programming model by reducing opportunities for race conditions and deadlocks [1]. We establish an event driven and multithreaded architecture to exploit the benefits of both system models. The projected software architecture achieves the following set of benefits:

• **Design flexibility:** Our architecture assigns the synchronization decision authority to the processes themselves. Hence, they are allowed to generate individual jobs using different synchronization semantics and will be notified on the outcome of IPC events job by job. This adds a good amount of design flexibility to software applications.

• **Unwanted blocking time elimination:** Due to admixing blocking and non-blocking IPC events at the application level, the applications can save execution time by eliminating unwanted waiting for message delivery and response in some specific cases.

• **Elimination of misinterpreted timeout by monitors:** The proposed architecture uses a thread pool concept that eliminates the overhead of creating numerous dynamic threads while handling IPC requests in a multiprocessing environment. Hence, our software architecture eliminates the possibility of misinterpreting timeout events due to message delivery/response.

• **Combining benefits of event driven and multithreaded concurrency:** The designed software middleware treats the IPC requests as discrete events. Hence, processes get immediate attention on IPC events. An event driven approach makes it easier to ensure protection against race conditions and deadlocks [6]. On the other hand, due to our multithreaded model, the benefit of execution concurrency is achieved.

• **Efficient CPU utilization:** The threads are scheduled by the middleware, and this scheduling is based on IPC event generations or completions. Such a scheduling ensures that idle threads will never spend unnecessary processor time cycles.

The remainder of our paper is composed as follows. In section 2, we define different IPC semantics and different programming models. In section 3, we construct a model for application controlled synchrony. This establishes a software architecture combining the multithreaded and event driven IPC paradigms and introduces the thread pool concept. In section 4, we describe our implementation and related data structures. In section 5, we conclude the paper.

2 IPC Semantics and Programming Models

A good interprocess communication facility is central to any distributed system and provides access to resources distributed over a network in a uniform manner. Generally, the primitives used for IPC are *message passing*, *remote procedure call* and *transaction communication*. In the case of message passing, a set of messages is communicated among coordinating processes. At the lowest level, message passing is the only means of communication in distributed systems [9]. But communication transparency is desirable by providing a higher level of abstraction. The remote procedure call (RPC) concept may be used to achieve this goal [9]. The RPC mechanism is built on top of a client/server model and is widely used. Finally, service oriented request/reply communication and multicast may be combined to form a third level, transaction communication. The transaction communication depicts a set of asynchronous request/reply messages without sequential constraints like transactions in a database [9]. All these primitives are mainly based on two different types of semantics; i.e. synchronous and asynchronous communication.

2.1 IPC Semantics and Monitors

The presence of monitors complicates the exploitation of synchronous/asynchronous communication. Three possible cases may arise [4]: (1) IPC monitors may change the identity of source and destination, (2) Monitors may wish to hide the timing of the destination's receipt of the request from a source, and (3) Monitors themselves may be of different trust levels. Hence, desirable aspects associated with IPC are reliability, error handling and timeout expiry handling. In addition, IPC monitors may redirect the IPC request in some cases [8].

2.2 Programming Models

An event driven programming model simplifies concurrency issues by reducing the risks for race conditions and deadlocks [6]. On the other hand, a multithreaded programming model ensures readability and maintainability of code [1]. The sources of conflation for task management and stack management are identified in [1]. The event driven programming model suggests that the gain in concurrency cannot be achieved without manual stack management, which is difficult to attain. However, with these two diagonally opposite programming paradigms, one can cautiously combine the advantages of both models [1].

3 Synchrony, Events, and Worker Threads

Usually, the issues related to IPC synchronization are controlled by the operating system, while the processes use this as a service. Modern OS microkernels implement synchronous IPC semantics [4]. But in some cases, allowing processes to decide about synchronization adds a lot of operating flexibility. In addition, it saves unnecessary blocking time in particular cases and helps a process to admix blocking and non-blocking semantics, as it needs during execution. Moreover, it is possible to combine the advantages of an event driven architecture with that of a multithreaded model to increase execution concurrency of multiple IPC requests on a timeshared basis. We use the thread pool concept to eliminate the overhead of dynamic thread creation and related timeout events.

3.1 Application Controlled Synchrony Model

Modern microkernels use synchronous IPC semantics where a source is blocked until the destination is ready or an error occurs. We have designed a modified architecture where the application or source process can decide, any time during execution, whether to get blocked until the destination process receives the message, or to proceed executing after the message is queued in the system. The calling process decides this while invoking an IPC event. Hence, the process may explicitly control IPC semantics. The task of the operating system will be to provide a uniform interface for message delivery and to report to the calling process about the result. The result may be success or failure. But the message synchrony decision will belong to the calling task.

3.2 Event Driven Architecture

As event driven and multithreaded concurrency models are diagonally opposite, we have tried to combine the advantages of both models. A compromising approach is cooperative threads management. The proposed software architecture is based on an event driven system model. The new approach to achieve cooperative thread management is to organize a system as a collection of event handlers. Hence, sending or receiving a message are two distinct events, and are attached to corresponding event handlers. Every process will have two distinct event handlers to get the result notification from its worker threads. Hence, we consider the "SendRequest" and "WaitResponse" operations as two distinct operations invoked by any process. Moreover, this will allow the IPC subsystem to process multiple blocking and non-blocking IPC events concurrently exploiting event driven multithreading. Our architectural model employs a three-tire middleware extension of a communication subsystem. The three different levels are as follows: level 1: the processes; level 2: the IPC subsystem; level 3: the thread pool. Hence, there exists two interfaces i.e., process to IPC subsystem interface (level 1-level 2) and IPC subsystem to thread pool interface (level 2-level 3). The interaction among the three levels is shown in Figure 1.

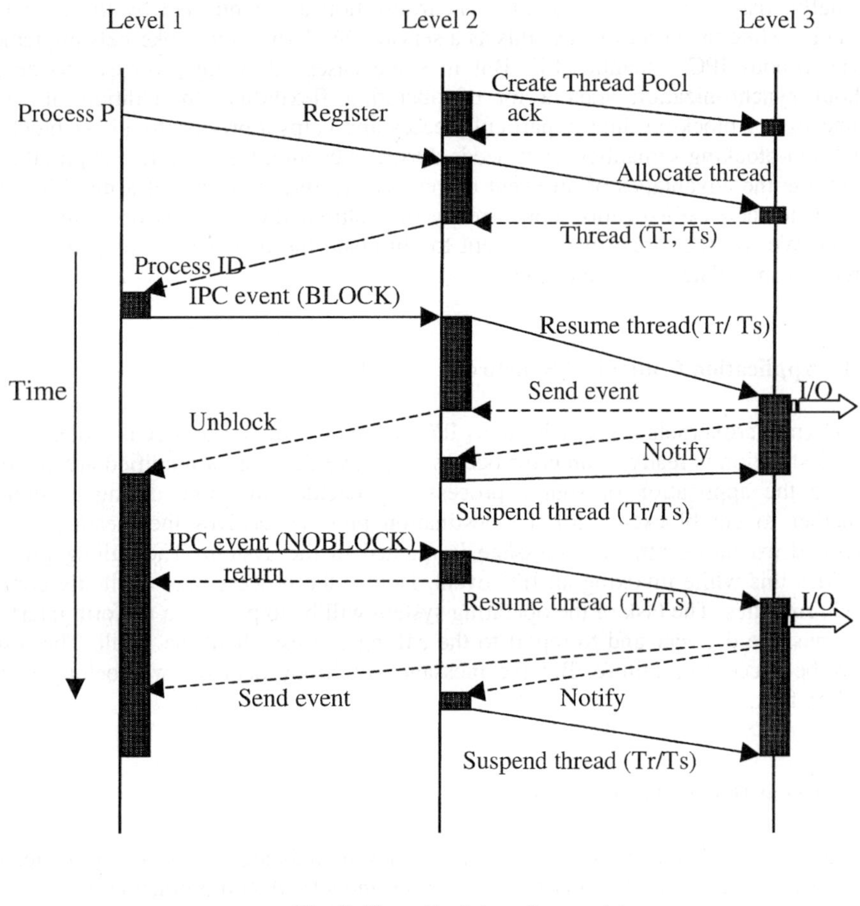

Fig. 1. Three-tire interaction model

During system initialization, level 3 creates a pool of suspended threads, and the thread stacks are initialized. After receiving confirmation from level 3 about successful creation of system resources (thread pool and thread stacks), level 2 creates its own data structures and becomes ready to handle IPC events. As shown in Figure 1, a process must register before requesting any IPC operation, and the process should nominate two event handlers. On successful registration, unique Process IDs will be returned to the individual processes for future use, and two threads will be allocated from the pool to each registered process. Registered processes belong to level 1 and may generate an IPC event explicitly indicating the type of event (BLOCK or NOBLOCK) as shown in Figure 1. Depending on the event types, level 2 schedules the threads to accomplish the requested service. On completion, the thread can notify the result either to level 2 or to level 1, depending on the type of IPC event. Thread allocation belongs to level 3, but scheduling allocated threads in various phases, and its status maintenance, belongs to the level 2. This event driven design ensures immediate thread scheduling upon the arrival of an IPC request, and prevents idle threads from gaining CPU resources. When a thread notifies level 2 about the end of a requested service, the scheduler in level 2 changes the thread state to suspended and marks it as idle. Hence, no idle thread uses any excess CPU cycles. In addition, our event based system model reduces the occurrence of race conditions and deadlocks. Detailed descriptions of the necessary data structures and algorithms are given in Section 4.

3.3 Concurrent Thread Pool

Our concurrent thread pool is a pool of threads in suspended state. For each send and receive operation, allocated working threads from the thread pool are responsible for completing the operations concurrently. Every thread in the pool has an individual thread stack. The thread pool is considered as a system resource containing suspended threads. When a process registers in the IPC subsystem, two working threads are allocated (by level 3) and the thread stacks get initialized. In addition, two job queues (one "JobQ" for "SendRequest" method and another "JobQ" for "WaitResponse" method) are allocated to each registered process (by level 2). In this model, a thread Tr acts on the receive "JobQ" of a process P and the other thread Ts acts on the send "JobQ" of the process P. When process P exits, Tr and Ts are returned to the thread pool in a suspended state. Our multithreaded event driven IPC system is capable of handling multiple requests concurrently for all processes without contention, and frees the system from the burden of dynamic thread creation, which may lead to misinterpreted timeouts in IPC monitors [4]. Additionally, allocated threads are efficiently scheduled to minimize consumption of unnecessary CPU resources.

4 Implementation Issues

We have implemented this application controlled synchrony in our event driven multithreaded execution model as a middleware on top of a sockets based communication subsystem of Windows OS. The data structures, interface definitions and algorithms of our experimental middleware are described below. The overall software architecture is presented in Figure 2.

4.1 Data Structures

The important data structures are described below.

• **Job queue (JobQ):** The JobQ data structure is used to queue the job requests for send or receive generated by processes. Every process has two distinct queues of finite length, one for "`SendRequest`" and one for "`WaitResponse`". The main members of JobQ are "`JobNo`", "`Status`" and "`Type`". "`JobNo`" represents the job sequence number assigned by the calling process. While "`Status`" and "`Type`" indicate the status of the job (serviced or pending) and nature of call (blocking or non-blocking as requested by the calling process) respectively.

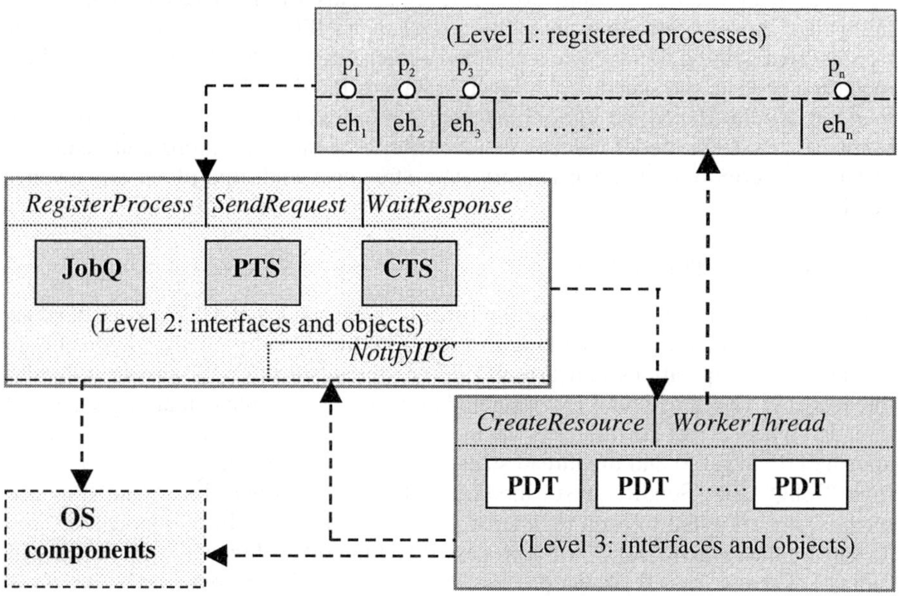

Fig. 2. Three-tire software architecture

- **Private data for thread (PDT):** Every working thread in the thread pool contains this on their individual thread stacks. One of the main members is "`ProcessStackNo`", representing the index in the process table. This field is used to get the entry points of the event handlers (eh) of the calling process. The member "`pThreadJobQueue`" holds a reference to the job queue (send or receive) on which the thread should act. Further, the member "`ThreadIndex`" gives the index for the associated thread. Finally, the member "`pThreadStateInfo`" holds the entry point in the IPC subsystem for this thread.

- **Common thread structure (CTS):** This data structure is used to register the context of every thread. The main members of this structure are "`ThreadHandle`" representing a handle to the thread and "`ThreadState`" representing the execution mode of the thread at any point of time. While the member "`ThreadIntifier`" is used to store the identifier assigned by the underlying OS and is used to access the thread from level 2 in the middleware.

• **Process table (PTS):** The execution context of every registered process is stored in this object. Among its members are "`ProcessID`", "`ProcessEventHandler`", "`pThreadAlloc`" and two JobQs. The member "`pSendJobQueue`" represents the job queue used by the "`SendRequest`" method and "`pRecvJobQueue`" represents the same for the "`WaitResponse`" method. The field "`ProcessID`" holds the distinct process ID assigned to any registered process. The "`ProcessEventHandler`" holds the event handler entry points for processes to notify the result of an IPC asynchronously. The member "`pThreadAlloc`" holds a reference to the global thread data of the threads attached to this process.

4.2 External Interface Definitions

The interface offered by our experimental middleware is given below.

• **`RegisterProcess()`**:This function is used to register a process in the IPC subsystem. It takes the entry points of process event handlers as the arguments and returns a unique "`ProcessID`".

• **`SendRequest()`**: This function is used by a process to send a message to another process. The "`Option`" argument of the function shall be "BLOCK" or "NOBLOCK" to indicate that it is a blocking or non-blocking call.

• **`WaitResponse()`**: This function is used to raise an exception, i.e. IPC event, by a registered process in order to receive/wait for a response message. The "`Option`" argument of the function shall be "BLOCK" or "NOBLOCK" to indicate that it is a blocking or non-blocking call.

4.3 Internal Interface Definitions

The main interlayer functional components are:

• **`CreateResource()`**:This function is responsible for resource allocation and initialization. It belongs to level 3. It returns "Success" or "Failure".

• **`WorkerThread()`**:This function describes the worker thread procedure. It belongs to level 3. It returns "Null".

• **`NotifyIPC()`**:This function is used by threads to notify level 2 about the end of a requested IPC service. In level 2, the thread scheduler gets invoked, and the thread is transferred into the suspended state and marked as idle.

4.4 Algorithms

The main algorithmic parts to accomplish our IPC system model are depicted in the following procedural steps.

```
//System initialization
Proc SysInit:
  begin
    Create_Process_Table;
    Create_Thread_Pool;
    Init_Thread_Stack;
    Regd._Callback_Level2;
    Suspend_All_Threads;
  end begin
```

```
//Process registration
Proc RegisterProcess:
  begin
    Update_Process_Table;
    Create_Job_Queues;
    Allocate_Thread;
    Init_Thread_Stack_Data;
  end begin
```

```
//Worker thread functions
Proc WorkerThread:
  begin
    Check_Private_Data;
    do
      Pick_From_Job_Queue;
      Invoke_IO;
        if(!Blocking)
          Send_Evt_Process;
        end if
      Clean_Up_Memory;
      Send_Evt_To_Level2;
    end do
  end begin
```

```
//SendRequest or
//WaitResponse events
Proc Send/Wait:
  begin
    Enqueue_Job;
    Activate_Thread;
      if(Blocking)
        Wait_Thread_Event;
      else
        Return_Result;
      end if
  end begin
```

5 Conclusions and Future Work

Our application controlled synchrony and event driven multithreaded IPC model enable concurrent message handling. As a pool of threads is created in the system beforehand, threads may be allocated to the processes on demand. This eliminates the overhead of creating threads dynamically while handling IPC requests. Hence, our proposed architecture eliminates the possibility for misinterpretation of timeout events due to non-availability of threads. In addition, our software architecture enables the processes to admix blocking and non-blocking IPC semantics in a flexible manner as the decision authority belongs to the processes. By admixing blocking and non-blocking IPC semantics, the processes save execution time by eliminating unwanted waiting for message delivery in some specific cases. Due to the event driven behaviour of the proposed model, immediate attention to IPC requests is achieved. In this software architecture, we have tried to combine the advantages of event driven systems and the multithreaded programming model. The event driven approach makes it easier to reduce the possibilities for race conditions and deadlocks. On the other hand, our architecture achieves the benefit of concurrent execution through multithreading. However, the thread scheduling mechanism relies on the middleware (level 2), and such scheduling is done on the basis of an IPC request or the completion notification of an IPC event from a worker thread. In this way we ensure that idle threads will never employ unnecessary processor cycles. When a process expires or exits, the workers threads previously allocated to this process returns to the thread pool.

The challenge of our execution model is that suspended threads preoccupy resources such as memory. Although such a waste is not dominant, processes may not use the entire thread population at any time. So, there may always be a few idle threads in the pool. On the other hand, if the system contains numerous processes, then the total thread population in pool may not be large enough to serve all the processes. We are currently investigating modifications to our present implementation to eliminate such difficulties.

References

1. Adya, A., Jon, H., Marvin, T., William, J. B., John, R. D., June 2002. Cooperative Task Management without Manual Stack Management, In the proceedings of the USENIX Annual Technical Conference, CA.
2. Colouris, G., Jean, D., Tim, K., 2002. Distributed Systems Concepts and Design, Addison-Wesley, Third edition, pp. 125–158.
3. Goscinski, A., 1991. Distributed Operating System, Addison-Wesley Publishing Company, pp. 204–207, 871–874.
4. Jaeger, T., Jonathan, E. T., Gefflaut, A., Park, Y., Kevin, J. E., Jochen, L., September 2002. Synchronous IPC Over Transparent Monitor, In 9th SIGOPS European Workshop, Denmark.
5. Jochen, L., Kevin, E., Sebastian, S., Hermann, H., Gernot, H., Nayeem, I., Trent, J., May 5–6, 1997. Achieved IPC Performance, In 6th Workshop on Hot Topics in Operating Systems (HotOS), Massachusetts.
6. John, O., January 1996. Why threads are a bad idea (for most purposes), In USENIX Technical Conference (Invited Talk), Austin, TX.
7. Kent, B. K., Muzio, C. J., Shoja, C. G., April 2001. Remote Transparent Execution of Java Threads, In proceedings of the HPC-2001, Seatle, pp. 184–191.
8. Liedtke, J., Clans & Chiefs, March 1992. In Architektur von Rechensystemen, Springer-Verlag, In English.
9. Randy, C., Theodore, J., 1997. Distributed Operating Systems & Algorithms, Addison-Wesley, pp. 48–49, 123–124.

ToCL: A Thread Oriented Communication Library to Interface VIA and GM Protocols

Albano Alves[1], António Pina[2], José Exposto, and José Rufino

[1] Instituto Politécnico de Bragança,
Campus Sta. Apolónia, 5301-857 Bragança-Portugal
albano@ipb.pt
[2] Universidade do Minho,
pina@di.uminho.pt

Abstract. In this paper we present ToCL a thread oriented communication library specially designed to fully exploit multithreading in a multi-networked cluster environment. ToCL provides a basic set of primitives to handle zero-copy message passing between application threads spread among cluster nodes. Large messages are fragmented and sent to remote threads as single messages using multiple low-level communication subsystems. The current implementation supports both Myrinet through GM and Gigabit Ethernet through VIA but we plan to extend it to other communication subsystems.

Keywords: multithreading, message-passing, intermediate-level library.

1 Introduction

With the advent of commodity SMP workstations and high performance SANs it became possible to do parallel computation at low cost. However, to fully exploit such power we need new programming models.

Multithreading and message-passing are two well-known techniques that may be combined to build appropriate platforms to use in a cluster of multi-processor machines.

1.1 Low-Level Communication Libraries

The increasing processors performance stressed the need of higher speed communication subsystems to achieve low-overhead calls to access the network.

User-level communication libraries are used to interface network adapters directly bypassing the operating system. Usually, these libraries provide a low-level interface for data exchange between processes executing in networked machines avoiding memory copies – zero-copy message transfers.

However, user-level API libraries are not intended to be used at the application level. To take advantage of the specificity of network adapters hardware, such as on-board processors, memory and DMA controllers, these libraries offer just a few basic facilities to application programmers.

P.M.A. Sloot et al. (Eds.): ICCS 2003, LNCS 2658, pp. 1022–1031, 2003.

GM [12] and MVIA [14] are two well-known low-level communication subsystems to exploit Myrinet and Gibabit/Fast Ethernet, respectively. FM [15], BIP [9], LFC [2] and MyVIA [5] also offer user-level interfaces to specific hardware.

1.2 High-Level Abstractions

If we want programmers to easily use libraries we need to improve, enrich and extend the facilities offered by low-level communication libraries. High-level abstractions and much richer interfaces are available from MPI [16], PM^2 [13] and PANDA [3], for example, but the use of these platforms forces programmers to learn specific programming models.

These high-level programming environments, among others, may be used to develop multithreaded applications, and inclusively support remote thread creation. However, application threads are not first class programmer entities involved in communication.

Traditionally the use of parallelism and high-performance computation has been directed to the development of scientific and engineering applications. However, the increasing use of multithreading on commercial and internet applications has shifted the market to new areas, where conventional programming models like MPI don't fit programmers' needs. In this context, the development and use of multithreading programming techniques in high-performance SMP cluster environments would benefit from the existence of a thread oriented communication library and higher level programming models.

2 ToCL Approach

ToCL is an intermediate-layer communication library designed to exploit high-performance networking hardware through available low-level communication subsystems. ToCL introduces the notion of thread oriented communication library.

Presently it provides a unified interface for GM and VIA but the overall design easies the integration of other low-level communication subsystems. The primary choice of GM and VIA [6] (mainly MVIA) is directly related with the research we are pursuing to take advantage of Myrinet and Gigabit high-performance technologies to build a scalable information retrieval environment.

2.1 Entities

ToCL uses three major entities to model applications executing in a cluster: hosts, processes and threads. Global identifiers are assigned to hosts and processes, to uniquely identify them within the context of a cluster. Threads may, also, have a global identifier that is compatible with the existence of a local identification in the context of a process.

ToCL entities have several attributes, as name, parent host (for processes), parent process (for threads), etc, and applications may locate them by using attribute-based queries. A distributed directory service is used to register and query entities present in the system. The current directory service prototype is based on a NFS shared database but more sophisticated systems may be considered, like [7] that uses LDAP fundamentals.

ToCL primitives offer remote spawning of processes and threads that register themselves into the directory through specialized enroll primitives. Processe registration is preceded by initialization of the existent low-level communication subsystems.

The attributes of entities, most of the times, remain unchanged during the overall application execution. To minimize the impact of the directory services, ToCL uses process local caches. The use of local identifiers for threads also contributes to minimize thread creation latency.

2.2 Communication

ToCL programs do not explicitly create communication end-points; processes and thread identifiers are used as message origin and destination, following PVM and TPVM approaches [8].

Threads in the same process share the communication facilities through port multiplexing. This approach introduces latency overheads but it has the advantage of not compromising scalability, given that the number of communicating entities (threads) is not limited by low-level communication end-points.

ToCL is based on POSIX threads (actually Linux Threads); it does not implement thread facilities. Asynchronous events from low-level communication subsystems are managed by using a few threads of control. More complex strategies to integrate communication management and thread scheduling are presented in [10, 11]. However, these strategies depend on the developing of specialized thread libraries, thus compromising applications portability and reutilization.

3 GM and VIA Basics

GM was developed by Myricom to take advantage of Myrinet hardware. VIA standard defines an architecture for the interface between general high performance network hardware and computer systems. MVIA is a well-known VIA implementation that makes possible to experiment a few Gigabit and FastEthernet network adapters.

Although distinct on the interface and protocols they use, both GM and VIA are low-level approaches to high-performance message passing.

3.1 Zero-Copy Messaging

Both VIA and GM use regions of registered memory to achieve zero-copy messaging. Registering a memory region comprises pining the associated memory pages and informing networking hardware (or communication library middleware) about buffer addresses. This means that applications must register data buffers before calling send primitives.

Message reception only succeeds if the communication library was previously notified to use pre-registered buffers. These buffers must be large enough to hold the incoming data, otherwise the message will be discarded. A single buffer may be used to store several incoming messages if applications explicitly inform the library, after each message reception, to reuse that buffer. To manage variable size messages, applications usually allocate huge buffers wasting memory.

3.2 Message Addressing

GM applications use ports to send/receive data. The number of ports is usually limited to eight per network adapter, so we just have a few valid destinations per host (assuming a network interface per host). Messages are addressed to remote ports by means of a pair ⟨node number, port number⟩. Node numbers correspond to network interfaces and are assigned by GM network mapping tools while a port number is returned when the application opens a port using a specific network interface.

VIA is a connection oriented communication protocol that uses pairs of VIs (Virtual Interfaces) to connect remote entities (processes or threads). To send messages to a particular destination, applications must create a VI and request a connection to the remote destination using the VIA network address. Messages are addressed to local VIs, however each local VI is related to one only remote VI. In MVIA, VIs are limited to 1024 per network interface. A full-connected multi-threaded application using h hosts and t threads per host would require at least $t^2 \times (h-1)$ VIs per host.

3.3 Sending and Reception

VIA uses descriptors for send and receive operations. Descriptors are registered memory regions containing information about the send/receive operations: buffer addresses, data length and other control information. On reception descriptors are processed according to a FIFO policy, thus making it hard to receive variable size messages. For example, notifying VIA to prepare reception of two different size messages requires knowledge about the order the messages will be sent.

GM library only needs to know the address and length of data. On reception, when multiple buffers are available they are used according to message needs, what makes it easy to deal with variable size messages.

Multithreaded applications may freely use VIA primitives – VIA is thread safe – but for GM it is necessary to use mutexes to handle concurrency.

4 ToCL Operation

ToCL offers a unique interface for the development of multi-networked and multithreaded cluster applications. It uses the basic functionality from the low-level communication libraries to not compromise the performance of the specific hardware and communication protocols.

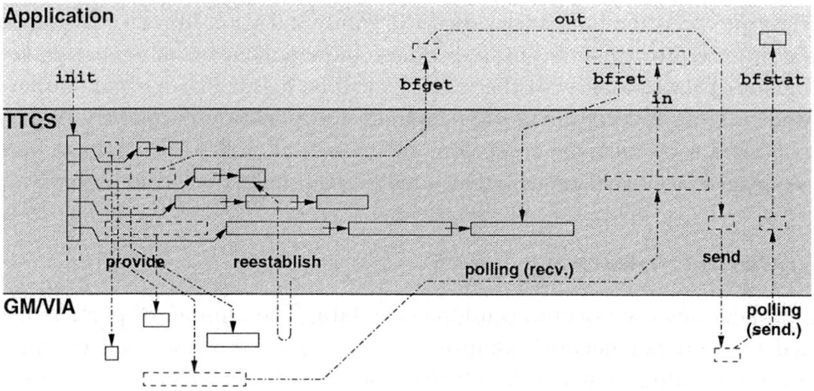

Fig. 1. Buffer management.

4.1 Buffer Management

ToCL manages a pool of pre-allocated registered buffers to ensure zero-copy communication. Because memory registering is a time consuming operation, buffers are created when ToCL initializes the underlying communication subsystems.

Applications must request buffers to store the data involved in a sending operation. To prepare itself for receptions, ToCL notifies the underlying communication subsystems to use pre-allocated buffers.

Figure 1 presents the buffer management mechanism used in ToCL.

Application Driven Operation. Applications must request specific buffers to store data before sending messages (fig.1:`bfget`). Because multiple low-level network libraries may be available, applications must select the target network (currently GM or VIA) when requesting buffers (by default the faster medium is used). The send operation selects the network that matches the buffer type.

To find out if a specific message was successfully transmitted, the status of the corresponding buffer must be checked (fig.1:`bfstat`). A specific buffer may then be reused for sending as long as the pending operation is terminated.

The receive operation returns to the application a library buffer (fig.1:`in`) that may also be reused to send data.

When a specific buffer (requested by the application or returned by a receive operation) is not needed anymore, the application should explicitly return it back to the ToCL library to be integrated in the buffer pool (fig.1:`bfret`). If a send buffer is intended to be used just once, the application may notify the library (before calling the send primitive) to automatically integrate it in the buffer pool after the sending completion.

Library Driven Operation. The ToCL library notifies GM and VIA of the availability of buffers for message reception (fig.1:provide).

When a message arrives (fig.1:polling (send.)), a buffer is used and the ToCL library must provide the communication sub-layer with a fresh buffer. If the buffer pool runs out of buffers, the ToCL library allocates and registers a new buffer to prevent message loss.

To avoid message discarding, whenever a message arrives before ToCL is able to provide a buffer, several buffers are provided in advance to the communication sub-layers. ToCL also exploits idle time to automatically reestablish buffer pool availability (fig.1:reestablish).

Buffer Sizes. For message sending the application requests appropriate size buffers according to data length. The ToCL library just adds a small extra space to the buffer to include message control data.

On message reception the library has to deal with variable length messages and buffer sizes are difficult to estimate.

A first possibility is to use buffers large enough to hold any possible message. This option requires large amounts of memory, unless data is copied from buffers to application memory and the same buffers are used to receive all messages.

ToCL choice is to define a maximum buffer size and to use buffers of 2^n bytes, to ensure that only a few different buffer sizes are handled. This way, for message reception, the library will provide the low-level communication subsystems, particularly VIA and GM, with at least one buffer for each allowed buffer size.

As an example, for a particular x bytes buffer request, a 2^y bytes buffer will be returned according to the formula $2^{y-1} \leq (x + msg.control) \leq 2^y$.

4.2 Message Dispatching

To ensure fair access to the multiple network facilities needed by application threads, ToCL uses a message dispatching mechanism that multiplexes the low-level communication subsystems.

Send and Receive Queues. Send and receive primitives executed by applications interact with the ToCL library through queues.

Messages are dequeued from the send queue and submitted to the low-level communication subsystems according to the buffer type. A FIFO queue is used for message sending, but other scheduling policies may be used to support priority messages.

ToCL uses a polling mechanism to monitor the low-level communication subsystems to detect the arriving of messages. New messages are enqueued into the receive queue and are indexed by source and tag. At application level the message interface may use the source and the tag of a message as a selection mechanism.

Low-level Transmission. ToCL communication model assumes that a thread can send a message to any other thread in a multithreaded multi-networked cluster environment, thus supporting thread oriented fine-grained communication and computation at application level. However, to deal with the expected

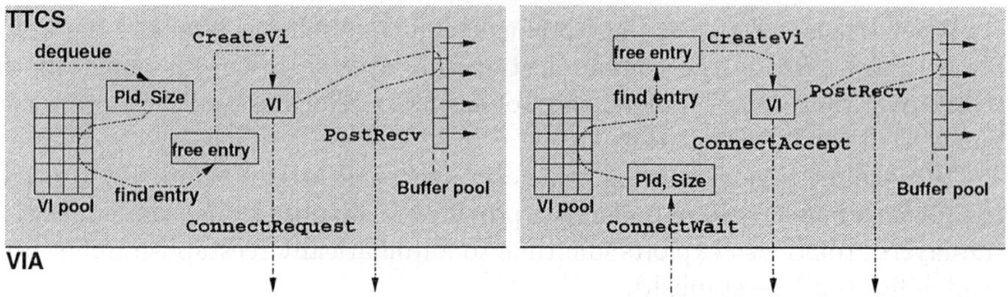

Fig. 2. Connection management for VIA.

large number of threads that applications may use, it creates one only low-level communication end-point for each process and low-level communication subsystem.

For each message dequeued from the receive queue it is then necessary to find the matching low-level subsystem end-point.

GM: Each process opens a port and announces its node and port identifiers by using a directory service. The directory service maps process global identifiers into GM ports. Local caches are used to minimize overhead.

GM is a connection-less protocol that allows any process to directly send a message to a target GM end-point. Messages to several destinations may be sent using the same port and messages from several origins may be received using one only port.

VIA: Each process announces itself using its network address and waits for connection requests. A network address is formed by the MAC address and a discriminator[3]. ToCL uses the global process identifier, provided by the directory service, as discriminator. The VIA network address allows the ToCL library to establish a connection to the right end-point and send the data (see below).

VIA Connections Details. Because connection establishment is a heavy operation it is mandatory to maintain persistent connections if we want to lower latency. The basic idea is to create a VI (virtual interface) and establish a connection to a specific remote process only once: the first time a message is sent to that process.

When using one only connection per remote process we have to deal with the following problem: how to receive variable size messages? In fact, notifying VIA to use a set of different size buffers for a particular VI is useless (see 3.3). To overcome this difficulty, ToCL uses a set of connections per remote process - one for each possible buffer size.

For each message to be sent, after dequeuing it from the send queue, the destination process identifier and the message size are used to find the matching

[3] Discriminators are similar to UDP or TCP ports but they may be any byte sequence.

connection (a VI entry) to the target process in the VI pool (fig.2[left]: find entry). In the absence of a connection to the target process, a new VI is created (fig.2[left]: `CreateVi`) and a connection request is issued to the remote process.

A connection request is addressed to the VIA network address registered in the directory service by the remote process. Apart from the request, the library also notifies VIA of the availability of buffers for future incoming messages, i.e. messages that will arrive after the connection is established.

For each process, ToCL uses a single thread to handle connection requests. Every time a request is detected (fig.2[right]: `ConnectWait`), a VI is created and stored in the VI pool. Note that connections are bi-directional, so that a VI may be used to send messages to the process that issued the connection request.

After notifying VIA of available buffers for message reception, ToCL also informs the requesting process that the connection was accepted (fig.2[right]: `ConnectAccept`).

Send and Receive. VIA uses descriptors to describe send and receive operations. Because both operations involve a buffer, ToCL creates a descriptor for each buffer and uses it as a buffer attribute.

After dequeuing a message, if the buffer type matches the VIA communication subsystem, ToCL uses the buffer descriptor to call the `VipPostSend` primitive. For GM, only the buffer address and the data length are needed to call the `gm_send` primitive. To handle sending success acknowledgments and set appropriately the buffer status flag, ToCL uses the GM callback mechanism or the VIA completion queue notifications.

To receive messages, ToCL polls periodically both communication systems; if a GM receive event is detected or if a descriptor is returned by the VIA completion queue then an enqueue operation is started.

4.3 Multi-Protocol Messages

ToCL has been designed to take advantage from the existence of multiple network adapters in cluster nodes. By using ToCL, applications may transparently exploit multiple low-level communication subsystems increasing communication bandwidth. These features are supported by a mechanism of message fragmentation that allows message fragments to travel independently through different communication media, reaching a destination end-point where they are reassembled in a solely message.

Fragment Dispatching. ToCL library distinguishes between two types of messages: short messages and long messages.

Short messages correspond to data stored in registered buffers and are transmitted at once using a single communication subsystem. The maximum size of short messages defaults to ≈32kbytes (VIA) and ≈64kbytes (GM).

Long messages are transmitted from regular user memory and require data copying.

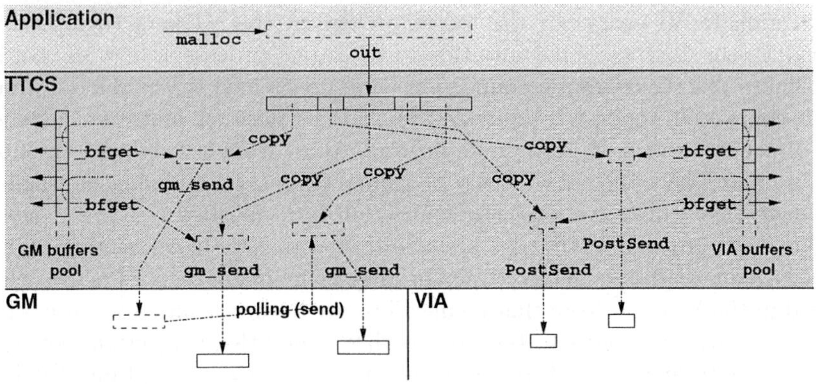

Fig. 3. Fragment dispatching.

After dequeuing a long message (whose buffer type is regular memory) the library first selects a communication subsystem, then searches for a free buffer in the buffer pool (fig.3: `_bfget`) and finally copies a message fragment to that buffer. Fragments are alternately sent using the available subsystems.

During fragment transmission, sending success acknowledgments may arrive (fig.3: polling send) and buffers used to send initial fragments may be reused to send other message fragments.

To make reassembling possible at destination, all fragments are tagged with the same message identifier along with fragment offsets and the originator process identifier.

5 Discussion

ToCL is an undergoing research being developed as a component of a major project aimed to design and to implement SIRe (a scalable information retrieval environment). SIRe basically will build a cluster architecture based on networked commodity workstations and high-performance networks to provide a large number of users with concurrent and efficient access to multiple text documents spread on cluster nodes.

Economics constraints and performance requirements lead us to develop ToCL as an intermediate-layer communication library, able to support a parallel and distributed multi-threading programming environment that can either scale up to provide higher performance or scale down to allow affordability or better cost effectiveness.

ToCL includes important features present in some of the most relevant communication libraries like Madeleine [4], an intermediate-level communication library used by PM^2 that supports multiple communication sub-systems, and PANDA which includes a specific layer to manage hardware heterogeneity. However, ToCL is unique in the use of existing POSIX threads libraries to built a fine-grained communication library where threads play the principal role.

A preliminary version of ToCL that exclusively supported GM, along with an application example and performance measurements, was presented in [1]. Performance tests were conduced using a small cluster environment consisting of four dual Pentium III (733MHz) machines, interconnected by Myrinet technology (LANai 9, 64bits/66MHz interfaces) .

More recently, a new communication library was devised, in order to support multiple communication subsystems. Presently a new suite of tests is being conduced to evaluate the overall design and performance. We plan to evaluate ToCL VIA support over Gigabit, and compare its performance to GM, by using another small cluster, consisting of four dual Athlon (1.8GHz) machines interconnected by Gigabit technology (SysKonnect SK-9821 interfaces).

References

1. A. Alves, A. Pina, J. Exposto, and J. Rufino. Scalable Multithreading in a Low Latency Myrinet Cluster. In *VECPAR'02*, 2002.
2. R. Bhoedjang, T. Rühl, and H. E. Bal. LFC: A Communication Substrate for Myrinet. In *4th Conf. of the Advanced School for Computing and Imaging*, 1998.
3. R. Bhoedjang, T. Rühl, R. Hofman, K. Langendoen, and H. Bal. Panda: A Portable Platform to Support Parallel Programming Languages. In *SEDMS IV*, 1993.
4. L. Bougé, J. Méhaut, and R. Namyst. Madeleine: Efficient and Portable Communication Interface for RPC-based Multithread Environments. In *PACT'98*, 1998.
5. Y. Chen, X. Wang, Z. Jiao, J. Xie, Z. Du, and S. Li. MyVIA: A Design and Implementation of the High Performance Virtual Interface Architecture. In *IEEE Int. Conference on Cluster Computing*, 2002.
6. Compaq Computer Corp., Intel Corporationnand & Microsoft Corporation. Virtual Interface Architecture Specification, 1997.
7. S. Fitzgerald, I. Foster, C. Kesselman, G. von Laszewski, W. Smith, and S. Tuecke. A Directory Service for Configuring High-Performance Distributed Computations. In *6th Int. Symposium on High Performance Distributed Computing*, 1997.
8. A. Geist, A. Beguelin, J. Dongarra, W. Jiang, R. Manchek, and V. Sunderam. *PVM: Parallel Virtual Machine. A User's Guide and Tutorial for Networked Parallel Computing.* Scientific and Engineering Computation. MIT Pres, 1994.
9. P. Geoffray, L. Prylli, and B. Tourancheau. BIP-SMP: High Performance Message Passing over a Cluster of Commodity SMPs. In *SC'99*, 1999.
10. Hansen, J. & Jul, E. Latency Reduction using a Polling Scheduler. In *Second Workshop on Cluster-Based Computing*, pages 27–31. ACM, 2000.
11. Langendoen, K., Romein, J., Bhoedjang, R. & Bal, H. Integrating Polling, Interrupts, and Thread Management. In *6th Symp. on the Frontiers of Massively Parallel Computing*, 1996.
12. Myricom. *The GM Message Passing System*, 2000.
13. R. Namyst and J. Méhaut. PM^2: Parallel Multithreaded Machine. A computing environment for distributed architectures. In *ParCo'95*, 1995.
14. National Energy Research Scientific Comp. Center. M-VIA: A High Performance Modular VIA for Linux. http://www.nersc.gov/research/FTG/via/index, 2002.
15. S. Pakin, M. Lauria, and A. Chien. High Performance Messaging on Workstations: Illinois Fast Messages (FM) for Myrinet. In *Supercomputing 95*, 1995.
16. M. Snir, S. Otto, S. Huss-Lederman, D. Walker, and J. Dongarra. *MPI - The Complete Reference.* Scientific and Engineering Computation. MIT Pres, 1998.

A Multi Dimensional Visualization and Analysis Toolkit for Astrophysics

Daniela Ferro[1], Vincenzo Antonuccio-Delogu[1], Ugo Becciani[1], Angela Germaná[1], Claudio Gheller[2], and Maura Melotti[2]

[1] INAF Astrophisical Observatory of Catania, via S.Sofia 78, Catania 95100, Italy,
dferro@ct.astro.it,
http://woac.ct.astro.it/
[2] CINECA, via Magnanelli 6-3,
Casalecchio di Reno BOLOGNA, Italy

Abstract. The AstroMD toolkit is a visualization and analysis software, specifically oriented to astrophysical data representation. It is included in the project Cosmo.Lab, financed by the European Community, which involves several European astrophysical institutions and CINECA. This tool gives a 3D graphic representation of the data exploiting all the available information and making use of the immersive visualization techniques. AstroMD uses the most advanced visualization technology, based on virtual reality, in order to build a leading edge instrument both for scientific research and for cultural dissemination and education. It is developed using the Visualization Toolkit (VTK) by Kitware (http://www.kitware.com), a freely available visualization library portable on several platforms. AstroMD is an open source completely free code which is freely available (http://cosmolab.cineca.it).

1 Introduction

Since the beginning of modern astronomy, the scientific community expressed a great interest in scientific visualization tools. A great improvement in this direction was determined by the introduction of modern CCD detectors to collect observational data in a digital format. Today, almost all the standard measures are digital and each measure can be generally considered as a collection of images forming a multi-dimensional data set. In many cases extensive image processing is required to obtain meaningful images. Useful scientific information can be obtained from the raw data only using a data reduction pipeline. The improvement of technology and the availability of super-computing multiprocessor systems, has led to a dramatic increase of the volume of data coming from numerical simulations. Today, astrophysical simulations produce many gigabytes of data which have to be efficiently visualized and analysed. Visualization is the most intuitive approach to the data and basic information can be obtained just "at a glance". Then the possibility of moving inside the data allows the scientist to focus on regions of interest and there to perform quantitative calculations. Therefore image processing tools are of fundamental importance in astronomy.

P.M.A. Sloot et al. (Eds.): ICCS 2003, LNCS 2658, pp. 1032–1041, 2003.

AstroMD is a new data visualization and analysis software specifically projected to deal with astrophysical data and can handle powerfully large datasets allowing both their graphical representation and analysis, responding to the requirements proposed by several research fields. The basic application on which the new software is applied focus on the dataset of the VIRMOS (observational galaxy catalogue) project, the data coming from observations of extragalactic radio sources and those obtained by cosmological simulations. Although these fields do not cover all the requirements of astronomy, they pose many typical problems that we expect to be solved by AstroMD and therefore represent a significant test-case. The solution to these problems was implemented following the suggestions and the indications of the research groups involved in the project and of a User Interest group. Astrophysical data have peculiarities that make them different from data coming from any other kind of simulation or experiment. Therefore they require a specific treatment. For example, cosmological simulations consider both baryonic matter (described by fluid-dynamics) and dark matter (described by N-body algorithms). Further components, like stars or different chemical species, can be introduced and followed in a specific way. These different species requires different types of visualization. Dark matter needs particles position or velocity rendering while baryons require mesh based visualization. Furthermore particle associated quantities, like the mass density or the gravitational potential, require their calculation and visualization on a mesh. Then simulated structures have a fully three-dimensional distribution. Therefore it is necessary to have a clear 3D representation and efficient and fast tools of navigation, selection, zoom and the possibility of improving the resolution and the accuracy of calculations in specific, user-selected, regions. Moreover evolution can change dramatically the properties of the simulated objects and the information that can be retrieved, therefore it is important to control efficiently sequences of time-frames.

The scientific goals of the project are the study of the evolution of field galaxies, the evolution of large scale structure, the evolution of galaxies in clusters and the search for distant quasars. For extragalactic radio sources AstroMD provides tools to guide the understanding of the emissivity, flow and field structure, a process which is currently limited by our ability to compute rapidly and visualize the results. In cosmological simulations, one tries to reproduce the formation, evolution and the properties of large scale structures of the universe.

2 The Software

In order to build a widely used product it was chosen to use a low cost software portable on a number of different platforms, the Visualization Toolkit (VTK) by Kitware [4]. VTK is an open source, freely available software system for 3D computer graphics, image processing, and visualization. It includes a C++ class library and several interpreted interface layers. VTK has been ported on nearly every Unix-based platform (e.g. Linux or IRIX) and PC's (Windows NT and Windows 98). The design and implementation of the library has been strongly influenced by object-oriented principles. The graphical model in VTK is at a higher level of abstraction than rendering libraries like OpenGL or PEX. This means it is much easier to create useful graphics and visualization applications. In VTK applications can be written directly in C++, Tcl, Java, or Python. Using these languages it is possible to build powerful, fast and portable applications. The built-in functionalities are controlled by a specific Graphic User Interface (Fig 2), written in incrTcl/Tk that contains the OO philosophy, slowly adding code as new functionality are added to the package. Each object can represent a reader, that allows to read data from a file or from a database, a filter, that allows to manipulate data, or a viewer, that allows to visualize the results. VTK supports a wide variety of visualization algorithms including scalar, vector, tensor, texture, and volumetric methods and advanced modelling techniques. It supports stereographic rendering and can be used for virtual reality visualization [?]. Furthermore, being easily extensible, the system allows ad hoc implementation of specific modules. All the features described above are integrated in the AstroMD package. Furthermore efficient manipulation and

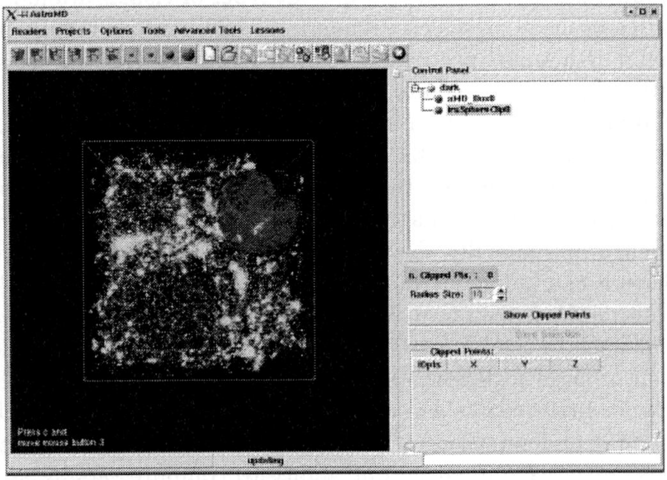

Fig. 1. The main GUI of AstroMD allows to visualize data (Render Window), to define variables and projects, to control variables and filters and to save results. The side of the cubic box is 50 Mpc wide. The sample consists of 15000 particles extracted from a N-body simulation of 16 millions of particles at redshift step z=0, showing the formation of several structures.

analysis tools, like smoothing of the particle masses on a mesh or calculation of the power spectrum and correlation functions, are parts of the basic functionalities. AstroMD has also stereographic rendering capabilities, which makes it usable for immersive visualization, presently implemented at the Virtual Theatre of CINECA. This completes and improve its capabilities in the representation of three dimensional data set. Sitting in the chairs of the theatre and wearing stereoscopic glasses viewers can experience a semi-immersive Virtual Reality experience. Nevertheless this tool is exploited on different platforms, from the very sophisticated virtual theatre, down to the personal workstation. AstroMD, in fact, is developed with particular care to the portability issues in order to make it usable on many different platforms and to allow a large diffusion and usage inside the scientific community and educational institutions.

2.1 Basic Functionalities

AstroMD is thought to work with large datasets and uses advanced visualization tools, therefore it requires a corresponding powerful computational system. It suggested, at least, to use a PIII 500 PC with 256/512 Mbytes RAM memory, 100 Mbytes of free disk space and a high level graphic card. The required software consists in: Open GL (MSWindows) or Mesa (Linux) visualization libraries, TCL-TK 8.2 (or later), and the Visual ToolKit (VTK), which is distributed together with the AstroMD package. AstroMD allows to treat both particles (unstructured data) and continuous fields discretely represented over a computational mesh (structured data). Data must be written as sequences of 3D coordinates in the case of particles and as sequences of scalar values (fortran or C order) in the case of meshes. The input data Formats presently accepted by AstroMD is the common unformatted C standard, the Raw Format, the TIPSY and the FITS Formats. Raw files are simply dump of the memory, written in a continuous sequence of x, y, z coordinates with no labels or other symbols in within. TIPSY is a visualization toolkit specifically designed to quickly display and analyse the results of N-body simulations (http://www-hpcc.astro.washington.edu/tools/tipsy/tipsy.html). Tipsy requires a specific data Format to work with, supported also by AstroMD. The basic data structure is an array of particle structures in three separate arrays for each of the types of particle used in the simulations: collisionless particles, SPH particles, star particles and their characteristic properties, as potential energy and temperature. Binary and ASCII files can be read. Data are visualized with respect to a cubic box which describes the computational region. A cubic or spherical sub-region can be selected interactively inside the parent box with a different spatial resolution, in order to focus on the most interesting regions. Data inside the sampler can be studied with the analysed tools or can be saved in specific files for a off-line analysis. Boxes can be translated, rotated, zoomed in and out with respect to selected positions. Colours and luminosities can be chosen by the user. Images of different evolutionary stages can be combined in order to obtain a dynamic view of the behaviour of the system. The opacity of the particles can be increased, so that low density regions are more easily detectable, or decreased, so that the details of the high density regions substructures are shown. Different particles species (e.g. dark matter and baryons) can be visualizes at the same time using different colours. Other particles related continuous quantities, like density fields, can be calculated as typical grid bades fields and visualized as isosurfaces or volumes (Fig. 2). Scalars fields can be visualized by isosurfaces or by volume rendering. The value of the isosurface can be selected on

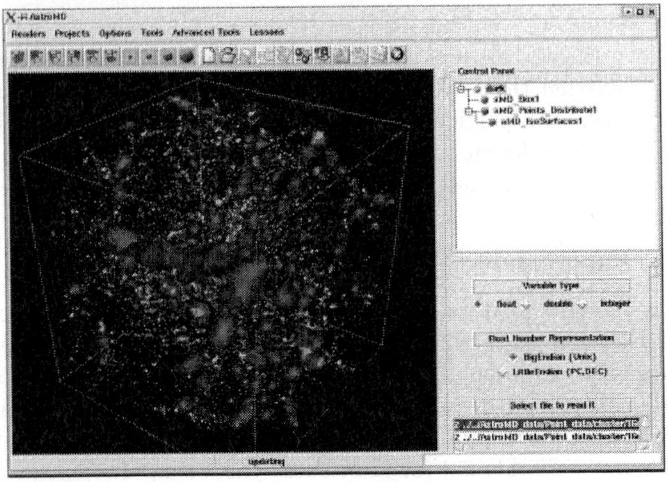

Fig. 2. Formation of clusters of galaxies in the universe, with overdensity of 200 with respect to the background density, during the evolution of a sample of 15000 particles in a cubic box of 50 Mpc.

the user interface. The volume rendering can be calculated using both the texture mapping and the ray tracing technique, depending on the specific hardware that is used (Fig. 3). Different time frames can be shown in a sequence. When particle-representation is used, the position of particles are interpolated between one time-step and the following. This improves the quality of the "animation" giving a fluidity otherwise unachievable. Both the single images and the whole sequence can be saved in bitmap or jpeg format to prepare an animation of the evolution. Enabling the steady-cam, the system can be rotated in Azimuth and Elevation during its evolution. Zoom-in and zoom-out possibilities are also offered. The whole set of particles can be visualized but it's also possible to use a sub-sample of the data, in order to get a faster and easier visualization. It was implemented a procedure which select randomly the sub-sample of data. Specific care has been devoted to avoid systematic errors in the selection procedure, so that the sample is still statistically significant.

2.2 Data Analysis Functionalities

Some built-in-tools, specifically directed to cosmological results of simulation, were implemented to allow an efficient manipulation and analysis of the data. At present the following functionality are implemented.

Particles mass density The mass density field associated to the particle distribution is calculated distributing the mass of each particle over the computational mesh by a eight points Cloud in Cell smoothing algorithm [?]. The computation can be done with the maximum accuracy using all the particles over a uniform

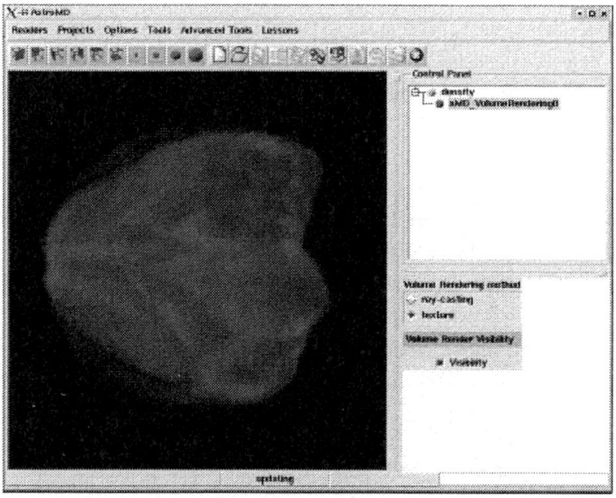

Fig. 3. Formation of clusters of galaxies in the universe, with overdensity of 200 with respect to the background density, during the evolution of a sample of 15000 particles in a cubic box of 50 Mpc.

high resolution mesh, but AstroMD allows the user to use only a sample of the whole set of particles, and the final result can be extrapolated to all particles of the simulation, reducing the CPU time consuming and the memory request. The smoothing of the masses can be performed generally using a coarse grid, that can be refined where high resolution is necessary.
The same tool can be used to calculate other fields related to quantities possibly associated to the particles, like, for example, the thermal energy density field or the X-ray luminosity field.

Gravitational field calculation Considering the mass density $\rho(x)$ defined over the computational mesh as above, the gravitational field can be calculated solving the Poisson equation

$$\phi(\boldsymbol{x}) \propto \nabla^2 \rho(\boldsymbol{x}), \tag{1}$$

where $\phi(\boldsymbol{x})$ is the gravitational potential, by a Fourier Transform procedure. The Poisson equation is transformed in its momentum space image using a FFT VTK built-in function. This reduces the equation to a much simpler algrebric operation

$$\phi(\boldsymbol{k}) \propto \frac{1}{|\boldsymbol{k}^2|} \rho(\boldsymbol{k}), \tag{2}$$

where $\phi(\boldsymbol{k})$ and $\rho(\boldsymbol{k})$ are the Fourier images of the potential and of the density and $|\boldsymbol{k}|^2$ is the square module of the wavenumber. Finally, the potential is transformed to the physical space using an inverse FFT.

Fourier decomposition, Power Spectrum and Correlation Function The quantity $\rho(\boldsymbol{k})$ is used to calculate the Power Spectrum $P(k)$ of the matter distribution, which is defined as the average value of the square norm of $\rho(\boldsymbol{k})$:

$$P(k) = \langle|\rho(\boldsymbol{k})|^2\rangle. \tag{3}$$

The Power Spectrum expresses the weight of each of the Fourier components of the mass distribution between k_{min} and k_{max} which represents the inverse of the size of the computational mesh and the Nyquist frequency [3]. The Power Spectrum is a powerful measure of the statistical properties of the distribution, together with the associated Correlation Function $\xi(r)$, which is its Fourier Transform. The two-point Correlation Function indicates the probability to find

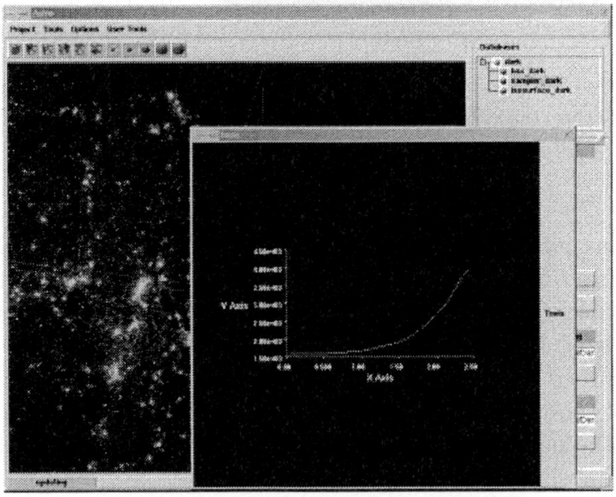

Fig. 4. Power Spectrum of the ΛCDM simulation sample.

a particle at a distance r from any other particle, and is usually used to analyze the clustering properties of a sample of discrete objects (particle, galaxies, galaxy clusters, etc) [1]. It was used HAM as his function estimator, because it is preferable at large scale, for samples with non uniform density [2]. The clustered zone is inserted in the sampler in which it is calculated the Correlation Function and the Power Spectrum. The plot 2D are shown in another window and is updated when the user points out to another clustered zone with mouse.

Minkowski Functionals The Minkowski Functionals provide a novel tool to characterize the large-scale distribution in the Universe [?]. They describe the Geometry, the Curvature and the Topology of a point set. Considering the set of points in 3D space, supplied by galaxies of a cluster of galaxies, and decorating each point with a ball of radius r, the tool measures the size, shape and

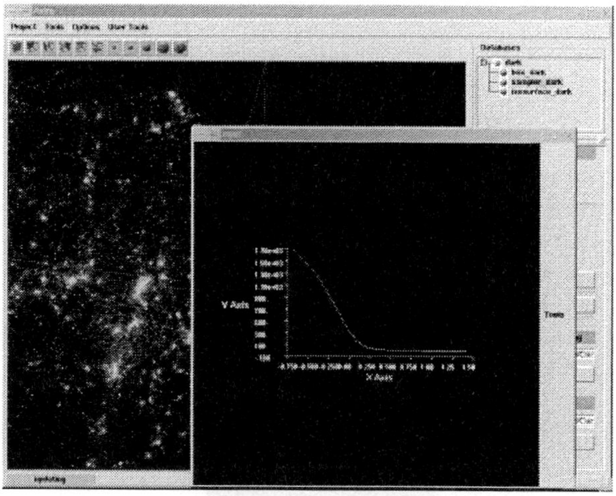

Fig. 5. Correlation Function of the ΛCDM simulation sample.

connectivity of the spatial pattern formed by the union set of these balls. These characteristics change with the radius r, which may be employed as a diagnostic parameter.

Friend of Friend Algorithm It was inserted a group finder, known as Friends-of-Friends [?] (http//hpcc.astro.washington.edu/tools/). A particle belongs to a FoF group if it is within some linking length of any other particle in the group. After all such groups are found, those with less than a specified minimum number of group members are rejected. User must set two parameters: the maximum distance among particles forming clustering and numMembers, the minimum number of clustered particles. At last, FoF cancels all groups whose members are less than numMembers. It was also implemented the calculation of the centre of mass of each group, the number of components of groups and the radius of each group. The graphical output contains the grouped particles. The centre of mass of each group identified by ball with radius as clustered group radius. In another window is displayed the plot of the fraction of grouped particles versus the number of components of the groups. When the parameters are changed by user, FoF updates all its outputs.

3 Conclusions

AstroMD represents the first experience of a tool of immersive visualisation and data analysis in astrophysics. It will be a valuable tool for the scientific groups, which will be able to interact efficiently with large amount of data, easily 'navigating' inside them, analysing their properties, calculating their statistical properties and reconstructing their three dimensional shapes and features. This will

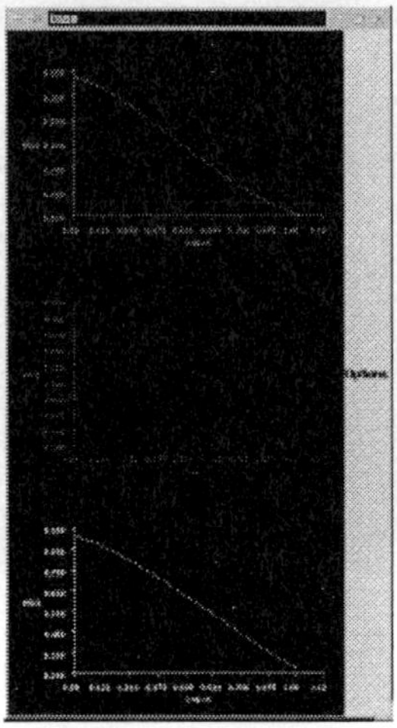

Fig. 6. Minkowski Functionals of a ΛCDM simulation.

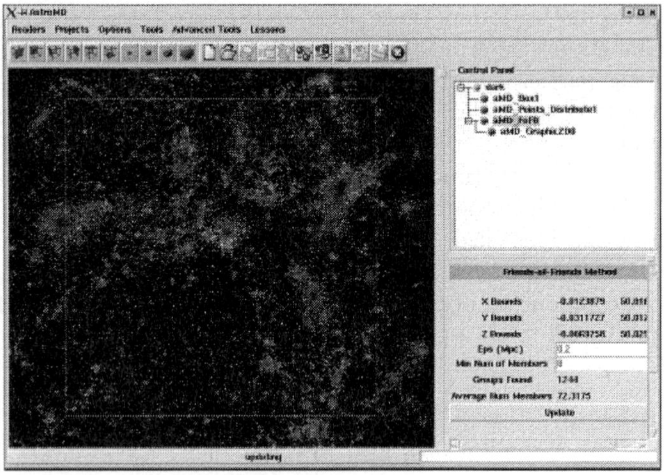

Fig. 7. FoF tool.

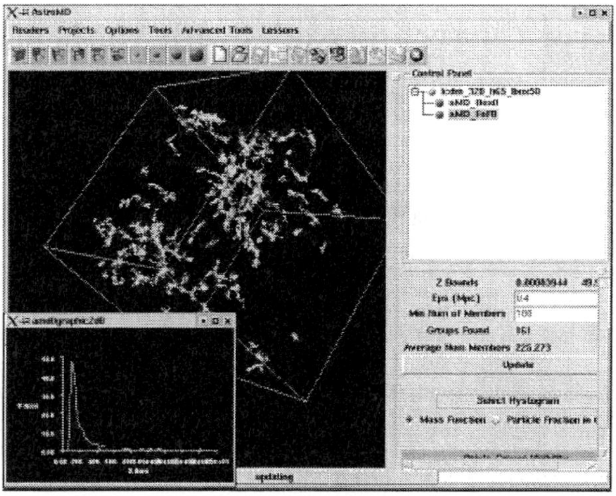

Fig. 8. FoF tool.

lead to a deeper understanding of the scientific problems and to their easier solution. The software developers and visualisation experts will have the possibility of strongly improving their knowledge in the involved techniques and their experience on the possible applications of 3D visualisation and immersive technology in the scientific research applications. Finally, the use of AstroMD for educational purposes will be a very effective way of presenting astrophysical problems to students, giving the possibility of observing objects and structures from a privileged and personalised.

References

1. Blanchard, A., Alimi, J.: Practical determination of the spatial correlation function. Astron. Astrophys. **203** (1988) L1–L4
2. Pons-Bordería, M.J., Martínez, V.J. et al: Comparing estimators of the galaxy correlation function. ApJ **523** (1999) 480–491
3. Jenkins, A., Frenk, C.S., Pearce, F.R. et al: Evolution of Structure in Cold Dark Matter Universes. Astrophys. J. **499** (1998) 20
4. Schroeder, W., Ken M., Lorensen B.: The Visualization Toolkit. Prentice Hall (1999)
5. Rabinowitz, P.: On subharmonic solutions of a Hamiltonian system. Comm. Pure Appl. Math. **33** (1980) 609–633

Error Correcting Codes with Mathematica

Igor Gashkov

Karlstad University, Department of Engineering Sciences, Physics and Mathematics 65188 Karlstad Sweden

Igor.Gachkov@kau.se

Abstract. The author (with Kenneth Hulth) got the idea to develop a non-standard, methodical-oriented course where hands-on sessions could add substantial understanding in the introduction of mentioned mathematical concepts. The package in MATHEMATICA in the field "Coding Theory" was developed for course " Error-Correcting codes with MATHEMATICA ", for students on advanced undergraduate level.

1 Introduction

Applications of the theories of Error-Correcting Codes have increased tremendously in recent years. Thus it is hardly possible today to imagine engineers working with data transmission and related fields without basic knowledge of coding/decoding of information. The possibilities of quantifying information with electronic equipment are developing rapidly, supplying the specialists working in communication theory with more sophisticated methods for circuit realization of concrete algorithms in Coding Theory. During preceding years courses in Coding Theory have been considered only for students on postgraduate level. This is due to the complexity of the mathematical methods used in most codes, such as results from abstract algebra including linear spaces over Galois Fields. With the introduction of computers and computer algebra the methods can be fairly well illustrated. The author has developed a course, 'Coding Theory in MATHEMATICA', using the wide range of capabilities of MATHEMATICA. The course was given at Jönköping University and Karlstad University, Sweden, on undergraduate level with a minimum of prerequisites. The hands on sessions were based on a package of application programs/algorithms, developed to illustrate the mathematical constructions, used in coding theory to encode and decode information. We will present some of our hands-on materials, which are used to construct Block Codes with means of algebraic methods. We first present the basic concepts of coding theory, such as, hamming distance, generator and parity check matrices, binary linear codes and group codes. We will then use some basic results from matrix algebra and linear spaces to construct code words from the information we want to send. Due to noise in the channel of transmission the received words may differ from the code words which were sent, and the very aim of coding theory is to detect and correct possible errors. In the linear codes (the code words having group structure) the encoding is accomplished by multiplying the

P.M.A. Sloot et al. (Eds.): ICCS 2003, LNCS 2658, pp. 1042–1051, 2003.

information word (seen as a vector) with a generator matrix, whereas the decoding process starts with multiplication of the received word with the parity check matrix. Within the cyclic codes it is preferable to work with generator- and parity check polynomials instead of matrices, and the code words here form a polynomial ring. With algebraic extensions of the field Z_p by irreducible (over Z_p) polynomials, the final step is taken into the BCH-codes, which have rich algebraic, structure. The application programs in the package support the learning processes by illustrating the algorithms in the mathematical constructions. The rich variety of manipulations in the algebraic structures, the possibility to vary the parameters in specific cases and the simplicity to construct concrete codes with MATHEMATICA should strengthen the understanding of the mathematical ideas used in coding theory.

2 Introduction to Coding Theory

As an introduction to the theory of Error-Correcting Codes we study the well-known Venn-Diagram with the three overlapping circles (Fig. 1)

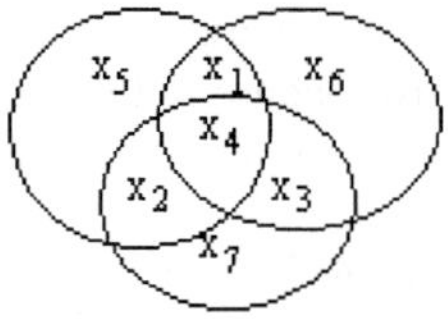

Fig. 1. Venn-Diagram with the three overlapping circles

We place 7 pieces x_i, where x_i is marked with the number (1) on one side and (0) on the other, on the respective i:th area of the figure. This could be performed in such a way, that the sum of the numbers in each circle is even, i.e. for each of the circles applies $\Sigma x_i = 0$ (mod 2). We will see that there are a total of 16 such possibilities, and if we write these $x = x_1x_2x_3...x_7$ we have created a code with 16 codewords. If we transmit such a codeword through a communication channel, it may happen, due to noise in the channel, that one (or several) of the x_i changes from 0 to 1 or vise versa. In our figure the error is easily found by checking the sum (mod 2) of the numbers in each of the circles. If we change x_i for exactly one i the conditions $\Sigma x_i = 0$ (mod 2) will be fulfilled. We thus see, that there do exist possibilities to correct errors in the transmission of information. We now turn to the figure again. The condition $\sum x_i = 0$ for each of the tree circles immediately gives (if we perform the operations modulo 2, i.e. within the field Z_2)

$$\begin{cases} x_1 + x_2 + x_4 + x_5 = 0 \\ x_1 + x_3 + x_4 + x_6 = 0 \\ x_2 + x_3 + x_4 + x_7 = 0 \end{cases} \Leftrightarrow \begin{cases} x_1 + x_2 + x_4 = x_5 \\ x_1 + x_3 + x_4 = x_6 \\ x_2 + x_3 + x_4 = x_7 \end{cases} \qquad (1)$$

It follows that in order to place the 7 pieces correctly, we simply place $x_1 \ldots x_4$ freely and then compute x_5, x_6 and x_7 in accordance with the equations. This means, that out of a total of 2^7 combinations of $x_1 \ldots x_7$ there will be 2^4 combinations satisfying our condition, i.e. from 2^7 *words* there will be 2^4 *codewords* . In order to find the codewords we could also proceed as follows: The word $v = (v_1 v_2 \ldots v_7)$ would be a codeword iff (1) is satisfied, which in matrix form means iff v satisfy

$$\begin{pmatrix} 1 & 1 & 0 & 1 & 1 & 0 & 0 \\ 1 & 0 & 1 & 1 & 0 & 1 & 0 \\ 0 & 1 & 1 & 1 & 0 & 0 & 1 \end{pmatrix} v^{tr} = \begin{pmatrix} 0 \\ 0 \\ 0 \end{pmatrix} \Leftrightarrow H\, v^{tr} = \mathbf{0} \qquad (2)$$

In order to check whether a word w received from the communication channel is a code word, we only compute $H\ w^{tr} = s^{tr}$, where s is called the syndrome, if $s^{tr} = \mathbf{0}$, then v is a code word. If exactly one error occurs in the transmission, then exactly one x_i would be false, and we write (as vectors) $w = v + e_i$ where v is a code word and e_i has a single nonzero bit, in the position i .
We compute $H\ v^{tr} = H\ (v^{tr} + e_i^{tr}) =$ (linearity!) $= H\ v^{tr} + H\ e_i^{tr} = 0 + H\ e_i^{tr} =$ the i:th column in H. One of these columns has to be in accordance with s^{tr} and so we correct the corresponding v_i. It can easily be proved that we would always find the error, provided we have exactly one false bit.
In general, if H is a binary matrix, the linear code with the *parity check matrix* H consists of all vectors v, satisfying $H\ v^{tr} = \mathbf{0}$. Usually H is an (n-k) x n matrix $H = [A\ \ I_{n-k}]$, with I_{n-k} the unit matrix (n-k) x (n-k) . With the information word $u = u_1 \ldots u_k$ we write the codeword $v = v_1 \ldots v_k v_{k+1} \ldots v_n$, where $v_i = u_i$ $1 \leq i \leq k$ and where $v_{k+1} \ldots v_n$ are the check symbols. We then have $H\ v^{tr} = \mathbf{0} \Leftrightarrow \mathbf{x} = \mathbf{u}G$, where $G = [I_k\ \ A^{tr}]$. G is called the generator matrix of the code. We have in our introductory example ended demonstrated the *Hamming Code* K[7,4,3], where the parameter 7 indicates the length and 4 the dimension (=number of information bits) of the code. The parameter 3 gives the *hamming distance* d of the code: the hamming distance d (x, y) between the words x and y equals the number of positions i, where $x_i \neq y_i$; d = min d (x, y) = minimum distance between any two codewords. d is easily found to be equal to the minimum *hamming weight* wt (v) of any codeword v, where wt(v) is the number of nonzero v_i.
The Hamming Code K [n, k, d] is characterized by (let m be the number of check bits, $m \geq 2$) $n = 2^m - 1$, $k = 2^m - 1 - m$, d=3 and is a perfect single-error-correcting code, meaning that the code has exactly the information needed for correcting one error.

3 The Package " Coding Theory"

The package "Coding Theory" is a file written in MATHEMATICA and will be read into MATHEMATICA with the commands.

```
In[1] :=<<CodingTheory.m
```

The package consists of two parts: one part with illustrative explanations, and one for scientific purposes. The illustrative part (commands starting with Show...) is considered to visualize the theoretical aspects of encoding / decoding, construct shift-register circuits etc. The command `?Show*` gives the following list of commands

```
In[2] := ?Show*

Out[2]=Show
ShowHammingCode
ShowBinaryGaloisField
ShowMeggittDecoder
ShowErrorTrappingDecoderBCHCode
ShowSystematicEncoderCyclicCode
...
```

The complete information about a command is received by using the command ? *Name.*

```
In[3] : = ? ShowHammingCode

Out[3]= ShowHammingCode[m,inf] shows the method of
encoding an information word inf into a hamming code
word of length 2^m - 1 .
```

We now use MATHEMATICA to construct Hamming codes. We first chose m = 3 which gives $n=2^3 - 1 =7$, i.e. the code K [7,4,3].

```
In[4]:= inf={1,1,1,0}; ShowHammingCode[3,inf]

Out[4]=
PARITY CHECK                GENERATOR
```

MATRIX= $\begin{pmatrix} 0 & 0 & 0 & 1 & 1 & 1 & 1 \\ 0 & 1 & 1 & 0 & 0 & 1 & 1 \\ 1 & 0 & 1 & 0 & 1 & 0 & 1 \end{pmatrix}$ MATRIX= $\begin{pmatrix} 1 & 1 & 0 & 1 & 0 & 0 & 1 \\ 0 & 1 & 0 & 1 & 0 & 1 & 0 \\ 1 & 0 & 0 & 1 & 1 & 0 & 0 \\ 1 & 1 & 1 & 0 & 0 & 0 & 0 \end{pmatrix}$

```
inf = {1,1,1,0}
v = inf * G  HAMMING CODE VECTOR v = {0,0,0,1,1,1,1}
```

We proceed by sending the code word v and let an error appear in the 5:th position. We thus decode the received word w = (0001011).

```
In[5]:= v={0,0,0,1,1,1,1};e={0,0,0,0,1,0,0};
w=Mod[v + e,2]; ShowDecHammingCode[3,w]

Out[5]=
```

PARITY CHECK MATRIX= $\begin{pmatrix} 0 & 0 & 0 & 1 & 1 & 1 & 1 \\ 0 & 1 & 1 & 0 & 0 & 1 & 1 \\ 1 & 0 & 1 & 0 & 1 & 0 & 1 \end{pmatrix}$ S = $\begin{pmatrix} 1 \\ 0 \\ 1 \end{pmatrix}$

```
POSITION OF ERROR = 5 RECEIVED WORD = {0,0,0,1,0,1,1}
DECODED WORD =  {0,0,0,1,1,1,1} inf ={1,1,1,0}
```

We now compute the parameters of the code.

```
In[6]:=
H = HammingMatrix[3];Length[H[[1]]] ;DimensionCode[H];
DistanceCode[H]
Out[6]=
7
4
3
```

4 Fundamental from Algebra

We need for the construction of the linear block-codes some prerequisites from algebra, such as structures of groups, cosets, rings and finite fields. The linear block codes are seen as subspaces of linear spaces over a finite field. Of fundamental importance are the following algebraic concepts:

- Irreducible polynomial : f (x) is irreducible over the field F, if f(x) cannot be factored as a product of two polynomials of degrees smaller than that of f(x).
- Primitive element: ↦ is a primitive element in the field F provided that every nonzero element in F is equal to some power of ↦.
- Minimal polynomial: If ↕ is an element of some extension of the field F, the minimal polynomial of ↕ (with respect to F) is the lowest-degree monic polynomial M (x) over F with M (↕) = 0. The minimal polynomial M(x) of ↕ ∈ GF(p^m) can be computed as
- M(x)= $(x-\beta)(x-\beta^{p})(x-\beta^{p^2})\ldots(x-\beta^{p^i})$, where $\beta = \beta^{p^{i+1}}$
- The Galois field GF(p^m) is defined as an algebraic extension of the field Z_p (p prime number) by an irreducible polynomial of degree m. If g(x) is a polynomial of degree m and irreducible over Z_p we construct GF(p^m) = Z_p / g(x) with p^m elements $\sum_{i=0}^{m-1} a_i\alpha^i$, $a_i \in Z_p$. The element α satisfies $g(\alpha)=0$.

```
In[7]:=
BIP = BinaryIrreduciblePolynomials[3,x]
Out[7] = {1 + x^2 + x^3 ,   1 + x + x^3 }

In[8]:=irr=BIP[[2]];ShowBinaryGaloisField[irr, x, b, b]
```

GF(8) is received by extending the field Z_2 by the irreducible polynomial $g(x) = 1 + x + x^3$ and observe that b is a primitive element in this extension field.

```
Out[8] =
```

Log	Vector	Pr. el.	Polynomial	Min. polynomial
$-\infty$	(0, 0, 0)	0	0	x
0	(1, 0, 0)	1	1	1 + x
1	(0, 1, 0)	b	b	$1 + x + x^3$
2	(0, 0, 1)	b^2	b^2	$1 + x + x^3$
3	(1, 1, 0)	b^3	1 + b	$1 + x^2 + x^3$
4	(0, 1, 1)	b^4	$b + b^2$	$1 + x + x^3$
5	(1, 1, 1)	b^5	$1 + b + b^2$	$1 + x^2 + x^3$
6	(1, 0, 1)	b^6	$1 + b^2$	$1 + x^2 + x^3$

We illustrate the multiplication of the elements $1 + b + b^2$ and $1 + b^2$ in GF(8):

```
In[9] := PolynomialMod[PolynomialMod[(1+b+b^2)*(1+b^2),
irrpol /. x - > b], 2]

Out[9] =      b + b^2
```

5 Cyclic Codes

The linear code K is called cyclic, if $v = (v_0, v_1, \ldots, v_{n-1}) \in$ K $\Leftrightarrow$ $w = (v_{n-1}, v_0, \ldots, v_{n-2}) \in$ K. If we then represent the vector $v = (v_0, v_1, \ldots, v_{n-1})$ with the polynomial $v(x) = \sum_{i=0}^{n-1} v_i x^i$, the cyclical shift of v to w corresponds to a multiplication of $v(x)$ with x. If $g(x)$ is a polynomial with lowest degree in K, it can be shown, that $g(x)$ is uniquely determined and generates K and

is thus called the *generator polynomial* of K $g(x)$ is a divisor of $x^n - 1$ and we have the following connection between generator matrix and generator polynomial:

$$G = \begin{pmatrix} g(x) \\ xg(x) \\ . \\ x^{k-1} g(x) \end{pmatrix} \tag{3}$$

In a cyclical code we now construct the code word $v(x)$ from the information word u in following steps:

- $u = (u_0, ..., u_k) \Rightarrow u(x) = u_k x^{n-1} + u_{k-1} x^{n-2} + ... + u_0 x^{n-k}$
- construct $r(x)$ as the remainder in $u(x) = q(x) \cdot g(x) + r(x)$
- let the code word be $v(x) = u(x) \cdot r(x)$

These algebraic operations could be performed by the *shift register circuit* (Fig. 2):

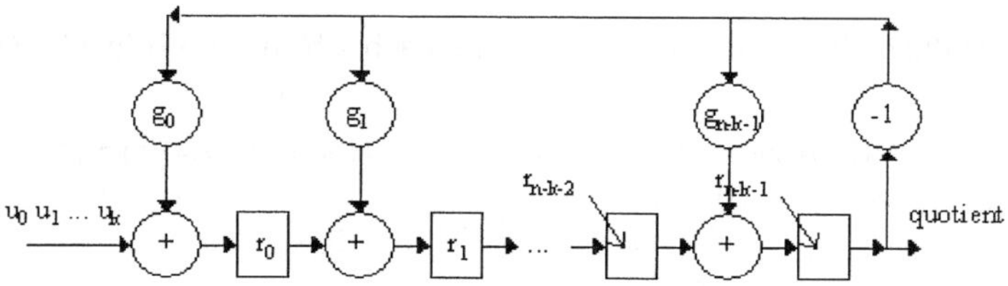

Fig. 2. Shift register circuit.

In MATHEMATICA we illustrate the procedure with the Hamming code K [7,4,3]: We first receive the generator polynomials to the code by the command:

```
In[10] : = gp=GeneratorPolynomials[7,x]
```

(This command gives all possible $g(x)$ of length 7)

```
Out[10] ={1, 1+x, 1+x+x^3, 1+x^2 +x^3, 1+x+x^2+x^4,
1+x^2+x^3+x^4, 1+x+x^2+x^3+x^4+x^5+x^6, 1+x^7}
In[11] : =       g=gp[[3]]
```

(We choose the third polynomial as $g(x)$ i.e. $g(x) = 1 + x + x^3$)

```
Out[11] = 1 + x + x^3

In[12] : = ShowCyclicCode[g,7,x]

Out[12] =

GENERATOR    POLYNOMIAL      g = 1 + x + x^3
```

GENERATOR MATRIX $\quad GM = \begin{pmatrix} 1 & 1 & 0 & 1 & 0 & 0 & 1 \\ 0 & 1 & 0 & 1 & 0 & 1 & 0 \\ 1 & 0 & 0 & 1 & 1 & 0 & 0 \\ 1 & 1 & 1 & 0 & 0 & 0 & 0 \end{pmatrix}$

PARITY CHECK POLYNOMIAL $h = 1 + x + x^2 + x^4$

PARITY CHECK MATRIX $\quad HM = \begin{pmatrix} 0 & 0 & 0 & 1 & 1 & 1 & 1 \\ 0 & 1 & 1 & 0 & 0 & 1 & 1 \\ 1 & 0 & 1 & 0 & 1 & 0 & 1 \end{pmatrix}$

From MATHEMATICA we then get the shift-register (systematic encoding)

```
In[13] : = ShowSystematicEncoderCyclicCode[g,7,x]

Out[13] =
```

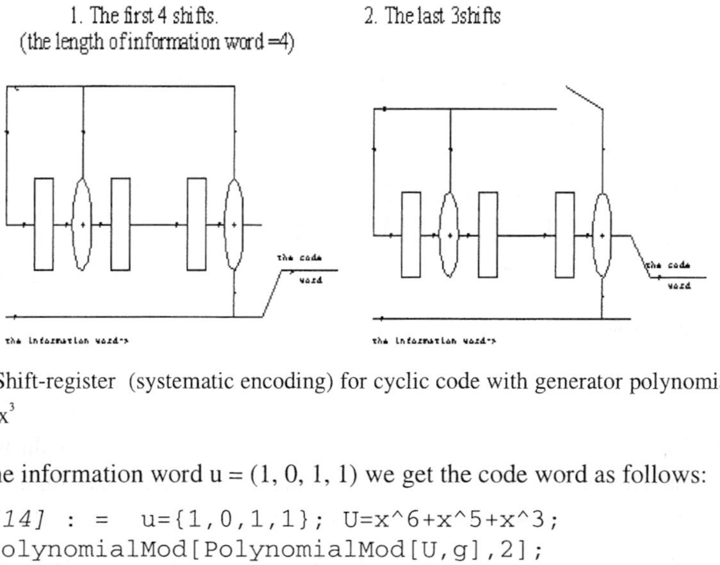

Fig. 3. Shift-register (systematic encoding) for cyclic code with generator polynomial $g(x) = 1 + x + x^3$

With the information word u = (1, 0, 1, 1) we get the code word as follows:

```
In[14] : =  u={1,0,1,1}; U=x^6+x^5+x^3;
r=PolynomialMod[PolynomialMod[U,g],2];
V=PolynomialMod[U-r,2] v=CoefficientList[V,x]
Out[14] = 1 + x^3  + x^5  + x^6 {1, 0, 0, 1, 0, 1, 1}
```

Using the package we get:

```
In[15] : = SystematicEncodeCyclicCode[g,u,7,x]
Out[15] =
THE INFORMATION POLYNOMIAL = x^3  + x^5  + x^6
```

THE CODE POLYNOMIAL = $1 + x^3 + x^5 + x^6$

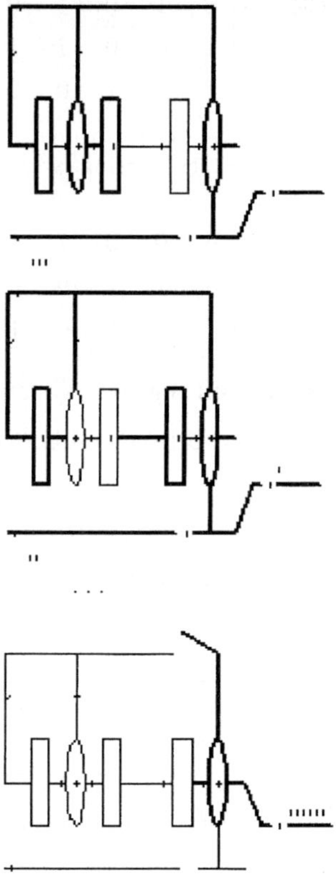

THE CODE WORD = (1, 0, 0, 1, 0, 1, 1)

From the information word u = (1, 0, 1, 1) we have thus got the code word $v(x) = 1 + x^3 + x^5 + x^6 = (1, 0, 0, 1, 0, 1, 1)$ which will be sent through the communication channel. Due to noise in the channel, the receiver might receive $w(x) = v(x) + e(x)$, where $e(x)$ is an error vector. We illustrate this in our example by letting $e(x) = x^4$, i. e $w(x) = 1 + x^3 + \mathbf{x}^4 + x^5 + x^6$. Our next aim is to decode the received word w = (1, 0, 0, 1, 1, 1, 1).

In general, when we receive a word w from the communication channel we use the concept of syndromes to reconstruct the information word. For a general linear code this requires the following steps:

- compute the syndrome $s^{tr} = H\, w^{tr}$
- divide the received words into cosets and fix a coset leader e with smallest hamming weight to each coset
- decode as v = w - e

Within the cyclic codes we construct the syndrome polynomial $s(x)$ from $w(x) = q(x) \cdot g(x) + s(x)$. For every $s(x)$ we chose an error vector $e(x)$ with smallest hamming weight and decode as $v(x) = w(x) - e(x)$. For the Hamming code K[7,4,3] with generator polynomial $g(x) = 1 + x + x^3$ we construct the syndromes and we get for error $e(x) = x^4$, syndrome $s(x)$ is $x^2 + x$.
We use MATHEMATICA to describe the methods of decoding. We consider (as in the example above) the Hamming code K [7, 4, 3], were we received the word $w(x) = 1 + x^3 + \mathbf{x}^4 + x^5 + x^6$. The following steps illustrate how the information word could be reconstructed:

```
In[16] : =V=1+x^3+x^5+x^6; e=x^4; W=V+e;
s=PolynomialMod[PolynomialMod[W,g],2]
Out[16] = x + x^2
```

We have got the syndrome polynomial and from the syndrome list above we recognize the error polynomial $e(x) = x^4$. An alternative way to decode the received word $w(x) = 1 + x^3 + \mathbf{x}^4 + x^5 + x^6$ is given in next steps:

```
In[17] : = w =CoefficientList[W,x]
Out[17] =   {1, 0, 0, 1, 1, 1, 1}
In[18] : =H = CyclicCode[g,7,x][[1]] Syndrome[H,w]
```

Out[18] = $\begin{pmatrix} 0 & 0 & 1 & 0 & 1 & 1 & 1 \\ 0 & 1 & 0 & 1 & 1 & 1 & 0 \\ 1 & 0 & 1 & 1 & 1 & 0 & 0 \end{pmatrix}$ s = $\begin{pmatrix} 1 \\ 1 \\ 1 \end{pmatrix}$

We see that this column is identical with the 5:th column in H, indicating that we have an error in the 5:th position.

References

1. MacWilliams, F. J., and Sloane, N. J. A. (1977) The Theory of Error - Correcting Codes. North - Holland, Amsterdam.
2. Adamek, J., Foundations of Coding , John Wiley & Sons, Inc 1991

Mobile Work Environment for Grid Users. Grid Applications' Framework

Michal Kosiedowski [1], Miroslaw Kupczyk [1], Rafal Lichwala [1], Norbert Meyer [1], Bartek Palak [1], Marcin Płóciennik [1], Pawel Wolniewicz [1], and Stefano Beco [2]

[1] Poznan Supercomputing and Networking Center,
ul. Z. Noskowskiego 12/14, Poznan, Poland
{miron, syriusz, meyer, palak, marcinp, pawelw}@man.poznan.pl
[2] DATAMAT S.p.A.,
Via Laurentina, 760 - I-00143 Rome, Italy
stefano.beco@datamat.it

Abstract. In this article we aim to describe a project developing of the Migrating Desktop infrastructure for mobile users. This functionality refers to the work environment of the users who very often change their location. The user interface called the Migrating Desktop, or grid desktop, is a very useful environment that accomplishes an integrated set of services and real applications, which could be run on the grid. We introduce a framework for improving the grid application launching. The possibility of usage of interactive application in developed environment is presented.

1 Introduction

What is needed for tomorrow is the proper remote and individual access to the resources, independently of the original location of the user. In the CrossGrid project (IST-2001-32243) [1] we introduce the Migrating Desktop [7]. It creates a transparent user work environment, independently of the system version and hardware. The Migrating Desktop would allow the user to access grid resources and his/her local resources from remote computers, like i.e. laptops. It will allow to run applications, manage data files, store personal settings (configuration definitions that characterise e.g. links to the user data files, links to applications, access to application portals and/or specialised infrastructure, as well as windows settings), independently of the localisation or the terminal type. In the paper we present our idea and the way of developing mechanisms for easier adjustment of the application which remains to be run on the grid. The Container and the Application Plugin paradigm will be discussed later on. We present the types of interactivity over the grid and the mechanisms for bridling it in our environment.

As an underlying layer we develop the middleware called the Roaming Access. This infrastructure is a set of modules and their interconnections hidden 'behind' the user interface. The Roaming Access and the Migrating Desktop features will not support the „moving users". It means that no special mechanisms to access the grid re

P.M.A. Sloot et al. (Eds.): ICCS 2003, LNCS 2658, pp. 1052–1058, 2003.

sources via mobile phones, PDAs (such as palmtops, organisers, etc.) will be considered within the confines of the given project. As a bottom line of the middleware we use the Globus toolkit [3]. These facilities give us important functionality like security policy, simple remote operations, user account mapping, etc. Some elements used in this work come from the DataGrid project [2] like the idea of the grid interactive job submission, the Virtual Organisation (VO) paradigm. The architecture of the project was fully detailed in the project documentation [5] and in the paper [6]. The components of the Roaming Access were presented in [7].

2 Application Framework

The Application Plugin and the Application Container are a general view of showing application specific input and output. These allow preparing the portal framework, which is independent of application type, so that we could easily extend it and prepare for next applications. It supports the batch and interactive application as well. It is worth mentioning that this framework will soon be integrated with the GVK (Grid Visualization Kernel) [10]. We would like to emphasize that the communication methods between the distributed nodes involved in the computations are not the point here. It could be MPI jobs and the HLA based as well [9].

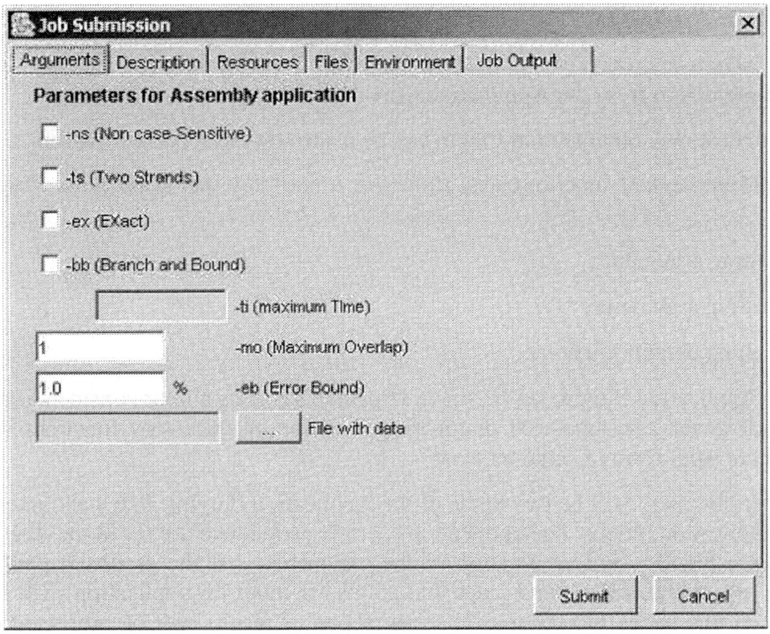

Fig. 1. The example of the Application Wizard – submission of the Assembly Application.

The Migrating Desktop provides a wizard that the user can use to specify the job details. This wizard (see Fig. 1.) simplifies the process of specifying parameters and limits, suggesting user default or last used parameters. That wizard consists of several panels. Two panels are reserved for the application specific plugin – the Arguments Panel and the Job Output Panel. Developers of the specific application should implement the contents of these panels in term of the Application Plugin.

The Application Container is a framework for the Application Plugin. It is an abstract class that is extended by the Application Plugin and extends the graphical component (JPanel) [8], which can be placed into different dialogs, panels etc. The Application Container can be characterised in the following way:

A generic abstract class that the Application Plugin will extend.

It is application independent.

It gives the Application Container API that can be accessed from the Plugin. They are represented by a set of methods.

It controls the Application Plugin by restarting, stopping it, catching error messages and invoking appropriate actions.

It can set or get parameters within the Application Plugin.

The Application Plugin is the content of the Application Container. It extends the Application Container class and implements a set of its abstract methods. It should provide and implement two panels of the Job Submission Wizard: the input arguments and the output specification. It should be able to visualise the output of application. The Application Plugin can be characterised in the following way:

It will have a reference to Application Container API, so that, if necessary, it could invoke a function from the Application Container.

The Application Container and Plugin can be placed on the separated servers.

It should implement functions that allow the control (by the Application Container) of:

- Setting arguments,
- Getting arguments,
- Getting default job name.

The Application Plugin can be a java JApplet class if its author considers it reasonable . However, the supported plugin should define all necessary functions (should extend the Application Container class).

Generally, there will be two kinds of the Application Plugins: batch and interactive work application plugin. The application specific parameters are set in the Arguments panel (see Fig. 1.). Pre-verification of the parameters and the graphical parameters setting (eg. operations on land maps that points are input for application) will be presented on this panel, too. Graphical visualisation of results will be presented on the Job Ouput panel, if necessary. The Interactive Job Plugin is a more complicated kind but it will contain all batch work plugin functionality. It shows the interactive stream of the job output. The interactivity is described later on.

3 Interactive Applications

A Grid Batch Job (GBJ) can currently be submitted using standard queuing software: the user submits the job providing a description by a Job Description Language file and waits for the end of the job asking for its status.

When the job is completed, the user will be able to request for output results downloading in his local machine.

When the user submits a Grid Interactive Job (GIJ), he/she needs the allocation of grid resources throughout his/her "interactive session". During the whole session a bi-directional channel is opened between the user client and the application programme on a remote machine. The entire job input and output streams are exchanged with the user client via this channel: the user sends input data to the job and receives output results and error messages. Details on the mechanism we foresee to manage GIJ's are in the following sections.

There is an additional group of applications, and they consist of zillions of chunk jobs persisting several minutes. Usually, the user – the decisive element - chooses the right way of running a job. Generally speaking, we can call such job an Application Interactive Job (AIJ). In this case, the Web Server automatically submits a GIJ that can be seen just as a container or, in other words, as a "shell" that will run on allocated grid resources. It opens a direct bi-directional connection between the Portal Server and allocated resources. The user chooses the AIJ that will run on allocated computing element(s), and the input data for the chosen application job.

In other words, we can say that:

the GIJ is the container. It is a piece of software that acts as a shell/interpreter/executor of an AIJ. It is a domain-dependant application. It sits on the needed resources (defined in accordance with the AIJ needs). It shall have a function capable to interact via standard streams with the Web Server. It will be submitted using a JDL via the Scheduling Agent.

the AIJ is the content. It is the real application that the user wants to be executed over the Grid. It is a domain-dependant application. It will be interpreted/executed by the GIJ. It will be submitted using whatever language is suitable for the application (e.g. C++ scripts) directly to the GIJ.

To give an example, the Web Server performs the GIJ submission after the user chooses a DATASET for his/her purposes within a list of DATASETS available for his/her Application Virtual Organisation. The submitted GIJ contains the process called the "Interactive Session Manager" (ISM) that handles the interactive session from the point of view of the application. ISM will run as close as possible to the Storage Elements hosting a physical copy of the selected DATASET.

Then the user "submits" one or more application scripts (the AIJ's). The user can choose the Application jobs among an available list of predefined applications or he/she can define one for his/her specific needs in an editor window.

A very important feature in interactive environments is the ability to recover from unexpected failures of e.g. workstation. There is no problem in case of batch jobs, but interactive applications need to be equipped with extra functionality. It would be better not to rewrite the existing application codes. However, it should be considerable to build the additional linkable library for the grid interactive session management. There is no such possibility in the existing load facilities (queuing systems, etc...).

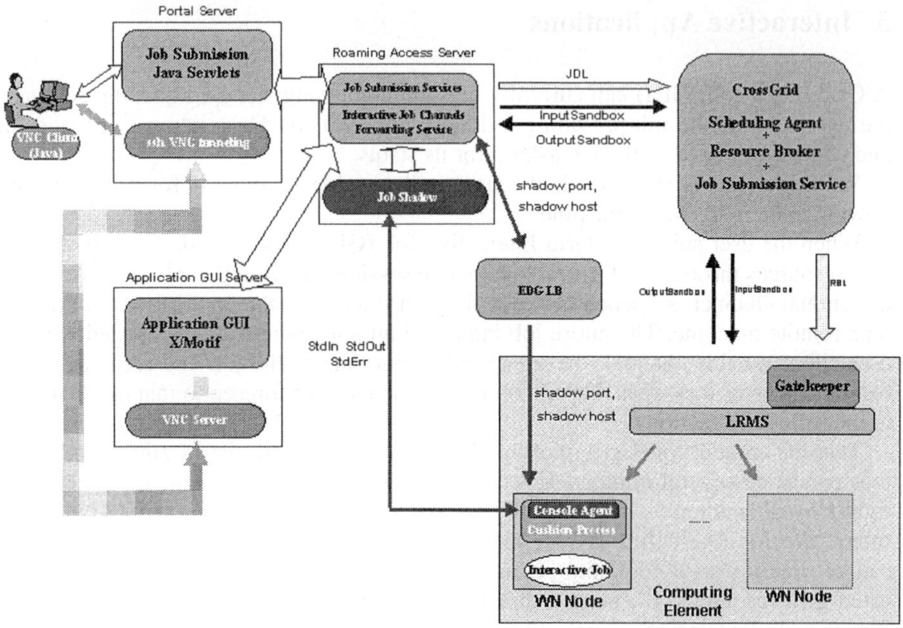

Fig. 2. How the interactive session works.

We present the solution that will be implemented for EU-CrossGrid (see Fig. 2.).

The interactive sessions [4] are handled by the Grid Console (GC) that is a system provided by Condor for getting mostly-continuous input/output from remote programs running on an unreliable network. A GC is a *split execution system* composed by two software components: an *agent* (Console Agent – CA) and a *shadow* (Console Shadow – CS or Job Shadow – JS, as it will be a slightly modified CS).
The *split execution system* is a special case of an *interposition agent*. An *interposition agent* transforms a program's operation by placing itself between the program and the operating system. When the program attempts certain system calls, the agent grabs control and manipulates the results.

In the *split execution system* an interposition agent (CA) traps some of the procedure calls of an application, and forwards them (via RPC) to a shadow process (CS) on another machine. Under this arrangement a program can run on any networked machine and yet execute exactly as if it were running on the same machine as the shadow. All the network communications are GSI-enabled.

The Console Agent runs on a Worker Node and it is a shared library that intercepts reading and writing operations on stdin, stdout, and stderr of the running job. When possible, the CA sends the output back to the CS. The shadow manages the input and output files according to the request of the agent.

If the output sending fails, CA will write it on the local disk instead. It does not matter why the input/output operation failed; the CA will keep the process running. At regular intervals it will attempt the network connection again. If the connection succeeds, it will transfer any buffered data to the shadow, and then resume normal operation.

The CA retries failed operations at regular intervals for a certain number of times, after which it will give up and kill the process. The number of times to retry and the number of seconds to pass between each retry are configurable.

The user submits his/her interactive job via the Portal Server (PS) and interacts with the remote machine where the job is running (Worker Node- WN). The Portal Server starts/stops the Job Shadow on the Submission Machine when the user opens/closes the interactive session. The PS writes in the Logging&Bookeeping (LB) and in the JDL file the ShadowPort (port number assigned to the shadow) and the ShadowHost (the host where the Job Shadow runs); this information is also used by the agent. In the JDL there will be an attribute called JobType set to "Interactive".

Before the submitted job goes in the running status, a script called *JobWrapper* prepares the environment for the job, transfers on the WN additional files within the job InputSandbox and starts the *Cushion Process*.

The *Cushion Process* is a process linked to the CA that handles the standard streams between the CA and the running job by using a named pipe. This process is needed to avoid dynamic linking between the user job process and the CA library. It also allows to avoid linking between a system library and a user process, as well as a possible failure that could appear if the user job works with other interposition agents (besides the CA).

For the standard X-Windows applications, the VNC [11] protocol will be used. It is foreseen to be integrated next year. The corresponding connections are shown in Fig. 2.

4 Summary

The Roaming Access and Migrating Desktop implement new generation tools for the end user according to the growing demands and needs. Nowadays it is not enough to have Internet or Grid access. It can be observed that the group of inexperienced users has increased significantly. On the other hand, the Grid and Internet populations are among the most mobile of all (moving between Virtual Organizations in an open network environment). Therefore, the Grid should be able to deliver a service allowing to keep and restore the users' working environment. The Job Wizard functionality is the response for easy and flexible work with the grid application, including interactivity. The mentioned functionality is a cutting edge in the grid utility. This is a goal that we approach by the delivery of the tool for the application programmers, and a uniform environment for the end-users.

The Desktop Portal Server extends the functionality of the Application Portal Server by providing a specialised advanced graphical user interface and a mechanism that allows the user to access all files stored on his/her personal machine available from other locations/machines. The mentioned functionality is developed within the CrossGrid project (IST-2001-32243, http://www.eu-crossgrid.org) and its first Prototype was released in February 2003.

References

1. Annex 1 to the "Development of Grid Environment for Interactive Applications" - EU-CrossGrid Project, IST-2001-32243, http://www.eu-crossgrid.org/CrossGridAnnex1_v31.pdf
2. Annex to the "Research and Technological Development for an International Data Grid" - EU-DataGrid Project, IST-2000-25182, http://eu-datagrid.web.cern.ch/eu-datagrid/1Y-EU-Review-Material/CD-1Y-EU-Review/3-DataGrid-TechnicalAnnex/DataGridAnnex1V5.3.pdf
3. Globus Toolkit 2.0, http://www.globus.org
4. Kupczyk, M., Lichwala, R., Meyer, N., et. al., Roaming Access and Portals: Software Requirements Specification", EU-CrossGrid Project, http://www.eu-crossgrid.org/Deliverables/M3pdf/SRS_TASK_3-1.pdf
5. Bubak, M., Malawski, M., Zajac, K., "Towards the CrossGrid Architecture", Proceedings 9th European PVM/MPI Users' Group Meeting, Linz, Austria, September/October 2002, LNCS 2474.
6. Bubak, M., Malawski, M., Zajac, K., "Current Status of the CrossGrid Architecture", Proceedings of the Cracow '02 Grid Workshop, December 11–14, 2002, Cracow, Poland
7. Kupczyk, M., Lichwala R., Meyer, N., Palak, B., Plociennik, M., Wolniewicz, P., "Roaming Access and Migrating Desktop", Proceedings of the Cracow '02 Grid Workshop, December 11–14, 2002, Cracow, Poland
8. Java Library, http://sun.java.com
9. High Level Architecture Run-Time Infrastructure RTI 1.3-Next Generation Programmer's Guide, https://www.dmso.mil/public/transition/hla/
10. Heinzlreiter, P., Kranzmueller, D., Volkert, J., "GVK – Visualization Services for the Grid", Proceedings Cracow '02 Grid Workshop, December 11–14, 2002, Cracow, Poland
11. Virtual Network Computing, http://www.uk.research.att.com/vnc/, University of Cambridge.

EC Transactions Use Different Web-Based Platforms

Whe Dar Lin

The Overseas Chinese Institute of Technology
Dept of Information Management,
No. 100, Chiao Kwang Road Taichung 40721, Taiwan

Abstract. A review of the methodological aspects of the EC transactions use different Web-based platforms is presented. Our design techniques for complex EC environments are present to avoid the viruses infected EC transaction files. When an executive service code is ordered it may get infected with some viruses before the signature is attached to it. The infected EC transaction files cannot be detected by signature verification and the origin of the infection order cannot be specified by different Web-based platforms. Applied our EC transactions use different Web-based platforms, these techniques lead to an EC framework with clients, agents, EC application servers and EC system management. The paper discusses how Web-based platforms can use signature method together with detection infected in the preprocessing step enable us to specify the origin of the infection side. The impact of our proposed method can improve Web-based platforms relationships between EC transaction agents.

1 Introduction

In EC era, there has been a general trend towards partnership Web-based platforms together with a reduction and consolidation of the supply based in order to have better EC application relationships with fewer Web-based platforms. It is an important step that an execution service code can be detected by checking the consistency of original transaction order with its accompanying digital signature [1], [2], [5], [6], [8]. The cooperative Web-based platforms approach to supplier is characterized by long term contracts, integrated key protocols, share marketing database, and a commitment to partner relationships.

We don't think this is a good architecture using in Web-based platform. [12][13][14] Some other efforts through the directory services, name services, library service to reduce complex about viruses impaction. This approach yields an efficient solution for simple and static point-to-point interaction in collaborative EC transaction order. Nevertheless, we need face dynamic communication commit from different Web-based platforms for multiple relationships between one another on supply chain management. We will choice the ID-based scheme [10], a large public-key file is not required because each user's public key is nothing but an identity and communication cost is low using in EC applications for different Web-based platforms.

Our proposed signature method is based on a public key scheme. We assume that special entities, such as transaction suppliers in supply chain are not reliable for

P.M.A. Sloot et al. (Eds.): ICCS 2003, LNCS 2658, pp. 1059–1068, 2003.

concerning the signature forgery or computer viruses. The rest of this paper is organized as follows. Sect. 2 gives some general assumptions about viruses and transaction supplier and Web-based platform. The proposed our security Web-based platform scheme will be described in Sect. 3 and Sect. 4 the correctness shall be discussion. Our proposed scheme security analysis is described in Sect. 5, and we discuss some implementation Web-based platform issue in Sect. 6. We conclude with some final remarks shall be stated in Sect.7.

2 Assumptions and Model

Here, we provide several assumptions about viruses and transaction supplier, and security in general using in our protocol. A protocol is a set of rules and conventions that define the communication framework between two or more EC transactions. After that, we shall describe the Web-based platforms.

2.1 General Assumptions about Viruses and Transaction Supplier

Our proposed scheme relies on the existence of a hash function h. Specifically,

Assumption 1: We assume there exist a function h such that:

On random input (r_i, m_i), it is difficult to generate (e_i) such that h(r_i, m_i)= (e_i). More generally, it is difficult to generate such (e_i) on input (r_i, m_i) and samples of signature on random messages signed with EC-based platforms.

Assumption 2: Web-based platform act according to the following conditions:

- They create transaction supplier honestly in marketing channels.
- They do not refuse to reply to the requests or questions from the EC agent.
- Viruses can infect and incubate warehouse files and EC order files on Web-based platforms.
- Viruses can damage both warehouse files and EC order files on Web-based platforms.

Our proposed automatic signature scheme of protection against viruses works upon the following security assumptions [1], [2], [3], [4], [9], [11]:

- Solving the discrete logarithm problem is difficult, so we use in cryptographic protocols.
- Inverse calculation of one-way hash functions is difficult that we use assumption is often made in cryptographic protocols.
- Distributed operating system in Web-based executes a verification program properly. This assumption is important because improper events must be ruled out in the verification procedure so that one can rely on the results of the verification.

2.2 Client-Server Transaction Model for Procurement

In this paper, the transaction supplier on the Web-based platform will reduce the amount of complete purchased work items and the turnaround time the deliver the order transaction services.

- Creates the service system library, good library, customer library and included files,
- Signs all library files following every subroutine, procedure and initial data structure,
- Calculates the fingerprints of the included files with a one-way hash function and creates the database of the fingerprints (fingerprint data file),
- Computes proxy integers and attaches them to the Web-based platform,
- Creates the signatures for the shops on Web-based platform and the fingerprint data file.

The included files and customer relation data are text files and are accessed during preprocessing. The transaction includes all processes from preprocessing to linking. The fingerprint of an included file is an output of a one-way hash function taking the included file as an input file. The fingerprint data file is composed of pairs of the names of the included files and their fingerprints. From this file, one can authenticate the validity of the included files.

3 Our Proposed Automatic Signature Scheme

In this section, we shall illustrate the structure of our proposed method. In our proposed protocol, The transaction supplier is denoted by u_m, the EC server u_s and the mobile user, client u_c.

They follow the steps below.
Step 1: The EC server u_s sends the request the Web-based platform R to the transaction supplier u_m.
Step 2: The transaction supplier u_m sends the Web-based platform R to the EC server u_s.
Step 3: The mobile user, client u_c sends the request the Web-based platform R for the transaction order P to the EC server u_s.
Step 4:The EC server u_s sends the executable transaction bill service code M to the client u_c.

3.1 The Initialization Phase

The system parameter are listed as follows:
(1)We use a prime p with $2^{511} < p < 2^{512}$, a prime divisor $q|p-1$, a generator g with order q over GF(p), and
(2) the security information of the transaction supplier u_m: x_m
(3) the public information of the transaction supplier u_m: y_m

$$y_m \equiv g^{x_m} \bmod p \quad (1)$$

(4) the security information of the EC server u_s: x_s
(5) the public information of the EC server u_s: y_s,

$$y_s \equiv g^{x_s} \bmod p \quad (2)$$

(6) the security information of the mobile user, client u_c: x_c
(7) the public information of the mobile user, client u_c: y_c,

$$y_c \equiv g^{x_c} \bmod p \quad (3)$$

3.2 The EC Server Requests the Delivery of a Web-Based Platform

When a server u_s requests the delivery of a Web-based platform named R. The server sends his secret key x_s along with four integers calculated from a random number k_1. The computation is as follows.

(1). Generate a random number k_1 satisfying $0<k_1<q-1$
(2). Compute

$$r_1 \equiv g^{k1} \bmod p \quad (4)$$

$$m_1 \equiv u_s \| R \bmod p \quad (5)$$

$$e_1 \equiv h(r_1, m_1) \bmod q \quad (6)$$

$$s_1 \equiv k_1 - x_s e_1 \bmod q \quad (7)$$

where h is a one-way function and || denotes concatenation.
(3). Send (u_s, R, e_1, s_1) with the request to the transaction supplier u_m.

3.3 The Transaction Supplier Sends the Web-Based Platform to the EC Server

When the transaction supplier u_m receives the integers (u_s, R, e_1, s_1), he verifies the server u_s requests and then delivers a Web-based platform C_R. The transaction supplier u_m verifies (u_s, R, e_1, s_1) by computing the following equations.

(1). compute

$$r^* \equiv g^{s1} y_s^{\ e1} \quad (8)$$

check whether the congruence e_1=h(r^*, $u_s \| R$) holds.
If it holds, the transaction supplier u_m does as follow.
(2). Generate a random number k_2 satisfying $0<k_2<q-1$
(3). Compute

$$r_2 \equiv g^{k2} \bmod p \quad (9)$$

$$t_m \equiv x_m + k_2 * r_2 \bmod q \tag{10}$$

$$m_2 \equiv C_R || t_m \bmod p \tag{11}$$

$$e_2 \equiv h(r_2, m_2) \bmod q \tag{12}$$

$$s_2 \equiv k_2 - x_m e_2 \bmod q \tag{13}$$

(4). Send the Web-based platform C_R and its signature (C_R, t_m, e_2, s_2) to u_s

3.4 The Client Requests to the EC Server to Send Service Code for an Order P

A client user u_c wants to get a service code for an order P, and the server creates an executable service code M and its signature. When the server u_s receives the integers (C_R, t_m, e_2, s_2), he verifies that a server u_m sends then accept the Web-based platform C_R. The server u_s verifies (C_R, t_m, e_2, s_2) by computing the following equations.

(1). Compute

$$r^* \equiv g^{s2} y_m^{\ e2} \tag{14}$$

check whether the congruence $e_2 = h(r^*, C_R || t_m)$ holds. If it holds, the server u_s does as the client wishes. In the process of calculating a signature, the client executes the following steps:

(2). Generate a random number k_3 satisfying $0 < k_3 < q-1$

(3). Compute

$$r_3 \equiv g^{k3} \bmod p \tag{15}$$

$$m_3 \equiv u_c || P \bmod p \tag{16}$$

$$e_3 \equiv h(r_3, m_3) \bmod q \tag{17}$$

$$s_3 \equiv k_3 - x_s e_3 \bmod q \tag{18}$$

(4). Send (u_c, P, e_3, s_3) with the request to the server u_s.

3.5 The Server Processes the Order and Sends the Execution Service Code to the Client

When the server u_s receives the integers (u_c, P, e_3, s_3), he verifies that a client u_c requests that the Web-based platform C_R processes the order P. The server u_s verifies (u_c, P, e_3, s_3) by computing the following equations

(1). compute

$$r^* \equiv g^{s3} y_s^{\ e3} \bmod p \tag{19}$$

check whether the congruence e_3=h(r^*, u_c||P) holds. If it holds, in the process of calculating a signature, the server processes the order P and executes the following steps:
(2). Generate a random number k_4, satisfying $0 < k_4 < q-1$
(3). Compute

$$r_4 \equiv Pg^{-k4} \mod p \tag{20}$$

$$s_4 \equiv k_4 - r_4 x_s \mod q \tag{21}$$

$$M \equiv C_R(P) \mod p \tag{22}$$

$$m_4 \equiv u_c || P || M \mod p \tag{23}$$

$$e_4 \equiv h(r_3, m_3) \mod q \tag{24}$$

$$t_s * h(M) \equiv t_m + r_2 x_s \mod q \tag{25}$$

$$r_2 \equiv g^{s2} y_m^{e2} \mod p \tag{26}$$

(4). Send (r_4, s_4, t_s, M) with the request to the server u_s.

The Web-based platform can be used only by the server u_s because the integer t_m is the proxy key of the transaction supplier u_m. However, if another user happens to know u_m's secret key, then he can use this u_m's Web-based platform and create a bad executable service code with this proper proxy. Then, the executable service code may damage someone's computer. The basic rule for this event is that the owner u_m of the Web-based platform storing the proxy integer key t_m is responsible for the damage so that the transaction supplier must hide his secret key.

3.6 Signature Verification

When a user executes a service code M made by u_s, the use verifies the signature with the transaction supplier's public key and server's public key. That he checks whether the signature satisfies the congruence such that

$$g^{t_s h(M)} \equiv y_m (r_2 y_s)^{r_2} \mod p \tag{27}$$

Since the signature (r_4, s_4, t_s, M) is automatically created, an infection of a generated executable service code can be detected in the above verification process. Using a signature can possibly be created for a contaminated executable service code, and in this case, the infection can be detected by the verification. We can determine who causes the injection.

4 Correctness of Our Scheme

Theorem 1:
1. The transaction supplier u_m can get the correctness request from the server u_s.
2. The server u_s can get the correctness from transaction supplier u_m.
3. The server u_s can get the correctness request from client u_c.

Proof:
(1) The transaction supplier u_m, receiving (u_s, R, e_1, s_1) from u_s can compute
$r^* = g^{s1} y_s^{e1} = g^{k1-xse1} y_s^{e1} = g^{k1-xse1} g^{xse1} = g^{k1} = r_1$
$h(r^*, u_s||R) = h(r_1, u_s||R) = h(r_1, m_1) = e_1$
Then we can be sure the signature of (u_s, R) is (e_1, s_1)
(2) The server u_s, receiving (C_r, t_m, e_2, s_2) from u_m can compute
$r^* = g^{s2} y_m^{e2} = g^{k2-xme2} y_m^{e2} = g^{k2-xme2} g^{xme2} = g^{k2} = r_2$
$h(r^*, C_r||t_m) = h(r_2, C_r||t_m) = h(r_2, m_2) = e_2$
Then we can be sure the signature of ($C_r||t_m$) is (e_2, s_2)
(3) The server u_s, receiving (u_c, P, e_3, s_3) from client u_c can compute
$r^* = g^{s3} y_c^{e3} = g^{k3-xce3} y_c^{e3} = g^{k3-xce3} g^{xce3} = g^{k3} = r_3$
$h(r^*, u_c||P) = h(r_3, u_c||P) = h(r_3, m_3) = e_3$
Then we can be sure the signature of (u_c, P) is (e_3, s_3)

Theorem 2:The server u_s can processes the correctness transaction order request from the client u_c.

Proof: The client u_c, receiving (r_4, s_4, t_s, M) from server u_s can compute
$$g^{t_s h(M)} \equiv y_m (r_2 y_s)^{r_2}$$
where $g^{t_s h(M)} \equiv g^{t_m + x_s r_2} \equiv g^{x_m + k_2 r_2 + x_s r_2} \equiv y_m r_2^{r_2} y_s^{r_2} \equiv y_m (r_2 y_s)^{r_2}$ Then we can be sure the signature of M is (t_s, M) and use the public key of transaction supplier and server.

5 Security Considerations

Theorem 3: The crackers want to reveal the secret key from public key is computing impossible.

Proof: The crackers want to from the public key $y_m \equiv g^{xm}$ mod p, $y_s \equiv g^{xs}$ mod p, and $y_c \equiv g^{xc}$ mod p to reveal x_m, x_s, and x_c, they have to solve the discrete logarithm problem.

Theorem 4: The malicious users want to forger the request from the proposed scheme is computing impossible.
(1) The transaction supplier u_m get the false receive from malicious users.
(2) The server u_s can get the false request from malicious users.

Proof:

(1) The transaction supplier can use (u_s, R, e_1, s_1) to verify $r^* = g^{s1}y_s^{e1} = g^{k1\text{-}xse1}y_s^{e1} = g^{k1\text{-}xse1}g^{xse1} = g^{k1} = r_1$, But the malicious users construct (u_s, R, e_1', s_1') without know x_s, the congruence, $r^* = g^{s1'}y_s^{e1'} = r_1$. They have to solve the discrete logarithm problem.

(2) The proof method is similar as part (1).

Theorem 5: The malicious users want to forger the signature from the proposed scheme is computing impossible.

(1) The server u_s can get the false Web-based platform from malicious user.

(2) The client u_c can get the false executable service code from malicious user.

(1) Proof: The server u_s the receive (C_r, t_m, e_2, s_2) from u_m can compute

$r^* = g^{s2}y_m^{e2} = g^{k2\text{-}xme2}y_m^{e2} = g^{k2\text{-}xme2}g^{xme2} = g^{k2} = r_2$

$h(r^*, C_r||t_m) = h(r_2, C_r||t_m) = h(r_2, m_2) = e_2$

We will sure the signature of ($C_r||t_m$) is (e_2, s_2)

Case I: The malicious users use the false (C_r', t_m, e_2, s_2) signature send to server u_s

$r^* = g^{s2}y_m^{e2} = g^{k2\text{-}xme2}y_m^{e2} = g^{k2\text{-}xme2}g^{xme2} = g^{k2} = r_2$

$h(r^*, C_r'||t_m) = h(r_2, C_r'||t_m) = h(r_2, m') = e' \neq e_2$

We will sure the signature of (C_r', t_m, e_2, s_2) is false signature.

Case II: The malicious users use the false (C_r, t_m', e_2, s_2) signature send to server u_s

$r^* = g^{s2}y_m^{e2} = g^{k2\text{-}xme2}y_m^{e2} = g^{k2\text{-}xme2}g^{xme2} = g^{k2} = r_2$

$h(r^*, C_r||t_m') = h(r_2, C_r||t_m') = h(r_2, m') = e' \neq e_2$

We will sure the signature of (C_r, t_m', e_2, s_2) is false signature.

Case III: The malicious users use the false (C_r', t_m', e_2', s_2') signature send to server u_s

$r^* = g^{s2'}y_m^{e2'} = r_2'$, $h(r_2', C_r'||t_m') = e_2'$

But the malicious users choose another Web-based platform C_r' and another proxy key t_m' without know x_m, the congruence equation that they have to solve the discrete logarithm problem.

(2) The proof method is similar as part (1).

Theorem 6: The computational complexity for an intruder to cryptanalyze a new session key in our scheme, after having received previous transaction order data, the public key, the old messages and the new message is as hard to cryptanalyze a plaintext in the ElGamal scheme when the order of g is a prime.

Proof. Let B_O correspond with the problem of breaking our scheme and B_E with breaking the ElGamal encryption scheme with the order of g is prime. There is an existence of a polynomial time B_E implies the existence of a polynomial time B_O.

We suppose it is possible to compute the discrete log of an output of G after seeing a sequence of all transaction orders. Without loss of generality we can assume success rate is an ε. We will use B_E by B_O as a procedure processing in computing feasible by a polynomial time. So, the input parameters for B_E is K_i=(r_i, s_i, t_j, M_k).

Now B_E computes H_I and h(r_I, m_I) which will be used as input parameters for B_O as follows. B_E chooses $x_1, x_2, \ldots, x_k \in Z_q$ at random and computes $y_1 \equiv g^{x_1}, y_2 \equiv g^{x_2}, \ldots, y_k \equiv g^{x_k}$ Then B_E has obtained a public message H_I=(p, q, g

$y_1, \ldots, y_{k-1}, y_k$). So B_E computes $y_i^{r_j} \equiv g^{x_i r_j}$ $for\ i = 1,2,\ldots,p, \quad j = 1,2,\ldots,q$. Thus, uses the H_I=(p, q, g $y_1, \ldots, y_{k-1}, y_k$) and K_i=(r_i, s_i, t_j, M_k) to B_O. In computing feasible condition for any queried transaction order by B_O, B_E computes by using H_I=(p, q, g $y_1, \ldots, y_{k-1}, y_k$) and sends h($r_I$, m_I) message to B_O. The EC transaction protocol will complete. Finally, B_E outputs the result of B_O, which is an $1/\varepsilon$ with non-negligible probability cases. We would find the discrete log in expected $1/\varepsilon$ steps. We prove under reasonable assumptions, that our proposed protocol is computationally secure even on public network within different Web-based platforms.

6 Performance

In this section, we shall calculate performance of our new method
I: The transmission cost on the delivery of the Web-based platform
In our method
Step 1: Send (u_s, R, e_1, s_1) with the request to the transaction supplier u_m.
Step 2:Send (C_R, t_m, e_2, s_2) with the response to the server u_s. Send the Web-based platform C_R and its signature (C_R, t_m, e_2, s_2) to u_s
II: The computation cost on verifying the signature of an executable program
In our proposed method verifies the signature.

$$g^{t_s h(M)} \equiv y_m (r_2 y_s)^{r_2} \bmod \quad p$$

Our method uses 2 modular exponentiation computing time and 1 hash function computing time.
III: The security model comparison in implement.
In our proposed method: The client checks if the signature satisfies the congruence

$$g^{t_s h(M)} \equiv y_m (r_2 y_s)^{r_2} \bmod \quad p$$

The server checks if the signature satisfies the congruence
e_3=h(r^*, u_c||P) and M ≡ C_R(P) mod p.

7 Conclusion

We proposed a protocol that developed to allow clients to get committing Web-based platform signature. In our signature scheme the public key of the Web-based platform owner is used for signature verification that a signature is created from both the public key of the transaction supplier and that of the server Web-based platform owner. In our proposed method, transaction suppliers no use any table to store EC Web-based servers' information that will more efficient and safety than any others EC Web-based systems. Without identify creator, our effective approach on the public keys lead to a verification procedure, and created signatures are checked relatively fast. The most nature extension to this novel protocol scheme is a server-based signature that integrated together with EC application package will allow client and the server to commit with one another.

References

1. T. ElGamal, "A public key cryptosystem and a signature ,scheme based on discrete logarithms," IEEE Trans. Inf. Theory, vol.IT-31, no.4, pp.469–472, 1985.
2. C,P, Schnorr, "Efficient identification and signatures for smart Cards," Advances in Cryptology-CRYPTO'89, LNCS 435, pp.239–252, Springer-Veriag, 1990.
3. R.C, Merklc, "A fast software one-way hash function," J. of Cryptology, vol.3, no.1, pp.43–58, 1990.
4. C. Park, "A Fiat-Shamir-like identification protocol without a highly reliable trusted center," Proc. the 1992 Symp. on Cryptography and lnf. Security, no.6D, 1992.
5. Y. Desmedt and Y. Frankel, "Multi-signatures for virus protection," Proc, the 5th lnt'l Computer Virus and Security Conf., 1992.
6. E. Okamoto, "Integrated security system and its application to anti-viral methods," Proc. the 6th Virus and Security Conf, 1993.
7. M. Cyirault, "Self-certified public keys," Advances in Cryptology-EUROCRYP'FO'91, LNCS 547, pp.490–497, Springer-Veriag, 1991.
8. K. Usuda, M. Mambo, T. Uyematsu and E. Okamoto "Proposal of an automatic signature scheme using a Web-based platform", IEICE Trans. Fundamentals E79-A (1) (1996) 94–101
9. K. Nyberg, R.A. Rueppel, "Message recovery for signature scheme based on the discrete logarithm problem," Designs, Codes and Cryptography, No 7, pp.61–81, 1996
10. A. Shamir, "Identity-based cryptosystem based on the discrete logarithm problem," Proc. CRYPTO'84, 1985, pp. 47–53.
11. D. W. Manchala, "E-Commerce Trust Metrics and Models," J. of IEEE Internet Computing, March 2000, pp. 36–44.
12. R. Sherwood, B. Bhattacharjee and A. Srinivasan, "A Protocol for Scalable Anonymous Communication," Proc. the IEEE Symposium on Security and Privacy, 2002, pp.1–12.
13. A. Ginige and S. Murugesan, "Web Engineering: An Introduction," J. of IEEE MultiMedia, January 2001, pp.14–18.
14. J. B. Lim and A. R. Hurson, "Transaction Processing in Mobile, Heterogeneous Database Systems," IEEE Trans. On Knowledge and data Engineering, Vol. 14, No. 6, 2002, pp.1330–1346.
15. Jan, J.K. and Whe Dar Lin :An Efficent Anonymous Channel Protocol in Wireless Communications, IEICE Trans. on Communications, Vol.E84-B, No.3, PP.484–491, 2001.
16. Whe Dar Lin, "Using Marketing Factors Associated with Web site in E-commerce," The International Conference on ASEMA, pp7~11, 2002.

MOIRAE – An Innovative Component Architecture with Distributed Control Features

Katia Leal[1], José Herrera[1], José M. Peña[1]*, and Ernestina Menasalvas[2]

[1] DATSI, Universidad Politécnica de Madrid, Spain
[2] DLSIIS, Universidad Politécnica de Madrid, Spain

Abstract. Today's distributed systems are very complex applications. Taming this complexity, providing both adaptabity and performance is still an open issue for distributed architectures. Component-based development is a key technique to design plugable elements for a distributed system. This article describes a new architecture for distributed component systems. This architecture defines each of its components as two different planes (i) operational actions and (ii) control policies. New issues in terms of adaptability and flexibility are available by this new architecture, called MOIRAE (Management of Operational Interconnected Robots using Agent Engines).

1 Introduction

Different component architectures has been proposed [1,6,5] in order to build complex distributed systems. MOIRAE architecture is an inovative approach for distributed environments with a strong emphasis on what has been called control and behaviour management. A MOIRAE application is designed under a tightly-coupled model called Policy/Mechanism (P/M) model. Using this architecture high flexible and adaptative systems can be designed to tackle performance-restrictive problems like distributed data mining scenarios, as we have presented before [4,3].

This article introduces a new development environment based on the MOIRAE architecture, MOIRAEToolKit. This environment allows developers to design and implement MOIRAE components in a easy manner.

The rest of the article is organized as follows. Section 2 presents the P/M model and its advantages in adaptative environments. In section 3 the MOIRAE architecture is reviewed. Section 4 introduces the new development environment called MOIRAEToolKit. The most important module of the MOIRAEToolKit, the Code Generator, is presented in the section 5. The article ends with a simple example, presented in section 6 followed by the conclusions and future lines shown in section 7.

* This work has been partially supported by UPM project #14494.

P.M.A. Sloot et al. (Eds.): ICCS 2003, LNCS 2658, pp. 1069–1078, 2003.

2 Policy/Mechanism Model

Complex applications like operating systems (OS) and database management systems (DBMS) have to provide both efficient resource management (performance) and extensibility with new requirements, services and tasks (flexibility). These applications are designed following a clear distintion between (i) operations and functions provided by the system and (ii) the rules under which these functions are managed. These rules decide when and what functions are used in order to perform user tasks, as well as which parameters and options configure them. These functions are called mechanisms and they are implemented for each of the simple operations the system provides. The policies are the separate descriptions of those management and configuration rules.

The Policy/Management model (P/M model) is a well-known technique in the design of complex systems. P/M models is the foundation of the MOIRAE reference architecture. This architecture extends how this technique has been used, making it applicable beyond the design phase. A MOIRAE component is an element that implements one or more mechanisms that can be managed by dynamic runtime-configurable policies.

3 MOIRAE Architecture

MOIRAE [2] is a reference architecture described by 4 different models:

① **Component Model**: Describes the atomic elements of the architecture. These elements are called *components*. This model defines each of the modules which the elements are divided into. Some of these parts have specific functionalities others depend on the operational function provided by the component.

② **Relationship Model**: Specifies the different communication methods among the components. This model defines relationship schemas depending on the cardinality of the members (unicast vs multicast), the scope of the relationship (public vs private) and the addressing methods (public vs anonymous).

③ **Architecture Model**: Combines Relationship and Component Models to describe how the architecture solves execution problems. This model provides a description of the cooperative work performed by the architecture.

④ **Control Model**: Defines how control decisions are performed. This model shows the mechanism used to solve control conflicts when they arise. Control Model is based on the concept of *control policies*.

3.1 Component Model

This Model describes the structure that every element takes as a skeleton. MOIRAE architecture provides control features based on the functionalities provided by each of the elements. These atomic elements are called *components*.

Following the M/P paradigm each component defines two kinds of functions, operational functions and control functions. Each of these groups of functions is implemented in a *plane.* The *Operational Plane* includes all the modules that provide the operational functions. The *Control Plane* provides the features necessary to control the other *plane* as well as to interact with other *Control Planes.*

The *Control Plane* of a MOIRAE component is a control agent itself. The elements inside of this plane are:

① *Event Sensor*: This element detects the abnormal situations that occurs inside the operational part of the component. These sensors are activated as asynchronous messages.
② *Actors*: They are a group of functions executed by the control plane to modify the operational plane. Using these functions the control part can execute actions, modify parameters or enable/disable operational functionalities.
③ *Sensors*: These elements are used when the control plane wants to scan the situations of the operational part. As a difference with the *Event Sersors*, these are synchronous information-gathering mechanisms.
④ *Control Interface*: This module allows the control plane to interact with the control planes of other components.
⑤ *Policy Engine*: It is the central module of the control part. This element manages the decisions taken by the control plane. When the control plane is summoned, either by a conflict (via *event sensor*) or by another component (via *control interface*), the *policy engine* is activated. This module may use the *sensors* to evaluate the status of the operational plane and also interact with other control components to decide the actions to be performed. These actions are either executed locally by the *actors* or submitted to another component using the *control interface.*

3.2 Relationship and Architecture Models

These two models describe the possible organization schemas of the components in the system. The *Relationship Model* defines the cardinality, scope and visibility of component relations. The *organization graphs*, interaction topologies of the components, are classified by the *Architecture Model.* Both Models are very important parts of the MOIRAE architecture, but they are not so related with the development environment MOIRAEToolKit. For a detailed description of this model see [2].

3.3 Control Model

The last Model of the architecture is the Control Model. This Model shows how control decisions are taken either locally or as a contribution of different control planes. When the control plane is activated, for example when a conflict is detected by the *event sensor*, the *policy engine* evaluates the alternatives to solve the problem. As a result, the control plane returns a sequence of actions to be performed to solve the conflict. For complex problems the control plane would be unable to achieve an appropriate solution by itself. Control Model specifies three different control actions that rule the cooperative solution of complex problems:

- **Control Propagation**: When a control plane is unable to solve a problem it submits the problem description (e.g.: the conflict) and any additional information to the control plane immediately superior in the hierarchy.
- **Control Delegation**: After receiving a control propagation from a lower element, the upper element may take three different alternatives:
  - ① If it is also unable to solve the problem it propagates up the conflict as well.
  - ② If it can solve the problem, it may reply to the original component with the sequence of actions necessary to solve the problem. This original component executes these actions.
  - ③ In the last situation it is also possible that the component, instead of replying with the sequence of actions the component may provide the p-DB information necessary to solve the problem in the lower component. This information could be used also in any future situation. This alternative is called *Control Delegation.*
- **Control Revoke**: This action is the opposite to the *control delegation* one. Using this control action any upper component in the hierarchy may delete information from the p-DB of any lower element. This action may be executed anytime and not only as a response of a *control propagation.*

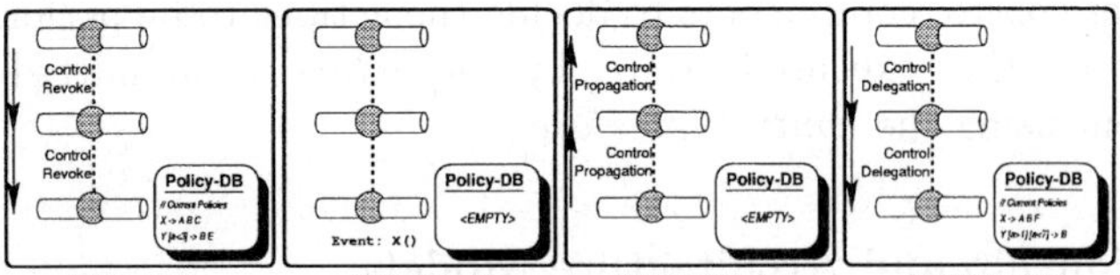

(a) Policy deletion *(b) Event detected* *(c) Propagation: event* *(d) Delegation: event*

Fig. 1. Policy management actions

4 MOIRAE ToolKit

MOIRAEToolKit is a collection of tools (as introduced by figure 2) to develop the components of the MOIRAE control architecture presented before. The figure shows the relationship among the different elements that are included in MOIRAE environment. These tools are:

- Analyser (*MOIRAE Parser*)
- Component Compiler
- Policy Verifier
- Document Generator

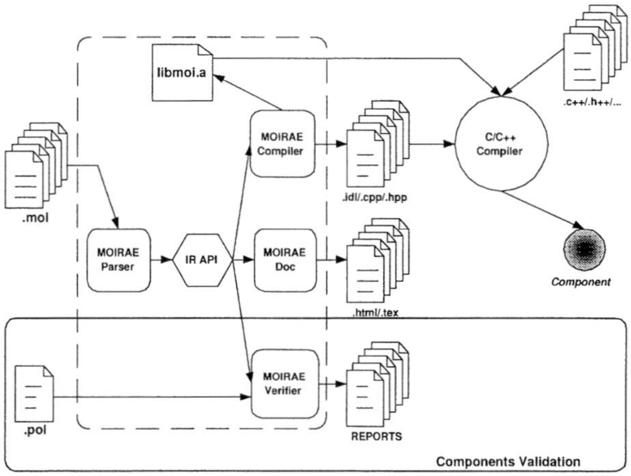

Fig. 2. MOIRAE tools.

4.1 Analyser (*MOIRAE Parser*)

This first module deals with lexical, syntactic and semantic (L/S/S) analysis of component description and implementation files (*.moi*). This element is the single frontend of most of the modules as its output is consumed by the different tools of this suit. This result is an internal structure called **IR** (*Internal Representation*). This structure is an abstract representation of the useful information provided by user through *.moi* file. This representation is a L/S/S-proven description of the defined component itself as well as the rest of the components it is related with. This module it is not actually a tool but it is linked with all the other modules of the systems in order to perform this common verification and analysis tasks.

4.2 Components Compiler

This module is basically a code generator tool. It takes the descriptive information represented by the IR and, depending on the library structure, it generates several high level language files, specifically in C++. Some of the classes generated by this tool, inherit from library classes. The others files depend directly on the options specified by user, such as the kind of middleware or the type of policy engine. This module is a key element of the MOIRAEToolKit because it provides part of the final code that will be compiled and linked when the actual component is built.

4.3 Policy Verifier

Policy Verifier is the second tool of the development environment. This module takes one or more component descriptions (IR structures) and a list of policy files. This module verifies whether the set of rules and procedures defined by the policies are able to solve all the possible conflicts generated by this group of components. As a result this module reports if the components can be properly controlled by the provided policies.

The verifications performed by this tool are approached in the following ways:

① Component local verification: implementation of all the functions declared.
② Component global verification: references to other components functions.
③ All events are covered by policies rules.
④ Conflicts and infinite loops detection.

4.4 Document Generator

The last tool of MOIRAE environment generates documentation based on HTML and LaTeX files, that describes services, commands and other component elements as well as it's implementation to the detail level defined in the *.moi* files.

This module reports all the information related to components in a more bright and compact way. In a future, this mechanism could be extended to generate component description tags to be processed by other development environment or registering services.

5 Component Code Generation

The source file of a component description is the *.moi* file. Using this file the Code Generator creates several C++ files that includes subclasses inherited from the MOIRAE component library classes. All the code generated deals with communication, control and registration operations. The actual operations (the tasks performed by the component) should be programmed as different files that will be compiled when the executable file is linked. A set of auxiliar files required for the compilation and execution of the component are also generated by this mechanism. These files are a default Makefile and several scripts to start/stop component execution. The *.moi* file has a source code according to the MOIRAE component description syntax. This description represents a component definition and it has to be provided by the developer.

The Code Generator module (presented above), uses the IR (Internal Representation) structure generated by the MOIRAE Parser. Figure 3 shows the complete sequence of the code generation procedure. In this figure, it can be seen that Code Generator uses IR class to generate the result set of files. Code Generator obtains component description from IR structure and generates subclasses that inherit from de MOIRAE library.

5.1 MOIRAE Library

MOIRAE library is a set of C++ classes each of which represents part of the structure and functionalities of a generic MOIRAE component. For example,

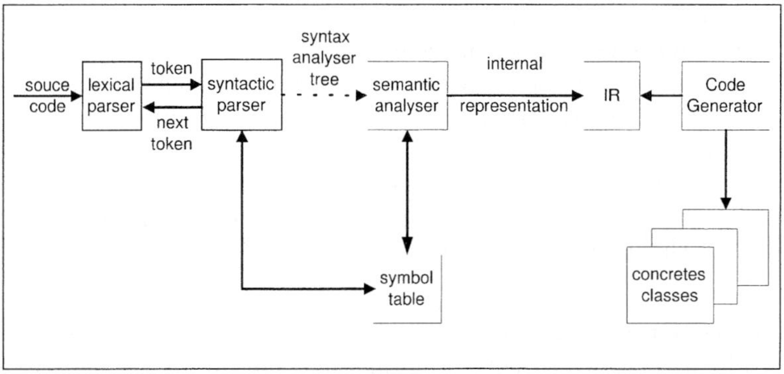

Fig. 3. Relationships between Code Generator, IR, and MOIRAE Parser

there is a class that represents the 'Operational Interface' of a MOIRAE component, as well as the 'Control Plane', 'Event Sensors' and so on. Using the attributes and methods contained in library classes, the features provided by a generic MOIRAE component can be obtained.

Code Generator creates a set of classes inherited from library classes, these new classes are intended to contain the specific functionalities of each of the components user specifies.

MOIRAE library includes the following classes:

- The different ways of communication between two agents (sockets or KQML), that is, between different control planes of several components.
- The different kinds of information that 'Policies Engine' and 'Control Plane' need to exchange.
- The different kinds of policies engines (reactive, deliberative ...)
- The basic skeleton of an operational or command procedures implementation, and so on ...

One of the advantages of the library is that it is independent from the final technology the user selects for the component. Component Compiler must generate code appropriate technology selected.

MOIRAE library presents different hierarchies for those elements who need a detailed level of particularization. It is the case of policy engines and communication mechanisms between agents. It does not matter the kind of policy engine, all of them work in the same way because they provide the same methods inherited from class at the top level.

5.2 Parser + IR

In this section we will explain the two auxiliar components used by Code Generator, MOIRAE Parser and IR structure.

MOIRAE Parser. MOIRAE Parser makes lexical, syntactic and semantic analysis of a MOIRAE component definition.

This component has two elements. The first one (called Scanner), performs lexical analysis of the MOIRAE file, this file has *.moi* extension. The second one, named Parser, contains the syntactic and semantic analyser.

In general, MOIRAE Parser has as an input file with *.moi* extension (the definition of the component to be implemented), and as an output, stores all useful data in the internal storage structure named IR.

IR (Internal Representation). The IR, is a intermediate storage structure, composed by lists, which stores all information from MOIRAE files. Each list is a set of instances of classes which representes concepts like services, commands, events or relationships of a MOIRAE component. IR is generated by MOIRAE Parser, and it is used as a bridge tool between Code Generator and Parser.

In addition, IR structure is requested by the rest of the MOIRAE tools. Once the MOIRE Parser has been executed the IR is taken as the validated representation of the component. An IR structure stores all information from the component definition files. This representation is not a file format rather than a memory structure to publish component description for other MOIRAEToolKit modules.

All the modules of the system are linked with the MOIRAE Parser code in order to call it to obtain an IR representation from a *.moi* file.

6 Example

This example shows how a component named `Account` is developed using the MOIRAEToolKit environment. `Account` component implements the typical functions that we can do with a bank account like deposit and withdraw an amount of money, and gets actual balance of the account.

The next code, represents a MOIRAE definition of the Account component, following MOIRAE syntaxis. As part of the design this component has a control event `redNumbers`, that the component will throw when the balance of account is negative.

```
// Set middleware CORBA_MICO
#option middleware = CORBA_MICO
component Account {
  // Put implementation class
  #option imple_modulo_op = "Implementation"
  events { // Events Declaration
     redNumbers() ; }
  services { // Services Declaration
     void deposit ( long amount );
     void withdraw( long amount ) throw (redNumbers);
     long balance (); }
}
```

The implementation class has the implementation of the three services (`deposit`, `withdraw` and `balance`). Once the *.moi* file is programmed, the code generator program is ran with the sentences shown in:

```
[jdoe@server] moirae Account.moi
[jdoe@server] make
[jdoe@server] Account myAccount accounts
```

When we typed the first sentence `moirae` program has generated the files shown in the next table.

Generated Files	
Files depending of CORBA middleware	
Account.idl *Main.cpp*	*Account_impl.hpp/cpp*
Files inherited from Moirae Library	
Component_Account.hpp/cpp	*ControlPlane_Account.hpp/cpp*
OperationalPlane_Account.hpp/cpp	*ControlInterface_Account.hpp/cpp*
OperationalModule_Account.hpp/.cpp	*PoliciesEngine_Account.hpp/cpp*
OperacionalInterface_Account.hpp/cpp	
EventSensor_redNumbers_Account.hpp/cpp	
Files depending of program language, C++	
Makefile	*MakeVars*

The Makefile generated is used by the `make` tool. The second sentence shows how the `Account` executable is built. The third sentence is needed to run the component. The first parameter specifics the name of the instance, the second parameter specifics the context name where the instances are registered. Now, *myAccount* instance is running on the server side. On the client side, we run a C++ program which uses the Account component.

```
[jwhy@client] Client myAccount accounts
```

This program uses a reference of *myAccount* object. The output of the client program is the set of next sentences: `Deposit: 700 euros`
`Balance: 700 euros`
`Withdraw: 600 euros`
`Balance: 100 euros`
`Withdraw: 150 euros`[1]
`Exception: MOIRAE`[2]

The trace message printed by the server is: `Throw redNumbers event`[3]

[1] At this moment `myAccount` component throws `redNumbers` event, and its control plane captures the event and deals with the procedures described by the policy rules. One of the possible actions proformed by the control plane could be user notification using a remote exception, other possible actions are accept the negative balance or report to higher level component about possible penalties.

[2] Client captures MOIRAE exception and prints the message "Exception: MOIRAE".

[3] The object server named `myAccount` throws `redNumbers` event: Control plane summoned.

7 Conclusion and Future Lines

MOIRAE is a new generic architecture for distributed component environments. One of the key elements of this new architecture is the possibility to deal with distributed control policies in order to provide hierarchical management facilites. Using simple, but flexible, distributed control mechanisms, it is possible to change during runtime the behaviour and control procedures the system is ruled by.

MOIRAEToolKit is a development environment designed to create MOIRAE components. Using its tools a new component can be defined using a description syntax. Code Generator tool uses this component description to create the code wrapper for the services described by the component. Developers only need to implement the remote services provided by the component, all the requirements in terms of communication, component registration and advanced control features are generated by the tool.

Policy Verifier and Document Generator are both tools from the same suit that checks policy rules against the component definitions and generates documents with component description.

At the moment, the current version of the MOIRAEToolKit provides only very simple control plane implementations. This part is the main development line of this project. A second line is the extension of the validation mechanisms in order to validate complex situations like event domain edge conditions.

Other minor issues are to extend the number of options that users can specify for their components. It would be a good idea to provide as much as possible options, about providing different middlewares, policies engines, communication methods and languages. At the present moment only C++ and CORBA objects are supported. This suit can be downloaded from HPDA project web site[4].

References

1. Object Management Group. Ccm tutorial. Document: omg/00-06-01, June 2000.
2. José M. Peña. Distributed Control Architecture for Data MiningSystems. PhD thesis, DATSI, FI, Universidad Politécnica de Madrid, Spain, June 2001. Spanish title: Árquitectura Distribuida de Control para Sistemas con Capacidades de Data Mining'.
3. José M. Peña, F. Javier Crespo, Ernestina Menasalvas, and Victor Robles. Parallel data miningexep erimentation using flexible configurations. LNAI, 2475:441–448, 2002.
4. José M. Peña and Ernestina Menasalvas. Towards flexibility in a distributed data mining framework. In Proceedings of ACM-SIGMOD/PODS 2001, pages 58–61, 2001.
5. Dale Rogerson. Inside COM: Microsoft's Component Object Model. Microsoft Press, 1997.
6. Sun Microsystems. Enterprise JavaBeans 2.0 specification. Whitepaper, Sun Microsystems, 1999.

[4] HPDA Project: `http://nova.ls.fi.upm.es/hpda`

Applying Computational Science Techniques to Support Adaptive Learning

Juan M. Santos, Luis Anido, Martín Llamas,
Luis M. Álvarez, and Fernando A. Mikic

E.T.S.E. Telecomunicacións
Campus Universitario S/N
36200, Vigo (Pontevedra), Spain
{jsgago, lanido, martin, lmsabu, mikic}@det.uvigo.es

Abstract. Adaptive Learning Systems offer customized learning experiences according to the actual student needs and capabilities. Effective student modelling, adequate representation of the knowledge domain and proper characterization of learning tools are key issues to provide high quality Adaptive Learning Systems. Most current systems are based on Artificial Intelligence techniques (e.g. fuzzy logic, neural networks, Bayesian networks, etc.) trying to reproduce human teaching behaviours by using a computational representation of expertise. This paper offers a survey on Adaptive Learning showing how Computational Science techniques are applied to instructional systems and identifying forthcoming trends for the future.

1 Introduction

Information and Communication Technologies (ICT) have been applied to the field of Education during the last years. Advantages of training supported, either completely or partially, by new technologies (commonly called *e-learning*) have been broadly discussed in many forums. Being aware of these advantages, many institutions both public and private have adopted e-learning in their training and instruction departments.

Web-based e-learning systems have progressively evolved from basic repositories of simple HTML documents to complex learning environments that include tools enabling different educational paradigms like “learning by doing” or “collaborative learning”. However, current commercial e-learning systems do not completely make use of all the potential that can be provided by New Technologies. Particularly, most of them do not offer students with learning experiences that are unique and tailored to their needs, interests, preferences, learning style and working environment in order to maximize the effectiveness of learning. Currently, few experimental platforms deal with personalisation and adaptation to each particular student. These platforms, called Adaptive Learning Systems (ALSs), are the result of the effort from researchers that combine two traditionally distinct areas: Instructional Science and Computational Science [1].

P.M.A. Sloot et al. (Eds.): ICCS 2003, LNCS 2658, pp. 1079–1087, 2003.

This paper presents a survey on Adaptive Learning, showing how Computational Science techniques and methods are applied to instructional systems and identifying forthcoming trends. Section 2 deals with the basic concepts involved in the field and presents a common abstract model for ALSs. In Section 3 some examples of currently available ALSs developed by different researching groups are discussed. Finally, Section 4 presents some conclusions and trends for the future.

2 A Conceptual Model for Adaptive Learning

According to [2], one of the main reasons to provide adaptation in Web-based e-learning systems is the great variety of users involved. In traditional educational environments, teachers apply a common pedagogical method regardless of the possible heterogeneity among their students. Most conventional systems assume that all the students in the same classroom have acquired the same level of knowledge so far and are prepared to acquire new knowledge at the same level. This is clearly arguable and definitely false in most Internet-based e-learning environments where students may have different background, availability to follow the courses and even a different language and culture.

This situation led researchers in the Artificial Intelligence area to develop Intelligent Tutoring Systems (ITS), which are software tools designed and programmed to "intelligently" reproduce human teaching behaviours by using a computational representation of expertise in instructional methods. For this, they need to model the knowledge domain in question and the learner skills and capabilities. Other aspects subject to adaptation include user interfaces, special requirements for disabled people, etc.

From a Conceptual point of view, an ALS is composed of the *Student Model*, the *Domain Model*, the *Adaptation Model*, the *Adaptation Engine*, and, in some modern systems, a *Environment Model* (c.f. Fig. 1):

- **Student Model:** The Student Model stores relevant information about a particular student and can include, depending on the specific system, his/her personal data, individual preferences, learning style, cultural facets, possible disabilities and background experience and current capabilities for a particular knowledge domain.
- **Domain Model:** The Domain Model is the repository for storing and structuring the learning contents and the overall knowledge on a particular domain: involved concepts, relevant properties from a pedagogical point of view to study, requirements, goals, etc.
- **Environment Model:** The Environment Model includes a description of the capabilities of the hardware devices and software applications used by the student in a particular learning session. It can be used to determine the most appropriate form of a resource to be delivered to each particular student's equipment. This model provides ubiquitous access [3] to the learning environment.

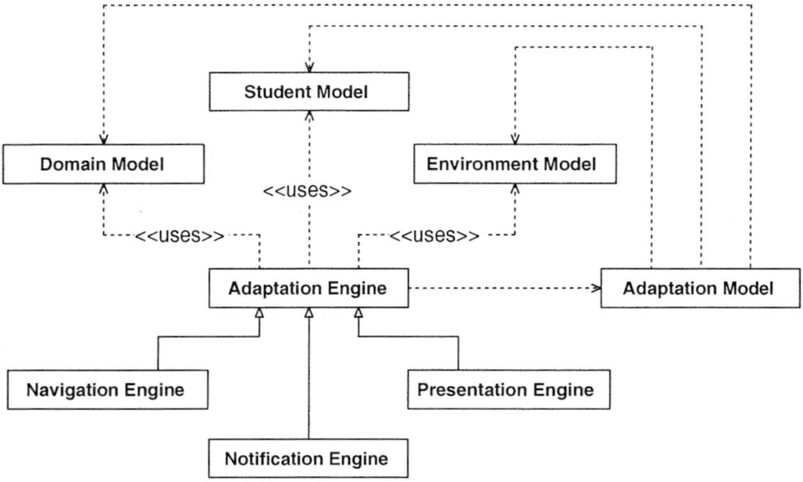

Fig. 1. Conceptual Model for Adaptive Learning Systems

- **Adaptation Model:** The Adaptation Model contains the specific inferential rules that define how the User Model, Domain Model, and Environment Model elements are combined to provide the actual adaptation.
- **Adaptation Engine:** The Adaptation Engine is the software process that combines information from the Student, Domain and Environment models and the actual learner interaction with the system in order to personalize the learning process for appropriate learning experiences. It is guided by the rules described in the Adaptation Model. In essence, it is an inference engine, i.e. an automatic control mechanism that applies the axiomatic knowledge present in the knowledge base to arrive at some conclusion: the adequate learning contents to be delivered to the learner.

3 Computational Science Techniques in ALS

The development of Adaptive Learning Systems is a highly complex activity. Architects of these sophisticated environments must address with complicated issues like effective student modelling, adequate representation of the knowledge domain, proper characterization of the learning environments or identification and design of appropriate and flexible pedagogical control rules. Many times, it is very difficult, or even impossible, to exactly deal with some of these problems. For example, how can be exactly represented the knowledge level of a student in a particular subject? or, even more, how can be represented such an abstract concept like the student motivation?. Futhermore, the computational complexity of the algorithms needed to automatically take decissions (e.g. which is the most appropiate next lesson for a particular learner) or "on-the-fly" adjust the different models from the user interactions with the system are difficult issues to deal with.

Fortunately, it is not necessary to design perfect models and algorithms to get useful adaptive outcomes. For example, it is often better an algorithm that recommends interesting paths in a course organization with a predictive accuracy of 80% over an algorithm that achieves 90% if the former requires considerably less CPU time. Modern Adaptive Learning Systems are almost always based on some Artificial Intelligence (AI) technique, like Fuzzy Logic, Neural Networks or, about all, Bayesian Networks.

Bayesian Networks (BNs) are directed acyclic graphs where nodes represent variables and arcs represent probabilistic dependence among variables. These networks, and several variations of them, have steadily been applied in ALSs [4]. They are usually used to represent the student model. Depending on the particular ALS, the variables can be asociated with "concepts", "problems", "abilities", etc, and they are linked by relationships between them, such as "part-of", "prerequisite-of", "evaluation-item-of", etc. A typicall usage setting is showed in Fig. 2.

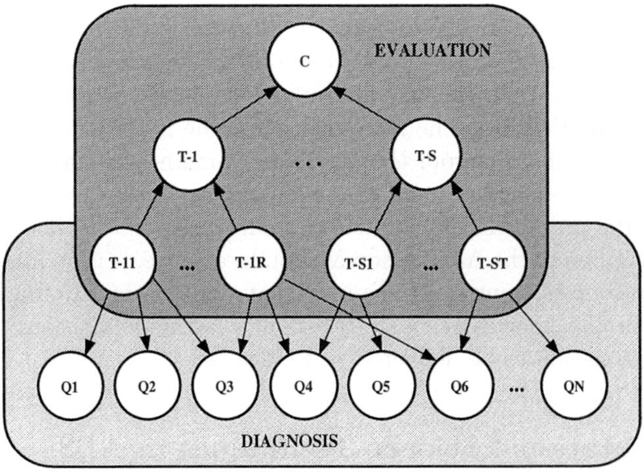

Fig. 2. Use of Bayesian Networks in Adaptive Learnin Systems

In the BN of the figure, upper nodes (*Knowledge Nodes*) are associated with the topics making up a course, arcs determine the relationship among the different topics and variables of each node indicate the likelihood of the student "knowing" that piece of the course. Bottom nodes (*Evidential Nodes*) are associated with the test items used to diagnose the knowledge level about the particular topics. Relationships among bottom and upper nodes represent the probabilistic influence of the test items on the Knowledge Nodes.

To characterize the network it is necessary to define its parameters, i.e., the set of discrete conditional probability distribution of each variable given its parents:

$$P(Xi|pa(Xi)), i = 1, ..., n$$

where Xi are the variables and $pa(Xi)$ represents the set of the parents of Xi. This set of probabilities defines the joint probability distribution for the entire network as:

$$P(X1, ..., Xn) = \prod_{i=1}^{n} (Xi|pa(Xi))$$

These networks are updated in real time, using some approximation algorithm, as the student interact with the learning platform. The information of the student given by the BN is used by the ALS to make different decisions, depending on the particular system. For example, if the BN determines that the student does not understand a topic, a short mini lesson explaining the involved concepts could appear on the user screen.

In the next sub-sections we briefly describe some developed systems, showing the different techniques or approaches used to deal with the different aspects of adaptation.

3.1 ELM Adaptive Remote Tutor (ELM-ART)

ELM-ART [5] is an intelligent interactive educational system to support learning programming in LISP. For a student, the system can be viewed as an interactive text-book with support for several types of adaptive navigation, interactive examples and automatic intelligent selection and evaluation of exercises (simple questions and programming problems to solve). It also supports additional functionalities, like a glossary of LISP constructs, a personal annotation tool, a portfolio viewer (where the student can see the set of all analyzed examples and all the solutions of solved problems) or several communication tools.

ELM-ART represents the domain knowledge (Domain Model) in terms of a conceptual network, where units to be learned are organized hierarchically into lessons, sections, subsections and terminal pages (units). Each unit is represented as an object containing slots for the text to be presented to the learner and for information that can be used to relate units to each other (prerequisites, related test items or inference links).

The Student Model is represented as an Overlay Model [6] (where the student' knowledge is considered as a subset of the expert' knowledge) arranged in four layers. The first layer, *Visited State*, describes whether the user has already visited a unit; the *Learner State* layer contains information on which exercises related to a particular unit the user has worked at and whether he/she successfully mastered them; the *Inferred State* layer describes whether a unit could be inferred as known via inference links from more advanced units the user already worked successfully; at last, the *Known State*, describes whether a unit has been marked by the learner as already known. Information in the different layers of the learner model is updated automatically and independently during each interaction with the system. Additionally, students are provided with a tool for inspecting and partially modifying his/her current model.

These models enable both the individual curriculum sequencing and the personalized visual annotation of links. ELM-ART also supports user interface personalization according to the preferences revealed by the students.

3.2 Intelligent Distributed Environment for Active Learning (IDEAL)

IDEAL [7] is an intelligent agent-based environment for active learning. The system consists of a number of specialized agents with different expertise. Each student is assigned a unique Personal Agent that manages the student's personal profile (Student Model) including knowledge background, learning styles, interests, courses enrolled in, etc. The Course Agents manage course materials and course-specific teaching techniques for a course. At last, the Teacher Agent interacts with a student and serves as an intelligent tutor of a course. The basic components of a teaching agent are a domain expert module (it creates exercises and questions according to the student's background and learning status, provides solutions and explains the concepts and solutions to cope with student's misconceptions), a pedagogical module (it is a rule-based production system that uses the student model and pedagogical knowledge to determine the appropriate actions) and a student modeller (it provides a model of a student based on his/her learning style, knowledge background and interests).

In IDEAL, the student level of expertise is inferred from the performance of students on exercises and quizzes using a Bayesian belief network. This model incorporates uncertainty to take into account the probability of slips (students sometimes miss questions that they should know) and lucky guesses (correct answers to questions of higher level than student's level). The measure of how well a skill is learned is represented as a probability distribution over skill levels, such as novice, beginning, intermediate, advanced and expert. The conditional probability of skill levels is as follows:

$$P(X = x_j|Q) = \frac{1}{P(Q)} * P(X = x_j) *$$

$$(1-s)^{\sum_{i=1}^{j}(q_{i+})} * s^{\sum_{i=1}^{j}(q_{i-})} * g^{\sum_{i=j+1}^{n}(q_{i+})} * (1-g)^{\sum_{i=j+1}^{n}(q_{i-})}$$

where X represents the skill levels; Q is the evidence vetor of n elements, in which each element q_i contains two numbers, e_{i+} and e_{i-}, corresponding to the number of correct and incorrect answers to questions at difficulty level i respectively; s is the probability of a slip; and g is the probability of a guess.

The system incorporate a proprietary approach to course content organization and delivery, which is developed based on smart instructional components, called *lecturelets*. Lecturelets contain both the learning resources (concept descriptions, examples, quizzes, etc.) and the instructions on how the resources should be processed or displayed.

3.3 Self-Paced and Adaptive Courseware (SAC) System

SAC [8] is a self-paced and adaptive courseware system developed at the Hong Kong Polytechnic University. This system provides dynamic navigational guidance to students taking online courses. It has been designed and implemented on a three-tier web application architecture which uses the Adaptive Hypermedia Application Model (AHAM) [9], a widely used reference model in Adaptive Hypermedia [10] systems.

The AHAM model is composed by tree sub-models:

- The Domain Model describes how the information is structured and linked together in term of concepts and concept relationships.
- The User Model is (conceptually) a table which associate each concept in the Domain Model with a set of attributes. Most of the implementations of this model include the attributes:
  - *Knowledge Value:* it indicates how much the user knows about the concept.
  - *Read:* It indicates whether the user read something (a fragment, a page or a set of pages) about the concept.

  A less common attribute would be *Ready To Read*, which indicates whether the user is ready to access this concept.
- The Teaching Model is a set of pedagogical rules which define how the domain model and the user model are combined to provide ways to perform the actual adaptation. These rules are used also to determine how to compute the attributes of the Domain Model. As an simplified example, the following rule:

  $$< \text{access(C2) AND C1.read=true} \Rightarrow \text{C3.ready-to-read=true, true} >$$

  expresses that when Concept2 is accessed and Concept1 is already read, the Concept3 can be accessed. The *true* parameter points out that more rules can be processed.

4 Conclusions and Future Trends

In spite of the great advantages reported by the researching community, as was mentioned in Section 1, at present, Adaptive Learning Systems have failed to gain widespread acceptance outside the researching laboratories. Thus, the many e-learning platforms that predominate in the current market (such as WebCT [11], IBM LearningSpace [12] or Blackboard Learning System [13]), used by thousands of educational providers, do not offer the possibilities of adaptation or personalization depending on the learner's characteristics, or they are very limited. Among the several reasons adduced by the scientists (c.f. [14]), one is widely cited and agreed:

Courses for ALSs are costly to build, needing among 100 and 1000 hours to produce an hour of instructional material.

Solving this problem will be an enormous breakthrough in ALS research. However we want to stand out another critical factor that will determine the final acceptance of these systems: Recently, several organizations and institutions have been thoroughly working towards the development of standards and recommendations aimed to solve the interoperability problems currently found in the e-learning domain [15]. It is foreseeable that near future educational platforms will be developed taken into account the results from this standardization process. Thus, it is essential that both ALS researchers to bear in mind this standardization process and standardization bodies to consider the issues implicated in ALS in order to guaranty the utilization of adaptation techniques.

First proposals from the e-learning standardization process are currently in scene and some manufacturers have begun to adopt them. However, they are not used at all in Adaptive Learning Systems. The reason of this fact resides, on the one hand, in the dedication of ALS researchers to solve particular teaching and pedagogical problems, whithout considering questions such as compatibility. On the other hand, current proposals present some troubles for using them in adaptive environments, being the most outstanding the lack of integration among them (see [16] for a more detailled discusion about this).

Acknowledgments. We want to thank "Ministerio de Ciencia y Tecnología" for their partial support to this work under grant "CORBALearn: Interfaz de Dominio guiada por Estándares para Aprendizaje Electrónico" (TIC2001-3767).

References

1. Jones, M., Greer, J., Mandinach, E., du Boulay, B., Goodyear, P.: Synthesizing Instructional and Computational Science. In: M. Jones & P. Winne (eds.): Adaptive Learning Environments: Foundations and Frontiers, pp. 383–401, Berlin: Springer-Verlag (1992)
2. Weber, G.: Adaptive Learning Systems in the World Wide Web. In: J. Kay (Ed.): User modeling – Proceedings of the Seventh International Conference (UM99), pp. 371–378. Vienna: Springer (1999)
3. Dillsus, D., Brunk, C.A., Evans, C. Gladish, B., Pazzani, M.: Adaptive Interfaces for Ubiquitous Web Access. In: Communications of the ACM, Vol 45, No 5. pp. 34–38 (2002)
4. Millán, E., Pérez-de-la-Cruz, J.L.: A Bayesian Diagnostic Algorithm for Student Modeling and its Evaluation. In: User Modeling and User-Adapted Interaction, vol. 12, pp. 281–330, Kluwer Academic Publishers (2002)
5. Weber, G., Brusilovsky, P.: ELM-ART: An adaptive versatile system for Web-based instruction. In: International Journal of Artificial Intelligence in Education, vol. 12, pp. 351–384 (2001)

6. Van Lehn, K.: Student modelling. In: Foundations of Intelligent Tutoring Systems. Hillsdale, NJ: Lawrence Erlbaum Associates Publishers, pp. 55–76 (1988)
7. Shang, Y., Shi, H., Chen, S.: An Intelligent Distributed Environment for Active Learning. In: ACM Journal of Educational Resources in Computing, vol. 1, issue 2, pp. 308–315 (2001)
8. Chan, A., Chan, S., Cao, J.: SAC: A Self-paced and Adaptive Courseware System. In: Proceedings of the IEEE International Conference on Advanced Learning Technologies (ICALT2001), Wisconsin, U.S.A., IEEE Press (2001)
9. De Bra, P., Houbent, G., Hongjing, W.: AHAM: A Dexter-based Reference Model for Adaptive Hypermedia. In: Proceedings of the 10th ACM Conference on Hypertext and Hypermedia, Darmstadt, Germany, ACM Press (1999)
10. Brusilovsky, P.: Adaptive Hypermedia. In: User Modeling and User-Adapted Interaction, vol. 11, issue 1–2, pp. 87–110, Kluwer Academic Publishers (2001)
11. WebCT web site at http://www.webct.com [last accesed January 1th 2003]
12. IBM' LearningSpace web site at http://www.lotus.com/products/learnspace.nsf/wdocs/homepage [last accesed on January 1th 2003]
13. Yaskin, D.: Blackboard Learning System (Release 6). Product Overview White Paper. Blackboard Technical Report. Online available at http://products.blackboard.com/cp/release6/LSR6WP.pdf
14. Woolf, B.P., Regian, J.W.: Knowledge-based training systems and the engineering of instruction. In: Training and retraining: A handbook for business, industry, government and the military, pp. 339–356, Macmillan Reference (2000)
15. Santos, J., Caeiro, M., Rodríguez, J., Anido, L.: Standardization in TelE-learning. A Critical Analysis. In: TelE-Learning. The Challenge for the Third Millenium. 17th IFIP World Computer Congress, pp. 321–328, Montreal (Canada). Kluwer Academic Publishers (2002)
16. Santos, J.M., Anido, L., Llamas, M., Rodríguez, J.S.: On the Application of the Semantic Web Concepts to Adaptive E-Learning. In: Shafazand, M. H.; Tjoa, A M. (Eds.) EurAsia-ICT 2002: Information and Communication Technology, Lecture Notes on Computer Science, Vol. 2510, pp. 536–543 (2002)

The Use of the Cooperative Solver SibCalc in Modeling of Complex Problems

Tamara Kashevarova and Alexander Semenov

Institute of Informatics Systems of the Russian Academy of Sciences
pr.ak. Lavrentieva, 6 Novosibirsk, Russia, 630090
{toma, semenov} @iis.nsk.su

Abstract. This paper describes the cooperative solver SibCalc that allows us to solve nonlinear algebraic problems with inexact and subdefinite data. Then, two examples of applying the solver to complex modeling problems are considered.

1 Introduction

In the mathematical modeling, we often need to study the processes described by complex models with incomplete and inaccurate data. As a rule, such models are non-closed and represent a set of equations, inequalities and logical expressions with integer and real variables. The values of the model parameters are often of a probabilistic nature or given by intervals of admissible values. Along with solving the problems with interval data, another difficult task of modeling is to solve the inverse problems: to find the model parameters, such that its complex indices should lie in some predefined ranges. Therefore, the development of effective and reliable methods for solving a problem with inaccurate and subdefinite data is an actual problem of modeling.

One of the approaches to solution of these problems is interval analysis [1], [2] successfully applied to solving a wide range of problems, both in interval and non- interval (classical) statement. However, this approach often proves to be not applicable to problems in non-traditional statement. At present, the artificial intelligence methods, such as constraint propagation [5], [9], are proposed to solve them. However, none of these methods can solve a complex problem completely, so the combination of such methods may be required. Effective application of several methods requires an appropriate software environment which allows us to manage the process of computation with the use of all accessible resources. One of such environments is the cooperative solver that we have created for investigation and solution of a wide range of modeling problems. The solver and the examples of applying it to specific problems of modeling are the theme of this work. In the first part we describe the basic methods and the software that we have developed on the basis of these methods. In the second part we give examples of applying the solver to design and financial-economic problems.

P.M.A. Sloot et al. (Eds.): ICCS 2003, LNCS 2658, pp. 1088–1097, 2003.

2 The Cooperative Solver SibCalc

The solver SibCalc [3], [7] is intended for solving the computational problems presented by the systems of algebraic equations and inequalities. Along with the problems in the "classical" statement (when the number of variables is equal to the number of equations), our solver makes it possible to solve arbitrary systems with the integer and real variables. Besides, it allows us to superpose additional conditions in the form of inequalities and logical relations. Another specific feature of SibCalc is its capability to work with inaccurate data given as intervals. Further we will describe some of the approaches used in the solver.

2.1 Interval Mathematics

The methods of interval mathematics that are now widely used possess a number of attractive properties, among which is the possibility to work with domains and to perform the guaranteed computations excluding loss of solutions and incorrect results caused by accumulation of roundoff errors. Let us give a review of the basic concepts and characteristics of interval mathematics. A finite connected subset of $\mathbf{R}$ denoted as $\mathbf{I} = [x_l, x_u]$, $x_l \leq x_u$, is called a real interval. x_l and x_u are called a lower and an upper boundary of the interval $\mathbf{I}$, respectively. The interval which does not contain any element is called empty. We will dwell on the case of closed intervals only, thus considering that the boundaries belong to an interval. If the lower boundary is equal to the upper one, the interval contains only one number and is called degenerate. Let $\mathbf{I_1}$ and $\mathbf{I_2}$ be two intervals. We define an interval extension of the arithmetic operations $\{+, -, \times, \}$ as follows:

$$\mathbf{I_1} \otimes \mathbf{I_2} = \{x = x_1 \otimes x_2 | x_1 \in \mathbf{I_1}, x_2 \in \mathbf{I_2}\}, \quad \otimes \in \{+, -, \times, /\}.$$

Note that for a nondegenerate interval $\mathbf{I}$, $\mathbf{I} - \mathbf{I} = [x_l - x_u, x_u - x_l] \neq [0, 0]$.

It is possible to analogously determine the interval extensions of functions. Let the function $f(x)$, $x \in \mathbf{R}$, be given. Then its interval extension will be $\mathbf{F(I)} = \{x = f(x) | x \in \mathbf{I}\}$ The interval extension of functions is, in its essence, the two-sided evaluation of the set of their values. As a rule, to obtain this evaluation, we make use of the **natural interval extension**, when all arithmetic operations are substituted by their interval extensions, while the interval values built in a special manner are used for the elementary functions. Computation of the natural interval extension usually results in the too wide boundaries. More precise but also more labor-consuming algorithms are applied to eliminate this effect.

One may notice that if the divisor in the operation of division contains zero, it will give the indeterminate result, which may lead to a loss of solutions. It is possible to avoid indeterminacy by introduction of the **extended interval arithmetic**. In this arithmetic, the symbol ∞ denoting "infinity" is introduced, and the intervals $[-\infty, a]$ and $[b, +\infty]$ denoting, respectively, all numbers smaller than a and all numbers larger than b are allowed. Then the interval $[-\infty, +\infty]$ denotes the set of all real numbers and can be considered as the result of division with the divisor containing zero. Infinite intervals also make it possible to construct interval extensions of functions with singular points and indefinite on some sets. Further on, only the extended interval arithmetic will be used.

Similar to the one-dimensional case, we can define multidimensional intervals (called interval vectors or boxes) and operations with them. A box is defined as a subset of $\mathbf{R}^n$ and is determined by a Cartesian product of n one-dimensional intervals $\mathbf{I}^n = [x_{l1}, x_{u1}] \times ... \times [x_{ln}, x_{un}]$.

The basic purpose of interval mathematics is to guarantee computations. This means that the mathematical result of any operation should lie within the interval presenting the result of this operation over the interval operands. It is reached by means of directed roundings, which allow us to obtain the machine-represented numbers greater or smaller than the result of machine operation (which generally can be not machine-represented). It is still more complicated to obtain the precise interval estimations of elementary functions, which are computed by expansion into a power series. Thus, to support interval computations, there is a need for effective implementation of the corresponding mathematical library, which ensures a precise and fast computation of the interval extensions of arithmetic operations and elementary functions.

The methods of interval mathematics allow us to effectively solve a number of problems, for example, the systems of nonlinear equations and the problems of global optimization [2]. In both cases, the interval version of Newton's algorithm is used. The algorithm makes it possible to find solutions in the form of boxes of the required width that contain the solution. Another extremely important property of this algorithm is its conclusiveness. The algorithm allows us to throw away domains which do not contain a solution (when intersection in the above formula gives an empty interval), as well as to prove that some domains contain solutions with guarantee; in some cases it proves that the solution is unique.

2.2 The Constraint Propagation Methods

Another approach, on which the computational part of the solver is based, is the constraint propagation. This technology, widely used for solving a broad class of problems, consists in rejection of domains that do not contain solutions a fortiori, and in a specially organized search in the remaining domains. Ideologically, this is close to the interval methods.

This approach was initially elaborated in the works on artificial intelligence for solving the problems over finite domains. The basic idea of the approach is establishment of local consistencies of various kinds. In the simplest case, only unary and binary relations connecting certain sets of variables are examined. The set of values of the variables belonging to a certain relation is called consistent, if the relation on this set of values is true. There are several forms of local consistencies (node-, arc-, path-, etc.), to obtain which the corresponding algorithms are proposed (for example, $AC-3$, $AC-4$, $PC-2$). These algorithms remove the values, which cause inconsistencies. Algorithms of this type are widely used in the problems of combinatory optimization, planning, scheduling, etc.

Another class of the constraint propagation methods is the algorithms over continuous domains. The intervals in these algorithms are considered as the domains of variables, and subintervals are removed during achieving local consistencies. Relations in the problems solved by this set of methods are equations and inequalities that are usually given as formulas or tables. In the constraint propagation methods on continuous domains, consistencies of special types - box-consistency, hull-consistency, bound- consistency - are determined. Consistency of every type is obtained by means of various algorithms, all of which use the methods of interval mathematics. So, the Newton one-dimensional interval method is used to establish the box-consistency, while for hull-consistency a partition of complex relations into the primitive (unary or binary) ones is used [6]. The methods of obtaining the local consistencies generally give similar solutions, but for different systems they may differ in computation time and widths of the boxes obtained. Besides, box-consistency can be used for real variables only, while achievement of hull-consistency can be also used for integer variables (assuming that a library for integer intervals is available), as well as for the relations on the variables of different types.

Another important component of the constraint propagation methods is the search for solutions in the obtained domains of smaller sizes. As a rule, the algorithms on the basis of the depth-first search (e.g., multilevel bisection), as well as a combination of these algorithms with the algorithms of achieving of local consistencies, are used for these purposes. To solve optimization problems, the branch and bound method is widely applied, as well as its specialized versions, such as the branch and prune method, and the methods based on local searches.

2.3 The Architecture and Capabilities of the Solver SibCalc

It is known that a really complex problem, especially in modeling, cannot be completely solved by only one method. Based on this, we have developed a cooperative solver SibCalc, which includes various methods and contains means for cooperative solving of problems. By **cooperative solving** we mean joint solution of the parts of an initial problem by different methods, where each method can transfer the results of its computations to other methods.

The solver SibCalc has its input language to record models of any complexity and a set of methods to solve the problems. An initial model is recorded in the language of model description. Then a translator converts the model from this language into **the universal internal representation**, which is then used practically by all methods of the solver and serve as a means of communication between the methods. The universal internal representation is a tree-like structure that is stored in the memory and has a clearly specified program interface.

The method library is a dynamic library of independent components which are used to organize the process of computations. All methods of the library have a unique interface, that simplifies the development and application of new methods. In particular, the interface allows us to the initial data, to obtain results, to launch a method, etc. Each method has a set of specific attributes, by which it is identified in the library of methods, and a set of variable attributes to control the method operation. At present, our library includes:

- algorithms of constraint programming over finite domains;
- algorithms of interval constraint programming over continuous domains;
- the Newton interval method;
- methods of solving systems of interval linear equations;
- methods of linear programming (an interval version of interior point method);
- a large set of search methods over continuous and finite domains, in particular, modifications of the branch and bound method for different domains;
- specialized methods of solving problems of nonlinear optimization;
- automatic differentiation;
- other special methods.

It should be noted that many methods use the interval library [4], specially developed for the solver, which implements arithmetic operations and computation of the interval extensions of elementary functions efficiently and with a high accuracy. It also includes the integer interval library.

Computation diagrams are used to define the ways of cooperative solving of problems. A computation diagram describes which methods will be used to solve a problem and how they interact with each other. So, it completely determines how to solve the problem, as it contains information on the set of methods in use, the order of their calls, the mode of data exchange between the methods. Methods are connected by a **channel** through which the data and control commands are transmitted. Channels can encapsulate a net transport protocol that automatically allows us to transfer the solver to the distributed architecture without any need for adaptation of the methods themselves and the algorithms. Computation diagrams can be created by users to solve specific problems, and diagram libraries can be developed to solve certain classes of problems.

3 Solution of the Modeling Problems

Let us consider application of the solver SibCalc to investigation and solution of problems that can appear in modeling. As mentioned before, SibCalc is a very powerful tool of work with complex non-closed models with inaccurate data, since it makes it possible not only to find the solutions, but also to perform qualitative research of the solution behavior. We consider application of SibCalc to two nonlinear mathematical models describing real-life problems.

3.1 The Problem of a Preliminary Design

The first example is the problem of a preliminary design. The first stage of a complex product design is a preliminary design, when the fundamental characteristics of the product are determined. At this stage, the majority of the parameters are either unknown at all, or given by a range of admissible values, and selection of values of some parameters gives us the precise values of other parameters. Parameters of models can be real (weight, size), integer (the number of passengers or engines) or enumerated (body type, wings type). Such models are usually non-closed and may contain logical expressions determining dependencies between the parameters (for example, the number of engines depends on the kind of empennage). Solving such systems is a complicated process of sorting a large number of various combinations, with no guarantee that the solution obtained will be optimal. The use of constraint programming methods over finite and continuous domains enables us to solve such problems with a rather high efficiency. In this example, the solver SibCalc was applied to solve mathematical models used at the stage of preliminary design of aircrafts. These models are sets of linear and nonlinear equations and inequalities, which describe the dependencies between different global parameters of the aircraft (takeoff weight, fuselage type, the number of wings, the number of pilots). The variables forming the models are of integer, real and logical types, the number of variables being not equal to the number of relations in the system. Below we give some relations of one model in the SibCalc input language (variables starting with I and N are integer, the others are real; the sign " $\rightarrow$ " indicates implication):

```
IE = 0 -> ST = SR * (2.5 + 0.5 * NM) and ND = 1 and SD = 0.13*SR
  and  SE = 0  and  SX = 0.7 * SR  and  vLX = vLR ;
IE = 1 -> QV = 215*(SX^1.5) / (vLX^0.5) * EL /100 and EV = 0.3 ;
CA = 3 * (vLF * (SS - SO) - 2 * VN) / (vLF ^ 3);
vLF ^ 2 = (2 * AM - SO - SS) / EF * 100;
vMC * vLX * SR *EL <> 0 ;  NM * PT > 0 ;
WVOIL = 1.15*SX*((15 + 0.15*SX)*(vMC * vLX/(SR*50*EL))^(2/3)+ 8);
WDER = ND * (27.5 * SD + 1.3 * SD ^ 2) ;
NS < 2 -> WSNA = 600 + 200 * NS ;
NS >= 2 -> WSNA = 400 + 300 * NS ;
```

The whole model contains 55 variable and 38 significant equations, thus having 17 degrees of freedom. The system is not partitionable, i.e. it is not possible to single out any group of equations with all variables being local but the system is weakly coupled. So, 19 variables occur only once, 12 variables occur only twice. As a rule, such subdefinite systems have infinite number of solutions; therefore it is usually necessary to find any first solution, else the problem of studying the system and finding the solutions with the required properties is posed. For example, it is required to find a solution, where some variables take the minimum values (e.g., takeoff weight), and other variables take the maximum values (e.g., quantity of passengers). To solve the second problem (the result can be also considered as the solution to the first problem), we have worked out a special methodology of localization of certain solutions, based on a combination of methods of constraint propagation and bisection. The dominant role in this methodology belongs to the strategy of selecting the sequence of variables in the algorithm of bisection, which ensures a reasonable compromise between the computation speed and the complexity of the selection algorithm. This strategy is based on the analysis of the systems for connectedness and on estimation of the speed of constraint propagation by the variables of the system. To determine connectedness of the system, the matrix of connectedness is created and formal subsystems are separated on the basis of the matrix. The variables which ensure faster constraint propagation are found after partition into subsystems, with global variables (included into more than one subsystem) being chosen first. Local variables (included into the given subsystem only) are additionally analyzed. The speed of change of interval solution for each variable is used to evaluate the speed of constraint propagation. As a result of this methodology, the system was decomposed into 3 submodels; after analysis of these submodels, a sequence of variables in the bisection algorithm was chosen. However, even after analysis of the system and selection of a "good" sequence of the selection of variables, much time is required for finding at least one solution in the case of the large number of degrees of freedom. To decrease the search time, we can fix the values of some variables and trace the solution behavior depending on the selected values. The question arises: how many variables, and which specifically, is it possible and necessary to assign in order to obtain the solution in a reasonable time. Since our model is described by a system of nonlinear algebraic equations, it is quite difficult to find out which variables are better to be considered as free. One of the approaches is to use the obtained partition of the model into subsystems and to analyze them for linearity and locality of variables. Based on such analysis, we have obtained that, to find the solution, it is sufficient to fix only six variables. Experiments have proved correctness of our approach, since with its help we were able to find required solutions in several seconds, while the solution of the system in its initial form gave no results after 72 hours of computations.

3.2 A Federal-Level Model of Materials and Finance

The federal-level model of materials and finance developed by a Russian economist V. Suslov [8] is studied in this paragraph. This material-financial model containing more than 350 nonlinear algebraic equations and inequalities and more than 320 variables considers inter-subject connections along all the main channels of movement of the material and financial resources: production and consumption of commodities and services, remuneration, for labor, taxes, state financing, etc. Various parameters and norms are used in modelling: indices of physical volume and indices of prices, materials consumption, funds consumption and labor expense, norms of reserves, tax rates, exchange rates, etc. This model is aimed at solving the problems of estimation and short-term forecast for financial-and-economic situation. This situation is represented in the model by a rather general closed system of macro-level indicators with respect to 7 sectors of economy:

- export-oriented sector
- production of other commodities
- distribution sector
- nonmaterial services to the population
- banking and finance
- government (including law enforcement and defense)
- population

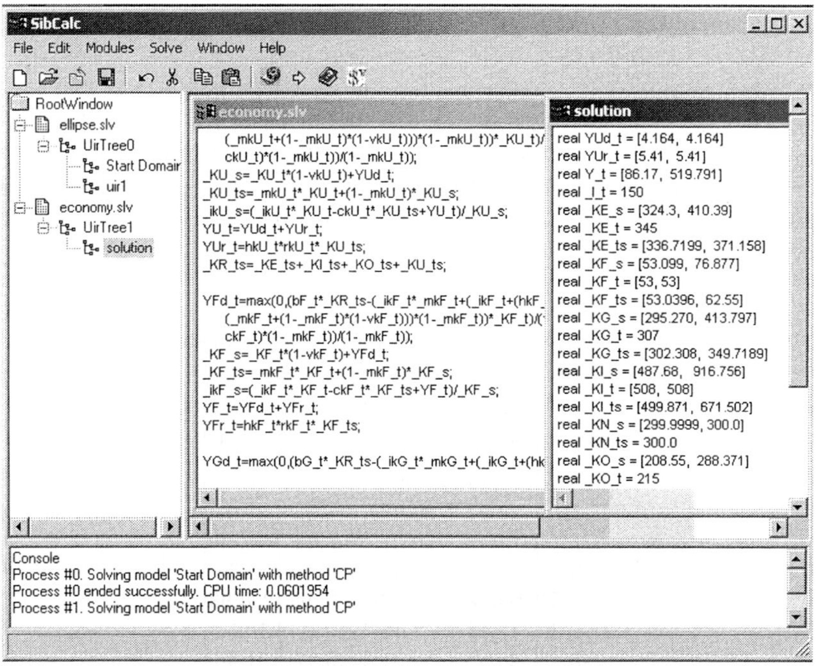

Fig. 1. The federal-level model in the SibCalc environment

Fig. 1 demonstrates a part of the model with its solutions in the SibCalc environment. In its purpose, the model is of a universal character and can be used for solving various problems, from traditional analysis of the structure and dynamics of "material" economy to investigation of the problems currently actual, such as inflation, budget deficit, tax rates, exchange rates, etc. The results of calculations enable us to reveal cause-effect dependencies between the financial and real indices, for instance, between the rates of rise in the prices, income of population, money supply, deposit and credit investments; between the indices of the financial state of the subjects of economy, of market conditions and the volumes of production and consumption. For many parameters of the model, only the limits of admissible values or economic evaluations can be determined, but not their precise values. Nonlinearity of the model and interval character of the parameters make it difficult to apply the classical computational methods to investigation of this model. At the same time, the interval approach gives us a possibility to find the optimal solutions with the model control parameters not completely defined. Application of the solver SibCalc makes it possible not only to solve traditional direct problems, but successfully perform computations for the reverse statement, e.g., to determine the indices of internal prices on the basis of different norms of the financial state of the subjects of economy and, first of all, their credit debts. Besides, with the help of the algorithms used in SibCalc, it is possible to formulate and to solve optimization problems.

3.3 Numerical Experiment

Let us perform one-year calculations for the economic model. For two levels of import (It = 180 and It = 150), we observe the levels of the budget deficit (def), unemployment (hNb), savings (sber) and profitability in the export-oriented sector (rentE), banking and finance (rentF) and in the government sector (rentG), depending on the minimal or maximal total amount of capital investment (Y). The calculation results are given in table 1.

Table 1. One-year results for eight parameters

	It = 150		It = 180	
	1	2	1	2
Y	86.1743	97.0288	86.1743	127.27
def	36.94	30.93	48.52	27.97
rentG	51.49	50.21	56.63	50.49
sher	-14.39	-7.51	-27.49	-2.56
rentF	196.30	214.69	175.94	234.93
rentE	18.13	-2.64	10.10	10.60
sber	28.58	30.51	23.97	31.46
hNb	11.69	6.09	25.87	4.67

In the table, column 1 presents the solution obtained for the minimal possible capital investment, and column 2 presents the solution obtained for the maximal possible capital investment. Based on the data presented in the table, we can draw the conclusions that give us the qualitative picture. With maximal total capital investment, import has no significant effect on the basic economic characteristics. However, the situation drastically changes if total capital investment is minimal. Thus, with increasing import, budget deficit is growing, savings are falling, and unemployment rate is jumping (more than twice). Furthermore, profitability in banking and finance declines and only profitability in the export sector is slightly growing (by 4%). Thus, computations with this economic model show that, with minimal capital investments, increase in import volume is destructive for the economy of the country. Application of a combination of SibCalc methods also allows us to conduct forecast computations with subdefinite data. Let us give the results of calculations according to this model, when the values of some indices are in ranges. Let the volume of production (production export-oriented sector) in the export-oriented sector vary from 200 to 300 (XE=[200, 300]), the total volume of investments vary from 86 to 90, with variations in the export sphere from 72 to 72.5, and net capital inflow (A) taking values of the interval [52690, 55090]. The computations show that in this case the volumes in the net capital inflow (A), non-material services (XU) and other production sectors (XI) will be: X = [105.9,106.3], XU = [100,100], XI = [659.6, 660.2].

4 Conclusion

In this work, we have considered the cooperative solver SibCalc implemented on the basis of the algorithms of interval mathematics and constraint propagation methods. We have demonstrated its application to solution of complex problems of modelling. The computational experiments carried out on a large set of problems showed that this approach is highly efficient and can solve a wide range of problems that cannot be solved by other methods. Moreover, the possibilities of the solver are not limited by these examples. Our investigations show that the methods here presented can be successfully applied to the sensitivity analysis of models. In this case, the possibility of working with interval data enables us to obtain qualitatively new results in this direction. Another possibility not considered in this work is parallelizing of calculations. Processing of large models may require much time and in some cases can prevent us from obtaining a solution in a reasonable time. Means of control of methods interaction, put into the computing schemes, make it possible to organize both parallel and distributed computations for large models, that allows us to substantially increase the size of the problems to be solved. These directions, along with the development of new methods to be incorporated into the solver, make part of our future work.

References

1. Alefeld, G., Herzberger, Ju.: Introduction in Interval Computations. Academic Press (1983)
2. Hansen, E.; Global Optimization Using Interval Analysis. Marcel Dekker (1992)
3. Kleymenov, A., Petunin, D., Semenov, A., Vazhev, I.: A model of cooperative solvers for computational problems. Proc. 4th Int. Conference PPAM'01, Naleczow, Poland, LNCS 2238, (2002) 797–802
4. Loenko, M.Yu.: Calculating the elementary functions with guaranteed accuracy. Programmirovanie, **27** (2001) 101–113 (In Russian)
5. Older, W., Vellino, A.: Constraint Arithmetic on Real Intervals. [in:] F. Benhamou and A. Colmerauer (editors): Constraint Logic Programming: Selected Research. MIT Press (1993) 175–195
6. Semenov A.L.: Solving Integer/Real Nonlinear Equations by Constraint Propagation. Technical Report N12, Institute of Mathematical Modelling, The Technical University of Denmark, Lyngby, Denmark (1994)
7. Semenov, A., Kashevarova, T.: Application of Constraint Programming Methods to Design Problems. Proc. ERCIM/Compulog Net Workshop on Constraints, Padova, Italy (200)
8. Suslov V.I. Measurement of the effects of interregional interaction: model, methods, results. Nauka, Novosibirsk, (1991) (In Russian)
9. Tsang, E.: Foundations of Constraint Satisfaction. Academic Press, Essex (1993)

Computational and Soft Skills Development through the Project Based Learning

Innokenti Semoushin, Julia Tsyganova, and Vladimir Ugarov

Ulyanovsk State University, 42 L. Tolstoy Str., 432970 Ulyanovsk, Russia
{SemoushinIV, TsyganovaJuV, UgarovVV}@ulsu.ru
http://staff.ulsu.ru/semoushin/

Abstract. In this paper we discuss some experiences gained by Department of Mathematical Cybernetics and Informatics, Ulyanovsk State University, Russia, on teaching computer science and mathematics courses for students majoring in Computational Science and Engineering (CSE).

1 Introduction

The demands for experts in Computational Science and Engineering (CSE) call into being different educational approaches such as new didactic environments [1], [2], new academic programs [3], [4], and new learning methods supported by technology [5]. Nevertheless, a building block to educate CSE professionals remains as before: highly developed skills of algorithmic thinking and computer programming. A promising development in university education involves the application of project-based learning as a curricular vehicle to develop students' computational talent. We define the Project-Based Learning (PBL) as a teaching paradigm stating that the student has perfectly understood a computational method not until he or she becomes able to make the computer to do all the work prescribed by the method. We express this by a piece of advice we give in jest to some of our students: "If you argue that you know the method well, please teach the computer to do the same".

2 Employers' Expectations

Training experts in CSE as well as in "Applied Mathematics" (AM), "Information Systems" (IS), "Information Technology" (IT) and some related fields assumes the profound study of computer programming. It is well known that contemporary employers expect not simply programmers but professionals having depth in one specific area of programming as well as breadth in computer science in general and in other fields relevant to information technologies, such as:

- Database Design and Management
- Data Communication (Computer) Networks

P.M.A. Sloot et al. (Eds.): ICCS 2003, LNCS 2658, pp. 1098–1106, 2003.

- System Programming (system management and control software in a variety of hardware and user environments)
- Application Programming (development of the problem-oriented applied programs)
- System analysis (finding the effective solution to the task and program specification development, on the basis of customer requirements)
- Etc.

Knowledge in these areas is requisite to carry out the main task – maintenance and development of information systems in industry and academic setting. One of the principal disciplines necessary to prepare experts for careers in CSE and IT is Programming. The purpose of this discipline is not only to develop in students the strong theoretical background but to teach the solid practical skills in computer programming. The students majoring in CSE must gain actual experience in solving problems through writing, entering, and running programs in order to develop their ability of structural algorithmic thinking. They also need to consider versatility of the discipline of Programming and its deep relation to many topics of mathematics: algebra, geometry, calculus, etc.

3 Traditional Approaches to Teaching Programming

Over the years many different techniques for teaching Programming and other related courses have been devised. Leading scientists, such as N. Wirth, E. W. Dijkstra, W. M. Turski and others paid much attention to the problem (see for example, [6], [7], [8], etc.).

Up to now, many educational courses in computer science are based on the methodological principles of teaching programming developed by Dr Niklaus Wirth, the famous Swiss scientist. Algorithmic language Pascal created by him, has been world wide practised, especially as the training programming language [6]. E. W. Dijkstra [7] has offered a method of formal program derivation from mathematical formulation of a problem, and some new approaches to decomposition of algorithms. Now Programming is the base of study at mathematical and engineering faculties of universities. The great experience has been accumulated, the various approaches to and methods of teaching this course differing on a goal orientation have been considered and approved. For example, the method suggested by Kaimin [9], is aimed at the developed student logic thinking and general computer literacy, allowing a student to work with ready-made applied software packages. The primary preparation method used at the Mechanics and Mathematics Faculty of Moscow State University (Russia) is aimed at establishing the base of general programming culture of the students [10]. This method is built upon developing the mini-projects to consider and make use of the basic propositions of programming. The issues of computational mathematics, numerical methods and other subjects of mathematics are not considered in this course. The material is based on concept of the "Executor", which can carry out "Actions" or "Instructions" over "Objects" to change their "States". The set of "Actions" refers to as "System instructions" of "Executor". This method gives the basic attention to consideration of algorithms, not especially becoming attached to the certain programming language.

4 Programming as a Sort of Engineering Design

4.1 Aiming at the Main Professional Qualities

Along with the other above mentioned methods, we consider the course "Programming" as a sort of engineering design. By definition of John R. Dixon [11]: the engineering design in essence is not the "art", but it represents activity which one can investigate and analyze, and master its bases during training... [11]. An expert in programming should correctly solve a lot of problem characteristic for design process while developing a software package. As a rule, the aim consists in the development of the software product able to carry out the job subject to a number of restrictions imposed on the product features. It is known that design is essentially iterative process consisting of successive stages on the way to the desired product some of which are to be repeated in order to improve or update the project or to correct the errors. The earlier corrections are made, the less amount of work and expenses will be spent for the whole project. On each design stage there exists a problem of a choice between many variants of the further work. This assumes decision making to satisfy one or other given criterion. In the designing process, it is very important to understand correctly the diverse requirements imposed on the future product. Consequently, along with the good knowledge of programming process, it is necessary to know well the subject area for which the project is developed. From the above brief reasoning about the design process it is clear that the modern programmer must be possessed of many qualities:

- High qualification in programming.
- Ability in doing analysis of a product on various design stages for revealing the elements having influence on the quality.
- The sufficient knowledge of the subject area for which the project is developed.
- Strong Computational Mathematics knowledge in designing solutions to mathematical and computer oriented problems.
- Thinking skills in making decisions under conditions of uncertainty in view of all essential factors [5].

The development of such qualities in prospective computer professionals is the basic task of the university courses "Computer Science" and "Programming".

4.2 The Basic Phase of Learning

As these courses are read at the first year of study and are basic, from the classical point of view it is necessary to keep elements of directive training.

Because programming is the practical discipline having, on the one hand, the strict language constraints and on the other hand, the wide variety of ways to implement the algorithm, the success on the initial stage of learning is possible only with the large amount of practical work. At this stage, it is important to gain the experience in simple algorithm implementation and acquire the working skills in one or other tool environment. In the same period, there must be introduced the common programming standards and elements of programming technology and program design techniques. The special attention should be given to the concept of the top-to-down program design and the algorithm decomposition methodology. As was shown by N. Wirth [6], there is some optimum set of algorithms whose realization allows to cover a necessary minimum of knowledge and skills of the initial level in the educational purposes while learning the chosen programming language. The set looks as follows:

- Base data structures and algorithms determined on them.
- Complex data structures: sequences, arrays, records, files.
- Search and sort algorithms on the given data structure.
- Recursive algorithms implementation.
- Dynamic data structures: lists, stacks, turns, trees.
- Program decomposition methods: procedures, functions, modules, objects.
- Algorithms on graphs.
- Interface realization.

The consecutive study of algorithms and their realization in particular language following the principle of increasing complexity of both data structures and algorithms determined on the given structures, allows to solve rather successfully the problem of training in students the basic programming concepts and skills.

4.3 Choosing the Subject Area for Algorithm Implementation

While developing assignments (or tasks) for practical algorithm implementation, a problem arises: what subject area is worthy to be chosen as a base for the assignment generation? Usually for that, the examples from various subjects are used: algebra, geometry, physics, chemistry, biology, mechanics, etc. The main requirement to the assignments is that the subject area should be familiar to the students. A variety of task themes has the advantage that it promotes expansion of student's mental outlook in application of algorithmization methods. However, it requires good correspondence between the students' training background and thematic orientation of the tasks.

Therefore the following programming training method is of interest: some subject area is chosen, its levels on increasing concepts complexity are defined, and the given subject area is offered as a basis for the tasks of programming course with a high degree of correlation on subjects and level of complexity of both the data structures, and algorithms. If the study of the chosen subject area will overtake a little the study of the corresponding programming sections, the parallel study method of the two disciplines will allow to raise a degree of mastering both programming and chosen subject area. As an example of a subject area it is possible to consider the course of "Numerical methods". A.P. Yershov [12] pointed out the strong, fundamental connection of the computer science and mathematics concepts:

Mathematics	Computer science
Algebraic system (structure)	Executor (robot, computer)
Carrier	Conditions
Element of the carrier	State of the conditions
Operation	Action changing conditions
Predicate	Question to the conditions
Signature	Instruction set
The protocol. A sequence of operations and predicates with their values + initial element	Behavior. A sequence of actions and questions to conditions with their answers, from the initial state
Predicate-precondition	Task condition
Predicate-postcondition	Task purpose
The available protocol realizing the appropriate predicates of pre- and post- conditions on the ends	The task solution. Behavior leading from the state satisfying the task condition to the state satisfying the task purpose
The program. Sub-recursive set including a set of feasible protocols	The code. The final instructions determining the behavior that leads to the end of each state satisfying the task condition

4.4 Our Suggestion: Mutually Supplementary Course Teaching

Mutual use of different disciplines methods was applied by several authors, for example [13], [14], [15]. While considering the Computational Mathematics or Graph Theory topics, the authors used elements of programming as an illustration of how to realize one or other method. However a crisp interrelation of the given courses was not determined. Our suggestion is that the methods of two courses should be thus interconnected that they supplement and support each other.

Let, for instance, the theme "Complex data structures: arrays" be studied in the course of "Programming" and the theme "Matrix algebra" of "Numerical methods" course have been considering. Then in the study of "Arrays", students may be given the tasks related to matrix operations: transposition, addition, multiplication, determinant calculation, and so forth. Realizing the tasks in the form of computer program, students will better mastering the "Matrix algebra". After that, it would be possible to pass to the next theme of study in "Numerical methods", for example, "Solving linear systems of equations". This theme gives the basis to form many assignments for the theme "Dynamic data structures" in "Programming". In the same manner, links between the other studied sections and courses can be introduced into the teaching and learning process.

5 Project-Based Learning

Most of the existing educational materials on Computational Mathematics provide main theoretical data and sometimes theoretical instructions on how to program a numerical method or algorithm. However, it seems to be inadequate to the end. We believe that the true understanding of a numerical method may be achieved if: (a) a student completes assignments related to a challenging programming project; (b) each project results in practical use of that particular method assigned for the student; (c) the student conducts a set of extensive computational experiments with the program he developed independently; and finally (d) frontal rating of the projects is carried out by the teacher together with the students.

Programming in itself is beneficial for student due to a number of reasons. First, it provides an opportunity to understand and learn a numerical method "from inside". This is quite different from utilizing ready-made software and significant for any creative professional. Second, it improves student's computer proficiency, as it requires keen programming. And finally, it develops general analytic and solution seeking performance and implants practical skill to attack and solve computationally oriented problems.

To organize Project-Based Learning while teaching numerical methods to a large audience requires to make programming assignments as varied as possible in terms of the methods' algorithmic significance rather than their initial data. However, the number of variations on every method is usually limited. In these conditions, finding as many as possible versions of every numerical method becomes a matter of great methodological importance for each teacher.

Organizing PBL means, also, that we should evaluate any laboratory programming project as a single study objective which possesses all the features of a completed software product. Among them are modular structure, convenient interface, efficient utilizing computer resources (memory and time), and possibility to implement a wide plan of computational experiments. This differs definitely from a widely used technique when the students work on one and the same ready-made software when they only enter their initial data and wait passively for a result. The approach we apply makes them perform valuable creative operations, stimulates each student's competitiveness, prevents cheating and helps to improve overall class performance.

A classic example of how to find as many as possible variant forms of a numerical method is the topic "Elimination and Matrix Inversion Methods". First of all, the teacher should systemize a set of Gauss and Gauss-Jordan elimination specific characteristics. They are: (1) direction of elimination of unknowns, (2) mode of access to the matrix entries, (3) mode of updating the active sub-matrix, (4) pivoting strategy etc. Then independence of these characteristics will result in a significant number of different variants of assignments on the same topic being studied.

Over the course on many years, our work is focused on the possible ways of applying PBL to teaching numerical methods in Linear Algebra, Least Squares, Optimal Filtering, Optimal Control, Linear Programming and Nonlinear Optimization. As a result, we recommend that teachers use textbooks, that offer a good choice of various project assignments. The first one [16] contains: Topic 1 - Elimination and Matrix Inversion Methods totalling 26 assignments, Topic 2 - Sparse System Solution totalling 48, Topic 3 - Cholesky Decomposition totalling 40, Topic 4 - Orthogonal Transformations totalling 28, Topic 5 - Simultaneous Least Squares totalling 28, Topic 6 - Sequential Least Squares and Optimal Filtering totalling 25 assignments. The second one [17] contains: Topic 7 - Simplex Method totalling 70, and Topic 8 - Nonlinear Optimization totalling 30 assignments.

5.1 Using the Modern Information Technologies through the PBL

Today, the functioning of a university is impossible without modern tools and technologies of access to information. Universities should pay special attention to the concept of mutually related courses within an educational environment. The development of an individual and self-realization of each student should be considered a priority. Modern information technologies give an opportunity to do so. They allow students to carry out academic tasks with the speed that is an optimum match for each student's talents, skills and character traits.

The mapping out of an educational process in this case consists of the creation of an informational environment.

An information environment assumes the presence of the following components:

1. Interconnected and complimentary academic programs focusing on modern information technologies.
2. A hardware plus software network environment with Internet connections.
3. Computer tools necessary for effective training, including multimedia technologies.
4. Resources for storage, accumulation and usage of data.
5. Organizational actions targeted at providing an effective interaction of all of the participants of the educational process within the modern information environment.

Within a given information environment, the concept of the PBL suggests that students come up with on their own projects to work on rather than use examples offered as a part of the lecture material. Such an approach considerably increases the level of mastering a course.

A great emphasis should be placed on the choice of topics for students' projects, as well as the methods of their realization. It is necessary to take into account both the studied section of the basic academic course, (for example, Numerical Methods) and the knowledge received within a mutually related course (for example, Programming).

The programming course of the Ulyanovsk State University, Ulyanovsk, Russia, uses the following training method based on Project-Based Learning. In the first semester, the students work on their projects, which involves using the studied programming language to program functions and sub-routines. Lab work within these courses is devoted to discussing various alternative methods of programming the same components in the presence of the project advisor.

The topics of assignments are based on the material that has not yet been covered, thus preparing the students for the material that will follow within the Numerical Methods course, for example:

1. Determining a number characteristics: parity, divisibility, and so on.
2. Determining calculation errors.
3. Processing final but unlimited sequences of numbers.
4. Processing one-dimensional and two-dimensional arrays of numbers.
5. Solving equations and systems of the equations.

The program components of every certain task are stored in the component library, i.e. in the module focused on the use of numerical methods.

It is possible to build modules on other subjects, such as computer graphics, graph algorithms, information compression algorithms, information metrics, text processing, sorting and searching, etc.

If a computer class has an Internet connection, then tasks, task specifications, the access to help material, and the demonstrational learning programs become available though the departmental web site, on which all of the educational material is stored. That allows students to use it both in class and at home for independent search. E-mail is a popular means of exchanging information. Teachers use demonstrational and test programs created on the basis of course video materials.

This cycle is completed by the design of the project based on the components created earlier. As s result, teachers reach the basic educational goal, namely the activization of the training process. This process means not only mastering the programming language itself, but also methods of task decomposition, program component design, project assembly and debugging, i.e. all of those elements that are essential for high-level knowledge of the subject.

Moreover, students simultaneously learn concepts and algorithms applied later in the Numerical Methods course that they effectively use for programming.

6 Conclusions

In this paper, leaning upon existing teaching methods and our own experience, we present the Project-Based Learning to prepare students for careers in Computational Science and Engineering.

References

1. Mori, P., Ricci, L.: Computational Science in High School Curricula: The ORESPICS Approach. In: Sloot, P.M.A., Kenneth Tan, C.J., Dongarra, J.J., Hoekstra, A.G. (eds.): Computational Science – ICCS 2002. Lecture Notes in Computer Science, Vol. 2331. Springer-Verlag, Berlin Heidelberg New York Barcelona Hong Kong London Milan Paris Tokyo (2002) 898–907
2. Anido, L., Santos, J.: An Online Environment Supporting High Quality Education in Computational Science. In: Sloot, P.M.A., Kenneth Tan, C.J., Dongarra, J.J., Hoekstra, A.G. (eds.): Computational Science – ICCS 2002. Lecture Notes in Computer Science, Vol. 2331. Springer-Verlag, Berlin Heidelberg New York Barcelona Hong Kong London Milan Paris Tokyo (2002) 872–881
3. Jeltsch, R.: CSE Program at ETH Zurich: Are We Doing the Right Thing? In: Sloot, P.M.A., Kenneth Tan, C.J., Dongarra, J.J., Hoekstra, A.G. (eds.): Computational Science – ICCS 2002. Lecture Notes in Computer Science, Vol. 2331. Springer-Verlag, Berlin Heidelberg New York Barcelona Hong Kong London Milan Paris Tokyo (2002) 863–871
4. Ruede, U.: Computational Engineering Programs at the University of Erlangen-Nuremberg. In: Sloot, P.M.A., Kenneth Tan, C.J., Dongarra, J.J., Hoekstra, A.G. (eds.): Computational Science – ICCS 2002. Lecture Notes in Computer Science, Vol. 2331. Springer-Verlag, Berlin Heidelberg New York Barcelona Hong Kong London Milan Paris Tokyo (2002) 852–857
5. Yip, W.: Generic Skills Development through the Problem-Based Lerning and Information Technology. In: Hamza, M.H., Potaturkin, O.I., Shokin, Yu.I. (eds.): Automation, Control, and Information Technology. Proc. of the IASTED International Conference – ACIT 2002. ACTA Press, Anahiem Calgary Zurich (2002) 102–107
6. Wirth, N.: Algorithms and Data Structure; ETH, Zurich (1986)
7. Dijkstra, E. W.: A Discipline of Programming. Prentice Hall, Englewood Cliffs (1988)
8. Turski, W. M.: Computer Programming Methodology. Rheine, London Philadelphia (1978)
9. Kaimin, V.: Course of Informatics: State, Methods and Perspectives. Informatics and Education **6** (1990) [In Russian]
10. Koushnerenko, A.G., Lebedev, G.V.: Programming for Mathematicians. Nauka, Moscow (1899) [In Russian]
11. Dixon, J. R.: Design Engineering: Inventiveness, Analisys and Decision Making. London Sydney (1966)
12. Yershov, A. P.: School Computerization and Mathematical Education. Programming **1** (1990) 3–15 [In Russian]
13. Lipski, W.: Kombinatoryka dla programistow. Warszawa (1982) [In Polish]
14. Ivanova, T. P., Pukhova, G. V.: Computational Mathematics and Programming. Prosveschenie, Moscow (1988) [In Russian]
15. McCracken, D.D., Dorn, W.S.: Numerical Methods and Fortran Programming. John Wiley and Sons, Inc., New York London Sydney (1965)
16. Semoushin, I.V., Kulikov, G.Yu.: Computational Linear Algebra: Collected Assignments for Laboratory Case Studies, Tests and Examinations. Ulyanovsk State University of Technology Publishers, Ulyanovsk (2000)
17. Semoushin, I.V.: A Practical Course in Optimization Methods. Part 1: Linear Programming. Ulyanovsk State University of Technology Publishers, Ulyanovsk (1999)

XML-Based Interface Model for Socially Adaptive Web-Based Systems User Interfaces

Janusz Sobecki

Department of Information Systems, Wroclaw University of Technology Wyb. Wyspianskiego 27, 50-370 Wroclaw, POLAND, sobecki@pwr.wroc.pl

Abstract. The differences among the population of the web-based systems users have very significant consequences in their interaction preferences. Many web-based systems offer different personalization mechanisms that enable users to adjust systems' interfaces to their personal needs, however, quite many users don't want to bother themselves doing it or are not able to make reasonable settings. In the paper the XML-based interface model for socially adaptive user interfaces is presented. This model enables personalization representation and consensus based interface reconciliation among the population of personalized interfaces.

Keywords: xml-based interface model, adaptive user interface, web-based system

1 Introduction

Web-based systems are nowadays used for many different reasons: to inform, to teach, to promote or to agitate. They also concern with many areas of people's interests. The user population of those systems is usually very differentiated and that's why it's members could have different information needs and interaction preferences. Today it is obvious for almost everyone that web-based systems are more appreciated by their users if their interfaces are well designed. But building a single user interface that is uniquely effective for each of these systems and for all the users is rather impossible [16].

The problems of user interfaces are addressed by many different scientific areas. Traditional approach has been worked out by the area of the Human Computer Interaction (HCI) and it suggests designing an interface according to the future user profile as well as to the system platform characteristic [9,14]. Obviously such solutions could not be fully applied in the today's web-based systems because of the differences that concern users and platforms they are using. The interface personalization is a commonly used solution in these circumstances [8].

The personalization enables to set three basic elements of the system: the system information content, the way the information is presented and how the information is structured [6]. For example in the very popular web portal — Lycos users are able to add and remove boxes with information (content), move them round the screen and change their sizes and colors (presentation) and making it the starting page (structure).

P.M.A. Sloot et al. (Eds.): ICCS 2003, LNCS 2658, pp. 1107–1116, 2003.

During the personalization process users are usually obliged to enter some demographic data about themselves, such as name, address, occupation and interests. These types of data are called the user data and are used to model the user. Usually all types of data concerning users that are gathered by the system could be also called the user profile. The user profiles could then be stored in the system supplier's servers or on the client side in cookies files in form of some entries [15]. The user profile could also contain the so-called usage data that may be observed directly from the users interaction with web-based systems, such as URL's of visited pages, links followed by the user and data entered by the user into the forms on the pages [4].

The alternative solution to the system personalization mechanism is offered by the implementation of interface agents. They help to automate very different user tasks and in this way help him or her to interact with many different information systems [5]. In this paper the problem of users interaction with web-based systems is addressed using another approach i.e. the social adaptive hypermedia methods [6].

2 The Architecture of the Socially Adaptive Web-Based Systems User Interfaces

In this chapter the general architecture of the socially adaptive web-based systems users interfaces will be presented only for short, in more details it has been presented in the following works [12,15,19].

The web-based system users may fall into three categories. First group is comprised of the users who have already used the system on the particular platform. Second group is made up by the users who have used the system already but on the other computer platform and in third group there are the users who have never used this particular system before.

In the first case, it is not much to do for the adaptation mechanism, because every user of the system, while using it, gathers experiences and acquires habits, so introduction of any automatic changes should be restricted to only such cases when the user asks for them or the computed usability function values are extremely low.

The second case, because of its dual character, is very problematic. On the one hand, the web-based system is already known for the user and the user is aware of his or her own information needs, how they are fulfilled and the user does not welcome any changes. On the other hand, different platforms have usually different characteristic of input and output devices. For example, for a typical PC we have an input by means of a 101 (or so) keys keyboard and a mouse, and color 1024x768 display as an output. For the modern handhelds, however, we have a touch screen and stylus together with some special keys as an input and maximum high color 320x240 display as an output. When we take into consideration other platforms such as: information kiosks, set-top boxes and navigation systems mounted in cars, we can encounter other specific differences in input and output devices characteristics. So in the adaptive interface construction for a new platform we should preserve as much as possible from the former platform, for example the information content, its structure but also the way of its presentation. In many cases, however, it is not possible, so it is suggested that at least these two former elements, included in the interface profile description, should be preserved. As for the latest element, i.e. the information presentation should be adaptively constructed as shown below.

In the third case, the interface should be automatically constructed for this particular user and this particular platform. It is assumed that web-based systems have a dynamic population of users located in a distributed environment. These web-based systems form a multi-agent system that communicates with their users. Agents collect both: the user data and the usage data of each particular system in form of the user profile that is transferred to the server. The users are able to personalize their systems. The quality of these settings is automatically measured by the appropriate interface usability measures [9,14]. The system personalization settings together with the usability measures values form the interface profile that is also sent to the server. Any novel user of the particular system is first classified according to the corresponding user profile [12], in this way the user is assigned to the particular class of users. Then the interface profiles belonging to the same class are used to create, by means of consensus methods [10,15], the interface profile for that particular user. The system quality measures and user classification will be described below. The problem of the personalization settings will be addressed in the following section.

The domain of the HCI [9,14] has worked out many methods for the interface evaluation but all of them are not straightforward and cannot be measured by a simple function. Quite often different empirical methods are used but also analytical methods or their mixture are applied. We must remember that in the socially adaptive web-based systems the usability factors values must be evaluated in an automatic manner. The usability is compound of several factors [9], such as: the speed of performance of the specified activity, incidence of errors, users' ability to correct errors, users' efforts to learn how to use the system, users' satisfaction with the system, etc. These factors, however, can be quite difficult to measure.

The user classification is a very well known problem that has been addressed for at least 40 years by the specialists from many fields, i.e. the information retrieval [3]. Nowadays, in the era of the e-economy and the Information Society, researches in this area are continued by many specialists from: HCI, user modeling and marketing [7].

The process of user classification results in grouping heterogeneous set of users $U = \{u_1, \ldots, u_n\}$ into disjoint subsets of the set U by some measure of similarity or distance. Finding an appropriate measure, as well as a user representation, are key problems of users classification. The process of classification could be computationally difficult, even with exponential complexity with respect to the number of elements to be classified [2,3], so practically other sub-optimal solutions have to be used. They are usually based on the selection of some initial partition as for example in the Dattola algorithm [3]. In the architecture proposed here for the centroides could serve, for example the profiles of randomly selected users who personalized the system interface by their own. Then for each vector (user profile) the similarity function between the vector and each centroide is determined. The vector is joined to the class with the centroid that has the highest value of the similarity function and also higher that assumed threshold, those below are assigned to the class of so-called isolated elements. Then for each class the centroides are recalculated and the process is repeated until any of vectors changes their class assignment. Finally all vectors from the class of isolated elements are assigned to the classes with the highest similarity function values.

3 XML-Based Interface Model

Because of the great differences among the users as well as the system platforms, the interface model to be used in the adaptive web-based systems should be very flexible in reflecting those differences in terms of some formalities that could be implemented in the chosen web technology. Firstly, the model should be easy to use and offer powerful tool for the system designer in the design of possibly wide range of interactive web-based systems [17]. Secondly, it should enable users of the web-based systems to personalize them, that mean to set them according to their information needs and interaction preferences. Finally, the model should enable determination of the representation of the set of interface profiles, that is the key element of the architecture proposed above.

In the following subsections the interface profile, the distance function between the interface profiles and the method for the representation determination in the set of the interface profiles will be presented.

3.1 Interface Profile

The interface profile is the element of the overall interface model, which task is to represent the actual hypermedia interface in the design, personalization and generation processes. The profile was first presented in the work [18]. In this paper a new modified version of the profile is presented.

To represent the interface we can use several different models known from the literature. We can use for example quite simple Pawlak [13] information system model, where the system S is a quadruple $S = \langle X, A, V, \rho \rangle$, X is a set of objects, A set of attributes describing these object, V set of attributes values and $\rho\colon X \times A \to V$ is an information function.

In many cases, however, it is more convenient to use multi-valued version of the Pawlak information system, where the information function is defined as follows: $\rho\colon X \times A \to 2^V \setminus \{\emptyset\}$. This is because of the nature of many attributes that describe an interface, i.e. topics of interest, sound types of different signals or other relevant information, which could be represent by set of values instead of an atom value. Moreover, in some cases the set of the attribute values could be ordered and repeated. For example the order of topics appearing on the screen could be quite important as well as the possibility to represent repeated values: template identifiers of some screen regions or soundtracks to be played in the background like in a jukebox. The case of ordered sets with repetitions seems to be the most general one, i.e. any other example: single value and subset of values could be treated as special cases of this one.

Nowadays object-oriented design is the most popular paradigm in the information system design. This is also the case in the user interface design. This type of structure could be then easily represented in form of a XML document [1]. XML has become recently the standard for exchanging any information among different applications or even their layers and can be directly implemented in wide range of user interfaces implementations.

Different attributes that describe user interfaces could have hierarchical dependencies and could be represented in an object-oriented manner or by a tree-like structure. Then attributes and their values are attached to the tree nodes and edges reflect dependencies among attributes and their values. In consequence we can model the every instance of a user interface as the following structure $I = \langle N, E, A, V, n_0, \sigma, \theta, \delta \rangle$, where

- N — is a set of nodes,
- $E \subset N \times N$ — is a set of edges between nodes, where the first node is called a father and the second a child,
- A — is a set of attributes,
- V — is a set of attribute values and $\Pi'(V)$ is a set of all ordered subsets with repetitions of values from V,
- n_0 — is the root node,
- $\sigma \colon N \to A \times \Pi'(V)$ — is a description function,
- $\theta \colon A \times \Pi'(V) \to 2^A$ — is the attribute mapping function,
- $\delta \colon A \times (V \cup \{\text{null}\}) \times (V \cup \{\text{null}\}) \to [0, 1]$ — is a distance function.

The structure to form a tree must have the following properties: there is only a single root node, there is only a single edge connecting two different nodes, all nodes except the root node have only one father each and all of them share the same predecessor i.e. the root node.

The description function σ assigns each node a pair attribute and ordered subset of their values. The attribute mapping function θ that assigns each pair: attribute and subset of their values the subset of the attributes that could be placed in the child nodes.

The distance function δ should be given by the system interface designer and fulfill all the distance function conditions but not especially all the metrics conditions [10]. The interface designer should specify the distance function between all pairs of atom values, together with the empty value, of all attributes. The distance function values could be enumerated or given in any procedural form. These values enable to find distances between ordered sets of values and also are necessary to determine the distance between any of two instances of the interface. The procedure for determining this distance can be found in [11], the values determined by this procedure falls in $[0, 1]$ interval and fulfils the distance function conditions.

3.2 Example of an Interface Profile

To represent the interface instance, as it was generated by the system or modified in the personalization process some elements for each web-based system should have been determined in advance, i.e. the set of attributes A, the set of their values V, the attribute mapping function θ, and distance function δ. These sets and functions are equal for every instance of the interface profile of that particular system. What differs one instance from another is: the set of nodes N, the set of edges E, the start node n_0, and the description function σ.

Let as consider a typical web-based information system, which goal is to promote, advertise and inform about a particular car model. It is obvious that this type of a system is intended for even hundreds of thousands of users from many different countries all over the world. These users may be of completely different age (teenagers, adults, elder people) and have different sex. So, it is essential that this type of system should be able to offer different interface profiles for different users, because of their different information needs and various interaction preferences.

It is possible to consider all the elements of the interface profile in the model presented above, i.e. the information content, its structure and presentation. However, in the following example only its presentation part is considered.

The sample system has the following parameters (see also Fig. 1):

- $A = \{$root, resolution, button_style, font_size, font_style, bckg_type, bckg_color, bckg_image, sound, sound_track, sound_effect, volume$\}$;
- $V = V_{\text{root}} \cup V_{\text{resolution}} \cup V_{\text{sound}} \cup V_{\text{sound\_effects}} \cup V_{\text{sound\_track}} \cup V_{\text{volume}} \cup V_{\text{bckg\_type}} \cup V_{\text{bckg\_color}} \cup V_{\text{bckg\_image}} \cup V_{\text{button\_style}} \cup V_{\text{font\_size}} \cup V_{\text{font\_style}}$, where
$V_{\text{root}} = \emptyset$, $V_{\text{resolution}} = \{640\text{x}480, 800\text{x}600, 1024\text{x}768, 1280\text{x}1024\}$,
$V_{\text{sound}} = \{\text{yes}, \text{no}\}$, $V_{\text{sound\_effects}} = \{\text{effect1}, \text{effect2}\}$,
$V_{\text{sound\_track}} = \{\text{track1}, \text{track2}\}$, $V_{\text{volume}} = \{0, 1, 2, 3, 4, 5\}$,
$V_{\text{bckg\_type}} = \{\text{color}, \text{image}\}$, $V_{\text{bckg\_color}} = \{\text{white}, \text{blue}, \text{black}\}$,
$V_{\text{bckg\_image}} = \{\text{stars}, \text{circles}\}$, $V_{\text{button\_style}} = \{\text{text}, \text{graphic\_static}, \text{animated}\}$,
$V_{\text{font\_size}} = \{1, 2, 3, 4, 5\}$, $V_{\text{font\_style}} = \{\text{arial}, \text{georgia}, \text{times}\}$;
- $\theta(\text{root}, \emptyset) = \{\text{resolution}, \text{sound}, \text{bckg\_type}, \text{button\_style}, \text{font\_size}, \text{font\_style}\}$, $\theta(\text{bckg\_type}, \text{image}) = \{\text{bckg\_image}\}$, $\theta(\text{bckg\_type}, \text{color}) = \{\text{bckg\_color}\}$, $\theta(\text{sound}, \text{no}) = \emptyset$, $\theta(\text{sound}, \text{yes}) = \{\text{sound\_track}, \text{sound\_effects}, \text{volume}\}$;
$\theta(\text{sound\_track}, \_) = \emptyset$, $\theta(\text{sound\_effects}, \_) = \emptyset$, $\theta(\text{volume}, \_) = \emptyset$,
$\theta(\text{button\_style}, \_) = \emptyset$, $\theta(\text{font\_style}, \_) = \emptyset$, $\theta(\text{font\_size}, \_) = \emptyset$;
- $\delta(\text{volume}, x, y) = |x - y| / |\max - \min|$, where max and min are maximal and minimal values accordingly;
$\delta(\text{volume}, x, \text{null}) = 1/2$, $\delta(\text{sound\_track}, \text{track1}, \text{track2}) = 0.7$,
$\delta(\text{sound\_track}, \text{track1}, \text{null}) = 1$, $\delta(\text{sound\_track}, \text{track2}, \text{null}) = 1$.

Because of the lack of a space the other values of the distance function will not be given. The information function values for this particular instance are given in the tree nodes presented in Fig. 1.

The tree structure shown in the Fig. 1 could be easily presented in form of XML document. This document is described by the interface profile structure formally given above in the section 3.1. It is also possible to use other formalisms to represent such structure, i.e. XML Schema [20]. XML Schema as DTD (Document Type Definition) is used to declare the XML document structure but is more powerful one. With XML Schema in contrary to the DTD it is possible to declare also types and patterns that could be use to declare attributes, their values (also in form of ordered subsets with repetitions of values) as well as information function σ, attribute mapping function θ and distance function values

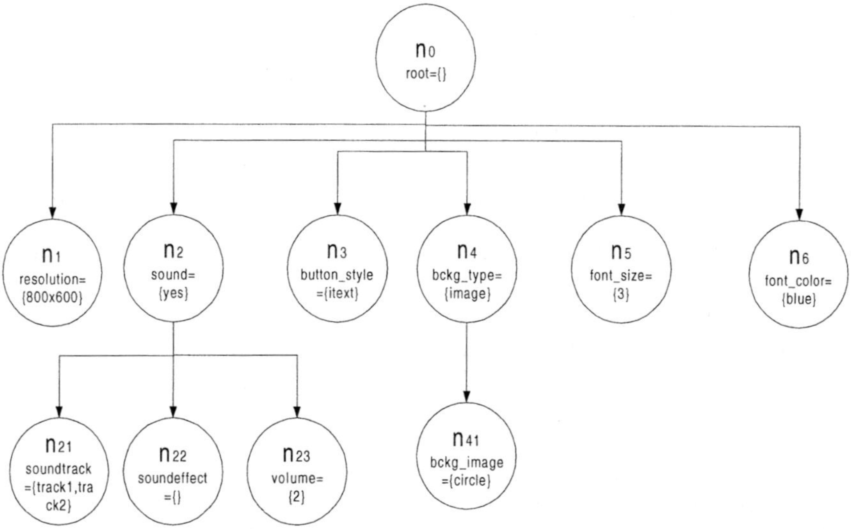

Fig. 1. Instance of the interface profile in form of the tree structure

δ. Unfortunately the XML Schema of the interface profile structure is outside the scope of this paper, here there is only the XML document that represents the interface instance and it will be given below.

```
<?xml version="1.0" encoding="ISO-8859-1"?>
<root>
  <resolution value="800x600"/>
  <sound value="yes">
    <sound_track value="track1 track2"/>
    <sound_effects value=""/>
    <volume value="2">
  </sound>
  <button_style value="text"/>
  <bckg_type value="image">
    <bckg_image value="circle"/>
  </bckg_type>
  <font_size value="3"/>
  <font_color value="blue"/>
</root>
```

3.3 Distance Determination among Interface Instances

The key element in the adaptive web-based system user interfaces model is its ability to measure the distance among different instances of user interfaces. Those instances may be given by the users themselves in the process of the interface personalization or determined by the adaptive procedures. In the interface model given above only the distance among attribute values are given, but the model assumes that attribute values may be in the form of ordered sets of values with repetitions, so it is necessary to determine the distance among them too. An example of this procedure could be found in work [18], this procedure is called **set_distance** in the program below. Having this it is possible to find the distance among the interfaces instances modeled by the tree structures t_1 and t_2. This distance is measured by the recursive function **tree_distance** described below, where the lv denotes the level of the node in the tree structure (equals 1 at start-point) and root(t) denotes the root node of the tree t. In the function construction it was assumed that when attributes in the roots of sub trees are different the distance between these sub trees should be equal, independently of all other values of this sub tree, the maximal value 1 (in the whole distance however, it is divided by the level of the root of the sub tree and the number of sub trees in this level). This assumption is a consequence of the fact that only comparable values could be compared. For example, there is no use to compare integer value of background color with the volume value. In the instance of the interface represented by a tree structure attributes and their values depend on the attributes and corresponding values that are placed higher in the tree so there is no use to compare them too.

```
function tree_distance(t1,t2,lv)
BEGIN
(a1,w1)=σ(root(t1))
(a2,w2)=σ(root(t2))
IF a1<>a2 THEN RETURN 1
ELSE IF θ(a1,w1)<> θ(a2,w2) THEN RETURN set_distance(w1,w2)
ELSE
BEGIN
c:=1/|θ(a1,w1)| /*θ(a1,w1)=θ(a2,w2) */
k:=lv+1
s:=Σ set_distance(w1,w2)
RETURN c*s+1/k*Σ tree_distance(t'1, t'2,k) /*where t'1
and t'2 are all sub-trees of the trees t1, t2 starting
from edges derived from the root node */
END
END
```

The value of distance determined by the above function falls in [0,1] interval and fulfils the distance function conditions.

3.4 Representation of the Set of Interface Instances

To solve the problem of determining the representation in the set of interface instances we should also have the values of usability function $u(j)$ assigned to each of the interface instance. Let assume that the usability function values falls also in the $[0, 1]$ interval and the value 0 denotes completely unusable interface and 1 denotes ideally usable interface. Then to find the interface instance for a particular user by means of consensus method, we must find the interface instance i that's the sum of the distances to all other instances j from the profile, i.e. those instances that are assigned to users belonging to the same (as our user) class C, multiplied with their utility function value $u(j)$ is minimal:

$$\min\left(\sum_{j\in C} u(j) * \mathsf{tree\_distance}(tr_i, tr_j, 1)\right). \tag{1}$$

This problem could be computationally difficult. We can have two approaches to solve it. First we can reduce the problem space only to those instances that belong to the set of all known instances (i.e. they were used already by other users from the same class of users) but the solution could be far from optimal one, or second, we can consider all possible interface structures that could be constructed out of the set of attributes A and values from the set V which are valid in the model for a particular system. However, in the works [10,11] it was proven that for some distance functions we could reduce the search space only to those attribute values that are already presented in the profile and this suffices to find satisfactory representation of the profile.

4 Conclusion

Adaptive user interfaces are becoming more and more popular among different web-based information services [6]. Unfortunately most of the solutions are commercial ones and quite often their methodologies are rather confidential. Considering the social perspective in the adaptive systems enables to follow those interface settings that are not only often used but also those that appeared to be the most effective, owing to application of usability measures.

In this paper the XML-model for adaptive interfaces has been presented. Using this model several web-based systems interfaces are being implemented already in Macromedia Flash. Macromedia Flash is very popular tool for multimedia web-based system implementation and its recent version MX is quite well suited for XML documents processing even without direct implementation of such mechanisms as XML Schema, DTD or XSL which the application developer has to implement in Flash or Action Script.

References

1. Bray T., Paoli J., Sperberg-McQueen C. M., Maler E.: Extensible Markup Language (XML) 1.0 (Second Edition), W3C. 6 October (2000) http://www.w3.org/TR/REC-xml.
2. Brown S. M. et al.: Using explicit requirements and metrics for interface agent user model for interface agent user model correction. In: Proceedings of the second international conference on Autonomous Agents, Minneapolis, Minesota, USA (1998) 1–7.
3. Dattola R. T.: A fast algorithm for automatic classification. Report ISR-14 to the National Science Foundation, Section V, Cornell University, Department of Computer Science (1968).
4. Fleming M., Cohen R.: User modelling in the design of interactive interface agents. In: Proc of the Seventh International Conference on User Modeling: (1999) 67–76.

5. Hedberg S. R.: Is AI going mainstream at last? A look inside Microsoft Research. IEEE Intelligent Systems **3/4** (1998) 21–25.
6. Kobsa A., Koenemann J., Pohl W.: Personalized Hypermedia Presentation Techniques for Improving Online Customer Relationships. The Knowledge Engineering Review **16** (2), (2001) 111–155.
7. Maglio P. P., Barrett R., Campbell C.S., Selker T.: SUITOR: an attentive information system. In: Proc. of the 2000 Int. Conf. on Intelligent User Interfaces (2000) 169–176.
8. Mobasher B., Cooley R., Srivastava J.: Automatic personalization based on Web usage mining. Comm of the ACM, **43** (8), (2000) 142–151.
9. Newman W. M., Lamming M. G.: Interactive system design. Addison-Wesley, Harlow (1996).
10. Nguyen N. T.: Conflict Profiles' Susceptibility to Consensus in Consensus Systems. Bulletin of International Rough Sets Society **5**(1/2): (2001) 217–224.
11. Nguyen N. T.: Consensus system for solving conflicts in distributed systems. Information Sciences **147** (2002) 91–122.
12. Nguyen N. T., Sobecki J.: Consensus problem of user classification in interactive systems. In: P. Grzegorzewski (et al., eds.): Soft Methods in Probability, Statistics and Data Analysis. Series Advances in Soft Computing. Physica Verlag Heildelberg (2002) 346–355.
13. Pawlak Z.: Information Systems — Theoretical Foundations. PWN Warsaw (1983).
14. Peerce J. et al.: Human-computer interaction. Addison-Wesley, Harlow (1996).
15. Sobecki J., Nguyen N. T.: Consensus-based adaptive user interface for universal access systems. In: Stephanidis C. (ed.) Proceedings of 9th Int. Conf. on Human-Computer Interaction and 1st Int. Conf. on Universal Access in Human-Computer Interaction. LEA London vol 3 (2001) 112–116.
16. Sobecki J.: One suits all — is it possible to build a single interface appropriate for all users? In: Grzech A., Wilimowska Z. (eds.) Proceedings of the 23rd Int. Scientific School ISAT, PWr Press Wroclaw (2001) 125–131.
17. Sobecki J.: Interactive multimedia information system planning. In: Valiharju T. (ed.), Digital Media in Networks. Perspectives to Digital World. University of Tampere (1999) 38–44.
18. Sobecki J.: Interface model in adaptive web-based system. In: Cellary W., Iyengar A. (eds.): Internet Technologies, Applications and Societal Impact. Kluver Academic Publishers, Boston (2002) 93–104.
19. Sobecki J.: Adaptive interface for hypermedia applications. In: Proceedings of VIPromCom-2002 — 4th EURASIP — IEEE Region 8 International Symposium on Video Processing and Multimedia Communications, 16–19 June 2002, in Zadar, Croatia: (2002) 147–152.
20. van der Vlist E.: Using XML Schema. O'Reilly & Assoc. Inc. (2000) http://www.zml.com/pub/a/2000/11/29/schemas/part1.html.

Author Index

Lecture Notes in Computer Science

For information about Vols. 1–2584

please contact your bookseller or Springer-Verlag

Vol. 2585: F. Giunchiglia, J. Odell, G. Weiß (Eds.), Agent-Oriented Software Engineering III. Proceedings, 2002. X, 229 pages. 2003.

Vol. 2586: M. Klusch, S. Bergamaschi, P. Edwards, P. Petta (Eds.), Intelligent Information Agents. VI, 275 pages. 2003. (Subseries LNAI).

Vol. 2587: P.J. Lee, C.H. Lim (Eds.), Information Security and Cryptology – ICISC 2002. Proceedings, 2002. XI, 536 pages. 2003.

Vol. 2588: A. Gelbukh (Ed.), Computational Linguistics and Intelligent Text Processing. Proceedings, 2003. XV, 648 pages. 2003.

Vol. 2589: E. Börger, A. Gargantini, E. Riccobene (Eds.), Abstract State Machines 2003. Proceedings, 2003. XI, 427 pages. 2003.

Vol. 2590: S. Bressan, A.B. Chaudhri, M.L. Lee, J.X. Yu, Z. Lacroix (Eds.), Efficiency and Effectiveness of XML Tools and Techniques and Data Integration over the Web. Proceedings, 2002. X, 259 pages. 2003.

Vol. 2591: M. Aksit, M. Mezini, R. Unland (Eds.), Objects, Components, Architectures, Services, and Applications for a Networked World. Proceedings, 2002. XI, 431 pages. 2003.

Vol. 2592: R. Kowalczyk, J.P. Müller, H. Tianfield, R. Unland (Eds.), Agent Technologies, Infrastructures, Tools, and Applications for E-Services. Proceedings, 2002. XVII, 371 pages. 2003. (Subseries LNAI).

Vol. 2593: A.B. Chaudhri, M. Jeckle, E. Rahm, R. Unland (Eds.), Web, Web-Services, and Database Systems. Proceedings, 2002. XI, 311 pages. 2003.

Vol. 2594: A. Asperti, B. Buchberger, J.H. Davenport (Eds.), Mathematical Knowledge Management. Proceedings, 2003. X, 225 pages. 2003.

Vol. 2595: K. Nyberg, H. Heys (Eds.), Selected Areas in Cryptography. Proceedings, 2002. XI, 405 pages. 2003.

Vol. 2596: A. Coen-Porisini, A. van der Hoek (Eds.), Software Engineering and Middleware. Proceedings, 2002. XII, 239 pages. 2003.

Vol. 2597: G. Păun, G. Rozenberg, A. Salomaa, C. Zandron (Eds.), Membrane Computing. Proceedings, 2002. VIII, 423 pages. 2003.

Vol. 2598: R. Klein, H.-W. Six, L. Wegner (Eds.), Computer Science in Perspective. X, 357 pages. 2003.

Vol. 2599: E. Sherratt (Ed.), Telecommunications and beyond: The Broader Applicability of SDL and MSC. Proceedings, 2002. X, 253 pages. 2003.

Vol. 2600: S. Mendelson, A.J. Smola, Advanced Lectures on Machine Learning. Proceedings, 2002. IX, 259 pages. 2003. (Subseries LNAI).

Vol. 2601: M. Ajmone Marsan, G. Corazza, M. Listanti, A. Roveri (Eds.) Quality of Service in Multiservice IP Networks. Proceedings, 2003. XV, 759 pages. 2003.

Vol. 2602: C. Priami (Ed.), Computational Methods in Systems Biology. Proceedings, 2003. IX, 214 pages. 2003.

Vol. 2603: A. Garcia, C. Lucena, F. Zambonelli, A. Omicini, J. Castro (Eds.), Software Engineering for Large-Scale Multi-Agent Systems. XIV, 285 pages. 2003.

Vol. 2604: N. Guelfi, E. Astesiano, G. Reggio (Eds.), Scientific Engineering for Distributed Java Applications. Proceedings, 2002. X, 205 pages. 2003.

Vol. 2606: A.M. Tyrrell, P.C. Haddow, J. Torresen (Eds.), Evolvable Systems: From Biology to Hardware. Proceedings, 2003. XIV, 468 pages. 2003.

Vol. 2607: H. Alt, M. Habib (Eds.), STACS 2003. Proceedings, 2003. XVII, 700 pages. 2003.

Vol. 2609: M. Okada, B. Pierce, A. Scedrov, H. Tokuda, A. Yonezawa (Eds.), Software Security – Theories and Systems. Proceedings, 2002. XI, 471 pages. 2003.

Vol. 2610: C. Ryan, T. Soule, M. Keijzer, E. Tsang, R. Poli, E. Costa (Eds.), Genetic Programming. Proceedings, 2003. XII, 486 pages. 2003.

Vol. 2611: S. Cagnoni, J.J. Romero Cardalda, D.W. Corne, J. Gottlieb, A. Guillot, E. Hart, C.G. Johnson, E. Marchiori, J.-A. Meyer, M. Middendorf, G.R. Raidl (Eds.), Applications of Evolutionary Computing. Proceedings, 2003. XXI, 708 pages. 2003.

Vol. 2612: M. Joye (Ed.), Topics in Cryptology – CT-RSA 2003. Proceedings, 2003. XI, 417 pages. 2003.

Vol. 2613: F.A.P. Petitcolas, H.J. Kim (Eds.), Digital Watermarking. Proceedings, 2002. XI, 265 pages. 2003.

Vol. 2614: R. Laddaga, P. Robertson, H. Shrobe (Eds.), Self-Adaptive Software: Applications. Proceedings, 2001. VIII, 291 pages. 2003.

Vol. 2615: N. Carbonell, C. Stephanidis (Eds.), Universal Access. Proceedings, 2002. XIV, 534 pages. 2003.

Vol. 2616: T. Asano, R. Klette, C. Ronse (Eds.), Geometry, Morphology, and Computational Imaging. Proceedings, 2002. X, 437 pages. 2003.

Vol. 2617: H.A. Reijers (Eds.), Design and Control of Workflow Processes. Proceedings, 2002. XV, 624 pages. 2003.

Vol. 2618: P. Degano (Ed.), Programming Languages and Systems. Proceedings, 2003. XV, 415 pages. 2003.

Vol. 2619: H. Garavel, J. Hatcliff (Eds.), Tools and Algorithms for the Construction and Analysis of Systems. Proceedings, 2003. XVI, 604 pages. 2003.

Vol. 2620: A.D. Gordon (Ed.), Foundations of Software Science and Computation Structures. Proceedings, 2003. XII, 441 pages. 2003.

Vol. 2621: M. Pezzè (Ed.), Fundamental Approaches to Software Engineering. Proceedings, 2003. XIV, 403 pages. 2003.

Vol. 2622: G. Hedin (Ed.), Compiler Construction. Proceedings, 2003. XII, 335 pages. 2003.

Vol. 2623: O. Maler, A. Pnueli (Eds.), Hybrid Systems: Computation and Control. Proceedings, 2003. XII, 558 pages. 2003.

Vol. 2625: U. Meyer, P. Sanders, J. Sibeyn (Eds.), Algorithms for Memory Hierarchies. Proceedings, 2003. XVIII, 428 pages. 2003.

Vol. 2626: J.L. Crowley, J.H. Piater, M. Vincze, L. Paletta (Eds.), Computer Vision Systems. Proceedings, 2003. XIII, 546 pages. 2003.

Vol. 2627: B. O'Sullivan (Ed.), Recent Advances in Constraints. Proceedings, 2002. X, 201 pages. 2003. (Subseries LNAI).

Vol. 2628: T. Fahringer, B. Scholz, Advanced Symbolic Analysis for Compilers. XII, 129 pages. 2003.

Vol. 2631: R. Falcone, S. Barber, L. Korba, M. Singh (Eds.), Trust, Reputation, and Security: Theories and Practice. Proceedings, 2002. X, 235 pages. 2003. (Subseries LNAI).

Vol. 2632: C.M. Fonseca, P.J. Fleming, E. Zitzler, K. Deb, L. Thiele (Eds.), Evolutionary Multi-Criterion Optimization. Proceedings, 2003. XV, 812 pages. 2003.

Vol. 2633: F. Sebastiani (Ed.), Advances in Information Retrieval. Proceedings, 2003. XIII, 546 pages. 2003.

Vol. 2634: F. Zhao, L. Guibas (Eds.), Information Processing in Sensor Networks. Proceedings, 2003. XII, 692 pages. 2003.

Vol. 2636: E. Alonso, D, Kudenko, D. Kazakov (Eds.), Adaptive Agents and Multi-Agent Systems. XIV, 323 pages. 2003. (Subseries LNAI).

Vol. 2637: K.-Y. Whang, J. Jeon, K. Shim, J. Srivastava (Eds.), Advances in Knowledge Discovery and Data Mining. Proceedings, 2003. XVIII, 610 pages. 2003. (Subseries LNAI).

Vol. 2638: J. Jeuring, S. Peyton Jones (Eds.), Advanced Functional Programming. Proceedings, 2002. VII, 213 pages. 2003.

Vol. 2639: G. Wang, Q. Liu, Y. Yao, A. Skowron (Eds.), Rough Sets, Fuzzy Sets, Data Mining, and Granular Computing. Proceedings, 2003. XVII, 741 pages. 2003. (Subseries LNAI).

Vol. 2641: P.J. Nürnberg (Ed.), Metainformatics. Proceedings, 2002. VIII, 187 pages. 2003.

Vol. 2642: X. Zhou, Y. Zhang, M.E. Orlowska (Eds.), Web Technologies and Applications. Proceedings, 2003. XIII, 608 pages. 2003.

Vol. 2643: M. Fossorier, T. Høholdt, A. Poli (Eds.), Applied Algebra, Algebraic Algorithms and Error-Correcting Codes. Proceedings, 2003. X, 256 pages. 2003.

Vol. 2644: D. Hogrefe, A. Wiles (Eds.), Testing of Communicating Systems. Proceedings, 2003. XII, 311 pages. 2003.

Vol. 2645: M.A. Wimmer (Ed.), Knowledge Management in Electronic Government. Proceedings, 2003. XI, 320 pages. 2003. (Subseries LNAI).

Vol. 2646: H. Geuvers, F, Wiedijk (Eds.), Types for Proofs and Programs. Proceedings, 2002. VIII, 331 pages. 2003.

Vol. 2647: K.Jansen, M. Margraf, M. Mastrolli, J.D.P. Rolim (Eds.), Experimental and Efficient Algorithms. Proceedings, 2003. VIII, 267 pages. 2003.

Vol. 2648: T. Ball, S.K. Rajamani (Eds.), Model Checking Software. Proceedings, 2003. VIII, 241 pages. 2003.

Vol. 2649: B. Westfechtel, A. van der Hoek (Eds.), Software Configuration Management. Proceedings, 2003. VIII, 241 pages. 2003.

Vol. 2651: D. Bert, J.P. Bowen, S. King, M, Waldén (Eds.), ZB 2003: Formal Specification and Development in Z and B. Proceedings, 2003. XIII, 547 pages. 2003.

Vol. 2653: R. Petreschi, Giuseppe Persiano, R. Silvestri (Eds.), Algorithms and Complexity. Proceedings, 2003. XI, 289 pages. 2003.

Vol. 2656: E. Biham (Ed.), Advances in Cryptology – EUROCRPYT 2003. Proceedings, 2003. XIV, 649 pages. 2003.

Vol. 2657: P.M.A. Sloot, D. Abramson, A.V. Bogdanov, J.J. Dongarra, A.Y. Zomaya, Y.E. Gorbachev (Eds.), Computational Science – ICCS 2003. Proceedings, Part I. 2003. LV, 1095 pages. 2003.

Vol. 2658: P.M.A. Sloot, D. Abramson, A.V. Bogdanov, J.J. Dongarra, A.Y. Zomaya, Y.E. Gorbachev (Eds.), Computational Science – ICCS 2003. Proceedings, Part II. 2003. LV, 1129 pages. 2003.

Vol. 2659: P.M.A. Sloot, D. Abramson, A.V. Bogdanov, J.J. Dongarra, A.Y. Zomaya, Y.E. Gorbachev (Eds.), Computational Science – ICCS 2003. Proceedings, Part III. 2003. LV, 1165 pages. 2003.

Vol. 2660: P.M.A. Sloot, D. Abramson, A.V. Bogdanov, J.J. Dongarra, A.Y. Zomaya, Y.E. Gorbachev (Eds.), Computational Science – ICCS 2003. Proceedings, Part IV. 2003. LVI, 1161 pages. 2003.

Vol. 2663: E. Menasalvas, J. Segovia, P.S. Szczepaniak (Eds.), Advances in Web Intelligence. Proceedings, 2003. XII, 350 pages. 2003. (Subseries LNAI).

Vol. 2665: H. Chen, R. Miranda, D.D. Zeng, C. Demchak, J. Schroeder, T. Madhusudan (Eds.), Intelligence and Security Informatics. Proceedings, 2003. XIV, 392 pages. 2003.

Vol. 2667: V. Kumar, M.L. Gavrilova, C.J.K. Tan, P. L'Ecuyer (Eds.), Computational Science and Its Applications – ICCSA 2003. Proceedings, Part I. 2003. XXXIV, 1060 pages. 2003.

Vol. 2668: V. Kumar, M.L. Gavrilova, C.J.K. Tan, P. L'Ecuyer (Eds.), Computational Science and Its Applications – ICCSA 2003. Proceedings, Part II. 2003. XXXIV, 942 pages. 2003.

Vol. 2669: V. Kumar, M.L. Gavrilova, C.J.K. Tan, P. L'Ecuyer (Eds.), Computational Science and Its Applications – ICCSA 2003. Proceedings, Part III. 2003. XXXIV, 948 pages. 2003.

Vol. 2670: R. Peña, T. Arts (Eds.), Implementation of Functional Languages. Proceedings, 2002. X, 249 pages. 2003.

Vol. 2674: I.E. Magnin, J. Montagnat, P. Clarysse, J. Nenonen, T. Katila (Eds.), Functional Imaging and Modeling of the Heart. Proceedings, 2003. XI, 308 pages. 2003.

Vol. 2675: M. Marchesi, G. Succi (Eds.), Extreme Programming and Agile Processes in Software Engineering. Proceedings, 2003. XV, 464 pages. 2003.

Vol. 2692: P. Nixon, S. Terzis (Eds.), Trust Management. Proceedings, 2003. X, 349 pages. 2003.

Vol. 2707: K. Jeffay, I. Stoica, K. Wehrle (Eds.), Quality of Service – IWQoS 2003. Proceedings, 2003. XI, 517 pages. 2003.